S0-BED-149

CONCISE ENCYCLOPEDIA OF

BUILDING & CONSTRUCTION MATERIALS

ADVANCES IN MATERIALS SCIENCE AND ENGINEERING

This is a new series of Pergamon Scientific reference works, each volume providing comprehensive, self-contained and up-to-date coverage of a selected area in the field of materials science and engineering. The series is being developed primarily from the highly acclaimed *Encyclopedia of Materials Science and Engineering*, published in 1986. Other titles in the series are listed below.

BEVER (ed.)
Concise Encyclopedia of Materials Economics, Policy & Management

BROOK (ed.)
Concise Encyclopedia of Advanced Ceramic Materials

CARR & HERZ (eds.)
Concise Encyclopedia of Mineral Resources

CORISH (ed.)
Concise Encyclopedia of Polymer Processing & Applications

EVETTS (ed.)
Concise Encyclopedia of Magnetic & Superconducting Materials

KELLY (ed.)
Concise Encyclopedia of Composite Materials

KIMERLING (ed.)
Concise Encyclopedia of Electronic & Optoelectronic Materials

SCHNIEWIND (ed.)
Concise Encyclopedia of Wood & Wood-Based Materials

WILLIAMS (ed.)
Concise Encyclopedia of Medical & Dental Materials

NOTICE TO READERS

Dear Reader
If your library is not already a standing order/continuation order customer to the series **Advances in Materials Science and Engineering**, may we recommend that you place a standing/continuation order to receive immediately upon publication all new volumes. Should you find that these volumes no longer serve your needs, your order can be cancelled at any time without notice.

ROBERT MAXWELL
Publisher at Pergamon Press

CONCISE ENCYCLOPEDIA OF
BUILDING & CONSTRUCTION MATERIALS

Editor
FRED MOAVENZADEH
MIT, Cambridge, MA, USA

Executive Editor
ROBERT W CAHN
University of Cambridge, UK

Senior Advisory Editor
MICHAEL B BEVER
MIT, Cambridge, MA, USA

PERGAMON PRESS
Member of Maxwell Macmillan Pergamon Publishing Corporation
OXFORD • BEIJING • FRANKFURT
SYDNEY • TOKYO

THE MIT PRESS
CAMBRIDGE
MASSACHUSETTS

Distributed exclusively in North and South America by The MIT Press, Cambridge, Massachusetts, USA

Published and distributed exclusively throughout the rest of the world by Pergamon Press:

U.K.	Pergamon Press plc, Headington Hill Hall, Oxford OX3 0BW, England
PEOPLE'S REPUBLIC OF CHINA	Pergamon Press, Room 4037, Qianmen Hotel, Beijing, People's Republic of China
FEDERAL REPUBLIC OF GERMANY	Pergamon Press GmbH, Hammerweg 6, D-6242 Kronberg, Federal Republic of Germany
AUSTRALIA	Pergamon Press Australia Pty Ltd., P.O. Box 544, Potts Point, N.S.W. 2011, Australia
JAPAN	Pergamon Press, 5th Floor, Matsuoka Central Building, 1-7-1 Nishishinjuku, Shinjuku-ku, Tokyo 160, Japan

First edition 1990

Library of Congress Cataloging in Publication Data
Concise Encyclopedia of building & construction materials / editor, Fred Moavenzadeh; executive editor, Robert W. Cahn. — 1st ed. p. cm. — (Advances in materials science and engineering; 4) Includes index.
1. Building materials—Dictionaries. I. Moavenzadeh, Fred, 1935– . II. Cahn, R. W. (Robert W.), 1924– . III. Title: Concise encyclopedia of building and construction materials. IV. Series.
TA402.C657 1989
624—dc20 89–8517

ISBN 0–262–13248–6 (MIT Press)

British Library Cataloguing in Publication Data
Concise encyclopaedia of building & construction materials.
1. Construction materials. 2. Building materials. I. Moavenzadeh, Fred, 1935– . II. Series.
624.1'8

ISBN 0–08–034728–2 (Pergamon Press)

Printed in Great Britain by BPCC Wheatons Ltd, Exeter

CONTENTS

HONORARY EDITORIAL ADVISORY BOARD

FOREWORD

In the short time since its publication, the *Encyclopedia of Materials Science and Engineering* has been accepted throughout the world as the standard reference about all aspects of materials. This is a well-deserved tribute to the scholarship and dedication of the Editor-in-Chief, Professor Michael Bever, the Subject Editors and the numerous contributors.

During its preparation, it soon became clear that change in some areas is so rapid that publication would have to be a continuing activity if the Encyclopedia were to retain its position as an authoritative and up-to-date systematic compilation of our knowledge and understanding of materials in all their diversity and complexity. Thus, the need for some form of supplementary publication was recognized at the outset. The Publisher has met this challenge most handsomely: both a continuing series of Supplementary Volumes to the main work and a number of smaller encyclopedias, each covering a selected area of materials science and engineering, will be published in the next few years.

Professor Robert Cahn, the Executive Editor, was previously the editor of an important subject area of the main work and many other people associated with the Encyclopedia will contribute to its Supplementary Volumes and derived Concise Encyclopedias. Thus, continuity of style and respect for the high standards set by the *Encyclopedia of Materials Science and Engineering* are assured. They have been joined by some new editors and contributors with knowledge and experience of important subject areas of particular interest at the present time. Thus, the Advisory Board is confident that the new publications will significantly add to the understanding of emerging topics wherever they may appear in the vast tapestry of knowledge about materials.

The appearance of Supplementary Volumes and the new series *Advances in Materials Science and Engineering* is an event which will be welcomed by scientists and engineers throughout the world. We are sure that it will add still more luster to a most important enterprise.

Walter S Owen
Chairman
Honorary Editorial Advisory Board

EXECUTIVE EDITOR'S PREFACE

As the publication of the *Encyclopedia of Materials Science and Engineering* approached, Robert Maxwell resolved to build upon the immense volume of work which had gone into its creation by embarking on a follow-up project. This project had two components. The first was the creation of a series of Supplementary Volumes to the Encyclopedia itself. The second component of the new project was the creation of a series of Concise Encyclopedias on individual subject areas included in the Main Encyclopedia to be called *Advances in Materials Science and Engineering*.

These Concise Encyclopedias are intended, as their name implies, to be compact and relatively inexpensive volumes (typically 300–600 pages in length) based on the relevant articles in the Encyclopedia (revised where need be) together with some newly commissioned articles, including appropriate ones from the Supplementary Volumes. Some Concise Encyclopedias will offer combined treatments of two subject fields which were the responsibility of separate Subject Editors during the preparation of the parent Encyclopedia (e.g., dental and medical materials).

At the time of writing, ten Concise Encyclopedias have been contracted and others are being planned. These and their editors are listed below.

Concise Encyclopedia of Advanced Ceramic Materials	Prof. Richard J Brook
Concise Encyclopedia of Building & Construction Materials	Prof. Fred Moavenzadeh
Concise Encyclopedia of Composite Materials	Prof. Anthony Kelly CBE, FRS
Concise Encyclopedia of Electronic & Optoelectronic Materials	Dr Lionel C Kimerling
Concise Encyclopedia of Magnetic & Superconducting Materials	Dr Jan E Evetts
Concise Encyclopedia of Materials Economics, Policy & Management	Prof. Michael B Bever
Concise Encyclopedia of Medical & Dental Materials	Prof. David F Williams
Concise Encyclopedia of Mineral Resources	Dr Donald D Carr & Prof. Norman Herz
Concise Encyclopedia of Polymer Processing & Applications	Mr P J Corish
Concise Encyclopedia of Wood & Wood-Based Materials	Prof. Arno P Schniewind

All new or substantially revised articles in the Concise Encyclopedias will be published in one or other of the Supplementary Volumes, which are designed to be used in conjunction with the Main Encyclo-pedia. The Concise Encyclopedias, however, are "free-standing" and are designed to be used without necessary reference to the parent Encyclopedia.

The Executive Editor is personally responsible for the selection of topics and authors of articles for the Supplementary Volumes. In this task, he has the benefit of the advice of the Senior Advisory Editor and of other members of the Honorary Editorial Advisory Board, who also exercise general supervision of the entire project. The Executive Editor is responsible for appointing the Editors of the various Concise Encyclopedias and for supervising the progress of these volumes.

Robert W Cahn
Executive Editor

EDITOR'S PREFACE

The largest consumer of materials in the world is the building and construction industry. The diversity of materials used for construction covers a broad spectrum, including at one extreme natural materials such as clay, stone, and timber, and at the other, all kinds of processed and synthetic materials. Several materials such as cement and concrete virtually have no other end user but the construction industry. The user of construction materials is also concerned with a wide diversity of properties, mechanical properties, durability, and fire resistance being among the most important. However, thermal properties, aesthetics in terms of color and texture, adhesion, friction, etc., are also of concern in the day-to-day use of construction materials.

This *Concise Encyclopedia of Building & Construction Materials*, for the first time, brings together a collection of articles that cover not only the very diverse materials used in construction, but also the major properties of concern to the construction industry.

The articles in the Encyclopedia are grouped into eight categories, each category or section being concerned with either a particular class of materials or a particular set of properties. Owing to the encyclopedic nature of the publication, the articles appear in alphabetical order, but an awareness of the underlying categories or sections into which the articles fall will help readers to better utilize the volume. These sections are:

1. Introduction, Economics, and Availability of Materials.
2. Building Materials, General
3. Mechanical Properties
4. Clay, Ceramics, Cement, Sand and Gravel
5. Glass
6. Metals
7. Polymers, Plastics, and Composites
8. Wood

In each of the last five sections, the user will find articles relating to production, properties, and applications of five major construction categories of construction materials.

There are 124 articles in this volume on building and construction materials, providing an excellent reference book for those involved in both the construction and the materials industries. The Encyclopedia can also be utilized as an excellent textbook or complementary reading reference book for undergraduate and graduate courses taught in civil engineering and architectural departments. The volume not only addresses the diversity of materials that have to be covered in such an encyclopedia, but also provides a select collection of referenced on each topic which can provide additional readings.

Over the past eighteen months in which I have been responsible for bringing this collection of articles together, I have had the pleasure of getting to know many dedicated and distinguished scholars who have generously contributed their time to make this a reality. My deepest gratitude goes to the authors of the articles, to Professor Michael Bever, Editor-in-Chief of the original Encyclopedia of Materials, and to Professor R.W. Cahn, Executive Editor of Advances in Materials Science and Engineering, of which this Concise Encyclopedia forms a part. I would also like to express my appreciation to Ms Patricia Vargas for her patience, unwavering attention and perseverance in this undertaking.

F. Moavenzadeh

GUIDE TO USE OF THE ENCYCLOPEDIA

This Concise Encyclopedia is a comprehensive reference work covering all aspects of building and construction materials. Information is presented in a series of alphabetically arranged articles which deal concisely with individual topics in a self-contained manner. This guide outlines the main features and organization of the Encyclopedia, and is intended to help the reader to locate the maximum amount of information on a given topic.

Accessibility of material is of vital importance in a reference work of this kind and article titles have therefore been selected, not only on the basis of article content, but also with the most probable needs of the reader in mind. An alphabetical list of all the articles contained in this Encyclopedia is to be found on p. xv.

Articles are linked by an extensive cross-referencing system. Cross-references to other articles in the Encyclopedia are of two types: in-text and end-of-text. Those in the body of the text are designed to refer the reader to articles that present in greater detail material on the specific topic under discussion. They generally take one of the following forms:

...as fully described in the article *Ceramics as Construction Materials*

...or on carbon fiber (see *Composite Materials: An Overview*).

The cross-references listed at the end of an article serve to identify broad background reading and to direct the reader to articles that cover different aspects of the same topic.

The nature of an encyclopedia demands a higher degree of uniformity in terminology and notation than many other scientific works. The widespread use of the International System of Units has determined that such units be used in this Encyclopedia. It has been recognized, however, that in some fields Imperial units are more generally used. Where this is the case, Imperial units are given with their SI equivalent quantity and unit following in parentheses. Where possible the symbols defined in *Quantities, Units, and Symbols*, published by the Royal Society of London, have been used.

All articles in the Encyclopedia include a bibliography giving sources of further information. Each bibliography consists of general items for further reading and/or references which cover specific aspects of the text. Where appropriate, authors are cited in the text using a name/date system as follows:

...as was recently reported (Smith 1988).

Jones (1984) describes...

The contributor's name and the organization to which they are affiliated appear at the end of each article. All contributors can be found in the alphabetical List of Contributors, along with their full postal address and the titles of the articles of which they are authors or co-authors.

The Introduction to this Concise Encyclopedia provides an overview of the field of composite materials and directs the reader to many of the key articles in this work.

The most important information source for locating a particular topic in the Encyclopedia is the multilevel Subject Index, which has been made as complete and fully self-consistent as possible.

ALPHABETICAL LIST OF ARTICLES

Acoustic Properties of Wood

Anyone who has heard a fine violin, xylophone, piano or almost any other stringed instrument being played has experienced some of the unique acoustic properties of wood. Principles based on these same properties are also used in the nondestructive evaluation of various strength properties of structural wooden members and wood-composite products and in tests for decay in existing utility poles and other "in-place" timbers. Investigations are also being carried out into the possibilities of using sound waves to identify various stages of wood drying either by moisture-content determination or by counting the number of acoustic emissions per unit time.

1. Velocity of Sound in Wood and Wood Products

The velocity of sound in wood and wood products can be measured either by using its natural or resonant frequency or by propagating stress waves by physical impact or ultrasonics. The velocity has been found to be influenced by wood species, moisture content, temperature and direction of wave propagation, (i.e., longitudinal, radial or tangential). Density does not significantly affect the velocity, but the ratio of the medium's elastic modulus E to its density ρ is important; for the case of rods, the velocity of sound v can be shown to be given by $v = (E/\rho)^{1/2}$.

James (1961) determined the speed of sound in clear Douglas-fir specimens subjected to longitudinal vibration at various moisture contents and temperatures using the specimen's resonant frequency. He calculated velocity from the expression $v = 2fl$ where f is the resonant frequency and l is the specimen length. Values from this study are given in Table 1; these show that as either moisture content or temperature is increased the velocity of sound decreases at an increasing rate.

Gerhards (1982) summarized the variables that affect sound velocity as determined by a number of researchers. Velocity was found to decrease substantially as grain angle increased from 0 to 90° with respect to the longitudinal axis. A decrease of 63% was found to occur over this range with the most apparent changes taking place at grain angles less than 20°. Velocity through knots was found to be lower, owing to grain curvature; however, when straight grain was present around the knot, the knot's effect was greatly reduced (Burmester 1965). Another important factor is advanced decay, which has been found to cause a 25% loss in sound velocity (Konarski and Wazny 1977).

The effect of advanced decay on sound velocity forms the basis for a commercial wooden-pole tester marketed under the name Pole-tek. This device uses a spring-loaded impact hammer and a sensing probe, which are placed on opposite sides of the pole to be tested. A comparison is made between sound-velocity values obtained for the test pole and a standard, decay-free pole; if the test pole gives a low value of sound velocity, incremental boring can be used to determine the actual extent of decay (Graham and Helsing 1979).

McDonald (1978), in a study utilizing ultrasonic wave propagation to locate defects in lumber, determined sound velocities in the longitudinal, radial and tangential directions of several wood species. Values for selected species are given in Table 2. These were obtained using a 400 V pulse of 1 μs duration at room temperature. If the radial and tangential values are averaged and compared with the longitudinal values, it can be seen that the longitudinal values are from 2.3 to 3.5 times greater than the transverse values. McDonald also found that a more precise detection system could be obtained if the ultrasonics were transmitted and received in water rather than in air. This results from better surface contact and hence better board adsorption of the induced stress wave.

Dunlop (1980), measuring the velocity of sound in particleboard, found a significant difference between values measured longitudinally (in the plane of the sheet) and those measured normal to this plane or in the transverse direction, as shown in Table 3. He also determined that the higher-density surface layers had different values than the core material as can also be seen in Table 3.

James et al. (1982) investigated the feasibility of monitoring stages in the kiln drying of lumber through changes in sound velocity per unit length as influenced by moisture content. Using ultrasonic pulses and correcting for effects of temperature, the velocity was shown to be well correlated with changes in moisture content. Moisture gradients influenced the readings, however, since the ultrasonic pulse would travel faster through the drier regions, tending to indicate lower moisture content than actually existed. It was found, however, that adjustment for this could be incorporated in the temperature-correction factors. The method shows promise but additional work is needed.

Skaar et al. (1980) made initial examinations in using acoustic emissions to monitor drying of red-oak lumber. Acoustic emissions are thought to originate from check development. The purpose of monitoring emissions is to control the severity of drying conditions to minimize degradation caused by checking.

As mentioned above, the velocity of sound is related to both its resonant frequency and its elastic modulus. Consequently, either the velocity of sound as

Table 1
Velocity of sound parallel to the grain at different temperatures as determined from resonant frequencies for clear Douglas fir (after James 1961)

Moisture content[a]	Sound velocity (m s^{-1}) −18 °C	27 °C	71 °C	93 °C	Percentage decrease in velocity[b]
7.2	5486	5359	5207	5131	6.5
12.8	5512	5334	5182	5080	7.8
16.5	5385	5182	4978	4851	9.9
23.7	5055	4877	4623	4420	12.6
27.2	4978	4775	4496	4293	13.8
Percentage decrease in velocity[c]	9.3	10.9	13.7	16.3	

[a] Oven-dry-weight basis [b] Over temperature range shown [c] Over moisture content range shown

Table 2
Velocity of sound as determined by ultrasonic wave propagation (after McDonald 1978)

Direction of sound propagation	Sound velocity (m s^{-1}) Douglas fir (dry)	Ponderosa pine (green)	Red oak (green)	Beech (dry)
Longitudinal	4350	4390	4110	6610
Radial	1980	1620	2040	1980
Tangential	1770	1460	1790	1770
Longitudinal: transverse ratio	2.3	2.9	2.9	3.5

Table 3
Velocity of sound in particle board (after Dunlop 1980)

Specimen	Density (kg m^{-3})	Sound velocity (m s^{-1}) Longitudinal	Transverse
18 mm Board	500	2200–2500	820
6 mm Skin	590	2200	920
8 mm Core	460	1800–1900	620

measured with longitudinal stress waves or resonant frequency as determined from free- or forced-vibration experiments can be used in the non-destructive evaluation of the elastic modulus of wood, which in turn can be linked to other mechanical properties.

Two acoustic properties related to sound velocity are the sound wave resistance, which is the product of velocity and density ($v\rho$), and the damping of sound radiation, which is determined essentially by the ratio of velocity to density (v/ρ). Sound wave resistance is important in sound propagation and the reflection of sound at media boundaries, whereas damping of sound radiation determines the energy loss from a vibrating body to surrounding media by radiation.

2. *Damping Capacity*

The damping capacity of vibrating wood results from energy dissipation through internal friction. A measure of damping capacity is logarithmic decrement δ, which is defined as the natural logarithm of the ratio of two successive decaying wave amplitudes (A_i and A_{i+1}) for a body set in harmonic motion and allowed to vibrate freely:

$$\delta = \ln(A_i/A_{i+1})$$

The values of logarithmic decrement are affected by moisture content, temperature, grain direction and frequency of vibration (Dunlop 1980, Pentoney 1955). The effects of each of these variables are nonlinear. In general, the minimum values for logarithmic decrement increase as moisture content increases and as temperature decreases. Values determined by James (1961) for clear Douglas-fir specimens, based on decay of its resonant frequency, are given in Table 4. These values were determined for free vibration in the longitudinal direction. Other researchers report values of logarithmic decrement parallel to the grain to be two to three times larger than values perpendicular to the grain (Dunlop 1981, Pentoney 1955).

Table 4
Logarithmic decrement values for clear Douglas fir at various moisture contents and temperatures (after James 1961)

Moisture content[a]	Specific gravity[b]	Logarithmic decrement (10^{-3})			
		−18 °C	27 °C	71 °C	93 °C
7.2	0.48	36.6	20.2	23.3	25.5
12.8	0.40	27.2	22.2	26.5	34.0
16.5	0.45	27.9	26.4	33.3	43.7
23.7	0.43	29.1	26.3	42.2	84.0
27.2	0.45	32.7	29.2	60.6	108.7

[a] Oven-dry-weight basis [b] Value based on oven-dry-weight and volume at specified moisture content

Dunlop (1981) investigated the feasibility of using damping capacity to determine decay in wood poles using a pulse-echo technique. He hypothesized that a more precise method for determining decay in wooden poles would be obtained by sending the signal in the longitudinal direction as opposed to the Pole-tek method which induces the pulse in the transverse direction. The results were promising but the method was not as precise as the Pole-tek technique, especially with respect to decay allocation.

3. Acoustic Properties Affecting Musical Instruments

The acoustic properties of wood make it a preferred material for sounding boards in musical instruments. The velocity of sound in dry wood parallel to the grain is about the same as in steel and most other metals. However, since the density of wood is much lower, it has low sound wave resistance and high damping of sound radiation. Low sound wave resistance facilitates resonance, and high damping of sound radiation combined with low damping capacity means that loss of the sound energy is consumed in internal friction and more is emitted as sound radiation to the surroundings—precisely what is desired of sounding boards in musical instruments. For example, values for sound wave resistance as given by Kollmann and Côté (1968) range from $20–37 \times 10^5$ N s m^{-3} for wood as compared with 395×10^5 and 258×10^5 N s m^{-3} for steel and cast iron, respectively. In Japan, it has been found that the best wood selected by experts in the piano industry for sounding boards had the lowest ratio of damping capacity to elastic modulus (Norimoto 1982).

The environmental conditions in which musical instruments are usually used are optimal for musical acoustic properties. Under these conditions (i.e., room temperature, wood at 8% moisture content), wave dissipation due to sound radiation is increased and damping due to internal friction is at a minimum.

4. Architectural Acoustics

Architectural acoustics involve two main considerations. The first is the acoustic quality of a room, which is dependent on its sound-absorption characteristics. The second is the isolation of sound in the source room to prevent transmission either to adjoining rooms or to floors below.

Sound absorption is necessary to keep the reflection or bounding of sound waves in a room to a minimum since such reflection or bounding adversely affects the clarity with which the source can be heard. Some reflection is necessary, however, to keep the room from becoming acoustically deadened. The sound absorption of a room is measured by its reverberation time; that is, the time for sound to diminish to one-millionth of its original intensity, equivalent to a reduction of 60 dB. This time is directly proportional to the volume of the room and inversely proportional to the sum of all objects and surfaces absorbing sound. Reverberation time is expressed as

$$t_R = 0.161\, V/\Sigma S_i a_i$$

where V is the room volume, S_i is the exposed surface area and a_i is the absorption coefficient of each object. An acoustically well-designed room, therefore, requires the correct amount of sound-absorbing materials. The optimum sound-absorption characteristics for a particular material result from a careful balance of its density, porosity, fineness of fibers, bulk elasticity and thickness. For sound to be absorbed, the surface porosity must be such that a sound wave will enter and be dissipated through internal heat produced by intermolecular friction between the air molecules and the absorbing structure. If the pore size is too small, the wave will be reflected; if it is too large, intermolecular friction will not take place, permitting the wave to pass through.

Owing to the porosity of the surface of normal wood, only 5–10% of sound waves are absorbed and 90–95% are reflected. The percentage absorbed is increased only slightly with a very rough surface.

Wood, therefore, is not a good absorber as acoustic materials should absorb at least 50% of the sound waves. Wood can, however, be used for acoustic materials either by product processing as in the manufacture of acoustic tiles or by the design of resonant panels.

Acoustic tile, manufactured from low-density fiberboard, absorbs sound since it contains open pores over at least 15% of its surface. Common fiber insulating board is not as porous and therefore is not a much better absorbent than solid wood. Other solid wood products such as extremely rough and porous particleboard may qualify as sound absorbers, but common particleboard and hardboard do not (Godshall and Davis 1969).

Well-designed resonant panels can absorb and control sound. In this method a panel is constructed to allow the panel skin to vibrate and transmit the sound wave to either a dead air space (at least 2.5 cm) or more efficiently to a sound-absorbing medium in that space such as fiber insulation. As expected, the thicker the air space or absorbing material, the higher the quality of absorbency.

Multifamily dwelling and commercial building codes require sound to be prevented from passing through partitions and floors into adjoining rooms. This is known as sound isolation and is measured as the transmission loss or drop in decibels from one room to the next. A drop of more than 45 dB at midfrequencies is considered necessary for party wall construction (Schultz 1969). In essence, a partition exhibiting a high transmission loss has the same objective as the resonant panels described above; this is to dissipate the vibrations of the wall surface set in motion by the sound waves before these vibrations can reach the wall surface in the adjoining room. If the two wall surfaces are rigidly attached, then both walls vibrate simultaneously and sound is transmitted. If they are allowed to vibrate independently, then the waves can be absorbed within the wall structure.

Jones (1973) has shown that an accurate prediction of the transmission loss for a structural system is dependent on knowledge of the sound frequency, partition mass and partition configuration. The effect of sound frequency has been categorized into three main regions (Schultz 1969). Region I consists of frequencies near the lowest resonance frequency; in this region, the transmission loss is dependent on partition stiffness. Region II consists of midfrequencies (250–1500 Hz); transmission in this region is controlled directly by the mass per unit area of the partition. The midfrequencies are of greatest concern since they are the most common and therefore are used for obtaining standard test values.

As frequencies increase above approximately 1500 Hz, region III is entered where transmission is again controlled by partition stiffness as in region I. This results from the partition "coincidence" frequency being reached, at which point the acoustic wavelength perfectly matches the flexural wavelength of the structure. The transmission loss is greatly reduced at this point (Jones 1973, 1976, 1978, 1979).

There are three main partition systems which use wooden components and are designed to decrease sound transmissions. The most simple involves placement of fiber insulation board between the studs and gypsum board to absorb the vibrations. The second system uses metal-resilient channels placed between the studs and gypsum board. The metal flanges in this system vibrate independently of each other, thereby helping to dissipate the wave. The third and most expensive partition system is a double stud wall with a 2.5 cm separation which allows each wall surface to vibrate independently of the other. All of these systems are enhanced by filling all air spaces with sound-absorbing insulation, using a double layer of gypsum board on the exterior to increase mass, and using adhesives to fasten components instead of nails.

Impacts from the floor above a room can be isolated by increasing the flooring density with foamed concrete or by using a "floating" floor design. The impact isolation for the foamed concrete is greatly improved by using a pad and carpet. The floating design is more complex, consisting of a glass-fiber interface between the flooring and subflooring and a heavy glass-fiber blanket between the joists and a resiliently suspended ceiling.

See also: Wood as a Building Material; Wood: An Overview

Bibliography

Bucur V 1984 Relationships between grain angle of wood specimens and ultrasonic velocity. *Catgut Acoust. Soc. J.* 41: 30–35

Burmester A 1965 Relationship between sound velocity and the morphological, physical, and mechanical properties of wood (in German). *Holz Roh- Werkst.* 23(6): 227–36

Dunlop J I 1980 Testing of particle board by acoustic techniques. *Wood Sci. Technol.* 14: 69–78

Dunlop J I 1981 Testing of poles by using acoustic pulse method. *Wood Sci. Technol.* 15: 301–10

Gerhards C C 1982 Longitudinal stress waves for lumber stress grading: Factors affecting applications: State of the art. *For. Prod. J.* 32(2): 20–25

Godshall W D, Davis J H 1969 *Acoustical Absorption Properties of Wood-Base Panel Materials*, USDA Forestry Service Research Paper FPL 104. US Department of Agriculture Forest Service, Madison, Wisconsin

Graham R D, Helsing G G 1979 *Wood Pole Maintenance Manual: Inspection and Supplemental Treatment of Douglas-fir and Western Red Cedar Poles*, OSU Forest Research Laboratory Research Bulletin 24. Oregon State University Forest Research Laboratory, Corvallis, Oregon

James W L 1961 Effect of temperature and moisture content on internal friction and speed of sound in Douglas fir. *For. Prod. J.* 9(9): 383–90

James W L 1986 *Effect of Transverse Moisture Content Gradients on the Longitudinal Propagation of Sound in Wood*, USDA Forest Products Laboratory Research

Paper FPL 466. US Department of Agriculture Forest Service, Madison, Wisconsin
James W L, Boone R S, Galligan W L 1982 Using speed of sound in wood to monitor drying in a kiln. *For. Prod. J.* 32: 27–34
Jones R E 1973 Improved acoustical privacy in multifamily dwellings. *Sound Vibr.* 7(9): 30–37
Jones R E 1976 How to accurately predict the sound insulation of partitions. *Sound Vibr.* 10(6): 14–25
Jones R E 1978 How to design walls for desired STC ratings. *Sound Vibr.* 12(8): 14–17
Jones R E 1979 Intercomparisons of laboratory determinations of airborne sound transmission loss. *J. Acoust. Soc. Am.* 66(1): 148–64
Koch P 1972 *Utilization of the Southern Pines*, Vol. I: *The Raw Material.* US Government Printing Office, Washington, DC
Kollmann F F P, Côté W A Jr 1968 *Principles of Wood Science and Technology*, Vol. I: *Solid Wood*, Springer, New York
Konarski B, Wazny J 1977 Relationships between ultrasonic velocity and mechanical properties of wood attacked by fungi (in German). *Holz Roh-Werkst.* 35(9): 341–45
McDonald K A 1978 *Lumber Defect Detection by Ultrasonics*, USDA Forestry Service Forest Products Laboratory Research Paper FPL 311. US Department of Agriculture Forest Service, Madison, Wisconsin
Norimoto M 1982 Structure and properties of wood used for musical instruments I. On the selection of wood used for piano soundboards. *J. Jpn Wood Res. Soc.* 28(7): 407–13
Pentoney R E 1955 Effect of moisture content and grain angle on the internal friction of wood. *Compos. Wood* 2(6): 131–36
Rudder F F Jr 1985 *Airborne Sound Transmission Loss Characteristics of Wood-Frame Construction*, USDA Forest Products Laboratory General Technical Report FPL 43. US Department of Agriculture Forest Service, Madison, Wisconsin
Schultz T J 1969 Acoustical properties of wood: A critique of the literature and a survey of practical applications. *For. Prod. J.* 19(2): 21–29
Skaar C, Simpson W T, Honeycutt R M 1980 Use of acoustic emissions to identify high levels of stress during oak lumber drying. *For. Prod. J.* 30(2): 21–22
Tanaka C, Nakao T, Takahashi A 1987 Acoustic property of wood I. Impact sound analysis of wood. *J. Jpn. Wood Res. Soc.* 33: 811–17

W. R. Smith
[University of Washington, Seattle, Washington, USA]

Adhesives for Wood

The first evidence of adhesive-bonded wood products can be traced back over 3500 years, to when the joining of wood was undertaken primarily for decorative purposes. The discovery of improved adhesives led to more sophisticated developments in wood composites and greater utilization of these products in structural applications. Until this century most of these adhesives were derived from animal, vegetable or mineral sources.

Projected limitations in the available harvest of wood resources and the overall decrease in wood quality emphasize the importance of adhesives and the science of adhesion in improving wood utilization by allowing smaller trees and present waste wood to be restructured into useful wood products. By proper choice of wood component, wood orientation, adhesive and manufacturing conditions, a variety of engineered composites are possible; these can have greater mechanical strength, decreased variability and improved stress-distributing properties compared to conventional solid wood.

This article outlines the mechanism of wood bonding and pertinent physical and chemical factors involved in achieving good bonds. Both wood-related and adhesive-related characteristics influencing the formation and performance of bonds in products such as laminated lumber, plywood, flakeboard and particleboard are considered.

1. *Nature of Adhesion*

Bond formation depends upon the development of physical and chemical interactions both within the bulk adhesive polymer and at the interface between adhesive and wood. Interactions within the adhesive accumulate to give cohesive strength while the forces between adhesive and wood provide adhesive strength. Both should exceed the strength of the wood, allowing substantial wood failure during destructive testing of high-quality bonds.

Most wood adhesives are liquids consisting of either 100% polymer material or more frequently an aqueous solution, emulsion or dispersion, wherein one half to one third of the mixture is polymer. Other solvents such as volatile alcohols, esters or aromatics may sometimes be used to replace all or part of the water. For more specialized bonding requirements, adhesives in a solid state (i.e., powders, hot melts) may be utilized with heat applied to transform the polymer into a flowable material.

The adhesive must accomplish three distinct stages bonding processes: it must wet the wood substrate, during flow in a controllable manner during pressing and finally set to a solid phase. Failure of the adhesive to accomplish any of these stages often results in reduced bond quality. The extent of wetting depends on the physical–chemical nature of both adhesive and wood surface. Solvent molecules and low-molecular-weight polymer molecules, being smaller and more mobile, tend to wet and penetrate into the wood substrate most rapidly. Adsorption and diffusion of these liquids into the wall often leads to swelling of the lignocellulose wood substance. The porous structure of wood and defect areas created by machining processes facilitate the transport of these materials into the wood. In contrast, higher-molecular-weight polymer material would wet and penetrate the wood substrate at a significantly slower rate, remaining largely as a concentrated polymer on the wood surface. Ultimately, the bonding area should consist of a steadily decreasing concentration of adhesive polymer as one proceeds from the surface to a few micrometers within the wood. Achieving this condition will depend on polymer physical properties, thermodynamic considerations and wood-surface properties. With solid adhesives, an intermediate heating stage which melts or softens the adhesive is necessary to accomplish wetting of the wood surface.

Optimum bond formation requires intimate contact between adhesive and wood substrates to ensure molecular interaction over a large area. This is accomplished using pressure and heat which causes viscous adhesive components to transfer and flow throughout the bonding sites while deforming the wood to achieve better contact between wood surfaces of different texture. Adhesive penetration into the wood structure must not be so great that inadequate amounts of adhesive remain in the glueline (starved glueline) or too little so that fibers and microdefects just below the wood surface are insufficiently bonded together.

Solidification and cure of the wood–adhesive system occurs either by solvent loss, polymerization and cross-linking reactions, or both. This process is time, temperature and pressure dependent with bond quality being a reflection of how well the final solidification stage is accomplished in the particular adhesive and wood substrate present.

2. *Mechanisms of Adhesion*

A number of theories have been proposed to explain bonding mechanisms in materials science. The unique heterogeneous characteristics of wood limit the relevance of any one particular theory, leading instead to wood bonding being rationalized in terms of a combination of mechanisms involving mechanical, physical and chemical parameters.

The ability of mechanical interlocking to occur readily is a function of the porous nature of most wood surfaces and the ease with which the adhesive can flow into these microcavities under the action of heat and external pressure. The mechanical anchoring of adhesive into wood occurs when the adhesive polymer hardens. This bonding mechanism is of obvious importance in hot melts and does increase the contact area available for bond formation. It cannot, however, account for the documented phenomena of bond-strength reduction with aging of wood surfaces and the limited correlation between adhesive penetration depth and bond-strength development.

Adsorption and diffusion mechanisms allow for intimate molecular contact between adhesive and wood substance over a large area. The accumulative physical interaction arising from attractive Van der Waals, hydrogen bonding and dispersion forces between adhesive polymer and wood components can result in strong bond formation which many consider a primary bonding mechanism. Complete wetting of the adhered surface enhances bonding potential. Both adhesive surface tension and wood surface tension have some bearing on wetting effectiveness. Adhesive spreading methods are additional factors of this mechanism since the mechanical energy input during glue application may overcome some surface force resistance to wetting.

Both adhesive and wood reactive functional groups allow for the high probability of covalent bond formation. In cross-linkable systems these chemical interactions can produce a strongly bonded wood–adhesive matrix capable of maintaining long-term stability under a variety of heat, moisture and stress conditions. While indications of covalent bonding have been found in model systems, conclusive evidence of their presence in wood-bonded products has not yet been made.

3. *Wood-Related Factors Affecting Adhesion*

Wood being a complex, three-dimensional tissue, has a distinct but variable structural and chemical makeup which must be allowed for in adhesive formulation. The heterogeneous character of wood is reflected in both its bulk and surface properties.

3.1 Bulk Features

Anatomical characteristics, permeability, density and moisture content are among the more important interrelated factors influencing wood-bond formation.

Fiber orientation influences the ease with which the adhesive polymer can wet wood while at the same time retaining sufficient polymer on the bonding surface to form a glueline interface. Rapid adhesive penetration is possible along the grain because of open lumen structures, while penetration across the grain is highly restricted by fiber walls. Anatomical differences between species will affect the permeability of wood to liquids and gases, thus influencing the degree to which adhesive can successfully intermix with wood fibers. In softwoods, fiber lumina are laterally interconnected by small capillaries which allow the ready passage of air and liquid to various depths below the wood surface. In hardwoods, many of these interconnecting routes are blocked or absent, resulting in more restricted flow of the adhesive into bonding sites. Further variations such as knots or grain distortions associated with tree-growth patterns can provide additional impediments to achieving bonding potential. Other structural features of importance are variations imposed by the presence of juvenile and mature wood, and earlywood and latewood effects within one piece of wood. These features are interrelated to wood density.

The density of wood relates directly to its strength and has primarily a physical influence on adhesion. The stronger the wood, the stronger the adhesive must be in order to utilize the full strength of the wood. Density variations are associated with the growth ring structure of wood, particularly in the ratio of fiber-wall to fiber-lumen volume. In dense wood, adhesive penetration is restricted while in lower density woods, care must be taken to prevent overpenetration. Where large differences in density can occur (i.e., earlywood–latewood zones) the adhesive must be designed to adapt adequately to both of these conditions. In general wood density in the range 250–450 kg m^{-3} (i.e., spruce, pine, fir, aspen) are easier to bond than woods of density greater than 600 kg m^{-3} (i.e., birch, oak, maple).

Wood physical properties are highly dependent on the interaction of wood density and moisture content which in turn affect gluing behavior. High-moisture-content conditions often result in increased adhesive flow and penetration while retarding polymer cure by limiting solvent loss into the wood or restricting condensation polymerization reactions. Excessive water at the bonding interface can also act as a barrier to bond formation especially with hot-melt systems. Typical mositure-content ranges for bonding are 2–16% although successful bond formation has been reported at up to 40% moisture content with some adhesives curing under ambient or slightly elevated temperature conditions. The likelihood of steam formation at higher moisture contents, with its destructive possibilities in the glueline, necessitates wood to be dried to 2–6% moisture content when using high-temperature-curing adhesives. Wood volumetric changes occurring between fiber saturation and oven-dry conditions result in directional shrinkage and swelling forces, directly proportional to density, which can place considerable stress on the glueline. Cured gluelines should partially maintain bond integrity with dimensional-change-induced stresses, even during continual cycling between wet and dry conditions because they may weaken the wood–adhesive interface considerably in high-density woods. Moisture and density also affect wood deformational properties requiring some adjustments in bonding pressure (i.e., lower density or higher moisture content requires less bonding pressure).

3.2 Surface Features

Both physical and chemical properties of the wood surface can profoundly affect bond-formation potential. Their primary influence impacts on the accessibility of adhesives to wood-fiber bonding sites. From a microscopic perspective, wood surfaces have a rough, physical appearance due to anatomy, in particular grain orientation and fiber size. Surfaces produced from transverse cuts (i.e., cut across the tracheids) expose end-grain areas whose large lumen volume often results in substantial over-penetration of adhesive. Consequently, direct bonding of end-grain surfaces is difficult and not commonly attempted. Cutting along tangential and radial surfaces inevitably excises some fiber walls, exposing different amounts and parts of fiber lumina. Surface topography may vary appreciably between species since lumina diameters may range from as small as 10 μm in fine-grained woods to as large as 200 μm in coarse-grained woods. Presence of resin ducts, vessel elements or knots give added variations to the wood surface. Machine operations during surface formation often result in further irregularities which can be associated with anatomical features. These can include saw-tooth abrasions, indentations from planer or shaper knives, lathe checks and roughness resulting from the peeling of veneer, torn and broken fibers caused during flake cutting and other splits or checks. Although these features may increase available bonding area, they hinder good bond formation by weakening the wood surface. Heat and pressure can be utilized to compress and plasticize the interface to improve contact in lower-density species but high-density woods must rely on smooth machining operations to achieve the intimate contact necessary for bonding. Mechanically weakened wood surfaces can be repaired during gluing by the action of adhesive penetration between and into the wood fiber.

Wood surfaces are chemically heterogeneous. Their composition does not necessarily correspond directly to the chemical makeup of bulk wood but rather is more a function of the conditions and methods of surface formation. Since wood fibers are composed of strands of oriented cellulose molecules intertwined within hemicellulose and lignin–polymer matrices, cleavage of a fiber wall may expose varying amounts of each chemical component at the wood surface layer. The carbohydrate components are highly polar, with fractions existing in both crystalline and amorphous forms and containing numerous functional groups available for reaction. Lignin, an amorphous polymer, offers additional reactive sites on its phenolic components. Both carbohydrate and lignin fractions are present in a range of molecular sizes and polarities which allow for bond formation with polar or nonpolar adhesives.

Extractives, although normally a minor component of the wood, can play a significant role in bond formation. These relatively low-molecular-weight materials are often highly mobile, nonpolar substances of limited reactivity which inhibit bond formation by blocking adhesive accessibility to reactive wood sites. This is accomplished by restricting wetting, altering the wood surface pH or forming complexes with adhesive or wood bonding sites.

Wood-surface history is an additional chemical feature affecting bonding potential. A smooth, freshly-cut wood surface dried at moderate temperatures offers the optimum conditions for bonding. As aging occurs, the surface is chemically modified at a rate depending on temperature and atmospheric conditions. Oxidation or contamination of the wood surface by nonpolar, mobile extractives and airborne pollutants are the primary phenomena occurring during this time. Oxidation reduces the number of active hydroxyl groups available for bonding. Under high-temperature drying conditions the chance of wood-surface oxidative degradation becomes highly probable. Sometimes chemical and thermal treatments can be utilized to change wood-surface chemistry and enhance bonding potential. Strong acids and oxidizing agents are common activators. They create new functional groups through depolymerization and ring cleavage reactions of lignin and carbohydrate molecules, or generate free radicals which can lead to bond formation by oxidative coupling. Resulting gluelines tend, however, to be brittle and have highly variable strength because wood can be degraded by many of the chemicals used.

4. Wood Adhesives

Adhesives can be classified as either natural or synthetic. Typical natural adhesives are carbohydrate or protein based, derived from either animal or vegetable sources, or maybe an inorganic material. Synthetic adhesives are produced by the controlled polymerization of various monomeric organic molecules. These synthetics can be further divided into thermosetting (most common for wood adhesives) and thermoplastic categories. They can function to produce structural or nonstructural bonds. The present discussion deals only with major adhesives used in the wood industry. The majority of these adhesives are formed by condensation polymerization processes involving the use of formaldehyde as a cross-linking agent. Details of their chemistry have been presented elsewhere (Skeist 1977).

Phenol–formaldehyde (PF) resin is the primary wood adhesive used in the production of strong, durable exterior products. These adhesives are low to moderate molecular weight thermosetting polymers, manufactured in aqueous solutions with acid or alkaline catalysts. By varying reaction time and temperature, catalyst type and amounts and ratio of reactants, a variety of different adhesives can be prepared. Resoles are a class of PF resins produced under alkaline conditions with a molar excess of formaldehyde to phenol. These reactions produce highly branched polymeric structures while remaining soluble in strongly alkaline solutions. Sufficient formaldehyde is present to allow a highly cross-linked network to be formed during cure. Novolacs are PF resins synthesized under mild acid or neutral conditions with a molar excess of phenol to formaldehyde. These polymers have a linear structure, relatively low molecular weight and maintain thermoplastic behavior unless additional formaldehyde is added when curing is undertaken. These are referred to as two-stage resin sytems. Plywood, oriented strandboard, particleboard and fiberboard intended for exterior service are manufactured using aqueous resoles tailored to give particular application, flow and cure properties for a specific product. A typical aqueous resole contains a phenol to formaldehyde to sodium hydroxide molar ratio of 1:(1.8–2.2):(0.1–0.8), with a resin-solids range from 35–50% and averaging 5–40 phenolic units linked together. These alkaline resoles require cure temperatures greater than 100 °C to attain adequate cross-linking. Fast cure at ambient temperatures can be achieved by addition of strong acids. Equipment corrosion and long-term acid degradation of the wood interface are, however, deterrents to this procedure.

For waferboard applications, the need for good adhesive distribution on large wafers and the limitations on moisture level in the glueline during high-temperature pressing require the use of powder PF resins. These are produced by spray drying or other solidification techniques.

Significant increases in phenolic resin reactivity are possible by copolymerizing with resorcinol to produce a phenol–resorcinol–formaldehyde (PRF) system. These are often used in specialized applications (i.e., glued laminated timbers, fingerjointed structures) where durable bonds and ambient cure conditions are needed. To ensure adequate pot life, the PRF is synthesized as a novolac system. Additional formaldehyde needed for cure is mixed in just prior to adhesive application. The high cost of resorcinol, relative to phenol, restricts most PRF applications to higher-value and large structural products.

Aqueous urea–formaldehyde polymers are low-cost, light-color, fast-curing adhesives with poor weather resistance. Their use is limited to products such as decorative plywood panels, particleboard subfloors or fiberboard for furniture or interior applications. Resins are prepared in molar ratios of formaldehyde to urea of (1–2):1 with the higher ratios usually utilized in solid wood or veneer bonding where higher strengths are required. Cure takes place at moderate acid conditions either by addition of acid salts or by use of the inherent acidity of wood species such as oak or cedar. Heat accelerates cure speed; polymer cross-linking, however, can be achieved at ambient temperatures. Although a thermosetting polymer, UF is sensitive to heat and moisture with formaldehyde slowly released as decomposition occurs. Requirements exist restricting formaldehyde emissions in wood products to low concentrations in air. To help control these emissions most particleboard is manufactured with formaldehyde to urea molar ratios less than 1.2:1. Alternatively, fortification of UF to improve its durability and strength is possible by copolymerizing with melamine. This modification adds increased chemical costs while requiring higher cure temperatures.

Polyvinyl acetates (PVAs) are thermoplastic, aqueous emulsions used primarily for furniture assembly and other non-structural applications. Bond strength develops from water loss into the wood from the adhesive, with a resultant increase in polymer concentration in the glueline. While providing excellent dry adhesion strength and good gap-filling properties, the tendency of PVAs to cold-flow or creep when subjected to sustained loads is a primary disadvantage. Modification of PVA with thermosetting phenolic resin results in cross-linkable PVAs with improved heat and moisture resistance but only moderate alterations in creep behavior.

Other adhesive systems such as epoxy, polyurethanes and polymeric isocyanates have limited applications to wood bonding at the present time. While these adhesives provide improved bond strength with no formaldehyde emissions, they have disadvantages of greater cost and incompatibility with water, which requires the use of organic solvents for applicator cleanup. Application of polymeric isocyanate in particleboard, waferboard and oriented strandboard have recently gained more favor due to their faster cure times and greater tolerance to moisture compared to conventional thermosets. Demand for durable, fireproof wood panels also has resulted in the application of inorganic cement adhesives in particleboard, although cure sensitivity to certain wood species and substantial increase in panel density limits their usage.

5. *Adhesive Factors Influencing Bond Formation*

Adhesive physical and chemical properties play as important a role as wood properties in determining bond performance. Strength, durability, working properties and cost are primary considerations governing adhesive selection. Interrelated polymer parameters which influence the overall performance of an adhesive include reactive functionality, molecular-weight distribution, solubility, rheology and cure rate.

Number and reactivity of adhesive functional groups determine how many active sites are available for further polymerization. These affect directly the reaction speed and degree of cure possible during bond formation. The functionality of the adhesive components must be greater than 2.0 to achieve a cross-linkable system.

Molecular weight and molecular-weight distribution is a polymer property affecting viscosity, flow and cure speed of the adhesive. Provided sufficient functional-group reactivity remains, higher-molecular-weight components can more rapidly attain cure but exhibit significantly greater solution viscosities and poorer flow properties than low-molecular-weight components. Optimum adhesive molecular-weight distribution, however, depends upon the wood composite. Solid wood products such as glued laminated lumber and plywood utilize a major portion of higher-molecular-weight components (i.e., $\bar{M}_N$ of 1500–3500) to avoid overpenetration of polymer into wood. Adhesives for waferboard and particleboard have their molecular weight distribution shifted to lower values (i.e. $\bar{M}_N$ of 500–1200). This allows efficient and homogeneous resin application while improving adhesive mobility to facilitate better resin transfer between wood particles during blending operations.

Polymer solubility or dispersibility in solution is an important factor affecting uniform application and cleanup of the adhesive. Improved solubility allows more concentrated adhesive solutions to be used which reduces the amount of solvent to be driven off during cure. Adhesive rheology and cure development, while related to polymer molecular weight and reactive groups present, are highly responsive to heat, pressure and wood moisture content. Various segments of the polymer molecule achieve mobility as plasticization occurs with moisture, or temperature increases to the glass transition (T_g) or softening temperature. Cure results in an increase of T_g. Often compromises are needed to ensure that heat and pressure are applied in a controlled sequence such that the adhesive adequately wets the wood surface and cures before resin overpenetration occurs.

6. *Bond Requirements for Composites*

Performance criteria for wood composites relate directly to product end use. This requires consideration of engineering strength needs, safety and short- and long-term response of the material to the service environment. Structural, exterior-grade products have the most demanding bond-quality requirements, since glueline failure could be catastrophic to these structures. For these situations glueline strength, durability and reliability must be assured, usually by extensive bond-quality testing programs. While conventional bond strength and short-term durability evaluation methodology is well established, there is at present only a limited understanding of the effect of long-term aging on bond integrity. Insight on how product dimensional change and sustained loading affect time-dependent performance properties of wood composites is developed by cyclic testing through simulated environments using variables of heat, moisture and loading stress. The nonlinear deterioration behavior of many adhesive bonds, together with our lack of knowledge of the aging process, are serious obstacles to assessing long-term performance of new wood adhesives.

For conditions where bond performance is less crucial, cost may be the overriding factor—not only in terms of adhesive price but also in terms of the impact adhesive costs have on overall production expenses. With more automated wood-composite manufacturing systems, compatibility of the adhesive with the assembly system often is the major consideration.

7. Future Directions

Continuing reductions in available harvest volumes and lower wood quality will lead to increased reliance on wood composites and the adhesives used to manufacture them. For commodity products, the importance of manufacturing efficiency will place greater emphasis on adhesives capable of tolerating variable wood properties and moisture contents while having rapid cure speed and reasonable cost. Adhesives will gain prominence in higher-valued wood composites by allowing products of varying shapes, improved engineering strength and greater aesthetic value to be more easily manufactured. Expanded use of continuous pressing techniques and heating methods such as steam pressing or microwave use will create added challenges for adhesive formulation development and wood-handling concepts. Improved knowledge of chemical factors influencing the reactivity of the wood-bonding surface would help enhance functional group reactivity, perhaps leading to reductions in amounts of adhesive required or possibly improved binderless wood composites. Advanced, high-strength composites are likely increasingly to utilize wood fiber or cellulose fiber as a directional, reinforcing matrix surrounded by adhesive polymers which may comprise up to 50% of the composite weight.

Future adhesive improvements will be concentrated in areas of durable, faster curing systems, capable of enduring moderate to long assembly times. Copolymer systems will increase in prominence with emphasis being directed toward multicomponent, cold-setting adhesives possibly applicable to on-site assembly. Isocyanates, because of their extensive reactivity and ability to bond at higher moisture contents are likely to be a component of these newer wood-bonding systems. Reactive hot melts, capable of rapidly curing to thermosetting polymers, is another development which will have significant impact on future bonding systems.

Bibliography

Collet B M 1972 A review of surface and interfacial adhesion in wood science and related fields. *Wood Sci. Technol.* 6: 1–42

Koch G S, Klareich F, Exstrum B 1987 *Adhesives for the Composite Wood Panel Industry.* Noyes Data Corporation, New Jersey

Kollman F F 1975 *Principles of Wood Science and Technology*, Vol. 2, *Wood Based Materials.* Springer, New York

Marian J E 1967 Wood, reconstituted wood and glue laminated structures. In: Houwink R, Salomon G (eds.) 1967 *Adhesion and Adhesives*, Vol. 2, *Applications.* Elsevier, New York

Oliver J F 1981 *Adhesion in Cellulosic and Wood-based Composites.* Plenum, New York

Pizzi A 1983 *Wood Adhesives Chemistry and Technology.* Dekker, New York

Skeist I 1977 *Handbook of Adhesives*, 2nd edn. Van Nostrand Reinhold, New York

Subramanian R V 1984 Chemistry of adhesion. In: Rowell R (ed.) 1984 *The Chemistry of Solid Wood*, Advances in Chemistry Series 207. American Chemical Society, Washington, DC

Wellons J D 1983 The adherends and their preparation for bonding. In: Blomquist R F, Christiansen A W, Gillespie R H, Myers G E (eds.) 1981 *Adhesive Bonding of Wood and Other Structural Materials*, Vol. 3. Clark C Heritage Series on Wood. Pennsylvania State University, University Park, Pennsylvania

P. R. Steiner
[Forintek Canada Corporation, Vancouver, British Columbia, Canada]

Adhesives in Building

The process of adhesion, by which a substance makes intimate contact with two surfaces and undergoes physical and chemical changes to bond them together, is complex and only partly understood. Nevertheless, within the last 40 years, age-old natural adhesives have been improved and a multitude of new adhesives has been chemically synthesized and now dominates the marketplace.

The construction industry is probably the largest user of adhesives. Adhesives increase the strength and stiffness of building components, effectively transfer and uniformly distribute stresses, reduce or eliminate mechanical fasteners and combine dissimilar materials, thin films, fibers and particles that could not be joined otherwise. Applications vary widely in structural capability, from the huge glued–laminated beams that support roof loads to non-structural applications where adhesives only support wall coverings, accessories and trim.

Considerably greater amounts of adhesives are used in the manufacture of building materials such as structural softwood plywood, particleboard, hardboard, decorative hardwood plywood, finger-jointed lumber, siding, glassfiber insulation, doors, sandwich panels and other factory-laminated products. For the most part, they are bonded with phenol–formaldehyde and urea–formaldehyde resin adhesives, although small amounts of melamine–urea–formaldehyde and melamine–formaldehyde are used. Millions of kilograms of these adhesives are used annually in the building products industry.

Although used in lesser amounts, adhesives employed to construct industrial and residential buildings constitute a very significant portion of the adhesives market. This group of adhesives is used to assemble building materials either at the construction site (where bonding conditions are sometimes difficult to control), or in assembly shops. Generally, construction adhesives differ in chemical composition and methods of bonding from adhesives used to manufacture building materials. It is this relatively large and diverse group of construction adhesives which is the primary focus of this article.

1. Adhesion

Adhesion is the state in which two surfaces are held together by interfacial forces which may be valence forces or interlocking action, or both. For adhesion to take place, the adhesive must be liquid and make intimate contact with the substrate. The more intimate the interfacial contact, where atoms and molecules approach electrostatic interaction, the more tenacious the bond. For strong bond formation, the surface must be free of contaminants and the adhesive must displace air and water molecules from the surface.

Wetting of the surface by the adhesive readily occurs when the contact angle between the edge of a drop adhesive and the solid surface is low. The contact angle approaches zero when the following conditions exist: the substrate possesses a relatively high surface attraction energy; the adhesive has a real affinity for the substrate; and the surface tension of the adhesive is low. If a drop of adhesive spreads to a thin film approaching zero contact angle, the adhesive will spread well and make intimate contact with the surface.

For maximum strength development on wood surfaces (the most common building material), an adhesive must penetrate to the most firmly attached fibers and interlock with the wood (see *Adhesives for Wood*). Low contact angle and viscosity are important in getting good capillary flow into crevices and pores. For the most part, the bonding surfaces of building materials are highly contaminated and replete with cracks and crevices. Such surface conditions cause gas pockets and blockages, which prevent complete wetting and introduce points for stress concentration. Furthermore, stresses develop in joints from dimensional changes brought about by the solvent, moisture, or gain and loss of heat during bond formation, and from external environmental service conditions. Even if building adhesives were capable of high levels of adhesion by valence forces, it seems likely that the many sources for interfacial blockages would nullify the potential for that kind of adhesion. In bonding building materials, perhaps the greatest contribution to joint strength can be made by preparing smooth and clean surfaces and using an adhesive with a high affinity for the particular substrate.

The process of adhesion is essentially completed in the transition from a liquid to a solid film. This transition can be a physical change, as in thermoplastic adhesives, or it can be a chemical change, as in thermosetting adhesives. In thermoplastic adhesives, the physical change to a solid film occurs by cooling of molten liquid on a cooler surface, or by loss of solvent through evaporation and diffusion. With thermosetting adhesives, the solid state is achieved by polymerization. Resin intermediates in partly reacted states may react with other intermediates or complete their reaction when heated, to form highly cross-linked structures.

2. Advantages of Adhesives

Nails, screws and bolts are still used more than adhesives to assemble building materials. However, adhesives offer advantages that fasteners cannot provide. For example, adhesives distribute stresses over the entire bonded area, thereby avoiding concentration of stresses at the fasteners. When all members are bonded all share the load. A laminated wood beam thus has much greater strength and stiffness than the combined laminates even when they are fastened together. Not only are these assemblies stronger and stiffer, but the designer can meet design requirements at a lower cost by using fewer pieces of lumber in smaller sizes and lower grades.

Adhesives reduce the number of fasteners that would otherwise be required. For example, 50–70% of the nails can be eliminated from stressed-skin panels by using adhesive. This reduces the cost of nails and can also drastically reduce nailing time and equipment needs.

Adhesives are advantageous when the appearance of fasteners is objectionable. Bonding of high-pressure laminations to kitchen counter tops is an example. Nail popping will mar the appearance of smooth surfaces such as gypsum wallboard or resilient tile floors. Adhesives eliminate most, if not all, fasteners.

Adhesives make it possible to eliminate defects in lumber, thereby increasing strength, grade and value. Short pieces of lumber can be finger-jointed into longer pieces of any length. Adhesives also permit randomizing of defects within a component, thereby avoiding a concentration of defects at one point, which could have disastrous effects on strength and stiffness.

Flexible adhesives have the ability to yield and distribute stresses during short-term dynamic loading. When used to bond modular or mobile homes for transporting, these adhesives can prevent damage to weaker materials such as gypsum wall-board and door and window frames.

3. Structural Performance

Adhesives in building construction have been categorized according to levels of expected structural performance. Five categories are presented here in descending order of structural integrity, with accompanying examples of applications and adhesives.

Prime structural is the most critical category because the adhesive is expected to contribute strength and stiffness for the life of the structure. The application is critical in the sense that failure could result in loss of life or serious damage to the structure. Examples of prime structural applications are glued-laminated beams, plywood and lumber I-beams and stressed-skin panels. The adhesives used are rigid types such as resorcinol, phenol–resorcinol and casein.

In semistructural applications, the adhesive is expected to impart stiffness to an assembly for the life of the structure. Failure of the adhesive would not cause any serious loss of structural integrity, because the load would be carried primarily by mechanical fasteners. An example of this category is the nailed–glued plywood floor system, where elastomeric construction adhesives are typically used.

Temporary structural adhesives contribute to the strength and stiffness of the structure, but the contribution is expected to be shorter than the life of the structure. If the adhesive should fail, damage might be significant but is repairable. Mobile and modular housing are examples where adhesives are used to resist severe dynamic racking, twisting and bending stresses that develop when transporting these large units over the highway. Poly(vinyl acetate) resin emulsions are used for this purpose in mobile homes, and elastomerics in modular units.

The secondary structural category comprises adhesives contributing strength and stiffness to an assembly but whose failure under loading would cause no significant damage to the structure. The failure could be easily recognized and repaired. An example is bonding of decorative panelling to gypsum board or furring strips with elastomeric adhesives.

The least critical category is nonstructural. Here, adhesives are expected to support the deadweight of lightweight substrates such as accessories and trim. If failure should occur, damage would be slight and repair would be inexpensive. Examples are bonding of counter tops, floor coverings, ceiling and wall tiles and moldings. The adhesives are usually contact-bond and elastomerics.

4. Types of Adhesives

4.1 Phenolic Resins

Resorcinol–formaldehyde and phenol–resorcinol–formaldehyde resin adhesives are used for critical, prime structural applications because they are exceptionally strong and durable and do not creep under high and long-term stresses. They are well suited for the slow laminating process because they tolerate long assembly times. They set at room temperature, thereby eliminating hot pressing of large, unwieldy beams and arches.

Resorcinol–formaldehyde resin is the product of a condensation reaction between resorcinol and formaldehyde. The reaction is carefully controlled to maintain a deficiency of formaldehyde to resorcinol, typically a 0.60–0.65:1 molar ratio. The reaction is completed at the time of use by adding formaldehyde. Resorcinol is very expensive, so ~15% of phenolic resin may be substituted and copolymerized to make a lower cost phenol–resorcinol resin. This adhesive still sets at room temperature with no sacrifice of strength and durability.

Resorcinolic adhesives are marketed as two component systems, that is, resin and hardener are separated and mixed just before use. In laminating plants, they are applied by powered roller–spreaders or extruders. In the field, a brush or hand roller is adequate. Open and closed assembly time varies according to reactivity, air and wood temperature, humidity, moisture content of wood and spread rate. Even though lumber is planed to clean and smooth both bonding surfaces, a pressure of 0.7–1.4 MPa is needed to ensure intimate contact between adhesive and substrate. The time for which a laminate remains under pressure may vary from 4 to 16 h.

Phenol–resorcinol adhesives with gap-filling capability are not commercially available, although there is a need for them in the construction industry. Conventional phenol–resorcinols are too thin to bridge gaps between rough, uneven plywood and lumber surfaces. Therefore, large presses, clamps, or extensive nailing must be used to apply enough pressure to achieve intimate contact between adhesive and substrate. Recent studies show that certain thickening agents can be incorporated into a phenol–resorcinol resin to make an extrudable, nonsagging adhesive that will bridge gaps of 1.5 mm or more. Instead of 100–150 mm, nails can be spaced 300–400 mm apart. Nails are needed only to hold the assembly together until the adhesive sets. The phenol–resorcinol has exceptional strength and durability. With added fibrous ingredients, its strength, rigidity and durability, even in thick bonds, exceeds that of plywood substrates.

A few years ago, desire to build houses rapidly spurred development of a fast-curing phenolic resin with gap-filling capability. Such a product was developed and used to assemble a few houses on a pilot scale but was never commercialized. The adhesive was made by reacting phenol–formaldehyde condensation products of low molecular weight with *m*-amino-phenol (and other di- and tri-substituted benzenes) to form a relatively stable phenol–formaldehyde *m*-aminophenol novolak-type polymer. This polymer was reacted with formaldehyde at the time of use. The cure rate could be varied between 20 s and 5 min. When the two components were mixed, the adhesive foamed and expanded to fill gaps between substrates. Sophisticated metering, mixing and application equipment was required, because the two components were highly reactive.

4.2 Polyurethanes

Polyurethane adhesives have also been developed for fast bonding of building components. They can be formulated for thin bondlines in laminating or for bridging gaps in construction joints. Two-component polyurethane systems are prepared from a pre-polymer of a polyol and a multifunctional polyisocyanate. When the two components are mixed, the cross-linking reaction can be quite rapid. By altering the molecular structure, that is, by providing different degrees of chain flexibility and controlling the number of sites for subsequent urethane formation, a wide variety of mechanical properties can be obtained, ranging from flexible to rigid. Isocyanates are very reactive with hydroxyl groups and develop excellent bonds on a variety of surfaces. These systems also require sophisticated metering, mixing and extruding equipment because of their reactivity.

4.3 Epoxy Resins

Epoxy resin adhesives are used very little in building construction. However, their strength capability on wood, glass, metal and masonry makes them highly versatile. They are capable of prime structural applications, but are actually used only for repairs, patching of voids in materials and accessory and trim attachment. Epoxies have excellent gap-filling capability. They set at room temperature and do not shrink on setting.

Epoxy resin is made by alkaline condensation of epichlorohydrin and bisphenol A. At the time of use, the resin is mixed with a hardener, usually a polyamine, and the reaction proceeds exothermically. Depending on formulation, the pot lifetime may be a few minutes to a few hours. The substrates must be held together until the adhesive sets, but close fitting of substrates is not essential, because of the gap-filling capability.

4.4 Casein

Casein was the most important moisture-resistant structural laminating adhesive before the introduction of synthetic resins such as resorcinol. A proteinaceous substance manufactured from skim milk, the casein molecule comprises a great number of different amino acids in its natural state. Its high molecular weight accounts for its colloidal properties and value as an adhesive. Water-resistant caseins are formulated as casein, calcium hydroxide and sodium salt. The proportions of these ingredients control the viscosity, working lifetime and water resistance. The casein–lime–sodium salt dry-mix is readily dispersed in water at the time of use and sets at temperatures above 4.5 °C with a clamp time of 4–8 h. It is easy to use in factories and at job sites because of its long pot lifetime, short assembly time and fair gap-filling capability. However, even the most water-resistant caseins are not suitable for exterior use. Nowadays casein is principally used to laminate structural beams and smaller components for interior service.

4.5 Vinyl Acetate Polymers (PVA)

Poly(vinyl acetate) emulsion, or white glues, are used in semistructural and temporary structural applications such as floors, walls and roofs of mobile homes. Although they have high strength, they are thermoplastic and yield under prolonged stress, high moisture and heat. Recently developed cross-linking poly(vinyl acetates) partly overcome these structural disadvantages.

Vinyl acetate polymers are formed by free radical addition polymerization of the monomer. They may also be copolymerized with other monomer species. Polymer chains can be tied together in two- or three-dimensional networks by cross-linking agents that lead to a higher degree of resistance to heat and moisture.

Most PVA adhesives that set at room temperature are sold as ready-to-use, one-component systems. Catalyzed systems consist of resin and catalyst which must be mixed before use. These also set at room temperature, or they may be set much faster with radiofrequency heating. PVA may be applied by brush or roller, but in mobile home construction it is applied in a bead from a squeeze bottle. PVA has some gap-filling capability, so reasonably strong joints can be made with only pressure from nailing. A bond strong enough for handling can be developed in <2 h on wood surfaces.

4.6 Elastomers

Elastomeric construction adhesives are a relatively new and important class of adhesives for job site and factory construction. They find use in practically all construction applications except the prime structural category. They are used extensively to bond plywood floors to joists and panelling to studs. Bonded floors are stiffer and stronger, and appeal to homeowners because they feel firm and do not squeak underfoot.

Elastomeric construction adhesives are not true elastomers; however, they are compounded with elastomers such as neoprene, styrene–butadiene, polyurethane, reclaimed and natural rubbers. Some compounds may be blends or copolymers. The elastomer usually constitutes 30–50% of the total solid ingredients. A neoprene construction adhesive may consist of neoprene, antioxidant, magnesium oxide, zinc oxide, phenolic resin and organic solvents. High loading with various fillers and extenders imparts mastic consistency for gap-filling capability. Most adhesives in this class set by loss of organic solvent or water. An exception is the polyurethane system, which sets chemically on reaction with moisture in the air. The setting time is generally 30 days for most organic solvent systems, although the polyurethane sets within 3 days.

Elastomerics are marketed as ready-to-use, one-part systems, packaged in caulking cartridges for hand extrusion or in bulk-size vessels for automatic extrusion in factories. An extruded bead adheres to vertical surfaces without sagging and bridges gaps between rough and poorly fitting lumber and plywood with only a minimum of pressure from nailing. These adhesives bond to both wet and frozen lumber surfaces although the temperaure must rise to 4.5 °C soon after bonding to allow the solvent to volatilize. They develop strong bonds, but usually not strong enough to make wood fail. Elastomerics are unusually tolerant of field conditions and have proved to be an excellent solution to many on-site bonding problems.

4.7 Contact-Bond Adhesives

Contact-bond adhesives are used in secondary structural and nonstructural applications, particularly where sheet materials are laminated to rigid substrates and close-fitting joints can be constructed. Examples are laminated high-density counter tops in kitchens and bathrooms and double-laminated gypsum wallboard.

Most contact-bond adhesives are made from neoprene, although styrene–butadiene, nitrile and natural rubbers are used. In making a typical neoprene contact, neoprene is mill-mixed with magnesium oxide, antioxidant and zinc oxide and then churn-mixed with *t*-butyl phenolic resin and an appropriate mixture of organic solvents to give 15–30% solids. The use of organic solvents in contact adhesives is diminishing because of the fire hazard and toxic fumes. Though an interest is developing in water-based systems, water-based contacts are slower drying and lack the rapid bond strength development and high ultimate strength that characterize solvent-based systems.

Contact adhesives are low-viscosity liquids that are spread in one or two coats on both substrates. They are allowed to air-dry to a tack-free state. The substrates are then pressed together, typically by roller pressure. Once in contact, the surfaces are stuck and cannot be repositioned. Setting takes place by loss of solvent, which is rapid on porous substrates but may take several days on nonporous substrates.

4.8 Mastics

A relatively large group of adhesives is used to bond wall and floor coverings. Generally, these are mastics able to bridge gaps between the flat covering materials and rough floor or wall surfaces. Adhesives for ceramic, metal and plastic wall tiles are rubber-based mastics in organic solvent or water, oil-based and resin-based mastics, and portland cement with resin binder. Mastics with organic solvent are not used with plastic tile. The adhesives may be applied to almost any sound, dry and clean surface by trowelling or buttering techniques. The tiles are simply pressed into the mastic while the adhesive is wet and within a few hours the bond develops by loss of solvent.

Adhesives for rubber, vinyl and cork tile flooring are rubber-based mastics with a water vehicle and resin-based mastics with alcohol. These adhesives are suitable for linoleum and vinyl sheet flooring. The flooring materials must be applied while the adhesive is still wet. Rigid asphalt and vinyl–asbestos tiles are bonded with asphalt emulsion, asphalt cutbacks and rubber-based systems with a water vehicle. The asphalt emulsion and asphalt cutback must be dry before the tiles are placed; however, the tiles must be placed before the rubber–water system dries.

4.9 Animal Products

For the most part, animal adhesives have been replaced by synthetics. A few craftsmen still use them to bond kitchen cabinets and built-ins. Animal adhesives are based on collagen—the fibrous protein found in bones, hides and connective tissue of animals. The adhesive is sold as a solid that must be dispersed in water and heated before use. On cooling, it sets to a hard film that makes a strong bond with some gap-filling capability. However, the bond has limited durability and will weaken on exposure to moisture and heat.

See also: Adhesives in Civil Engineering; Structural Adhesives

Bibliography

Bikerman J J 1961 *The Science of Adhesive Joints*. Academic Press, New York, pp. 1–258

Blomquist R F, Christiansen A W, Gillespie R H. Myers G E (eds.) 1983 *Adhesive Bonding of Wood and Other Structural Materials*, Vol. III. Educational Modules for Materials Science and Engineering (EMMSE) Project, Materials Research Laboratory, Pennsylvania State University, University Park, Pennsylvania, pp. 1–436

Gillespie R H, Countryman D, Blomquist R F 1978 *Adhesives in Building Construction*, Agricultural Handbook 516. US Department of Agriculture, Forest Service, Washington, DC, pp. 1–160

Pizzi A (ed.) 1983 *Wood Adhesives Chemistry and Technology*. Marcel Dekker, New York, pp. 1–364

Selbo M L 1975 *Adhesive Bonding to Wood*, Technical Bulletin 1512. US Department of Agriculture, Forest Service, Washington, DC, pp. 1–122

Skeist I (ed.) 1977 *Handbook of Adhesives*. Van Nostrand Reinhold, New York, pp. 1–921

US Department of Agriculture Forest Service 1974 *Wood Handbook: Wood as an Engineering Material*, Agricultural Handbook 72. US Department of Agriculture, Forest Service, Washington, DC, Chap. 9, pp. 1–14

Wake W C 1982 *Adhesion and the Formulation of Adhesives*. Applied Science, London, pp. 1–332

C. B. Vick

[US Department of Agriculture Forest Service, Athens, Georgia, USA]

Adhesives in Civil Engineering

Civil engineering concerns that branch of technology associated with large-scale structures and works:

dams, roads, railways, harbor and coastal installations, large buildings and structures, earthworks and foundations. Such works have been carried out by man for centuries and we have much evidence of the success and ingenuity of our forefathers. Initially, they used stone or primitive bricks (sometimes with mortar for bonding and sealing), wood and earth. More recently, Portland-cement concrete (usually called simply concrete) and structural steel have become the principal materials used, although advantage is taken of natural rock and earth infill where possible for cost reasons alone.

Since concrete is less strong in tension than in compression (by a factor of about 10) it is normally used so that the tensile loading is small, or is reinforced with steel bar or mesh (see Fig. 1(a)). A bond or adhesion between the steel and concrete is essential for the two materials to act as a composite. While shrinkage of the concrete on setting and hardening contributes some grip, the load transfer is enhanced by using twisted or specially-deformed bars. In steel beam–concrete slab construction, composite action is achieved by welding or otherwise attaching steel studs to the steel section in the form of shear connectors as shown in Fig. 1(b). However, in neither of these applications is there any attempt to use *structural adhesives* for the load transfer mechanism. It is the intention of this article to show what is available in structural adhesives for civil engineering applications, to discuss the problems which are likely to be faced, and to illustrate these by some actual applications.

It should be pointed out that there are two topics which will not be covered here. First, polymeric materials, together with pitches and tars, have been used as sealants, but these are not directly load-bearing

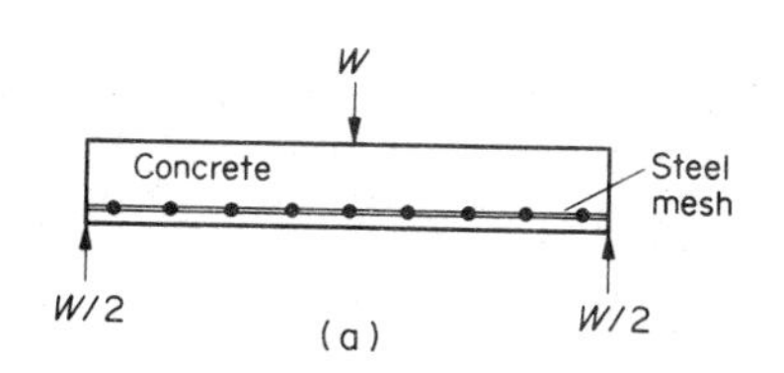

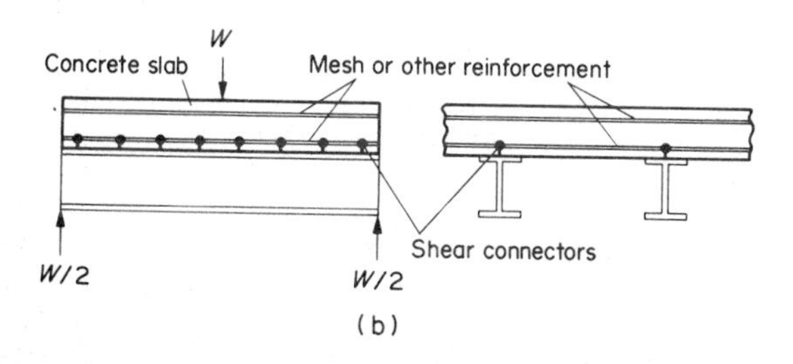

Figure 1
Load transfer between steel and concrete: (a) bar or mesh reinforcement for a simple slab; and (b) shear stud reinforcement

applications. Second, there has recently been considerable interest in polymer concretes. In these, the aggregates are bound together with synthetic resins or polymer modified cements. These polymer concretes can show a two-fold increase in strength (compressive, tensile or flexural) over ordinary concrete. The polymer used is often a polyester resin. For further information, the reader is referred to the excellent article by Hewlett and Shaw (1977).

1. Adhesive Materials Available

The chemistry of structural adhesives is complex and there are many variations on apparently similar formulations; the engineer who intends to use a structural adhesive is therefore well advised to consult the adhesives manufacturer about the chemical compatibility of the different types. Fortunately, these many variants fall into two basic forms: polyesters and epoxies. Generally speaking, epoxies are about three or four times as expensive as polyesters, but provide significantly improved performance. Because of the problems of heat provision and dissipation and shrinkage stresses arising during curing and cooling of hot-cured resins, cold-curing adhesives are often preferred in civil engineering applications. In special cases, heating strips might be laid in the adhesive, but the expense involved would rarely be justified.

Both epoxies and polyesters may be formulated so that they can be used for laminating glass or carbon fiber composites. In this form, the resins need to flow so as to wet the fibers completely. Thus, these resins would have very poor gap-filling properties for applications where two ill-fitting surfaces have to be bonded and where gravity would cause the adhesive to drain out. Fortunately, it is possible to use various thixotropic agents to overcome this problem. Also, fillers such as china clay, talc and powdered metals can be used to provide the necessary gap-filling properties. In addition, these fillers can reduce the cost of the product, since they are often cheaper than the resins. However, fillers, although providing some stiffening action, sometimes weaken the cured resin.

Resins, particularly epoxies, can be flexibilized by various polymer additives. This generally reduces the strength and modulus slightly and considerably increases the strain to failure, but reduces the creep resistance significantly. Of these effects, the gain in strain to failure usually outweighs the slight losses in the other mechanical properties.

Typical shear stress–strain curves of structural adhesives are shown in Fig. 2. These are meant to illustrate the range of available properties rather than to identify proprietary materials. It should be pointed out that the strength of adhesives is usually markedly poorer in tension than in compression or shear. (In this respect, they bear some similarity with concrete.) Typically, Young's modulus is in the range 1–7 GPa

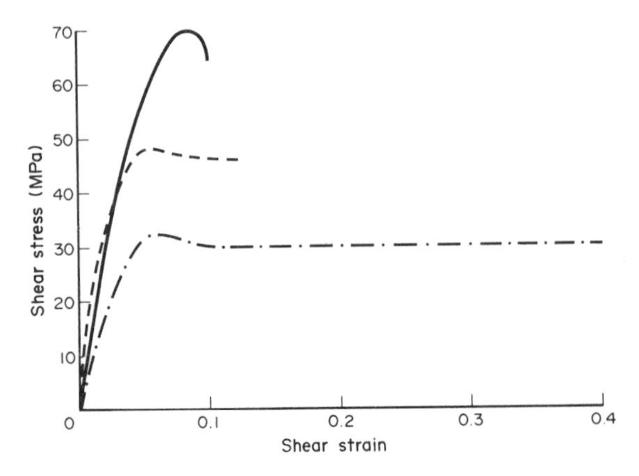

Figure 2
Shear stress–strain curves for some epoxy adhesives: ——, unmodified; – – –, filled, toughened; —·— · rubber-modified (failure strain, 0.6)

and Poisson's ratio is ~0.4. The shear modulus is typically 0.3–2 GPa.

2. *The Mechanics of Adhesive Joints*

A considerable amount of mathematical analysis has been carried out on the type of overlap joints commonly used in aircraft and similar construction where thin sheets are bonded. Figure 3 shows that the shear stress is nonuniform, peaking at the ends of the joint, and that there are also considerable transverse tensile stresses at the ends. These transverse tensions are the usual cause of failure in this type of joint. Even when monolithic adherends are used, these stress concentrations still occur, particularly in the transverse direction. Adams and Peppiatt (1974) showed that at the ends of the joint a tensile stress ~3.6 times the average applied shear stress exists, and that this stress concentration is due to the high stress gradients at the joint ends.

It is generally recommended that load-bearing adhesives be used in shear rather than tension (they are rarely likely to fail in compression), owing to their low tensile strength. The same applies to concrete. But consider the case of a steel plate bonded to concrete and loaded in shear, as shown in Fig. 4(a); this is a common structural application of adhesives. Fracture will be initiated at the corner A and the crack will extend to the free surface of the adhesive at B and to the adhesive–concrete interface at C. There will now exist a large transverse tensile stress at C. If the lower adherend were also steel (or similar material with a high tensile strength), the crack would propagate through the adhesive near the steel–adhesive interface, as shown in Fig. 4(b), to D. However, with a concrete adherend, the concrete itself may fail in tension, the crack now propagating to E (Fig. 4(c)).

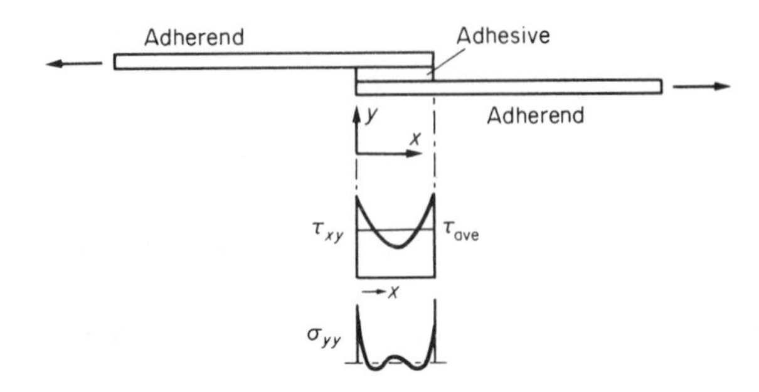

Figure 3
Stress distributions in a simple overlap joint

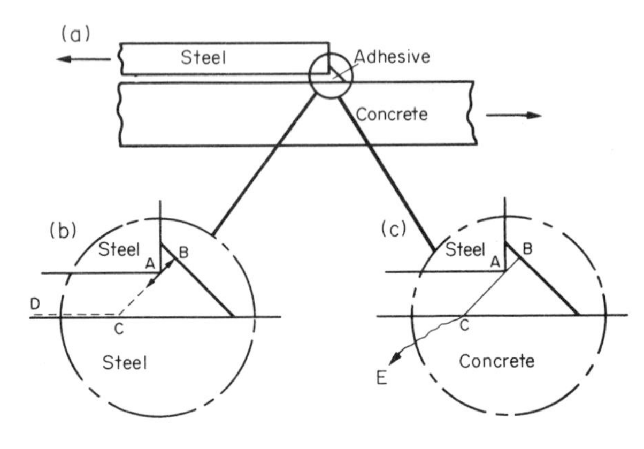

Figure 4
Failure in steel plate bonded to concrete, loaded in shear

Conversely, if the upper adherend were concrete, as shown in Fig. 5, it is likely that this too would crack under the tensile loading, leading to a series of microcracks before the adhesive fails and the cracks again run into the lower adherend.

One method of overcoming this problem is to use a flexibilized adhesive which yields considerably before it fractures. In this way, the stresses are redistributed, leading to a reduction in the stress concentrations in the concrete. Unfortunately, such flexibilized adhesives tend to creep readily and may not find favor in long-term applications, especially in hot locations.

3. *Technical Problems*

To the technical difficulties which face the mechanical or aeronautical engineer in using structural adhesives must be added a number of others which are specific to civil engineering. Principally, these are related to the large scale of the works. An excellent document which details the various precautions and tests and covers many aspects of using adhesives in civil engineering site work has been issued by the Fédération

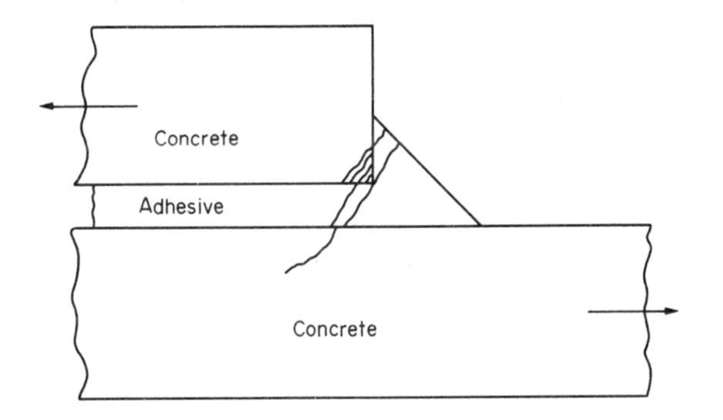

Figure 5
Failure in concrete bonded to concrete, loaded in shear

Internationale de la Précontrainte (1982). It is essential for engineers intending to use adhesives on site.

3.1 Surface Preparation

Good bonding requires good surface preparation. Adhesives will not normally give a satisfactory bond strength if the adherend surface is dirty, oily, lichen-covered or otherwise contaminated. Grit blasting or bush hammering of both steel and concrete is usually satisfactory for removal of surface contamination and laitance. However, long-term protection of the steel is recommended to avoid corrosion at the adhesive–adherend interface. Any protective layer (such as is formed by galvanizing) must have a bond to the steel and adhesive at least as good as that of the untreated steel to the adhesive, otherwise it becomes a weak link in the chain. Also, it is important that the surfaces are not wet. If there is any moisture present, this should be dabbed away with a clean cloth.

In the aircraft industry, a primer is often applied to the surfaces prior to bonding. This acts as an intermediate compound, providing good adherence to both the adhesive and the substrate and protecting the latter from further contamination. This practice does not appear to be used in civil engineering, but it might be advantageous in allowing the use of different adhesive systems, such as one which does not bond well to concrete but otherwise has excellent mechanical and environmental properties. This could be a profitable idea for both adhesive manufacturers and civil engineers to follow.

3.2 Environmental Effects

The adhesive must not be unduly affected by the presence of ambient humidity or likely service temperatures. Fire is an additional hazard which must be given serious consideration where adhesives carry critical structural loads. In addition, the presence of chemicals, such as those of the alkaline environment associated with concrete or due to oil and similar hydrocarbon spillages, must not cause degradation of the adhesive.

3.3 Duration of Application

The components to be bonded are large, requiring a substantial amount of adhesive which has to remain active until the whole surface has been covered. Premature cure or skinning would be disastrous as it would severely reduce the ultimate bond strength.

3.4 Mixing

It is recommended that where two-component adhesives are used these should be of two distinct colors so that good mixing can be *seen* to have taken place. Power mixing is recommended and a 350 W electric-drill on slow speed should be adequate for a 5 kg mix. High mixing speeds can lead to entrainment of air and may also cause the mix to overheat and cure prematurely.

3.5 Temperature of Application

In cold environments (5–20 °C) a fast-reacting adhesive system should be used, while in hot environments (>40 °C) a slow-reacting system is essential. It is best to consult the adhesive manufacturer to ensure that a properly cured bond is achieved.

3.6 Bonding Thickness and Gap Filling

In aerospace applications, bondlines down to 0.02 mm are possible, but in civil engineering the thinnest practical bondline is in the region of 1 mm, owing to lack of fit of the components and the difficulty of removing the excess adhesive (squeezing out to the edges). Normally, the excess adhesive squeezed out is removed by a scraper before it has gelled.

The gap is often vertical, such as when bonding precast segmental bridge sections, and it is essential that the adhesive does not run out. There are various tests for thixotropy, a property which indicates the ability of a liquid to hold to a vertical surface without running. Such a test in described in ASTM D 2730. This property is a function not only of the adhesive but also of the ambient temperature. If problems are encountered, it may be necessary to caulk the edges of the gap or otherwise prevent the adhesive from running out.

3.7 Application

The adhesive may be applied to either or both of the surfaces to be joined. The use of a trowel or stiff brush is usually all that is required, although special tools are sometimes used.

4. Examples of Use

The following examples of the use of adhesives in civil engineering are not meant to be exhaustive; indeed, this technology is finding ever wider applications in construction.

4.1 Segmental Constructions

Since the 1960s there has been increasing interest in the technique of segmental construction, using precast concrete units made under factory-controlled conditions, maybe remote from the construction site. The sections are then erected and post-tensioned on the site. To achieve a good match, each segment is match-cast against the end of its neighbor, using a release agent to aid separation. This technique has been widely used not only for bridge building but also for many other structures, such as tunnels, stadia and lighthouse foundations.

Initially, the gaps between the sections were large (>250 mm), using concrete cast in situ to make the joint. Stressing of the segments had to be delayed until the mortar had hardened sufficiently. In 1963, at Choisy-le-Roi in Paris, a bridge over the Seine was constructed using match-cast segments in which the joint was made using an epoxy resin. Match-casting allowed for very small glue lines (about 2 mm thick in this case). This was the first use of adhesives in this way in a major construction. Since then, many other bridges have been built in this fashion in Europe and the Americas. The functions of an adhesive in such constructions are:

(a) to act as a continuum to transmit shear, compressive and (in some cases) tensile stresses between the adjacent concrete sections;

(b) to develop strength (i.e., to cure) quickly so as to speed the erection of the segments;

(c) to act as a lubricant to facilitate the location of the mating surfaces; and

(d) to provide a watertight seal to protect the prestressing bars.

Two notable recent bridges in the UK in which adhesive-bonded segments have been used are the Byker viaduct, which is S-shaped in plan and some 800 m long (part of the Metro scheme at Newcastle upon Tyne), and the M180 motorway bridge at Scunthorpe. The Byker viaduct has the distinction of being the first to be built in the UK using cantilever construction and precast segments with epoxy-resin-bonded joints. The bonded construction allowed the use of woven fiberglass to provide joint thickening where needed, so as to correct for minor misalignments. The M180 bridge at Scunthorpe is interesting in that adhesives were successfully used even during severe (for the UK) wintry conditions. It was found necessary to warm the mating surfaces before applying the adhesive so as to remove all traces of frost. Also, space heaters were installed inside the hollow sections to help cure the adhesive. In an excellent technical account of the use of epoxy adhesives in the M180 bridge, Moreton (1981) recommends that heating must be used, at least to keep the concrete above 0 °C, and warns that allowance must be made for the possibility that the adhesive may not fully cure. It may be possible to insert a series of resistance wires to provide direct electrical heating of the adhesive and local concrete. These wires could either be withdrawn before cure or left in place; in the latter case, they will not significantly weaken the bond if they are non-reactive with the adhesive or the concrete.

Other notable structures in which adhesive-bonded segmental construction has been used include the roof of the Sydney Opera House and the large sugar silos in Durban, South Africa. A breakwater was formed around the Ekofisk caisson (North Sea oil installation) by bonding concrete segments with an epoxy adhesive.

4.2 Externally-Bonded Reinforcement

A technique which has found substantial application in recent years has been the strengthening and stiffening of reinforced concrete structures by externally bonded steel plates. In particular, this has been applied to bridges in many parts of the world. Several reasons have prompted the use of this technique:

(a) to correct faulty design;

(b) to correct faulty workmanship;

(c) to increase the bridge capacity to cope with increased traffic loads; and

(d) to counter deterioration of the steel reinforcement, especially that subjected to salt-spray exposure.

This technique was pioneered in France and South Africa during the period 1965–75. A particularly interesting example in France is on the Autoroute du Sud (Bresson 1972). The Japanese were quick to recognize the possibility of uprating many bridges for which the traffic loading (especially heavy commercial vehicles) had caused a serious problem. By 1975, at least 240 bridges had been strengthened in this way. More recently, several motorway bridges in the UK have been similarly strengthened.

The reinforcing plates may be applied to the underside of bridges, where the concrete is in tension, both to stiffen and to strengthen the structure. They can also be applied to the webs of concrete beams to increase the shear stiffness. The benefits of using bonded steel plates instead of additional beams or props are that the application can be carried out with a minimum loss of headroom and a minimum disruption to other traffic. However, the long-term durability of bonded joints used in this type of construction has not been established.

The Quinton interchange on the M5 motorway in England was strengthened during 1975 and the work has been monitored (Raithby 1980). Over 1000 steel plates, 250 mm wide, 6.5 mm thick and up to 3.5 m long, were bonded to the underside of the bridge. In some cases, two or three plates were bonded together to increase the thickness. Each plate was suspended from bolts located in the lower surface of the deck slab and then wedged into position so that the excess adhesive was squeezed out. To allay fears of detachment of the plates, the bolts remain, although they take virtually none of the structural loads, which are all carried in shear by the adhesive. In addition, the free ends of the plates are secured by small clamps bolted into the concrete; this is used to counteract the most likely cause of failure, local peeling at these free ends.

In 1977, two bridges on the M25 motorway at Swanley, UK, were strengthened by adding steel plates to the underside of the deck. Additional plates were bonded to the top of the deck over the supports to counteract tensile loads.

External plating has also been used to strengthen floors in buildings (again headroom advantages are obvious). An eight-story prefabricated building has been strengthened by bonding steel plates to the structure so as to withstand statistically rare, but possible, high wind loads (Hewlett and Shaw 1977, Shaw 1971).

Although bonded external reinforcement, used wisely, can be a very useful maintenance tool, there are as yet no codes of practice or standards (Mander 1981). Caution should therefore be exercised and the designer should consult those who already have experience. Also, the long-term effects should be carefully monitored.

4.3 Fixings in Masonry and Concrete

Resin-bonded anchors can be used in masonry, rock or concrete. These consist of a glass tube which contains a cold-setting polyester or epoxy resin, together with a separate internal tube containing the hardener. After drilling, the tube is placed in the hole and a rotary percussion drill is attached to the bolt. The bolt is then quickly driven into the tube so that the glass components are broken and the resin and hardener are mixed. Excellent adhesion is thus obtained between the bolt and the substrate when the resin has cured. Internally threaded sockets can also be bonded in by this method, thus leaving a demountable fixing. A variety of proprietary fixings are available for tensile loads of up to 52 kN (12 000 lbf) in K350 (35 MPa, 5000 lbf in $^{-2}$) concrete or shear loads of 39 kN (9000 lbf). Fast curing is essential if overhead work is contemplated. For this reason, polyester resins are often favored and a strong bond can be achieved in a matter of minutes.

4.4 Structural Repairs to Concrete

Internal repairs to concrete are carrried out by injecting adhesives into cracks (Army Construction Engineering Research Laboratory 1973). Generally the adhesive must have a low viscosity to penetrate narrow cracks; a viscosity in the region of 1 Pa s is often recommended, so filled adhesives are impractical. To avoid running of the resin, it is usual to close the crack either by taping or by caulking with tar or a hot thermoplastic polymer. The function of the adhesive is partly to bond the cracked surfaces together, thus restoring the monolith (provided that a good bond can be achieved), and partly to act as a sealant which prevents environmental attack on the steel reinforcement.

External repairs to spalled concrete structures tend to be partly cosmetic as well as structural. Typically, the defective concrete is cut away and the area is grit blasted. Any steel reinforcement thus exposed is treated with a rust inhibitor. The surface is then built up using a resin–cement grout cast on to the structure behind a thin, preformed section of glass-reinforced plastic.

4.5 Resin-Bonded Nonskid Surfaces

Since the mid-1960s, there have been many successful applications of nonskid surfaces, especially for roads. These consist of an epoxy resin, sometimes mixed with coal tar, overlaid with finely divided sand. Such surfaces have been used for pedestrian areas as well as bridges, highway intersections and pedestrian road crossings.

4.6 Cable Anchors

Cables are widely used in civil engineering for carrying substantial tensile loads. The decks of suspension bridges are usually carried by cables and in offshore technology there is an increasing use of cable-stabilized structures. The end attachments for these cables have to be treated carefully, for it is here that the maximum stress concentrations occur. Commonly, the ends of the cable are splayed into a conical socket which is then filled with molten zinc for large (structural) ropes or white metal for stranded wire ropes. When the cable is put under tension, the conically formed end is pulled into the conical socket, producing a compressive load between the socket and the formed end by wedge action. This compressive load results in a large interfacial compressive stress between the wires and the zinc or white metal. It is argued that, even in the absence of a good interfacial bond, this frictional force is sufficient to retain the cable end in the socket.

Recently, the molten metal has been replaced in some applications by a thermosetting polymer. By using a cold-setting resin which has a relatively low viscosity, excellent filling can be achieved without having to heat the socket and cable end. Polyester resins are normally used, since these tend to give a quicker cure, although the use of epoxy systems has also been reported. The rapid-curing polyesters exhibit an exothermic reaction. To reduce this effect, which can lead to overheating of the resin and to thermal shrinkage cracks, fillers such as silica can be used which both reduce the volume of resin and increase the heat capacity of the mix. Although epoxies generally give a better bond to steel than do polyesters, it has been argued that the system will work with only a small bond strength, owing to the cure shrinkage and to the wedging action under load. However, there must be some bond strength and action, otherwise the wires would simply pull out before the wedging action could begin.

Cables for which this system has been used range from 2 to 150 mm in diameter (2000 t breaking load). The big advantage of using resins compared with molten metals is that the application may be carried out on site with little danger and modest equipment. This is particularly useful when operating offshore with large diameter cables.

4.7 Steelwork

When an adhesive is introduced between the faces of a bolted steel connection, it is reported (Mander 1981) that both the static and the dynamic (fatigue) load-carrying capacities are increased.

Prevention of buckling failure by welding ribs and stiffeners to beams, plates and diaphragms can lead to distortion and fatigue. It is possible that bonding could replace welding. This would also have the advantage that dissimilar materials, such as aluminum or carbon-fiber-reinforced plastics, could be used which might provide useful weight savings—particularly if the stiffening had to be added as an afterthought.

See also: Adhesives in Building; Structural Adhesives

Bibliography

Adams R D, Peppiatt N A 1974 Stress analysis of adhesive-bonded lap joints. *J. Strain Anal.* 9: 185–96

Bresson J 1972 Reinforcement par collage d'armatures du passage inferieur du CD126 sous l'Autoroute du Sud. *Ann. Inst. Tech. Batim. Trav. Publics.* Suppl. 297

Fédération Internationale de la Précontrainte 1982 *Proposal for a Standard for Acceptance Tests and Verification of Epoxy Bonding Agents for Segmental Construction*

Fuller J D, Kriegh J D 1973 *A Guide for Pressure Grouting Cracked Concrete and Masonry Structures with Epoxy Resin.* Army Construction Engineering Research Laboratory Technical Report AD-755 926. National Technical Information Service, Springfield, Virginia

Hewlett P C, Shaw J D N 1977 Structural adhesives used in civil engineering. In: Wake W D (ed.) 1977 *Developments in Adhesives—I.* Applied Science, London, Chap. 2, pp. 25–75

Mander R F 1981 Use of resins in road and bridge construction and repair. *Int. J. Cem. Compos. Lightweight Concr.* 3: 27–39

Moreton A J 1981 Epoxy glue joints in precast concrete segmental bridge construction. *Proc. Inst. Civ. Eng.* Part 1, 70: 163–77

Raithby K D 1980 *External Strengthening of Concrete Bridges with Bonded Steel Plates*, Report No. 612. Transport and Road Research Laboratory, Crowthorne, UK

Shaw J D N 1971 The use of epoxy resin impregnation for deteriorated concrete structures. In: Guy I N (ed.) 1971 *Advances in Concrete.* Concrete Society, London

R.D. Adams
[University of Bristol, Bristol, UK]

Advanced Ceramics: An Overview

Ceramics make up one part of the portfolio of materials available for use and development by man and are in this respect complementary to the metals and polymers. The fabrication and application of ceramics are among the oldest technological skills, and modern products which bear a close and direct relationship to these ceramics include pottery, structural clay products and clay-based, heat-resistant (refractory) materials, all of which exploit the availability and properties of a particular group of nonmetallic, inorganic raw materials, the clays. These products constitute the traditional ceramics.

At a much later stage in historical development, there appeared a range of products which were similarly formed from nonmetallic, inorganic solids but which relied on very considerable raw material modification and refinement, or even on the synthesis of entirely new compositional systems, in order to provide properties matched to more specific and exacting requirements. Examples of these materials are the simple binary oxides such as alumina (Al_2O_3), magnesia (MgO) and zirconia (ZrO_2), or the ternary oxides such as the spinels (compounds based on $MgAl_2O_4$) and the perovskites (those based on $CaTiO_3$). There are also more complex oxide systems synthesized to satisfy an exact property requirement, as for example the ionically conducting material $Na_{3.1}Zr_{1.55}Si_{2.3}P_{0.7}O_{11}$. In addition, there are groups of nonoxide ceramics such as the carbides (SiC, silicon carbide or carborundum, is an important example), the nitrides (Si_3N_4, silicon nitride), the borides, the silicides, the halides and other such categories of solid compound. With the possibility of mixtures between these types (the sialons, a set of materials based on combinations of Si_3N_4 with Al_2O_3 and other oxides, are the prime example), it is apparent that the variety of such ceramic systems becomes very wide. In recognition of the differences that lie between these systems and the classical or traditional clay-based systems, a number of attempts have been made to establish a distinctive term to describe them. No usage has won general acceptance and the forms "technical," "special," "engineering," "fine," and as here "advanced" will all be encountered. The distinction, despite its occasional convenience, should not be overemphasized; there is much to be gained from transfer of technology and experience from the "traditional" to the "advanced" sectors and vice versa.

1. Applications

The applications for which advanced ceramics have been developed and proposed are many. A first division can, however, be made into categories of materials with electrical and electronic functions, those with mechanical function at ambient temperatures and those with mechanical function at elevated temperatures.

The first application of ceramics in the electrical context was as an insulator, and electrical porcelains and aluminas sustain an important role in this connection. In the period since 1940, however, a great diversification in function has occurred with innovations in materials development and in the associated solid state theory taking place in concert. Thus, the range of magnetic materials based on the ferrimagnetism of the spinel ferrites and the range of dielectric phenomena (notably ferroelectricity and piezoelectricity) encountered in the perovskite titanates have sprung from a close linking of solid state scientists and engineers with the materials design of ceramics. The products stemming from this dramatic development now form an established industry characterized by rapid innovation and an increasing sophistication of materials specification.

Mechanical applications at ambient temperature are largely based on the combination of wear resistance, hardness and corrosion resistance which many ceramics (notably alumina and silicon carbide) provide. Such applications include wear parts in medical engineering (total prostheses for hip joints), in process plant (pump components and valve faces, lining for pipework) and in mechanical engineering (bearings and valves). Alongside the many advantages which ceramics bring to these uses, the principal disadvantage and impediment to more rapid exploitation has been that, while the materials can display great strength (the attainment of 1 GPa associated with high-strength steels is no longer seen as exceptional), the mode of fracture is most commonly brittle, resulting in sudden and complete failure of the component. The reluctance of engineers to welcome ceramics more enthusiastically is mainly caused by the severity of the design problems posed by this attribute.

Despite the great advances being made elsewhere, mechanical applications at elevated temperatures are seen as perhaps the key target for the realization of the promise of advanced ceramics. The properties that lie behind this promise include wear resistance, hardness, stiffness, corrosion resistance and relatively low density. The main attraction, however, is the refractoriness of ceramics, that is, the high melting point and the retention of mechanical strength to high temperature. Some materials of interest for engineering applications are listed together with their melting (or sublimation) temperatures in Table 1.

The significance of these properties can already be seen in one successful application, namely in the use of ceramic tool tips (aluminas and sialons) for the high-speed machining of metals. The conditions of operation of the tip are arduous (high tip temperature, high mechanical loads, severe impact conditions in intermittent machining) and the success of ceramics in this sector is one of the most convincing demonstrations of their eventual suitability for a wider range of applications in high-temperature engineering. Of these, undoubtedly the most significant in terms of eventual scale is that of components for heat engines (diesels and turbines), the objective being to allow a raising of the operating temperature and, with it, engine efficiency and fuel economy.

The great attraction of this development has been long recognized, and from earlier experience it is known that ceramics pose severe problems in such applications, the main ones being susceptibility to thermal shock (the tendency to sudden fracture on the part of ceramics exposed to sudden cooling is familiar even in the domestic environment) and to sudden brittle fracture when subjected to impact. The reasons for thinking that successful development of engine components will eventually result are the large levels of current financial support, the pressing requirement for the resulting economies, the dramatic advances in materials quality, and in particular the fact that the origins of the associated problems are now better recognized. A key feature will be the quality and refinement that it is possible to bring to the processing involved in the fabrication of the ceramics.

In summary, therefore, the advanced ceramics have won a place in everyday use through their successful exploitation in a wide variety of electrical and magnetic applications and in a growing range of mechanical applications of which tool tips are perhaps the most established. A large question remains, however, in estimating the extent to which they will be capable of entering mainstream mechanical engineering as highly stressed components in high-temperature engineering systems.

Table 1
Melting or sublimation temperatures of some materials used in engineering applications

Material		Melting point (°C)
Aluminum oxide (alumina)	Al_2O_3	2054
Zirconium oxide (zirconia)	ZrO_2	2770
Silicon carbide	SiC	2650
Silicon nitride	Si_3N_4	1900[a]
Nickel	Ni	1453
Aluminum	Al	660

[a] Sublimation temperature

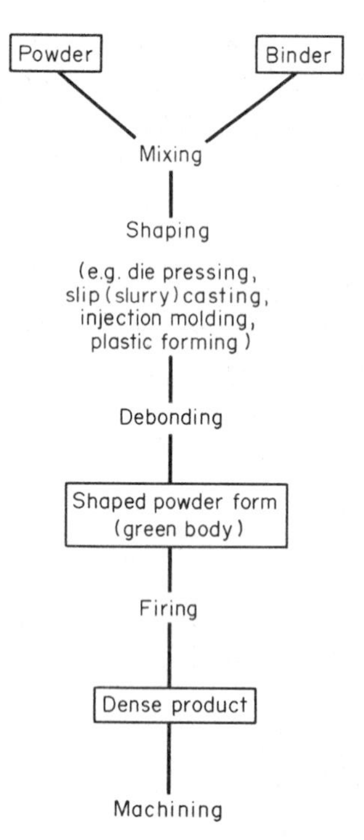

Figure 1
Basic flow chart for the fabrication of ceramics

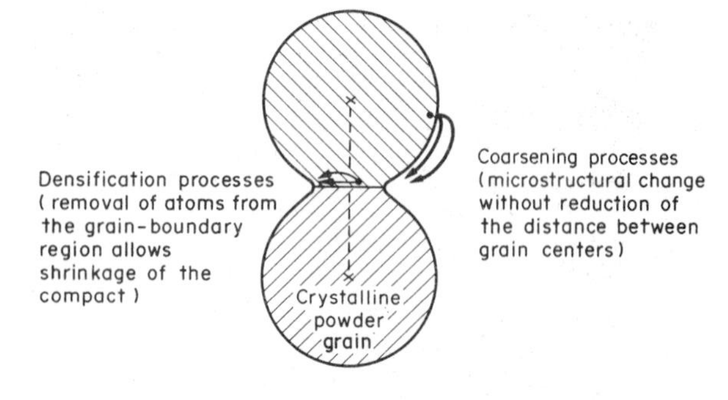

Figure 2
Schematic indication of the distinction between densifying and nondensifying microstructural changes resulting from atom movement during the firing of ceramic powders

2. *Fabrication*

One consequence of the high melting point of ceramics is that they are generally fabricated from powders. The shortest flow chart then follows the pattern shown in Fig. 1. In considering this pattern, it is convenient to split the processing into two parts, namely those processes that come before the shaped powder form (processing before firing) and those that come after (firing). In this section, some of the considerations of special concern to ceramics for mechanical engineering applications are reviewed.

Until recently, the changes that happen during firing have received the bulk of the attention given to ceramics processing, and the understanding of these changes is now considerable. In the most idealized form, solid-state sintering, the powder consists of individual crystal grains which remain solid throughout the heat treatment. The system is heated to temperatures where diffusion of the atoms in the solid state can occur. Atom movements (Fig. 2) can then bring about a reduction in the surface energy of the system either by densification (movement of atoms from the grain-boundary region to the solid–gas interface) or by coarsening (movement of atoms between different regions on the solid–gas interface). Since the former process results in a reduction in the particle–particle distance, shrinkage of up to 20% occurs and porosity is removed as required for component fabrication (Fig. 3). The coarsening process consumes the surface energy, which provides the driving force for densification, without at the same time reducing the level of porosity (Fig. 4).

The rates of these processes are enhanced by higher fabrication temperatures and by fine particle size, and densification is further enhanced if pressure can be applied to the system during heating, thus increasing the driving force for shrinkage and pore elimination. Coarsening is reduced if the powder particles in the initial compact are all of similar size. The consequence is that the firing process commonly requires high temperatures (for example, 1400–1650 °C for Al_2O_3, 1400–1700 °C for ZrO_2 and 1800 °C for sinterable Si_3N_4 are typical), fine particles (from 1 μm down to 30 nm is the commonly explored range), and uniform particles.

A common difficulty that can occur in the heat treatment of powder preforms, for example, with silicon nitride and silicon carbide, is that coarsening processes are relatively favored with the result that the heating of the pure powders is in itself an ineffective means of producing dense components. One solution here is to use an additive which produces a liquid phase between the grains at the firing temperature; this then acts as a rapid diffusion path for material from the grain boundary and densification can occur, particularly if assisted by pressure. Magnesium oxide (5 wt%) acts this way to permit sintering of silicon nitride powders (Fig. 5).

A second problem arises from the use of fine powders in the search for the enhanced activity capable of yielding high-density components. As powder sizes go below 1 μm, there is increased difficulty in powder handling and increased tendency for particles to interact with one another giving rise to the formation of agglomerates which then act to all intents and purposes in the same way as a coarse powder with a size corresponding to that of the agglomerate. The benefits of fine powders have been confirmed (zirconium oxide can be sintered to full density at 1100 °C in place of the

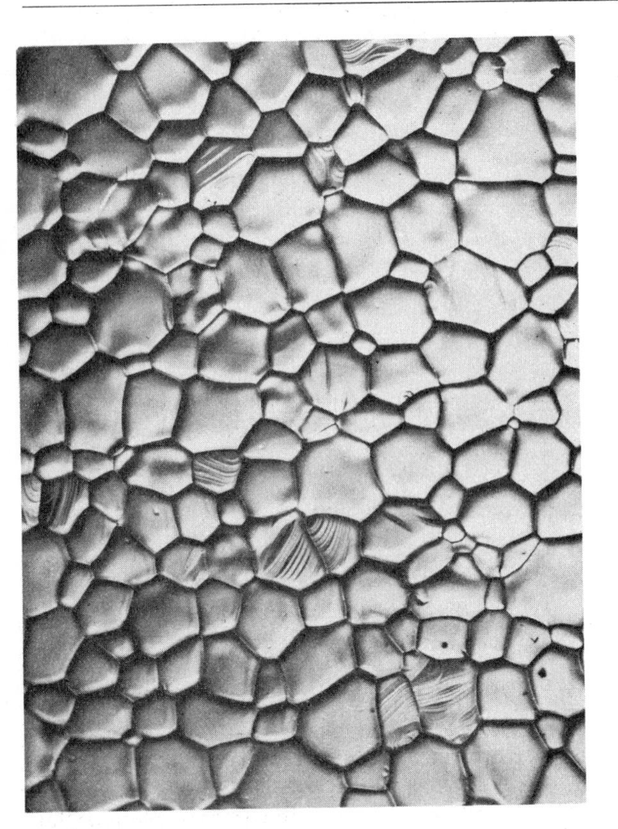

Figure 3
The surface of an alumina ceramic; all porosity has been removed during the firing of the powder. The microstructure consists of the crystalline grains and the boundaries (interfaces) between them. (Courtesy of E W Roberts)

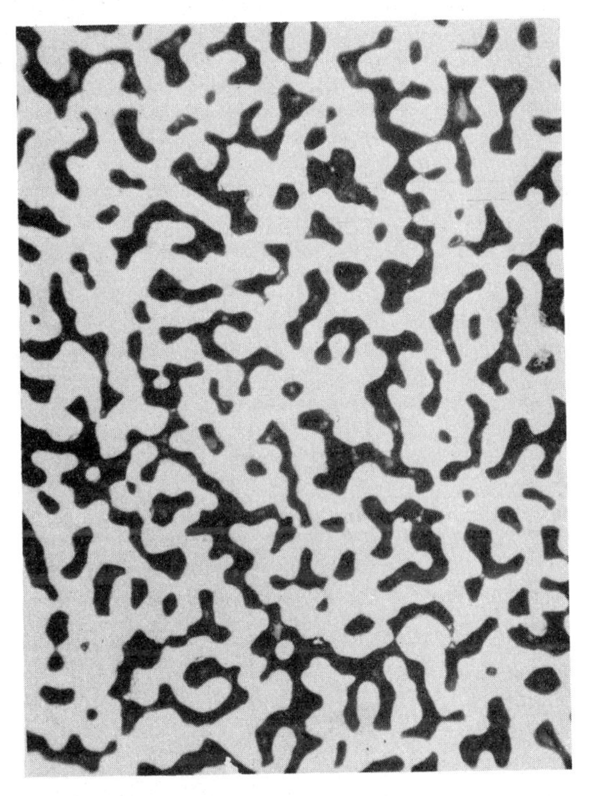

Figure 4
The sintering of silicon results in the formation of a continuous network of solid material (white) and porosity (black); this microstructural change is not accompanied by any shrinkage. (Courtesy of C F Paine)

more normal 1700 °C if a genuine particle size of 30 nm can be achieved) but the precautions needed to avoid agglomerate formation in powders of such size are demanding.

A final problem is that conventional firing can only yield results as good in terms of uniformity as the quality of powder compact produced by the earlier processing. If severe variations in packing density occur in the compact, then corresponding inhomogeneities and faults will occur in the finished product. Two general approaches can be adopted. In the first, removal of suspected flaws is attempted in a final processing step; thus, in hot isostatic pressing (hipping), partly sintered components of reasonable density are heated under an imposed gas pressure to remove the last trace of porosity. The development of sophisticated nondestructive-evaluation methods capable of confirming that all porosity has been removed is an important element in this approach and is currently an area receiving much research attention. A second approach sets out with the objective of making the ideal powder compact during the processing before firing. This approach requires meticulous powder preparation by controlled nucleation and growth from solution followed by structural assembly achieved by careful balance of interparticle forces within the solution; the resulting fault-free arrangement of powder particles in the system prior to firing offers a promising basis (Fig. 6) for the fabrication of optimal microstructures in the final product.

The present position with respect to processing can be summarized by noting that the target of development programs is that of achieving convenient, practical and reproducible fabrication procedures. The center of activity has if anything now moved into the prefiring stage where powder preparation and handling methods are being developed with a view to reducing the maximum fault size within the material. At present, it appears that prior to the eventual sophistication illustrated in Fig. 6, very substantial improvements can be made by applying to the fabrication of ceramics the clean room conditions and other refined production engineering methods long familiar in semiconductor-grade work.

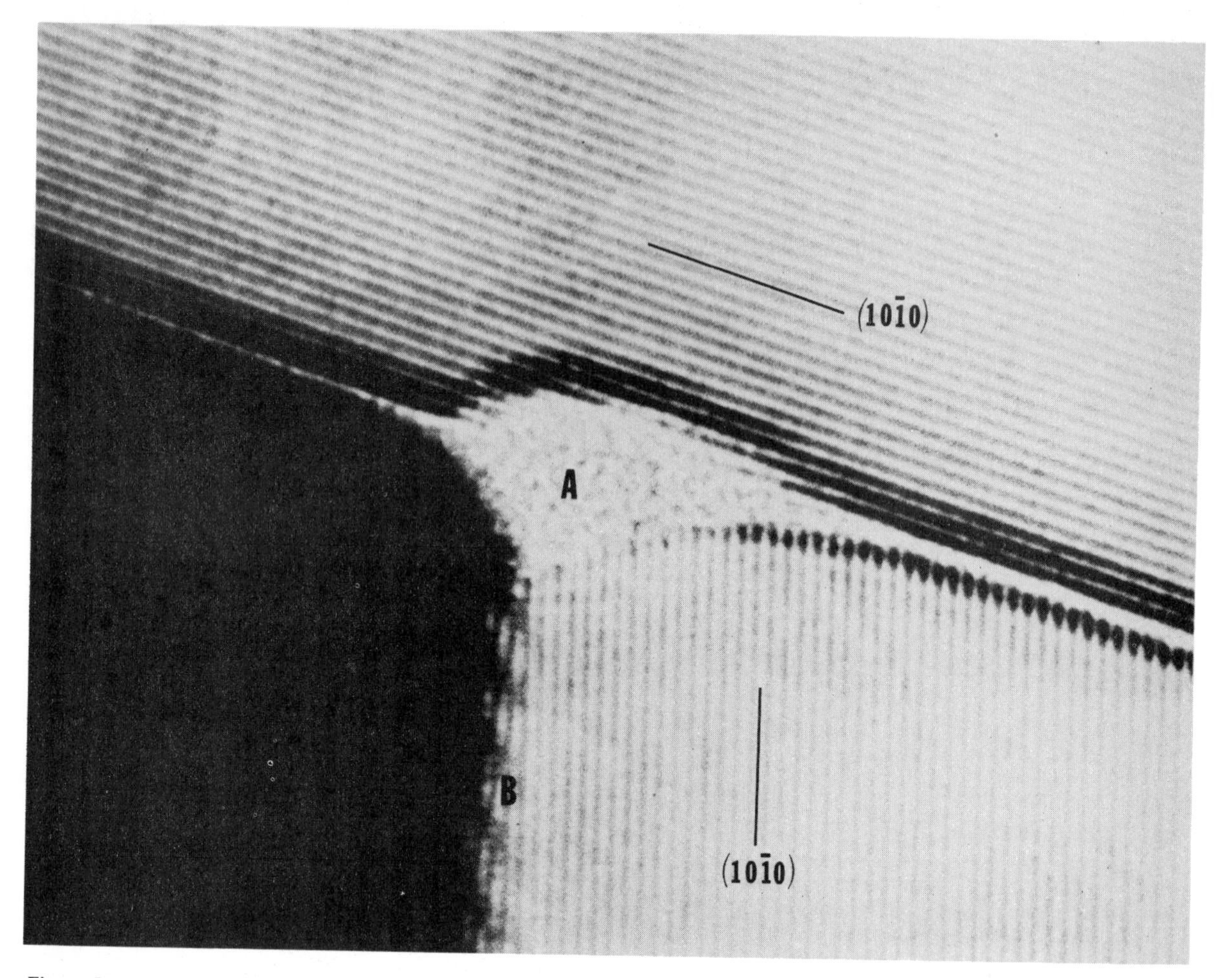

Figure 5
The existence of a continuous grain-boundary phase between the crystalline grains of Si_3N_4 hot pressed with MgO additive can be seen in high-resolution electron microscopy. The film is some 8 Å thick. (Courtesy of D R Clarke)

3. Microstructures

The processing methods that are adopted determine, as noted earlier, the eventual microstructure of the resulting ceramic, that is, the nature, quantity and distribution of the structural elements or phases making up the material. Thus, if entirely solid-state sintering processes are used and are fully successful, simple single-phase structures (Fig. 3) consisting of crystalline grains and the interfaces between them can be produced. Less complete densification can result in the persistence of a residual pore phase between (Fig. 7) or within the grains. The use of a liquid phase sintering aid results in an additional phase at the grain boundaries. Complex microstructures can result where reactions occur during the firing process, or where multicomponent compositions are used (Fig. 8).

A central argument in advanced ceramics, as elsewhere in materials science and engineering, is that many of the properties of the resulting component depend upon the microstructure of the material. The microstructure-independent properties that stem directly from the interatomic bonding of the material such as the melting point, the stiffness and the thermal expansion coefficient are by definition less susceptible to improvement by microstructural control, and due attention must be paid to these at the time of materials choice. (Silicon carbide is selected for its great hardness, stiffness and refractoriness and silicon nitride for similar advantages coupled with an even lower thermal expansion coefficient which minimizes dimensional changes and internal stresses arising from imposed temperature gradients.) In contrast, many of the properties most critical for exacting engineering use vary strikingly with the microstructure.

Taking first the example of mechanical strength, classical arguments of brittle fracture suggest that fracture strength is determined by the term $K_{Ic}/\sqrt{c}$,

Figure 6
The extent of sphere ordering that can be achieved with uniform submicrometer TiO_2 spheres offers a basis for the preparation of unfaulted structures prior to the firing stage. (Courtesy of E A Barringer and H K Bowen)

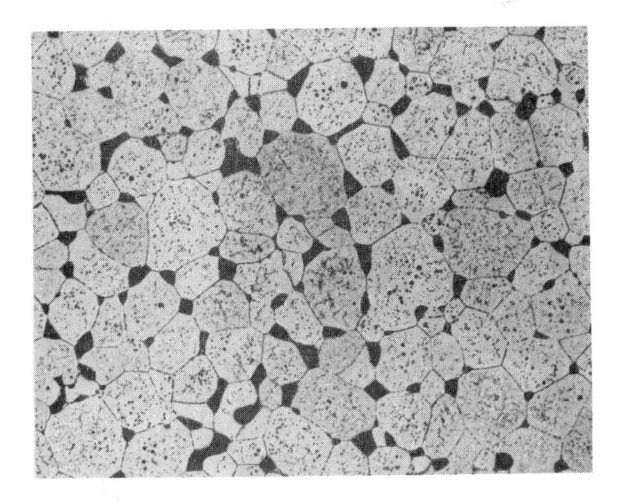

Figure 7
The presence of residual porosity (dark phase) is a common feature of solid-state sintered ceramics. (Courtesy of S P Howlett)

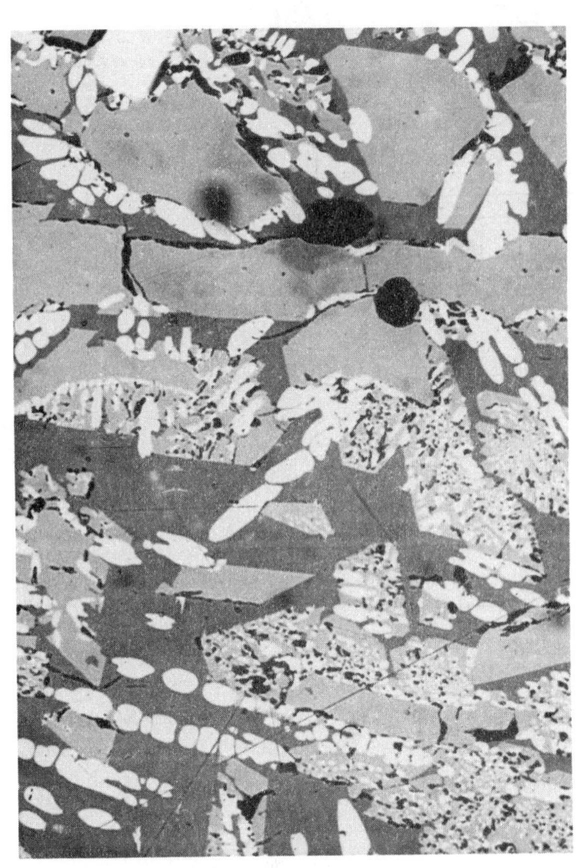

Figure 8
A multiphase ceramic microstructure comprising zirconia, alumina, glass and porosity as phases in order of decreasing lightness. (Courtesy of E W Roberts)

where K_{Ic} is the fracture toughness (related to the energy required to extend a crack of given size through the material) and c is the size of the fault from which the crack initiates, for example, from the porosity shown in Fig. 7.

The dependence on c and on structural quality has the direct consequence that any variability in fault size from one piece to another in the components from a production run results in variability in strength. There is therefore a great requirement for precise reproducibility in production methods and in particular for the elimination of fault-producing accidents such as can result from the use of insufficiently clean processing conditions.

In considering the avenues available for strength improvement in terms of microstructural refinement, two complementary approaches can be identified: the first is to reduce the size of fault that is to be found in the component by improving the processing technology; the second is to increase the energy required to extend a given crack in the material by developing toughening mechanisms.

A recent example of the significance of c, the fault size, is to be found in the development of macrodefect-free (MDF) cement. Conventional cements contain pores (faults) of sizes around 1 mm; the use of processing developments to reduce the pore size to a few micrometers (Fig. 9) results in strength enhancement from 5 to 150 MPa when tested by bending; this dramatic improvement emphasizes the benefits to be gained from such microstructural engineering.

The second approach, that of increasing the fracture toughness, draws attention to one of the clearest

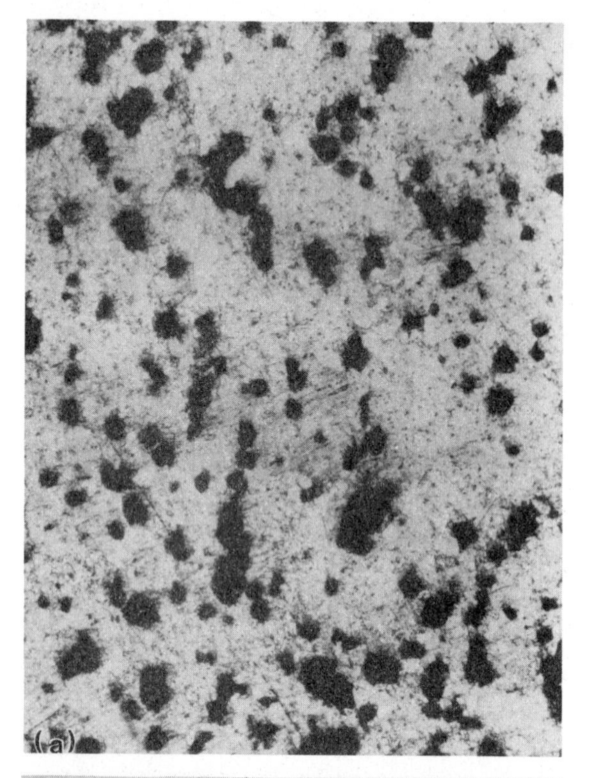

Figure 9
The comparison of (a) conventional and (b) macrodefect-free cements shows the striking absence of large pores in the latter. (Courtesy of J D Birchall)

distinctions between the ductile metals and the brittle ceramics, namely that the metals are capable of absorbing 100 000 times more energy when a crack grows through the material. Much benefit is accordingly to be expected from the development of mechanisms capable of absorbing energy during the process of crack extension in ceramics.

An example of such an approach is that of the fiber-reinforced composites; another example that has achieved considerable recent success is the transformation-toughening mechanism as represented by the zirconia ceramics (Fig. 10). Transformation-toughened materials such as the currently developed fine-grain-size zirconias are now known with bend strengths reaching 2 GPa (equivalent to high-strength engineering steels) and with projected strengths of 3 GPa.

Taking as a second example the ability of the material to resist creep, a well-recognized consequence of the need to sustain loads at high temperature without undergoing progressive distortion is that the material should not contain any phase which becomes liquid at high temperatures and which consequently allows high creep rates, by sliding of lubricated surfaces (Fig. 5) between the grains of the material for example. The use of grain-boundary engineering to remove such films either by crystallization to form high-melting-point compounds or by reaction with the grain material (as proposed for sialon compositions) is an area of active development.

To summarize, a central aspect of the development of advanced ceramics has been the recognition of the close link between the microstructure and the resulting properties and hence of the types of microstructure required for specific applications. The gap between the

Figure 10
The strength of alumina (gray phase) is doubled by the addition of 15 vol% zirconia (white phase). The zirconia particles change their crystal structure and expand as a crack passes through the material and the resulting strains between particles and matrix impede crack growth. (Courtesy of N Claussen)

level of sophistication now sought in the precision and reliability of fabrication and that commonly available has become a basis for major development effort.

4. *Areas of Development*

The pattern of development of advanced ceramics has been characterized by a steadily increasing awareness of each of the linking factors between processing and microstructure and between microstructure and properties. With this awareness has grown an increased ability to produce components well matched to applications and hence to exploit ceramics over a wide range of specialized functions. This increased specialization has resulted in more careful compositional control (notably in the levels of purity required in raw materials) and in an increasingly deliberate use of microstructural design (porosity, pore distribution; grain size, grain aspect ratio; crystallinity of boundary phases; inclusion size, composition and location). The benefits that have sprung from this progression have been notable and can be expected to continue.

An important area of future development will accordingly lie in effecting such improvements and in then making them attainable in a reproducible manner and at a reasonable cost. With respect to fabrication, the great sensitivity of ceramics to the existence of microstructural inhomogeneities and flaws will ensure sustained levels of attention to flaw elimination: approaches such as powder refinement and improvement in the forming steps are a sensible extension of existing trends. It may be wondered, however, if the continuing quest for consolidation methods other than firing may not eventually prove fruitful. The advantages of removing the apparatus of powder preparation, forming, firing, hipping and nondestructive testing that the classical methods seem to require are ever apparent.

In compositional developments one can expect minor adjustments to existing systems such as in the type of additive selection employed in the grain-boundary engineering described earlier. It also seems reasonable as in the sialons and in the alkali ion conductors that new phases will be synthesized and that property improvements and even property discoveries will occur. In microstructural engineering, the lessons of recent years, in which such clear improvements in strength and toughness have resulted from the exploitation of the zirconia transformations, show the reserves of property enhancement that can still be explored. The fact that ceramics remain at the level of marginal exploitation for many applications emphasizes the significance that such property developments may have.

The overall picture therefore is of a group of materials, surprising in their diversity of character and impressive in their range of application. The breakthrough into the level of widespread acceptance in mechanical engineering is perhaps the step that will confer the degree of familiarity that is still lacking when comparisons are made with the metals. As noted earlier, this breakthrough is dependent on the achievement of standards of fabrication that will not be easy to meet. Whether the low-cost fabrication of multiple complex shapes in ceramics can be performed with sufficient reliability and precision to win the markets which the properties of these materials promise for general engineering is perhaps the most significant and intriguing of current questions in materials science.

Bibliography

Many helpful reports on work relating to the processing, properties and applications of ceramics are to be found in a group of series, the units of which are published at more or less regular intervals. These include:

Advances in Ceramics. American Ceramic Society, Columbus, Ohio
Materials Science Research. Plenum, New York
Proceedings of the British Ceramic Society. Institute of Ceramics, Stoke
Progress in Ceramic Science. Pergamon, New York
Science of Ceramics. Published by a sequence of European ceramic societies

Other references:

Claussen N 1984 Transformation-toughened ceramics. In: Kröckel H, Merz M, van der Biest O (eds.) 1984 *Ceramics in Advanced Energy Technologies.* Reidel, Dordrecht, pp. 51–86
Creyke W E C, Sainsbury I E J, Morrell R 1982 *Design with Non-Ductile Materials.* Applied Science, London
Davidge R W 1980 *Mechanical Behavior of Ceramics.* Cambridge University Press, Cambridge
Evans A G 1982 Structural reliability: A processing dependent phenomenon. *J. Am. Ceram. Soc.* 65: 127–37
Hlavac J 1983 *Technology of Glass and Ceramics.* Elsevier, Amsterdam
Kenney G B, Bowen H K 1983 High tech ceramics in Japan: Current and future markets. *Bull. Am. Ceram. Soc.* 62: 590–96
Kingery W D 1960 *Introduction to Ceramics.* Wiley, New York
Kingery W D, Bowen H K, Uhlmann D R 1976 *Introduction to Ceramics.* Wiley, New York
McColm I J 1983 *Ceramic Science for Materials Technologists.* Leonard Hill, Glasgow
Pampuch R 1976 *Ceramic Materials.* Elsevier, Amsterdam
Richerson D W 1982 *Modern Ceramic Engineering.* Dekker, New York

R. J. Brook
[University of Leeds, Leeds, UK]

Architectural Design and Materials

The type, availability and physical properties of different materials have undoubtedly had significant effects on the design of buildings and other man-made

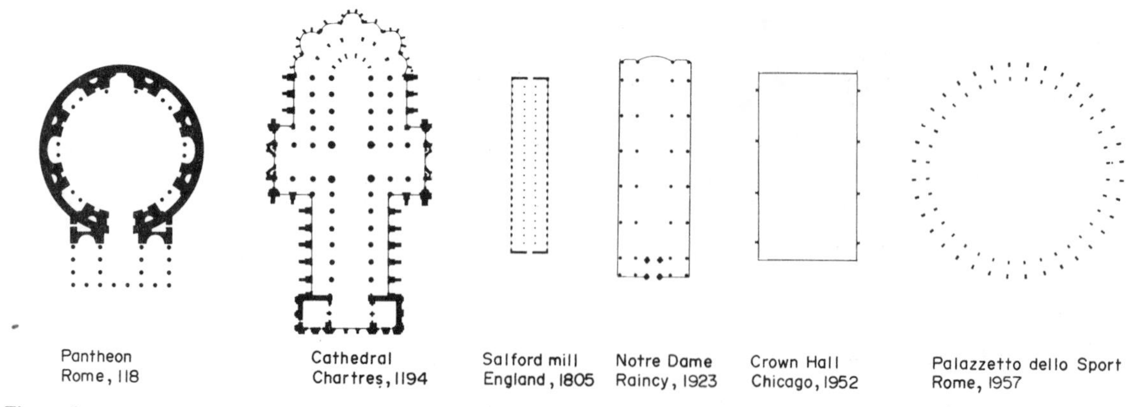

Figure 1
The reduction in plan area of vertical support systems from Roman to modern times

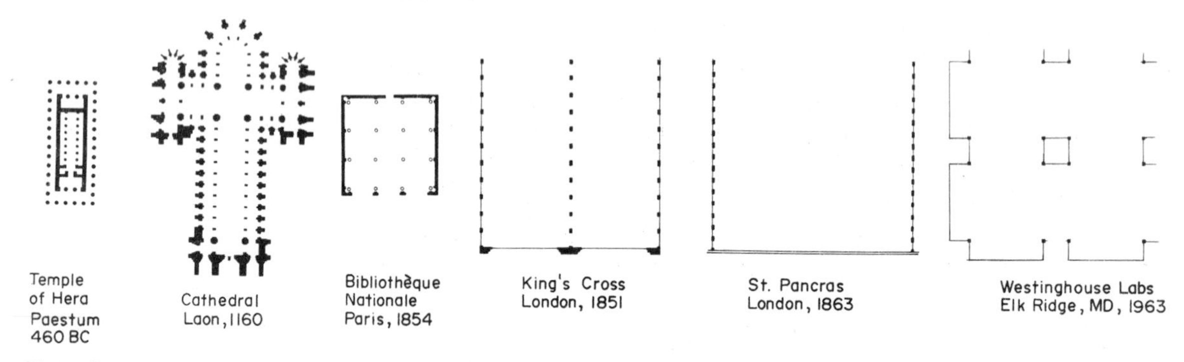

Figure 2
The development of long-span structures

environments. During the past several hundred years in particular, materials-related innovations and developments have aided society in meeting its rather demanding building needs, in both quantity and types of facilities. Possible building plan organizations and spaces, and configurations, sizes and appearances of buildings have all been affected by these developments. The capacity to provide and maintain controlled internal building environments has also changed. In many instances, these effects have been dramatic; in others, subtle and difficult to trace.

1. Materials-Related Innovations in Design

Some of the more obvious materials-related changes in building designs are evident in Figs. 1, 2 and 3. Figure 1 illustrates how the dimensions of vertical support elements in buildings (e.g., columns, load-bearing walls) have decreased over the years, a process that profoundly affected the architecture of the times. For example, the contrast in the vertical support patterns of two comparable domed structures, the Pantheon (Rome, 118AD) and the Sports Palace (Rome, 1957) by Nervi, is quite remarkable. Much of the site plan of ancient buildings such as the Pantheon was devoted to massive walls or columnar elements. With changes in the types and properties of materials available for construction, particularly increases in unit strengths and stiffnesses, the total amount of space in plan required for vertical support elements has decreased with time. The floor plan of the Crown Hall in Chicago (1952) shown in Fig. 1 illustrates the extent to which this trend has taken place. Only a small fraction of the total floor area is occupied by columns.

Changes have also occurred in horizontal support systems. Figure 2 illustrates how open spans became greater. In early buildings, the creation of large open spaces necessitated the use of a limited number of specific shapes (e.g., domes). During the nineteenth century, innovative long-span systems consisting of trusses, rigid arches, cables and other forms made possible by new developments in materials, offered new solutions to spanning large, open areas. Where long-span structures were used, the overall form and organization of buildings changed. The form of St. Pancras railroad station in London (1863) would have been undreamed of by the builders of the cathedral at Laon (1160).

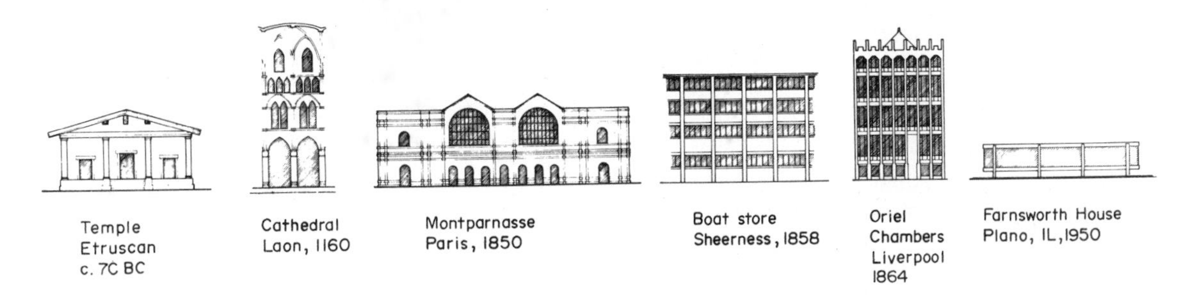

Figure 3
Changes in facades: the development of the distinction between structural and nonstructural elements

Materials-related innovations have also affected the facades and overall visual characteristics of buildings. The massive vertical support elements of early buildings and limited span capabilities of horizontal elements dictated relatively small openings in exterior walls. As Fig. 3 illustrates, increasing long-span options, linked with rapid developments in glass technology, facilitated the larger openings in exterior walls. For example, a series of long-span structures is seen clearly in the facade of the Montparnasse in Paris (1850) shown in Fig. 3. Materials-related innovations have led to the now commonplace distinctions made in exterior surfaces between structure and enclosure. The image of many modern buildings is now that of enclosure rather than structure, a trend carried to its extreme in many of the glass-enclosed skyscrapers of today.

Materials-related innovations were not the sole cause of significant changes in architecture. As with any history, the history of architecture is far too involved and complex to ascribe major developments to one cause only. However, changes have been possible because of, and frequently were encouraged by, developments in the field of materials. More importantly, developments related to the emerging field of materials have caused designers to think differently about shaping objects to meet building needs. Consequently, it is useful to look at the larger impact of materials-related innovations on the history of architecture.

2. *Impact of Materials on Architecture*

In tracing the impact of materials on architecture, it is crucial to focus on the periods between the seventeenth and eighteenth centuries, when mechanical theory was postulated and the mechanical properties of materials were under investigation. Before then, the materials influenced the design and construction of buildings mainly through the historic role of buildings as climatic modifiers and by adapting existing technologies to locally available materials.

The form and construction of many early buildings, particularly indigenous housing, were responsive to the prevailing climatic, technical and socioeconomic conditions. In particular, the climate has influenced the general shapes, plan organizations and overall massing of building complexes; designers have tried to mitigate adverse environmental conditions. Climatic concerns also influenced the design of specific elements (e.g., walls and roofs), the shapes of openings, and the choice of materials for construction (especially with respect to their thermal characteristics and weather responses). For example, buildings in hot, dry climates are often compact, with massive wall and roof elements (typically brick or stone), and small window and door openings. Because of their thermal transmission and storage characteristics, the massive enclosure elements offer relatively cool interiors during hot days and warm interiors during cool nights. Buildings in warm humid climates, by contrast, are often characterized by dispersed forms that are constructed of light materials with limited thermal storage capacities. Elements are arranged to encourage constant flowthrough ventilation. Fortunately, in both of these examples, the local natural materials are well suited to construction of climatically responsive buildings.

Material and building responses in more complex climatic conditions are less clear than the examples cited, and the available materials do not always meet the needs. Consequently, the development of active rather passive means of controlling interior environments and the development of new materials are encouraged in these instances.

The arrangement of structural and enclosure elements in early buildings was also influenced strongly by the mechanical properties of the materials and the technologies for producing and assembling the materials. In buildings using high-mass structural and enclosure elements, forms such as arches, domes and corbels have evolved. The resulting states of internal force and the complicated construction procedures were uniquely compatible with the low tensile strengths and the limited dimensions of the materials (e.g., brick or stone). In other regions, the technology used to join and assemble light, but relatively long, linear elements (e.g., wood members) dictated the architecture of the buildings, as in traditional Japanese construction.

The design of most building elements was based on experience; rules were empirical rather than theoretical. By the sixteenth century, some codes had been formulated for proportioning—at least for certain structural forms used in major buildings. Many of these rules, however, were also influenced by abstract geometrical principles of proportion. These principles, which arose from visual, cultural and symbolic rationales, also affected the selection of specific materials. The clearly drawn distinctions in current building practices among various design objectives (e.g., structural, visual) were blurred and not well understood at that time.

The rapid development and evolution of mechanical theory in the seventeenth and eighteenth centuries, coupled with experiments by Hooke and others investigating the mechanical properties of materials, altered the design practices of the time. These developments, which provided an alternative approach for selecting materials and for sizing and shaping building elements, were based on the inherent states of force, or, more generally, on the action of any physical phenomenon that could be isolated and described. Along with this new approach to materials, "efficiency" evolved as a conscious design criterion for shaping, sizing, and selecting material properties for building elements. The related concept of "optimum design solutions" developed as well. Although all this is familiar today, it represented a radical method of design at the time.

For a while, these concepts provided a whole new direction in architecture (beyond the design of specific elements). Optimality and efficiency became conscious criteria in the design of buildings, as is evidenced by the proposal by LeRoy (1787) for a new hospital in Paris, in which the actual building shape was dictated by contemporary ideas about medicine and air dynamics; these considerations departed significantly from previous criteria for building form. The elaborate computations made by Bentham for the multipurpose Panopticon Building reveal that design concepts having their roots in the theory of mechanics had extended into problems of space allocation and activities. Many French institutions, such as the Académie Royale d'Architecture (which studied building materials in 1678—including the roles played by the sun, wind and even the moon on the aging and deterioration of stones used in buildings) and the Académie des Sciences, disseminated the new approach into the design practices of the time.

In the late seventeenth century, other theories of architecture soon led to the development of two new French institutions, the École des Beaux Arts and the École Polytechnique. The Polytechnique taught technical skills and the Beaux Arts those of aesthetic concerns, beginning a split into the engineering and architectural professions that is still apparent today. To a large extent, materials-related concerns began to assume lesser importance in the development of architectural forms.

Advances in technology and related developments in materials continued and blossomed with the advent of the industrial revolution. The rise of the industrial state and the subsequent changes in society created needs not only for new materials, but also for new building types (e.g., large train sheds). Significantly, in the nineteenth century, the concept of repetitively produced products made its way into architecture. For many designers, a fundamental change of perspective took place. No longer could buildings or parts of buildings be conceived of only as handcrafted artifacts, but as products. New materials, and their associated production processes, induced this conceptual shift.

3. Types of Architectural Material

3.1 Iron and Steel

Iron was the first "new" material that changed a structure's appearance. At first, it was used as a slender support for balconies, as in St. Anne's Hospital, Liverpool (1770–72), where cast iron columns were used. These supports, like most cast iron columns to follow, were ornamented in a traditional design style, without consideration of the inherent nature of iron. Cast iron columns and beams were soon used in factories, as in Strutt's calico mill, in Derby, England (1792–93). They took up little floor space, were relatively inexpensive and were thought to be fireproof. However, the role of cast iron was visually played down or ignored and was never expressed on the exterior of buildings.

In contrast, iron was displayed unashamedly in bridge construction, such as at Ironbridge (1779). Labrouste left all the interior ironwork exposed in his Bibliothèque Ste. Geneviève. The Coal Exchange in London had a grand interior court of cast iron, using ornamentation made possible by the flexibility of cast iron. These buildings, while using iron on a grand scale inside, did not express it on the exterior.

The opposite is true with cast iron facades dating from the 1840s on through to the end of the nineteenth century. Often, a cast iron facade was put onto a brick and timber row house when the house was converted for retail or commercial uses. These facades, with their large windows and thin spandrels, greatly raised interior illumination. Bogardus built the first all-iron building in the USA, his New York factory, in 1849. All-iron buildings became popular with retail merchants, especially in Paris, because of their light, highly ornamented members, elegant stairways and interior courts.

Early examples of structures whose style was dictated by materials, and not traditional architecture, were the train sheds of the UK. Early train sheds were considered "functional" and were designed by engineers, not architects. Architects designed traditional entrance facades for them, as at King's Cross station in London (1851). The ironwork in the appended sheds was unornamented, functional, and spanned long distances in a simple and beautiful manner. But fully enclosed buildings (except greenhouses) were slow to follow this example. One of the first to employ an unornamented iron and glass enclosure, with no masonry or wood (except for flooring), was the Crystal Palace of 1851 in London. Although many "crystal palaces" were built in the next half-century, other building types did not approach this bare, minimalist attitude until Behrens and Gropius built factories in Germany just prior to World War I. Train sheds and exhibition halls were building types that came into being with the increased use of iron and, later, steel.

Established building types, such as churches and residential buildings, for many decades did not express any use they may have made of iron or steel. Commercial buildings slowly began to reveal their structure, as iron and steel became more important. The Owen warehouse was an early example of "honest" construction—the structure is clearly evident on the elevations.

A subsequent and important innovation was the skyscraper, especially in Chicago and New York. Skyscrapers were feasible only with the development of the iron, and soon thereafter, the steel skeleton. Later buildings used reinforced concrete. Large office buildings were still usually clad in masonry, but window area increased and ornamentation decreased. The PSFS building in Philadelphia, designed in 1929, was one of the first steel skyscrapers in the "international style." Its plain banding indicated the floor levels, strip windows allowed maximum fenestration, and vertical ribs expressed the vertical supports and upward thrust of the building. In the early 1950s, Seagram's Building and Lever House, both in New York City, took structural expression to an extreme, using no masonry cladding and only steel and glass on the exterior of the building. This became an oft-repeated formula, a supposedly "pure" form of building derived from the structure.

In the last few decades, several new forms made possible by the use of steel have been explored, including cable structures, space frames and doubly curved forms (e.g., spheres) made of various types of bar networks. Cable structures have long existed, but have not been explored as a building form until fairly recently. Although James Bogardus submitted a proposal for the New York Exhibition of 1853 in which the roof of a circular cast iron building, 213 m in diameter, was to be suspended from radiating chains anchored to a central tower, the building was not constructed. The pavilions of the Nijny-Novgorod Exhibition designed by Shookhov in 1896 were the first actual modern applications. Subsequent structures include the locomotive roundhouse pavilion at the Chicago World Fair in 1933 and the livestock judging pavilion built at Raleigh, North Carolina, in the early 1950s. Since then, a number of other significant cable-supported buildings have been constructed.

Three-dimensional forms also can be made of assemblies of short, rigid steel bars. Structures of this type were first used extensively in the nineteenth century. The Schwedler dome, consisting of an irregular mesh of hinged bars, was introduced in Berlin in 1863 when Schwedler designed a dome with a span of 40 m. Other, more recent structures have bars that are placed on curves generated by the medians and parallels of revolution. Some of the largest domes in the world (e.g., the Houston Astrodome and the Louisiana Superdome) follow this latter scheme.

3.2 Concrete

The efficient production of concrete and the subsequent widespread use of reinforced concrete members introduced another series of developments. A whole new, twentieth-century vernacular of architecture is evident in the designs of Le Corbusier and others. Reinforced concrete was rarely used in buildings until the end of the nineteenth century. At first, it was just used for floor slabs, where it was unseen, as iron had been initially. By the turn of the century it was being used in houses, factories (notably the AEG factory, Germany) and churches (e.g., St. Jean-de-Montmartre, Paris). But the concrete was detailed and incised with coursing to give it the appearance of masonry. Even Perret's apartment house in Rue Franklin in Paris, which expressed its structural skeleton, did so while covering the concrete with terracotta.

The use of concrete increased in Europe as this century progressed. Concrete was a readily available, inexpensive material, whose pure and plain qualities appealed to Futurists and Modernists. They saw it as a revolutionary material indicative of progress and the machine age; because it was inexpensive, they used it as an egalitarian material for mansions and workers' housing. Loos, Le Corbusier and others used it unashamedly in housing schemes, where it was painted, but not disguised to resemble any other material. This "style" became quickly adapted by "modern" architects who applauded its ostensible "purity" and exported it to all corners of the earth.

Reinforced concrete also made possible the extensive use of flat plates or slabs. These systems influenced the architecture of many buildings by eliminating the hierarchical structure evident in the deck, beam and column arrangements characteristic of buildings made of other materials.

The plasticity of concrete was rarely explored before World War II. The penguin pool at Regent's

Park Zoo in London is a notable exception. Like steel, concrete freed the building envelope from any structural elements, as at the Johnson Wax Tower in Racine, Wisconsin. Concrete could also assume nonlinear forms very easily. A rare early example is the airplane hangar at Orly Airport built in the 1920s. Nervi, Maillart and others explored concrete's form-shaping possibilities.

The development of prestressing and post-tensioning led to not only new shapes for building elements (e.g., the long-span single tee), but also whole new complete building systems made of planar or volumetric elements.

3.3 Timber

Timber, an ancient building material, was used in different ways beginning in the nineteenth century. Long serving as structure (in the form of heavy post and beam assemblies) and enclosure (shingles, clapboards), the use of wood was revolutionized by Augustine Taylor, a carpenter from Hartford, Connecticut. In 1833, the balloon-frame of construction that he developed utilized a light framework of studs and joints, and changed the way most wood buildings were designed and built. Technological innovations, such as mechanical means of sawing timbers and new ways of connecting pieces (e.g., nails), made the balloon frame possible. The structure could be assembled quickly by relatively unskilled carpenters working with hand tools only.

The truss—a structural innovation making large spans possible—was also explored initially using timber members. The covered truss bridges of New England and other regions of the USA led to the sophisticated steel truss designs so familiar today.

New glues and special fabrication processes led to the widespread use of plywood. For the first time, relatively large surface-forming sheets of an easily workable material were available for general use. These adhesives also facilitated the development of glued laminated structures, which made it possible to increase the length and strength of wood members and to shape them in ways undreamt of by seventeenth-century carpenters.

3.4 Other Materials and Composites

An interesting materials-related effect on building design and construction entails making walls and other enclosure elements out of layers of different materials. Walls in early buildings were often constructed of a single material that served the multiple roles of moisture, thermal, weather and fire barriers and also provided the structural support.

Many building designs depend on the specific properties of the single material chosen for use. With the advent of new materials and improved understanding of the functions of exterior walls, different materials were introduced as layers in walls—each serving specific roles. Insulation was introduced as a thermal barrier, membranes as vapor barriers, gypsum wall board as enclosure and fire barriers, and studs as structural elements. Each material was designed to serve its designated role optimally, in marked contrast to earlier building practices.

Many new materials-related investigations focus on developing single elements that are capable of serving multiple roles. Sandwich construction, typically layered and bonded, is one result of these efforts, as is the development of composite materials. Materials with a limited range of properties are combined with other materials with different properties, producing composites with their unique properties. Included in this category are various reinforced plastics, glass fibers and woven materials that are coated or mixed with other materials. Many are composed of randomly oriented, short fibers that are made by chopping various types of rovings into short lengths and embedding them into binding material.

Major developments have occurred in other materials as well. From its initial use as the tip element on the Washington Monument, for example, aluminum has found its way into most buildings constructed today. Various types of plastics and elastomeric materials are commonplace.

Plastics have assumed a primary role in building construction. They offer advantages such as formability, strength, toughness, lightness, light transmission and low costs. End-use applications range from primary and secondary building elements (walls, roofs, doors and windows) to various applications as finishes and parts of electrical and mechanical systems. Primary building elements are reinforced, commonly with glass fibers, to increase their strength and stiffness. Plastic-coated fabrics, for example, are commonly employed in air-supported buildings and as covers for storage areas, radomes and tents. Secondary uses include parts of windows and doors and coatings, adhesives and sealants.

In summary, it is clear that materials-related innovations have significantly affected the design of buildings. The course of architecture has undoubtedly been altered by these developments. The majority of our future buildings will not be designed and built as they were in the past, because they will incorporate new materials along with the ones in current use.

See also: Steel for Construction; Wood as a Building Material; Plastics and Composites as Building Materials; Indigenous Building Materials; Building Systems and Materials

Bibliography

Collins P 1967 *Changing Ideals in Moden Architecture.* Faber and Faber, London

Dietz A 1969 *Plastics for Architects and Builders.* MIT Press, Cambridge, Massachusetts

Fitch J M 1972 *American Building: The Environmental Forces That Shape It*, 2nd edn. Houghton Mifflin, Boston, Massachusetts

Giedion S 1941 *Space, Time and Architecture: The Growth of a New Tradition.* Harvard University Press, Cambridge, Massachusetts

Huntington C, Mickadeit R E 1975 *Building Construction: Materials and Types of Construction*, 4th edn. Wiley, New York

Merritt S 1975 *Building Construction Handbook*, 3rd edn. McGraw-Hill, New York

Straub H 1952 *A History of Civil Engineering: An Outline from Ancient to Modern Times.* Leonard Hill, London

Tzonis A 1980 *Aspects of Mechanization in Design: The Rise, Evolution and Impact of Mechanics in Architecture*, Publication Series in Architecture No. a-7907. Harvard University Press, Cambridge, Massachusetts.

D. L. Schodek
[Harvard Graduate School of Design, Cambridge, Massachusetts, USA]

B

Binders: Clay Minerals

The clay minerals most widely used as binders are sodium and calcium montmorillonite, and kaolinite. Binders based on these minerals are employed in the foundry, pelletizing and ceramics industries because of their plasticity when mixed with water and their fine particle size. Kaolinite ($Al_2O_3{\cdot}2SiO_2{\cdot}2H_2O$) is the major clay mineral constituent of the rock kaolin, and the montmorillonites are the chief constituents of bentonite.

1. Kaolin

In addition to the pigment grade of kaolin used as a binder in ceramics and other applications, binders based on other kaolinitic materials (ball clays, fire clays and plastic underclays) are also available. Ball clay is a secondary clay commonly characterized by the presence of organic matter, high plasticity, high dry strength, long vitrification range and a light color when fired. Kaolinite is the principal mineral constituent of ball clay. Ball clays are mined in the USA, the UK, France, Germany, India and South Africa. Fire clay is a kaolinitic clay that is not white-burning and has a pyrometric cone equivalent above 19, that is, it has a fusion point above 1520 °C. The term fire clay excludes most kaolin and ball clays because they become white on burning. Plastic underclay is a nonrefractory kaolinitic clay found immediately below coal beds of carboniferous age.

In ceramic applications, kaolins, ball clays, fire clays and plastic underclays are used as binders. Determination of which is the best binder for a particular application involves matching the properties of each with the specifications required in the final product. The specifications normally considered are modulus of rupture, plasticity, pyrometric cone equivalent, casting rate, fired color and shrinkage. Other tests may be required for certain products.

The modulus of rupture (MOR) is the fracture strength of a material under a bending load. The MOR measurement is made on a long bar of either rectangular or circular cross section, supported near its ends; a load is applied to the central part of the supported span. Plasticity, the ease with which a mass can be deformed and retain the deformed shape without any elastic rebound, is measured on a plastigraph or with a liquid-limit tester. The casting rate is important; fine-grained bodies cast more slowly than coarse ones. Viscosity of the casting slip must be carefully controlled, as if the slip is too viscous it will not properly fill the mold or drain cleanly. The fired color is important in many instances; for example, in whiteware bodies the fired color must be white or cream. This is a visual test and the the test specimen, fired to a predetermined temperature, is compared with a standard. Shrinkage is also important, as ceramic articles undergo shrinkage at several different points in the manufacturing sequence. During drying, articles will shrink by varying amounts, depending on their composition and the amount of water present. Further shrinkage occurs during firing; therefore it is important to know both the drying and firing shrinkage.

The particle size of kaolin products is generally controlled, as a coarse kaolin has very different ceramic properties to a fine kaolin. In general, the finer the particle size the higher the MOR, the plasticity and the shrinkage, and the lower the casting rate. A particular grade of kaolin can be matched with the properties needed in the final product. In castings a coarse kaolin is preferred, and in whiteware a plastic kaolin of fine particle size that fires white is preferred. The Georgia kaolins have pyrometric cone equivalent values ranging from 33 to 35 (i.e., fusion points of 1745–1785 °C indicating a highly refractory clay.

Ball clays are used where high plasticity, high dry strength and a long vitrification range are required along with a white or light cream color. Fire clays are used in refractory applications when fired color is not important. Plastic underclays are used when high plasticity is required and the fired color does not have to be white.

2. Bentonite

The most important binder mineral for use in the foundry and pelletizing industries is bentonite, the chief mineral of which is sodium or calcium montmorillonite. The montmorillonites are extremely plastic over a wide moisture range and can meet specifications for moisture content, gelling index, binding properties and liquid limit. The later is a measure of the ability of montmorillonites to hold water without flowing.

Sodium montmorillonite clays have the ability to hold moisture and are not easily dehydrated when subjected to heat. They give a very durable mold or pellet, and much less clay can be used to bind the foundry sand or the iron ore in the pellet. Sodium montmorillonites also have a high resistance to expansion defects, good refractoriness and a low moisture requirement. The hot and dry strengths and thermal stability are preferred for the production of steel and heavy iron castings in a clay-bonded foundry sand. Therefore the steel casting industry uses sodium montmorillonites (Wyoming bentonite).

Table 1
Comparison of foundry binding properties of typical sodium and calcium montmorillonites

Property	Calcium montmorillonite	Sodium montmorillonite
Green compression strength	high	slightly lower
Bond development	rapid	moderate
Hot compression strength	moderate	high
Dry compression strength	low	high
Durability	low	high
Resistance to expansion defects	low	high
Thermal stability at casting temperatures	poor	very good
Sand flowability	very good	moderate
Shakeout characteristics	easy	poor

Calcium montmorillonites are called nonswelling bentonites or sometimes sub-bentonites. They give very rapid bond development, a high green compression strength, moderate hot strength, low dry compression strength, low durability and very good sand flowability. Because the majority of iron foundries want a low dry strength and low durability to achieve good shakeout characteristics, calcium montmorillonites are in common use. The best nonswelling bentonites are mined in Mississippi and Texas in the USA, and in the Federal Republic of Germany, Italy and Greece.

Blends of sodium and calcium montmorillonites are used in many foundries to achieve desired binding properties. Table 1 shows a summary of the properties of sodium and calcium montmorillonites with respect to foundry use. Some producers of calcium montmorillonites activate the clay by changing the calcium cation to sodium through treatment with soda ash or occasionally with a sodium polymer such as sodium polyacrylate. Some calcium bentonites respond very well to this treatment and give considerably improved properties. Although a soda ash-treated calcium montmorillonite is often called sodium montmorillonite after treatment, the properties at high casting temperatures and the durability are not as good as those of natural sodium montmorillonite.

Specifications for montmorillonites used as binders in the foundry industry are outlined by the Steel Founders Society of America and the American Foundrymen's Society. The tests include moisture content, pH, calcium oxide content, liquid limit, thermal stability at casting temperatures, and green, dry and hot strengths. Specifications for montmorillonites used for pelletizing taconite-type iron ores have not been standardized, and several tests are used as green pellets must be capable of withstanding handling, compaction and drying, and the dry pellets must be even stronger. Mixtures of bentonite, iron ore and water are tested for wet drop strength, wet compression strength, plastic deformation and dry compression strength. The final choice of a suitable bentonite for pelletizing appears to be largely logistical with locational advantage and transportation costs playing a major role. In the case of foundries, although the locational advantage and transportation costs are important, the performance characteristics play the major role in the selection of the binder.

Bibliography

American Foundrymen's Society 1963 *Foundry Sand Handbook*, 7th edn. American Foundrymen's Society, Chicago, Illinois

Norton F H 1968 *Refractories*, 4th edn. McGraw-Hill, New York

Patterson S H, Murray H H 1983 Clays. In: Lefond S J (ed.) 1983 *Industrial Minerals and Rocks*, 5th edn. American Institute of Mining, Metallurgical and Petroleum Engineers, New York, pp 585–651

Steel Founders Society of America 1965 *Tentative Specifications for Western Bentonite*, SFSA Designation 13T-65. Steel Founders Society of America, Cleveland, Ohio

H. H. Murray
[University of Indiana, Bloomington, Indiana, USA]

Bond Strength: Peel

A peel test is often employed to assess the strength of an adhesive bond. In order to use this geometry, at least one of the two bonded components (adherends) must be sufficiently flexible so that the moment arm m of the peel force about the advancing peel front (Fig. 1a) remains constant as peeling proceeds. This feature distinguishes a peel test from a cantilever-beam cleavage experiment, in which the moment arm m continually increases as the separation front progresses (Fig. 1b). Typical peel specimens consist of a pressure-sensitive adhesive tape adhering to a rigid substrate or two thin films bonded together. There are generally three components: the two adherends and a layer of adhesive between them. The mechanical properties of

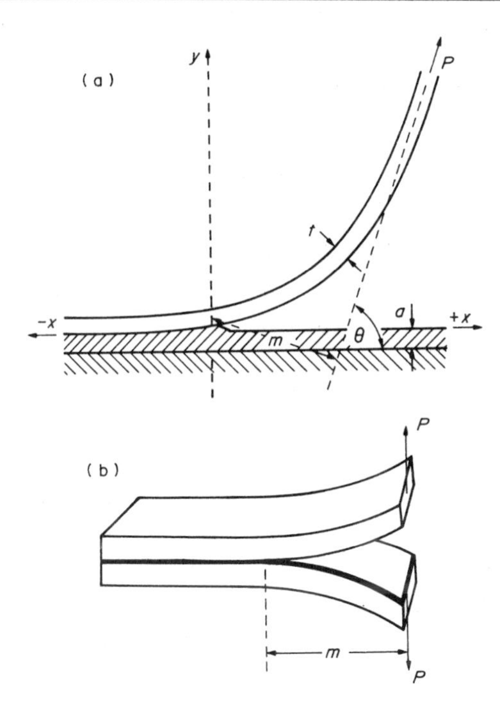

Figure 1
Peel test (a) contrasted with cantilever-beam cleavage test (b)

all three layers affect the resistance to peeling apart, as described later.

1. Peel Tests

The strength of a suitable specimen is usually determined at peel angles θ of 90° (Fig. 2a) or 180° (Fig. 2b) or in a T-peel configuration (Fig. 2c). The bond strength is characterized by the peel force P per unit width of bonded surface when a constant stripping velocity v is imparted to the free ends. Once failure has started, the peel force generally fluctuates about an average value as peeling proceeds. Failure may take place by separation at either interface or within one of the three component layers if the layer itself tears apart instead of detaching. It is quite important to note the site of failure when testing an adhesively bonded joint. For example, if failure occurs within the adhesive layer, it is misleading to associate the measured peel strength with interfacial properties of the adhesive and adherends.

Scientifically, there are two simplifying advantages of the peel test compared with other methods, such as butt tensile or lap shear tests (Gent and Hamed 1977a). Failure proceeds at a controlled rate and the peel force is a direct measure of the work of detachment W_a (for interfacial failure) or the work of cohesive fracture W_c (for cohesive tearing). The values of W_a and W_c are, respectively, the energy which must be expended to detach a unit area of interface and the energy expended per unit area of material torn.

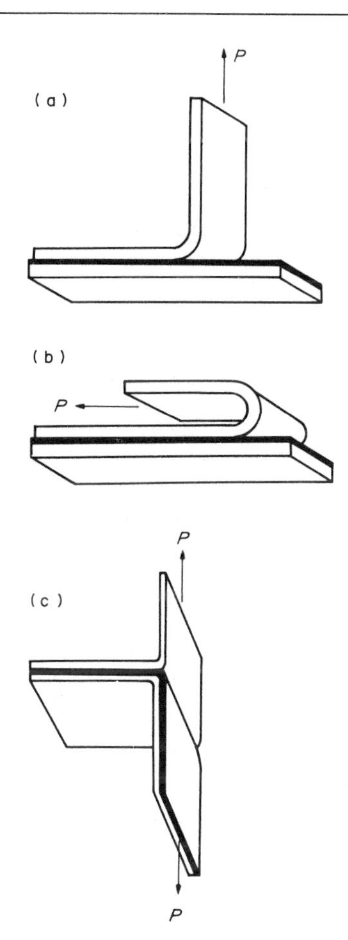

Figure 2
Common peel test geometries: (a) 90° peel; (b) 180° peel; (c) T-peel

Two quite different approaches to the analysis of peeling rupture have been employed. The first sets up an energy balance for peeling an adhering strip (Lindley 1971). The second involves calculations of the stresses set up in the adhesive layer during peeling (Kaelble 1959, 1960).

2. Energy Considerations in Peeling

During steady-state peeling, conservation of energy requires that:

work done by peel force = work of detachment + strain energy imparted to newly detached strip (1)

For peeling a unit length of bonded interface apart, Eqn. (1) becomes:

$$P(1+\varepsilon-\cos\theta)=W_a+tE_{strain} \qquad (2)$$

where P is the peel force per unit width, t is the thickness of the peeled strip, ε and E_{strain} are, respectively, the tensile strain and energy density imparted to the detached strip and θ is the peel angle. Adhering strips are generally thick enough to detach without stretching significantly, at least at large peel angles, so that ε and E_{strain} are relatively small. In this case, Eqn. (2) can be simplified and the peel force becomes:

$$P = W_a/(1-\cos\theta) \quad (3)$$

Thus the peel force P provides a direct measure of the work W_a of detachment.

There are several contributions to W_a. The first is the intrinsic work $W_{a,0}$ required to break the interfacial bond. This is generally quite small, ranging from about 0.1 $J\,m^{-2}$ for van der Waals bonding between nonpolar materials up to about 100 $J\,m^{-2}$ for covalently bonded polymeric materials. Other contributions come from energy dissipated within the adhesive and adherend layers as they are deformed during peeling. If these layers were deformed reversibly, without dissipating any mechanical energy, the peel force would be small, reflecting only the intrinsic interfacial strength $W_{a,0}$. However, for typical adhesives this criterion of perfectly elastic deformations is not met and W_a is correspondingly much greater than $W_{a,0}$, being of the order of 10^4 $J\,m^{-2}$ or more for tough, ductile adhesive and adherend layers.

Because energy dissipated in deforming the adhesive and adherends and in bending the layers apart on separation depends generally upon the peel angle θ, the thickness of the various layers, the rate of peeling, the test temperature and the dissipative properties of the various components, the value deduced for W_a is found to depend upon all these factors also, even though the intrinsic interfacial strength $W_{a,0}$ is in principle independent of them.

The exact way in which the various components of W_a add together has not yet been worked out. They are clearly not independent of each other. If there is no intrinsic adhesion, no deformation energy is required to bring about separation. A factorization principle has been suggested (Gent and Schultz 1972):

$$W_a = W_{a,0}(1+H) \quad (4)$$

where H is a numerical factor representing the dissipative properties of the adhesive and adherends. However, no method of calculating H from first principles has yet been devised.

3. Stress Analysis

An elastic beam (adhered layer) on an elastic foundation (adhesive) has been employed as a model of a peeling strip. Simple bending relations are used to calculate the peel force. It is assumed that the tensile stress σ in the adhesive layer is constant across the width of the bond and through its thickness. The stress at a distance $-x$ into the bond is obtained (Kaelble 1959, 1960) as

$$\sigma = \sigma_0(\cos Bx + K\sin Bx)\exp Bx \quad (5)$$

where

$$B = \left(\frac{Yw}{4EIa}\right)^{0.25} \quad (6)$$

and

$$K = \frac{Bm}{Bm + \sin\theta} \quad (7)$$

in which σ_0 is the maximum tensile stress at $x=0$, Y is Young's modulus of the adhesive, E is Young's modulus of the adherend, a is the adhesive layer thickness, w is the bond width, I is the moment of inertia of the adherend layer cross section, m is the moment arm of the peel force and θ is the peel angle. Equation (5) reveals that the cleavage stress in the adhesive layer is a highly damped harmonic function of distance $-x$ from the peel front, involving alternating regions of tension and compression. However, it is strictly valid only for adhesive joints made of linearly elastic materials and for values of $K \approx 1$ (i.e., $Bm \gg \sin\theta$), since the derivation is based on small bending theory. Under these circumstances,

$$P = a\sigma_0^2/2Y(1-\cos\theta) \quad (8)$$

It is instructive to compare Eqn. (8) with the result obtained previously from energy considerations, Eqn. (3). The predicted dependence of peel force on peel angle is identical in both equations.

4. Effect of Thickness of the Adhesive Layer

By combining Eqns. (3) and (8), it is seen that

$$\sigma_0 = (2YW_a/a)^{0.5} \quad (9)$$

Thus, the maximum tensile stress set up in a linearly elastic adhesive is predicted to be inversely proportional to the square root of the adhesive layer thickness. The stress at the point of failure would be expected to be rather independent of the adhesive layer thickness, but it should be noted that Eqn. (9) applies only to the mean tensile stress throughout the adhesive layer thickness and not to the local stress set up at the line of detachment.

In practice, it is found that the peel force increases as the thickness of the adhesive layer is increased, as shown in Fig. 3. This is due to energy dissipation within the bulk of the adhesive. As larger volumes of adhesive are subjected to deformation per unit area of detachment, so the total work expended in peeling increases. At sufficiently large thicknesses, however, the peel force becomes independent of thickness. The highly stressed region around the line of detachment is now small in comparison with the total thickness of

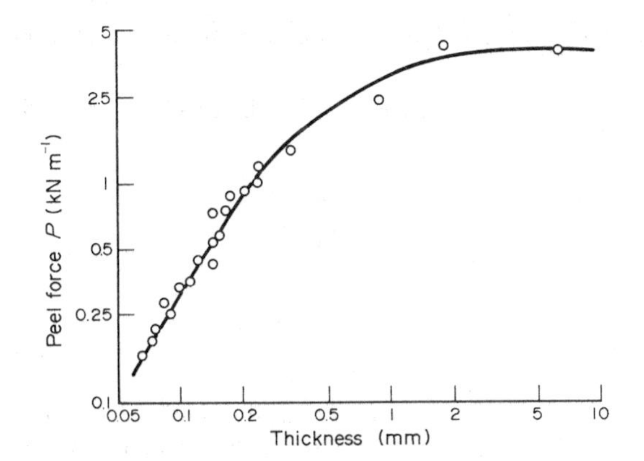

Figure 3
Dependence of peel force upon adhesive layer thickness for an elastomer detaching from a polyester substrate (after Gent and Hamed 1978)

the adhesive layer and the major dissipation process no longer involves the entire layer of adhesive.

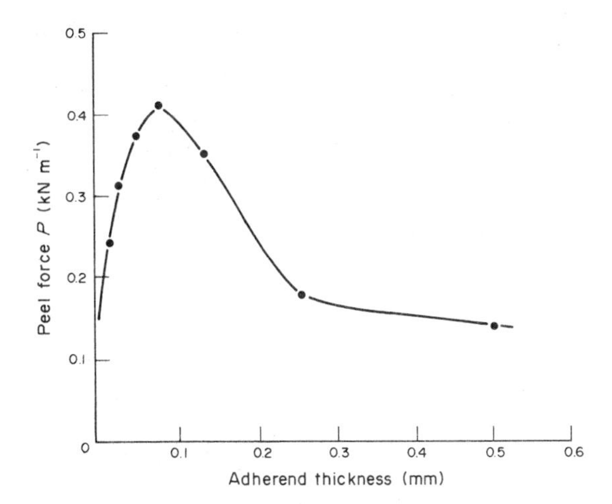

Figure 4
Effect of thickness of backing on peel force for a pressure-sensitive adhesive tape stripped from a rigid substrate (after Satas and Egan 1966)

5. Effect of Thickness of an Elastic–Plastic Peeling Layer

Generally, the peel force passes through a maximum with thickness of the peeling layer (Fig. 4). This is due to energy dissipated in bending the layer away from the interface (Gent and Hamed 1977b). For a sufficiently thick peeling strip, the bending stresses set up are small and the strip remains substantially elastic as peeling proceeds. Thus, little energy is dissipated within it. However, if the strip is sufficiently thin, the bending stresses may be large enough to cause plastic deformation during peeling and W_a is then augmented by the work expended plastically in the strip. For an elastic–plastic adherend, peeled at 180°, the maximum contribution to the peel force due to yielding, $P_{y,\max}$, when the entire strip undergoes plastic yielding, is calculated to be

$$P_{y,\max} = t\sigma_y/4 \tag{10}$$

where σ_y is the yield stress.

The critical thickness t_c of the adherend, above which no plastic yielding occurs, is calculated as

$$t_c = 12EP_0/\sigma_y^2 \tag{11}$$

where P_0 denotes the peel force per unit width in the absence of plastic yielding.

Thus, sufficiently thick strips do not undergo plastic yielding and hence there is no contribution to the observed peel strength from this cause. At the other extreme, a very thin strip undergoes complete plastic yielding; however, the total energy dissipated in this way is small because the thickness t is small. Hence, the additional peel force contribution due to plastic yielding is small (Eqn. (10)). As the thickness of the strip is increased, more energy is dissipated in yielding and the peel force rises proportionately. Eventually, at a large enough thickness, the bending stresses set up in the strip are insufficient to cause the entire strip to undergo yielding and the contribution to the peel force from this mechanism diminishes, eventually reaching zero again when the strip thickness exceeds t_c. Thus, the peel force for both infinitesimally thin and thick adhering layers is equal to P_0.

6. Rate and Temperature Effects in Pressure-Sensitive Adhesives

Pressure-sensitive adhesives consist of soft elastomeric semisolids. Their peel strength depends strongly upon the rate of peel and the test temperature, as shown for a model system in Fig. 5. At low rates, the peel force increases with rate, and failure takes place entirely within the adhesive layer, which fails by flowing apart. At a critical rate of peel (which depends upon the test temperature) an abrupt transition takes place to interfacial separation, at much smaller peel forces. This transition occurs when the rate of deformation at the peel front is so high that the adhesive molecules are unable to disentangle and flow apart like a liquid but remain together as a coherent elastic solid. Although the local stress required to disentangle the molecules at low rates is relatively small, the work expended in ductile flow is large and the peel force (which measures the work of separation) is correspondingly high. In the elastic state, the work of separation is mainly expended at the interface and is then relatively small.

The rate (and temperature) at which the abrupt transition occurs depend upon the molecular weight of the polymeric adhesive and its segmental mobility.

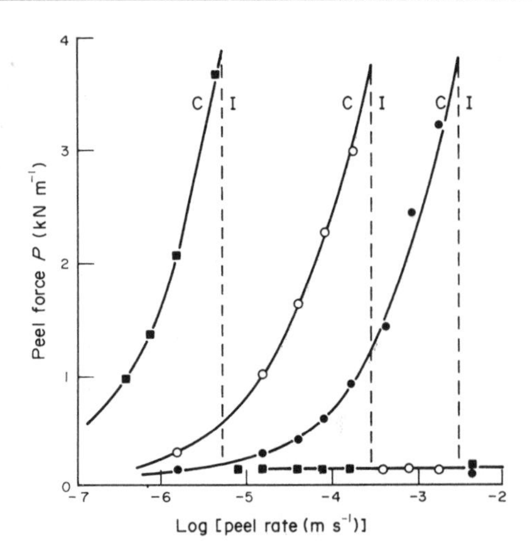

Figure 5
Peel force as a function of peel rate and test temperature for an elastomer adhering to a polyester substrate: ■, −5 °C; ○, 10 °C; ●, 23 °C; C and I denote cohesive and interfacial failure modes (after Gent and Petrich 1969)

The peel strength above the critical rate depends upon the factors discussed previously. Below the critical rate, the peel strength is primarily a measure of the work of extending a viscoelastic liquid to the point of rupture.

Bibliography

Gent A N, Hamed G R 1977a Peel mechanics of adhesive joints. *Polym. Eng. Sci.* 17: 462–66
Gent A N, Hamed G R 1977b Peel mechanics for an elastic–plastic adherend. *J. Appl. Polym. Sci.* 21: 2817–31
Gent A N, Hamed G R 1978 Adhesion of elastomers with special reference to triblock copolymers. *Plast. Rubber Mater. Appl.* 3: 17–22
Gent A N, Petrich R P 1969 Adhesion of viscoelastic materials to rigid substrates. *Proc. R. Soc. London, Ser. A* 310: 433–48
Gent A N, Schultz J 1972 Effect of wetting liquids on the strength of adhesion of viscoelastic materials. *J Adhes.* 3: 281–94
Kaelble D H 1959 Theory and analysis of peel adhesion: Mechanisms and mechanics. *Trans. Soc. Rheol.* 3: 161–80
Kaelble D H 1960 Theory and analysis of peel adhesion: Bond stresses and distributions. *Trans. Soc. Rheol.* 4: 45–73
Lindley P B 1971 Ozone attack at a rubber–metal bond. *J. Inst. Rubber Ind.* 5: 243–48.
Satas D, Egan F 1966 Peel adhesion and pressure sensitive tapes. *Adhes. Age* 9: 22–25

A. N. Gent and G. R. Hamed
University of Akron, Akron, Ohio, USA]

Brittle Fracture: Linear Elastic Fracture Mechanics

All materials tend to fracture if stressed severely enough. Some materials fracture much more easily than others, and are thereby said to be "brittle." Most brittle of all are solids with covalent–ionic bonding, glasses and ceramics; by contrast, metals and polymers have the quality of "toughness" required for general engineering applications. In many areas of technological development the question of brittleness is a vitally important factor in design strategy. Essentially, the problem is to contain the growth of cracks. Linear elastic fracture mechanics (LEFM) is the formalism developed to address this problem. Fundamentally rooted in the laws of classical mechanics and thermodynamics, LEFM provides a sound theoretical basis for characterizing resistance to fracture. The generality of the methodology is such as to accommodate an unlimited diversity of crack configurations, and to allow for materials evaluation under different conditions of testing. At the same time, the underlying assumption of a linear continuum is restrictive in the insight that may be gained into the actual physical processes of crack extension; LEFM describes the mechanics, not the mechanisms, of fracture.

1. Griffith's Theory of Fracture

The starting point for present-day fracture mechanics is the classic paper of Griffith (1920). Griffith considered a continuous, isotropic, homogeneous elastic body containing an internal through-the-thickness planar crack of length $2a$, with applied loading at the external boundary (Fig. 1). Treating the body as a reversible thermodynamic system, he formulated a requirement for crack extension in terms of the equilibrium condition

$$dU/da = d(U_M + U_S)/da = 0 \quad (1)$$

where the total system energy U consists of a mechanical component U_M (energy stored elastically in the body less work done by applied tractions in displacing the boundary) and a surface component U_S (energy associated with creation of new crack surfaces). For $da > 0$, $dU_M < 0$ and $dU_S > 0$ always, so Eqn. (1) represents a balance between mechanical energy release (driving force) and surface energy increase (resisting force). In this interpretation, the crack extends or contracts reversibly according to whether the total energy differential in Eqn. (1) is negative or positive.

The remainder of Griffith's paper deals with various aspects of the strength of brittle solids. The first step was to derive an expression for the equilibrium of a crack system loaded remotely in uniform tension. For this purpose Griffith modelled the crack as an elliptical cavity of infinitesimal semiminor axis and unit thickness, making use of an earlier stress analysis by Inglis to obtain $U_M = -\pi a^2 S_a^2/E$ for the mechanical energy

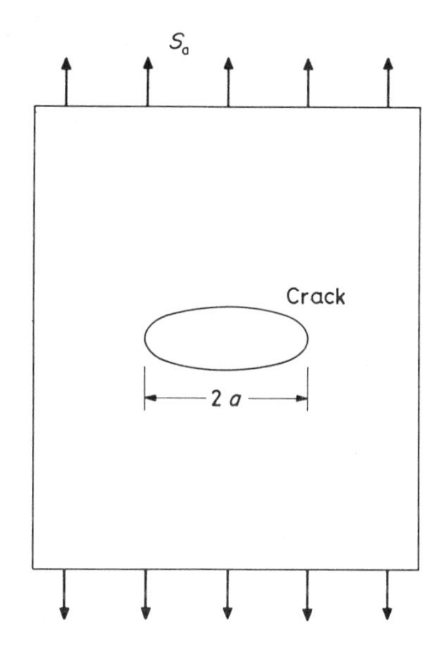

Figure 1
Internal crack system subjected to loads applied at outer boundaries. The configuration shown is that considered by Griffith

term, where S_a is the applied tensile stress and E is Young's modulus (the minus sign indicating an energy reduction relative to the uncracked state). Writing $U_S = 4a\gamma$ for the surface energy term, with γ the free surface energy per unit area (there are two new surfaces created), Eqn. (1) yields the critical stress

$$S = (2E\gamma/\pi a)^{1/2} \tag{2}$$

It is readily shown that Eqn. (2) represents an unstable equilibrium, that is, the crack grows catastrophically when $S_a > S$.

Griffith next introduced cracks of measurable lengths (millimeter-scale) into glass test pieces, which he stressed to failure to establish that $Sa^{1/2}$ is constant. He then calculated γ from Eqn. (2), and confirmed this to be of the order of the true surface energy of glass ($\sim 1\,\mathrm{J\,m^{-2}}$). Extrapolation of the results to atomic-scale crack sizes gave $S \approx E/10$ as an estimate of the theoretical limiting strength for the glass structure.

On testing glass test pieces in their as-received state, Griffith found strengths in the region $S \ll E/10$. He concluded that the glass must ordinarily contain flaws, submicroscopic microcracks or other stress-concentrating defects, effective crack size ~ 1–$10\,\mu\mathrm{m}$. Only in exceptional circumstances (e.g., untouched, freshly drawn fibers) could the theoretical strength be approached.

2. *Formalism of LEFM*

Griffith's two major contributions, the energy-balance concept and the flaw hypothesis, remain the cornerstone of brittle fracture theory. However, to be a practical tool in structural analysis a more general mathematical framework is necessary. The work of Irwin (1958) provides this framework. Several important fracture parameters, relating to crack propagation at different levels with respect to the crack tip, emerge from the theoretical development.

2.1 Crack Extension Force

Following Griffith's revelation that it is the mechanical energy release which drives a brittle crack, Irwin defined a generalized crack extension force:

$$G = -dU_M/dA \tag{3}$$

where A is the crack area. The object of the fracture mechanics approach is to express G in terms of characteristic applied loading and crack-dimension variables, and elastic constants, for a given fracture specimen. Taken in conjunction with an appropriate fracture criterion, this then allows for prediction of the crack history to failure.

The most direct route to an evaluation of G is via the mechanical energy term U_M. An instructive example is a body with an internal crack, subjected to a tensile force P applied at some remote point. For any fixed crack area A, the body behaves as a linear elastic spring with a compliance $\lambda = u/P$, where u is the load-point displacement. In an incremental crack growth dA, the compliance changes by an amount $d\lambda = \lambda(du/u - dP/P)$. The corresponding mechanical energy change during this extension is the elastically stored energy, $d(\frac{1}{2}Pu)$, less the work done in displacing the load point, $P\,du$, that is, $dU_M = -\frac{1}{2}(P\,du - u\,dP)$. This gives the crack extension force in terms of the compliance characteristics,

$$G = \tfrac{1}{2}P^2\,d\lambda/dA = \tfrac{1}{2}\,(u^2/\lambda^2)\,d\lambda/dA \tag{4}$$

The energetics of crack expansion are therefore expressible in terms of the load–deflection response of the system.

It may be noted in the previous example that G is independent of the manner of loading; for example, the same result is obtained at constant P (deadweight loading), and constant u (fixed-grip loading). The latter is a special case, in that dU_M is determined entirely by the amount of strain energy released (the applied load doing zero work). For this reason G is often referred to as the fixed-grip strain energy release rate.

2.2 Stress Intensity Factor

Another extremely powerful fracture parameter which derives from the Irwin approach is the stress intensity factor. This parameter derives from a consideration of the near stress field about a sharp, slitlike crack (Fig. 2). Assuming the slit walls to be free of tractions, analysis of the elasticity equations gives simple solu-

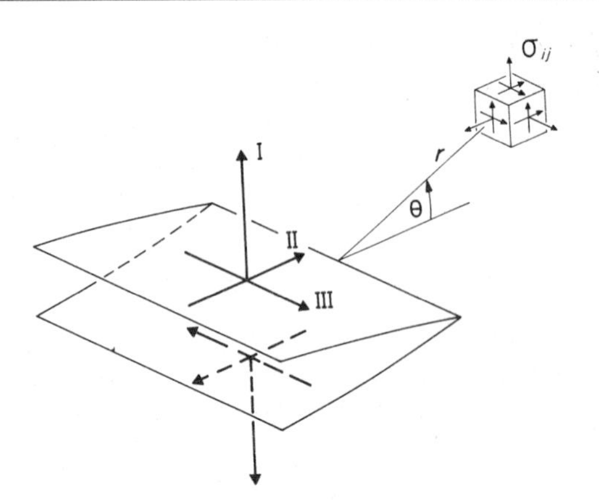

Figure 2
Slitlike crack tip, showing stress field coordinate system; arrows designate displacement vectors for the three fracture modes

tions for the stresses,

$$\sigma_{ij} = K f_{ij}(\theta)/(2\pi r)^{1/2} \tag{5}$$

where r and θ are polar coordinates about the crack tip. The coordinate-dependent terms in Eqn. (5) determine the distribution of the field; the factor K likewise determines the intensity—hence the terminology. It can be readily shown that the stress intensity factor depends on the magnitude of the applied loading, and is a function of the characteristic crack size. In this context it is noted from Eqn. (5) that K has dimensions of stress × distance$^{1/2}$.

A major advantage of the stress intensity factor as a fracture parameter is its property of additivity (a product of the principle of superposition for linear elastic deformations) for a given mode of crack extension. Three such modes, identified by their (mutually orthogonal) crack-surface displacements, can be defined (Fig. 2): I, an opening mode, where tensile forces act to separate crack walls normally; II, a sliding mode, where shear forces produce equal and opposite displacements in a direction perpendicular to the crack front; and III, a tearing mode, similar to II but with the shear direction parallel to the crack front. It is common to designate the operative mode by an appropriate subscript notation, thus K_{I}. In the case of brittle fracture, however, cracks tend to propagate almost exclusively under mode I conditions, and on this understanding the subscript is often conveniently omitted.

By considering the virtual opening and closure of an incremental crack element, the strain energy release may be calculated in terms of the stress components in Eqn. (5). Such an operation shows the G and K parameters to have a certain equivalence. For mode I fracture

$$G = K^2/E \tag{6}$$

Similar terms may be added to the righthand side to account for any shear loading, indicating that for different modes it is the crack extension force rather than the stress intensity factor which is superposable.

2.3 Crack Opening Displacement

The stress intensity factor concept takes us closer to the region where fracture processes actually operate, namely the crack tip. However, there is an inbuilt limit as to how close we can get, as the stress field described by Eqn. (5) contains a singularity at $r \rightarrow 0$. An approach which purports to overcome this difficulty is that which gives direct consideration to the response of an element of material immediately ahead of the crack tip. It can be shown that

$$G = 2\int_0^u \sigma(u)\,du \approx \sigma u \tag{7}$$

where σ is the normal stress acting on the element and u is the corresponding crack opening displacement. Thus, with knowledge of an appropriate constitutive relation $\sigma(u)$ for the crack element, the driving force for fracture may be expressed exclusively in terms of the parameter u.

It is at this point that LEFM becomes inadequate, as it can be demonstrated that the requisite function $\sigma(u)$ in Eqn. (7) is necessarily nonlinear. This will be taken up again in Sect. 6.

3. *Fracture Criteria*

The preceding formalism provides parameters which characterize the driving force for crack propagation. The next step is to determine the conditions under which the driving force becomes sufficient to cause extension. It is convenient to subdivide the various criteria into three categories.

3.1 Equilibrium Fracture

The Griffith energy-balance concept lays the foundation for defining the intrinsic resistance of a material to fracture under equilibrium conditions. Such conditions are generally realized in vacuum or inert environments, or at low temperatures. Crack extension occurs when the driving force attains a critical level, $G_C = 2\Gamma$, where Γ is the energy required to create unit area of new crack surface (again two surfaces are created). In the Irwin interpretation, Γ may contain a dissipative component in addition to the reversible surface energy term γ, thus opening the way for extension of the concept to materials of varying degrees of brittleness. In accordance with Eqns. (6) and (7), the equilibrium condition may equally well be expressed in terms of a critical stress intensity factor, $K_C = (2\Gamma E)^{1/2}$, or crack-tip opening displacement, $u_C \approx 2\Gamma/\sigma_C$. These critical quantities may be regarded as material toughness parameters.

For highly brittle materials, typical values of the material constants are $\Gamma \approx 5\,\mathrm{J\,m^{-2}}$, $K_C \approx 1\,\mathrm{MPa\,m^{1/2}}$ ($E \approx 100$ GPa). Inserting $\sigma_C \approx E/10$ as the critical crack-tip stress, we obtain $u_C \approx 0.5$ nm for the crack-tip opening displacement; that is, the crack is atomically sharp, consistent with a bond-rupture process of fracture (Lawn et al. 1980, Lawn 1983). With less brittle materials, values of the toughness parameters may be markedly larger, by several orders of magnitude in the case of most metals and polymers.

3.2 Kinetic Fracture

Cracks can often extend at subcritical stress levels, especially in the presence of reactive chemical environments. A major characteristic of this type of fracture is its time dependence. The crack extends at a certain rate which depends sensitively on the magnitude of the applied stress. Other important variables are the concentration of chemically active species c (either within the environment or the material structure itself), and temperature T. Consequently, the crack extension condition has its roots in chemical kinetics, and is generally expressed in terms of a crack velocity function

$$v = da/dt = v\,(K, c, T) \tag{8}$$

in the region $K < K_C$.

Kinetic fracture is not well understood at the fundamental level. In brittle materials the process involves some form of chemically assisted bond rupture. In some cases transport processes can become a dominant factor. The form of Eqn. (8) is thus generally complex, and usually has to be determined empirically. The situation is even more obscure with nonbrittle materials, where dissipative processes can play an important role, especially in cyclic loading.

3.3 Dynamic Fracture

Once the equilibrium condition is exceeded, that is $K > K_C$, the crack is free to extend dynamically, at a velocity controlled by the inertia of the crack walls. If the equilibrium is a stable one, the crack grows only as quickly as the load is applied. If unstable, the crack accelerates rapidly toward a terminal velocity governed by the speed of sonic waves in the solid. According to a suggestion by Mott, the mechanics of dynamic fracture may be accommodated within the energy-balance framework by adding a kinetic energy term; thus, $U = U_M + U_S + U_K$. Differentiation with respect to crack area then gives

$$G - 2\Gamma = dU_K/dA \tag{9}$$

so that the net crack driving force is expressible as an energy dissipation rate. If $U_K(A)$ is an increasing function, the crack propagates without limit, and catastrophic failure ensues; if it is a decreasing function, the crack slows down, and ultimately arrests.

4. Testing Methods

Various laboratory test pieces have been designed to measure fracture parameters for prospective engineering materials. The basic idea is to follow the response of a contrived crack system to specified conditions of loading, for given environment, temperature, and so on.

One method of widespread usage, especially in the evaluation of ceramic materials, involves adoption of specimens with reasonably simple geometry, thereby allowing for analytical solution of the elasticity equations in terms of well-defined boundary conditions. A large number of crack configurations have been subjected to such analysis, and standard solutions, generally expressed in stress intensity factor notation, are available in handbook compilations. For example, the solution for a crack of characteristic dimension a in uniform tension S_a is of the form

$$K = YS_a(\pi a)^{1/2} \tag{10}$$

where Y is a crack geometry term; inserting $Y = 1$ for an internal through crack, and invoking Eqn. (6) in conjunction with the equilibrium condition $G_C = 2\gamma$ for ideally brittle fracture, it is then a straight-forward matter to rederive the Griffith formula, Eqn. (2). It should be noted that $K(a)$ is an ever-increasing function, so that the crack, once beyond the equilibrium point, is unstable, as indicated in Sect. 1. In many cases, it is preferable to work with a test piece for which $K(a)$ is a diminishing function, the greater configurational stability making for a more controlled experiment, particularly in subcritical crack growth studies. Specimens with constant K are of special interest, for then one may, in principle, determine fracture parameters without having to follow the crack propagation explicitly. The proper specimen for any given study will, of course, be determined by the actual properties under investigation—the study of crack arrest, for example, demands a specimen with both unstable and stable equilibrium crack lengths—and by material considerations such as microstructure.

An alternative approach is to determine the availability of energy to the crack system by directly monitoring the load–deflection curve. There are several variants. One method is to predetermine an appropriate compliance function $\lambda(A)$ by measuring the elastic response of dummy specimens containing cracks of different, fixed area A, and then using Eqn. (4) to calculate G. Another method is to contrive a geometrical configuration in which the crack can be driven stably through the entire test piece without extraneous energy dissipation (e.g., at supports), so that the area under the load–deflection curve (work to fracture) may be identified exclusively with the total surface energy of the crack produced. Techniques in this category are less restrictive on crack geometry, an advantage bought at the expense of reduced amenability to theoretical stress analysis.

5. *Strength*

Linear elastic fracture mechanics provides the means for quantifying the factors which control the strength of brittle materials. Given a specimen with a dominant flaw of effective linear dimension $a' = Y^2\pi a$, insertion of $K = K_C$ into Eqn. (10) gives the instability stress

$$S = K_C/(a')^{1/2} \qquad (11)$$

as a modified form of Eqn. (2). Strength is thereby seen as controlled by two parameters: one intrinsic, the material toughness K_C, the other extrinsic, the nominal flaw size a'. Equation (11) is applicable under a wide range of test or service conditions. In essentially inert environments the flaw remains stationary up to the point of instability, in which case S defines the maximum stress level the given material can withstand. For design purposes, it becomes a matter of ensuring that this level is never attained (either by direct mechanical loading or by virtue of some thermal or electrical response). In reactive environments, on the other hand, flaws may grow subcritically prior to failure, in which case it becomes more meaningful to interpret Eqn. (11) as specifying the maximum flaw size a' that a material can sustain at a fixed stress. This in turn leads to the notion of a finite lifetime for a brittle component, an evaluation of which may be obtained via integration of Eqn. (8) in conjunction with Eqn. (10).

The key to high strength therefore rests with maximizing toughness and minimizing flaw size. These are topics which strictly fall outside the scope of continuum mechanics, and are taken up elsewhere in the Encyclopedia (see *Brittle Fracture: Micro-mechanics*). However, there are two points which should be mentioned here in connection with the design aspect of strength. The first concerns the scale of the underlying micromechanical processes, especially as reflected in the flaw size. With highly brittle ceramics and glasses this scale is small, typically < 100 µm, so that direct observation of flaw evolution is not usually practicable. This is the realm of high resolution microscopy techniques, as illustrated by the example in Fig. 3. The second point relates to the vast diversity of processes that can lead to flaw generation, with consequent strong variability in flaw dimension a and geometry Y. As a direct result of these two difficulties, a class of strength theories based on probabilistic concepts has arisen, in which flaw populations are taken to be distributed in accordance with some empirical statistical function. Although it gives little physical insight into the actual processes of brittle fracture, this approach is attractive to the engineer who seeks only a simple mathematical formula for predicting prospective strengths of components.

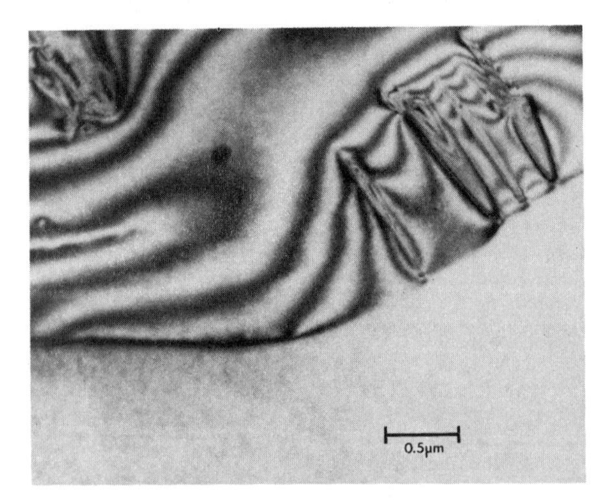

Figure 3
Transmission electron micrograph of part of a flaw in a brittle solid (silicon). The flaw has the nature of a microcrack, revealed by virtue of diffraction contrast at the residual interface (courtesy of B J Hockey)

6. *Mechanics* vs *Mechanisms*

Although LEFM derives from the basic laws of classical physics, and has great power in the analysis of brittle fracture systems, it has important limitations. Brief allusion was made to these limitations earlier in the article. Put simply, LEFM loses its validity as we attempt to view events closer to the vital crack-tip region where physical separation actually occurs. The stress-field singularity at $r \to 0$ in Eqn. (5) exemplifies the problem. Hooke's law has been assumed to hold everywhere in the crack system; in reality, materials tend to a maximum in their intrinsic stress–strain behavior, so that the essential mechanisms of fracture are nonlinear. The linear formalism is accordingly inadequate for describing how cracks propagate. The same formalism nevertheless survives as a useful means of determining when cracks propagate, as in most brittle fractures the nonlinear separation processes are effectively confined to a small region (small compared with crack and specimen dimensions) about the tip. Several models based on this "small-scale zone" principle have been proposed in an attempt to incorporate the essence of nonlinearity into the analysis, for example, via the elemental stress–displacement function $\sigma(u)$ in the crack opening displacement parameter of Eqn. (7).

Insofar as truly brittle fracture is concerned (i.e., where cracks proceed via sequential rupture of cohesive bonds), LEFM is limited also by its continuum basis. Science contains many examples, even in classical physics, where the discrete nature of matter can have a profound influence on properties. The lattice crack models of Thomson (1973) and others, based on the representation of solids as assemblages of point masses (atoms) and linking nonlinear springs (bonds), are no exception. A major prediction from these models is that the intrinsic surface energy resistance to fracture is not a smooth function of crack area, but contains perturbations at the atomic level. Crack propagation is thus seen as a process of activation over an array of local energy barriers. From such a description, it is possible to obtain first-principles derivations of the fundamental mechanical relations which govern crack extension, for example, the crack velocity function in Eqn. (8). Again, the more macroscopic driving-force parameters such as G and K remain unaffected by the detailed nature of these highly localized crack-tip events, and the atomic modifications can accordingly be accommodated, via the surface energy resistance term, into the LEFM framework (Lawn and Wilshaw 1975).

A third factor which requires attention is the reversibility implicit in the elasticity-based formulation. It has already been indicated (Sect. 3) that the fracture surface energy can contain a dissipative component; indeed, that this component can be dominant in the energetics of crack propagation (notably in metallic and polymeric materials). Dissipative processes can occur in all materials, but in the case of intrinsically brittle fractures these processes necessarily operate away from the crack tip. The picture is that of an atomically sharp crack triggering some localized energy absorption "sinks" (e.g., isolated dislocations and second-phase particles) within its advancing near field (see *Brittle Fracture: Micromechanics*). Such sinks are confined to a region close to the crack tip where the stresses are high, within a so-called process zone. One way of incorporating this phenomenon into LEFM is to regard the process zone as shielding the crack tip from the remotely applied loading. The local field intensity experienced by the tip is accordingly reduced in relation to that evaluated from the remote boundary conditions, by an amount which is determined by micromechanical considerations. With this qualified stress intensity factor (or equivalent crack-extension force), the basic rules of LEFM (e.g., the fracture criteria of Sect. 3) remain intact.

See also: Mechanics of Materials: An Overview

Bibliography

Griffith A A 1920 The phenomena of rupture and flow in solids. *Philos. Trans. Roy. Soc. London, Ser. A.* 221: 163–98

Irwin G R 1958 Fracture. In: Flugge S (ed.) 1958 *Handbuch der Physik*, Vol. 6. Springer, Berlin, pp. 551–90

Lawn B R 1983 Physics of fracture. *J. Am. Ceram. Soc.* 66: 89–91

Lawn B R, Hockey B J, Wiederhorn S M 1980 Atomically sharp cracks in brittle solids: An electron microscopy study. *J. Mater. Sci.* 15: 1207–23

Lawn B R, Wilshaw T R 1975 *Fracture of Brittle Solids*. Cambridge University Press, Cambridge

Thomson R M 1973 The fracture crack as an imperfection in a nearly perfect solid. *Annu. Rev. Mater. Sci.* 3: 31–51

B. R. Lawn
[National Bureau of Standards,
Washington, DC, USA]

Brittle Fracture: Micromechanics

Fracture in brittle solids results from discrete flaws, activated by an applied or concentrated stress. The flaws are created either during the fabrication process (e.g., inclusions, voids or large grains generated during sintering) or by surface interactions (e.g., surface cracks induced by machining or projectile impact). A recognition of these flaws, a comprehension of their mode of formation and an evaluation of their influence on the failure stress are essential constituents of designs and procedures that permit the reliable use of brittle materials. This article examines the formation of flaws and describes the principal relationships between the flaw type and size and the ultimate failure stress.

The failure-initiating flaws invariably exhibit a size and spatial distribution that necessitates a probabilistic approach to failure. Some of the essential statistical features of the brittle fracture process are thus included here. Particular emphasis is placed on relations between the statistical parameters and the physical characteristics of the important failure-initiating flaws.

Brittle materials are frequently used in preference to other materials because of their high-temperature capabilities. Components are often, therefore, subject to thermal transients which generate relatively large thermal stresses and induce failure. This thermal shock process is of special importance to the use of brittle solids and so a description of thermal failure is given.

1. Failure Behavior

Fracture in brittle solids can be considered to occur either by the direct activation of noninteracting, preexistent flaws or by flaw generation and coalescence processes, as shown in Fig. 1. The former mode of failure generally pertains to solids with a fine-scale microstructure at low to intermediate temperatures ($\lesssim 1100\,°C$), and thus encompasses most of the high-strength brittle solids. A reasonable understanding of this failure characteristic has evolved in recent years. This article is primarily concerned with this mode of failure.

The failure of materials with coarser microstructures often involves the generation of subcritical microcracks. These microcracks form during application of stress by virtue of the existence of very large,

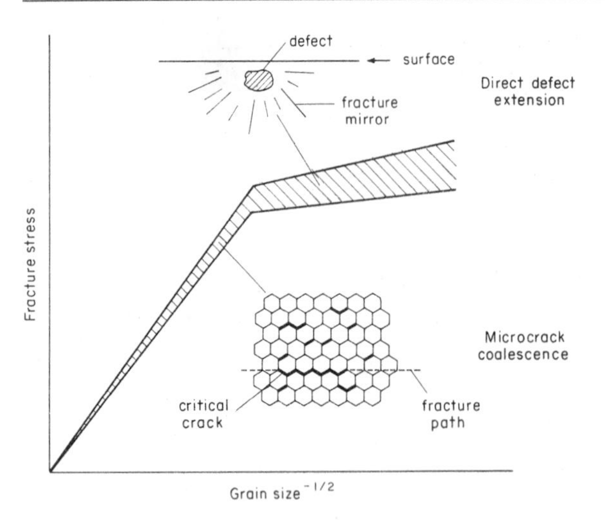

Figure 1
Failure of brittle solids: in typical fine-grained materials fracture occurs from preexistent flaws and the strength is relatively grain size insensitive, whereas fracture in coarser-grained materials usually proceeds by the coalescence of grain-boundary-located microcracks induced by the applied stress

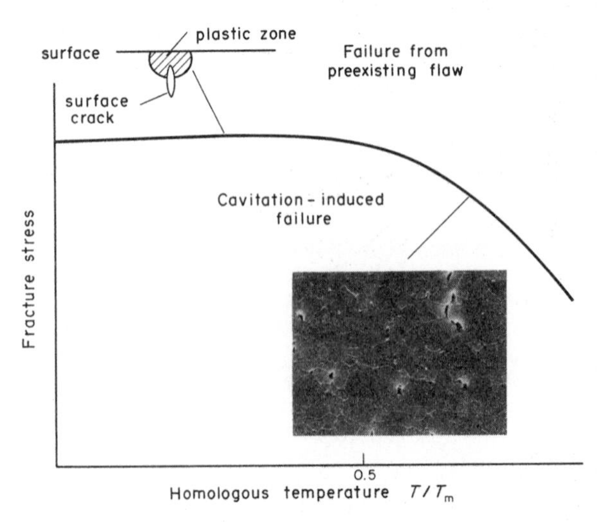

Figure 2
Fracture stress in brittle solids as a function of homologous temperature: fracture at low to intermediate temperatures occurs from preexistent flaws (e.g., surface cracks), whereas fracture at high temperatures occurs by cavity growth and coalescence at grain boundaries

localized residual stresses resulting from thermal contraction mismatch or inherent anisotropy of the material. The failure process involves the formation and coalescence of such microcracks, in accord with statistical concepts. Preliminary analyses have been developed, but a full understanding has yet to emerge. No specific consideration of this mode of failure is given here.

Crack formation can also occur at elevated temperatures in conjunction with the onset of viscous deformation. The most commonly encountered failure process involves diffusive cavitation at grain boundaries (Fig. 2). The process exhibits the same general phenomenology as creep rupture in ductile solids, but the failure details are specific to the microstructure.

2. Flaw Mechanics

In those classes of brittle solids that fail by the extension of preexistent flaws, three principal flaw categories can be distinguished: surface cracks, fabrication defects and environmentally induced defects. The typical origins of these flaws are examined here and their respective roles in the failure process are described. Study of the influence of flaws on failure involves (a) considerations of the expected flaw size and its relation to the dominant extrinsic and intrinsic variables, and (b) analysis of the dependence of the failure stress on the flaw size and other pertinent material parameters. The importance of comprehending the failure details is illustrated by the wide range of fracture strengths that obtain for different flaws of the same size in Si_3N_4 (Fig. 3). These details include some understanding of the local or residual stresses in the flaw vicinity and the existence of complementary flaws in adjacent regions.

2.1 Surface Cracks

Hard particles subject to a force P can plastically penetrate the surface of brittle solids. The localized plastic deformation creates residual stresses which motivate the formation of surface cracks (Fig. 4). Particle penetration, leading to crack formation, usually occurs during surface finishing operations such as machining, grinding or polishing. Similar plastic penetration occurs when a hard particle impinges on a surface. Specific surface crack populations should thus be expected on surfaces exposed to impacting, eroding or abrading particles.

The crack formation process can be considered to result from outward forces exerted by the plastic zone on the surrounding elastic material (Fig. 4). These outward forces develop because volume is conserved in plastic flow and accommodation of the indentation volume causes expansion of the plastic zone. Analysis of cracking subject to central forces imposed by the plastic zone indicates a crack radius a given by

$$a = \lambda (E/H)^n \hat{P}^{2/3} K_{Ic}^{-2/3} \cot^{2/3}\Psi \quad (1)$$

where $\hat{P}$ is the maximum normal force exerted on the penetrating particle, K_{Ic} is the fracture toughness, E is Young's modulus, H is the hardness, Ψ is the angle

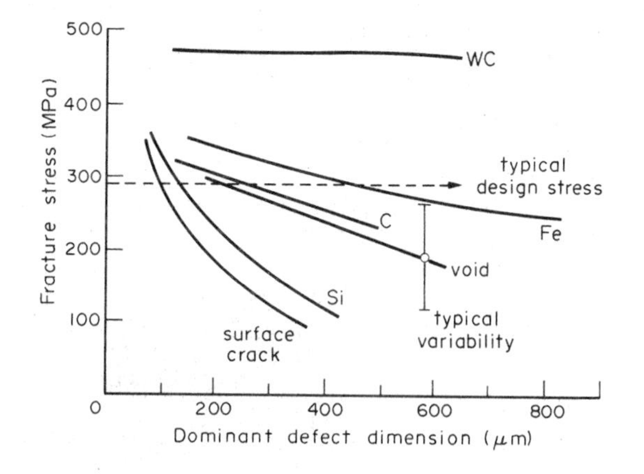

Figure 3
Relations between the fracture strength and the defect size for various types of defect in Si_3N_4

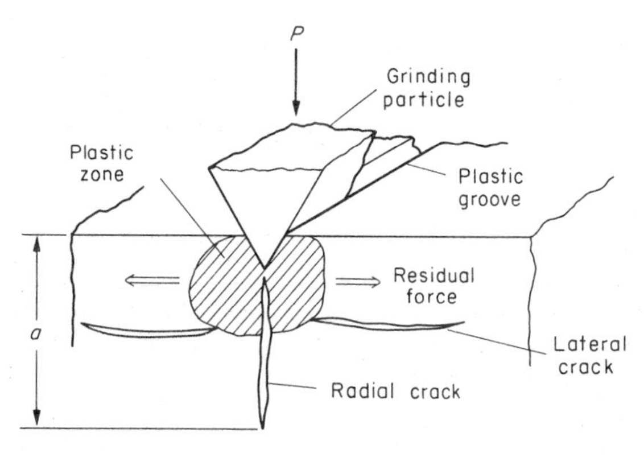

Figure 4
A schematic indicating the mode of formation of surface cracks during grinding. The primary formation period occurs when (plasticity-induced) residual "forces" develop after the particle has moved away from the region of interest

included by the penetrating particle, λ is a material-independent constant ($\sim$0.017) and n is an exponent ($\sim$1/3). The normal force is thus the important external variable, as dictated by the machining or impact conditions, while the fracture toughness is the material parameter of principal significance. (E/H is not substantially material dependent in brittle solids.)

The magnitude of a surface crack (which evidently relates to the failure stress) directly influences the scattering of impinging acoustic waves. In consequence, surface cracks can be characterized using acoustic surface waves, and hence acoustic techniques provide a basis for relating the crack dimensions to the failure strength in a practical way.

Analysis of failure from surface cracks must, in general, recognize the continuing existence of the residual stress that created the crack (unless the specimen has been subject to annealing treatments which eliminate this source of stress). In the presence of residual stress the crack extends immediately upon application of an external stress, and a period of stable crack extension precedes failure (Fig. 5). The failure stress σ_f under these conditions is given by

$$\sigma_f = (3/4^{4/3})\, K_{Ic}\,(a_0/\pi\Omega)^{-1/2} \qquad (2)$$

where Ω depends on the crack geometry (for a semicircular crack $\Omega \approx 5.8/\pi^2$) and a_0 is the initial crack radius. Effective lower-bound predictions of failure from surface cracks have been demonstrated using Eqn. (2).

2.2 Fabrication Defects

The fabrication defects encountered as failure origins include large voids, large grains and inclusions. The formation of these flaws can be related to the initial stages of fabrication, particularly powder processing. Materials produced by other means—for example, melt processes (silicon, glass) and vapor deposition—are less susceptible to fabrication flaws.

The influence of fabrication flaws on failure is incompletely understood at present, but some of the important crack responses are summarized in Fig. 6. Local stresses can develop within or around fabrication flaws, because of differences in thermal contraction (equilibrium exists at elevated temperatures where rapid diffusion occurs) or in elastic modulus. These local stress concentrations are believed to activate imperceptible microcracks that exist around the flaw and thus produce a detectable crack that subsequently (or concurrently) initiates the failure. The microcracks are statistically distributed throughout the material (or inclusion) volume and at the interface.

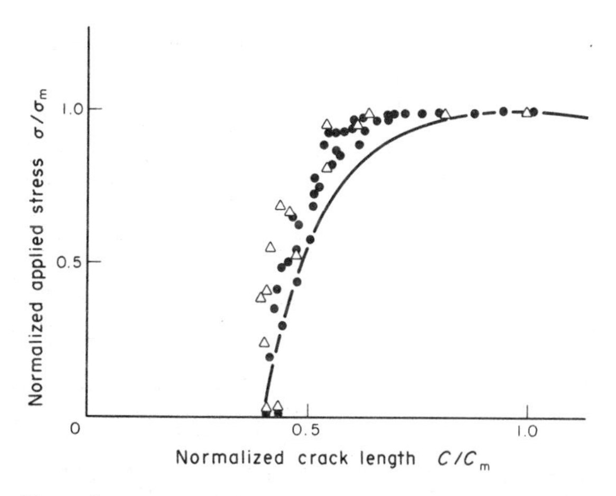

Figure 5
The change in crack length that occurs during the application of load to a specimen containing a surface crack: ●, Knoop; △, Vickers

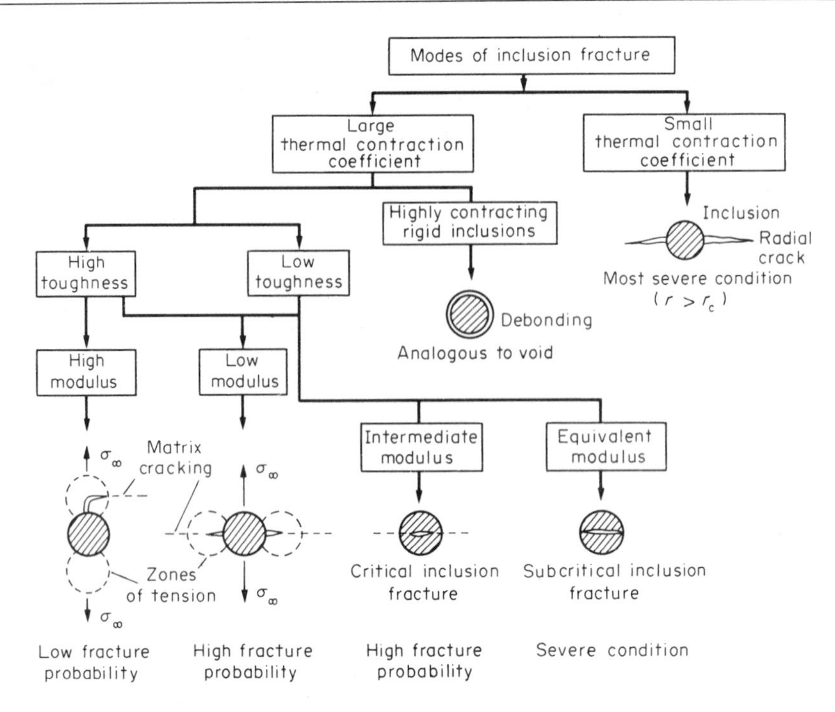

Figure 6
A schematic indicating the various crack responses that can develop in the presence of inclusions. The important parameters are the thermal contraction, the fracture toughness, the elastic modulus and the statistical distribution of microflaws in the vicinity of the inclusion

In consequence, the fracture process exhibits relatively wide statistical variability.

Models that provide a reasonable correlation between observed and theoretical fracture behavior are exemplified by fracture analyses conducted for large voids. Such analyses invoke a statistical microcrack distribution that interacts with the concentrated stress in the vicinity of the equatorial plane. A probabilistic failure relation can then be deduced, using weakest link statistics. The utility of the resultant expression for relating failure to voids of different size has been validated by experiments conducted on Si_3N_4. However, more complex void-induced failure processes have been encountered in other systems and a general description of void failure has yet to emerge.

Fracture from inclusions is much more complex. The first distinguishing feature is the tendency for cracking due to thermal contraction mismatch (Fig. 6). If the thermal expansion coefficient of the inclusion is appreciably lower than that of the matrix, radial matrix cracks can initiate when the defect exceeds a critical size. This situation can produce severe strength degradation, but such behavior is unusual for structural materials, which must have an intrinsically low thermal expansion coefficient in order to resist thermal shock. Alternatively, if the expansion coefficient of the inclusion exceeds that of the matrix, several possibilities can exist. Highly contracting, high-modulus inclusions tend to detach from the matrix and produce a defect comparable in character to a void. Inclusions that are more compliant, or exhibit smaller relative contractions, remain attached to the matrix. Thereupon, several modes of failure are possible, as exemplified by the results for several types of inclusion in Si_3N_4 (Fig. 3). The expected failure mode depends upon the elastic modulus and fracture toughness of the inclusion in relation to the corresponding values for the matrix. For example, when the inclusion has a larger toughness than the matrix, fracture initiates within the matrix, usually from microflaws located within (or adjacent to) the interface. If the bulk modulus of the inclusion exceeds that of the matrix, the tensile stresses (in a direction suitable for continued extension of the crack due to the applied stress) are confined to a relatively small zone near the poles of the inclusion (Fig. 6). For this case, the fracture probability is expected to be relatively small, as exemplified by the high fracture stress for WC inclusions in Si_3N_4 (Fig. 3). Alternatively, when the modulus of the inclusion is small, the maximum tensile stresses occur in the matrix near the equatorial plane. Fracture initiates within this region of the matrix, essentially irrespective of the inclusion toughness. The fracture condition is thus comparable to that for a void. This case is expected to be an important one, because inclusions are often porous (following high-temperature mass transport driven by thermal contraction anisotropy) and hence of low effective modulus.

2.3 Environmentally Induced Defects

Flaws can be created by interactions with the environment, particularly at elevated temperatures. Oxidation can, for example, generate pits (beneath the primary oxidation layer) that behave in a manner analogous to inclusions. The flaw creation process is rather poorly comprehended at this juncture. Two extremes of behavior have been encountered. Internal oxidation accompanied by an appreciable increase in the volume of the oxidation-prone phase can lead to extensive radial fracture and, in some cases, complete destruction of the component. More typically, a volumetric defect subject to minimal local stress develops with time, up to a maximum size, which is presumably related to the size of the local region prone to rapid oxidation (e.g., the size of an oxidation-susceptible inclusion).

2.4 Relative Role of Dominant Flaws

Most brittle materials can be expected to contain at least two populations of flaws: surface cracks and fabrication defects (Fig. 7a). The different spatial character of these two flaw populations signifies a variable relative role of each population in the fracture response, observed with different components or test configurations (see Sect. 3). Strength tests are frequently conducted in flexure, which imposes the largest tensile stresses on the specimen surface. Hence, fractures that occur in flexure tests generally originate from surface cracks, and flexural strengths often reflect the surface crack population (Fig. 7b). Conversely, simple tensile tests frequently induce failure from fabrication defects, and hence tensile strengths relate primarily to the presence of inclusions, voids or large grains.

The different spatial character of the important flaw populations also determines the failure mode experienced by components. For example, thermal stresses usually exhibit a sharp maximum at the surface, and hence thermal shock failures are most sensitive to the surface crack population. In other situations (e.g., rotational stresses in turbine disks) the fabrication flaws dominate failure.

Environmentally induced flaws can also be encompassed within this framework. The fabrication flaw population remains invariant, while the smaller surface cracks become consumed by the smaller surface cracks become consumed by the surface chemistry changes, and an environment-related flaw population emerges. The relative influence of each population on the observed mechanical strength depends upon the time of exposure, the stress-state volume condition and the specific character of the flaw distributions. Often, however, the low-strength extreme of the environment-induced population is truncated. In consequence, the environmentally related population dominates the strength of small test specimens, but has little effect on the mechanical performance of large components.

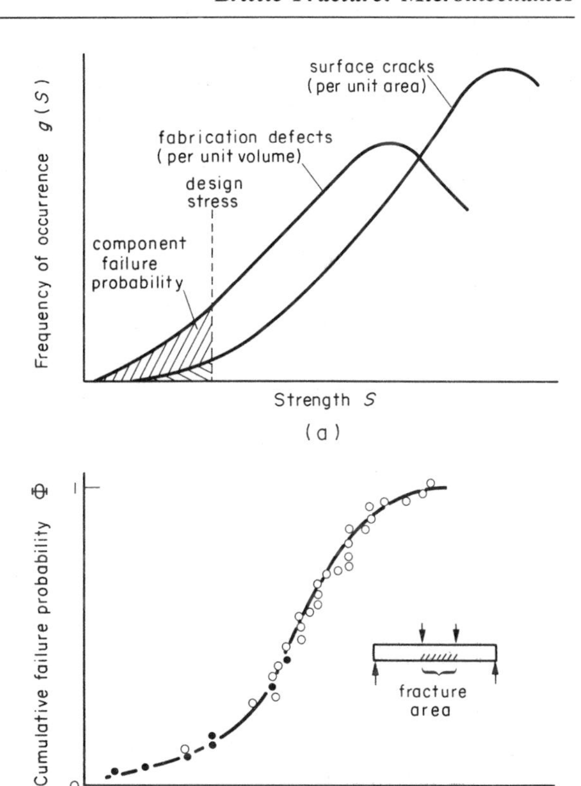

Figure 7
Some flaw population characteristics. (a) There are generally at least two flaw populations with distinct statistical characteristics. (b) Flexural strength tests usually sample the surface crack population (○) at the higher strength levels and the fabrication defect population (●) in the lower strength range, giving a convolved cumulative failure probability (based on indiscriminate order statistics)

3. *Statistical Effects*

The flaws that preexist in brittle solids can be characterized by a statistical distribution that encompasses both the flaw size and the volume sampled. Explicit statistical relations can be written for situations where the flaws behave independently, that is, flaw coalescence does not occur before failure. These relations are based upon the weakest link concepts that permit the survival probability of a volume element ΔV subject to a stress $\sigma(x, y, z)$ to be written as

$$1-\phi(\sigma, \Delta V)=\exp\left[-\int_{\Delta V} dV \int_0^{\sigma(x,\,y,\,z)} g(S)dS\right] \quad (3)$$

where ϕ is the failure probability and $g(S)dS$ is the number of flaws in unit volume with a fracture strength between S and $S+dS$. (It is convenient to express the result in terms of "flaw strength" rather than flaw size, although Eqn. (3) could be converted into a flaw size function by applying the relations outlined in Sect. 2.) The quantity $g(S)$ can be determined for each flaw population directly from fracture results, obtained using tensile or flexure tests. It is imperative, of course, that the failure-initiating flaw be identified for each test result. An alternative (and more common) procedure is to assume a functional form for $g(S)$ and to derive the statistical parameters of the function from the experimental results. A function with adequate versatility (the Weibull assumption) is

$$\int g(S)dS=[(\sigma-\sigma_{\mu})/\sigma_0]^m \quad (4)$$

where m is the shape parameter, σ_0 is the scale parameter and σ_μ is an offset parameter (all of which have no physical significance).

Once $g(S)$ has been determined from experimental results for each of the important flaw populations (e.g., see Fig. 7a), the fracture probability of components can be predicted by inserting $g(S)$ into Eqn. (3) and integrating. Extrapolation beyond the data range leads to considerable uncertainty and data must, therefore, be acquired at strength levels approaching the design stress in order to obtain useful predictions. This can be achieved by conducting experiments, such as tensile tests and spin tests, on specimens with a large, relatively uniformly stressed volume. Also, it must be realized that the statistical predictions are only valid whenever the flaw populations are invariant. This invariance can be compromised if there are appreciable batch-to-batch variations in the fabrication-induced flaw populations, or if the surface finishing is erratic. When such situations are likely to occur, statistical failure prediction should only be used for approximate estimation of size and shape effects. More definitive predictions of failure must rely upon proof testing and/or quantitative forms of nondestructive evaluation.

The weakest link assumption, which is a condition for the well-developed statistical procedures discussed above, is not always valid. It is generally most valid in fine-grained materials, which fracture in the desired manner. However, materials with a coarser microstructure often exhibit fracture by microcrack coalescence (see Sect. 1). The fracture of these materials cannot be described using weakest link concepts. Appropriate statistical methods are at a preliminary stage of development.

4. Crack Extension Resistance

The operational strengths of brittle materials usually scale with the fracture toughness, as reflected in relations between the fracture strength and the flaw dimensions (e.g., Eqn. (2)). Also, the extent of strength degradation (e.g., from projectile impact or thermal shock) is known to be suppressed by high values of fracture toughness. It is of substantial consequence, therefore, to comprehend the influence of microstructure on fracture toughness.

The fracture resistance of a brittle solid can appreciably exceed the basic cleavage resistance of its microstructural constituents. The additional toughness is associated with crack–microstructure interactions. At the most direct level, crack deflection can be induced by microstructural anisotropy. The stress intensification of the deflected crack is reduced and an increase in toughness ensues. This is a widely encountered toughening mechanism that accounts (experimentally) for toughness levels up to three times the fundamental toughness (i.e., three times the toughness of a polycrystal subject to transgranular fracture). The magnitude of this toughness has not been rigorously analyzed, because of the complexity associated with the crack advance process, involving a spectrum of fracture resistance at the crack front (with both in-plane and out-of-plane components). Solutions for the deflected crack, integrated over the crack front, provide an approximate measure of the deflection effect, although experimentally observed trends are not fully accounted for by oversimplified analyses of this type.

Additional toughening is attributed to stress-induced microstructural changes: most notably, martensitic transformations, mechanical twinning and microcracking. These stress-activated processes occur predominantly in the vicinity of a crack tip and usually tend to shield the tip region from the applied stress (Fig. 8a). The extent of the shielding, and hence of the toughness increase, depends on the microstructural details.

4.1 Transformation Toughening

The net dilational and shear strains that accompany the martensitic transformation of a region of material near a crack tip invariably result in a change in the local stress immediately ahead of the crack. The local stress intensity factor $K_{\mathrm{I}}^{\mathrm{local}}$ thus differs from the applied value K_{I}^{∞} (Fig. 8c) by an amount that depends upon the stress–strain behavior of the transformation region (Fig. 8b). In brittle solids, the stress–strain relation exhibits a linear elastic character before and after transformation, with an essentially constant slope, reflecting the relatively small changes in elastic moduli that accompany the transformation. The transformations are also hysteretic, a phenomenon that enables the transformed region to be retained following a crack advance. An appreciable reduction in $K_{\mathrm{I}}^{\mathrm{local}}$ commences when the crack advances into the transformed zone, resulting in R-curve behavior. The magnitude of the maximum increase in K_{Ic} may be expressed in terms of the dilational component of the transformation strain by

$$\Delta K_{\mathrm{Ic}}\approx 0.3\varepsilon^{\mathrm{T}}Eh^{1/2}V_{\mathrm{f}} \quad (5)$$

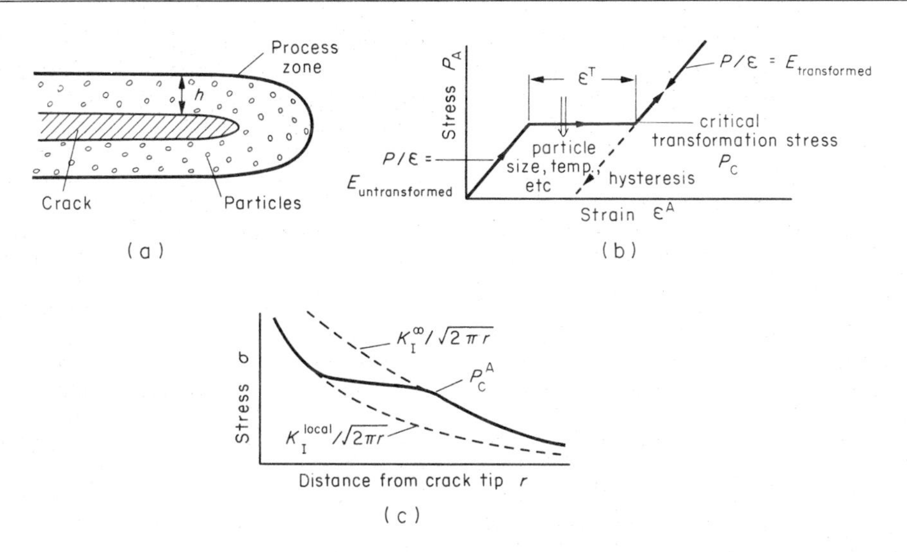

Figure 8
Schematic indicating the crack shielding effect of a process zone: (a) The shape of the process zone. (b) The stress–strain curve (of the process zone) required for shielding. (c) The magnitude of the shielding expressed in terms of the crack-tip stress field

where ε^T is the unconstrained dilational strain, E is Young's modulus, h is the zone height and V_f is the volume concentration of material subject to transformation. An additional increment in toughness would pertain in the presence of a net shear strain.

The zone height h is a particularly important parameter. It is determined by the transformation stress P_c (analogous to the yield stress in plastic zone formation) such that $h \approx (1/2\pi)(K_{Ic}/P_c)^2$. The stress P_c is known to depend upon temperature, chemical composition and particle size. The effects reside in the influence of the dominant variables on the net thermodynamic driving force for transformation and/or on the nucleation process. The relative importance of the driving force with respect to the nucleation has not yet been fully ascertained.

4.2 Microcrack Effects

Crack shielding can also occur in the presence of a process zone of microcracks. The shielding in this instance again resides in the dilatation of the microcracked zone. The propensity for microcracking exhibited by a material depends upon the scale of the dominant microstructural entity, such as the grain size. Specifically, larger-grained materials are usually more prone to microfracture. This effect originates from residual stresses created by thermal contraction anisotropy or mismatch. Microstructural effects on the fracture toughness can, therefore, often be attributed (at least qualitatively) to the incidence of microfracture.

4.3 Crack Size Dependence

Another factor affecting fracture toughness of brittle solids is crack size. In some instances, the scale of the cracks that initiate failure is comparable to the microstructural dimensions. The effective fracture toughness is then intermediate between the cleavage value and the polycrystalline level. The variation in toughness with crack length is not yet comprehended, but experimental results suggest that the polycrystalline value is assumed when the crack radius becomes about an order of magnitude larger than the average microstructural dimension. Interpolation permits a reasonable estimation of toughness at intermediate crack lengths. Fracture from very small flaws is determined by the single-crystal toughness. Such behavior is encountered in coarse-grained materials. Flaws of intermediate relative size result in a flaw-size-independent fracture stress attributed to the stable growth of preexistent cracks to a uniquely determined critical size. This mode of failure occurs frequently in conventional ceramics. Finally, in fine-grained materials the flaws are generally much larger than the grain size and fracture is dictated by the polycrystalline level of toughness. The desirability of a fine-scale microstructure is thus evident.

5. Thermal Fracture

The fracture of brittle solids subject to thermal stress occurs in accord with the same criterion that dictates fracture in the presence of mechanical stress, in that the stress intensity factor, derived from both the stress level and the crack size, exceeds the critical value. However, because thermal stresses can exhibit rapid spatial and temporal variation and because stresses arise in response to local deviations in thermal conductivity, thermal fracture exhibits several special features. Some of these special characteristics are described in this section.

5.1 Surface Cracks

The most deleterious surface cracks are those nearly normal to the surface, that is, along the heat flow axis. These cracks do not offer a significant impediment to the heat flow and the only appreciable source of stress intensification is that which derives from the temperature gradient along the heat flow axis. The thermal fracture problem can thus be fully characterized by dimensionless groupings similar to those that determine the thermal stress:

(a) $K/\alpha E \Delta T r^{1/2}$, the dimensionless stress intensity factor, which dictates the incidence of fracture

(b) hr/k, the Biot number β, which measures the importance of the surface heat transfer relative to the internal heat conduction

(c) $\kappa t/r^2$, the Fourier number, which represents the rate of energy storage to the rate of thermal conduction

(d) $x/(\kappa t)^{1/2}$, a measure of the depth of penetration of the thermal effect

(e) $h(\kappa t/k)^{1/2}$, a measure of the ratio of the surface heat flux to the heat conducted and of the region of thermal penetration

(f) x/r, H/r, dimensionless locations

Here α is the thermal expansion coefficient, E is Young's modulus, ΔT is the imposed temperature differential, r is a specimen dimension, h is the heat transfer coefficient, k is the thermal conductivity, κ is the thermal diffusivity and t is a characteristic time. Typical variations in the normalized stress intensity factor for surface cracks are plotted in Fig. 9, indicating the role of several of the dimensionless parameters. The stress intensity factor exhibits a characteristic initial increase, followed by a maximum and then a decline at large crack lengths (or at long times). The existence of a peak reflects the strain-controlled nature of thermal fracture, which permits the effective stress over the crack surface to diminish as the crack extends (by virtue of the increase in compliance with increase in crack length). Crack arrest is thus a generally encountered phenomenon in thermal fracture, occurring when $K \approx K_{Ic}$. Stress intensity curves, such as Fig. 9, deduced for the appropriate thermal environment, coupled with a statistical characterization of the surface crack population (see Sect. 3), provide an adequate basis for predicting and interpreting thermal fracture from surface cracks.

5.2 Fabrication Defects

The thermal fracture from fabrication defects differs from that attributed to surface cracks by virtue of additional stresses that derive from inhomogeneity in

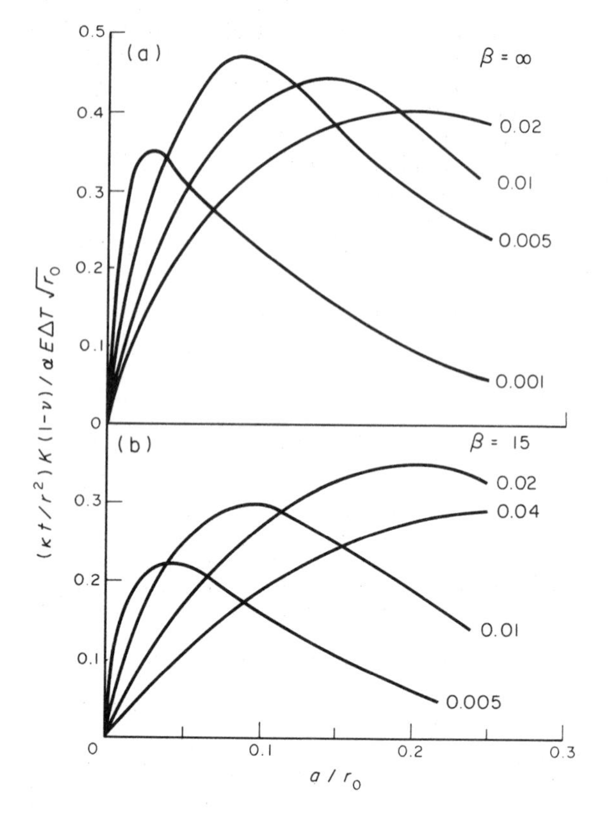

Figure 9
The normalized stress-intensity factor plotted as a function of the relative crack length for various values of the dimensionless time $\kappa t/r^2$ for two choices of the Biot number β

the thermal conductivity. Most fabrication defects exhibit thermal conductivities that differ from the matrix. The displacement gradient across the defect thus differs from that across the matrix and a thermal strain develops. This strain superimposes on that generally computed from the temperature gradient. This additional stress causes the thermal fracture condition to deviate from that predicted using the strength distribution (characterized by mechanical tests) and the generalized thermal stress field.

6. Summary

The micromechanics of fracture in brittle solids has been advanced in recent years to a level wherein most of the significant fracture phenomena have been identified. However, a detailed micromechanics interpretation of the various fracture processes only exists in certain cases. A comprehension of brittle fracture involves aspects of fracture mechanics and fracture statistics, elasticity and plasticity.

As outlined in this article, it is important to recognize the existence of multiple flaw populations as typical failure sources, and to understand the failure process associated with each significant population that exists in a specific material. This comprehension necessitates the formation of a direct link with the material microstructure, and with the methods of fabrication, machining and surface finishing. Then, the ultimate quantification of failure resides in an ability to conduct the mechanics with an adequacy that permits a unique correlation with experimental results.

See also: Brittle Fracture: Linear Elastic Fracture Mechanics; Mechanics of Materials: An Overview

Bibliography

Bradt, R C, Lange F F, Hasselman D P H, Evans A G (eds.) 1974–1982 *Fracture Mechanics of Ceramics*, Vols. 1–6. Plenum, New York

Davidge R W 1979 *Mechanical Behaviour of Ceramics*. Cambridge University Press, Cambridge, UK

Emery A F 1980 Thermal stress fracture in elastic–brittle materials. In: Hasselman D P H, Heller R D (eds.) 1980 *Thermal Stresses in Severe Environments*. Plenum, New York, pp. 95–121

Evans A G 1982 Structural reliability: A processing-dependent phenomenon. *J. Am. Ceram. Soc.* 65: 127–37

Evans A G, Langdon T G 1976 Structural ceramics. *Prog. Mater. Sci.* 21: 171–441

Evans A G, Meyer M E, Fertig K W, Davis B I, Baumgartner H R 1980 Probabilistic models of defect initiated fracture in ceramics. *J. Non-Destr. Eval.* 1: 111–14

Lawn B R, Evans A G, Marshall D M 1980 Elastic/plastic indentation damage in ceramics: The median/radial crack system. *J. Am. Ceram. Soc.* 63: 576–81

Lawn B R, Wilshaw T R 1975 *Fracture of Brittle Solids*. Cambridge University Press, Cambridge, UK

Rice R W 1977 Microstructure dependence of mechanical behavior of ceramics. In: MacCrane R K (ed.) 1977 *Properties and Microstructure*, Treatise on Materials Science and Technology Vol. 11. Academic Press, New York, pp. 199–381

A. G. Evans
[University of California, Berkeley, California, USA]

Building Systems and Materials

The notion of building systems in the modern sense of the term has its origin in the radical changes brought about by the industrial revolution and the subsequent effects on materials, processes and design in construction (Giedion 1969).

During recent decades, two distinct but related approaches to the development of industrialized building systems have emerged which may be labelled as the "open" versus the "closed" system approach to industrialized construction (Fig. 1). Along with the mechanization and rationalization of traditional construction practices and processes, the evolution of

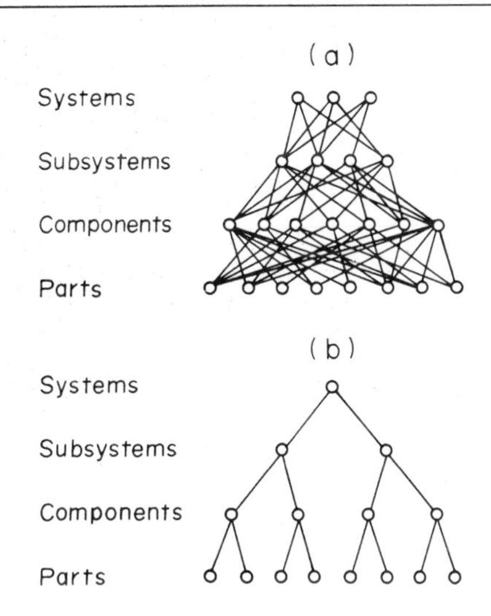

Figure 1
Comparison between (a) open and (b) closed systems

open and closed systems was accompanied by progressive industrialization and prefabrication of components, elements and structural systems as well as complete, whole building systems (Sullivan 1980).

In addition to the above, new procedures of planning, programming, design, procurement, production, transport, mechanical methods of assembly, and efficient management of total system packages have radically changed the organizational structure of many firms as well as whole sectors of the construction industry. These changes in organization have led also to changes in professsional roles and, to some extent, the emergence of entirely new professions, such as construction management, system engineering, planning and programming, value engineering, life-cycle costing, benefit–cost analysis, to name but a few which have not only affected the efficiency of the industrialized building sector, but also conventional construction methods (Brandle 1974).

Finally, the change from almost total reliance on prescriptive norms and standards for material and process control to the use of performance-based standards for new product development, has brought the practices of the construction industry closer to those of the manufacturing sector, including the need to stimulate innovation by organized scientific research and development in both the private and public sectors. Modular and dimensional coordination, pre-engineered, functional and technical compatibility requirements, along with the application of the methods of operations research, and the so-called "systems approach," have led to a gradual shift from conventional and/or traditional use of materials to scientifically and professionally managed industrial mass production of precoordinated components and elements in the factory, capable of being assembled without significant modification on site (see Fig. 2).

1. Evolution and Typology of Building Systems

A distinction should be made between traditional construction systems, which refer to the conventional process of assembling traditional building elements on site by cutting, fitting, bending and so on, and prefabricated and/or industrialized building systems, which are delivered either as fully compatible, pre-engineered elements or as total building packages. The term "building system" should not be confused with so called "system building," which deals with the management of the total construction process and includes all phases of the building process, from production to final erection. The stages in the evolution, from craft-based construction to industrialized building, are shown in Fig. 3.

Furthermore, it is important to recognize that industrialization is not necessarily synonymous with prefabrication, since prefabrication on or off site may be realized without the methods of industrialization. The term prefabrication covers any process, including conventional ones, that coordinates the fabrication of parts, components, assemblies and other building elements, either on or off site, to be incorporated in the assembly of the finished building without additional need for processing, shaping or modification, except for connections to be made at predetermined joints or junctions where elements meet in a dimensionally and/or geometrically predetermined and coordinated manner. Another way to express this difference is to distinguish between a building that has been "assembled" rather than "constructed".

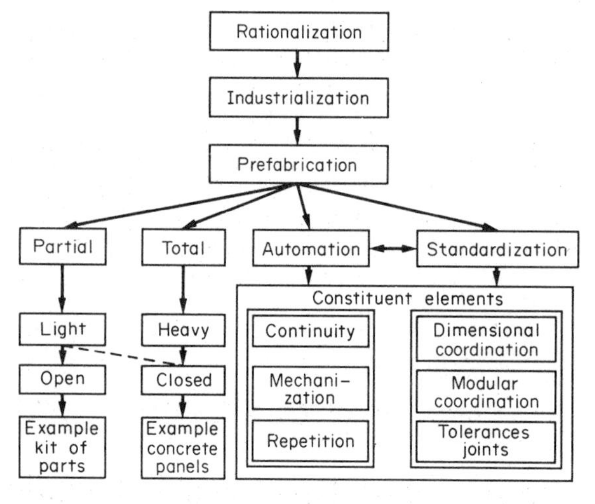

Figure 2
Structure and constituent elements of industrialized and prefabricated systems development

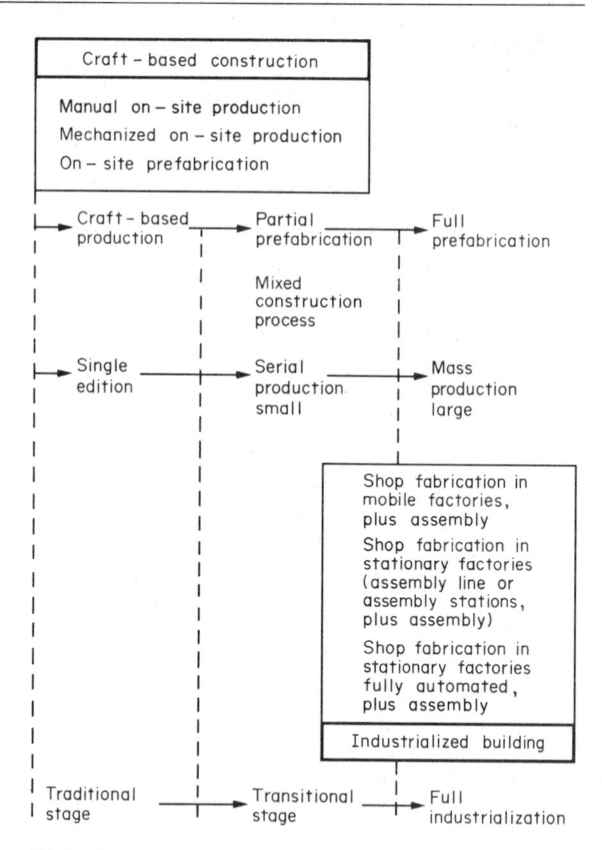

Figure 3
Stages in the evolution of industrialization of building

2. Types of Building Systems

The combination of new materials with new processes of management has led to the evolution of both "type-specific" and "material-specific" building systems. Classification may be predicated on either category, but is usually based on type rather than material, for the simple reason that most building systems are capable of being realized by more than one material (see Sect. 3), whereas systems depending for their design on a single material are rare. Thus, the conventional way of classifying building systems is usually based on the type of structural support system, that is, panel systems, box or volumetric (space-enclosing) systems, skeleton or frame systems, and hybrids of these (Schmid and Testa 1969).

Generically, any one of the preceding systems may be conceived as either "open" or "closed". Essentially, in a closed system, parts and components are peculiar to that system alone and are not interchangeable or transferable to another system. Such systems are commercially known as "proprietary" systems and are generally producer sponsored.

In open systems, parts and components can be arranged to form a reasonable number of alternative combinations of plans and/or geometrical forms. Components are usually of different origin, and their use is not confined to a single system. Open systems are generally user sponsored.

Conventional systems are not open in the above sense, but may contain as "catalogue" items ranges of open components which may be mutually compatible, but which usually require some additional reprocessing or manipulation on site or during assembly not accounted for in their design.

In general, compatibility can be achieved by dimensional and/or modular coordination, spatial and/or technical norms and standards, geometrical control of joints and interfaces, and so-called "bypassing" systems. In bypassing systems of coordination, diverse elements, components or subsystems are allowed to meet within a considerable latitude of allowable joint tolerance, which allows elements to "bypass" each other by means of adaptable joints and connections.

3. *Materials*

Developments in prefabrication and industrialization in construction have been matched by improvements in materials and methods, along with the introduction of entirely new materials. Advances range from the modest to the exotic.

Cement is one of the most common materials in prefabricated panel and box systems, especially outside the USA (see *Cement as a Building Material*). New types of cement, such as controlled-set and expansive cements, have led to improved methods of controlling quality and preventing shrinkage and cracking during curing. Additives, used to shorten curing time and to control setting in varying temperature conditions, have also improved control of production and scheduling. Experimental cements, possessing the ability of self-prestressing by controlled shrinkage, have been tested in projects, such as Operation Breakthrough (US Department of Housing and Urban Development 1970), and promise to simplify prestressing and/or posttensioning by eliminating the need for the special equipment currently used to induce tension mechanically (Dietz 1971).

Industrialization, as related to concrete, may be divided into five basic categories:

(a) Precast open structural components, such as T, double-T, U, H and L sections, delivered to the site from the factory or precast on site. To increase span, and to reduce the depth and area of a given section, post- or pretensioning methods of reinforcement may be used.

(b) Vertical structural elements of various designs and sizes. Basically, these may be divided into single-story and multistory types, and may be combined with precast elements of category (a).

(c) Precast facade elements of various designs, either seated on or suspended from the structural frame, which may be fabricated from reinforced concrete or steel. The use of such elements is often predicated on aesthetic rather than utilitarian considerations, and thus they are usually rather heavy, increasing the load on the structure, while their structural potential remains unused.

(d) Prefabricated, reinforced-concrete closed panel and box (volumetric) systems, where large space-enclosing elements may be produced either on or off site for large-volume projects, usually housing.

(e) Miscellaneous precast elements, such as culverts, pipes, lintels, road curbs, dividing strip elements and fence posts.

Steel technology has also made considerable progress in producing new varieties of high-strength, as well as highly ductile, steels—including the change from riveting to bolting by means of high-stress bolts. More efficient welding methods in the shop allow for higher quality, and more economical methods, of prefabrication off site and for development of lightweight skeleton systems for various building systems. Contumescent paints and sprayed-on fireproofing have made steel compatible with concrete systems where high fire resistance is required.

Steels, with the ability to develop self-sealing rust surfaces, have led to new design solutions for exteriors. New designs for space frames, staggered trusses and cable-supported structures, usually prefabricated off site and assembled without waste on site, have made it possible to cover vast spaces for spectator events; examples include football stadia and Olympic sports facilities (see *Steel for Construction*).

Modern industrialized metal systems may be classified as follows:

(a) Industrialized structural components (usually steel), other than conventional standard steel sections, including tapered beams, rigid frames, trusses, post and beam systems, and space frames.

(b) Covering systems for facades, interiors, roofs, overhangs and other miscellaneous cladding purposes. These may be flat panels or formed elements made of steel, aluminum or composite (sandwich) construction with other materials. The range of finishes is considerable (e.g., painted, sprayed, anodized, enamelled and galvanized).

(c) Combinations of (a) and (b), comprising open components, kits of parts and fully coordinated (closed) functional packages for various building types.

(d) Core unit systems, such as kitchens and bathrooms, often combined with other materials, such as plastics and fiberglass.

New glues and epoxies have transformed the range of wood usage and have led to the controlled structural use of plywood, laminated beams, arches and trusses (see *Wood as a Building Material*). The use of plastic monomers and other chemical impregnation compounds has made wood more durable, while at the same time enhancing its dimensional stability and hardness (Dietz 1971), thus leading to factory-based manufacturing processes of various wood construction systems.

The evolution of tinted, photochromic glass has led to the use of this material for sunlight control as a replacement for shading devices. The uses for highly tempered glass have expanded as its strength has increased. It can now be employed as a structural material at reasonable cost in the preassembly of glazed windows and door fittings, among other products made of glass (see *Glass as a Building Material*). Safer transfer from the factory to the building site has also resulted.

Plastics continue to proliferate, both with new uses and as substitutes for traditional materials such as wood and glass. Their ability to be molded or extruded into practically any shape desired has made plastics the prefabricated material par excellence. Their use is not confined to finishing products alone, but has been expanded into other applications as well, such as piping, prefabricated kitchen and bathroom core units, hardware and insulation.

Industrialized systems utilizing polymers and plastics are by and large confined to nonstructural uses (even though structural applications are feasible and exist) such as:

(a) volumetric (three-dimensional) elements and shapes, making up partial or complete functional units or elements, such as bathrooms and kitchens;

(b) profiles, shells and other molded or formed shapes or sections to be incorporated into prefabricated panels, elements or volumetric units, or to be combined with conventional supports or materials;

(c) fully integrated functional units, such as small houses, portable toilets, kiosks and similar mobile units; and

(d) domes and related geodesic structures, either as components, kits of parts or as fully assembled units.

New sealants, high-strength mortars and joint connecting devices have made the assembly of prefabricated elements and/or components easier, at the same time assuring better protection against the intrusion of rain, air, insects and dust, as well as improving the structural integrity of the completed assembly by their contributions to strength and stability.

The greatest advances and innovations have been made in the development of composite materials (see *Plastics and Composites as Building Materials*). Their main characteristic is polyvalent performance; that is, the ability to combine the advantages of their constituent properties with a resulting overall performance which surpasses that of its parts. Sandwich and/or composite panels, made up of various layers of metal, polyurethane, lightweight concrete, gypsum, reinforced fiberglass—to name but a few—combine high structural strength with good insulating, acoustic and fire-resisting properties, while at the same time reducing overall weight and assembly time.

More exotic methods, such as filament winding, developed by the aerospace industry, and related glass-fiber-reinforced plastic techniques, are finding increasingly more applications in construction as their cost becomes competitive with other materials.

In general, it may be said that these improved and new materials may be considered as the most suitable candidates for a truly industrialized approach to building construction, since they offer controlled and often superior performance, precise tolerance control, dimensional stability, and ease of assembly without the need of additional cutting, patching or other alterations once they have been delivered to the site. Given dimensional and functional coordination and/or compatibility, the development of such open components will lead to more flexible and adaptable open systems, a trend which is emerging in all industrialized countries and which promises to revolutionize the construction sector throughout the world, including the developing countries.

See also: Architectural Design and Materials

Bibliography

Banham R 1960 *Theory and Design in the First Machine Age*. Architectural Press, London

Borrego J 1968 *Space Grid Structures—Skeletal Frameworks and Stressed Skin Systems*. MIT Press, Cambridge, Massachusetts

Brandle K 1974 *The Systems Approach to Building—A Core of Information*, Associated Schools of Architecture Learning Package. University of Utah, Salt Lake City, Utah

Dietz A 1969 *Plastics for Architects and Builders*. MIT Press, Cambridge, Massachusetts

Dietz A 1971 *Industrialized Building Systems for Housing*. MIT Press, Cambridge, Massachusetts

Giedion S 1969 *Mechanization Takes Command*. Oxford University Press, Oxford

Hornbostel C 1978 *Construction Materials*. Wiley–Interscience, New York

Koncz T 1970 System building with large panels. *Manual of Precast Concrete Construction*, Vol. 3. Bauveriog, Wiesbaden

Morris A E J 1978 *Precast Concrete in Architecture*. The Whitney Library of Design, Watson-Guptil, New York
Otto F (ed.) 1969 *Tensile Structures*. MIT Press. Cambridge, Massachusetts
Schmid T, Testa C 1969 *Systems Building: An International Survey of Methods*. Praeger, New York
Sullivan B J 1980 *Industrialization in the Building Industry*. Van Nostrand Reinhold, New York
US Department of Housing and Urban Development 1970 *Operation Breakthrough*. US Government Printing Office, Washinton, DC

E. Dluhosch
[Massachusetts Institute of Technology, Cambridge, Massachusetts, USA]

C

Calcium Aluminate Cements

Hydraulic cements consisting predominantly of monocalcium aluminate, $CaAl_2O_4$ (or CA in cement industry nomenclature—see *Portland Cements, Blended Cements and Mortars*), are referred to variously as calcium aluminate cements, aluminous cements or high-alumina cements. A major use of CA cements is as the binder in refractory concretes. Structural uses in low-temperature load-bearing applications are at present discouraged because of strength reductions due to the "conversion" reaction.

1. Composition and Manufacture

The composition of CA cements varies widely from low- to high-purity types (Table 1). The C–A–S phase diagram indicates that $C_{12}A_7$, C_2AS, C_2S and CA_2 are the only phases which can coexist with CA at equilibrium. Most of the iron is usually in a ferrite solid solution of composition between C_4AF and C_6AF_2. Ferrous iron compounds are present in cements produced under reducing conditions.

The raw materials for manufacturing the lower-purity CA cements usually consist of relatively pure limestone and low-silica bauxite. High-purity (white) CA cements are formed from the purest raw materials—limestone and Bayer alumina. After grinding and proportioning of the raw materials, the mixture is sintered or fused between 1350 and 1500 °C under either oxidizing or reducing conditions. In reductive fusion, molten iron can be separated from the molten clinker, and the CA clinker produced tends to have a low iron content. The final clinker compounds (Table 2) in high-purity cements are formed during the nonequilibrium heating cycle of a sinter process, while in lower-purity cements they are usually formed under equilibrium conditions during the cooling of a melt.

The cooled clinker is made into cement by grinding it to a specific surface area in the range of 250–1800 $m^2\,kg^{-1}$. The range of physical properties of CA cements is shown in Table 3. CA cement concrete can develop a 24 h compressive strength comparable with a 28 day strength in Portland cement concrete.

2. Reactions with Water

The phases in CA cements that react most readily with water are CA, $C_{12}A_7$ and CA_2. Although C_3AH_6 is the stable hydrate, the main hydration product of the anhydrous calcium aluminates at temperatures $<21\,°C$ is the metastable CAH_{10}:

$$CA + 10H \rightarrow CAH_{10} \quad (1)$$

Some AH gel is also present. Above 21 °C, the main products are C_2AH_8, C_3AH_6 and AH_3. If β-C_2S or C_2AS is present in CA cements, their hydration is believed to be slow and to produce either a calcium silicate hydrate similar to that in hydrated Portland cements or gehlenite hydrate. Also, hydration products of the ferrite phases are similar to phases produced by ferrites in Portland cements.

3. Uses

Uses for CA cements (Table 4) exploit the refractory nature of these cements and their hydration products, rapid strength development, and high chemical resistance. White CA cements are used for the most demanding refractory applications. Aggregates used with CA cements in refractory concretes must also have refractory properties. They include blast furnace slags, expanded clay or shale, vermiculite, crushed firebrick, calcined flint clay or bauxite and tabular alumina. Combinations of proportioned cements and aggregates are referred to as castables. They are commercially available in dry packaged form and classified by their modulus of rupture, bulk density and volume stability at high temperature.

Table 1
Typical oxide compositions of calcium aluminate cements (%)

Cement type	SiO_2	Al_2O_3	CaO	Total Fe as Fe_2O_3	MgO	SO_3	Alkalis
Low purity	4.5–9.0	39–50	35–39	7–16	0.2–1.0	0.1–2.5	0.1–0.3
Intermediate purity	3.5–6.0	55–66	31–36	1–3	0.2–0.7	0.1–0.6	0.1–0.3
High purity	0.0–0.2	70–80	17–27	0.0–0.4	0.0–0.4	0.0–0.5	0.1–0.6

Table 2
Phases identified in calcium aluminate cements[a]

CA	β-C_2S	α-A
CA_2	C_4AF	$C_{12}A_7$[b]
C_2AS	$C_4A_3\bar{S}$	CT[b]

[a] For details of the nomenclature of these phases, see *Portland Cements, Blended Cements and Mortars* [b] Minor amounts only

Table 3
Physical properties of calcium aluminate cements

Property	Range
Color	Dark gray to white, depending on the quantity of iron and its oxidation state
Density (kg m^{-3})	3000–3250
Bulk density (kg m^{-3})	1500
Fineness (m^2 kg^{-1})	250–1800
Set time, initial. Vicat (ASTM C191)[a] (h)	0.5–9

[a] American Society for Testing and Materials (1982)

4. The Conversion Reaction

"Conversion" is the name given to the change from the metastable hexagonal CAH_{10} in CA cement concretes to stable cubic C_3AH_6 and AH_3 (gibbsite). The reaction occurs slowly at low ambient temperatures and more rapidly at higher ambient temperatures under conditions of high relative humidity:

$$3CAH_{10} \rightarrow C_3AH_6 + 2AH_3 + 18H \qquad (2)$$

This reaction increases the porosity of the solid concrete and can cause a loss in strength. These occurrences can be minimized by using a low water–cement ratio and curing at $>30°C$ in the first 24 h after placement.

Conversion occurs in hydrated cements used for refractory applications. These reactions do not cause detrimental effects in concrete used under high-temperature conditions, because the low-temperature strength is still adequate and ceramic bonds form at high temperatures.

Table 4
Some uses of calcium aluminate cement mortars and concretes

Refractory
- Insulating walls and furnace linings
- Heat-resistant floors and foundations
- Linings for pipes and chemical reactors
- Rocket exhaust deflector pads

Nonrefractory
- Quick-setting and high 24 h strength concretes
- Grouting
- Insulators for power lines
- Cementing of bore holes
- Decorative panels and terrazzo flooring
- Jointing and repointing in railroad tunnels and sewers
- Chemically resistant floors and tanks
- Chimney (stack) linings (where condensation occurs)

5. Chemical Resistance

CA cement mortars and concretes are usually more resistant than Portland cement to dilute acids (pH > 3.5) and sulfate solutions, but not to alkali hydroxides. Possible reasons for acid resistance are that the hydration of CA cements does not produce much calcium hydroxide and that hydrated alumina acts as a barrier to acid reaction. To explain sulfate resistance, it has been suggested that any calcium sulfate aluminate hydrates which form do so through solution rather than by a topochemical reaction, such as is believed to lead to expansion in sulfate attack of Portland cement concretes. The aggregates selected for use with CA cement in these applications must have a comparable chemical resistance

See also: Cement as a Building Material; Cements, Specialty; New Cement-Based Materials; Portland Cements, Blended Cements and Mortars

Bibliography

American Concrete Institute 1978 *Refractory Concrete*, ACI Publication SP-57. American Concrete Institute, Detroit

American Concrete Institute 1979 *Refractory Concrete*, ACI State-of-the-Art Report 547R-79. American Concrete Institute, Detroit

American Society for Testing and Materials 1984 *Annual Book of ASTM Standards*, Sect. 15, Vol. 15.01, *Refractories; Carbon and Graphite Products; Activated Carbon.* American Society for Testing and Materials, Philadelphia, Pennsylvania

Lea F M 1971 *The Chemistry of Cement and Concrete.* Chemical Publishing, New York

Neville A M, Wainwright P J 1975 *High Alumina Cement Concrete.* Construction Press, Hornby

Robson T D 1962 *High Alumina Cements and Concretes.* Wiley, New York

Robson T D 1964 Aluminous cements and refractory concretes. In: Taylor H F W (ed.) 1964 *The Chemistry of Cements*, Vol. 2. Academic Press, New York, pp. 3–35

G. J. Frohnsdorff and J. E. Kopanda
[National Bureau of Standards,
Washington, DC, USA]

Carbon-Fiber-Reinforced Plastics

To utilize the high stiffness and strength that carbon fibers possess it is necessary to combine them with a matrix material that bonds well to the fiber surface and which transfers stress efficiently between the fibers. Fiber alignment, fiber content and the strength of the fiber–matrix interface all influence the performance of the composite. Furthermore, highly specialized processing techniques are necessary which take account of the handling characteristics of carbon fibers, which are different from those of other reinforcements. It is the processibility of polymers that make them particularly attractive for use in a whole range of composite materials. For example, one form of carbon-fiber-reinforced plastic (CFRP) contains continuous carbon fibers, possibly in the form of a unidirectional or woven mat, which are impregnated by a thermosetting resin such as an epoxy. Prepreg material of this type can then be layed up to ensure the appropriate orientation of the fiber before curing of the resin is undertaken. Alternatively, if fiber orientation does not have to be so rigorously defined, then the processibility of thermoplastic polymers means that with the reinforcement in the form of short fibers (typically 0.2–10 mm), a molding compound of fiber and thermoplastic can be injection-molded to produce directly finished components.

From the perspective of materials performance, the potential of CFRP is enormous: for example, a civil aircraft constructed, where possible, entirely of CFRP instead of aluminum alloy would have a total weight reduction of 40%; a communications satellite built from CFRP can carry more telecommunications channels than permitted by previously specified materials; and a new generation of tactical aircraft can be planned, which will have a versatility regarded as impossible prior to the advent of CFRP, e.g., forward-swept wings for low stalling speeds, and very small turning circle radii. However, the high cost of the fiber and the lack of cost-effective manufacturing processes are still inhibiting the more general use of CFRP, as is the newness of the art of composites design.

1. Matrix Materials

1.1 Thermosets

Examples of thermoset resins commonly used for carbon-fiber composites are listed in Table 1, together with their advantages and disadvantages. Long shelf life is an important advantage for prepregs, while low viscosity is vital to processing techniques such as pultrusion, filament winding and rapid resin injection molding (RRIM). Cure of the resin should not take a long time or require extreme temperatures, while a high value of the glass transition temperature (T_g) (the temperature of the glassy–rubbery transition) is necessary if the composite is to be used at elevated temperatures. Carbon-fiber–epoxy composites represent about 90% of CFRP production. As can be seen in Table 1 the properties of an epoxy are dependent on the curing agent used and the functionality of the epoxy. The attractions of epoxy resins are that (a) they polymerize without the generation of condensation products which can cause porosity, (b) they exhibit little volumetric shrinkage during cure, which reduces internal stresses, and (c) they are resistant to most chemical environments.

The major disadvantages of epoxy resins are their flammability and their vulnerability to degradation in aggressive media. Phenolic resins are, however, quite stable in both respects, although as structural composite materials, phenolic CFRPs are quite uninteresting, as their fatigue and impact properties are very

Table 1
Examples of thermosetting resins used for CFRP

Resin	Advantages	Disadvantages
Epoxides		
DGEBA (aliphatic amine curing agent)	very low viscosity; room-temperature cure	very low T_g; short shelf life; flammability
DGEBA (aromatic diamine curing agent)	medium viscosity; long shelf life	medium T_g; difficult cure
TGMDA (aromatic diamine curing agent)	high T_g; long shelf life; toughness	high viscosity; difficult cure; moisture uptake; flammability
Novalac (phenolic)	nonflammability; high T_g	difficult cure; shrinkage; brittleness
PMR 15 (polyimide)	very high T_g; low moisture uptake; high strength	difficult cure; thermal fatigue

DGEBA, diglycidylether of bisphenol A
TGMDA, tetraglycidylmethylenedianiline

poor. As indicated earlier, the prospect of using CFRPs as a substitute for light metal alloys is particularly attractive but there are problems to be overcome at high temperatures. Using polyimides, peak service temperatures of ~400 °C can be reached. However, conventional polyimides present significant processing problems. Toxic solvents have to be evaporated and, during curing, condensation products have to be vented from the tooling. Cure cycles are also prohibitively long. PMR 15 is a polyimide developed to surmount the above problems. Three monomer reactants are mixed in an alcohol, the solution being used to impregnate the fiber. The alcohol is then evaporated, during which time a thermoplastic prepolymer is formed. This prepreg can then be cured using tooling identical to that used with epoxies, although much higher temperatures (~380 °C) are required.

Another compromise using polyimide chemistry to enhance CFRP performance is to form copolymers with epoxy resins. In this way flame retardancy and reduced moisture absorption are realized, but there is no substantial increase in high-temperature performance. Polybismaleimides represent a very good compromise in many respects and should command the applications arena between the epoxies and the polyimides.

1.2 Thermoplastics

A considerable number of thermoplastics are currently being used in CFRP; they may be either semicrystalline or amorphous. Semicrystalline thermoplastics offer greater solvent resistance, but are generally more difficult to process.

In load-bearing applications, carbon fibers are most widely used in combination with nylon 66 which gives the highest strength and modulus over a fairly wide temperature range. A thermoplastic that is currently of considerable interest to fabricators looking to mass-produce articles is poly-1,4 butanediol terephthalate (PBT). It has a melting point of 224 °C and, if a low molecular weight grade is used, it can be injection molded at 250 °C. It crystallizes very rapidly so very short cycle times can be employed. CFRP based upon PBT exhibits a heat distortion temperature of 200 °C with adequate resistance to creep up to 120 °C.

The prospects of using high-temperature engineering thermoplastics, examples of which are shown in Table 2, with continuous-fiber reinforcements instead of thermosetting resins, look particularly exciting. The advantages of thermoplastics are lower production costs, improved toughness and impact resistance and lower susceptibility to moisture uptake, which is found to reduce the value of T_g for epoxy matrices. Additional benefits are the ease of repair of the composites and the recovery of waste or used material.

In Table 3 the properties of injection-molded nylon 66, polyethersulphone (PES) and polyether ether ketone (PEEK), each with 40 wt% of carbon fiber are compared. The nylon 66 compound exhibits superior properties at normal ambient temperature. It is with properties such as creep modulus (stress to induce 0.1% strain within 100 s) and tensile strength at elevated temperatures that the high-temperature thermoplastics reveal their merit.

2. Fiber–Matrix Interphase

The properties of the material between the fiber reinforcement and the plastic matrix have a strong influence on the mechanical properties of the CFRP. To produce a high-modulus composite requires maximizing adhesion between the fiber and matrix. However, if the polymer matrix is brittle (e.g., epoxies) there is a corresponding reduction in impact strength as mechanisms of energy dissipation such as debonding and fiber pullout become suppressed. If a tough polymer matrix is to be used (e.g., PEEK), where under impact the principal mechanism of energy dissipation is yielding within the polymer, good adhesion between the fiber and matrix is required.

Unfortunately this region of the composite, sometimes known as the interphase, is extremely complex. Constituents of the interphase might include a modified fiber surface, the treatment being undertaken to promote better adhesion, or a coating of polymer present to ensure compatibility with the matrix polymer. In addition, the presence of the fiber surface may

Table 2
High-temperature engineering thermoplastics

Generic name (trade name; manufacturer)	Heat distortion temperature at 1.82 MPa (°C)
Polycarbonate (Lexan, Merlon; GE, Mobay)	133
Polyphenylene sulfide (Ryton; Phillips)	137
Polyether ether ketone (Victrex PEEK; ICI)	148
Polysulfone (Udel; Union Carbide)	174
Polyetherimide (Ultem; GE)	204
Polyethersulfone (Victrex PES; ICI)	204

Table 3
Properties of three thermoplastics reinforced with carbon fiber (CF)[a]

Property	Material		
	40% CF–nylon 66	40% CF–PES	40% CF–PEEK
Density (gm cm^{-3})	1.34	1.52	1.45
Tensile strength (MPa)			
at −55°C			275
24°C	246	176	227
50°C	168	168	208
100°C	108	140	165
140°C	75	112	129
180°C			72
Tensile elongation (%)	1.65	1.1	1.35
Tensile creep modulus (GPa)			
at 24°C	26.2	22.1	26.5
120°C	11.5	21.4	23.9
160°C	9.0	21.1	10.5
200°C		17.5	6.6
Flexural strength (MPa)	413	244	338
Flexural modulus (GPa)	23.4	16.8	20.8
Compressive strength (MPa)	240	216	
Compressive modulus (GPa)	20.2	17.2	

[a] "Grafil" product, reference RG40

affect the cure of thermosets, resulting in chemically different polymer, or may affect the morphology of semicrystalline thermoplastics.

2.1 Surface Treatment

Surface-treatment processes are now established as an integral part of fiber manufacture. Most are either wet oxidative treatments using solutions of, for example, sodium hypochlorite, sodium bicarbonate or chromic acid, or dry treatments in which the oxidizing agent is ozone. The precise action of the treatment is undoubtedly complex, but is currently thought to include the introduction of chemically active sites onto which the polymer can graft, as well as chemical cleaning, an increase in rugosity of the fiber surface and the removal of any weak surface layer on the fiber.

A plethora of techniques has been proposed to measure fiber–polymer adhesion. Such techniques are experimentally tedious but, more significantly, the assumptions necessary to deduce adhesive strengths from the various types of measurement make them only comparative. The techniques include measurement of the composite short-beam interlaminar shear strength, the distribution of fiber lengths in a fully fragmented single fiber, the stress required to pull out a fiber embedded in the polymer and the stress required for a bead of polymer to become detached from a single fiber.

2.2 Surface Coating and Sizing

Polymer coatings and sizes are frequently applied to fibers immediately after manufacture. The light application of resin binds individual filaments together and maintains a compact bundle for processing into an intermediate product or final composite form. Size compositions applied to the surface can afford protection against filament damage and provide lubrication for a smooth delivery of fibers in continuous processing. The sizing or coating may be bulk polymer or resin specially functionalized to improve fiber wetting and adhesion while still retaining compatibility with the bulk polymer.

3. *Properties of CFRP*

The main points to appreciate when considering a carbon-fiber composite as compared with a conventional material such as a metal alloy are that the composite is heterogeneous and anisotropic. By heterogeneous it is meant that the properties vary from point to point within the material and by anisotropic it is meant that the properties depend on the direction in which they are measured. These two factors are important when considering not only mechanical properties (such as stiffness and strength) but also the physical, electrical and thermal characteristics of the composite.

The physical and mechanical properties of a CFRP are strongly dependent on the properties of the particular fiber used. There are a variety of fiber types available and the properties of fibers are being advanced continually by the development of new grades (see Table 4).

Table 4
Types of commercially available carbon fibers

Carbon fiber type	Tensile strength (GPa)	Elastic modulus (GPa)	Breaking strain (%)
High strength (PAN)	3.5	240	1.5
High strain (PAN)	5.5	240	2.2
Intermediate modulus (PAN)	4.5	285	1.6
High modulus (PAN)	2.5	350	0.7
Ultrahigh modulus (PAN)	2.0	450	0.5
Isotropic (pitch)	1.0	35	2.5
Ultrahigh modulus (pitch)	2.0	550	0.4

The section below outlines some of the main features of the mechanical, thermal and electrical properties of CFRP.

3.1 Mechanical Properties

CFRP containing continuous aligned fibers (which would typically have a volume fraction of 60%) are essentially elastic to failure (when loaded in tension parallel to the fibers) and exhibit no yield or plasticity region. There is, however, a slight nonlinearity in the stress–strain curve, with the stiffness increasing slightly with applied strain, as a result of improved orientation of the graphitic planes within the fibers. Although the failure strains in tension for CFRP may seem small when compared with those of other structural materials, the predictably elastic behavior under load allows a high proportion of the ultimate strength to be utilized in practice. Consequently, strain levels at useful working stresses are comparable with those of metals, alloys and glass-fiber composites. The properties of unidirectional CFRP are compared with similar performance characteristics of other commonly used structural materials in Table 5. A range of material properties typical of those required for design calculations is given in Table 6. These data confirm the anisotropy of the unidirectional material (which is much more marked in CFRP than in glass or aramid composite); the stiffness and strength of the materials are very much lower when they are loaded perpendicular to the fibers than when they are loaded parallel to the fibers.

It should also be noted that the unidirectional compression strength of the lamina is only about two-thirds of the tensile strength. In early carbon-fiber composite materials (from the early 1970s) the tensile strength and compression strength were approximately equal, with the failure mode in compression being shear failure. Improved fiber tensile properties have been achieved with a reduction in fiber diameter and, during this time, more ductile resins have come into favor. These factors have tended to promote fiber instability or microbuckling failure when unidirectional materials are loaded in compression. Hence the compression strength of the material is not a true measure of the fiber strength, but more of the ability of the matrix to support the fiber against buckling, and the integrity of the fiber–matrix interface. As a result of these developments, compressive performance of CFRP can be a limiting design condition.

There are very few practical applications for CFRP which require only unidirectional reinforcement. Laminates are assembled from unidirectional layers stacked at various orientations to one another. Layers are included at 90° to give adequate transverse properties and at 45° to give good shear properties. A laminate construction of 0°/90° ±45° plies has negligible anisotropy when loaded in-plane and is known as quasi-isotropic. Typical data for such a laminate

Table 5
Some properties of fiber-reinforced composites (55–70 vol% fiber) and metals

Material	Density ($gm\,cm^{-3}$)	Ultimate tensile strength (GPa)	Young's modulus (GPa)	Specific tensile strength (GPa)	Specific Young's modulus (GPa)
CF–Epoxy:					
high strength	1.5	1.9	130	1.27	87
high modulus	1.6	1.2	210	0.94	119
E-glass–epoxy	2.0	1.0	42	0.50	21
Aramid–epoxy	1.4	1.8	77	1.30	56
Steel	7.8	1.0	210	0.13	27
Titanium	4.5	1.0	110	0.21	25
Aluminum L65	2.8	0.5	75	0.17	26

Table 6
Typical performance data for unidirectional laminates

Property	High strength	High modulus
0° Tensile strength (GPa)	1.90	1.15
0° Tensile modulus (GPa)	137	200
Elongation (%)	1.4	0.6
Interlaminar shear strength (MPa)	90	70
0° Flexural strength (GPa)	1.80	1.10
0° Flexural modulus (GPa)	130	190
0° Compression strength (GPa)	1.25	0.70
0° Compression modulus (GPa)	140	210
Density ($g\,cm^{-3}$)	1.55	1.63
90° Tensile strength (MPa)	50	35
90° Tensile modulus (GPa)	10	7

Table 7
Performance data for multidirectional laminates with 60 vol% fiber and 0°/90°/±45° laminate construction

Laminate	Tensile strength (GPa)	Tensile modulus (GPa)
High strength	0.6	50
High modulus	0.35	75

are shown in Table 7. As a result of the marked anisotropy, CFRP can experience large stress concentration factors at cutouts or mechanically fastened joints. Given that the material is predominantly elastic to failure this might be expected to present serious practical problems. Fortunately, certain damage mechanisms (such as matrix cracking parallel to the fibers and delamination cracking between layers) can occur in laminates so that the stress concentrations do not have the deleterious effects that might be expected. Even so, the designer must take care when dealing with cutouts or joints. Bonded joints must also be designed carefully to avoid failure of the laminate by interlaminar (through-thickness) failure.

The range of basic mechanical properties of CFRP is enhanced by excellent resistance to creep rupture and fatigue. Creep rupture times at equivalent stress levels are several orders of magnitude greater for CFRPs compared with those for glass fiber-reinforced composites (GRP), and also significantly better than for aluminum. CFRP is also capable of operating well under conditions of alternating stress. For high strength fibers, S–N fatigue curves are much flatter for CFRP compared with those for GFRP or aluminum, and a higher ratio of working stress to ultimate stress can be utilized. However, the newer fiber–matrix systems are not yet characterized fully and there are signs that, with these materials, fatigue may be more of a consideration than in the past. In passing it should be noted that fatigue failure in composite materials is not caused by the initiation and growth of a dominant crack as it is in metallic materials; instead there is an accumulation of damage (matrix cracks, delamination, fiber breaks) which may result in a lowering of the stiffness and the residual strength. Much more work is needed to develop damage-tolerant design for CFRP composites.

The impact resistance of CFRP is generally considered low, and in fact the post-impact compression strength is considered to be a design limiting factor. There is, however, scope for the modification of impact performance and fracture toughness by fiber hybridization with glass or aramid reinforcements leading to structures with much greater damage tolerance. Moreover, impact performance will improve as strain-to-failure of the fiber continues to increase.

3.2 Thermal Properties

CFRP is thermally conductive along the direction of the fibers. Conductivity perpendicular to the fibers is much less as it is dominated by the polymer matrix. The ability to dissipate heat is thought to contribute to the very good fatigue properties of CFRP. The conductivity of fibers increases with graphite content, and highly graphitized fibers, such as high-modulus fibers, have thermal conductivity values of $34\,W\,m^{-1}\,K^{-1}$ in the longitudinal direction which approach the value for steel ($50\,W\,m^{-1}\,K^{-1}$).

Coefficients of thermal expansion for carbon-fiber composites are dependent on the fiber type and on

Table 8
Typical thermal expansion data for CFRP with 60 vol% fiber

CFRP	Coefficient of thermal expansion[a] (10^{-6} K^{-1})	
	Longitudinal	Transverse
High modulus–epoxy	−0.25– −0.60	20–65
High strength–epoxy	0.30– −0.30	20–65

[a] over the temperature range 100–400 K

temperature. The fibers themselves have small negative coefficients of thermal expansion and this leads to the possibility of designing composites with almost zero overall thermal expansion. This dimensional stability, which can be available over extremes of temperature, is very attractive in applications such as antennas for aerospace vehicles. As with the thermal conductivity, the coefficient of thermal expansion is anisotropic—the value measured perpendicular to the fibers is dominated by the polymer matrix and is usually much higher (see Table 8) than the value measured parallel to the fibers. This can have important practical consequences in that residual stresses are left trapped in a laminate after it has cooled from the processing temperature. These stresses can be large enough to promote matrix cracking, especially in systems such as PMR-15 which, as described previously, is processed at high temperatures. For such systems to achieve wider use this problem has to be overcome either by toughening the resin further, or by further development of a damage-tolerant approach to design.

3.3 Electrical Properties

CFRP is electrically conductive along the direction of the fiber. As with thermal conductivity, electrical conductivity increases with the level of graphitization of the fiber.

There are practical applications which make use of the electrical properties of carbon fiber, notably to provide resistance heating effects (in nonmetallic tooling, for example) or static charge dissipation in composite structures.

The electrochemical behavior of graphite can lead to problems when carbon-fiber composites are in contact with metal alloys, as they may be in an aircraft structure, for example. Graphite is cathodic to most metal alloys and this can lead to appreciable corrosion (especially in mechanically fastened assemblies) unless adequate precautions are taken.

4. Forms of Carbon Fibers

Continuous-filament fibers represent the primary production route for manufacture, and other products are derived from this form by secondary processes. The conversion of fibers into some conveniently handleable intermediate product plays an important part in the effective utilization of the fibers as a reinforcement. Specific products are needed to meet the individual processing requirements of different fabrication techniques.

4.1 Continuous-Filament Tow

Continuous fibers are grouped together in tow bundles, each containing a specified number of individual parallel filaments. The tows are produced in different bundle sizes, ranging in discrete steps from 400 to 320 000 filaments per tow. The main sizes used are in the 3000–12 000 range; finer tows are employed only for specialized applications. Heavier tows offer some advantages for rapid volume conversion to composite and can lead to economies of scale for both fiber production and composite manufacture by certain processes.

Twisted tows, containing typically 15 turns per meter, arise from processes employing twisted precursor fibers. The presence of twist reduces the efficiency of the fiber reinforcement but can enhance the handleability of the tow, aiding conversion to secondary products such as woven fabrics and braids.

4.2 Woven Fabrics

Carbon fibers can be woven into a variety of unidirectional and bidirectional fabrics capable of being processed efficiently into composite materials. The weave construction (e.g., plain, satin, twill) determines the handle of the fabric and controls its ability to conform to contoured shapes.

4.3 Preimpregnated Products

Forms of preimpregnated materials (prepregs) include continuous tows, unidirectional tapes and sheets, and woven fabrics. A resin is applied to the fibers using a solvent, melt or film impregnation technique. The resins used are specially formulated with a latent cure, so that at room temperature the cure reaction proceeds only very slowly and the material can be handled and stored at ambient temperature for a reasonable length of time. Curing temperatures are typically between 100 and 180 °C. A choice can be made from the wide range of commercial prepregs to suit the end use and fabrication technique employed.

4.4 Nonwoven Fabrics

Carbon-fiber mats, felts and papers manufactured by conventional nonwoven processes from randomly distributed short fibers (about 6 mm in length) provide multidirectional reinforcement but convert inefficiently into composite form. New products containing longer fibers (about 25 mm) are now being developed, exhibiting good handling characteristics and generating improved composite performance.

4.5 Molding Materials

Short-length fibers chopped from continuous tows are widely used for the manufacture of molding compounds in both thermoplastic and thermosetting matrices. The application of special resin sizes and coatings facilitates processing into molding materials and other products such as papers and felts.

Thermoplastic molding compounds are supplied as granules containing short-length (0.2 mm) fiber reinforcements for processing on conventional injection-molding equipment.

Sheet molding compounds (SMCs) and dough molding compounds (DMCs) are already well established with glass fibers. The use of a small specific quantity of carbon-fiber reinforcement is envisaged in SMC with careful placement of aligned continuous fibers for optimum reinforcing effect at minimum additional cost. Carbon-fiber-reinforced DMCs in epoxy and polyester systems achieve excellent stiffness but strength levels achieved so far are disappointing.

Aligned short fibers in the form of prepreg sheets (ASSM) are available from special processes developed in the UK and Germany. The products can be molded to generate high performance in composites with highly curved or contoured shapes by virtue of the aligned fiber reinforcement coupled with the ability to promote controlled fiber movement for mold conformity.

5. *Fabrication Techniques*

Fabrication techniques have an important bearing on the properties obtained from carbon-fiber composites, and efficient processes must be devised for handling the fibers to ensure the maximum contribution to mechanical performance. The manufacture of GRP provides some useful experience although two major points of difference arise. First, the handling characteristics of the two fibers are different; secondly, the aim in the manufacture of carbon-fiber composites is to obtain the best possible mechanical performance, requiring the fiber to be incorporated in the precise direction and in the exact amount necessary to meet the stresses applied.

The basic principles of the manufacture of carbon-fiber composites are similar to those used for GFRP (see *Glass-Reinforced Plastics: Thermosetting Resins*), although changes in the actual techniques are necessary in most cases. Processes unique to CFRP production are also being developed.

5.1 Compression Molding

Three suitable material forms containing carbon-fiber reinforcement are available: DMCs, SMCs, and preimpregnated tapes, sheets and fabrics.

Certain considerations which affect plant and mold tool design and process procedures are necessary for the manufacture of CFRPs. The thermal conductivity and low thermal-expansion characteristics of carbon fibers introduce the need for carefully controlled heating and cooling cycles and special attention to molding shrinkage (and expansion) to achieve dimensional accuracy in the final part. SMCs usually comprise hybrid reinforcements, mainly glass, with a small amount of carbon included for specific stiffening. The effectiveness of the carbon fiber is influenced by careful placement for maximum reinforcing effect and for minimum risk of misalignment caused by resin and glass-fiber flow during molding. Alignment accuracy, parallelism of mold-tool platens and additional considerations arise with the use of prepreg materials.

5.2 Autoclave and Vacuum Bag Molding

The production of high-quality composites from prepreg materials (tapes and fabrics) is achieved by the consolidation on a single-sided tool of a laminated preform. The method makes use of a flexible bag or blanket through which pressure is applied to the prepreg layup on the mold. The mold is supported on a platen inside an autoclave which is pressurized with gas heated to the required cure temperature.

Unidirectional and woven materials can be molded equally successfully and the choice depends on the end-use requirements and the layup techniques employed. Highly automated procedures for the precise placements of prepreg layers in laminate preforms are now well developed and are being introduced on a production basis in aircraft-component manufacture. Computer-controlled tape-laying and ply-profiling machines are a major step towards a fully automated integrated composite-manufacturing scheme.

Current trends are towards prepregs with simplified cure cycles and zero resin-bleed characteristics. The careful use of viscosity modifiers in resin formulation offers greater flow control which is important for large or thick composite moldings in which significant temperature variations may exist.

Vacuum molding represents a simplified version of the autoclave technique where the pressure chamber is substituted by an oven and consolidation is achieved by virtue of the vacuum drawn inside the flexible membrane. The relatively low net pressure applied requires the resin to have easy flow characteristics.

5.3 Filament Winding

The filament-winding process is appropriate for the production of hollow carbon-fiber components which have an axis of rotation. Excellent mechanical properties are achieved by virtue of the accurate positioning of the fibers and the high volume fractions possible in the composite. The process is highly automated and can be programmed to lay down specific reinforcement patterns to comply with complex design requirements in multistressed components.

Carbon fibers can be processed successfully by both wet-winding and prepreg-winding techniques. Special attention to the gradual buildup of winding tensions and to the choice of smooth-surfaced materials for guide points is necessary in order to promote even, high-speed delivery of carbon fibers with minimum risk of abrasion and filamentation.

5.4 Mandrel Wrapping

Prepreg wrapping is used widely in the manufacture of sports goods, for the production of parallel and taped tubular shapes. Special purpose-built equipment is available for semiautomated production. Mechanical performance is determined by the layup pattern of the tube, although simple configurations are required for successful volume throughput. Unidirectional tapes and woven fabrics are equally applicable. Good material handleability is essential; unidirectional prepregs are often combined with a light scrim support to improve transverse properties.

5.5 Pultrusion

Pultrusion is a well-established technique for the production of profiles of glass-reinforced polyester composites and offers scope for CFRP manufacture. The process is simple, continuous, can be highly automated and does not require a high level of financial or technological investment. The use of carbon-fiber reinforcement with high-performance systems, such as epoxies, presents a completely new set of considerations for a continuous molding operation. The interacting effects of thermal expansion and resin shrinkage impose strict requirements for the tool design both in terms of dimensional accuracy and material of construction.

High-quality, high-performance pultruded stock is commercially available. Hybrid reinforcement configurations, new matrix materials (e.g., thermoplastics offering postformability) and methods for introducing off-axis reinforcement are being developed to extend the capabilities of the process.

5.6 Wet Layup

The contact molding method is simple and is a low-cost method, but is prone to variability in quality and performance. The method is therefore not attractive for the widespread use of carbon fibers. However, the addition of a small proportion of carbon fibers as a specific reinforcement for GFRP structures can be both cost- and performance-effective. For example, unidirectional woven tapes strategically incorporated into GFRP laminates promote maximum stiffness for minimum added cost, and nonwoven fabrics applied as surfacing tissues for GFRP structures provide chemical resistance and electrical conductivity.

5.7 Other Methods

Thermoplastic molding compounds are processed by injection molding. Long-fiber reinforcements, combined with polyethersulfone, polyether ether ketone and other polymers can be converted to composites by thermoforming operations.

6. *Applications*

Applications for CFRPs fall generally into three broad categories: aerospace (aircraft, space and satellite structures), sports goods, and engineering and industrial equipment. The overriding application is that of metals substitution for improved mechanical performance and weight reduction. The stiffness-to-weight ratio for high-modulus CFRPs is the highest available for any structural material and hence carbon-fiber composites are applied where stiffness, strength and light weight are essential.

7. *Future Perspective*

Today, about 4000 t of carbon fiber per year are consumed worldwide. Compared to glass fiber used in composite materials, estimated at five to six million tonnes per year, this is indeed a modest market. The latter is now a mature product and advances at the approximate rate of world economic growth. Before carbon fiber arrives at this phase of its history, it must first establish a concrete role in the engineering consciousness and achieve an economic scale of production. Carbon fiber costs US$30 a kilogram at 1988 prices. Potentially, using polyacrylonitrile (PAN) produced on a commercial scale, it could reduce in price to US$10 a kilogram. On the basis of modulus, volume for volume, this would make CFRP less expensive than GFRP. If the early promise of using pitch as a precursor to carbon fiber should be fulfilled, the cost expectations would exceed this prediction by far.

The most important factor influencing the future of CFRP is the cost of energy. By the late 1980s, earlier anxieties about energy supplies had abated, but ultimately the driving force in the materials economy will be the energy cost of materials plus the energy cost of their use. CFRP offers considerable advantages in both respects compared to conventional materials.

Bibliography

Anon 1981 *Processing and Uses of Carbon Fiber Reinforced Plastics.* VDI, Düsseldorf

Anon 1983 Fiber composites; design, manufacture and performance. *Composites* 14(2): 87–139

Anon 1986 *Carbon Fibers, Uses and Prospects.* Noyes Data Corporation, Park Ridge, New Jersey

Clegg D W 1986 *Mechanical Properties of Reinforced Thermoplastics.* Elsevier, London

Goodman S 1986 *Handbook of Thermoset Plastics.* Noyes Data Corporation, Park Ridge, New Jersey

Kelly A, Mileiko S T 1983 *Fabrication of Composites.* North Holland, Amsterdam

P. J. Mills and P. A. Smith
[University of Surrey, Guildford, UK]

Cellular Materials in Construction

Many materials have a cellular structure, made up of an interconnected network of struts or plates. Such cellular materials occur in nature as wood, cancellous bone and cork. Recently, man has begun making his own cellular materials, in the form of honeycomb-like materials and foams. Although initially honeycombs and foams could only be made from a limited range of materials, both can now be made from a wide range of polymers, metals, ceramics and glasses. Micrographs of several natural and man-made cellular materials are shown in Fig. 1.

The cellular structure of these materials gives rise to unique properties which make them useful in many applications. Because they can undergo large deformations at relatively low loads, they are widely used in the packaging industry for providing protection from impacts. Their low volume fraction of solids and small cell size give them a low thermal conductivity, making them ideal thermal insulators. And their low weight is exploited in structural sandwich panels, where they are used as a lightweight core to separate two stiff, strong faces; the resulting sandwich gives an efficient member for resisting bending and buckling loads.

In this article, the structure and mechanical behavior of cellular materials are reviewed, and their use in construction is described.

1. Structure and Mechanical Behavior

Several features characterize the structure of a cellular material. The volume fraction of solids in the material can be described by its relative density: the density of the cellular material divided by that of the solid cell wall material from which it is made. The cells may extend in one plane only (as they do in honeycombs), so that the normals to all of the cell walls lie in a single plane, producing a "two-dimensional" cellular structure. Alternatively, they may extend in any direction

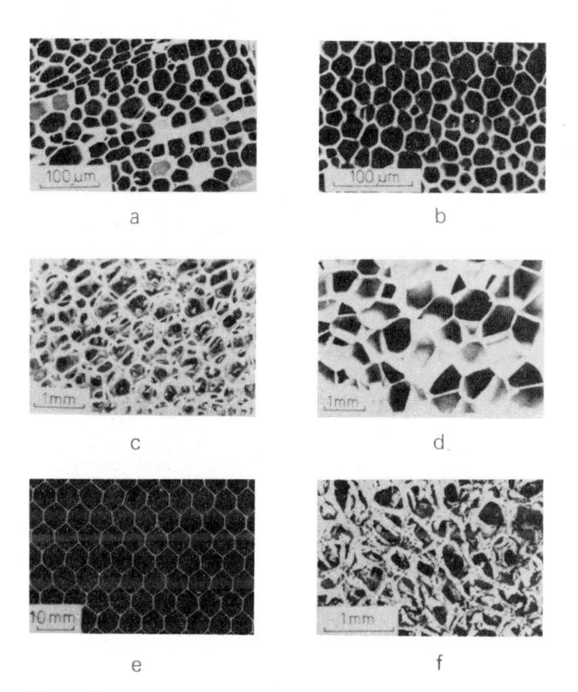

Figure 1
Micrographs of cellular materials: (a) balsa wood, (b) cork, (c) polyurethane foam, (d) polyethylene foam, (e) aluminum honeycomb, (f) copper foam

(as they do in foams), giving a less restricted, "three-dimensional" structure. The cells may be open or closed, depending on whether or not the faces of the cells are empty or covered by thin membranes. The cells may be geometrically anisotropic, with different mean intercept lengths in different directions. Anisotropy is often the result of the rise of the foaming gas in the liquid matrix which produces cells which are elongated in the rise direction and roughly equiaxed in the plane normal to it. The mechanical properties of foams are often anisotropic as a result of anisotropy in the cell geometry. Each of these structural features is shown in Fig. 1.

A typical stress–strain curve for a cellular material is shown in Fig. 2. There are three distinct regimes of behavior. At relatively low loads, the material is linear elastic. At some critical level of load, the cells begin to collapse by elastic buckling, plastic yielding or brittle fracture, depending on the nature of the cell wall properties. Cell collapse then progresses at a roughly constant load, producing a roughly horizontal stress plateau, until, at relatively large strains (typically about 0.8), the opposing cell walls begin to meet and touch. The stress then rises steeply as the material densifies (Maiti et al. 1984).

Each of these regimes of behavior is related to the mechanism by which the cells deform within that regime. Observation of model honeycomb materials and of foams in a scanning electron microscope reveals that the initial linear elasticity is produced by

Table 1
Property–density relationships for cellular materials

Property	Open celled foams	Honeycombs In-plane	Honeycombs Axial
E	$C(\rho/\rho_s)^2 E_s$	$C(\rho/\rho_s)^3 E_s$	$C(\rho/\rho_s) E_s$
ν	C	C	C
G	$C(\rho/\rho_s)^2 E_s$	$C(\rho/\rho_s)^3 E_s$	$C(\rho/\rho_s) E_s$
σ^*_{el}	$C(\rho/\rho_s)^2 E_s$	$C(\rho/\rho_s)^3 E_s$	$C(\rho/\rho_s)^3 E_s$
σ^*_{pl}	$C(\rho/\rho_s)^{3/2} \sigma_{ys}$	$C(\rho/\rho_s)^2 \sigma_{ys}$	$C(\rho/\rho_s)^2 \sigma_{ys}$
σ^*_{cr}	$C(\rho/\rho_s)^{3/2} \sigma_{fs}$	$C(\rho/\rho_s)^2 \sigma_{fs}$	$C(\rho/\rho_s) \sigma_{fs}$
K^*_{Ic}	$C(\rho/\rho_s)^{3/2} \sigma_{fs}\sqrt{\pi l}$	$C(\rho/\rho_s)^2 \sigma_{fs}\sqrt{\pi l}$	$C(\rho/\rho_s) \sigma_{fs}\sqrt{\pi l}$

Note: 1. The constants of proportionality are different for each property and each case (foams, honeycomb-in-plane and honeycomb-axial). 2. E_s, σ_{ys} and σ_{fs} are the Young's modulus, yield strength and modulus of rupture, respectively, of the solid cell-wall material. 3. The characteristic length of the cell size is l

bending in the cell walls. Cell collapse is caused by elastic buckling in elastomeric honeycombs and foams, by plastic yielding in materials which yield, and by brittle crushing in materials which fracture (Gibson and Ashby 1982, Maiti et al. 1984). The strain at which the honeycomb or foam begins to densify depends on its relative density.

The tensile behavior of cellular materials is similar to the compressive behavior with the following two exceptions. Elastomeric honeycombs and foams do not buckle so that the stress plateau between the linear elastic regime and densification is eliminated in tension. And the tensile stress–strain curves of brittle cellular solids is terminated by brittle fracture at the end of the linear elastic regime.

This stress–strain behavior can be characterized by several mechanical properties. The linear elastic regime is described by a set of elastic moduli such as the Young's modulus E, the Poisson ratio ν, and the shear modulus G. The plateau stresses corresponding to elastic buckling, plastic yielding and brittle crushing are σ^*_{el}, σ^*_{pl} and σ^*_{cr}, the limiting strain for densification is ε_D, and the tensile fracture toughness is K_{Ic}. By modelling the structure of cellular materials and the mechanisms by which they deform, relationships between the properties of the cellular material and its relative density, its solid cell wall properties, and its geometry can be found. Table 1 summarizes these relationships for both honeycombs and foams. A more detailed analysis can be found in Maiti et al. (1984).

2. *Applications of Cellular Materials in Construction*

2.1 Lightweight Structural Sandwich Panels

Structural panels made up of two stiff, strong faces separated by a lightweight core are known as sandwich panels. The core acts to separate the faces, increasing the moment of inertia of the panel, with little increase in weight, producing an efficient structure for resisting bending and buckling loads. Usually, the core is made of a lightweight cellular material. In aerospace applications, aluminum or paper–resin honeycomb cores are preferred for their extremely low weight. In building applications foam cores, which are typically somewhat heavier than honeycomb cores, are preferred as they offer low thermal conductivity in addition to relatively low weight. Typical face materials include fiber-reinforced composites, steel, aluminum and plywood.

The technology for producing sandwich panels was first developed in the aerospace industry, beginning with the balsa core–plywood face sandwiches of the World War II Mosquito aircraft (Allen 1969). Modern aerospace sandwich components are now made almost exclusively with honeycomb cores and fiber-composite faces bonded together with advanced adhesives. Sandwich panels are now made routinely and are reliable for components such as the rotor blades of helicopters, wing-tip sections, tail sections and flooring panels. This technology is now well developed and is being transferred to the construction industry.

The main use of sandwich components in the construction industry is in building panels. Several manufacturers are now producing wall, roof and door panels using sandwich construction. Different combinations of face and core materials are used; typically the faces are made of steel, aluminum or waferboard and the cores are made of foamed polyurethane, foamed polystyrene beadboard or foamed glass. The foam core gives excellent thermal insulation while the faces provide the structural stiffness and strength of the panel. The panels are used in housing and in low-rise, nonresidential buildings and as fascia panels in commercial buildings.

The reduced weight of sandwich construction is also useful in applications where the component must be lifted into place by crane or helicopter. For example, a recent EPA ruling required the addition of roofing covers to exposed water distribution reservoirs. Domed covers made with fiberglass-reinforced plastic faces and a paper–resin honeycomb core were assembled adjacent to the reservoir sites and then lifted into place by crane. The low weight of the sandwich reduced the installation costs. In another application, modular sandwich lighthouse structures with fiber-reinforced plastic faces and balsa cores were installed using helicopters; again, the low weight of the units reduced installation costs.

Sandwich construction is also used for portable structures, such as the portable, folding bridges being developed by the US Army. The deck of these bridges is doubly sandwiched, with a top face of high-strength aluminum and a bottom face of a graphite-epoxy–aluminum composite. Between these two faces is a webbed core which is itself a sandwich with aluminum faces and a honeycomb core. The low weight of the sandwich structure allows the bridge to be transported easily. The bridges are hinged so that they fold during transport and can be deployed by a hydraulic actuator for use.

The weight of sandwich panels can be minimized by appropriate optimization analyses. Wittrick (1945), Ackers (1945) and Kuenzi (1965) have developed minimum weight analyses for pin-ended columns subject to an end load. Huang and Alspaugh (1974) and Ueng and Liu (1979) have minimized the weight of a sandwich panel subject to multiple constraints. More recently, Gibson (1984) and Demsetz and Gibson (1987) have found the core density as well as the face and core thickness which minimize the weight of a sandwich beam or plate for a given stiffness. These optimization analyses are useful in applications where the weight of the component is critical.

2.2 Wood

Wood is still one of the most widely used structural materials throughout the world. The present world production is about 10^9 t per year, roughly equal to that of iron and steel (Dinwoodie 1981). In North America more wood products are used for construction than all other construction materials combined (Bodig and Jayne 1982).

The cellular structure of wood is shown in Fig. 1 (a). Softwoods used in construction have two main types of cells: tracheids and parenchyma. The tracheids are the long, narrow cells (aspect ratio about 100:1) which make up the bulk of the tree. They conduct fluids and nutrients within the tree and give it its structural support. The parenchyma cells, found in the rays, store sugars.

At a finer level, the walls of the cells have a fiber-composite structure with partially crystallized cellulose molecules forming stiff, strong, fiber-like chains helically wound in a matrix of amorphous lignin and semicrystalline hemicellulose. Each cell wall is made up of four layers, each with a different orientation of cellulose fibers.

Mechanically, wood behaves very much like other cellular materials. Stress–strain curves for several types of wood in uniaxial compression along and across the grain are shown in Figs. 3 and 4. These curves show that wood is highly anisotropic: it is much stiffer and stronger when loaded along the grain than across it. Part of this anisotropy results from the different mechanisms by which wood deforms when loaded along and across the grain, and part comes from the anisotropy in the cell wall properties. The general shape of the curves resembles that of Fig. 2.

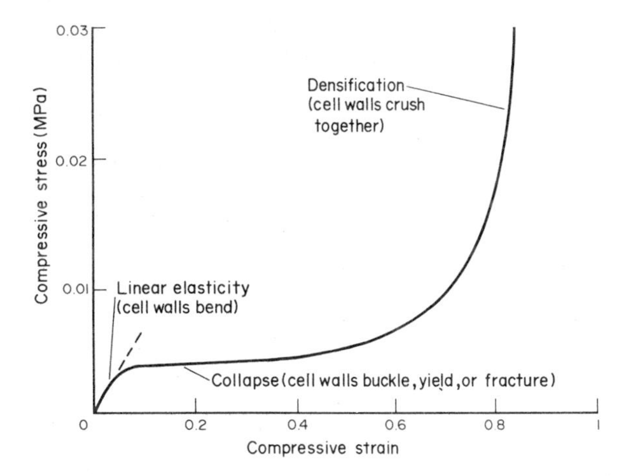

Figure 2
A typical stress–strain curve for a cellular material in uniaxial compression

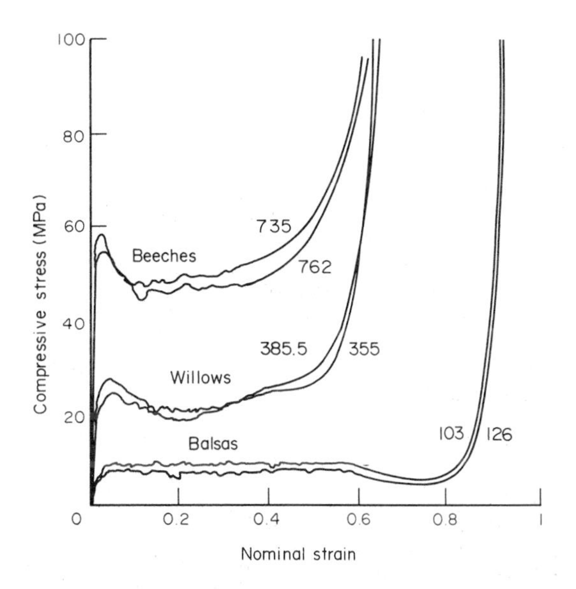

Figure 3
Stress–strain curves for wood loaded along the grain. Figures by curves indicate density (kg m^{-3}); axial compression = $1.25–1.5 \times 10^{-5} s^{-1}$

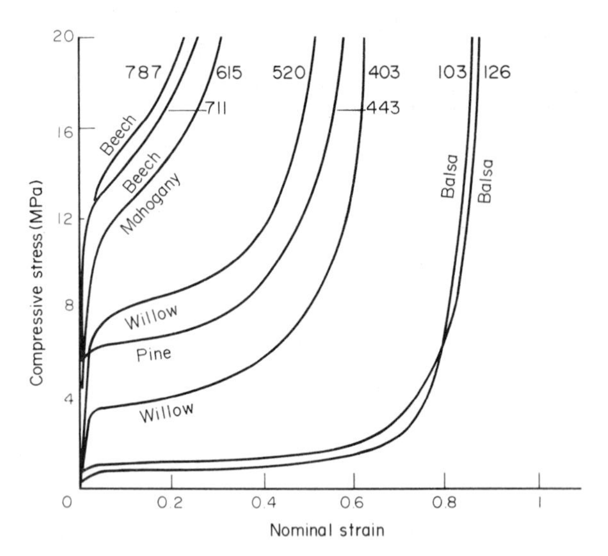

Figure 4
Stress–strain curves for wood loaded across the grain. Figures by curves indicate density ($kg\,m^{-3}$); tangential compression $= 1.6 \times 10^{-3}\,s^{-1}$

The mechanical behavior of wood can be modelled by analyzing the mechanisms by which it deforms (Easterling et al. 1982). When loaded across the grain, the cell walls bend, in the same way that a honeycomb does when loaded in the plane of the honeycomb. Because of this, the Young's moduli and compressive strength of woods loaded across the grain depend on the cube and the square of the relative density of the wood (see Table 1). When loaded along the grain, the cell walls compress uniaxially, as a honeycomb does if loaded normal to the plane of the hexagonal cells. Because of this the Young's modulus along the grain depends linearly on the relative density of the wood. Finally, the axial compressive strength of the wood is related to yielding in the cell walls, again giving a linear dependence on relative density. These results are plotted in Figs. 5 and 6, along with data for wood. The modelling begins to explain the dependence of the properties of wood on the loading direction and the relative density of the wood. The tensile fracture toughness of wood has been explained in a similar way; see, for example, Ashby et al. (1985).

The mechanical properties of wood tend to be variable, due to the presence of defects in the wood, such as nonparallel grain and knots. Recently, glue-

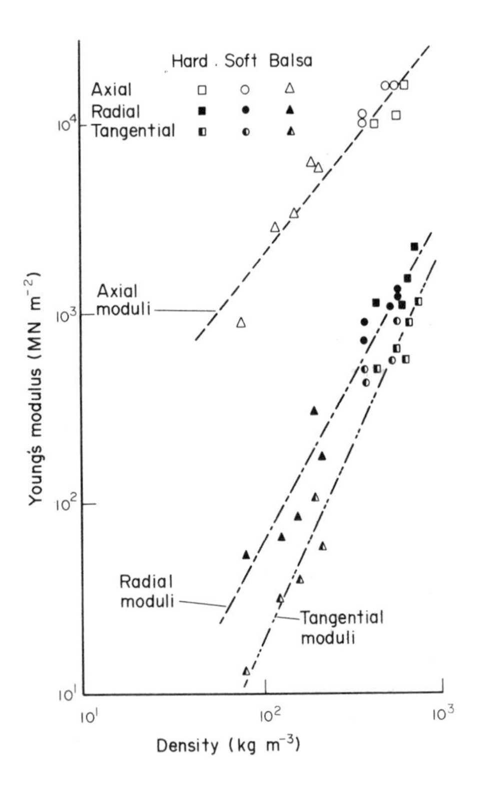

Figure 5
Young's modulus for woods, plotted against density. The lines indicate the theorectical dependencies

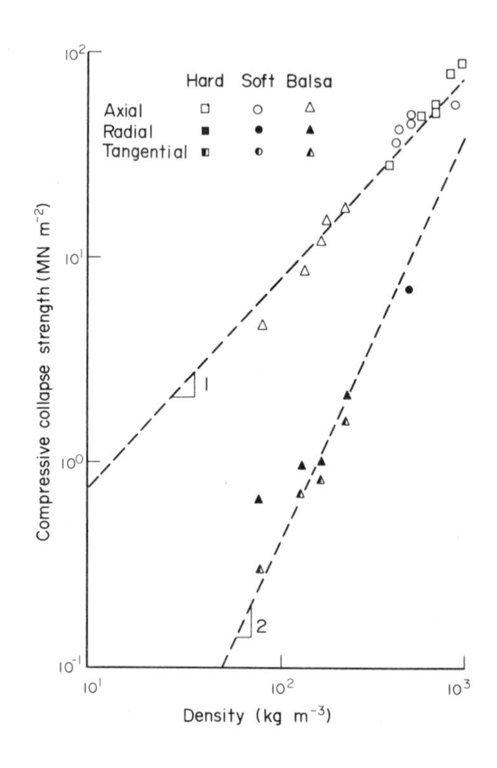

Figure 6
Compressive strength for woods, plotted against density. The lines indicate the theoretical dependencies

laminated wood members (glulam) have been developed to minimize the effect of such defects. Glue-laminated members are made by gluing together thin strips of wood, from which the defects have been cut out. The properties of such members are more uniform than those of dimensioned lumber. Glulam sections also have the advantage that they can be curved and built up to any size. The efficiency of wood members can also be improved by using them in sandwich constructions. Typically, plywood faces are used in conjunction with honeycomb, foam or balsa cores.

2.3 Thermal Insulation

The thermal conductivity of cellular solids is typically low, partly because of their low volume fraction of solids (about 10%) and partly because of their small cell size (usually between about 0.04 mm and 1 mm). This makes them ideal thermal insulators; one of the most prevalent applications of foams is in thermal insulation.

In Arctic construction, icing can be a problem. Foams are sometimes used as thermal insulation to reduce icing. For example, polyethylene foam sheets have recently been used to insulate railway tunnels outside Anchorage, Alaska, and in British Columbia, Canada. Near the portals of the tunnels ice build-up on the crown and sidewalls of the tunnels and on the track caused severe icing problems. Insulation shields, made of polyethylene foam faced with galvanized steel plates, installed at the portals of the tunnels, reduced icing to a tolerable level.

Foam sheet has also been used to insulate foundations from permafrost in the Arctic. In this case, the foam is used to prevent the permafrost melting from the heat of the building; melted permafrost can cause excessive sliding and settlement of the foundations.

Bibliography

Ackers P 1945 The efficiency of sandwich struts utilizing calcium alginate core, R & M 2015. Aeronautical Research Council, Farnborough, UK

Allen H G 1969 *Analysis and Design of Structural Sandwich Panels.* Pergamon, Oxford

Ashby M F, Easterling K E, Harrysson R 1985 The fracture toughness of woods. *Proc. R. Soc. London, Ser. A.* 398: 261–80

Bodig J, Jayne B A 1982 *Mechanics of Wood and Wood Composites.* Van Nostrand Reinhold, New York

Demsetz L A, Gibson L J 1987 Minimum weight design for stiffness in sandwich plates with rigid foam cores. *Mater. Sci. Eng.* 85: 33–42

Dinwoodie J M 1981 *Timber: Its Nature and Behaviour.* Van Nostrand Reinhold, New York

Easterling K E, Harrysson R, Gibson L J, Ashby M F 1982 On the mechanics of balsa and other woods. *Proc. R. Soc. London, Ser. A.* 383: 31–41

Gibson L J 1984 Optimization of stiffness in sandwich beams with rigid foam cores. *Mater. Sci. Eng.* 67: 125–35

Gibson L J, Ashby M F 1982 The mechanics of three-dimensional cellular materials. *Proc. R. Soc. London, Ser. A.* 382: 43–59

Huang S N, Alspaugh D W 1974 Minimum weight sandwich beam design. *AIAA J.* 12: 1617–18

Kuenzi E W 1965 Minimum weight structural sandwich. US Forest Service, Research Note FPL-086. Forest Products Laboratory, Madison, Wisconsin

Maiti S K, Gibson L J, Ashby M F 1984 Deformation and energy absorption diagrams for cellular solids. *Acta Metall.* 32: 1963–75

Ueng C E S, Liu T L 1979 Least weight of a sandwich panel. In: Craig R R (ed.) 1979 *Proc. ASCE Engineering Mechanics Division 3rd Specialty Conf.* University of Texas at Austin, pp. 41–44

Wittrick W H 1945 A theoretical analysis of the efficiency of sandwich construction under compressive end load, R & M 2016. Aeronautical Research Council, Farnborough, UK

L. J. Gibson
[Massachusetts Institute of Technology, Cambridge, Massachusetts, USA]

Cement as a Building Material

Cement is the binder used to make concrete, mortars, stuccos and grouts for all types of building and construction. Although cement is a generic term and can be applied to many inorganic and organic materials, by far the most widely used and most versatile cement is Portland cement. This article is primarily concerned with Portland cement, but a brief description of other inorganic cements is also given. Unlike gypsum plaster, Portland cement is a hydraulic cement; that is, it sets and hardens under water and hence can be used safely in all structures in contact with water.

The ancestry of Portland cement can be traced back to Greek and Roman times, but the year 1824 is generally considered to mark the origin of modern Portland cements when the name "Portland" was used in a patent by Joseph Aspdin to emphasize the similarity of concrete made with his cement to Portland limestone, which was a favored building material at the time. By the 1860s the Portland-cement industry was flourishing in the UK and Europe, but the first patent in the USA was not taken out until 1870 by David Saylor. Today about 70 Mt are used annually in the USA.

1. Manufacture

Portland cement is commonly made from limestone and clay, two of the most widespread and abundant materials. The principal constituents are thus lime

(CaO) and silica (SiO_2) which, when subjected to high temperatures (~ 1450 °C), form the hydraulic calcium silicates that are the active compounds in cement. Other calcareous materials (such as muds, marls and shell deposits) or argillaceous materials (such as slates, shales and coal ashes) are used in some cases.

The overall manufacturing process is shown schematically in Fig. 1. The various steps are: (a) quarrying and crushing the raw materials; (b) proportioning and fine grinding of the crushed materials; (c) burning in a rotary kiln at high temperature; and (d) grinding the resulting clinker to a fine powder with the addition of a small amount of gypsum. Although wet milling and proportioning with the aid of slurries (the wet process) is traditional and is found in older plants in the USA, the trend is towards dry grinding and proportioning (the dry process). Although it was made possible by improvements in grinding mills, in particle-size classification, and in dust collection, the adoption of the dry process has been motivated largely by energy efficiency. In the wet process, large amounts of energy are required to evaporate the water, and long kilns must be used to attain large throughputs of materials (Fig. 2). This increases the heat losses through the kiln. In addition, the dry process lends itself to the use of more-efficient heat exchangers, such as the suspension preheater, in which the incoming kiln feed draws heat from the exit gases in a series of cyclones. As a result, dry-process rotary kilns are much shorter than those used in the wet process (Fig. 2), heat losses are much reduced, and throughputs are greater. Further savings can be realized by precalcining the hot feed in a flash furnace (precalciner) before it enters the kiln.

Typical energy consumptions in cement manufacture are given in Table 1. Average overall energy

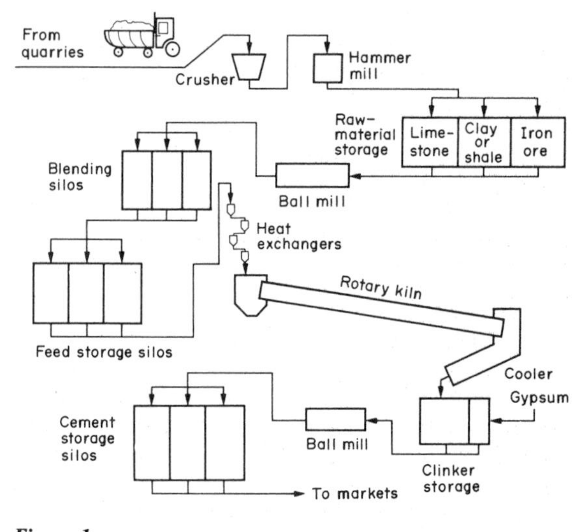

Figure 1
Schematic representation of a modern dry-process cement plant

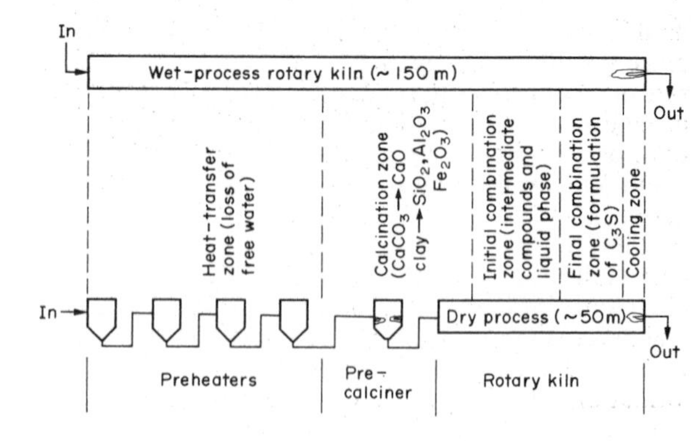

Figure 2
Comparison of kiln reactions in the wet and dry processes

Table 1
Average energy consumption in cement manufacture ($MJ\ kg^{-1}$)

Kiln operation	Wet process	Dry process: Long kiln	Dry process: Preheater
Burning process	8.0	6.7	5.6
Electrical energy	1.5	1.2	1.1

consumptions ($MJ\ kg^{-1}$) for countries with cement industries employing various processes are as follows: Japan, 4.5; FRG, 4.2; USA, 6.0. Theoretically, a value of 1.7 $MJ\ kg^{-1}$ is calculated from the basic chemical processes involved; in practice, modern plants now have energy consumptions approaching 3.5 $MJ\ kg^{-1}$. The average figure for the USA is relatively high because of a preponderance of older plants built when energy was inexpensive.

As calcination of the raw materials progresses, chemical combination takes place in the hotter zones of the rotary kiln (see Fig. 2). Formation of intermediate compounds by sintering occurs first, until at 1335 °C a quarternary liquid phase appears with a composition close to that predicted from the C–S–A–F phase diagram, where C = CaO, S = SiO_2, A = Al_2O_3 and F = Fe_2O_3 in conventional cement chemists' notation. (Other symbols employed include $\bar{S} = SO_3$, H = H_2O and K = K_2O.) The amount of liquid formed depends on the amount of A and F present in the raw feed and on the kiln temperature. Generally, about 25% of the total mix is liquid and hence the process is neither a complete fusion or a pure solid-state sintering—it is called clinkering. The liquid phase allows chemical combination to go to completion in a relatively short time. Much C_2S has formed by solid-state reactions before the liquid appears but its formation is completed in the liquid. The final reaction between C_2S and lime to form C_3S takes place wholly within the liquid at 1450–1500 °C. The rate of combination depends on the rate of dissolution of lime in the clinker liquid, which is dependent on both particle size and temperature.

On leaving the hottest zone, the clinker cools to about 1100–1200 °C. Then it is air-quenched as it exits the kiln. However, cooling is usually too rapid to allow true equilibrium to be maintained. The rate of cooling also controls the crystal growth of the calcium silicates and the crystallization of the liquid phase to form C_3A, C_4AF and additional C_2S; slow cooling results in the formation of larger and less-reactive crystals. The structure of a clinker can be studied by optical microscopy and used to predict clinker performance. After cooling, the clinker may be stored for future grinding or it may be ground immediately. Finish grinding takes place in ball mills equipped with particle-size classifiers that avoid over-grinding. At this time, small quantities of gypsum are added to control the early hydration reactions of C_3A and thus optimize cement performance.

2. *Composition*

Portland cement consists of four major clinker compounds, C_3S, C_2S, C_3A and C_4AF, together with the gypsum added during grinding. The relative portions of these compounds can be changed to optimize specific properties of the cement (discussed in more detail below). Table 2 shows typical compound compositions, which are calculated assuming (incorrectly) ideal stoichiometries and chemical equilibrium in the kiln (the Bogue calculation).

All the clinker compounds form solid solutions with small amounts of every element that is in a clinker. In addition to the four major oxides, MgO, K_2O and Na_2O are ubiquitous impurities in most raw materials. Impure C_3S is called alite; it reacts more rapidly with water than does pure C_3S. C_2S exists as four major polymorphs: γ, β, α' and α. At ambient temperatures, the nonhydraulic γ polymorph is thermodynamically stable but, in cement, impurities stabilize the hydraulic β polymorph. Impure C_3A contains considerable quantities ($\sim$ 10 mol%) of iron, which may make it less reactive than the pure compound. High levels of sodium substitution modify the crystal structure from orthorhombic to cubic with a corresponding decrease in reactivity.

C_4AF does not have an exact stoichiometry; continuous solid solution between A and F allows its composition to vary from about C_6A_2F to C_6AF_2. For this reason C_4AF is known as the ferrite phase. Although in many cements the composition of the ferrite phase is close to C_4AF, in those that have very low C_3A contents, the composition is closer to C_6AF_2. As the iron content increases, the reactivity of the ferrite phase decreases. It is the ferrite phase that gives cement its color: when MgO is present in the ferrite phase, it becomes black and the cement gray. If the MgO content is low, or if the ferrite phase contains appreciable ferrous iron (due to a reducing atmosphere in the kiln), the cement is browner. In addition, crystalline MgO (periclase) is present in most clinkers, typically in the range 2–4 wt%. There are limits placed on the amount of MgO, because its slow hydration can cause disruptive expansions in the hardened paste. In clinkers that are produced using high sulfur coal as a fuel, alkali sulfates, such as $K\bar{S}$ or $C_2K\bar{S}_3$, sometimes occur; these compounds may affect setting behavior and can lower the 28-day strength.

Cements can be classified according to the ASTM types given in Table 2. These classifications are based mostly on performance, not composition, although in general cements of the same type have broadly similar compositions. About 90% of all Portland cement made in the USA is type I. When higher strength is required at early ages (1–3 days) a type III can be used. In addition to having a higher C_3S content, it is also more finely ground so that its overall hydration is faster. Type IV cement can be used to avoid thermal cracking in mass concrete, by lowering the heat of hydration and hence the internal temperature of the mass. In the presence of high concentrations of sulfates in ground waters or soils, the hydration products of C_3A may undergo further reactions that result in disruptive expansion (see Sect. 3). The resistance of concrete to sulfate attack can be improved by lowering the C_3A content in a type V cement. Type II cement has properties intermediate between those of type I and types IV or V.

Aside from the five ASTM types of Portland cement, other special cements are available. Blended cements, containing reactive siliceous materials (pozzolans), or

Table 2
Typical compound composition of Portland cement

ASTM type	Description	C_3S	C_2S	C_3A	C_4AF	$C\bar{S}H_2$
I	General purpose	55	20	10	7	5
II	Moderate sulfate resistant	45	30	7	12	4
III	High early strength	60	15	10	7	5
IV	Low heat of hydration	30	45	5	12	3
V	Sulfate resistant	40	35	3	15	3

hydraulic slags mixed with type I, are used in most countries as alternatives to types IV and V. On hydration, they form more C–S–H and less calcium hydroxide. Expansive cements are modified Portland cements that counteract the effects of drying shrinkage by forming large amounts of ettringite to generate restrained expansion. Some special cements use ettringite to provide high early strength. Masonry cements and oil-well cements are Portland cements, modified to optimize the special requirements of these applications. Calcium aluminates (high-alumina cements) are used principally for refractory concretes. They cannot be used for structural purposes because of potential strength loss during service. Magnesium oxychloride (Sorel) cements can be used only for interior applications because they do not resist moisture.

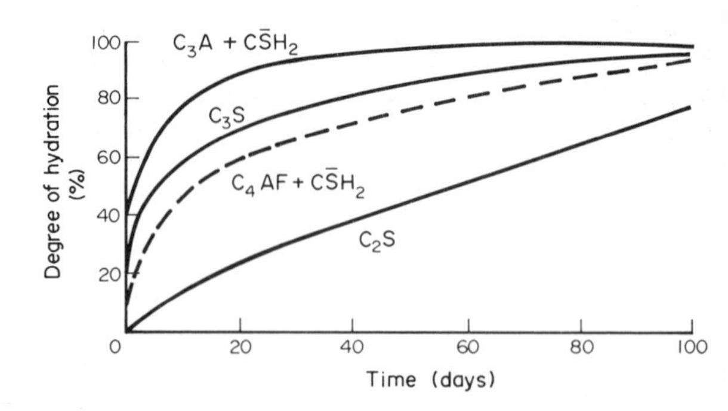

Figure 3
Rate of hydration of clinker compounds

3. Hydration

The setting and hardening of concrete is the result of complex chemical reactions between cement and water, that is, hydration of the paste fraction. The overall reactions are summarized in Table 3.

Both alite and belite form the same hydration products: an amorphous calcium silicate hydrate ($C_3S_2H_4$) and crystalline calcium hydroxide (CH). (The designated stoichiometry of the calcium silicate hydrate is only an approximation, since its composition is variable, so it is usually referred to as C–S–H. This implies no definite stoichiometry or regular crystal structure.) The rate of hydration (Fig. 3) and the heat of hydration of the two silicates are quite different. The silicates are the components responsible for the strength of cement pastes, but in the first seven days the alite content controls the development of strength. The hydration of alite has an initial induction period that keeps concrete fluid for 1–2 h and allows it to be handled and placed conveniently.

In the presence of gypsum, C_3A and the ferrite phase form calcium sulfoaluminate hydrates. The first product is the trisulfate form, ettringite, but when all gypsum has been consumed ettringite converts to the monosulfate form. C_3A or the ferrite phase contribute little to strength, but are still important. When $C_4A\bar{S}H_{12}$ is exposed to an external source of sulfate ions, it will convert back to ettringite with concomitant expansions.

Interactions occur between the clinker compounds when they are hydrating in cement. For example, belite hydrates faster in the presence of alite, C_3A and the ferrite phase compete for the gypsum during their hydration, and gypsum accelerates the hydration of alite. During its formation, C–S–H can incorporate considerable quantities of alumina, sulfate and Fe_2O_3 into solid solution. Thus less sulfoaluminates are formed during hydration than theory predicts—perhaps as little as half the expected amount.

It is important to know the effect of temperature on the hydration process because often concretes are cured at elevated temperatures. During the first 24 h, the hydration of cement is very sensitive to temperature; apparent activation energies of 40 kJ mol^{-1} have been reported for cement or alite. Thus, at low temperatures the rate of hydration, and hence of strength development, is very much reduced, while at high temperatures the reaction is accelerated. At about 90 °C the calcium sulfoaluminate hydrates become unstable, while above 100 °C, C–S–H will convert to crystalline compounds. In the temperature range of autoclaved concrete products (150–200 °C under

Table 3
Hydration of clinker compounds

Compound	Reaction	Heat of reaction (kJ mol^{-1})
Alite	$2C_3S + 7H \rightarrow C_3S_2H_4 + 3CH$	−114
Belite	$2C_2S + 5H \rightarrow C_3S_2H_4 + CH$	−43
C_3A	$C_3A + 3C\bar{S}H_2 + 26H \rightarrow C_6A\bar{S}_3H_{32}$	−367
	$2C_3A + C_6A\bar{S}_3H_{32} + 4H \rightarrow 3C_4A\bar{S}H_{12}$	
	$C_3A + C\bar{S}H_2 + 10H \rightarrow C_4A\bar{S}H_{12}$	
Ferrite	$3C_4AF + 4CSH_2 + 5OH \rightarrow 12[C_4(A,F)SH_{12}] + 2(A,F)H_3$	−403

pressurized steam), C–S–H rapidly transforms to α-C_2SH. This results in low strength because the transformation increases porosity, but when finely divided silica is added, C–S–H converts to 11 Å tobermorite, which has good cementing properties.

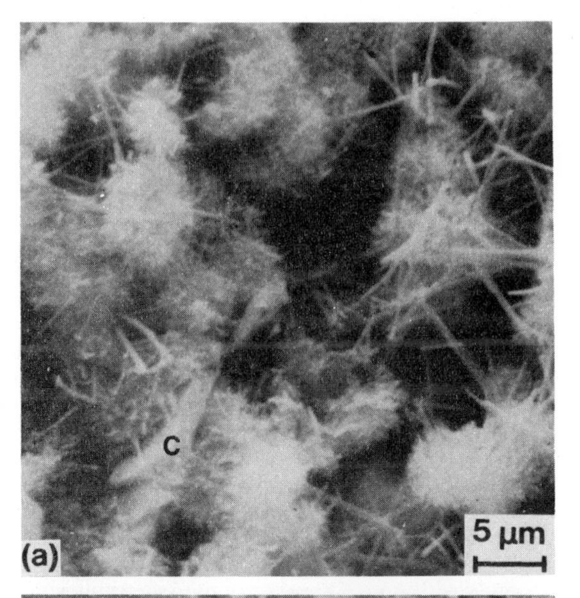

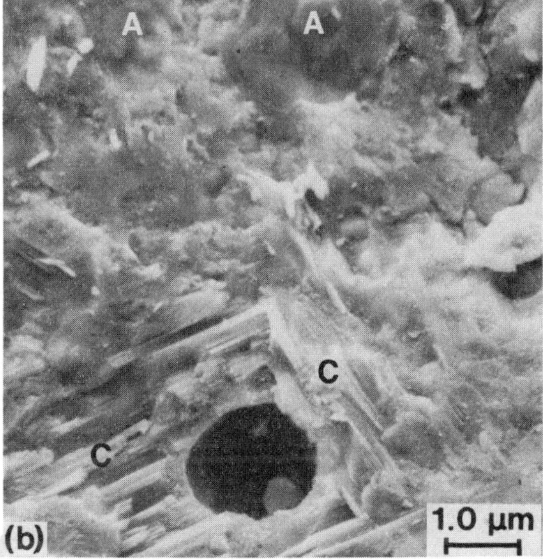

Figure 4
Microstructure of hardened cement paste. (a) After one-day curing, showing C–S–H and ettringite needles in the pore space. A small calcium hydroxide crystal C is seen to the lower left. (b) After 64 days curing, massive calcium hydroxide crystals C have filled most pore space, while unhydrated residues A can be seen embedded in C–S–H. The large hole is an entrapped air void

4. Structure of Hardened Cement Paste

The transition of a hydrating cement paste from a fluid to a rigid material is a continuous process resulting from progressive hydration of the calcium silicates. Ettringite may control rheological properties, but does not form a continuous matrix because C_3A and the ferrite phase are minor components. Ettringite crystallizes mostly at the surface of the aluminate grains but also grows in the pore space (Fig. 4a). Under special conditions crystallization of larger amounts of ettringite can cause premature stiffening because of physical interlocking of the needle-like crystals. The continuous matrix of hardened paste is formed by C–S–H that coats the surface of calcium-silicate grains (Fig. 4a), which eventually merge as hydration proceeds (Fig. 4b). Calcium hydroxide grows in the water-filled voids between the grains. The approximate composition of the paste is given in Table 4.

C–S–H is amorphous colloidal material with high surface area and small particle size; these features can be seen only with an electron microscope. It has a large volume of intrinsic microporosity (~25 vol. %), which lowers its intrinsic strength and causes volume instability on drying. To a large extent, the material properties of concrete reflect the properties of C–S–H. In contrast, massive crystals of calcium hydroxide can be readily seen, even by optical microscopy. Calcium hydroxide contributes to strength because it fills voids, but since it is highly soluble and very alkaline it can reduce the durability of concrete in certain circumstances. The calcium sulfoaluminate hydrates play a relatively minor role in hardened pastes. All properties of concrete are affected by the amount of capillary porosity, which is controlled by the water–cement ratio. A high water–cement ratio increases the porosity, thereby lowering strength, increasing permeability (hence decreasing durability), and increasing drying shrinkage and creep.

Table 4
Composition of hardened paste (water–cement weight ratio, $w/c = 0.5$)

Component	Volume percent	Remarks
C–S–H	55	Continuous matrix; high intrinsic porosity; amorphous
CH	20	Large crystals (0.01–1 mm)
Calcium sulfoaluminate hydrates	10	Small crystals (1–10 µm)
Capillary porosity	~15	Depends on the amount of water used

See also: Concrete as a Building Material; Portland Cements, Blended Cements and Mortars

Bibliography

Barnes D (ed.) 1984 *Structure and Performance of Cement.* Applied Science, Barking, UK

Bye G C 1983 *Portland Cement: Composition, Production and Properties.* Pergamon, Oxford

Copeland L E, Kantro D L 1969 *Proc. Int. Symposium Chemistry of Cement. II: International Process Engineering in the Cement Industry.* Cement Association of Japan, Tokyo, pp. 387–421

Duda W H 1977 *Cement Data Book*, 2nd edn. Bauverlag, Wiesbaden

Francis A J 1977 *The Cement Industry 1796–1914: A History.* David and Charles, Newton Abbot, UK

Ghosh S N (ed.) 1983 *Advances in Cement Technology.* Pergamon, Oxford

Lea F M 1971 *The Chemistry of Cement and Concrete*, 3rd edn. Chemical Publishing, New York

Mindess S, Young J F 1981 *Concrete.* Prentice-Hall, Engelwood Cliffs, New Jersey

Skalny J P, Young J F 1980 *Proc. Int. Congress Chemistry of Cement*, Vol. 1, Sect. 2, Paper 1. Editions Septima. Paris, pp. 3–45

Soroka I 1979 *Portland Cement Paste and Concrete.* Macmillan, London

Young J F 1983 *Hydraulic Cements for Concrete.* American Concrete Institute, Detroit, Michigan

J. F. Young
[University of Illinois at Urbana, Urbana, Illinois, USA]

Cements, Specialty

There are many commercially available inorganic cements. Certain types, such as ASTM type I Portland cements, are commodities manufactured in enormous quantities; others, such as colored Portland Cements and dental cements, have more limited markets and must be considered specialties. This article describes some special inorganic cements (Table 1) not covered in the articles *Portland Cements, Blended Cements and Mortars* and *Calcium Aluminate Cements.* They are inorganic powders which can be mixed with a liquid, usually water, to form a fluid or plastic mixture which will set and harden. These cements are usually, though not necessarily, mixtures of crystalline inorganic phases such as silicates and aluminates of calcium.

1. Modified Portland Cements

Portland cement clinkers manufactured for use in Portland cements complying with ASTM specifications may also be used in making specialty cements which either do not have to meet the ASTM specification or have to meet additional requirements. For example, pipe, block and expansive cements, and cements for cement–asbestos products, are used in the factory fabrication of concrete products in processes which require special equipment and curing conditions. In contrast, plastic, gun-plastic and masonry cements are used mostly on the construction site by home builders who require special mortars for the construction of walls.

1.1 Asbestos Product Cements

Cement–asbestos products are made on machinery similar to that used in papermaking. For pipe manufacture, a slurry of cement, ground quartz and asbestos fibers is dewatered by filtration through a moving felt and the solid is then transferred to a rotating mandrel upon which a thick layer can be built up. The product formed is cured with either low- or high-pressure steam. The special cements used help to optimize the manufacturing operations. They are made from ASTM type I or type II Portland cement clinkers, particular attention being given to the particle size distribution and the form and quantity of the calcium sulfate used as a set retarder.

1.2 Block Cements

Block cements are Portland-type cements for the manufacture of concrete blocks. Since the strength and other performance characteristics of the blocks can be evaluated easily, the cements do not necessarily have to comply with ASTM specifications. As a result, small quantities of materials not permitted under ASTM specifications may be added to block cements. The main ingredients are usually type I or type II Portland cement clinkers, and a small quantity ($<10\%$) of limestone, fly ash or blast furnace slag. Block cements tend to be more finely ground than ASTM type I and type II cements. They give concrete mixes which can easily be compacted and released from the mold in the block pressing operation; they also give rapid strength development under the steam curing conditions used.

1.3 Colored Cements

Colored cements are usually Portland cements. Cements with earth tones are sometimes produced from clinkers manufactured under slightly reducing conditions. The color depends particularly on the transition metals present and their oxidation states. The raw materials may be enriched with manganese for dark cements or with titanium in a patented process for buff cement. Colored cements are also obtained by intergrinding Portland cements with pigments. For bright colors or a clear black, white cement is used with pigments which must be stable in the alkaline environment of concrete. To obtain slight changes in color, gray cements are sometimes interground with additives such as pozzolans or carbon black. It is often difficult to obtain hardened concrete with a completely uniform appearance; also, carbonation of calcium hydroxide on the surface, or continuing slow reactions and the migration of soluble

Table 1
Some specialty inorganic cements

Cement type	Typical use
Modified Portland cements	
Asbestos product	cement–asbestos product manufacture
Block	concrete block manufacture
Colored	architectural concrete and terrazzo
Expansive	self-stressing reinforced concrete
Gun plastic (see also plastic)	gun-applied stucco
Hydrophobic	storage in damp conditions prior to use
Jet set (see regulated set)	
Masonry	masonry mortar
Nonstaining	mortar for limestone blocks
Oil well	cementing of oil well casings
Pipe	concrete pipe manufacture
Plastic	stucco
Rapid hardening	rapid-hardening concrete
Regulated set	rapid-hardening nonstructural concrete
Self-stressing (see expansive)	
Shrinkage-compensated	crack-free concrete slab
Waterproofed (see hydrophobic)	
White	white architectural concrete
Other calcium silicate cements	
Hydraulic fly ash	sub-bases
Lime-pozzolan	sub-bases
Natural	mass concrete, masonry mortar
Slag	mass concrete
Supersulfated (see slag)	mass concrete
Other inorganic cements	
Ammonium phosphate	patching of concrete highways
Barium or strontium silicate or aluminate	concrete for aggressive environments
Magnesium oxy-salt	flooring, terrazzo
Silicate dental	dental cement
Sodium hexametaphosphate	bonding of furnace linings
Zinc phosphate	dental cement

salts to the surface (efflorescence), may lead to substantial changes in appearance with time (see Sect. 1.7).

1.4 Expansive Cements

Expansive cements are similar to shrinkage-compensated cements (Sect. 1.13) but give higher levels of expansion. Self-stressing cements are expansive cements for use in reinforced concretes in which the reinforcing steel is to be stressed by the expansion of the concrete. Though still under development, self-stressing cements based on mixtures of monocalcium aluminate and calcium sulfate hydrates have been used in the USSR. The applicable ASTM standard for expansive cements is ASTM C806.

1.5 Hydrophobic Cements

Normal Portland cements stored in bags may deteriorate and become lumpy if the bags are exposed to high relative humidity. In an attempt to protect against this deterioration, cements are sometimes ground with small quantities of a water-repellent material such as stearic acid. The products are called hydrophobic cements.

1.6 Masonry Cements

Masonry cements are used in mortars for the laying of clay bricks and concrete blocks. The mortars should not normally be stronger than the units to be bonded. When mixed with water and sand, masonry cements give mortars with "buttery" characteristics which aid the mason; they also retain water well when in contact with the porous masonry units. The most commonly used standards for masonry cements are those defined in ASTM standard C91. The cements are made from type I or type II Portland cement clinkers and they usually contain other ingredients which help achieve the desired properties at the lowest cost. The additions may include hydrated lime, fly ash, clays, ground limestone and small quantities of air-entraining agents and organic flow modifiers such as soluble cellulose derivatives.

1.7 Nonstaining Cements

Nonstaining cements are Portland cements formulated for use in mortars which may be in contact with limestone blocks. Because discoloration of limestone may result from reaction of soluble alkali salts with the pyrite which is sometimes present, non-staining cements have low alkali contents.

1.8 Oil Well Cements

Oil well cements are used for cementing oil well casings to surrounding rock. The cements, in the form of fluid slurries, are pumped down the well through the steel casing and up the annular space between the casing and the rock. They must have predictable behavior and their slurries should flow easily without premature setting under the high pressures and temperatures in the well. Cements similar to ASTM type II or type V Portland cements are used for cementing shallow wells, but even more slowly reacting cements are required for very deep wells. These may be based on specially prepared, high-C_2S Portland cement clinkers or on mixtures of Portland cements with pozzolans or quartz. Standards for oil well cements are published by the American Petroleum Institute. They emphasize uniformity of the cements and their behavior with retarding and flow-modifying additives.

1.9 Pipe Cements

Pipe cements are for concrete pipe manufacture. They are made from ASTM type I or type II Portland cement clinker, but they are less finely ground than the ASTM cements and the quantity of gypsum added is usually smaller. Pipe cements are formulated to give a relatively dense concrete product in the centrifugal pipe spinning process. They give good strength development under steam curing.

1.10 Plastic and Gun-Plastic Cements

These cements are manufactured chiefly for use in surface renderings on wall of wooden frame houses and other buildings. They are used most extensively in California. Their mortars adhere well to vertical surfaces and they can be spread easily; further, they are formulated so as to be able to be applied in warm, dry weather without cracking excessively. Gun-plastic cements are formulated for use with a mortar gun. To aid pumping, they usually contain small quantities of fibrous or platy minerals, such as short fiber asbestos, or certain clays or mica. Concern about health hazards has fostered searches for substitutes for asbestos, such as clay and mica. The cements frequently contain small quantities of organic materials to modify the flow properties and improve water retention.

1.11 Rapid-Hardening Cements

Whereas high-early-strength Portland cements, ASTM type III, are intended for use when rapid strength development is required, even more rapid hardening and strength development are sometimes desired. Cements with very high early strength and ultrarapid hardening can be manufactured, using one or more of (a) a special Portland cement clinker containing large quantities of C_3S and C_3A, (b) fine grinding and (c) water-reducing or accelerating admixtures.

1.12 Regulated-Set Cements

To meet perceived needs for short setting times (< 60 min) and high early strengths, regulated-set cements were developed by the Portland Cement Association (PCA). Potential uses include substitution for gypsum plaster as, for example, in roof decks. The PCA patent (US Patent No. 3,628,973) covers compositions based on a special Portland-type clinker with a halogen-containing phase, either $C_{11}A_7.CaF_2$ or $C_{11}A_7.CaCl_2$. The clinker is ground with a carefully controlled mixture of calcium sulfate hydrates. The setting time depends upon the quantities of calcium sulfate hydrates added and the quantity of the halogen-containing phase. Regulated-set cement is not now being produced in the USA, though it is manufactured in Japan under the name "jet cement" and in West Germany. Uses include castable fillers for the walls and doors of fireproof safes and the patching of damaged concrete.

1.13 Shrinkage-Compensated Cements

The drying of hardened mortars and concretes is always accompanied by at least a small amount of shrinkage which may cause cracking. To minimize the cracking, shrinkage-compensated cements were developed. They are formulated to expand a little during hardening, the expansion being approximately equal and opposite to the drying shrinkage. Shrinkage compensation can lead to a reduction in the number of cracks which develop on drying if the concrete is at least lightly restrained internally by some reinforcing steel. This can be important in large slabs.

Shrinkage-compensated cements include high-C_3A Portland cements to which carefully controlled mixtures of calcium sulfate hydrates have been added, and Portland cements to which small quantities of expansive components have been added. The expansive components are usually one of four types:

(a) a special clinker containing a substantial quantity of the mineral $C_4A_3\bar{S}$, together with most of the usual Portland cement clinker minerals;

(b) a mixture of monocalcium aluminate and calcium sulfate or its hydrates;

(c) periclase (magnesium oxide); and

(d) hard-burned lime (calcium oxide).

Some of the expensive ingredients and their uses are covered by patents. The ASTM standard specification for shrinkage-compensated cements is ASTM C806.

1.14 White Cements

The usual gray color of Portland cements is attributable to the compounds of iron and other transition metals which they contain. Using raw materials which consist almost exclusively of CaO, SiO_2 and Al_2O_3, with MgO, Na_2O and K_2O as the major impurities, a very pale green or brown Portland cement clinker can be obtained. It can be ground with gypsum to give a white Portland cement. The manufacture of white cement clinker requires special equipment to avoid contamination by coloring elements, to provide for the high clinkering temperatures required and for quenching the clinker rapidly in a reducing atmosphere to obtain the whitest product. White Portland cement is used for architectural purposes such as precast decorative panels and terrazzo.

2. Other Calcium Silicate Cements

2.1 Hydraulic Fly Ash

Whereas low-lime fly ash has no significant cementing ability unless mixed with a source of lime, some high-lime fly ash types are cements in their own right. The increasing use of coals from the western USA which produce high-lime fly ash offers the potential for use of this ash for cementing applications.

2.2 Lime–Pozzolan Cements

Lime–pozzolan mixtures containing 5–10% of lime are well known as cements for use in soil stabilization. The reactions in lime–pozzolan–water mixtures form calcium silicate hydrates similar to those formed by Portland cements.

2.3 Natural Cements

Before Portland cements were discovered, natural cements were made from the product formed by heating naturally occurring cement rocks to high temperatures. Cement rocks have compositions similar to the feed used in the manufacture of Portland cement clinker but, because natural cements are made at lower temperatures than Portland clinkers, the mineralogical compositions of the products are significantly different. In particular, natural cements seldom contain C_3S, though they may contain much free lime and β-C_2S. Natural cements have a long history of use as masonry cements and they may be blended with Portland cements to make acceptable structural cements for some purposes. The manufacture of natural cements in the USA has now been all but discontinued. When the properties of natural cements are needed, as in the restoration of historic buildings, formulations based on Portland cements can give comparable properties.

2.4 Slag Cements

These cements consist mostly of ground granulated blast furnace slag with a small amount of activator—an additive which accelerates setting and strength development. In slag cements, the activator is usually Portland cement or a hydroxide of calcium, sodium or potassium, which give the aqueous phase a high pH. The activator may also be a sulfate such as gypsum or anhydrite. Supersulfated cements are slag cements which contain a higher quantity of anhydrite than is necessary for activation; significant amounts of ettringite ($C_3A.3C\bar{S}.32H$) occur in the hardened product. Slag cements and supersulfated cements tend to develop strength more slowly than Portland cements, but the concretes produced from them are usually highly sulfate-resistant.

3. Other Inorganic Cements

3.1 Ammonium Phosphate Cements

Many rapid-hardening mixtures of acidic phosphate solutions with metal oxide powders or aggregates are known, and many have been patented. As an example, mortars of magnesia with ammonium dihydrogen phosphate solutions are used for patching of concrete highways.

3.2 Barium or Strontium Silicate or Aluminate Cements

Cements consisting mostly of dibarium or distrontium silicate are sometimes made in small quantities for use where particularly high sulfate resistance is needed. Whereas the calcium hydroxide formed from the hydration of Portland cements can be leached out in sulfate soils and waters, barium and strontium cements form highly insoluble barium and strontium sulfates under these conditions. Cements consisting mostly of barium or strontium aluminates have also been used in small quantities in applications requiring high chemical resistance.

3.3 Magnesium Oxy-Salt Cements

The most common noncalcareous inorganic cements are magnesium oxy-salt cements. The best known is Sorel cement. It is formed by mixing lightly burned magnesium oxide with a concentrated (20%) solution of magnesium chloride. The cement hardens by the precipitation of hydrated magnesium oxychlorides to form a hard, white product. Since the product is not very resistant to water, it is used only in applications where exposure to water is not expected, as in terrazzo and other interior floors of buildings. Magnesium oxysulfate cements are also used commercially.

3.4 Sodium Hexametaphosphate Cements

Solutions of sodium hexametaphosphate are mixed with dead-burned magnesia powder to form a cementing material for use in the bonding of magnesite brick furnace linings. The materials are cured for ~24 h at 120 °C.

See also: Calcium Aluminate Cements; Cement as a Building Material; New Cement-Based Materials; Portland Cements, Blended Cements and Mortars

Bibliography

American Petroleum Institute 1984 *Specification for Materials and Testing for Well Cements*, 2nd edn. American Petroleum Institute, Dallas, Texas

American Society for Testing and Materials 1984 *Annual Book of Standards*, Sect. 4, Vol. 04.01, *Cement; Lime; Gypsum*. American Society for Testing and Materials, Philadelphia, Pennsylvania

Lea F M 1970 *The Chemistry of Cement and Concrete*, 3rd edn. Chemical Publishing Company, New York

Ramachandran V S, Feldman R F, Beaudoin J J 1981 *Concrete Science: Treatise on Current Research*. Heyden and Son, Philadelphia, Pennsylvania

G. J. Frohnsdorff
[National Bureau of Standards, Washington, DC, USA]

Ceramics as Construction Materials

Ceramics—materials whose essential ingredients have inorganic or nonmetallic bonds—are among the oldest and most enduring of construction materials. Brick and tile have a history of at least 100 000 years' use, and there is evidence that sheet glass was cast for windows as long ago as the time of Christ. Very early applications of cements and concrete-like materials took place in the Middle East thousands of years ago, and the ancient Egyptians set stone with mortars derived from both gypsum and lime. The Greeks and Romans added volcanic ash or ground burnt clay tiles (pozzolanic minerals) to render lime cements hydraulic (i.e., mixed with water to cure, and resistant to water when hardened); and, by further adding sand, broken tiles and brick, and crushed stone, the Romans produced a concrete, the technology and quality of which were not surpassed until the eighteenth century.

Today ceramics encompass a wide selection of materials with many different uses in construction. These materials can be grouped within three general categories: structural clay products (e.g., brick, tile, porcelain fixtures, sewer and drain pipe), glass, and cement-based products (e.g., mortar, plaster, wallboard, concrete). While the raw ingredients and basic products of construction ceramics have remained largely the same for centuries, many advances and improvements have been introduced to gain better performance, reduce costs, and extend the range of applications of these products. Thus, for example, in addition to its predominant use in windows, glass is also applied in construction as glass blocks, curtain walls and, as fibers, in insulation and for reinforcing plastics. Similarly, with advances in its chemistry and methods of production, delivery and placement, concrete has become the most widely used construction material (by weight, constituting about 75% of all construction materials used annually), with many different applications in buildings, roads, tunnels, foundations, linings and subaqueous structures. Most ceramic products used in construction are composed of materials containing silica (SiO_2) and thus are included in the segment of the ceramics industry known as the silicate industries.

1. Entering the Modern Era

Many of the advances in both structural clay products and glass products have been in their production processes. Structural clay products have seen improvements in the form of more efficient use of plant and ease of installation. From the early centuries of the last millennium when window glass was made by blowing a long cylinder of glass, cutting it open and unrolling it, the glass industry has developed increasingly better methods of producing flat glass, leading to the current float glass method wherein a sheet of glass is produced by floating molten glass on a bed of molten metal (see *Float Glass Process*). Other improvements have affected the materials themselves or produced wholly new products. For instance, through more careful selection of constituents and control in firing, bricks and tiles have been made stronger and, when necessary, lighter.

Modern advances in cementitious materials occurred in the eighteenth and nineteenth centuries, beginning in England and spreading to other European countries and the USA. It was during this period that interest in hydraulic cement and concrete was revived, accompanied by a growing understanding of both the raw materials and the manufacturing processes needed to produce higher quality cements. In a series of experimental developments begun by Smeaton in 1756 (with the use of a mortar derived from clayey limestone and a pozzolanic material), improvements in cements were gained by investigators such as Parker, Frost, Aspdin and Johnson in the first half of the nineteenth century. A patent for "Portland cement" manufactured by heating finely ground limestone and clay was issued to Aspdin in 1824, but the first cement factory was established by Frost (to produce his "British cement") in 1825. Subsequent discoveries by Johnson improved the technology of cement production and stressed the importance of higher, controlled temperatures in heating the raw materials to produce cement clinker. By the latter half of the nineteenth century, cement industries were established in England, Belgium, Germany, and France; in the 1890s the US industry grew very rapidly. The twentieth century has seen further improvements in the scientific understanding of the chemistry of cement and of concrete batching and placement, in the development of new cements and additives to alter the characteristics of cements, and in the development of new products for construction. Today concrete and other cementitious products are used virtually worldwide in construction.

2. *Structural Clay Products*

2.1 Industry Characteristics

Over the past 30 years, the importance of these products, as measured by their intensity of use (the ratio of the quantity consumed to the GNP), has shown contrasting trends. On the one hand, the intensity of use of brick has steadily declined in the face of competition from other exterior building materials while the intensity of use of tile has greatly increased due to a surge in popularity. In 1982, 4.8 billion clay bricks and about $2.75 \times 10^7 \text{m}^2$ ($3.05 \times 10^8 \text{ft}^2$) of floor and wall tile were shipped from US plants. Altogether the value of structural clay products (brick, tile and pipe) shipped in that year from the USA amounted to about $1.2 billion.

The raw materials for clay products are abundant and available. Often production plants are situated close to the raw material source (or the market) to avoid transporting clay products long distances, since transportation costs of the relatively inexpensive but heavy products would constitute a major share of the final product cost. Energy costs are also important to this industry since the processing of clay products requires preparation of the clay powder using mechanical means as well as firing of the end product at fairly high temperatures.

Common brick is probably the most important product of the brick and structural clay tile industry. Glazed floor and wall tile and clay sewer pipe are also major products of the industries providing clay products to construction. Clay products have encountered much competition from other materials. For instance, brick must compete with siding materials like wood and aluminum which are easier to install; wall and floor tiles, popular for bathrooms, compete with plastic tiles and fiberglass reinforced showerstalls and tub-units; and clay sewer pipe must compete with concrete, cast iron and plastics. A recent countertrend is the increased popularity of ceramic floor and wall tiles. For this product, as well as other clay products, competition exists internationally. The USA is a net importer of these products which include floor and wall tile and unglazed brick.

2.2 Materials Properties

Structural clay products are composed of a combination of three types of minerals. The first is a hydrated clay which is composed of layers of silicon bonded to oxygen (with a general chemical formula of $(Si_2O_5)_n$) and aluminum bonded to oxygen and hydroxide ions ($AlO(OH)_2$). Typically, this alumina layer will be bonded to one or two silica layers to form one layer. Between these thicker layers the bonding is very weak (van der Waals bonds) which gives these clays a platelike character and highly directional properties. The second constituent is a fluxing material, like feldspar ($K_2(Na_2)O \cdot A_{12}O_3 \cdot 6SiO_2$). The third component is an inert filler like quartz.

Within this general composition there are certain variations and additives which make a particular material suitable for particular products. For instance, clays for building and paving bricks are usually rich in fluxes, or materials that lower the firing temperature of the clay. Ceramics used for wall and roof tiles are usually made from low-melting clays whereas facing ceramics can be made from low- or high-melting clays. Certain minerals, such as chamotte or sand can lower the plasticity of the clay and decrease shrinkage on firing. Combustible materials (like wood shavings, ashes or fuel slag) can be added as extenders which burn out, producing a lighter, more porous product.

Sometimes structural clay products, like floor and wall tiles, are glazed. A glaze is a smooth, glossy coating that may be applied by painting, spraying or dipping. Glazing materials can be varied to achieve a certain color, gloss or texture.

Once the clay mixture has been formed and fired, the product consists of a crystalline phase embedded in a glassy (noncrystalline) matrix. The crystals are strongly bonded by ionic–covalent bonds, the strength of which gives rise to many of the properties attributed to traditional ceramic products.

The strong bonding of structural clay products makes them fairly strong (up to 138 MPa (20 ksi) compressive strength for facing bricks) but very brittle since planes of atoms do not easily slide past each other under a load to produce deformation. The water absorption of bricks is preferred to be 5–8% though it can be as high as 25–30% for bricks or tiles. Traditional ceramic products are electrical and thermal insulators, again due to the strong bonding at the microstructural level. For this reason also, ceramics have good abrasion, chemical, and heat resistance.

2.3 Processing and Manufacturing

The processing of structural clay products follows that of traditional ceramic products, which can be divided into four steps: powder preparation, forming, heat treatment and finishing. Powder preparation includes the raw material selection and the powder production. Once the raw materials (clays, fluxes, etc.) have been selected, usually from whatever local sources are available, they are ground and sieved to form a powder.

Forming of the product can take place in a variety of ways. Sewer and drain pipe can be extruded through a shaping die. Bricks can be formed in sand-coated wooden molds or extruded into long strips which are subsequently cut to size. Floor and wall tile are formed by dry pressing them into molds. Sanitary ware is formed by slip casting. In this process a slurry of clay particles in water is poured into a porous plaster of paris mold. The mold absorbs water from the slurry and a layer of compacted clay particles is deposited inside the mold. For hollow pieces this is allowed to proceed until the desired part thickness is achieved and then the remaining slurry is drained out. For solid pieces the slurry remains in the mold until all of the water is absorbed.

Heat treatment occurs when the formed piece is baked or sintered in a kiln to turn the loosely bonded powder into a dense body. Finishing procedures usually entail grinding to the desired shape.

2.4 Typical Products

Structural clay products fall into several categories. Building brick comes in various colors and shapes. Some are solid, others are hollow or perforated to decrease weight or enhance insulation properties. Brick-laying is a time- and labor-intensive process so prefabricated panels of brick are now manufactured which can be a full story high. Sand-lime brick is often used in Europe, though less so in the USA. Other brick products include paving brick, glazed brick and firebrick. Terracotta, a product made from clay and grog (prefired clay, to reduce shrinkage), is an exterior finish that can be produced in different shapes; a glaze is also applied for a smooth finish.

Several types of tiles are available both for indoor and outdoor exposure. These also appear in a wide variety of colors and shapes. Quarry tile is a red or brown floor tile usually about 20 cm (8 in.) square. A rectangular hollow tile is used for flue linings. Roofing tile, common in the Western USA, is attractive and durable but rather expensive. Drainage tile is used to convey water away from, for example, building foundations.

Other types of structural clay products are pipes and porcelain fixtures. Sewer pipes must be produced from a good grade of clay and glazed on their interior to avoid absorption. Sanitary ware, made from a combination of ball-clay, china clay and other clays, is often included under construction materials, since it is an integral part of building and housing construction. Sanitary ware includes all of the ceramic bathroom fixtures: sink, toilet, and tub.

3. Glass

3.1 Industry Characteristics

The flat glass industry in the USA is fairly concentrated with five firms producing almost all of the flat glass. In 1985 the flat glass industry in the USA shipped about 3.15×10^8 m^2 (3.5×10^9 ft^2), amounting to more than one billion dollars' worth of glass. One third of the shipments of this product goes to the construction market with most of the remainder used by the automotive market.

In the early 1980s the USA maintained a positive trade balance in flat glass although imports rose sharply in 1982 due to a growing domestic demand. Some competition is coming from the use of plastic, especially in residential skylights, two thirds of which were plastic in 1984. On the other hand, nonresidential skylights maintain a large and growing share of the demand for glass skylights. In addition, glass has been very popular as a design medium both for exterior architecture and interior design, entailing applications like glass curtain walls and glass blocks.

3.2 Materials Properties

Glass building products are composed of sand (SiO_2), soda (Na_2O) and lime (CaO). Various other oxides like potash and boric oxide are often used. In addition, coloring and opacifying agents can be added to give a wide variety of looks to common glass. When molten glass cools to its solid state the molecules do not fall into a crystalline arrangement, as is the case in other types of ceramics, but rather maintain their unordered or amorphous arrangement.

Various properties of glass make it very suitable as a window material. Its transparency allows natural light into a building. It is a thermal insulator so it can be used in fairly thin sheets (or as part of an insulating system of several sheets). Its insulating properties and ability to be processed into fibers has also led to its incorporation into insulating systems for buildings and pipes. Glass does allow some solar radiation to enter a building so windows have been developed which can reflect this heat, preventing it from raising the temperature inside the building in hotter months. Glass is fire resistant although, unless specially laminated, it can transmit radiation from fires, thereby possibly spreading the fire. Window glass is very durable and easy to maintain.

3.3 Processing

The main process used for making flat glass, from which windows are made, is the float process. In this process molten glass is poured onto a pool of molten tin. At one end of the tin bed the glass layer is pulled off by rollers producing a continuous sheet of glass of uniform thickness and free from distortions. Older, less common methods of producing flat glass are to pull the glass vertically from a furnace (sheet glass) or to draw a thicker layer between rollers and then grind and polish to form plate glass. Glass can also be drawn into tubing or fibers.

3.4 Products

Flat glass comes in many varieties used for windows, skylights and curtain walls. Clear transparent glass is the standard, though various colors and degrees of opacity can be achieved. Laminated glass is composed of two layers of glass bonded with an elastomeric layer to make penetration into the glass difficult. Insulating glass comprises two or more sheets of glass with sealed spaces between them. Tempered glass is strengthened by heating and rapid cooling and sometimes by chemical means. When broken, tempered glass breaks into small, relatively harmless pieces. Wired glass is also available. It consists of a wire mesh embedded in the sheet of glass. Low emissivity (or "Low-E") glass is coated to reflect radiant heat to reduce heat loss in the winter and heat gain in the summer.

Other glass products used in construction include fibers and process-related equipment. Fibers are used for insulation and to reinforce plastics. Process equipment includes glass pipe and glass-lined steel pipe, pumps and pipe fittings.

4. Cement and Concrete

4.1 Industry Characteristics

The primary cementitious material with respect to ceramics in construction is Portland cement, which accounts for over 90% of the total production of hydraulic cement in the USA. From the start of US cement production in the late 1800s, production capacity of Portland cement clinker grew to a peak of over 90 Mt in 1975, but has stabilized since then. This capacity is provided by 58 firms operating 164 Portland cement plants (as of 1980), and although the long-term trend is toward fewer plants in the USA, the number has not changed much since 1974.

The production of Portland cement is energy-intensive: costs of fuel for heating raw materials to form clinker, and for electricity to grind and pulverize raw materials and clinker, account for 30–50% of the costs of cement production. The Europeans and Japanese have led the industry worldwide in applying new, more efficient production technology. Challenges facing the industry in the USA, apart from the need to adopt this new technology, include the difficulty in raising the capital needed for costly plant modernization, antipollution and other government regulations, energy availability, and the matching of production capacity to shifting and fluctuating construction demand.

The primary use of Portland cement is in concrete, and several other industries contribute to the production of this construction material; e.g., crushed rock and sand and gravel firms which supply the aggregates, and chemical companies and other sources which provide admixtures, coatings, sealants and substitutes for natural aggregate. The major consumer of Portland cement in the USA is the ready-mix concrete industry, comprising over 6000 firms which alone consume more than half of the cement shipped. The concrete masonry industry (bricks, blocks, etc.) accounts for between 5 and 6% of consumption; and other concrete products, almost 8%. The remainder is used by concrete site plants, with about 1.5% being used by other industries. Among segments of construction output, housing accounts for 30–33% of cement consumption, with the remainder distributed as follows: industrial/commercial, 21–23%; public buildings, 7%; public works, 17–18%; transportation, 17–19%; and miscellaneous uses, 3–4%. These data encompass new construction, maintenance, and repair of facilities, and include both batch concrete and concrete products. Besides concrete, Portland cement is also used in construction in mortar and grout.

The gypsum industry is another major dealer in cementitious products related to construction. One major product is drywall or gypsum board, used primarily in housing construction, although nonresidential building markets are also being developed. The growing volume of building repairs, alterations and additions offers another attractive opportunity for this material. Although the primary constituent of drywall is gypsum, the major costs are associated with the paper covering and the energy of production. Other major products of this industry are plasters, including plaster of paris, wall plaster and hard-finish plaster.

4.2 Properties and Processing of Cement

Portland cement is primarily a hydrated calcium silicate. The raw material for calcium is typically limestone, chalk, marl or seashells. The silica is provided by shale, clay, slate, slag, sand or iron ore. Other ingredients, such as alumina and iron, are also included in cement from readily available sources. The constituent rocks and minerals are carefully ground, proportioned, and mixed (using wet or dry processes). The ground material is burned in a kiln at about 1500 °C, causing the formation of clinker. The clinker is then pulverized (with any additional chemicals, like gypsum) to form Portland cement.

The ultimate properties of concrete are influenced (but not determined) by the chemistry of the cement. Unlike sintered ceramic products (e.g., from clay), the hardened paste in concrete results from a chemical reaction of hydration. Portland cement encompasses four primary compounds: dicalcium silicate ($2CaO.SiO_2$, abbreviated C_2S), tricalcium silicate (C_3S), tricalcium aluminate ($3CaO.Al_2O_3$, or C_3A), and $4CaO.Al_2O_3.Fe_2O_3$, or C_4AF. The short-term and long-term characteristics of the resulting product depend upon the relative amounts of these four compounds and how they react during hydration. For example, a greater proportion of C_3S will produce a higher early-strength cement (since C_3S hydrates relatively quickly), but more C_2S results in higher ultimate strength (through a greater production of the hydration product calcium silicate hydrate, or C–S–H). C_3A hydrates rapidly, and is therefore useful in cements that must set up quickly; however, the speed of this reaction is so rapid that it must be retarded by the addition of gypsum during cement manufacture. The hydration reaction is complex, produces heat (which is desirable in some applications, undesirable in others), is influenced by time, temperature, amount of water, and other chemicals present, and yields a time-varying series of compounds whose structures and mechanisms in influencing the properties of the hydrated cement paste are still only incompletely understood.

Several types of cement are available, depending upon what characteristics are desired in concrete during placement, over the short term, and ultimately. Briefly, these are as follows: Type I—general use; Type

II—general use where moderate sulfate resistance and moderate heat of hydration are needed (C_3A and C_3S are limited in such cements, because of their more rapid hydration); Type III—high early strength cement; Type IV—low heat of hydration cement; and Type V—high sulfate resistance cement (C_3A is limited, since its hydration product is unstable in sulfate environments).

There are variations of these cement types to include air entrainment for frost resistance. Not all these cements are manufactured or used widely, however, since it is possible to achieve some of these same objectives more economically and with specific control by using admixtures. In the USA, 90% of the cements produced are of Types I and II; the remainder are Type III and special cements such as the following: expansive cements (used to counteract shrinkage cracking or to induce compressive stresses (chemically prestressing) in the concrete); rapid setting and hardening cements; oil-well cements (used to cement the well casing to the surrounding rock, requiring fluidity during pumping and placement, but rapid hardening thereafter); and calcium aluminate cements or high alumina cements (in which strength is derived from aluminates rather than silicates). In addition, Portland cements may be blended with blast furnace slag or pozzolans for lower heat of hydration and greater economy and durability.

4.3 Concrete

Concrete is the product that results from the combination of cement, coarse and fine aggregate, and admixtures. Because of its economy, durability, versatility, and use of readily available (or synthetic) constituent materials, concrete has found wide application in virtually all types of construction. Most of the concrete used in construction uses hydrated Portland cement as its bonding agent, although polymer concrete (having a polymerized monomer as its bonding matrix) is also used. While the type of Portland cement influences the characteristics of the concrete at all stages of its life, other factors are important as well; e.g., any admixtures that are included, the size, shape, and porosity of the aggregate, and conditions prevalent during placement of the material.

Important improvements to the workability, control, strength, and durability of concrete may be achieved by using admixtures. Admixtures are materials other than cement, water and aggregate that are included in the concrete. Several types of admixtures have been developed, for example, in the following areas: air entrainment (for frost protection, increased durability, improved workability); plasticizers or water-reducers (to provide higher strength, more economical use of cement, or greater workability); accelerators or retardants (to control the rate of set for reasons of weather, job requirements, methods of placement (e.g., rapid set needed when spraying shotcrete), etc.); and pozzolans (materials which react with the free lime during hydration to form additional cementitious compounds, saving in cement and providing a more durable concrete). Admixtures often fulfill more than one function simultaneously, and if used in combination, may react with each other. Therefore, proposed mixes involving admixtures are tested beforehand for their effect on all concrete properties.

The tensile strength of concrete is about one-tenth its compressive strength. Therefore, concrete is usually reinforced with steel bars or wire mesh to withstand tensile loads. Concrete may also be prestressed (i.e., loaded prior to service in compression, to offset loads that would otherwise cause tension). Fiber-reinforced concrete includes short, discontinuous fibers throughout the cement paste; although strength is not increased, the resulting concrete is tougher. The coarse and fine aggregate bound by the cement matrix influence the weight, elastic modulus and dimensional stability of the concrete. Where higher strength-to-weight ratios are desired, lightweight aggregates (either natural or artificial) are used. Conversely, heavy aggregates are used where a dense concrete is needed (e.g., for radiation protection). Through careful control of constituent materials and production, higher strength concretes are continually being developed (from $35\,MN\,m^{-2}$ in the 1950s to $76–96\,MN\,m^{-2}$ available in the 1980s).

5. *Advances in Ceramic Construction Products*

Ceramic materials are well-suited to the concept of engineering construction materials to meet the requirements of the design since their macrostructural properties can be well controlled through careful selection of raw materials and precise formation of microstructure in the processing steps. This control of manufacture is what distinguishes advanced or high-performance ceramics from traditional ceramics, many of which have been discussed in this article. The use of such concepts to fashion materials is beginning to occur in construction. For example, computer simulation models are being developed and applied to understand better the internal structure, physics and chemistry of concrete, to improve strength, durability and placement during construction. In other cases innovative products have been developed. One example is a high-performance ceramic tile manufactured for use in heavily trafficked, high-maintenance areas on highways and bridges. The product is made from a high alumina (87%) powder which is cold pressed and sintered at 1470 °C. Another example is an opaque glass-ceramic panel developed in Japan as an exterior facing for buildings. (A glass-ceramic is a glass which has been allowed to crystallize.) The panel looks like marble but performs better than marble or granite in weather resistance. It is strong, nonabsorbent, fire-, chemical- and wear-resistant and relatively easy to use and maintain.

Bibliography

Budnikov P P 1964 *The Technology of Ceramics and Refractories*. MIT Press, Cambridge, Massachusetts
Jones J T, Berard M F 1972 *Ceramics: Industrial Processing and Testing*. Iowa State University Press, Ames, Iowa
Kingery W D et al. 1976 *Introduction to Ceramics*. Wiley, New York
Mehta K P 1986 *Concrete: Structure, Properties, and Materials*. Prentice-Hall, Englewood Cliffs, New Jersey
Newton F H 1974 *Elements of Ceramics*. Addison-Wesley, Reading, Massachusetts
Phillips C J 1960 *Glass: Its Industrial Applications*. Reinhold, New York

M. J. Markow and A. M. Brach
[Massachusetts Institute of Technology,
Cambridge, Massachusetts, USA]

Chemically Modified Wood

Chemically modified wood is wood which has been treated by chemical processes to change the biological, chemical, mechanical or physical properties of the natural wood. This article primarily discusses modifications involving chemical reactions with the wood components and their effect on wood properties, as well as selected nonreactive treatments.

1. Reactive Modification of Wood

By far the most abundant functional group in wood is the hydroxyl group, and it follows that reactive chemical modification of wood depends principally on reagents which react readily with these groups. The distribution, accessibility and reactivity of these hydroxyl groups vary throughout the wood depending on the component (cellulose, hemicellulose or lignin), degree of order (crystalline or amorphous regions) and type of hydroxyl group (primary, secondary or phenolic). These variations influence the distribution of the reaction products in the modified wood.

Virtually every type of reagent capable of reacting with hydroxyl groups has been applied to the chemical modification of wood. These include acid anhydrides, inorganic acid esters, acid chlorides, chloroethers, aldehydes, lactones, reactive vinyl compounds, epoxides and isocyanates. Both monofunctional and difunctional reagents have been studied.

Since the modified wood must still possess the basic desirable properties of untreated wood, the reagents must be capable of reaction with wood under relatively mild conditions. Extremes of acid or alkaline conditions or temperature which might degrade the wood must be avoided. Penetration is favored by small polar molecules. With reagents which form a by-product in the reaction with hydroxyl groups an additional processing step for its removal may be necessary. Among the most successful modifying agents are acetic anhydride, acrylonitrile, epichlorohydrin, formaldehyde, methyl isocyanate, β-propiolactone and propylene oxide.

2. Properties of Chemically Modified Wood

The changes in wood properties which are most apparent after chemical modification are an increase in resistance to biological deterioration and an improvement in dimensional stability towards variations in humidity. Under the most favorable modification conditions, these improvements are achieved with no sacrifice of mechanical properties or with even a slight increase in strength. However, there are unavoidable increases in weight and cost.

Chemical modification of the hydroxyl groups in wood affects wood properties through two mechanisms. One relates to the increased steric influence of the larger substituent compared with that of the small hydroxyl hydrogen. The other involves the reduced hydrogen-bonding capability of the modified wood and its resultant lower hygroscopicity. Both mechanisms contribute to the observed improvements in properties.

Biological deterioration of wood, whether by fungi or by the symbiotic protozoa in termites, depends on enzymatic hydrolysis of the polymeric wood carbohydrates to simple sugars. Enzymatic action is extremely stereospecific, so the introduction of substituents on the hydroxyl groups prevents hydrolysis. The substitution of one hydroxyl per anhydroglucose unit in cellulose provides biological resistance. Decay fungi also need a high moisture content in the wood to thrive. Reduction of wood hygroscopicity through chemical modification also retards biological deterioration.

In contrast to conventional treatments of wood to impart biological resistance, chemical modification of wood provides treatments which are effective, nontoxic and nonleachable. As such they are certainly more environmentally acceptable. However, the high degree of modification required for effectiveness results in a rather expensive treatment.

Dimensional stabilization of wood by chemical modification takes advantage of the larger size of the substituent groups through a mechanism known as bulking. The larger volume of these groups occupies space in the swollen cell wall and prevents the cell wall from shrinking in response to loss of moisture. In effect, the wood has been preswollen and is used in this condition. Furthermore, the substituent groups are less hygroscopic, so less water is adsorbed at any specific humidity level. The stabilizing groups are permanently attached to the wood structure.

Chemical modification of wood can also provide stabilization by mechanical restraint at the molecular level by the formation of cross-links between cellulose chains or microfibrils which prevent the spreading and swelling of the cellulose structure by water. The most effective cross-linking agent is formaldehyde which can provide a high degree of dimensional stability (90%) at a weight gain of only 7% compared with bulking agents which require 20–30% weight gain for 70–75% stabilization. However, cross-linking with formaldehyde causes serious embrittlement of the wood as well as loss of abrasion resistance.

Chemical modification without cross-linking provides dimensional stability with no loss in toughness. In contrast, polymer impregnation often causes severe embrittlement because the disordered regions of the cellulose chains are embedded in a rigid matrix which restricts chain motion and permits failure at low levels of absorbed energy. In chemical modification the cellulose chains are still free to absorb energy by vibration without chain failure occurring at low impact stresses.

Since the strength properties of wood are higher at low moisture contents the lower equilibrium moisture content of the less hygroscopic chemically modified wood at any given humidity results in higher strength values. For example, chemically modified pine wood does not fail first under compressive stress in the upper part of the beam during bending strength determinations like untreated pine, but rather in tension at the lower surface, thus allowing higher values of fiber stress at the proportional limit and of rupture modulus. The lower moisture content of the modified wood provides sufficient increase in crushing strength to bring about this change in mode of failure.

In a specific example of chemical modification, wood acetylated with uncatalyzed acetic anhydride to a weight gain of about 20% shows reductions of swelling in high humidity of 70–80%. Acetylated pine becomes resistant to wood-destroying fungi and termites at a weight gain of about 18%. Acetylated wood distorts and checks less than untreated wood during weathering. The impact strength of acetylated wood is not decreased by the treatment, whereas wet compressive strength is doubled. Hardness and work to proportional limit are materially increased. The elastic modulus remains unchanged. Density is increased and equilibrium moisture content is reduced.

3. *Nonreactive Chemical Modification*

Chemicals which do not react with the wood components can also be used to bring about changes in wood properties. The bulking mechanism for dimensional stabilization is operative for any chemical which penetrates the cell wall. Water-soluble inorganic salts and organic compounds, although effective bulking agents, are hygroscopic and cause finishing problems and reduction in strength. Polyethylene glycols are very effective water-soluble bulking agents, and at higher molecular weights (~ 1000) are somewhat less hygroscopic than the small molecules. They can be applied to green wood by diffusion and used to stabilize gunstocks, carvings and turnings.

The use of bulking agents which are solvent-soluble but water-insoluble can avoid the hygroscopicity problem entirely. Cellosolve (ethylene glycol monoethyl ether) can be used as an intermediary solvent because it is soluble in water in all proportions; it is also used as a solvent for waxes, oils and resins. Replacement of the water in swollen wood by cellosolve followed by replacement of the cellosolve with molten waxes or oils affords a high degree of dimensional stability. However, the process is very slow.

Plasticization of wood can be accomplished by chemical treatment, most notably by the use of ammonia in gaseous or liquid form. The use of amines, amides and urea permits permanent plasticization by retention of the chemical or temporary plasticization followed by its removal.

Some natural woods are self-lubricating, such as lignum vitae which was long used for marine bearings. More common species of wood may be made self-lubricating by impregnation with waxes or lubricating oils. Internal lubricity improves machinability. Cutting properties of cedar pencil stock are improved by impregnation with paraffin wax or polyethylene glycol.

The resistance of wood to degradation by various chemicals such as acids and alkalis can be enhanced by preventing access of the reagents to the wood fibers by means of barriers. Filling the pores of wood with resistant materials such as waxes, tars, asphalt and sulfur improves chemical resistance. Still higher resistance to acids, even in the presence of organic solvents or at elevated temperatures can be obtained by polymerization of the thermosetting phenol–formaldehyde resins within the cell wall. This treatment does not resist alkali. Formation within the wood of an acid- and alkali-resistant resin by polymerization of furfuryl alcohol provides broad-spectrum chemical resistance to everything but oxidizing agents.

See also: Wood: An Overview

Bibliography

Goldstein I S, Loos W E 1973 Special treatments. In: Nicholas D D (ed.) 1973 *Wood Deterioration and Its Prevention by Preservative Treatments*, Vol. 1. Syracuse University Press, New York, pp. 341–71

Rowell R M 1975 Chemical modification of wood: Advantages and disadvantages. *Proc. Am. Wood-Preserv. Assoc.* 71: 41–51

Rowell R M 1983 Chemical modification of wood. *For. Prod. Abstr.* 6(12): 363–82

Rowell R M, Youngs R L 1981 *Dimensional Stabilization of Wood in Use*, Research Note FPL-0243. US Forest Products Laboratory, Madison, Wisconsin

I. S. Goldstein
[North Carolina State University, Raleigh, North Carolina, USA]

Composite Materials: An Overview

The term composite originally arose in engineering when two or more materials were combined in order to rectify some shortcomings of a particularly useful component. For example, cannons which had barrels made of wood were bound with brass because a hollow cylinder of wood bursts easily under internal pressure. The early clipper sailing ships, which were said to be of composite construction, consisted of wood planking on iron frames, with the wood covered by copper plates to counter the attack of marine organisms on the wood.

A composite material can be defined as a heterogeneous mixture of two or more homogeneous phases which have been bonded together. Provided the existence of the two phases is not easily distinguished with the naked eye, the resulting composite can itself be regarded as a homogeneous material. Such materials are familiar: many natural materials such as wood, are composites; so are automobile tires, glass-fiber-reinforced plastics (GRP), the cemented carbides used as cutting tools, and paper—a composite consisting of cellulose fibers (sometimes with a filler, often clay). Paper is essentially a mat of fibers, with interfiber bonding being provided by hydrogen bonds where the fibers touch one another.

It is sometimes a little difficult to draw a distinction between a composite material and an engineered structure, which contains more than one material and is designed to perform a particular function. The combination is usually referred to as a composite material provided that it has its own distinctive properties, such as being much tougher than any of the constituent materials alone, having a negative thermal expansion coefficient, or having some other property not clearly shown by any of the component materials.

In many cases, the dimensions of one of the phases of a composite material are small, say between 10 nm and a few micrometers and under these conditions that particular phase has physical properties rather different from that of the same material in the bulk form; such a material is sometimes referred to as a nanocomposite.

The breaking strength of fibers of glass, graphite, boron and pure silica, and of many whisker crystals, is much greater than that of bulk pieces of the same material. In fact, it may be that all materials are at their strongest when in fiber form. In order to utilize the strength of such strong fibers, they must be stuck together in some way: for example, a rope is appropriate for fibers of hemp or flax, and indeed carbon and glass fibers are often twisted into tows. However, for maximum utilization, a matrix in which to embed such strong fibers is required, in order to provide a strong and stiff solid for engineering purposes. The properties of the matrix are usually chosen to be complementary to the properties of fibers; for example, great toughness in a matrix complements the tensile strength of the fibers. The resulting combination may then achieve high strength and stiffness (due to the fibers), and resistance to crack propagation (due to the interaction between fibers and matrix).

Nowadays, the term "advanced composite" means specifically this combination of very strong and stiff fibers within a matrix designed to hold the fibers together. This type of composite combines the extreme strength and stiffness of the fibers and, owing to the presence of the matrix, shows much greater toughness than would otherwise be obtainable.

1. Prediction of Physical Properties

An important consideration, both for engineering design of a composite material and for scientific understanding of its properties, is how the overall properties of the composite depend upon those of the individual constituents.

Properties of composite materials can be considered under two headings: (a) those that depend solely upon the geometrical arrangement of the phases and their respective volume fractions, and not at all upon the dimensions of the components; and (b) those that depend on structural factors such as periodicity of arrangement or the sizes of the pieces of the two or more component phases. For (a) to be true, the smallest dimension of each phase must usually be greater than 10 nm, and sometimes much greater than this.

1.1 Properties Determined by Geometry—Additive Properties

The geometrical arrangement of the phases in a composite material can often be described in simple terms (Fig.1). The simplest physical property of a composite—namely its density ρc—is given by the volume-

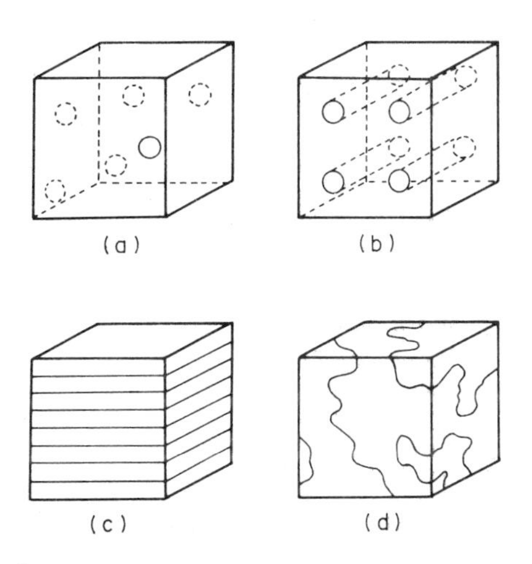

Figure 1
Composite geometries: (a) random dispersion of spheres in a continuous matrix; (b) regular array of aligned filaments; (c) continuous laminae; and (d) irregular geometry (after Hale 1976)

weighted average of the densities of the components:

$$\rho c = \rho_1 V_1 + \rho_2 V_2 \qquad (1)$$

where V is the volume fraction and subscripts 1 and 2 refer to the components. If there are no voids $V_1 + V_2 = 1$. More complicated physical properties, e.g., those described by a second-rank tensor, relate two vectors: either a solenoidal vector and an irrotational vector (as with magnetic or electric susceptibility), or else a flux vector and the gradient of a scalar function (as with diffusivity, and electric and thermal conductivity). The relations derived for these properties for a composite material in terms of the same property in each of the submaterials or components are all formally identical. They have usually been discussed in terms of the dielectric constant ε and, because of this, the general attempt to predict the overall physical properties of a multiphase composite from the same properties of the individual phases is sometimes called the dielectric problem. Attempts to devise exact, generally applicable theoretical expressions for the dielectric constants of mixtures of unknown phase geometry are futile. However, the dielectric constant ε_c of a mixture of two phases must lie between certain limits, whatever the geometry. If some knowledge of the phase geometry is available, still closer bounds can be set on the value of ε_c; the same is true of other properties. There is therefore a good deal of utility in the search for bounds on the properties. For a few special geometries, exact solutions can be obtained. Approximate solutions are also often obtained for two-phase composites when the volume fraction of one of the components is low. Such solutions are often extended to nondilute situations by use of a self-consistent scheme, in which each particle of the minor component is assumed to be surrounded not by the other component but by average composite material.

For example, as given by Hashin and Shtrikman (1962), the effective dielectric constant of an isotropic composite must lie in the range

$$\varepsilon_1 + \frac{V_2}{[(\varepsilon_2 - \varepsilon_1)^{-1} + (V_1/3\varepsilon_1)]} \leqslant \varepsilon_c \leqslant \varepsilon_2 + \frac{V_1}{[(\varepsilon_1 - \varepsilon_2)^{-1} + (V_2/3\varepsilon_2)]} \qquad (2)$$

where $\varepsilon_2 > \varepsilon_1$.These are the best possible bounds obtainable for the dielectric constant of an isotropic two-phase material if no structural information, apart from the volume fractions, is available.

Figure 2 compares the predictions of those bounds for the cases $\varepsilon_1/\varepsilon_2 = 0.1$ and $\varepsilon_1/\varepsilon_2 = 0.01$, with the most elementary bounds which can be rigorously derived for an isotropic composite. These elementary bounds correspond to simple Voigt (polarization in parallel) and Reuss (polarization in series) estimates. The Voigt estimate is of the form of Eqn. (1), while the

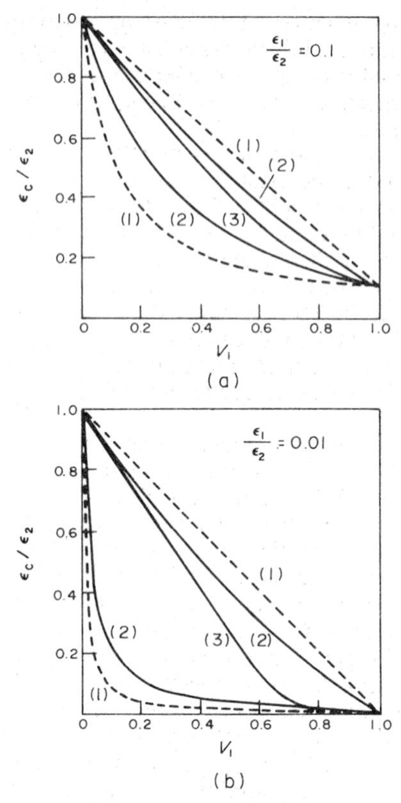

Figure 2
Dielectric constant of composite materials: (a) $\varepsilon_1/\varepsilon_2 = 0.1$ and (b) $\varepsilon_1/\varepsilon_2 = 0.01$. (1) Bounds from Eqns. (1) and (3); (2) Hashin and Shtrikman bounds for arbitrary geometry (Eqn. (2)); (3) self-consistent approximation for spheres in a continuous matrix (after Hale 1976)

Reuss estimate gives the composite dielectric constant as

$$1/\varepsilon_c = V_1/\varepsilon_1 + V_2/\varepsilon_2 \qquad (3)$$

If ε_1 and ε_2 do not differ greatly, the bounds given by Eqn. (2) are very close together, but for $\varepsilon_1/\varepsilon_2 = 0.01$ they are far apart (such as in Fig. 2); therefore, in many practical cases, better bounds are required.

In the foregoing discussion, the properties of the individual components or submaterials within the composite are assumed to be identical to the same properties measured in a piece of the submaterial outside the composite. In addition, the composite is taken to be statistically homogeneous in the sense that, if small elements of the material were extracted, these would have the same physical properties as the whole sample. In addition, implicit assumptions are made that there are no voids, that space-charge effects and polarization are absent (in the case of electrical conductivity) and that there is no discontinuity in temperature at the interface (in the case of heat flux).

The prediction of the elastic properties of particulate composites is discussed in the article *Particulate Composites*.

Thermal expansion coefficients of composites involve both the elastic constants and the thermal expansion coefficients of the individual phases. An important function of reinforcing fillers and fibers in plastics is the reduction and control of thermal expansion. For example, with dental filling materials, a difference in thermal expansion between the filling material and the tooth can lead to a marginal gap, and hence composite filling materials are designed to have a thermal expansion coefficient very close to that of the tooth.

With some components, especially at low fiber volume fractions, the transverse thermal expansion coefficient of, for example, a glass-fiber–epoxy-resin composite, can be greater than that of the matrix. This effect is particularly noticeable with fibers of high modulus and low expansion coefficient (e.g., boron or carbon) in a low-modulus matrix.

Fibers are often used in laminated arrangements because properties parallel to the fibers are very different from properties perpendicular to the fibers. The effective in-plane thermal expansion coefficients for angle-ply laminates (in which the fibers are arranged at plus and minus an angle ϕ to a particular direction) show that in such laminates a scissoring or lazy-tongs type of action can occur, and, with appropriate values of ϕ, can lead to a zero or even negative thermal expansion coefficient in one direction.

1.2 Special Property Combinations—Product Properties

Composites can, in principle, be thought of as materials which produce properties unobtainable in a single material: for example, by combining a piezoelectric material with a material showing magnetostriction, the composite should show a magnetoelectric effect; that is, an applied magnetic field would induce an electric dipole moment; or a material could be produced in which an applied magnetic field produced optical birefringence, by coupling a material which shows strain-induced birefringence with one showing magnetostriction.

This idea led van Suchtelen (1972) to classify such effects as product properties of composites and so now there are considered to be two different types of physical property of composite materials.

The first is the type discussed so far in Sect. 1.1. These are sum or additive properties where the composite property is related to that same property of each of the components and so depends on the geometry of arrangement of the components. The geometry of arrangement includes of course the volume fraction. Examples are elastic stiffness, relating applied stress and measured strain, or electrical resistance, relating applied electric field and measured current density, or the simple example of mass density. The value of the physical property of the composite in general lies between those of the (two) components.

There is a subclass of additive properties in which the value of the property of the composite can lie well outside the bounds set by the values of the property of the components; examples are Poisson's ratio involving the ratio of two compliance coefficients under the action of a single applied stress, or acoustic wave velocity, which depends on the ratio of elastic modulus to density. Here, because the elastic modulus and the density can follow quite different variations with volume fraction, e.g., elastic modulus following Eqn. (3) and density following Eqn. (1), the acoustic wave velocity of the composite can lie well outside the value for either component.

A further, but less obvious, example is the thermal expansion coefficient, which depends on the thermal expansion coefficients and the elastic constants of the two phases. Newnham (1989) distinguishes this last set of properties and calls them combination properties (distinct from sum or additive properties) when the composite property lies outside the bounds of the same property of the two or more constituents. However, since a value lying outside the range of the constituents' properties can in principle occur for all composite properties (even mass density) it seems best to view all cases in which the same property of the composite and of the components is considered as cases of additive or sum properties.

Product properties are those cited at the beginning of this section: the property of the composite depends specifically on the interaction between its components. Any physical property can be considered as the action of a physical quantity X resulting in physical quantity Y giving the X–Y effect. Product properties are those in which an X–Y effect in submaterial 1 produces a Y–Z effect in submaterial 2, producing in the composite an X–Z effect. A good example is the one cited of a magnetoelectric effect in a composite material having one magnetostrictive and one piezoelectric phase. Application of a magnetic field produces a change in shape in the magnetostrictive phase which then stresses the piezoelectric phase and hence generates an electric field. In this case the coupling is mechanical but the coupling could also be electrical, optical, magnetic, thermal or chemical (Van Suchtelen 1972).

Viewed in this way, sum properties are those in which an X–Y effect in submaterial 1 and the same X–Y effect in submaterial 2 combine to give an X–Y effect in the composite.

Table 1 classifies some physical properties or phenomena according to the input–output parameters X and Y. A small selection of possible product properties is given in Table 2. Newnham (1989) gives practical examples of these effects.

Table 1
Matrix classification of some physical properties or phenomena in materials according to the type of input and output parameters (after van Suchtelen 1972)

	Input parameter (X)					
Output parameter (Y)	Mechanical (force/ deformation) (1)	Magnetic (field/ polarization) (2)	Electrical (field/ polarization, current) (3)	Optical and particle radiation (light or particle flux) (4)	Thermal (temperature, temperature gradient, heat current) (5)	Chemical (chemical composition, chemical composition gradient) (6)
Mechanical (force/ deformation) (1)	elasticity	magnetostriction magnetoviscosity (suspension)	electrostriction Kirkendall effect electroviscosity (suspension) indirect: thermal expansion		thermal expansion	osmotic pressure
Magnetic (field/polarization) (2)	piezomagnetism	magnetic susceptibility	superconductors galvanic deposition of ferromagnetic layer direct generation of magnetic field	photomagnetic effect	thermomagnetism ferromagnetic material at $T \simeq T_c$ (+ magnetic field)[a]	dependence of T_c on ferromagnetic composition
Electrical (field/ polarization, current) (3)	piezoelectricity piezoresistivity	magnetoresistance (+electric current)[a] Hall effect (+electric current)[a] ac resonance induction of voltage	dielectric constant, dielectric polarization Hall effect (+magnetic field)[a]	photoconductivity photoemission photoelectromagnetic effect (+magnetic field)[a] ionization	thermoelectricity ferroelectrics at $T \simeq T_c$ (+electric field)[a] temperature-dependent resistivity (+electric current)[a]	dependence of T_c on ferroelectric composition

Optical and particle radiation (light or particle flux) (4)	stress birefringence triboluminescence	Faraday effect magnetooptic Kerr effect deflection of charged particles	electroluminescence laser junctions refractive index Kerr effect absorption by galvanic deposits cold emission of electrons	refractive index fluorescence scintillation color-center activation	thermoluminescence	chemoluminescence
Thermal (temperature, temperature gradient, heat current) (5)	heat of transition of pressure-induced phase transition piezoresistivity and Joule heating	adiabatic demagnetization Nernst-Ettingshausen temperature gradient effect (+ electric current)[a] magnetoresistance effect + Joule heating (+ electric field)[a]	dissipation in resistance Peltier effect Nernst-Ettingshausen temperature gradient effect (+ magnetic field)[a]	absorption	thermal conductivity	reaction heat
Chemical (chemical composition, chemical composition gradient) (6)	pressure-induced phase transition		electromigration galvanic deposition	light or particle stimulated reactions (photosensitive layers)	Soret effect (temperature gradient)[a] phase transition change of chemical equilibrium	

[a] Indicates the parameter in parenthesis is essential as a second input

Table 2
Product properties of composite materials (after van Suchtelen 1972)

X–Y–Z (Table 1)	Property of phase 1 (X–Y)	Property of phase 2 (Y–Z)	Product property (X–Z)
1–2–3	piezomagnetism	magnetoresistance	piezoresistance phonon drag
1–2–4	piezomagnetism	Faraday effect	rotation of polarization by mechanical deformation
1–3–4	piezoelectricity	electroluminescence	piezoluminescence
1–3–4	piezoelectricity	Kerr effect	rotation of polarization by mechanical deformation
2–1–3	magnetostriction	piezoelectricity	magnetoelectric effect
2–1–3	magnetostriction	piezoresistance	magnetoresistance spin-wave interaction
2–5–3	Nernst–Ettingshausen effect	Seebeck effect	quasi-Hall effect
2–1–4	magnetostriction	stress-induced birefringence	magnetically induced birefringence
3–1–2	electrostriction	piezomagnetism	electromagnetic effect
3–1–3	electrostriction	piezoresistivity	coupling between resistivity and electric field (negative differential resistance, quasi-Gunn effect)
3–4–3	electroluminescence	photoconductivity	
3–1–4	electrostriction	stress-induced birefringence	electrically induced birefringence light modulation
4–2–1	photomagnetic effect	magnetostriction	photostriction
4–3–1	photoconductivity	electrostriction	photostriction
4–3–4	photoconductivity	electroluminescence	wavelength changer (e.g., infrared–visible)
4–4–3	scintillation	photoconductivity	radiation-induced conductivity (detectors)
4–4–4	scintillation, fluorescence	fluorescence	radiation detectors, two-stage fluorescence

1.3 Properties Dependent on Phase Dimensions and Structural Periodicity

In Sects. 1.1 and 1.2 the physical properties of the components or submaterials have been assumed to be unaltered by the incorporation of the components into the composite. This is usually not the case. The effects of phase changes or chemical reactions during fabrication are of importance in a wide range of composites. Excessive shrinkage of one component can result in high internal stresses, which may lead to premature failure or even preclude successful fabrication. Matrix shrinkage also has an important indirect effect on the mechanical behavior of fiber composites, since the resulting internal stresses can determine the frictional forces at the fiber–matrix interface.

The dimensions and periodicity of a composite structure also have an important effect on the properties when they become comparable with, for example, the wavelength of incident radiation, the size of a magnetic domain or the thickness of the space-charge layer at an interface. The dimensions are also particularly important when the properties of a material depend on the presence of defects of a particular size (e.g., cracks). The strength of a brittle solid containing a crack depends on the square root of the length of the crack. Since very small particles cannot contain long cracks and because the surface region of most solids behaves differently from the interior (a small fiber or sphere contains proportionately more surface material than a large piece), the breaking strength depends on the dimensions of the piece tested. This is particularly important when considering mechanical strength and toughness of composites.

Many of the optical effects which depend on structural dimensions also depend in a very complex way on other factors, and optical effects produced by dispersion of a second phase within a material are often used to determine the distribution of that phase (e.g., determination of the molecular weights of polymers or of the size of crystal nuclei in glasses). A composite material containing aligned elongated particles of an optically isotropic material in an optically isotropic matrix may exhibit double refraction as a straightforward consequence of the relationship between dielectric constant and refractive index.

In a ferromagnetic material at temperatures below the Curie point, the electronic magnetic moments are aligned within small contiguous regions—the magnetic domains—within each of which the local magnetization is saturated. Magnets of very high coercivity can be obtained by using very fine particles: a particle with a diameter of between 10^{-1} and 10^{-3}μm would be magnetized almost to saturation, since the formation of a flux-closure configuration would be energetically unfavorable. Magnetization reversal cannot take place in a sufficiently small single-domain particle by boundary displacement; it must occur by a single jump against the magnetocrystalline anisotropy and the shape factor. Fine-particle permanent magnets of rare-earth alloys such as $SmCo_5$, enveloped in an inert matrix such as tin, have been developed which have superior properties; for example, the energy product can be up to 79.6×10^3 A m^{-1}. The preparation of these fine-particle magnets shows that attention to small size alone is not enough, and that care must be taken to prevent domain-wall nucleation at the surfaces of the particles.

Considerable attention has also been paid to rod-type eutectics as permanent magnetic materials. If the permanent magnet is long and thin, the demagnetization energy can be very large as a consequence of the shape anisotropy, even in the absence of high magnetocrystalline anisotropy. This is the concept of the elongated single domain. In the Bi–Mn, Bi system, very high coercive forces e.g., 79.6×10^3 A m^{-1} can be obtained but, because the volume fraction of magnetic material is low, so is the energy product.

Composite principles are also used in the construction of powerful electromagnets containing filamentary superconducting composites: for example, high-performance electromagnets producing field densities of 16 T with a Nb–Sn (intermetallic) winding, or 8 T with Nb–Ti. One of the problems here is instability in the interaction between the superconductor and the magnetic field, which causes quenching of superconductivity. When this occurs, a very large current must be carried by other means, and one solution to this problem is to surround the conductor with, say, ten times its volume in copper. The copper can then cool the superconductor and return it to the superconducting state. The energy dissipated is proportional to r^2, where r is the radius of the wire, and the cooling is proportional to r. The ratio of surface area to cross-sectional area per unit length varies as $1/r$ and therefore small wires are more effectively cooled.

The theory of how instability is produced also shows that it can be almost eliminated by ensuring that the diameter of the individual superconducting filaments is reduced below approximately 20–50 μm. This size effect arises because instability occurs whenever a small rise in temperature reduces the shielding current (proportional to the superconducting current), causing a flux change which in turn leads to heating, and so on. The energy liberated per flux jump is smaller if the superconducting element is narrower, and the time to cool the superconductor from outside is smaller if the diameter is smaller. Both effects produce an upper limit to the diameter of the superconductor for effective stabilization. For ac applications it is also necessary to reduce the eddy-current losses in the copper. This can be done by using a material with a high electrical resistivity to decouple the wires.

To produce the very highest magnetic fields attainable in the laboratory (50–100 T), pulsed-field systems are used. This requires the conductor to show high mechanical strength as well as very good electrical conductivity; a composite has the ideal geometry for producing this combination of properties.

2. *Advanced Fiber Composites*

The best known and most striking example of a modern composite material is the fiber composite, designed for high strength, high stiffness and low weight.

The breaking strength and the stiffness of typical specimens of a number of modern fibers are shown in Fig. 3. Steel wire and glass provide the strongest materials and pitch-based carbon fibers the stiffest. The strength and stiffness of the same fibers divided by their density and by the acceleration due to gravity is shown in Fig. 4. The resulting physical unit is a length and, as far as the strengths are concerned, represents (intriguingly) the greatest length of a uniform rod which could in principle be picked up from one end on the surface of the Earth without breaking.

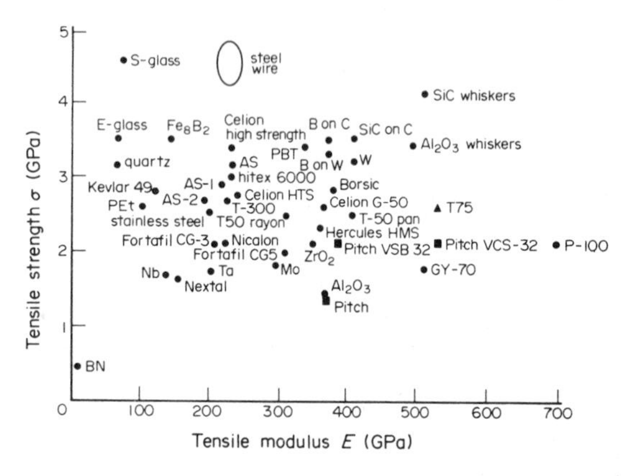

Figure 3
Tensile strengths and stiffness of a variety of fibers

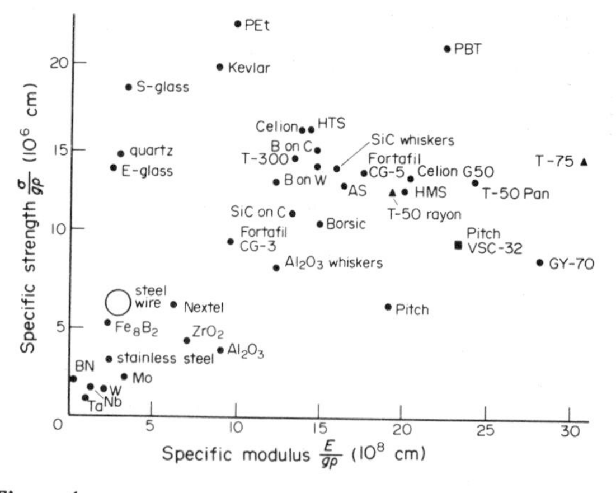

Figure 4
Specific strength and stiffness of a variety of fibers. The specific value is the value of the property divided by $g\rho$ where ρ is the density and g the acceleration due to gravity

Quantities such as specific strength and stiffness are often quoted in units such as strength divided by specific gravity, in which case units of stress are used, or else as stress divided by density, the resulting unit being a velocity squared. In the case of stiffness the velocity represents that of longitudinal sound waves in the material. In Fig. 4 a unit of specific stiffness, $E/\rho g$ (or of specific strength $\sigma/\rho g$) of 10^7 cm is equivalent to a value of stiffness E/ρ (or of strength σ/ρ) of 9.8×10^9 $cm^2\, s^{-2}$ (E is Young's modulus and σ is the tensile strength).

Figure 3 shows that the stiffest materials are ceramics composed of elements from the first rows of the periodic table (B, C, SiC, Al_2O_3, etc.). This is because these elements contain the highest possible packing of strong covalent bonds directed in three dimensions. Such materials also show a high melting point, a low coefficient of thermal expansion and a low density. All of these are very desirable engineering properties. The strongest materials are the glasses and the metals.

In all structural applications, the weight of the structure necessary to bear a given load and to do so with a minimum elastic deflection is important, and so strong fibers are compared on the basis of strength divided by density. The fibers shown in Fig. 3 are compared in this way in Fig. 4. When this is done the stiffest materials on a weight-for-weight basis remain the ceramics but the strongest are the plastics (PET, PBT, Kevlar) and the glasses. The metals make a very poor showing on both stiffness and strength. However, fibers cannot be used directly and must be bound together within a matrix.

A set of aligned fibers in a pliant matrix is only useful as an engineering material under direct tensile forces parallel to the fibers. The salient properties are represented by those of a pocket handkerchief which is stiff in the direction of the warp and weft (parallel to the fibers) but shears easily. Hence the properties of fiber composites are very directional (see Fig. 5), being very strong and stiff parallel to the fibers, but rather weak in shear parallel to the fibers (because this property depends principally upon the shear properties of the matrix), and very weak indeed in tension perpendicular to the fibers. In order to overcome this, the fibers are arranged in laminae, each containing parallel fibers, and these are stuck together (Fig. 5) so as to provide a more isotropic material with a high volume loading of fibers. Figure 6 shows the specific strengths and stiffness values of laminated forms of fibrous composites made of some of the fibers where properties are shown in Figs. 3 and 4. The advantage over conventional isotropic constructional materials represented by the metal alloys is seen to be considerable but much less marked than in Figs. 3 and 4.

Alternatively, the fibers may be randomly arranged in a plane or in three dimensions; such arrangements

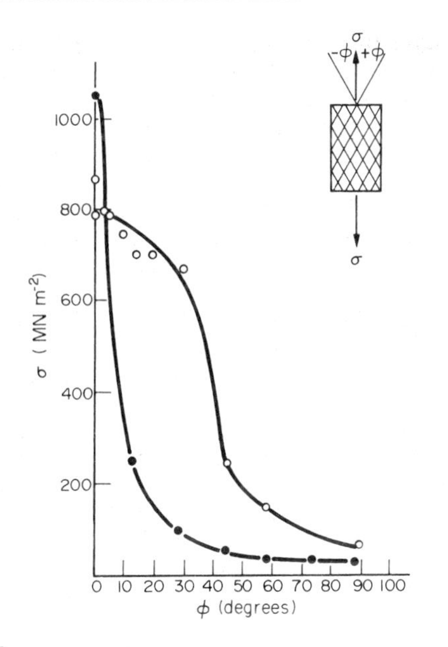

Figure 5
Measured variation of tensile strength with angle ϕ for specimens consisting of a number of alternate layers of fibers. The fibers in each layer are parallel and continuous. Alternate layers are at $+\phi$ and at $-\phi$ to the tensile axis. Open circles represent data for a volume fraction of 40% of silica fibers in aluminum. Full circles represent data for a volume fraction of 66% of E glass fiber in an epoxy resin

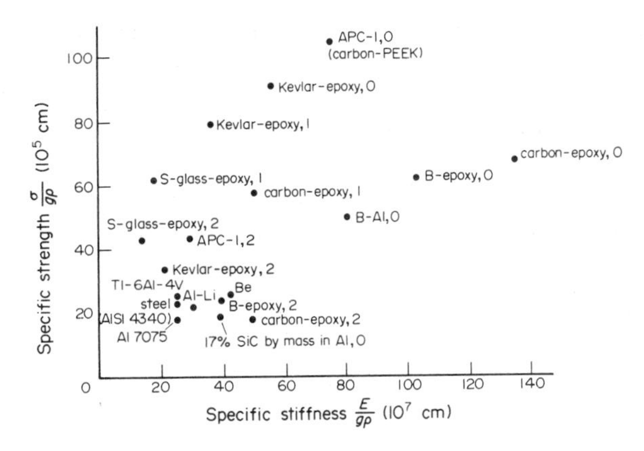

Figure 6
Specific strength and stiffness of some isotropic materials and of fiber composites. The designation 1 on the composites means the following arrangements of fibers : 50% at 0°; 40% at ±45° and 10% at 90° to the stress; 2 denotes balanced laminates with equal proportions at 45°, 90° and 135°; O indicates aligned fibers in the specified matrix. The volume fraction of fibers in the various composites are not the same in the different systems. They vary between 40 and 60%. The metal alloys are those without such designation

limit the obtainable fiber packing density. Fibers are also often woven into mats before incorporation into the composite because this aids the handling of fiber arrays.

When carbon- or other fiber-reinforced plastics are made with woven fabric rather than nonwoven material, distortion of the load-carrying fibers parallel to the applied stress reduces the tensile strength further and reduces stiffness and toughness. When, however, woven fabric is oriented at 45° to the load direction these properties compare favorably with those for nonwoven ± 45° material. Indeed, the residual strengths after impact can be greater. At present, between one third and one half of the market of carbon-fiber-reinforced plastics (CFRP) consists of woven fabric; the rest consists of aligned material.

The anisotropy of the properties of advanced fiber composites represents a completely new feature in engineering design. The nonisotropy can be varied; in fact it is possible now to place the fibers precisely where they are needed in a structure to bear the loads. In fact it is also possible, for instance, to vary the amount and stacking of the fibers in various directions along the length of a beam so as to vary the stiffness and torsional rigidity of the material.

This is precisely what is needed in helicopter rotor blades, and it is expected that within a few years' time all rotor blades except perhaps those for heavy-lift helicopters will be made from GRP or combinations of GRP and CFRP; despite the fact that such composites were not applied to helicopter blades until mid-1960s.

The AV 8B, the latest version of the Harrier vertical takeoff and landing aircraft, employs composites in torque boxes, auxiliary flaps, the forward fuselage and cone structure, the horizontal stabilizers and in the ammunition and gun pods.

It would be impossible to make use of the inherent advantages of anisotropic materials such as CFRP without the invention of computer-aided methods. The reiterations and the number of variables involved require computer programs both to relate the properties of the individual laminae to the properties of the fibers and to take into account the interactions in terms of bending–twisting coupling between the various laminae. Initially, of course, these complications were viewed as a great disadvantage of the material, since the engineer was not happy with anisotropy. Now the problem of how to use it and make a virtue, not so much out of necessity but of supposed vices, has been solved. The latest Grumman X29 aircraft makes use of bending–twisting coupling so that, as the loads on an aircraft wing increase, the structure can deform but remain tuned to the aerodynamic requirements. The material possesses almost a form of gearing, so that it changes its shape automatically without the need for sensors and levers.

3. *Applications of Composites*

High-performance composites use fibers in order to attain the inherent properties of the fiber, coupling this with a judicious choice of matrix so that toughness and impact damage are not lost.

Strong fibers, whether they be stiff or not, have the great advantage of restraining cracking in what are called brittle matrices. It is this use of fibers which is often referred to when showing how the ancients employed composite materials. Quotations from the Bible are made, such as that from the Book of Exodus, Chapter V, verses 6 *et seq.*, which refer to the difficulty of making of bricks without straw. Fibers restrain cracking because they bridge the cracks. The principle is very simple. It has been used in asbestos-reinforced cement and the principle of reinforced concrete is not too far from the same idea. The principle is illustrated in Fig. 7. A crack passing through a brittle material may not enter the fibers but leaves a crack straddled by fibers so that a material which would normally have broken with a single crack now has to be cracked in a large number of places before the fibers themselves fail or pull out. This is utilized in fiber-reinforced cement, which is based on the same principles as reinforced concrete, but since the fibers are very much finer than reinforcing bars and are often not visible to the naked eye, a homogeneous material is effectively produced which can show quasi-plastic deformation.

This judicious use of fibers enables us to think nowadays in terms of using brittle materials for construction purposes, because if they are cracked the fibers will render the cracks harmless. Building panels and pipes, of course, are made of such material. Glass-reinforced cement is becoming commonplace. However, a much greater prize appears ahead. Metals have limited high-temperature capability. The ceramic materials described above as providing the best fibers also provide the materials of highest melting point. For the construction of prime movers at high temperature these will have to replace metals. They cannot do this as monolithic pieces because, as we have seen, these would be easily broken. However, they can resist cracking if they contain fibers. The composite principle then leads to suggesting that even fibers of the same material as the matrix may give marked resistance to cracking and fracturing. Such is true of carbon–carbon composites which are used in brakes for Concorde and in certain ballistic missile and reentry vehicle applications. Of course carbon oxidizes and so would have to be protected if exposed to hot air. In contrast, therefore, a good deal of interest centers on the use of refractory oxide glasses such as cordierite or a material of almost-amorphous silicon nitride or silicon carbide for use at high temperature, containing fibers again of silicon nitride or silicon carbide. These will restrain cracking and give an engineering material usable for thousands of hours at temperatures in excess of 1100 °C. As yet these are not commercially available but aircraft engine manufacturers are making and testing pieces of ceramics, sometimes unreinforced, for aerospace applications; but the main thrust of research is towards reinforced materials.

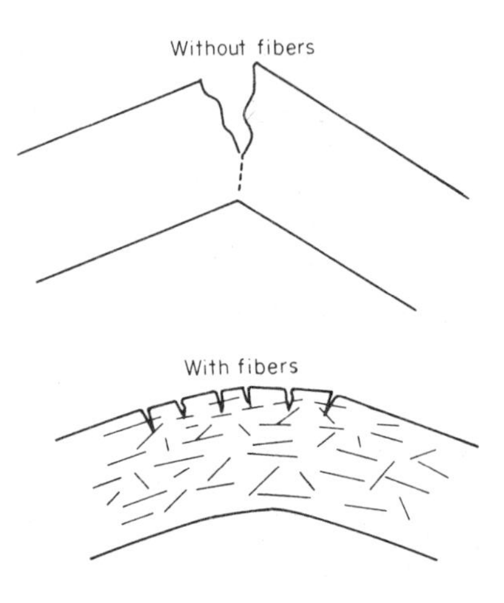

Figure 7
Demonstration of how cracks are prevented from running in a brittle material because of the fibers in their path

Ceramic materials are of widespread abundance and are all essentially cheap. If they can be fashioned by simple chemical means, such as reacting acids with alkalis, they are classed as phosphate-bonded materials. These again have great potential for replacing, say, sheet steel in normal household use, again provided cracking can be restrained. Fibers will do this and, if the material is not exposed to high temperature, glass fibers or even textile fibers may be employed.

4. *Fibers*

The very strong and stiff fibers comprise the following: asbestos (a naturally occurring fiber), boron and carbon of varying degrees of graphitization, made in principle from a number of starting materials but in practice mainly two: PAN (polyacrylonitrile) and pitch. Inorganic glasses based on silica yield many fibers. Those made in the form of continuous filaments are most important for composite materials. Many linear polymers such as polyparabenzamide or simple polyethylene can be processed to form very stiff and strong fibers. The ways of doing so are very varied. Fibers of Al_2O_3 are important for high temperature use in filtration and other applications because of their chemical inertness. Whiskers or tiny (~ 1 μm diameter) fibers are known of most materials. They are often very strong and stiff and possibly will have uses as agents producing toughness, stiffness or wear resistance when introduced into other materials.

Historically the first materials designed specifically by chemists to hold fibers in a composite material were the organic thermosetting resins.

5. *Examples of Composite Materials*

Many natural products such as wood, shells and manufactured derivatives such as paper and plywood are of course composites, and the understanding of the physical properties of these materials both illuminates our understanding of the properties of wholly synthetic composites and, more frequently, is itself enhanced from our knowledge of the behavior of modern composite materials.

Glass-reinforced plastic (GRP) is by far the most commonly-used commercial composite in terms of volume. It may be fabricated from thermosetting or thermoplastic resins. Carbon-fiber-reinforced plastics have been developed rapidly during the 1970s and 1980s; based initially on thermosetting resins they have recently begun to employ thermoplastics. The newer thermoplastic polymers such as polyether ether ketone (PEEK), polysulfone or the modified polyamides have advantages over the thermosetting resins: they can withstand higher strain before failure, they have lower moisture absorption and they can be shaped when heated and repaired without requiring a lengthy and intricate cycle of cure. They also have, in effect, an infinite storage life at room temperature. The newer thermoplastic polymers are, however, expensive.

There is no reason why a fibrous composite need contain fibers of only one chemical type. Hybrids are possible containing more than one type of fiber. Nor is it necessary that fiber and matrix be chemically different: carbon–carbon composites in which strong stiff carbon fibers are incorporated into a matrix of quasiamorphous carbon produced by pyrolysis, chemical vapor deposition or by other means have a number of important uses and great potential for use at higher temperatures. It is the potential of ceramics to displace metals for use at high temperatures (> 1000 °C) because of the ceramics' greater oxidation resistance and lower density, that primarily causes interest in the use of these materials in composites containing fibers. Ceramic materials used alone are brittle and hence produce fragile components. One way, perhaps the only way, to produce very high fracture toughness into them is to incorporate fibers. Fiber-reinforced ceramics possess great potential but have few uses at present. However, the incorporation of fibers to increase toughness has been commercially realized in fiber-reinforced cements.

Under some conditions metals such as aluminum or magnesium may be better vehicles for carrying fibers, and hence for making a useful fiber composite, than polymer matrices. These metals have been reinforced with a large number of fibers because, as a matrix, they show advantages over all thermosetting resins and over some thermoplastic resins: they have high thermal conductivity, little hydrothermal degradation and dimensional stability, and they are not susceptible to radiation damage or low-temperature brittleness. Their higher melting point can be an advantage. Some metals, e.g., tungsten, are extremely stiff (tungsten has a Young's modulus of 350 GPa) and possess very high melting points besides being well known as fibers. If they can be protected from oxidation, they offer the possibility of being used as a reinforcement of another metal (e.g., nickel-base alloy), or of a ceramic for high-temperature use.

The idea of making a composite for engineering use has arisen in recent years from the desire to utilize very stiff (and strong) fibers. The fibers cannot be used alone and therefore need a matrix. The need to manufacture the two component materials separately and then combine them could be avoided if fibers were made in situ in a matrix. Here ingenuity is at a premium since the resulting composite, if it has adequate properties, is likely to be more cheaply produced than if each component is made separately. It may well also be that the two components in the final product are so chemically related that the stability of each in the presence of the other is ensured. Processes (in situ composites) used with metals and inorganic materials include plastic deformation, eutectic solidification, precipitation and the recently discovered method of great promise, melt oxidation. The latter is used to form ceramic fibers or particles either alone or within a metal matrix. The corresponding processes in organic polymer systems are based on copolymerization and extrusion.

Although the great majority of composites referred to are advanced composites, the increasing use of stiff fibers and the interest in these enhances interest in particulate composites where a stiff and thermally resistant phase of other than the fibrous form is used in the composite. Some of these composites have been known for years and have found use as cutting tools, wear resistance parts, and so forth. They are described in the article *Particulate Composites*.

6. *Properties of Composites*

The science of the physical properties of composite materials relates the properties of the composite to those of the individual constituents and of the interface between them. Most of the simple physical properties such as density, elastic modulus and thermal expansion coefficient can be calculated. The elastic properties of laminates which consist of thin layers (lamellae) stacked upon one another, within each layer the fibers being parallel, is an important topic for design. Another form of arrangement of nonparallel fibers is that of the woven fabric.

An engineer may wish to choose a particular arrangement of fibers of given properties within a given matrix in order to attain or to approach a particular set of properties within the composite. For the simpler properties, where there is confidence in the prediction of a property from consideration of the properties of the constituents, such as for elastic modulus or for thermal expansion coefficient, attainable combinations of properties can be displayed graphically, leading to the concept of a structure-performance map.

Properties such as strength and toughness of composite materials are not as well understood as the simpler elastic properties, because in many cases the modes of failure under a given system of external load are not predictable in advance. An added complication is the fact that the breaking strength of individual fibers within a population shows a spread of values. This variation is described generally by what are called Weibull statistics. Since the breaking strength of a composite will depend not only on the highest breaking strength of the individual fibers but also on how the individual breaks are arranged in space within the composite, the accurate prediction of the breaking strength requires a complicated statistical theory even when time-dependent effects are ignored. Despite the complications of arriving at an exact prediction of the strength, advanced composites are used nonetheless and their strength measured. Because of the anisotropy of elasticity and strength these quantities must be measured in special ways. Methods of achieving reliable measurements of strength and what simple rules there are for variation of strength with orientation, fiber length, and so on, are dealt with in the article *Composite Materials: Strength*.

In engineering practice, the static strength of a composite in the presence of notches and in areas where stress is concentrated due to geometric form is at least as important as that of simple test pieces used in the laboratory. In practice, composite materials must be joined to one another and to other structural members. Here every effort is made to minimize stress-concentrating effects such as abrupt changes of elastic moduli and of geometrical section or form.

The ability of a material to remain serviceable when containing cracks, or when cracks arise during service, depends upon its toughness. It is by no means clear that measures of toughness or crack resistance, such as the concept of a critical stress intensity factor derived from linearly elastic mechanics are applicable to composite materials. Some progress can be made in elucidating the primary mechanisms responsible for the energy required to break a composite material. These are the sliding friction between fibers and matrix if broken fibers pull past one another, the deformation within the fiber and matrix, and the formation of subsidiary cracks. These energy-dissipating mechanisms depend on the variables of volume fraction, the diameter of the fibers, etc., and so representative diagrams indicating the relative importance of these can be constructed.

This description of the mechanical properties of composite materials has so far made no mention of time-dependent properties; in particular, strength has been described almost as a static property. Strength under oscillating stress and variation of deformation of the composite with time are aspects of a general investigative analysis of the failure of composite materials which has generally been called damage mechanics. It has been shown that emphasis on the strain range to which a given load subjects a composite can simplify greatly the interpretation of the results. During cyclic testing of a composite material, cracks in one of the constituents occur usually within the matrix. The same occurs in a static test where it is called multiple fracture. It is necessary then to understand the physical properties of a material containing a large number of cracks.

The creep deformation, though not the creep rupture, of aligned fibrous composites with continuous fibers stressed parallel to these cracks is, on the other hand, relatively well understood and can be modelled in terms of the constitutive equations of deformation for the fibers and matrix deforming in parallel. Discontinuous-fiber composites can also be dealt with to some extent and hence deformation normal to the fibers can be at least partly understood.

The long-term stability of composite materials under load depends on chemical factors, oxidative stability of the two components, and so on, particularly at high temperatures. For most composites based on resin matrices, water absorption at room and slightly elevated temperatures is of great importance. Epoxy resins, for example, absorb water and this alters their glass transition temperature. If the fibers are susceptible to water attack, as glass is, the ingress of water leads to attack on the reinforcing fibers.

Before composite materials enter service, and while they are in engineering service, their properties must be evaluated in a nondestructive manner. The principal objective of nondestructive evaluation is to provide assurance on the quality and structural integrity of a particular component. This can be achieved directly by using a nondestructive evaluation (NDE) technique or indirectly by monitoring or controlling the fabrication process. The latter is really only an extension of the usual procedures of process control necessary to achieve a consistent product. But NDE can also assist in optimizing fabrication procedure by, for example, monitoring the local state of cure. Since composite materials are essentially arrays of fibers assembled into place, the possibility arises of using the fibers themselves as monitors of their performance in service or alternatively doing this by incorporation of other types of sensors. The monitoring of fracture within a glass fiber or of the failure of resin adhesion to the surface of a fiber of glass is quite possible using visible light, and other fibers may be interrogated using other wavelengths of radiation.

These ideas raise intriguing possibilities of monitoring performances in situ by NDE so that the material itself becomes perceptive and can communicate "how it feels." These ideas may be of great importance for the use of advanced materials generally, not just for composites (Kelly 1988).

7. Engineering Applications

It is already apparent that the applications of composite materials, even if mention is not made of those based on naturally occurring composites such as wood, are very wide, ranging from a simple GRP tank or boat hull to a sophisticated aeroplane wing aeroelastically designed, or to another primary aircraft structure such as a helicopter rotor-blade. Glass reinforced plastics are used in the building industry, in marine applications in boats and minehunters, in building, transport and the electrical industry.

In many applications of composite materials it has been corrosion resistance which has determined the success in substituting for a traditional material. It is mainly in leisure goods, medical materials and in aerospace, where performance requirements overcome the considerations of cost, that the newer advanced composites are gaining ground. Consideration of the manufacturing methods will often dictate which composite is to be used, because an important principle of composite materials is that almost any desired combination of physical properties can, within certain limits, be obtained.

The manufacturing methods, of course, depend upon the feedstock, and again the GRP industry has developed most of the methods employed. Glass is available in many forms, as rovings, chopped fibers or fabrics. The fibers may be laid up and then impregnated with resin, or guns may be used to spray liquid resin or fibers onto a suitably shaped core. Alternatively, preimpregnated arrays of fibers (prepregs) are used which in their simplest form consist of parallel tows, rovings or aligned individual fibers spread out to produce a uniform distribution of fibers within the thickness of the sheet, which is impregnated with resin to produce a material of controlled volume fraction. The resin is partially cured (B-staged) to a slightly tacky condition. Sometimes this must be refrigerated until used. Sheet, dough and bulk molding compounds are variants of this, generally using discontinuous fibers.

All manufacturing methods aim to avoid the entrapment of air or the formation of voids since these represent gross defects. Articles may be made by hand or spray placement, by press molding using heated matched male/female tools under pressures of 3–7 MPa, or by vacuum molding using a flexible membrane to obviate the need for a press. If higher density and lower void content are required, molding is done in an autoclave using pressures of, say, 1–2 MPa at an elevated temperature, $\sim 100\,^{\circ}\mathrm{C}$ or so.

Alternatively, fibers may be enclosed in a mold and the resin injected, the resin being precatalyzed so that it cures. In reaction injection molding, two fast-reacting components (usually based on urethanes) of initial low viscosity are pumped into the mold. This method gives low cycle times (1–2 minutes).

In the process of pultrusion, continuous fibers are passed through a bath of resin and the impregnated fibers passed through a heated die so that resin cures. This process is good for the production of rod, sheet, tube and bar forms. The most accurate positioning of fibers necessary for some high-performance applications is attained by filament winding, in which fibers impregnated with resin are wrapped onto a former mandrel which is withdrawn after the resin is cured. Modern filament winding machinery can produce complicated nonaxisymmetric shapes as a result of the application of robotics. The process is admirably suited to computer control and this will become more widespread in the future.

The extrusion process of reinforced polymer systems is essentially a form of polymer processing involving the incorporation of discontinuous reinforcing fibers such as chopped glass-fiber or natural cellulose.

All of these more conventional methods are used with high-performance composites but the increasing interest, and some use, of higher melting point thermoplastic matrices (such as polyether ether ketone) containing carbon fibers, is leading to the development of forming methods in the solid state by pressing, drawing and extrusion comparable to conventional metal-forming methods.

Bibliography

Bunsell A R (ed.) 1988 *Fibre Reinforcements for Composite Materials*. Elsevier, Amsterdam

Christensen R M 1979 *Mechanics of Composite Materials*. Wiley, Interscience, New York

Dhingra A K, Lauterbach H G 1986 Fibers, engineering. In: *Encyclopedia of Polymer Science and Engineering*, Vol. 6, 2nd edn. Wiley, New York, pp. 756–802

Gerstle F P Jr 1985 Composites. In: *Encyclopedia of Polymer Science and Engineering*, Vol. 3, 2nd edn. Wiley, New York, pp. 776–820

Hale D K 1976 The physical properties of composite materials. *J. Mater. Sci.* 11: 2105–41

Hannant D J 1978 *Fiber Cements and Concretes*. Wiley, New York

Hashin Z, Shtrikman S 1962 A variational approach to the theory of the effective magnetic permeability of multiphase materials. *J. Appl. Phys.* 33: 3125–31

Hull D 1981 *An Introduction to Composite Materials*. Cambridge University Press, Cambridge

Kelly A 1988 Advanced new materials; substitution via enhanced mechanical properties. *Advancing with Composites, International Conference on Composite Materials*. CUEN, Naples, pp. 15–23

Kelly A, Macmillan N H 1986 *Strong Solids*, 3rd edn. Clarendon, Oxford

Kelly A, Rabotnov Y N (eds.) 1983 *Handbook of Composites*, Vols. 1–4. North Holland, Amsterdam

Maddock B J 1969 Superconductive composites. *Composites* 1: 104–11
Newnham R E 1989 Nonmechanical properties of composites. In: Kelly A (ed.) 1989 *Concise Encyclopedia of Composite Materials*. Pergamon, Oxford.
Suchtelen J van 1972 Product properties: a new application of composite materials. *Phillips Res. Rep.* 27: 28–37
Suchtelen J van 1980 Non-structural applications of composite materials. *Ann. Chim. Fr.* 5: 139–53

A. Kelly
[University of Surrey, Guildford, UK]

Composite Materials: Strength

Composite materials made from aligned fibers embedded in a plastic matrix are orthotropic; that is, their properties vary symmetrically about orthogonal axes. The characterization of strength therefore requires the measurement of a minimum of five independent strength parameters. Assuming transverse isotropy, the principal parameters are the tensile and compressive strengths measured parallel to, and at right angles to, the fibers and the shear strength (measured parallel to the fibers). The longitudinal tensile strength (i.e., the tensile strength parallel to the fibers) is determined largely by the strength properties of the fibers themselves whereas the longitudinal compressive strength is generally governed by fiber stability. The transverse and shear strengths are limited by the relatively low strength properties of the matrix and fiber–matrix bond so they are usually an order of magnitude lower than the longitudinal properties.

In practical applications, the relatively poor transverse and in-plane shear properties can be overcome by the use of multidirectional laminated composites. The strength properties of multidirectional laminates can be determined from those of the unidirectional plies using models based on elastic lamination theory. Relatively simple models will often give results which are sufficiently accurate for initial design, but for high accuracy, effects such as material nonlinearity, the development of noncatastrophic damage and the presence of hygrothermal strains must be included. In practice, therefore, it is often more convenient to make direct measurements of the strength of multidirectional composites.

1. The Effect of Fiber Orientation on Strength

The tensile strength of a unidirectional composite loaded at an angle to the fiber direction may be related to the longitudinal and transverse tensile strengths (σ_{1T} and σ_{2T}) and in-plane shear strength (τ_{12}) as illustrated in Fig. 1. The failure mode, and hence the controlling strength parameter, depends on the angle between the fiber and loading axes. For compressive loading the effect of fiber orientation follows a similar pattern, the strength and failure mode being governed by the longitudinal and transverse compressive strengths and the shear strength.

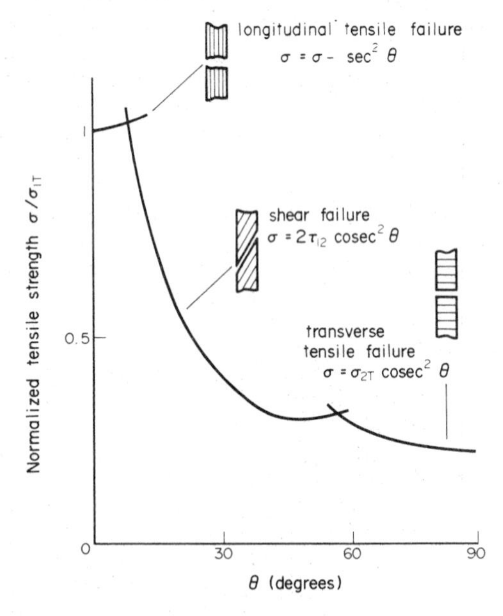

Figure 1
The effect on tensile strength of fiber orientation

This description of the effect of fiber orientation is based on the theory of maximum principal stresses which makes the reasonable assumption that failure in any one mode is not influenced by the magnitude of the stress in any other mode. The corresponding theory of maximum principal strains, which is based on similar assumptions concerning the independence of principal strains, results in slightly different curves due to Poisson's ratio effects. Both of these theories have the advantage of simplicity, but they indicate abrupt changes in the relationship between strength and fiber orientation at the points of transition between failure modes. Failed specimens corresponding to the transition points exhibit features of both modes suggesting that they are not entirely independent. Moreover, since a single continuous curve is generally preferred for analysis and computation, a variety of interaction theories have been proposed. Most of these theories are modifications of the various yield criteria developed for homogeneous materials and are therefore used purely empirically. The differences between the various strength predictions are often relatively small compared to the scatter in experimental data so that many of these theories have been used successfully and none is universally preferred.

2. Variation of Strength with Fiber Volume Fraction

The variation with fiber volume fraction of the longitudinal tensile and compressive strength and the in-

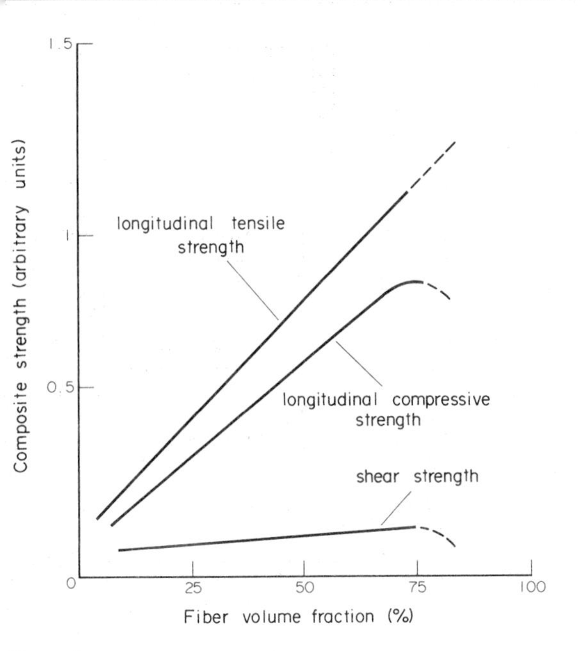

Figure 2
The variation of strength with fiber volume fraction

plane shear strength of a typical fiber reinforced plastic is illustrated in Fig. 2. Over much of the range (including the great majority of practical composites) the strength parameters vary in direct proportion to the fiber volume fraction. Thus, the longitudinal tensile strength, for example, may be defined by the equation

$$\sigma_{1T} = \sigma_f V_f + \sigma_m(1 - V_f)$$

where V_f is the fiber volume fraction, σ_f is the effective fiber strength and σ_m is the stress on the matrix at the time of failure of the fibers. The value of σ_f depends on the complex relationships between the properties of both fibers and matrix and, for engineering purposes, it is usually deduced from the measured strength of the composite at known fiber volume fraction. The failure strains of the fibers and matrix are rarely equal so that, for brittle fibers in an elastic matrix, the value of σ_m is given by $\sigma_f E_m/E_f$, where E_m and E_f are the matrix and fiber moduli, respectively. Alternatively, the yield strength may be more appropriate for a matrix capable of yielding. In many practical composites the contribution of the matrix is comparatively small so that the expression

$$\sigma_{1T} \simeq \sigma_f V_f$$

may be used with little error.

The upper limit to composite tensile strength is determined largely by the maximum fiber volume fraction which may be achieved without loss of integrity due to inadequate continuity of the matrix phase. This will depend on the cross-sectional shape and degree of alignment of the fibers but, in general, fiber volume fractions in excess of about 70% are rarely achieved without excessive fiber damage.

In longitudinal compression, failure can occur by shear across both fibers and matrix but, in general, it occurs by instability of the fibers. In either case, compressive strength varies approximately in proportion to fiber volume fraction and may be described by equations similar to those for tensile strength above. However, the loss of composite integrity at high fiber volume fraction has a more profound effect on compressive strength, since it permits premature instability of the fibers. The transverse and shear strength properties depend primarily on the strength of the matrix and fiber–matrix interface and exhibit only minor increases in strength with fiber volume fraction. These properties are also limited by the loss of composite integrity at higher fiber volume fractions.

3. *Other Factors Affecting Composite Strength*

In composites containing aligned discontinuous fibers, the stress in each fiber is not uniform along its length but builds up from either end. The tensile strength of such composites depends on the relative magnitudes of the mean fiber length and the critical fiber length, which may be defined as the minimum length which will allow the development of sufficient stress to fail the fiber at its midpoint. For advanced fiber reinforced plastics the critical fiber length could be as much as 50 fiber diameters, and only composites containing fibers very much longer than this will exhibit strengths approaching that of the corresponding continuous fiber composite. As fiber length is reduced, the reinforcement efficiency decreases until, when fiber length is less than the critical length, failure will no longer involve fracture of the fibers and will be governed by fiber pull-out.

Poor fiber alignment also substantially reduces reinforcement efficiency in discontinuous fiber composites. By examination of the effects of fiber orientation for continuous fibers, illustrated in Fig. 1, it is evident that the contribution to composite strength from fibers more than a few degrees off-axis is greatly reduced. This effect together with the effect of finite fiber length means that, although reinforcement efficiencies in excess of 90% have been observed, values of 50% to 70% are perhaps more typical of aligned discontinuous-fiber composites.

The effects on compressive strength of both fiber length and fiber alignment are broadly similar to the effects on tensile strength. However, while fiber length will have virtually no effect on transverse or shear properties, poor fiber alignment will tend to increase the transverse tensile and compressive strengths and also the shear strength.

A particularly important factor affecting the strength of both continuous- and discontinuous-fiber composites is the occurrence of microstructural defects and, in particular, the occurrence of voids. In fiber-reinforced plastics, a small fraction of voids can have a quite disproportionate effect on those strength properties which depend on the integrity of the matrix phase. As an example, Fig. 3 shows the effect of matrix void content on the compressive strength and shear strength of a typical glass-fiber-reinforced plastic. In practice, void content in fiber-reinforced plastics may be limited to less than about 5% by quite simple manufacturing techniques, and the more sophisticated methods such as those used in the production of aerospace components enable void content to be reduced to 0.5–1%. However, even at these low levels, the effects of void content are not negligible and can contribute significantly to material strength variability.

Finally, it is important to note that environmental effects, particularly temperature and moisture absorption in fiber-reinforced plastics, can substantially affect composite strength. Detailed discussion of these effects is beyond the scope of this article but since both moisture and temperature tend to soften plastic matrices and may, in some systems, degrade the fiber–matrix bond the effects are most marked on compressive and shear strength properties.

4. Strength of Multidirectional Laminates

Unidirectional composites are often too anisotropic for practical applications, but this may be overcome by the use of laminated materials in which the fiber plies are oriented in two or more prescribed directions. Such laminates can be made most efficient in terms of strength-to-weight ratio if the applied loads are carried by direct stresses in the fibers. Thus, a proportion of the plies should lie parallel to each of the principal tensile and compressive loads whereas the remainder should be oriented at $\pm 45°$ to carry the shear loads.

Varying the orientation and stacking sequence of the plies gives an infinite family of laminates. Many of these laminates will exhibit phenomena not observed in isotropic materials including, for example, coupling between bending, stretching and twisting deformations. However, there is rarely any particular advantage in using more than four fiber directions and except for a few specific applications in which coupling effects can be exploited to advantage, structural laminates will generally contain only 0°, 90° and $\pm 45°$ plies assembled in sequences which eliminate coupling effects and exhibit orthotropic behavior, that is, the in-plane properties are symmetric about axes at right angles. The strength properties of such laminates clearly depend on the proportions of fiber in each direction and Fig. 4 shows a typical carpet plot of tensile strength in the 0° fiber direction for a carbon fiber composite. It may be seen that the most important factor is the proportion of fiber lying parallel to the applied load. Except for very low proportions of 0° fiber, the relative proportions of $\pm 45°$ and 90° fiber have very little effect on tensile strength. Compressive strength varies with layup in a manner which is generally similar but, because the ply stacking sequence can affect the stability of the load bearing 0° plies, the use of carpet plots such as that shown below for tensile strength should be treated with caution. The in-plane shear strength of the 0°, 90°, $\pm 45°$ family of laminates is unaffected by the relative proportions of 0° and 90° plies but increases linearly with the proportion of $\pm 45°$ plies as shown in Fig. 5.

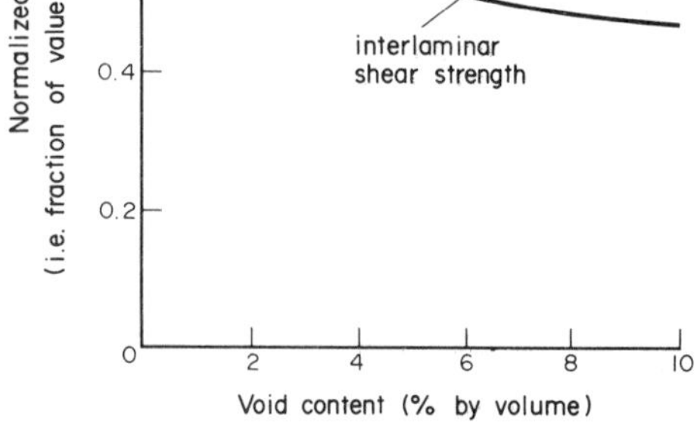

Figure 3
The effect of voids on composite strength

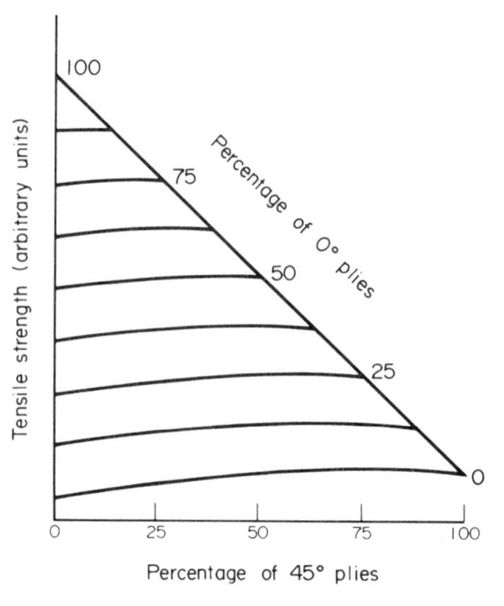

Figure 4
Tensile strength of $0°/\pm 45°/90°$ family of laminates

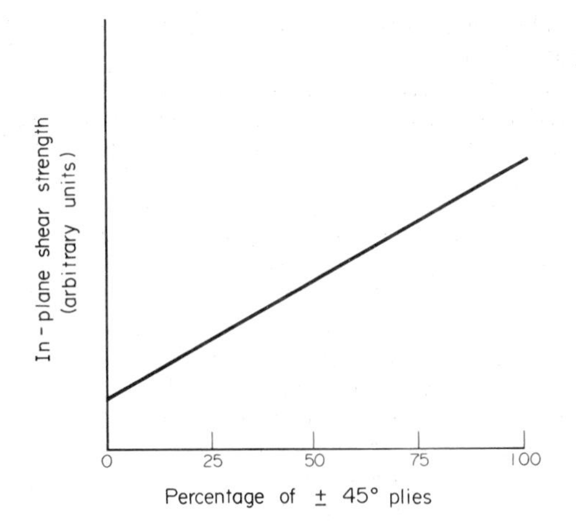

Figure 5
In-plane shear strength of 0°/±45°/90° family of laminates: (a) tensile test specimen; (b) compressive test specimen; (c) rail shear test specimen; (d) short-beam shear test specimen

The most simple model for the prediction of strength properties of multidirectional laminates makes the assumption that the transverse and shear load-carrying capacity of a unidirectional ply is negligible. Thus, for example, the tensile strength along a given axis is given by the product of the longitudinal tensile strength of the unidirectional material and the fraction of fiber plies which are parallel to that axis. By resolving in-plane shear stress into tensile and compressive stresses in the ±45° plies, the shear strength may also be expressed as a simple function of the unidirectional strength properties. By comparison with Figs. 4 and 5, it may be seen that this approach would give conservative strength values of varying accuracy. Nevertheless, the extreme simplicity of such a model makes it particularly attractive for initial design and optimization of composite laminates for specific applications.

More accurate analytical models of the strength of multidirectional laminates are based on the constitutive relationships used to describe the elastic behavior of a unidirectional ply, and to relate that behavior to the elastic properties of multidirectional laminates on the basis of strain compatibility between the plies (see, for example, Ashton et al. 1969). To determine a particular strength property, the first step is to calculate the laminate strains for unit applied load and thence to calculate the longitudinal, transverse and shear stresses and strains in each of the plies. The margin of strength for each mode of loading in each ply may then be calculated, whence it is possible to determine the minimum level of applied load at which one of the plies will fail.

At this point the remaining plies must absorb some additional load and this may be modelled by modification of the stiffness matrix. Recalculation of the ply stresses and strains will indicate whether any further failures will occur immediately or whether further load may be sustained. By an iterative procedure, a point will eventually be reached at which the laminate can carry no additional load.

When using models such as that described above, it is generally assumed that the strength properties of the unidirectional material may be applied to each of the plies within the laminate. This assumption is not always valid since, for example, the strain at which cracks form in 90° plies has been shown to depend on ply thickness. However, for most purposes this assumption is not unreasonable. The major differences between the various models of this type lie in the assumptions concerning the load carrying capability of a failed ply and hence the manner in which the stiffness matrix must be modified.

For accurate strength prediction it will often be necessary to include thermal- and moisture-induced strains into the calculations, and in some cases it may also be necessary to include material nonlinearity. This latter effect may be modelled most accurately by an iterative incremental procedure in which the stiffness properties are adjusted according to the nonlinear stress–strain relationships. Even relatively simple models of the strength of multidirectional laminates require considerable computation so that the great majority are computer based and iterative procedures do not present any particular difficulty.

5. *Measurement of Composite Strength*

The high degree of anisotropy and the wide range of failure mechanisms exhibited by fiber composites give rise to a number of novel problems in test specimen design. Moreover, as a result of differences in the relative magnitudes of the various properties, it is not possible to specify a unique set of specimens which will give satisfactory results for all fiber composites. Various standards have been proposed by, for example, the American Society for Testing and Materials (ASTM) and the British Standards Institution (BSI), and the use of such standards is to be encouraged. However, it is important to understand the principles governing the design of such specimens to ensure that the specimen and test method selected for a particular purpose will result in failure by a mechanism appropriate to the intended use of the data. The principles of design and the consequent form of typical specimens are outlined in the following paragraphs.

For unidirectional fiber composites, the very high ratio of longitudinal tensile or compressive strength to shear strength leads to two major problems in the measurement of longitudinal properties. First, since load must be transferred into the body of the specimen predominantly by shear, large shear areas are required at the end fittings to generate adequate longitudinal load. Moreover, any waisting necessary to ensure failure in a region of uniform stress must be very gradual to avoid unwanted shear failures. A further problem arises from the very high stiffness and generally linear-elastic behavior of these materials. Any slight bending of the specimen or misalignment of the load can give rise to significant bending stresses and lead to premature failure. The transverse failing strain is often even lower than the longitudinal failing strain so that sensitivity to bending effects is even greater in the measurement of transverse strength properties.

The problems mentioned above may be largely overcome using specimens of the general form shown in Figs. 6a,b which are based on specimens designed for carbon-fiber-reinforced plastics. For tensile specimens, load is applied by friction grips bearing on thin aluminum alloy or glass-fiber-reinforced plastic tabs bonded to the specimen. The specimen is waisted in thickness which allows a significant reduction in cross-sectional area to be achieved within a reasonable length while meeting the requirement for a shallow waisting curve. This also reduces the bending stiffness

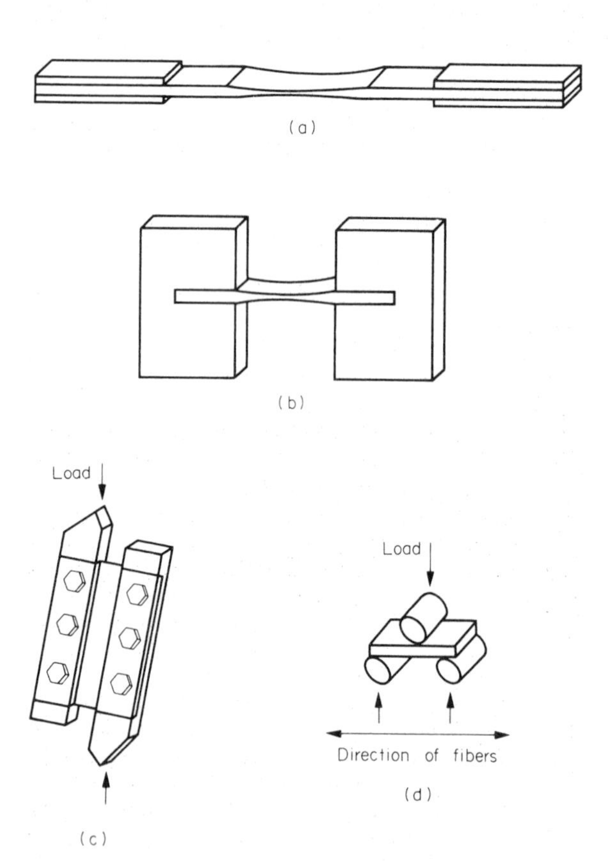

Figure 6
General configuration of test specimens for unidirectional fiber reinforced plastics

in the plane in which test machine misalignment is most likely.

For compression testing, Euler instability is a major problem and the need to shorten the free length conflicts with the need for a shallow waisting curve. For low-modulus materials, fixtures to prevent buckling and to provide rotational restraint of the ends are necessary. However, for high-modulus fibers such as carbon it may be possible to manage without antibuckling guides if the end tabs are replaced by slotted aluminum blocks (Fig. 6b) which provide an encastré end condition. A further difficulty results from the fact that any restraint of Poisson expansion by the end fittings can lead to artificially high results for very short specimens. Transverse compression specimens require less waisting since transverse isotropy leads to lower stress concentrations at the end fittings. Thus, in spite of the low transverse stiffness, stability may often be achieved using a short specimen with slotted end blocks.

The main problem associated with the measurement of in-plane shear strength of unidirectional composites is that of ensuring failure in a region of pure shear. Applying torsional load to a thin-walled tube is the most effective technique, but for flat laminates some other technique is required. The rail shear test (Fig. 6c) is generally preferred for unidirectional fiber composites. Although not entirely free from stress concentrations, it provides a reasonable measure of shear strength if the fibers are at right angles to the loading rails and can therefore sustain the direct loads occurring along the short free edges. Alternatively, the shear properties of the unidirectional material can be derived from a tensile test on a $\pm 45°$ laminate using a specimen similar to that described for multidirectional laminates below. It may easily be shown that the shear strength is equal to half the tensile strength of the $\pm 45°$ specimen.

The most simple method of evaluating shear strength is the short beam shear test (Fig. 6d). This test actually measures interlaminar shear strength near the midplane of the laminate and since there are also direct stresses due to the bending of the beam and stress concentrations due to the loading rollers this test is far from ideal. Nevertheless, its great simplicity makes it an attractive test method which finds widespread use, particularly for quality control purposes.

The measurement of tensile and compressive strength of multidirectional laminates is usually effected using unwaisted specimens as shown in Fig. 7a. The fact that some of the off-axis fibers run into the end fitting tends to have a local reinforcing effect and the majority of failures occur near the center of the specimen. For specimens containing plies at $\pm\theta°$ the length to width ratio (L/W) should be chosen to ensure that no $\theta°$ fibers run from one end fitting to the other. This may not be possible for very small values of θ and results from such specimens should be treated with caution. For compressive loading, a wide range of

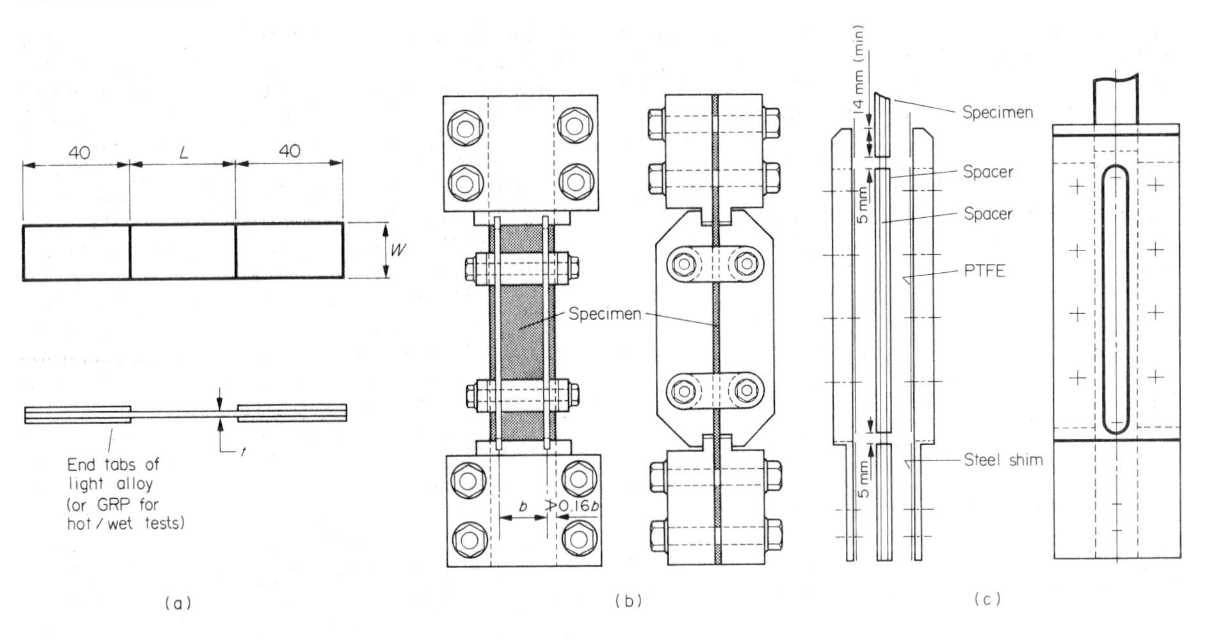

Figure 7
Multidirectional test specimen and typical antibuckling guides

antibuckling guides has been successfully used and typical examples are illustrated in Figs. 7b,c. Measurement of in-plane shear strength should be carried out by torsion of a thin-walled tube. However, since multidirectional tubular specimens are expensive and difficult to manufacture, the rail shear test has been used even though the stress concentrations in the vicinity of the short free edges vary with fiber layup.

In summary, it must be recognized that fiber composite materials exhibit a much wider range of failure mechanisms than isotropic materials. The specimen design principles outlined above are based on the need to initiate particular failure mechanisms in regions of uniform stress. It is clearly important to examine all failed specimens to ensure that failure occurred at the correct location and through a mechanism appropriate to the intended use of the data.

6. *Acknowledgement*

This article is Crown Copyright and is reproduced with the permission of the Controller of Her Majesty's Stationery Office.

Bibliography

American Society for Testing and Materials 1988 *Annual Book of ASTM Standards.* ASTM, Philadelphia, Pennsylvania

Ashton J E, Halpin J C, Petit P H 1969 *Primer on Composite Materials: Analysis.* Technomic, Stamford, Connecticut

British Standards Institution 1987 *Catalogue.* BSI, Milton Keynes

Ewins P D 1974 Techniques for measuring the mechanical properties of composite materials. In: *Composites: Standards, Testing and Design Conf. Proc.* IPC Science and Technology Press, Guildford

Holister G S, Thomas C 1966 *Fibre Reinforced Materials.* Elsevier, London

Jones R M 1975 *Mechanics of Composite Materials.* McGraw-Hill, New York

Kelly A, Macmillan N H 1986 *Strong Solids*, 3rd edn. Clarendon, Oxford

Sih G C, Skudra A M 1986 Failure mechanics of composites. In: Kelly A, Rabotnov Y N (eds.) 1986 *Handbook of Composites*, Vol. 3. Elsevier, Amsterdam

R. T. Potter
[Royal Aircraft Establishment, Farnborough, UK]

Concrete as a Building Material

Portland cement concrete is used in many kinds of construction such as hydraulic structures, transportation facilities, and residential, commercial and industrial buildings. It thus competes directly and successfully with all other major construction materials (see Table 1). Concrete consists of aggregates (sand and gravel or crushed rock) held together by a matrix of hardened cement paste (the binder), which is formed by reacting Portland cement with water.

Table 1
Typical properties of construction materials

Material	Density ($kg\,m^{-3}$)	Tensile strength (MPa)	Weight–strength ratio ($MPa\,kg^{-1}$)	Elastic modulus (GPa)	Elongation at failure (%)	Energy requirement ($GJ\,m^{-3}$)	Annual volume used in US construction ($10^6\,m^3$)
Aluminum alloy	2800	300	9	70	~20	360	~0.3
Mild steel	7800	300	26	210	25	300	
High-strength steel	7800	1000	8	210	10		6
Glass	2500	60	42	65	0.1	50	
Softwood[a]	350	~100	3.5	~12	~1.0	170	170
Plastic (polystyrene)	1000	~50	20	~3	2.0		
Rock	2600	10–25	100–260	35–70	0.1	3	0.3
Concrete	2300	~3	~770	~30	0.1	3	200

[a] Clear wood, parallel to grain

As compared with other construction materials (Table 1), concrete shows relatively poor performance: it is weak and brittle in tension, has a high weight–strength ratio and shows volume instabilities under load. It is, however, economical in terms of both money and energy. Concrete is made from locally obtained aggregates and cements that are manufactured within 150–300 km of most construction sites. Although cement manufacture has a high energy demand, cement is only a small fraction of the concrete. A variety of complex shapes and surface finishes can be cast on-site or fabricated under controlled conditions. Concrete is also fire resistant and provides energy-efficient buildings because of its thermal properties; it does not require protective coatings for long service lifetime.

1. Constituent Materials

Cements used in concrete are discussed elsewhere (see *Cement as a Building Material*). Aggregates make up 75–85% of concrete (by weight or volume) and hence are important constituents of concrete. They are well-graded gravels or crushed rock with sizes in the range 20–40 mm to as small as 0.15 mm. It is convenient to separate aggregates into coarse and fine (sand), with an arbitrary division at 4.75 mm (No. 4 sieve). No particular geological type of rock is favored over any other; either natural gravels and sands or crushed rock and manufactured sands may be used. Aggregates should be strong and dense (low porosity) and have a uniform shape; rocks that tend to cleave in certain directions should be avoided.

Most available water supplies are suitable for making concrete; those to avoid are waters rich in dissolved salts (e.g., seawater), waste waters with a high content of soluble organic materials, and turbid waters. If there is any doubt about the suitability of water, or if there is no alternative source, it should be tested in a concrete sample before use.

Other materials added to concrete are collectively called admixtures, the most common categories being set-modifying admixtures, water-reducing agents, air-entraining agents and mineral admixtures. The first three categories are water-soluble materials, which are added in relatively small quantities (< 1% by weight of the cement). Set-modifying admixtures are used to decrease or extend the time of setting (and early strength development), water-reducing agents reduce the amount of water required to make workable concrete, and air-entraining agents incorporate a finely dispersed air void system into the concrete to improve its frost resistance.

Mineral admixtures are solids that are usually added in much greater quantities (> 10% by weight of the cement). The most important of these are pozzolans and hydraulic slags. Pozzolans are siliceous or aluminosiliceous materials that are often used to replace part of the cement, to lower the heat of hydration, or to improve durability. Reactive silica in pozzolans reacts with the calcium hydroxide in the cement paste to form additional calcium silicate hydrate. Pozzolans can be natural materials (e.g., diatomaceous earth or volcanic tuffs) or waste materials (e.g., fly ash). Hydraulic slags are generally quenched glassy slags from the steel industry. In the presence of Portland cement the slags are "activated" and react with water to form cementitious products. Thus, slags can also be used to replace part or even most of the cement, in order to lower the heat of hydration or improve durability.

2. Mix Design, Handling and Curing

Concrete must be made to suit the particular application for which it is intended. Three overall requirements must be met: (a) the workability of the fresh concrete; (b) the strength of the hardened concrete; and (c) the durability of the hardened concrete.

The first criterion is important because concrete is usually cast on-site into temporary molds (formwork). The ease with which concrete can be handled during casting is known as its workability. The workability depends on a number of factors, the most important being the shape and texture of the coarse aggregate, and the amount and fineness of the sand. Different degrees of workability can be selected by varying the amount of paste used. On the construction site workability is generally measured using the slump test, although other tests can be used. The second requirement is determined primarily by the water–cement weight ratio (w/c). This controls the porosity of the cement paste and hence its strength. The third criterion is also satisfied by w/c, although certain precautions must also be observed. For example, an air-entrainment agent must be used when resistance to freezing and thawing is needed, and a sulfate-resistant cement should be used when the concrete will be in contact with sulfates.

The essence of any mix design procedure is to satisfy all three criteria simultaneously by a suitable selection of w/c, cement content (hence paste content) and weight ratio of sand to gravel. Several recognized procedures are available which can be modified to suit specific circumstances. In the USA the American Concrete Institute's method is widely used (ACI 1977b). Once the proportions have been selected, trial batches of concrete are mixed and tested to ensure compliance with all criteria and specifications (Mindess and Young 1981).

On most modern construction jobs concrete is mixed at a central batch plant by a concrete supplier (ready-mixed concrete). During transportation the concrete is gently agitated to prevent segregation of the constituents, until it is placed in formwork at the job site and consolidated to form void-free masses. Large volumes of concrete can be conveniently cast in this manner. Concrete can be transported in a variety of ways: crane and bucket, buggies, chutes or conveyors, or pipelines (pumping). The method chosen will depend on the layout of the job site and the relative economics. In the forms, concrete is usually consolidated by high-frequency vibrators. Concrete behaves like a thixotropic liquid with a definite yield stress; under vibration the yield stress is exceeded and the concrete flows readily, thereby filling the forms and expelling air voids. During placement and consolidation, excessive segregation of the constituents should not occur, but water, the lightest constituent, always separates to some degree, a phenomenon known as bleeding.

Once in place, concrete must be kept moist to allow hydration to occur and strength to develop. The length of this curing period depends on the strength required and the ambient temperature; concrete should be kept in the forms until it is strong enough to carry existing loads. Usually structural concrete should be cured for seven days or until it reaches 70% of its specified strength. Longer times should be used if the temperature is below 20 °C or if cements with low rates of strength gain are used. Concretes can be steam-cured to obtain rapid strength gain.

3. *Properties of Hardened Concrete*

3.1 Strength and Fracture

The compressive strength of concrete is normally in the range 15–35 MPa, but concretes have been produced with strengths as high as 85 MPa commercially and over 200 MPa in the laboratory. These high strengths are obtained by specifying a low w/c to give low-porosity, high-strength pastes, and also by using strong, low-porosity aggregates. Concrete is a brittle material with low tensile strength and it fails catastrophically in tension; the tensile strain at failure is only 0.1%. In compression, concrete is only slightly more ductile, with the strain at failure about 0.3%. This value changes very little with strength (Fig. 1).

When concrete is loaded, the aggregates (relatively high elastic modulus) and the cement paste (low elastic modulus) respond to the load differently. Different amounts of strain in the two components cause a differential stress at the interface and hence premature bond failure under low applied stresses. At higher stresses cracks propagate through the paste but may be arrested by aggregate particles. Thus, concrete undergoes multiple cracking as it approaches failure, and a curvilinear stress–strain curve results. Less microcracking occurs in high-strength concrete, and consequently high-strength concrete will fail catastrophically in compression as well as in tension. Progressive microcracking causes concrete to fail when subjected to sustained loads greater than 75% of its ultimate strength. Under loading–unloading cycles (dynamic fatigue) failure will eventually occur at even

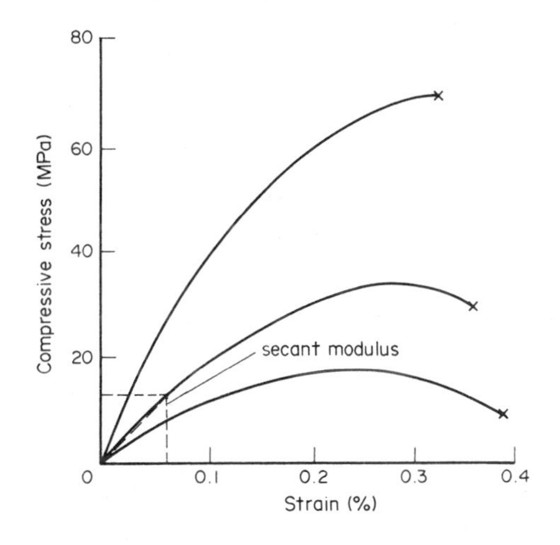

Figure 1
Stress–strain curves for concretes of different strength

lower stresses; typically 50% of the static strength after 10^6 cycles.

The parameter affecting concrete strength is the w/c of the paste. To estimate f_c (compressive strength), one of two formulae is used; either Abram's equation:

$$f_c = A/B(w/c)^{1.5} \quad \text{(MPa)} \tag{1}$$

where A (~ 100) and B (~ 4) are empirical constants, or Powers' equation:

$$f_c = A\left[\frac{0.68\alpha}{0.32\alpha + (w/c)}\right] \quad \text{(MPa)} \tag{2}$$

where A (~ 285) is an empirical constant and α is the fraction of cement that has hydrated (~ 0.85 after 28 days' moist curing). The effect of the aggregate on the strength of concrete is important only in lightweight or high-strength (>40 MPa) concretes. Air entrainment will lower strength because it increases porosity. In the preceding expressions it can be allowed for by replacing w/c by $(w+a)/c$, where a is the weight of water that would occupy the volume of entrained air in the concrete.

The strength of concrete in tension and flexure can be estimated from its compressive strength. The tensile strength is usually slightly less than 1/10 of the compressive strength; the flexural or shear strengths are slightly less than 1/5. The exact relationship depends on the size and shape of the aggregate, the age of the concrete, and the kind of test used. Typical values are given in Table 2. The elastic modulus of concrete can be calculated from the moduli of the aggregate and paste and their concentrations, using simple composite theory. The modulus E_p of cement paste depends on its w/c:

$$E_p = 32(1.30 - w/c)^3 \quad \text{(GPa)} \tag{3}$$

The secant modulus at 40% of ultimate stress is usually used.

As it dries, concrete shrinks (Table 2). If this movement is restrained, then tensile stresses that can cause cracking are developed. This will occur while concrete is still in the plastic state if excessive evaporation of moisture is not prevented. Hardened concrete will invariably crack on drying and control joints should be used to localize these cracks so that they can be sealed. Drying shrinkage is a paste phenomenon and is restrained by the aggregate:

$$\varepsilon_c = \varepsilon_p(1-A)^n \tag{4}$$

where ε_c, ε_p are the shrinkage strains of concrete and paste, respectively, A is the aggregate content, and n is a constant that depends on the aggregate stiffness. The shrinkage of paste is decreased by lowering the w/c, increasing the time of moist curing and raising the curing temperature. The rate of shrinkage is dependent on the surface–volume ratio of the concrete member and the ambient relative humidity—parameters that control the rate of moisture loss. Shrinkage strains can be estimated using empirical formulae that are provided in building codes.

Concrete also experiences stress-induced, time-dependent deformations that are called creep. Loading and drying usually occur simultaneously and are synergistic, that is, the total strain is greater than the sum of the individual creep and shrinkage strains. Creep is also a paste phenomenon and the factors that affect shrinkage affect creep in a similar manner. In addition, creep is proportional to the applied stress if this is not greater than 50% of the ultimate strength; specific creep (creep per unit stress) is used to compare the creep of different concretes. Formulae used in building codes to estimate creep also include drying shrinkage.

3.2 Other Properties

The thermal properties of concrete are given in Table 3. The coefficient of thermal expansion is not a unique value; it depends on the kind of aggregate used, as well as on w/c, age and moisture content. The thermal conductivity of concrete depends primarily on its density. A low-density, highly porous concrete has a low thermal conductivity, no matter whether the porosity resides in the aggregate or in the paste. The specific heat of concrete is independent of aggregate type, since most rocks have about the same specific heat. Concrete has a good fire resistance because the low thermal conductivity and high specific heat decrease the rate of temperature rise in the concrete.

Table 2
Typical properties of structural concrete

Classification	Typical w/c	Compressive strength (MPa)	Elastic modulus (GPa)	Tensile strength (MPa)	Flexural strength (MPa)	Ultimate specific creep (10^6 MPa^{-1})	Density (kg m^{-3})	Ultimate drying shrinkage (%)
Low strength	0.6	17	20	1.5	3	160	2300	0.08
Medium strength	0.5	35	28	3	6	100	2300	0.075
High strength	0.4	70	36	6	12	30	2300	0.07
Lightweight	$\sim$0.5	28	12	2.5	5	150	1800	0.1
Cement paste	0.3	140	32	12	25	$\sim$300	2100	0.4
Cement paste	0.6	70	14	6	12	$\sim$600	1800	0.8

Table 3
Thermal properties of concrete and its constituents

Classification	Coefficient of linear expansion ($10^{-6}\,K^{-1}$)	Thermal conductivity ($W\,m^{-1}\,K^{-1}$)	Specific heat ($J\,kg^{-1}\,K^{-1}$)	Thermal diffusivity ($10^{-6}\,m^2\,s^{-1}$)
Normal weight	7.5–13	1.5–3.5	800–1200	0.6–1.9
Lightweight	7.5–13	0.5–1.1	800–1200	0.3–0.6
Cement paste				
$w/c=0.4$	18–20	1.3	1450	0.4
$w/c=0.6$	18–20	1.0	1600	0.3
Granite	7–9	3.1	800	0.3
Quartzite	11–13	4.3		
Mild steel	11–12	120	460	0.3

Also, heat is consumed by evaporation of water. The high temperature gradient, combined with drying shrinkage, will cause surface spalling, but the bulk of the concrete will remain structurally sound if its temperature remains below 300 °C. If it is cooled slowly and not subjected to thermal cycles, concrete can be used at very low temperatures.

Concrete exhibits good attenuation of penetrating γ rays and neutron fluxes because of its high water content, and special aggregates can enhance these properties. Dry concrete is also a good insulator (resistivity about $10^{13}\,\Omega m$) but wet concrete is semi-conducting (resistivity $10^6\,\Omega\,m$).

3.3 Durability

High-quality concrete is very durable without the need for protective coatings. However, concrete is inherently susceptible to freezing and thawing and must be protected by air entrainment. Also, concrete is susceptible to several agents that cause deleterious expansions and cracking. Sulfates will attack the cement paste, chlorides will corrode the reinforcement, and alkali (sodium and potassium) salts will attack certain aggregates. The first two agents come from external sources and they can be repelled by lowering the permeability, which is controlled by w/c of the concrete. At $w/c=0.4$ the permeability of well compacted concrete is about $10^{-12}\,m\,s^{-1}$; this increases about 20 times at $w/c=0.6$. Using sulfate-resistant cements, or adding pozzolans or slags, will help control sulfate attack. Corrosion of reinforcing steel is enhanced by chlorides in deicing salts or seawater, but can be minimized in many cases by using adequate concrete cover. However, prestressed concrete should never be allowed to come in contact with chlorides because of the risk of stress corrosion.

Certain aggregates that contain reactive silica can react with the sodium and potassium present in Portland cements. Rocks containing volcanic silica glass or opaline quartz are most susceptible, and are found mainly in the southern and western USA. Such rocks should be avoided if at all possible, otherwise the use of low-alkali cements or pozzolans will help alleviate the problem.

4. Special Concretes

Although most concretes used in construction have properties close to those given in Table 1, special concretes with quite different properties can be made (see Table 4). For example, ultralightweight concretes are made primarily for insulating purposes, while heavyweight concretes are used in radiation shielding. Concrete can be combined with other materials to offset some of the deficiencies of plain concrete (Table 4). Fiber reinforcement can improve its post-cracking strength and ductility. The fibers, which are added at the mixer, arrest tensile cracks in the concrete matrix; thus, catastrophic failure is prevented since tensile stresses are transferred to the fibers. Steel fibers are used most widely, while in some countries glass fibers are replacing asbestos fibers. Plastic fibers (e.g., polypropylene) absorb large amounts of energy and increase resistance to impact loads.

Polymer latexes can also be added to fresh concrete. As the concrete dries, polymer films form which restrict tensile crack propagation and thereby improve ductility. The use of low w/c and the tendency for polymer films to line the surfaces of pores reduce permeability and improve durability. Hardened concrete can also be impregnated with a suitable monomer to fill the empty pore space. After polymerization in situ a strong but brittle composite with a compressive strength as high as 150 MPa is obtained. Polymer-impregnated concrete does not shrink or creep appreciably because moisture is removed, and it is very durable because the pore space is now filled with solid polymer. Surface impregnation of in-situ structures greatly improves their durability.

See also: Calcium Aluminate Cements; Cement as a Building Material; Cements, Specialty; Portland Cements, Blended Cements and Mortars

Table 4
Properties of special concretes

Classification	Compressive strength (MPa)	Elastic modulus (GPa)	Flexural strength (MPa)	Tensile strain at failure (%)	Relative abrasion resistance	Density (kg m^{-3})
Normal weight (structural)	35	28	6	0.1	1	2300
Heavyweight	35	28	6	0.1	1	>3200
Ultralightweight (insulating)	3	13	0.5	0.1	1	<750
Fiber-reinforced (steel fibers)	40	28	25	~1	1.5	2300
Latex-modified (acrylic)	30	~16	13	0.35	3.0	2300
Polymer-impregnated (methyl methacrylate)	135	~45	17	0.15	3.3	2400

Bibliography

American Concrete Institute 1977a *Polymers in Concrete*. American Concrete Institute, Detroit, Michigan
American Concrete Institute 1977b *Recommended Practice for Selecting Properties for Normal and Heavyweight Concrete*, ACI 211 1–77. American Concrete Institute, Detroit, Michigan
American Society for Testing and Materials 1980 *Significance of Tests and Properties of Concrete and Concrete-Making Materials*, STP 169B. American Society for Testing and Materials, Philadelphia, Pennsylvania
Biczok I 1967 *Concrete Corrosion and Concrete Protection*. Chemical Publishing, New York
Concrete Manual: A Manual for the Control of Concrete Construction, 8th edn., 1975. US Government Printing Office, Washington, DC
Hannant D J 1978 *Fiber Cements and Fiber Concretes*. Wiley, London
Illingworth J R 1972 *Movement and Distribution of Concrete*. McGraw-Hill, Maidenhead
Lea F M 1971 *The Chemistry of Cement and Concrete*, 3rd edn. Chemical Publishing, New York
Malhotra V M (ed.) 1980 *Progress in Concrete Technology*. Canadian Government Publishing Center, Hull, Quebec
Mindess S, Young J F 1981 *Concrete*. Prentice-Hall, Englewood Cliffs, New Jersey
Murdock L J, Brook K M 1979 *Concrete Materials and Practice*, 5th edn. Edward Arnold, London
Neville A M 1981 *Properties of Concrete*, 3rd edn. Pitman, London
Neville A M, Dilger W H, Brooks J 1983 *Creep of Plain and Structural Concrete*. Construction Press, Lancaster
Popovics S 1979 *Concrete-Making Materials*. McGraw-Hill, New York
Rixom M R (ed.) 1977 *Concrete Admixtures: Use and Applications*. Construction Press, Lancaster
Rüsch H, Jungwirth D, Hilsdorf H K 1983 *Creep and Shrinkage: Their Effect on the Behavior of Concrete Structures*. Springer, New York
Short A, Kinniburgh W 1978 *Lightweight Concrete*, 3rd edn. Applied Science, London
Soroka I 1979 *Portland Cement Paste and Concrete*. Macmillan, London
Tattersall G H, Banfill P F G 1983 *Rheology of Fresh Concrete*. Pitman, London
Woods H 1968 *Durability of Concrete Construction*. American Concrete Institute, Detroit, Michigan

J. F. Young
[University of Illinois at Urbana, Urbana, Illinois, USA]

Construction Materials: Crushed Stone

Crushed stone is a prosaic but nearly indispensable construction material. About 3 billion tonnes are produced annually worldwide, excluding the centrally-planned-economy countries, from a great variety of rock types. This immense amount of material is used for a multitude of purposes, ranging from low-cost bulk aggregate for road building and repair—its largest use—to specialty products, such as fillers and coatings, which command high prices but are a comparatively small part of the total production.

Crushed stone must have physical properties that suit the requirements for each use. For example, crushed stone for road building must be tough enough to withstand the crushing stresses and wear of modern motorized vehicles, a property found in many different rock types, and rock used for whiting must be very white and bright, properties found in only a few types of rock. Thus, for any particular use there may be many types of suitable rock; while to cover the whole spectrum of usage, many different rock types are needed.

The crushed-stone industry seems nearly ubiquitous. Large reserves of stone are widespread, and even where modern efficient machinery is used in processing and transporting the stone, bulk stone cannot be hauled more than 40 or 50 km by truck before transporting the stone becomes more costly than producing it. Numerous mines therefore have been opened within short distances of each other to serve local needs. Because most, if not all, countries are self-sufficient in crushed stone for the same reasons, there is little international trade in crushed stone for aggregate.

1. Definitions

Crushed stone is rock that has been mined and crushed to sizes that meet various consumer requirements. It is also commonly described as an aggregate: i.e., a mass or an assemblage of sized rock fragments that does not denote any particular rock type, but that is commonly composed of only one or two rock types. Crushed stone can be bound in cement or in asphalt, or can be used loose for many products, such as roadstone, fill, ballast, riprap and agricultural limestone. This last use and others depend on the chemical properties as well as the physical properties of crushed stone, but it needs only to be durable and reasonably inert for most physical uses.

Classifications used by industry for types of crushed stone are somewhat ambiguous geologically, and include many rock types (Table 1). Most crushed stone is classified as limestone, which includes dolomite, a closely related carbonate rock. Limestone and dolomite are commonly intermixed and are described by geologists as dolomitic limestone (limestone that contains between 10% and 50% of the mineral dolomite) or calcareous (limey) dolomite (dolomite that contains between 10% and 50% of the mineral calcite). High-calcium limestone is more than 95% calcium carbonate ($CaCO_3$), ultrahigh-calcium limestone is more than 97% $CaCO_3$, and high-magnesium dolomite is more than 95% calcium magnesium carbonate ($CaMg(CO_3)_2$).

Granite includes all coarse-grained light-colored igneous rocks. Sandstone includes impure sandstones and most siliceous rocks with a grained texture, and may include the metamorphic rock, quartzite. But the commercial term quartzite includes true quartzite, a metamorphosed sandstone, and may also include the sedimentary rock, orthoquartzite, which is sandstone that is strongly cemented with quartz. Similarly, the commercial term marble includes true marble, a metamorphosed limestone, and may include sedimentary crystalline limestone that is dense and will take a polish. All fine-grained dark igneous rocks are commonly included in the classification traprock.

Table 1
Selected rock types used for crushed stone

Rock group	Rock type	Commercial classification
Igneous	basalt gabbro diorite andesite	traprock
	granodiorite quartz diorite quartz monzanite syenite granite	granite
Metamorphic	gneiss	
	slate	slate
	marble	marble or limestone
	quartzite	quartzite or sandstone
Sedimentary	orthoquartzite	
	sandstone graywacke conglomerate	sandstone
	coral shell	shell
	dolomite limestone chalk marl	limestone or marble

2. Rock Types and Uses

Of the estimated 2.6 billion tonnes of crushed stone produced in developed market economy countries in 1983, 0.8 billion tonnes, or more than one-quarter, was produced in the USA. Limestone dominated the crushed-stone market in the USA; it accounted for nearly three-quarters of the production and was followed by granite, traprock, sandstone, shell and marble. Limestone was also a large part of the 78 million tonnes of crushed stone produced in Canada, and more than one-third of the 247 million tonnes produced in the UK. Other large producers of stone were the Federal Republic of Germany, Australia, France and Japan.

Limestone and dolomite, which are included in the same classification, have close geologic affinities. Dolomite is commonly a replacement for limestone, or an alteration of limestone to dolomite by the addition of magnesium. This chemical reaction takes place under natural conditions through geologic time and is accompanied by a molecular reduction in volume of about 11%, which may result in a dolomitic rock that is porous or vuggy.

Much limestone that is mined was first deposited in widespread sheets in ancient marine seas, although patchy deposits may be mined in some areas. Some strata were tilted or folded by large-scale earth movements through long periods of time. High-quality limestone or dolomite strata range from a few feet to hundreds of feet in thickness, and are commonly interbedded with strata of various rock types of varying quality. Extensive layers of limestone and dolomite may provide a widespread source of crushed stone that has consistent physical characteristics.

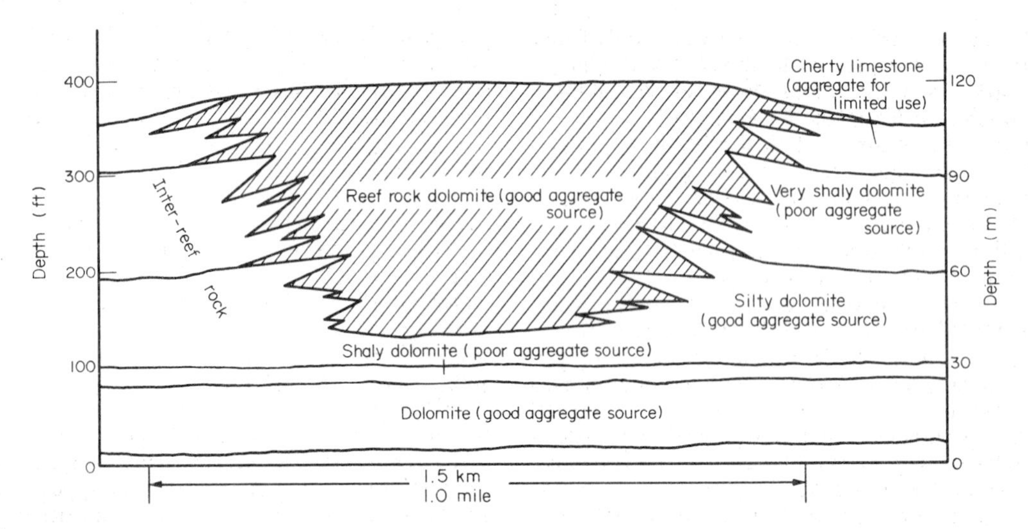

Figure 1
Cross section showing sources of crushed-stone aggregate in typical reefal and inter-reefal rocks of Silurian age in the Great Lakes area

Reefal deposits, which support many quarries in the Great Lakes area of the central USA and are a source of high-quality limestone and dolomite in other areas, also grew in some of the ancient seas and may be over a hundred meters thick. Reefs of Silurian age in the Great Lakes area range in size from 30–60 m in diameter and less than that in thickness, to more than 1.5 km in diameter and 120 m thick. These reefs are surrounded by inter-reef limestones and dolomites that are only partly suitable for aggregate (Fig. 1).

Crushed limestone and dolomite have many uses. More than 30 uses were reported in a canvass by the US Bureau of Mines (Tepordei 1987), and more than 70 were described by Lamar (1961). Secondary uses, such as those for calcined lime, increase the number of uses several fold. By far the greatest use of crushed limestone and dolomite in the USA is for roadstone and coverings, which accounted for more than 142 million tonnes of the 649 million tonnes used in 1985. It was followed in usage by concrete aggregate, which exceeded 79 million tonnes, and cement, which was about 84 million tonnes.

The construction of a major interstate-highway system in the USA during the 1960s and 1970s used large amounts of crushed stone, mostly limestone and dolomite. But road building has decreased, and road repair and other construction needs have increased.

Crushed limestone with less than about 5% magnesium carbonate is mined for cement making in many localities, although the restrictions on its composition make suitable deposits much less common than those for ordinary aggregate. Demand and exploration for relatively scarce new sources of high-calcium limestone, ultrahigh-calcium limestone, high-magnesium dolomite for calcined lime (includes calcined dolomite), fluxstone and many other chemical uses have increased in recent years.

Uses for crushed marble are nearly identical to those of limestone, with particular uses in terrazzo, as high-purity filler, and for decorative crushed stone because of its exceptional whiteness and brightness in some deposits. True marble is a metamorphosed limestone that is found in geologic terranes that have been subjected to great pressure and heat. The geographic distribution of marble is limited, and the amount used for crushed stone is further limited because of its greater value as dimension stone. Most marble that is crushed for aggregate comes from waste piles and discards at dimension-stone quarries and mills. Only 2.2 million tonnes was produced in the USA in 1985; that amount is less than 1% of the limestone used that year. An undetermined amount, probably quite small, is produced with and reported as limestone.

Chalk, which may also be classified as limestone, is produced in considerable amounts in the UK for cement making. A small amount from England and France is used as a mild abrasive. The high surface area and purity of chalk in some deposits make it an excellent raw material for stack-gas desulfurization processes, but its limited geographic distribution may keep it from becoming an important material for this purpose.

Marl, a calcareous material deposited especially in freshwater lakes, bogs and marshes, is an earthy substance that may be high-purity calcium carbonate, but more commonly contains considerable clay and silt impurities. It is crushed in comparatively small amounts for cement manufacture and soil conditioning.

Shell, which is produced in several southern states in the USA, is used mostly for construction and road aggregate, and in lesser amounts for cement making and lime manufacturing. Shell is derived mainly from fossil mounds of oyster shell, mostly by dredging, which limits its market area. Even so, 8.3 million tonnes was produced in the USA in 1985.

Although the commercial classification of granite includes the geologic rock types of granite gneiss, syenite, quartz monzanite, quartz diorite and diorite, most of the granite used for crushed stone has a composition in the family of granitic rocks that includes true granite and granodiorite.

Granite is a common igneous rock found in many places in folded and intruded mountain ranges and in shield areas. This distribution makes reserves of the stone practically inexhaustible in areas of widespread igneous terrane.

Fresh, unweathered granites are exceptionally durable and have low porosity, hardly any detectable loss on freeze-and-thaw tests, exceptionally high hardness, and high compressive strength. Because of the favorable physical properties of granite, nearness to market is more important for locating quarries than variation in physical properties.

In the USA, most of the granite used for crushed stone is for concrete and roadbase aggregate, which consumed nearly 45 million tonnes in 1985. Granite was also used for construction aggregate and roadstone, concrete aggregate and bituminous aggregate. But total consumption in the USA for 1985 was about 132 million tonnes, an amount comparable to only 20% of the limestone produced.

Major physical uses of crushed granite parallel those of crushed limestone. Open-pit quarries for crushed granite are similar to limestone operations, except that equipment for producing crushed granite must take into account its high strength and abrasiveness. Most granites have preferred partings. This is of little consequence in open-pit quarries, but it is important for granite produced for dimension stone (see *Construction Materials: Dimension Stone*).

Nearly all traprock is basalt and diorite, although all dark fine-grained igneous rock falls within this classification (Table 1). Seventy-one percent of the 76 million tonnes of traprock used in the USA in 1985 was for construction aggregate.

Basalts and diorites occur mostly in flows of intrusive or extrusive sheets, sills and dikes that range in thickness from 15 m to more than 300 m. Weathered zones or vesicular layers may be undesirable, but great quantities of fresh, unweathered stone are usually available in areas where traprock is found. Traprock is the heaviest crushed stone of the above categories, which increases its cost of transportation somewhat, but most traprock is tough and durable and makes an excellent aggregate. Its dark color may make it less desirable for aesthetic reasons.

The commercial classification sandstone includes graywacke, a tough sandstone that contains a considerable amount of impurities, and conglomerate, a poorly sorted pebbly sandstone. Since quartz is hard but brittle, sandstones tend to lack the toughness of limestone or granite, but will give satisfactory service for most aggregate uses if they are well cemented. Crushed sandstone is used for aggregate, ballast, fluxstone, and as a source of silica for cement among other uses, but in much smaller amounts than limestone; US consumption in 1985 was only 21 million tonnes. The quartz minerals in sandstone are abrasive and can cause rapid wear of processing equipment. Quartzite serves well as an aggregate and is crushed to sand sizes for high-silica uses.

3. Sampling and Testing

Physical characteristics of rock types are affected by their mineralogy, texture, cementation and fabric. Physical properties differ among rock types, and even among similarly classified rocks. Impurities or minor constituents can be a major cause of this variance. Thin shale laminae, for example, which form only a small percentage of the rock in which they are found, can make the stone unsuitable for many purposes. Conversely, impurities may actually increase the durability and usefulness of some stone. For example, a fine-grained limestone that contains some disseminated quartz silt or sand may have more skid-resistance qualities or may be more durable than the same limestone without these "impurities."

Accurate sampling and testing when particular physical qualities are required for specific uses cannot therefore be overemphasized. Sampling and testing are inexpensive procedures when compared with the cost of opening a mine and building a crushing plant. Sampling patterns and techniques are nearly as varied as the rock types sampled, but the sample should be as representative of the rock being tested as it can be.

The number of samples needed to characterize a stone deposit adequately depends on the geologic variability and the intended use of the stone. Limestone that has physical properties near the minimum required for a particular use will require more samples to account for the variability than a limestone that is well above the minimum. If a sampling pattern for a few test holes reveals variability of physical properties that might make the deposit unsuitable for its intended use, additional holes will be needed to define the limits and location of suitable reserves.

Rock cores from drill tests will normally yield representative samples, although the core must be large enough to provide enough sample for testing, about 5 cm or more in diameter for most layered deposits. Thick, nearly homogeneous stone (e.g., some thick calcareous or dolomitic reefs) is characterized by fewer samples than thinly layered deposits, which may require many samples to detect adequately significant changes in physical properties.

Changes in the composition and gross texture of the stone obviously require additional sampling, whether from cores or from outcrops of the stone. Changes in color may not be indicative of significant changes in physical properties, but are always suspect, and differently colored rocks should be sampled separately.

Sampling stone that contains sporadic impurities presents problems. One common example is limestone or dolomite containing chert. The chert, especially if chalky and weathered, is deleterious in cement and bituminous mixtures. Chert can be present in bands or as discrete lenses or nodules. Commonly, the bands are not continuous, and a drill-hole test may penetrate between nodules or through gaps in the bands. A much better estimate of the amount of chert in a deposit can be had from quarry or outcrop exposures, although this is also not without its dangers, because the amount of chert in a limestone or dolomite may vary from one part of a quarry to another.

Because of its sporadic occurrence, then, any chert at all in a core makes that rock suspect for certain uses of aggregate, and a more precise quantitative measure of the amount of chert in stone may wait until the quarry or mine is opened and operating. A regional knowledge of the character of the rock and the amount of chert is of great help in the initial site evaluations.

Many tests for crushed-stone aggregate in the USA have been standardized by the American Association of State Highway and Transportation Officials (AASHTO) and the American Society for Testing and Materials (ASTM) (Table 2). Most states in the USA have additional specifications and tests for local needs. Standards in the UK are issued by the British Standards Institution, and international standards are promoted by the International Federation for the Applications for Standards.

The most common physical tests include those for determining specific gravity, absorption , resistance to freezing and thawing (soundness), and resistance to abrasion (durability).

A knowledge of specific gravity is needed to determine the volume occupied by aggregate in concrete. Absorption measurements will indicate the tendency of crushed stone to absorb water from concrete mixes, which affects the strength of the concrete. Bonding between the particles and the cement is also affected by the water absorbed, and the amount of absorption of hydrocarbons is important in asphalt mixes. The standard AASHTO T-85 test for these properties for coarse aggregate is simple and inexpensive (Table 2). The maximum allowable absorption for many purposes is 5%.

Soundness properties are measured in part by subjecting crushed-stone particles to a series of freezing and thawing cycles to simulate seasonal changes and the action of ice and water on aggregate. Clay minerals, shaly laminae and finely fractured particles are particularly susceptible to these processes. Sedimentary stone containing abundant clay minerals or fine fractures frequently fails the freeze-and-thaw test, which has limits for most uses of about 10–20% loss on the AASHTO T-103 test.

A more rigorous test for soundness is the sodium sulfate or magnesium sulfate test (AASHTO T-104), which simulates freezing and thawing by alternately soaking and drying the material in a saturated solution. As the particles dry, crystals of sodium or magnesium sulfate form and expand, which causes disintegration of the aggregate. This test is fast, but does not simulate actual field weathering as well as the more time-consuming freeze and thaw test. Some specifications, therefore, allow retesting of aggregate by freezing and thawing if the material fails the sodium sulfate test.

The grinding and crushing action that takes place when vehicles pass over loose aggregate or macadam roads, or when heavy trains flex railroad beds, is difficult, if not impossible, to reproduce in a laboratory. For lack of better terms, this resistance to grinding and crushing might be called durability or strength, but it is commonly by a test called resistance to abrasion—somewhat of a misnomer—that some measure of the property is taken.

AASHTO Test T-96, resistance to abrasion, subjects aggregate to at least part of this grinding and crushing action by rotating it in a cylindrical drum in which steel spheres are dropped on the aggregate every revolution. This test has been widely accepted in the USA for many years, although the relationship of the results of the test to the actual wear and tear of aggregate in roads remains unclear.

Research to understand problems with aggregate and to develop better methods of testing aggregate for roads and for many other uses is conducted in many countries, particularly Germany, Australia, the UK and the USA. This research is conducted mostly by governmental organizations, although universities and private organizations sponsor some research.

Of continuing interest is the work on antiskid aggregates from two different lines of approach: (a) determining kinds of aggregate that polish the most on road surfaces, and (b) determining rock types that have the greatest natural antiskid properties. A few standardized tests have been developed for measuring these properties. Some success has been achieved by using limestone, a rock that by itself may polish but that may contain impurities such as angular sand or chert particles, which provide an increase in friction in surface courses.

For a surface aggregate to remain rough through extensive use and to continue to provide good traction for a long time, the particles on the surface should wear at different rates and continue to present unworn particles to traffic. Such requirements may open a new market for aggregate containing impurities that have been rejected in the past, but much work must be done before practical and effective aggregates and mixes can be found or developed.

Table 2
Selected tests standardized by the American Association of State Highway and Transportation Officials (AASHTO) and the American Society for Testing and Materials (ASTM) for crushed-stone aggregate

AASHTO designation	ASTM designation	Purpose	Procedure
Sampling			
T 2	D 75	Sampling stone, slag, gravel, sand and stone block	Sets forth procedures for sampling sources of aggregate at the deposit, at the plant and at the delivery point
T 248	C 702	To reduce field samples of aggregate to testing size	Describes the use of mechanical splitters and splitting by quartering
Particle size			
T 27	C 136	Determining particle-size distribution of aggregates	Crushed-stone aggregate is passed through a nest of sieves and the percentage retained on each sieve is calculated
Specific gravity and absorption			
T 84	C 128	Determining bulk and apparent specific gravity and absorption for aggregate	A weighed sample is introduced into a pycnometer, water added, and weighed. Calculations are from mass of pycnometer, amount of sample and mass of water
T 85	C 127	Determining bulk and apparent specific gravity and absorption for aggregate greater than 4.75 mm (No. 8 sieve)	Sample is dried, weighed, soaked in water and reweighed
Durability			
T 3	D 289	Testing resistance to abrasion by use of the Deval machine	Rock cubes are cut from aggregate particles, weighed and rotated in a cylinder of the Deval machine. Amount of loss of material below a 4.75 mm sieve is measured
T 96	C 131	Testing resistance of aggregate to abrasion and wear by use of the Los Angeles machine	A weighed and sieved sample is tumbled in a cylindrical drum containing loose steel spheres. The amount of sample that is reduced below a 1.70 mm sieve is calculated
T 103		Testing resistance of aggregate to freezing and thawing in water	Sized and weighed aggregate is frozen in water and thawed for a specified number of cycles. Amount of loss of sample less than certain size is calculated
T 104	C 88	Testing resistance of aggregate to disintegration by solutions of sodium sulfate or magnesium sulfate	A sized and weighed sample is soaked in the solution, removed, and dried for a number of cycles. Loss below certain size is determined

Considerable research also continues on better testing for durability and resistance to impact, and on the role of particle shape in bound or loose mixtures.

4. Mining and Processing

Sources of crushed stone have historically been so plentiful that early mining was simply a matter of removing stone from an outcrop or small open pit and crushing it by hand at the quarry or at the site of use. Only since the early and mid-1900s has widespread use been made of explosives and machinery to remove and process large amounts of stone from centralized quarries.

Although modern quarry operations are still relatively simple compared with some other industrial operations, important matters must still be considered. Stone, which has been cleared of overburden, must first be blasted from the quarry face. The method of blasting will partly determine the size and shape of aggregate particles and the amount of fines; the character of the rock itself will account for much of its shape. Sedimentary rock containing laminae of soft material, for example, may result in thin, elongate or bladed particles as compared with a homogeneous rock, which will produce stronger, more nearly cubical particles.

Stone that is reduced as near as possible to the desired product size by blasting is less expensive to produce than stone requiring undue mechanical crushing after blasting. Blasting for efficient production of rock fragments that will require minimal further processing includes judicious spacing and arrangement of blast holes, selection of correct types of explosive, designed loading of blast holes, and calculated timing of sequential blast.

Broken rock from the quarry face is loaded into trucks by shovels or front-end loaders and transported to a primary crusher, which is usually as close as is practical to the quarry face. Crushing equipment can be selected to efficiently process the type of stone being mined, produce the size and shape of the aggregate particles needed, and minimize the amount of waste sizes. A jaw crusher, for example, may produce a slightly elongate particle but usually with less fines than an impact crusher, which may produce more equant particles and efficiently handle large amounts of some materials. Cost considerations of different types of crushers, or the ability of a crusher to efficiently process different rock types, can be more important to an operator than the slight differences in particle shapes or amounts of waste stone.

The crushed stone is screened and conveyed to stockpiles containing various sizes of aggregate, and any oversize particles that were screened out previously are crushed again in a secondary crusher. The stone is loaded into trucks, weighed and transported to the site of use. Rail, ship and barge transportation require additional loading and unloading facilities, and are not economical unless the sources of stone are scarce and transporting the stone long distances can be justified.

Some of the largest operations are captive quarries producing stone for major industries, such as cement, steel and burned lime, but many more quarries produce stone for local construction needs, largely for road building and repair. Individual quarries may produce only a few tens of thousands of tonnes of aggregate per year, to more than a million tonnes per year near metropolitan areas.

Complicated quarrying can be caused by unusual or complex geologic structures. In areas where the stone strata are essentially flat lying, the quarry may extend over a large area if the stone section is relatively thin. Thick sections of suitable stone will encourage the quarrier to deepen his pit, especially in areas where additional land is high priced or unavailable because of environmental or other reasons.

Tilted or folded beds may necessitate production from irregular or deep, narrow quarries. Deep quarries may have greater water influx, and overburden thickness can increase dramatically in a short horizontal distance in tilted beds. Tunnel mines into the sides of quarries with thick overburden can be an economical solution.

Although stone reserves in many areas are essentially inexhaustible, lack of available land, intensive surface use of land, zoning regulations, thick overburden or environmental considerations may make underground mining attractive to crushed-stone producers. Underground mining, though commonly more expensive than open-pit mining, is becoming more common and has some advantages compared with open-pit mining. Underground mines can be operated on a year-round basis, and many environmental problems of surface mines can be eliminated or moved underground. Extensive removal of overburden is not required, and steeply dipping rock strata can be mined.

Tunnel mines into the sides of deeper quarries allow continued mining of high-quality stone in areas of high surface use, with a minimum increase in the cost of mining since in some mines much of the same equipment can be used in large underground rooms, and a new processing plant may not be needed, keeping costs down.

Combined mining of sand and gravel and crushed stone at the same mine site has become more common in the last decade. Sand and gravel pits near population centers may have bedrock floors suitable as a source of crushed stone at a depth shallow enough to be quarried after the sand and gravel have been removed. Intense use of surface lands or limited sand and gravel reserves may restrict horizontal expansion of the operation and make the bedrock the only source of additional stone reserves.

A fortunate set of geologic circumstances, however, must be present for combined mines to be practical.

Sand and gravel must be sufficient to allow economical removal of all overburden from the bedrock, which then must be suitable for use as crushed-stone aggregate (see *Construction Materials: Sand and Gravel*).

5. *Environmental Concerns*

Mining is a heavy industry that will generate dust, noise, blasting vibrations and heavy traffic near surface operations. Underground mines are much less visible but also can affect the environment in significant ways.

The amount of environmental disturbance caused by a surface quarry depends mainly on its size and activity. The perception of its effect may be greatly influenced by its location: a mine near a metropolitan area is very noticeable, but a mine in sparsely settled country may attract little if any notice from nearby inhabitants. The impact of time and size on environment is illustrated by many small quarries that were opened to supply aggregate for rural-road construction during the early 1900s in the midwestern USA. Some of these quarries were active for only a few weeks and most of them have now been reclaimed by nature. They are overgrown with vegetation or filled in by stream sediments and blend well into the countryside. Although there are many more of these old quarries than there are active quarries today, they have practically no effect on the environment and in fact may be difficult to recognize in places.

The large centralized quarries of today, however, are impossible to miss, and much planning is required to make them acceptable, both as active quarries and as abdandoned ones. Obvious steps can be taken to erect barriers of vegetation or fences to block the sight of unsightly equipment, and dust- and noise-suppressing equipment is helpful. Planning for eventual rehabilitation of a large quarry may be more complex, but the uses for the quarry can be numerous and range from water reservoirs in some old pits to drive-in theaters in dry pits. But quarries do have limitations for rehabilitation. Limestone quarries may not be suitable for waste disposal because of possible groundwater contamination through solution channels or porous formations. Water influx is common in quarries, and the expense of pumping the water out may make them unsuitable for many uses; but a water-filled quarry may have much utility and may increase the value of the site for industrial, residential or recreational development.

Underground mines remove many of the environmental disturbances of an open-pit quarry to underground quarters, but some surface installations still remain with some of the same environmental problems as surface mines. When an underground mine is abandoned, its utility depends on such factors as accessibility, roof stability, room size and amount of water influx. If a tunnel mine in competent rock is above the water table and has a direct outlet to the surface from the side of a hill, the possibilities near an urban area for using the mine for offices, storage, underground factories, or many similar uses are excellent.

Underground mines served by a shaft may present difficulties, although a deep dry mine with no roof problems may have utility similar to that mentioned above. Mines below the water table may serve as a reservoir for water near areas of concentrated population.

Underground mines in durable materials may last indefinitely without roof collapse, and examples of such old mines can still be seen; but roof collapse, especially in shallow mines, can cause serious subsidence of surface land. The subsidence above room-and-pillar mines is variable and can be disruptive.

Planning for sequential land use solves many environmental problems. Much valuable stone that could have been mined by open-pit methods has been covered by industrial complexes or the buildings and streets of expanding cities and towns. Zoning areas for quarrying stone as the first land use and for reclamation of the sites for recreation or residential construction as the intermediate or the final use would have allowed exploitation of much stone that is now lost to man's use.

See also: Construction Materials: Fill and Soil; Construction Materials: Granules; Construction Materials: Industrial Minerals; Construction Materials: Lightweight Aggregate

Bibliography

Harben P W, Bates R L 1984 *Geology of the Nonmetallics.* Metal Bulletin, Worchester Park

International Road Federation 1980 *World Survey of Current Research and Development on Roads and Road Transport.* International Road Federation, Washington, DC

Lamar J E 1961 *Uses of Limestone and Dolomite*, Illinois Geological Survey Circular 321. Illinois Geological Survey, Urbana-Champaign, Illinois

Rooney L F, Carr D D 1971 *Applied Geology of Industrial Limestone and Dolomite*, Indiana Geological Survey, Bulletin 46. Indiana Geological Survey, Bloomington, Indiana

Tepordei V V 1985 Crushed stone. *Mineral Facts and Problems*, US Bureau of Mines Bulletin 675. USBM, Washington, DC, pp. 757–68

Tepordei V V 1987 Crushed Stone. *Minerals Yearbook 1985*, Vol. 1. US Bureau of Mines, Washington, DC

C. H. Ault
[Indiana Geological Survey, Bloomington,
Indiana, USA]

Construction Materials: Dimension Stone

Dimension stone is a comprehensive term for rock that is cut, split or worked to form masses or units of desired shapes and sizes for specific uses, or that is selected from naturally occurring material to meet requirements of shape and size. The term does not cover crushed, broken or ground stone (see *Construction Materials: Crushed Stone*). Most dimension stone is used for building, but other uses are numerous and include monuments, markers, paving, flagging, curbing, whetstones, grindstones, surface plates, beds for billiard and pool tables, refractory units, chalk boards, electrical panels, laboratory sinks and furniture, curling stones, and a host of ornamental objects and industrial items.

1. Building Stone

Virtually every variety of rock that is firm enough to hold together until it can be put into its intended place has been used, somewhere and at some time, as building stone, but only five lithologic types of rock are produced commercially to any appreciable extent. These are granite, limestone, marble, sandstone and slate. Several of the terms as they are used in the stone industry have greater latitude than the strict petrologic definition would permit.

All of the three main generic rock types—igneous, sedimentary and metamorphic—are utilized as dimension stone. Igneous rocks are those that have cooled from a melt; they are made up largely of silicate minerals and free silica in the form of quartz. Those that cool at depth are called intrusive rocks; they range in texture from visible grain size to coarse crystals. Extrusive igneous rocks cool much more rapidly and are generally fine-grained. Some igneous rocks, called porphyries, consist of a fine-grained ground mass containing separate coarser crystals of one or more of the minerals.

Sedimentary rocks are deposited by the geological agencies of transportation (running water, marine currents, wave action or the wind) or by chemical precipitation.

Metamorphic rocks are formed from preexisting igneous rocks and sedimentary rocks, and from earlier to more advanced stages of metamorphic rocks, by pressure or heat (commonly both) within the earth's crust.

1.1 Granite

In geological terms, granite is an intrusive igneous rock composed of discrete, visible, crystals of quartz and the potassium and sodium feldspars, with a minor permitted content of biotite and muscovite mica, and of certain other accessory minerals, such as augite. Within the dimension stone industry, however, igneous rocks of wider variety are classed as granite. They include: quartz–feldspar mixtures in which calcium feldspars are fairly abundant or predominant; igneous rocks that have little or no free quartz; others that have a high, rather than accessory, content of biotite, augite and other ferromagnesian minerals; and porphyries in which the matrix is below visible grain size.

1.2 Limestone

A sedimentary rock that consists largely of the carbonate minerals calcite or dolomite, or a mixture of the two, is a limestone. The color is generally light, ranging from white or near-white through gray or buff or tan, but limestones may be dark because of mineral impurities. In texture, the limestones range from those that are so fine-grained as to lack visible particles, through the sand sizes, to coarse material; and in compactness they may range from dense to very open texture with sizeable interstices, vugs and cavities. Bedding may be thin, thick or massive, and this controls the architectural use to which they may be put. Thin-bedded stone, in which nature has determined one dimension of the building unit, is usable for flagging, for vertical application and for masonry patterns emphasizing horizontal directions. Massive stone, from which sizeable blocks may be quarried, is necessary for large masonry units, columns and sculpture.

1.3 Sandstone

A sedimentary rock composed of grains within the range of sand size (0.0625–2 mm) is a sandstone. Most of the sandstones used for building and other dimension purposes are composed largely of quartz grains, but the rock is still a sandstone if a high proportion of the grains are feldspar or other minerals. The cementation holding the sand grains together may be calcareous, dolomitic, siliceous or ferruginous. Siliceous sandstones, given equal degree of cementation, are harder than those that have carbonate cement, and in consequence they are more difficult to work but more durable. Sandstones range more widely in color than do the limestones. Many of those used in building are light, ranging from near-white to gray and tan, but highly colored varieties have dominated the building sandstone trade at various times and places. Examples are the "brownstone" of the Eastern Seaboard of the USA, red sandstones from the upper Midwest to the New York region of the USA, red and dark-gray sandstones in Scotland, and red sandstone in India. A wide range of colors, mostly in the warm hues, is produced in northeast-central Ohio, at Crossville, Tennessee, and at various locations in the western states. In some of these products, the same billet of finished stone contains a range of colors.

1.4 Marble

Geologically, marble is a metamorphic rock formed by recrystallization of limestone or dolomite through some combination of heat and pressure, and many of the marbles utilized in classical structures of antiquity and up to the present day fit this definition, but by custom and precedent various rocks of other origin and somewhat different composition are classified as trade marbles. They include limestones and dolomites that will take a polish because they are microcrystalline or have recrystallized because of processes other than dynamic or thermal metamorphism, including alteration by groundwater. The term encompasses travertine, a generally banded hotspring or cavern deposit that typically has vesicular openings arranged in layers, and "onyx" marble, which is very fine-grained and is generally considered to have been formed by cold-water precipitation of calcium carbonate. Finally, the serpentine marbles are truly metamorphic and consist of that mineral, with or without accompanying carbonate minerals and chlorite or biotite; some contain flecks of magnetite. A necessary characteristic of all the trade marbles, except travertine, is that they take a polish; the solid part of the travertine structure does polish, but the voids prevent a mirror finish. Of all the building stones, the marbles present the widest variety of colors, in some instances as a single pervasive color and in others as a wide range distributed in flecks, bands, pods or aggregates. Some of these deposits of true metamorphic marbles are most elusive and difficult to explore and to develop, as color, texture and composition may change drastically in a short distance within the quarry area. Thus, an unusual marble may be literally "worked out" with little prospect of finding more—this is true to a greater extent than in the case of other main types of building stone.

1.5 Slate

The fifth major category of dimension stone is slate which is metamorphic. It is microcrystalline rock produced by dynamic metamorphism of clays and shales, causing the flat, micaceous particles to be oriented in planar fashion, with their long axes perpendicular to the direction of pressure. This gives rise to slaty cleavage which permits the rock to be split into sheets. The original bedding of shale may show faintly, or prominently, across the cleavage surfaces of slate, and for most dimension stone uses this is an undesirable characteristic. Slate of this kind is termed "ribbon" slate.

2. *Nature of Building Stone Use*

Within the last century, a vast change has taken place in the manner in which stone is used for building. Through most of the nineteenth century stone was used largely as a structural material to carry the weight of the building. The walls were bearing walls, and the load of the roof and floors was supported by the exterior walls, and by interior masonry walls, commonly of brick. Beams, joists and rafters were carried on the exterior and interior masonry walls. In these circumstances a great volume of stone was used. To erect a tall building it was necessary for the walls to be very thick at the base, although they could be progressively narrowed upward as the load decreased. Fenestration and entryways were awkward and cumbersome because of the wall thickness. The advent of the elevator permitted buildings to be ever higher, and it was necessary to devise new methods of building. The resulting development was the skeletal framework, in which structural steel or reinforced concrete, or a combination of the two carried the load of the building, and stone came to be used mainly as cladding. Under these circumstances the stone could be appreciably thinner, and needed to be to reduce weight and unneccessary bulk. Originally, much of the stone cladding was still fairly thick, but it has become progressively thinner, in some instances so thin as to cause problems with anchoring the stone to the walls and with wind load. These problems have been increased by a trend towards panels of larger area, in many instances extending floor-to-floor, and even to multifloor panels. In single-story and other low-rise residential and commercial buildings, stone is used as veneer over frame and block buildings. These developments have resulted in the use of a lower volume of stone relative to the amount of space enclosed.

3. *Quarrying*

Dimension stone is quarried by numerous methods that range from primitive to advanced. At most sites of open-cut removal, soil or other unconsolidated material overlies the bedrock, and at some it is also necessary to remove rock overburden to gain access to the stone to be quarried. Soil cover is generally removed by bulldozers or other earth-moving machinery. Hydraulic sluicing has been used on occasion, and may still be applied in places, although environmental protection controls, where they exist, act as deterrents. Rock overburden is commonly quarried away by methods similar to those employed for the desired stone if blasting is likely to damage the material beneath, as is frequently true for limestone and marble. Explosives in drill holes are used for loosening unusable rock overburden in many granite and sandstone quarries. Even where the stone that is sought is the most shallow bedrock, the surface to some variable depth may be weathered, jointed or creviced to an extent that precludes efficient recovery and necessitates removal as waste. Most building stone is obtained from surface quarries, but underground removal is feasible, particularly for limestone and marble. A room-and-pillar pattern is used. Recently developed belt saws and chain saws are capable of horizontal as well as verticle cutting; they are called "gallery" saws, and they hold promise for more efficient underground extraction. Subsurface quarrying already enjoys the advantages of stabilized year-long working conditions, improved ability to follow desired types of stone without removal of excessive amounts of overburden and reduced impact on the surface environment.

One method of quarrying employs a channelling machine that moves along a track, cutting a slot, typically about 5 cm wide, along one or both sides. The cutting is carried out by multiple cutter bars, sheaved together, and edged and hardened at their lower ends. These bars, actuated by a cam, move successively upward and downward, deepening the channel, as the machine moves back and forth by some amount ranging from a small fraction of a centimeter in hard stone to several centimeters in very soft stone.

A second method of separating stone from the bedrock is by use of wire saws. A power-driven, endless, helical wire is drawn initially over the surface of the ledge, and later in the slot that it has cut, carrying a mixture of water and abrasive. Quartz sand or crushed chert (flint) is sufficiently harder than limestone or marble to serve as the cutting agent, but for granite a harder abrasive, such as silicon carbide, is necessary. The wire is under tension, exerting constant pressure on the stone at the base of the slot, and feeding over sheaves that slowly descend or advance as the slot deepens. A single length of wire, typically 300 m or more, is actuated by a power unit and is so rigged over an assembly of pulleys and posts that it may perform many cuts at different places in the quarry simultaneously. Wire sawing is used principally for vertical cutting, but it can be used for undercutting in the horizontal direction.

A variant of the wire saw, generally used in shorter increments, carries "beads," impregnated or coated with diamond and spaced along the wire.

Especially in the limestone industry, either the channelling machine or the wire saw permits the separation of long "cuts" about 1.2 m in width and ranging in depth from about 2 to 4 m. At either end of such a cut, crosscuts are made by either channeller or wire saw, and a substantial tonnage of stone is thus freed from bedrock, except at the base where, unless a natural separation is present, sleeved wedges or points may be driven into closely spaced pneumatic drill holes until the cut breaks. Alternatively, the base may be freed by undercutting with a wire saw or chain saw. Then the cut, in some instances as long as 20–25 m, is tipped forward (turned down) by crane, derrick or other heavy equipment, or increasingly by distention of inflatable air bags that are lowered into the channel or slot behind the cut, onto a series of "pillows" made up of chunks of stone that crush on impact and cushion the fall, reducing or eliminating breakage. Once on its face, the cut is sectioned into mill blocks of desired size by using wedges and slips in pneumatic drill holes.

An alternative to channelling or wire sawing, but a system that still permits the turning down of cuts, is the use of a quarry bar, a massive assembly along which a drill may slide and be locked into successive positions that permit the drilling of closely spaced holes with webs of stone between them—the web then being removed wth a spade-like broaching bar, also actuated by a pneumatic drill, leaving a ribbed vertical surface.

Chain saws of various types may be utilized either to cut channels or to slice a quarry floor into sections of desired width. A single chain saw, its teeth generally armored with tungsten carbide, diamond or other hard materials, may travel on a track to cut a single channel, or multiple saws may be ganged to create thinner units and supplant a substantial part of the later mill work that is generally required. Chain saws are more effective in softer stone than in harder material. They are most commonly used for fairly soft limestones, although they will function in firm, well-consolidated stone, the speed of cutting diminishing with increased hardness. They may be used for sandstone, slate and marble.

Belt saws, which substitute tough, mechanically driven belts for toothed chains, are a relatively recent development of impressive effectiveness. The belts are cored with multiple steel wires that pass through spaced metallic segments impregnated to some depth with diamond.

For both granite and sandstone, it is possible to drill holes 30 cm or more apart in the same plane and achieve a fairly flat surface of breakage by detonating charges in them.

With granite, and no other common type of dimension stone, it is possible to cut a channel or slot by a process called jet piercing, which utilizes a very high-temperature flame attained by compulsion of fuel oil or diesel oil in a high-pressure discharge of compressed air. The cutting action results from thermal shock, the coefficient of thermal expansion being different in the various minerals that constitute the granite, with the result that a shower of rock flakes is constantly driven off in advance of the flame tip.

Hydraulic piercing, achieved through use of a water jet at very high pressure is at an early stage of development but is successful in rock as obdurate as granite and may further supplant other methods for various parts of the quarrying process as the technology improves.

All the foregoing methods apply to situations in which stone must be actually cut from a continuous body, but it should be noted that much sedimentary dimension stone is taken up in its natural bedding thickness, one of the dimensions having been determined by nature, leaving only the need to free the stone in the other two directions by one or another of the methods previously described. Similarly, marble and slate, and especially granite, may have the vertical dimension of the blocks established by sheeting, roughly following the earth's surface, occasioned by reduction of pressure as the rocks formed at depth were unloaded by erosion following uplift.

4. Milling

Building stone may be used as it comes from the quarry or in the form in which it is collected as field stone, stream shingle or boulders, but most is processed to a minor or major degree in a building stone mill.

Mill blocks of granite, marble, sandstone or limestone are generally trimmed or slabbed in a gangsaw, which slices the block to the desired thicknesses with parallel steel blades that travel back and forth, deepening the cuts with each pass. The steel blades may carry an abrasive (sand, crushed chert, steel shot or silicon carbide) in a water or slurry mixture to do the cutting, or they may have toothed diamond or tungsten carbide inserts. The various gangsawing methods yield different finishes, sand or toothed blades being the smoothest. Alternatively, the blocks may be slabbed or trimmed by wire saws including those that use spaced diamond-impregnated beads as the cutting agents, or by circular saws.

The slabs from such sawing procedures are commonly trimmed to nearer final to nearer final size by diamond-toothed circular saws. In the softer varieties of stone the principal face of the unit, the edges and moldings may be planed to the final finish and shape. Carved details and parts of the work initiated by the planer (such as the ends of rabbeted grooves) must be finished by hand. Columns, balusters and any turned work are done on a lathe. Concealed features, such as anchor slots and holes for hanging the units, and relieved areas for conduit and framework are provided by stone cutters. Stone carvers execute Ionic, Corinthian or compound capitals, decorative details, rose windows, tracery, sculpture, friezes and inscriptions.

A high proportion of building stone is sawed or otherwise cut in some way, but in much of the industry the term "cut stone" refers to units that are completely dimensioned for installation at a specific location in the finished structure. Numerous units that appear to be identical after placement are actually different in their accommodations for anchorage. Cut stone is installed by stone setters, a separate craft.

Stone supplied in modular course heights and cut to length on the job (in which case wall pattern is determined by the mason) is not, under such a restricted definition, considered to be cut stone.

Finishes range from rock face, which means a broken irregular surface, through sawed, planed, honed and polished surfaces. Broken but fairly planar surfaces result from splitting sawed slabs with a hydraulic shearing device, either straight-bladed or with articulated teeth, called a "guillotine." Flame finishes for granite are achieved through the thermal shock from a high-temperature jet flame passed over the surface, causing spalling. Grooved, ribbed and other textured surfaces are created by routing tools in planers, or by hand work. Batted or scabbled surfaces may be imparted by hand tools such as scabbling picks. Bush-hammered finishes are created by hand or mechanical work.

5. Specifications and Tests

Since rocks are inhomogeneous and naturally formed materials, varying in such aspects as mineralogy, texture, cementation and moisture, and since test specimens cannot be prepared and tested uniformly, they exhibit variations in physical properties, such as resistance to abrasion, absorption, compressive strength, rupture modulus and specific gravity.

Technical bodies and trade organizations have established standards to assure the quality of building stone, and testing procedures have been developed for determining whether materials meet stated requirements. Among the most widely recognized are those issued by the American Society for Testing and Materials (ASTM). ASTM specifications for various rock types include ANSI (American National Standards Institute)/ASTM specifications for limestone, sandstone, granite, marble and slate. Some are restricted to interior or exterior use, and some are for special uses, such as slate for blackboards or roofing. Comparable specifications and tests, in general based on the same physical characteristics, exist in various other countries.

6. Selection of Stone

Aesthetic considerations are more important than any others in the selection of building stone by architects for most purposes. Exceptions are surfaces subject to substantial wear from foot traffic; surfaces vulnerable to attrition and impact, such as base courses; interior base courses likely to be affected by washing and chemicals; and exterior paving that may be damaged by salt or other deicing compounds.

Persons familiar with building stone are frequently asked what guidance the standards, tests and specifications offer for selection of stone for specific construction projects, and the answer is that this is not their purpose, although they may be helpful. Examples of their usefulness are to be found in the choice of stone for stair treads, flooring and paving where the stipulation of minimum resistance to abrasion may be important, depending on the amount of traffic that may be expected. Absorption is a consideration where exposure to liquids, to spillage and soiling, or to freezing are likely to be factors. For the most part, however, the physical characteristics expressed in the standards are principally assurance to the consumer that the stone purchased falls within the satisfactory range for the type of stone that has been selected. For example, minimum compressive strength is not specified for, say, the various categories of limestone or granite to guarantee that the stone will not crush in use, as most construction practice does not subject the stone to vertical pressures that will pose a problem in this respect. For example, Currier (1960) mentioned that the load on the stone at the base of the Washington Monument, a 169 m high load-bearing column, is only 4140 kPa; hollow load-bearing concrete block used in the same way requires compressive strength of only 4830–6890 kPa, and hollow block that is not load bearing (much building stone is not load bearing in use) has only 2070 kPa minimum. However, the specified minimum compressive strength for one category of limestone is 27 600 kPa, and the minimum for granite is 131 000 kPa. Compressive strength that falls below the normal range for the type of stone being tested indicates possible flaws, or at least peculiarities, that suggest the possibility of performance unsatisfactory in ways other than compressive failure.

The ultimate test of whether stone is satisfactory for any given purpose is performance, and various building stones have withstood the test of time, in many cases for centuries, in all types of buildings and in bridges and monuments distributed through a wide diversity of climates and environments. Such difficulties as have been encountered are attributable in a high proportion of the cases to placement in unsuitable situations (below grade, in constantly wet locations or subject to salt action), to faulty construction procedures (improper anchorage or lack of expansion space), and to the effects of incompatible associated materials (e.g., leaching of alkalis from mortar or concrete), rather than to any inherent frailty in the stone.

See also: Building Systems and Materials; Construction Materials: Crushed Stone; Construction Materials: Fill and Soil; Construction Materials: Granules; Construction Materials: Industrial Minerals; Construction Materials: Lightweight Aggregate; Construction Materials: Sand and Gravel

Bibliography

American Society for Testing and Materials 1984 *Annual Book of ASTM Standards*, Sect. 4, Vol. 4.08, *Soil and Rock: Building Stones*. American Society for Testing and Materials, Philadelphia, Pennsylvania, pp. 1–40, 78, 79

Ashurst J, Dimes F G 1977 *Stone in Building: Its Use and Potential Today*. Architectural Press, London

Bowles O 1939 *The Stone Industries: Dimension Stone, Crushed Stone, Geology, Technology, Distribution, Utilization*, 2nd edn. McGraw-Hill, New York

Bowles O 1960a Dimension stone. In: Gillson J L (ed.) 1960 *Industrial Minerals and Rocks: (Nonmetallics Other than Fuels)*, 3rd edn. American Institute of Mining, Metallurgical, and Petroleum Engineers, New York, pp. 321–37

Bowles O 1960b Slate. In Gillson J L (ed.) 1960 *Industrial Minerals and Rocks: (Nonmetallics Other than Fuels)*, 3rd edn. American Institute of Mining, Metallurgical, and Petroleum Engineers, New York, pp. 791–98

British Standards Institution 1957 *Glossary of Terms for Stone Used in Building*, BS 2847. British Standards Institution, London

Currier L W 1960 *Geologic Appraisal of Dimension-Stone Deposits*, US Geological Survey Bulletin 1109. US Geological Survey, Reston, Virginia

Patton J B 1974 *Glossary of Building Stone and Masonry Terms*, Indiana Geological Survey Occasional Paper 6. Indiana Geological Survey, Bloomington, Indiana

Patton J B, Carr D D 1982 *The Salem Limestone in the Indiana Building Stone District*, Indiana Geological Occasional Paper 38. Indiana Geological Survey, Bloomington, Indiana

Power W R 1983 Dimension and cut stone. In: Lefond S J (ed.) 1983 *Industrial Minerals and Rocks: (Nonmetallics Other than Fuels)*, 5th edn., Vol. 1. American Institute of Mining, Metallurgical, and Petroleum Engineers, New York, pp. 161–81

Winkler E M 1973 *Stone: Properties, Durability in Man's Environment*. Springer, New York

J. B. Patton
[Indiana Geological Survey, Bloomington, Indiana, USA]

Construction Materials: Fill and Soil

Fill is any man-made deposit of natural soil, rock products or waste materials. Fills can be carefully constructed to form dense, stable embankments, or they can be random accumulations of soils, rocks, organic materials and even trash and garbage. To the geologist, pedologist and agronomist, soil is the very superficial layer of the earth's crust which is capable of supporting plant life. To the civil engineer, interest in the soil goes much deeper because of construction activities which must take place on or in soil, such as foundations and tunnels, or which use soil, such as dams, embankments and levees. Thus, soils in an engineering sense are the sediments and other unconsolidated accumulations of materials found above the bedrock, which are produced by the physical and chemical disintegration of rocks and which may or may not contain organic or even man-made materials.

Even though soils often possess less than desirable properties compared with other civil engineering materials such as concrete and steel, they are often used as engineering materials because they are so economical. But they must first be stabilized, reinforced or otherwise changed to improve their engineering properties.

1. Soil Formation and the Nature of Soil Constituents

Soils are primarily the products of weathering in the earth's crust. Physical (mechanical) weathering causes disintegration of rocks by alternate freezing and thawing, temperature changes, erosion, glaciation and the activities of plants and animals. Chemical weathering decomposes rock minerals by oxidation, reduction and carbonation. In general, chemical weathering is much more important than physical weathering in soil formation. Soils can be weathered in place (residual), or they may be transported by water, wind, glaciers or gravity. Soils may consist of purely mineral matter or they may also contain organic material. Some cumulose soils such as peat and muck may be almost entirely organic. Because soils consist of discrete particulate materials, they have voids and usually the voids contain both air and water.

The predominant minerals in rock and therefore in soils are silica, mica, feldspar, calcite and dolomite, ferromagnesium minerals, iron oxides and the clay minerals. The latter are basically hydrous aluminosilicates plus other metallic ions which result from chemical weathering especially of feldspars, micas and the ferromagnesium minerals. They are very small plate-shaped crystals, with a diameter of the order of 1 μm or less. The most important clay minerals in soils are kaolinite, montmorillonite, illite and chlorite. Detailed descriptions of clay minerals are given in Holtz and Kovacs (1981) and Mitchell (1976), and in the article *Binders: Clay Minerals*.

2. *Soil Texture and Classification*

The texture of a soil is related to its appearance or "feel," and it depends on the relative sizes and shapes of the particles as well as on the distribution of those sizes. Soils with particles larger than about 0.05 mm, which is about the smallest grain that is visible to the naked eye, are coarse grained and they appear coarse textured. Examples are boulders, cobbles, gravels and sands. Engineering properties of these soils are strongly dependent on grain size and distribution and grain shape. Fine-grained and fine-textured soils are silts and clays. The term clay is applied to soils which contain clay minerals in sufficient quantities to affect their engineering behavior. The behavior of fine-grained soils, especially clays, has very little relation to texture, grain size, shape or grain size distribution. Their specific surfaces are relatively large and thus their behavior is more strongly affected by electrochemical than by gravitational forces. Clays have plasticity and cohesiveness, and the presence of water significantly affects their engineering behavior. Silts, although fine grained, are granular, nonplastic and cohesionless.

Soil classification based on soil texture and plasticity is very useful for correlation with engineering properties. Many systems have been proposed for specific engineering purposes, for example, highways, earth dams and airfield construction.

3. *Grain Size and Grain Size Distribution*

The range of particle sizes in soils spans at least eight orders of magnitude. Sizes can range from boulders or cobbles of several centimeters in diameter down to ultrafine-grained colloidal particles. Even though the

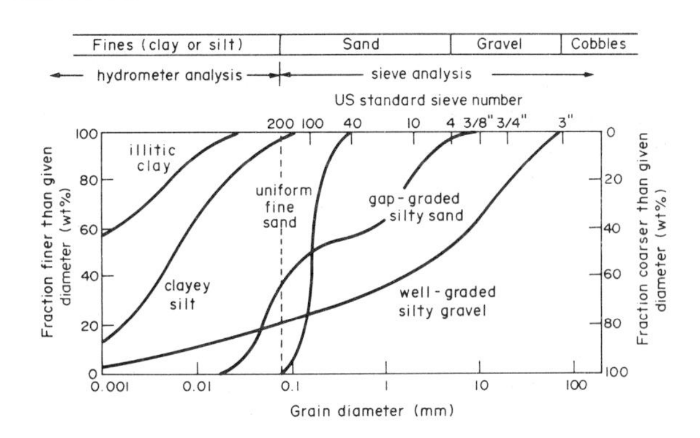

Figure 1
Typical grain size distribution curves

individual particles may have shapes very different from spheres, mean diameters of the particles are determined either by mechanical (sieve) analysis or some procedure using Stokes' law (American Society for Testing and Materials 1986). Typical cumulative grain size distribution curves are shown in Fig. 1.

4. *Atterberg Limits of Fine-Grained Soils*

The presence of water in the voids is probably the most important factor affecting the engineering behavior of fine-grained soils. Not only is the water content important but it is also useful to compare the water content with certain standards of engineering behavior. Atterberg limits indicate different boundaries of mechanical response in terms of water content, and they are very useful in the classification of fine-grained soils because they are good indices of soil behavior (Holtz and Kovacs 1981). It is possible for a soil to have behavior ranging all the way from a brittle solid to a liquid depending only on the water content (Fig. 2). At water contents (w) greater than the liquid limit (LL), soils can support essentially no static shear

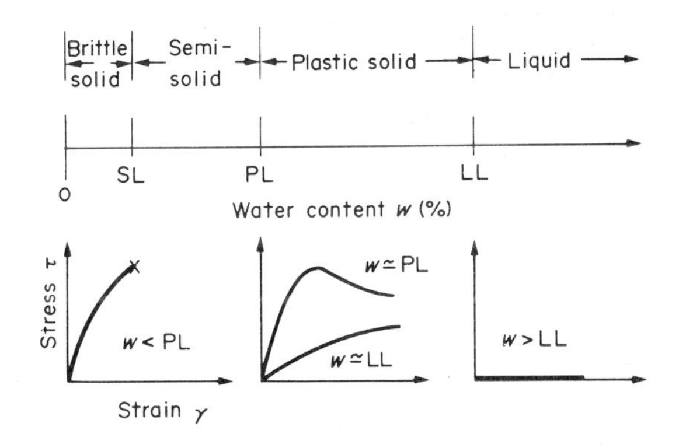

Figure 2
Behavior states and stress–strain response of soils as a function of water content

stress. For water contents less than the liquid limit but greater than the plastic limit (PL), soils behave as if they were plastic or Bingham solids.This range is called the plasticity index, or PI = LL − PL. At water contents below PL, soils behave as if they were semisolid or even brittle. Below the shrinkage limit (SL), no further decrease in volume occurs as drying continues. Standard tests for Atterberg limits are given by the American Society for Testing and Materials (1986).

5. *Soil Structure and Fabric*

The structure of a soil mass controls to a very large extent its engineering behavior, and this is true for both natural as well as man-made soil deposits. Soil structure includes both the geometric arrangement or fabric of the soil particles and their interparticle forces. For granular soils, since the surface activities and interparticle forces are small, structure is identical to fabric. In general, the fabric of these soils is single grained. Fabrics range from loose with a low bulk density and open structure to dense with close packing and a high density. It is also possible for the grains to have a bulked or honeycomb structure which can support some static load but collapses if it is vibrated. Large masses of loose saturated sands may liquefy if subjected to dynamic loading such as earthquakes.

With the development of the scanning electron microscope, fabrics of fine-grained soils have been found to be extremely complex. Because of their large surface activities, they always seem to be aggregated or flocculated into basic fabric units (domains) and fabric assemblages (clusters and peds) with pore spaces in between.

The macrostructure, including peds, joints, fissures, silt and sand seams, root holes and varves, often controls the engineering behavior of soil masses. For example, the strength of a soil is significantly less along a crack or fissure than through the intact material, while the permeability is greater because water flow is concentrated in joints and fissures. Engineering problems involving stability, settlement or drainage must carefully consider soil macrostructure. Although the structure of natural deposits is destroyed by excavation, soils in earth fills can have rather complex fabrics due to the construction process (Mitchell 1976).

6. *Engineering Properties of Soils*

Important engineering properties of soils include permeability, compressibility, modulus and shear strength. For fills and embankments, the shrink–swell and frost susceptibility characteristics also must be considered (Holtz and Kovacs 1981). Permeability determines how water flows through the soil pores, and is a function of the texture, grain size and distribution, density, and soil structure. The physical and chemical interaction of the pore water and air with the solid particles in the soil structure causes both natural and man-made soil deposits to have very complex mechanical behavior. The response of soils to boundary stresses depends on the rate of load application, permeability and soil structure. For example, fine-grained saturated soils loaded rapidly will develop pressures in the pore water that strongly affect both their modulus and strength (Holtz and Kovacs 1981). The developed pore pressures can be either negative (dilatancy) or positive, depending on the density and stress history of the soil. Because soils are in general nonlinear, nonconservative, anisotropic, viscoelastic–plastic materials, it is impossible to formulate exact mathematical descriptions of their stress–strain–time behavior. Consequently, for engineering purposes great simplifications are made, and designs are often based on empirical rules of thumb, precedence and the experience of the engineer. Empirical correction of "safety" factors must be applied to account for real material response and the unknowns imposed by the geological conditions and the construction process. Therefore, the engineering of soil materials is more of an art than a science (Terzaghi and Peck 1967, Winterkorn and Fang 1975).

7. *Stabilization and Compaction*

Because their engineering properties are usually poor, the economic use of soils in fills requires some type of improvement or stabilization.

Stabilization can be thermal, electrical, chemical or mechanical (compaction). The first two are only occasionally used. Chemical stabilization includes the mixing or injecting of materials such as Portland cement, lime, asphalt and other bituminous materials, calcium chloride, sodium chloride and even industrial wastes into soils (Winterkorn and Fang 1975, American Society of Civil Engineers 1978, 1987).

Mechanical stabilization, called compaction, is the densification of soils by the application of some type of mechanical energy. Densification and particle rearrangement occur because air is expelled from the soil voids without significant change in the water content. Soil strength is increased, compressibility and permeability decreased and detrimental effects of shrinkage, swelling and frost action often can be ameliorated. Desirable engineering properties of compacted fills can be achieved by proper consideration of the soil type and the most appropriate compaction technique and equipment (United States Bureau of Reclamation 1974).

Cohesionless soils are most efficiently compacted by vibration. Particle rearrangement and densification occur due to cyclic deformations produced by the vibrating equipment. Equipment used for compacting granular fills includes vibrating plates, tampers, motorized rollers and large weights dropped from a crane.

With increasing cohesion the efficiency of vibration as a densification agent significantly decreases because even slight bonding or cohesion between the particles interferes with particle rearrangement. In the field, common compaction equipment includes hand-operated tampers, sheepsfoot rollers and rubber-tired rollers. Rubber-tired rollers are most effective for compacting sandy and silty soils with relatively low cohesion. Sheepsfoot rollers are better for more cohesive soils. Most efficient compaction takes place if the soils are laid down and compacted in relatively thin (15–30 cm) layers or lifts.

8. *Theory of Compaction of Fine-Grained Soils*

The compaction of fine-grained soils is a function of dry density, water content, soil type and compactive effort which is a measure of the mechanical energy applied to the soil mass (Holtz and Kovacs 1981). In the field, compactive effort is the number of passes or coverages of a roller of a certain type and mass. In the laboratory, a hammer is dropped several times on a soil sample of known volume, or a tamper kneads the soil by applying a given pressure for a few milliseconds which is supposed to simulate the effect of field compaction equipment.

The process of compaction can be illustrated by the common laboratory compaction or Proctor test (American Society for Testing and Materials 1986). Several samples of the same soil, but at different water contents, are compacted with the same compactive effort. When the bulk dry densities of each sample are plotted against water content, a characteristic parabolic curve is obtained (Fig. 3). The curve is unique for a given soil type, method of compaction and compactive effort. The water content corresponding to the maximum dry density is called the optimum water content. Typical values of maximum dry density are 1.6–2.0 Mg m^{-3} with an outside range of about 1.3–2.4 Mg m^{-3}. Optimum water contents range between 10 and 20% with a maximum range of 5–40%.

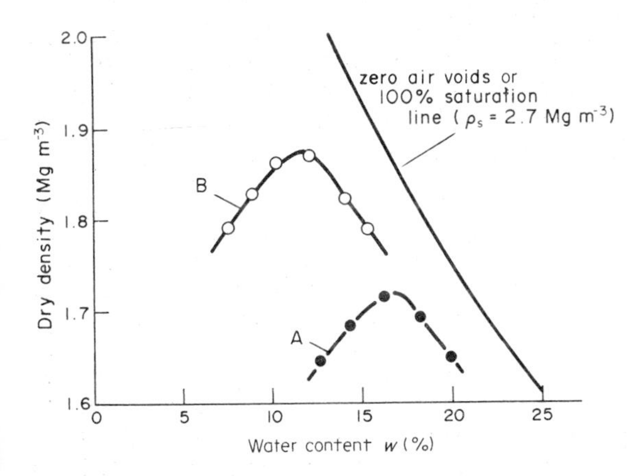

Figure 3
Typical standard Proctor (A) and modified Proctor (B) laboratory compaction curves

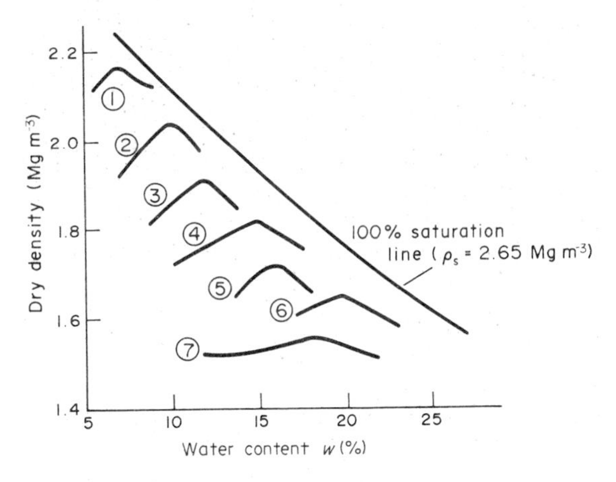

Figure 4
Water content/dry density relationships for seven soils compacted by the standard Proctor method: (1) well-graded silty sand; (2) well-graded clayey sand; (3) medium graded clayey sand; (4) sandy silty clay; (5) clayey silt; (6) fat clay; (7) poorly graded sand

Even at very high water contents and high compaction energies, the compaction curve never quite reaches the 100% saturation line. However, more energy will increase the maximum dry density and decrease the optimum water content (curve B, Fig. 3).

Different soils will have different shaped compaction curves even when compacted by the same compactive effort (Fig. 4). Well-graded soils will have a higher dry density than uniform soils, and, as plasticity increases, the maximum dry density tends to decrease. Different types of compaction, that is, impact, static or kneading, produce different soil structures and therefore somewhat different shaped compaction curves (Holtz and Kovacs 1981).

9. *Properties of Compacted Cohesive Soils*

The structure, and thus the engineering properties, of compacted cohesive soils depends on the type of compaction, compactive effort, soil type and water content. When soils are compacted dry of the optimum water content, the structure is always flocculated and the engineering properties are essentially independent of the type of compaction. However, such soils are much more sensitive to changes in water content because of environmental factors. The type of compaction has a significant effect on soils compacted wet of optimum. As the water content increases, the soil fabric becomes increasingly more oriented as clusters of particles tend to align themselves along zones of movement caused by the compacting device.

Table 1
Comparison of soil properties: dry of optimum versus wet of optimum compaction

Property	Comparison
1. Structure:	
A. Particle arrangement	Dry side more random
B. Water deficiency	Dry side more deficient; thus imbibes more water, swells more, has less (negative) pore pressure
C. Permanence	Dry side structure more sensitive to change
2. Permeability:	
A. Magnitude	Dry side more permeable
B. Permanence	Dry side permeability reduced much more by permeation
3. Compressibility:	
A. Magnitude	Wet side more compressible in low pressure range, dry side in high pressure range
B. Rate	Dry side consolidates more rapidly
4. Strength:	
A. As molded:	
a. Undrained	Dry side much higher
b. Drained	Dry side somewhat higher
B. After saturation:	
a. Undrained	Dry side somewhat higher if swelling is prevented; wet side can be higher if swelling is permitted
b. Drained	Dry side about the same or slightly greater
C. Pore water pressure at failure	Wet side higher
D. Stress–strain modulus	Dry side much greater
E. Sensitivity	Dry side more likely to be sensitive
5. Shrinkage	Wet side greater
6. Swell	Dry side much greater

Table 1 compares different soil properties wet and dry of optimum.

See also: Construction Materials: Crushed Stone; Construction Materials: Dimension Stone; Construction Materials: Granules; Construction Materials: Industrial Minerals; Construction Materials: Lightweight Aggregate; Construction Materials: Sand and Gravel

Bibliography

American Society for Testing and Materials 1986 *Annual Book of ASTM Standards*, Vol. 04.08, *Soil and Rock; Building Stones*. American Society for Testing and Materials, Philadelphia, Pennsylvania

American Society of Civil Engineers 1978 *Soil Improvement: History, Capabilities, and Outlook*. Committee on Placement and Improvement of Soils, Geotechnical Engineering Division, American Society of Civil Engineers, New York

American Society of Civil Engineers 1987 *Placement and Improvement of Soils: A Ten Year Update*. American Society of Civil Engineers, New York

Harr M E 1977 *Mechanics of Particulate Media: A Probablistic Approach*. McGraw-Hill, New York

Holtz R D, Kovacs W D 1981 *An Introduction to Geotechnical Engineering*. Prentice-Hall, Englewood Cliffs, New Jersey

Mitchell J K 1976 *Soil Behavior*. Wiley, New York

Perloff W H, Barron W 1976 *Soil Mechanics*. Ronald Press, New York

Rowe P W 1972 The relevance of soil fabric to site investigation practice. *Géotechnique* 22(2): 195–300

Scott R F 1963 *Principles of Soil Mechanics*. Addison-Wesley, Reading, Massachusetts

Terzaghi K, Peck R B 1967 *Soil Mechanics in Engineering Practice*, 2nd edn. Wiley, New York

US Bureau of Reclamation 1974 *Earth Manual*, 2nd edn. US Bureau of Reclamation, Engineering and Research Center, Denver, Colorado

Winterkorn H F, Fang H-Y 1975 *Foundation Engineering Handbook*. Van Nostrand Reinhold, New York

R. D. Holtz
[Purdue University, West Lafayette, Indiana, USA]

Construction Materials: Granules

Granules as used in construction (specifically roofing materials) are defined as natural or synthetic mineral grains smaller than gravel and larger than sand, used primarily to form a protective and decorative coating on the weather-exposed surface of composition roofing.

1. Historical Development

Roofing granules in today's market are the products of a long technological evolution. Primitive constructions of burlap saturated with tar appeared as early as 1780. Dusting of talc and mica on the coated surfaces to keep the roofing from sticking to itself in storage, represents an early improvement to the original construction product. Sometimes pigmentation was added with the dusting material for initial color. Exposure to the sun caused the asphalt to become brittle and it eventually cracked and leaked. This led to the first important use for crushed mineral products on asphalt-coated shingles and provided the impetus for the subsequent growth of the asphalt roofing industry. When a layer of crushed mineral was applied on the exposed surface the asphalt coating was protected from ultraviolet light for long periods of time.

Screened waste from slate quarries was the most common granule used in the early asphalt roofing products after the significant technological advances of the 1900s. Records show that coarse slate screenings were used for asphalt roofing as early as 1906. As the asphalt roofing industry further developed, other naturally colored rocks were used in roofing surfaces. These included sand, quartz, mica, feldspar, talc, glass, slag, basalt, granite, rhyolite and oyster shells. In addition, a number of manufactured granular mineral materials, such as crushed brick and tile, fired clay and white porcelain, were used. By 1910, mineral-surfaced asphalt roofing was becoming better known as an economical, long-lasting serviceable product.

The first artificially colored roofing granule was produced in 1914, by running molten slag into a solution of alkaline metal silicate which was absorbed by the porous slag. Various colors were obtained by adding pigments to the silicate coating and by subsequently heat treating the granules. In 1922 another technique was developed in which green slate granules were saturated with copper sulfate solution and heated to 760 °C to form copper oxide. Many other granule coloring processes were patented by 1970 and probably the most significant were the sodium silicate coatings introduced about 1975.

2. *Specifications and Manufacture*

A rock deposit suitable for roofing granules must meet the following requirements.

(a) Resist weathering: the weathering properties of a roofing granule are of primary importance. If it does not adequately withstand the elements of nature, the granule will not provide the necessary protection to the asphalt matrix, nor will it give the color performance to the roofing shingle.

(b) Accept coloring: after crushing, many mineral granules do not have the surface characteristics necessary for granule base—for example, quartz with its conchoidal fracture is very difficult to coat. Also, the mineral constituents must be chemically inert to processing chemicals. Many rocks spall in the heating process because of their thermal expansion characteristics and the presence of molecular water. This condition alters the grade and leaves uncolored surfaces on the granule. To maintain acceptable standards of quality, it is of the utmost importance that the rock base be of uniform quality and homogeneous in color as well as in other physical characteristics.

(c) Low porosity: high porosity allows the penetration of moisture and the seepage of water into the asphalt shingle, causing it to blister and/or crack. It must be completely opaque to ultraviolet light. The actinic rays of the sun rapidly deteriorate asphalt roofing. The quality of toughness is especially important during the manufacturing cycle where unavoidable attrition by conveyors, chutes, screens and elevators could severely reduce the size of the original grains.

The production of granules involves exploration, plant site location, quarrying, crushing, screening and coloring. Detailed geological explorations, laboratory evaluations and marketing considerations determine the plant location. Accessibility to mass transportation routes, water, power and labor sources are important economic factors to be considered. Quarrying operations are similar to those of rock aggregates. Roofing granules are produced principally in open-pit operations. After removal of overburden, drilling and blasting are followed by hauling the rock to the primary crusher. Crushing is carried out by either jaw or gyratory crushers. Depending on the quarry and weather conditions, a washing operation might be necessary. Secondary crushing is usually performed by cones, rolls or hammer mills in circuit with vibrating screens. In the modern granule plant, as illustrated in Figs. 1 and 2, elaborate dust control systems are necessary, not only to protect employees' health but also to comply with local, State and Federal laws (in the USA). The crushed and screened granules are transported to the coloring section of the plant, where they are mixed with pigments and other coating materials in rotary mixers. The coated granules are then fed into a rotary kiln for the final heat treatment to mature the bond. Finally, the granules are discharged from the kiln into a rotary cooler or similar device. Rescreening on vibratory screens is usually necessary for quality control. All modern granule plants are supported by elaborate quality control departments in order to meet the stringent requirements imposed by rigid product specifications.

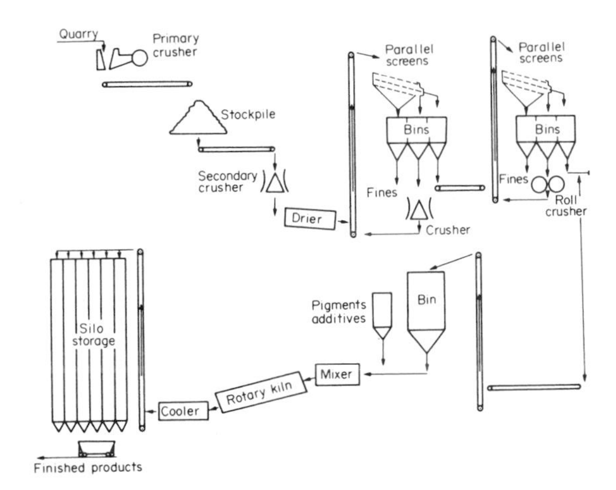

Figure 1
Flowsheet for a typical roofing granule plant (after Jewett et al. 1975)

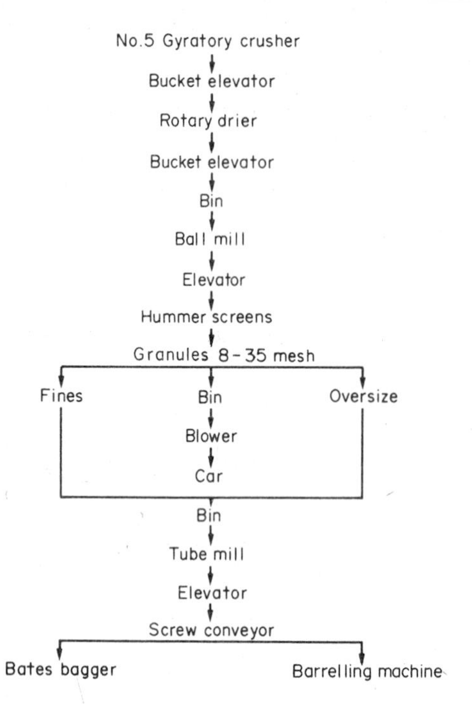

Figure 2
Flowsheet of a mill for manufacturing slate granules

Recent advances have been made in synthetic granules technology. Methods have been developed for the production of granules, organic and inorganic, for use in chemical and pollution technologies with controlled morphology, particle size, particle shape and flow characteristics, and extreme inertness to various liquid and gaseous media in purification applications.

3. Uses

The principal use of granules is in asphalt roofing production. With the increasing importance of the quality and color of roofing materials, ceramic-coated artificially colored granules have become the dominant mineral surfacing on asphalt roofing products.

Today the utilization range of synthetic granules, organic and inorganic, goes far beyond roofing shingles, covering areas of pollution control, industrial filtration and the production of high-technology ceramics. An example of the use of granules in pollution control technology involves the various filtering devices such as granular bed filters used in industrial cleaning processes. Mathematical models have been developed describing the filtration efficiency and pressure drop across granular bed filters. Clarification of mine waters by filtration through granular charges is well known.

Many ceramic processes employ granulated raw materials and raw material mixes in the production of final ceramic ware. Such is the case in the refractories industry. Sintered granular materials enable the ceramic technologist to produce refractory wares of high density, which is necessary for improved high-temperature mechanical properties and resistance to slag attack.

See also: Construction Materials: Crushed Stone; Construction Materials: Dimension Stone; Construction Materials: Fill and Soil; Construction Materials: Industrial Minerals; Construction Materials: Lightweight Aggregate; Roofing Materials

Bibliography

Bowles O 1922 *The Technology of Slate,* US Bureau of Mines Bulletin No. 218. US Government Printing Office, Washington, DC, pp. 87–103

Bowles O 1934 *The Stone Industries,* 1st edn. McGraw-Hill, New York, pp. 461–64

Brown J A 1960 Granules. In: Gillson J L (ed.) 1960 *Industrial Minerals and Rocks (Nonmetallics Other Than Fuels),* 3rd edn. American Institute of Mining, Metallurgical and Petroleum Engineers, New York, pp. 443–54

Henry E C 1948 Utilization of slate waste. *J. Can. Ceram. Soc.*17: 27–31

Jewett C L, Collins R C, Weaver L W, McSheer T 1983 Roofing granules. In: Lefond S J (ed.) 1983 *Industrial Minerals and Rocks (Nonmetallics Other Than Fuels),* 5th edn., Vol. 1. American Institute of Mining, Metallurgical and Petroleum Engineers, New York, pp. 203–12

Johnstone S J 1954 *Minerals for the Chemical and Allied Industries.* Wiley, New York, pp. 461–64

Josephson G W 1949 Granules. In: Dolbear S H (ed.) 1949 *Industrial Minerals and Rocks (Nonmetallics Other Than Fuels),* 2nd edn. American Institute of Mining, Metallurgical and Petroleum Engineers, New York, pp. 404–16

Larabee D M 1968 Roofing granules. *Mineral Resources of the Appalachian Region,* US Geological Survey Professional Paper 580. US Government Printing Office, Washington, DC, pp. 252–54

Saxena S C 1980 *Mechanistic Modeling of Granular Bed Filters for the Removal of Particulate Matter from High-Temperature High-Pressure Gas Streams: A Critical Assessment,* Special Publication DOE/METC/SP-80/24. US Department of Energy, Washington, DC

Sherrington P J, Oliver R 1981 *Granulation.* Heyden, New York

Subramaniam A F 1967 Charnockites and the granulites of southern India: A review. *Dan. Geol. Foren.* 17: 473–93

Tohata H 1979 Development of granulation technology. *Kemikaro Enjiniyaringu* 24(1): 76–81

R. S. Kalyoncu
[US Bureau of Mines, Tuscaloosa, Alabama, USA]

Construction Materials: Industrial Minerals

Natural industrial minerals used for construction are crushed stone, sand and gravel, dimenson stone, fill and soil, and granules. All except granules are used either alone or with a paste or a binder, such as cement or asphalt, in the construction of, for example, highways, airports, railroads, dams, homes and public buildings. Granules have a specialized use in making asphalt roof materials. Construction materials have so many applications, although often disguised, that the average person fails to recognize their true importance. In 1980, the USA, the leading consumer of construction materials, produced more than 2 billion tonnes of sand and gravel and crushed stone—about 9 tonnes per person. Except for dimension stone and granules, construction materials are generally produced near the places where they are used. In 1980, sand and gravel and fill and soil were produced in all 50 states; crushed stone was produced in 48. Because dimension stone and granules require special properties not found widely distributed, they have more restricted production.

The natural industrial minerals used for construction are characterized by low unit cost and high unit weight. Because they are generally consumed in highly populated areas, places inhospitable to mining operations, they must sometimes be transported from surrounding areas. In some places the cost of transportation to market exceeds the cost of production. Processing natural construction materials is generally simple and, if needed at all, consists of breaking, crushing, washing or screening to size. In general, as processing increases for a product, the cost increases. Dimension stone, for example, may have a high unit cost when it requires cutting, carving or polishing to meet specifications.

Although chemical properties of a rock may be important for some construction applications, physical properties are far more important. In some construction projects, the material is used to occupy space or to provide the base on which other materials are used. The material must therefore be competent yet chemically inert in the environment in which it is used. For example, when rock is crushed for aggregate it should not split along cleavage planes or thin partings but form equidimensional fragments; the aggregate should be hard, low in porosity, strong and free of reactive or nondurable materials. Government agencies that consume large amounts of materials generally set specifications for use, but these agencies may instead follow the recommendations of an association or a society that has established specifications for using particular products. Some users have their own specifications, which vary according to availability and cost of the materials.

See also: Construction Materials: Crushed Stone; Construction Materials: Dimension Stone; Construction Materials: Fill and Soil; Construction Materials: Granules; Construction Materials: Lightweight Aggregate; Construction Materials: Sand and Gravel

Bibliography

Brobst D A, Pratt W P (eds.) 1973 *United States Mineral Resources*, US Geoglogical Survey Professional Paper 820. US Geological Survey, Washington, DC

Lefond S J (ed.) 1975 *Industrial Minerals and Rocks*, 4th edn. American Institute of Mining, Metallurgical and Petroleum Engineers, New York

US Bureau of Mines 1980 *Mineral Facts and Problems*. US Government Printing Office, Washington, DC

D. D. Carr
[Indiana Geological Survey, Bloomington, Indiana, USA]

Construction Materials: Lightweight Aggregate

Lightweight aggregates are classified as those minerals, natural rock materials, rock-like products and by-products of manufacturing processes which are used as bulk fillers in concrete building blocks, lightweight structural concrete, precast concrete structural units, road surfacing materials, plaster aggregates and insulating fill. Other uses include soil conditioners, architectural wall covers and suspended ceilings.

Lightweight aggregates for concrete masonry units, according to the American Society for Testing and Materials (1986), consist of "aggregates of low density used to produce lightweight concrete, including: pumice, scoria, volcanic cinders, tuff and diatomite; expanded or sintered clay, shale, slate, diatomaceous shale, perlite, vermiculite or slag; and end products of coal or coke combustion."

Materials used for lightweight aggregate may be classified into four general groups:

(a) natural lightweight aggregate materials such as pumice, scoria, tuff, breccia and volcanic cinders;

(b) manufactured structural lightweight aggregates prepared by pyroprocessing clay, shale or slate in rotary kilns or on travelling-grate sintering machines;

(c) manufactured insulating ultralightweight aggregates prepared by pyroprocessing perlite, vermiculite and diatomite; and

(d) by-product lightweight aggregate prepared by crushing and sizing foamed and granulated slag, cinders and coke breeze.

The last group of lightweight materials, with the exception of slag, has become less significant with time and will not be addressed in this article.

The principal properties sought in materials which are used as lightweight aggregates vary according to the end use. Lightweight aggregates are normally valued for the following characteristics:

(a) lightweight, thermal and acoustical insulation properties;

(b) high fire resistance, toughness and substantial compressive strength;

(c) low water absorption and good resistance to freezing and thawing;

(d) low shrinkage on drying and minimal thermal expansion;

(e) good bonding properties with cement; and

(f) chemical inertness, good elastic properties and good abrasion resistance.

The list of the varieties and the uses of materials as lightweight aggregates is extensive. Most lightweight aggregates and aggregate materials are bulky and low in unit value. Usually they cannot be transported great distances in their final form yet still remain competitive with alternative building materials. Some, which have an economical transportation means available to them, service international markets. Special materials, such as perlite, vermiculite and diatomite, service both national and international markets. This is owing to their relatively high unit value. They are normally shipped prior to pyroprocessing, which expands these materials to a much larger size.

Pumice, scoria and expanded slag may be used extensively and in substantial quantities in some areas. These may not be available economically in other locations. The more specialized the end use, the greater the market area served. The greater the unit value of these materials, the greater the geographical distribution and use.

1. Naturally Occurring Lightweight Aggregates

Naturally occurring lightweight aggregates, such as pumice, scoria, tuff and volcanic cinders, are products of explosive volcanic eruptions. Pumice and pumicite are produced by the violent expansion of dissolved gases in a viscous silicic lava such as rhyolite or dacite. The relatively high viscosity and the volatile content of this type of lava are an explosive combination. During formation the pressure of dissolved gases is suddenly released and the expansion of the gases generates frothy masses of expelled lava. These masses cool very quickly in the atmosphere to form fragments of glass containing innumerable bubble-shaped cavities. Each bubble cavity or cell is sealed off from its neighboring cell by a thin membrane of glass, making the permeability of pumice very low. Pumicite consists of grains, flakes, threads and shards of glass finer than 4 mm in diameter.

Scoria, the mafic counterpart of pumice, is a rusty-red to black pyroclastic material, generally of andesitic to basaltic composition that accumulates in volcanic cinder cones. The vesicles of scoria tend to be coarser and more variable than those of pumice. The cell walls are thicker and the texture is finer grained and more compact. Scoriaceous material less than 25 mm in particle diameter is termed volcanic cinders.

The cellular nature of pumice, pumicite, scoria and volcanic cinders provides these materials with two main properties of value: they are lightweight and have insulating ability. These materials are used primarily in the construction industry as aggregates in lightweight structural concrete, as a plaster aggregate and as loose-fill insulation in structures. Coarse pumice aggregate is used extensively in the manufacture of concrete blocks and masonry products where its low specific gravity reduces the weight of concrete blocks, which makes handling easier. Pumice also minimizes the bulk of foundations and supporting columns in multistory construction, reducing the reinforcement required in the suspended floors. It has high insulation and good acoustical properties, is fireproof and resists condensation, mildew and vermin. Block pumice ranges in size from 5 cm to more than 1.5 m in diameter. Larger blocks may be sawed and used architecturally for wall coverings and suspended ceilings. Scoria is used as a decorative stone in landscaping. Other uses include material for road surfacing, ballast and roofing granules. In 1985 crushed volcanic cinder and scoria sold or used by producers in the USA totalled 2.4 million tonnes for all uses. Use as lightweight aggregates consumed 371 870 t.

Commercial production of volcanic lightweight materials tends to be concentrated in two main zones—the circum-Pacific belt and a belt that spreads from the Mediterrean Sea, along the Himalayas and into the East Indies. The 1985 production in the USA of pumice and pumicite was 368 242 t; it was concentrated in the western USA with 22 mines operated by 21 companies in eight states.

In 1985 world production of pumice and related volcanic materials declined by 4% from the 1984 production level to 11 million tonnes. The combined output of Greece and Italy accounted for 65% of the total world production. France, Spain, the USA and Yugoslavia produced 63% of the remaining 35%.

These materials are normally mined by open-pit methods. The exact method of extraction depends upon the degree of consolidation of the raw materials. In 1984, US pumice was priced at US\$8.98 ton^{-1}. Imported pumice from Greece was quoted at US\$6.47 ton^{-1} for use in concrete masonry products.

2. Manufactured Lightweight Aggregates

Manufactured lightweight aggregates are produced by either the rotary-kiln or sintering-grate processes. Clays, shales and slate used in the manufacture of brick, tile, sewer pipe and other structural clay products are frequently satisfactory sources for manufactured lightweight aggregates. These materials must contain certain impurities, such as iron oxides, carbon, sulfides, alkalis or alkaline earths, to be suitable for consideration as a lightweight aggregate. When the shale, clay or slate is heated to incipient fusion the impurities decompose and generate gases, which are trapped in the viscous glassy exterior of the particle. The rotary-kiln process produces rounded particles which have a glassy impervious coating and a fine internal pore structure with unconnected pores. This process produces an aggregate that is very low in weight and also low in absorption.

The sintering process uses similar materials that have been mixed with a proper amount of combustible materials, such as coke, coal or petroleum residue. In this process the material is fired on a sintering grate, where it partially melts; the added combustibles burn out leaving voids. Some sintered materials will expand slightly. The product from this process is then crushed to specification size. The exposed pore structures of the angular particles make the absorption higher than that of rotary-kiln fired aggregates. When used in concrete, sintered aggregate requires more cement to obtain a specific compressive strength than other aggregates because the capillary action of the pore structure tends to draw additional amounts of the cement into the void.

In 1985, consumption of clay, shale and slate in the manufacture of lightweight aggregate increased by 8% from that consumed in 1984. The largest category of use was in the production of concrete block (2.2 million tonnes), and the second largest use was in structural concrete (1.2 million tonnes). Highway surfacing, recreational and horticultural uses consumed the balance (544 200 t). In the USA the average price for all uses in 1985 was US$6.11 ton^{-1}.

3. Manufactured Ultralightweight Aggregates

3.1 Perlite

Perlite is a glassy volcanic rock with a rhyolitic composition containing 2–5% combined water. True perlite is characterized by a system of macroscopic and microscopic concentric onion-like structures, produced as a result of shrinkage during cooling. Commercially, the term includes any volcanic glass that will expand or "pop" when heated quickly, forming a lightweight frothy material.

Perlite occurs naturally as steep-sided block-type lava domes restricted in area which are formed by the extrusion of highly viscous magmas. Perlite deposits are restricted to Tertiary age or younger deposits of rhyolitic composition. Rhyolitic glasses are unstable and devitrify with age into a microcrystalline aggregate of quartz and feldspar. They are also susceptible to hydrothermal alteration, forming clay minerals and zeolites.

Perlite differs from other volcanic glasses principally by its combined water content. Perlite is formed by the secondary hydration of obsidian after its implacement. The combined water exists in two forms—molecular and hydroxyl. The ratio of these two kinds of water is different for perlite of different origins. It is the volatilization of the combined water that provides the energy to "pop" the perlite into a frothy material.

When grains of crushed perlite are brought to the point of incipient fusion, the contained water is converted to steam and the grains swell into light, fluffy cellular particles. Expansion takes place in either a stationary vertical furnace or a horizontal rotary furnace. Some preheating may be utilized to drive off excessive surface water on particles. The temperature at which expansion takes place lies in the range 760–1090 °C. A volume increase of 10–20 times the original value is normal.

Expanded perlite has a low thermal conductivity, high sound absorption and a high resistance to heat. More than half of the perlite produced goes into the construction industry as aggregate in insulation board, plaster and concrete. It is also used as a rooting medium and soil conditioner.

In 1985, crude perlite produced in the USA was mined by 12 operations in six western states. New Mexico was the major producer with 86% (557 800 t) of the total production. Other sources were in Arizona, California, Colorado, Idaho and Nevada. In 1985, expanded perlite was produced by 64 plants in 31 states. Perlite, because of its lightweight characteristics and limited distribution, can be shipped long distances prior to expansion and still remain competitive with other materials. The average price of processed perlite in 1985 was US$33.85 ton^{-1}. Greece and the USSR are other major producers of perlite. The major deposits in Greece are located on the islands of Milos and Kos. The deposits in the USSR are located in Armenia, Kazakhatan and Buryat. Smaller deposits are located in Turkey, Italy, Japan, Australia, New Zealand and South Africa.

3.2 Vermiculite

The term vermiculite is used to classify or define a small group of minerals that have chemical affinities with montmorillonite, chlorite and biotite. Physically, vermiculites are similar to the micas, having the typical cleavage that allows splitting into thin flexible flakes. The mineral is an alteration product of biotite and other micas. The three broad types of deposits are:

(a) those formed within large ultramafic intrusive masses, such as pyroxenitic plutons cut by syenite or alkalic granite and by carbonatitic rocks and pegmatites;

(b) those associated with ultramafic intrusive bodies, such as dunite and unzoned pyroxenite, and peridotite cut by pegmatite and syenitic or granitic intrusives; and

(c) those formed from ultramafic metamorphic rocks such as amphibole schist in contact with pyroxenite or peridotite and cut by pegmatite.

Vermiculite has a hardness of 2.1–2.8, a specific gravity of 2.5, and a bronze to yellowish-brown or black color. The unit cell of vermiculite contains a layer of water. In rapid heating (to above 870 °C), the water flashes into steam, forcing the layers apart at right angles to the plane of cleavage. Some of the resulting particles have a worm-like appearance; hence, the name "vermiculite," which is the Latin for "little worm." Pure vermiculite can expand to 30 times its original volume, but generally averages 8–12 times.

In 1985, 284 800 t of vermiculite concentrate valued at $32.4 million was produced in the USA. In the same year, world production of vermiculite concentrate was estimated to be more than 498 850 t. More than 90% of world production comes from Libby, Montana and Palabora, South Africa. Other noted deposits occur in the Blue Ridge Mountains area of North Carolina, the Green Springs area of Louisa County, Virginia, and the Enoree district of South Carolina. Other countries with significant vermiculite production are Japan, Brazil and Argentina.

See also: Construction Materials: Crushed Stone; Construction Materials: Dimension Stone; Construction Materials: Fill and Soil; Construction Materials: Granules; Construction Materials: Industrial Minerals; Construction Materials: Sand and Gravel

Bibliography

American Society for Testing and Materials 1986 *Annual Book of ASTM Standards*, Part 16, *Lightweight Aggregates for Concrete Masonry Units*. ASTM, Philadelphia, Pennsylvania

Harben P W, Bates R L 1984 *Geology of the Nonmetallics*. Metal Bulletin, New York, pp. 64–76, 282–84

Kadey F L Jr 1983 Perlite. In: Lefond S J (ed.) 1983 *Industrial Minerals and Rocks*, 5th edn. American Institute of Mining, Metallurgical and Petroleum Engineers, New York, pp. 997–1015

McCarl H N 1983 Lightweight aggregates. In: Lefond S J (ed.) 1983 *Industrial Minerals and Rocks*, 5th edn. American Institute of Mining, Metallurgical and Petroleum Engineers, New York, pp. 81–95

Peterson N V, Mason R S 1983 Pumice, Pumicite, and volcanic cinders. In: Lefond S J (ed.) 1983 *Industrial Minerals and Rocks*, 5th edn. American Institute of Mining, Metallurgical and Petroleum Engineers, New York, pp. 1079–84

Strand P R, Steward O F 1983 Vermiculite. In: Lefond S J (ed.) 1983 *Industrial Minerals and Rocks*, 5th edn. American Institute of Mining, Metallurgical and Petroleum Engineers, New York, pp. 1375–81

US Bureau of Mines 1987 *Minerals Yearbook 1985*. USBM, Washington, DC

B. H. Mason
[The France Stone Co., Waterville, Ohio, USA]

Construction Materials: Sand and Gravel

Sand and gravel are among the more important materials of the construction industry, and therefore are essential resources among the heavily populated, industrialized nations of the world. Sand and gravel are used for road base, fill, road surfacing, and for aggregate in both bituminous and Portland cement mixes. Sand is used alone in mortar and plaster.

Sand and gravel are widely distributed throughout the world, but because they are high-bulk, low-volume materials they are almost invariably mined near places where they are used, and except in border areas they are unimportant in international trade.

World production of sand and gravel was estimated by the US Bureau of Mines in 1977 to be about 9000 Mt per year; most of that production was concentrated in Europe and North America. Heavily industrialized nations with large populations, like the USA and the USSR, produce as much as 1000 Mt of sand and gravel per year. Other industrialized nations with smaller populations, such as the UK, France and the Federal Republic of Germany, produce 100–200 Mt per year. Nonindustrialized nations generally have negligible production of sand and gravel for construction.

1. Definitions

The definitions of terms relating to sand and gravel that are used throughout the industry in the USA are those proposed by the American Society for Testing and Materials (ASTM C125-76). Sand is defined as fine aggregate resulting from natural disintegration and abrasion of rock or processing of completely friable sandstone. As fine aggregate the material must pass the 3/8 in. (9.5 mm) sieve and almost entirely pass the No. 4 (4.75 mm) sieve. Manufactured sand is the product resulting from crushing rock, gravel or sand.

Gravel is coarse aggregate resulting from natural disintegration and abrasion of rock or processing of weakly bound conglomerate. Coarse aggregate must be predominantly retained on the No. 4 sieve. Crushed gravel is the product resulting from artificially crushing gravel and having substantially all fragments with at least one face resulting from fracture.

2. Composition

Gravel may be composed of many rock types that differ in their physical properties. Igneous rocks form from molten materials. Those that are intruded deep within the earth's crust cool slowly and form a coarsely crystalline texture. Gravels derived from these rocks tend to be durable, tough and chemically inert. Molten material extruded onto the earth's surface forms finely crystalline igneous rocks. Some of these rocks may contain minerals that can produce undesirable chemical reactions within concrete, but most of them are acceptable as aggregate. Certain porous volcanic types are finding increased use as lightweight aggregate.

Metamorphic rocks have been altered from their original state by high temperature and pressure well below the earth's surface. Many metamorphic rocks produce high-quality aggregate, but schists and gneisses with high mica content deteriorate rapidly in gravel deposits, and slates contain cleavage planes that reduce the durability of the rock.

Sedimentary rocks are deposited on the earth's surface by water, wind, chemical action or organic activity. Chemical rocks include salt, gypsum and anhydrite; all are unsuitable for use as aggregate because of their water solubility. Chert is a special type of chemical rock that forms as an inclusion in a host rock. Certain types of chert have long been known to produce severe problems in aggregate because of their potential to cause deleterious chemical reactions with the alkalis in Portland cement. Other types of chert are not susceptible to dissolution in Portland cement, and because chert has a high resistance to abrasion, these types may be considered high-quality aggregate.

Organic rocks include coal, which is a deleterious component of aggregate because of its low strength, and limestone and dolomite, which, except for certain argillaceous and siliceous varieties, are acceptable for use as aggregate.

Sedimentary rocks deposited by water or wind include conglomerates, sandstone, siltstone and shale. Highly cemented varieties of these rocks are usually acceptable, but many of them are too porous or their individual grains are not sufficiently cemented together to produce durable rock.

Sand is produced when rocks are more completely broken down, many of them to their individual mineral components. Sands tend to be composed predominantly of quartz, because it is the most abundant mineral in many rock types and is more resistant to chemical and mechanical breakdown than most other minerals. Quartz is a desirable component of fine aggregate because of its hardness, durability and relative insolubility. Besides quartz, sands may contain labile minerals, mineral aggregates, chert, coal or mica, which reduce the quality of the aggregate.

3. Origin and Occurrence

3.1 Introduction

Sand and gravel in lower latitudes accumulate in rivers, on beaches or as residuum. The deposits tend to be lithologically simple, reflecting the sorting and abrading action of transport mechanisms and the limited size of the source area from which the materials were derived. Besides, in humid low-latitude regions chemical weathering may be so intense that all but the most resistant rocks are destroyed.

Higher-latitude deposits originated directly or indirectly from the glaciers that covered large portions of the mid and upper latitudes during the Pleistocene epoch (Ice Age). Glaciers originated in subpolar regions, moved south hundreds of kilometers, and incorporated into their sediment load all rock types represented in the bedrock over which they passed. Sorting action is minimal, and chemical weathering does not take place during glacial transport. Crushing and plucking rather than abrasion are the dominant mode of mechanical breakdown. For these reasons the sediment load carried by glaciers ranges in size from clay to boulders, and tends to consist of angular clasts that may include rock types that are soft or susceptible to rapid chemical degradation.

Fortunately, glacial sand and gravel deposits are almost invariably the result of the concentration of the coarse-grained fraction of the glacial load by running water. Depending on the mode of origin, sorting and abrasion have altered the original material to varying degrees. Fluvial transport in ice-contact deposits such as kames and eskers is limited, but where sediments are carried away from the glacier, as in outwash fans and valley trains, the resulting deposits have the characteristics of fluvial sediments.

The geologic setting and mode of origin control not only the mineralogy of sand and gravel deposits but also the size and shape of the deposits and the amount and distribution of waste materials. These are the dominant factors affecting the quality and quantity of the aggregate and the cost of extracting it.

3.2 Nonglacial Deposits

Sand and gravel suitable for aggregate are found in many active and inactive stream channels. Although local bedrock outcrop may contribute deleterious materials to nearby deposits, the washing and abrading action of streams tends to produce clean deposits of durable material. The deposits may occur in channels or bars of the present river or on adjacent terraces that formed when the river flowed at a higher level (Fig. 1).

Alluvial fans occur in mountainous regions, where streams abruptly decrease in gradient as they emerge from mountains onto basins or plains. Coarse material is deposited near the mountain front, while finer-grained sediments are carried out onto the basin or plain. The composition of the deposit reflects the bedrock composition in the adjacent uplands. Poor-quality aggregate can result, because transport distances in many places are too short to allow abrasion of deleterious material. And as the fan builds, stream positions shift rapidly, causing a complex arrangement of sand and gravel and fine-grained waste material (Fig. 1).

Beach deposits consist of clean sand and gravel that tend to be relatively free of deleterious materials due to the abrading action of waves. The deposits tend to be thin, but many of them extend along the coast for

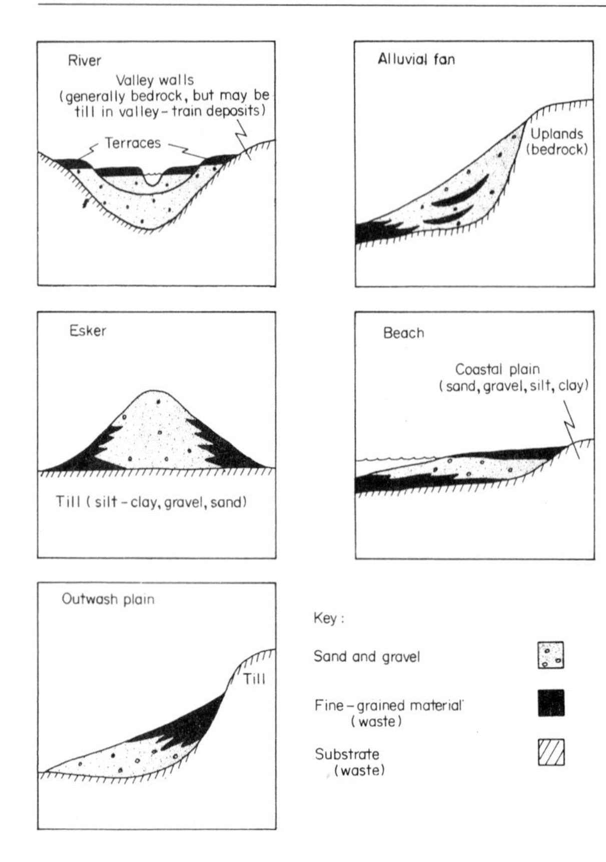

Figure 1
Cross sections of selected types of sand and gravel deposits showing their shapes and the relationships between waste and minable material

great distances, and they may extend both landward and seaward of the actual shoreline (Fig. 1). In nonglaciated areas with wide coastal plains they may be an important source of aggregate.

Residual deposits result from the concentration of the more durable products of deeply weathered rock. These deposits may show all stages of transport, from the near in situ concentration at the site of weathering to some stream transport. Extensive deposits of late Tertiary age cover wide areas. But many of these deposits consist of chert that may be generally unsuitable for use as aggregate without beneficiation, and many of these clasts have well-developed iron oxide coatings that can adversely affect the cement–aggregate bond. In some areas, such as the southeastern USA, residual deposits may be the only local source of aggregate.

3.3 Glacial Deposits

Ice-contact deposits are usually small, irregularly shaped, and vertically and laterally inhomogeneous. Wide variations in grain size within a deposit make it difficult to produce a uniform product from them, and interbedded fine-grained waste material can seriously impede mining operations (Fig. 1). Because of short transport distances, gravels tend to be angular, and in many places consist of labile clasts that adversely affect the quality of the aggregate.

Deposits that are carried beyond the glacial margin are more laterally and vertically uniform, and are therefore much easier to exploit than ice-contact deposits. Outwash plains are broad, thick deposits similar to alluvial fans in that local accumulations of fine-grained waste material can hinder mining operations (Fig. 1). Valley-train deposits occur along the rivers that drained glaciers, and are similar in many ways to nonglacial fluvial sediments. They differ from other fluvial deposits in the wide variety of rock types that may be included.

Use of depositional models to predict the location and characteristics of sand and gravel deposits in glacial terrains has increased recently due to advances in understanding how glaciers deposit these materials (Eyles 1983).

4. Mining and Processing

Mining operations range in size from those producing a few thousand tonnes per year of pit-run (unprocessed) material, to those producing several million tonnes per year and consisting of large, complex plants able to process a wide variety of construction products. Production has tended to be concentrated in these large plants, but because of steadily mounting transportation costs and improvements in the design of small plants, especially wheel-mounted portable processors, smaller plants nearer places of use may become more common.

Sand and gravel may be dredged from natural bodies of water, such as lakes and rivers. On land they may be dredged from wet pits where the deposit extends below the water table, or they may be excavated from dry pits where the water table is below the pit floor or where dry conditions are maintained by pumping.

There are two steps in producing sand and gravel for aggregate. The material is excavated, and then it is processed according to the character of the deposit and the intended use of the material produced (Fig. 2).

In rivers, lakes or oceans, sand and gravel are excavated by floating dredges, some of which have complete processing plants aboard. The material may be excavated by dragline, or by suction, bucketladder or cutterhead dredges. It is then either processed aboard or transferred to barges for transport ashore. Excavation is similar for wet pits on land, except that draglines operate from the shore and dredges are usually tethered to the shore, and the sand and gravel are immediately transferred to land-based processing plants by floating pipelines. In a dry pit, sand and

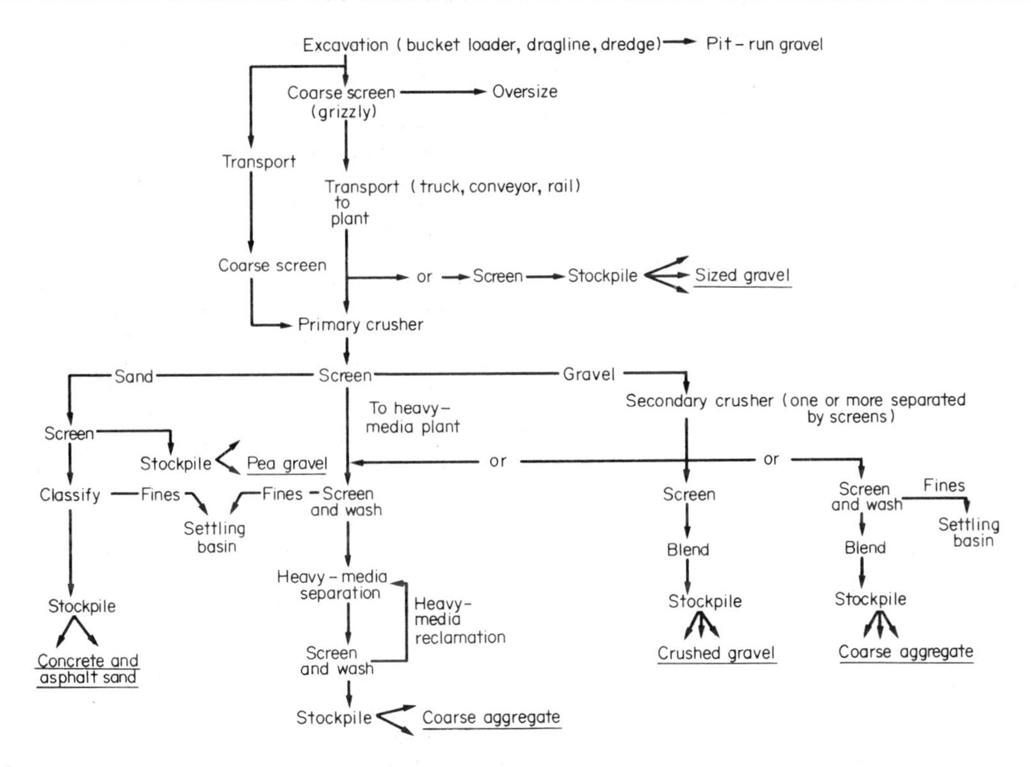

Figure 2
Flow chart showing the various stages in mining and processing sand and gravel. The processed material becomes more valuable as it moves through the system increasing its capability to meet more stringent specifications

gravel may be mined by front-end loaders, draglines or power shovels.

After excavation, the sand and gravel are usually run through a coarse screen to remove particles too large to be used in the plant. The resultant material can then be stockpiled as pit-run gravel to be used for fill, or it can be sent on to the processing plant by truck or rail. Where material is transported to the plant by conveyor belt, the coarse screen is usually combined with a primary crusher that reduces boulders to sizes that can be carried on the belt.

At the plant the aggregate is first screened to separate sand from gravel. Sand is fed to the sand plant, where secondary screening may remove pea gravel, and finer sand is size graded, usually in wet classifiers. Very fine sand is trapped in cyclones, and material which passes a No. 200 screen (0.074 mm) is passed in slurry form to a settling basin, where the water is reclaimed or drained away.

Gravel-size material can enter one of two lines. In one line the material is fed into a crusher and then screened to produce a size-graded material suitable for use as road base or fill. In the other line, the material is crushed, screened and washed, producing a material that can be used for concrete or bituminous aggregate. After screening, the material can be stockpiled, or the various sizes can be blended to produce the different classes of aggregate.

In some plants, beneficiation of the raw material must take place. The most common type of beneficiation is heavy-media separation, whereby lightweight deleterious materials, such as coal, shale and chert, are floated off in mixtures of water, magnetite and ferrosilicon. Other types of beneficiation include specialized screens to remove flat particles, differential impactors to reduce soft particles to fines that may be washed away, and elastic fractionation, which relies on the difference in elasticity between hard particles and soft deleterious particles to effect separation.

Sand and gravel operations, like all extractive industries, place a heavy burden on the environment. They release large amounts of fine particulate matter into the air (or water in dredging operations), and noise levels near pits can be high. The industry uses different measures to ameliorate this burden. Pit roads can be periodically sprayed with water or chemical dust suppressants. Cyclones and settling ponds can be used to trap fine-grained waste material, and if water supply is critical, wash plants can be fitted with settling tanks, where effluent from the plant is treated with flocculants that cause fine particles to settle out rapidly so that wash water can be quickly recycled. Fences and berms can be used to cut noise levels, and landscaping can be used imaginatively not only to reduce noise levels but also to improve the appearance of the operation.

Many abandoned pits are near expanding population centers, and the large areas they occupy have considerable value in land-short urban areas as potential residential or commercial developments or as recreational areas. The industry is becoming increasingly aware of the value of successfully reclaimed sand and gravel pits, and some firms now routinely plan their mining strategy to minimize the adverse environmental aspects of the operation and to maximize the potential of the reclaimed land.

5. *Quality*

For sand and gravel to be suitable aggregate, they must be capable of being bound together by cementing agents to produce a durable material able to withstand stress for a long time. Asphalt or Portland cement are the most common cement materials, but certain sulfur compounds are potential asphalt substitutes. The aggregate must be clean, durable and chemically inert, and should have the proper shape. The aggregate and the cement should also have similar thermal characteristics. A series of rather stringent specifications, based on the performance records of various constituents of aggregate, has been written by the American Society for Testing and Materials (ASTM) to assess the probable performance of untested aggregate from new sources (Table 1). Most other specifications written by government agencies and private firms are based on these tests. Other countries have their own specifications and tests, but most of these regulations differ only slightly from those written by the ASTM.

Besides promoting uniformity in testing and specifications, the ASTM also acts as a clearing house for information dealing with the physical properties of materials. The organization in Europe that performs similar functions is the Réunion Internationale des Laboratoires d'Essais et de Recherches sur les Matériaux et les Constructions (RILEM).

The aggregate must first be clean—i.e., free of very fine particles, mica and organic matter. There also should be no coatings of such materials as clay, calcium carbonate, iron oxide or manganese that might inhibit the bonding of cement to particle. Fines can be determined by sieving the potential aggregate on a No. 200 (0.074 mm) sieve. In general, no more than 5% of a −200 mesh material will be allowed in aggregate. Excessive amounts of organic matter can be determined by a chemical colorimetric test, and mica and surface coatings can be determined by petrographic examination.

The aggregate must also be durable and free of deleterious particles that may affect the strength of the concrete. Deleterious materials, such as coal and lignite, can be identified by petrographic analysis, and soft or poorly lithified particles can be identified by the scratch-hardness test. The amount of these deleterious materials allowable in an aggregate is strictly limited. Among the tests for measuring the durability of an aggregate are the Los Angeles or Deval Abrasion tests, which determine the resistance of an aggregate to abrasion. The resistance to weathering is determined by the freeze–thaw test, or by the magnesium or sodium-sulfate soundness test. In both of these tests, particles are subjected to cycles of saturation and either freezing or drying. Crystals of ice or sodium or magnesium sulfate grow in the particle and cause accelerated splitting, flaking or crumbling. Many factors, including alumina (clay) content, carbonate mineralogy, specific gravity and hardness, contribute to the performance of rocks in the test, but the two with the greatest effect are the presence of fractures and laminae, and the absorption of water.

Another property of aggregate which is receiving increasing attention is its ability to withstand polishing under road conditions. It is desirable for aggregate, especially in bituminous mixes, to present a rough surface to increase the antiskid properties of the road. Such aggregate may contain hard minerals that resist polishing, such as quartz or chert, or softer rocks, such as siliceous limestone, containing harder impurities that continue to present fresh, rough surfaces as they wear down. Two tests currently in use measure the skid-resistance properties of aggregate under simulated and actual road conditions.

Other tests measure the properties of the aggregate which affect the behavior of the concrete during its preparation and while it is in use. The specific gravity is measured to calculate the volume occupied by the aggregate in Portland cement, and bulk specific gravity is required to exceed 2.50 for concrete aggregate in most specifications. Water absorption is measured, because the particle–cement bond is adversely affected when water is absorbed by aggregate particles in concrete. A limit of 5% water absorption is usually placed on concrete aggregate. The moisture content of the aggregate must also be known for proper preparation of concrete, and the total evaporable water and surface moisture, especially of fine aggregate, are routinely measured.

The shape and grain-size distribution of the aggregate are carefully controlled because these properties affect the void ratio and flowability of the concrete. Shape, as determined during petrographic examination, is especially important in determining the quality of aggregate. In concrete, rounded equant clasts are the most desirable because highly angular or flat particles reduce the workability of poured concrete, especially when it is used in forms for walls. In asphalt, however, crushed gravel and manufactured sand are preferred, because the asphalt bonds better to angular surfaces and angular clasts pack together

Table 1
Quality-control tests frequently performed on sand and gravel used for aggregate

ASTM[a] designation	Test	Purpose
		Sampling
D75-71	Sampling stone, slag, gravel, sand and stone block	Describes the methods used to take representative samples at the deposit or plant or on delivery
C702-75	Reducing field samples of aggregate to testing size	Describes methods used to reduce field samples to an appropriate size for various tests
		Deleterious materials
C117-76	Amount of material finer than 0.075 mm (No. 200 screen) in aggregate	Outlines the procedure for determining the total quantity of material finer than a standard No. 200 sieve
C40-73	Organic impurities in sands for concrete	Describes the method used to determine injurious organic compounds through a colorimetric test, and warns whether further tests may be necessary
C87-69	Effect of organic impurities in fine aggregate on strength of mortar	Determines the effect of organic impurities on mortar strength by means of a compression test on mortar of plastic consistency
C142-71	Clay lumps and friable particles in aggregates	Determines approximate amount of clay lumps and friable particles in natural aggregates through disaggregation and washing
C123-69	Lightweight pieces in aggregates	Determines the approximate percentage of lightweight pieces in aggregate by means of heavy-liquid separation
C586-69 C289-71 and C227-71	Potential alkali reactivity of concrete aggregates (rock-cylinder method (C586-69), chemical method (C289-71), or mortar-bar method (C227-71))	Determines the potential reactivity of aggregate with alkalis in Portland cement by rock-cylinder expansion, by chemical reactivity, or mortar-bar expansion
		Durability
C131-76	Abrasion of small-size coarse aggregate by use of the Los Angeles machine	Describes procedure used in testing resistance to abrasion of coarse aggregate smaller than 37.5 mm ($1\frac{1}{2}$ in.) in the Los Angeles machine
C535-69	Abrasion of large-size coarse aggregate by use of the Los Angeles machine	Describes procedure used in testing resistance to abrasion of coarse aggregate larger than 37.5 mm ($1\frac{1}{2}$ in.) in the Los Angeles machine
C88-76	Soundness of aggregates by sodium sulfate or magnesium sulfate	Determines the resistance of aggregates to disintegration by saturated solutions of sodium or magnesium sulfate to judge soundness of aggregate subjected to weathering under freeze-thaw conditions
C235-68	Scratch hardness of coarse aggregate particles	Identifies and determines the quantity of soft or poorly lithified particles in coarse aggregate on basis of scratch hardness
C295-65	Petrographic examination of aggregates for concrete	Outlines procedures to be used in examining aggregate by optical microscopes to determine physical and chemical properties and to describe and classify constituents, especially those that bear on the quality of the material
E274 and E303	Skid resistance of paved surfaces by using a full-scale tire (E274) and measuring surface frictional properties using the British portable tester (E303)	Determines the skid resistance of aggregate in pavement under actual or simulated road conditions
		Other properties
C136-76	Sieve analysis of fine and coarse aggregates	Determines the particle-size distribution of fine and coarse aggregate by using sieves with square openings
C128-73	Specific gravity and absorption of fine aggregate	Determines bulk and apparent specific gravity (to calculate aggregate displacement in concrete) and absorption of fine aggregate
C127-77	Specific gravity and absorption of coarse aggregate	Determines the bulk and apparent specific gravity (to calculate aggregate displacement in concrete) and absorption of coarse aggregate
C70-73	Surface moisture in fine aggregate	Determines in the field the surface moisture in fine aggregate by water displacement; may also be used for coarse aggregate with appropriate changes in method
C29-76	Unit weight of aggregate	Determines the unit weight of fine, coarse or mixed aggregates

[a] American Society for Testing and Materials 1979 *Annual Book of Standards*, Pt. 14. ASTM, Philadelphia, Pennsylvania

better. This combination produces a road surface less subject to flow under heavy loads.

The potential of chemical reactivity between aggregate particles and Portland cement must be considered. Most silicate and carbonate rocks are relatively inert. Some silicate rocks, chiefly opaline and chalcedonic chert, some volcanic and zeolitic rocks, and even some siliceous carbonate rocks can react with alkalis in cement and form alkali–silica gels. These gels can absorb water and expand, and therefore cause expansion failures in concrete. Among the properties of the deleterious rock types that determine the potential reactivity are a low degree of crystallinity and a large surface area due to high microporosity. Other factors also affect the reactivity, and the degree to which these factors combine can be determined by measuring the expansion of rock cylinders or mortar bars that have been saturated in alkaline solutions, or by measuring the degree to which siliceous minerals are dissolved in alkaline solutions.

The tests described above do not measure the actual behavior of the aggregate in cement, but instead measure those properties that are believed to influence performance. Because the behavior of aggregate in cement is a function of the complex interaction of many physical and chemical properties, the tests may be misleading for some aggregates. The only true test of the quality of aggregate is its actual performance in use.

See also: Concrete as a Building Material; Construction Materials: Industrial Minerals

Bibliography

American Society for Testing and Materials 1978 *Significance of Tests and Properties of Concrete and Concrete-Making Materials*, ASTM Special Technical Publication 169B. American Society for Testing and Materials, Philadelphia, Pennsylvania

Bates R L 1969 *Geology of the Industrial Rocks and Minerals*. Dover Publications, New York

Dunn J R 1983 Aggregates—sand and gravel. In: Lefond S J (ed.) 1983 *Industrial Minerals and Rocks*, 5th edn., Vol. 1. American Institute of Mining, Metallurgical and Petroleum Engineers, New York, pp. 96–110

Evans J R 1978 *Sand and Gravel*, Mineral Commodities Profiles MCP-23. US Department of the Interior, Washington, DC

Eyles N (ed.) 1983 *Glacial Geology: An Introduction for Engineers and Earth Scientists*. Pergamon, Oxford

Goldman H B, Reining D 1975 Sand and gravel. In: Lefond S J (ed.) 1975 *Industrial Minerals and Rocks*, 4th edn. American Institute of Mining, Metallurgical and Petroleum Engineers, New York, pp. 1027–42

Lenhart W B 1960 Sand and gravel. In: Gillson J L (ed.) 1960 *Industrial Minerals and Rocks*, 3rd edn. American Institute of Mining, Metallurgical and Petroleum Engineers, New York, pp. 733–58

Neville A 1973 *Properties of Concrete*. Wiley, New York

Orchard D F 1962 *Concrete Technology*, Vol. 1, *Properties of Materials*. Contractors Record, London

Schellie K L 1977 *Sand and Gravel Operations—A Transitional Land Use*. National Sand and Gravel Association, Washington, DC

G. S. Fraser
[Indiana Geological Survey, Bloomington, Indiana, USA]

Cork

Cork is a natural material with a remarkable combination of properties. It is light yet resilient; it is an outstanding insulator for heat and sound; it has exceptional qualities for damping vibration; has a high coefficient of friction; and is impervious to liquids, chemically stable and fire resistant. These attributes have made cork commercially attractive for well over 2000 years.

1. Origins, Chemistry and Structure

Commercial cork is the outer bark of *Quercus suber* L., the cork oak. The tree, first described by Pliny in AD 77, is cultivated for cork production in Portugal, Spain, Algeria, France and parts of California. As Pliny said, the cork oak is relatively small and unattractive; its only useful product is its bark which is extremely thick and which, when cut, grows again.

Cork occupies a special place in the history of microscopy and of plant anatomy. When Robert Hooke perfected his microscope around 1660, one of the first materials he examined was cork. What he saw led him to identify the basic unit of plant and biological structure, which he called "the cell." His book, *Micrographia* (Hooke 1664), contains careful drawings of cork cells showing their hexagonal-prismatic shape, rather like cells in a bee's honeycomb. Subsequent studies by Lewis (1928), Natividade (1938), Eames and MacDaniels (1951), Gibson et al. (1981), Ford (1982) and Pereira et al. (1987) confirm Hooke's picture, and add to it the observation that the cell walls themselves are corrugated like the bellows of a concertina (Fig. 1), a structure which gives the cork extra resilience. The cells themselves are considerably smaller than those in most foamed polymers: the hexagonal prisms are between 10 and 40 μm in height and measure between 10 and 15 μm across the base, with a cell-wall thickness of between 1 and 1.5 μm. There are between 4×10^4 and 2×10^5 cells per cubic millimeter.

The cell walls of cork are made up of lignin and cellulose, with a thick secondary wall of suberin and waxes (Eames and MacDaniels 1951, Sitte 1962, Esau 1965, Zimmerman and Brown 1971, Pereira 1982). All trees have a thin layer of cork in their bark. *Quercus suber* is unique in that, at maturity, the cork forms a layer several centimeters thick around the trunk of the tree.

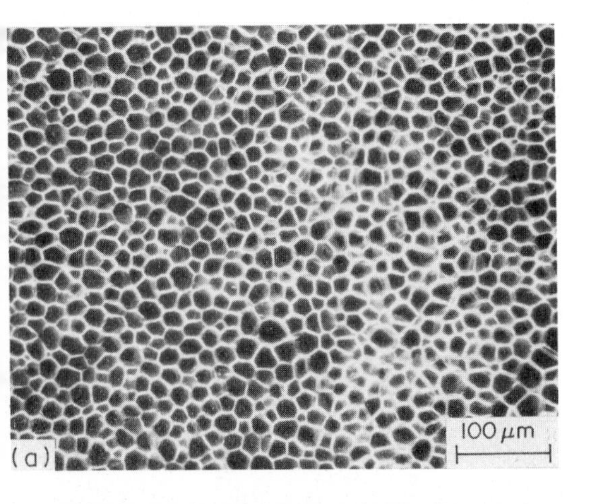

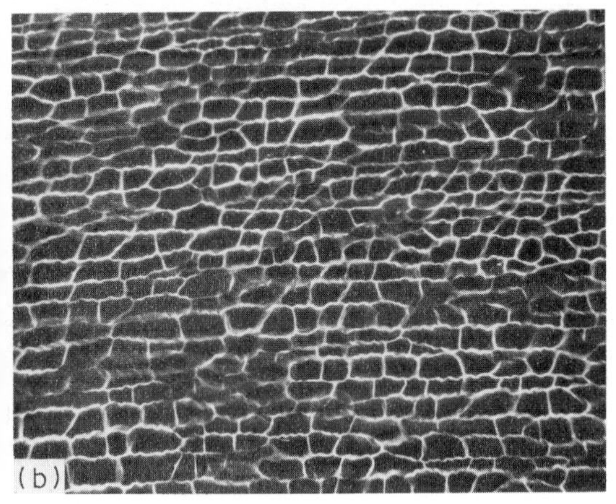

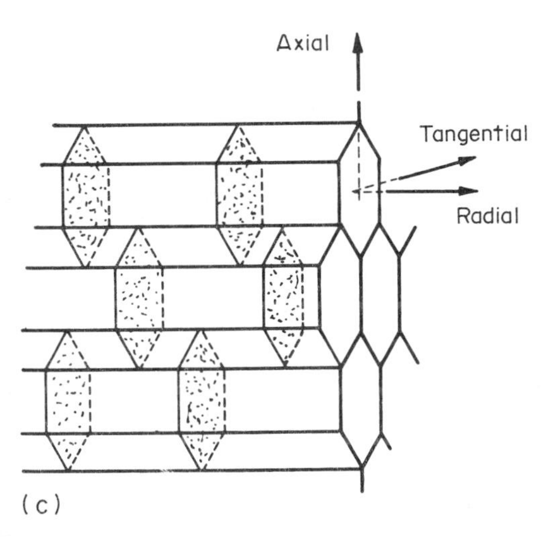

Figure 1
Cells in cork: (a) a section normal to the radial direction of the tree, (b) a section normal to the axial direction of the tree (i.e., the direction parallel to the trunk). The cells (c) are hexagonal prisms

2. *Properties*

Many of the properties of cork derive from its cellular structure (Gibson et al. 1981, Gibson and Ashby 1988). When cork is loaded in compression or tension, the cell walls bend, and this bending allows large elastic change of shape. For this reason, the elastic moduli are low (Tables 1, 2)—roughly one hundred times lower than those of the solid of which the cell walls are made. When compressed beyond the linear-elastic limit (Fig. 2) the cell walls buckle, giving "resilience," meaning a large nonlinear elastic deformation. The prismatic shape of the cells makes cork anisotropic. Young's modulus E_1 measured in a direction parallel

Table 1
Average properties of cork

Density	120–240 kg m^{-3}
Young's modulus	4–60 MN m^{-3}
Shear modulus	2–30 MN m^{-2}
Poisson's ratio	0.15–0.22
Collapse strength (compression)	0.5–2.5 MN m^{-2}
Fracture strength (tension)	0.8–3.0 MN m^{-2}
Loss coefficient	0.1–0.3
Coefficient of friction	0.2–0.4
Thermal conductivity	0.025–0.028 J mK^{-1}

Table 2
Anisotropic properties of cork (density 170 kg m^{-3})

Young's modulus E	20 ± 7 MN m^{-2}
Young's modulus E_2, E_3	13 ± 5 MN m^{-2}
Shear moduli G_{12}, G_{21}, G_{13}, G_{31}	2.5 ± 1 MN m^{-2}
Shear moduli G_{23}, G_{32}	4.3 ± 1.5 MN m^{-2}
Poisson's ratio ν_{12}, ν_{13}, ν_{21}, ν_{31}	0.05 ± 0.05
Poisson's ratio ν_{23}, ν_{32}	0.5 ± 0.05
Collapse strength, compression, σ_1^*	0.8 ± 0.2 MN m^{-2}
Collapse strength, compression, σ_2^*, σ_3^*	0.7 ± 0.2 MN m^{-2}

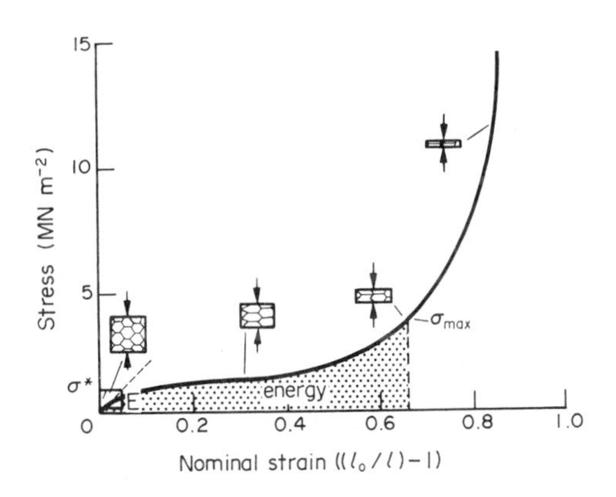

Figure 2
A compression stress–strain curve for cork, showing the modulus E and the collapse stress σ^*

to the prism axis (the "radial" direction of Fig. 1) is almost twice as great as E_2 and E_3, measured in directions normal to this axis (the "axial" and "tangential" directions); the shear moduli, too, are anisotropic. Most anisotropic of all is Poisson's ratio, which, for compression parallel to the prism axis, is almost zero. This means that compression in this direction produces little or no lateral spreading—a property useful in gaskets, and in corks which must seal particularly tightly. The wide range of values for a given property derives from the range of density of natural cork (which depends on growing conditions and the age of the tree) and on its moisture content.

Cork is a particularly "lossy" material, meaning that, when deformed and then released, a considerable fraction of the work of deformation is dissipated. This fraction is measured by the loss coefficient (Table 1); for cork it has a value between 0.1 and 0.3, giving the materials superior vibration and acoustic-damping properties. The high coefficient of friction, too, derives from the ability to dissipate energy. When an object slides or rolls on a cork surface, the cells deform and then recover again as the slider passes; the energy loss gives friction.

The cellular structure gives cork a particularly low thermal conductivity. Heat is conducted through a cellular solid by conduction through the cell walls, by conduction through the gas within the cells and by convection of this gas (Gibson and Ashby 1988). The cells in cork are sufficiently small that convection is suppressed completely, and the thin cell walls contribute little to heat transfer; the thermal conductivity is reduced to a value only slightly above that of the gas contained within the cells themselves.

Cork is remarkable for its chemical and biological stability. The presence of suberin, waxes and tannin give the material great resistance to biological attack and to degradation by chemicals. For this reason, it can be kept in contact with foods and liquids (such as wine) almost indefinitely without damage. And for reasons which are not fully understood, cork is fire-resistant: it smoulders rather than burns.

3. Uses

For at least 2000 years, cork has been used for—among other things—"floats for fishing nets, and bungs for bottles, and also to make the soles for women's winter shoes" (Pliny, AD 77). Few materials have such a long history or have survived so well the competition from man-made substitutes. We now examine briefly how the special structure of cork has suited it so well to its uses.

3.1 Bungs for Bottles and Gaskets for Woodwind Instruments

Connoisseurs of wine agree that there is no substitute for corks made of cork. Plastic corks are hard to insert and remove, they do not always give a good seal, and they may contaminate the wine. Cork corks are inert and seal well for as long as the wine need be kept. The excellence of the seal derives from the elastic properties of the cork. It has a low Young's modulus E and, more importantly, it also has a low bulk modulus K. Solid rubber and solid polymers above their glass transition temperature have a low E but a large K, and it is this that makes them hard to insert into a bottle and gives a poor seal when they are in place.

One might expect that the best seal would be obtained by cutting the axis of the cork parallel to the prism axis of the cork cells; the circular symmetry of the cork and of its properties are then matched. This is correct: the best seal is obtained by cork cut in this way. But natural cork contains lenticels—tubular channels that connect the outer surface of the bark to the inner surface, allowing oxygen into, and CO_2 out from, the new cells that grow there. The lenticels lie parallel to the prism axis (the radial direction); a cork cut parallel to this axis will leak. This is why almost all commercial corks are cut with the prism axis (and the lenticels) at right angles to the axis of the bung (Fig. 3a).

A way out of this problem is shown in Fig. 3b. The base of the cork, where sealing is most critical, is made of two disks cut with the prism axis (and lenticels) parallel to the axis of the bung itself. Leakage is prevented by gluing the two disks together so that the lenticels do not connect. The cork, when forced into the bottle, is then compressed (radially) in the plane in which it is isotropic, exerting a uniform pressure on the inside of the neck of the bottle.

Cork makes good gaskets for the same reason that it makes good bungs: it accommodates large elastic distortion and volume change, and its closed cells are

Figure 3
Ordinary corks (a) have their axis at right angles to the prism axis to prevent leakage through the lenticels (dark lines). A laminated cork (b) makes better use of the symmetry of the cork structure

mpervious to water and oil. Thin sheets of cork are ised, for instance, for the joints of woodwind and brass nstruments. The sheet is always cut with prism axis and lenticels) normal to its plane. The sheet is then sotropic in its plane, and this may be the reason for utting it so. But it seems more likely that it is cut like his because the Poisson ratio for compression down he prism axis is zero. Thus when the joints of the nstrument are mated, there is no tendency for the heet to spread in its plane and wrinkle.

.2 Floor Covering and the Soles of Shoes

Manufacturers who sell cork flooring claim that it etains its friction even when polished or covered with oap, and experiments (Gibson et al. 1981) confirm his.

Friction between a shoe and a cork floor has two origins. One is adhesion: atomic bonds form between he two contacting surfaces, and work must be done to break them if the shoe slides. Between a hard slider and a tiled or stone floor this is the only source of riction; and since it is a surface effect, it is destroyed by a film of polish or soap. The other source of friction is lue to anelastic loss. When a rough slider moves on a cork floor, the bumps on the slider deform the cork. If cork were perfectly elastic no net work would be done: he work done in deforming the cork would be recovered as the slider moves on. But if the cork has a high loss coefficient (as it does) then it is like riding a bicycle through sand: the work done in deforming the material ahead of the slider is not recovered as the slider passes on, and a large coefficient of friction appears. This anelastic loss is the main source of riction when rough surfaces slide on cork; and since it lepends on processes taking place below the surface, not on it, it is not affected by films of polish or soap. The same thing happens when a cylinder or sphere rolls on cork, which therefore shows a high coefficient of rolling friction.

3.3 Packaging and Energy Absorption

Many of the uses of cork depend on its capacity to absorb energy. Cork is attractive for the soles of shoes and flooring because, as well as having good frictional properties, it is resilient under foot, absorbing the shocks of walking. It makes good packaging because it compresses on impact, limiting the stresses to which the contents of the package are exposed. It is used as handles of tools to insulate the hand from the impact loads applied to the tool. In each of these applications it is essential that the stresses generated by the impact are kept low, but that considerable energy is absorbed.

Cellular materials are particularly good at this. The stress–strain curve for cork (Fig. 2) shows that the collapse strength of the cells (Table 1) is low, so that the peak stress during impact is limited. Large compressive strains are possible, absorbing energy as the cells progressively collapse. In this regard, its structure and properties resemble polystyrene foam, which has replaced cork (because it is cheap) in many packaging applications.

3.4 Thermal Insulation

Trees are thought to surround themselves with cork to prevent loss of water in the hotter seasons of the year. The two properties involved—low thermal conductivity and low permeability to water—make it an excellent material for insulation against cold and damp (the hermit caves of southern Portugal, for example, are lined with cork). It is for these reasons that crates and boxes are sometimes lined with cork. And the cork tip of a cigarette must appeal to the smoker because it insulates (a little) and prevents the tobacco getting moist.

3.5 Indentation and Bulletin Boards

Cellular materials densify when they are compressed or indented. So when a sharp object, like a drawing pin, is stuck into cork, the deformation is very localized. A layer of cork cells, occupying a thickness of only about one quarter of the diameter of the indenter, collapses, suffering a large strain. The volume of the indenter is taken up by the collapse of the cells so that no long-range deformation is necessary. For this reason the force needed to push the indenter in is small. Since the deformation is (nonlinear) elastic, the hole closes up when the pin is removed.

Bibliography

Eames A J, MacDaniels L H 1951 *An Introduction to Plant Anatomy*. McGraw–Hill, London

Emilia Rosa M, Fortes M A 1988 Stress relaxation and stress of cork. *J. Mater. Sci.* 23: 35–42

Esau L 1965 *Plant Anatomy*. Wiley, New York, p. 340

Ford B J 1982 The origins of plant anatomy: Leeuwenhoek's cork sections examined. *IAWA Bull.* 3: 7–10

Gibson L J, Ashby M F 1988 *Cellular Solids: Structure and Properties*. Pergamon, Oxford, Chap. 12

Gibson L J, Easterling K E, Ashby M F 1981 The structure and mechanics of cork. *Proc. R. Soc. London, Ser. A.* 377: 99–117

Hooke R 1664 *Micrographia*. Royal Society, London, pp. 112–21

Lewis P T 1928 The typical shape of polyhedral cells in vegetable parenchyma. *Science (N.Y.)* 68: 635–41

Natividade J V 1938 What is cork? *Bol. Junta Nac. da Cortica (Lisboa)* 1: 13–21

Pereira H, Emilia Rosa M, Fortes M A 1987 The cellular structure of cork from *Quercus suber* L. *IAWA Bull.* 8: 213–18

Sitte P 1962 Zum Feinbau der Suberinschichten in Flaschen Kork. *Protoplasma* 54: 55–559

Zimmerman M H, Brown C L 1971 *Trees Structure and Function*. Springer, Berlin, p. 88

M. F. Ashby
[University of Cambridge, Cambridge, UK]

Corrosion and Oxidation Study Techniques

Textbooks and journals reveal such a plethora of methods for the study of corrosion that some guidance is necessary for the uninitiated to discover the optimum method for their needs. This article is intended to form such a guide and for this reason a classification has been adopted (Table 1) which groups methods according to purpose rather than according to common features in the methods. This article is concerned with mechanistic studies. Mechanistic studies imply the ability to distinguish transport processes or phase boundary reactions down to the atomic scale. However, measurements are made with all the benefits of a laboratory setting and equipment may be very sophisticated.

The first choice which must be made is between environments, especially whether it is aqueous (see *Corrosion of Metals: An Overview*) or gaseous. When metals corrode in the gas phase it is rare for significant losses of material to arise from vaporization. Thus it is usually possible to assess the extent of corrosion by the mass or thickness of corrosion product formed. Many methods in the literature are variations on this basic measurement. Samples may also be transferred readily to the electron microscope or spectrometer for identification of composition or morphology. In contrast, when metals corrode in an aqueous media, transport or loss of material in soluble form must be expected and it is usually impossible to gauge the extent of reaction by the mass of products on the surface. Moreover, it is usually difficult to interpret the results of electron microscopy or of spectrometry because of the effects of a vacuum on the normally immersed surface. Reaction rates must be measured directly from metal losses or inferred from the use of the electrochemical parameters of electropotential or exchange current density. The different ways in which these can be determined give rise to the variety of

Table 1
Classification of corrosion and oxidation study techniques[a]

Type	Thickness (m)	Aqueous	Gaseous
Kinetics	10^{-9}	Scratching electrode	Quartz piezobalance XPS/AES
	10^{-6}	Linear polarization resistance	Microbalance Volumetric measurement AES depth profile
	10^{-3}	Weight loss Ac impedance Rotating disk	Thermobalance
	Local	Artificial pits and crevices Constant strain	Metallographic examination
Chemistry	10^{-9}	Differential reflectance Ellipsometry	SIMS Electron diffraction AES/XPS
	10^{-6}	EXAFS Mössbauer Raman	XRD EPMA STEM Laser optical and Raman spectroscopy
	10^{-3}	Infrared Coulometric reduction	Taper sections XRF Spark source mass spectrometry
	Local	Microelectrode	AES Analytical TEM
Morphology	10^{-9}	Ellipsometry	Holography TEM
	10^{-6}	Light microscopy	Markers RBS/RNR SEM
	10^{-3}	Flexure	Flexure and vibration
	Local	SEM	SEM

[a] AES, Auger electron spectroscopy; EPMA, electron probe microanalysis; EXAFS, extended x ray absorption fine structure spectroscopy: RBS, Rutherford backscattering; RNR, resonant nuclear reactions; SEM, scanning electron microscopy; SIMS, secondary ion mass spectroscopy; STEM, scanning transmission electron microscopy; TEM, transmission electron microscopy; XPS, x ray photoelectron spectroscopy; XRD, x ray diffraction; XRF, x ray fluorescence

electrochemical methods listed for corrosion studies in aqueous environments.

The investigator must secondly decide whether the need is for kinetic, chemical or morphological data and which thickness range (nm, μm or mm) is a significant rate of wastage in the given situation. In some corrosion problems, such as pitting, breakaway or stress-corrosion cracking, the thickness is irrelevant as the phenomenon is purely local; methods for such studies are highlighted separately.

1. Gaseous Corrosion

1.1 Kinetics

In the laboratory, the rate of corrosion or oxidation of a metal or alloy is most simply measured in terms of its weight gain in a given interval of time. Analytical balances with a sensitivity of ± 0.1 mg are readily available and this corresponds to an oxide thickness of ~ 1 nm on a sample of 1 cm^2 in area. Thus gravimetric methods can be used throughout the entire thickness range. To avoid unintended thermal cycles it is usual to suspend the sample in the reaction chamber from an external balance so that a continuous readout of sample weight is available under isothermal conditions. Such devices are termed thermobalances. Traditionally, thermobalances are open to the environment and only the atmosphere within the reaction chamber is controlled. The need for a free-moving suspension to pass through an orifice in this chamber makes it impracticable to evacuate the chamber. Care must be taken to compensate for buoyancy and heating-up times which may give a significant zero point error. Thus, for work in the nanometer range a small balance totally enclosed within the envelope of the reaction vessel is used. The enclosure can then be evacuated as a standard state at the commencement and end of the test exposure. Such a balance is usually termed a microbalance and is useful also in working with toxic and high- or low-pressure environments or with atmospheres which would condense at room temperature. At a subnanometer level of thickness, a quartz piezobalance may be used. At such trivial levels of reaction the investigator is likely to be interested less in corrosion than in adsorption phenomena. One problem with complex gas atmospheres containing several potential reactants (e.g., O_2, C, Cl_2 or S_4) is that of determining their reacting proportions. Analysis of the product after exposure yields the gross uptakes, but on a continuous or semicontinuous basis, analysis of the atmosphere must be used. This measurement gives rise to a series of volumetric methods in which the change in partial pressure of a component in a known isothermal volume of gas is monitored.

Position-sensitive measurements of the extent of oxidation can be made by determination of the corrosion product thickness . This measurement is often made intermittently by taking metallographic samples, but in the case of submicrometer films can be monitored continuously by changes in interference colors. In the laboratory, measurement of shifts in phase angle and intensity between incident and reflected light permits the information available to be expanded by the techniques of ellipsometry.

The results of thickness measurements, x, may be extrapolated in time t by use of an appropriate rate law. Important examples are (a) parabolic, $x^2 = k_1 t$, which applies when diffusion across the corrosion product is rate controlling; (b) linear, $x = k_2 t$, which applies when the corrosion product offers no impedance to reaction; and (c) logarithmic, $x = k_3 \ln(1 + ct)$, when a passive layer of limiting thickness is reached. The coefficients k_1, k_2, k_3 and c are all constant at a given temperature. The data can be interpolated and, with care, extrapolated in temperature by the Arrhenius relationship, $k . k_1 \exp(-Q/RT)$, where k is the specific reaction rate constant, Q is the activation energy, R is the gas constant and T is the absolute temperature. The value of Q may indicate the nature of the controlling transport step; for example, ~ 200 kJ mol^{-1} indicates solid state diffusion. The value is reduced by short circuit diffusion, along grain boundaries for example, or surface diffusion.

1.2 Chemistry

The chemical nature of the corrosion product is fixed by both thermodynamic and kinetic constraints. The composition of the outermost surface tends towards thermodynamic equilibrium with the gas phase, while the inner surface is in equilibrium with elements in the alloy. It is the activity gradients between these equilibrium positions which give rise to diffusion across the layer and hence to the segregation of distinct phases between them. The relative stabilities of the oxides, sulfides or chlorides commonly of concern in gaseous corrosion are determined by their free energies of formation and may be predicted for a given temperature by the use of an Ellingham diagram.

There is now an excellent range of methods available for investigation of the chemical structure of the corrosion product. They are all vacuum techniques involving large fixed equipment. For the earliest stages of film growth various forms of ion spectroscopy may be used, such as secondary-ion mass spectroscopy (SIMS) which is available in scanning or imaging forms, or ion scattering spectroscopy (ISS). This group of techniques has sensitivity to monolayers of material and can be especially valuable because of the ability to detect hydrogen. They are reflection techniques and hence the surface film can be examined in place on the metal, but the mass spectra may need specialist interpretation and quantitative evaluation is still difficult to achieve.

The most useful method available for analysis of films in the nanometer range of thickness is x ray excited photoelectron spectroscopy (XPS or ESCA), which yields a composition within ±5% together with firm assignation of the bonding state of most species, including oxygen and carbon. Analysis for hydrogen is not possible and spatial resolution to ~200 μm is only just becoming available. This is, however, an area of current development and localized analysis of thin films is best undertaken by Auger electron analysis (AES), which is often combined with scanning electron microscopy (SEM) for good visual control of the area analyzed. Identification of crystal structure in this thin film region must be by electron diffraction. On very flat single crystal surfaces, low-energy electron diffraction (LEED) is useful, but generally reflection high-energy electron diffraction (RHEED) would be used. Conventional transmission patterns from films removed from the surface, by methanolic iodine solution for example, may still be the best method.

For films of several μm in thickness, the x ray electron probe microanalyzer (EPMA) is of exceptional value and widely available in the form of an energy-dispersive x ray detector (EDXA) mounted on a scanning electron microscope. Special facilities are needed for elements below sodium in the periodic table, and carbon is the practical limit. Valence states are not available by EDXA and XPS can be used on fractured or spalled interfaces within thick scales, often to great effect. Structural information within this range can be obtained by x ray diffraction (XRD) in addition to electron diffraction methods.

Very thick films can be examined in scans across metallographic sections by EPMA, or in total by the methods of bulk analysis, x ray fluorescence (XRF) or spark source mass spectroscopy.

Investigation of local compositional variations in thin films and particularly for light elements (below Na) is best achieved by AES and compositional variations in thick films can be followed by a method in which AES is operated in sequence with oxide removal by an energetic ion beam to yield a depth profile. A complementary method for this form of analysis is the examination of thin sections of oxide scale in the analytical transmission electron microscope (ATEM). Special provision can be made with some of these vacuum methods for corrosion processes to be studied in situ using heated gas cells. Alternatively, laser-excited optical spectroscopy or Raman spectroscopy, which do not require a vacuum, can be used.

1.3 Morphology

Morphological changes during growth of an oxide scale are important to both the corrosion scientist and the engineer. These are shape and profile changes induced by preferred crystallinity, directional growth or locally accelerated attack. Such phenomena frequently give rise to stress. Thus important experiments in which inert markers are placed on the surface indicate that oxidation of elements in the first half of the periodic table (< Ga) occurs by diffusion of metal ions across the corrosion product. The subsequent drift of the oxides towards the metal surface creates stress in the film, depending on the local geometry. This may cause the oxidation rate of small coupons to differ from that of large fabrications. The uptake of oxygen or the distribution of dopant elements are conveniently measured using high-energy ion beams by Rutherford backscattering (RBS) or resonant nuclear reactions. Stress is measured by unilateral oxidation of strip or disk specimens followed by measurement of flexure. Young's moduli are measured during oxidation by frequency changes in a vibrating strip; they may differ from those of non-growing oxide. The morphological structure of scales can be observed on metallographically prepared specimens using optical or SEM methods.

2. Aqueous Corrosion

2.1 Kinetics

Whilst electrochemical measurements form the basis for the study of aqueous corrosion, the emphasis depends on the kinetic region of importance. The initial period of reaction is important for the light it throws on the rate of passivation of new surface as it would be exposed during the fracture of metals. This rate, in turn, determines whether a crack continues to propagate or whether, instead, the crack becomes blunt under the action of electrochemical dissolution.

The problem of instrumentation for work in this region is that of response rate, since records of events occurring over millisecond periods are required. The major design feature is the means provided to expose new surface within the electrolyte at a controlled moment. In polarography this is done by dropping a liquid metal, mercury, from an orifice; in the case of normal metals it can only be done mechanically, for example by straining or scratching the electrode. The potential of the surface relative to a standard cell then indicates the period during which metal is exposed, the current flow through the electrode gives the instantaneous rate of reaction, which diminishes as the film heals, and the charge transferred indicates the thickness of the film formed. In a useful variant of the technique a circular scratch is described by a stylus on a rotating disk.

The vast body of studies on aqueous corrosion is concerned not with the passivation stage but with either the steady loss of metal from unfilmed or partly filmed surfaces or the loss of passivity of a surface under chemical or electrochemical action. In the standard three electrode test cell, the potential of the testpiece or working electrode relative to the standard electrode is measured by a very high-impedance electrometer as a function of the current flow between the working electrode and a counter electrode. When a potentiostatic circuit is used, the current flow is varied automatically until a preset potential is reached and held, whereas in a galvanostatic circuit it is the potential which is varied. Commercial electronic devices are available to perform these functions and make it possible to scan the sample's potential in a stepwise, continuous or repetitive pattern.

When the external current is passed through the electrode, its potential changes as a new balance between the anodic and cathodic reactions is established. At high currents (100 or 1000 times the corrosion current), the reaction appropriate to the direction of the current is sufficiently dominant for the testpiece to behave as a single electrode. Its potential V then varies logarithmically with the measured current I according to the Tafel equation:

$$V = \beta \log I/I_0$$

This relation gives a basis for the extrapolation of the electrode current through the mixed potential range to the rest potential and hence for determining the corrosion current. Unfortunately, polarization of the testpiece by currents great enough to reach the Tafel region may cause side reactions inappropriate to the corrosion process being studied. Hence methods have been introduced to permit estimation of corrosion rates using measurements within the mixed electrode region. In the linear polarization method, use is made of the proportionality between current and potential close to the rest potential (± 20 mV). The slope of this line is known as the polarization resistance R_p and

$$R_p = -(\beta/4.6\ I_0)$$

The assumption of linearity over the range allows I_0 to be determined using three or four point measurements, which can be ideal for monitoring. More recently, computer-based iterative methods have been introduced which permit the corrosion current to be determined from the nonlinear regions either side of the rest potential; these will become powerful tools in laboratory work.

As the corrosion rate increases, accumulation of corrosion products as surface layers becomes increasingly evident and the reaction rate is controlled by mass transport rather than by electron transfer. Each transport step has its own characteristics of path resistance and relaxation time and these can be separated by the techniques of ac electrochemistry. In these methods an ac ripple is imposed on the dc current flowing through the testpiece so that the current–frequency spectrum can be analyzed and hence the transfer paths described by a "circuit diagram" of resistors and capacitors. The methods are of especial value for studies on the mechanism of failure of painted and coated surfaces.

The corrosion rate of immersed surfaces frequently depends on hydrodynamic parameters because of the importance of mass transfer through the boundary layer. Such situations can be modelled and controlled in the laboratory by means of electrodes in the form of rotating disks or cylinders, using equations for boundary layer thickness developed originally by Levich.

Locally high corrosion rates, as in crevices or in pits, are of particular concern in aqueous corrosion and in this situation the overall current flow through the electrode can be irrelevant or even misleading. Standard forms of crevice are produced by tightening testpieces together at a controlled torque while pits can be simulated using the relevant metal as the lead in a pencil-type electrode. In the latter instance, prior dissolution of the metal core produces the artificial pits and charges them with the appropriate solution. Special methods are used to prevent salt precipitates in the pits, but thereafter pitting corrosion rates can be studied in isolation from general corrosion. The most serious form of local corrosion is that which occurs conjointly with applied stress which may be static (stress-corrosion cracking) or cyclic (corrosion fatigue). Early methods for the study of stress-corrosion cracking utilized a static strain, imposed by constraining the test pieces in a "C" jig for example, but more modern methods use a constant strain rate applied by means of a tensometer, which is found to predict actual failure more closely.

2.2 Chemistry

Surfaces are capable of being examined in a multitude of ways once they are removed from the test medium but the validity of this approach must be checked against conclusions drawn from in situ examination by specialist methods. For the earliest phase of growth, ellipsometry and differential reflectometry have been employed. Thin (5 nm) passive films can be examined by EXAFS (extended x ray absorption fine structure), by Mössbauer spectroscopy, or by laser Raman spectroscopy in the presence of at least thin films of water. Thicker films can be examined in situ by infrared spectroscopy or electrochemically by examination of the coulometric reduction spectrum. The study of the chemistry of local corrosion by in situ methods is very difficult, but microelectrodes can be inserted into crevices and pits using the techniques traditionally employed by microbiologists.

2.3 Morphology

The thickness and texture of deposits formed in an aqueous environment can be very important in operating plant, because of their effect on heat transfer and, through friction factors, on pressure drop. Such overall effects may be easier to monitor in situ than they are to examine in detail. Early stages of roughening can be studied by ellipsometry or holography, and light microscopy can be employed for the study of surface texture on the micrometer scale. The flexure methods employed in dry oxidation to measure stress are valid, but hydrodynamic disturbance associated with the cyclic methods may make these unacceptable.

3. Conclusion

The reader will have discovered that corrosion methods cover the whole breadth of techniques for chemical and physical analysis. It is hoped that this review will have placed the groups of techniques into their conventional context and will thus have helped the reader to make an efficient literature search. No scientist should, however, be dissuaded by this review from combining techniques with imagination in order to best suit his own needs.

See also: Investigation and Characterization of Materials

Bibliography

Azzam R M, Bashara N M 1977 *Ellipsometry and Polarised Light*. North-Holland, Amsterdam

Birks N, Meier G H 1983 *Introduction to High Temperature Oxidation of Metals*. Edward Arnold, London

Frankenthal R P, Kruger J (eds.) 1978 *Passivity of Metals*. Electrochemical Society, Princeton, New Jersey

Hearle J W S, Sparrow J T, Cross P M (eds.) 1972 *The Use of the Scanning Electron Microscope*. Pergamon, Oxford

Mansfeld F 1976 The polarization resistance technique for measuring corrosion currents. In: Fontana M G, Staehle R W (eds.) 1976 *Advances in Corrosion Science and Technology*, Vol. 6. Plenum, New York, pp. 163–262

Mulvey T, Webster R K 1974 *Modern Physical Techniques in Materials Technology*. Oxford University Press, Oxford

Parkins R N 1979 Development of strain-rate testing and its implications. In: Ugianski G M, Payer J H (eds.) 1979 *SCC—The Slow Strain Rate Technique*, STP 665. American Society for Testing and Materials, Philadelphia, Pennsylvania, pp. 5–25

Rapp R A (ed.) 1970 *Techniques of Metals Research*, Vol. 4, Pt. 2. Interscience, New York

Seah M P, Briggs D (eds.) 1983 *Practical Surface Analysis by Auger and X-Ray Photoelectron Spectroscopy*. Wiley, Chichester

Shreir L L (ed.) 1976 *Corrosion*, 2nd edn. Newnes–Butterworths, London

J. E. Castle
[University of Surrey, Guildford, UK]

Corrosion of Iron and Steel

Iron and steel are the most common of all engineering metals and are used in a far greater tonnage than all the other metals added together. Many industrial processes involve the handling of strong chemicals and iron and steel are usually only suitable for such applications when protected with surface coatings or when highly alloyed, as in the case of stainless steels. These are specialized applications since the main environments to which iron and steel are exposed are the natural media of air, water and soils. The rate of corrosion of unalloyed iron and steel is primarily determined by the aggressivity of the environment; variations in metal composition and microstructure have a very much smaller effect.

1. The Mechanism of Corrosion

The mechanism of corrosion of iron and steel, commonly known as rusting, involves dissolution at anodic sites on the metal surface, and the reaction can be considered by the following simple equation:

$$Fe \rightarrow Fe^{2+} + 2e^{-} \qquad (1)$$

Natural environments are usually approximately neutral (pH = 7) or slightly alkaline and in these circumstances the balancing cathodic reaction is

$$O_2 + 2H_2O + 4e^{-} \rightarrow 4OH^{-} \qquad (2)$$

The essential point is that corrosion can only occur when both oxygen and water are present simultaneously to enable the cathodic reaction to consume the electrons generated during the anodic dissolution of the iron. Normally, the ferrous ions and hydroxyl ions combine to form ferrous hydroxide, which is subsequently oxidized to ferric hydroxide. This is the basic component of hydrated rust although other metal oxides and salts, such as sulfates, may also be present. The physical characteristics of rust are such that it has a light, friable, brittle nonadherent nature and repeatedly spalls away from the metal surface. Accordingly, it provides little or no protection and rusting continues at a rate dependent upon the environment to which the steel is exposed.

2. Atmospheric Corrosion

In the interior of most buildings and in clean, rural districts the rate of corrosion of iron and steel is very low when the relative humidity is less than 60%. However, rusting is accelerated as the time of wetness rises and as the air becomes contaminated with industrial pollutants or, in marine locations, chlorides. The most damaging industrial pollutant is sulfur dioxide, which reacts with moisture to produce dilute sulfurous and sulfuric solutions, known as acid rain. Although the naturally occurring chloride salts are usually neutral, they are highly conductive and assist the electrochemical reactions involved in corrosion. In addition, compounds such as sodium chloride are hygroscopic and absorb moisture from the atmosphere so that rusting can occur at low relative humidities. Typical rates of corrosion for unalloyed irons and steels in rural, industrial and marine locations are shown in Table 1.

The chemical composition of iron and steel has very little effect on corrosion resistance unless a few percent of alloying elements are intentionally added for this purpose. These steels are known as the weathering steels and the alloy content is about 2%, made up by small additions principally of copper, nickel and chromium. The rust that forms on these steels is adherent and has a granular appearance; the cracks and pores tend to be filled with insoluble salts which effectively block the access of oxygen and water to the

Table 1
Atmospheric corrosion rates for bare iron and steel

Environment	Site	Corrosion rate (μm per year)
Rural—very dry and clean	Khartoum, Sudan	3
Rural/urban—very dry	Delhi, India	8
Rural—temperature	Godalming, UK	48
Suburban	Berlin, Germany	53
Urban	Teddington, UK	70
Industrial, Europe	Motherwell, UK	97
Industrial, America	Pittsburgh, USA	109
Heavy industrial	Sheffield, UK (1945–65)	135
Marine/rural	Sandy Hook, USA	84
Marine/industrial	Corgella, South Africa	114
Marine—tropical surf	Lagos, Nigeria	619

steel surface. The protective rust takes several years to develop and forms most rapidly (2–3 years) in locations which have substantial dry seasons. In very wet areas and in atmospheres containing a high concentration of chloride ions the rust films may never stabilize. However, these steels are used in many countries of the world and comparative corrosion rates of mild steel and weathering steel for locations in the UK are shown in Fig. 1. Because weathering steels are not usually painted, their big advantage is the elimination of the costs and inconvenience of maintenance. Although used for steel-framed buildings, their principal application is for bridges over roads, rivers and railways. In these locations maintenance would be particularly expensive, due to difficult or dangerous access, and would involve closing motorways, railway tracks and other installations, which is usually undesirable.

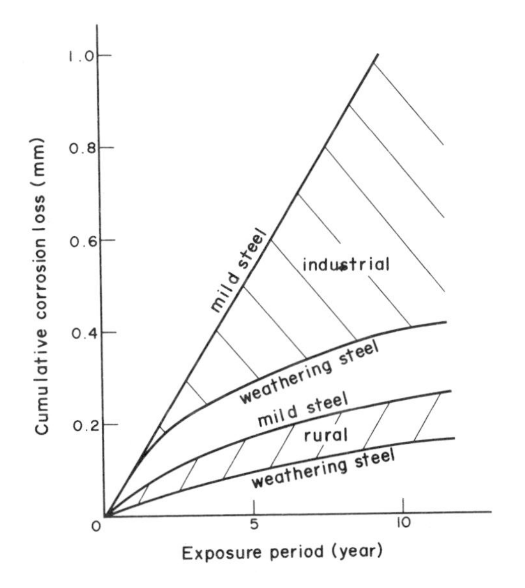

Figure 1
Atmospheric corrosion of weathering and mild steels

3. Fresh Water

As in air, the chemical and microstructural compositions of the various irons and steels have little effect on corrosion rates in fresh water. Even the addition of small amounts of copper, nickel and chromium do not have the beneficial effects observed in atmospheric exposure. However, the water composition is of great importance and corrosion rates are found to depend on such factors as conductivity, pH, hardness, type and amount of dissolved solids, dissolved gases (especially oxygen) and organic matter. Natural waters only vary in pH between about 5 and 9 and this has little effect on corrosion, but corrosion can be virtually prevented by making the water alkaline artificially ($pH > 10$). Corrosion can be slowed if the rust is not removed from the metal surface and this can be assisted by the entrainment of hard-water salts in heated systems or on structures under cathodic protection. Conversely, continuous exposure of bare metal by erosion or abrasion in fast-flowing water containing solid particles will greatly increase the rate of corrosion of iron and steel. Brackish waters have a high conductivity, which accelerates corrosion, while polluted waters, with low dissolved oxygen contents, are less aggressive unless microbial attack by sulfate-reducing bacteria is present. For these reasons it is very difficult to assign a typical corrosion rate for iron and steel in fresh water but 0.04 mm per year is generally accepted as a reasonable rate for unpolluted, oxygenated fresh water. It is interesting to note, and not often realized, that this rate of corrosion is no greater than the atmospheric corrosion rate in a temperate climate.

4. Seawater

Again, the rate of corrosion of iron and steel in seawater is little affected by the metal composition and, as the open sea is of similar composition around the world, corrosion rates are remarkably constant. Even temperature has little effect on immersed iron and steel because the increased rate of the corrosion reaction at higher temperatures is offset by the reduced solubility of oxygen, which depresses corrosion rates. Of course, local coastal factors, such as pollution, can greatly alter the extent of rusting but the main variable in seawater corrosion is the position of the iron or steel with respect to the waterline. Figure 2

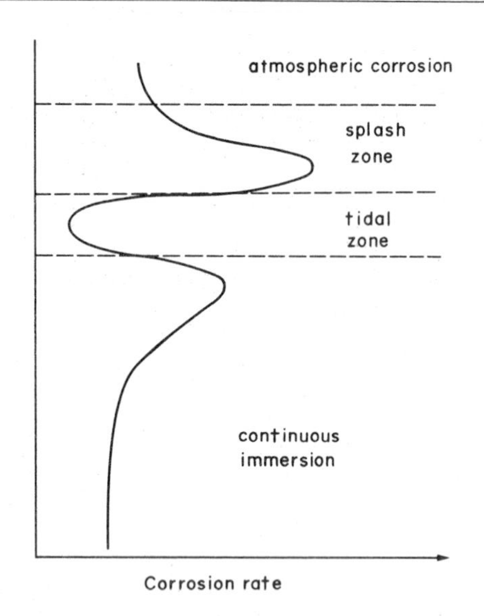

Figure 2
Schematic corrosion profile for steel piling in seawater

shows schematically the change in corrosion rate for bare steel piling in seawater.

The highest corrosion rate, 0.0–0.2 mm per year, is at the splash zone where the water is highly oxygenated and the wave action prevents protective rust films from forming. The intertidal zone corrodes much more slowly, ~0.05 mm per year, but there is usually a second peak in corrosion rate at the low-water mark, particularly if there is any abrasive action from floating fenders. In the fully immersed zone the corrosion rate is controlled by oxygen availability and also benefits from the galvanic protection that occurs naturally from the relatively anodic tidal and splash zones. A typical rate of 0.05 mm per year is only slightly greater than that observed in fresh water. Where marine fouling builds up it acts as a barrier which minimizes rusting, and if cathodic protection, is (or has been) applied this barrier is fortified with a calcareous layer of magnesium and calcium carbonates. Above the splash zone the corrosion rapidly decreases to the normal marine atmospheric rate.

Where maintenance is feasible, for example on ships, rusting is usually prevented by a combination of painting and cathodic protection, whereas permanently submerged structures, such as offshore oil platforms or marine piling, sometimes rely on cathodic protection alone or the use of thick steel to delay, rather than prevent, failure by corrosion.

5. *Soils*

Soils are complex and their aggressivity is determined by factors such as retained wetness, oxygen content, dissolved salts, pH and organic matter content; minor variations of iron and steel composition are inconsequential. Sandy soils are drier than clays, and peaty soils are acidic. Oxygen-free soils, such as river and sea beds, produce virtually no corrosion (0.02 mm per year), even when waterlogged, but if sulfate-reducing bacteria are present, then pitting can be a problem. Generally, the extent of rusting in undisturbed soils is very low, much lower than atmospheric corrosion rates and driven piles have very long lives. However, disturbed soils such as backfill are more corrosive, and buried steel structures such as pipelines are usually coated and given cathodic protection.

6. *Corrosion Prevention*

The main methods of minimizing the rusting of iron and steel rely on stifling the cathodic reaction shown in Eqn. (2) (see *Corrosion Protection Methods* and *Corrosion Protective Coatings for Metals*).

Rusting in air can be prevented by reducing the relative humidity to less than 60% by air conditioning and by minimizing atmospheric pollution. This is obviously only feasible in enclosed spaces but during transport, iron and steel components can be protected by the use of packaging impregnated with vapor-phase inhibitors, which slowly release volatile, organic compounds to inhibit rusting. For aqueous environments, such as cooling systems, dissolved inhibitors such as potassium chromate, sodium nitrite and sodium benzoate are used and are particularly effective in closed, recirculating systems. Inhibitors are often added together with antifreeze solutions. A further, and very effective, method is to remove the oxygen from the water by deaeration and chemical means in plants such as boiler feed water systems.

If the cathodic reaction cannot be stifled by the removal of water or oxygen, then the anodic dissolution reaction can be prevented by forcing electrons into the iron or steel by cathodic protection. This can be done by electrically connecting the structure to a more active metal, such as zinc, or by impressing a direct current by which the steel is made the cathode, the inert anode being a corrosion-resisting metal, such as platinized titanium. To minimize the rate of dissolution of the sacrificial anode, or to reduce the requirement for impressed current, the iron or steel structure is usually also coated. The potential of the structure is controlled at about 850 mV $Cu/CuSO_4$; lower potentials are dangerous as hydrogen evolution can occur and cause hydrogen embrittlement. Cathodic protection can only be operated in the presence of an electrolyte and is, therefore, only used for submerged or buried structures.

Clearly, another common method of corrosion prevention is to isolate the steel from the environment by means of organic or metal coatings. Paints with inhibitive primers are the most frequently used coatings but coating with sacrificial metals (e.g., hot dip galvanizing with zinc), and the deposition of a noble metal (e.g., electroplating with chromium) is often carried out.

Finally, iron and steel rusting can be minimized or prevented by alloying to prevent the formation of loose, friable, iron-rich corrosion products. Low-alloy, weathering steels have been referred to, and by increasing the chromium content to 12% or more, thin chromium-rich oxides form in preference to rust. These high-alloy steels often also contain nickel and molybdenum and are known as the stainless steels.

See also: Corrosion of Metals: An Overview

Bibliography

Burns R M, Bradley W W 1967 *Protection Coatings for Metals.* Reinhold, New York

Butler G, Ison H C K 1966 *Corrosion and Its Prevention in Waters.* Hill, London

Evans U R 1963 *An Introduction to Metallic Corrosion.* Arnold, London

Glasstone S 1948 *Text-Book of Physical Chemistry.* Macmillan, London

Iron and Steel Institute 1949 *Corrosion by Industrial Waters and Its Prevention,* ISI Special Report No. 41. ISI, London

LaQue F L, Copson H R 1963 *Corrosion Resistance of Metals and Alloys.* Reinhold, New York

Morgan J H 1959 *Cathodic Protection.* Hill, London

Shrier L L (ed.) 1976 *Corrosion,* 2nd edn., Vols. 1 and 2. Newnes-Butterworth, London

Uhlig H H 1971 *Corrosion and Corrosion Control.* Wiley, London

R. A. E. Hooper
[Arthur-Lee and Sons, Sheffield, UK]

Corrosion of Metals: An Overview

Corrosion is the degradation or destruction of a material that occurs when it reacts with its environment. Only the degradation of metals in aqueous and atmospheric environments is discussed here.

Studies in the UK and the USA have shown the economic consequences of metallic corrosion to be very serious. These studies found that economic losses resulting from metallic corrosion were at least 4% of the gross national product. The consequences of corrosion are, however, not limited to economic losses, but also involve health (as in metallic implants in the body), safety (as in the failure of vehicles and structures) and technology (as a developmental barrier to new technologies).

1. Principles

Corrosion in an aqueous environment and in an atmospheric environment (which also involves thin aqueous layers) is an electrochemical process. It is necessary, therefore, to describe the electrochemical principles that underlie corrosion.

1.1 Corrosion Cells

A metal surface exposed to a conducting aqueous electrolyte usually becomes the site for two types of chemical reaction. There is an oxidation or anodic reaction that produces electrons as, for example,

$$Fe \rightarrow Fe^{2+} + 2e^- \qquad (1)$$

and a reduction or cathodic reaction that consumes the electrons produced by the anodic reaction, such as

$$O_2 + 2H_2O + 4e^- \rightarrow 4OH^- \qquad (2)$$

$$O_2 + 4H^+ + 4e^- \rightarrow 2H_2O \qquad (3)$$

$$2H^+ + 2e^- \rightarrow H_2 \qquad (4)$$

The cathodic reaction represented by Eqn. (2) is usually the most relevant to corrosion in natural environments, since such corrosion occurs at nearly neutral pH values. However, Eqns. (3) and (4) must also be considered for the acidic environments encountered in industrial processes or for the confined volumes (pits, crevices) where the pH can reach acidic values because of hydrolysis reactions such as

$$Fe^{2+} + 2H_2O \rightarrow Fe(OH)_2 + 2H^+ \qquad (5)$$

which produce H^+ ions, the concentration of which can become large as the H^+ ions cannot readily move out from the confined volumes.

Because these anodic and cathodic reactions occur simultaneously on a metal surface, they create an electrochemical cell of the type shown in Fig. 1. The sites where the anodic and cathodic reactions take place, the anodes and the cathodes of the corrosion cell, are determined by many factors. Firstly, they need not be fixed in location: they can be adjacent or widely separated. For example, if two metals are in contact, one metal can be the anode and the other the cathode, leading to galvanic corrosion (see Sect. 2.6) of the more anodic metal. Variation over the surface of oxygen concentration in the environment can result in the establishment of an anode at those sites exposed

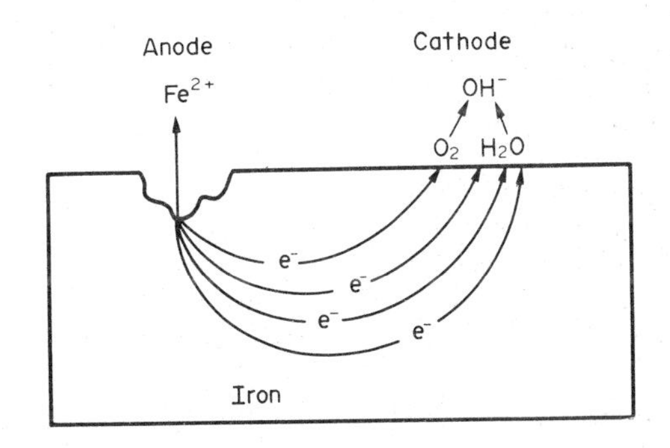

Figure 1
The electrochemical cell set up between anodic and cathodic sites on an iron surface undergoing corrosion

to the environment containing the lower oxygen content. This causes corrosion due to differential aeration. Similar effects can occur because of variations in the concentration of metal ions or other species in the environment. These variations arise because of the orientation of the corroding metal. A vertical metal surface can suffer corrosion because gravity can affect the concentration of certain environmental species near its bottom. Finally, variations in the homogeneity of the metal surface, due to the presence of inclusions, different phases, grain boundaries, disturbed metal and other causes, lead to the establishment of anodic and cathodic sites.

The process occurring at the anodic sites is the dissolution of metal as metallic ions in the electrolyte or the conversion of these ions to corrosion products such as rust. This is the destructive process called corrosion. The flow of electrons between the corroding anodes and the noncorroding cathodes forms the corrosion current, the value of which is determined by the rate of production of electrons by the anodic reaction and their consumption by the cathodic reaction. The rates of electron production and consumption, of course, must be equal or a buildup of charge would occur.

1.2 Thermodynamics and Corrosion Tendencies

A driving force is necessary for electrons to flow between the anodes and the cathodes. This driving force is the difference in potential between the anodic and cathodic sites. This difference exists because each oxidation or reduction reaction has associated with it a potential determined by the tendency for the reaction to take place spontaneously. The potential is a measure of this tendency.

A very useful way of studying the relation of potential to corrosion is through the use of an equilibrium diagram developed by Pourbaix which plots potential versus pH, a parameter also of great importance in corrosion processes. The simplified Pourbaix diagram for iron is shown in Fig. 2. The potential plotted is the potential of iron as measured versus a standard hydrogen reference electrode. The Pourbaix diagram enables the determination by means of potential and pH measurements of whether a metal surface is in a region of immunity where the tendency for corrosion is nil, in a region where the tendency for corrosion is high, or in a region where the tendency for corrosion may still exist but where there is also a tendency for a protective or passive film to exist. Such a film can drastically affect the rate of corrosion, and, in some cases, practically stop it.

Lines (a) and (b) on Fig. 2 represent Eqns. (2) or (3) and (4), respectively. They are a measure of the concentration of oxygen or hydrogen in solution and show the variation of potential with pH for solutions in equilibrium with 1 atm of either oxygen or hydrogen. Lines drawn parallel to (a), but lower, would give the variation of potential with pH for a solution

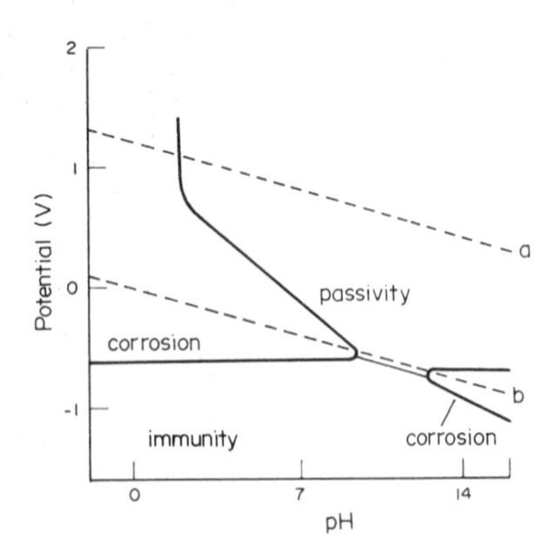

Figure 2
The potential–pH (Pourbaix) equilibrium diagram for iron

containing less oxygen than that represented by (a). Thus, the potential of a specimen can give some indication of the oxygen concentration in the environment. The differences in oxygen concentration between different sites on a metal lead to the differential aeration corrosion mentioned above (Sect. 1.1). By means of the Pourbaix diagram, through potential and pH measurements, some idea can be gained about the tendency of a metal to be in an immune, passive or corroding regime and also to estimate the influence of limited or ready access of oxygen on the corrosion process.

1.3 Kinetics and Corrosion Rates

Obtaining corrosion tendencies from the determination of potentials is necessary but not sufficient to ascertain whether a given metal or alloy will suffer corrosion under a given set of environmental conditions. This is so because, while the tendency for corrosion may be strong, the rate of corrosion may be very low and corrosion may not present any problems.

To determine the corrosion rates of a metal surface a current is applied to the surface by means of an auxiliary electrode and the change in potential of the surface is measured. When the potential of the metal surface is changed as a result of a flow of current to it, it is said to be polarized. If it is polarized in a positive direction (potential increases), it is said to be anodically polarized; a negative direction signifies that it is cathodically polarized. The degree of polarization (the degree of potential change as a function of the amount of current applied) is a measure of how the rates of the anodic and the cathodic reactions are retarded by environmental (concentration of metal ions in solution) and the surface-process (adsorption, film formation, ease of release of electrons) factors. The former is called concentration polarization, the latter activation polarization. The variation of potential as a function of current (a polarization curve)

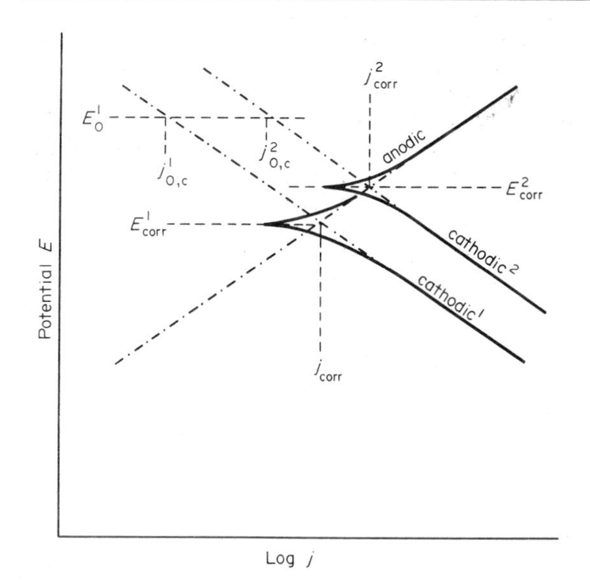

Figure 3
Schematic illustration of potential versus log (current density) curves (polarization curves) produced by making the metal surface measured an anode and then a cathode. Two cathodic reactions are shown, the one with the higher cathodic exchange density $j^2_{0,\,c}$ giving the higher corrosion current j^2_{corr}. The extrapolations of the anodic and cathodic curves (–·–·) produce the Evans diagrams (see Fig. 4)

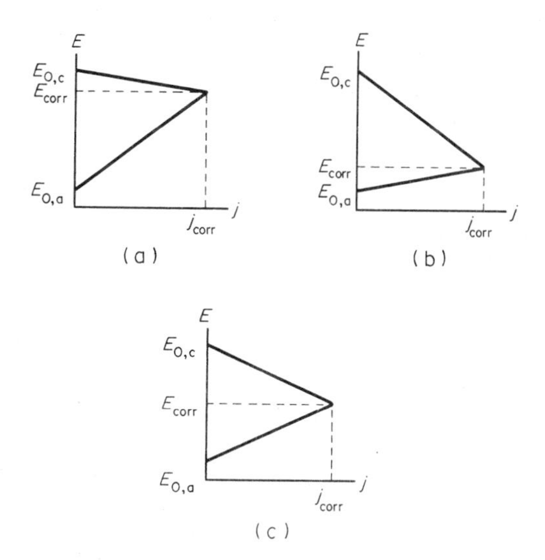

Figure 4
Evans diagrams (produced from the extrapolations shown in Fig. 3) illustrating the three main ways in which the corrosion rate is controlled: (a) control by the reactions on the anode; (b) control by the reactions on the cathode; and (c) control by a mixture of both

allows the study of the effect of concentration and activation processes on the rate at which anodic or cathodic reactions can give up or accept electrons. Hence, polarization measurements allow the determination of the rate of the reactions that are involved in corrosion—the corrosion rate.

Figure 3 shows anodic and cathodic polarization curves representing anodic reactions like Eqn. (1) or cathodic reactions like Eqns. (2–4). The potential E is plotted as a function of the logarithm of the current density j. When the corrosion reactions are controlled by activation polarization, as is often the case, polarization behavior follows the Tafel equation:

$$\eta = a + b \log j \qquad (6)$$

where η is the overpotential, the difference between the reversible (equilibrium) potential, $E_{0,\,a}$ or $E_{0,\,c}$ for the reaction under consideration (anodic or cathodic) and the measured potential E; b is the slope of the linear part of the polarization curve and a is the intercept at the reversible potential. The constant a is proportional to $\log j_0$, where j_0 is called the exchange current density. The values of the constants a and b in Eqn. (6) depend on the metal and the environment under consideration. It is possible to extrapolate the anodic and cathodic linear portions of the polarization curves to E_{corr}, the corrosion potential, where, under ideal conditions, they should intersect. The value of the current at their intersection is j_{corr}, the rate of corrosion. These linear portions can be extrapolated further until the anodic line reaches the reversible potential (the potential where the metal is in equilibrium with its ions) of the anodic reaction and the cathodic line, the reversible potential of the cathodic reaction. Figure 3 also shows the profound effect that the exchange current density for the cathodic reaction $j_{0,c}$ can have on the corrosion rate j_{corr}: the larger $j_{0,c}$ the larger the corrosion rate.

Parts of the extrapolations shown in Fig. 3 produce the diagrams shown in Fig. 4. These are called Evans diagrams. The slopes b in Eqn. (6) of the anodic or cathodic polarization curves, determine whether the anodic, cathodic or both reactions control the rate of the corrosion process. Evans diagrams are used extensively by corrosion scientists and engineers to evaluate the effect of various factors on corrosion rates.

1.4 Passivity and Breakdown of Passivity

As discussed above, both thermodynamic (corrosion tendency) and kinetic (corrosion rate) considerations are crucial in determining the extent of corrosion of a metal surface. A crucial factor controlling the rate is the existence of the phenomenon of passivity for certain metals and alloys, such as stainless steels and titanium.

What is passivity? A great deal of research, resulting in considerable controversy, has been undertaken to provide the answer to this question. For the purposes of this discussion it is sufficient to characterize passivity as the conditions existing on a metal surface because of the presence of a protective film that markedly lowers the rate of corrosion, even though from thermodynamic (corrosion-tendency) considerations active corrosion is expected.

An effective passive film must keep the corrosion current on a metallic surface at a low enough value such that the extent of corrosion damage is at an acceptable level. To remain effective, such a film must resist the breakdown processes leading to the forms of localized corrosion that are major sources of corrosion failures: pitting, crevice, intergranular and stress corrosion. The breakdown processes cause the disruption of the passive film and thus expose discrete bare sites on the metal surface to an environment in which the tendency for attack is very high. Two types of breakdown process exist: chemical and mechanical. The mechanism of chemical breakdown is poorly understood. Among the mechanisms proposed for chemical breakdown are desorption of the species (e.g., oxygen) responsible for the passive film by a damaging species, and a chemomechanical rupture of the passive film by mutual repulsion of adsorbed damaging species. It should be noted that each of these mechanisms involves a damaging species. Unfortunately, one of the major species causing breakdown of passivity is the chloride ion, a damaging species abundantly available in nature. The other kind of breakdown, mechanical breakdown, occurs when the passive film is ruptured as a result of stress or abrasive wear.

Fortunately, competing with the breakdown processes there is a repair process, repassivation. Thus, an effective alloy for resisting localized corrosion would be one whose surface not only forms a passive film that resists the processes leading to breakdown but also is capable of repassivating at a rate sufficiently high that once breakdown has occurred exposure to a corrosive environment is minimal.

2. *Types of Corrosion*

There are a number of types of corrosion, many (but not all) of which depend on the passive-film breakdown and repair processes just described.

2.1 Pitting Corrosion

The initiation of a pit occurs when a chemical breakdown process exposes a discrete site on a metal surface to damaging species such as chloride ions. The sites where pits nucleate are poorly understood, but some possible locations are at surface compositional heterogeneities (inclusions), scratches or places where environmental variations exist. The pit grows if rapid repassivation does not prevent the formation of a high local concentration of metal ions produced by dissolution at the point of initiation. If the rate of repassivation is not sufficient to choke off the pit growth, two new conditions develop. First, the metal ions produced by the breakdown process are precipitated as solid corrosion products (such as the $Fe(OH)_2$ in Eqn. (4)) which usually cover the mouth of the pit. This covering over of the pit restricts the entry or exit of ions in or out of the pit and allows the buildup of hydrogen ions through a hydrolysis reaction of the type given by Eqn. (4). Also, the chloride or other damaging-ion concentrations increase because a diffusion of these ions becomes necessary to maintain charge neutrality in the pit. In the highly acidic and highly concentrated damaging-ion and metallic-ion solution in the pit, repassivation becomes even more difficult and the rate of pit growth accelerates. The pit is the anode of a corrosion cell, and the cathode of the cell is the nonpitted surface. Since the surface area of the pit is considerably smaller than the cathodic surface area, all of the anodic corrosion current flows to a very small surface area. Thus, the anodic current density becomes quite high, and penetration of a metal structure with only a few pits can be rapid.

2.2 Crevice Corrosion

Crevice corrosion results when a portion of a metal surface is shielded in such a way that the shielded portion has limited but, nevertheless, some connection with the surrounding environment. Such surrounding environments contain damaging corrosion species, usually chloride ions. A typical example of crevice corrosion is the crevice that often forms in the area between a gasket and a metal surface. The environment that eventually forms in such a crevice is similar to that formed under the precipitated corrosion product cover in a pit. Similarly, a corrosion cell is formed with the unshielded surface as the cathode, the crevice interior acting as the anode because of its lower oxygen concentration compared with the surrounding medium. This combination of being the anode of a corrosion cell and existing in an acidic high-chloride environment where repassivation is difficult makes the crevice interior subject to corrosive attack.

2.3 Stress-Corrosion Cracking, Hydrogen Damage and Corrosion Fatigue

Stress-corrosion cracking (SCC) is a form of localized corrosion which produces cracks in metals by the simultaneous action of a corrodent and sustained tensile stress. Because the corrosion results from the application of stress, the breakdown of the passive layer on a metal surface in SCC is generally ascribed to mechanical causes. A number of mechanisms have been proposed to explain SCC.

Corrosion fatigue has a similar mode of failure to SCC. It mainly differs from SCC in that it involves situations where the stress is applied cyclically rather than as a static sustained tensile stress. With both corrosion fatigue and SCC, the processes of film mechanical breakdown, repassivation, adsorption of damaging species, hydrogen embrittlement and electrochemical dissolution are to a greater or lesser extent involved in determining susceptibility.

2.4 Intergranular Corrosion

Almost all metals used in practical devices are made up of small crystals or grains whose surfaces join the surfaces of other grains to form grain boundaries. Such boundaries, or the small regions adjacent to these boundaries, can under certain conditions be considerably more reactive (by being more anodic) than the interior of the grains. The resulting corrosion at the boundaries is called intergranular corrosion. It can result in a loss of strength of a metal part or the production of debris (grains that have fallen out).

A major cause of intergranular corrosion is sensitization, which occurs especially in 18%Cr–8%Ni stainless steels (austenitic stainless steels) as the result of improper heat treatment. Such heat treatment depletes the chromium content in areas adjacent to the grain boundaries so that the alloy near the grain boundaries has a composition close to that of an ordinary steel containing nickel. Thus, at the grain boundaries a less corrosion-resistant and more anodic alloy results from sensitization, and intergranular corrosive attack can occur. The driving force for such corrosion is the difference in potential between the cathodic grain interiors and the anodic grain-boundary regions.

2.5 Uniform Corrosion

Uniform corrosion, unlike the localized types of corrosion described thus far, is generally predictable because it results from electrochemical reactions that occur uniformly over the entire surface of a metal part; it is thus capable of being measured more reliably than localized corrosion. The damage caused by uniform corrosion, sometimes called wastage, is usually manifested in the progressive thinning of a metal part until it virtually dissolves away or becomes a delicate lace-like structure.

2.6 Galvanic Corrosion

When two metals have different potentials in a conducting electrolyte, the more anodic metal usually suffers galvanic corrosion. Galvanic corrosion differs from the other forms of corrosion described previously in that the anodic and cathodic sites of the corrosion cell exist because of the electrical contact of two different metals, rather than because of differences existing on the surface of the same metal.

The value for the differences in potential between two dissimilar metals is usually obtained from a listing of the standard oxidation potentials for various metals for reactions of the type given in Eqn. (1). Since the standard oxidation potentials assume bare metal surfaces in a standard solution containing their ions at unit activity, they are not reliable guides to the corrosion tendencies of the anodic member of a corrosion cell produced by two coupled dissimilar metals which invariably have films on their surfaces and are exposed to nonstandard environments.

Another factor besides the coupling of two different metals that can lead to galvanic corrosion is a difference in temperature at separated sites on the same metal surface. Such a situation leads to thermogalvanic corrosion. This kind of corrosion can be encountered in heat-exchanger systems where appreciable temperature differences are common.

2.7 Selective Leaching

When one component of an alloy is removed selectively by corrosion resulting in an increase in the concentration of the remaining components, selective leaching occurs. This type of corrosion is also called parting and, after the most common example, the selective leaching out of zinc from brass, dezincification. For other alloys undergoing selective leaching, names such as dealuminumification and denickelification are also used. Another important and somewhat misleading name for a form of selective leaching suffered by gray cast iron, when iron is selectively leached out leaving behind a graphite matrix, is graphitization.

2.8 Erosion and Fretting Corrosion

Erosion and fretting corrosion result from a disruption of protective passive films by erosive or abrasive processes. In the case of erosion, protective films are removed by the rapid movement of a corrosive fluid, liquid or gas, with or without solid abrasive particles. Film removal by fretting corrosion results from the abrasion that can occur when multicomponent devices rub together under load. The repeated removal of protective oxide films by fretting corrosion produces corrosion-product debris.

3. Corrosion Control and Design

To prevent or control the types of corrosion described above, the proper utilization of the present understanding of the principles underlying corrosion processes has been necessary. Out of this understanding have come a number of corrosion-control measures that have been incorporated into more-effective designs for structures and mechanisms resistant to corrosion damage.

The major corrosion-control measures are as follows.

3.1 Corrosion-Resistant Alloys

Most of the alloys developed to be corrosion resistant contain constituents, such as chromium, which produce more-effective protective films that resist breakdown and self-repair rapidly.

3.2 Coatings

Rather than depend on the naturally formed barrier that a corrosion-resistant alloy provides, an artificial barrier to the corrosive environment may be applied in the form of a coating on a metal surface (see *Corrosion Protective Coatings for Metals*). The most widely used artificial barriers are organic coatings, especially paints. Paints, in addition to providing polymeric film barriers, contain pigments that infuse corrosion-inhibitive chemicals into the micro-environments that develop at the metal–coating interface. Other organic coatings include lacquers, waxes and greases. Some metallic coatings such as chrome plate provide a protective barrier because they are more corrosion resistant than the metals they protect. On the other hand, a metallic coating such as zinc coating works because of the greater tendency of the zinc to corrode than the substrate metal (usually steel), and thereby protect it. Finally, ceramic coatings such as porcelain enamels and glasses are used to protect metal surfaces by providing chemically resistant barriers, usually oxides, which are more stable than metals.

4. Cathodic and Anodic Protection

Cathodic protection provides corrosion control by making the structure to be protected the cathode of a corrosion cell (see *Corrosion Protection Methods*). In doing so, the potential of the structure is usually brought into or near the immunity region shown in the Pourbaix diagram in Fig. 2. This is achieved either by impressing a current using a power supply and nonreactive anode or by using a sacrificial anode (zinc, magnesium or aluminum) which supplies a current as it corrodes. Anodic protection makes the structure to be protected an anode and an applied current brings the structure's potential into the passivation region of the Pourbaix diagram.

5. Corrosion Inhibitors

This protection scheme seeks to alter the environment in which a given metal must be used by adding chemicals—corrosion inhibitors—that make the environment less corrosive. The inhibitors act either by adsorbing on metallic surfaces and thereby producing barriers to the environment or by keeping the environment from becoming more corrosive by providing a buffering action.

Effective design for corrosion control uses the many corrosion prevention measures described above as well as assuring that structures and parts are not designed in ways that promote corrosion.

Bibliography

Fontana M G, Greene N D 1978 *Corrosion Engineering*, 2nd edn. McGraw-Hill, New York

Frankenthal R P, Kruger J (eds.) 1978 *Passivity of Metals*. Electrochemical Society, Pennington, New Jersey

Pourbaix M J N 1966 *Atlas of Electrochemical Equilibria in Aqueous Solutions*. Pergamon, Oxford

Pourbaix M J N 1973 *Lectures on Electrochemical Corrosion*. Plenum, New York

Scully J C 1966 *The Fundamentals of Corrosion*. Pergamon, Oxford

Shreir L L (ed.) 1976 *Corrosion*, 2nd edn., Vols. 1 and 2. Newnes-Butterworths, London

Staehle R W, Brown B F, Kruger J, Agrawal A (eds.) 1974 *Localized Corrosion*. National Association of Corrosion Engineers, Houston, Texas

Uhlig H H (ed.) 1948 *The Corrosion Handbook*. Wiley, New York

Uhlig H H, Revie RW 1985 *Corrosion and Corrosion Control: An Introduction to Corrosion Science and Engineering*, 3rd edn. Wiley, New York

West J M 1980 *Basic Corrosion and Oxidation*. Ellis Horwood, Chichester

J. Kruger
[Johns Hopkins University, Baltimore, Maryland, USA]

Corrosion Protection Methods

The structural metal most often used in present technology is steel. The protection of steel in a corrosive environment will therefore be used here as an example to illustrate corrosion protection methods. In fact, these methods apply to most metals with very few exceptions.

The conditions under which iron is corroded or protected in aqueous environments can be schematically illustrated in a Pourbaix diagram (Fig. 1). When the oxidizing conditions and the pH of a certain environment intersect in the corrosion region (say at point A in Fig. 1), iron can be expected to be attacked.

Obviously, the methods designed to protect the metal should depend on changing the conditions so that the intersection point is removed from the corrosion region. There are three directions that may be followed on the diagram to achieve this end:

(a) decreasing the oxidation potential to the immunity region;

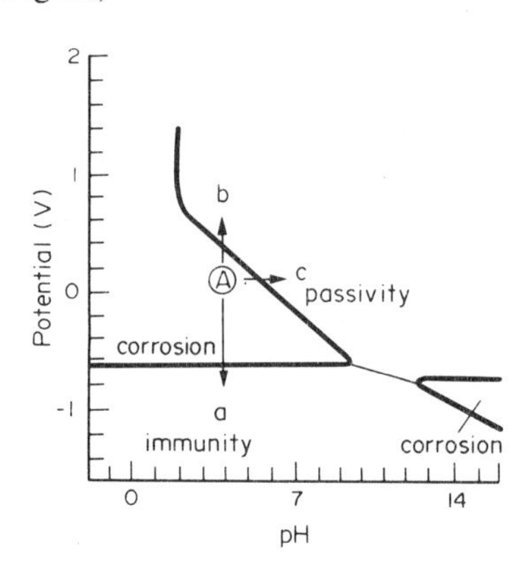

Figure 1
Corrosion-protection directions in the potential–pH (Pourbaix) diagram for iron

(b) increasing the oxidation potential to the passivity region; and

(c) increasing the pH to the passivity region.

Alternatively the metal may be separated from its environment, or metals of higher corrosion resistance may be used for construction. These five approaches essentially cover the numerous methods for the protection of metals.

1. Decreasing Potential

1.1 Cathodic Protection

When applicable, cathodic protection is by far the most reliable of all corrosion protection methods since it is designed to bring a metal into a thermodynamically stable regime (direction a in Fig. 1). In this method, the structure to be protected is made the cathode; its potential is lowered with respect to an auxiliary anode embedded in the corroding medium (usually wet soil, or seawater for ships and offshore structures). The polarization between the structure and the anode can be produced by means of a dc power supply, a method called impressed-current cathodic protection (ICCP). The anode in such a system is made of an inert material such as carbon or platinum coated over titanium, niobium or tantalum. Alternatively, the anode may be of a metal which is more active than steel (e.g., magnesium, aluminum or zinc). This electrode is directly joined to the structure to be protected, and the method is termed sacrificial-anode cathodic protection (SACP) since the active metal anode is corroded when coupled to the protected structure, thus releasing the anodic current to the system. The choice between ICCP and SACP is based on economic considerations, on the availability of dc current supplies and on the geometry of the structure.

Cathodic protection is widely used for the protection of ship hulls and offshore structures, as well as underground water and fuel tanks, and, most notably, pipelines. The potential of a steel structure in most natural environments should be -0.850 V with respect to a $Cu/CuSO_4$ reference electrode for full cathodic protection. The combination of cathodic protection with organic coatings usually makes the system more economical. Certain metal coatings such as zinc, aluminum and cadmium provide cathodic protection for the underlying steel by virtue of their active nature. In these cases, the coating acts as a sacrificial anode.

1.2 Lowering Oxidant Concentration

The oxidation potential of a structure can be brought down towards the immunity region of the Pourbaix diagram (Fig. 1) by depletion of the supply of oxidant. The most notable case is that of steam boilers. The methods usually employed involve reduction of the dissolved oxygen content in boiler water by physical deaeration or chemical oxygen scavenging.

Physical deaeration is attained by passing the water in jets into a lower-pressure ambient so that gas solubility in the water decreases. In this way, both air and carbon dioxide are released from the water, resulting in an increased pH value, which is a desirable effect in itself in controlling water-side corrosion. Chemical scavenging is performed by reducing agents. The most commonly used agents are sodium sulfite and hydrazine, the latter usually being used for higher-pressure boilers.

2. Protection by Oxidation-Potential Increase

2.1 Anodic Protection

In contrast to cathodic protection, the metal is brought to a higher potential (direction b in Fig. 1) in the anodic protection method. The metal becomes passive by the formation of a protective layer. This is achieved by means of a potentiostat, a power supply designed to keep a metal at a prescribed potential. This potential is set in the passivity region of the Pourbaix diagram. The method is not as widely used as cathodic protection, and is usually limited to metals whose passive layers are sufficiently reliable, such as stainless steel. It is, however, used to some extent in the phosphoric acid industry. Two major considerations are involved in the application of this method. First, it is important to ensure that all of the system is in the passive range. This requires paying special attention to remote parts such as branching piping, and ensuring that the structures to be protected are not just brought to a higher potential within the active range (which could lead to more severe corrosion). The other main consideration is a precise knowledge of the presence of certain ions (e.g., Cl^-) which can lead to extensive pitting at potentials above the "critical pitting potential." This sets an upper limit to the potential, the requirement being that it be kept below the critical pitting potential, which depends on the aggressive-ion concentration. With stainless steel, the potential should also be kept below the transpassive range of potential over which the protective Cr_2O_3 layer is oxidized to soluble hexavalent chromium.

2.2 Oxidizing Additives

The use of oxidizing inhibitors is quite common in many cooling-water systems. Chromates are used extensively for this purpose. Their main action is similar to the electrical increase of oxidation potential in anodic protection (direction b in Fig. 1). The normal concentration of chromates in cooling water is of the order of 400 ppm, but sometimes an initial boost is recommended with a concentration of about 900 ppm. Usually an increase in pH to about 9 is desirable, thus moving the system into the passive range. A risk involved in the use of this kind of corrosion inhibitor results from allowing the inhibitor concentration to fall below the prescribed value, leading to local attack or pitting. Chromates have also become less attractive in recent years owing to health hazards discovered to be associated with their use. Oxidizing additives are very useful in primary layers in anticorrosive paints. Lead oxide used to be a major constituent in primary coatings (red lead), but increasing awareness of the danger of lead poisoning has led to a decline in the use of lead-base paints. Zinc dichromate has largely replaced red lead but, as noted, chromium itself presents health problems.

3. pH Increase

Increasing the pH brings about passivity by a horizontal shift in the Pourbaix diagram (direction c in Fig. 1). For steel, a safe pH is considered to be above 9.2. Increasing the pH is a common practice in many systems that lend themselves to such a change in acidity. The most common use is for industrial cooling water where the pH increase is usually achieved by adding sodium hydroxide. Another case is steam boilers; here, the alkaline addition is usually sodium phosphate. The reason why this compound is used rather than the cheaper alkali hydroxides is the need to prevent the pH from reaching very high levels when evaporation occurs in a leaking system, a situation not uncommon in large steam systems. An increase of pH to above 10 may result in "caustic embrittlement" of steel at high temperature and pressure. On the other hand, phosphates act as buffers and prevent such a local increase in pH.

Protection under coatings by a local increase of pH is attained with the use of red lead which, apart from being oxidizing, is also alkaline.

4. Separating Metal from Environment

4.1 Paints and Organic Coatings

Coating the metal with insulating layers would be the best protection method if such layers were completely impervious and prevented the corrosive environment from coming into contact with the metal. Such coatings do not exist and provisions must be made to provide for the permeability and defects in these coatings. For this reason, more than one layer of coating is usually applied. This decreases but does not eliminate the probability that defects are continuous from the outer part of the coating system to the metal.

Most protective systems consist of a base layer containing an inhibitive additive, usually an oxidant such as zinc chromate, covered by polymer-based insulating layers. The oxidizing agents and other inorganic compounds such as paint pigments are typically embedded in the polymer matrix (see *Paints, Varnishes and Lacquers*).

4.2 Protection by Metallic Coatings

Metal coatings are used to separate a structural metal from a corrosive environment. Such metal coatings can be classified into two groups—those that are more active than the base metal, and those that are more noble. For the protection of steel, the first group includes coatings such as zinc, aluminum, and to some extent, cadmium. These coatings act as sacrificial anodes at sites of discontinuity and thus afford cathodic protection to the metal. The lifetime of a coated structure is often related to the coating thickness, which determines the amount of sacrificial material.

The group of coatings more noble than the substrate includes such metals as chromium, nickel and copper. The protection afforded by such coatings depends entirely on their continuity: at holes in such coatings enhanced corrosion of the base metal usually occurs. These coatings are used in many applications in spite of the risks involved because they impart special properties such as luster, hardness and electrical conductivity.

4.3 Corrosion Inhibitors

The separation of metals from the environment can be achieved by the adsorption of inert material on the metal surface in the form of thin, sometimes monomolecular, layers. Chemicals which form such layers are usually referred to as corrosion inhibitors. Such inhibitors can form the separating barrier by becoming attached to a metal surface through an ionic group. Examples of such inhibitors are derivatives of benzomethyl sulfonate and similar organic compounds. Other inhibitors, such as polyphosphate, act by limiting crystallization of minerals, such as calcium carbonate in water, to laminar growth on the metal surface. A third group of inhibitors act by oxidizing the surface and forming passive layers. These include oxidizing inhibitors like chromates.

4.4 Passivation

Sometimes a pretreatment of metals is performed prior to their use by immersion in oxidizing agents such as nitric acid and chromates. This treatment is termed passivation. It differs from the inhibition described above in that the oxidizing agent is not present during the service life of a part and the passive layer which is acquired during the pretreatment cannot be repaired if damaged during service. The passivation treatment is therefore useful only for mildly corrosive environments, but is extremely useful in extending shelf life of products prior to their use.

4.5 Anodic Oxidation

Anodic oxidation protection is most commonly used for aluminum and its alloys, but some other metals such as titanium are treated similarly. In this method, the metal is oxidized electrochemically in an electrolyte bath prior to exposure to the corrosive environment. Typical thicknesses of such oxide films are

5–30 μm. The electrolytic bath is usually sulfuric, chromic or phosphoric acid but other electrolytes are sometimes used. The oxide film formed in an acid solution is porous and the pores have to be sealed before the metal can be used in a corrosive medium. The sealing is usually done by immersion in boiling water. Thin nonporous coatings can be formed by anodizing in a neutral electrolyte.

5. *Protection by Use of Alloys*

It is often economically feasible to replace the corrosion-susceptible metal of a structure by one of its alloys which is more resistant in the specific environment. Many alloying additions extend the passive region to pH values or potentials which are in the corrosion range for the original metal. Stainless steels are the most important example. Additions of chromium to iron in excess of 9% bring about a passive behavior over a wide range of conditions.

In the common environment of aerated fresh water, iron is usually in its active state and stainless steels in the passive state. This is also the case in many acids used in the chemical industry where stainless steels play a major role as construction materials. Small alloying additions are sometimes used in stainless steels to make them resistant under some particular conditions; for example, additions of 2–4% molybdenum enhance their resistance to pitting in chlorides and 0.5–1% titanium or niobium inhibit sensitization in heat-affected zones near weld seams, a condition that promotes intergranular corrosion.

See also: Corrosion Protective Coatings for Metals; Corrosion and Oxidation Study Techniques; Corrosion of Metals: An Overview

Bibliography

Fontana M G, Greene N D (eds.) 1978 *Corrosion Engineering*, 2nd edn. McGraw-Hill, New York

Pourbaix M 1973 *Lectures on Electrochemical Corrosion*. Plenum, New York

Shreir L L (ed.) 1976 *Corrosion*, 2nd edn. Newnes-Butterworths, London

Uhlig H H, Revie R W 1985 *Corrosion and Corrosion Control: An Introduction to Corrosion Science and Engineering*, 3rd edn. Wiley, New York

J. Yahalom
[Technion-Israel Institute of Technology,
Haifa, Israel]

Corrosion Protective Coatings for Metals

Combinations of metals (alloys) can generally be fabricated with physical properties that meet the demands of various applications, but the resulting alloys often lack the required chemical properties and are not resistant to many of the environments to which they may be exposed. The solution to this problem is a corrosion-resistant coating that provides the necessary resistance to a harsh environment and a metallic substrate that provides the necessary physical properties. It is the purpose of this article to outline some of the principles by which a coating protects a metal from corrosion at temperatures of 0 to 200 °C. Three general classes of coatings are discussed: cementitious coatings, metallic coatings and organic coatings.

1. *Cementitious Coatings*

These coatings are based largely on three bonding systems: alkali silicate cements, Portland cement and calcium aluminate cements. The absence of holes in the coating, good adhesion to the substrate, and resistance to cracking during thermal cycling are essential requirements for corrosion protection.

All cementitious coatings are alkaline in nature. Water or aqueous solutions that penetrate the coating leach alkali from the cement such that the pH at the cement–metal interface is nonacidic. Structural steels are resistant to corrosion when the pH is mildly alkaline and the naturally occurring oxide on the steel retains its integrity under these conditions. Thus, corrosion protection of steels by cements is a consequence of the near neutral or alkaline nature of the aqueous environment that penetrates through the pores in the coating.

2. *Metallic Coatings*

Metallic coatings are widely used for corrosion protection of metals and alloys because of the many methods by which the coatings can be applied. Among the major methods are electrodeposition, chemical deposition, hot dipping, spraying, condensation from the vapor, mechanical cladding and sputtering.

Metallic coatings protect a substrate metal by two major mechanisms: simple screening of the substrate from the environment, and sacrificial (galvanic) protection. Examples of the first mechanism include the use of gold and palladium coatings to protect copper connectors in electronic devices, and copper coatings on steel wire. Examples of the second mechanism include galvanized steel (zinc on steel), tinplate (tin on steel) and aluminized steel (aluminum on steel). Sacrificial protection is the favored type of protection mechanism under those circumstances where the coating is expected to be damaged during service. Coatings based on barrier protection alone fail to provide protection at those places where the substrate becomes exposed because of scratches, impact or mechanical deformation.

The mechanism of sacrificial protection can be understood from the electrochemical mechanism of corrosion. The corrosion process can be visualized as consisting of two half-reactions: (a) the anodic electron-generating reaction, which in the case of iron may be described as $Fe \rightleftharpoons Fe^{2+} + 2e^-$; and (b) the cathodic electron-consuming reaction, which under atmospheric conditions is usually $H_2O + \frac{1}{2}O_2 + 2e^- \rightleftharpoons 2OH^-$. Since metals are excellent electrical conductors, the anodic and cathodic half-reactions may occur at significant distances from one another when a conductive liquid is available to complete the electrical circuit. Rain plus contaminants on the surface of the metal provide the necessary conductive liquid during outdoor exposure. Thus, the anodic reaction will occur on that portion of the metal where the resistance to the formation of the metal ion is a minimum, and the cathodic reaction will occur where the conditions are such that the oxygen reduction reaction occurs most readily.

In the case of galvanized steel, the zinc surface serves as the anode and any exposed steel surface serves as the cathode. The zinc thus corrodes and the steel is protected. The steel is further protected by the secondary effect that the iron oxide on the steel surface is inert in the mildly alkaline environment generated by the cathodic reaction. The distance over which protection may be maintained is a function of the conductive nature of the liquid which overlays the separate steel and zinc regions; a highly conductive liquid permits good throwing power and protection over centimeter distances.

When metal coatings are formed by dipping into a bath of molten metal, as for example in the preparation of tinplate and galvanized steel, chemical interactions may occur at the metal–metal interface. Intermetallic compounds of zinc and iron form during the manufacture of galvanized steel and the intermetallic compound $FeSn_2$ forms during the manufacture of tinplate. The $FeSn_2$ layer is very resistant to chemical attack and greatly reduces the area of the steel substrate that becomes exposed after partial dissolution of tin. Tinplate depends for its performance on the fact that in the absence of oxygen the tin serves as the anode and any exposed steel serves as the cathode. The rate-controlling step in the overall corrosion reaction is the cathodic half-reaction, which occurs very slowly on tin and $FeSn_2$, and relatively rapidly on steel. Thus, a low corrosion rate occurs so long as the steel substrate is shielded from the environment by either tin or $FeSn_2$.

One of the more widely used coatings in outdoor exposure is chrome plate, which finds extensive application in the automotive industry. Chrome plate consists of a relatively thick layer of nickel on steel and a very thin layer of chromium on the nickel. The nickel undercoat is applied in two, or sometimes three, steps. The first layer, nearer the steel substrate, is electrodeposited to give a semibright deposit, and the second layer is electrodeposited from a plating bath in which sulfur-containing additives are included. The resulting nickel deposit is lustrous and contains a considerable amount of sulfur. The top layer is a thin electrodeposit of chromium which maintains the brightness of the second nickel layer. The chromium-plating bath is designed so that it deposits the chromium with many microcracks invisible to the naked eye. This three-layer electrodeposit provides corrosion protection to the steel in an unusual way. The chromium layer functions as the cathode since it has access to the oxygen in the air, and the intermediate layer of sulfur-containing nickel becomes the anode. The microcracked nature of the chromium provides access of the environment to the nickel layer immediately below the chromium, and thus the anodic reaction is distributed over a very large area. The sulfur-bearing nickel layer corrodes laterally without penetration to the steel because the nickel layer nearer the steel is inactive as an anode compared with the sulfur-bearing nickel layer. The corrosion products of nickel are soluble and are washed away by rain. Rust spots are not observed when the system is functioning optimally because there is no penetration to the steel.

3. Organic Coatings

Organic coatings—lacquers, varnishes and paints—are complex mixtures of polymers, fluid carriers, fillers, pigments, corrosion inhibitors and additives for special purposes. They protect metal surfaces from corrosion by a combination of barrier action and the presence in the coating of inhibitors that function when the coating is wet. Coatings that contain high concentrations of zinc flakes also provide corrosion protection, particularly of steel, by a combination of sacrificial protection and the formation of zinc corrosion products that serve as corrosion inhibitors.

The appraisal of new coating systems and a comparison of the relative performance of different coatings is best done under service conditions, and many outdoor test facilities have been constructed to test coatings underground, in desert, industrial and marine environments, and in areas where varying amounts of rainfall and different temperature regimes are experienced. However, considerable emphasis is placed on the performance of coatings in accelerated tests, since times for deterioration under service conditions are often measured in terms of many years.

Coatings that are expected to be exposed only to atmospheric conditions are generally tested in a device known as a weatherometer, in which the deteriorating medium is a combination of ultraviolet light, moisture and elevated temperature. Coatings that might be used near the seashore or in environments where salt is present are tested in a salt spray cabinet for many hours. The exact experimental conditions are often determined by the user of the coating or by industrial trade associations. The performance of coated-steel test panels is determined by an evaluation of blistering, the number of rust spots and the area covered by rust. The ability of the coating system to resist deterioration after it has been damaged is appraised by scratching an "X" in the coating through to the substrate before the panel is inserted into the salt spray cabinet. The performance is characterized by the distance from the scratch that the coating loses adherence from the substrate. Modifications of these tests to appraise coatings for service on the exterior of automobiles include repeated exposure to salt spray and to freeze–thaw conditions.

Coatings that are expected to be exposed to moisture at high temperatures are appraised in a hot-water test. The coated metal is exposed to hot water at 80–100 °C for times ranging from one hour to twenty days. The adherence of the coating to the substrate is determined after exposure to the hot water and signs of corrosion are sought beneath the coating. The mechanism by which coatings lose adherence in the test is not known.

Coatings for pipelines, for ships or for other structures that use an applied cathodic current to reduce corrosion are tested in accelerated tests to determine their ability to resist deterioration under these conditions. A hole is made in the coating and the coated metal is immersed in an electrolyte and is made the cathode against an inert counter-electrode by means of an external power source. A predetermined voltage is applied and the system is allowed to remain in this condition for many days. At the end of an agreed time, the system is removed from the electrolyte and the distance from the hole where the coating has lost adherence to the substrate is determined. Good coating systems exhibit very little loss of adherence.

The major driving force for cathodic delamination at low driving potentials, or for the delamination that occurs during corrosion in the presence of air, is the cathodic reaction, $H_2O + \frac{1}{2}O_2 + 2e^- \rightleftharpoons 2OH^-$. At very negative cathodic potentials or in the absence of oxygen, the alternative cathodic reaction, $2H^+ + 2e^- \rightleftharpoons H_2$, occurs. Both reactions have in common the consequence that the alkalinity in the very small liquid volume under the coating increases as the reaction occurs. Values of the pH as high as 14 in liquid volume beneath a coating have been reported. This high pH may cause dissolution of the thin oxide at the metal–organic coating interface or it may result in attack on the polymer near the interface. Which of these two delaminating mechanisms dominates is determined by the resistances of the interfacial oxide and the polymer to alkaline attack. The value of the pH at the point where the delamination is actively occurring is determined by the rate at which the cathodic reaction occurs, the shape of the delaminating front, the rate of diffusion of OH^- ions away from the delaminating front, and any buffering reactions that may occur.

The five main rules for minimizing cathodic delamination are: (a) bare metal, or superficially oxidized metal, should be avoided at the coating–substrate interface; (b) there should exist at the coating–substrate interface a layer which is a very poor conductor of electrons; (c) the interfacial layer at the coating–substrate interface should be a poor catalyst for the cathodic reaction; (d) the boundary between the substrate and the coating should be rough in order to provide a tortuous path for lateral diffusion; and (e) the interfacial region should be resistant to alkaline attack.

Many organic coatings are formulated so as to provide corrosion protection to the metal on which they are used. The agents that are incorporated in the coating to perform this function are known as inhibitors, of which there are four major classes: oxidizing inhibitors, organic compounds, metallic cations, and nonoxidizing inorganic and organic salts.

The most widely used oxidizing inhibitor is chromate. Commercial metals are often exposed to chromate solution before shipment to avoid corrosion during transit or storage. Phosphate pretreatments for corrosion protection are often followed by a chromate treatment. Metals are often treated with chromate before the application of an organic coating, and chromates are used in coatings to provide long-term corrosion protection. Chromate is an effective corrosion inhibitor for iron, zinc, aluminum, cadmium and tin; it is effective in aerated and deaerated solutions and in acid, neutral and alkaline environments. Chromate is an effective inhibitor for several reasons, the most important of which is the low solubility of Cr_2O_3 in mildly acidic media. Upon exposure to chromate solution, the surface of the inhibited metal develops a coating of chromium hydroxide or chromium oxide. Any chromate incorporated in the oxide film serves as an oxidizing agent and allows self-repair of the film when it is chemically deteriorated. Much effort is being devoted to finding substitutes for chromate inhibitors in paints because of environmental hazards associated with their use.

Organic corrosion inhibitors function primarily by adsorption and insulation of the metal from the environment. Evidence for the adsorption of the organic compound on the metal comes from ellipsometric measurements, electrochemical studies, solution depletion studies, radiotracer studies, spectroscopic measurements and studies of the double layer. Circumstantial evidence comes from corrosion inhibition data that are interpretable in terms of adsorption isotherms. Corrosion potential and corrosion rate measurements in the presence and absence of the inhibitor are the criteria used to determine whether the inhibitor affects primarily the anodic or cathodic half-reactions or both equally.

Lead oxides and salts are widely used as corrosion inhibitors in organic coatings. When the coating is wet, a small amount of lead is dissolved, probably as $PbOH^+$, and is transported by diffusion to the metal surface. The $PbOH^+$ serves as a buffering agent to assist in the precipitation of ferric oxide or hydrated ferric oxide near the site where corrosion occurs.

Oxygen is essential in corrosion inhibition by lead because it oxidizes any ferrous ions to ferric ions. Precipitation of ferric oxide or hydrated ferric oxide occurs because the hydrolysis reactions are driven to the right by the consumption of H^+ by the buffering reaction $PbOH^+ \rightleftharpoons Pb^{2+} + OH^-$. The hydrolysis reactions involving Fe^{3+} may be written as

$$Fe^{3+} + H_2O \rightleftharpoons FeOH^{2+} + H^+$$

$$FeOH^{2+} + H_2O \rightleftharpoons Fe(OH)_2^+ + H^+$$

$$Fe(OH)_2^+ + H_2O \rightleftharpoons Fe(OH)_3 + H^+$$

Inorganic and organic salts such as borates, acetates, carbonates and benzoates are considered to function in the same manner as lead salts because they also are only effective in the presence of oxygen. The inhibitors are buffering agents and cause the precipitation of the ferric oxide by driving the hydrolysis reactions to the right by the consumption of H^+.

See also: Corrosion and Oxidation Study Techniques; Corrosion of Metals: An Overview; Corrosion Protection Methods

Bibliography

Carter V E 1977 *Metallic Coatings for Corrosion Control.* Newnes-Butterworths, London

Federation of Societies for Paint Technology (1964 onwards) *Federation Series on Coatings Technology.* Federation of Societies for Paint Technology, Philadelphia, Pennsylvania

Leidheiser H Jr (ed.) 1981a *Corrosion Control by Organic Coatings.* National Association of Corrosion Engineers, Houston, Texas

Leidheiser H Jr 1981b Mechanism of corrosion inhibition with special attention to inhibitors in organic coatings. *J. Coat. Technol.* 53 (678): 29–39

Leidheiser H Jr 1983 Towards a better understanding of corrosion beneath organic coatings. *Corrosion* 398: 189–201

Pludek V R 1977 *Design and Corrosion Control.* Wiley, New York

H. Leidheiser Jr
[Lehigh University, Bethlehem, Pennsylvania, USA]

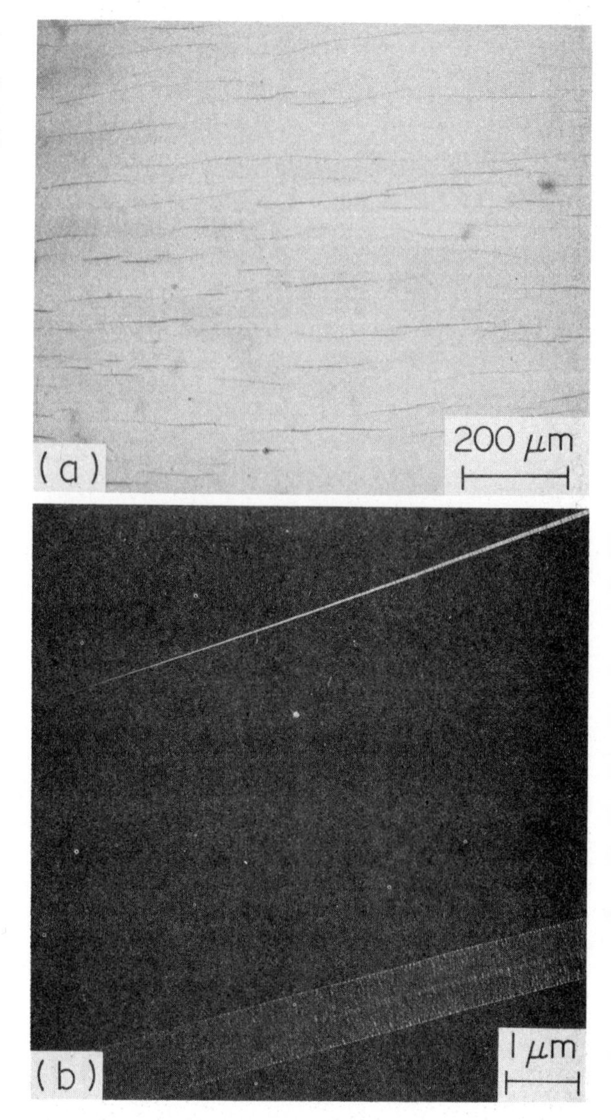

Figure 1
(a) Optical micrograph of crazes in a thin film of polystyrene. (b) Transmission electron micrograph of the typical craze microstructure in such a film

Crazing and Cracking of Polymers

When exposed to tensile stresses glassy polymers exhibit a unique form of plastic deformation called crazing (Fig. 1a). Crazes are thin, planar defects which strongly reflect and scatter light in the clear plastic, and which grow along the plane of maximum principal stress. They look like cracks, but transmission electron microscopy (Fig. 1b) shows that they consist of tiny fibrils ($\sim$6 nm in diameter) in which the polymer molecules are oriented along the fibril direction, and which span from one surface of the craze to the other. Between the fibrils is an interconnected network of voids.

1. Craze Stresses

The craze microstructure shown in Fig. 1b is typical of that of crazes grown in air in many glassy polymers; the fibrils enable the craze to support tensile stresses. Figure 2 shows the tensile stresses supported by a craze as a function of distance from the center of the craze to its tip. The center of the craze is typically

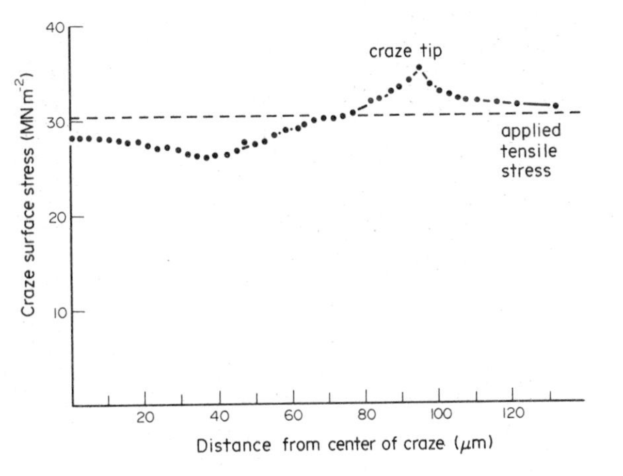

Figure 2
Profile of the tensile stress exerted by the craze fibrils on the craze surfaces in polystyrene (after Lauterwasser and Kramer 1979)

0.5 μm wide, the craze tapering to less than 10 nm in width at the tip. There is a moderate stress concentration at, and just behind, the craze tip.

2. *Craze Growth Mechanisms*

It is observed that as crazes expand in area they increase in width. Although it was originally thought that the craze tip advances by nucleation of isolated voids ahead of the tip, recent stereo-transmission electron microscopy studies have shown that the tip advances by breaking up into a series of void fingers which advance ahead of the main craze front (Donald and Kramer 1981). The fibrils develop from the polymer webs between these void fingers. The fingerlike growth is thought to be produced by Taylor meniscus instability.

It was also originally thought that crazes widen by creep of the polymer fibrils—fibrils formed just behind the craze tip would thus become increasingly extended toward the center of the craze. However, recent measurements of the local craze fibril volume fraction v_f from mass–thickness contrast of crazes on transmission electron microscope image plates show that fibril creep does not contribute significantly to the widening of isolated air crazes. (Since the fibril deformation takes place at constant volume, the fibril extension ratio $\lambda_{craze} = 1/v_f$.) Instead, more polymer is drawn into the fibrils from the craze boundaries as the craze widens, so that λ_{craze} is roughly constant along the craze. The craze extension ratio is highest just behind the craze tip, where the craze stress is high (Fig. 2), and in the midplane of the craze (the so-called midrib, which forms by fibril drawing just behind the growing craze tip).

3. *Craze Microstructure and Fibril Breakdown*

When cracks develop in glassy polymers they invariably originate by breakdown of the craze fibril structure to form large voids (Rabinowitz and Beardmore 1972). These voids grow slowly in size until a crack of critical length is formed within a craze. The crack then propagates rapidly, breaking the craze fibrils as it grows. The natural flaw in a glassy thermoplastic is therefore usually not a static preexisting crack, but a crack which develops over time as a result of craze nucleation, growth and breakdown.

An important determinant of craze fibril stability is the craze fibril volume fraction v_f or its inverse λ_{craze}. Usually, for production of tough materials, brittleness may be decreased by increasing the craze extension ratio λ_{craze}. For crazes, however, λ_{craze} is already quite large, and since the true stress σ_t in the craze fibrils is magnified by a factor of λ_{craze} over the stress S on the craze surfaces ($\sigma_t = S\,\lambda_{craze}$), increasing λ_{craze} usually

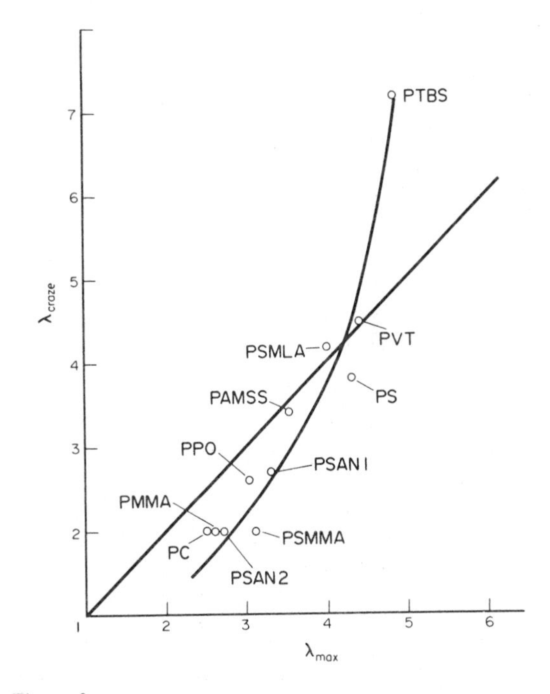

Figure 3
Craze fibril extension ratios against the computed maximum extension ratio of the entanglement network for various homopolymers and copolymers: PTBS, poly-*t*-butylstyrene; PVT, polyvinyltoluene; PSMLA, styrene–maleic anhydride copolymer; PAMSS, α-methylstyrene–styrene copolymer; PPO, poly(2,6-dimethylphenylene oxide); PMMA, poly(methyl methacrylate); PC, polycarbonate (bisphenol A polycarbonate, "Lexan"); PSAN, styrene–acrylonitrile copolymer (1, 24% acrylonitrile; 2, 66% acrylonitrile); PSMMA, styrene–methyl methacrylate copolymer (65% methyl methacrylate)

increases the rate of craze fibril breakdown and void nucleation within the craze. Plastic deformation by crazing does not ordinarily contribute much to the macroscopic plastic strain because the total volume of crazed matter is so small; an increase in λ_{craze} thus increases the apparent brittleness of such polymers by accelerating craze breakdown to form cracks.

In turn λ_{craze}, which can be considered to be a natural draw ratio of the craze fibrils, is determined by the molecular entanglement network of the polymer glass. It depends strongly on the entanglement molecular weight M_e (or the chain contour length l_e between entanglements) as established from measurements of the elasticity of the polymer melt (on the rubbery plateau) and increases dramatically from low- to high-l_e polymers. These values may be compared with λ_{max}, the maximum extension ratio of the entanglement network computed from l_e and the entanglement mesh size d, which is established from M_e and random coil dimensions. These λ_{max} values assume no slippage of entanglements and no chain scission; the entanglement network is treated as permanently cross-linked. As shown in Fig. 3, the values of λ_{craze} follow λ_{max} for the low-l_e polymers, but rise to, and significantly above, λ_{max} for the high-l_e polymers (Donald and Kramer 1982). This comparison (and other independent evidence) suggests that λ_{craze} is largely determined by λ_{max} of the entanglement network but that there is a significant contribution to λ_{craze} from chain breakage in the higher λ_{craze} values of the high-l_e polymers. This contribution probably arises from the higher true stresses in the craze fibrils. In view of the connection between a high λ_{craze} and rapid fibril breakdown it is not surprising that most of the low-λ_{max} polymers are often considered ductile engineering thermoplastics (e.g., polycarbonate, PC), whereas the high-λ_{max} polymers (e.g., polystyrene, PS) are considered brittle (Kramer 1983, Kramer 1984).

The effects of prior molecular orientation and chemical cross-linking of the glass on craze microstructure should be largely based on the changes they produce in λ_{max}. Chemical cross-linking, for example, should decrease λ_{max} and thus λ_{craze} when the cross-link density reaches high enough levels, and experiments indicate that this is so (Henkee and Kramer 1984). Uniaxial molecular orientation (e.g., by stretching the polymer quickly above its glass-transition temperature T_g) should decrease λ_{max} along the molecular orientation axis M but increase λ_{max} perpendicular to this axis. Crazes in a highly oriented polystyrene grown by applying tension parallel to M have $\lambda_{craze} = 2.2$, whereas crazes in the same polystyrene grown by applying tension perpendicular to M have $\lambda_{craze} = 20$. For comparison, the λ_{craze} value of crazes grown in unoriented polystyrene is approximately 4. The large change in craze microstructure (and fibril breakdown kinetics) produced by molecular orientation is a fundamental reason for the large fracture anisotropy in uniaxially oriented polymers.

4. *Crazing in Rubber-Modified Polymers*

It is possible to exploit the high microscopic ductility (high λ_{craze}) of the high-λ_{max} polymers, although normally too few crazes are formed in such polymers to produce sizeable macroscopic strains before a crack reaches critical size within one of the crazes. Ideally, it is desirable to nucleate many crazes; profuse nucleation can be promoted by embedding rubber particles (1–10 μm) in the glassy polymer. Figure 4 shows the typical rubber particle microstructure in high-impact polystyrene (HIPS), an important example of this class of materials. When this material is pulled in tension a very dense array of crazes develops (Fig. 5). There are so many crazes that the material turns white (shows stress whitening) owing to strong scattering of light.

Well-bonded and highly occluded rubber particles act to reinforce, as well as nucleate, crazes and do not give rise to premature craze breakdown and crack formation. Macroscopic plastic strains of 50% or more due to crazing can be achieved in such materials.

5. *Environmental Stress Crazing and Cracking*

Perhaps the most important drawback to the use of glassy polymers, rubber-modified or not, for engineer-

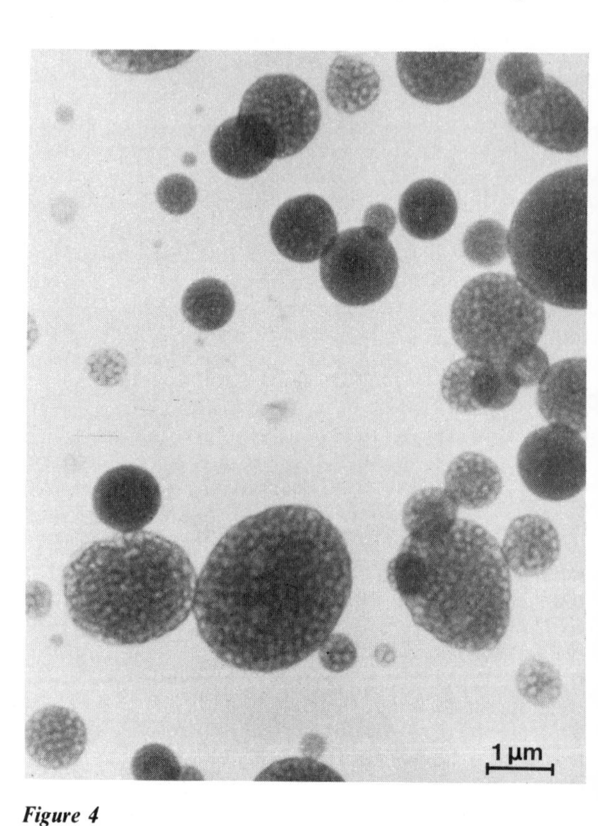

Figure 4
Transmission electron micrograph of the rubber particle morphology (stained with OsO_4) in HIPS

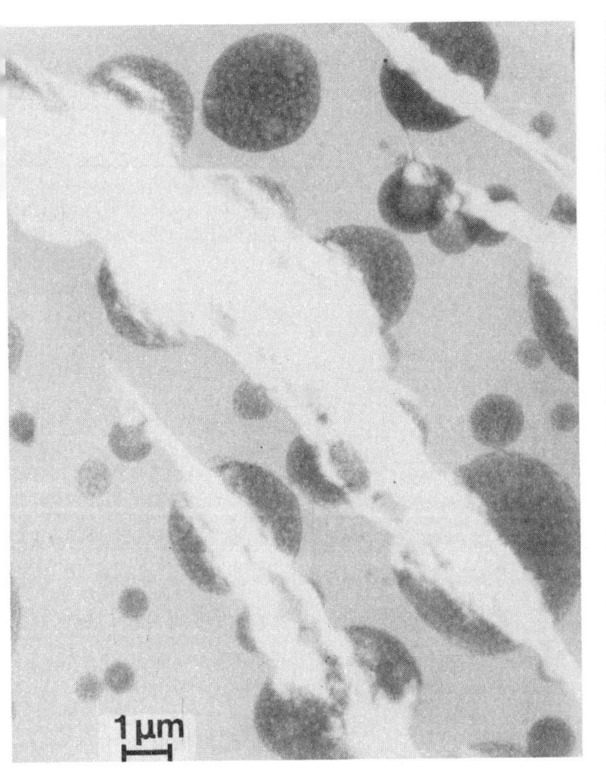

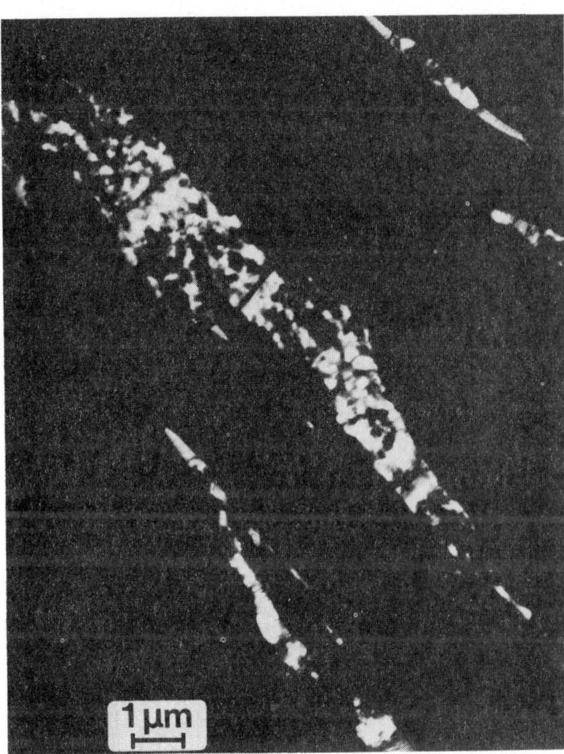

Figure 5
Crazes growing from and through rubber particles in a thin film of HIPS under stress. The craze fibrils are not visible in the left image because of the long exposure necessary to reveal the rubber particles

ing applications is their sensitivity to various organic environments and their susceptibility to environment-assisted cracking in a wide variety of situations. Here again, crazing is the important microscopic process. It can occur at much lower stresses in the presence of certain environments and crazes break down to form cracks more easily. The most important effect of the environment is the plasticization (depression of T_g) of the craze fibrils. Ordinarily many liquid organic media cannot diffuse sufficiently quickly into polymer glasses to cause much change in the bulk mechanical properties. However, the fibrils in a craze have diameters less than 100 nm and absorb these liquids rapidly to the equilibrium solubility limit. The resulting drop in T_g of the polymer makes it easier to plastically draw the polymer into fibrils at the craze surfaces. At the same time, molecular disentanglement in the swollen glass fibrils can proceed more rapidly, hastening the ultimate breakdown of craze fibrils to form voids. Some liquids (e.g., *n*-heptane in polystyrene) can even depress the fibril T_g to below room temperature, in which case a very weak craze is formed. Such environments are commonly called stress-cracking environments since the precursor crazes may be converted so rapidly to cracks that they are difficult to detect optically.

6. *Summary*

In summary, crazes have a highly drawn fibril and interconnected-void microstructure which is strongly load bearing. The craze tip advances by the meniscus instability mechanism and the craze thickens by drawing more material into the polymer fibrils from the craze surfaces. The draw ratio in the fibrils and the fibril volume fraction are primarily controlled by the entanglement network of the polymer glass; higher-draw-ratio (lower fibril volume fraction) crazes break down rapidly to form large voids and cracks. Although too few crazes ordinarily nucleate to produce a macroscopic plastic strain, if craze nucleation is encouraged by adding rubber particles, large strains due to crazing are possible. Environmental crazing and cracking occur principally due to plasticization of the craze surfaces and fibrils by the environment; strongly plasticized fibrils can break down very rapidly to form cracks.

See also: Mechanics of Materials: An Overview; Polymers: Structure, Properties and Structure–Property Relations

Bibliography

Donald A M, Kramer E J 1981 The mechanism for craze-tip advance in glassy polymers. *Philos. Mag. A.* 43: 857–70

Donald A M, Kramer E J 1982 Effect of molecular entanglements on craze microstructure in glassy polymers. *J. Polym. Sci., Polym. Phys. Ed.* 20: 899–909

Donald A M, Kramer E J, Bubeck R A 1982 The entanglement network and craze micromechanics in glassy polymers. *J. Polym. Sci., Polym. Phys. Ed.* 20: 1129–41

Farrar N R, Kramer E J 1981 Microstructure and mechanics of crazes in oriented polystyrene. *Polymer* 22: 691–98

Henkee C S, Kramer E J 1984 Crazing and shear deformation in crosslinked polystyrene. *J. Polym. Sci., Polym. Phys. Ed.* 22: 721–37

Kambour R P 1973 A review of crazing and fracture in thermoplastics. *J. Polym. Sci., Macromol. Rev.* 7: 1–154

Kramer E J 1979 Environmental cracking of polymers. In: Andrews E H (ed.) 1979 *Developments in Polymer Fracture.* Applied Science, Barking, pp. 55–120

Kramer E J 1983 Microscopic and molecular fundamentals of crazing. *Adv. Polym. Sci.* 52/53: 1–56

Kramer E J 1984 Craze fibril formation and breakdown. *Polym. Eng. Sci.* 24: 761–69

Lauterwasser B D, Kramer E J 1979 Microscopic mechanisms and mechanics of craze growth and fracture. *Philos. Mag. A.* 39: 469–95

Rabinowitz S, Beardmore P 1972 Craze formation and fracture in glassy polymers. CRC *Crit. Rev. Macromol. Sci.* 1: 1–45

E. J. Kramer
[Cornell University, Ithaca, New York, USA]

Creep

Creep includes a wide range of physical and practical phenomena, ranging from effects of strain rate on brittle fracture in steel structures and relaxation of prestressed steel-reinforcing bars, at temperatures near atmospheric, to creep of metals and ceramics in high-temperature boilers or chemical reactors. What distinguishes creep from plasticity is its time-dependent deformation that remains on release of load; what distinguishes creep from linear viscoelasticity is the usual cessation of deformation on unloading and the nonlinear dependence of deformation on stress history. As with plasticity, creep strains reduce stress concentrations, but often increase strain concentrations. Applications often require strains of 1% or less, with varying loads and temperatures over a 10–100 year life. Predicting such service requires evolutionary equations for the parameters of the metallurgical and dislocation structures that govern creep, as well as the flow equations for strain rate in terms of those parameters. Available data are usually at constant stress and temperature for much shorter times, especially for new alloys.

In this article, some physical bases of creep are introduced for critical insight into the creep equations often used. Solutions with those equations, in common with all continuum statics, must also satisfy the strain-displacement and equilibrium equations (see *Small Strain, Stress and Work*). Some solutions and their general character are discussed, along with sources of analytical, bounding, numerical and experimental methods for further solutions.

1. The Physical Bases of Creep in Solids

Many mechanisms of rate dependence in solids depend on the existence and motion of vacancies (missing atoms) in the crystal structure. After showing that these can be dealt with using classical, not quantum, mechanics, we consider how vacancies and hence metal ions diffuse and produce creep, either directly or by changes in the dislocation and metallurgical microstructure. The combined thermal- and stress-activated breakaway of dislocations from pinning points provides a third mechanism for the insight needed to judge the effects of changing structure and the approximations of the steady-state equations discussed later. The treatment is less detailed where references are readily available.

1.1 Applicability of Classical Statistical Mechanics

Quantum mechanics states that a mode of vibration, with frequency ν, can be excited only in increments of internal energy Δu proportional to ν:

$$\Delta u = h\nu \tag{1}$$

where h is Planck's constant (6.62×10^{-34} J s). In classical thermal motion, each degree of freedom of the ensemble of particles has an average energy of $kT/2$, where k is Boltzmann's constant (1.381×10^{-23} J K^{-1}). An atom vibrating along a line has two degrees of freedom, corresponding to its displacement and velocity, giving a total energy kT. To excite a mode, the temperature must be above its Debye temperature T_D:

$$kT > h\nu \quad \text{or} \quad T > h\nu/k \equiv T_D \tag{2}$$

The modes to be considered in classical statistical mechanics are those satisfying Eqn. (2) out of such conceivable modes as the linear or rotational vibrations an atom in a solid, the orbital motion of an electron, end-over-end or spin modes of molecules in a gas, or the relative motion of the atoms in a molecule. In most crystals, the highest excited frequency below the melting point is translation of individual atoms with neighbors oscillating out of phase. For translational modes, the Debye frequency ν_D can be estimated as that associated with sound waves of length $2b$ (with b the closest atomic spacing) passing through the crystal at a sound velocity c. c can be approximated from the modulus of elasticity E and the density ρ as $c=(E/\rho)^{1/2}$, and for many solids with primary bonds is about 5000 m s^{-1}. Thus for a typical interatomic spacing of $b = 0.25 \times 10^{-9}$ m, $\nu_D \approx c/2b \approx 10^{13}$ s^{-1}. Then for most solids with primary bonding, T_D from Eqn. (2) is about room temperature, above which the atomic motion can be treated with classical mechanics. A notable exception is diamond, with light carbon atoms and very stiff (high energy) bonds.

Alternatively, the Debye temperature can be found from the specific heat of a solid, which arises from the energy in the three degrees of freedom, each with translation and velocity. In terms of the energy per mole u and the universal gas constant $R = kN_A = 8.314$ J K^{-1} mol^{-1}:

$$u = 6\frac{kT}{2}N_A = 3RT; \quad c_v = 3R = 25 \text{ J K}^{-1}\text{ mol}^{-1} \qquad (3)$$

where N_A is the Avogadro constant. At lower temperatures, fewer modes are excited, and the specific heats drop, as calculated from quantum mechanics.

1.2 Thermal Equilibrium of Vacancies

At high-enough temperatures, thermal equilibrium requires that vacancies (vacant lattice sites) exist. Their migration can itself produce strain. They also promote strain by helping dislocations to climb around obstacles. Finally, they provide a mechanism for diffusion, which leads to changes in metallurgical structure, in turn allowing strain by dislocation motion.

A crystal lattice expanded to contain a vacancy has a higher enthalpy due to the stretched interatomic bonds and due to the expansion of the lattice by the vacancy formation volume $\Omega_v \approx b^3$ under pressure p. These give an enthalpy of formation of a vacancy of $h_f = u_f + p\Omega_v$. Vacant lattice sites also introduce a large number of alternative configurations and, therefore, a rise in entropy s. At high-enough temperatures this rise in entropy can offset the rise in enthalpy, giving a decrease in the Gibbs free energy $g_f \equiv h_f - Ts_f$. Thermodynamic equilibrium occurs at a minimum g_f, requiring a definite concentration of vacancies, as follows. At the atomic level, entropy changes are primarily due to changes in the number of alternative configurations of a state. In terms of Boltzmann's constant k and the number of distinguishable configurations W,

$$S_2 - S_1 = k \ln W_2/W_1 \qquad (4)$$

With N atoms and n vacancies ($N+n$ total sites), $W_2 = (N+n)!/(N!n!)$, and the excess Gibbs free energy of the crystal is

$$G = nh_f - kT \ln\left(\frac{(N+n)!}{N!n!}\right) \qquad (5)$$

Using Stirling's approximation $\ln N! \approx N \ln N$, and assuming $n/N \ll 1$, gives the equilibrium vacancy concentration in terms of h_f, either per vacancy or per mole:

$$c_v \equiv \frac{n}{N+n} = \exp(-h_f/kT) = \exp(-h_f/RT) \qquad (6)$$

Equation (6) has been confirmed, and h_f measured, with several physical and metallographic techniques.

1.3 Thermal Activation; Vacancy Diffusion

The atomic-scale point of view not only gives the equilibrium concentration of vacancies, but it also accounts for their approach to equilibrium by migration (diffusion). The rate of approach is often more important to the engineer than the equilibrium state itself. After all, the equilibrium state of a bridge is rust throughout the world!

To see how vacancies migrate, consider a polycrystal under uniaxial tension. At a grain boundary under a normal stress σ_{nn}, an ion will tend to be pulled into the boundary, leaving a vacancy in the crystal lattice. The enthalpy of formation of the vacancy is reduced by the tensile stress σ_{nn} acting on the volume of the vacancy b^3:

$$h_f(\sigma_{nn}) = h_{f0} - \sigma_{nn}b^3 \qquad (7)$$

Because vacancies can be introduced easily from grain boundaries, their nearby concentrations will be the equilibrium values. Thus from Eqns. (6) and (7), the concentration of vacancies will be higher near boundaries under transverse tension than near boundaries parallel to the tension. The vacancies will tend to diffuse from regions of high concentration to those of low. The diffusion of vacancies around a corner in one direction gives creep by the diffusion of ions in the other direction.

The creep rate follows from considering adjacent layers of higher and lower vacancy concentrations. More vacancies will jump from the higher to the lower, as an ion jumps from the lower to the higher vacancy layer. Jumping is impeded by the need for an ion to squeeze between others. The probability that an ion will have the activation free energy g_a required for a jump is given by kinetic theory:

$$p = \exp(-g_a/kT) \qquad (8)$$

The rate of activation of an individual event is this probability times the attempt frequency, approximately the Debye frequency ν_D. (Actually the jumping is a many-body problem, but $p\nu_D$ is a useful approximation.) Also, if the free energy for activation is expressed in terms of the activation enthalpy and entropy, $g_a = h_a - Ts_a$, and the approximation (from quantum mechanics) made that the ratio s_a/k is small compared with unity for large T/T_D, then the rate of activation of an individual event is approximately

$$\dot{R} \approx \nu_D p = \nu_D \exp(-g_a/kT) = \nu_D \exp(-h_a/kT)$$
$$\exp(-s_a/k) \approx \nu_D \exp(-h_a/kT) \qquad (9)$$

Vacancy diffusion through a grain occurs at a rate

proportional to the product of the vacancy concentration gradient by the activation rate. Thus vacancy diffusion (and self-diffusion of ions) are proportional to $\exp[(h_D = h_f + h_a)/RT]$. h_D correlates with the melting point: $h_D/RT_m = 16$–18 for Al, Cu, Au, Ni, Pt, Fe, and W; Pb is an exception at 22. See Nowick (1986) and LeClaire (1986) for more on the diffusion in metals and alloys, and Atkinson (1986) for ceramics.

1.4 Mechanisms of Thermally Activated Deformation

For insight into creep and its dependence on the structural changes that evolve with deformation and time, as well as equations that may be useful in predicting 10–100 year service of new alloys from necessarily short-time data, consider three of a number of mechanisms that are reasonably well understood (Frost and Ashby 1982, Owen 1986, Owen and Grujicic 1986, Martin 1988, and Langdon 1988).

The diffusional flow discussed above gives a macroscopic creep strain rate in terms of the grain diameter d and the applied uniaxial stress σ:

$$\dot{\varepsilon} = 42\nu_D \frac{b^2}{d^2} \frac{\sigma b^3}{RT} \exp[-(h_f + h_a)/RT] \qquad (10)$$

(Below 0.6–0.9 of the melting point T_m, with typical grain sizes, grain-boundary diffusion is more active than the bulk diffusion of Eqn. (10).) Dislocation climb and glide can occur when dislocations have been blocked by obstacles (see *Plasticity*). The line energy of dislocations is too large for dislocation loops to be in thermal equilibrium, as are vacancies, or for thermal motion to activate dislocation multiplication by bowing out loops. But vacancies can migrate to the extra half plane of an edge dislocation, allowing it to climb over an obstacle. An applied stress then makes the dislocation sweep out a large area. The vacancy migration gives a thermal activation factor, which is multiplied by a pre-exponential factor that accounts for the effect of applied stress σ, dislocation structure s, and temperature T on the resulting dislocation motion:

$$\dot{\varepsilon} = \dot{\varepsilon}_p(\sigma, s, T)\exp[-(h_f + h_a)/RT] \qquad (11)$$

Dislocation breakaway from pinning points can be activated from smaller and weaker obstacles than those requiring climb. Assuming that the force–distance relation (a) rises gradually from zero, and (b) is locally parabolic near its maximum gives the activation enthalpy for breakaway as a function of the applied shear stress σ and a strength parameter s:

$$h_b = h_{max}\left[1 - \left(\frac{\sigma}{s}\right)^p\right]^{3/2} \qquad (12)$$

The tail on the interaction curve requires $p \leqslant 1$, typically 0.75. The dislocation and metallurgical structures both affect the activation enthalpy through s. Sometimes the stress affects h through an activation volume, so that $h = h_{max} - \sigma\Omega_a$. In either case, with high enough stresses the process is stress-activated without the need for thermal motion, and rate-independent plasticity is approached.

1.5 Determining Steady-State Equations from Data

The effects of temperature in the pre-exponential factor of Eqns. (10) and (11) are relatively weak, compared with the effect through the exponential function. Also, the temperature dependence of the activation enthalpy function h (e.g., through $E(T)$) is small compared with the effect of RT. Furthermore, both effects are offset by the temperature dependence of the strength parameter s through the temperature dependence of the shear modulus $G(T)$. Then Eqns. (10)–(12) take on the form

$$\dot{\varepsilon} = \dot{\varepsilon}_p(\sigma, s)\exp[-h(\sigma, s)/RT] \qquad (13)$$

In dislocation breakaway h is a strong function of stress σ and structure s. The wide variety of pinning strengths and particle spacings means that there can be a whole spectrum of enthalpy functions, of which only one or a few are active at any one stress and temperature. The validity of Eqn. (13), with its single enthalpy function, must be experimentally verified.

Likewise, there is a spectrum of structural (strength) parameters s. These require a set of evolutionary equations to describe their changes with increments of strain and time in terms of their current values and the temperature. Assuming hardening a linear function of $\dot{\varepsilon}$ gives

$$\dot{s} = H(s, T)\dot{\varepsilon} - \dot{R}(s, T) \qquad (14)$$

In general, both H and $\dot{R}$ can be of either sign due to strain hardening or recovery and aging, overaging, or recrystallization. These changes are especially important (and complex) in transient creep arising from changes in stress and temperature, discussed below.

If a steady-state structure (both dislocation and metallurgical) can be assumed, s becomes a unique function of σ and T, and Eqn. (13) takes on the form

$$\dot{\varepsilon} = \dot{\varepsilon}_p(\sigma, T)\exp[-h(\sigma, T)/RT] \qquad (15)$$

Special cases of Eqn. (15) are found by plotting data in various ways. If $\ln\dot{\varepsilon}$ vs. $1/T$ for constant stress gives straight lines, the temperature effect disappears from $\dot{\varepsilon}_p$ and h. If lines of constant stress all extrapolate to the same point at $1/T = 0$, h contains all the stress effect and there is none in the pre-exponential term. If the lines of constant stress all have the same slope, h is independent of stress, and the stress effect appears only in the pre-exponential term. Further, if cross-plotting $\ln\dot{\varepsilon}$ vs. $\ln\sigma$ at constant temperature gives a straight line, the slope is the exponent m of the commonly used power-law creep, in terms of constants $\dot{\varepsilon}_0$ and σ_0:

$$\dot{\varepsilon} = \dot{\varepsilon}_0\left(\frac{\sigma}{\sigma_0}\right)^m \exp(-h/RT) \qquad (16)$$

The assumption of constant m is often only mathematically convenient: Evans and Wilshire (1985,

p. 112) quote a ferritic steel at 838 K for which m varied continuously from 17 at $\sigma = 300\ \mathrm{MN\,m^{-2}}$, $\dot{\varepsilon} = 10^{-4}\ \mathrm{s^{-1}}$, to $m = 1.2$ at $\sigma = 35\ \mathrm{MN\,m^{-2}}$, $\dot{\varepsilon} = 10^{-11}\ \mathrm{s^{-1}}$ (1% creep in 30 years). There may have been a stress effect in the activation enthalpy, distorting these exponents. In general, it is hard to find data accurate enough to determine whether the stress effect is in the pre-exponential term, the activation enthalpy, or both. In extrapolating short-term data to long-term service, especially necessary with new alloys, both forms should be tried.

Deformation mechanism maps show the regimes of temperature and stress in which various mechanisms predominate. To compare different materials, the shear stress should be normalized by dividing it by the shear modulus, and dividing the temperature by the melting point (Fig. 1). Lines of constant strain rate are plotted. Interest may range from the $10^{10}\ \mathrm{s^{-1}}$ in front of a crack running in a pipe to the $10^{-11}\ \mathrm{s^{-1}}$ rate for 1% creep in 30 years. In this 1%Cr–Mo–V steel, the dominant mechanisms at progressively higher strain rates below the phase transition are grain boundary diffusion (see Eqn. (10)), power-law creep (Eqn. (16)), power-law breakdown (Eqn. (16) with $\sinh h\sigma$ for σ), and a very narrow region of obstacle-controlled glide (Eqn. (12), with the pre-exponential term proportional to σ^2). Note the hazard of predicting creep in the diffusion regime by extrapolating data from the power-law creep regime. Overaging due to precipitate growth by diffusion is possible after unspecified times above 550 °C, and would shift the map. Oxidation near 600 °C would interact with fatigue due to load changes.

Rate-modified temperature plots are often convenient because the effects of strain rate are small compared with those of temperature and can sometimes be included in the temperature effect (McGregor and Fisher 1946; later known as the Larson–Miller parameter; see also McClintock and Argon 1966). Take the logarithm of Eqn. (15), solve for temperature, introduce a reference strain rate $\dot{\varepsilon}_R$, and rearrange to obtain the rate-modified temperature. If the logarithm of $\dot{\varepsilon}_p$ is nearly constant over the region of interest, define $a \equiv 1/\ln(\dot{\varepsilon}_p/\dot{\varepsilon}_R)$, and assume h depends only on stress. Then

$$T_{rm} = T(1 - a \ln \dot{\varepsilon}/\dot{\varepsilon}_R) = ah(\sigma)/R \tag{17}$$

Find a by applying Eqn. (17) to pairs of points at the same stress, but different temperatures and strain rates.

T_{rm} is useful when the stress effect is in the activation enthalpy, and when an added parameter such as structure is to be plotted, as in tensile tests or transient effects discussed below. At high temperatures the activation enthalpy is more nearly constant, and the Dorn or Zener–Hollomon temperature-modified time, $t \exp(-h/RT)$, is more satisfactory, although it leads to very large numbers.

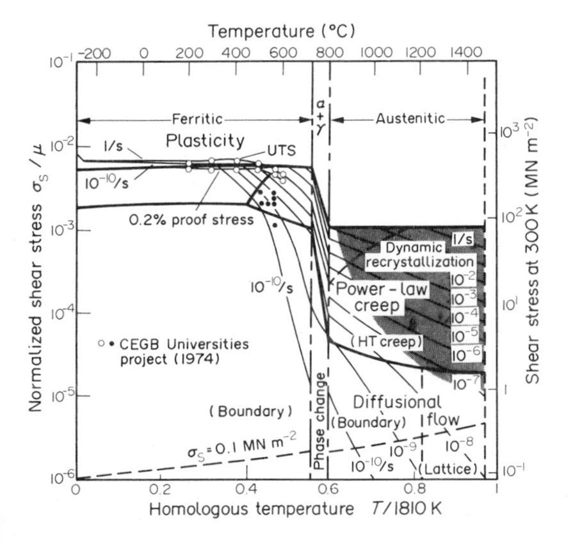

Figure 1
A Cr–Mo–V steel of grain size 100 μm, showing data (after Frost and Ashby 1982)

1.6 Transient Creep

Transient or initial creep is important because loads and temperatures may change in service, leaving residual stresses that lead to stress reversals, and because in many applications the typical 1% of primary creep is an important part of the total allowable. There is such a wide variety of mechanisms that there can hardly be a general correlation.

(a) Stress-drop tests show that more than one structural parameter s is required, for stress decrements can be found that give an initial negative creep rate which later turns positive (Evans and Wilshire (1985, pp. 122–3). Thus the idea of a "back-stress" or "rest-stress" as an inherent strength parameter is an approximation, not fundamental.

(b) Formulations based on the strain-hardening and recovery of a strength parameter show promise at high temperatures in pure metals where the dynamic recovery due to straining is small (e.g., Argon and Bhattarchaya 1987). At lower temperatures and higher stresses the spectrum of the strengths of dislocation nets is distorted by small strain changes or reversals (see McCartney and McLean 1976, Fig. 3). Then the steep stress–strain curve on load reversals is due more to exhaustion of easily moved dislocations rather than to hardening of a structural parameter s.

(c) Any power-law relation is a nonanalytic function and hence of doubtful physical validity.

(d) Frost and Ashby (1982, pp. 177–120) discuss transients in pure metals, including that transient in the diffusion creep regime required to establish equilibrium vacancy gradients.

(e) Sometimes the establishment of a steady-state dislocation structure requires an initially nearly strain-free time before steady-state creep, i.e., a negative primary creep.

(f) The dislocation mechanisms of creep and plasticity merge continuously into each other, so rate effects will gradually appear in the plastic phenomena of cyclic hardening and softening, cyclic creep under mean stress, and relaxation under mean strain.

(g) Residual stresses may promote primary creep.

(h) Cycling may facilitate changes in dislocation and metallurgical structure, including those due to radiation, thus maintaining transient creep over long periods.

With the wide variety of mechanisms and service conditions, the best practice is to run tests covering the range of the service conditions (fortunately not as long a time as for most creep tests) and choose a sufficiently accurate and convenient formulation.

One such condition is the rate-modified temperature, which allows plotting the effect of structure under nonsteady conditions, as represented by extension at constant strain rates. The references of Sect. 1.5 showed satisfactory correlation of strain rates varying by a factor of 100, up to true strains of 0.5 (50%), even with an inversion of the temperature effect due to changing regimes from carbon being fixed to being dragged along with the dislocations. It would be interesting to apply the concept to tests of cyclic plastic strain at various rates and amplitudes. From Eqn. (17) the activation enthalpy reaches that of iron at about $T_{rm} = 540$ K; above that the physical validity of T_{rm} is lost because of the assumption that the pre-exponential term, through its logarithm in a, is a constant with no stress effect.

2. Stress–Strain Relations for Creep

For an isotropic polycrystal under multiple stress components, in the steady-state creep regime, the relations for creep are those for plasticity with plastic strain replaced by creep strain rate. The uniaxial Eqns. (12) in (13) with (14), or (10), (15), or (16) for steady state now give the dependence of an equivalent strain rate:

$$\bar{\dot{\varepsilon}}^{c2} \equiv \sum_i \sum_j \tfrac{2}{3}\dot{\varepsilon}^c_{ij}\dot{\varepsilon}^c_{ij} = \tfrac{2}{3}(\dot{\varepsilon}^{c2}_{11} + \dot{\varepsilon}^{c2}_{22} + \dot{\varepsilon}^{c2}_{33}) + \tfrac{4}{3}(\dot{\varepsilon}^{c2}_{23} + \dot{\varepsilon}^{c2}_{31} + \dot{\varepsilon}^{c2}_{12}) \quad (18)$$

on the equivalent stress, defined in terms of either the stress deviators

$$\sigma'_{ij} \equiv \sigma_{ij} - \delta_{ij}\Sigma_k \sigma_{kk}/3$$

or the actual stress components:

$$\bar{\sigma}^2 \equiv \sum_i \sum_j \tfrac{3}{2}\sigma'_{ij}\sigma'_{ij} = \tfrac{1}{2}[(\sigma_{11}-\sigma_{22})^2 + (\sigma_{22}-\sigma_{33})^2 + (\sigma_{33}-\sigma_{11})^2] + 3[\sigma^2_{23} + \sigma^2_{31} + \sigma^2_{12}] \quad (19)$$

The components of the strain rate are given in terms of the stress deviators

$$\dot{\varepsilon}^c_{ij} = \tfrac{3}{2}\sigma'_{ij}\frac{\bar{\dot{\varepsilon}}^c}{\bar{\sigma}} \quad (20a)$$

or the stress components:

$$\dot{\varepsilon}^c_{11} = [\sigma_{11} - \tfrac{1}{2}(\sigma_{22}+\sigma_{33})]\,\bar{\dot{\varepsilon}}^c/\bar{\sigma}, \ldots,$$
$$\dot{\gamma}^c_{23} = 3\sigma_{23}\bar{\dot{\varepsilon}}^c/\bar{\sigma}, \ldots \quad (20b)$$

After unloading or a change in ratios of stress components, until the subsequent strains reach 0.2% to 1%, these steady-state equations no longer hold, as discussed in Sect. 1.6 and in the analogous Bauschinger effect (see *Plasticity*).

The above strain rate vs. stress relations must be added to the elastic strains, and to any plastic strains used to model transient creep, to obtain total strains to satisfy the strain-displacement relations and any velocity boundary conditions. The stresses must satisfy equilibrium and stress boundary conditions. Together, these give a complete continuum mechanics solution.

3. Character of Solutions

Power-law creep is analogous to incompressible power-law elasticity and deformation theory plasticity (stresses determine total plastic strain, not strain increments). In the limit for a creep exponent of unity, the equations become those of linear incompressible viscosity, which are mathematically analogous to linear incompressible elasticity. In the other limit as $m \to \infty$, the equations are analogous to nonhardening plasticity. It is therefore often worthwhile to interpolate between solutions in elasticity and in nonhardening plasticity as a guide and as a check for solutions in creep (see *Elasticity; Plasticity*).

As the creep exponent increases from unity, strain-rate concentrations become stronger and farther-reaching (St Venant's principle weakens). Uniqueness also weakens, so necking and buckling intensify. For more insight, consider first axial tension and compression, then bending, and finally more general results. Often for simplicity the transient creep (t) is assumed to occur quickly enough to be accounted for by time-independent power-law plasticity, during loading, and the secondary creep rate (s) by a power law:

$$\dot{\varepsilon}^c = \frac{d}{dt}\left[\left(\frac{\sigma}{\sigma_t}\right)^{m_t}\right] + \frac{1}{\tau}\left(\frac{\sigma}{\sigma_t}\right)^{m_s} \tag{21}$$

3.1 Axial Loading

For a constant tensile load, the time to reduce the cross-section from A_0 to A (called the rupture time t_R if $A = 0$) is nearly independent of elastic and transient strains. In terms of the initial secondary creep rate from Eqn. (21),

$$t(A) = \frac{1}{m_s \dot{\varepsilon}(\sigma_0)}\left[1 - \left(\frac{A}{A_0}\right)^{m_s}\right]; \quad t_R = \frac{1}{m_s \dot{\varepsilon}(\sigma_0)} \tag{22}$$

Only if the reduction of area is less than $\approx 25\%$ does the second of Eqns. (22) give times much shorter than the rupture time; if so, a scalar damage parameter is sometimes introduced into the flow and evolutionary equations.

Extension at constant head rate produces more gradual necking in creep than plasticity, because reaching a load maximum in one section does not entirely stop straining in the others. The resulting increased inhomogeneity in strain rates makes interpretation difficult, without diametral measurements of strain.

Relaxation of stress at constant total strain is important in prestressed reinforcement and in bolted joints. The transient creep rate is reduced by the falling stress (after prior hardening) but "steady-state" creep is not affected. Assuming the transient creep occurs immediately underestimates the required tightening stress σ_1, but allows solving for the initial tightening strain in terms of the final stress σ_f. First find

$$\frac{\sigma_1}{\sigma_f} = \left(1 + \frac{(m_s - 1)(\sigma/\sigma_t)^{(m_s-1)} t/\tau}{\sigma_t/E}\right)^{1/(m_s-1)} \tag{23}$$

Then the imposed strain must be

$$\varepsilon_1 = \frac{\sigma_1}{E} + \left(\frac{\sigma_1}{\sigma_t}\right)^{m_t} \tag{24}$$

Creep buckling depends strongly on the initial eccentricity of the load, e_0. An approximate model consists of a cantilever column of total length L, with a length l from the built-in end consisting of two flanges with a separation h and total cross-sectional area A. For $e_0 < h/2$, both flanges are loaded according to the tangent modulus rule for plasticity. At constant load, the inner flange will be subject to increasing stress and hence creep, with the outer subject to decreasing stress so that creep becomes less than the power-law prediction. After $e > h/2$, the stress reverses and the creep rate becomes greater than power-law creep due to a Bauschinger effect. For $e_0 > h/2$ the creep rates in the two flanges are always of opposite sign, and hence Eqn. (21) is applicable. For the initial deflection, these give an implicit equation that can be solved, and for the ensuing steady-state creep rate it gives a differential equation for numerical integration. Odqvist (1966) found the time t_c to eccentricity e from any initial eccentricity e_i after loading and transient creep, for pure power-law creep with $m_s = 3$ (even with varying stresses along both flanges extending the length of the column). In terms of the steady-state creep rate $\dot{\varepsilon}_0$ at the axial load alone, and the elastic strain ε_E^e due to pure axial loading at the Euler buckling load,

$$t_c = \frac{\varepsilon_E^e}{6\dot{\varepsilon}_0} \frac{\ln[1 + (h/e_i)^2]}{\ln[1 + (h/e)^2]} \tag{25}$$

3.2 Bending, Residual Stress and Beam-Bar Transitions

In pure creep bending of a beam b wide by h deep, at a rate of curvature $d^2\dot{v}/dx^2 = \dot{v}''$, the strain rate varies linearly with the distance v from the centroid: $\dot{\varepsilon} = y\dot{v}''$. For steady-state power-law creep the stress is S-shaped. For $y \geqslant 0$, $\sigma = \sigma_t(y\dot{v}''\tau)^{(1/m_s)}$. The resulting stress in terms of bending moment turns out to be

$$\sigma = (1 + 1/(2m_s)) \frac{M y^{1/m_s}}{b(h/2)^{2+1/m_s}} \tag{26}$$

If the beam is initially loaded elastically to M, the stress is linear, and the peak stress is higher than in steady-state creep. The resulting initial rates of peak strain and curvature are higher. Thus a structural member can have transient creep even if the material does not. In fact, part of the transient creep of the material itself comes from its microstresses (another part is dislocation rearrangement). The relaxation of the peak stresses towards their steady-state creep stress is very rapid, but that in the low-stress regions is very slow for typical creep exponents of $m_s = 3$–10.

The residual stress on unloading is found, as in plasticity, by adding the elastic stress due to an equal and opposite moment (in the absence of transient creep). The subsequent relaxation of residual stress will be very slow, because even the peak stresses are low. Completely reversing the initial moment gives a much higher transient than even the initial one. This is the creep analogue, in the structure, of the Bauschinger effect. Again, part of the effect observed in the material itself is due to microstresses.

As in plasticity and elasticity, the same part of a structure may be modelled in different ways at different loads or times. A bar built-in between two rigid supports under a transverse load may initially be modelled as quasi-steady creep bending, but later as two bars under tension creep. Likewise, a plate can become a shell and then a membrane.

3.3 General Results

Power-law creep exhibits the small-strain uniqueness and at least some of the locality of stress concentrations typical of elasticity, but superposition does not apply. The stress-load and strain-displacement parts of a solution to plane problems often depend on the creep exponent, not on the magnitudes of creep rates. This extends even to power-law plasticity, and is illustrated by the relations $\dot{\varepsilon} = y\dot{v}''$ and Eqn. (26) of the bending problem.

Near the tips of rapidly loaded cracks, the stress and strain fields are initially those set by the elastic stress intensity factor K. Higher loads give the power-law strain hardening fields governed by the scalar J. With time, a field of transient creep may spread out from the tip, and finally a field of steady-state creep overwhelms the earlier fields. If crack growth is slow enough for this field to dominate, it still sets (through a scalar C), a similitude between laboratory and service growth rates. Within that zone, a singularity solution for a growing crack is useful if it characterizes the local fields outside the fracture process zone associated with the microstructure (see McClintock and Bassani 1981, Kanninen and Popelar 1985). Load reversals (fatigue) have large effects.

4. Sources of Solutions

4.1 Bounds to the Deformation Rate or Load

For a body under a load P and a load-point velocity $\dot{u}$ giving a power $P\dot{u}$, the creep analogue of the principle of minimum energy in nonlinear elasticity states that for any stress distribution σ_{ij} in equilibrium with the load P, and for equivalent stresses and strain rates given by Eqns. (18) and (19) with an associated flow rule leading to Eqn. (20),

$$P\dot{u} \leqslant \int \bar{\sigma}(\sigma_{ij})\dot{\bar{\varepsilon}}(\sigma_{ij})\,\mathrm{d}V \qquad (27)$$

An upper bound to the steady-state deformation rate for a given load is found by introducing Eqn. (21) to obtain an integral in terms of the stress field, normalized by its equilibrium load:

$$\dot{u} \leqslant \frac{\sigma_t}{P\tau}\int\left(\frac{\bar{\sigma}}{\sigma_t}\right)^{m_s+1}\mathrm{d}V = \frac{1}{\tau}\left(\frac{P}{\sigma_t}\right)^{m_s}\int\left(\frac{\bar{\sigma}}{P}\right)^{m_s+1}\mathrm{d}V \qquad (28)$$

A lower bound to the load for a given deformation rate can be found by solving Eqn. (28) for P:

$$P \geqslant \sigma_t\left[\frac{\dot{u}\tau}{\int(\bar{\sigma}/P)^{(m_s+1)}\,\mathrm{d}V}\right]^{1/m_s}\mathrm{d}V \qquad (29)$$

Stress fields for the bounds may be based on elastic or plastic solutions, or on constant stress patches. As with the limit load theorems of plasticity, these equations require deflections and crack growth small enough for the resulting stress changes to be negligible. Bounds have also been given for elastic-creeping and plastic-creeping materials, for creep at points away from the load, and for varying loads and temperatures (e.g., Leckie and Ponter 1970, Ponter 1974).

Analogous theorems for rigid-plastic power-law plasticity, modelling monotonic loading by total strains rather than strain increments, seem to be known but seldom mentioned.

4.2 Analytical Solutions

Where available, a closed-form solution carries much information about the effects of variables, as well as a check for numerical methods. There are relatively few closed-form solutions and correspondingly few books on the mechanics of creep. Finnie and Heller (1959) discuss tubes under internal pressure combined with bending or torsion, plate and shell buckling, and rotating disks. Odqvist (1966) discusses or includes the nonlinear elastic analogue, the power theorems used to determine bounds, and plates and membranes. Hult (1966) emphasizes transient creep through uniaxial stresses and pressure vessels. Kachanov (1967) introduces the bound theorems and a scalar damage simulating diffuse cracking. Applications include torsion, plates, shells and rotating disks.

4.3 Finite-Element Solutions

The advantages of the finite-element method are better matching of flow and evolutionary equations, including elastic strains and rate effects (see Chan et al. 1985, Krempl and Yao 1987). Beyond the problems with strain-hardening plasticity, stability of direct (explicit) integration requires time steps small enough so that the creep strain increment is less than the local elastic strain by a factor equal to the creep exponent: $\dot{\varepsilon}^c\delta t < \varepsilon^e/m_s$. In practice various schemes are used to reduce solution time. See, for example, Owen and Hinton (1980), compendia such as Pittman et al. (1984) and Nakazawa et al. (1987), and journals such as *International Journal of Numerical Methods in Engineering, Computer Methods in Applied Mechanics and Engineering* and the *Journal of Plasticity*.

4.4 Experimental Methods

Extensometer measurements in creep are difficult because of the indentation of gauge points into the surface, and the need to design extensometer arms to minimize effects of temperature gradients. If a protective atmosphere allows scribing fine lines, laser illumination may be needed to maintain contrast against the radiant light. Furthermore, surface displacements may not be indicative of plane strain conditions beneath.

As noted above, model tests of loads and displacements may be useful if temperatures are uniform and the exponents are the same in the model material and the prototype.

Bibliography

Argon A S, Bhattarchava A K 1987 Primary creep in nickel: Experiments and theory. *Acta Metall.* **35**, 1499–514

Atkinson A 1986 Diffusion in ceramics. In: Bever M B (ed.)

1986 *Encyclopedia of Materials Science and Engineering.* Pergamon, Oxford, pp. 1175–80
Chan K S, Lindholm U S, Bodner S R, Walker K P 1985 A survey of unified constitutive theories. In: Thompson R L 1984 *Nonlinear Constitutive Relations for High Temperature Application—1984*, Conference Publication 2369. NASA Cleveland, Ohio, pp. 1–23
Evans R W, Wilshire B 1985 *Creep of Metals and Alloys.* Institute of Metals, London
Finnie I, Heller W R 1959 *Creep of Engineering Materials.* McGraw-Hill, New York
Frost H J, Ashby M F 1982 *Deformation Mechanism Maps.* Pergamon, Oxford
Hult J A H 1966 *Creep in Engineering Structures.* Blaisdell, Waltham, Massachusetts
Kachanov L M 1967 *The Theory of Creep* 1, 2. Nat. Lend. Libr. Sci. Tech., Boston Spa
Kanninen M F, Popelar C H 1985 *Advanced Fracture Mechanics.* Oxford University Press, New York
Krempl E, Yao D 1987 The viscoplasticity theory based on overstress applied to ratchetting and cyclic hardening. In: Rie K-T (ed.) 1987 *Low Cycle Fatigue and Elasto-Plastic Behavior of Materials.* Elsevier/Applied Science, London, pp. 137–48
Langdon T G 1988 Creep of metals and alloys: Factors governing creep resistance. In: Cahn R W (ed.) 1988 *Encyclopedia of Materials Science and Engineering*, Suppl. 1. Pergamon, Oxford, pp. 93–98
Leckie F A, Ponter A R 1970 Deformations bounds for bodies which creep in the plastic range *J. Appl. Mech.* **37**, 426–30
LeClaire A D 1986 Diffusion in crystalline metals: Phenomenology. In: Bever M B (ed.) 1986 *Encyclopedia of Materials Science and Engineering.* Pergamon, Oxford, pp. 1186–93
MacGregor C W, Fisher J C 1946 A velocity modified temperature for the plastic flow of metals *J. Appl. Mech.* **13**, A11–A16
Martin J L 1988 Creep of metals and alloys: Microstructural aspects. In: Cahn R W (ed.) 1988 *Encyclopedia of Materials Science and Engineering*, Suppl. 1. Pergamon, Oxford, pp. 98–104
McCartney L N, McLean D 1976 Some relations describing creep. *J. Mech. Eng. Sci.* **18**, 39–45
McClintock F A, Argon A S (eds.) 1966 *Mechanical Behavior of Materials.* Addison-Wesley, Reading, Massachusetts
McClintock F A, Bassani J L 1981 Problems in environmentally-affected creep crack growth. In: Nemat-Nasser S (ed.) 1981 *Three-Dimensional Constitutive Relations and Ductile Fracture.* North-Holland, New York, pp. 123–45
Nakazawa S, William K, Rebelo N (eds.) 1987 *Advances in Inelastic Analysis* AMD 88, PED 28. American Society for Mechanical Engineers, New York
Nowick 1986 Diffusion in crystalline metals: Atomic mechanisms. In: Bever M B (ed.) 1986 *Encyclopedia of Materials Science and Engineering.* Pergamon, Oxford, pp. 1180–86
Odqvist F K 1966 *Mathematical Theory of Creep and Creep Fracture.* Clarendon Press, Oxford
Ohtani R, Ohnami M, Inoue T (eds.) 1988 *High Temperature Creep-Fatigue, Current Japanese Materials Research* 3. Elsevier/Applied Science, London
Owen W S 1986 Plastic deformation: Creep mechanisms. In: Bever M B (ed.) 1986 *Encyclopedia of Materials Science and Engineering.* Pergamon, Oxford, pp. 3536–40
Owen W S, Grujicic M 1986 Plastic deformation: Thermally activated glide of dislocations. In: Bever M B (ed.) 1986 *Encyclopedia of Materials Science and Engineering.* Pergamon, Oxford, pp. 3540–43
Owen D R, Hinton E 1980 *Finite Elements in Plasticity: Theory and Practice.* Pineridge, Swansea
Pittman J F, Zienkowicz O C, Wood R D, Alexander R R (eds.) 1984 *Numerical Analysis of Forming Processes.* Wiley, New York
Ponter A R 1974 General bounding theorems for the quasi-static deformation of a body of inelastic material, with applications to metallic creep. *J. Appl. Mech.* **41**, 947–52

F. A. McClintock
[Massachusetts Institute of Technology,
Cambridge, Massachusetts, USA]

D

Design and Materials Selection and Processing

Producing a design is the critical link in the chain of events that starts with a creative idea and ends with a successfully marketed product. Important stages between these endpoints are detailed design, materials selection, materials processing (manufacturing), assembly and quality control. There are important materials decisions to be made in all of these areas of the design process, but most emphasis must be given to the steps of materials selection and materials processing. Indeed, as Fig. 1 illustrates, the steps in the design processes are highly interrelated. A poorly chosen material can add significantly to the processing cost, yet a material selected only to optimize processing might fail in service.

1. Design Performance Requirements

The performance or functional requirements of a design must first be expressed in terms of physical, mechanical, thermal, electrical or chemical properties. Materials science is chiefly concerned with understanding how to control the structure of materials. When structure can be controlled through composition control, thermal treatment and processing, then the material properties which determine performance can be improved. An important aspect of materials selection is deciding just which properties are most important for the successful performance of the design. One logical way to approach this problem is to establish the failure modes of the design, either by simulated service testing or proof testing, or by fault-tree logic models (Marriott and Miller 1982). For most failure modes two or more material properties interact to control the material behavior. Moreover, the service conditions met by materials in many designs are more complex than the test conditions ordinarily used to measure material properties. For example, the stress level is not likely to be constant but will fluctuate with time in a non-sinusoidal way. Or the material may be subjected to a complex superposition of environments such as an alternating stress (fatigue) at high temperature (creep) in a highly oxidizing environment (corrosion).

In addition to technical properties which define performance, careful consideration should be given to the properties which determine producibility. These may be difficult to define quantitatively and will be more experience based. However, as Fig. 1 emphasizes, the materials selection step cannot be divorced from its processing. The decision whether to choose a zinc-based die casting or an injection-molded nylon polymer for a pump part may be more a question of the manufacturing process than the material properties.

The typical materials selection problem involves balancing the decision among many candidate materials for which several properties are important, but not necessarily of equal weight. A decision based on a weighted property index is an appropriate technique (Dieter 1983). However, many factors other than material properties enter into the materials selection decision. The paramount factor is cost. Once the performance requirements have been achieved, materials selection comes down to buying properties at the best available price. Closely related to cost is availability. Other considerations are the ability to obtain the material in the required forms, the tolerances and size limitations on these products, and the variability of the properties of the material.

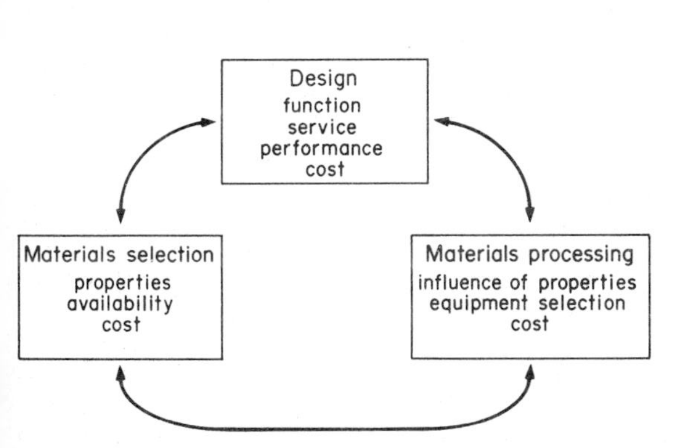

Figure 1
The interrelations of design, materials selection and processing to make a final product

2. Materials Selection Process

It can therefore be concluded that materials selection is a type of engineering decision-making that is carried out in the presence of multiple constraints without a clear-cut objective function. Since there is usually not a single property to optimize, the materials selection process involves many trade-offs and compromises. However, it can be approached within the same framework that is used for engineering problem solving.

The steps in the materials selection process are as follows:

(a) Analysis of the materials requirements. This involves determination of the conditions of service and environment that the product must withstand, and their translation into critical material properties.

(b) Screening of candidate materials. This involves comparing the needed properties with a large materials property database to select the few materials that look promising for the application. To do this, it is necessary to establish screening properties which have an absolute lower (or upper) limit. No trade-off beyond this limit is tolerable. In other words, should the material be evaluated further for the application?

(c) Selection of candidate materials. The materials which pass the screening tests must be analyzed with respect to trade-offs of product performance, producibility, cost and availability to select the best material (or small set of best materials) for the application. Techniques like the weighted property index, value analysis, or benefit/cost analysis may be helpful in this decision.

(d) Development of design data. The critical material properties for the selected material(s) must be determined experimentally to obtain statistically reliable measures of the material performance under the specific conditions expected to be encountered in service. Not all material selection decisions utilize this step. For a routine application it may be sufficient to rely on the values of properties in published specifications, but for applications where reliability is of great importance, design data properties must be determined. Often the expenditure of funds for this phase of materials selection is a trade-off between using a more costly material with properties which exceed those needed for the application.

3. Process Selection

Selecting the best manufacturing process is not an easy task. Rarely can a product be made by only one method, so that several competitive processes generally are available. As in all of engineering design, cost is a very important factor. Therefore, the selection of the optimum process can be made only after the costs of manufacturing by the competing processes have been evaluated (Ostwald 1980). A key item in the cost of manufacture is the cost of the raw material and the yield of good parts from the process. The influence of the processing method on the properties of the final product must also be taken into consideration, since this determines the performance of the part in service. A more detailed listing of the factors that influence the selection of a manufacturing process is as follows:

(a) selection of material,

(b) quantity of pieces required,

(c) complexity of geometric shape,

(d) size and section thickness,

(e) dimensional accuracy (tolerances),

(f) surface finish,

(g) available equipment,

(h) delivery date, and

(i) cost of manufactured product.

Generally, the selection of the material determines the processing methods which can be used. Some materials are too brittle to be plastically deformed, and others are too reactive to be cast, or have poor weldability. Once beyond this broad generalization, the selection of material has its greatest impact on manufacturing cost. The greater the workability or machinability of the material, the lower the loss due to defective parts and the lower the unit processing cost. However, the workpiece should be slightly larger than the finished part to allow for removal of surface defects or to permit finishing to size by machining. Often the manufacturing process dictates the use of extra material such as sprues and risers in castings, or flash in forgings. At other times extra material must be provided for purposes of handling, positioning, or testing the part.

Steels, aluminum alloys and many other engineering materials can be purchased from the supplier in a variety of metallurgical conditions other than in the annealed state. Examples are quenched and tempered steel, solution-treated, cold-worked and aged aluminum alloys, and cold-drawn and stress-relieved brass rod. It may prove more economical to have the metallurgical strengthening produced in the workpiece by the supplier than to heat-treat it separately after it has been manufactured.

A basic factor in selecting the manufacturing process is the quantity of pieces required. For each process under consideration there will be a break-even point in terms of quantity, beyond which one process will be more economical than the others. Those processes which require a large investment in tooling or special machinery usually are not economical for small production runs but result in the lowest unit cost for large quantities of product. Processes which require extensive preparation of tooling are not suitable for short delivery schedules.

Different manufacturing processes vary in their limitations to produce complex shapes and parts with minimum dimensions. Limitations on how large a part can be made may be imposed by the available processing equipment. However, more often the limitation on size is concerned with the minimum part thickness that can be obtained. The wall thickness of a casting may be limited by the fluidity of the molten metal or the thickness of a forging by the very high die pressures generated when the plan area-to-thickness ratio becomes very large.

Each manufacturing process has an inherent ability to produce parts within a certain range of dimensional accuracy and a certain surface finish. When the design requirements demand tolerances and/or surface finishes outside of this range, then the process should be changed, or costly finishing steps (e.g., grinding, honing and lapping) will have to be added to the processing sequence. Manufacturing costs increase exponentially as tolerances are reduced. Any tolerance less than ± 0.125 mm should be scrutinized carefully to see whether it is absolutely necessary for the design.

While it is possible to produce almost any design, it does not follow that production costs will be reasonable or acceptable. The design should be made with one or more production processes in mind. The designer should work cooperatively with manufacturing engineers to ensure that the design details are consistent with the characteristics of the process. There is a growing literature on this subject (Trucks 1974). Special emphasis should be given to:

(a) minimizing the number of production steps;

(b) standardizing materials and basic parts; and

(c) designing to ensure ease in locating, orienting, feeding, holding and transferring parts to facilitate the assembly step.

Greater integration between the design, materials selection and materials processing steps is one of the major achievements which will lead to increased manufacturing productivity.

See also: Selection Systems for Engineering Materials: Quantitative Techniques

Bibliography

Dieter G E 1983 *Engineering Design: A Materials and Processing Approach.* McGraw–Hill, New York, Chap. 6

Farag M M 1979 *Materials and Process Selection in Engineering.* Applied Science, London

Marriott D L, Miller N R 1982 Materials failure logic models: A procedure for systematic identification of material failure modes in mechanical components. *Trans. J. Mech. Eng. Design* 104: 626–34

Ostwald P F (ed.) 1980 *Manufacturing Cost Estimating.* Society of Manufacturing Engineers, Dearborn, Michigan

Schey J A 1977 *Introduction to Manufacturing Processes.* McGraw–Hill, New York

Trucks H E 1974 *Designing for Economical Production.* Society of Manufacturing Engineers, Dearborn, Michigan

G. E. Dieter
[University of Maryland, College Park, Maryland, USA]

Design with Polymers

Successful manufacture of plastics products requires a combination of sound judgement and experience. Experience and familiarity with polymers and their properties, and knowledge of mold design, mold construction methods and procedures subsequent to molding are essential. Early in the design process, there should be consultation and rapport between product designer, material supplier, mold builder and processor. Discussions should be continued throughout the design stage to identify the appropriate manufacturing processes, their variables and the influence on the product. This should be continued through material selection and prototype evaluation.

Often, products which are to incorporate plastics components are designed before final selection of the polymer materials. Because of this, compromises may be required and the processor should review the design for manufacturing technique and available equipment to be able to produce the plastic part within the required economic bounds. Consultation with specialists ensures that the most effective material and processes are selected, that processing limitations and variables have been identified and that a testing procedure for ensuring product quality has been developed. Participation by specialists in product assurance, maintenance, packaging, materials and environmental requirements will also contribute to a successful design.

1. Basic Requirements for Designing with Plastics

It must first be determined what the material must do and the environmental conditions to which it will be exposed. The design must also take into account economic factors, production rates, and special feature requirements. For each product, an initial description of performance must be established. Since the resulting part design and material selection are generally arrived at through many design compromises, it is important to distinguish between true requirements and those which are only desirable. The general considerations in the design process can be grouped as follows.

(a) Configuration
- (i) basic use
- (ii) size and shape
- (iii) weight

(b) Structural or mechanical properties
- (i) static, dynamic or cyclic forces
- (ii) load magnitudes and duration
- (iii) strength and stiffness

(c) Nonstructural or physical properties
- (i) thermal, electrical, optical and other physical properties
- (ii) color, surface finish and identification

(iii) fire resistance and material compatibility
(iv) permanence

(d) Environmental factors
(i) temperature and humidity range in storage and use
(ii) weathering and water resistance
(iii) chemical and biological resistance
(iv) service life and degradation characteristics
(v) recyclability

(e) Production and cost
(i) total quantities and production rates
(ii) secondary operation and special features
(iii) cost restrictions
(iv) standardization

2. *Materials Selection*

There are hundreds of distinctly different plastics materials, the result of combining generic types, fillers, reinforcements and manufacturing processes. Designing with plastics demands knowledge of specific material behavioral factors which include creep, strength, stiffness, water absorption, dimensional stability, fatigue, toughness, chemical resistance and stress-crack resistance. For this reason the selection process for a plastic material requires expertise in materials technology.

2.1 Initial Screening

The initial screening of materials is performed to select the type of plastics to be examined as possible final candidates. This choice is based on inherent materials properties not subject to enhancement by design. These "go or no-go" requirements, which may include transparency, chemical resistance and/or temperature resistance, immediately limit the types and numbers of materials to be considered. The materials list may also be narrowed down by considering part shape and general strength requirements.

Screening should also consider processability, that is, ease of production. Limitations of the manufacturing process and its tooling cover such factors as parting line, draft, thicknesses, tolerances, undercuts, inserts and subsequent joining techniques.

2.2 Final Selection

Choosing the right material is probably the most important decision to be made in designing with plastics. The selection of candidate materials can range from a simplified choice based on a well established precedent to a complex assessment of key properties for the full range of available plastics.

A key element in the selection process is the analysis of part design requirements and the translation of these into meaningful material property requirements. Be they simple or complex, the requirements must be described in terms of all important properties, namely, strength and resistance to impact, temperature, stress, time and chemicals.

A problem faced in the selection of materials resides in the property data that are available. It is difficult for the uninitiated to assimilate the large volume of available information and to adapt the data to a specific need, without a thorough understanding of polymer materials. A fundamental understanding is required of what properties are useful in design. In cases where there is no material property that relates directly to a part requirement, a qualitative judgement must be made, followed by testing to evaluate performance. The result of the selection process should be a list of candidate materials, with the associated cost factors, worth pursuing in the prototype evaluation phase.

3. *Change in Material Properties during Processing*

When designing any plastics product, a number of variables must be kept in mind. For example, the strength of high-impact plastics containing long glass fibers or rag stock fillers generally remains high when the materials are compression molded. On the other hand, values may drop by as much as 50% when the materials are transfer, plunger or injection molded, because gating can break up filler fibers. Certain sections become weaker per unit of thickness than other sections. Geometric design of the product may offset any advantage to be gained by compression molding. If a high-impact material is selected, design of the part may produce sections no stronger than if they had been molded of the least expensive, low-impact material. Figures 1–4 and Table 1 summarize various processing considerations and should assist in the selection of materials and processes.

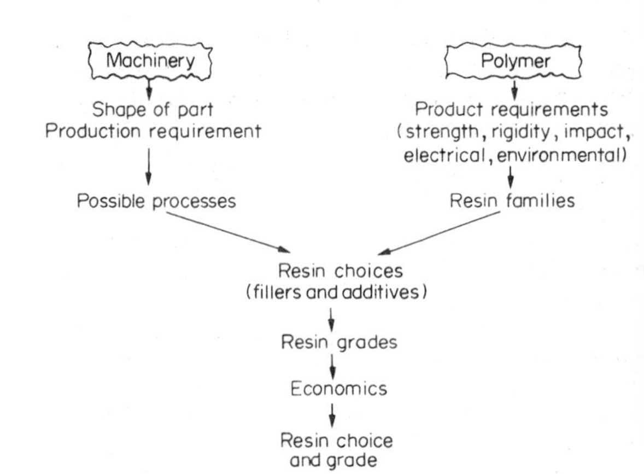

Figure 1
Choice of polymer and manufacturing process (courtesy of Dr R. Progelhof, New Jersey Institute of Technology)

Process	Shape limitation	Factor limiting maximum size	Complex shapes	Controlled wall thickness	Open hollow shapes	Closed hollow shapes	Very small items	Plane area > 1 m^2	Inserts	Molded in holes	Threads
Blow molding	hollow thin wall	platten			○	○		○			○
Calendering	sheet	width of roll		○				○			
Casting		mold	○	○					○	○	
Compression molding		platten	○	○	○				○	○	○
Extrusion: film	sheet or tube	die		○				○			
profile		die		○							
sheet	sheet	width of roll		○				○			
Filament winding	surface of revolution			○		○		○			
Hand or spray layup	large thin wall	mold		○	○			○	○	○	
Injection molding: solid		platten	○	○	○		○		○	○	○
foam		platten	○		○			○	○	○	○
reactive		platten	○		○			○	○	○	○
Machining			○	○	○		○				○
Melt flow stamping		platten		○							
Pultrusion	linear	die		○							
Rotational molding	hollow				○	○		○	○	○	○
Thermoforming	thin wall	platten			○			○			
Transfer molding		platten	○	○	○		○		○	○	○

Figure 2
Processing limitations to part design: ○ process compatible with part type (courtesy of Dr R. Progelhof, New Jersey Institute of Technology)

4. Shrinkage and Tolerance

The actual dimensions of the final part will be different from those of the mold. The net shrinkage, or growth in some instances, depends on various factors, such as the stock temperature and the molding pressure. Two different forms of shrinkage must be considered when designing plastics products—the initial shrinkage which occurs while the part is forming, called "mold shrinkage," and that which occurs in the part after a period of time, called "postshrinkage" ("aftershrinkage").

Some materials are more stable than others during aging, regardless of their initial shrinkage. Commonly used thermoset plastics, such as a glass-filled phenolic compound, may have a mold shrinkage of 0.6% but no postshrinkage. On the other hand, an α-cellulose-filled melamine may have a mold shrinkage of 0.7–0.9% but show an aftershrinkage of as much as 0.6–0.8%; thus, the total shrinkage may be as high as 1.3–1.7%. In thermoplastic materials, there are large variations in shrinkage; for example, 30% glass-filled polycarbonate has 0.2% shrinkage, whereas the value is 2.5–3.5% for unmodified polyethylene.

The product design should allow for as broad a tolerance as possible. If it has been determined that the molded part must be postcured, stress-relieved or baked, allowance must be made for the additional shrinkage that could occur during these treatments. Figures 5 and 6 indicate the tolerances required for various polymers as a function of dimension.

5. Product Evaluation by Prototyping

The prototype phase of a design project provides the first opportunity to evaluate product performance. Prototype evaluation presupposes the fabrication of a part or product by the manufacturing process that will be used in production. A prototype could be made by machining or other simplified modelmaking techniques. However, this practice should be avoided when detailed engineering tests are to be performed. Parts manufactured by a method other than the

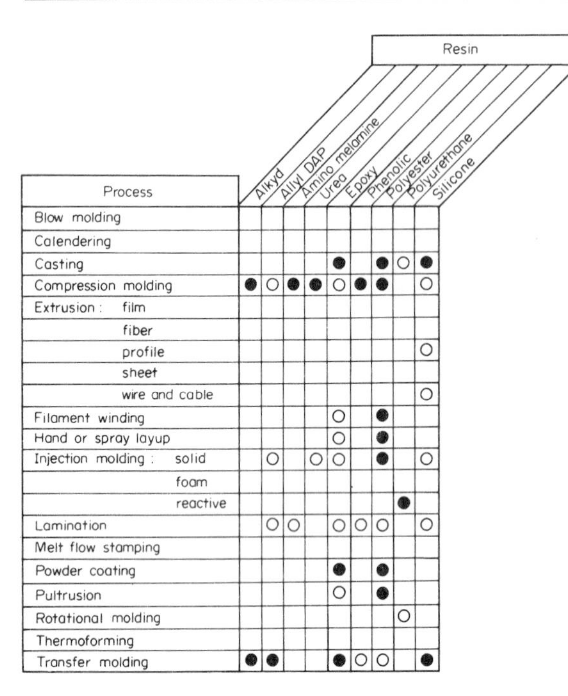

Process	Alkyd	Allyl DAP	Amino melamine	Urea	Epoxy	Phenolic	Polyester	Polyurethane	Silicone
Blow molding									
Calendering									
Casting					●		●	○	●
Compression molding	●	○	●	●	○	●	●		○
Extrusion: film									
fiber									
profile									○
sheet									
wire and cable									○
Filament winding					○		●		
Hand or spray layup					○		●		
Injection molding: solid		○		○	○		●		○
foam									
reactive								●	
Lamination		○	○		○	○	○		○
Melt flow stamping									
Powder coating					●		●		
Pultrusion					○		●		
Rotational molding								○	
Thermoforming									
Transfer molding	●	●			●	○	○		●

Figure 3
Process suitability for thermosetting resins:
● primary choice; ○ secondary choice
(courtesy of Dr R. Progelhof, New Jersey Institute of Technology)

manufacturing process can provide very misleading results, since material properties can be greatly affected by the fabrication process.

Prototype evaluation should also be representative of the final design with respect to the specific material, the configuration and the use of special manufacturing techniques such as ultrasonic welding. It is essential, for instance, when testing colored parts for impact, to have the dye incorporated, since the presence of the dye may have a significant effect on impact resistance or other mechanical properties.

Testing of the final product under actual end use conditions is the most meaningful form of evaluation. Although this may not be practicable, due to cost or induced hazards in the event of failure, all performance requirements are encountered and a fairly complete assessment can be made. When end use testing is not practicable, simulated service conditions must be employed in a manner which duplicates the end use conditions as closely as possible. Tests may include temperature cycling, load cycling, electrical performance, environmental exposure, static and impact strength tests and others.

Since testing for the service life of the item is often impracticable under conditions of actual service, accelerated testing can be used to evaluate performance. It has been found to be advantageous to initiate long-term testing as early as possible, concurrent with accelerated tests, and to continue the long-term tests indefinitely.

Another useful approach to plastics part evaluation is that of "overtesting" or testing to failure. Such testing may include temperature cycling beyond the required limits, drop tests from greater heights and at lower temperatures, application of induced stresses until part failure occurs, and other destructive tests. These tests tend to establish the ultimate performance limits of the product and provide an estimate of the factor of safety for the ultimate user.

In the final design phase, consideration is normally given to specifications to control production. These specifications usually include raw material identification, part dimensions, tolerances, color, finish and other pertinent appearance features. In many cases an engineering test is needed to assure product performance. This test would be based on critical product requirements and the means of measurement, and may involve the measurement of material properties. Generally, quality assurance procedures should be developed from the prototype testing phase.

The prototype phase affords the opportunity to assess the influence of the manufacturing process variables on product performance. Plastic fabrication processes subject the material to rather severe conditions involving elevated temperature, high pressure and high shear-flow rates. These conditions, coupled with variable cycle times, other machine variables and procedures subsequent to molding, can result in internal stresses which can lead to product failure. Knowledge of these variables and their influence on the material will contribute significantly to a successful design. The evaluation of process variables can also be used to determine the effectiveness of the product assurance procedures intended for use as a measure of product integrity.

6. Design Specifications

The completion of a design must be followed by development of appropriate engineering drawings. These drawings detail dimensions and tolerance limits, materials, surface finish, color and marking, as well as special requirements and features as needed for production control. Design finalization should also include technical documentation of the entire integrated product design process, describing the requirements, the design, material selection and product evaluation. Such reports can be the basis for future reference with respect to production, product improvement and maintenance.

6.1 Considerations of Product Assurance

Hardware or products are produced chiefly on the basis of a requirements package. The package contains drawings and specifications detailing the requirements for each part and assembly. In essence, it

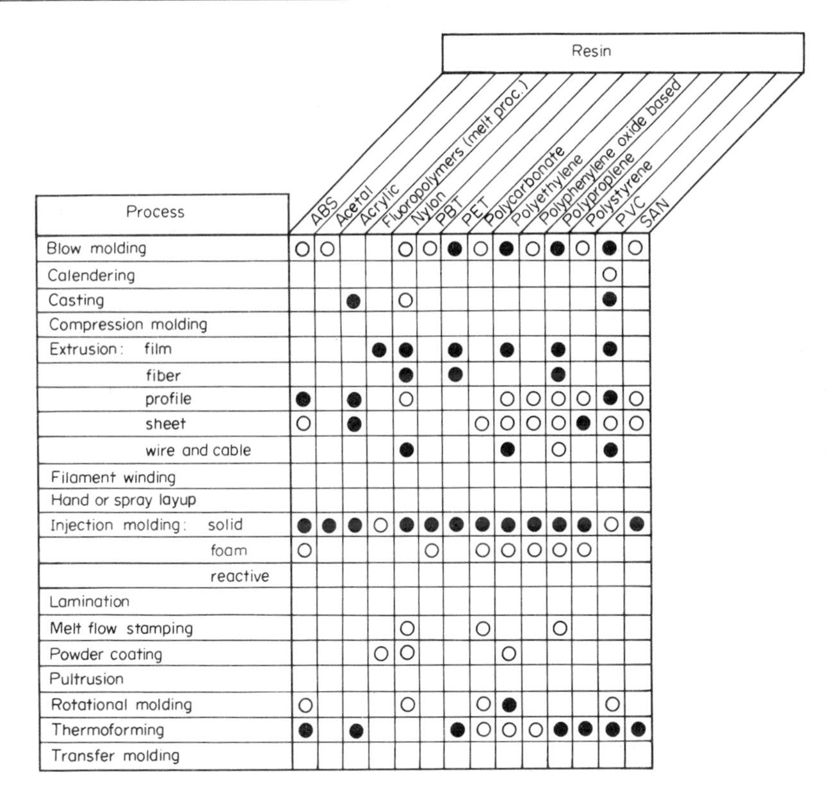

Process	ABS	Acetal	Acrylic	Fluoropolymers (melt proc.)	Nylon	PBT	PET	Polycarbonate	Polyethylene	Polyphenylene oxide based	Polypropylene	Polystyrene	PVC	SAN
Blow molding	○	○			○	○	●	○	●	○	●	○	●	○
Calendering													○	
Casting			●		○								●	
Compression molding														
Extrusion: film				●	●		●		●		●		●	
fiber					●		●				●			
profile	●		●		○				○	○	○	○	●	○
sheet	○		●					○	○	○	○	●	○	○
wire and cable					●				●		○		●	
Filament winding														
Hand or spray layup														
Injection molding: solid	●	●	●	○	●	●	●	●	●	●	●	●	○	●
foam	○					○		○	○	○	○	○		
reactive														
Lamination														
Melt flow stamping					○			○			○			
Powder coating				○	○				○					
Pultrusion														
Rotational molding	○				○			○	●				○	
Thermoforming	●		●				●	○	○	○	●	●	●	●
Transfer molding														

Figure 4
Process suitability for thermoplastics: ● primary choice; ○ secondary choice (courtesy of Dr R. Progelhof, New Jersey Institute of Technology)

contains the product assurance considerations, which include quality control and performance specifications. The emphasis placed on the design concept is to develop more effective performance specifications.

6.2 Standards and Specifications

Quality provisions are frequently cited by reference to specifications or standards, particularly for materials acceptance. Use of available documents must be scrutinized to ensure that conformance to the specifications also results in compliance with product requirements.

6.3 Quality Control

The term "quality control" is a general term implying the control of product quality, either through product measurement or in the control of a manufacturing process. The requirements package sets the general standards by which the product is accepted. This includes initial material purchase, dimensions, tolerances, surface finish, color and general workmanship. These requirements, although a measure of acceptability, may not completely reflect the true quality of the item. For instance, the dimensional stability cannot be assessed accurately unless the item is exposed to an elevated temperature as expected in use.

6.4 Performance Specifications

Performance specifications are defined as those methods used to evaluate performance, the basis of which is related to the function of the component.

Performance specifications are extremely important in designing with plastics, to ensure that the production techniques employed do not hinder performance. Testing can be either destructive or nondestructive. It may employ existing methods or novel ones unique to the product. Tests may be applied to individual components, subassemblies or total assemblies as appropriate. In all, performance specifications are the most meaningful form of product evaluation.

7. Secondary Process Operations in Designing

There are numerous types of secondary process operations, which fall into the categories of finishing, joining and decorating. Many are quite useful in achieving design features not readily possible with

Table 1
Processing cost indication (courtesy of Dr R. Progelhof, New Jersey Institute of Technology)

Cost per unit part	Process
Very low	calendering film extrusion RIM solid injection molding wire and cable extrusion
Low	flow molding[a] foam injection molding melt flow stamping profile extrusion rotational molding[a] sheet extrusion thermoforming transfer molding
Medium	blow molding[a] compression molding powder coating pultrusion rotational molding[a]
High	casting filament winding hand or spray layup machining

[a] Varies with size

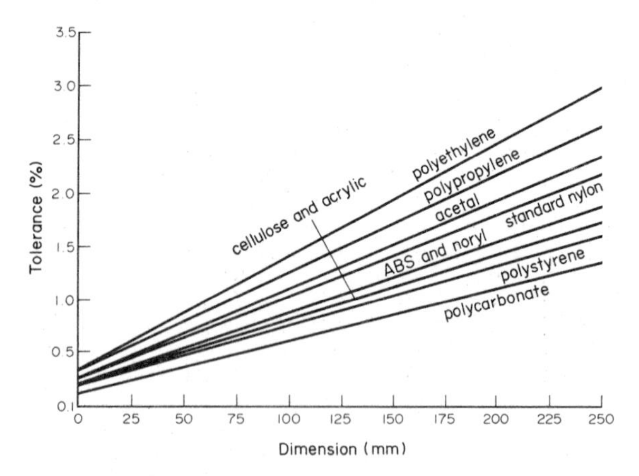

Figure 5
Dimensional tolerances for thermoplastics (no glass reinforcement)

other materials. Their adaptability to design is important in controlling product uniformity.

7.1 Finishing

Finishing operations include degating, deflashing and machining, which are required after the part has been fabricated. Although degating and deflashing are very

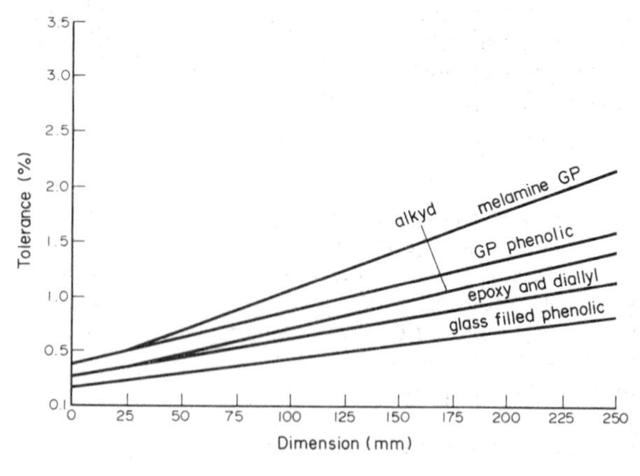

Figure 6
Dimensional tolerances for thermosetting resins

simple procedures, machining of plastics warrants greater attention. Techniques used to machine metals are generally applicable to plastics but must be modified for effective performance. Such modifications are necessary because plastics are of lower stiffness, lower thermal conductivity, higher thermal expansion and lower softening point. Each type of polymer has unique properties and as a result it can be assumed that each has different machining characteristics.

7.2 Joining

The joining of plastics is generally accomplished by mechanical methods, adhesive bonding or welding. Each requires different design configurations, shape and size, which must be considered early in product development, to ensure evaluation in prototype testing.

Mechanical means of joining include screw fasteners, inserts, snap fits, hinge mechanisms and staking. These may be used for joining plastics to plastics or to metals. Because of potential induced stress and thermal considerations, mechanical joining cannot be treated routinely and must be given careful consideration in design.

The field of adhesive bonding is a highly specialized area. Adhesives for plastics fall into five basic groups: solvent cements, bodied adhesives, monomeric cements, elastomeric adhesives and reactive adhesives. There is no universal adhesive or cement that will bond all types of plastics materials. Also, it should be noted that certain plastics materials cannot be joined with adhesives, due to their chemical makeup. Although a very cost-effective means of joining plastics, adhesive bonding requires a good deal of knowhow for effective utilization.

Welding processes for plastics include ultrasonic assembly, hot gas welding and heat sealing. Each process is an art in itself and employs specific design configurations for effective utilization.

The many techniques for joining plastics afford the designer great latitude in product development to facilitate assembly, incorporate unique features and reduce cost.

7.3 Decorating

Whether applied for functional or aesthetic reasons, decorating is frequently required in the manufacture of plastic parts. The methods include painting, printing, coloring, metallizing and hot stamping. Many of the basic molding processes are also adaptable to decorating, in that raised or lowered letters, two-color molding and in-mold decorating are practicable. To meet product requirements effectively these secondary processes and their variations must be compatible with the material and the design.

8. Use of Computers in Part Design

The computer provides a useful tool both for improving original designs and for upgrading existing designs by incorporating different materials. Component designers are faced with designing the lightest possible component meeting performance specifications set by the systems designer and providing optimal cost–benefit balance. Development and evaluation of various design concepts in terms of cost and performance must be iterated until this "best" design is achieved. Faced with many parameters, designers need a technique for evaluating many alternatives.

8.1 Computer-Aided Engineering

Computer-aided engineering (CAE) is a systematic process of applying a computer to each design application concern. CAE organizes design processes into unique elements (design, analysis, test, drafting, documentation and control, and related manufacturing), to which computer tools are applied to gain information and make better decisions. With CAE, performance of alternative materials and concepts is investigated via "computer prototypes." By using computer models rather than hardware prototypes, CAE can substantially shorten a design cycle, improve product, and reduce cost. The CAE method is compared favorably with the CAD/CAM and manual methods in Fig. 7.

8.2 Materials Considerations

Past lightweight component design efforts reveal two characteristic materials problems. First, theoretical equations are used to evaluate relative material effectiveness without consideration of overall component design, an approach satisfactory only for simple structures or properties. The second problem occurs when alternative materials are substituted directly in a design.

See also: Polymers: Technical Properties and Applications

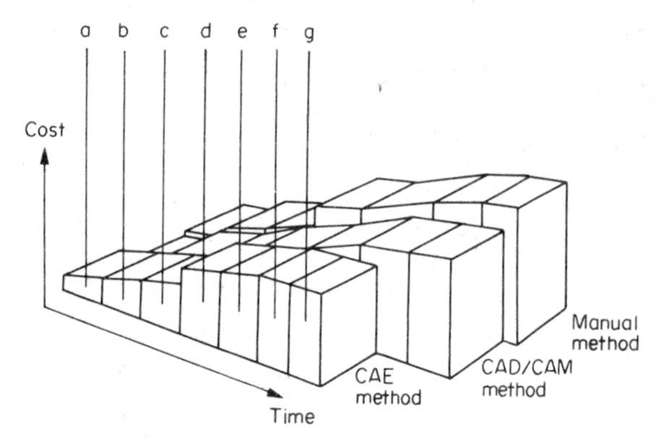

***Figure** 7*
Illustration of the benefits of computer-aided engineering in part design: (a) system simulation/concept development, (b) design guided by CAE analysis, (c) CAE detail design, (d) first prototype built, (e) preproduction engineering, (f) production prototype, (g) release to manufacturing (courtesy of Structural Dynamics Research Corporation)

Bibliography

Grafton P 1980 Design considerations for plastics components. *Plastics Mechanical Performance and Physical Testing*, PIA graduate course. Plastics Institute of America, Hoboken, New Jersey

Hagan R S, Thomas J R 1965 *Plastics Part Properties and Their Relationship to End-Use Performance*. American Society for Testing and Materials, Philadelphia, Pennsylvania

Hagan R S, Thomas J R 1979 Environmental effects on plastics parts and properties. *Plastics Mechanical Performance and Physical Testing*, PIA graduate course. Plastics Institute of America, Hoboken, New Jersey

Maddux K C 1983 Use of computers in plastics part design. *Applications of Computer Aided Engineering in the Injection Molding Industry*, PIA graduate course. Plastics Institute of America, Hoboken, New Jersey

Maddux K C, Amgas G D 1984 *Lightweight Component Design via Computer-Aided Engineering*, SAE paper No. 790984. Society of Automotive Engineers, New York

Nardone J 1981 Engineering plastic products. *Plastics Product Design*, PIA graduate course. Plastics Institute of America, Hoboken, New Jersey

O'Toole J L 1981 Applications and design properties of plastics. *Plastic Mechanical Performance and Physical Testing*, PIA graduate course. Plastics Institute of America, Hoboken, New Jersey

Progelhof R C 1981 Process technology and product assurance. *Plastics Product Design*, PIA graduate course. Plastics Institute of America, Hoboken, New Jersey

Starita J 1978 Getting meaningful impact data for product design. *Plastics Mechanical Performance and Physical Testing*, PIA graduate course. Plastics Institute of America, Hoboken, New Jersey

A. Spaak
[Plastics Institute of America, Hoboken, New Jersey, USA]

Design with Wood

Wood is a remarkable design material. It has generally high strength-to-weight ratios, requires only limited amounts of energy in preparation for end use and can be made more durable by various treatments. It is aesthetically pleasing, comes from a renewable resource and its countless successful applications testify to its recognized value. In many places in the world, wood is the main structural material used in residential and other light-frame construction. Heavy construction now makes increased use of highly refined laminated timber products that are available in a wide variety of sizes and shapes, and are suitable for all types of structures.

The design of main load-bearing systems is usually covered by law and requires certification of engineering competence on the part of the designer. However, this is not always necessary for the design of other objects or systems. The design process must be comprehensive and cover all attributes desired in the finished product. The primary consideration is the expected lifetime of the product. A wooden toy may be designed to lead a short rigorous life, whereas a building is designed to last for many years. There are other considerations if the design object is a structure such as a building. For example, it must be able to carry the expected loads, provide adequate fire safety, provide the desired interior environment, and resist all expected agents of degradation during the expected life. The structure may need to meet acoustic, vibrational or other special requirements, all of which must be met within the constraints of economy. Nonstructural wood systems share many of these demands and will probably have further special requirements such as surface finish performance.

1. Design Considerations

Although all design can be recognized as a combination of art and science, this is particularly true with wood design because the available scientific knowledge is frequently incomplete, and it is therefore necessary to resort to knowledge accumulated through years of experience. For example, to know the materials the designer must develop a subjective familiarity with wood products including accurate mental pictures of the species, its appearance and the range of variation that can be expected. Commercial wood-product specifications must be clearly understood since some designs may require special sorting or manufacture or both, and this can narrow sources of availability and increase cost.

Details of successful and unsuccessful construction must be absorbed from observation and the experience of others who can point out important features. Some aspects of the design may be quite demanding on wood characteristics such as freedom from warp, swelling or shrinkage, and the occurrence of defects such as knots. The designer must be aware of these matters when defining material specifications because a particular species or species group may be desirable or certain grades most suitable. All of this is necessary because wood is a variable biological product and continues to behave like a living material during service, changing size according to environmental conditions. Even within a species and grade, wood products will vary significantly in appearance, and in physical, chemical and biological properties.

As a natural material, wood is subject to attack by natural forces; these must be arrested to give the required life-span to the design. With the proper knowledge, long-term service can be readily obtained from wood. There is considerable overlap between structural and nonstructural design, with possible change of priorities in factors relating to safety and appearance.

2. Moisture and Temperature Considerations

In service all wood contains some moisture, the amount varying with the temperature and humidity of the environment. Moisture-content changes can cause shrinkage or swelling, and this must be anticipated in the design. In fine wood products, the moisture content must be controlled from the time of manufacture of parts through fabrication and after the wood product is put in service. Many surface finishes retard moisture flow in and out of the wood, thus offering a degree of protection against environmental change in service. Special treatment can stabilize wood against shrinkage and swelling, but it is expensive. The control of moisture in construction wood products is not usually as stringent, but serious problems can occur. Undue restraint on a wood structural member experiencing a moisture change can cause failure in the material. Thus close attention must, for example, be given to structural connector spacing, particularly in large members, to preclude the development of tensile failures perpendicular to grain owing to shrinkage restraint of the connectors. Constrained wood parts may buckle or create other unsightly defects in response to a moisture increase caused by a normal environmental change.

Wood engineering manuals and design specifications offer rules and advice on adjusting design stresses and elastic moduli for common end use situations. Design for particular conditions of moisture and temperatures should be undertaken with great care. When conditions of high stress are added to unusual moisture–temperature conditions, experienced advice is mandatory and, if this is not available, tests should be carried out in a simulated environment. The influences of cycling moisture content between high and low levels and/or cycling temperature on wood strength are not well covered by available research but indications are that these cycles can cause strength degradation.

Temperature as an isolated variable also influences the strength of wood, with strength decreasing as temperature increases. This effect is more pronounced at higher wood moisture contents. In the range of temperatures normal to the human environment, engineering adjustments of permissible stresses are generally omitted. If temperatures exceed 65 °C for extended periods or are extremely low, it is necessary to consult the literature available on the subject.

Higher wood moisture contents short of full saturation provide opportunities for growth of fungi which destroy the material. Decay will stop when the moisture content falls, but will begin again with the return of higher moisture. In building construction, ground proximity or contact, leaks, rainwater standing on horizontal exposed surfaces, and condensation are the primary offenders in creating moisture levels conducive to fungal growth (see *Wood: Decay During Use*).

Where there is any chance of intermittent moisture accumulation, dual priority in wood design must be to keep the wood dry while also providing backup ventilation. Insulated walls, attic spaces and crawl spaces are examples of danger zones in buildings that must be ventilated. Horizontal, weather-exposed wood surfaces are to be avoided because water can more easily spill over into poorly ventilated cracks in addition to saturating the horizontal surface. Effective drainage on the exterior of the structure is necessary for prevention of decay. Some structural situations, such as salt storage buildings and certain types of animal housing, produce high-moisture environments hostile to wood, thus necessitating preservative treatment.

Vapor barriers are commonly used in conjunction with insulation in walls, floors and ceilings to control condensation. Condensation can reduce the effectiveness of the insulation and can cause decay and paint problems. No vapor barrier can be expected to be perfect and allowance for ventilation to remove trapped condensation is necessary in good design.

Wood is often exposed to the weather for architectural or other reasons, but this is a questionable practice in many geographical locations. If successful local examples exist, they should be checked out as to species of wood, fastening and geometry of design. Weathering effects from alternate wetting and sunlight combine to create surface degradation and other potential cosmetic defects.

Although dry wood will last indefinitely, a stable temperature–humidity environment is even more conducive to long satisfactory service. Wood that remains completely submerged in water will not decay, although animal life such as marine borers may attack and destroy the wood.

3. Thermal and Electrical Properties

Wood is usually classified as a thermal insulator but does not perform as well as commercial insulation since its insulation qualities decrease with increasing moisture content and are lower in woods of higher density. Wood is also classified as an electrical insulator: its electrical resistance changes rapidly with increasing moisture content within the normal service range (i.e., near oven-dry to 20% moisture content). At higher moisture values resistance loss is smaller and erratic. In the event that thermal or electrical considerations must be included in the design, available literature should be consulted with regard to the specific situation.

4. Insect Protection

Insects can attack wood and this fact must be considered in both the design and the selection of materials. Information concerning the local incidence of insect problems and their control must be appraised early in the design procedure. The many areas of the world where wood is a historic building material indicate that local insect problems can be either avoided or economically controlled. In other areas, insect threats can be very serious and may be the major consideration of design.

5. Fire Considerations

The fire safety of wood construction is a matter of concern that is important but sometimes overrated. Uncontrolled contents of buildings and combustible decorative materials frequently cause fire risks that overshadow safety features incorporated into the basic structure. In the USA, sprinklered wood construction is widely accepted and provides for a reduction in risk from the total fire load including contents. The experimental designer must, however, always keep fire in mind, as highly efficient structures using slender or thin wood parts may not be practical because of potential fire problems. Fire tests and qualification of new construction systems can be very expensive, and hence proposed new constructions that cannot be generically related to already approved construction should be developed with this potential cost in mind (see *Wood and Fire*).

6. Chemical Treatment

Wood can be chemically treated to change its properties in response to special environment needs (e.g., to resist decay and insects, to retard burning, or to resist the effects of moisture). The designer should be careful in selecting treated wood, taking into consideration problems of handling, possible toxicity of treatment materials and the likelihood of interaction with other components of design such as corrosion of fasteners.

7. Acoustic and Vibrational Considerations

Acoustics and vibration in wood-building construction have been the subject of increased research in recent years, with some results becoming regulatory in the USA. Acoustic and vibrational problems are, of course, not confined to wood but occur with all structural materials. As a material, wood has a favorable vibration damping coefficient, and its acoustic performance varies according to the application (see *Acoustic Properties of Wood*). The methods of study underlying current acoustic/vibration recommendations for construction have, of necessity, been mainly empirical. This is because the total wall, floor, ceiling or building represents a complex unit that is generally intractable from an analytical standpoint. The available design information on acoustics relates to specific constructions including fastener details. New constructions could require testing for code approval.

8. Strength–Load Duration

The strength properties of wood and connections are time dependent at constant moisture and temperature, subject to complex relationships. Contemporary engineering methodology for dealing with this aspect of design is relatively simple because the present base of research findings is limited. Simplified factors tied to available research and backed by experience are used to adjust allowable stresses and elastic moduli for design purposes. Many common structural designs use prescribed loads that have a relatively low probability of occurrence and a limited period of expected duration at full design load. This situation, quite typical in ordinary building design, is relatively straightforward and backed by an extensive successful history. In such applications, tabulated allowable mechanical properties can be adjusted for the expected load duration and compared with design-load stresses with good assurance of satisfactory performance over the design lifetime. The load duration adjustments used in such a case are specified by building codes and can be found in wood engineering handbooks.

9. Importance of Structural Connections

If established practices of connection design as found in the handbooks and manuals are followed, and if loading and environment are not unusual, tabulated design values for connectors can be expected to yield good results. Large joints must be carefully designed, keeping the need for allowable wood movement in mind and giving special attention to the possibility of development of high tensile stresses perpendicular to the grain as a result of shrinkage. Shear stresses within heavy timber joints can also become critical and are given careful attention by the experienced engineer.

Adhesive connections are a natural choice for wood assembly (see *Adhesives for Wood*), but practical considerations frequently preclude their use. Many wood adhesives require high-quality machining and fitting of parts. Most have rigid temperature requirements and all require considerable expertise and discipline in their application. For structural purposes, adhesive usage is confined, with few exceptions, to prefabrication of parts. The difficulty in inspecting finished adhesive connections requires good assembly quality control along with subsequent destructive test sampling. The possibility of joint stresses arising from moisture- or temperature-related size changes must also be kept in mind. If the changes occur slowly, some relief can be expected as a result of stress relaxation. Finally, the designer must clearly understand the costs of suitable adhesive joining and balance these against the resultant gain in product performance.

Mechanical connectors such as nails, bolts, screws, shear plates, timber connector rings and similar traditional devices have a long service record. With the exception of timber connector rings, completed joints fabricated with these mechanical devices are relatively easy to inspect and simple field assembly is usually possible under quite adverse field conditions. The toothed metal-plate connector, although not well suited to field assembly, is a relatively new and efficient device that has found wide acceptance in light-frame building component construction. It has also found more recent use as a connector for upholstered furniture frames. This newer connector, available in a variety of sizes, thicknesses and tooth patterns, has given rise to a significant industry in the USA that produces the plates and provides extensive engineering design services using sophisticated computer systems. These services, including the required practical expertise, are readily available in most countries.

Advances in metal plate and other new connectors, coupled with the development of computer systems for complex analyses, have fostered the development of new, longer-span structural components consisting of smaller lumber parts. Much of this progress has been made possible through the use of computer-based techniques of structural analysis that emerged in the 1960s and have since been adapted into improved design techniques for trusses, frames and composite beams. Special programs can treat floor and wall systems whereas others deal with furniture. These tools are available to the designer but do not eliminate the requirement for the consideration of the important fundamental aspects of wood use. The computer system only analyzes a model; the model is created by the designer and is thereby limited by the designer's insight and knowledge of materials, connections, structure and loads. As an example, the modelling can accommodate the behavior of a partially rigid mechanical wood connection but the user of the system must supply the quantitative inputs that describe the partial rigidity.

See also: Wood: An Overview; Wood as a Building Material

Bibliography

American Institute of Timber Construction 1985 *Timber Construction Manual*, 3rd edn. Wiley, New York

Avery-Phares 1980 *The Woodbook*. Avery-Phares, San Francisco, California

Hoyle R J Jr, Woeste F E 1988 *Wood Technology in the Design of Structures*, 5th edn. Iowa State University Press, Ames, Iowa

Percival D H, Suddarth S K 1981 Structural wood systems. *Educ. Modules Mater. Sci. Eng.* 3: 291–311

US Forest Products Laboratory 1987 *Wood Handbook: Wood as an Engineering Material*, USDA Handbook No. 72. US Government Printing Office, Washington, DC

S. K. Suddarth
[Purdue University, Lafayette, Indiana, USA]

Dislocations

Once it was recognized that metals are usually aggregates of crystals, it became possible to calculate their theoretical shear strength. This turned out to be about $G/30$, where G is the shear modulus, in total disagreement with the observation that the most perfect single crystals available would shear plastically under resolved shear stresses of $G/50\,000$ or less. After a long period of development, which has been recorded by Taylor (1965) and Orowan (1965), it became clear (Orowan 1934, Polanyi 1934, Taylor 1934) that the mechanism of slip was the passage across the glide plane of line defects of the crystal lattice called dislocations. These line defects (Fig. 1) are simply the boundaries of areas in which the material on one side of the glide plane has slipped by an interatomic spacing $\boldsymbol{b}$ (the Burgers vector) with respect to the material on the other side. Outside the boundary the atoms on one side of the glide plane are still in their original good register with those on the other side; inside the boundary the register is again good, since the relative displacement brings an atom into the position previously occupied by another atom.

Once dislocations are moving in the crystal, they multiply by the Frank–Read mechanism (Fig. 2). There are thus more dislocations available to produce plastic deformation, but at the same time their motion is hindered by their collisions with other dislocations, especially those on intersecting glide planes. Crystal whiskers, even of soft metals such as tin or of organic materials such as metaldehyde, that are free from dislocations show elastic limits of several percent elongation, comparable with the theoretical values. Dislocations are not merely theoretical constructs, but

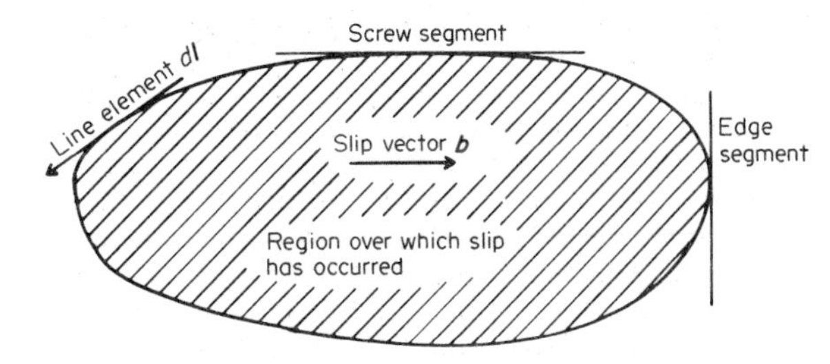

Figure 1
A dislocation line showing edge and screw segments: the slip plane is the plane of the figure, and the slip vector $\boldsymbol{b}$ lies in this plane

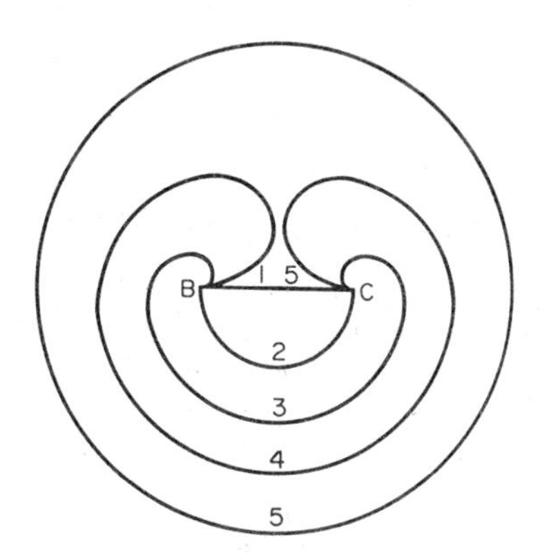

Figure 2
The Frank–Read mechanism. A segment BC of dislocation lies in the glide plane, which is the plane of the paper. At B and C it turns out of the plane, and is held fixed. Under an applied stress the mobile segment BC moves from 1 to 2, then 3, 4 and finally to 5. It now consists of a large loop together with the original segment BC, which can again repeat the sequence 1, 2, 3, . . .

may be observed directly by transmission electron microscopy or x ray topography, or optically if they are "decorated" with strings of precipitates. Their intersections with the surface may be marked by crystal growth spirals or etch pits. They can be seen moving in thin foils in the electron microscope. Their motion in bulk insulators and in thin metal samples can be studied by nuclear magnetic resonance (Hackelöer et al. 1977), in metals by resistance fluctuations (Bertotti et al. 1979), and in both by observation of acoustic emissions (Wadley et al. 1980). The fiftieth anniversary of the papers of Orowan, Polanyi and Taylor was marked by four international conferences. Their reports summarize the present state of dislocation theory.

1. Geometry of Dislocations

A dislocation is characterized by its Burgers vector, which is most satisfactorily defined as follows: choose arbitrarily the positive direction of an element dl of the dislocation line, and let u be the displacement of any point in the crystal from its position in some unstrained state of the crystal. Then the Burgers vector b of the dislocation is given by

$$b = -\int (\partial u/\partial s)\, ds \tag{1}$$

where the integral is taken once round a circuit s, called the Burgers circuit, which encircles the dislocation line with a right-handed screw. (The finish-to-start right-hand (FS–RH) rule yields the opposite sign.) The Burgers vector is constant along a dislocation, which cannot end in the lattice, and if several dislocations meet in a node the sum of their Burgers vectors vanishes (Frank's sum rule) if all positive line directions are taken outwards from the node.

If the dislocations are numerous, so that they may be taken as continuously distributed, the dislocation density tensor α_{ij} is defined as the sum of the b_j components of the dislocations threading unit area normal to the i axis. If the long-range strains in the lattice are negligible and the components $\tilde{\omega}_j$ of the lattice rotation are small, the lattice curvature tensor $\kappa_{ij} = \partial\tilde{\omega}_j/\partial x_i$ is related to α_{ij} by

$$\kappa_{ij} = \alpha_{ij} - \tfrac{1}{2}\delta_{ij}\alpha_{mm} \tag{2}$$

where δ_{ij} is the Kronecker symbol and the Einstein summation convention is used so that $\alpha_{mm} = \alpha_{11} + \alpha_{22} + \alpha_{33}$. If the strains are small but the rotations are large, the combination of elastic and plastic distortions produces a lattice which in the terminology of differential geometry has torsion but no curvature. The source function for the elastic strains is the Moriguti–Kröner elastic incompatibility tensor η^{E}_{ij} defined by

$$\eta^{E}_{ij} = -\varepsilon_{ikr}\varepsilon_{jls}\partial^2 e_{rs}/\partial x_k \partial x_l \tag{3}$$

where ε_{ikr}, the Levi–Civita tensor, is equal to $+1$ when i, k and r are an even permutation of 1, 2, 3, is equal to -1 for an odd permutation and is equal to zero otherwise; e_{rs} is a tensor component of the elastic strain.

2. Elastic Fields of Dislocations

The elastic fields of dislocations are calculated in a number of standard works (Cottrell 1953, Friedel 1964, Nabarro 1967, 1979, Hirth and Lothe 1982, Teodosiu 1982). The formulae given below apply to edge and screw dislocations in an isotropic medium. Poisson's ratio ν appears in the formulae for an edge dislocation, but not in those for a screw. In most real crystals, the distortion of the elastic field produced by elastic anisotropy introduces changes quite as large as those indicated by the presence of the term ν in these formulae. There are also effects due to the nonlinearity of the elastic forces, of which the principal one is the introduction of an effective volume of the core of the dislocation roughly equal to one atomic volume for each atomic length of the dislocation.

For an edge dislocation of Burgers vector $(b, 0, 0)$ lying along the negative z axis, the displacements are

$$u_x = \frac{b}{4\pi(1-\nu)}\frac{xy}{(x^2+y^2)} - \frac{b}{2\pi}\tan^{-1}\frac{x}{y} \tag{4}$$

$$u_y = -\frac{(1-2\nu)b}{8\pi(1-\nu)}\ln\frac{x^2+y^2}{b^2} + \frac{b}{4\pi(1-\nu)}\frac{y^2}{(x^2+y^2)} \tag{5}$$

and the tensor components of strain are

$$e_{xx} = -\frac{b}{4\pi(1-\nu)}\frac{y(x^2-y^2)}{(x^2+y^2)^2} - \frac{b}{2\pi}\frac{y}{(x^2+y^2)} \tag{6}$$

$$e_{xy} = \frac{b}{4\pi(1-\nu)}\frac{x(x^2-y^2)}{(x^2+y^2)^2} \tag{7}$$

$$e_{yy} = \frac{b}{4\pi(1-\nu)}\frac{y(3x^2+y^2)}{(x^2+y^2)^2} - \frac{b}{2\pi}\frac{y}{(x^2+y^2)} \tag{8}$$

$$e_{xx} + e_{yy} = -\frac{b}{2\pi}\frac{1-2\nu}{1-\nu}\frac{y}{x^2+y^2} \tag{9}$$

$$e_{rr} = -\frac{b}{4\pi}\frac{(1-2\nu)}{(1-\nu)}\frac{\sin\theta}{r} \tag{10}$$

$$e_{r\theta} = \frac{b}{4\pi(1-\nu)}\frac{\cos\theta}{r} \tag{11}$$

$$e_{\theta\theta} = -\frac{b}{4\pi}\frac{(1-2\nu)}{(1-\nu)}\frac{\sin\theta}{r} \tag{12}$$

$$e_{rr} + e_{\theta\theta} = -\frac{b}{2\pi}\frac{(1-2\nu)}{(1-\nu)}\frac{\sin\theta}{r} \tag{13}$$

and the stresses, which are derived from an Airy stress function

$$\chi = -Dy\ln(r/b) \tag{14}$$

with

$$D = bG/2\pi(1-\nu) \tag{15}$$

are

$$P_{xx}=-Dy(3x^2+y^2)/(x^2+y^2)^2 \qquad P_{rr}=-Dr^{-1}\sin\theta \tag{16, 17}$$

$$P_{xy}=Dx(x^2-y^2)/(x^2+y^2)^2 \qquad P_{r\theta}=Dr^{-1}\cos\theta \tag{18, 19}$$

$$P_{yy}=Dy(x^2-y^2)/(x^2+y^2)^2 \qquad P_{\theta\theta}=-Dr^{-1}\sin\theta \tag{20, 21}$$

and

$$P_{rr}+P_{\theta\theta}+p_{zz}=-2(1+\nu)Dr^{-1}\sin\theta \tag{22}$$

Since $p_{rr}=p_{\theta\theta}$, the principal axes of stress and of strain lie at 45° to the radius vector.

For a right-handed screw dislocation with Burgers vector $(0, 0, -b)$ lying along the $0z$ axis, the only component of displacement is u_z, and the displacement depends only on x and y, being given by

$$u_z=(b/2\pi)\tan^{-1}(y/x) \tag{23}$$

The only independent tensor components of strain which do not vanish are

$$e_{xz}=-\frac{b}{4\pi}\frac{y}{(x^2+y^2)} \tag{24}$$

and

$$e_{yz}=\frac{b}{4\pi}\frac{x}{(x^2+y^2)} \tag{25}$$

The stresses are

$$p_{xz}=-\frac{bG}{2\pi}\frac{y}{(x^2+y^2)} \tag{26}$$

and

$$p_{yz}=\frac{bG}{2\pi}\frac{x}{(x^2+y^2)} \tag{27}$$

In cylindrical coordinates the only nonvanishing components are

$$u_z=b\theta/2\pi \tag{28}$$

$$e_{\theta z}=b/4\pi r \tag{29}$$

and

$$p_{\theta z}=bG/2\pi r \tag{30}$$

If the Burgers vector of a dislocation makes an angle θ with its line, then in the isotropic approximation its elastic energy per unit length is

$$E(\theta)=(b^2G/4\pi)[\cos^2\theta + (1-\nu)^{-1}\sin^2\theta]\ln(R/r_0) \tag{31}$$

where R is an outer cutoff radius usually taken to be the distance to the nearest dislocation or to the free surface, and r_0 is an inner cutoff radius of the order of b. Whether the isotropic approximation is used or not, the line tension $T(\theta)$ which resists a bowing out of a dislocation aligned along θ is not $E(\theta)$, because a segment in a direction of low E rotates into directions of high E, and vice versa. It is found that

$$T=E+d^2E/d\theta^2 \tag{32}$$

The general elastic theory of dislocations in a continuous isotropic medium was worked out by Volterra (1907). He considered a multiply connected body such as a torus made singly connected by a cut (Fig. 3). The faces of the cut are subjected to a relative displacement, any surplus material is pared away, and gaps are filled with new material. A theorem due to Weingarten shows that this operation of paring and welding can only be carried out in a way which leaves the strains everywhere finite and twice differentiable if the initial displacement of the cut surfaces was one possible for a rigid body, that is, composed of three components of rotation and three of displacement. If in addition the crystal lattice is to be continuous across the seam, the rotation and displacement must be symmetry elements of the lattice. The rotations would have to be large (at least $\pi/6$), and the dislocation would have such a high energy that it would not be formed in a solid. The translational symmetry displacements are the displacements $\boldsymbol{b}$ from one atom to an equivalently situated atom, and these are the dislocations which move during plastic deformation. The displacement $\boldsymbol{b}'$ from one atomic site to a crystallographically inequivalent site generates a partial dislocation. The motion of a partial dislocation through the lattice produces stacking faults.

The elastic field of a dislocation in a bounded medium is not the same as that in an infinite medium, because the tractions on the free surface must be

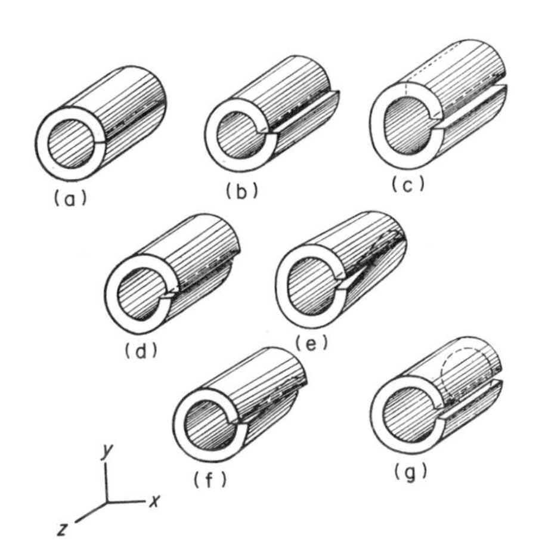

Figure 3
The six simple orders of dislocation made by Volterra's cutting and welding procedure. The positive direction of the dislocation line runs into the paper in the $-z$ direction. (a) Original cut cylinder, (b) dislocation of type b_x, (c) type $-b_y$, (d) type b_z (right-handed), (e) type $-d_{yz}$, (f) type $-d_{zx}$, (g) type d_{xy}

relaxed. In the case of a cylinder containing a screw dislocation parallel to its axis, only one component of traction has to be annulled; the problem is formally analogous to a problem in electrostatics, and can be solved by the method of images or by conformal representation. In a case such as that of an edge dislocation with its line parallel to a plane boundary, or lying in a circular cylinder, two components of traction have to be annulled. This cannot be done by means of a simple image dislocation; solutions are, however, available.

3. *Strain Produced by the Motion of Dislocations and the Force Tending to Move Them*

Consider a rectangular crystal of sides p, q, r, containing N sources of dislocation loops per unit volume, and subjected to a shear stress τ (Fig. 4). If a single loop of dislocation spreads right across its glide plane, the relative displacement of the opposite faces of the crystal is b. If it spreads over its glide plane, of area pq, only far enough to make a loop of area A, the mean relative displacement of the opposite faces is Ab/pq, and the shear strain is $\gamma = Abpqr$. There are $Npqr$ such loops in the volume pqr, and the total plastic strain is thus given by

$$\gamma = NAb \tag{33}$$

Now consider a straight edge segment XY moving across the glide plane. The forces on the opposite faces are $pq\tau$, and when the dislocation moves right across the glide plane by a distance p the external forces do work $bpq\tau$. The force tending to move the dislocation is thus $bq\tau$, and the force f per unit length is given by

$$f = b\tau \tag{34}$$

In the general case, we consider an element $\boldsymbol{dl} = dl_i$ of a dislocation with Burgers vector $\boldsymbol{b} = \boldsymbol{b}_t$ lying in a crystal subjected to a stress $p = p_{kt}$. Then the Peach–Koehler–Weertman formula tells us that the total force $\boldsymbol{dF} = dF_m$ on the dislocation element (which may have a component normal to the glide plane) is given by

$$\boldsymbol{dF} = \boldsymbol{dl} \times (p'\boldsymbol{b}) \quad \text{or} \quad dF_m = \varepsilon_{mik} dl_i p'_{kt} b_t \tag{35}$$

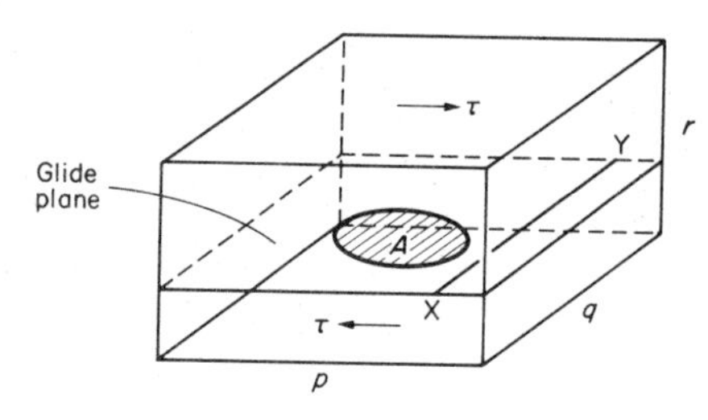

Figure 4
A dislocation loop spreading to cover an area A of a glide plane

where $p' = p'_{kt}$ is the deviatoric stress defined by

$$p' = p - \tfrac{1}{3}\mathrm{l}\,\mathrm{tr}\,p \quad \text{or} \quad p'_{kt} = p_{kt} - \tfrac{1}{3}\delta_{kt} p_{mm} \tag{36}$$

The component $\boldsymbol{df}$ of this force acting in the glide plane normal to $\boldsymbol{dl}$ in such a way as to expand the dislocation loop is given in terms of the unit vector:

$$\boldsymbol{n} = [\boldsymbol{b} \times \boldsymbol{dl}]/|[\boldsymbol{b} \times \mathrm{d}\boldsymbol{l}]|$$

or

$$n_i = \varepsilon_{ijk} b_j dl_k / |\varepsilon_{rst} b_s dl_t| \tag{37}$$

by

$$\boldsymbol{df} = [\mathrm{d}\boldsymbol{l} \times \boldsymbol{n}](\boldsymbol{npb}) \quad \text{or} \quad df_i = \varepsilon_{ijk} dl_j n_k n_r p_{rs} b_s \tag{38}$$

These formulae also apply if the stress p in the neighborhood of an element of a dislocation is produced by other dislocations or by distant parts of the same dislocation. As an example, consider an edge dislocation with its line parallel to the negative z axis and Burgers vector $(b, 0, 0)$ free to glide on the plane $y = y_0$ while the dislocation described by Eqns. (4–22) is anchored at $x = 0$, $y = 0$. The stress component which causes it to glide is, from Eqn. (38), p_{xy}. Equation (18) shows that this vanishes, and the mobile dislocation is in equilibrium, when $x = 0$ or $x = \pm y_0$. The latter positions are stable, and an edge dislocation dipole has two equivalent unsymmetrical equilibrium configurations.

4. *Arrays of Dislocations*

Consider a crystal which contains a wall of edge dislocations with their lines parallel to the negative z axis and Burgers vectors $(b, 0, 0)$. The lines meet the xy plane at the points $(0, nb/\psi)$, where $n = \ldots -1, 0, 1 \ldots$ and $0 < \psi \ll 1$. The wall then divides the crystal into two blocks tilted with respect to each other by an angle ψ about a line parallel to the dislocation lines (a tilt boundary). The stress field of the array of dislocations decays in proportion to $\exp(-2\pi\psi|x|/b)$. This kind of wall (sub-boundary) may be produced by the process of polygonization.

Now consider the case in which the axis of relative rotation of the two blocks is perpendicular to the boundary. A wall of screw dislocations produces both a relative rotation and a shear strain which tend to constant values at large distances from the wall; a relative rotation without long-range strain is produced by a crossed grid composed of two walls of screw dislocations of the same strength and helicity (a twist boundary). In the general case of one subgrain rotated with respect to the other by a small rotation $\boldsymbol{\omega}$, the density of dislocations in the wall may be explored by taking an arbitrary vector $\boldsymbol{V}$ lying in the plane of the wall. If $c_i(\boldsymbol{V})$ is the number of dislocations of Burgers vector $\boldsymbol{b}_i$ which is cut by $\boldsymbol{V}$, then

$$\sum_i c_i(\boldsymbol{V})\boldsymbol{b}_i = \boldsymbol{V} \times \boldsymbol{\omega} \tag{39}$$

for every V. The state of lowest energy of the dislocation wall is usually not a rectangular grid but a distorted hexagonal mesh, and this structure is observed experimentally. The stress field of the boundary is essentially confined to a slab of thickness b/ψ, and at any point within this slab the stress field is dominated by that of the nearest dislocation. The total length of dislocation in unit area of the wall is of order ψ/b. It follows that the surface energy W of the small-angle boundary is given by

$$W=(\alpha bG/4\pi)\ln(\psi_0/\psi) \tag{40}$$

where α is determined exactly by the elastic properties of the crystal but ψ_0 depends on the core structure of the dislocation. This relation is well confirmed by experiment for $\psi<\psi_0$.

It is also possible (Schober and Balluffi 1970) to analyze the structure of a grain boundary in which the orientation lies close to one in which a moderately high density of lattice sites in the crystals on either side of the boundary can be brought into coincidence by a relative translation without any distortion or rotation of the lattices. A small deviation from this orientation is then accommodated by grain-boundary dislocations with small Burgers vectors representing the smallest vectors in the displacement-shift complete lattice, which is made from all vectors joining sites in one lattice to sites in another when the two are in a position of best coincidence. These grain-boundary dislocations can be seen with an electron microscope.

Similarly, when two crystals with the same orientation, but slightly different lattice spacings, are joined, the interface has its lowest energy if patches of each crystal face are slightly strained to produce a good register between the lattice sites on either side of the interface. The long-range stresses arising from these strains are cancelled by the stresses of a grid of misfit dislocations which form the boundaries of the patches. In the important case of a thin layer deposited epitaxially on a thick substrate (van der Merwe 1978), the proportion of the misfit taken up by elastic strain in the deposited layer gradually decreases and the proportion taken up by misfit dislocations gradually increases as the thickness of the layer increases. Dislocations may also provide the mechanism of martensitic phase transformations.

A third type of dislocation array is produced if dislocations are emitted from a source but held up by an obstacle at a distance L from the source. Dislocations accumulate on the glide plane, piling up towards the obstacle, until they produce a stress at the source equal and opposite to the applied shear stress τ on the glide plane. In the case of pure edge dislocations, the total number n which can accumulate in the glide plane is given asymptotically by the formula

$$n=\tau L/2D \tag{41}$$

The positions of the individual dislocations may be determined by the method of Eshelby, Frank and Nabarro (1951), or more readily when n is large by that of Leibfried (1951) and Head and Louat (1955). For large n, the glide plane packed with dislocations is simply thought of as a lubricated crack in the stressed solid. The displacement u of one side of the crack relative to the other side is calculated as a function of distance x along the crack by ordinary elasticity theory, and the number of dislocations per unit length of crack $D(x)$ is given by

$$bD(x)=du/dx \tag{42}$$

If there is a source at $x=0$ and an obstacle at $x=L$, it is found that

$$D(x)=[\tau x/\pi D(L-x)]^{1/2} \tag{43}$$

The total shear stress on the glide plane produced by the applied shear stress τ and the dislocation pile-up is given by

$$\begin{aligned}\tau(x)&=\tau[x/(x-L)]^{1/2} \quad (x<0 \quad \text{or} \quad x>L)\\ &=0 \quad (0<x<L)\end{aligned} \tag{44}$$

There are large forward stresses ahead of the obstacle, but no large reverse stresses. A simple virtual-work argument shows that the force exerted on unit length of the obstacle is $nb\tau=b\tau^2L/2D$.

In covalent and ionic crystals and in alloys of low stacking-fault energy, but not in pure copper, arrays of dislocations having distributions close to that of Eqn. (43) are sometimes seen (Fig. 5). When their observed spacings are compared with those given by the exact theory (Fig. 5), the agreement can be very satisfactory (Bilby and Entwisle 1956). However, the spacings of like dislocations moving in procession across a glide plane are more often nearly uniform and are largely determined by other interactions.

5. *Dislocations in Crystals*

A dislocation of Burgers vector $\boldsymbol{b}$ will not usually glide readily on any plane including $\boldsymbol{b}$, although cases of such "pencil glide" are important. Glide is confined to a small number of low-index crystal planes, often those most closely packed with atoms, called glide planes. Thus in the face centered cubic structure a dislocation with $\boldsymbol{b}=[a/2, a/2, 0]$ glides readily on $(1\bar{1}1)$ or $(11\bar{1})$, and sometimes on (001), but not on other planes. If the dislocation is gliding on $(1\bar{1}1)$, it can cross slip onto $(1\bar{1}\bar{1})$. However, this cross slip requires the dislocation, which is extended on the $(1\bar{1}1)$ plane into two partial dislocations, to constrict until at least part of it lies in pure screw orientation along a line parallel to [110] before extending again in $(1\bar{1}\bar{1})$. The constricted region has an increased energy. When a dislocation lies predominantly in one glide plane, but has an element in a parallel glide plane, with short segments which do not lie in these planes, the short segments are called jogs. If the jogs lie in a cross-slip plane they may be able to move and are then called glissile jogs. Jogs in

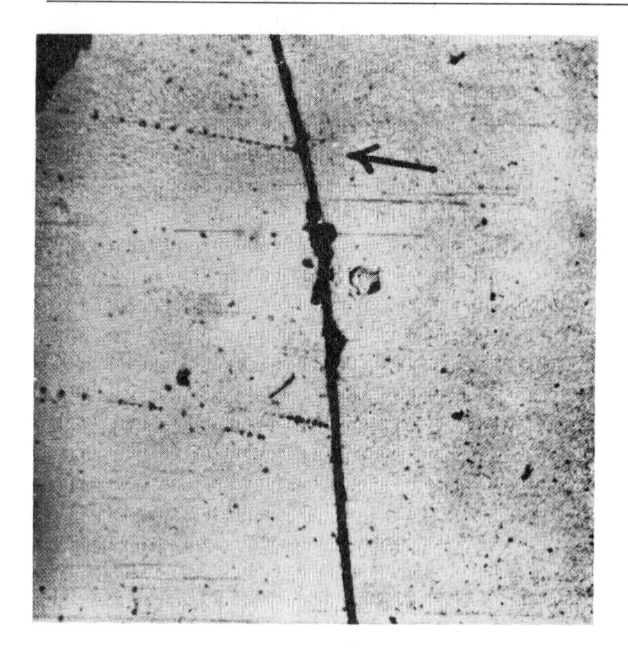

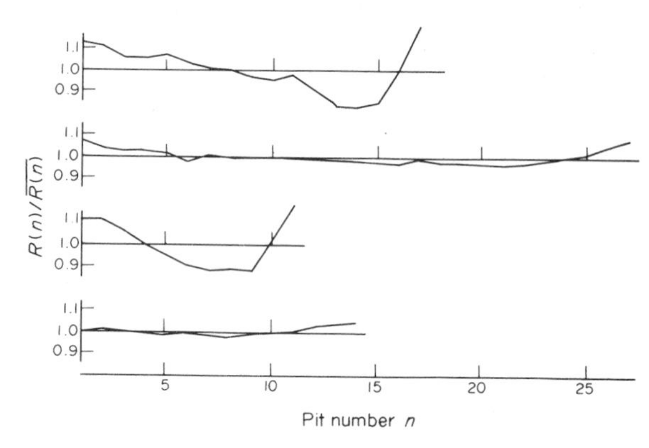

Figure 5
An array of dislocation etch pits in α-brass. For four such arrays the ratio of the observed distance of the nth pit from the head of the procession to the calculated distance is plotted as a function of n

screw dislocations, particularly if the jogs themselves are extended, move with difficulty, and are called sessile jogs. The length of a jog may be as little as one interatomic spacing. By definition, the line of a jog does not lie in the glide plane of the jogged dislocation. A kink, on the other hand, is an element of the dislocation line which lies in the glide plane, but not along the line of the dislocation when the Peierls potential compels that line to run predominantly along a low-index direction in the glide plane. When two dislocations cut through one another, each acquires a jog or kink equal in length to the Burgers vector of the other. A jog in an edge dislocation is always mobile, but a jog along $[1\bar{1}0]$ in a screw dislocation with Burgers vector along $[110]$ can move only in the $[110]$ direction, while the dislocation as a whole may be moving in the $(1\bar{1}1)$ plane in the $[11\bar{2}]$ direction, perpendicular to the possible motion of the jog. Thus the jog cannot travel with the dislocation (i.e., it is sessile).

6. *Moving Dislocations*

When a screw dislocation moves in an isotropic medium with a uniform velocity v, its elastic field contracts in the direction of motion by the factor $(1-v^2/c^2)^{1/2}$, where c is the speed of transverse waves. Its total energy increases by the reciprocal of this factor, the increase involving both kinetic energy and an increase in the elastic strain energy which arises from the contraction of the elastic field. The dependence of energy on velocity may be used to define a mass per unit length m which, like the line tension, depends logarithmically on inner and outer cutoff radii as

$$m=(\rho b^2/4\pi)(1-v^2/c^2)^{-1/2}\ln(R/r_0) \approx \tfrac{1}{2}\rho b^2 \qquad (45)$$

where ρ is the density of the medium. The moving dislocation also has a pseudomomentum per unit length of mv, which is of the same nature as the crystal momentum of a phonon. The corresponding results for an edge dislocation are more complicated, involving the speed of longitudinal waves as well as that of transverse waves.

The motion of the dislocation is always damped. The Peierls–Nabarro force must be considered, but there are many other mechanisms (Al'shitz and Indenbom 1975). The dominant mechanism is usually the scattering of phonons by the moving dislocation, which can occur by two processes. In one, the scattering arises because of nonlinearities in the core of the dislocation. The scattering cross section for phonons of wavelength λ is of order $\gamma^2 b^2/\lambda$, where γ is Grüneisen's constant (numerically about 2). The drag coefficient B relating the resistive force per unit length F to the dislocation velocity v by $F=Bv$ is given approximately by

$$B=bE/\phi c \qquad (1<\phi<10) \qquad (46)$$

where E is the energy density of phonons, proportional to T at high temperatures. Especially at low temperatures, this static mechanism is dominated by the flutter mechanism of scattering. Here, the shear component of the phonon causes the dislocation to vibrate and hence radiate crystal momentum. At low temperatures this effect leads to a drag $B \propto T^3$. In metals, there is also a contribution to B from the scattering of electrons. This is independent of temperature (Kaganov et al. 1974) and usually dominates the contribution of phonon scattering below about 15 K. It is suppressed if the metal becomes superconducting, and is restored if superconductivity is suppressed by a magnetic field.

7. Dislocations, Vacancies and Interstitial Atoms

The glide of a dislocation in its glide plane does not remove atoms from lattice sites. It is clear from Fig. 6a that the climb of an edge dislocation out of its glide plane by a single interplanar distance requires the absorption of a vacancy or the creation of an interstitial atom for each atomic length of the dislocation. A sessile jog which is dragged along by the rest of the dislocation is in fact climbing, and the motion of a jog of monatomic height produces a line of vacancies or interstitial atoms.

Just as dislocations can act as sources or sinks for point defects, so point defects can aggregate to form dislocations. Figures 6b and 6c show how it is possible for vacancies to aggregate on a plane to form a disk-shaped void. The faces of the disk fall together and join to form a perfect crystal, and the edge of the disk becomes a loop of dislocation with a Burgers vector which does not lie in the plane of the loop. In a close-packed crystal, the faces do not come together to form a perfect crystal, but a stacking fault is formed, while the dislocation bounding the fault is a sessile Frank partial dislocation. The glide of a Shockley partial dislocation across the loop removes the stacking fault and renders the dislocation bounding the loop perfect.

8. Other Physical Effects of Dislocations

Dislocations were introduced into the science of materials because they are the mechanism of some of the most important types of plasticity, and they were found to have a controlling influence on crystal growth in some important circumstances. They may also have a considerable influence on other material properties. As centers of elastic strain, they act as nuclei of precipitation in supersaturated solid solutions. Except in very special conditions, clean dislocations cannot be observed optically, but the atmospheres of point defects round dislocations are often visible. Dislocations have a profound influence on processes of radiation damage (Brailsford and Bullough 1981).

A high density of dislocations produces a small decrease in the Debye temperature, and the vibrations of the dislocation segments introduce a term in the specific heat which is linear in T and may amount to 1.9% of the electronic specific heat (Bevk 1973). Dislocations reduce the thermal conductivity of insulators, the resistance at low temperatures being dominated by flutter, and being especially strong in the transverse components of heat pulses (Anderson and Malinowski 1972, Roth and Anderson 1979). Resonant scattering by dislocation segments which are pinned at their ends can be detected.

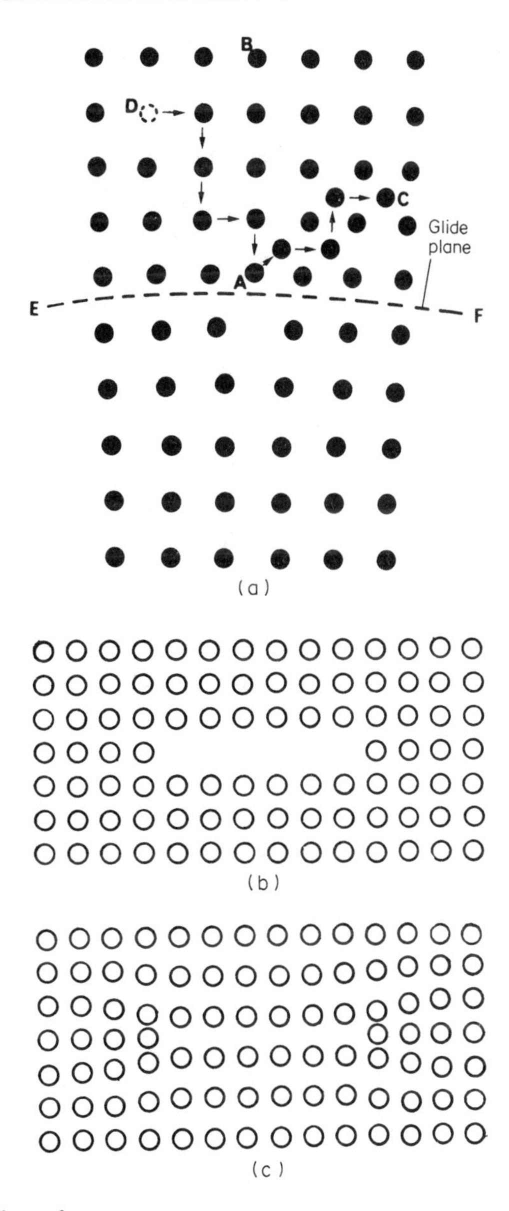

Figure 6
(a) Climb of an edge dislocation: either the atom A diffuses to an interstitial position such as C, or a vacant site such as D diffuses to A. The line of the dislocation is perpendicular to the plane of the figure.
(b) Vacancies collecting to form an internal crack.
(c) The sides of a crack falling together

Dislocations contribute to the electrical resistivity of a strained metal. The observations suggest a scattering width of about b, while the best calculations lead to about a tenth of this value, a discrepancy which has not been explained. In semiconductors they act as effective recombination centers for electrons and holes, thereby reducing the lifetime of minority carriers and degrading the performance of devices.

The small long-range strains produced by isolated dislocations deflect conduction electrons through small angles and make only a very small contribution to the electrical resistivity. However, the de Haas–van Alphen oscillations in the magnetic susceptibility depend on the fitting of an integral number of electron wavelengths into an orbit, an effect which is readily perturbed by small strains. While the theory is not yet entirely clear, the experimental evidence (e.g., Chang and Higgins 1975) shows that dislocations have a far larger effect on the de Haas–van Alphen oscillations than do point defects which produce the same change in the electrical resistivity.

Dislocations, like vacancies, provide regions in the lattice of a metal which are depleted of positive ions and therefore attract positrons. These positrons then annihilate with conduction electrons which have a lower density and lower velocities than atomic core electrons. The positron lifetimes are thus increased, and the energies and angular correlations of the γ ray pairs produced are altered (see Hautojärvi 1979).

9. Disclinations

While the rotational dislocations permitted by Weingarten's theorem and illustrated for small angles of rotation in Fig. 3(e–g) do not occur in crystalline solids, they appear in nematic and cholesteric liquid crystals, which are composed of elongated molecules which have a preferred orientation (the director) for their long axes. In the state of lowest energy the director has the same orientation in all parts of the liquid, but the energy required to produce configurations in which the orientation of the director varies from place to place is not large. If the total rotation on going round a circuit is an integral multiple of π radians, the pattern of molecular axes joins up continuously (except perhaps along the line of the dislocation). A rotational dislocation of this kind is usually called a disclination.

Bibliography

Al'shitz V A, Indenbom V L 1975 Dynamic dragging of dislocations. *Sov. Phys. Usp.* 18: 1–20

Anderson A C, Malinowski M E 1972 Interaction between thermal phonons and dislocations in LiF. *Phys. Rev. B* 5: 3199–3210

Bertotti G, Celasco M, Fiorillo F, Mazzetti P 1979 Application of the current noise technique to the investigation on dislocations in metals during plastic deformation. *J. Appl. Phys.* 50: 6948–55

Bevk J 1973 The effect of lattice defects on the low-temperature heat capacity of copper. *Philos. Mag.* 28: 1379–90

Bilby B A, Entwisle A R 1956 Dislocation arrays and rows of etch pits. *Acta Metall.* 4: 257–62

Brailsford A D, Bullough R 1981 The theory of sink strengths. *Philos. Trans. R. Soc. London, Ser. A* 302: 87–137

Chang Y K, Higgins R J 1975 de Haas–van Alphen effect study of dislocations in copper. *Phys. Rev. B* 12: 4261–81

Cottrell A H 1953 *Dislocations and Plastic Flow in Crystals.* Clarendon, Oxford

Eshelby J D, Frank F C, Nabarro F R N 1951 The equilibrium of linear arrays of dislocations. *Philos. Mag.* 42: 351–64

Friedel J 1964 *Dislocations.* Pergamon, Oxford

Hackelöer H J, Selbach H, Kanert O, Sleeswyk A W, Hut G 1977 Determination of the velocity of mobile dislocations by nuclear spin relaxation measurements. *Phys. Status Solidi B* 80: 235–43

Hautojärvi P J (ed.) 1979 *Positrons in Solids,* Topics in Current Physics Vol. 12. Springer. Berlin

Head A K, Louat N 1955 The distribution of dislocations in linear arrays. *Aust. J. Phys.* 8: 1–7

Hirth J P, Lothe J 1982 *Theory of Dislocations.* McGraw-Hill, New York

Kaganov M I, Kravchenko V Ya, Natsik V D 1974 Dislocation dragging by electrons in metals. *Sov. Phys. Usp.* 16: 878–91

Leibfried G 1951 Verteilung von Versetzungen im statischen Gleichgewicht. *Z. Phys.* 130: 214–26

Nabarro F R N 1967 *Theory of Crystal Dislocations.* Clarendon, Oxford

Nabarro F R N (ed.) 1979 *Dislocations in Solids.* North-Holland, Amsterdam

Orowan E 1934 Zur Kristallplastizität III: Über den Mechanismus des Gleitvorganges. *Z. Phys.* 89: 634–59

Orowan E 1965 Dislocations in plasticity. In: Smith C S (ed.) 1965 *The Sorby Centennial Symposium on the History of Metallurgy.* The Metallurgical Society (AIME), New York, pp. 359–77

Polanyi M 1934 Über eine Art Gitterstörung, die einen Kristall plastisch machen könnte. *Z. Phys.* 89: 660–64

Roth E P, Anderson A C 1979 Interaction between thermal phonons and dislocations in LiF. *Phys. Rev. B* 20: 768–75

Schober T, Balluffi R W 1970 Quantitative observation of misfit dislocation arrays in low and high angle twist grain boundaries. *Philos. Mag.* 21: 109–23

Taylor G I 1934 The mechanism of plastic deformation of crystals: Part I: Theoretical. *Proc. R. Soc. London, Ser. A* 145: 362–87

Taylor G I 1965 Note on the early stages of dislocation theory. In: Smith C S (ed.) 1965 *The Sorby Centennial Symposium on the History of Metallurgy.* The Metallurgical Society (AIME), New York, pp. 355–58

Teodosiu C 1982 *Elastic Models of Crystal Defects.* Springer, Berlin

van der Merwe J H 1978 The role of lattice misfit in epitaxy. *CRC Crit. Rev. Solid State Mater. Sci.* 7: 209–31

Volterra V 1907 Sur l'équilibre des corps élastiques multiplement connexes. *Ann. Sci. Ec. Norm. Supér. Ser. 3* 24: 40–517

Wadley H N G, Scruby C B, Speake J H 1980 Acoustic emission for physical examination of metals. *Int. Met. Rev.* 25: 41–64

Reports of Conferences Celebrating the Fiftieth Anniversary of Crystal Dislocations

Loretto M H (ed.) 1985 *Dislocations and Properties of Real Materials.* Institute of Metals, London

Suzaki H et al. 1985 *Yamada Conference on Dislocations in Solids.* University of Tokyo Press, Tokyo

The 50th Anniversary of the Introduction of Dislocations, 1983. Metallurgical Society (AIME), New York

Veyissiere P, Kubin L, Castaing J (eds.) 1984 *Dislocations 1984*. Editions du CNRS, Paris

F. R. N. Nabarro
[University of The Witswatersrand, Johannesburg, South Africa]

Ductile Fracture: Micromechanics

The fracture toughness of a material is conventionally assessed in terms of the critical value of some parameter characterizing the crack tip field at the initiation of unstable crack growth. In plane strain under small-scale yielding conditions, that is, the plastic zone is small compared with the crack length and uncracked ligament, the critical value of the linear elastic stress intensity factor K_{Ic} is generally determined at the onset of crack extension and can be referred to as the "toughness" (see *Brittle Fracture: Linear Elastic Fracture Mechanics*). With appreciable nonlinearity in the load–displacement curve, however, the (crack initiation) toughness is measured in terms of the critical value of an energy release rate parameter J_{Ic} (referred to as the J-integral). Alternatively the opening displacement at the crack tip, δ_{Ic}, may be used. These parameters are explicitly related for linear elasticity in terms of the plastic flow stress σ_0 and the elastic (Young's) modulus E:

$$J_{Ic} \equiv \frac{K_{Ic}^2}{E} = \frac{1}{\alpha}\delta_i \sigma_0 \tag{1}$$

where α is a proportionality factor of order unity, dependent upon the yield strain $\varepsilon_0(=\sigma_0/E)$, the work hardening exponent n and whether plane stress or plane strain conditions are assumed.

Although in "brittle" structures, catastrophic failure or instability is effectively coincident with this onset of crack extension, in the presence of sufficient crack tip plasticity, crack initiation is generally followed by a region of stable crack growth. Under elastic–plastic conditions, such subcritical crack advance has been macroscopically characterized in terms of crack growth resistance curves (Fig. 1). Crack growth toughness is now assessed in terms of the tearing modulus T_R:

$$T_R \equiv \frac{E}{\sigma_0^2}\frac{dJ}{da} \tag{2}$$

Whereas crack initiation toughness values (e.g., K_{Ic}, J_{Ic}) are by far the most widely measured and quoted, it has been noted in high-toughness ductile materials, for example, that stable ductile crack growth can occur at J values some 5–10 or more times the initial J_{Ic} value prior to instability (e.g., Fig. 2). Furthermore, microstructural influences on fracture resistance would appear to be enhanced in the crack growth regime, compared with initiation behavior (Fig. 2; see further explanation in Sect. 2).

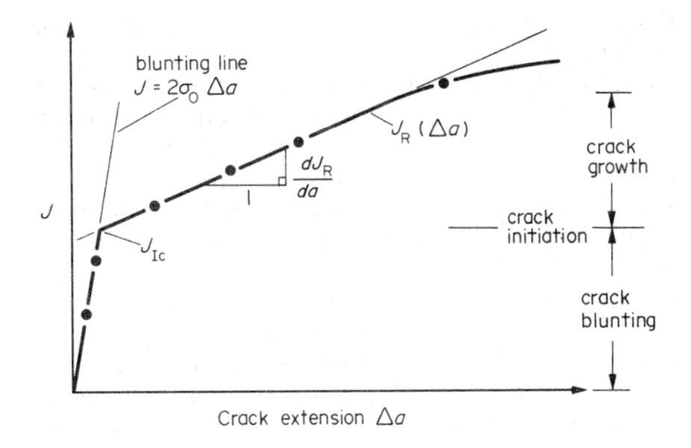

Figure 1
$J_R(\Delta a)$ resistance curve, showing definition of $J_i = J_{Ic}$ at initiation of crack growth where the blunting line intersects the resistance curve; ● experimental data

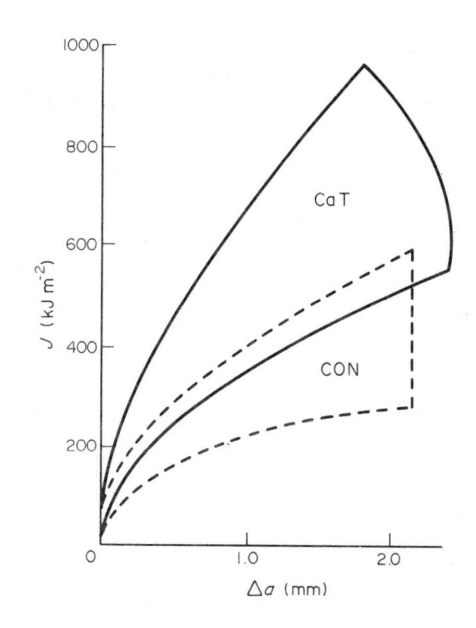

Figure 2
Experimental data showing $J_R(\Delta a)$ resistance curves for several heats of A516 grade 70 plain carbon steel plate ($\sigma_0 \sim 260$ MPa) (after Wilson 1979)

The purpose of this article is to examine the relationship between crack initiation and (quasi-static) crack growth toughness at both macroscopic and microscopic levels and in particular to identify the role of microstructure. Continuum and local models for the initiation and continued propagation of cracks are considered in terms of near-tip stress and deformation fields and macroscopic or microscopic fracture criteria. Specifically, model expressions for crack initiation and crack growth toughness are presented which indicate a relationship between J_{Ic} and T_R.

1. Crack Initiation Toughness

Full comprehension of a fracture process requires that microscopic models for specific fracture mechanisms be constructed. Such models are generally referred to as "micromechanisms." Unlike the continuum approach, this requires a microscopic model for the particular fracture mode, which incorporates a local failure criterion and consideration of salient microstructural features, as well as detailed knowledge of both stress and deformation fields. However, physical fracture processes, and consequently the local failure criterion and characteristic microstructural dimensions, vary substantially with fracture mode, as Fig. 3 illustrates for the four classical fracture morphologies: microvoid coalescence, quasicleavage, intergranular cleavage and transgranular cleavage.

In view of the specificity of such models to particular fracture mechanisms for particular microstructures, a complete microscopic–macroscopic characterization of toughness has been achieved in only a few simplified cases. For example, for slip-initiated transgranular cleavage fracture (Fig. 3d) in ferritic steels, Ritchie et al. (1973) have shown that the onset of brittle crack extension (at $K_I = K_{Ic}$) is consistent with a critical stress model in which the local tensile opening stress σ_{yy} directly ahead of the crack must exceed a local fracture stress σ_f^* over a microstructurally significant characteristic distance $x = l_0^*$, as depicted in Fig. 4a. This model for the cleavage fracture toughness implies:

$$K_{Ic} \propto [(\sigma_f^*)^{(1+n)/2n}/(\sigma_0)^{(1-n)/2n}]\, l_0^{*0.5} \tag{3}$$

In mild steels, with ferrite–carbide microstructures, the characteristic distance was found to be of the order of the spacing of the void-initiating grain boundary carbides, that is, typically about two grain diameters (d_g), although different size scales were found when the analysis was applied to other materials. The model has been found to be particularly successful both in quantitatively predicting cleavage fracture toughness values in a wide range of microstructures and in rationalizing the influence on K_{Ic} of such variables as temperature, strain rate, neutron irradiation and warm prestressing. Somewhat similar microscopic models involving a critical stress criterion have been suggested for other fracture modes, including intergranular cracking (Fig. 3c) in temper-embrittled steels and during hydrogen-assisted fracture.

For initiation of ductile fracture by microvoid coalescence (Fig. 3a), McClintock (1969), Rice and Johnson (1970) and Rice and Tracey (1969) considered the criterion that the critical crack tip opening displacement must exceed half the mean void-initiating particle spacing (i.e., $2\delta_i \approx l_0^* \approx d_p$), based on the notion that in nonhardening materials this takes place when the void sites are first enveloped by the intense strain region at the crack tip. This model implies that

$$\delta_{Ic} \approx (0.5 \text{ to } 2)\, d_p \tag{4a}$$

or

$$J_{Ic} \approx \sigma_0 l_0^* \tag{4b}$$

However, it is unusual to find the fracture toughness increasing directly with increasing strength. This problem is overcome by the approach of McClintock (1977) and Mackenzie et al. (1977), who alternatively have utilized a stress-modified critical strain criterion. Here, the local equivalent plastic strain $\bar{\varepsilon}_p$ must exceed a critical fracture strain $\bar{\varepsilon}_f^*$ (specific to the relevant stress state), over a characteristic distance l_0^* comparable with the mean spacing d_p of the void-initiating particles, as shown schematically in Fig. 4b. Then, by following the approach of Ritchie et al. (1973), assuming that $\bar{\varepsilon}_f^*$ is proportional to $1/r$, the ductile fracture toughness becomes

$$\delta_{Ic} \approx \bar{\varepsilon}_f^* l_0^* \tag{5a}$$

or

$$J_{Ic} \approx \sigma_0 \bar{\varepsilon}_f^* l_0^* \tag{5b}$$

or

$$K_{Ic} \approx \sqrt{E' \sigma_0 \bar{\varepsilon}_f^* l_0^*} \tag{5c}$$

Unlike the critical crack tip opening displacement (CTOD) criterion, the stress-modified critical strain criterion now implies that J_{Ic} for ductile fracture is proportional to strength times ductility, which is more physically realistic and permits rationalization of the toughness-strength relation for cases where microstructural changes which increase strength also cause a more rapid reduction in the critical fracture strain.

Analyses by Rice and Tracey (1969) and McClintock (1969) can be used to estimate the critical fracture strain $\bar{\varepsilon}_f^*$. For the rate of void expansion in the triaxial stress field ahead of a crack tip in a non-hardening material, Rice and Tracey suggest that, in terms of the void radius r_p

$$\frac{dr_p}{r_p} = 0.28 d\bar{\varepsilon}_p \exp(1.5\sigma_m/\bar{\sigma}) \tag{6}$$

where σ_m is the hydrostatic stress and $\bar{\sigma}$ is the equivalent stress. For an array of void-initiating particles of diameter D_p and mean spacing d_p, setting the initial void radius to $D_p/2$ and integrating to the point of ductile fracture initiation gives an expression for the fracture strain $\bar{\varepsilon}_f^*$ as:

$$\bar{\varepsilon}_f^* \approx \frac{\ln(d_p/D_p)}{0.28 \exp(1.5\sigma_m/\bar{\sigma})} \tag{7}$$

The earlier analysis by McClintock for a strain-hardening material (of exponent n) containing cylindrical holes similarly suggests:

$$\bar{\varepsilon}_f^* \approx \frac{\ln(d_p/D_p)(1-n)}{\sinh[(1-n)(\sigma_a^\infty + \sigma_b^\infty)/(2\bar{\sigma}/\sqrt{3})]} \tag{8}$$

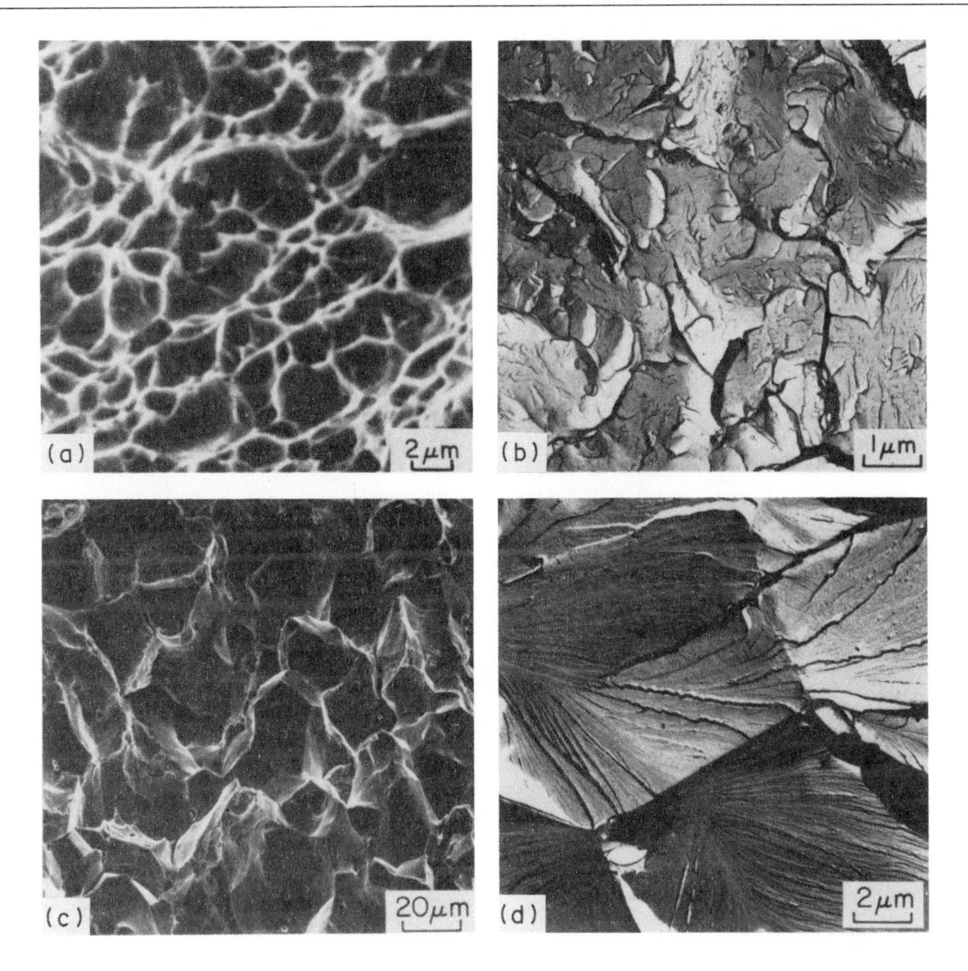

Figure 3
Scanning and transmission electron micrographs of the classical fracture morphologies, showing: (a) ductile fracture via microvoid coalescence involving the nucleation, growth and final linkage of voids formed by cracking or interface decohesion of second phase particles; (b) transgranular cracking by quasicleavage induced primarily by the fracture of matrix particles; (c) intergranular cracking involving a generally brittle separation along grain boundaries due to local embrittlement at these interfaces; (d) transgranular cleavage where brittle fracture along low energy "cleavage" planes results primarily from the cracking of grain boundary precipitates (fractographs courtesy of A W Thompson)

where σ_a^∞ and σ_b^∞ are the transverse stress components.

Both analyses consider the fracture strain to be limited by the simple impingement of the growing voids and thus tend to overestimate $\bar{\varepsilon}_f^*$ by ignoring strain localization. However, they correctly suggest a dependence of $\bar{\varepsilon}_f^*$ on stress state $(\sigma_m/\bar{\sigma})$, strain hardening (n) and purity (d_p/D_p). For example, a large effect of stress state (i.e., triaxiality) on fracture strain is predicted, such that $\bar{\varepsilon}_f^*$ would be expected to be reduced by an order of magnitude between conditions of unnotched plane strain and the condition ahead of a sharp crack. Increased strain hardening, however, can enhance $\bar{\varepsilon}_f^*$, particularly at high triaxiality, but the benefits of increased purity (i.e., increased hole spacing d_p) are pronounced only at low D_p/d_p ratios due to the logarithmic terms in Eqns. (7, 8). For example, reducing the volume fraction of inclusions from 10^{-2} to 10^{-6} would increase $\bar{\varepsilon}_f^*$ only by a factor of 2.

More recently, Thompson and coworkers have suggested a local means of evaluating $\bar{\varepsilon}_f^*$ through use of the fracture surface microroughness M, defined in Fig. 5. The basis for this approach is the recognition that the ratio of void height h to the diameter D_p of the initiating particle is a measure of the local fracture strain, such that

$$\bar{\varepsilon}_f^* \approx \ln(h/D_p) \qquad (9)$$

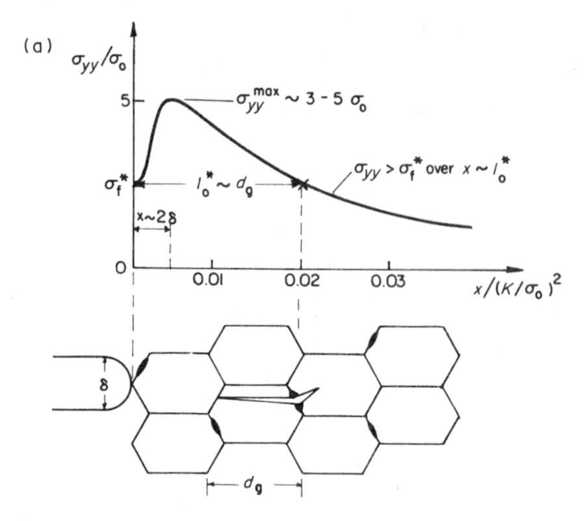

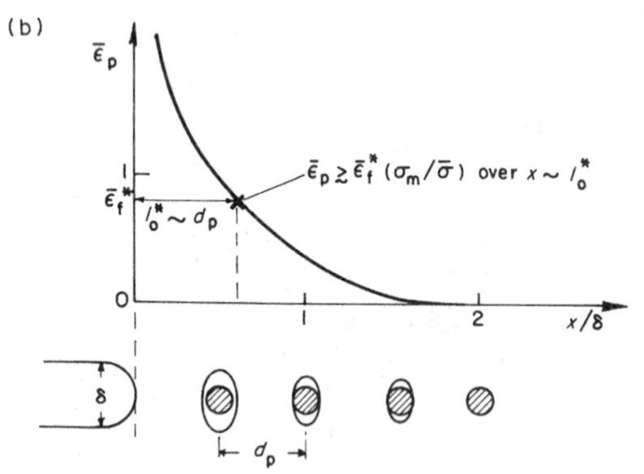

Figure 4
Schematic idealization of microscopic fracture criteria pertaining to (a) critical stress-controlled model for cleavage fracture; (b) critical stress-modified critical strain-controlled model for microvoid coalescence

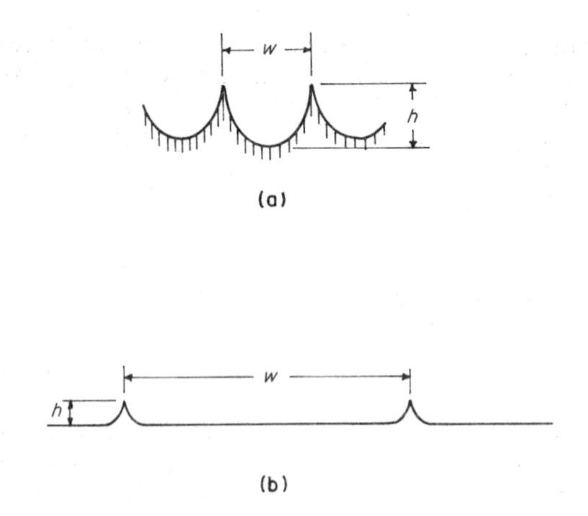

Figure 5
Definition of fracture surface roughness, $M = h/w$, for (a) microvoid coalescence; (b) other locally ductile fracture modes, such as quasicleavage (after Thompson 1982)

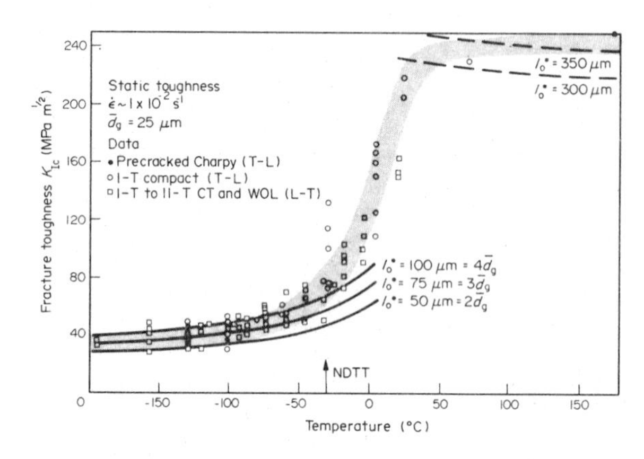

Figure 6
Comparison of experimentally measured fracture toughness K_{Ic} data for crack initiation in SA533B-1 nuclear pressure vessel steel ($\sigma_0 \sim 500$ MPa) with predicted values based on the critical stress model for cleavage on the lower shelf, and on the stress-modified critical strain model for microvoid coalescence on the upper shelf

or, in terms of M and volume fraction f_p of void-initiating particles:

$$\bar{\epsilon}_f^* \approx \tfrac{1}{3}\ln\left(\frac{M^2}{3f_p}\right) \tag{10}$$

Thus, Eqn. (5) would be written as

$$J_{Ic} \approx \frac{\sigma_0}{3}\ln\left(\frac{M^2}{3f_p}\right) l_0^* \tag{11}$$

The success of these microscopic models for crack initiation toughness can be appreciated from Fig. 6, where the critical stress model for cleavage and the stress-modified critical strain model for ductile fracture are utilized to predict respectively the lower and upper shelf toughness in ASTM A533B-1 nuclear pressure vessel steel. Whereas the characteristic distance l_0^* for cleavage fracture is ~2–4 times the grain size (essentially the bainite packet size), for ductile fracture l_0^* was found to be ~5–6 times the average major inclusion spacing d_p.

2. Crack Growth Toughness

A critical strain-based microscopic criterion for ductile crack growth was first proposed by McClintock (1958) for mode III (antiplane shear) crack extension under elastic–perfectly-plastic conditions and involved the attainment of a critical shear strain γ_f^* over some characteristic radial distance $r = l_0^*$ into the plastic zone. Applying this local criterion for $\gamma_p > \gamma_f^*$ over distance $r = l_0^*$ both for crack initiation and for crack growth yields estimates for the critical plastic zone sizes at initiation as

$$r_{yi} = l_0^* \frac{\gamma_f^*}{\gamma_0} \tag{12a}$$

and at instability as

$$r_{yc} = l_0^* \exp\left[\sqrt{\frac{2\gamma_f^* - 1}{\gamma_0}} - 1\right] \tag{12b}$$

where γ_0 is the shear yield strain. On the assumption that the critical fracture strains and distances are identical for initiation and growth, stable crack growth would occur with J increasing from an initiation value J_i to a steady-state value J_{ss}, where $dJ/da \to 0$, such that

$$\frac{J_{ss}}{J_i} = \frac{\gamma_0}{\gamma_f^*} \exp\left[\sqrt{2\left(\frac{\gamma_f^*}{\gamma_0}\right) - 1} - 1\right] \tag{13}$$

Eqn. (13) implies that the potential for stable crack growth increases dramatically as γ_f^* becomes large compared with the yield strain γ_0.

The concept of the attainment of a critical strain over some characteristic dimension directly ahead of a growing crack is not so amenable for the nonstationary crack opening (mode I) case, since the regions of intense strain are directly above and below the crack plane. Accordingly, Rice and his coworkers have proposed several alternative local failure criteria for initiation and continued growth of plane strain tensile cracks, all involving the notion of a geometrically similar crack profile very near the tip. Rice and Sorensen (1978) stated the crack growth criterion in general fashion by requiring that a critical opening displacement δ_p be maintained at a small distance l_p^* behind the crack tip. The local criterion, $\delta = \delta_p$ at $r = l_0^*$, then yields

$$\frac{\delta_p}{l_0^*} = \frac{\alpha}{\sigma_0}\frac{dJ}{da} + \beta\frac{\sigma_0}{E}\ln\left(\frac{e r_y'}{l_0^*}\right) \tag{14}$$

where α and β are constants, e is the natural logarithm base ($= 2.718$) and r_y' is the plastic zone size.

By comparing Figs. 5 and 7, it is apparent that the left-hand side of Eqn. (14), the ratio of local microscopic parameters δ_p/l_0^*, can be identified with the fracture surface microroughness $M = h/w$, for microvoid coalescence and possibly other modes (Fig. 7). Subsequently, Rice et al. (1980) have rephrased this geometrically similar near tip profile criterion to remove reference to the local microscopic parameters δ_p and l_0^*, by noting that

$$\delta = \beta r \frac{\sigma_0}{E} \ln\frac{\rho}{r} \quad \text{as} \quad r \to 0 \tag{15}$$

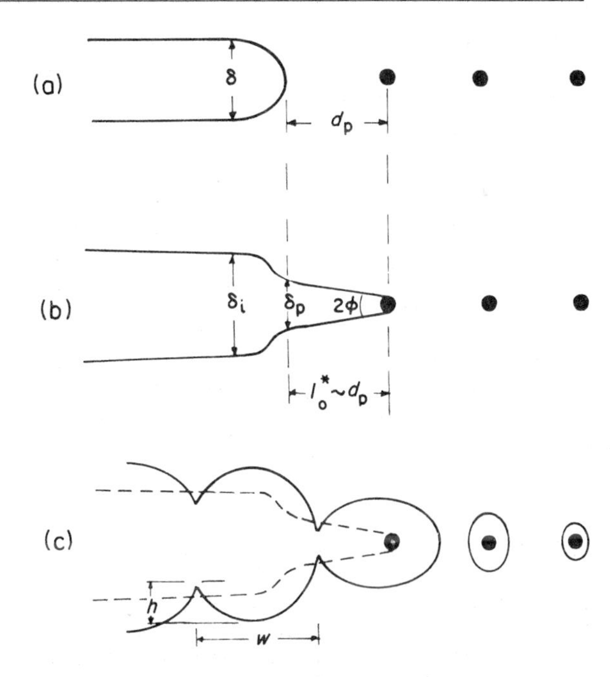

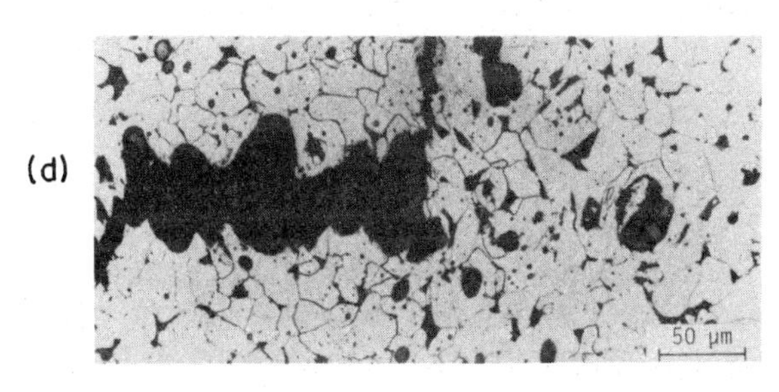

Figure 7
Idealization of stable crack growth by microvoid coalescence, showing (a) blunted crack tip; (b) crack growth to next inclusion based on constant crack tip opening angle (ϕ) or on critical crack tip opening displacement (δ_p) distance ($l_p^* \approx d_p$) behind the crack tip; (c) morphology of resulting fracture surface relevant to the definition of fracture surface microroughness ($M = h/w$); (d) fractographic section through ductile crack growth via coalescence of voids in free-cutting mild steel (micrograph courtesy of J. F. Knott)

where

$$\rho \equiv r_y' \exp\left\{1 + \frac{\alpha}{\beta}\left(\frac{E}{\sigma_0^2}\right)\frac{dJ}{da}\right\} \equiv r_y' \exp\left(1 + \frac{\alpha}{\beta} T_R\right)$$

with $\beta = 5.6$ for a Poisson ratio $\nu = 0.3$. Since the parameter ρ fully characterizes the near tip crack profile, Rice et al. (1980) proposed $\rho =$ constant as a criterion for continued growth, such that

$$\rho = \frac{sEJ_{Ic}}{\sigma_0^2} \exp\left(1 + \frac{\alpha_{ssy}}{\beta} T_0\right) \tag{16}$$

where $\alpha_{ssy} \approx 0.58$ and T_0 is the initial value of the tearing modulus, given by

$$T_0 = \frac{E}{\alpha_{ssy}\sigma_0}\frac{\delta_p}{l_0^*} - \frac{\beta}{\alpha_{ssy}}\ln\left(\frac{esEJ_{Ic}}{l_0^*\sigma_0^2}\right) \quad (17)$$

This implies a general growth criterion:

$$T_R \equiv \frac{E}{\sigma_0^2}\frac{dJ}{da} = \frac{\alpha_{ssy}}{\alpha}T_0 - \frac{\beta}{\alpha}\ln\left(\frac{r_y'}{sEJ_{Ic}/\sigma_0^2}\right) \quad (18)$$

which provides the differential equation governing plane strain tensile crack growth with increase in J from initiation at $J_i = J_{Ic}$ to the steady state value J_{ss} where a plateau in the $J_R(\Delta a)$ curve is reached (i.e., as $dJ/da \to 0$). For small-scale yielding with $\nu = 0.3$, this gives

$$\frac{J_{ss}}{J_{Ic}} = \exp\left(\frac{\alpha_{ssy}}{\beta}T_0\right) = \exp(0.1\,T_0) \quad (19)$$

Although the "constant" criterion for continued mode I crack extension has been found to be consistent with experimental $J_R(\Delta a)$ measurements (in deeply cracked bend tests on AISI 4140 steel), the approach is essentially not microscopically based. A more physically realistic approach is to consider a local fracture criterion, similar to the mode III case, where crack advance is consistent with the attainment of a critical accumulated plastic strain γ_f^* within a microstructurally characteristic radial distance l_0^* from the crack tip. By analogy with mode III, the local criterion $\gamma_p > \gamma_f^*$ over radial distance $r = l_0^*$ is applied for both mode I initiation and continued growth and instability. This yields estimates for the critical plastic zone sizes for mode I crack initiation as

$$r_{yi} \approx l_0^*\left(\frac{1}{\pi}\frac{\bar{\varepsilon}_f^*}{\varepsilon_0}\right) \quad (20a)$$

and for instability as

$$r_{yc} \approx l_0^* \exp\left[\frac{0.6(1+\nu)}{2-\nu}\frac{\bar{\varepsilon}_f^*}{\varepsilon_0}\right] \quad (20b)$$

It should be noted that, similar to the mode III expressions, the critical plastic zone size for the growing crack is an exponential function of the ratio of fracture to yield strain, rather than a direct function for crack initiation. However, unlike mode III, there is no square root dependence in the exponential term. Although the numerical constants are only approximate, this implies that for mode I cracks

$$\frac{J_{ss}}{J_i} \approx \pi\frac{\varepsilon_0}{\bar{\varepsilon}_f^*}\exp\left(0.6\frac{\bar{\varepsilon}_f^*}{\varepsilon_0}\right) \quad (21)$$

Similar to the mode III case, a comparison of the microscopically based relationships for crack initiation and instability in mode I clearly shows that microstructural changes in a material which increase the fracture ductility and decrease the yield strength (i.e., lower ε_0) can have a much larger (beneficial) effect on crack growth toughness as opposed to crack initiation toughness. This can be appreciated by comparing experimental J_{Ic} and $J_R(\Delta a)$ data. For example, Fig. 2 shows $J_R(\Delta a)$ resistance curves for A516 grade 70 steels with different steelmaking processes used to control the inclusion content. It is apparent that the effect of controlling the volume fraction and shape of oxides and sulfides through additional calcium treatments (CaT), compared with conventional vacuum degassing (CON) only, becomes progressively more significant with increasing crack extension. According to the simple modelling analysis described above, this can result simply from the different strain distributions for the stationary and running crack rather than from any change in the local fracture criteria

3. *Relationship Between T_R and J_{Ic}*

Conventionally, correlations between various toughness parameters have been made in purely empirical fashion through regression analysis of experimental data. For example, the many (often dimensionally incorrect) expressions purporting to define relationships between K_{Ic} and Charpy V-notch impact energy have been obtained exactly in this manner. However, the failure criteria reviewed above for both initiating and growing cracks provide an ideal physical basis for examining the relationship between crack initiation and growth toughness parameters without recourse to such purely empirical procedures. Specifically, the model of Rice et al. (1980) of the geometrically similar very near crack tip profile of an extending crack within the asymptotic mode I deformation field of the nonstationary flaw permits a logical correlation of J_{Ic} and the tearing modulus T_R, as previously noted by Shih et al. (1980a). Following Rice et al., the general expression for T_R, when evaluated under small-scale yielding conditions, simplifies to

$$T_R = T_0 - \frac{\beta}{\alpha_{ssy}}\ln\left(\frac{J}{J_{Ic}}\right) \quad (22)$$

Under fully plastic conditions, where the "plastic zone" dimension r_y' is considered to saturate with full yielding at some fraction of the uncracked ligament ($r_y' \approx b/4$), Eqn. (18) becomes

$$T_R = \frac{\alpha_{ssy}}{\alpha_{fp}}\left[T_0 - \frac{\beta}{\alpha_{ssy}}\ln\left(\frac{b/4}{sEJ_{Ic}/\sigma_0^2}\right)\right] \quad (23)$$

where for fully yielded conditions $\alpha_{fp} \approx 0.51$.

Since the parameters α_{ssy}, α_{fp}, β and s have all been determined to a fair degree of accuracy by finite element computations, the major problem in utilizing Eqns. (22, 23) to relate J_{Ic} and T_R reduces to interpreting quantitatively the microscale parameters δ_p and l_0^*. Dean and Hutchinson (1980) suggest that the ratio δ_p/l_0^*, normalized with respect to the yield strain, should exceed 100 for intermediate strength steels.

When δ_p/l_0^* can be related to the microroughness parameter M, this implies M values >0.16 (for 350 MPa yield strength) to 0.50 (for 1000 MPa yield strength). These values are consistent with available experimental data. Rice and Sorensen (1978), on the other hand, equate l_0^* with the fracture process zone size, which for microvoid coalescence is taken to be of the order of the spacing d_p of the void-initiating particles. Since from the analysis of Rice and Johnson (1970) the CTOD at initiation, δ_i, should be in the range 0.5–2.0d_p, it has been suggested that δ_p be regarded as an independent empirical parameter and that l_0^* be taken to be in the region $(0.5\text{–}2.0)\delta_i$. Ordinarily both α_p and l_0^* would be regarded as material parameters, so this implicit assumption of $\delta_p/l_0^*=1$ seems artificially restrictive. In fact, in terms of the fracture surface microroughness, it implies M values of order unity, which appears to be an upper limit for fully plastic conditions unless particles are rare.

To simplify the functional form of the expressions between J_{Ic} and T_R, it is to be noted that the additional CTOD at the advancing crack tip δ_p must be smaller than the CTOD δ_i to cause initiation at the original crack tip (consistent with studies which show far less crack tip blunting associated with the extending, as opposed to the initiating, crack). That is:

$$\delta_p=\frac{\lambda\alpha J_{Ic}}{\sigma_0} \tag{24}$$

where λ is of the order of, yet less than, unity. Incorporating Eqn. (24) into Eqns. (22, 23) gives, for small-scale yielding,

$$T_R=\gamma\left(\frac{J_{Ic}E}{l_0^*\sigma_0^2}\right)-\frac{\beta}{\alpha_{ssy}}\ln\left[es\left(\frac{J_{Ic}E}{l_0^*\sigma_0^2}\right)\right]-\frac{\beta}{\alpha_{ssy}}\ln\left(\frac{J}{J_{Ic}}\right) \tag{25}$$

and for large-scale yielding,

$$T_R=\frac{\alpha_{ssy}}{\alpha_{fp}}\left[\lambda\left(\frac{J_{Ic}E}{l_0^*\sigma_0^2}\right)\right]-\frac{\beta}{\alpha_{ssy}}\ln\left[es\left(\frac{J_{Ic}E}{l_0^*\sigma_0^2}\right)\right]+\frac{\beta}{\alpha}\ln\left[\frac{4sl_0^*}{b}\left(\frac{J_{Ic}E}{l_0^*\sigma_0^2}\right)\right] \tag{26}$$

Using the most recently reported values for the parameters α_{ssy}, α_{fp}, β, e and s from finite element computations, and taking realistically $\lambda\approx0.2$, the variations in T_R values with nondimensional J_{Ic} values predicted from Eqns. (25, 26) are shown in Fig. 8 for conditions of small-scale yielding and full plasticity. For the former, curves are shown at the onset of growth, that is, at $J/J_{Ic}=1$ when $T=T_0$, and when J values are an order of magnitude larger than the initiation value. For full yielding, the T_R versus J_{Ic} expression is evaluated for both 30 mm and 120 mm uncracked ligaments, which represent typical values of b in 25 mm and 100 mm thick precracked compact specimens. However, it is clear that in each case these logarithmic terms have only a marginal effect in the range of values quoted.

Using experimental toughness data for a wide variety of steels ranging from low to high strength (i.e., σ_0/E values from 0.002 to 0.006), Eqns. (25, 26) can be seen from Fig. 8 to provide an excellent basis for comparison of J_{Ic} and T_R, except perhaps at very high tearing modulus values, exceeding 300 or so. It should be noted, however, that in the construction of Fig. 8, the characteristic dimension l_0^*, which was used as a fitting parameter, was assigned a value of 130 μm. Although the basis for this is arbitrary, in keeping with the physical idealization of ductile crack growth, it is apparent that this size should be comparable with the mean inclusion spacing in these pressure vessel steels, such that l_0^* is of the order of d_p. The spacing of all inclusions in such steels is far below 130 μm, but d_p may correspond to the largest, earliest-initiating inclusions. If these have a volume fraction of 2%, for example, their size for a 130 μm spacing would have to be ~25 μm, a large but not unreasonable value.

In this connection, it is important to recognize that d_p must refer, not to a metallographic spacing of all particles in the material, but to the spacing of those particles which are "effective" in the fracture process. The effective particles may include not all which debond from the matrix in crack extension, but only those which nucleate early enough in the strain to fracture to drive the necking, shear localization, or both, of the interparticle ligaments. Thus, both experimental data and analytical predictions show a trend of virtually a linear increase in crack growth toughness T_R with increase in the non-dimensional crack initiation toughness J_{Ic}.

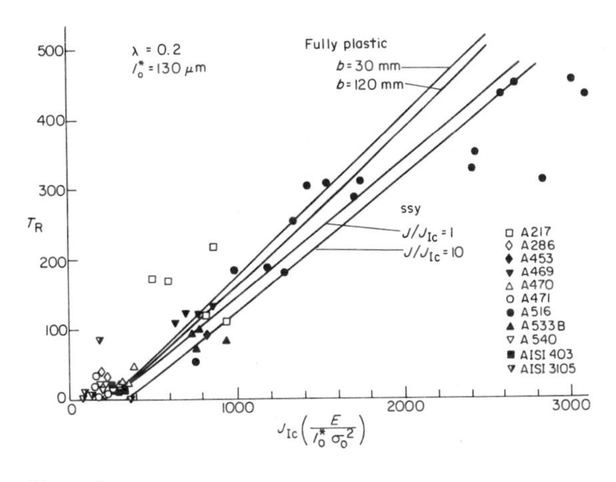

Figure 8
Variation of crack initiation toughness J_{Ic} with crack growth toughness T_R, showing a comparison of theoretical predictions, for both small-scale yielding (ssy) and fully plastic conditions, with experimental toughness data for steels (after Ritchie and Thompson 1985)

The foregoing implicitly assumes that material characteristics determine d_p and δ_p (for a given crack tip stress state), while l_0^* would correspond to a particular fracture micromechanism in that material. It is tempting to guess that l_0^* would typically be a small, integer multiple of a well defined microstructural dimension, but in fact the general case may be a wide range in relative size scales, depending on whether the fracture is locally controlled, as seems to be the case in microvoid coalescence, or is partly regionally controlled, as in blocky fractures, ductile intergranular fractures and quasicleavage. At the present level of understanding, it appears more appropriate to attempt to measure d_p and regard l_0^* as a fitting parameter which both depends on the particular fracture micromechanism and also must exhibit a size scale consistent with that micromechanism and the relevant microstructural features.

4. Conclusion

In this article, the distinction between fracture toughness associated with crack initiation and associated with subsequent slow crack growth has been examined on the basis of differences between the stress and deformation fields local to the crack tip regions of stationary and nonstationary mode I cracks. By considering a critical strain micromechanical model for void coalescence, it was found that metallurgical factors which specifically influence yield strength (σ_0) and local fracture ductility ($\bar{\varepsilon}_f^*$) can have a far greater influence on crack growth toughness (e.g., T_R) than on crack initiation toughness (e.g., J_{Ic}). This result is principally a consequence of the weaker strain singularity ahead of a slowly growing tensile crack and is analogous to early results of McClintock (1958) for mode III. Furthermore, by considering a similar micromechanical model for crack growth based on the attainment of a geometrically similar crack tip profile, a relationship between the tearing modulus T_R and the crack initiation fracture toughness J_{Ic} was described and found to provide a good fit to experimental toughness data on a wide range of steels.

Such analyses highlight the significance of the nonstationary crack tip fields in the modelling of subcritical crack growth where the presence of crack tip plasticity and associated plastic zones in the wake of the crack tip can lead to lower strains at the same nominal driving force (i.e., same K_I or J) than that predicted by the currently used stationary crack analyses. For strain-controlled fracture, this effect results in resistance curve behavior in which an increasing nominal driving force (i.e., increasing K_I or J) must be applied to sustain crack extension, even though the failure criterion can remain unchanged. Furthermore, the enhanced role of microstructure in influencing resistance to crack growth, compared with crack initiation, which follows from this modified strain distribution for slowly moving cracks, implies that fracture toughness, when assessed in terms of crack growth parameters such as the tearing modulus T_R, becomes far more amenable to metallurgical control.

See also: Mechanics of Materials: An Overview

Bibliography

American Society for Testing and Materials 1983 Standard test method for plane strain fracture of metallic materials, E399-83. *1983 Annual Book of ASTM Standards*, Section 3. American Society for Testing and Materials, Philadelphia, Pennsylvania, pp. 170–553

American Society for Testing and Materials 1983 Standard test method for J_{Ic}, a measure of fracture toughness, E813-81. *1983 Annual Book of ASTM Standards*, Section 3. American Society for Testing and Materials, Philadelphia, Pennsylvania, pp. 762–80

Begley J A, Landes J D 1974 *Fracture Toughness*, ASTM STP 514. American Society for Testing and Materials, Philadelphia, Pennsylvania, pp. 170–86

Curry D A 1980 Predicting the temperature and strain rate dependences of the cleavage fracture toughness of ferritic steels. *Mater. Sci. Eng.* 43: 135–44

Dean R H, Hutchinson J W 1980 *Fracture Mechanics, 12th Conf.*, ASTM STP 700. American Society for Testing and Materials, Philadelphia, Pennsylvania, pp. 383–405

Drugan W J, Rice J R, Sham T L 1982 Asymptotic analysis of growing plane-strain tensile cracks in elastic ideally plastic solids. *J. Mech. Phys. Solids* 30: 447–73

Green G, Knott J F 1975 On effects of thickness on ductile crack growth in mild steel. *J. Mech. Phys. Solids* 23: 167–83

Hermann L, Rice J R 1980 Comparison of theory and experiment for elastic–plastic plane–strain crack growth. *Met. Sci.* 14: 285–91

Hutchinson J W, Paris P C 1979 *Elastic–Plastic Fracture*, ASTM STP 668. American Society for Testing and Materials, Philadelphia, Pennsylvania, pp. 37–64

Kameda J, McMahon C J Jr 1980 Solute segregation and brittle fracture in an alloy steel. *Met. Trans. A* 11A: 91–101

Knott J F 1973 *Fundamentals of Fracture Mechanics*. Butterworth, London

Mackenzie A C, Hancock J W, Brown D K 1977 On the influence of state of stress on ductile fracture initiation in high strength steels. *Eng. Fract. Mech.* 9: 167–88

McClintock F A 1958 Ductile fracture instability in shear. *J. Appl. Mech.* 25: 582–88

McClintock F A 1969 Crack growth in fully plastic grooved tensile specimens. In: Argon A S (ed.) 1969 *Physics of Strength and Plasticity*. MIT Press, Cambridge, Massachusetts, pp. 307–26

McClintock F A 1977 In: Jaffee R I, Wilcox B A (eds.) 1977 *Fundamental Aspects of Structural Alloy Design*. Plenum, New York, pp. 147–72

Paris P C, Tada H, Zahoor A, Ernst H 1979 *Elastic–Plastic Fracture*, ASTM STP 668. American Society for Testing and Materials, Philadelphia, Pennsylvania, pp. 5–36

Parks D M 1976 Interpretation of irradiation effects on the fracture toughness of a pressure vessel steel in terms of crack tip stress analysis. *J. Eng. Mater. Tech.* 98: 30–36

Rice J R 1976 In: Erdogan F (ed.) 1976 *The Mechanics of Fracture*. American Society for Mechanical Engineers, Warrendale, Pennsylvania

Rice J R, Drugan W J, Sham T-L 1980 Elastic–plastic analysis of growing cracks. *Fracture Mechanics, 12th*

Conf., ASTM STP 700. American Society for Testing and Materials, Philadelphia, Pennsylvania, pp. 189–221
Rice J R, Johnson M A 1970 In: Kanninen M F, Adler W G, Rosenfield A R, Jaffee R I (eds.) 1970 *Inelastic Behavior of Solids.* McGraw–Hill, New York, pp. 641–72
Rice J R, Sorensen E P 1978 Continuing crack-tip deformation and fracture for plane-strain crack growth in elastic–plastic solids. *J. Mech. Phys. Solids* 26: 163–86
Rice J R, Tracey D M 1969 On the ductile enlargement of voids in triaxial stress fields. *J. Mech. Phys. Solids* 17: 201–17
Ritchie R O, Geniets L C E, Knott J F 1973 *The Microstructure and Design of Alloys, Proc. 3rd Int. Conf. Strength of Metals and Alloys*, Vol. 1. Institute of Metals and Iron and Steel Institute, London, pp. 124–28
Ritchie R O, Knott J F, Rice J R 1973 On the relationship between critical tensile stress and fracture toughness in mild steel. *J. Mech. Phys. Solids* 21: 395–410
Ritchie R O, Server W L, Wullaert R A 1979 Critical fracture stress and fracture strain models for the prediction of lower and upper shelf toughness in nuclear pressure vessel steels. *Met. Trans. A* 10A: 1557–70
Ritchie R O, Thompson A W 1985 On macroscopic and microscopic analyses for crack initiation and crack growth toughness in ductile alloys. *Met. Trans. A* 16A: 237–48
Shih C F 1981 Relationships between the *J*-integral and the crack opening displacement for stationary and extending cracks. *J. Mech. Phys. Solids* 29: 305–26
Shih C F, Andrews W R, Wilkenson J P D 1980a Characterisation of crack initiation and growth with applications to pressure vessel steels. In: Miller K J, Smith R F (eds.) 1980 *Mechanical Behavior of Materials, Proc. 3rd Int. Conf.*, Vol. 3. Pergamon, Oxford, pp. 589–601
Thompson A W 1983 The relation between changes in ductility and in ductile fracture topography: Control by microvoid nucleation. *Acta Met.* 31: 1517–23
Thompson A W, Ashby M F 1984 Fracture surface microroughness. *Scripta Met.* 18: 127–30
Vassilaros M G, Joyce J A, Gudas J P 1983 In: Lewis J C, Sines G (eds.) 1983 *Fracture Mechanics, 14th Symp.*, Vol. 1, *Theory and Analysis*, ASTM STP 791. American Society for Testing and Materials, Philadelphia, Pennsylvania, pp. I-65–I-83
Wilson A D 1979 The influence of inclusions on the toughness and fatigue properties of A516-70 steel. *J. Eng. Mater. Tech.* 101: 265–74

R. O. Ritchie
[University of California, Berkeley, California, USA]

E

Economics of Materials, Energy and Growth

The principal determinants of the relations observed between a country's materials usage, its energy consumption and the production of final goods demanded are access to natural resource endowments and the structure of its economy. Both assume at any time the country's level of capital investment and stage of industrialization are given, as well as the degree of free trade linking its economy to the rest of the world through imports and exports. Over time, changes occur in technologies and in other dynamic factors, so that the structure of the world economies will change. For this reason, a historical review would appear to offer the simplest approximation to and best insights into the changing patterns of materials and energy use as a developed economy grows. This article employs the Leontief structure of the post-war US economy and data on corresponding national accounts from 1947 to 1985 for measures of materials and energy input and final output values. It then looks at future changes and implications, dividing the discussion into three parts. The years from World War II through to 1972 represent a period of unparalleled growth in which increasing free trade in materials and fuels helped to establish stable prices and rather consistent patterns of use among these sectors and the rest of the economy. Beginning in 1973, these patterns were disturbed as both world trading conditions and the organization of the international markets changed significantly. A surge of cartel activity and an increasing number of restrictions on open market exchange reversed the trends toward trade liberalization and growth that prevailed earlier. New technologies appeared, some of them influenced by the threat of resource scarcities. From 1973 to 1985 these changes illustrate the more dynamic factors underlying the structure of the economy, inducing important changes in some materials supplies and demands. The period ahead, through the turn of the century and beyond, may make these changes more pronounced. Some of the implications for policies on materials, energy and growth for the future are explored in this final part.

It is important at the outset to establish the data and the model employed. Statistics on individual consuming and producing industries are taken from the national accounts in the *Survey of Current Business* and the *Census of Manufactures*, both its subject series on fuels and electric energy consumed and its industry series on mineral industries. These reports provide data for the model in the following form. They publish aggregate values of gross industrial output (x), the array of input values as interindustry flows (x_{ij}) from producing to consuming sectors, which form the transactions matrix of the economy (X), and the net final output values (y) of all goods produced in the economy during the given year. The value of the vector of final goods demand y, called Gross National Product, is equal to the sum of values added in the input industries (v). The input–output framework of the economy then permits the construction of input–output relations among the economy's industries (a_{ij}). These are economic analogues to the process engineering estimates of specific materials use rates, and give the array of technical coefficients, $A = X_{ix}^{-1}$, describing the structure of the economy in terms of three accounting identities:

$$x = (I - A)^{-1} y = Ly \qquad (1)$$

$$x = Xi + y \qquad (2)$$

$$x = X^{\mathrm{T}} i + v \qquad (3)$$

where I and i are the identity matrix and unit vector, respectively; x is the vector of 81 sectors defined at the two-digit level of the US national accounts; y is the vector of final demands, and T indicates transpose. The inverse matrix L is the "Leontief matrix" of the economy, defining its structure, so called after its originator, Wassily Leontief.

In the first equation, the national accounts distinguish the gross output of each materials or energy industry (x_i) required to produce its own product (y_i) and the rest of the final demands (y_j). These demands are net of the interindustry contributions ($-x_{ij}$) given by the inputs from the economy as a whole in Eqn. (2). When the vector of input prices and final goods prices is known, the value of demands y is clearly the same as the value of the aggregate inputs v which go to make it up. This identity is shown in Eqn. (3). In Section 1, data from these identities are summarized in Tables 1–3 through to 1972, providing an overview of the materials- and energy-use rates and responses to changes which occurred in market prices and technologies over the 25-year first period. After 1972, a surge of restrictions on open markets left shortages in many supply sectors and excess supplies in others, suggesting an alternative accounting in which the interindustry flows, Ax, are defined proportional to the selling industry's output, rather than to the output of the purchasing industry. The resulting output–input coefficients are written $\boldsymbol{a}_{ij} = x_{ij}/x_i$ to distinguish them from Leontief's input–output coefficients ($a_{ij} = x_{ij}/x_j$). This modifies Eqn. (1) to become:

$$x^{\mathrm{T}} = v^{\mathrm{T}}(\boldsymbol{I} - \boldsymbol{A})^{-1} = v^{\mathrm{T}} G \qquad (1')$$

where G is the array of output–input coefficients. The importance of the energy sector to the economy is

Table 1
The relationship of minerals and energy sectors to the gross national product (GNP), 1958–1972

Industry classification (BEA 2-digit number)	Value added (millions of current dollars) 1958[a]	1963[b]	1967[c]	1972[d]
Materials processing machinery and equipment				
Construction and mining (45); materials handling (46); and metalworking (47) machinery and equipment	3619	5139	7896	9273
Electrical wiring (53); and industrial (55) machinery and equipment	3611	4285	6467	7945
Metal containers (39); heating, plumbing (40); screw machine (41); and other fabricated metal (42) products	8231	10186	15550	19839
	15461	19610	29913	37057
Basic metals manufacturing and mining				
Nonferrous primary metals (38); and mining (6)	3322	4605	6326	7247
Primary iron and steel (37); and mining (5)	8102	10928	12893	15435
	11424	15533	19219	22682
Basic nonmetal manufacturing and mining				
Stone and clay products (36); and mining (9)	4606	5717	6346	9040
Chemicals and selected products (27)	4674	6887	8540	10510
	9280	12604	14886	19550
Energy				
Petroleum refining and related industries (31)	3608	5100	6890	7547
Electricity, gas, water and utilities (68)	9914	13874	17712	25281
Coal mining (7)	1604	1540	1844	3189
Crude petroleum and natural gas (8)	6671	6926	8611	11725
	21797	27440	35057	47742
Total energy and minerals sectors	57962	75187	99075	127031
Total final demand	447334	590389	795388	1182765
Energy and minerals fraction of total demand	0.13	0.13	0.13	0.11

[a] National Economics Division Staff 1965 The transactions table of the 1958 input–output study and revised direct and total requirements data. *Surv. Curr. Bus.* 45(9) [b] National Economics Division 1969 Input–output structure for 1963. *Surv. Curr. Bus.* 49(11) [c] Interindustry Economics Division 1974 The input–output structure of the US economy: 1967. *Surv. Curr. Bus.* 54(2) [d] Ritz P M Roberts E P Young P C 1979 Dollar-value tables for the 1972 input–output study. *Surv. Curr. Bus.* 59(4)

revealed more clearly and analyzed in this fashion in Sect. 2, Tables 4 and 5. In Sect. 3, the implications for the growth of the US and world economies are briefly discussed.

1. Materials and Energy Growth in a Free Trading World: 1945–1972

It took more than a decade after World War II for Europe and Japan to restore their industrial structure, and for the less developed countries to re-establish their primary exports of materials and fuels. The period 1945–1972 saw an unprecedented and continuous expansion in the world in materials trade and use, which took place under relatively open market trade conditions. In the US national accounts defined by Eqns. (1) and (2) at the two-digit level, the domestic mineral and energy sectors comprise 20 of the total of 78 industries. The tables exclude the trade accounts. Materials are defined by the metals numbered 6, 38, 37, 5; the basic nonmetals net of glass and polymers numbered 36, 9, 27; and the materials processing industries numbered 46, 45, 47, 55, 53, 40, 41, 39, 42. Energy is defined by the fossil fuel extractive and refining industries and electricity numbered 31, 68, 7, 8. Twenty-five percent of the structure thus accounts for about 13% of the national product. This share is small and falling (Table 1). Energy or mining alone accounts for less than 4%. For perspective, it is interesting to note that mining was more important in the structure of the 1929 economy than it is in today's much larger and more complex structure.

The influence of energy in the interindustry flow matrix is broadly spread across the economy. This implies that when international ore and energy markets are functioning freely, the lack of ample energy resource endowment will mean little to economic growth. This case is confirmed in the period outside the US by the successful revival of countries without domestic endowments of these resources, such as Japan and much of Europe.

Unlike energy, chemicals and metals processing industries are clustered in the L matrix. To appreciate the importance to the economy of such metals and processing complexes, it is necessary only to observe that together the firms of each group form separable parts of the structure. They are "strategic" because of their forward linkages and the fact that they are found either in an economy functioning together or not at all. Their important role is confirmed again by the experience of post-war growth in Germany, Japan and the US among developed countries. The underdeveloped nations attempted less successfully to stimulate industrialization by building protected materials sectors.

In any case, the interaction between materials industries and energy at both initial and advanced industrialized stages of an economy can be seen summarily in Tables 2 and 3. These give account values for the midpoint (1958) and the end (1972) of the period. The absolute volume of the flows (X) and the Leontief coefficients (in parentheses, A) are also shown. In 1958 these specific flows summed to \$57 billion of inputs required of a \$441 billion economy. By comparison, in 1972, \$127 billion of materials and energy were required for a \$1183 billion economy. The tables confirm a declining share of US materials and energy in the value product. This is especially notable in principal metals industries (e.g., steel). In part this decline is due to the shift to lighter material technologies, such as aluminum or plastic materials, that is, a change in A. In part the decline is due to larger import components of certain metals in fabricated or end-product forms now shown in the tables, for example, steel strip and automobile imports. The relative shares of energy in the materials and processing industries remained constant over the period, despite the significant transformation of most of the relevant technologies. Finally, the economy as a whole grew faster than its materials sectors and this helped to produce declining value added shares in the US.

While the energy coefficients were relatively large for a number of the materials industries, only the exceptional industry, such as aluminum, glass or paper, shows a coefficient high enough to warrant its relocation were fuel prices to suddenly rise. This result stands in contrast to the energy sector's own consumption in the refining, petrochemicals and electric power industries, and suggested to economists at the end of the period of cheap energy in 1973 that higher OPEC tariffs and the rising costs of environmental control would not have severe or lasting impacts on growth or levels of demand outside the energy sector itself. As the period closed, the government sought to mitigate impacts with energy regulations.

2. Materials and Energy in a Restricted Trading World: 1973–1983

Beginning with the 1973 OPEC cartel restrictions on oil exports, a decade of restricted materials and energy supply followed. Energy prices quadrupled. Further, many nontariff barriers to world materials trade surfaced as negotiations lowered tariff walls. The increased distortions in world markets turned economists to the consideration of the problems of materials and energy interrelations in a regime of regulated trade.

The analysis of such a world of manipulated shocks suggests a model which places emphasis on the production relations less in terms of final demands than in terms of the supply availabilities of inputs, as would be typical of technologies forced to favor energy substitutions or materials conservation. Redefining Eqn. (1), using Eqn. (2) and A, yields in Eqn. (1′) the input value added of $v^{\mathrm{T}}G$, where G is a matrix of supply side multipliers relating unit changes in primary inputs to

Table 2
Mineral and energy sectors, services and the costs of energy, 1958 (in 1958 US$ million at producers' prices for energy sector accounts and major interindustry transactions with Leontief coefficients of direct requirements)

		Energy sector account (inputs)[a]						
		Petroleum refining and related industries (31)	Electric, gas, water and sanitary services (68)	Coal mining (7)	Crude petroleum and natural gas (8)	Total energy inputs	Total gross output of industry	Sum of energy sector technical coefficients
Basic metal[b]	Primary nonferrous metal manufacturing (38)	42 (0.0042)	217 (0.0215)	12 (0.0012)		271	10097	0.0269
	Nonferrous metal ores mining (6)	8 (0.0062)	30 (0.0227)	1 (0.0008)		39	1319	0.0296
	Primary iron and steel manufacturing (37)	147 (0.0076)	466 (0.024)	507 (0.0261)		1120	19391	0.0578
	Iron and ferroalloy ores mining (5)	10 (0.0080)	17 (0.0137)	5 (0.004)		32	1245	0.0257
Basic non-metals	Stone and clay products (36)	85 (0.0112)	213 (0.028)	63 (0.0083)		361	7609	0.0474
	Stone and clay mining and quarrying (9)	5 (0.0031)	39 (0.024)	3 (0.0018)		47	1624	0.0289
	Glass and glass products (35)	7 (0.0032)	80 (0.0363)	3 (0.0014)		90	2206	0.0408
	Paper and allied products except containers (24)	121 (0.0115)	189 (0.018)	73 (0.0069)		383	10508	0.0364
	Forestry and fishery products (3)	19 (0.013)	c			19	1451	0.013
	Miscellaneous agricultural products (2)	898 (0.0384)	172 (0.0074)	1		1071	23395	0.0458
	Plastics and synthetic materials (28)	56 (0.0131)	41 (0.0096)	23 (0.0054)		120	4272	0.0281
	Chemicals and selected chemical products (27)	658 (0.0543)	288 (0.0238)	61 (0.005)	24 (0.002)	1031	12107	0.0852
	Chemical and fertilizer minerals mining (10)	5 (0.0089)	24 (0.0426)	c	1 (0.0018)	30	563	0.0533

Energy sector	Petroleum refining and related industries (31)	1243 (0.0691)	262 (0.0146)	10 (0.0006)	9291 (0.5163)	10806	17997	0.6004
	Electric, gas, water, sanitary services (68)	245 (0.0121)	3380 (0.1666)	546 (0.0269)	1166 (0.0575)	5337	20289	0.2630
	Coal mining (7)	30 (0.0109)	65 (0.0236)	471 (0.1711)		566	2752	0.2057
	Crude petroleum and natural gas (8)	52 (0.0048)	78 (0.0072)	c	242 (0.0223)	372	10852	0.0343
Services	Federal government enterprises (78)	6 (0.0015)	73 (0.0178)	47 (0.0114)		126	4105	0.0307
	Transportation and warehousing (65)	1518 (0.0445)	147 (0.0043)	27 (0.0008)		1692	34119	0.0496
	State and local government enterprises (79)	40 (0.0084)	376 (0.0786)	79 (0.0165)	20 (0.0042)	515	4784	0.1077
	Hotels, personal and repair services excluding auto (72)	144 (0.0118)	235 (0.0193)			379	12171	0.0311
	Automobile repair and services (75)	27 (0.0034)	151 (0.0191)	10 (0.0013)		188	7913	0.0238
	Medical and educational services and nonprofit organizations (77)	70 (0.0031)	418 (0.0184)	c		488	22703	0.0215
	Wholesale and retail trade (69)	727 (0.0076)	1916 (0.0201)	4	c	2647	95250	0.0278
	Maintenance and repair construction (12)	375 (0.0222)	25 (0.0015)			400 (0.0237)	16875	(0.0474)
Energy accounts	Gross raw sales	6538	8902	1946	10744	28130	345597	
	% of total sales	36	44	71	99			
	Net product (total final demand)	8836	8929	631	−12			

[a] The two-digit industrial classifications are as given in Bureau of Economic Analysis 1965 *Surv. Curr. Bus.* 45(9) [b] In parentheses below transactions are given the technical coefficients, that is, the amount of inputs required from each industry at the top of the column to produce one dollar's worth of the output of the industry on the left [c] Less than $500 000

Table 3
Mineral and energy sectors, services and the costs of energy, 1972 (in 1972 US$ million at producers' prices for energy sector accounts and major interindustry transactions with Leontief coefficients of direct requirements)

		Energy sector account (inputs)[a]						
		Petroleum refining and related industries (31)	Electric, gas, water and sanitary services (68)	Coal mining (7)	Crude petroleum and natural gas (8)	Total energy inputs	Total gross output of industry	Sum of energy sector technical coefficients
Basic metal[b]	Primary nonferrous metal manufacturing (38)	112 (0.0048)	618 (0.0265)	11 (0.0005)		741	23353	0.0363
	Nonferrous metal ores mining (6)	25 (0.0113)	78 (0.0353)	c		103	2208	0.0466
	Primary iron and steel manufacturing (37)	173 (0.0049)	1223 (0.0346)	763 (0.0216)		2159	35380	0.061
	Iron and ferroalloy ores mining (5)	13 (0.01)	82 (0.0633)	4 (0.0031)		99	1295	0.0764
Basic non-metals	Stone and clay products (36)	147 (0.0096)	500 (0.0325)	93 (0.006)		740	15374	0.0481
	Stone and clay mining and quarrying (9)	86 (0.0295)	106 (0.0364)	2 (0.0007)		194	2911	0.0666
	Glass and glass products (35)	20 (0.0035)	223 (0.0395)	2 (0.0004)		245	5642	0.0434
	Paper and allied products except containers (24)	223 (0.0115)	579 (0.0298)	109 (0.0056)		911	19425	0.0469
	Forestry and fishery products (3)	49 (0.0218)	1 (0.0004)			50	2245	0.0222
	Miscellaneous agricultural products (2)	877 (0.0257)	248 (0.0073)	1		1126	34163	0.033
	Plastics and synthetic materials (28)	80 (0.0084)	208 (0.022)	43 (0.0045)	10 (0.0011)	341	9468	0.036
	Chemicals and selected chemical products (27)	279 (0.0108)	1184 (0.0458)	110 (0.0043)	57 (0.0022)	1630	25869	0.063
	Chemical and fertilizer minerals mining (10)	13 (0.0264)	69 (0.1402)	1 (0.002)		83	492	0.1687

Energy sector	Petroleum refining and related industries (31)	2332 (0.0761)	638 (0.0208)	35 (0.0011)	14981 (0.4886)	17986	30664	0.5866
	Electric, gas, water, sanitary services (68)	1370 (0.0235)	11382 (0.195)	2308 (0.0395)	3686 (0.0631)	18746	58383	0.3211
	Coal mining (7)	78 (0.0143)	110 (0.0202)	656 (0.1206)		844	5439	0.1552
	Crude petroleum and natural gas (8)	66 (0.0039)	235 (0.014)		596 (0.0356)	897	16730	0.0536
Services	Federal government enterprises (78)	51 (0.0053)	241 (0.0252)	171 (0.0179)		463	9568	0.0484
	Transportation and warehousing (65)	2610 (0.0341)	580 (0.0076)	1	32 (0.0004)	3223	76618	0.0421
	State and local government enterprises (79)	249 (0.089)	1463 (0.5227)	138 (0.0493)		1850	2799	0.661
	Hotels, personal and repair services excluding auto (72)	325 (0.0107)	902 (0.0298)	8 (0.0003)		1235	30291	0.0408
	Automobile repair and services (75)	181 (0.0074)	150 (0.0061)			331	24551	0.0135
	Medical and educational services and nonprofit organizations (77)	429 (0.0051)	1728 (0.0204)	21 (0.0002)		2178	84900	0.0257
	Wholesale and retail trade (69)	1543 (0.0071)	3280 (0.0150)			4823	218236	0.0221
	Maintenance, repair and construction (12)	1137 (0.0312)	65 (0.0018)			1202	36417	0.033
Energy accounts	Gross raw sales	12468	25893	4477	19362	62200	772421	
	% of total sales	41	44	82	116[d]			
	Net product (total final demand)	12905	24406	771	−2634			

[a] The two-digit industrial classifications are as given in Bureau of Economic Analysis 1974 *Surv. Curr. Bus.* 54(2) [b] In parentheses below transactions are given the technical coefficients, that is, the amount of inputs required from each industry at the top of the column to produce one dollar's worth of the output of the industry on the left [c] Less than $500 000 [d] Due to positive net import balance

the sellers' gross output. Table 4 from Giarratani (1976) applies such an interindustry supply model to energy issues showing that when total supply multipliers are calculated and ranked by size, the extractive energy sectors become very crucial. Coal and crude petroleum and natural gas rank ninth and eleventh in the top twenty of the eighty-two sectors in the structure. The importance of the materials and energy supply is reinforced by the finding that ten of the top twenty are basic fuels, metals and nonmetals industries. True, this framework also confirms that the materials sectors do not have large impacts on fuel production. Nevertheless, the direct and indirect multiplier effects or "elasticity" of energy impacts on supply are very high, approximately 1.4 using the Bureau of Economic Analysis (BEA) 1974 input–output matrix. In confirmation of this analysis Table 5 shows four of the five largest major industrial users of purchased fuels and energy are chemicals, fuel products, stone, clay, glass and primary metals. In 1977 these absorbed 62 vol% of the purchased fuels sold to all industries.

Clearly, then, restrictions of the free access to fuels have a very large impact on the materials and processing sectors as well as on the use of energy itself, and rises in energy prices such as those which have occurred since 1973 can be expected to force greater changes in the competitiveness and technologies of the materials producers than on other industries.

3. Some Implications for the Future: 1989–2000 and Beyond

Development economists have long noted the various stages of economic growth and their effect on the distribution of sectoral output and final demands. More recent studies of materials and energy demands confirm that materials sectors may experience rising and then declining shares as they mature. The contrasting Japanese and US consumption rates for materials also offer examples of how the dynamic growth of support industries as well as the shift to service sectors over time may affect materials-use rates at different stages of maturity. Exchange rate distortions will affect such trends. However, the disparate rates of growth over the past twenty years in lighter materials such as plastics or aluminum in contrast to heavier metals such as steel (all relatively intensive in energy use) show clearly that technological changes and time permit no easy generalizations to laws of demand about materials consumption rates. Indeed, differential growth in productive capacities and switches in international trade flows are subject to myriad real and monetary influences. These increase the difficulty of relating trends in materials propensities to consume primarily to stages of growth or the composition of final demands. Economists dispute the basic claim that a country's growth in income per capita can be disassociated from either materials intensive or energy

Table 4
Total supply multipliers of the twenty top-ranked US sectors (GNP), 1974[a,b,c]

Sector number		Total supply multipliers	Rank
5	Iron and ferroalloy ores mining	4.0106	1
6	Nonferrous metal ores mining	3.9546	2
67	Radio and TV broadcasting	3.5115	3
10	Chemical and fertilizer mineral mining	3.4993	4
28	Plastic and synthetic materials	3.2469	5
38	Primary nonferrous metal manufacturing	3.2116	6
4	Agriculture, forestry and fishery services	3.1879	7
27	Chemical and chemical products	3.1438	8
7	Coal mining	3.1285	9
37	Primary iron and steel manufacturing	3.1158	10
8	Crude petroleum and natural gas	3.0927	11
9	Stone and clay mining, and quarrying	3.0496	12
24	Paper and allied products, except containers	2.9487	13
3	Forestry and fishery products	2.9433	14
30	Paints and allied products	2.9382	15
16	Broad and narrow fabrics	2.9080	16
21	Wooden containers	2.8502	17
50	Machine shop products	2.8491	18
25	Paperboard containers and boxes	2.8068	19
20	Lumber and wood products, except containers	2.8012	20

[a] Range of total supply multipliers: the largest is iron and ferroalloy ores mining, value 4.0106; the smallest is medical, educational services and nonprofit organizations, value 1.0922 [b] Source: Giarratani 1976 *Environ. Plann.* 8(4): 44
[c] Using the Leontief interindustry matrix for 1967, BEA

Table 5
Purchased fuels and electric energy consumed by major industry groups, 1977[a,b]

SIC Code	Major group	Quantity (10^{12} J)	Total cost (million dollars)
	All industries	12929.0	33379.6
20	Food and kindred industries	952.4	2537.8
21	Tobacco products	20.8	63.9
22	Textile mill products	339.2	1136.6
23	Apparel and other textile products	65.6	285.8
24	Lumber and wood products	227.7	726.5
25	Furniture and fixtures	52.7	198.3
26	Paper and allied products	1308.4	2960.4
27	Printing and publishing	92.3	421.7
28	Chemicals and allied products	2978.5	6448.7
29	Petroleum and coal products	1303.4	2445.4
30	Rubber and miscellaneous plastics products	272.3	962.4
31	Leather and leather products	23.1	82.6
32	Stone, clay, and glass products	1251.8	2587.4
33	Primary metal industries	2539.4	7043.6
34	Fabricated metal products	395.3	1348.7
35	Machinery, except electrical	339.6	1268.0
36	Electric and electronic equipment	249.3	986.7
37	Transportation equipment	389.9	1402.6
38	Instruments and related products	78.4	272.5
39	Miscellaneous manufacturing industries	49.1	199.9

[a] Source: *Census of Manufacturers* 1978, Subject Series, Fuels and Electric Energy Consumed, pp. 4–9 [b] In Btu 10^9 and $\$10^6$ for two-digit SIC sectors

intensive industrial growth. Some conjecture that the ability to support ever-increasing service sector values may require continuing comparative advantages and strengths in the materials and processing industries. Notice is given to the fact that the metals, materials and chemical industrial complexes are decomposable in the X and A matrices of most economies.

Thus, a country tends either to have the entire set of closely linked metals industries or to lack them entirely. This implies that in the dynamics of growth there are technological or other externalities and economies defined in these industries at the microlevel which are less discernible at the macrolevel. Such a finding supports concern over the decline of important materials and energy processing sectors in developed countries such as Germany, France and the United States, and prompts the adoption of industrial policies that restrict free trade.

In the materials sector some countries maintain high subsidies or nontariff barriers to trade, and these restrictions combined with aggressive export policies can warp the growth of domestic materials sectors in other countries more open to free trade. The implications of such asymmetries in trade and restrictions in the materials and energy markets are significant for the economic analysis of commercial strategies in both the private and public policy arenas.

On the energy side, the dynamics of energy substitution and conservation have been debated by economists as avidly as the materials substitution issues. The most recent input–output analyses of world materials use and forecasts suggest that trade restrictions, depletion and technological changes can effect notable differences in observed regional propensities to produce given materials, not to mention major differences in final investment and consumption demands. The range of such predictions based on energy models and different scenarios in the United States forecast of materials use is very great. It is important to note that the applied models of the sort most useful to the analysis of materials, energy and economic policy are largely static. This reflects a large degree of ignorance about the dynamics of technical innovation and engineering design and the influence of such changes in the future on the maintenance of national comparative advantages or national economic growth. A simple example from automotive engineering is illustrative. The rise in gasoline prices and concerns over exhaust emissions motivate current research on new automotive engines by the industry, which portends a significant shift toward higher-temperature combustion. This would greatly increase fuel efficiency and reduce some residuals emissions, but would require a shift away from conventional engine materials. The switch would favor the use of ceramic materials with higher heat resistance than currently used metallic materials. The expected result would reduce weight and improve performance while lowering operating costs. The elimination of conventional metals in the course of such system redesign, however, cannot be predicted by comparing static input–output coefficients at the two-digit industry level or by observing the trend in aggregate cost indices employed by economists to measure technical progress. Finally, such switches cannot become industrially feasible without a vital metals, materials and tool manufacturing complex operating within the economy.

See also: Materials Economics, Policy and Management

Bibliography

Ghosh A 1958 Input–output approach to an allocative system. *Econometrica* 25: 58–64
Giarratani F 1976 Application of an interindustry supply model to energy issues. *Environ. Plann.* 8: 447–54
Leontief W 1966 *Input–Output Economics.* Oxford University Press, New York

R. Newcomb
[University of Arizona, Tucson, Arizona, USA]

Elasticity

Idealizing the deformation of materials into several pure modes for which analytical solutions are available gives insight into practical problems, as discussed in the article *Time-Independent Plasticity.* Four such modes are considered in this and the articles *Plasticity, Creep* and *Linear Viscoelasticity.* For the elastic idealization, the deformation is assumed to be small enough and applied for a short enough time that it disappears immediately on release of load and restoration of temperature. Even such small, elastic deformation can cause failure by misalignment, jamming, vibration or buckling. Avoiding yielding (permanent deformation) often limits elastic strains to 0.1–1.0%. Such elastic strains, applied thousands or millions of times, can cause fracture by the initiation and growth of cracks. Finally, elasticity characterizes the overall behavior of some structures that fracture under a single load because of cleavage fracture mechanisms or microscopically ductile hole-growth mechanisms, with large enough cracks in large enough structures.

For these applications, the strains and rotations are small compared with unity. The resulting definitions of strain in terms of displacements, along with the equations of equilibrium in terms of stress gradients (see *Small Strain, Stress and Work*) apply to these problems. Small-strain elasticity is distinguished from the inelastic phenomena (discussed in other articles) by the relations between stress and strain. These are needed, along with the boundary conditions, to specify solutions to problems in continuum mechanics. The physical basis for the relations and the parameters involved, the general character of the solutions to problems of interest, and where to find solutions by analytical, approximate, numerical or experimental methods are described.

1. Linear Elastic Relations for Changes in Strain, Stress and Temperature

Strictly speaking, linear elasticity deals only with changes from some reference state because there may be initial (residual) stress or strain, as in a rolled steel bar.

1.1 Isotropic Materials

A material which has the same behavior in any direction, such as glass or a polycrystal with random orientations of the individual crystals, is called isotropic. Linearity and symmetry arguments then lead to expressions for the change in strain ε due to changes in temperature T in terms of a modulus of elasticity E, a Poisson's ratio ν, a nearly constant coefficient of linear thermal expansion α and a stress σ:

$$\left.\begin{aligned}\Delta\varepsilon_{xx}&=\frac{\Delta\sigma_{xx}}{E}-\frac{\nu\Delta\sigma_{yy}}{E}-\frac{\nu\Delta\sigma_{zz}}{E}+\alpha\Delta T\\ \Delta\varepsilon_{yy}&=-\frac{\nu\Delta\sigma_{xx}}{E}+\frac{\Delta\sigma_{yy}}{E}-\frac{\nu\Delta\sigma_{zz}}{E}+\alpha\Delta T\\ \Delta\varepsilon_{zz}&=-\frac{\nu\Delta\sigma_{xx}}{E}-\frac{\nu\Delta\sigma_{yy}}{E}+\frac{\Delta\sigma_{zz}}{E}+\alpha\Delta T\end{aligned}\right\}\quad(1a)$$

(Subscripts correspond to components of displacement and gradient; see *Small Strain, Stress and Work.*) Frequently the initial, or reference state can be taken as zero. Then the operator Δ can be dropped from Eqns. (1) except in the thermal expansion terms. Poisson's ratio ν is the negative of the ratio of lateral to axial strain under uniaxial stress.

The changes in shear components of strain in an isotropic material are found by symmetry and linearity arguments to be independent of changes in normal stress or thermal expansion. In terms of a shear modulus G,

$$\begin{aligned}\Delta\varepsilon_{yz}&\left(=\frac{\Delta\gamma_{yz}}{2}\right)=\frac{\Delta\sigma_{yz}}{2G}\\ \Delta\varepsilon_{zx}&=\frac{\Delta\sigma_{zx}}{2G}\\ \Delta\varepsilon_{xy}&=\frac{\Delta\sigma_{xy}}{2G}\end{aligned}\quad(1b)$$

(The quantity γ is the engineering shear strain.) Isotropy implies that rotating the axes through 45° should give the same result, requiring that $G=E/[2(1+\nu)]$.

The ratio of dilational stress $\sigma\equiv\Sigma_k\sigma_{kk}/3$ to strain $\varepsilon\equiv\Sigma_k\varepsilon_{kk}$ is the bulk modulus $K=E/[3(1-2\nu)]$. The value of K is also the negative of the pressure per unit volumetric strain:

$$K \equiv \frac{\sigma}{\varepsilon} = -\frac{p}{\Delta V/V} \tag{2a}$$

In terms of stress and strain deviators, $\sigma'_{ij} \equiv \sigma_{ij} - \sigma\delta_{ij}$ and $\varepsilon'_{ij} \equiv \varepsilon_{ij} - (\varepsilon/3)\delta_{ij}$ (see *Small Strain, Stress and Work*), the deviatoric components of small elastic strain, both normal and shear, are

$$\varepsilon'_{ij} = \frac{\sigma'_{ij}}{2G} \tag{2b}$$

Equations (2) are simpler and more physical than Eqns. (1). For instance, for rubber the bulk modulus is similar to that of liquid polymers, and only a factor of 10 less than that of stiff glassy polymers. The shear modulus of rubber, on the other hand, is several orders of magnitude less than that of glassy polymers. Equations (2) are also useful for finding stress in terms of strain by expressing stress in terms of its dilational and deviatoric values, introducing Eqns. (2) for the dilatation and strain deviator, and converting back to strains:

$$\sigma_{ij} = 2G\varepsilon_{ij} + (K - \frac{2}{3}G)\varepsilon_{kk}\delta_{ij} \tag{3}$$

1.2 Anisotropy

If the behavior of a material differs with direction, it is called anisotropic. Sometimes the shape of the part makes it convenient to choose coordinate axes which show no material symmetry. Then, imposing any component of strain produces all components of stress as reactions. Crystallographers use a notation in which the six independent components of stress and strain are each numbered 1 through 6:

$$\begin{aligned}
\sigma_{11} &= C_{11}\varepsilon_{11} + C_{12}\varepsilon_{22} + C_{13}\varepsilon_{33} + C_{14}\gamma_{23} \\
&\quad + C_{15}\gamma_{31} + C_{16}\gamma_{12} \\
\sigma_{22} &= C_{21}\varepsilon_{11} + C_{22}\varepsilon_{22} + \ldots + C_{24}\gamma_{23} + \ldots + C_{26}\gamma_{12} \\
\sigma_{33} &= C_{31}\varepsilon_{11} + C_{32}\varepsilon_{22} + \ldots + C_{36}\gamma_{12} \\
\sigma_{23} &= C_{41}\varepsilon_{11} + C_{42}\varepsilon_{22} + \ldots + C_{46}\gamma_{12} \\
\sigma_{31} &= C_{51}\varepsilon_{11} + C_{52}\varepsilon_{22} + \ldots + C_{56}\gamma_{12} \\
\sigma_{12} &= C_{61}\varepsilon_{11} + C_{62}\varepsilon_{22} + \ldots + C_{66}\gamma_{12}
\end{aligned} \tag{4a}$$

The constants C are variously called components of elastic stiffness, elastic constants or elastic moduli. The general anisotropic stress–strain relation (4a) can be expressed more concisely in terms of the tensor components of strain and the stiffness components C_{ijkl} as

$$\sigma_{ij} = \sum_{k=1}^{3}\sum_{l=1}^{3} C_{ijkl}\varepsilon_{kl} \tag{4b}$$

Term-by-term comparison of Eqns. (4) and the symmetries $\sigma_{ij} = \sigma_{ji}$ and $\varepsilon_{ij} = \varepsilon_{ji}$ show that, for example, $C_{24} = C_{2223}$. Equality of cross-derivatives of a strain energy function and the above symmetries of stress and strain give relations such as $C_{61} = C_{16}$, making the tensor of elastic stiffness symmetric and reducing the number of independent components to 21.

The components of stiffness C_{ijkl} transform with coordinate axes according to the rule for fourth-order tensors in terms of the direction cosines:

$$C_{i'j'k'l'} = \sum_{i=1}^{3}\sum_{j=1}^{3}\sum_{k=1}^{3}\sum_{l=1}^{3} C_{ijkl} l_{i'i} l_{j'j} l_{k'k} l_{l'l} \tag{5}$$

When both elastic and plastic deformation are present, the strains turn out to be additive. It is then easier to work with expressions giving the components of strain in terms of those of stress. A general linear relation for strain in terms of stress and temperature change involves coefficients of elastic compliance:

$$\begin{aligned}
\varepsilon_{11} &= S_{11}\sigma_{11} + S_{12}\sigma_{22} + \ldots + S_{14}\sigma_{23} + \ldots + \alpha_1\Delta T \\
\varepsilon_{22} &= S_{21}\sigma_{11} + S_{22}\sigma_{22} + \ldots + S_{24}\sigma_{23} + \ldots + \alpha_2\Delta T \\
\varepsilon_{33} &= S_{31}\sigma_{11} + \ldots + \alpha_3\Delta T \\
\gamma_{23} &= S_{41}\sigma_{11} + \ldots + S_{44}\sigma_{23} + \ldots + \alpha_4\Delta T \\
\gamma_{31} &= S_{51}\sigma_{11} + \ldots + \alpha_5\Delta T \\
\gamma_{12} &= S_{61}\sigma_{11} + \ldots + \alpha_6\Delta T
\end{aligned} \tag{6a}$$

or

$$\varepsilon_{ij} = \sum_{k=1}^{3}\sum_{l=1}^{3} S_{ijkl}\sigma_{kl} + \alpha_{ij}\Delta T \tag{6b}$$

The tensor components of compliance S_{ijkl} transform as fourth-order tensors and have the same symmetries as the components of stiffness C_{ijkl}. The components of thermal expansion α_{ij} transform as do the components of strain. The tensor and crystallographers' components of compliance are related by equations such as

$$\begin{aligned}
S_{1111} &= S_{11}, \qquad S_{1122} = S_{12} \\
S_{1123} &= \frac{1}{2}S_{14}, \qquad S_{2331} = \frac{1}{4}S_{45}
\end{aligned} \tag{7}$$

An orthotropic material such as wood or rolled steel sheet, referred to its three mutually perpendicular planes of symmetry, has just nine independent components of stiffness or compliance: six relating normal components of stress and strain, and three relating shear. For wires or rolled rods with x_3 axial, the independent components of compliance are further reduced in number to five by $S_{22} = S_{11}$, $S_{23} = S_{13}$, $S_{44} = S_{55}$, and $S_{66} = 2S_{11}(1 - S_{12}/S_{11})$. Cubic crystals have just three independent elastic constants S_{11}, S_{12} and S_{44}. For cubic crystals, Eqn. (5) can be simplified to give a component of compliance in a new direction, $S_{i'j'k'l'}$, in terms of the corresponding (possibly zero) component in the crystallographic direction S_{ijkl} (where $i' = i$, $j' = j$ etc.):

$$\begin{aligned}
S_{i'j'k'l'} &= S_{ijkl} \\
&\quad + (S_{11} - S_{12} - S_{66}/2)\left[\left(\sum_{m=1}^{3} l_{mi'} l_{mj'} l_{mk'} l_{ml'}\right)\right. \\
&\quad \left. - \delta_{i'j'}\delta_{j'k'}\delta_{k'l'}\right]
\end{aligned} \tag{8}$$

1.3 Values of Elastic Constants and their Physical Basis

Table 1 gives typical room-temperature values of the elastic constants E, G, K, v, and α, as well as the density ρ, which is used in designing for minimum weight, against vibration, and in finding the speed of sound. Within a given category, properties vary roughly in order of listing of the materials. The constants for polymers vary widely with crystallinity; the values given are for polyethylene, but cover the ranges of polystyrene, polymethylmethacrylate, nylon and polycarbonate. Polymers are also strongly affected by plasticizers (which soften them) and fillers, such as the chopped glass fibers used in some molded plastics (see Table 1). Finally, polymers are rate-dependent (see *Linear Viscoelasticity*).

In solids with metallic or covalent bonds, the elastic constants ultimately depend on the quantum-mechanical interactions of electrons with each other and with ions containing inner shell electrons. In view of the scarcity and complexity of calculations from the electronic structure, it is often useful to think of the interatomic forces as arising from a very stiff repulsive potential, associated with deformation of the ionic core, added to a much softer attractive potential associated with electrostatic or van der Waals forces. The dilatational stress is proportional to the derivative of the potential energy with respect to volume or to radius. Starting from the equilibrium radius r_0, the stress increases until the cohesive strength is reached, at the point of inflection of the potential energy curve. The bulk modulus is proportional to the slope of the stress–strain curve at the equilibrium spacing, and hence to the minimum radius of curvature of the potential.

The potential can be expressed approximately in terms of a binding energy per unit mass u_0 and power-law potentials with attractive and repulsive exponents n_a and n_r, respectively:

$$u = -u_0 \frac{n_r n_a}{n_r - n_a}\left[\frac{1}{n_a}\left(\frac{r_0}{r}\right)^{n_a} - \frac{1}{n_r}\left(\frac{r_0}{r}\right)^{n_r}\right] \tag{9}$$

Because the shapes of the interatomic potentials are similar, with repulsive exponents n_r of the order of 10 and attractive exponents n_a of 1–4, a number of useful conclusions can be drawn from the potential model.

(a) The equilibrium spacing r_0 is largely determined by the repulsive potential and hence by the size of the ionic core. The size of the core increases only slowly with the atomic number because the inner electron shells are compressed by the outer ones.

(b) The strain at the cohesive strength is of the order of 0.1, small compared with unity, but large compared with the typical strains of 0.001–0.01 at which elastic behavior in most solids is terminated by plastic flow or fracture. For such small elastic strains, the bulk modulus is nearly constant and the elastic stress–strain relations are nearly linear.

(c) The cohesive strength is typically about 0.05 of the bulk modulus.

(d) The bulk modulus is roughly proportional to the binding energy per unit volume ρu_0:

$$K/\rho u_0 = 2{:}4$$

This ratio holds for a number of liquids as well as solids (except for methanol and water, for which $K/\rho u_0 = 0.9$). Thus information on the binding energy (or heat of sublimation or vaporization) can give a rough estimate of the compressibility of, for example, oil in a hydraulic system.

Table 1
Room-temperature values of elastic properties for nominally isotropic materials and fibers (values vary in order of listing of materials)

	Young's modulus E ($GN\,m^{-2}$)	Shear modulus G ($GN\,m^{-2}$)	Bulk modulus K ($GN\,m^{-2}$)	Poissons ratio v	Coefficient of linear thermal expansion ($10^{-6}\,°C^{-1}$)	Density ρ ($10^3\,kg\,m^{-3}$)
Diamond, tungsten carbide, tungsten, alumina	1100–350	450–140	600–200	0.18–0.3	1–6	3.5–19
Steels	190–220	73–82	140–170	0.26–0.3	9–18	7.7–8.0
Aluminum, glass	79–66	26–30	79–37	0.34–0.2	24–4	2.9–2.2
Fibers:						
graphite, Kevlar, glass	400–70					1.4–2.6
Polymers	0.1–3.7	0.03–1.4	0.4–5	0.46–0.3	220–50	0.9–1.4
(10–35% glass fiber)	3.4–10				15–60	1.2–1.5
Rubber	0.0008–0.004	0.0003–0.0014	1.4	0.50	39–60	1.0–1.3
Foam rubber	$10\text{–}100 \times 10^{-6}$	4.40×10^{-6}	$10\text{–}100 \times 10^{-6}$	0.2–0.4		0.01–0.1

The potential model of Eqn. (9) is also useful in more qualitative ways. The thermal expansion of materials can be pictured by considering the potential well around the equilibrium spacing. At absolute zero, the atoms reside at the bottom of the potential well, with energy, $-u_0$. With increasing temperature, the atoms oscillate about the bottom of the potential well. Since the potential well is asymmetric, with a shallower slope for radii larger than r_0, the atoms return more slowly from larger volumes than from smaller ones. Thus, on average, atoms spend more time at the larger volume, giving an average increase in specific volume at temperatures above absolute zero.

The decrease in bulk modulus with temperature can be pictured in a similar way. At larger average equilibrium spacings, the curvature of the potential well is less, so the bulk modulus is less. The resulting change in the elastic constants with temperature can be normalized by dividing the bulk and shear moduli by their values at absolute zero, K/K_0 and G/G_0, and by dividing the temperature by the melting point temperature T/T_m. The normalized moduli drop nearly linearly with normalized temperature, from about unity at the Debye temperature (nearly room temperature for most solids, which have values of E/ρ close to those of aluminum or steel) to $K/K_0 \approx 0.8$ and $G/G_0 \approx 0.5$ just before melting.

Central-force potentials such as Eqn. (9), in crystal structures in which the strains between all atoms are those of the crystal as a whole, follow Cauchy's relations between the elastic constants. For isotropy this means that Poisson's ratio is exactly 0.25. From Table 1, the central-force model does not hold for most materials (but it nearly does for ionic solids such as common salt). Foams and glasses, which have lacy structures on the macroscopic and molecular scales, respectively, have low values of Poisson's ratio $\nu \approx 0.2$. When $\nu = 1/3$, the relationship $K = E/[3(1-2\nu)]$ shows that $K = E$. When the distortional or shear modulus drops toward zero, as for rubber and the softer polymers, $E \to 0$, but the bulk modulus drops only toward that of liquids, so Poisson's ratio $\nu \to 0.5$.

The distortional stiffness of rubber arises not from interatomic forces, but from the entropy of its thermally moving long-chain molecules, with thousands of molecules between the cross-links that form a permanent network. (Were it not for the cross-linking, rubber would be a liquid, and its bulk modulus is comparable to that of organic liquids at room temperature.) Random thermal motion tends to kink the chains, so the distance between cross-links is some average random spacing. Applying a distortion stretches this distance to a longer, less likely configuration. The resistance to such distortion can be derived by using the relation between entropy and the probability of a configuration, the central limit theorem of statistics, and a random walk model of the positions of the links in a chain segment between cross-links. The result is given in terms of the molecular weight M_c of the chain segment between cross-links, the gas constant per mole R, the absolute temperature T, and the mass density ρ:

$$G = \frac{\rho R T}{M_c} \qquad (10)$$

In contrast to crystals, the shear modulus of rubber increases with temperature. Hard rubber has a higher volume fraction of cross-linking agent, decreasing the average molecular weight between cross-links.

Although the physical basis of the bulk modulus of liquids is similar to that of solids, the resistance to distortion drops to zero. The distortional or deviatoric components of stress depend only on the rate of distortion (see *Creep; Linear Viscoelasticity*).

While the components of elasticity have been calculated from quantum mechanics for some metals, the best values are still empirical, usually obtained from the sound velocity.

The minimum size limit to homogeneous elasticity is the atomic dimension for single crystals, but for polymers, composites or textiles with molecular, fiber or yarn diameter d_f, the minimum size for homogeneous elasticity is some fraction of the shear lag length, $\approx d_f(E_{chain}/G_{transverse})^{1/2}$.

The limiting elastic strain before yield or fracture is typically much less than the ideal cohesive strain calculated from Eqn. (9); most structural solids yield or fracture at a maximum strain of $\varepsilon \approx 0.001$–$0.01$, while a few very strong fibers reach $\varepsilon = 0.03$–0.04. Rubber's behavior becomes nonlinear at about $\varepsilon = 0.1$.

2. Characteristics of Solutions

The equilibrium and strain-displacement equations, which hold for all materials, together with the linear elastic stress–strain relations, provide as many equations as unknowns. In the typical three-dimensional case, the boundary conditions required for a solution consist of one normal and two shear components of either traction or displacement at each point. More general friction or compliant boundary conditions (e.g., load-distributing clamping bars across the ends of a sheet under tension) involve $3N$ independent relations between one normal and two shear components on one or more of the N boundary elements.

2.1 Uniqueness

Uniqueness implies that if insight can be used to find a solution, it will be the only solution. Small-strain linear elasticity is unique with certain limitations.

First, if the deflection of the body affects the stress distribution, not only must the boundary conditions be applied to the deformed shape of the body, but also the solutions may not be unique. For example, a column may buckle to either side. (Mathematically, two or more solutions become possible: the solution bifurcates.) Note that buckling may be unstable under given loads, but stable under given displacements.

Second, only the increments of stress or strain due to increments of applied load are unique; different residual stresses give different final stresses for the same load.

Third, the in-plane components of stress are identical for plane strain and plane stress for all isotropic materials subject to the same load boundary conditions, if there is no net load on any closed boundary (thus excluding loaded rivet holes). This greatly broadens the utility of the tests on models in experimental mechanics.

2.2 Linearity and Superposition

Small-strain elasticity is linear in the sense that multiplying all components of the stress, strain, displacement and boundary conditions by the same constant provides another solution. Furthermore, the sum of the stresses etc., of two such solutions is also a solution. Thus the effects of the loads on a structure can be determined one at a time and then scaled and superimposed, simplifying the study of various combinations of loads that occur in practice. This summing is valid for linear elasticity and viscoelasticity, but not for plasticity and creep.

2.3 St Venant's Principle and Stress Concentrations

Mathematically, the elasticity equations are elliptical, which means that all boundary conditions must be satisfied before any part of the solution can be found. In turn, the stress distribution at any one point is at least somewhat affected by the boundary conditions at all points of the surface. In practice, the effects may be slight, according to St Venant's principle, which states that, if a set of self-equilibrating forces is applied to a small region of a body, say of extent d, then the effect of these forces on the stress is important over a region no more than $2d$ or $3d$ in extent. It is this principle that allows the stress-concentration factors to be described separately from the general geometry of the structure in which they occur, and that allows the stress distribution around a crack to be described in terms of stress-intensity factors that are simply scalar quantities, rather than requiring a separate and distinct stress distribution for each external shape of the boundary and each kind of applied load. For instance, in an infinite plane solid with an applied load normal to an elliptical hole of half length a and tip radius ρ, the stress concentration, defined as the ratio of maximum local to net applied stress, is

$$K_t \equiv \frac{\sigma_{\text{loc}}}{\sigma_{\text{app}}} = 1 + 2\left(\frac{a}{\rho}\right)^{1/2} \qquad (11)$$

For other shapes of holes or notches the stress concentrations can be roughly estimated from Eqn. (11) by taking a to be a typical minimum notch or ligament dimension and the factor multiplying $(a/\rho)^{1/2}$ to be as low as 0.3 for mild stress concentrations in bending or torsion. Roark and Young (1975) and Petersen (1974) give specific data. For a discussion of stress intensity factors associated with sharp cracks, see *Brittle Fracture: Linear Elastic Fracture Mechanics*. For stress intensity factors from loads and geometries, see Tada et al. (1985), Rooke and Cartwright (1976), or Murakami (1987).

2.4 Wave Propagation

Within an elastic isotropic material, small-strain plane waves can propagate either as shear waves or as dilatational (more precisely, uniaxial strain) waves, with velocities

$$c_1 = \left(\frac{G}{\rho}\right)^{1/2}, \qquad c_2 = \left(\frac{E(1-\nu)}{\rho(1+\nu)(1-2\nu)}\right)^{1/2} \qquad (12)$$

Rayleigh waves travel along the surface of a body at velocities of roughly c_1. See Kolsky (1953) for a discussion of waves and their effects.

3. Sources of Solutions

Solutions for many problems are worked out by Timoshenko and Goodier (1970), for example. For many further results, see Roark and Young (1975).

For practical approximations, upper bounds to the stiffness of a body, $K = dP/du_P$, can be obtained by assuming displacement fields satisfying any displacement boundary conditions and finding the strain energy per unit volume

$$e\rho_0 \equiv \int \sum_i \sum_j \sigma_{ij}\, d\varepsilon_{ij}$$

from the stress–strain relations Eqns. (3) or (4). Similarly, lower bounds to the stiffness can be found by assuming stress fields satisfying any stress boundary conditions and equilibrium,

$$\sum_j \frac{\partial \sigma_{ij}}{\partial x_j} + \rho b_i = 0$$

and finding the complementary energy per unit volume

$$e_c\rho_0 \equiv \int \sum_i \sum_j \varepsilon_{ij}\, d\sigma_{ij}$$

from the strain–stress relations, Eqns. (1) or (6):

$$\left.\begin{aligned} e\rho_0 &= \frac{1}{2}\sum_i \sum_j \sum_k \sum_l C_{ijkl}\varepsilon_{kl}\varepsilon_{ij} \\ e_c\rho_0 &= \frac{1}{2}\sum_i \sum_j \sum_k \sum_l S_{ijkl}\sigma_{kl}\sigma_{ij} \end{aligned}\right\} \qquad (13)$$

Since the energy and co-energy are $Ku_P^2/2$ and $P^2/2K$, the stiffness lies within the range

$$K_{\text{lb}} = \frac{P^2}{\int e_c\rho_0\, dV} \leqslant K \leqslant K_{\text{ub}} = \frac{\int e\rho_0\, dV}{u_P^2} \qquad (14)$$

For the stiffness of a polycrystal from single crystal constants, or for a hole or a crack in a bar, rough bounds can be obtained from uniform displacement or stress fields, or from classical solutions for the net section.

Energy methods are also used for series and numerical solutions, since the least upper bound is the exact solution. See, for example, Sokolnikoff (1956) and, for the finite element method (FEM), Reddy (1984). The boundary element method (BEM, see Brebbia et al. 1984) allows the solution of plane problems, without supplying an interior mesh, by the automatic superposition of the combination of the Green's functions that will satisfy the general applied boundary conditions.

Experimental mechanics (experimental stress analysis) is often useful. In addition to the classical methods of photoelasticity, strain gauges and brittle coatings (including flaking rust), recent developments include laser-generated interferometric, Moiré, holographic and speckle patterns (e.g., Jones and Wykes 1983, Kobayashi 1987).

Bibliography

Brebbia C A, Telles J C F, Wrobel L C 1984 *Boundary Element Techniques, Theory and Applications in Engineering.* Springer, Berlin

Gurtin M E 1981 *An Introduction to Continuum Mechanics.* Academic Press, New York

Hull D 1981 *An Introduction to Composite Materials.* Cambridge University Press, Cambridge

Jones R, Wykes C 1983 *Holographic and Speckle Interferometry.* Cambridge University Press, Cambridge

Kobayashi A S (ed.) 1987 *SEM Handbook on Experimental Mechanics.* Prentice-Hall, Englewood Cliffs, New Jersey

Kolsky H 1953 *Stress Waves in Solids.* Oxford University Press, Oxford

McClintock F A, Argon A S (eds.) 1966 *Mechanical Behavior of Materials.* Addison-Wesley, Reading, Massachusetts

Murakami Y (ed.) 1987 *Stress Intensity Factors Handbook.* Pergamon, Oxford

Petersen R E 1974 *Stress Concentration Factors.* Wiley, New York

Reddy J N 1984 *An Introduction to the Finite Element Method.* McGraw-Hill, New York

Roark R J, Young W C 1975 *Formulas for Stress and Strain,* 5th edn. McGraw-Hill, New York

Rooke D P, Cartwright D J 1976 *Compendium of Stress Intensity Factors.* HMSO, London

Sokolnikoff I S 1956 *Mathematical Theory of Elasticity,* 2nd edn. McGraw-Hill, New York

Tada H, Paris P C, Irwin G R 1985 *The Stress Analysis of Cracks Handbook.* Paris Productions, St. Louis, Missouri

Timoshenko S P, Goodier J N 1970 *Theory of Elasticity,* 3rd edn. McGraw-Hill, New York

F. A. McClintock
[Massachusetts Institute of Technology, Cambridge, Massachusetts, USA]

Electrogalvanized Steel

In response to a strong demand for improvement in the corrosion resistance of coated steel mainly for automotive applications, zinc-electroplating technology has made great progress since the late 1970s, both in quality performance and in its manufacturing process. That is, the quality performance of coated products were enhanced by the development of alloy electroplated steel, including zinc–cobalt, zinc–nickel, zinc–iron alloy electroplated steel and others; the electrogalvanizing process was innovated by the development of high productivity and energy-efficient electroplating processes which remarkably increased the current density from approximately 30 A dm^{-2} to 100–200 A dm^{-2}.

1. Progress of Zinc Alloy Electroplated Steel

Investigations of zinc alloy electroplating on steel strip have been conducted since the early 1970s. A zinc–cobalt alloy containing a small amount of alloying elements (Ariga and Kanda 1980, Matsudo et al. 1980) was the first found to be corrosion-resistant and thus attracted strong attention in the field of zinc electroplating. The zinc–cobalt alloy contains 0.3–3.0 wt% cobalt and a very small amount of chromium or molybdenum. The alloy exhibited between two and six times the corrosion resistance of pure zinc in a salt-spray test. Since the alloying elements are at low concentrations, the phase of the alloy is η phase, which is the same as that of pure zinc.

Subsequently, zinc–nickel alloy electroplating was developed. Although this had been investigated for a long time (Averkin 1964, Dini and Johnson 1977), it was the latter half of the 1970s when electroplating of zinc–nickel alloy began to be applied to continuous steel strip (Hirt 1980, Shibuya et al. 1980). Zinc–nickel alloys contain 10–15 wt% Ni. They attain a higher corrosion resistance than the zinc–cobalt alloys, that is, four to eight times the corrosion resistance of pure zinc, in accordance with a higher concentration of alloying element. The alloys are composed of a γ-phase ($NiZn_3$ or Ni_5Zn_{21}) intermetallic compound, thus resulting in an increase in hardness, a decrease in elongation and expansion, and improvement in both weldability and corrosion resistance after painting. In addition to excellent corrosion resistance, zinc–nickel alloys are the most easily manufactured of zinc-alloy coatings. Consequently, zinc–nickel electroplated steel became a representative product in the field.

At almost the same time as zinc–nickel alloys were being developed, research and development of zinc–iron alloys was also being conducted (Fukuzuka et al. 1980, Watanabe et al. 1982, Toda et al. 1984). The idea of producing zinc–iron alloys by means of electroplating aimed at attaining the same quality as that of "galvannealed" steel (which proved to be suitable as a substrate for painted steel) but without deteriorating its mechanical properties. The properties of zinc–iron alloys vary with iron concentration. Without painting, the corrosion resistance of zinc–iron alloys differs little from that of pure zinc. However, the corrosion resistance after painting is significantly improved in zinc–iron alloys containing 10–20 wt% iron. Since zinc–iron alloys are mainly composed of δ phase ($FeZn_7$), an intermetallic compound in the range 10–20 wt% iron, weldability is improved as in zinc–nickel alloys. Accordingly, zinc–iron alloys containing ~ 15 wt% iron are electroplated on steel strip to attain a high corrosion resistance after painting, and weldability. Zinc–iron alloy electroplated steel has also been massively produced as a corrosion-resistant steel for automobiles. In the range of α phase (iron $\geqslant 50$ wt%), zinc–iron alloys exhibit the properties of iron, and thus result in an improvement in paintability, including uniform appearance and wet adhesion of cathodically electrodeposited paint film. The latter has been widely used as primer for automobile paint systems. Such an iron-rich zinc–iron alloy, 0.5 µm thick, is used as an upper layer on zinc–iron alloys containing ~ 15 wt% iron, when improved paintability is required (Toda et al. 1984, Adaniya et al. 1985). Although zinc–iron alloys provide various kinds of useful properties, zinc–iron alloy electroplated steel is commercially produced by only a few makers because of some difficult problems involved in its manufacture.

Zinc–manganese alloys are the most recently developed alloys. Those containing 40–50 wt% manganese exhibit 40 times the corrosion resistance of pure zinc and excellent corrosion resistance after painting (Sagiyama et al. 1986, Urakawa et al. 1986). As a result of its well-balanced properties, zinc–manganese alloy electroplated steel has a possibility of being adopted as a corrosion-resistant steel. However, industrial manufacture of the product poses more difficult problems than those of zinc–iron alloy electroplated steel. Furthermore, a composite coating containing aluminum in a matrix of zinc was investigated. The state of the development of zinc alloy electroplating reached by the late 1980s is summarized in Table 1.

In recent years, dispersion coatings which disperse inorganic substances such as SiO_2 and Al_2O_3 in the zinc deposit have been actively investigated (Yamazaki et al. 1984, Hiramatsu et al. 1985, Umino et al. 1987) in Japan to develop new products following the alloy electroplated steels. These new products are expected to be manufactured in the mid-1990s, and will possibly compete with zinc alloy electroplated steel.

2. *Industrial Electroplating Process*

With respect to manufacturing technology, great progress was made in high-speed electroplating and zinc alloy electroplating.

High-speed electroplating technology is needed to increase production and to produce heavily coated products for automotive use. It can be achieved by increasing the number of cells and by applying a high current density. The latter is obviously desirable when facility cost and operation are considered. There are several problems in applying high current density, but the most fundamental problem is the limiting current density i_l which is determined by the thickness of the diffusion layer δ as shown in Eqn. (1), when electrolyte composition is fixed:

$$i_l = ZFD\ C^0/\delta \tag{1}$$

where Z is the valency of the metal ion, D is the diffusion coefficient of the metal ion, C^0 is the concentration of the metal ion and F is the Faraday constant. To increase i_l, δ must be reduced by increasing the flow rate of the electrolyte as shown in Eqn. (2) (Schlichting 1968); a high flow rate of electrolyte is essential to increase i_l:

$$\delta = k(\nu X/U_0)^{1/2} \tag{2}$$

where U_0 is the flow rate of electrolyte, ν is the viscosity, X is the length of electrode and k is a constant (~ 5).

The relationship between flow rate and available current density (equivalent to limiting current density) in industrially used baths, including sulfate baths, chloride baths and mixed baths, is exemplified in Fig. 1 (Fukuda et al. 1981). The available current density increases in proportion to the flow rate in each bath. The highest available current density is derived from a chloride bath, when the flow rate is kept constant.

Table 1
Zinc alloy electroplated steels developed in Japan (1987)

Alloy	Typical compositions (manufacturers)
Zn–Co	0.3–3 wt% Co, plus 0.1–0.5 wt% Mo (Toyo Kohan) or plus 0.05 wt% Cr (Nippon Kokan)
Zn–Ni	13 wt% Ni (Sumitomo Metal, Kawasaki Steel, Nippon Kokan, Kobe Steel), plus 0.3 wt% Co (Nippon Steel)
Zn–Fe	10–25 wt% Fe (Nippon Kokan, Nippon Steel, Kobe Steel, Kawasaki Steel, Sumitomo Metal)
Zn–Al	10–15 wt% Al in powder form (Kawasaki Steel)
Zn–Mn	30–80 wt% Mn (Nippon Kokan)

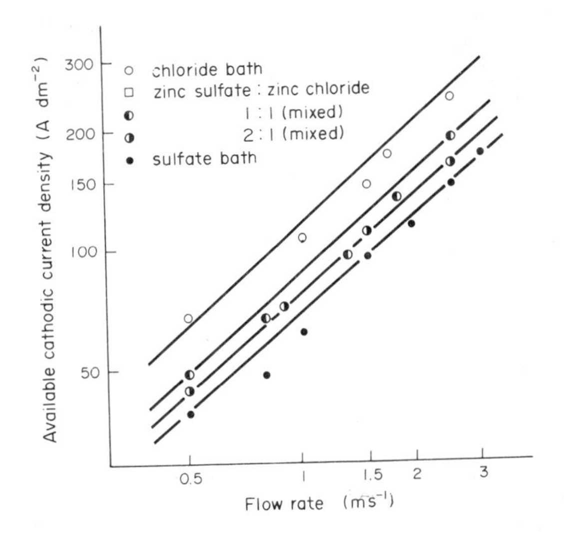

Figure 1
The available cathodic current density compared with the flow rate of the solution in industrially used baths

The second problem involved in high-current-density operation is avoiding too great an increase in the bath voltage. Since the bath voltage is determined by the distance between the anode and the cathode and also the specific electric conductivity of the bath, shortening the distance between the former and using a high value for the latter leads to a lowering of the bath voltage. Improvement in cell structure is needed to shorten the distance between the anode and the cathode. The structure of new cells developed for high-current-density operation is shown in Fig. 2 (Adaniya 1985). The specific electric conductivity of the bath significantly differs between types of bath. That of a chloride bath is generally two to four times that of a sulfate bath (see Table 2). Therefore, use of a chloride bath is very effective in lowering the bath voltage, thus saving electric power consumption.

Another point to be considered concerning high-current-density operations is the selection of anode. Anodes are classified as either soluble or insoluble. In conventional electrogalvanizing lines, where comparatively low current density is applied, only soluble anodes have been used. However, when a high current density of more than 100 A dm^{-2} is applied, adjustment of the anode–cathode distance and anode exchange are frequently needed, owing to the rapid consumption of anodes, in order to keep the anode–cathode distance short and constant. In such a case, manual anode exchange is no longer effective and automatic handling of the anode or mechanical handling at least is needed. From the viewpoint of anode handling, the use of an insoluble anode is reasonable. However, a few items must be taken into account when an insoluble anode is used. First, a huge dissolving tank must be installed to supply zinc ions.

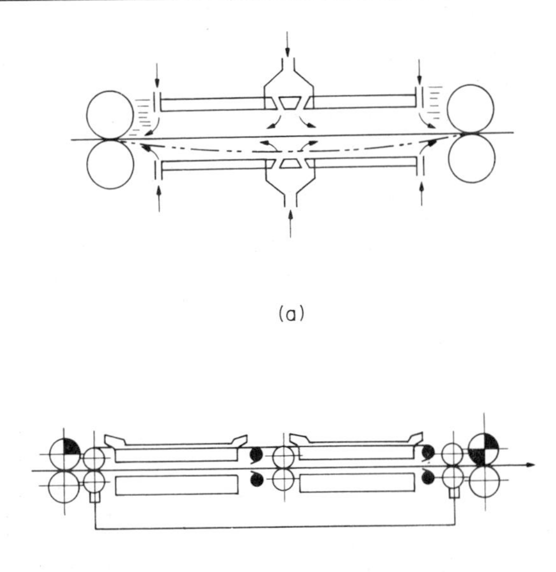

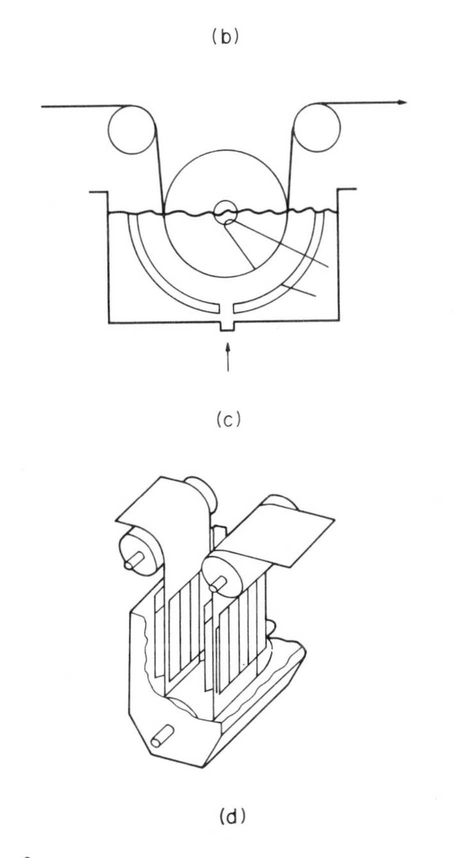

Figure 2
Cells for electrogalvanizing which were developed in the mid-1980s: (a) liquid cushion type (Nippon Steel); (b) horizontal flow type (Nippon Kokan); (c) radial cell (USX, Kawasaki Steel); (d) gravitel cell (Ruthner)

Table 2
Specific electric conductivities of various electrogalvanizing baths, at a temperature of 50 °C and at pH 4

Type of bath	Specific electric conductivity ($m\,S\,cm^{-1}$)
Sulfate bath ($ZnSO_4$ 1.5 mol)	110
Mixed bath ($ZnSO_4$ 1.5 mol, $ZnCl_2$ 0.8 mol)	130
Chloride bath I ($ZnCl_2$ 2.2 mol, NH_4Cl 5.7 mol)	490
Chloride bath II ($ZnCl_2$ 2.2 mol, NH_4Cl 1.9 mol)	220

Second, the adverse effect of metal ions such as lead ions dissolving slightly from an insoluble anode (lead alloy) on the quality of the product must be considered. In addition, their effect on the increase in bath voltage and nonuniform current distribution due to the evolution of oxygen gas must be eliminated. Furthermore, an insoluble anode cannot be used in a chloride bath of a high specific electric conductivity owing to the evolution of chlorine gas. These advantages and disadvantages must be considered when selecting the type of anode. In the most recently developed cells shown in Fig. 2, both types of anode are adopted: (a) and (d) are equipped with insoluble anodes while (b) and (c) are equipped with soluble anodes.

In the case of the electroplating of alloys such as zinc–iron, zinc–cobalt and zinc–nickel, further items must be taken into account in addition to those described for pure zinc electroplating. First, the method of adding alloy elements and controlling bath composition must be considered. If iron, cobalt or nickel is added in the form of sulfate or chloride, anions are accumulated in the bath during plating, thus resulting in an imbalance of the bath composition. Therefore, alloy elements must be added by dissolving metals or in the form of a compound such as an oxide, hydroxide or carbonate. Another method of supplying metal ions into the bath is to use soluble anodes, including the combination of a zinc anode and iron, cobalt or nickel anodes, or an anode of alloy to be plated. However, alloy anodes are not yet used in the commercial line.

Second, the upper limit of impurity concentration in the bath must be controlled. Metal cations such as Sn^{2+}, Ti^{4+} and Fe^{3+} prevent codeposition of alloys (Fukushima et al. 1983b) and Fe^{3+}, Pb^{2+}, and so on, deteriorate the quality performance of the product. Therefore, the concentration of such impurities must be kept low. In zinc–iron alloy electroplating, Fe^{3+} ions are generated in large quantities when an insoluble anode is used. The countermeasure for Fe^{3+} generation is still a serious problem.

Current density and the flow rate of plating solution are restricted in alloy plating, which determines the limitations of manufacturing conditions. Zinc and metals of the iron group anomalously codeposit (Fukushima et al. 1983a). Accordingly, not only bath composition, but also current density and relative velocity of strip to plating solution must be kept constant to obtain a constant composition of alloy (Adaniya et al. 1985). This restricts production efficiency of the line on which electroplated strips of various widths are manufactured.

Cathode-current efficiency is also a consideration affecting production efficiency and electric power cost. Compared with pure zinc electroplating, alloy electroplating is generally conducted at a low pH, as this accelerates the dissolution of metals to be plated and enhances the uniformity of alloy coatings. Moreover, a large amount of hydrogen gas evolves on the cathode because of the low hydrogen overpotential of

Table 3
Main specifications of electrogalvanizing lines introduced in the mid-1980s in Japan

Company (Works, Line No.)	Start-up	Production capacity ($kt\,year^{-1}$)	Line speed ($m\,min^{-1}$)	Type of cell	No. of cells	Type of electrolyte	Type of anode	Rectifier capacity (kA)
Kawasaki (Chiba, 1)	1982	272	120	radial	7	chloride	soluble	310
Nippon Kokan (Fukuyama, 3)	1983	327	150	horizontal	11	sulfate plus chloride	soluble	550
Nippon Steel (Nagoya, 1)	1983	327	200	horizontal	16	sulfate	insoluble	640
Sumitomo (Kashima, 1)	1984	327	200	vertical	14	sulfate	insoluble	672
Nippon Steel (Kimitsu, 2)	1985	240	150	horizontal	8	sulfate	insoluble	320
Nisshin (Hanshin, 1)	1986	131	100	horizontal	3	sulfate	insoluble	120

Table 4
Main specifications of electrogalvanizing lines introduced in the mid-1980s in the US (and the USX line of 1977)

Company (works)	Start-up	Production capacity (kt year^{-1})	Line speed (m min^{-1})	Type of cell	No. of cells	Type of electrolyte	Type of anode	Rectifier capacity (kA)
USX (Gary)	1977	329	183	radial	18	chloride	soluble	1008
National (GLX, Detroit)	1985	329	198	vertical	20	sulfate	soluble	1000
Armco (Middletown)	1986	206	92	vertical	16	sulfate	insoluble	750
Double Eagle (Rouge Steel, MI)	1986	576	183	radial	37	chloride	soluble	2072
LTV (Cleveland)	1986	321		vertical		sulfate	insoluble	
PFM/Beth./In. (Walbridge, OH)	1986	163		vertical		sulfate	insoluble	

alloy elements. Therefore, the pH should be determined from overall standpoints of alloy electroplating.

Tables 3 and 4 show the main specifications of electrogalvanizing lines (EGLs) constructed recently in Japan and the USA. Almost all the EGLs are constructed so as to be capable of producing alloy electroplated steels. It should be noted that cell structure and the type of anode are different, even though the EGLs are built for the same purpose. This fact may be understood to mean that EGLs still have both advantages and disadvantages. A new engineering approach surpassing the present one is expected to appear in the future.

Bibliography

Adaniya T 1985 Electroplated steel sheets. *Tetsu to Hagane* 71(3): 520–22

Adaniya T, Hara T, Sagiyama M, Honma T, Watanabe T 1985 Zinc–iron alloy electroplating on strip steel. *Plat. Surf. Finish.* 72: 52–56

Ariga K, Kanda K 1980 A consideration on the corrosion resistance of Zn-Co based Co-electrogalvanized steel sheet. *Tetsu to Hagane* 66(7): 797–806

Averkin K A 1964 Electrodeposition of zinc-nickel alloys. In: Averkin K A (ed.) 1964 *Electrodeposition of Alloys*. Israel Program for Scientific Translations, Jerusalem, pp. 102–15

Dini J W, Johnson H R 1977 Electrodeposition of zinc–nickel alloy coating. *Sandia Lab. (Tech. Rep.)* Oct: 77–8511

Fukuda S, Ohkubo Y, Abe M, Watanabe T, Honma T, Tonouchi A 1981 High speed electrogalvanizing. *Nippon Kokan Tech. Rep. Overseas* 31: 37–44

Fukushima H, Adaniya T, Higashi K 1983a Zinc–iron-group metal alloy plated steel. *J. Met. Finish. Soc. Jpn.* 34(9): 446–51

Fukushima H, Akiyama T, Kakihara K, Adaniya T, Higashi K 1983b Effect of the inorganic impurities in bath on the composition of electrodeposited Zn–Fe-group metal alloys. *J. Met. Finish. Soc. Jpn.* 34(8): 422–27

Fukuzuka T, Kajiwara K, Miki K 1980 The properties of zinc–iron alloy electroplated steel sheets. *Tetsu To Hagane* 66(7): 807–13

Hiramatsu M, Kawasaki H, Kusano F 1985 Zinc and zinc-silica composite coatings having superior adhesion property. *Proc. 71st Annual Meeting of the Metal Finishing Society of Japan* pp. 32–33

Hirt T A 1980 Continuous strip plating of nickel-zinc alloy. *3rd American Electroplaters Society Continuous Strip Plating Symp.* Annapolis, Maryland

Matsudo K, Adaniya T, Ohmura M, Ogawa M 1980 Development of corrosion resistant electrogalvanized steel including small amounts of Co and Cr. *Tetsu To Hagane* 66(7): 814–20

Sagiyama M, Urakawa T, Adaniya T, Hara T 1986 Zinc–manganese alloy electroplated steel for automotive body. *Technical Paper Series* 860268. Society of Automotive Engineers, New York

Schlichting H 1968 *Boundary-Layer Theory*. McGraw-Hill, New York, pp. 24–28

Shibuya A, Kurimoto T, Korekawa K, Noji K 1980 Corrosion-resistance of electroplated Ni–Zn alloy steel sheet. *Tetsu to Hagane* 66(7): 771–78

Toda M, Kojima H, Morishita T, Kanamaru T, Arai K 1984 Development of two-layered Zn–Fe alloy electroplated steel sheet—New coated steel sheet for automotive body. *Technical Paper Series* 840212. Society of Automotive Engineers, New York

Umino S, Yamato K, Kimura H, Ichida T 1987 Effects of Co^{2+} and Cr^{3+} on codeposition of Al_2O_3 in zinc electroplated steel. *Tetsu to Hagane* 73(5): 36

Urakawa T, Sagiyama M, Adaniya T, Hara T 1986 Corrosion-resistance and paintability of Zn–Mn alloy plated steel sheets. *Tetsu to Hagane* 72(SI): 968–75

Watanabe T, Ohmura M, Honma T, Adaniya T 1982 Iron–zinc alloy electroplated steel for automotive body. *Technical Paper Series* 820424. Society of Automotive Engineers, New York

Yamazaki F, Wada K, Shindo Y 1984 Investigation of Zn-SiO_2 and Zn-Ni-SiO_2 electro-codeposited steel sheet. *Tetsu to Hagane* 70(13): 145

T. Adaniya
[Steel Research Center, Kanagawa-Ken, Japan]

Energy Conservation: Insulation Materials

Thermal insulation is the resistance to heat transfer through a material and can be expressed in terms of degrees temperature difference across a slab of the material per unit of heat passing through unit area per unit time. A thermal insulator is a material or an assembly of materials that provides a thermal resistance *R*, measured in degrees temperature difference across the slab per unit of heat through a unit area per unit time, exceeding 2.0 per inch of thickness. The fundamental relation of the unidirectional flow of heat by conduction through a homogeneous solid under steady-state temperature conditions is related to the thermal conductivity of the substance, designated *k*. This *k* factor is determined by testing the actual material in accordance with the "Standard Test Method for Steady-State Thermal Transmission Properties by Means of the Guarded Hot Plate" (ASTM C-177-76(3)) or by means of the "Heat Flow Meter Standard Test Procedure" (ASTM C-518-76(3)). Under cycling conditions, a portion of the heat gain at one surface is absorbed and stored and a fraction is lost at the opposite side. This heat storage results in a temperature rise, which is a function of the mass and the specific heat of the material. When determining the rates of heat flow through composite constructions under transient conditions, the heat storage capacity of each individual element must be included.

1. Factors Affecting the Properties of Thermal Insulating Materials

Most insulating materials rely on still air to provide the thermal resistance. Wet materials are poor insulators. Moisture migrates to the colder surface and reduces the still air space by filling it with water, which is about 25 times less effective as an insulator than still air. When moisture vapor reaches a cold surface, liquid forms if the temperature is low enough to remove the latent heat of liquefaction. If the cold surface is far enough below the freezing temperature to remove the latent heat of fusion, the vapor turns directly into ice and the thermal performance of the insulation system is proportionally reduced.

The thermal conductivity of various insulating materials increases with increasing temperature, as shown in Fig. 1. As the density of the insulating material is increased, the still air space is reduced and the conduction due to the contact between particles or fibers increases. However, the higher the density of the insulating material, the less radiant energy transmission occurs, since still air provides little resistance to radiant energy transmission. The shorter the wavelength of radiant energy generated as the temperature is increased, the greater the radiant heat flow. The heat flow through thermal insulation due to solid conduction increases at a slower rate than that due to radiation. For high-temperature applications, therefore, materials of higher density are used to minimize radiant energy transmission. The pores of such materials may be filled with a gas of lower thermal conductivity than that of still air to further reduce heat flow.

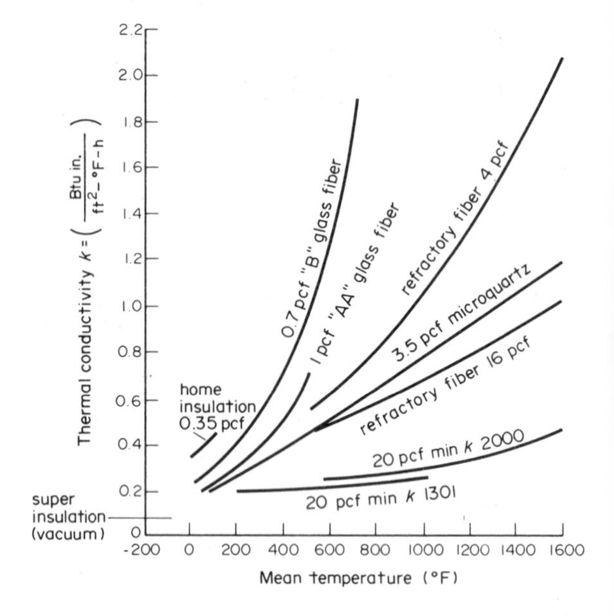

Figure 1
Thermal conductivity of various insulating materials as a function of temperature

Heat flow can also result from convection that takes place at surfaces where a film can form and encapsulate minute quantities of gas. Such convection at insulation jackets is affected by gas motion and is generally more significant than the convection within the thermal insulation pack. The rate of heat flow between a surface and the air in contact with it can be expressed in terms of the temperature difference and the film coefficient of heat transfer. The radiant heat flows through space from a substance at higher temperature to a substance at lower temperature and is determined by the difference in absolute temperatures to the fourth power and the emissivity of the substance. Within insulation systems, the overall heat flow by radiation is the fraction reflected by the particles or fibers and is dependent upon the absorbance and emittance of the particles or fibers that make up the insulation. Figure 2 demonstrates the radiant heat transfer between two 1 ft square radiantly black surfaces, one at 100 °F and the other at the temperatures indicated along the abscissa. The bottom curve demonstrates the heat transfer resulting from the placement of 1 in. of insulation between the plates.

The transfer of thermal energy from one region or potential to another resulting from conduction, convection and radiation can be combined to determine

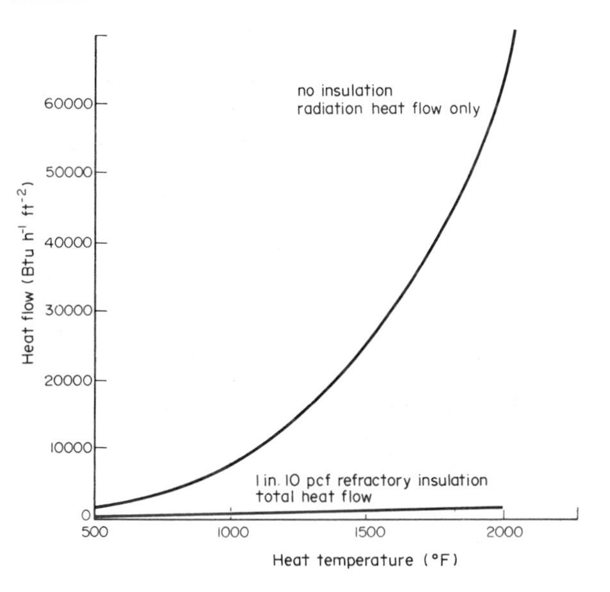

Figure 2
Heat flow as a function of hot face temperature with the cold face at 100 °F

Table 1
Examples of insulating materials with their maximum or minimum service temperatures

Insulating material	Service temperature (°C)
High-temperature applications	
alumina–silica fibers	1265 (max.)
potassium titanate fibers	1200 (max.)
diatomaceous silica	1040 (max.)
mineral fibers	1000 (max.)
perlite	875 (max.)
Intermediate-temperature applications	
hydrous calcium silicate	650 (max.)
glass fibers	540 (max.)
cellular glass	425 (max.)
cellulose fibers	95 (max.)
Low-temperature applications	
plastic foams	−140 (min.)
polystyrene	−130 (min.)
poly(vinyl chloride)	−40 (min.)
cellular rubber	−40 (min.)
cellular glass	−245 (min.)
mineral fibers	−270 (min.)
evacuated foil and fiber mats	−270 (min.)

the total heat flow. In addition to the thermal resistances of the materials in a composite system, the surface or film conductances are added. The surface coefficient of conductance is variable because it is affected by radiation, conduction and convection as dicussed above.

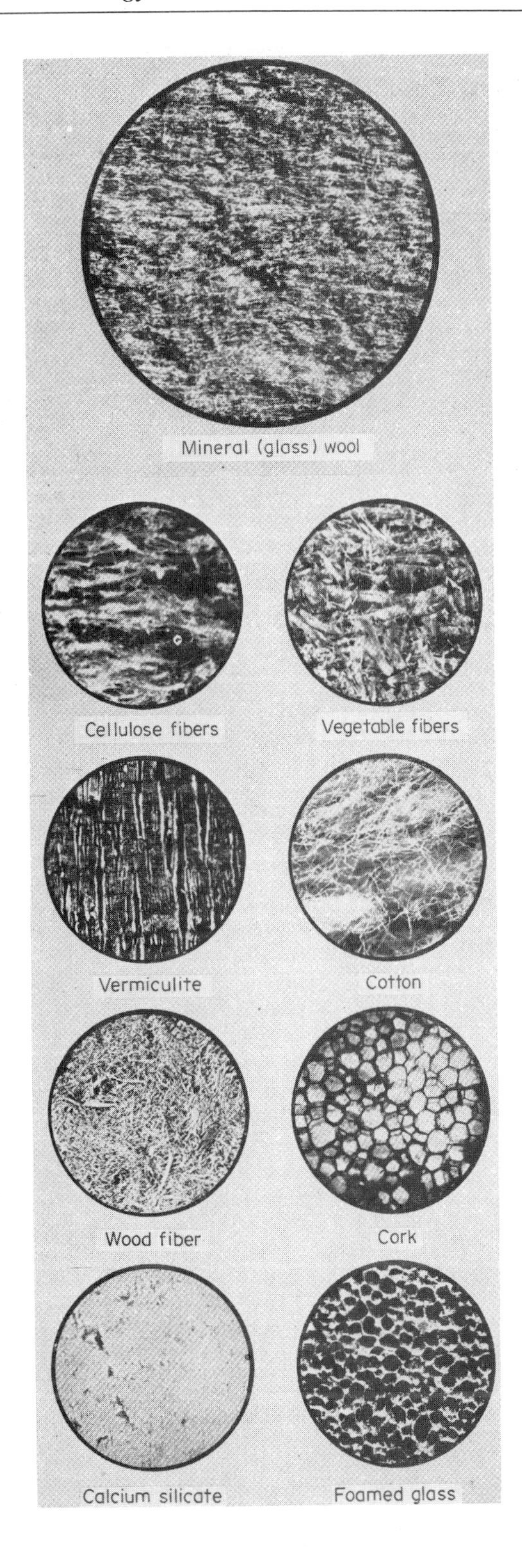

Figure 3
Photomicrographs of various insulating materials

2. *Principal Types of Insulating Materials*

The principal types of insulating materials for high-, intermediate- and low-temperature applications are given in Table 1. Photomicrographs of selected insulating materials are shown in Fig. 3.

Thermal insulation may be in the form of blankets, blocks, boards, cement, fitting covers, loose-fill, pipe, reflective roof, underground or multilayered mats or systems, and may be made from one or more materials combined to meet specific service requirements. All the properties desired may not be available in a single material, so that some high thermal resistance may have to be sacrificed to balance some other property, such as toxicity hazards, fire or corrosion resistance, strength, cost or resiliency. Ablative insulation provides an example of where materials are combined to meet specific service requirements. It comprises a combination of materials which melt and/or gasify so that the heat of gasification and/or liquefaction reduces the heat transfer through the backing medium at very high temperatures.

Bibliography

American Society for Testing and Materials (issued annually) *Standard Definition of Terms Relating to Thermal Insulations–Term No.58: Thermal Insulation*, ASTM C 618. ASTM, Philadelphia, Pennsylvania

American Society for Testing and Materials (issued annually) *Annual Book of ASTM Standards*, Sect. 4: *Constructions*, Vol. 04.06, *Thermal Insulations and Environmental Acoustics*. ASTM, Philadelphia, Pennsylvania

American Society of Heating, Refrigeration and Air Conditioning Engineers *ASHRAE Handbook of Fundamentals*. ASHRAE, Atlanta, Georgia

Close P D 1947 *Thermal Insulation of Buildings*. Reinhold, New York

Vershoor J D, Greebler P 1952 Heat transfer by gas conduction and radiations in fibrous insulations. *ASME Trans.* 74: 961–68

D. E. Morganroth
[Perrysburg, Ohio, USA]

Epoxy Resins

A very large number of chemically different polymers are sold under the generic term epoxy resins. These polymers are in general noncrystalline and thermosetting: that is, the molecular chains are arranged in cross-linked, three-dimensional networks. The polymers are formed by reacting an epoxy monomer, containing two or more reactive epoxide groups, with any of a large number of curing agents. The epoxy monomer which is by far the most common is based upon the reaction product of epichlorohydrin and bisphenol A (Eqn. (1)).

The most popular liquid epoxy for casting or potting applications has a molecular weight per epoxide group of approximately 190, and is a mixture of epoxy monomers with $n = 0, 1, 2, 3$, and with $n_{\text{average}} \approx 0.2$. The pure product, with $n = 0$, is crystalline; the advantage of having several n values present is that the product is a liquid with a viscosity of approximately 19 Pa s at 25 °C.

Higher average values of n result in higher molecular weight per epoxide group and higher viscosity, as shown by the commercial products in Table 1. These epoxy monomers are known as DGEBA (diglycidyl ether of bisphenol A) resins. The DGEBA resins enjoy a favored position due to their good economics, and their good temperature resistance and mechanical properties which are imparted by the stiff bisphenol A moiety. In addition to the DGEBA resins, many other types are available; a few examples are given in Table 2.

Since epoxy groups are quite reactive, they can be readily linked with other multifunctional compounds (such as diamines), or, in the presence of certain catalysts or initiators, the epoxy groups can be induced to react with one another to form chains. In either case, the nonepoxy compound is called a curing agent. A very wide range of curing agents is available,

Table 1
Properties of commercial DGEBA epoxy monomers, with increasing value of n

Molecular weight per epoxide group	Melting point (°C)	Viscosity at 25 °C (Pa s)
184	liquid	13
190	liquid	19
255	35–40	solid or very viscous liquid
310	40–45	solid or very viscous liquid
355	55–68	solid
650	75–85	solid

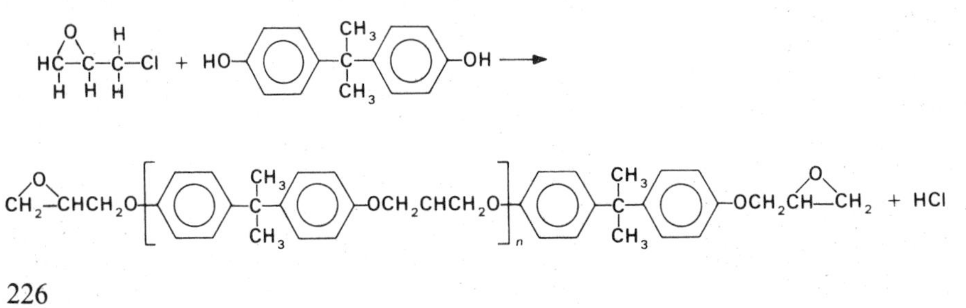

$$\underset{\text{(epoxide ring: O bridging first two C)}}{HC(H)\text{–}C(H)\text{–}C(H)_2\text{–}Cl} + HO\text{–}C_6H_4\text{–}C(CH_3)_2\text{–}C_6H_4\text{–}OH \longrightarrow$$

$$\overset{O}{CH_2\text{–}CH}CH_2O\text{–}\left[C_6H_4\text{–}C(CH_3)_2\text{–}C_6H_4\text{–}OCH_2CHCH_2O\right]_n\text{–}C_6H_4\text{–}C(CH_3)_2\text{–}C_6H_4\text{–}OCH_2\overset{O}{CH\text{–}CH_2} + HCl \qquad (1)$$

Table 2
Properties of sample epoxy resins

Type	Typical molecular weight	Typical viscosity at 25 °C (Pa s)
Polyglycidylether of phenol–formaldehyde novolac	175	1.7
Aliphatic resin: triglycidyl ether of glycerol	150	0.135
Aliphatic resin: diglycidyl ether of linoleic dimer acid	430	0.4–0.9
Tetrafunctional polyglycidyl ether of tetraphenylene ethane	220	solid

resulting in a wide range of mechanical properties, temperature resistance, chemical resistance and processability. A few of the more commercially important curing agents are briefly discussed below.

Primary amines react with epoxy groups as shown in Eqn. (2). The reaction proceeds between primary amino groups and epoxy groups such that linear chains are formed. The secondary amino hydrogens also react with epoxy groups, but at a slower rate (Eqn. (3)). The result is a cross-linked, three-dimensional network. If the rate of primary amine reaction is very much faster than the reaction rate of the secondary amino hydrogens, the polymer may solidify and the reaction may appear to stop. At this stage, heating permits limited flow processing, and further heating causes cross-linking to occur. This linear polymerization, followed later by cross-linking, is called B staging and is particularly important in the manufacturing of polymer composites.

$$H_2N{-}R{-}NH_2 + H_2C{-}\overset{O}{\widehat{\ \ }}{-}\underset{H}{C}{-}R'{-}\underset{H}{C}{-}\overset{O}{\widehat{\ \ }}{-}CH_2 \longrightarrow H_2N{-}R{-}\overset{H}{N}{-}\underset{H}{\overset{H}{C}}{-}\overset{OH}{C}{-}R'{-}\underset{H}{C}{-}\overset{O}{\widehat{\ \ }}{-}CH_2 \quad (2)$$

$$\sim R{-}\overset{H}{N}{-}\underset{H}{\overset{H}{C}}{-}\overset{OH}{C}\sim + \underset{H}{HC}{-}\overset{O}{\widehat{\ \ }}{-}\underset{H}{C}\sim \longrightarrow \sim R{-}N(HCH{-}C(OH){\sim}){-}\underset{H}{\overset{H}{C}}{-}\underset{OH}{\overset{H}{C}}\sim \quad (3)$$

Aliphatic di-, tri- and tetraamines react very rapidly, and can be used for the ambient-cure systems utilized in the two-tube consumer epoxy adhesives; a particulate filler is often added. For greater temperature resistance, more rigid structures are common, such as the aromatic diamines methylene dianiline and metaphenylene diamine. The aromatic amines are cured in a two-step temperature sequence with a final temperature of ~150 °C or more, and they are used where such elevated temperatures are feasible. Amine toxicity and possible mutagenicity considerations are leading to a shifting use of various amines.

Aliphatic polyamide systems cure in a few hours to a few days at room temperature, by the scheme for secondary amines given in Eqn. (3), and there are property-related advantages to starting with this type of high-molecular-weight curing agent. Tertiary amines facilitate epoxide ring opening and thereby catalyze homopolymerization of epoxy compounds with functionality >2. Tertiary amines also catalyze the reaction between hydroxyl groups and epoxy groups.

Curing can also occur through the action of acid curing agents—e.g., Lewis acids such as BF_3, anhydrides, organic acids and phenols. Anhydride cures are the most common, owing to their relatively low viscosity and good pot life (available time for processing after mixing). Cyclic anhydrides react with water or the hydroxyl groups in the DGEBA to form organic acids, which then react with the epoxy groups (Eqn. (4)). If the initial reaction of the anhydride was with water, the second acid group is now available for polymerization. If the initial reaction was with a DGEBA hydroxyl group, there is the basis for a cross-link.

$$\sim\overset{O}{\overset{\|}{C}}{-}OH + \underset{H}{\overset{H}{}}C{-}\overset{O}{\widehat{\ \ }}{-}C\sim \longrightarrow \sim\overset{O}{\overset{\|}{C}}{-}O{-}\underset{H}{\overset{H}{C}}{-}\overset{OH}{C}\sim \quad (4)$$

Lewis acids, such as BF_3 or its complexes, serve as catalysts for ring opening, resulting in homopolymerization of the epoxy resin. Lewis acids often cause very rapid reaction at room temperature, with hardening times of $\lesssim 2$ min being common. Salts which decompose above room temperature (such as the monoethylamine salt of BF_3) have the effect of retarding the reaction until the salt dissociation temperature is reached.

In general, curing agents are either nucleophilic as with amines, or electrophilic, as with BF_3 and organic acids. At elevated temperatures, nucleophilic curing agents such as amines or carboxylates are often capable of generating anions, which are then capable of initiating an ionic polymerization.

Bibliography

Kakuichi H (ed.) 1969 *Epoxy Resins*. Dekker, New York

Lee H, Neville L 1967 *Handbook of Epoxy Resins*. McGraw–Hill, New York

May C A, Tanaka Y (eds.) 1973 *Epoxy Resins: Chemistry and Technology*. Dekker, New York

Ranney M W 1977 *Epoxy Resins and Products: Recent Advances*, Chemical Technology Review No. 97. Noyes Data Corporation, Park Ridge, New Jersey

Sherman S, Gannon J, Buchi G, Howell W R 1980, Epoxy resins. In: Mark H F, Othmer D F, Overberger C G, Seaborg G T (eds.) 1980 *Kirk-Othmer Encyclopedia of Chemical Technology*, 3rd edn., Vol. 9. Wiley–Interscience, New York, pp. 267–90

J. P. Bell
[University of Connecticut, Storrs, Connecticut, USA]

F

Failure Analysis: General Procedures and Applications to Metals

This review of failure-analysis guidelines and procedures is concerned with the failure analysis of metallurgical components. The same guidelines apply in general to components made of ceramic or polymeric materials.

The two main goals of a failure analysis are (a) to find the primary causes of the failure, and (b) to define a corrective action which will prevent repetition of the same failure.

A failure analysis requires the integration of the work of experts in different disciplines of engineering such as mechanics, physics, chemistry, metallurgy, fracture, manufacturing, quality control and statistics. Depending upon the degree of complexity of the failure under investigation, and also depending upon the confidence level required of the analysis, a failure analysis may take one hour, one day, one week or even one month to be completed. The time-scale available for a given failure analysis will control each of the steps to be followed. The guidelines given below are in principle time and cost independent.

1. General Procedures

It is important to make sure that all the information needed to perform a detailed failure analysis is collected very quickly from all possible sources. This information includes the manufacturing, processing and service history of the failed component, as well as the reconstruction of the sequence of events resulting in the failure. The different types of failure which are likely to be encountered are very varied and a listing of the main types is given in Table 1.

1.1 Reconstruction of Failure

In the case of failure of a complex system such as the explosion of a petrochemical plant or the crash of an aircraft, it is important to establish unambiguously the location of the initial failure source. This wreckage analysis is assigned to reconstruction experts. The word reconstruction assumes that the experts who have a practical knowledge of the different functions of a component are able to establish the failure sequence. For instance, if the tail section of an aircraft is found a mile and a half away from the crash site, it will be concluded that it is likely that the tail section failed first, prior to impact of the aircraft. The same is true of a helicopter blade: if a single blade is found away from the main crash site, it is a sure indication that the blade attachment or the blade itself led to the crash. The reconstruction work may take a fair amount of time. One of the many findings derived from the reconstruction work should be the clear identification of the primary fracture path and of all the secondary fracture paths. The identification of a primary fracture path leads to the determination of the origin of the fracture and subsequently to the cause of the failure.

Table 1
Types of failure

Ductile failures	Wear-fretting failures
Brittle failures	Liquid-metal embrittlement failures
Fatigue failures	Gas embrittlement failures
Stress-corrosion failures	High-temperature failures
Corrosion failures	

This can be done through the reconstruction of the sequence of the accident and also by fractography. A rapid visual evaluation of different but characteristic fracture features will give the local direction of crack propagation. Examples of such features are chevron markings, beach marks, crack branching and shear lips (see Sect. 2.2). The identification of the primary fracture path and of all the secondary fracture paths is critical because it will guide the next steps in the failure analysis, and much valuable time could be wasted if the primary fracture path is not correctly identified.

The reconstruction experts will also analyze in detail all of the available data on the operation of the component prior to the failure.

1.2 Component History

At the same time as the reconstruction experts are doing their work, it is important to review the history of the failed part. This will include the following factors.

(*a*) *Design criteria.* A component is usually designed on the basis of engineering design guidelines, codes and specifications. The original design and the modifications brought to the design of a part over many years of service should be reviewed. It is important to note that a design code serves only as a guideline for minimum or typical performance. A good designer will add a factor of safety to the design parameters recommended by the code. Furthermore, a code is usually based on a body of technical knowledge which may be outdated by the time the part is manufactured or put into service. Thus the fact that a component may have met a code at a given time does not ensure that it is fail-safe. The design codes used in different countries or by different manufacturers may have to be compared to assess the technical soundness of the design criteria used at the time of the design.

(*b*) *Materials selection.* During failure analysis, the alloys used in the fabrication of a component should be carefully reviewed. For a given part, it is not uncommon to see marked modifications in the alloys used for the initial fabrication of the part. For instance, a part made with air-melted steel in the early 1970s would now be made with vacuum arc-melted steel in order to ensure better performance in fatigue and stress-corrosion applications as well as greater toughness. A change of alloy does not necessarily mean that the old alloy was unsuitable for the part at the time of the design, it just implies a higher degree of safety. The selection and purchase of an alloy is usually based on a set of specifications which are agreed upon between the vendor and the buyer of the material. All the specifications should be reviewed to see whether the material specifications were adequate, up-to-date and complete, and whether the materials used in the failed component met the specifications.

(*c*) *Process history.* A detailed review of the processing history of the key component involved in the primary fracture path is a critical part of failure analysis. This review will include an evaluation of all the manufacturing steps as well as all the quality-control tests performed on the failed part. A detailed understanding is required of all the different correlations and interactions between manufacturing procedures and variables and the structure and properties of the part.

(*d*) *Actual service history.* This is one of the most important pieces of information needed for each failure analysis. The detailed service history of the failed component and a reconstruction as far as possible of the loads, displacements, temperature and environment "service envelope" versus time is needed. It is not very common to have a log book giving a recorded history of the component, and thus in general it will always be difficult to know the exact service history of the failed part.

In most cases a component is designed within a narrow service envelope so that a mean set of service conditions can be obtained with a typical standard deviation. It is usually also possible to go back to a similar component to review the service conditions, or even to set up a new record of the current operating conditions.

(*e*) *Statistics of failures of the component.* Up-to-date statistics giving the number of components which have failed in a similar manner should be sought. These can be obtained from trade associations, or for example in the case of an aircraft from the Federal Aviation Authority, and in the case of a ground vehicle, from the Department of Transportation. Other sources of failure statistics are the manufacturers and the users. At the same time as the statistics of failures are collected, it is necessary to review the cause of each previous failure. The modifications implemented following each previous failure need to be reviewed.

1.3 Failure Stages

The failure history of a component can be divided into four stages.

(*a*) *Local degradation of the component.* This stage can take the form of wear, corrosion, fretting corrosion, pitting, oxidation, local microplasticity and other degradation processes.

(*b*) *Crack initiation.* This is the stage at which a microfissure or microcrack is initiated. This microcrack may have the depth or length of one, two or three grains ($< 100\,\mu m$) and may be difficult to find optically, but it can be detected with modern nondestructive test methods. The microcracks formed as the result of the material degradation stages may or may not grow into longer cracks. There are many components in which numerous microcracks can be generated without further crack growth. In this case, one of the questions to be answered is whether the component containing microcracks can remain in service for a finite lifetime without any probability of failure.

(*c*) *Slow crack growth.* Once initiated, in general, a microcrack will grow by ductile tearing, fatigue, creep, stress corrosion or liquid metal embrittlement. These processes of slow crack growth usually take place during a fraction of the lifetime of the component. One of the most important problems in each failure analysis is to identify the fraction of component life spent in growing a crack from a size equal to two or three grains to the final stage of fast crack propagation. A knowledge of the crack-length–time history can be used to define a quality control inspection period. It can also be used to assess the risk of leaving the other components in service.

(*d*) *Fast crack propagation.* A growing crack will eventually reach a critical length which, when combined with local stresses, leads to rapid crack propagation by brittle fracture, ductile tearing or fully plastic collapse. This process of fast crack propagation occurs over a very small time scale, usually of the order of less than a second.

1.4 Incipient Failure

The processes of degradation and crack growth may occur at any time during the life of a component. Incipient failure may occur in the following ways.

(a) During manufacture. Here, failure may be caused by forging cracks, forging laps, quench cracks, hydrogen flakes, solidification defects and shrinkage porosity, among others.

(b) During quality control or proof testing. Some quality-control procedures are such that the components are subjected to vibratory stresses or to large peak stresses during proof testing. It is not uncommon to see cracks generated during quality-control handling. For instance, stress corrosion cracks in pressure vessels made of high-strength steels have been initiated during pressurized proof testing with water.

(c) In prototype tests. Very often a prototype component is tested during its manufacture. The prototype testing program is designed to provide a component which leads all the other parts in service time by simulating service loads in an accelerated time frame. If the prototype fails during testing, the failure analysis should be especially painstaking in order to plan the corrective action and to derive a reliable prediction of the probability of failure of all the other components in service.

(d) Failure of the part in service. Most failure analyses deal with parts which fail in service, for which a decision/action will be needed very quickly. In this case the reliability of the failure-analysis conclusion is very important to the decision process.

1.5 Decision/Action Following Failure Analysis

Following failure analysis, and on the basis of the understanding obtained by the failure-analysis team, the following decisions/actions are possible.

(a) Do nothing. This is not an uncommon approach when the cause of failure is not clearly understood, or when the failure is blamed on improper use of the component.

(b) Proceed with a detailed and careful quality-control inspection of all the components in service. Such a decision requires careful planning of the steps to be taken, i.e., locations to be inspected, samples to be taken, time intervals for inspection. There will always be a conflict between the cost of the quality-control inspections, including the downtime cost of the component, and the degree of reliability of the inspections.

(c) Perform accelerated tests. Following the failure analysis, it may be critical to verify the results of the analysis by performing a series of accelerated tests on a component similar to the failed one, or on a smaller scaled-down part, taking into account the size effect. With modern mechanical testing equipment (e.g., servohydraulic machines) it is possible to simulate very rapidly a set of loads, displacements or strains on a given component. Furthermore, it is also easy to reproduce the size of the flaw which may have led to the failure. Using linear-elastic and elastic–plastic fracture mechanics, correlations can readily be established for the cracking rates and failure probabilities between a crack in the test component and a crack in the real component.

(d) Repair the parts. The conclusions of the failure analysis may be such that repairs of all the components in service are feasible and can be performed quickly, without any detrimental effects. However, it should always be remembered that a "quick fix" might be the source of other failures if it is not carried out properly and tested carefully.

(e) Cost versus probability of failure. In each case, the decision/action recommendations are based on an evaluation of the costs of the action to be taken. This is part of the field of risk analysis and risk assessment. Depending upon the risk of further failures, it must be decided either (i) to rework the failed component, (ii) to take all the other components out of service or (iii) to leave them in service. Removing all the components from service is a drastic action which may be necessary if the probability of failure is high. In this case a redesign of the component will be needed. An alternative choice is to change the service conditions to ensure a safe life.

2. Metallography and Fractography

2.1 Chemical Analysis and Metallography

For a metallic component it is crucial to know its overall chemistry, and also the chemical gradients near the origin of the crack. It is also recommended to characterize fully the macrostructure of the part in order to take into account the gradients of microstructure. For instance, the orientation of the different processing planes in a forging (e.g., short transverse plane, long transverse plane) or the directions of solidification in a casting should be identified. The flow lines of the grains in a forging are very important because they introduce a strong mechanical anisotropy. A detailed understanding of the influence of the processing history on the macro- and microstructure of the material and of the component is required to account for the fracture behavior. The microstructure at the location of the initiation site should be determined without entirely destroying the fracture features which may be needed for further investigation. The microstructure analysis includes grain size, grain shape, inclusions (size, shape and volume fraction), voids and chemical gradients.

2.2 Macrofractography

A macrofractographic analysis should be carried out in parallel with the stress analysis. It should include the following steps.

(*a*) *Fracture path versus load direction.* It is important to know the load directions near to and at the location of the fracture origin and the relative orientation of the fracture plane with respect to the loading direction. In fracture-mechanics terminology the modes of crack-tip opening are referred to as mode I (normal opening), mode II (shear opening) and mode III (antiplane-strain opening). The cracking plane is directly related to the cause of cracking. For instance, for shear overload it is a plane of maximum shear stress. On the other hand, a stress-corrosion or a fatigue crack propagates in a plane normal to the maximum-principal-stress axis. Textbooks in applied mechanics give the principal-stress directions for simple loading conditions (e.g., bending, torsion, internal pressure) for different structural components.

(*b*) *Brittle fracture versus ductile fracture.* A macrofractographic analysis gives an evaluation of the degree of plasticity associated with the fracture path in the component. It is important to differentiate between a fully plastic fracture and a very localized brittle fracture where there is little or no plastic deformation (see Sect. 3).

(*c*) *Fracture markings.* The identification of different fracture markings is critical to the macrofractographic analysis. Fracture origins are identified by tracing back all the directions of crack growth. An origin is often characterized by a difference in texture, roughness and color, indicating a long exposure to oxidizing or corrosive environments. Its location is usually associated with a high-stress or high-strain region such as the edge of a hole, the fretting marks near a bearing surface, machining defects and processing pores.

In many components having the shape of a sheet or a plate, there will be a unique propagation direction of the crack. In this case the fracture surfaces have a series of markings, referred to as chevron markings. These are initiated along the center-line of the plate and run to the edges. The chevrons point in the direction away from the crack propagation direction (see Fig. 1). Chevron markings are extensively used in failure analysis of large or small components to trace the local directions of crack propagation and thus to find the fracture origins. There are some alloys for which the chevron markings may not always be clearly defined; this is the case for clear steels or some alloys which contain only a limited volume fraction of inclusions and which do not show a gradient of inclusion content from the center of the plate to its edges. In these alloys there will always be some "river" fracture lines normal to the crack front, similar to chevron markings. In most cases these lines can also be used to identify the main directions of crack propagation.

Beach marks represent the successive positions of the crack front as a function of time. They represent markers in the history of the progression of the crack front and are due to transient and/or cyclic variations in loads, corrosion and temperature. Beach marks are extensively used to reconstruct the life history of the component by correlating specific beach marks with particular events during service. Whenever it is difficult to understand the sequence of beach marks, it is recommended that similar beach marks be produced

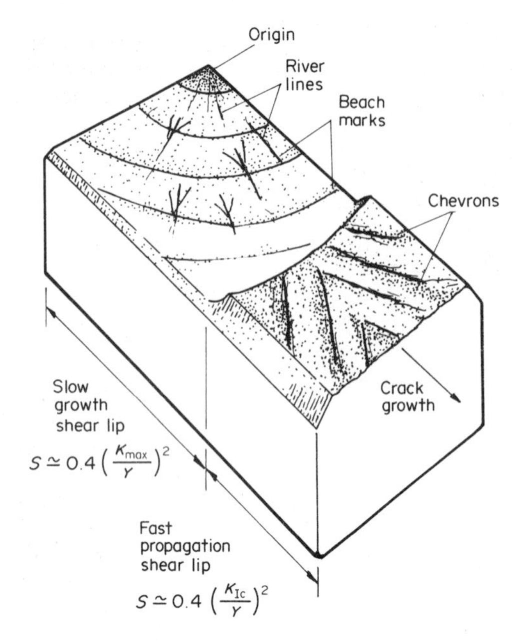

Figure 1
Sketch of typical fracture features including origin, river lines, beach marks, chevrons and shear lips

in an accelerated time frame on the fracture surfaces of a test component or test bar subjected to a history similar to the service history.

Shear lips represent the near-surface zone which is in the pseudo-plane-stress condition and where fracture proceeds along planes at $\pm 45°$ to the plane of the surface (Fig. 2). The width or thickness of the shear lips is very imporant because it gives a measure of the toughness of the component during crack growth. The empirical correlation between fracture toughness and shear-lip width is given by

$$W_{\text{shear lip}} = 0.4\,(K_{Ic}/Y)^2 \qquad (1)$$

where K_{Ic} is the fracture toughness index and Y is the yield stress. When the failure of a plate is such that the fracture proceeds under a plane-stress condition, the secondary cracks resulting from the change of orientation of the shear lips can also be used to trace the direction of crack propagation. This fracture feature is very useful when chevron markings are not readily visible. In the case of slow crack growth, the shear-lip width gives a measure of the size of the plastic zone at the crack tip which can in turn be related to the local stress-intensity factor or to the J parameter (see Sect. 3.) and then to the crack propagation rates.

Compression spalls occur in components that are broken in bending. The bending fracture of a part leads to fracture from the tensile side to and through the compression side; branching of the crack in the compression region is referred to as a compression

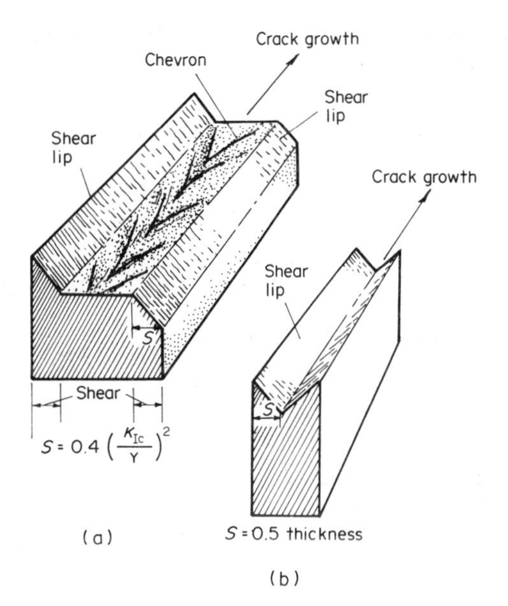

Figure 2
Sketch of typical fracture features for (a) plane-strain and (b) plane-stress fractures

spall. This fracture marker is used to find the direction of crack propagation under conditions of bending. The origin is usually 180° away from the compression spall.

The texture of fracture surfaces, that is, the roughness and the color, gives a good indication of the interactions between the fracture path and the microstructure of the alloy. For instance, at low stress, a fatigue fracture is typically silky and smooth in appearance. Stress-corrosion fractures show extensive corrosion features and corrosion beach marks. A discontinuous ductile fracture shows some stages of crack-tip blunting, crack arrest and "pop-in." All the features related to the texture of the fracture surfaces can, if necessary, be duplicated in a prototype or coupon fracture test.

(*d*) *Reproducibility of macrofractographic features in accelerated tests.* It is, in general, quite easy to reproduce the macroscopic fracture features of a component with a test sample of the same material subjected to an accelerated test program. This is very important when the correlation between the microstructure and the fracture features of a material is not well understood. All the fracture features mentioned above can be readily reproduced in an accelerated test program.

2.3 Microfractography

Assuming that there is a clear understanding of all the macrofractographic features, a need often exists for a high-magnification electron-microfractographic analysis to provide further information about the micromechanisms involved in the material failure (see Table 2). The scanning electron microscope, with or without plastic replicas, is an ideal tool for this type of work. The guidelines for microfractographic analysis are given in detail in the American Society for Metals Fractography Handbook. Figure 3 gives a sketch of the fracture features associated with different fracture conditions.

Great care should be exercised not to take the microfracture features out of context, that is, at too

Table 2
Micromechanisms of fracture

Transgranular crystallographic cleavage
Interface decohesion
Intergranular brittle decohesion
Microvoid initiation, growth and coalescence
Rupture by single shear or by alternating shear
Slip plane shear decohesion by stage 1 fatigue
Transgranular or intergranular fatigue cracking with fatigue striation
Transgranular or intergranular decohesion by stress corrosion cracking
Mixed shear and cleavage modes on atomic scale
Creep cavity nucleation and growth

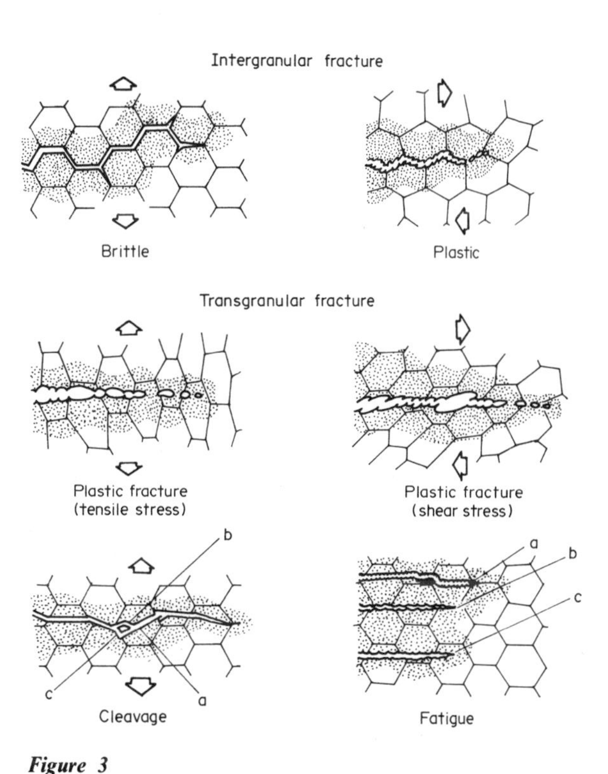

Figure 3
Sketch of intergranular and transgranular fracture paths and associated fracture features. The hexagonal cells represent the grain size

high a magnification, or to view artifacts which may be irrelevant and misleading. It is important to avoid conclusions based on a single high-magnification SEM fractograph. In cases where an unknown feature could be a key to the overall fractographic analysis, it is necessary to perform a duplicate test on the same material under similar conditions to provide a clean and undamaged fracture surface for a comparative fractographic analysis. It is only when all the fracture features have been reproduced under controlled conditions that conclusions can be drawn from the microscopic fractographic observations. As an illustration of the problems that may arise from fractographic analyses, a case occurred in which pearlite lamellae etched by seawater on the fracture surfaces in a low-carbon steel were mistaken for fatigue striations. The pearlite spacing was assumed to represent the fatigue crack growth per cycle, and a detailed but worthless fracture mechanics analysis was based on that erroneous interpretation. In the case of fatigue failures, measurements of fatigue striation spacings (when the striations are well-defined) provide a powerful measure of the crack growth rates for a given crack length, and the loading history. These data should be used in the reconstruction of fracture sequences.

3. Mechanical Analysis

A failure analyis is not complete without a detailed study of the mechanical conditions which led to failure. There are certain steps which may be taken to perform this analysis. First, however, the mode of fracture must be determined. There are two different types of fracture: fully plastic and elastic–plastic.

Fully plastic fracture is typically the cause of failure of a component made of a low-strength material such as pure aluminum, pure copper or pure brass. A fully plastic fracture can be due to a buckling instability or to an overload. There is always extensive yielding of the component over its entire cross section. A plastic-limit analysis will give an approximate value of the critical load at failure. This estimated load should be compared with the actual service loads.

In elastic–plastic fracture, plastic deformation is limited to the region around the crack tip and crack flanks. An elastic–plastic crack is initiated in a region of localized plastic strain. The crack initiation criterion can be verified by using the following procedures.

(a) Estimate the loads and displacements applied to the part, taking into account the directions as well as the intensities of these parameters.

(b) Calculate the local stress at the sites of stress concentration, taking into account, if needed, the residual stresses. This can be done either by calculating the stress concentration factor at the root of the notch or by using finite element analysis. Note that an elastic analysis will in most cases be too simplistic. K_t, the elastic stress concentration factor, can be obtained from tables or from handbooks, and the K_t can then be converted following Neuber's equation into a stress concentration and a strain concentration factor:

$$(K_t)^2 = K_{\text{stress conc.}} \times K_{\text{strain conc.}} \qquad (2)$$

The Neuber analysis is quite accurate, and the strain concentration factor gives a first estimation of the total local strain existing at the root of the notch. The finite element analysis gives the stress and strain gradients around the notch. This information is used to predict whether a crack will or will not grow. Following the notch analysis, it shoud be possible to predict whether a crack will be initiated either by low-cycle fatigue, by stress-corrosion cracking or by creep deformation. A fracture criterion will be needed to predict the initiation of a crack at the root of a notch or at the stress concentration site. Several textbooks on fracture provide the needed guidelines (*Metals Handbook* 1974, 1975).

3.1 Fracture Mechanics Analysis of Crack Growth

Assuming that it has been clearly demonstrated that the stress and strain conditions at the initiation sites are such that a crack will be initiated, the next step is to predict the rates at which the crack will grow. A crack may grow under conditions of ideally linear elastic fracture mechanics, where the plastic-zone size at the tip of the crack is very small compared with the crack length. On the other hand, if there is extensive plasticity near the crack tip, the crack growth conditions are referred to as elastic–plastic conditions. In that case the analysis of crack growth becomes more difficult. Finally, a crack may grow under a condition of fully plastic yielding.

In order to estimate the rate of crack propagation, one should have access to a handbook of fracture mechanics data which will provide the crack growth rate data for fatigue (da/dn), creep crack growth (da/dt), stress corrosion cracking (da/dt) and fracture toughness (K_{Ic} or J_{Ic}). The data given in the handbooks are never directly applicable to the failure analysis at hand, but they can be interpolated or extrapolated depending upon the need. Correlations between the mechanical properties of the material and the crack growth data are also needed to derive rapidly the magnitude of the crack growth rates.

3.2 Crack Growth Reconstruction

Having considered fractography, fracture mechanics and crack growth, the crack growth history of the failure can be reconstructed. This is best done by estimating the time and/or the number of cycles it takes to grow a crack from a length a_0, of the order of a few grains at the initiation site, to a final fracture length a_f, under the service loads or conditions reported. The analysis should quickly verify, within a factor of 2 to 4, whether the failure history is plausible. This calculation can be done by direct integration or by iteration with a computer.

Following this first approximation, it is necessary to perform a second calculation to take into account all the secondary factors such as residual stress, corrosion effects and microstructural variations which may account for marked differences in crack growth rates.

3.3 Accelerated Test Plan

It may be crucial, in order to verify the validity of the crack growth analysis, to perform a series of accelerated mechanical tests to obtain the rates of crack growth and the fracture conditions for the material or component at hand. With the servohydraulic equipment available, this can be done within a very short time. Accelerated test results provide valuable information to clarify and complete the failure analysis.

3.4 Estimated Costs versus Risk of Failure

In order to be able to provide some guidelines on the decision or action to be taken with respect to the other components in service, it is necessary to have a complete review of the analysis of the time spent in crack growth in the failed component and in other components in service. Once this analysis is performed it should be possible to define a final set of crack-detection inspection intervals to ensure safe operation and planned retirement from service of cracked components.

The final step in a failure analysis is to review the probability of failure and to estimate the costs of failure versus the cost of quality-control inspections and the cost of a fracture control plan. This is somewhat beyond the scope of the failure analyst, but the reliability of the conclusions arrived at in the failure analysis is very important in the risk-analysis task.

4. Conclusions

A report should be prepared which takes into account all the facts that have been clearly established. This may take the form of a preliminary report, in order to have it reviewed and coordinated by the team of experts who have participated in the analysis. A preliminary report will give everyone an opportunity to verify that a consensus of opinion has been reached and that there is no misunderstanding about the facts on which the analysis is based. The final analysis report should not only include the results of analysis but should also set guidelines for a short-term solution to avoid a repetition of the failure, as well as a long-term solution to the fracture problem. Typically, the final report would include the following sections:

(a) description of the failed component;

(b) service conditions at the time of failure;

(c) wreckage reconstruction;

(d) service history;

(e) design, manufacturing and quality-control history;

(f) mechanical and metallurgical conditions of failure;

(g) summary of causes and sequence of failure; and

(h) recommendation for prevention of similar failures and guideleines for corrective action for similar components in service.

See also: Failure Analysis of Ceramics; Failure Analysis of Plastics; Mechanics of Materials: An Overview

Bibliography

American Society for Metals 1974 *Metals Handbook*, 8th edn., Vol. 9, *Fractography and Atlas of Fractographs*. ASM, Metals Park, Ohio

American Society for Metals 1975 *Metals Handbook*, 8th edn., Vol. 10, *Failure Analysis and Prevention*. ASM, Metals Park, Ohio

Hutchings F R, Unterweiser P M (eds.) 1981 *Failure Analysis: The British Engine Technical Reports*. American Society for Metals, Metals Park, Ohio

Nauman F K 1983 *Failure Analysis: Case Histories and Methodology*. American Society for Metals, Metals Park, Ohio

Pomey G, Rabbe P 1968 *Ruptures de fatigue de pièce de machines classification et analyse*. Societé Francais de Metal, Dunod, Paris

R. Pelloux
[Massachusetts Institute of Technology, Cambridge, Massachusetts, USA]

Failure Analysis of Ceramics

Failure analysis entails examination of a fractured component or test specimen and correlation of the observed features with design, with analytical and instrumentation data and with prior fabrication information. The aim is to elucidate the cause of the failure. It is critical in component application to determine whether the failure was the result of a design deficiency, a material deficiency (arising from the fabrication process) or an abnormal off-design condition. Examination of the fracture surfaces and surrounding regions of a ceramic component can aid in this determination.

The techniques for failure analysis pertinent to glasses and fine-grained ceramics are relatively well established. These materials exhibit distinctive topographical features in the vicinity of the fracture origin which can be observed by a combination of macroscopic and microscopic methods. The essence of these procedures is briefly described in this article. The corresponding analysis for large-grained ceramics, especially those ceramics with elastic and thermal expansion anisotropy (such as Al_2O_3), is appreciably more difficult. The indications are more subtle and experience with a specific material is often required.

1. Morphology of Fracture

The fracture of ceramic components typically arises either from defects introduced during processing or machining, or from service (especially impact) damage. The specific fracture-initiating defect responsible for the failure of a component can usually be located, and its origin identified, by recognizing the various morphological features by the following procedures.

The first step is to reassemble the fragments whenever possible, to determine, macroscopically, distinctive crack patterns that suggest the position or cause of failure initiation. A frequently encountered fracture pattern is illustrated in Fig. 1a. The fracture origin is invariably located near the center of the planar central section, between O and O′ in Fig. 1a. This feature develops because, as the crack accelerates from the origin, it begins to bifurcate (Fig. 1b). This typically occurs when the stress intensity K (quasistatically defined) exceeds a critical value. The bifurcations occur repeatedly from both crack tips, resulting in substantial fragmentation of material in the plane beyond the fracture origin, outside OO′ (Fig. 1). Consequently the loss of small fragments is not detrimental to the failure analysis, because such fragments do not contain the failure origin. The larger fragments, on both sides of the planar initiation zone, are now separated from the assemblage and can be exposed to closer scrutiny. (Occasionally a highly defective component separates into only two fragments. In such instances, the general region of the failure origin cannot be so readily located. This can also be true of failure caused by thermal shock.)

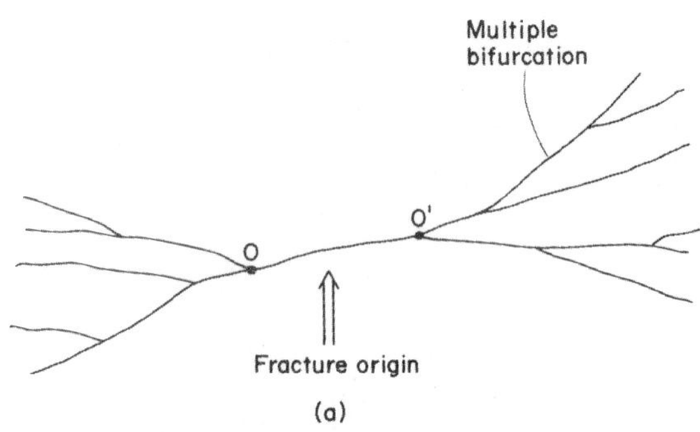

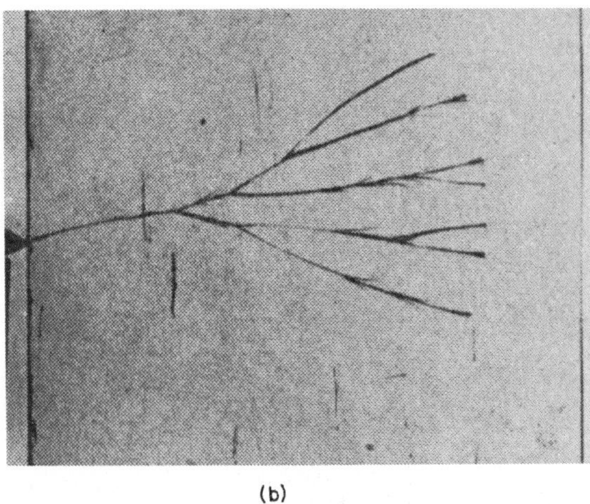

Figure 1
(a) Schematic illustrating a typical crack morphology in the vicinity of the origin, and (b) crack bifurcation in glass from an edge-initiated failure (Schardin 1958)

In some instances, additional or alternative diagnostic surface features are observable at the macroscopic level. These include:

(a) fractures intersecting (or radiating from) a single point or small area;

(b) a "witness mark," consisting of a debris-coated or burnished area indicating abnormal localized surface distress during service;

(c) fractures extending along or parallel to machining grooves;

(d) a fracture extending inwards from an edge of a component;

(e) a fracture which appears to be associated with a contour or discontinuity in the component configuration;

(f) a fracture intersecting a hole in the component (e.g., a bolt hole, feed-through hole); and

(g) a fracture intersecting or passing through a point of attachment.

Often the fundamental cause of failure can be deduced from such macroscopic features. For instance, fracture radiating from a point can often result from localized impact or some other form of mechanical interference. A fracture at a hole, edge, contour or joint can suggest an unacceptable design with excessive stress concentration. Alternatively, an abnormal defect may be involved in the fracture process. To distinguish between a stress concentration and an abnormal defect as the primary cause of failure, the fracture surfaces adjacent to the suspected region of fracture must be investigated.

The examination of the fracture surfaces thus constitutes the second step in failure analysis. This phase is achieved visually, by low-magnification optical microscopy and sometimes by scanning electron microscopy (SEM). The examination is predicated on the recognition that, as a crack initiates and propagates through a glass or ceramic, distinguishing features are created on the fracture surface. Some of these features are illustrated in Fig. 2.

The features of interest are referred to as river markings. These markings radiate toward the fracture origin but terminate in a relatively featureless, essentially circular zone before converging on the actual origin. This featureless central zone has been referred to as the "mirror" region in the glass literature while the zone containing the river markings has been termed the "mist and hackle" region. The river markings are created while the crack is moving dynamically. The dynamic crack tip stress field causes the crack to deflect onto parallel planes, in adjacent regions of the front. When these planes overlap, steps are created. These steps are the source of the river markings. The crack velocity in the mirror region is too low to cause crack bifurcation and deflection. Consequently the crack remains relatively planar and the associated fracture surface is comparatively featureless. The perimeter of the mirror zone coincides, in fact, with the initial bifurcation point OO′ on the surface crack pattern (Fig. 1a).

Once the river markings have been identified and their center traced (from the point of intersection of the radial lines), the actual failure origin can be sought by inspection at higher magnification of the center of convergence. This phase of the investigation is important, because the nature of the fracture origin establishes whether failure was caused by a design deficiency, a material deficiency or an abnormal service condition. The appearance of a variety of fracture origins and the implications regarding the probable cause of failure are as follows.

The facility with which the origin can be identified is contingent on the type of origin involved. If the

Figure 2
(a) Schematic illustrating river markings on a fracture surface; (b) failure origin in silicon nitride near the surface

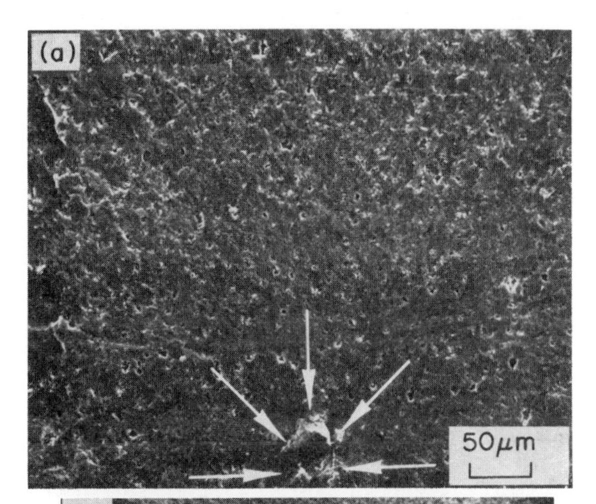

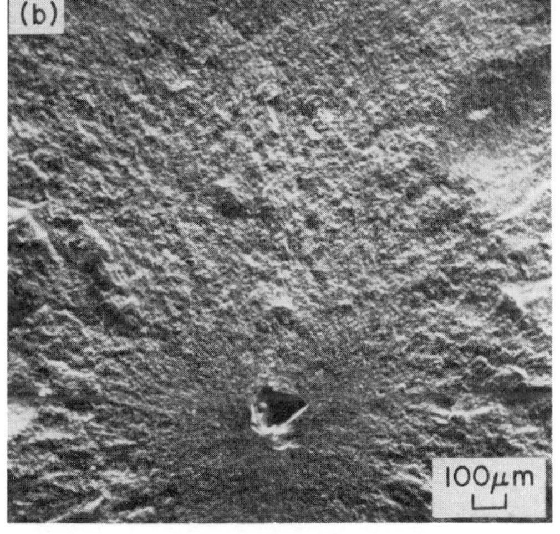

Figure 3
Scanning electron micrographs of (a) an Si inclusion and (b) a large pore as failure origins in Si_3N_4

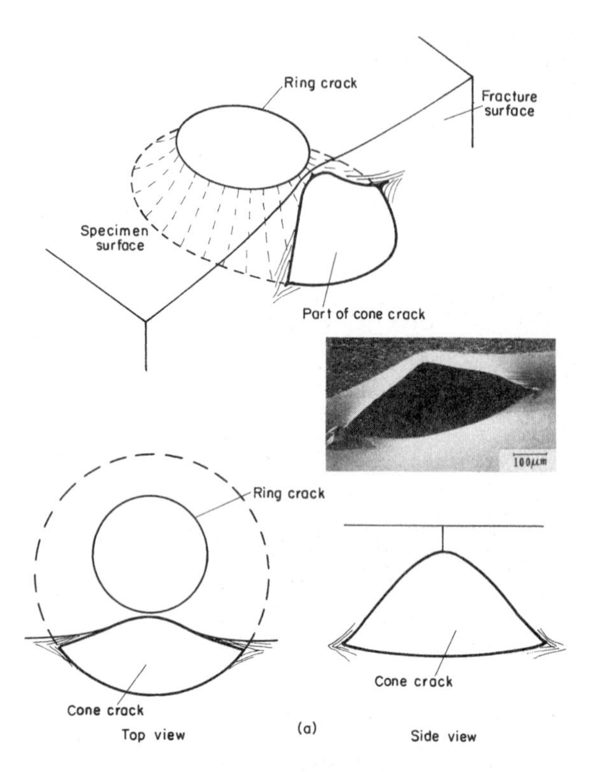

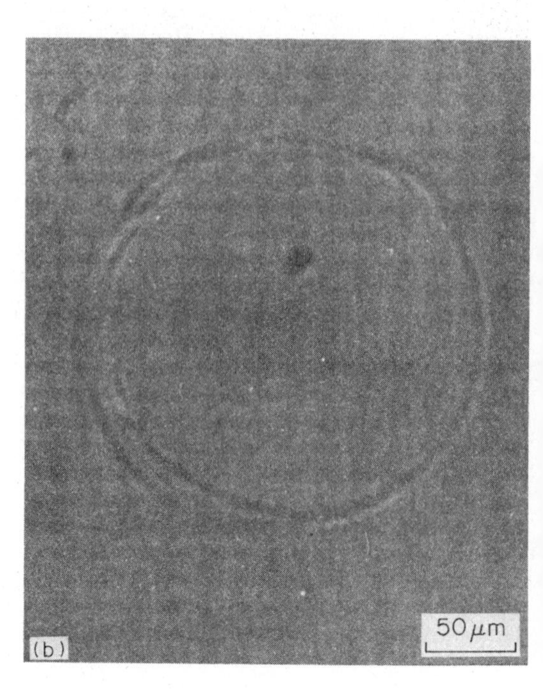

Figure 4
(a) Partial Hertzian cone crack revealed by the fracture process, and (b) ring crack observed on the surface

material is defective or if the component has been subject to impact damage, the origin tends to be relatively large (compared with typical microstructural features) and is easily distinguished. Common, readily detected defects (Fig. 3) include large pores or pore clusters, inclusions, isolated large grains or preexisting cracks (cracks present prior to fracture). Pores derive from deficiencies in the fabrication process. The size, shape or distribution of the pores often provides insight regarding the specific processing step which initiates the pores and thus has implications for material improvements. Inclusions typically result from contamination during the fabrication process. Chemical analyses of the inclusion by an electron microprobe technique can determine its likely origin. Isolated large grains originate from inadequate powder quality, inhomogeneous mixing or insufficient control of the densification parameters. The size, shape and crystal structure of the large grains can indicate the specific source of such defects.

Preexisting cracks may originate in several ways. Each source produces a crack or crack network having distinct characteristics visible on the fracture surface adjacent to the fracture origin. Sources of cracks include laminations formed during pressing, and shrinkage separations formed during drying or binder removal.

Some failures result from damage produced in the material during handling or during operation. Impact and oxidation/corrosion-induced damage are two examples of defects that can be readily detected by examination of the fracture origin. In the case of impact, some of the characteristics of the impacting body can also be deduced. Two principle types of impact damage occur: that induced by blunt projectiles (e.g., spheres or soft projectiles) and sharp projectile damage. Blunt projectile impacts elicit elastic response of the ceramic and result in the creation of Hertzian cone cracks. Such cracks can be readily identified at the failure origin as a cone partially exposed by the final fracture process (Fig. 4a). A thin circular ring on the surface is also generally associated with this type of origin (Fig. 4b). In contrast, irregular projectiles may cause plastic deformation of the ceramic and result in an array of radial cracks (Fig. 5). Evidently only one of these cracks causes the final fracture. Consequently, when this type of damage is the source of failure, several radial cracks remain on the surface around the fracture origin and can be readily identified by inspection of the surfaces. In this

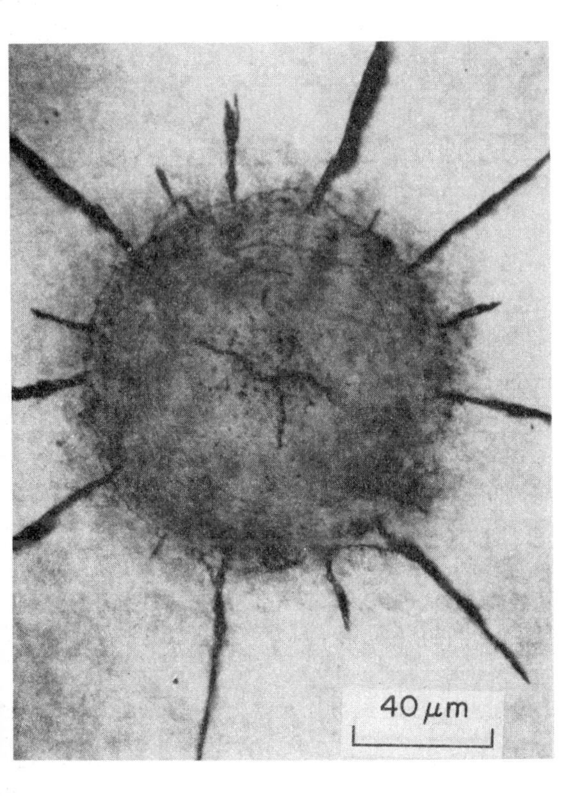

Figure 5
Radial cracks caused by the impact of an irregular projectile

instance, fracture surface observations are of minimal utility, because the profile of the specific radial crack that caused the failure is difficult to distinguish, being morphologically similar to the fracture surface created during final failure.

Oxidation and corrosion-related damage processes include surface recession, pitting and reaction layers usually visible on the fracture surface. An example of a fracture which has initiated at a surface pit caused by oxidation/corrosion is illustrated in Fig. 6.

The fracture surfaces discussed so far are relatively easy to diagnose. Some fracture surfaces are much more difficult to evaluate, because the features are less distinct. This is the case for large-grained ceramics, low-strength ceramics, bimodal microstructure ceramics and fractures initiated by machining damage, contact stress and thermal shock.

Failure from machining damage is probably the most difficult to distinguish. Machining generates small, subsurface cracks (Fig. 7). The cracks develop along failure surfaces similar to those associated with the final, unstable failure. Consequently, fracture surface observations do not readily reveal the profile of the critical machining flaws. Relatively subtle distinguishing features must be sought. For example, in Si_3N_4, machining damage exhibits a larger proportion of transgranular fracture than the subsequent load-induced stable growth. The machining flaws can thus be most effectively identified using optical microscopy with oblique illumination, by taking advantage of the

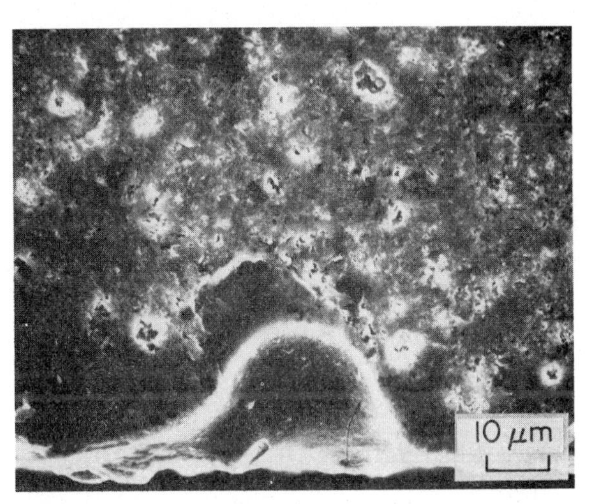

Figure 6
Failure from an oxidation-induced surface pit in Si_3N_4

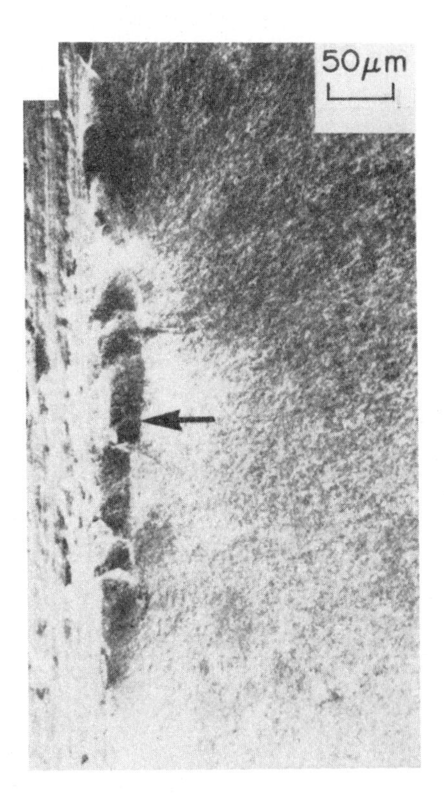

Figure 7
Scanning electron micrograph of machining-induced cracks in MgF_2 (courtesy of R. W. Rice)

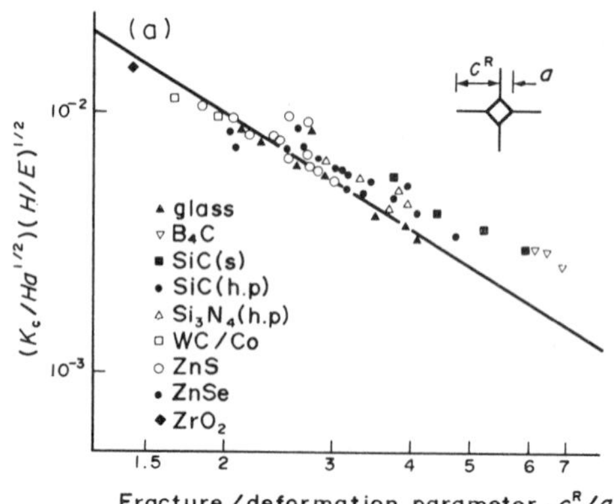

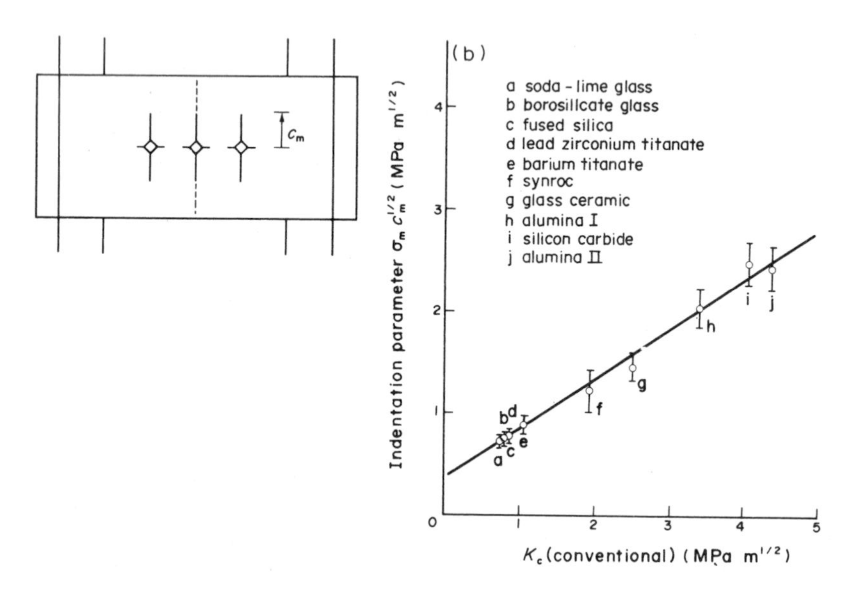

Figure 8
Synopsis of the indentation technique for determining the toughness on small fragments of the failed ceramic component: (a) direct method; (b) indirect method

greater reflectivity of the transgranular segments of the failure surface. Scanning electron microscopy is of minimal utility, because the transgranular regions are not readily discerned at the low magnification needed to locate the critical flaws. In other materials, changes in trajectory between the machining flaws and the failure surface can be used to define the flaw profile, but unambiguous identification is difficult.

2. *Analysis of Failure*

In addition to flaw identification, measurements of various dimensions at the failure origin can provide knowledge of the local stress that caused the failure. In particular, the radius r_m of the mirror (Fig. 2a) is uniquely related to the fracture stress σ_f by

$$\sigma_f = \lambda K_c r_m^{-1/2} \tag{1}$$

where K_c is the fracture toughness and λ is a material-independent constant ($= 3.5 \pm 0.3$). Consequently, an independent measurement of the toughness, in conjunction with determination of r_m, allows the fracture stress to be determined *a posteriori*.

Furthermore, the toughness can be conveniently ascertained on a fragment of the actual failed component, using the indentation technique (Fig. 8). The estimate of the failure stress obtained in this manner is probably the least ambiguous, because it relies on the measurement of characteristic dimension of the failure crack (i.e., the bifurcation length), which in turn is apparently determined by a characteristic stress intensity factor $K \approx 4.5K_c$. However, additional complexity arises when the component is highly defective. The mirror radius then becomes a significant fraction of the component dimension in the fracture plane and λ in Eqn. (1) is modified due to the effect of the compliance on the stress intensity factor K. The modified λ can be computed from the pertinent stress intensity factor solutions, some of which are available in handbook form.

Separate measurements of the dominant flaw dimension can also be informative. Clearly a relationship exists between flaw size and strength. In principle, therefore, the flaw size should provide an independent estimate of the failure stress. However, such measurements are of limited utility, because the strength–flaw size relationship depends on the flaw type (Fig. 9). Furthermore, unique relationships have not been developed for many important flaw types, notably pores, low-strength inclusions and large grains. Nevertheless, flaw size measurements can be advantageous in the analysis of impact damage or failures associated with machining damage. In particular, cracks generated by sharp projectile impact and machining are subject to stable growth on load application, due to plasticity-induced residual stress. Consequently the crack dimension at instability is straightforwardly related to the stress. For example, a circular crack with instability radius a_m relates to the stress by

$$K_c = (2/\pi^{1/2})\sigma a_m^{1/2} \qquad (2)$$

Measurement of the radius a_m when clearly delineated on the fracture surface thus provides the requisite estimate of the failure stress.

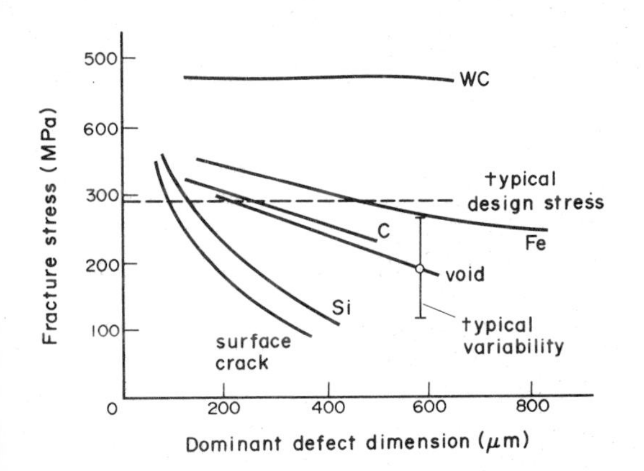

Figure 9
Effects of flaw type on the strength–flaw size relation for Si_3N_4 (curves labelled WC, Fe, C and Si are for inclusions of these substances)

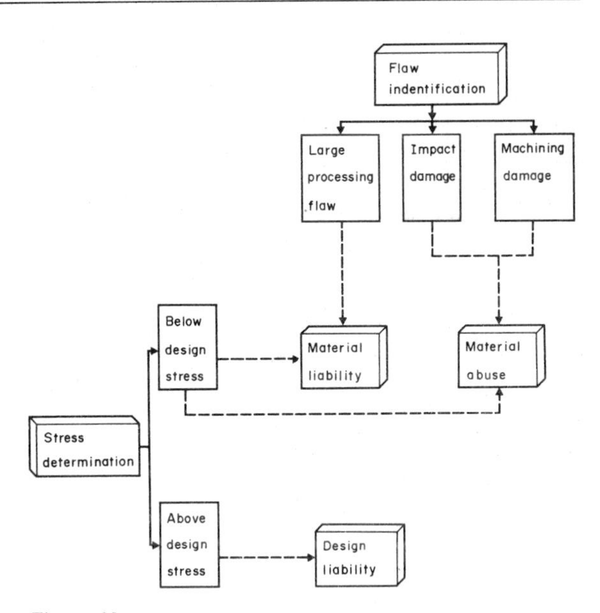

Figure 10
Procedure for assigning liability based on the two principal steps in the failure analysis: flaw identification and failure stress evaluation

3. Diagnosis of Failure

The information obtained in accordance with the above procedures may be assembled and used to specify the cause of failure, thereby permitting recommendations for avoiding a recurrence of the failure. The unambiguous identification of the failure-initiating flaws provides direct evidence of the defective nature of the material. In particular, a large processing flaw at the origin would be a tentative basis for classifying the material as defective. Conversely, a flaw attributed to impact or machining would indicate that the material had been abused after processing. Furthermore, evaluation of the stress at which the fracture occurred establishes whether the material satisfies the strength specification afforded by the supplier and, additionally, whether the component was subject to an overload condition (i.e., by comparison with the design stress). These implications are summarized in Fig. 10.

The recommendations based on the preceding diagnosis are self-evident, with liabilities assigned to the material manufacturer, to the handling or installation engineer or to the design engineer. Corrective action based on these findings can then be instituted.

See also: Failure Analysis: General Procedures and Applications to Metals; Failure Analysis of Plastics

Bibliography

Evans A G, Charles E 1976 Fracture toughness determinations by identification. *J. Am. Ceram. Soc.* 59: 371–72

Kirchner H P, Gruver R M 1976 In: Bradt R C, Hasselman D P H, Lange F F (eds.) 1976 *Fracture Mechanics of Ceramics*, Vol. 1. Plenum, New York, p. 309
Lawn B R, Evans A G, Marshall D B 1980 Elastic/plastic indentation damage in ceramics: The median/radial crack system. *J. Am. Ceram. Soc.* 63: 574–81
Rice R W 1978a In Herman H (ed.) 1978 *Treatise on Materials Science and Technology*, Vol. 11. Academic Press, New York, p. 199
Rice R W 1978b In: Palmour H, Davis R F (eds.) 1978 *Processing of Crystalline Ceramics*. Plenum, New York, p. 303
Rice R W 1984 *Fractography of Ceramic and Metal Failure*, ASTM STP 827. American Society for Testing and Materials, Philadelphia, Pennsylvania
Richerson D W 1983 *Modern Ceramic Engineering: Properties, Processing and Use in Design*. Dekker, New York
Schardin H 1958 In: Averbach B L, Filbeck D K, Hahn G T, Thomas D A (eds.) 1958 *Fracture*. Wiley, New York, p. 304

A. G. Evans
[University of California, Berkeley, California, USA]

D. W. Richerson
[Garrett Corporation, Phoenix, Arizona, USA]

Failure Analysis of Plastics

Failure analysis of plastics is frequently an investigation of the occurrence of brittle failure in a material which is generally accepted as ductile. Failures do occur in the ductile mode but their appearance can usually be explained satisfactorily on the basis of simple mechanical properties, such as the yield stress. Account must be taken, however, of the effect of variables, including the time of loading and temperature, on such properties, since the load-bearing capability of plastics in general, and thermoplastics in particular, decreases with time and with increasing temperature.

In this article, factors which may induce brittleness in otherwise ductile plastics are identified, and relevant examples are discussed.

1. Ductile Failure

1.1 Ductile Failure in Air

This type of failure is found in pressure vessels and pipes, but is perhaps observed most frequently in polyethylene film: overloaded garbage sacks extend several hundred percent locally until the seriously thinned polymer can no longer support the load and breaks. In more critical engineering applications, the appropriate design calculations must be made, involving stress (or strain) level, time under load and temperature and also environmental effects, if any. A ductile failure is characterized by the occurrence of cold drawing, often accompanied by stress whitening.

1.2 Influence of Liquids on Ductile Failure

The mechanical properties of plastics which absorb appreciable quantities of specific liquids can be affected adversely; for example, the stress at yield of high density polyethylene is reduced by 10% if the polymer is immersed in gasoline. This polymer is suitable, nevertheless, for the manufacture of fuel tanks for road vehicles. The ductile failure stress for polyamides (nylon) is reduced even more seriously by the absorption of water, but the toughness is improved.

1.3 False Yielding of Particulate Structure

Polymers in which the particulate structure is preserved after melt shaping (especially poly(vinylchloride) and polytetrafluoroethylene), and some polymers containing particulate fillers, may suffer premature ductile failure at an unexpectedly low stress. This is ascribed to shear failure of the links between the particles, and is illustrated in Fig. 1.

2. Factors which May Induce Brittle Failure

2.1 Materials Factors

(*a*) *Choice of polymer*. If brittle failure constitutes a hazard which should be minimized, it is reasonable that one should choose a plastic with a low propensity to brittle failure; or, if the choice is between two brittle materials, the one with the higher strength under the operating conditions should be chosen. Brittle failures produce smooth surfaces, with characteristic zones generated by an accelerating crack in the material. These zones are shown in Fig. 2.

(*b*) *Wrong grade of polymer*. Within a family of plastics, there are frequently many different "grades:"

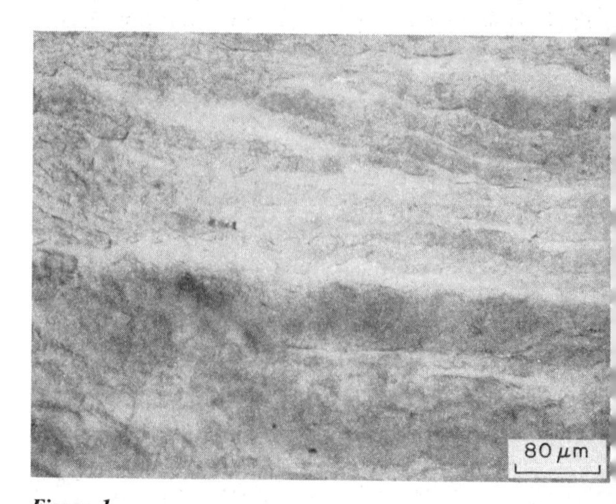

Figure 1
Thin section of polytetrafluoroethylene after tensile failure, showing interparticle separation. Normal light micrograph

Figure 2
Fracture surface of polystyrene. Reflected light, differential interference contrast

these may be based on polymers of different molecular chain length (often identified as "molecular weight") or polymers with different additives to improve particular properties (e.g., impact modifiers). The incidence of brittleness is increased in polymers of lower molecular weight, as illustrated by the cracking of a strip cut from a pressure vessel which was inadvertently manufactured from an injection-molding grade of propylene copolymer, instead of the higher-molecular-weight blow-molding grade. The failure in Fig. 3 resulted from 100 h loading at 100 °C, while a similar specimen of the blow-molding grade remained intact.

(*c*) *Inclusion of hard particulate fillers.* Embrittlement is an inevitable consequence of adding particulate fillers, which are of higher modulus (hardness), to a ductile plastics material; the elongation at break decreases rapidly with increasing filler content. The effects are particularly disastrous if the particle size exceeds 10–20 μm, or when the polymer matrix–filler particle interface is weak.

2.2 Design Factors

(*a*) *Poor design and stress concentrations.* Stress concentrations cause engineers many problems, yet designers insist on their inclusion in many products. The example cited here in Fig. 4 is an automobile coolant reservoir tank, which incorporated a concave section to accommodate an electric cable. Proof testing, and stress analysis by a finite elements method, both highlighted this feature as a serious stress concentrator when the vessel was pressurized. Sharp corners are even more widespread examples of stress concentrators which can induce brittle failure.

(*b*) *Poor mold or die design.* Features of mold design which lead most frequently to unexpected failure are

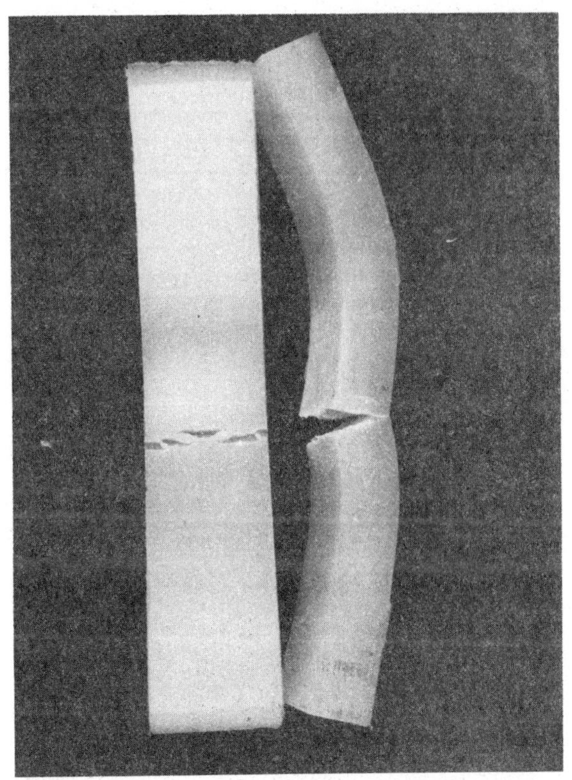

Figure 3
High-temperature stress cracking in low-molecular-weight polypropylene

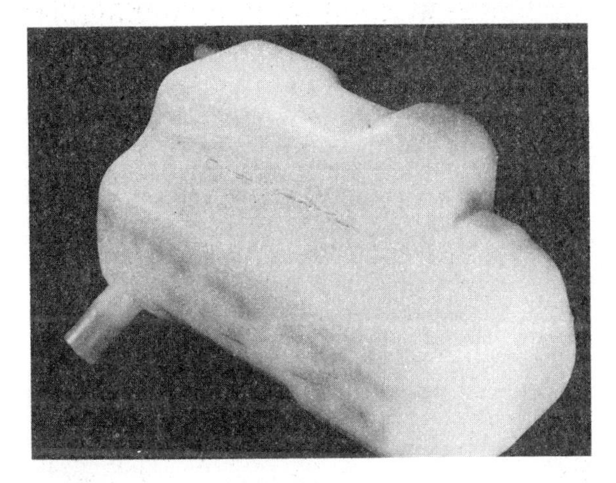

Figure 4
Stress cracking in polypropylene

associated with a lack of understanding of the behavior of polymer melts, particularly where multiple flow paths are involved. There are numerous examples of such failures; for instance, a washing machine tub which is fed by eight gates located circumferentially around the central hub. The requirement that the multiple flows join together leads to eight lines of weakness along the hub, providing ample explanation of the cracks which develop in service. Another common feature which encourages failure is an abrupt change in thickness.

(*c*) *Unsatisfactory assembly—welds*. Many plastics products are fabricated by joining several subcomponents; the methods employed are many and include solvent welding, adhesives and hot welding. There are many variants of the last, which is illustrated here by hot plate welding. In this process the two parts to be joined are brought into contact with a "hot plate," maintained at a temperature above the softening or melting temperature of the plastic materials. After a suitable time, the hot plate is removed and the softened parts brought into contact and kept under a steady pressure until the polymer has cooled sufficiently to harden. Variables such as hot plate temperature, time of heating, pressure of the components on the hot plate, pressure exerted at the welding stage and the time of welding all affect the finished weld, as does the texture of the original components. Even for the optimum weld there are regions of weakness associated with the demarcation between polymer which was melted during the welding and the original texture, and with the presence of welding beads, made up of material which was forced out of the weld. An ideal weld should retain about 50% of the softened polymer in the weld, with the other 50% being forced into the beads, carrying any contaminant or degraded polymer at the junction out of the weld, into the beads. Figure 5 shows a far from perfect weld, made at too high temperature, or too high welding pressure, or both. Noteworthy are the sharp notches formed by the welding beads, leading into the weakened region at the boundary of the weld. This component failed catastrophically at the weld during proof testing.

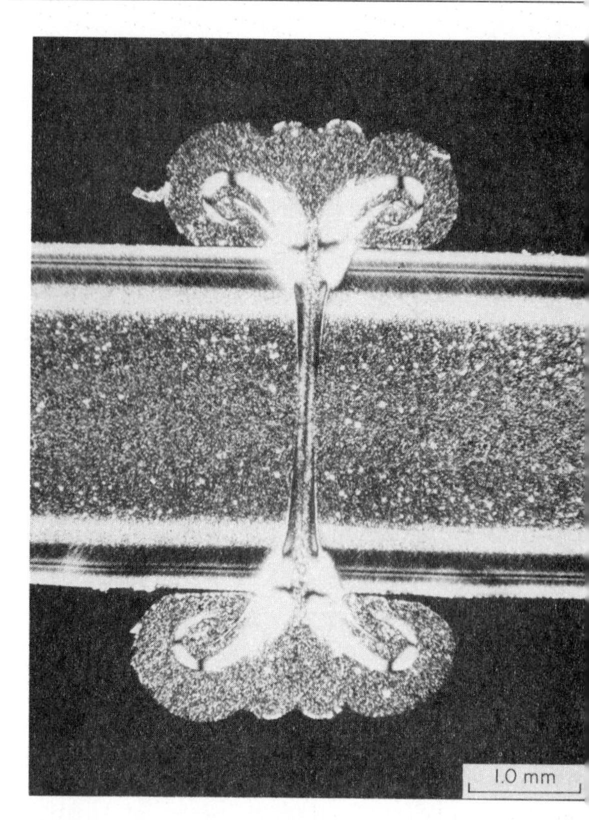

Figure 5
Unsatisfactory polypropylene weld. Crossed polars

2.3 Processing Faults

(*a*) *Too low temperature—inhomogeneous melt*. One of the important considerations in the shaping of thermoplastics is that the melt should be homogeneous with respect to consistency and temperature. If the melt is prepared at too low temperature, with a minimum amount of shear, it is possible that unmelted granules pass through the melting zone into the shaping process and are incorporated in the final product. For crystalline plastics particularly, such inclusions in the finished product frequently initiate brittle failure, as illustrated in Fig. 6, which shows such a granule of unmelted acetal polymer "contaminating" an injection modling.

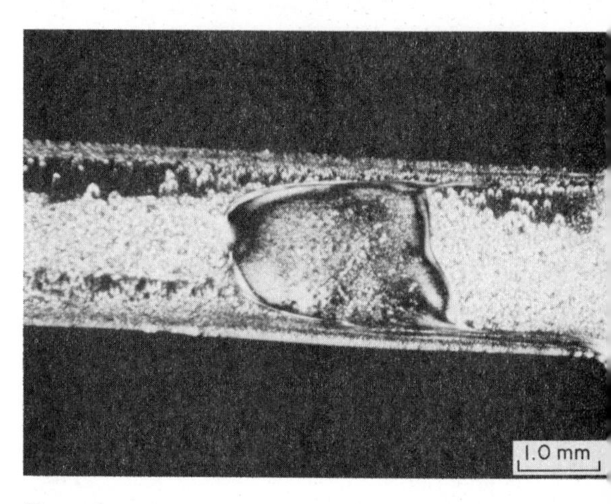

Figure 6
Unmelted granule in acetal injection molding. Crossed polars

(*b*) *Too high temperature—degraded polymer*. Many plastics are unstable at elevated temperatures, degrading by mechanisms which include depolymerization (unzipping) and random chain scission; oxidation during processing is a further powerful route for degradation. Figure 7 shows infrared spectra of two regions of a 250-liter container, shaped by rotational casting of medium-density polyethylene; both spectra

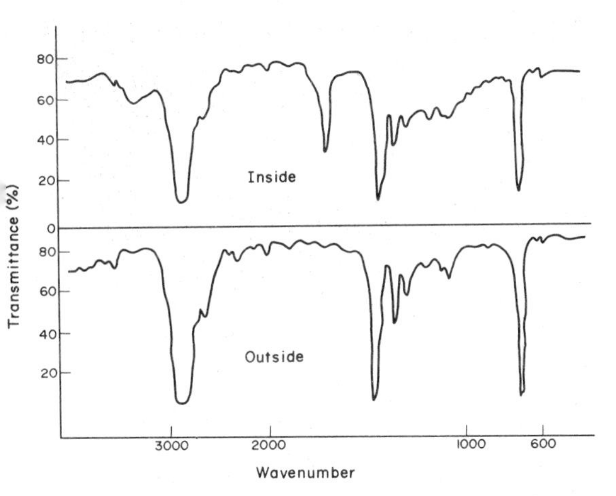

Figure 7
Infrared spectra of oxidized and normal polyethylene

Figure 8
Degradation and splitting at an interface in poly(vinylchloride). Fluorescence microscopy

are from the base of the container; that labelled "outside" had been in contact with the mold surface, whereas that coded "inside" had free access to air during the molding operation. The latter spectrum shows evidence of heavy oxidation, indicating that an unreasonably high temperature, possibly with a long molding cycle, had been employed. This product split across the base when abused by impact blows. Degraded regions in poly(vinylchloride) (PVC) can be located using fluorescence, as shown in Fig. 8.

(*c*) *Pressure too low—surface defects.* The shaping process of extrusion blow molding utilizes gas pressure—usually air—to blow a hollow extrudate to conform to the shape of a mold. During the cooling stage the molding seeks to shrink and to part from the mold. Should this occur, the surface quality of the molding is impaired, and this can contribute to failure. Figure 9 illustrates the surface texture of blow moldings: micrographs (a) and (b) were direct observations of the surface by scanning electron microscopy, whereas micrographs (c) and (d) were obtained by a surface replication technique, using carboxymethyl cellulose, and photographed in normal light. The latter technique is applicable when nondestructive examination is required.

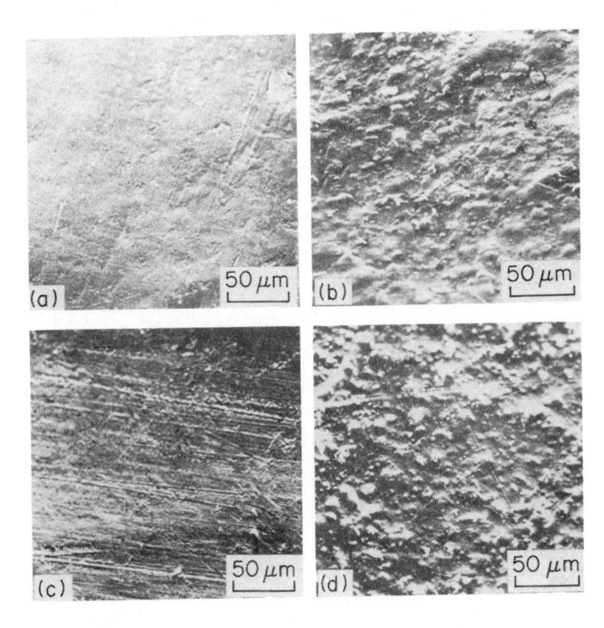

Figure 9
Micrographs (a) and (c) are of the surface of a satisfactory blow molding; (b) and (d) are from a blow molding at too low pressure

(*d*) *Inadequate dispersion of additives.* This is one of the most common causes of the embrittlement of ductile plastics and often involves fillers or pigments, which are frequently hard particulate solids. Sometimes the dispersion is so poor that mere visual examination is sufficient to pinpoint this as a prime cause of brittle failure, but it is more usual to cut thin sections from the faulty product (3–10 μm thickness) and examine these under normal light illumination in a microscope. Figures 10 and 11 are micrographs which illustrate two aspects of poor dispersion, both of which have led to brittle failure in service. Figure 10 shows typical pigment smears which constitute a plane of weakness; the pigment is carbon black, which is a difficult material to disperse satisfactorily. Figure

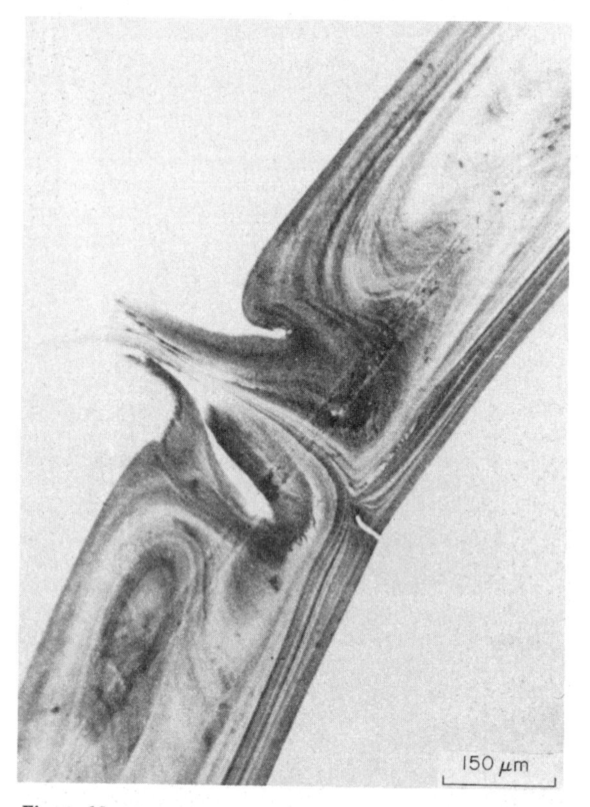

Figure 10
Section of carbon-black-filled polyethylene. Normal light

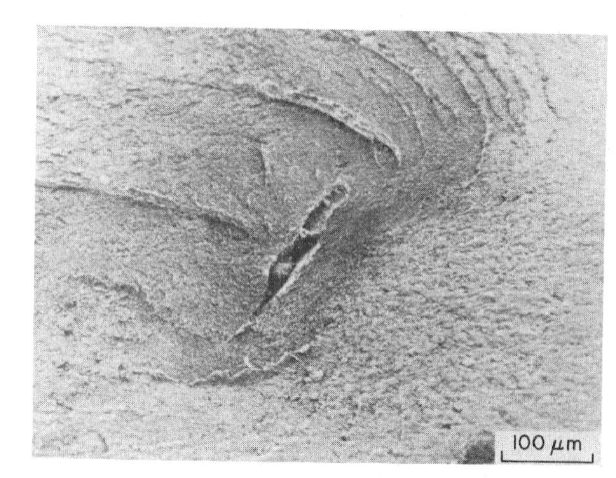

Figure 11
Pigment agglomerate in PVC. Scanning electron micrograph

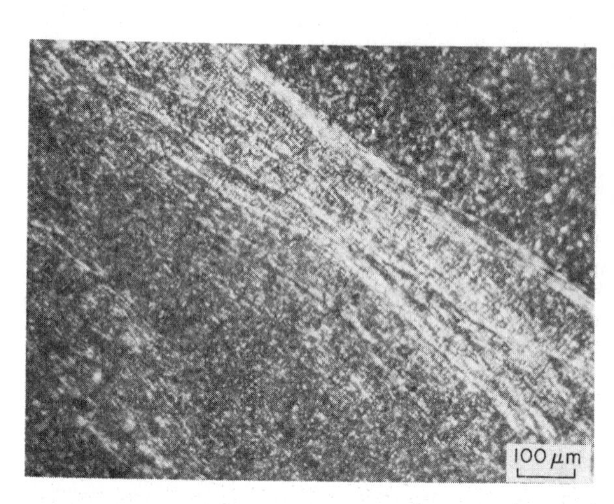

Figure 12
Low-density polyethylene inclusions in polypropylene. Crossed polars

11 is titanium dioxide in PVC, demonstrating in this case the difficulty of breaking down pigment agglomerates, the large particle acting as a stress concentrator; a crack can be seen originating at the particle. Streaks of high pigment concentration passing through the entire thickness of the product may give rise to high gas and vapor permeability.

(*e*) *Incompatible masterbatch polymer.* The practice of incorporating additives by means of concentrates in a carrier polymer, a masterbatch, is growing. It allows the accurate dispensing of additives which may be present at very low concentration in the final product. In principle, the polymer in the masterbatch should be similar to the bulk polymer to which it is added, but the use of low-molecular-weight polymers facilitates the mixing of the masterbatch. Low viscosity impedes dispersion, however. There is an increasing tendency to regard masterbatches as "universal," with some manufacturers recommending the use of "alien" polymer masterbatches. Thus low-density polyethylene is employed as a masterbatch carrier for incorporating additives in polypropylene. The incompatibility of these two polymers, coupled with poor mixing, results in heterogeneous texture which, in critically stressed regions, leads to premature failure. Such an inclusion is shown in Fig. 12; the micrograph, taken with crossed polars, shows very clearly the undispersed low-density polyethylene. This contributes to failure in two ways: first, by acting as a stress concentrator, and second, by being a region of low-molecular-weight polymer, which is weaker than the higher-molecular-weight matrix polymer.

(*f*) *Particulate contamination.* This type of failure is similar to that resulting from inadequate dispersion of additives which are incorporated deliberately, except that contaminants are included accidently, often as a consequence of poor working practice. Such inclusions act as stress concentrators, and are effective at sizes of 5–10 μm and upwards. They are thus visible in a light microscope, and this method, together with x ray dispersive analysis, is used in their identification. Contaminants are frequently inorganic, often minerals, but may be alien polymer used elsewhere in the plant.

(*g*) *Voids*. Should they occur, voids are obvious stress concentrators, and thereby contribute to premature brittle failure. The origin of voids is frequently associated with poor processing, and their appearance is promoted by parts of thick section. Figure 13 is a micrograph taken in polarized light showing voids in an injection molding of acetal copolymer, giving rise to a crack.

(*h*) *Unsuitable cooling conditions*. Blow moldings are often cooled only by their contact with a cold mold which, to reduce cycle time and thereby increase productivity, is chilled in a temperature marginally higher than the dew point of the surrounding air. For artifacts of considerable thickness, 4 mm and upwards, the cooling time may be a minimum of 2 min, imposed by the low cooling rate of the polymer at the inner surface. Such a slow cooling rate allows the growth of a very coarse texture, exemplified here by an automobile coolant reservoir tank, blow molded in polypropylene. Figure 14 is a scanning electron micrograph of the inner surface of the tank in a region where the thickness was 6 mm, implying a very low cooling rate when the cooling route is by natural convection. Cracks can be seen developing between the spherulites, which have grown to large size, approximately 150 μm.

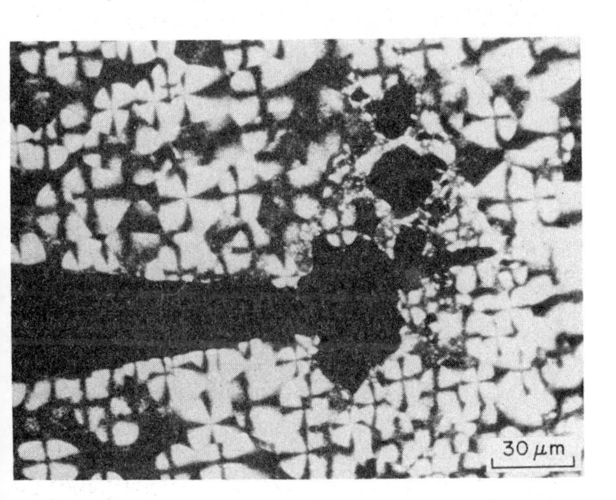

Figure 13
Voiding and crack in injection molded acetal polymer. Crossed polars

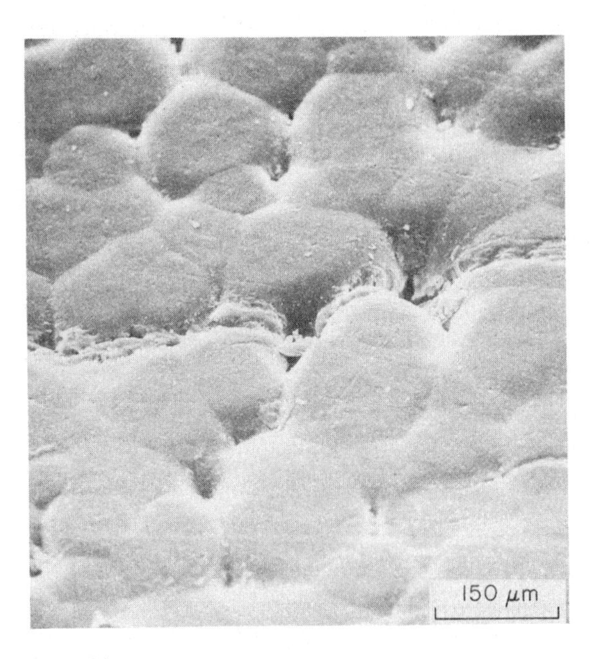

Figure 14
Slow-cooled inner surface of blow-molded polypropylene tank. Scanning electron micrograph

(*i*) *High residual orientation*. High shear rate shaping processes, such as injection molding, inevitably give rise to orientation of the polymer chains; further, the mold is used to cool and solidify the plastic, so that the polymer in contact with the mold surface is solidified rapidly, before there is opportunity for the orientation to relax. This region of high molecular alignment is a fruitful source of cracks, particularly when the injection molding is subsequently remelted in part, as in welding. Figure 15 is a micrograph, taken with crossed polars, of a weld joining two sections of subcomponents injection molded in high-molecular-weight, high-density polyethylene. The very high residual orientation of the original texture can be seen, together with the very confused morphology this leads to in the weld, which itself was near optimum for these subcomponents. To improve the quality of the injection molding, it is necessary to change the processing conditions, increasing cycle times; however, this reduces the profit to the molder, and therefore this change is avoided.

(*j*) *Imperfect internal welds*. When two flow paths meet head on, as they do after passing round an obstacle, a weak region is generated unless they integrate perfectly, which they do only rarely. Similar problems can arise when flows from multiple gates come together, so that internal welds are common features in injection moldings. Attention must be focused on minimizing their effect, both at the mold

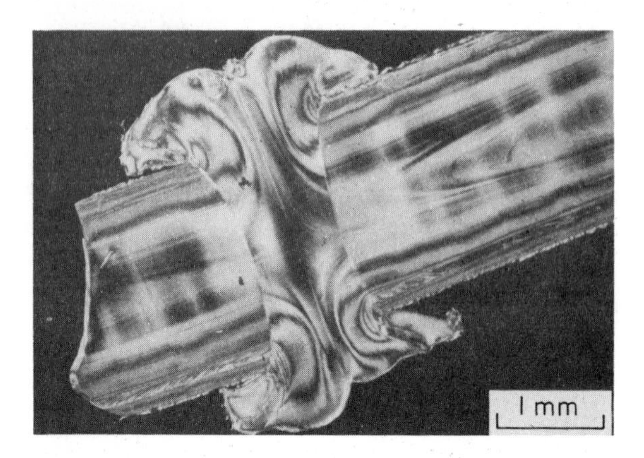

Figure 15
High residual orientation in high-molecular-weight, high-density polyethylene

design stage and in subsequent processing. Internal welds can be readily identified by microscopic examination of a thin section, usually between crossed polars, and exemplified here by a molding in acetal polymer (Fig. 16). Remedies at the mold design stage include thickening of the section at the expected meeting point, allowing greater opportunity for the mixing of the two flows.

(*k*) *Faults with gates and runners*. The gate is a unique feature in any molding, since it is the region where packing the mold to compensate some of the shrinkage in cooling is likely to lead to a textural hiatus, and consequently a region of stress concentration. This is demonstrated in Fig. 17, which is the gate area of an injection-molded nylon sealing ring. Cracks which initiated brittle failure were generated in this region.

(*l*) *Orientation of fillers, especially fibers*. Anisotropic fillers, and especially fibers, react to flow by turning their major axes into a perferential direction, which depends on the nature of the flow (e.g., whether it is converging or diverging). The effect on the properties of the composite is to make them anisotropic, and in extreme cases this can promote failure.

(*m*) *Morphological interfaces*. Most thermoplastics products are of nonuniform texture, a feature which is often caused by the different rates of cooling experienced. Thus, when a cold mold is used in either injection or blow molding, the product has a quenched zone of fine texture adjacent to the mold, with the interior, which is cooled more slowly, exhibiting a coarser texture. The change in texture (and density) may be abrupt and the mismatch appears to give a plane of weakness from which cracks may generate. Figure 18, which is a section taken from blow-molded polypropylene, shows this phenomenon.

Figure 16
Internal weld in molded acetal polymer

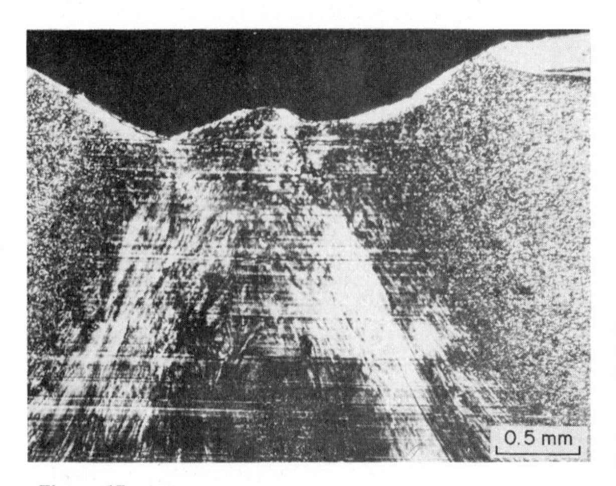

Figure 17
Gate region of nylon injection molding. Crossed polars

2.4 Service Factors

(*a*) *Prolonged loading*. The possibility of brittle failure is increased by longer times under load, the failure occurring apparently at low deformation, although

Figure 18
Thin section of blow-molded polypropylene. Crossed polars

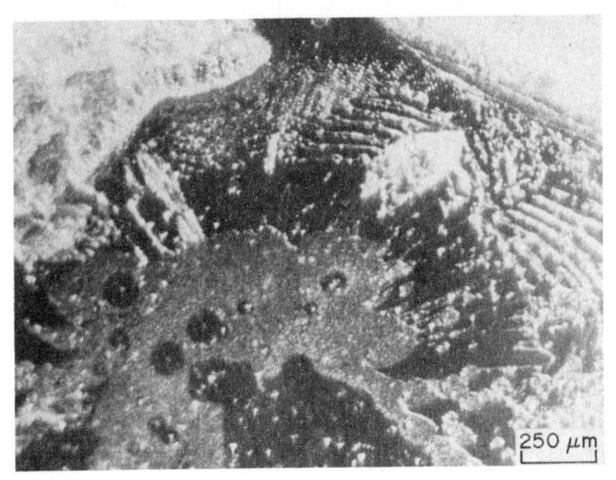

Figure 19
Fatigue crack in polystyrene

microscopic examination reveals evidence of partial ductility. The classic example of brittle failure brought about by long-term loading can be found in the proof testing of high-density polyethylene pipes; failure of the pipes under internal pressure is characteristically ductile at short times, but at long times, particularly at high temperature, brittle failures at low strain are encountered.

(*b*) *Cyclic loading*. This mode of stressing, often known as "fatigue," or "dynamic fatigue," leads to embrittlement, and is a cause of failure of some materials which are otherwise very durable and can withstand almost unlimited impact abuse. Thus, although acrylonitrile–butadiene–styrene (ABS) is generally regarded as a very tough plastic, finding application where impact stresses are likely, for example, as a cover for an air-cushion grass cutter, it is rapidly embrittled on being subjected to cyclic (alternating) stresses. The illustration in Fig. 19 is a fatigue crack in polystyrene, the micrograph being taken in the dark field, reflected light mode. Examination of the fracture surface reveals the periodic undulations, which record the progress of the crack with each cycle of stress.

(*c*) *Thermal/photochemical degradation.* Most polymers will degrade if heated to sufficiently high temperature, which is usually, however, above the melting or softening point. This behavior, therefore, mainly concerns the processor. Light energy, and especially ultraviolet radiation, is potentially more damaging and, at ambient temperatures, is the more usual cause of degradation. The requirements for photochemical reaction are that the radiation quanta are absorbed by the substrate, and that the quanta contain sufficient energy to cause reaction to occur. Some PVC compounds are susceptible to photochemical attack which reduces the molecular weight of the polymer, embrittling it. Further, volatile products are evolved, decreasing the volume, the two effects together giving severe surface cracking, which is exemplified in Fig. 20, a micrograph obtained by scanning electron microscopy.

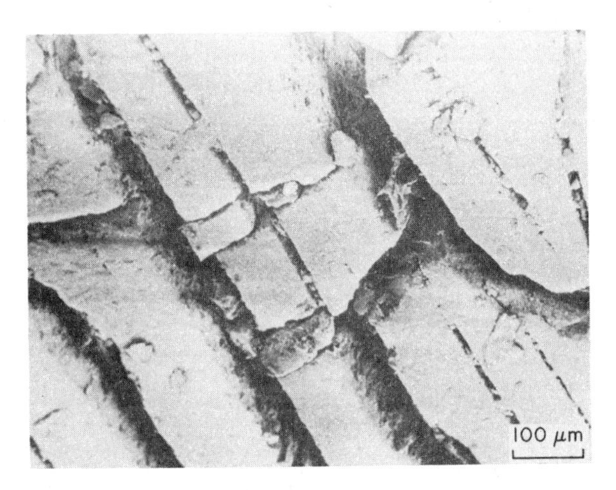

Figure 20
Typical surface cracking of PVC

(*d*) *Thermal/photochemical oxidation.* As with degradation in service, again the most insidious effects are photochemical in origin; natural weathering causes deterioration in many unprotected plastics, although stabilized grades of most are available and give satisfactory lifetimes. An example is the use of propylene–ethylene sequential copolymer in crates for storing and transporting beers and soft drinks. When not in use, such crates are frequently stored outside and subjected, therefore, to weathering. Suitably protected grades can withstand such treatment for many years, but unprotected grades degrade and embrittle by photochemical oxidation in a year or two. Oxidized polypropylene residues can be found in the surface layers of such exposed crates; reprocessing them spreads and activates the oxidized products, resulting in products which are far less stable to further thermal or photochemical attack.

(*e*) *Attack by environment.* It is the responsibility of the designer to choose materials for a product which are resistant to the environment in which the product is expected to operate. These "rules" are generally followed, except in the case of water, which is often not considered to be an active environment. Several important plastics react with water, however, particularly at high temperatures, although deterioration at temperatures near the ambient is not unknown. Susceptible polymers include polyamides and polyesters, including polycarbonate, which is taken here as an example.

Polycarbonate (PC) is a tough transparent plastic, with a softening point of 145–150 °C, which implies that it can be sterilized by boiling water or steam techniques. Further, it absorbs only a small amount of water, even after prolonged immersion, and the mechanical properties are affected only marginally by water content. However, water causes degradation of PC by hydrolysis; thus, a baby's feeding bottle, for which PC seems the ideal material, giving lightweight and tough transparent products, embrittles after a few weeks' use, because of degradation by hydrolysis resulting from the repeated sterilization. A similar mechanism degrades nylon when the common practice of boiling completed moldings in water to stabilize the water content, and thereby the dimensions and properties, is followed.

(*f*) *Environmental stress cracking.* Although plastics are generally resistant to chemical attack, and can be loaded over long periods without failure, the combination of certain environments with stress, applied externally or generated internally, or both combined, can induce brittle failure in particular plastics. Polypropylene is a plastic favored for the manufacture of automobile coolant reservoir tanks, although the ethylene glycol added to the water as antifreeze is a mild stress cracking agent for the polymer. This can be appreciated by reference to Fig. 21, which shows samples taken from the same regions of three similar tanks. Sample (a) is the control, which was not subjected to any test, while samples (b) and (c) were subjected to the same proof test (identical pressure, time and temperature). For (b) the tank had been filled with a 50:50 mixture of ethylene glycol and water, while for (c) the tank contained water only. The deterioration observed in (b) is typical of this type of failure.

(*g*) *Abuse of product.* Most products are designed, and expected, to withstand limited abuse; toys provide many examples where maltreatment is almost inevitable. Hammer blows frequently simulate the main type of abuse; the example in Fig. 22 was hit repeatedly with a heavy hammer, in an attempt to prove that the vessel was brittle! By counting the number of scars which did not cause failure, it can be deduced that it was the twelfth blow which precipitated the failure. Fortunately, there are not so many people about with this destructive urge, so this level of abuse need not be catered for. Nevertheless, the designer of a product must try to foresee the types of abuse most likely to be encountered and to design against failure under such circumstances; economics will inevitably curtail what can be achieved.

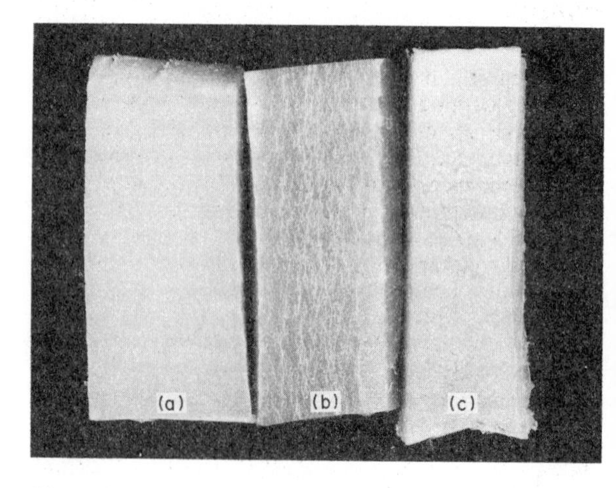

Figure 21
Environmental stress cracking of polypropylene in ethylene glycol–water mixture. Temperature 115 °C

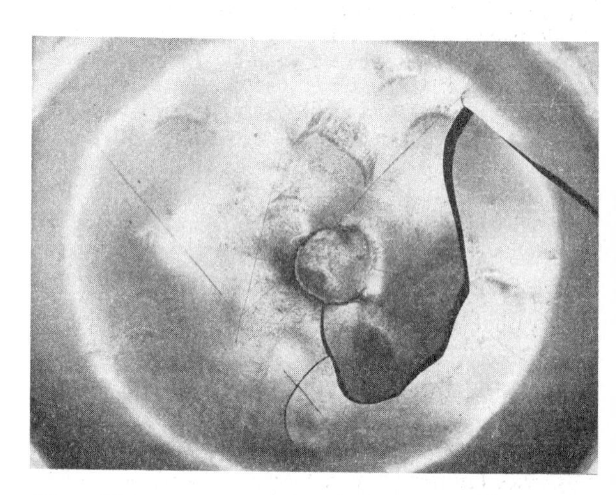

Figure 22
Much abused base of large polyethylene container

3. Concluding Comment

The foregoing may seem to be a very full and possibly alarming catalogue of the failures that can, and indeed have, affected plastics products, since all those cited have been failures in commercial products. On the pessimistic side, the list is probably incomplete, as further contributors to failure are being identified as more is learned about plastics. On the credit side, many of the known types of failure can now be prevented.

See also: Failure Analysis: General Procedures and Applications to Metals; Mechanics of Materials: An Overview

Bibliography

Birley A W, Scott M J 1982 *Plastics Materials: Properties and Applications.* Blackie, Glasgow

Brown R (ed.) 1984 Polymer Testing—A Special Issue. *Meas. Tech. Polym. Solids* 4 (2–4)

Engel L, Klingele H, Ehrenstein G, Schaper H 1981 *Atlas of Polymer Damage.* Wolfe Science, London

Haslam J, Willis H A, Squirrel D C M 1972 *Identification and Analysis of Plastics.* Butterworth, London

Rodriguez F 1982 *Principles of Polymer Systems.* McGraw–Hill, New York.

Tadmor Z, Gogos C G 1979 *Principles of Polymer Processing.* Wiley, New York

D. A. Hemsley and A. W. Birley
[University of Technology, Loughborough, UK]

Fatigue Crack Growth: Macroscopic Aspects

Fracture by the progressive growth of incipient flaws under cyclically varying loads, that is, metal fatigue, must now be considered as the principal cause of in-service failures in engineering structures and components. However, the process of fatigue failure itself consists of several distinct processes including initial cyclic damage (cyclic hardening or softening), formation of an initial "fatal" flaw (crack initiation), macroscopic propagation of this flaw (crack growth) and final catastrophic failure or instability. The purpose of this article and the article *Fatigue Crack Growth: Mechanistic Aspects* is to focus attention primarily on the macroscopic crack propagation stage of fatigue failure, and to attempt to relate the microscopic (metallurgical) and macroscopic (continuum mechanics) descriptions which have been presented for this process.

The physical phenomenon of fatigue was first seriously considered in the mid-nineteenth century when widespread failures of railway axles in Europe prompted Wöhler in Germany and Fairbairn in England to conduct the first systematic investigations into the fracture of materials under cyclic stresses around 1860. In 1917 the first reported observations of corrosion fatigue were made by Haigh concerning the effect of seawater on the failure of steel cables. However, the main impetus for research directed at the crack propagation stage of fatigue failure, as opposed to mere lifetime calculations, did not occur until the mid-1960s when the concepts of linear elastic fracture mechanics and so-called "defect-tolerant design" were first applied to the problem of subcritical flaw growth. This approach recognizes that all structures are flawed, and that cracks may initiate early in service life and propagate subcritically. Lifetime is then assessed on the basis of the time or the number of loading cycles for the largest undetected crack to grow to failure, as might be defined by an allowable strain, or limit load or fracture toughness (K_{Ic}) criterion. Implicit in such analyses is that subcritical crack growth can be characterized in terms of some governing parameter (often thought of as an effective "crack driving force") which describes local conditions at the crack tip yet may be determined in terms of loading parameters, crack size and geometry. Linear elastic and elastic–plastic fracture mechanics have, to date, provided the most appropriate methodology for such analyses to be made, and consequently, considerable effort has been directed towards defining parameters which uniquely describe stress and strain fields at the crack tip over length scales characteristic of the local fracture mechanisms involved.

The concept of directly applying fracture mechanics to subcritical faitgue crack growth was first suggested by Paris et al. (1961) in their famous "Rational Analytic Theory of Fatigue." Despite difficulties in finding a technical journal which would publish the manuscript, their proposal of correlating fatigue crack propagation rates (da/dN) with the stress intensity factor (K_I) has remained the basis of the defect-tolerant fatigue design approach ever since.

The current widespread acceptance of the fracture mechanics approach to fatigue, as utilized in defect-tolerant design philosophies, has meant that a major emphasis in fatigue research in recent years has been directed towards the study of macroscopic crack propagation, particularly with regard to the collection of engineering materials data on cyclic crack growth rates. The macroscopic aspects of such fatigue crack propagation behavior are reviewed in this article.

1. Fracture Mechanics Characterization

For cracks subjected to cyclically varying loads, maximum and minimum stress intensity factors, K_{max} and K_{min}, defined at the extremes of the cycle, are assumed to prescribe crack growth. According to the original analysis of Paris and his co-workers, the crack growth increment per cycle in fatigue (da/dN) can be described in terms of a power law function of the range of K_I, given by the alternating stress intensity $\Delta K(=K_{max} - K_{min})$, thus:

$$\frac{da}{dN}=C\Delta K^m \quad (1)$$

where C and m are scaling constants for a particular material–heat-treatment condition. It is important to note here that such an asymptotic continuum mechanics characterization does not necessitate detailed quantitative microscopic models to be known for the individual fracture events, and in view of the complexities of these processes on the microstructural scale, this must be regarded as fortunate, at least for a macroscopic description of fatigue crack extension.

One of the principal limitations in this approach (and, in fact, the criterion that K_I or ΔK are valid descriptions of crack tip fields) is that a state of small-scale yielding (ssy) exists, as dictated by the size of the crack tip plastic zone in relation to the overall dimensions of the body. For a cyclically-stressed elastic–perfectly-plastic solid, plastic superposition of loading and unloading stress distributions can be used to compute the extent of plastic zones ahead of a fatigue crack (Fig. 1). On loading to K_{max}, a monotonic or maximum plastic zone is formed at the crack tip of dimension:

$$r_{max} \approx \frac{1}{2\pi}\left(\frac{K_{max}}{\sigma_y}\right)^2 \quad (2)$$

where σ_y is the yield strength. However, on unloading from K_{max} to K_{min}, superposing an elastic unloading distribution to maximum extent $-2\sigma_y$, results in a region ahead of the crack tip where residual compressive stresses of magnitude $-\sigma_y$ exist. This region is known as the cyclic plastic zone size r_Δ and is approximately one fourth the size of the monotonic zone, i.e.:

$$r_\Delta \approx \frac{1}{2\pi}\left(\frac{\Delta K}{2\sigma_y}\right)^2 \quad (3)$$

where, strictly speaking, σ_y is now the cyclic yield strength. The correlation of K_I (or ΔK) to crack extension by fatigue will thus be a valid approach provided that r_{max} and r_Δ are small compared with the in-plane dimensions of crack length and ligament depth.

In situations where the above conditions are not met, such as for fatigue crack growth under large-scale yielding at high stress intensity ranges or for very short cracks, alternative macroscopic descriptions have been proposed based on elastic–plastic (nonlinear) fracture mechanics. One such approach is based on the J integral where, despite difficulties in its precise meaning as applied to cyclically stressed nonstationary cracks in regions of elastic unloading and nonproportional loading, some authors have proposed a power-law correlation of the fatigue crack growth rate under elastic–plastic conditions to the range of J, i.e.,

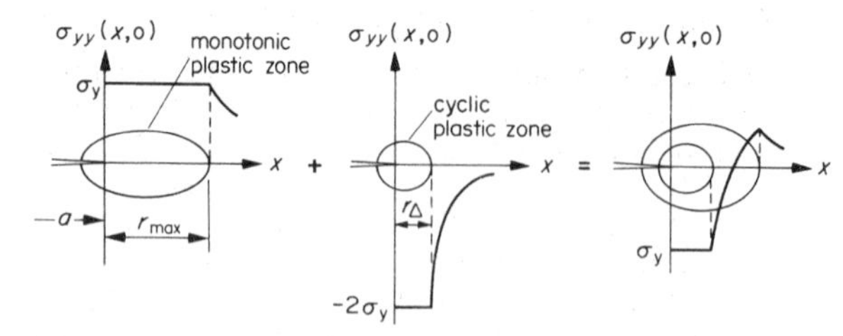

Figure 1
Plastic superposition of loading and unloading stress distributions for an elastic–perfectly-plastic solid of flow strength σ_y during fatigue crack propagation used to estimate the extent of the maximum plastic zone (r_{max}) and cyclic plastic zone (r_Δ). $\sigma_{yy}(x,0)$ is the maximum principal stress directly ahead of the crack (after Rice 1967)

$$\frac{da}{dN} \propto \Delta J^{m'} \quad (4)$$

as discussed in Sect. 4.

An alternative approach to elastic–plastic fatigue crack growth, which is not necessarily limited by the problems associated with the nonlinear elastic definition of J (i.e., based on deformation theory of plasticity), can be derived using the concept of crack tip opening displacement (CTOD), δ_t. For proportional loading:

$$\delta_t = Jd(\varepsilon_y,n)/\sigma_y \quad \text{(elastic–plastic)}$$
$$\propto K_I/\sigma_y E' \quad \text{(linear elastic)} \quad (5)$$

where E' is the appropriate elastic modulus and d is a proportionality factor dependent upon the yield strain ε_y, the work hardening exponent n, and whether plane strain or plane stress is assumed. Since, like J, δ_t can be taken as a measure of the intensity of the elastic–plastic crack tip fields, it is feasible to correlate the rate of fatigue crack growth to the range of δ_t, the cyclic crack tip opening displacement (ΔCTOD), as

$$\frac{da}{dN} \propto \Delta\text{CTOD} \quad \text{(elastic–plastic)}$$
$$\propto \frac{\Delta K_I^2}{2\sigma_y E'} \quad \text{(linear elastic)} \quad (6)$$

Approaches based on J and δ_t are fundamentally equivalent for proportional loading situations, and can be converted to K_I-based analyses for linear elastic behavior. However, unlike J and K_I, the CTOD approach perhaps offers more physical insight into the mechanisms of crack advance since it can be more readily related to the physical crack tip failure processes involved.

2. *Experimental Measurement*

Fatigue crack propagation rate data are generally measured using standard fracture mechanics-type specimen geometries, such as the compact tension, edge-notched bend, edge-notched tension, center-cracked sheet test-pieces, and so forth. Starting from a mechanically sharpened crack, specimens are subjected to a cyclically varying load ΔP, generally under constant load control (increasing stress intensity K_I), and the increase in crack length is monitored as a function of time or number of cycles N at constant frequency. Numerous crack monitoring techniques have been employed, such as optical procedures using calibrated travelling microscopes or high-speed photography, electrical procedures using eddy current probes or resistivity measurements (electrical potential techniques), compliance procedures using strain gauges or crack mouth displacement gauges, and ultrasonic or acoustic emission detectors. Data are used to construct crack length a versus number of cycles N plots, which are then differentiated, either graphically or numerically, to determine the rate of crack growth $(da/dN)_{ai}$ for each crack length a_i (Fig. 2). Corresponding to each crack length a_i, the value of $(\Delta K)_{ai}$ is computed from the applied loads and the relevant K_I calibration for the particular test geometry, such that the data are finally presented in the form of log–log plots of da/dN versus ΔK (Fig. 2). Empirical expressions, of the form of Eqn. (1), are then numerically fitted to the data to define the crack growth relationship (i.e., $da/dN \propto \ln(\Delta K)$) for the particular combination of material, testing conditions and environment in question. These procedures have now been standardized by the American Society for Testing and Materials in their ASTM Standard Test Method E647-81 for constant-load-amplitude fatigue crack growth rates above 10^{-8} m per cycle.

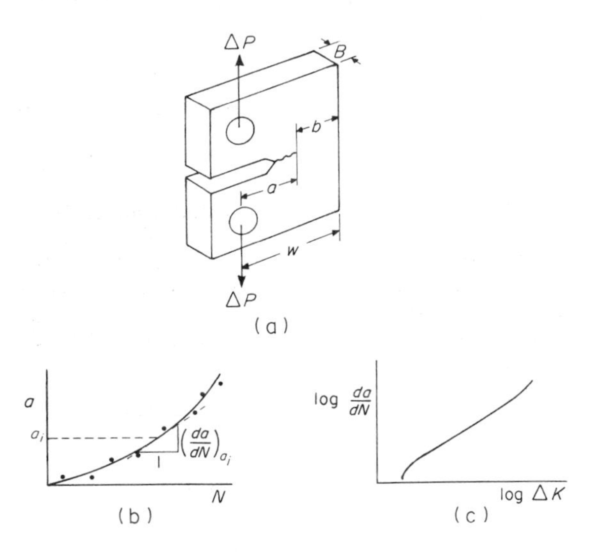

Figure 2
Schematic illustration showing procedures for measuring fatigue-crack propagation rates: (a) compact specimen stressed under cyclic loads ΔP, (b) crack length a versus number of cycles N curve differentiated to give growth rate (da/dN) at particular crack length a_i and (c) resulting log–log plot of da/dN vs alternating stress intensity ΔK

To determine crack growth behavior at lower growth rates, i.e., in the so-called near-threshold regime where stress intensities approach a threshold ΔK_{th}, below which cracks remain dormant or propagate at experimentally-undetectable rates (Fig. 3), it is generally more realistic to monitor growth rates under load-shedding (decreasing K_I) conditions, in order to minimize transient residual stress effects. Threshold levels are approached using a procedure involving successive load reduction followed by crack growth. Measurements of da/dN are taken at each load level, over typical increments of 1–1.5 mm increase in crack length, after which the load range is reduced by not more than 10%, and the same procedure followed. Larger reductions in load are liable to give premature crack arrest from retardation effects owing to residual plastic deformation (see Sect. 6). The increments, over which measurements of growth rate are taken, are chosen to represent distances at least four times larger than the maximum plastic zone dimension generated at the previous load level. The threshold ΔK_{th} for no crack growth is then measured as the value of ΔK where the growth rate is infinitesimal, which is generally operationally defined in terms of minimum growth rate, say 10^{-11} m per cycle, calculated from the accuracy of the crack monitoring technique and the number of cycles elapsed. For example, using electrical potential crack length measurements where an accuracy of 0.1 mm on absolute crack length can be conservatively assumed, detecting no crack extension in say 10^7 cycles could be employed to define the threshold as the stress intensity range to give a growth rate not exceeding 10^{-11} m per cycle. Following ΔK_{th} measurement, the load may be increased in increments and procedures again used to measure da/dN. In this way, low growth rate data are monitored under decreasing and increasing K_I conditions, and provided that care is taken to minimize transient effects caused by too severe load changes, growthrate behavior should be similar. A typical test plot of da/dN versus ΔK for fatigue crack propagation in an ultrahigh strength steel, representing data spanning the entire range of growth rates from threshold levels to final failure determined by these procedures, is shown in Fig. 3.

Such fatigue crack growth measurement procedures can be easily automated using a suitable crack monitoring technique and a computer-controlled testing machine. This is particularly relevant for near-threshold testing under load-shedding conditions, where a

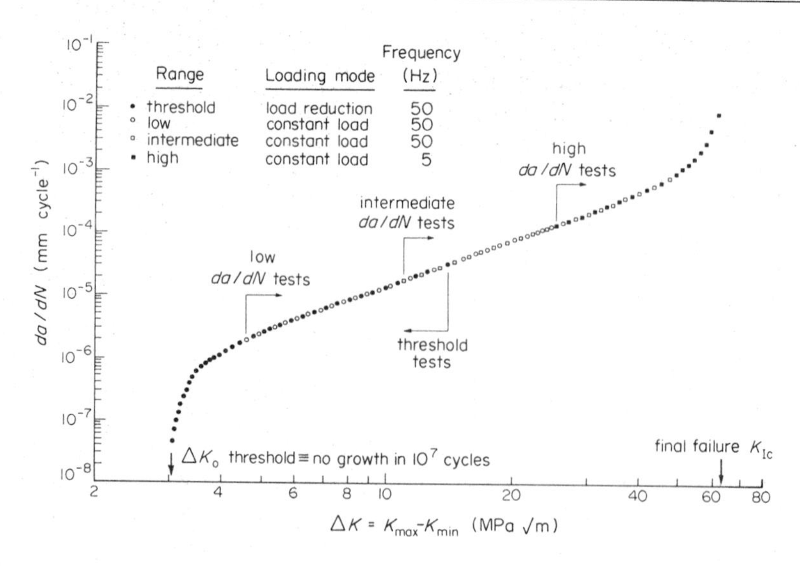

Figure 3
Typical test conditions for obtaining fatigue crack propagation data spanning entire range of growth rates from threshold levels (ΔK_{th}) to final failure (defined by the fracture toughness K_{Ic}). da/dN vs ΔK data measured for an ultrahigh-strength 300-M steel ($\sigma_y = 1700$ MPa)

programmed constant decrease in the normalized K_I gradient, i.e., $|\Delta K^{-1}.\ d\Delta K/da|$, is utilized. Such procedures are currently being standardized by the ASTM.

3. Fatigue Crack Growth Under Linear Elastic Conditions

As described above, the application of linear elastic fracture mechanics and related small-scale crack tip plasticity has provided a basis for describing the phenomenon of fatigue crack propagation, where the crack growth increment per cycle (da/dN) is found to be principally a function of the stress intensity range ΔK through power-law expressions of the form $da/dN = C\Delta K^m$ (Eqn. (1)). This basic relationship provides an adequate engineering description of behavior at the mid-range of growth rates, typically $\sim 10^{-5}$ to 10^{-3} mm per cycle. At higher growth rates, however, as K_{max} approaches instability conditions, defined by the fracture toughness (K_{Ic}, K_c) or the limit-load, Eqn. (1) often underestimates the propagation rate, whereas at lower (near-threshold) growth rates it is generally conservative, as ΔK approaches ΔK_{th}. In general, the sigmoidal variation of growth rates with ΔK can be characterized in terms of three regimes, governed by different primary mechanisms of failure, as shown schematically for ambient environments in Fig. 4. From this figure, it is apparent that crack growth behavior can be dependent upon a number of different variables (to an extent dependent upon the growth rate

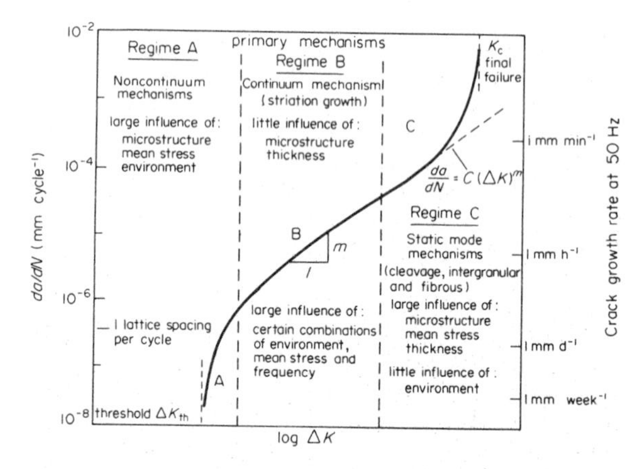

Figure 4
Schematic variation of fatigue-crack propagation rate (da/dN) with stress-intensity range (ΔK), showing regimes of primary growth rate mechanisms, and effects of several major variables on crack-growth behavior

regime), such as mean stress (characterized by the load ratio $R = K_{min}/K_{max}$), microstructure, frequency, environment, test-piece thickness and so forth. Mechanisms describing such kinetics of fatigue crack extension, together with the specific influence of such variables, are discussed in detail in the article *Fatigue Crack Growth: Mechanistic Aspects.*

4. Elastic–Plastic Fatigue Crack Growth

The majority of fatigue crack propagation data is obtained in conditions where the scale of local crack tip plasticity is small compared with the in-plane test-piece dimensions of crack length and ligament depth. As described in Sect. 1, under such circumstances the conditions of small-scale yielding exist and characterization of the rate of crack growth in terms of a linear elastic stress intensity factor is appropriate. There are, however, several situations where crack growth is accompanied by more extensive plasticity such that linear elastic analyses become invalid. These situations can arise for crack propagation at high stress intensity ranges (i.e., regime C in Fig. 4), for initial crack growth from notches where the crack is advancing within the strain field of the notch, and for the growth of so-called short cracks where local crack tip plastic zone sizes are comparable with crack lengths (see Sect. 5).

Early approaches to characterizing such elastic–plastic fatigue crack growth were based on the plastic strain range ($\Delta\varepsilon_p$) where the rate of crack extension was empirically approximated by expressions of the form:

$$\frac{da}{dN} \approx (\Delta\varepsilon_p)^2 a \tag{7}$$

However, the difficulties associated with measurement of a local parameter such as $\Delta\varepsilon_p$ in the vicinity of the crack prompted Dowling (1976) to propose a more global approach relying on elastic–plastic (nonlinear) fracture mechanics. Based on the concept that crack tip stress and deformation fields in an elastic–plastic strain hardening solid can be described by the Hutchinson–Rice–Rosengren (HRR) singularity and thus uniquely characterized by the J-integral, Dowling suggested that fatigue crack propagation in the presence of extensive plasticity could be correlated to the range of J, that is, ΔJ, which under linear-elastic conditions would be equal to $\Delta K^2/E'$. The success of this approach is evident from Fig. 5 where both linear elastic and elastic–plastic crack growth data in nuclear pressure vessel steel are shown to closely correspond, even though plastic zone sizes above $\sim 10^{-3}$ mm per cycle are of the order of 25 mm, that is, comparable with overall test-piece dimensions. However, the fundamental validity of this approach has often been questioned since it appears to violate the basic assumption in the definition of J that stress must be proportional to current plastic strain. This follows because J is defined from the deformation theory of plasticity (i.e., nonlinear elasticity) which does not allow for the elastic unloading and nonproportional loading effects inherent in crack extension. However, despite these limitations the use of ΔJ does provide a good fit to fatigue crack propagation data over a very wide range of growth rates, as shown in Fig. 5, and therefore must still be regarded as a feasible, yet perhaps empirical, characterization parameter for cyclic crack advance.

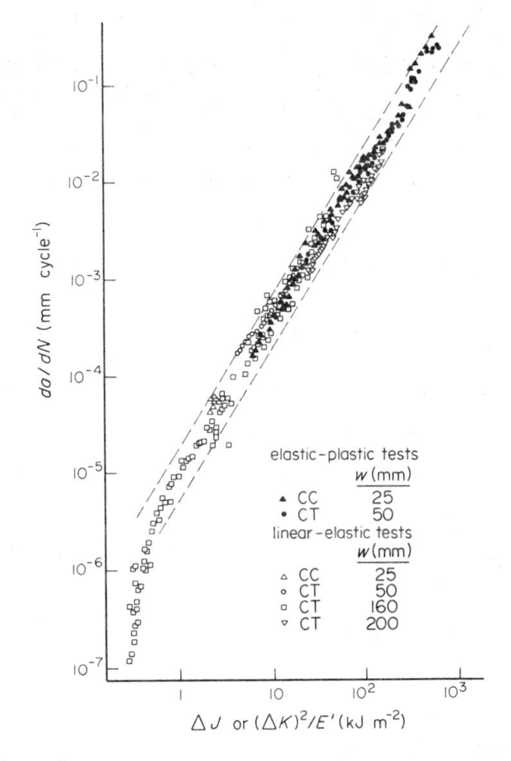

Figure 5
Fatigue crack propagation rates (da/dN) under linear elastic and elastic–plastic conditions in A533B nuclear pressure vessel steel ($\sigma_y = 480$ MPa) as a function of ΔJ for test geometries of compact (CT) and center cracked (CC) specimens of varying width w. Note ΔJ is equal to $\Delta K^2/E'$ under linear-elastic conditions (after Dowling 1976)

An alternative approach to elastic–plastic fatigue crack growth, which is not subject to these nonlinear elastic restrictions, is to correlate crack extension to the range of crack tip opening displacement ΔCTOD. This is appealing since it permits a clearer physical interpretation of the macroscopic crack propagation data in terms of the microscopic fracture events involved, although, somewhat surprisingly, it has not been extensively utilized in the literature, possibly because of difficulties in measurement and interpretation.

5. Short Fatigue Cracks

Current defect-tolerant design codes have been developed to estimate the service life of cyclically loaded structures in terms of the subcritical growth rates of incipient flaws utilizing the existing database of laboratory-determined fatigue crack propagation results.

However, the majority of such data has been determined from test-piece geometries containing cracks of the order of 25 mm or so, whereas many defects encountered in service are far smaller than this, for example, in turbine disk and blade applications. In the relatively few cases where the behavior of such short cracks has been studied, it has been found, almost without exception, that at the same nominal driving force, the growth rates of short cracks are greater than (or equal to) the corresponding growth of long cracks. Thus, a potential exists for nonconservative defect-tolerant predictions in components where the growth of short flaws represents a large proportion of the life.

Cracks may be considered short if their length is: (a) comparable to the microstructural size-scales (e.g., of the order of a grain size); (b) comparable to the scale of local plasticity (e.g., $\sim 1\ \mu m$ in ultrahigh strength materials ($\sigma_y \sim 2000$ MPa) to ~ 1 mm in low strength materials ($\sigma_y \sim 300$ MPa)); or (c) simply physically small ($< 0.5-1$ mm).

For the growth of cracks small compared with the scale of plasticity, at intermediate growth rates (i.e., regime B in Fig. 4), the use of elastic–plastic fracture mechanics, specifically involving ΔJ, has shown a one-to-one correspondence between short and conventional long crack data. However, this is only one aspect of a very complex problem, since the kinetics of short flaw crack growth may be influenced by microstructural and environmental factors in ways markedly different from long flaws. In particular, it has been shown that short cracks can propagate at stress intensities below the long crack threshold ΔK_{th}, both from smooth surfaces and when emanating from notches. Such cracks then advance at progressively decreasing growth rates before arresting or joining the long crack propagation curve (Fig. 6). Further details concerning the microscopic aspects of the short crack problem can be found in the bibliography and in the article *Fatigue Crack Growth: Mechanistic Aspects*.

Short crack studies are currently of high priority in fatigue research since the problem represents a breakdown in the similitude concept ingrained in conventional fracture mechanics analyses. Extreme care, therefore, must be taken in the application of existing (long crack) fatigue crack propagation data for defect-tolerant lifetime calculations in components containing small defects, since without making allowances for the possible faster propagation rates and lower thresholds associated with such short flaws, predictions may be dangerously nonconservative.

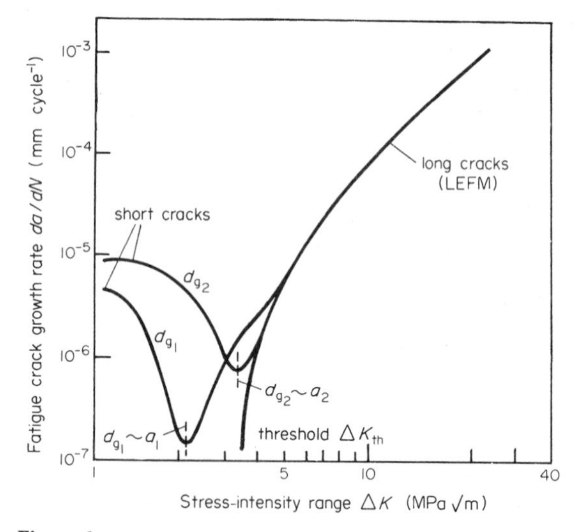

Figure 6
Data showing the growth rate behavior of microscopically short cracks, as a function of grain size (d_g), in 7075-T6 aluminum alloy ($\sigma_y = 515$ MPa). Note how short cracks propagate below the long crack threshold (ΔK_{th}) at progressively decreasing growth rates, with the minimum velocity occurring when the crack length is of the order of the grain size (i.e., $a \sim d_g$) (after Lankford 1982)

6. *Variable Amplitude Loading*

Since the majority of components in service are subjected to variable amplitude, spectrum or random loading, rather than constant amplitude loading, an appreciation of the influence of periodic overloads and block loading sequences on fatigue crack propagation is clearly of practical significance. Classically, the approach taken for variable amplitude loading in fatigue design is to use the Palmgren–Miner relationship, where for a body subjected to n_i number of cycles of different stress amplitudes $\Delta\sigma_{a_i}$, with $i = 1$ to k,

$$\sum_{i=1}^{k} n_i/N_i = 1 \qquad (8)$$

where N_i is the number of cycles to failure had all the stress amplitudes been of magnitude $\Delta\sigma_{a_i}$ (computed from S/N diagrams). This approach, which is utilized for total life prediction and thus includes both crack initiation and crack propagation stages, predicts that the cycles to failure will be decreased if a large single positive (spike) overload is applied during constant amplitude cycling. However, as shown schematically in Fig. 7, the effect of such an overload on mode I crack propagation rates is to initially retard crack growth, sometimes to the point of arrest, and thus to increase the propagation life. Hence it is apparent that the application of Miner's rule (Eqn. (8)) to crack propagation behavior under variable amplitude loads is often inappropriate.

For crack growth studies, however, the above approach can be utilized for narrow band random loading where each cycle does not differ significantly from its predecessor. Here interaction effects can be considered small and it is reasonable to assume that each cycle causes the same amount of crack extension as if it were applied as part of a sequence of constant

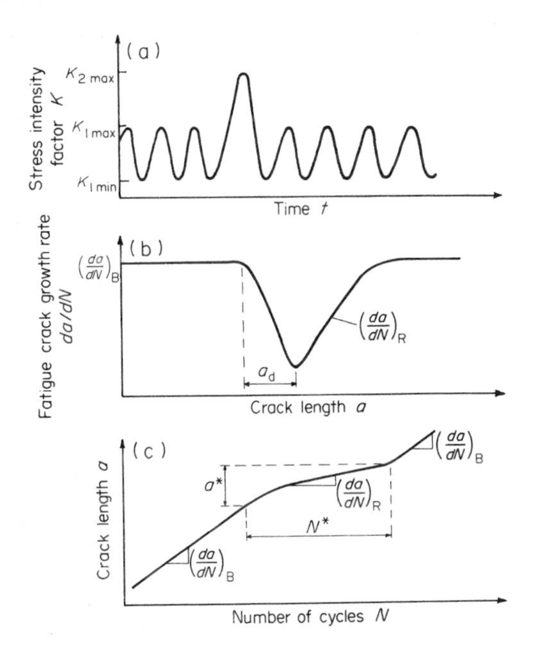

Figure 7
Schematic illustration showing the transient deceleration in growth rates during mode I fatigue crack propagation resulting from the application of a single positive (spike) overload

amplitude loads. For broad band loading, on the other hand, cyclic crack extension has been estimated on the basis that the load, or stress intensity, ranges are constant at their root mean square value. However, since loading sequence interaction effects are more relevant here, this procedure often yields conservative predictions for mode I crack advance.

The nature of these interaction effects is shown in Fig. 7. The application of a single positive (spike) overload of sufficient magnitude (generally $>50\%$ of the baseline ΔK) can result in significantly retarded crack growth over a crack length increment a* which is of the order of, or greater than, the overload plastic zone size, whereupon crack growth rates return to their original baseline level. Sometimes cracks overloaded in this fashion totally arrest, where presumably the effective stress intensity actually experienced at the crack tip is reduced below ΔK_{th}. However, the onset of the retardation or crack arrest often does not coincide with the actual overload cycle but is delayed until the crack has grown some distance into the overload plastic zone (so-called delayed retardation). For fully reversed overload cycles, the application of an underload (compressive overload) prior to a single overload does not appear to have much effect on subsequent retardation, whereas an underload following an overload tends to minimize retardation. Block loading sequences are also known to produce large transient effects on fatigue crack propagation. For example, high–low block loading sequences, where the stress or stress intensity range is reduced from one constant level to a lower one, can result in an initial retardation or crack arrest which is actually much greater than for an equivalent single overload.

The origin of these interaction effects is associated with mechanisms of crack tip blunting and branching, premature crack closure and the fact that the overloaded crack is growing into the residual stress field of the overload plastic zone. In defect-tolerant design, however, the role of variable amplitude cycles is incorporated through the use of more empirical models, such as those of Wheeler and Willenborg. The Wheeler model assumes that the extent of crack growth retardation following the overload cycle is related to the distance the retarded crack has transversed into the overload plastic zone, whereas the Willenborg approach computes an effective stress intensity, based on the nominal K_{max} being reduced due to the presence of residual stresses within this plastic zone. However, since in many materials the onset of retarded crack growth does not coincide with the point of application of the overload cycle (i.e., delayed retardation), both models are somewhat artificial physically. Nevertheless, they are utilized practically in the development of more complex crack growth relationships (based on Eqn. (1)) which are in widespread use for defect-tolerant design procedures, particularly for airframe components.

7. *Design Philosophies*

The conventional approach to design against fatigue failure during a finite lifetime has classically involved the use of S/N curves (representing the life N resulting from a stress amplitude S), suitably adjusted to take into account the effects of mean stress (using Goodman diagrams), effective stress concentrations at notches (using fatigue-strength reduction factors), variable-amplitude loading (using the Palmgren–Miner cumulative-damage law), multiaxial stresses, environmental effects, and so forth. In materials that show a well-defined fatigue limit, such as steels in benign atmospheres, infinite life might be expected when applied alternating stresses are below this limit, with due consideration of the factors listed above. This approach essentially represents design against crack initiation, since near the fatigue limit, especially in smooth specimens, this may occupy the major portion of the lifetime ($\sim 80\%$). Crack initiation here may represent either the nucleation of a microcrack that becomes the "fatal" macrocrack by continued propagation or the coalescence of several microracks to form the macrocrack—whichever is the critical event for the particular microstructure in question. This approach, although simply empirical in nature, is still in widespread use for design.

For safety-critical applications, especially with welded and riveted structures, there has been a growing awareness that the presence of defects in a material below some certain size must be assumed and taken into account at the design stage. Under such circumstances, the integrity of a structure will depend on the lifetime spent in crack propagation since the crack-initiation stage can be assumed to be small, and conventional S/N analyses, where remaining lifetime is based on the number of cycles both to initiate and to propagate the crack, may dangerously over-estimate the life. Such considerations have led to the widespread adoption of the defect-tolerant fatigue design approach based on a fracture mechanics description of crack propagation behavior. In this approach, maximum possible initial defect sizes are estimated using nondestructive inspection or proof-testing procedures so that fatigue lifetimes can be computed as the number of cycles for this defect (assumed to be a crack) to propagate to some critical size at failure (defined by the fracture toughness, limit load or allowable strain criteria, for example). This analysis involves integration of the crack-growth law for that material (e.g., of the form of Eqn. (1)) starting from some small initial crack size, suitably modified to take into account mean-stress effects, variable-amplitude loading, environmental data and so on, as required. It must be appreciated, however, that successful application of this approach in design for finite life demands periodic nondestructive-testing evaluation both to define initial defect sizes and to detect cracks before they reach critical dimensions—a procedure that is feasible for aircraft structures, large pressure vessels, power-plant shafts and rotors, large turbines and so forth. However, from economic considerations alone the defect-tolerant approach is unlikely ever to be generally adopted. For many applications, including such everyday commodities as washing machines and automobiles, analyses using S/N data will remain the principal design assessment criteria.

In certain applications, where large numbers of small-amplitude, high-frequency stress reversals are involved for lifetimes in excess of 10^{10} to 10^{12} cycles (beyond the range of available data used to construct S/N curves), the defect-tolerant approach has been utilized in design for infinite fatigue life based on no crack growth. In such cases, initial defect sizes are estimated and a design stress is chosen such that the stress intensity factor range (ΔK) is maintained below the crack propagation threshold (ΔK_{th}). Although overly conservative for many applications, this approach has proved essential for the design of turbine blades and in the operation of gas circuitry in nuclear-reactor systems subject to high-frequency (~ 1300 Hz) vibrational stresses from the gas circulators ("acoustic fatigue"), where growth rates as minute as 10^{-7} mm per cycle, if allowed, would cause failure in a small fraction of the required life. Accordingly, calculations must err on the side of pessimism, and the size of permissible defects must be such that the overwhelming majority of the stress cycles produce ΔK ranges that are less than ΔK_{th} and presumably do not cause growth. This threshold concept can, however, lead to difficulties if the size of the "threshold defect" at the design stress amplitude is so small that uncertainty in the value of ΔK_{th} arises from the short crack problem and further that nondestructive inspection cannot be depended upon to detect slightly larger cracks with 100% reliability. Here, other means, such as proof-testing, realistic inspection and repair schemes, must be sought to achieve the required durability.

8. Conclusion

This article attempts to provide a basic framework for understanding the macroscopic aspects of fatigue crack propagation in metals and alloys from a continuum mechanics viewpoint. Further details may be obtained with reference to the Bibliography.

Although much progress has been made in the analysis of fatigue with the continuum mechanics characterization of cyclic growth rates through the application of linear elastic and elastic–plastic fracture mechanics, particularly with the widespread use of defect-tolerant design codes, many uncertainties still exist with mechanistic interpretations. A comprehension of the micromechanisms of fatigue crack growth, aside from the obvious benefits for the design of new and improved fatigue-resistant alloys, is vital to the definition of the parameters that govern the growth of short cracks, near-threshold crack growth, and crack extension under variable amplitude loading. These issues are described from both continuum and metallurgical viewpoints in the article *Fatigue Crack Growth: Mechanistic Aspects*.

See also: Mechanics of Materials: An Overview

Bibliography

Dowling N E 1976 *Cracks and Fracture*, ASTM STP 601. American Society for Testing and Materials, Philadelphia, Pennsylvania, p. 19

Elber W 1971 *Damage Tolerance in Aircraft Structures*, ASTM STP 486. American Society for Testing and Materials, Philadelphia, Pennsylvania, p. 230

Frost N E, Pook L P, Denton K 1971 A fracture mechanics analysis of fatigue crack growth data for various materials. *Eng. Fract. Mech.* 3: 109–26

Hudak S J 1981 Small crack behavior and the prediction of fatigue life. *J. Eng. Mater. Technol.* 103: 26–35

Johnson H H, Paris P C 1968 Subcritical flaw growth. *Eng. Fract. Mech.* 3: 1–45

Kocańda S 1978 *Fatigue Failure of Metals*. Sijthoff and Noordhoff, The Hague

Paris P C, Erdogan F 1963 A critical analysis of crack propagation laws. *J. Basic Eng.* 85: 528–34

Paris P C, Gomez M P, Anderson W E 1961 Rational analytic theory of fatigue, *Trend Eng.* 13: 9

Rice J R 1967 *Fatigue Crack Propagation*, ASTM STP 415.

American Society for Testing and Materials, Philadelphia, Pennsylvania, p. 247
Rice J R 1968 In: Liebowitz H (ed.) 1968 *Fracture: An Advanced Treatise*, Vol. 2. Academic Press, New York, p. 191
Ritchie R O 1979 Near-threshold fatigue-crack propagation in steels. *Int. Met. Rev.* 20: 205–30
Suresh S, Ritchie R O 1984 In: Davidson D L, Suresh S (eds.) 1984 *Fatigue Crack Growth Threshold Concepts*. Metallurgical Society (AIME), Warrendale, Pennsylvania, p. 227
Wheeler O E 1972 Spectrum loading and crack growth. *J. Basic Eng.* 94: 181–6

R. O. Ritchie
[University of California, Berkeley, California, USA]

Fatigue Crack Growth: Mechanistic Aspects

Micromechanical modelling of fatigue crack propagation behavior to develop predictive growth rate relationships from first principles still is an imprecise science. This is due to the complex nature of the fatigue process, involving striation formation, additional static fracture modes at high growth rates, the three-dimensional and noncontinuum nature of low growth rates, environmental interactions, crack closure effects and so forth. In fact, no single model for fatigue crack propagation currently embodies all these salient factors.

McClintock (1963), however, concluded that fundamentally there were two limiting mechanisms by which a fatigue crack could grow. If the crack advanced by a mechanism of damage accumulation or renucleation over the volume of the crack tip plastic zone, a fourth-power dependence of crack growth rates (da/dN) on the stress intensity range (ΔK) would result, whereas if crack growth occurred by kinematically irreversible crack tip plastic deformation based on the instantaneous crack tip opening displacement range (ΔCTOD), a second-power dependence would result.

Experimentally determined growth rate behavior (typically between 10^{-6} and 10^{-3} mm per cycle) tends to fall between these two extremes. Experiments involving (a) interrupted crack propagation tests, where the fatigue damage was annealed out and the test restarted at the same ΔK, and (b) variable amplitude loading, where the striation spacings following the retardation period after single spike overloads were found to be identical to the pre-overload spacings, have tended to favor the instantaneous CTOD models.

While perhaps valid over the range of data on which they are based, the majority of these theoretical fatigue crack growth models suffer from the fact that they generally provide poor agreement with measured propagation rates and furthermore are unable to account for such factors as near-threshold behavior, crack closure effects and the influence of aggressive environments. This is evident from the fact that the crack growth relaionships most often used in engineering defect-tolerant design and lifetime calculations are based on mere empirical curve fits to the relevant experimental crack propagation data. Perhaps the safest approach to this problem has been to develop semiempirical relationships based on physical models for crack growth and then to determine the constants in the model by fitting to relevant experimental data.

1. General Characteristics of Fatigue Crack Growth

The application of linear elastic fracture mechanics and related small-scale crack tip plasticity has provided a basis for describing the phenomenon of fatigue crack propagation. The crack growth increment per cycle (da/dN) is found to be principally a function of the stress intensity range ΔK through power-law expressions of the form

$$\frac{da}{dN} = C\Delta K^m \tag{1}$$

where ΔK is defined as the difference between the maximum and minimum stress intensities in the cycle (i.e., $\Delta K = K_{max} - K_{min}$). This basic relationship provides an adequate engineering description of behavior in the mid-range of growth rates, typically $\sim 10^{-5}$–10^{-3} mm per cycle.

At higher growth rates, however, as K_{max} approaches instability conditions, defined generally by the fracture toughness (K_{Ic}, K_c) or the limit load, Eqn. (1) often underestimates the propagation rate, whereas at lower (near-threshold) growth rates it generally is conservative, as ΔK approaches a threshold stress intensity range (ΔK_{th}) below which cracks remain dormant or propagate at experimentally undetectable rates. In general, the sigmoidal variation of growth rates with ΔK can be characterized in terms of three regimes, governed by different primary mechanisms of failure, as shown schematically in Fig. 1. It should be noted that the form of the relationship illustrated in Fig. 1 may be changed markedly in the presence of active environments (i.e., in corrosion fatigue) and elevated temperatures (i.e., in creep fatigue).

1.1 Mid-Range of Growth Rates (Regime B)

In the mid-range of growth rates, typically $\sim 10^{-6}$–10^{-3} mm per cycle, where the simple power-law relationship of Eqn. (1) applies (regime B in Fig. 1), fatigue failure is generally characterized by a transgranular ductile striation mechanism, as shown in Fig. 2a. Such striations represent local crack growth increments per cycle and it has been suggested that they occur by a mechanism of opening, advancing and blunting of the crack tip on loading, followed by

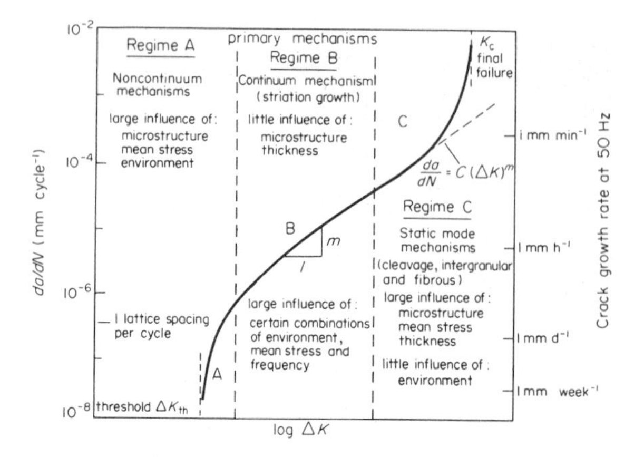

Figure 1
Schematic variation of fatigue crack propagation rate da/dN with stress intensity range ΔK, showing regimes of primary growth rate mechanisms

resharpening of the tip on unloading. An example of this mechanism is shown in Fig. 3, although the precise details of this process are still uncertain.

Several theoretical models for such macroscopic striation crack growth (often termed stage II crack propagation) have been proposed which rely on the fact that, where plastic zones are sufficiently large compared with microstructural dimensions, plastic blunting at the crack tip is accommodated by shear on two slip systems at roughly 45° to the crack plane. As such sliding-off is largely irreversible, new crack surface can be created during cyclic crack advance by simultaneous or alternating slip on these two systems. An example of the latter process is shown in Fig. 4, based on observations of fatigue crack tips in copper single crystals subjected to bending or push-pull loading. The observations that striations are formed less readily, and that cracks propagate far slower, in vacuo often are considered to be consistent with such models, since due to the lack of crack surface oxidation, the irreversibility of crack tip shear should be less efficient.

Regardless of the precise details of the mechanism, such crack tip sliding-off models predict that the increment of crack advance per cycle should be proportional to the cyclic crack tip opening displacement (ΔCTOD). From fracture mechanics considerations, the fatigue crack growth rate for the striation mode of propagation should then be given by

$$\frac{da}{dN} \propto \Delta\mathrm{CTOD} = \beta \frac{\Delta K^2}{2\sigma_y E'} \qquad (2)$$

where σ_y is the cyclic flow stress, E' the appropriate elastic modulus and β a proportionality constant reflecting the efficiency of crack tip blunting (taken theoretically to be in the region 0.5–1). Equation (2) predicts that growth rates in regime B should be dependent on ΔK^2, and this certainly is consistent with most data in this range, which show the power-law exponent m to be in the region 2–4.

Furthermore, the inverse dependence of da/dN on elastic modulus has also been verified, as shown in Fig. 5 for a wide range of metallic systems. However, the inverse dependence on cyclic yield strength is rarely seen experimentally (Fig. 6). Moreover, in absolute terms, experimentally measured fatigue crack growth rates (under small-scale yielding conditions) invariably indicate a rate of incremental crack extension which is a far smaller fraction of ΔCTOD, particularly at lower growth rates (Fig. 7). Numerous explanations have been advanced for the lack of direct correlation between da/dN and ΔCTOD (including the minimal role of strength), although none is entirely satisfactory. The discrepancy appears to be related to a number of factors, including the nonuniformity of crack advance in three dimensions, the occurrence of other fracture modes, the role of crack closure and the micromechanisms of crack tip blunting. The latter factor specifically refers to the degree of strain localization at the tip as a function of its opening, and has been used to rationalize the absence of a yield strength effect of da/dN by assuming $\beta \propto \sigma_y/E'$.

Aside from the strong dependence on modulus, rates of fatigue crack extension by this striation mechanism are largely insensitive to metallurgical microstructure in the mid-growth rate range, as shown in Figs. 6 and 7. In Fig. 6, data are plotted for a wide range of ferrous alloys representing an order of magnitude change in yield strength from simple ferritic to tempered martensitic structures. It is apparent that where crack growth occurs via a striation mechanism (indicated by the letters), growth rates vary by a factor of < 10. Furthermore, propagation rates in this regime appear relatively independent of the mean stress, characterized by the load ratio $R = K_{min}/K_{max}$. Data for a 2.25%Cr–1%Mo and pressure vessel steel in Fig. 8 reveal that growth rates between $\sim 10^{-6}$ and 10^{-3} mm per cycle are almost totally insensitive to changes in load ratio from 0.05 to 0.75, although it should be noted that strong mean stress effects are present at lower and higher (not shown) growth rates. Behavior in the mid-growth regime, in the absence of environmental interactions, is also essentially independent of changes in cyclic frequency, wave shape and whether deformation is in plane stress or plane strain.

Such load ratio, frequency and wave shape effects are, however, important in the presence of certain active environments. Under such circumstances, fatigue crack growth rates in this regime become environmentally assisted through additional contributions to crack advance from such mechanisms as active path corrosion (metal dissolution) and hydrogen embrittlement. An example of such corrosion fatigue behavior can be seen in Fig. 9 for a bainitic 2.25%Cr–1%Mo steel tested in air and hydrogen. Below a critical frequency, sufficient to permit the

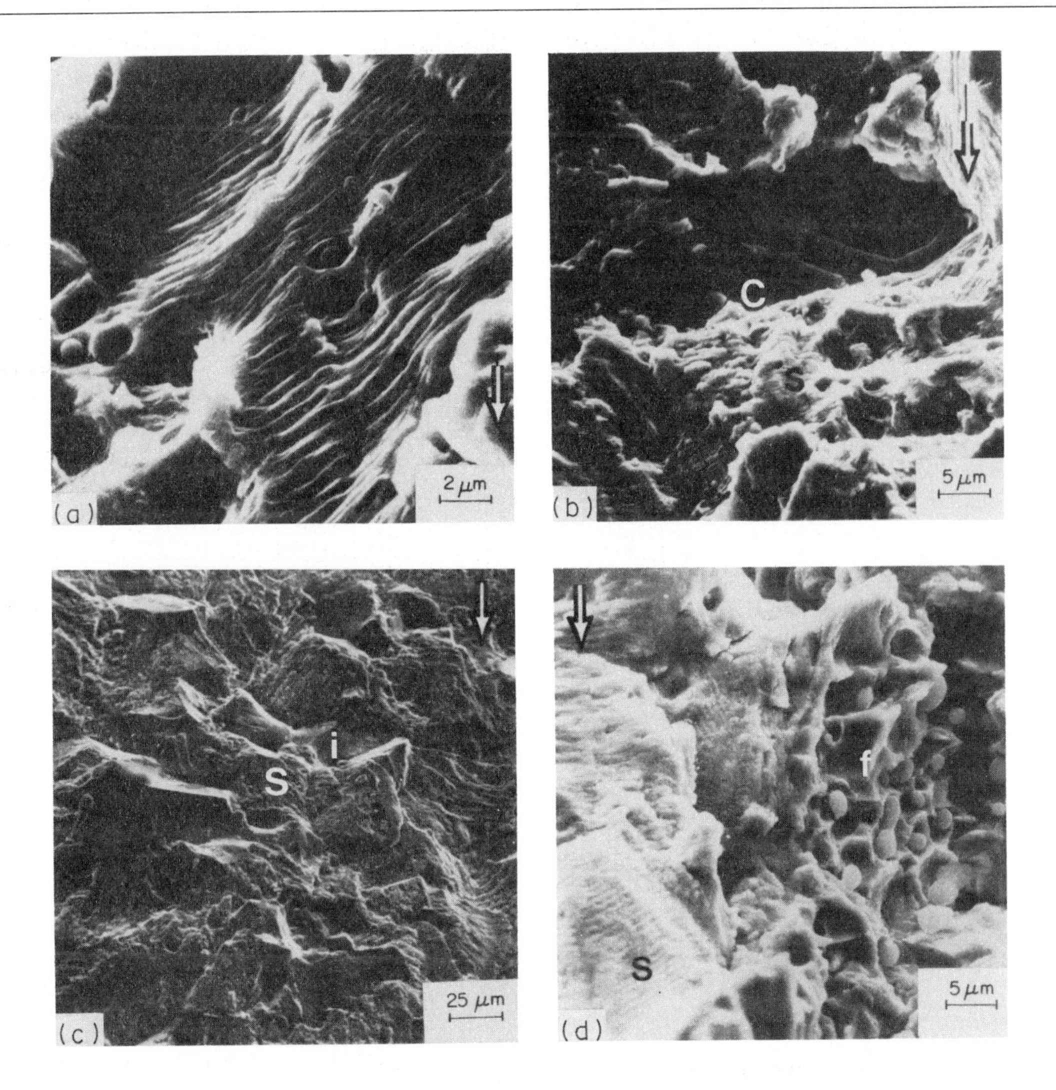

Figure 2
Fractography of fatigue crack propagation at intermediate (regime B) and high (regime C) growth rates in steels tested in moist room air at $R = 0.1$: (a) ductile striations in mild steel at $\Delta K = 30\,\mathrm{MPa\,m^{1/2}}$; (b) additional cleavage cracking (C) during striation growth (S) in mild steel at $\Delta K = 19\,\mathrm{MPa\,m^{1/2}}$; (c) additional intergranular cracking (i) in low-alloy Ni–Cr steel at $\Delta K = 15\,\mathrm{MPa\,m^{1/2}}$; (d) additional microvoid coalescence and fibrous fracture (f) with striations (S) in AISI 316L stainless steel at $\Delta K = 30\,\mathrm{MPa\,m^{1/2}}$ (arrows indicate general direction of crack growth)

chemisorption of hydrogen at the crack tip, fatigue crack growth rates are seen to be markedly accelerated in hydrogen (above $\sim 10^{-5}$ mm per cycle) compared with air, the onset of the acceleration being a function of the load ratio. In this instance, hydrogen-assisted crack propagation often is coincident with a change in fracture mode from predominantly transgranular to intergranular and can be attributed to hydrogen-induced degradation of material ahead of the crack tip. Analogous behavior has been reported in water, hydrogen sulfide and other corrosive environments, although the resulting corrosion fatigue crack growth behavior observed tends to be very specific to the particular material, environment and test conditions in question.

1.2 High Growth Rate Regime (Regime C)

At high fatigue crack growth rates, typically above $\sim 10^{-3}$ mm per cycle, where K_{max} approaches K_{Ic} or limit load failure (regime C in Fig. 1), Eqn. (1) generally underestimates measured growth rates, due to the occurrence of brittle fracture mechanisms (so-called

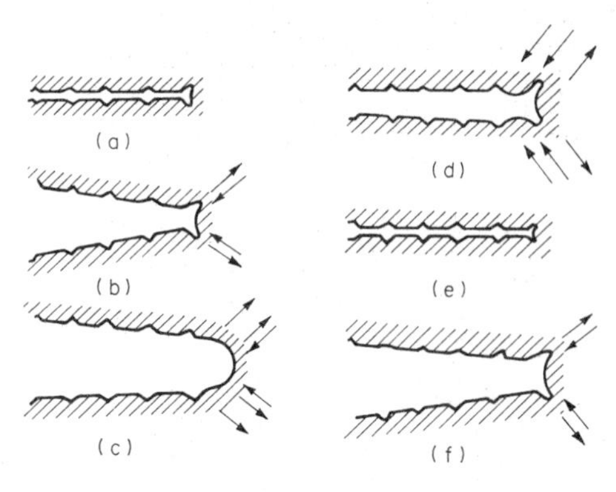

Figure 3
Diagrammatic representation of the sequence of formation of ductile fatigue striations according to the plastic blunting model (Laird 1967), involving sequential blunting, crack extension and resharpening

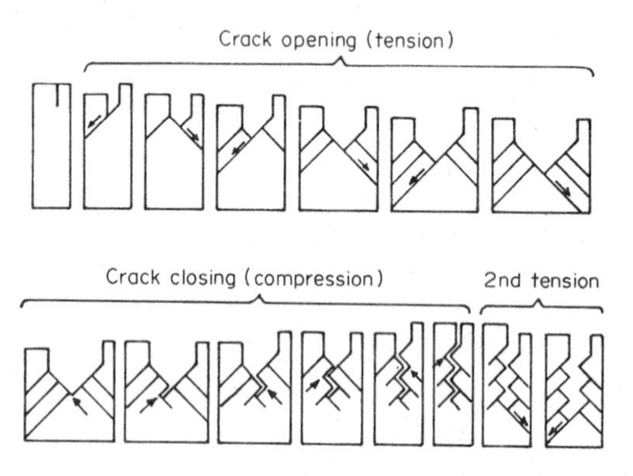

Figure 4
Model for the alternating slip processes involved in fatigue crack extension during a loading cycle (after Neumann 1974)

static modes) which replace or are additional to striation growth. Such static modes include cleavage, intergranular cracking and microvoid coalescence (Fig. 2), and their presence results in growth rate behavior which is markedly sensitive to microstructure, mean stress and testpiece thickness. Since both cleavage and intergranular cracking are stress-controlled fracture mechanisms and microvoid coalescence is dependent upon the nature of the hydrostatic stress, their occurrence is promoted by high mean stresses (high R values), thick specimens (promoting plane strain conditions) and low toughness microstructures (e.g., when $K_{max} \rightarrow K_{Ic}$). This is apparent in Fig. 10, where high growth rate data for a low-toughness (temper-embrittled) low-alloy steel can be seen to be markedly dependent upon load ratio. Here, brittle intergranular cracking was found to be responsible for the marked

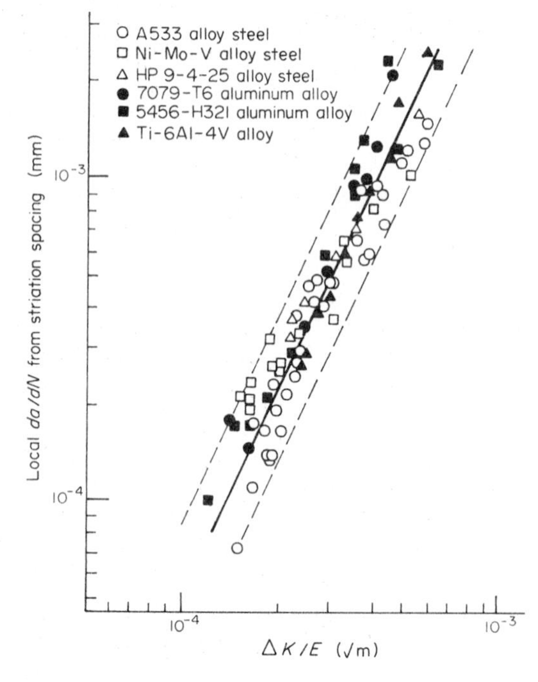

Figure 5
Experimental data showing inverse dependence of local fatigue crack propagation rate da/dN, measured from striation spacings, on elastic modulus E (note how normalizing ΔK with respect to E yields similar mid-growth rate behavior for all materials) (after Bates and Clark 1969)

acceleration in growth rates with increasing load ratio as K_{max} approached K_{Ic}. However, because crack velocities are so fast, there is generally insufficient time per cycle for strong corrosion effects to be active, and behavior in this regime is considered to be largely insensitive to environment.

1.3 Low Growth Rate (Near-Threshold) Regime (Regime A)

At ultralow fatigue crack growth rates, typically below $\sim 10^{-6}$ mm per cycle, approaching the threshold stress intensity range ΔK_{th} (regime A in Fig. 1), Eqn. (1) generally overestimates measured growth rates and behavior becomes markedly sensitive to mean stress, microstructure and environment. In additon, at such near-threshold levels, the scale of local plasticity approaches that of the microstructure and measured propagation rates become less than an interatomic spacing per cycle, indicating that crack extension is not occurring uniformly over the entire crack front. This deviation from continuum crack growth mechanisms is evident in Fig. 7 from the large discrepancy, at near-threshold stress intensities, of measured growth rates from predictions based on the crack tip opening displacement model, Eqn. (2).

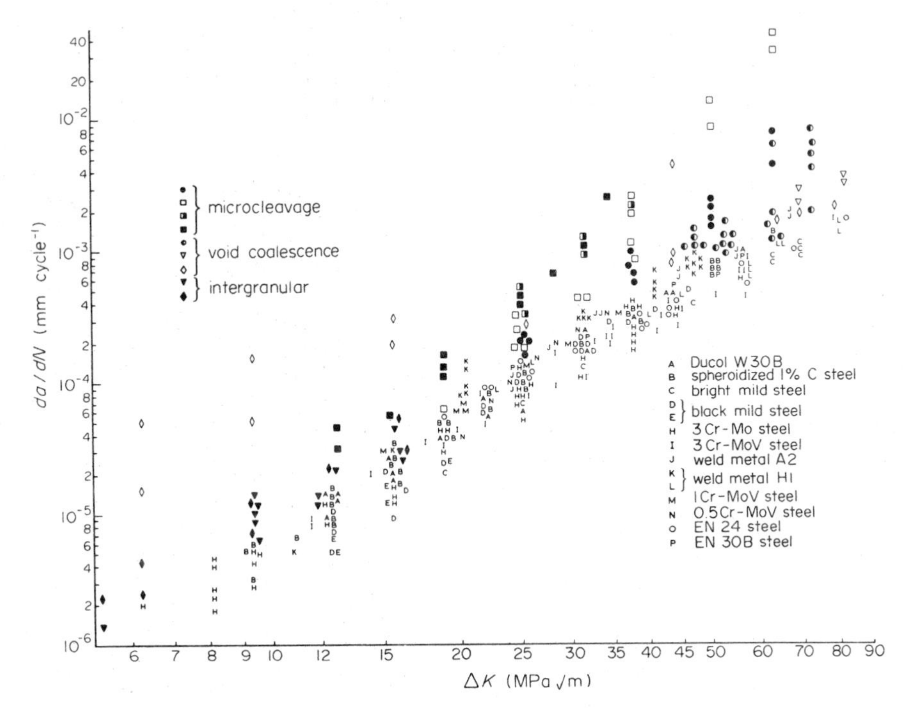

Figure 6
Fatigue crack propagation rate da/dN as a function of ΔK for a wide range of ferrous alloys ($\sigma_y \sim 200$–2000 MPa). Note little change in intermediate propagation rates where mechanism of growth involves striations (after Lindley et al. 1976)

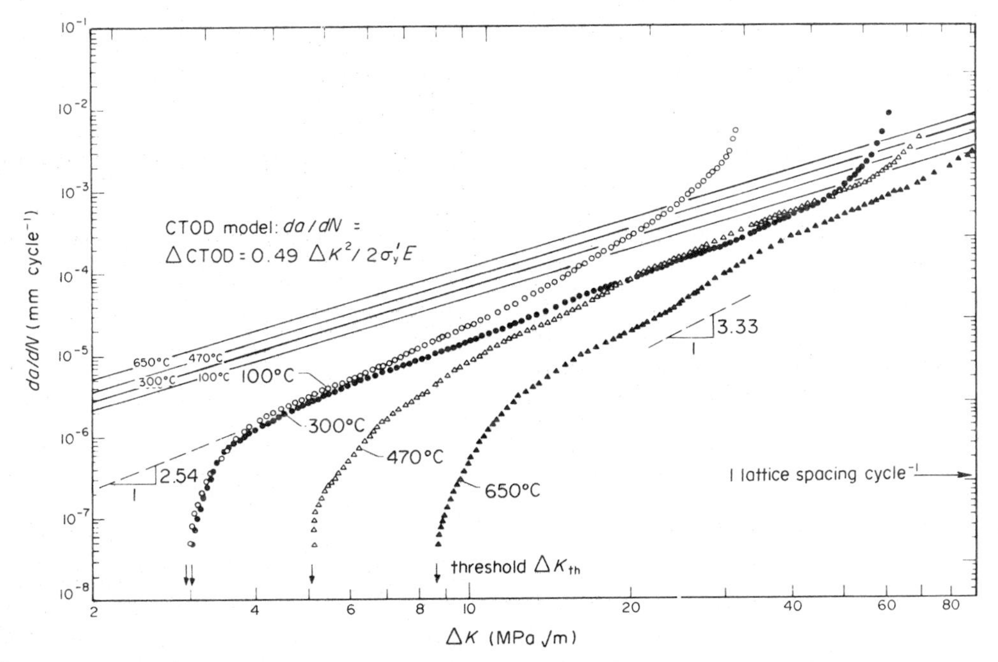

Figure 7
Variation of fatigue crack propagation rate da/dN in moist air at $R = 0.05$ and 23 °C with ΔK for ultrahigh-strength 300-M martensitic steel austenitized at 870 °C, oil-quenched and tempered between 100 and 650 °C to vary tensile strength from 2300 to 1190 MPa, respectively, showing smallest effect of strength level in the mid-growth rate regime. Solid lines indicate predictions based on the CTOD model for crack growth, Eqn. (2)

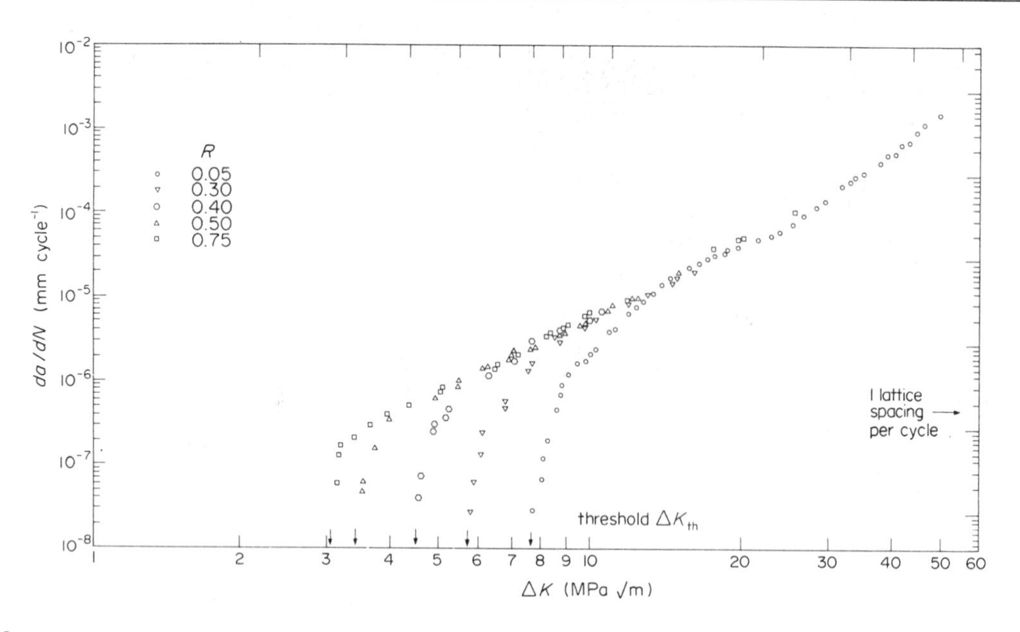

Figure 8
Influence of load ratio R on fatigue crack propagation rate da/dN as a function of ΔK in a bainitic 2.25%Cr–1%Mo pressure vessel steel SA 542 class 3 ($\sigma_y = 500$ MPa) at 50 Hz (after Ritchie and Suresh 1983)

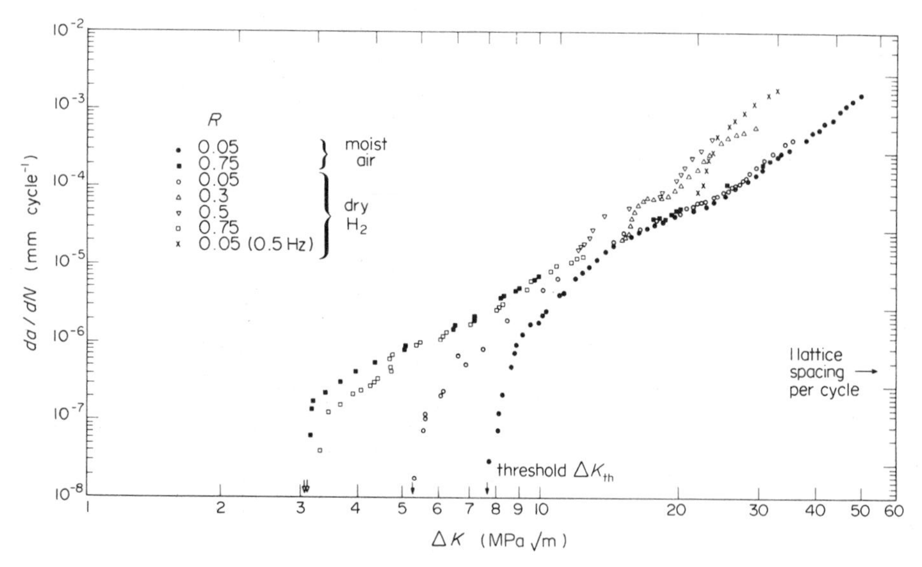

Figure 9
Influence of load ratio R and environment on fatigue crack propagation in bainitic 2.25%Cr–1%Mo steel SA 542 class 3 ($\sigma_y = 500$ MPa, frequency $= 50$ Hz), tested in moist air and dry hydrogen at ambient temperature gas (after Ritchie and Suresh 1983)

Mechanistically, comparatively little is known about near-threshold crack propagation, although fractographically surfaces tend predominantly to be transgranular with evidence of isolated intergranular or "cleavage-like" facets (Fig. 11). Such faceted-type crack advance, sometimes referred to as "microstructurally sensitive" or "crystallographic" fatigue, is observed at ultralow growth rates, where maximum plastic zone sizes are typically less than the grain size. The low restraint of the elastic surroundings on cyclic

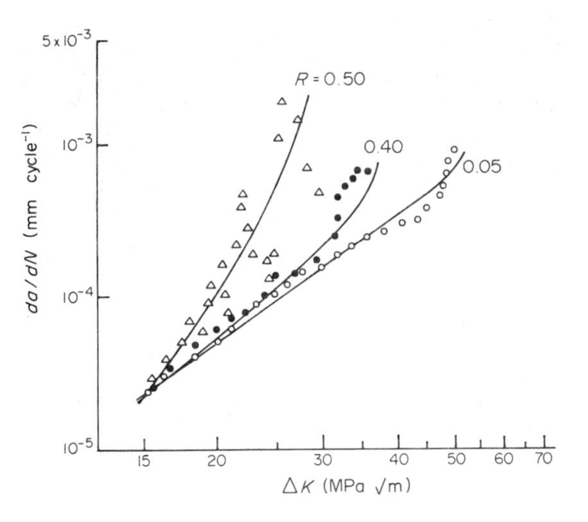

Figure 10
Influence of load ratio R on fatigue crack propagation at high stress intensities (regime C) for a low-toughness Ni–Cr En 30A steel ($\sigma_y = 735$ MPa), tested in ambient temperature moist air (after Ritchie and Knott 1973)

slip will promote single slip at the crack tip on the slip system with the highest critical resolved shear stress, resulting in a slip-band cracking mechanism (stage I growth) with associated mode II + mode I displacements. This gives rise to a faceted or zigzag fracture path, unlike ductile striation growth at higher ΔK levels (stage II growth), where fracture morphologies are macroscopically more planar (Fig. 12).

This transition in fracture mode follows from the fact that, at higher stress intensities, the crack tip plastic zone encompasses many grains and the maintenance of slip-band cracking in a single direction becomes incompatible with a coherent crack front. Thus, the accompanying increase in cyclic plasticity activates further slip systems, leading to the noncrystallographic mode of striation crack advance by alternating or simultaneous shear. Since the single shear type crack advance is accompanied by significant mode II crack tip displacements in addition to mode I, this combination of faceted fracture morphologies and local shear displacements at near-threshold levels has particular importance in the development of fatigue crack closure effects, as discussed in detail below.

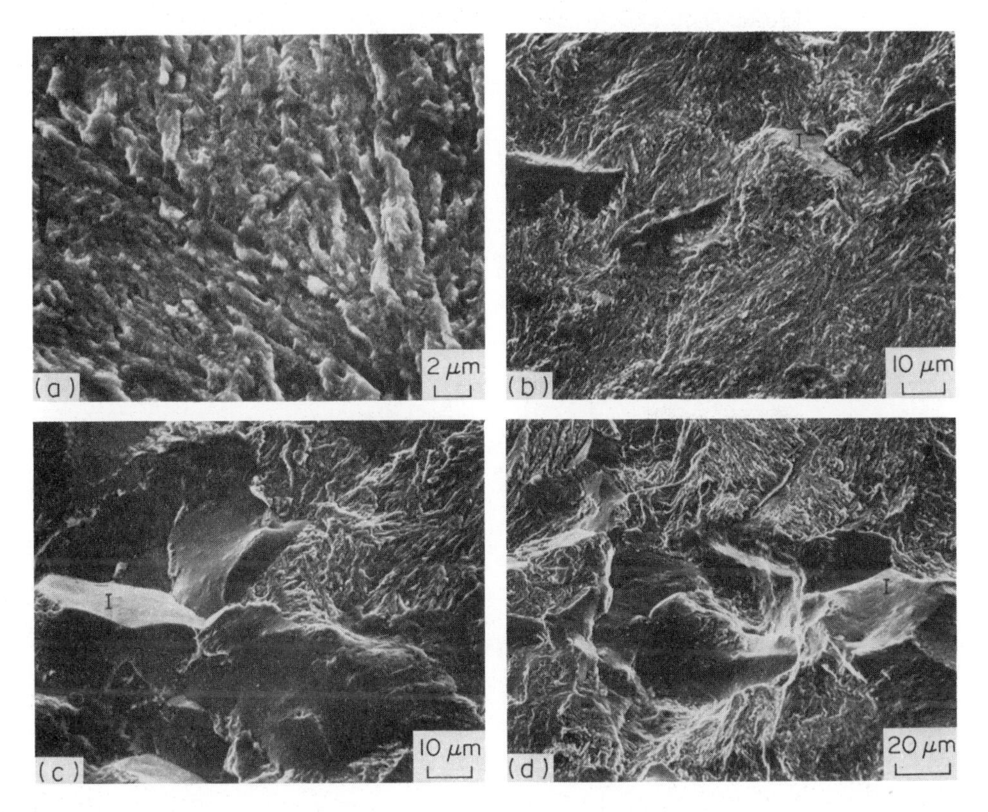

Figure 11
Fractography of near-threshold fatigue crack propagation (regime A) in high-strength martensitic 300-M steel ($\sigma_y = 1500$ MPa) tested in moist ambient temperature air: (a) transgranular fracture mode close to threshold of $\Delta K = 5.1$ MPa m$^{1/2}$ segments of additional intergranular facets at $\Delta K =$ (b) 5.2, (c) 7.6 and (d) 11 MPa m$^{1/2}$ ($R = 0.05$) (arrows indicate general direction of crack growth)

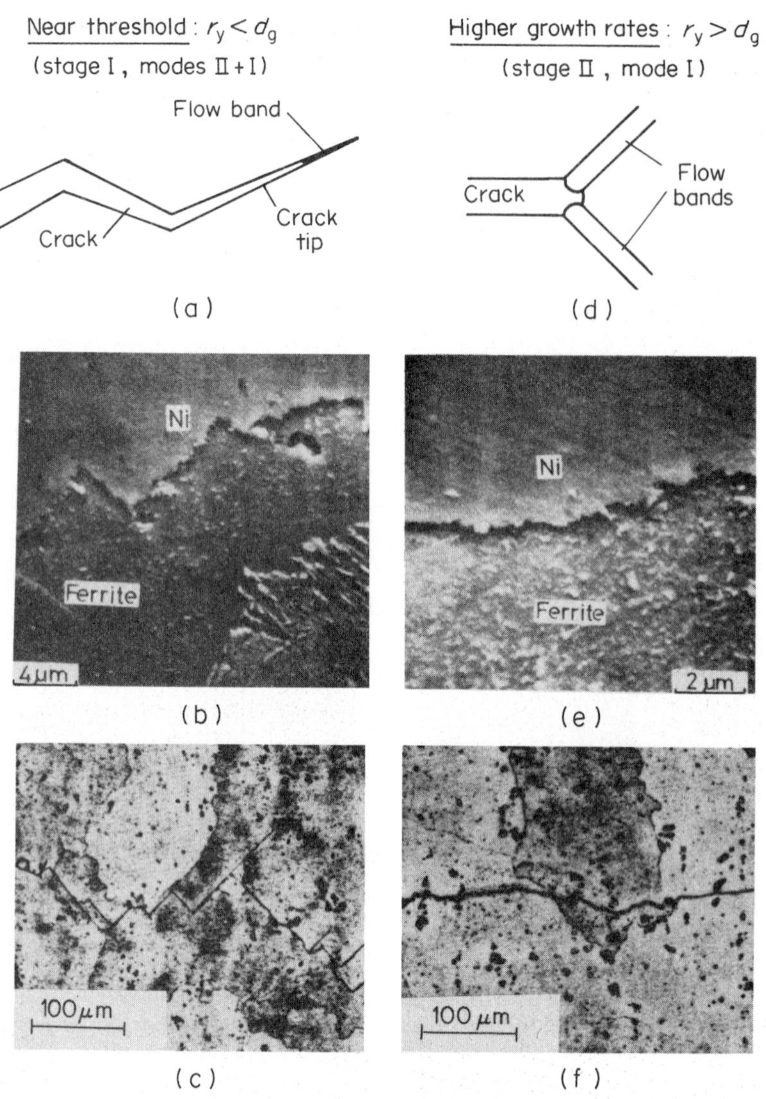

Figure 12
Crack opening profiles and resulting crack path morphologies corresponding to (a–c) near-threshold (Stage I) and (d–f) higher growth rate (Stage II) fatigue crack propagation: (b, e) nickel-plated fracture sections of fatigue crack growth in 1018 steel (afer Minakawa and McEvily 1981); (d, f) metallographic sections of crack growth in 7075-T6 aluminum alloy (after Louwaard 1977) (r_y is maximum plastic zone size, compared with average grain diameter d_g)

In terms of macroscopic growth rate behavior, near-threshold propagation rates are strongly sensitive to mean stress, unlike behavior in regime B. With rising load ratios, growth rates are increased significantly and threshold ΔK_{th} values are reduced, as shown in Fig. 8. Similarly, in contrast to regime B behavior, near-threshold growth rates are markedly higher with increasing strength level in steels (Fig. 7) and generally slower in coarser-grained microstructures (Fig. 13). Furthermore, the nature of the environment plays a prominent role in this regime, as illustrated in Fig. 9: near-threshold growth rates in dry hydrogen are significantly faster (at $R = 0.05$) than in moist air.

In general, however, environmental interactions with cyclic crack extension at ultralow growth rates are particularly complex, since in lower-strength steels, for example, seemingly aggressive environments such as water and wet hydrogen result in marginally slower propagation rates compared with moist air, whereas crack propagation rates in inert gases, such as dry helium, appear to be substantially faster (Fig. 14). It is pertinent here, though, that the marked effects of

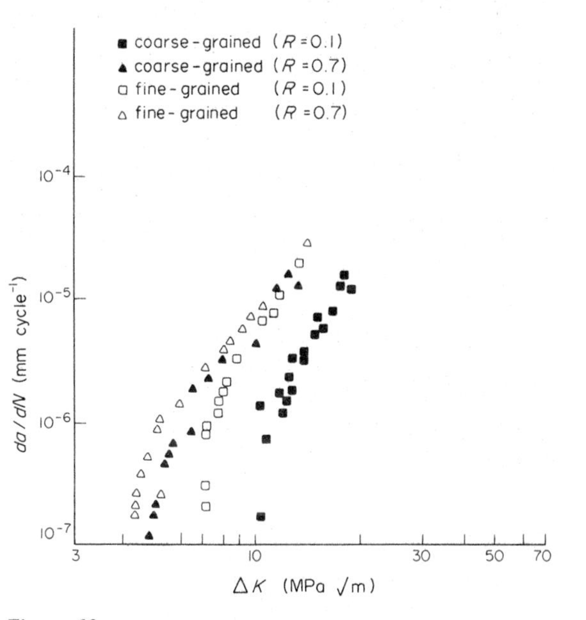

Figure 13
Fatigue crack propagation in a fully pearlitic hot-rolled eutectoid rail steel ($\sigma_y = 410$ MPa), tested in moist ambient temperature air at $R = 0.1$ and 0.7 and a frequency of 30 Hz, showing lower near-threshold growth rates and higher threshold ΔK_{th} values in coarser-grained microstructure ($d_g = 130$ μm) than in finer-grained structure ($d_g = 25$ μm) (note that major influence of grain size is at low load ratios, with little effect at $R = 0.7$) (after Gray et al. 1981)

strength level (Fig. 7), grain size (Fig. 13) and environment (Figs. 9, 14) all predominate at low mean stresses and are largely absent at high load ratios, behavior which is indicative of major contributions from crack closure, as discussed in the following section.

2. *Fatigue Crack Closure*

Crack closure, or more precisely plasticity-induced crack closure, can be considered as arising from interference between crack surfaces in the wake of the crack tip, resulting from the constraint of surrounding elastic material on the residual stretch in material elements previously plastically strained at the tip. This process is aided by the fact that the crack is growing into a reversed plastic zone in which a state of residual compressive stress exists. Since the crack cannot propagate while it remains closed, the net effect of this closure is to reduce the nominal stress intensity range ΔK to some lower effective value ΔK_{eff} actually experienced at the crack tip:

$$\Delta K = K_{max} - K_{min} \qquad \text{(no closure)}$$
$$\Delta K_{eff} = K_{max} - K_{cl} \qquad \text{(with closure)} \qquad (3)$$

where K_{cl} is the stress intensity to close the crack ($\geqslant K_{min}$).

This phenomenon was demonstrated effectively by Elber (1971) for crack propagation at high stress intensities in aluminum alloys by mounting surface strain gauges of very small gauge length (so-called Elber gauges) at the tips of fatigue cracks. The resulting elastic compliance curves are illustrated schematically in Fig. 15, where it is apparent that a fatigue crack of length a initially "opens" as if no crack were present before following the expected compliance slope for a physically open flaw of length a. The transition point in this curve thus defines the physical opening or closing load and hence the value of K_{cl}. Subsequently, other experimental measurements of closure loads have been performed using crack mouth opening gauges, back-face strain gauges, electrical potential measurements and ultrasonic techniques, although the data from the different procedures are not always totally consistent.

The closure concept has proved to be extremely useful in explaining, at least qualitatively, many aspects of fatigue crack propagation behavior, including the existence of a threshold, the role of variable amplitude loading and the observations of mean stress-sensitive crack growth. For the latter case, since as the mean stress or load ratio is raised, the crack can be considered to remain open for a larger proportion of the cycle, it follows that, at a fixed nominal ΔK, a higher R value results in a higher ΔK_{eff} and hence a faster growth rate. However, once the minimum stress intensity K_{min} exceeds K_{cl}, the crack remains open during the entire loading cycle, so that above the critical load ratio R_{cr}, closure effects become insignificant as $\Delta K_{eff} = \Delta K$, and further increases in the mean stress have little influence.

One of the more serious objections to this concept is that plasticity-induced closure is most prevalent under essentially plane stress conditions, yet in the mid-range of growth rates where such conditions often exist, mean-stress effects are generally minimal (Fig. 8). Furthermore, the major influence of mean stress is usually found at the ultralow growth rates associated with near-threshold fatigue, where invariably plane strain conditions are present. Recent experimental compliance measurements on near-threshold cracks have confirmed that very significant effects of closure are present in plane strain, although it is difficult to rationalize these effects solely in terms of the plasticity arguments of Elber. Accordingly, several "microscopic" mechanisms of crack closure have been identified recently, based on the role of crack surface corrosion deposits and fracture surface roughness or morphology. These additional closure mechanisms are relevant specifically to the near-threshold regime, since it is principally here that the size of the corrosion debris and the roughness of the fracture surface are comparable with the crack tip opening displacements.

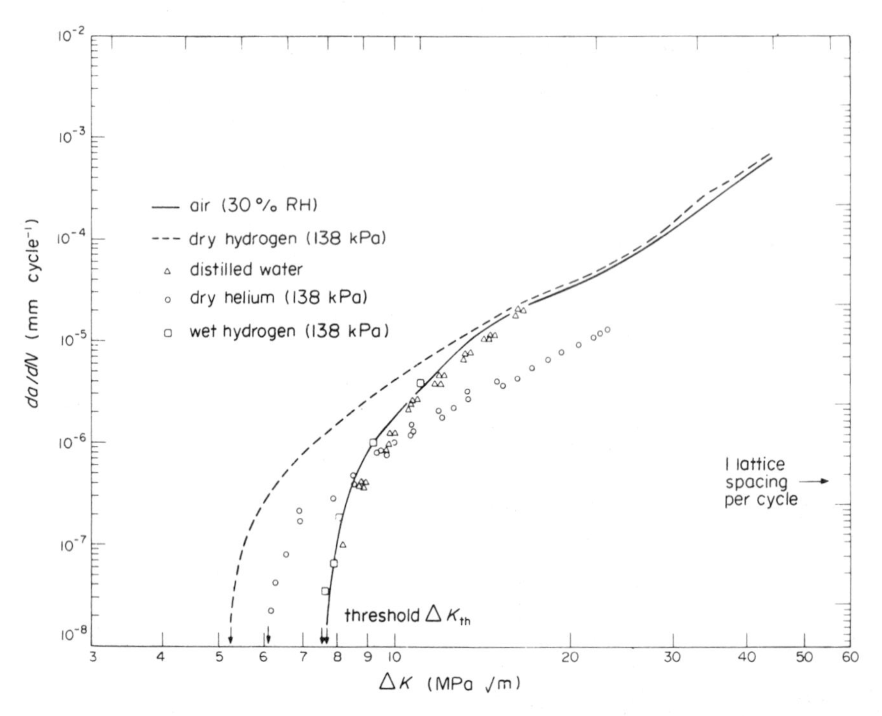

Figure 14
Fatigue crack propagation in bainitic 2.25%Cr–1%Mo, SA 542 class 3 ($\sigma_y = 500$ MPa, frequency = 50 Hz, $R = 0.05$ and ambient temperature), showing influence of environment at near-threshold and mid-growth rate regimes (note that above ~ 10^{-5} mm per cycle, growth rates in hydrogen and air are similar at this high frequency, whereas growth rates in inert helium gas are slower; in contrast, below ~ 10^{-6} mm per cycle, near-threshold growth rates are fastest in dehumidified environments—dry hydrogen and dry helium—and slowest in moist environments—wet hydrogen, moist air and water) (after Suresh et al. 1981)

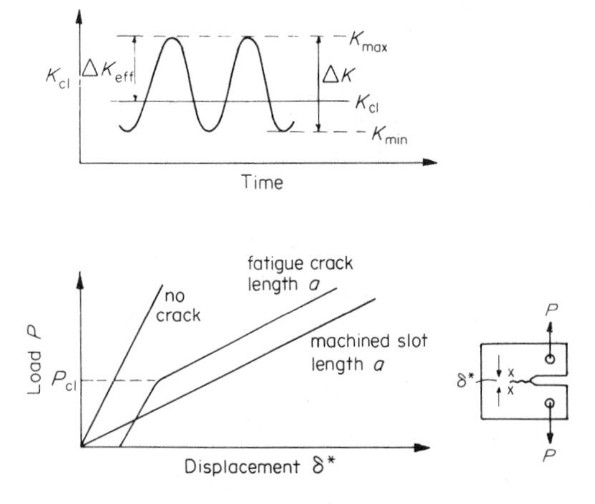

Figure 15
Procedures for experimentally demonstrating plasticity-induced crack closure, showing elastic compliance curves for uncracked test piece, test piece containing a finite width slot of length *a*, and test piece containing fatigue crack of length *a* (note how fatigue cracked specimen does not appear to indicate presence of crack until above the closure load) (after Elber 1971)

2.1 Oxide-Induced Crack Closure

Oxide-induced crack closure arises from the fact that during near-threshold fatigue crack propagation at low load ratios in moist environments, corrosion deposits of a thickness comparable with the size of crack tip opening displacements can build up near the crack tip, thus providing a mechanism for enhanced closure so that the crack becomes "wedged closed" at stress intensities above K_{min} (Fig. 16). Such oxide films (Fig. 17) are thickened at low load ratios by a mechanism of fretting oxidation—a continual breaking and reforming of the oxide scale behind the crack tip due to a smashing together of the crack surfaces as a result of plasticity-induced closure aided by the strong mode II displacements characteristic of near-threshold crack growth.

For example, in lower-strength steels tested in moist air at $R = 0.05$, oxide films have been observed near ΔK_{th} with a maximum thickness some 20–40 times the limiting thickness of films formed naturally on metallographically polished samples exposed to the same environment for similar periods. The closure effect promoted by such deposits has been substantiated by Auger measurements of oxide thicknesses (Fig. 18),

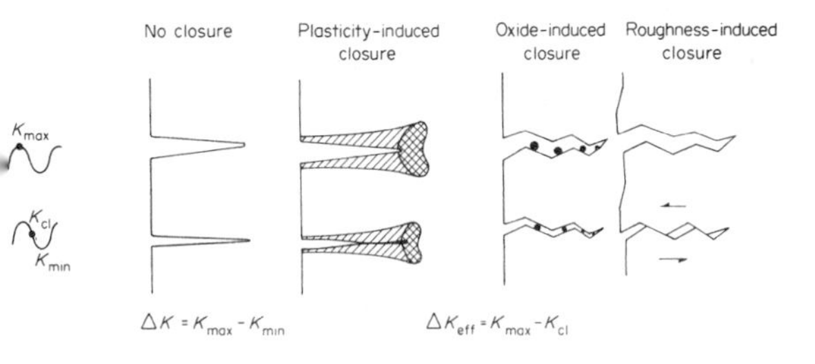

Figure 16
Schematic illustration of mechanisms of crack closure arising from plasticity, oxide debris and fracture surface roughness

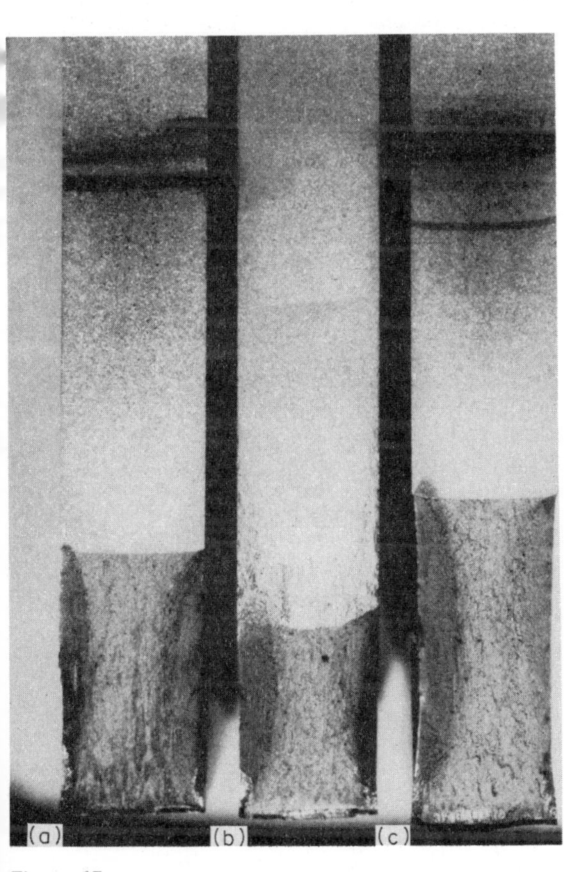

Figure 17
Bands of visual corrosion deposits from close to ΔK_{th} during fatigue crack propagation in a HP 9-4-20 (9%Ni–4%Co) steel tested in moist air: (a, c) near-threshold tests ($da/dN < 10^{-6}$ mm per cycle); (b) mid-growth rate test ($>10^{-5}$ mm per cycle) (all tests at low load ratios of $R = 0.05$–0.10)

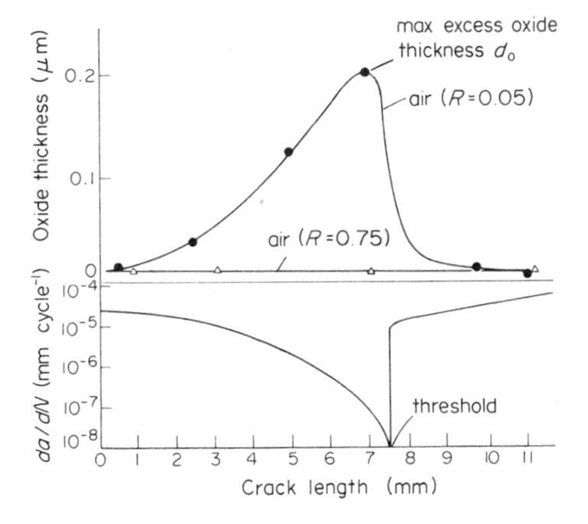

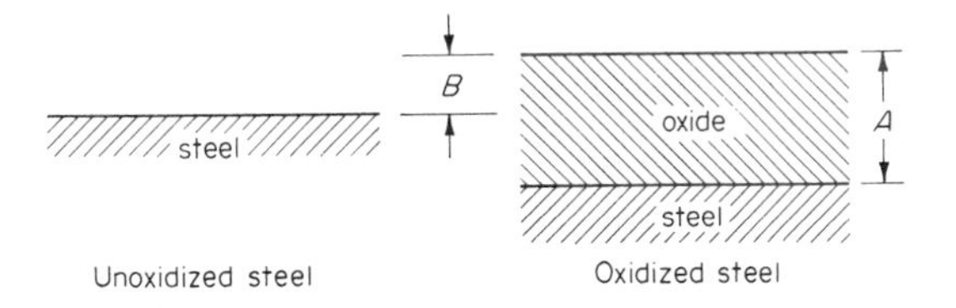

Figure 18
Measurements of fatigue fracture surface oxidation products in lower strength steel (2.25%Cr–1%Mo steel, SA 542 class 3), showing oxide film thickness at ambient temperature and a frequency of 50 Hz as a function of crack length (measured from notch) and crack growth rate da/dN during typical load-shedding near-threshold testing procedure: growth rates measured in moist air and dry hydrogen at $R = 0.05$ and 0.75, and oxide data generated from Ar^- sputtering analysis using Auger spectroscopy (note enhanced oxidation close to threshold ΔK_{th} at low load ratios compared with minimal excess oxidation at high load ratios) (after Suresh et al. 1981)

direct experimental closure measurements using ultrasonics and simple modelling studies involving rigid wedges inserted into elastic cracks. Since the formation of enhanced crack surface oxide deposits is far less predominant in dry, oxygen-free environments, at high load ratios (where plasticity-induced closure effects are minimal) and in higher-strength materials (where fretting oxidation is less effective), the resulting concept of oxide-induced crack closure provides an attractive explanation of many of the somewhat surprising effects of microstructure and environment on near-threshold behavior (outlined above).

First, the occurrence of oxide-induced closure can be associated directly with the large load ratio effects observed for near-threshold growth and the value of ΔK_{th} (Fig. 8), since enhanced crack face oxidation cannot occur (except in very oxidizing environments) at high load ratios (above R_{cr}), as the crack is sufficiently open to prevent fretting oxidation. This is consistent with observations of a substantial reduction in load ratio effects in the mid-range of growth rates where oxide-induced closure effects are minimal.

Second, the reported reduction in threshold values and increase in near-threshold crack propagation rates in steels with increasing strength level (Fig. 7) is also in accordance with this mechanism, since oxide formation inside near-threshold cracks is increasingly restricted in higher-strength materials (presumably because plasticity-induced closure is reduced and fretting oxidation is less predominant). The fact that the influence of strength level on near-threshold behavior is far less apparent at high load ratios is totally consistent with this.

Third, and most important, the concept of oxide-induced crack closure provides a very feasible explanation for the complex role of environment at near-threshold levels. Environmentally influenced near-threshold fatigue crack propagation can be interpreted in terms of competition between two concurrent processes: crack closure due to corrosion debris (which decreases growth rates) and hydrogen embrittlement or active path corrosion arising from the production of hydrogen vis-a-vis the oxidation process (which increases growth rates). In lower-strength steels ($\sigma_y < 1000$ MPa), for example, the susceptibility to hydrogen embrittlement is relatively low, particularly at the high frequencies ($\geqslant 50$ Hz) commonly employed for near-threshold measurements, and thus the retarding influence of oxide-induced closure appears to dominate overall behavior. Hence, faster near-threshold crack growth rates are to be expected in dry hydrogen or helium, compared with the moist environments of air, wet hydrogen and water, because behavior in the dry atmospheres is retarded less by closure (Figs. 9, 14).

Furthermore, these differences should become insignificant at high load ratios, since closure phenomena will be ineffective. This is shown by the similar near-threshold growth rates and ΔK_{th} values for 2.25%Cr–1%Mo steel tested at $R = 0.75$ in dry hydrogen and moist air, compared with the large effect of environment for tests at $R = 0.05$ (Fig. 9). Thus, because of this important influence of closure mechanisms at low stress intensity ranges, where cyclic CTOD values approach the size of the oxide scales, the nature and characteristics of environmentally influenced crack propagation behavior at near-threshold levels is often radically different from the more customary corrosion fatigue behavior observed at higher growth rates.

Finally, the concept of oxide-induced closure may provide one important rationale for the very existence of a fatigue crack propagation threshold. Comparisons between the cyclic crack tip opening displacements and the maximum thickness of the excess oxide scale indicate that there is a one-to-one correspondence at the threshold (Fig. 19). Since the crack is unable to propagate once it is totally wedged closed with debris, this result suggests one physically appealing explanation for the threshold—an explanation which is consistent with data which indicate that near-threshold growth rates in vacuo, where presumably little oxide debris are generated, show no well defined threshold and actually can exceed growth rates in air at very small crack tip displacements.

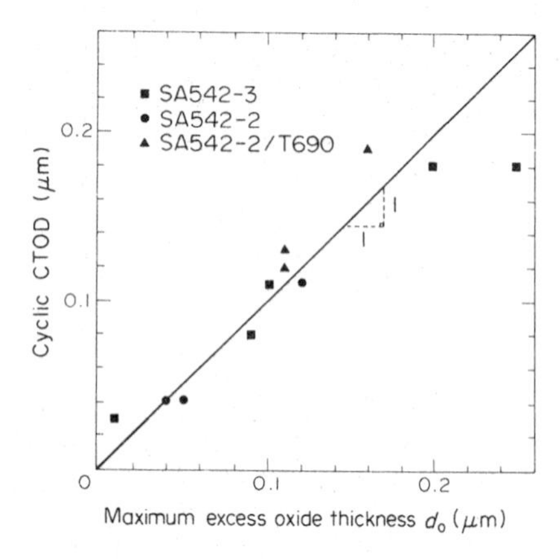

Figure 19
One-to-one correspondence of maximum excess crack surface oxide thickness and cyclic crack tip displacement ΔCTOD at the threshold (i.e., at ΔK_{th}, $d_0 \sim \Delta$CTOD), for a range of lower-strength 2.25% Cr–1%Mo steels ($\sigma_y = 290$–770 MPa) tested in air, water, hydrogen and inert gas environments (after Ritchie and Suresh 1983)

2.2 Roughness-Induced Crack Closure

A further mechanism of crack closure of specific relevance to near-threshold behavior is that induced by fracture surface roughness or morphology (Fig. 16). Such roughness-induced closure can arise if the fracture surface roughness is comparable in size with crack tip opening displacements and if significant mode II displacements exist, such that the crack can become wedged closed at discrete contact points along the crack faces (Fig. 20). Like oxide-induced closure, this mechanism is most prevalent at low stress intensity ranges, where maximum plastic zone sizes typically are smaller than microstructural dimensions. In this regime, crack advance proceeds by a single shear mechanism, with associated mode I plus mode II displacements, resulting in serrated fracture paths (Fig. 12) which enhance the closure effect.

Such roughness-induced closure provides an important contribution to the role of microstructure in influencing near-threshold crack growth and in fact has wide significance for a large range of engineering materials (e.g., titanium alloys which form little oxide debris). For example, with regard to the well known observations of improved near-threshold crack growth resistance in coarser-grained materials (e.g., Fig. 13), coarse microstructures tend to promote

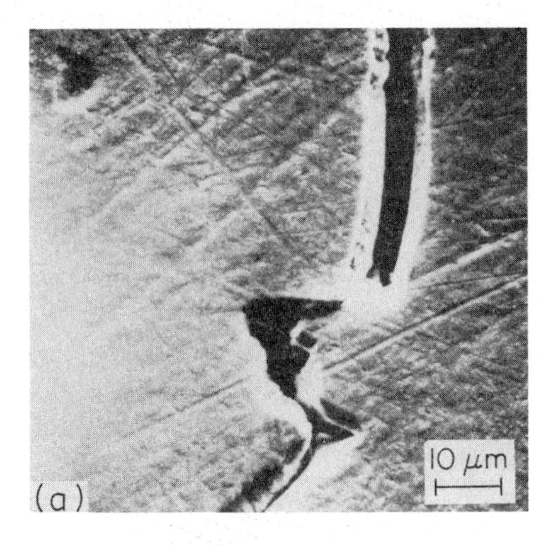

Figure 20
Example of crack closure due to near-threshold fracture morphologies from replicas of a fatigue crack in α-titanium ($\Delta K \approx 9\,\mathrm{MPa\,m^{1/2}}$, $R = 0.1$), taken at minimum load fatigue cycle at (a) 9.2 mm and (b) 14.5 mm from crack tip (note contact between faceted crack surfaces at discrete points, aided by the mode II crack tip displacements, leading to roughness-induced crack closure) (after Walker and Beevers 1979)

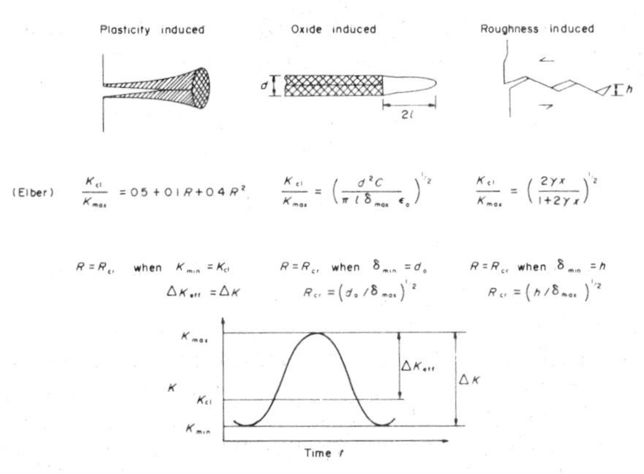

Figure 21
Summary of the various mechanisms of fatigue crack closure and their respective analytical expressions (δ = crack tip displacement; ε_0 = yield strain; d_0 = total oxide thickness at ΔK_{th}; C = constant ($\sim 1/32$); x = ratio of mode II to mode I crack tip displacements; γ = nondimensional roughness factor, defined as the ratio of the height h of fracture surface asperity to its width w; R_{cr} = critical load ratio above which closure effects are minimal, at given ΔK) (after Ritchie and Suresh 1983)

rougher fracture morphologies and further delay the transition from a serrated (single shear) to a more planar (alternating or simultaneous shear) fracture surface to higher ΔK values when the maximum plastic zone size is of the order of the grain size. Thus, coarse microstructures generate greater closure, making effective ΔK values lower, at the same nominal ΔK, than in finer-grained microstructures. This explanation is entirely consistent with the fact that at high load ratios where closure effects are minimal, the effect of grain size on near-threshold behavior is reduced markedly (Fig. 13).

Simple analytical expressions for both oxide-induced and roughness-induced crack closure mechanisms are summarized in Fig. 21. Both mechanisms predominate at near-threshold levels, but it is worth noting that their effect is not necessarily unique to this regime. Given a sufficient degree of faceted fracture morphology, such as in the case of "crystallographic" fatigue crack propagation in nickelbase superalloys, or substantial crack face oxidation, which may occur for elevated temperature creep or fatigue crack growth, such mechanisms may be active at the larger ΔCTOD values associated with cyclic crack growth at higher ΔK levels. Their effect, however, is maximized at near-threshold stress intensities, due to the single shear (mode I + mode II) character of the crack extension mechanisms there and to the small magnitude of the associated crack tip opening displacements.

3. *Short Fatigue Cracks*

As outlined in the article *Fatigue Crack Growth: Macroscopic Aspects*, short fatigue cracks have been observed experimentally to show crack growth kinetics which can be markedly different from those commonly observed for long cracks (i.e., typically

~25 mm in length in commonly used geometries such as the compact specimen). There are several factors which define a short crack, namely cracks which are (a) of a length comparable with the scale of microstructure, (b) of a length comparable with the scale of local plasticity and (c) simply physically small (< 0.5–1 mm).

The first of these cases, the microstructurally small crack, represents a continuum mechanics limitation to current analysis procedures. Although much metallurgical research has focused on this topic, it is generally not a major problem in the engineering sense, since microcracks of this size are not readily detectable by nondestructive testing and their formation and growth are generally classified in terms of the crack initiation stage of an engineering-sized flaw. It is important to note, however, that metallurgically the microstructures which show optimum resistance to such microcrack growth (or crack initiation) do not necessarily guarantee optimum resistance to the propagation of long cracks. A classic example of this is the well known fact that the smooth bar fatigue limit in steels (which represents an effective threshold for crack initiation) increases with increasing strength level, whereas the crack propagation threshold ΔK_{th} decreases. This poses something of a dilemma in the selection and design of alloys, such that the choice of a material for a given fatigue-sensitive application often depends critically upon whether design is based on crack initiation (i.e., using *S–N* curves) or on crack propagation (i.e., using defect tolerance).

The majority of short crack research has focused on cracks which are comparable in size with the scale of local plasticity. This case arises when the crack length is of the order of the plastic zone size or when cracks emanating from notches are growing in the strain field of the notch, and represents a linear elastic fracture mechanics limitation to current design methodologies. An example of this is given in Fig. 22, where the threshold conditions for no fatigue failure are shown as a function of crack size. For long cracks, the threshold condition is simply given, in fracture mechanics terms, as one of a constant threshold stress intensity range ΔK_{th}, whereas for very small cracks the data approach the smooth bar fatigue limit, or endurance strength $\Delta\sigma_e$, of a constant threshold stress. It is readily apparent that extending the linear elastic fracture mechanics analysis of a constant ΔK_{th} to small crack sizes will result in dangerously nonconservative predictions. The transition between long and short crack behavior here is often known as the intrinsic crack size a_0 and is given by

$$a_0 \approx \frac{1}{\pi}\left(\frac{\Delta K_{th}}{\Delta\sigma_e}\right)^2 \tag{4}$$

where $\Delta\sigma_e$ is the fatigue limit corrected for the appropriate load ratio. Substitution of typical values into Eqn. (4) reveals a_0 to be of the order of 1 μm in

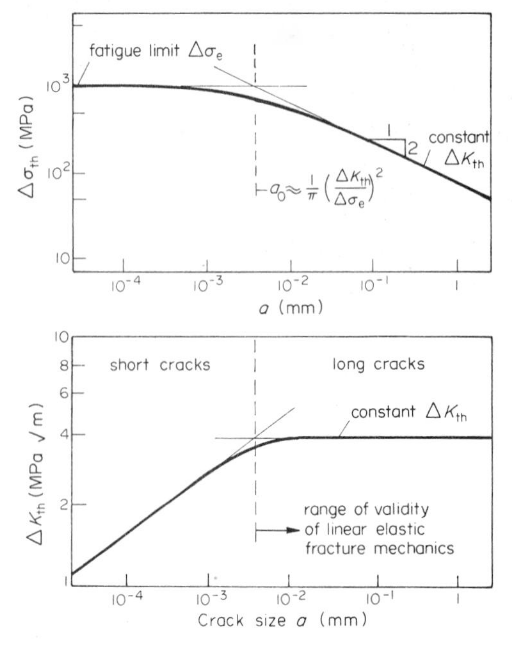

Figure 22
Variation with crack size a of threshold stress range $\Delta\sigma_{th}$ and threshold stress intensity range ΔK_{th} for no crack growth in 300-M steel ($\sigma_y = 1400$ MPa) at $R = 0$

ultrahigh-strength materials ($\sigma_y \sim 2000$ MPa) and ~1 mm in low-strength materials ($\sigma_y \sim 300$ MPa), indicating that this problem is most predominant in lower-strength alloys.

Allowing for the geometrical factors associated with notches, the discrepancy between long and short crack behavior in this case has been traced largely to inappropriate fracture mechanics characterization. Whereas the use of the stress intensity range ΔK may be valid for long cracks, when crack sizes approach the dimension of the plastic zone, conditions of small-scale yielding no longer apply and analyses based on a linear elastic K_I essentially are meaningless. Accordingly the growth of such short cracks can be treated using elastic–plastic fracture mechanics, so that when compared on the basis of ΔJ or ΔCTOD, the propagation rates of long and short cracks show a significantly closer correspondence (Fig. 23).

The third type of short crack, the physically short flaw, perhaps is the most dangerous, since it may be considered "long" in terms of both continuum mechanics and linear elastic fracture mechanics analyses, yet it may still show faster propagation rates than corresponding long cracks at the same ΔK level. Two reasons have been advanced for such behavior. The first is associated with crack closure. Since the origin of the differences in behavior resulting from a contribution from crack closure arises from the fact that such

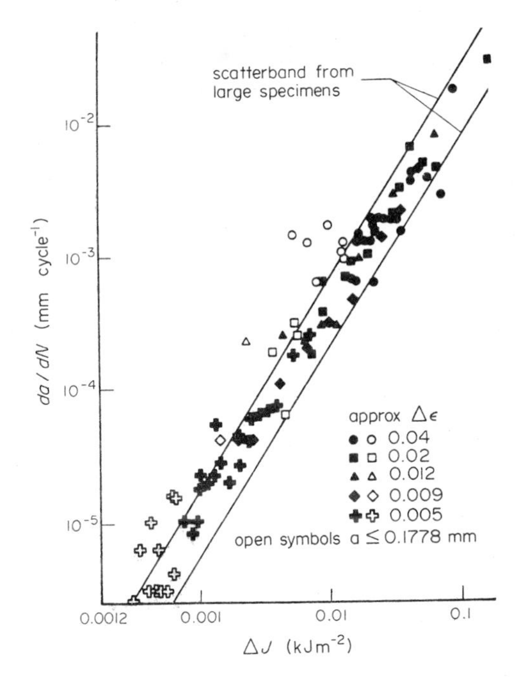

Figure 23
Comparison of fatigue crack propagation rates da/dN in A533B nuclear pressure vessel steel ($\sigma_y = 480$ MPa) obtained from experiments with long cracks ($a \sim 25$ mm) and short cracks ($a < 0.18$ mm) (after Dowling 1978)

closure effects predominate in the wake of the crack tip, and since, by definition, short cracks possess only a limited wake, it is to be expected that in general such cracks will be subjected to less closure. Thus at the same nominal ΔK, short cracks may experience a larger effective ΔK compared with the equivalent long crack. This may arise from a restricted envelope of prior plastic zones in the wake of the tip, less corrosion debris or less roughness-induced closure generated on the smaller crack surface. In fact the latter effect has been reported: experimental measurements of the crack opening displacements (at zero loads) of small surface cracks in titanium alloys clearly have shown far less closure associated with the short flaw. Thus at equivalent nominal ΔK levels, the short crack would be expected to propagate faster (and show a lower threshold ΔK_{th}) than the corresponding long crack because the effective stress intensity range experienced at the crack tip would be greater.

The second factor is associated with electrochemical effects and is relevant to the growth of physically short flaws in aqueous corrosive environments. Experiments by Gangloff (1981) on high-strength 4130 steel tested in NaCl solution revealed that corrosion fatigue crack propagation rates of short cracks (<0.8 mm) were more than two orders of magnitude faster than corresponding rates of long cracks (>25 mm) at the same ΔK level, although behavior in inert atmospheres was essentially similar. A complete understanding of this phenomenon is as yet lacking, but preliminary analysis indicated that the effect could be attributed to different local crack tip environments in the long and short flaws, principally resulting from the influence of crack length on crack surface reactions and on the solution renewal rate at the crack tip region.

4. *Variable Amplitude Loading*

Mode I fatigue cracks subjected to broad band variable amplitude loading spectra show transient crack growth behavior in the form of local accelerations or decelerations, depending upon the magnitude and sequencing of the applied loads (see *Fatigue Crack Growth: Macroscopic Aspects*). For example, the application of a single positive (spike) overload of sufficient size (generally $>50\%$ of the baseline ΔK) can result in significant retardations in crack growth over crack lengths comparable with the overload plastic zone size. Furthermore, low–high block loading sequences can yield local accelerations, whereas high–low sequences can produce local decelerations (and sometimes even arrest). The origin of these interaction effects is currently uncertain, but mechanisms involving crack tip blunting, residual compressive stress fields and fatigue crack closure concepts have been suggested which appear, at least qualitatively, to rationalize the data.

Recent analysis of the post-overload retardation effect by Suresh (1982) have suggested that during the overload cycle, the crack tip profile becomes blunted, often with one or two 45° branch cracks subsequently emanating from the tip. The reduction in effective stress intensity range resulting from this blunted and branched crack, coupled with the fact that the crack is now propagating into the compressive residual stress field of the overload plastic zone, acts to retard crack growth to near-threshold levels and in certain cases to arrest it. Furthermore, the retardation can be enhanced during the initial post-overload region because the crack is experiencing effective stress intensities ranges in the near-threshold regime, such that the process of crack advance becomes influenced by additional (near-threshold) crack closure mechanisms (see Sect. 2).

Specifically, the branched morphology of the overloaded crack together with the single shear faceted nature of the subsequent (near-threshold) crack growth can promote significant roughness-induced crack closure, evidenced by severe abrasion between the crack faces in the post-overload zone (region C in Fig. 24). It is also apparent from Fig. 24 that the stretch zone (region B) created at the application of the overload is free from abrasion marks, which suggests that the initial blunting of the crack tip minimizes the effect of plasticity-induced crack closure (due to the

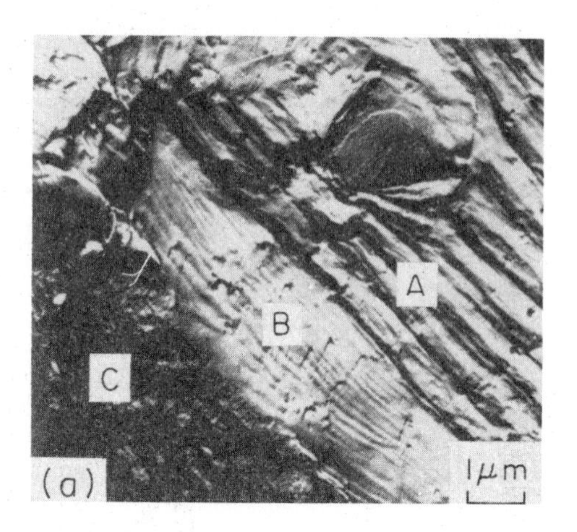

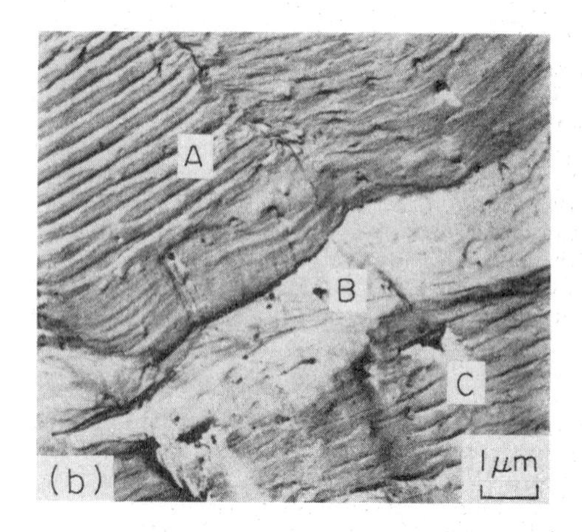

Figure 24
Transmission electron micrographs showing fracture surface appearance of a 2024-T3 aluminum alloy ($\sigma_y \approx 380$ MPa) during a single positive (spike) overload sequence: the crack is propagating from the top right-hand corner to the lower left-hand corner and indicates ductile striations in the pre-overload zone (A), a relatively featureless stretch zone (B) at the overload cycle and a post-overload zone (C) showing severe signs of crack surface abrasion (after Hertzberg 1976)

residual plastic displacements created during the overload cycle) in contributing to the retardation in crack growth. Similar arguments involving crack tip blunting and branching, residual compressive stress and crack closure mechanisms have been advanced to rationalize transient crack growth behavior following spectrum and block loading sequences.

One final point of interest with regard to the influence of variable amplitude loading is that for fatigue crack growth in antiplane shear (i.e., in mode III rather than in mode I), the response to single overload cycles and high–low block loading sequences is almost exactly opposite to that observed for mode I cracks. This difference in the transient growth rate behavior of mode III and mode I cracks follows from the fact that, in mode III, cracks are not influenced by such mechanisms as crack tip blunting or branching and fatigue crack closure (other than rubbing between sliding fracture surfaces), which so profoundly affect the behavior of mode I cracks. Accordingly, analyses of the role of variable amplitude loading sequences on mode III crack growth can be based more on the Miner's rule approach, in that the transient behavior can be rationalized simply in terms of the damage accumulated within the reversed plastic zones for each individual load reversal.

5. Conclusion

This article has attempted to provide a simplified basis for understanding the physical mechanisms associated with fatigue crack propagation in metals and alloys from both microscopic (metallurgical) and macroscopic (continuum mechanics) viewpoints. Although fatigue represents one of the major causes of failure in engineering service, a complete understanding of how a fatigue crack actually propagates has not yet been reached. Good progress has been made with the continuum mechanics characterization of cyclic crack growth rates and fracture mechanics analyses are now in widespread use for defect-tolerant design codes. Similarly, an understanding of the role of microstructure in improving the resistance to fatigue crack growth has emerged to the point where alloy design guidelines exist for the production of alloys with optimum resistance to fatigue failure.

However, much work remains in the definition of mechanisms associated with environmentally influenced crack growth, with the effect of variable amplitude loading and particularly with the problem of the short flaw. These problems demand an interdisciplinary approach to fatigue research involving applied mechanics, materials science and surface chemistry studies, and clearly offer substantial opportunities for further investigations, both of a fundamental nature and to provide reliable engineering data needed in the design and maintenance of fatigue-critical structures.

See also: Mechanics of Materials: An Overview; Fatigue Crack Growth: Macroscopic Aspects

Bibliography

Beevers C J 1977 Fatigue crack growth characteristics at low stress intensities of metals and alloys. *Met. Sci.* 11: 362–67

Elber W 1971 *Damage Tolerance in Aircraft Structures*, ASTM STP 486. American Society for Testing and Materials, Philadelphia, Pennsylvania, p. 230

Fine M E, Ritchie R O 1979 *Fatigue and Microstructure.* American Society for Metals, Metals Park, Ohio, p. 245
Gangloff R P 1981 The criticality of crack size in aqueous corrosion fatigue. *Res. Mech. Lett.* 1: 299–306
Hertzberg R W 1976 *Deformation and Fracture Mechanics of Engineering Materials.* Wiley, New York
Laird C 1967 *Fatigue Crack Propagation,* ASTMP STP 415. American Society for Testing and Materials, Philadelphia, Pennsylvannia, p. 131
Lindley T C, Richards C E, Ritchie R O 1976 Mechanics and mechanisms of fatigue crack growth in metals: A review. *Metall. Met. Form.* 43: 268–80
McClintock F A 1963 In: Drucker D C, Gilman J J (eds.) 1963 *Fracture of Solids.* Interscience, New York, p. 65
McEvily A J 1983 *Quantitative Measurement of Fatigue Damage.* American Society for Testing and Materials, Philadelphia, Pennsylvania, p. 283
Minakawa K, McEvily A J 1981 On crack closure in the near-threshold region. *Scr. Metall.* 15: 633–65
Neumann P 1974 New experiments concerning the slip processes at propagating fatigue cracks—I. *Acta Metall.* 22: 1155–65
Paris P C, Erdogan F 1963 A critical analysis of crack propagation laws. *J. Basic Eng.* 85: 528–34
Pelloux R M N 1969 Mechanisms of formation of ductile fatigue striations. *Trans. Am. Soc. Met.* 62: 281–85
Ritchie R O 1982 In: Bäcklund J, Blom A, Beevers C J (eds.) 1982 *Fatigue Thresholds,* Vol. 1. Warley, UK p. 503
Ritchie R O, Knott J F 1973 Mechanisms of fatigue crack growth in low alloy steel. *Acta Metall.* 21: 639–48
Ritchie R O, Suresh S 1983 Behavior of short cracks in airframe components. *Proc. 55th Specialists Meeting of AGARD Structural and Materials Panel,* AGARD Vol. CP328. North Atlantic Treaty Organization, Advisory Group for Aerospace Research and Development, Paris, p. 1.1
Starke E A, Lütjering G 1979 *Fatigue and Microstructure.* American Society for Metals, Metals Park, Ohio, p. 205
Suresh S, Zamiski G F, Ritchie R O 1981 Oxide-induced crack closure: An explanation for near-threshold corrosion fatigue crack growth behavior. *Metall. Trans. A* 12: 1435–43
Tomkins B 1968 Fatigue crack propagation—An analysis. *Phil. Mag.* 18: 1041–66
Weertman J 1979 *Fatigue and Microstructure.* American Society for Metals, Metals Park, Ohio, p. 279
Wei R P, Speidel M O 1971 In: Devereux O, McEvily A J, Staehle R W (eds.) 1971 *Corrosion Fatigue.* National Association of Corrosion Engineers, Houston, p. 433
Wei R P, Simmons G W 1982 Fatigue—environment and temperature effects. *Proc. 27th Sagamore Army Materials Research Conf.* AMMRC, Watertown, Massachusetts

R. O. Ritchie
[University of California,
Berkeley, California,
USA]

Fatigue: Engineering Aspects

Wide use has been made of metals and their alloys in engineering components and structures over the past two centuries. A major factor has been the ability of metals to withstand tensile and torsional loads, thus enabling rotating machinery and lightweight structures to be developed. The cyclic and transient tensile loading of such components and structures has revealed a major failure process—fatigue. The physics of this process, which can produce failure after many cycles under a repeated fraction of the yield load, is now well understood. It concerns the initiation and subsequent growth of cracks (see *Fatigue Crack Growth: Macroscopic Aspects; Fatigue Crack Growth: Mechanistic Aspects*). This article describes the important aspects of fatigue as seen by the engineer who has to provide some assurance that structures will have adequate fatigue resistance. Understanding and quantitative description of the process is of great help in underpinning engineering judgement of this process, so some reference is made where appropriate to basic principles.

1. Behavior of Materials Under Cyclic Loading

Tensile or compressive straining of a metal induces dislocation movement and multiplication, with the eventual development of discrete dislocation substructures. Cyclic straining, either by repeated tension or alternate tension and compression modifies such substructures and leads to hysteretic behavior in the stress–strain plane (Fig. 1). For many metals and alloys, a unique relationship exists between the applied cyclic stress range $\Delta\sigma$ and cyclic plastic strain range $\Delta\varepsilon_p$, often of power law form:

$$\Delta\sigma = k\,\Delta\varepsilon_p^{n'} \quad (1)$$

This is the cyclic analogue of the static tensile stress–strain curve. However, there can be significant differences between monotonic and cyclic stress–strain

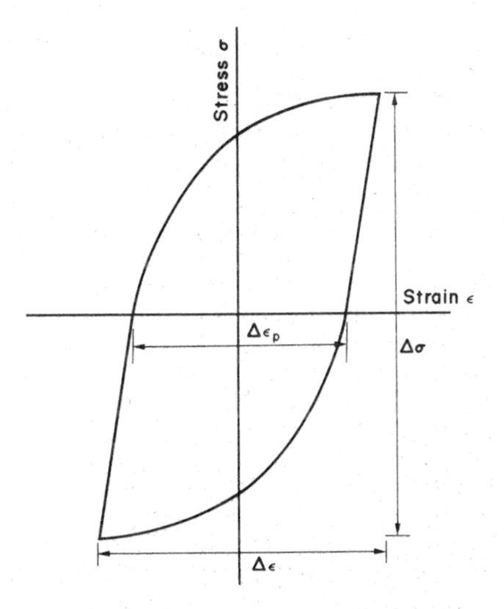

Figure 1
Typical cyclic stress–strain hysteresis loop

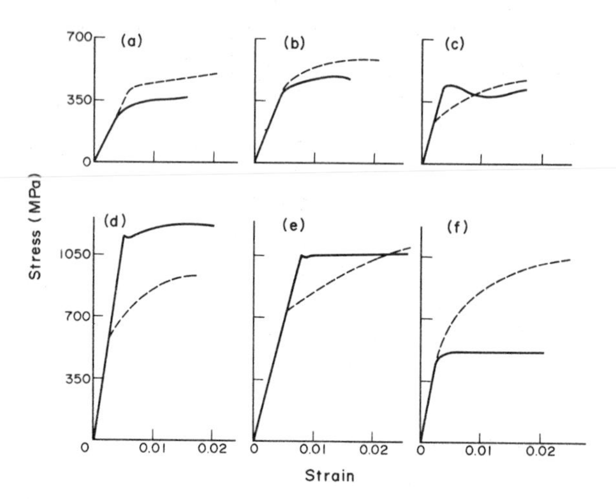

Figure 2
Monotonic and cylic stress–strain curves for (a) 2024-T4 Al alloy, (b) 7075-T6 Al alloy, (c) Man-ten steel, (d) SAE 4340 (350 BHN) steel, (e) Ti-811 titanium alloy, (f) Waspaloy A nickel alloy (--- cyclic stress, —— monotonic stress)

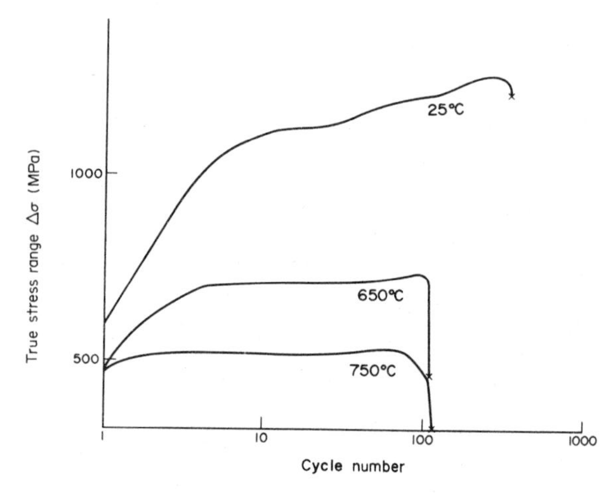

Figure 3
Variation of true stress range $\Delta\sigma$ with cycles, 20% Cr–25% Ni/Nb stainless steel ($\Delta\varepsilon_p = 0.4$) for various test temperatures

curves, as shown in Fig. 2 for a range of engineering alloys at ambient temperature. Both cyclic "hardening" and cyclic "softening" behavior are observed.

The cyclic stress–strain curve is usually described in terms of the steady-state stress and strain ranges, arrived at after a number of cycles. Steady state is often defined in terms of half-life values. Figures 3 and 4 show extreme hardening and softening behavior for two high-temperature alloys. The initial hardening period is associated microstructurally with the gener-

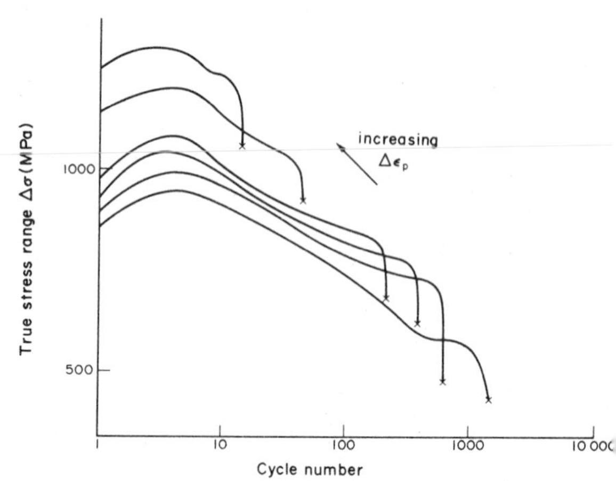

Figure 4
Variation of true stress range $\Delta\sigma$ with cycles, aged PE16, test temperature 750 °C

ation of a stable dislocation substructure. In this sense it is analogous to primary creep. The rapid softening always observed towards the end of life is caused by the presence of macrocracks and is not solely a material effect. Surface crack initiation is often noted to occur at the end of the initial hardening period. Precipitation-hardened alloys may exhibit dramatic cyclic softening as shown in Fig. 4. This is thought to be due to precipitate shearing by dislocation channelling. Cold-worked alloys also exhibit cyclic softening. The cyclic-stress-induced generation of dislocation substructures can occur at stress levels below general yield and in some alloys this is thought to be the basis of the material's fatigue strength (fatigue limit), via the link between such substructures and crack initiation. Other alloys (e.g., austenitic stainless steels) can sustain significant cyclic plastic straining beyond general yield without initiating cracks, because the dislocation substructures generated are irregular. The fatigue strength of an alloy is therefore a material structure sensitive property.

2. Endurance Relationships

2.1 Material

The basic endurance relationships for a material are expressed in terms of the applied cyclic stress/strain range ($\Delta\sigma$, $\Delta\varepsilon$) or amplitude versus endurance (cycles to failure, N_f). In terms of elastic and plastic components these are often of power law form,

$$\Delta\varepsilon_e \mathrm{N}_f^{\alpha'} = A_1 \qquad (2)$$

$$\Delta\varepsilon_p N_f^{\alpha} = A \qquad (3)$$

and the curves are shown schematically in Fig. 5. Equation (2) represents the well known *S–N* or

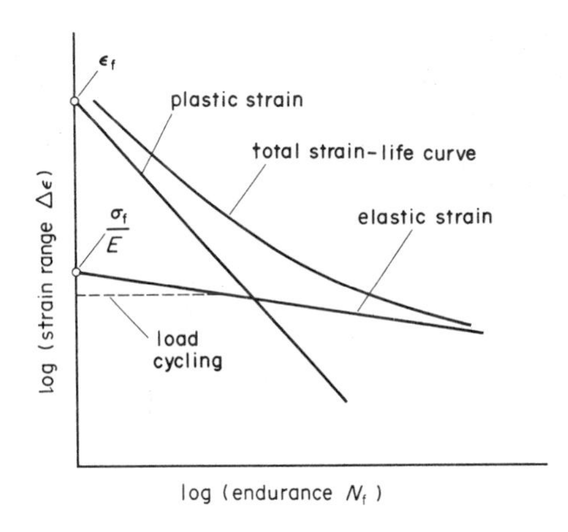

Figure 5
Lifetime as a function of elastic, plastic and total strain amplitude

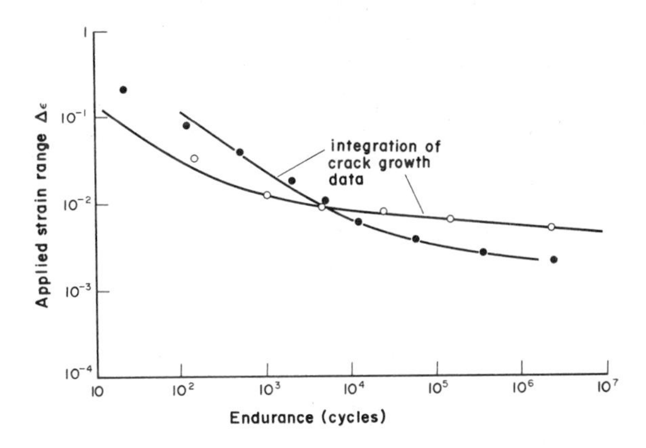

Figure 6
Strain endurance curves for two ferritic steels: ○, En 25 steel, 200 °C temper; ●, mild steel

Basquin relationship whereas Eqn. (3) is the Coffin–Manson law. It is worth noting that above general yield, the Basquin relation holds only for strain cycling conditions as load cycling leads to excessive accumulation of plastic deformation (ratchetting) and tensile rather than fatigue failure. For materials where crack initiation is rapid, these relationships are effectively integrated crack-growth laws between an initial crack size a_0 (which may be only a few micrometers) and final crack size a_f of several millimeters. Figure 6 shows such an integration compared with endurance data for mild steel and a high-strength low-alloy steel. Increase in material yield strength tends to raise the fatigue strength, which depends on crack initiation, but leads to a decrease in lifetime at high strain levels. The latter is caused by an increase in crack growth rate, and reduction in a_f because of reduced toughness.

The material endurance curve is also influenced by other factors. An increase in tensile mean stress tends to reduce fatigue strength by promoting crack initiation, whereas a compressive mean stress marginally increases it. A rise in temperature also tends to reduce fatigue strength as well as increasing crack growth rates, thus leading to a dropping of the whole fatigue curve. Multiaxial straining induces different effects in the initiation and propagation dominated regions. Where initiation is by dislocation processes alone, described by a "cyclic" yield point, it may be represented by a typical multiaxial yield criterion such as von Mises, expressed in cyclic terms. The crack growth region, however, is described by a more complex criterion, dependent on both shear and normal stresses or strains (Brown and Miller 1973). If "natural" crack initiation is overcome by some other process, then the fatigue strength can be drastically reduced, to a level determined solely by crack growth considerations. Examples of such reductions are encountered in corrosion fatigue, where surface pitting induces initial surface "cracks," wear-induced fatigue and fatigue of materials with mechanical surface damage (e.g., scratches, machining marks).

2.2 Components

Endurance relationships for components often reflect the sensitivity of fatigue life to surface damage. Welded joints, for example, contain microflaws (e.g., slag, porosity, undercut) which act as already initiated cracks. Figure 7 shows the endurance data for fillet and butt welded joints in a structural steel as a function of the nominal stress at the weld location. A comparison is made with plain material behavior. The typical scale of initial flaws in the welded joints producing such dramatic effects on life and strength is of order 0.1–0.5 mm. The fillet weld curve is of interest in that it represents an integration of the Paris linear elastic fracture mechanics (LEFM) crack growth law (see *Fatigue Crack Growth: Macroscopic Aspects*). This is of the form

$$da/dN = C\Delta K^m \quad (4)$$

where ΔK is the cyclic stress intensity factor ($\propto \Delta\sigma a^{1/2}$) and C and m are material constants. For structural steel, m is ~ 3. Integration of Eqn. (4) then gives an S–N curve of the form

$$\Delta\sigma N_f^{-1/3} = \text{constant} \quad (5)$$

consistent with the curve shown in Fig. 7. The fatigue limit would be related to the LEFM threshold for crack growth ΔK_{th}, based on the initial crack size and making due allowance for weld residual stresses. It should be noted that this value is less sensitive to material yield strength than the material fatigue strength. In fact, higher strength materials tend to have

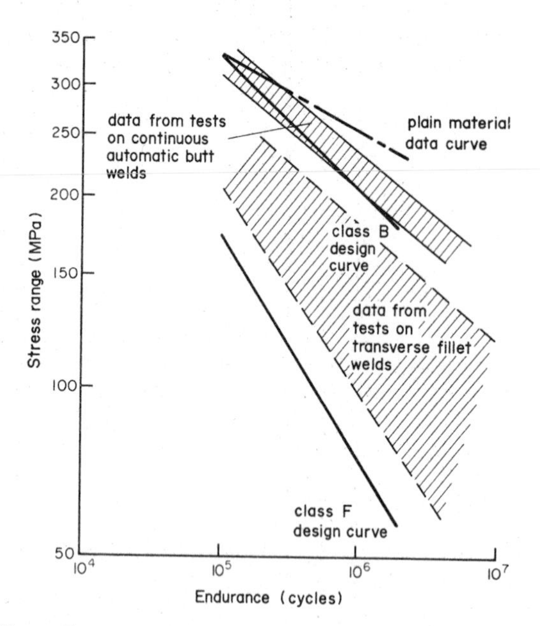

Figure 7
Weldment fatigue endurance data and design curves (after Gurney and Maddox 1973)

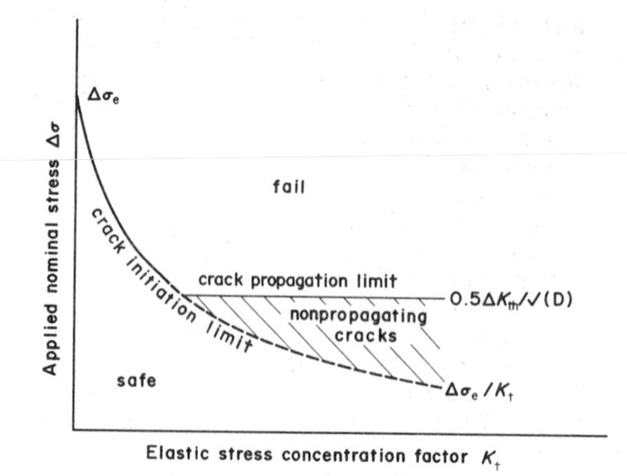

Figure 8
Variation of fatigue strength of notched components wtih K_t

lower ΔK_{th} values (Lindley 1981). For butt welds, the lower initial stress condition at the weld toe means that initial flaws do not behave as cracks from the outset and some cycles are spent initiating cracks from them, giving a higher fatigue strength and shallower S–N slope than that observed for fillet welds.

If flaws are not of a sharp crack profile, they behave as blunt notches which have different stress concentrating effects dependent upon the notch root radius ρ. The notch is the strongest feature controlling fatigue life in components, whether it is on a macro- or microscale. This has been recognized for many years and careful determination of the fatigue strength reduction due to notches is a major element in the design of rotating machinery. A notch stress concentration is characterized by two dimensions, ρ and the notch depth D. Under elastic conditions, the stress concentration factor K_t is given by

$$K_t = 1 + 2(D/\rho)^{1/2} \tag{6}$$

and the static stress field at a notch root,

$$\sigma = K_t \sigma_a [\rho/(\rho + 4x)]^{1/2} \tag{7}$$

where σ_a is the nominal applied stress and x the distance from the root. The fatigue strength of a notched component varies with K_t as shown in Fig. 8. The actual fatigue strength $\Delta\sigma_f$ is defined by two limits. At low values of K_t, the limit is determined by crack initiation, where

$$\Delta\sigma_f = \Delta\sigma_e/K_t \tag{8}$$

where $\Delta\sigma_e$ is the plain material fatigue strength. At higher values of K_t, cracks are readily initiated and the actual limit is defined by the crack growth threshold ΔK_{th}, as

$$\Delta\sigma_f = 0.5[\Delta K_{th}/(D+a)^{1/2}] \approx 0.5(\Delta K_{th}/D^{1/2}) \tag{9}$$

In the region between the two limits, nonpropagating cracks are observed.

3. Design Assessment of Fatigue

The design of an engineering structure or component is related to its primary function but it involves consideration of issues pertinent to likely service history, one of which may be fatigue. Whether this is a major or minor issue, it is a matter of design assessment rather than a design basis. When the assessment of a detailed design shows fatigue as a potential problem, the designer has several options. These include reduction of cyclic stresses, improvement of component details and modification of service history. However, the simple expedient of using stronger material to raise fatigue strength seldom improves overall fatigue behavior.

In considering fatigue, the designer usually adopts one of two philosophies: "safe life" or "damage tolerance." A third philosophy, "fail safe," is really a subset of damage tolerance, although historically its development, primarily for aircraft structures, preceded the more generalized concept. The safe-life philosophy relies on the provision of a valid endurance curve, which is factored to provide a design curve. The endurance curve may be for the material (e.g., as in Fig. 6) or a critical structural detail (e.g., as in Fig. 7), expressed in terms of a relevant stress range. The design factor is chosen to provide an adequate margin on failure taking account of considerations such as accuracy of stress analysis, relevance of testpiece to structure and inherent scatter of test data. In addition, the factor is increased for safety related structures, for example, nonredundant structures whose catastrophic fatigue-induced failure is unacceptable.

Although the safe-life approach is that most widely used for fatigue assessment, it has an inherent uncertainty in the applicability of the test data to an actual structure or component, whose initial state, particularly with regard to flaws, is unknown. This has led to increased development and use of the damage-tolerance approach. This philosophy, which has become feasible with the advent of fracture mechanics, relies on knowledge and awareness of acceptable or "tolerable" levels of fatigue damage in the structure. The fail-safe condition is a special case of this approach, where the structure has sufficient redundancy to tolerate partial failure which can be defined as damage. Any initial failure is then to a safe state involving stress redistribution. However, it must be possible to detect this change before further damage leading to a critical unsafe state occurs. Examples of fail safe are in some airframe structures, offshore structural joints and leak-before-break in pressure boundary components (pipes, vessels). Damage tolerance is a wider concept, but it always relies on damage or flaw detection. Nondestructive examination and fracture mechanics methodology are the basis for the development and utilization of the approach. Damage tolerance is used in the assessment of nuclear pressure vessels and aeroengine components.

The main features of the damage-tolerance approach have been outlined elsewhere (see *Fatigue Crack Growth: Macroscopic Aspects*). In practice, limitations are currently found in several areas.

(a) Detection and accurate measurement of initial flaws or cracks induced in service.

(b) Determination of the stress intensity factor range ΔK for a crack in its specific geometric location. This includes consideration of crack shape.

(c) Determination of a relevant material crack growth curve. This should cover the range of interest, which often includes the threshold and tearing regions, noting any dependence on mean stress. It may also include an environmental factor.

(d) Determination of a_f.

For many structures, limitations of knowledge in these areas mean that unduly conservative assumptions are made.

The safe-life approach also has limitations which include the ability to cope adequately with the following.

(a) Stress concentrations.

(b) Stress gradients—"size" effects are most often related to stress gradients.

(c) Load history—the variable load history experienced by most components and structures still provides one of the major uncertainties in fatigue assessment. The linear Palmgren–Miner rule is still the most widely used rule for coping with variable fatigue loads. It can be derived from a crack growth analysis, but mechanics/material interactions under variable amplitude conditions lead to nonlinearities (see *Fatigue Crack Growth: Mechanistic Aspects*). Developments in waveform analysis and simulative testing are producing some more realistic rules for specific applications.

(d) Stress state—mean and multiaxial stresses.

4. Fatigue Interactions

Major uncertainties in the estimation of fatigue behavior are introduced if time-dependent effects are present which can interact with the fatigue crack initiation and growth process. Corrosion was mentioned earlier as a factor which can seriously erode fatigue limits based on crack initiation. Crack growth rates can also be strongly influenced by an environment, gaseous or liquid. Figure 9 shows an example of such an influence for a pipeline steel tested in a water environment. The effect is accentuated by a high mean stress (R ratio) and low cyclic frequency.

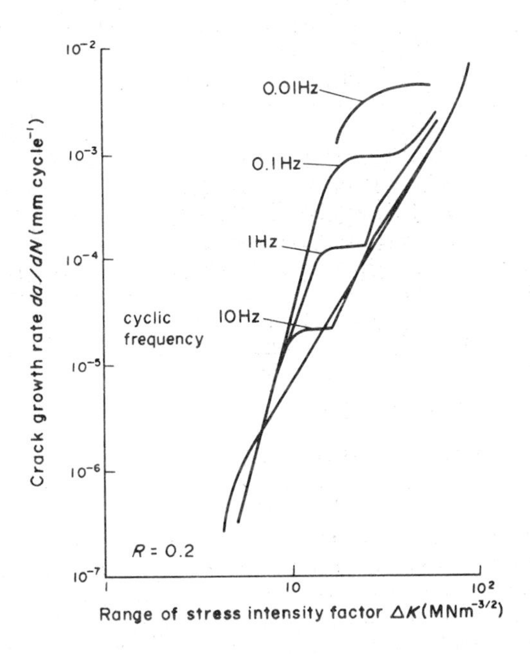

Figure 9
The influence of a seawater environment on fatigue crack growth in a C–Mn steel (after Vosikovsky 1975)

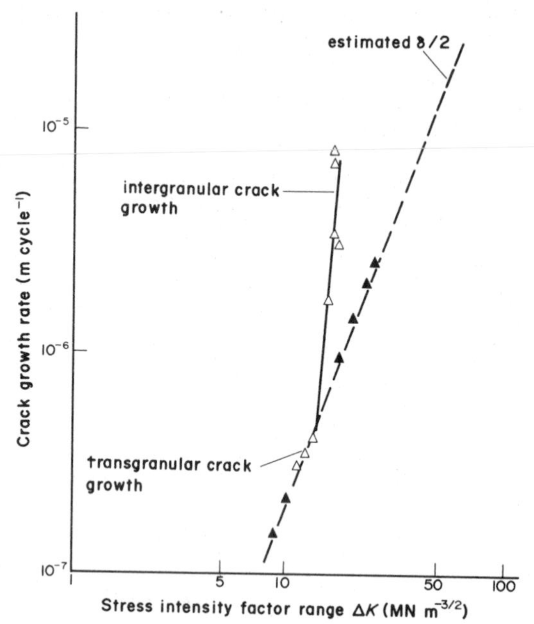

Figure 10
Effect of creep on fatigue crack growth in type 316 stainless steel at 625 °C, $R = 0.6$ (▲, sawtooth cycle, 10^{-12} Hz; Δ, tensile dwell cycle, 10^{-4} Hz) (after Lloyd and Wareing 1979)

At elevated temperature, creep-induced cavitation damage can interact with a growing fatigue crack to produce accelerated crack growth by effectively reducing the K_{max} level for tearing. Figure 10 shows this for an austenitic stainless steel where a crack is growing under a load cycle involving long dwell periods at maximum tensile load.

See also: Failure Analysis: General Procedures and Applications to Metals; Mechanics of Materials: An Overview

Bibliography

Brown M W, Miller K J 1973 A theory for fatigue failure under multiaxial stress–strain conditions. *Proc. Inst. Mech. Eng.* 187: 65
Gurney T R, Maddox S J 1973 A reanalysis of fatigue data for welded joints in steel. *Weld. Res. Int.* 3(4): 1
Lindley T C 1981 Near threshold fatigue crack growth. *Proc. ASFM3 Ispra.* Applied Science, London, p. 167
Lloyd G J, Wareing J 1979 Stable and unstable fatigue crack propagation during high-temperature creep—fatigue in austenitic steels. *J. Eng. Mater. Technol.* 101: 274
Vosikovsky O 1975 Fatigue crack growth in an X-65 line pipe steel at low cyclic frequencies in aqueous environments. *J. Eng. Mater. Technol.* 97: 298

B. Tomkins
[United Kingdom Atomic Energy Authority,
Warrington, UK]

Ferrous Physical Metallurgy: An Overview

Ferrous physical metallurgy is the branch of physical metallurgy that is concerned with the design and processing of steels or cast irons with special properties. The design process consists of selecting alloying additions and processing procedures to produce microstructures that result in the desired properties. The alloying elements of importance include major alloying elements (C, Mn, Si, Ni, Cr, Mo, Al, B) and microalloying elements (Nb, Ti, V); in many cases, control of the impurity elements (S, P, O, N, H) is also a critical issue. The processing procedures include melting, casting, hot and cold rolling, annealing, thermomechanical treatment and heat treatment. The microstructural elements include grain size; dislocation density; crystallographic texture; the size, amount, distribution and chemical composition of the individual phases; and possible segregation effects of alloying elements at grain or interface boundaries. The properties that are subject to control may be physical, chemical, mechanical, or manufacturing-related (e.g., weldability, formability).

The design of steels or cast irons with special properties thus involves a complicated interplay between alloying, processing and microstructural elements. Although scientific principles can be used to investigate the effect of single alloying additions on a single property (e.g., Mn on solid solution hardening), multiple alloying additions and complex microstructures generally require that simplistic assumptions (e.g., additive strengthening factors) or empirically developed regression equations be used in predicting various steel properties. In addition, in many cases, several alternative alloying/processing/microstructure paths may be available to achieve the same properties, and then economic factors may control the final design decision.

In this article, the major steel alloying additions are discussed, as is the classification of steels into various types, based on their chemical composition, their properties or their end uses. The various microstructural elements in steel are then described, as well as how processing procedures are used to control them. The article then covers those steel properties that are difficult to control and, finally, the different types of cast irons are described. For more in-depth information on these subjects, the reader should consult the bibliography at the end of the article.

1. Alloying Elements in Steel

1.1 Major Alloying Elements

Major alloying elements in steels consist of C, Mn, Si, Ni, Cr, Cu, Mo, Al and B.

(*a*) *Carbon.* Carbon is the principal alloying element in steels and is present in amounts between 0.02 and

2%. Its principal functions are to increase hardness and strength in hot-rolled steels and also to increase hardness and strength in quenched and tempered steels. C (and N) segregation to dislocations or precipitates from α-Fe also affects steel properties by strain-aging or quench-aging reactions. Ductility and weldability decrease with increasing C content. The importance of C as an alloying element results in the Fe–C phase diagram serving as the basis for the discussion of steel microstructures.

(*b*) *Manganese.* Manganese has several functions. In small amounts (0.2%) it reacts with the S impurity to form MnS inclusions and prevents formation of low melting FeS which causes hot-shortness. In larger amounts (1.0–1.5%) Mn contributes to solid-solution hardening, decreases ferrite grain size in hot-rolled plate steels, and increases the hardenability of quenched and tempered steels.

(*c*) *Silicon.* Silicon in small amounts (0.2%), is used as a deoxidizer, reacting with dissolved O in the liquid steel to form SiO_2. In intermediate amounts (0.5–1.0%) it contributes to solid-solution hardening. In larger amounts (1–3%) it is used to increase the electrical resistivity and decrease core loss of magnetic steels. Steels containing more than 3% Si may remain ferritic to very high temperatures because of the strong ferrite-stabilizing effect of this element.

(*d*) *Nickel.* Nickel has several functions. In small amounts (0.5–1.0%) it contributes to solid-solution hardening in hot-rolled plate steels and increases the hardenability of quenched and tempered steels. In larger amounts (3–9%), Ni is used to lower the ductile-to-brittle transition temperature of plate steels. Ni is also added in amounts up to 10% in stainless steels to stabilize the austenitic phase at room temperature.

(*e*) *Chromium.* Chromium is added in small amounts (0.2–2%) to increase the hardenability and to promote secondary hardening in quenched and tempered steels. In large amounts (10–20%), Cr improves corrosion and oxidation resistance, and is the principal alloying element in many high-temperature or stainless steels.

(*f*) *Copper.* Copper additions in small amounts (0.2–0.5%) are used to improve the atmospheric corrosion resistance of plate steels. In larger amounts (1.0–1.5%) Cu results in precipitation hardening after quenching and aging.

(*g*) *Molybdenum.* Molybdenum additions (0.1–0.2%) are used to increase the hardenability and to promote secondary hardening in quenched and tempered steels.

(*h*) *Aluminum.* Aluminum additions (0.01–0.04%) are used primarily as a deoxidizer to react with O impurities and prevent gas evolution during solidification. Al also reacts with N impurities to form a fine dispersion of AlN particles that restrict austenite grain growth during reheating.

(*i*) *Boron.* Boron additions (0.0003–0.003%) are used to improve hardenability in quenched and tempered steels by segregation to austenite grain boundaries which inhibits nucleation of high-temperature decomposition products.

1.2 Microalloying Elements

Microalloying elements in steels consist of additions of Nb, Ti or V in amounts generally less than 0.2%. Nb, Ti and V all react with C and N to form a fine dispersion of NbCN, TiCN or VCN, respectively. These fine dispersions of precipitates have several important functions. First, they contribute to strength by precipitation hardening effects. Secondly, they inhibit austenite grain growth during reheating of steels into the austenite temperature range because of their pinning action. Lastly, they inhibit recrystallization of the austenite during hot rolling and are necessary to achieve optimum properties during controlled rolling. In some tool steels, these elements may be added in larger amounts (1–2%) to form large volume fractions of hard carbides that increase wear resistance.

1.3 Impurity Elements

Impurity elements in many steels must be controlled to achieve optimum properties. The most important impurity elements are S, P, O, N and H. These elements originate from many sources including the iron ore or scrap used in steelmaking, the ferroalloy additions used for major alloying elements, the reaction of the liquid steel with air, the furnace linings, or the slag during melting or casting operations.

(*a*) *Sulfur.* Sulfur is present in all steels in amounts ranging from 0.001 to 0.04%. S is generally deleterious because of the formation of MnS inclusions which act as nucleation sites for initiating ductile fracture. As a result removal of S as an impurity can greatly increase the tensile ductility or toughness of many steels. However, in some free machining steels, S is added as an alloying element (0.1–0.2%) to promote machinability, since ductile fracture is not a critical issue in these steels.

(*b*) *Phosphorus.* Phosphorus is present in all steels in amounts between 0.001 and 0.04%. Its effect is generally considered to be deleterious because of segregation to grain boundaries and the promotion of brittle intergranular failure. However, in some sheet steels P is considered beneficial because of its solid solution hardening effect. Also, in motor lamination steels, P increases the ease of cold-punching operations.

(*c*) *Oxygen.* Oxygen occurs as an impurity in the form of oxide inclusions since its solubility in solid steel is negligible. Its concentration is sensitive to melting practice but varies from 20 to 200 ppm. Because high inclusion contents lead to lower ductility or toughness, high O contents are always considered detrimental.

(*d*) *Nitrogen.* Nitrogen may occur in the form of precipitates as dissolved interstitial N. In the form of AlN precipitates, N has a beneficial effect in restricting austenite grain growth. In the form of TiCN, VCN or NbCN precipitates, N has a beneficial effect in strengthening. In solution, N interacts with dislocations to cause strain aging, and generally leads to a decrease in ductility and toughness. However, N can be added as a deliberate solid solution strengthening or austenite stabilizing element in some renitrogenized sheet steels or in stainless steels. Total amounts range from 0.001 to 0.1%.

(*e*) *Hydrogen.* Hydrogen is a particularly deleterious impurity because of its rapid rate of diffusion to internal defects where it recombines to form molecular hydrogen and generates a high internal pressure that can cause blistering, flaking or embrittlement. It is present in extremely small amounts of the order of 1–100 ppm. Special vacuum degassing treatments or slow cooling treatments may be used to lower the hydrogen content.

2. *Steel Types*

The major types of steels include carbon, alloy, high-strength low-alloy (HSLA), tool, stainless, heat-resisting, and electrical or magnetic steels (see *Steels: Classification*). These distinctions are partly based on alloying additions and partly on end use or properties. Additional subtypes may exist within these steel types based on special applications, the method of manufacture or the shape of the product. For instance, there are hot-rolled structural and plate steels; heat-treated alloy steels; cryogenic steels; wire, rod and bar steels; tubing and pipe steels; ultrahigh-strength steels; or cold-rolled sheet steels. The major types are briefly discussed below.

(*a*) *Carbon steels.* Carbon steels are by far the most important type of steel in terms of total tonnage. They are the low-cost choice for many applications including structural beams; wire, rod and bar products; hot-rolled plates; and hot- and cold-rolled sheet products. Also, the microstructure of carbon steels serves as a basis for understanding the properties of other steel types, since alloying additions only have a minor effect on the amount and distribution of the various phases. C is the only major alloying addition in these steels, although they may contain small amounts of Mn, Si and Al as deliberate alloy additions. These steels are generally separated into subtypes of low-carbon, medium-carbon and high-carbon steels. Low-carbon steels contain 0.02–0.2% C; medium-carbon steels contain 0.30–0.50% C; and high-carbon steels contain 0.60–1.2% C.

(*b*) *Alloy steels.* Alloy steels are steels that derive improved properties from the addition of alloying elements or the presence of larger amounts of Mn and Si than are ordinarily present in carbon steels. The most important subtype of these steels are heat-treated alloy steels in which the higher alloy contents promote increased hardenability. Typical alloy additions include Mn, Ni, Mo, Cr, V and B. The importance of these steels has led to the designation of various grades according to SAE or AISI codes. Carbon contents of these grades vary from 0.10 to 0.80% C with the strength increasing and the ductility or toughness decreasing as the carbon content increases. In the heat-treated condition, strength levels can vary from 100 to 250 ksi (690–1723 MPa).

(*c*) *High-strength low-alloy steels.* These are steels with yield strengths greater than 40 ksi (276 MPa) and with alloy additions designed to provide improved strength, toughness, formability or corrosion resistance. These steels are designed for applications where higher strengths are required than those that can be obtained in simpler hot-rolled carbon steels. These higher strength levels can be achieved by microalloying additions of Nb, Ti, or V; by additions of Ni, Cr and Mo; by additions of Cu; or by various combinations of these alloying additions. Also, these steels can be processed by hot rolling; by controlled rolling; by quenching and tempering; or by quenching and aging treatments. Although higher in cost than carbon steels, this type of steel is finding increasing usage because of weight savings. Total tonnage amounts are about 10% of that of carbon steels. Important applications include structural and bridge, pipeline and automotive body components. Generally, the carbon content of these steels is low (0.04–0.2% C) to ensure weldability during fabrication of structural components.

(*d*) *Tool steels.* Tool steels are either carbon or alloy steels capable of being hardened and tempered. Their use is for mechanical fixtures for cutting, shaping, forming and blanking of materials. AISI designates seven grades: (a) high-speed tool steels; (b) hot-work tool steels; (c) cold-work tool steels; (d) shock-resisting tool steels; (e) mold steels; (f) special purpose tool steels; and (g) water-hardening tool steels. Bearing steels are another important subtype of tool steels. Tool steels contain 0.3–2.5% C, and depending on the grade, may contain large amounts of Cr, W, Mo, V and Co. The heat treatment of these steels is sophisticated because of their high carbon content and the need to avoid cracking during quenching and subsequent tempering. With the higher C and alloy content steels, hard carbides of Cr, W, Mo or V may be formed to increase wear resistance. Also, secondary hardening reactions during tempering help to promote hot hardness in high-speed tool steels.

(*e*) *Stainless steels.* Stainless steels are steels that contain substantial amounts of chromium to increase corrosion and oxidation resistance. The actual value of Cr that qualifies a steel as a stainless steel is not fixed. Some AISI standards use 5.5% Cr as a minimum whereas generally in excess of 10% Cr is required to achieve substantial corrosion resistance. Steels containing more than 14% Cr are always classified as stainless steels. Stainless steels are generally separated into subtypes of ferritic, martensitic, austenitic and precipitation-hardening. Ferritic grades contain 17–25% Cr and 0.1–0.2% C and generally are not heat-treatable. Martensitic grades contain 12–17% Cr and 0.15–1.2% C and are generally quenched and tempered to high strengths. Austenitic grades contain 17–25% Cr and 7–20% Ni with 0.15–0.25% C. Additions of Ti or Nb may be added to these grades to tie up C as TiC or NbC and prevent sensitization of these steels by precipitation of Cr_7C_3 or $Cr_{23}C_6$ at grain boundaries. Precipitation-hardening grades contain 16–17% Cr, 4–7% Ni, and additions of Cu or Al to promote precipitation hardening by precipitation of Cu or Ni_3Al intermetallic compounds.

(*f*) *Heat-resisting steels.* These are steels that have increased high-temperature strength, and better oxidation resistance compared to carbon steels. They generally contain increased Cr levels for better oxidation resistance. Also, additions of Mo are used to induce precipitation of alloy carbides during high-temperature exposure to improve creep strength. Carbon contents are generally in the range 0.1–0.2% C to ensure weldability.

(*g*) *Electrical and magnetic steels.* These are steels used in motor and transformer applications. There are three general types: low-carbon steels, non-oriented Si steels and oriented Si steels. Low-carbon steels are hot-rolled steels that contain less than 0.08% C, 0.25–0.75% Mn, and sometimes additions of P to make cold-punching of the laminations for motors easier. They are the cheapest grade of electrical steels and have the highest core loss. Non-oriented silicon steels contain between 0.8 and 3.5% Si (Si+Al) to increase the electrical resistivity of the steel and decrease core loss. Carbon contents are less than 0.025% or lower to minimize the amount of cementite particles which tend to pin domain walls during the magnetization cycle. No attempt is made to produce a crystallographic texture in these steels, but their core loss is appreciably lower than that of low-carbon grades. Oriented Si steels contain up to 3% Si and undergo a complex cold-rolling, annealing and decarburization cycle to produce a coarse-grained, ultralow-carbon steel with most of the grains having a (110) [100] texture. Such steels have the lowest core loss obtainable because the high Si content increases the electrical resistivity and the texture is such that the easy direction of magnetization ⟨100⟩ lies in the direction of the magnetic field.

3. Steel Microstructures

The various microstructural elements that can be controlled in steels to improve properties include grain size; dislocation density; crystallographic texture; the size, amount, distribution and chemical composition of the individual phases; and the possible segregation effects of certain elements at grain or interface boundaries.

(*a*) *Grain size.* Grain size of steels has an important influence on many properties but particularly on strength and toughness. As the grain size is decreased, the strength increases and the ductile-to-brittle transition temperature decreases. As a result, refinement of the grain size is one of the most important strengthening mechanisms in steel products because it also increases toughness. Ferrite grain size is indirectly controlled by the austenite grain size since after reheating into the austenite range for normalizing or hot-rolling, the steel subsequently transforms during cooling by nucleation of ferrite at the austenite grain boundaries. Many metallurgical processes such as normalizing, accelerated cooling or controlled rolling are thus designed to result in a refinement of the austenite grain size since this results in a refinement of the ferrite grain size. Quenching and tempering operations that produce martensitic transformations also result in marked refinement of the microstructure because each austenite grain is subdivided into many small martensite plates or units.

(*b*) *Dislocation density.* Dislocation density increases as the steel is cold worked so that the strength of the cold-worked steel products is higher than that of hot-rolled products. The increase in strength is proportional to the square root of the dislocation density. The stored energy in the dislocations also acts as the driving force for the recrystallization process in annealing operations. Transformation of austenite to martensite in quenching and tempering operations produces a high dislocation density (10^{12} cm^{-2}) which is largely responsible for the high strength of heat-treated steels.

(*c*) *Crystallographic texture.* The crystallographic texture of polycrystalline steels may be controlled by proper cold-rolling and annealing treatments. In deep-drawing operations, strong {111}⟨110⟩ textures are beneficial because the resulting plastic properties give an increased resistance to thinning of the steel sheet and delay fracture until higher plastic strains are reached. In electrical steels, the {110}⟨100⟩ texture is beneficial because the easy direction of magnetization of steel is in the ⟨100⟩ direction.

(*d*) *Individual phases.* Individual phases in steels contribute to many of the steel properties. Although the matrix phase is always bcc or fcc Fe, a large number of separate phases can be precipitated from the matrix during heat treatment or solidification including carbides (Fe_3C or alloy carbides), aggregates of carbides and other phases (pearlite, bainite, or tempered martensite), or inclusions (MnS, Al_2O_3, AlN). All of these individual phases influence the properties as their size, distribution, spacing or chemical composition are changed. For instance, the strength of HSLA steels is increased as the interparticle spacing of microalloy carbides (NbCN, VCN) is decreased. Also, the strength of eutectoid carbon steels is controlled by the spacing between the carbide (Fe_3C) lamellae of the pearlite. On a coarser scale, the strength of hot-rolled carbon steels is determined by the volume fraction of pearlite. The ductility and toughness of steels is also controlled to a large extent by the volume fraction, size and shape of inclusions. In addition, the strength of the Fe matrix depends on the solid-solution hardening effects of the individual alloying elements. In some ways, steels can be considered as composite materials with the overall properties resulting from the combined action of the matrix and precipitated phases (see *Steels: Classification*).

(*e*) *Segregation.* Segregation of alloying or impurity elements to grain boundaries plays an important role in many embrittlement processes in steel. Segregation of Sb, P and Sn impurities in heat-treated alloy steels during tempering in certain temperature regimes leads to severe embrittlement, referred to as temper embrittlement. Certain alloying additions such as Mo help to alleviate this condition by slowing the rate of diffusion of the impurities to the grain-boundary regions. Other alloying elements like Mn increase the susceptibility to such embrittlement. Segregation of hydrogen to grain boundaries may also play a role in hydrogen embrittlement phenomena. The driving force for such segregation is the lowering of strain energy of the impurity species or a lowering of the grain boundary interfacial free energy.

On a coarser scale, segregation of alloying elements occurs during solidification of steel ingots on both a microscale (dendrite arm spacings of about 100 μm) and a macroscale (top to bottom of ingot). Microsegregation can lead to significant differences in transformation behavior. For instance, hot-rolling deforms regions of Mn segregation into bands. During slow cooling of hot-rolled low-carbon steels, pearlite forms in the high Mn regions of these bands. Such "banding" phenomena are responsible for the banded microstructure exhibited by hot-rolled plate steels. Macrosegregation of C, Mn and other alloying elements from top to bottom of an ingot can lead to variation in properties from head to tail of a rolled plate or sheet. The origin of such segregation is the difference between solubilities of elements in solid and liquid Fe, so that as freezing progresses, many alloying elements are rejected from the solid Fe and the liquid Fe phase becomes enriched in alloying element. Increased casting speeds, as in continuous casting or in rapid solidification processing, can minimize or prevent such segregation, whereas slow casting speeds, as in ingot mold casting, maximize such segregation.

4. Steel Processing

Steels, like other metals, must be melted, cast and then forged, rolled, extruded or drawn into useful shapes. Subsequently these shapes or parts formed from them may be heat treated. Many of these steps prior to heat treatment can have a significant effect on properties.

(*a*) *Melting.* Melting practices can be utilized to help remove impurity elements or to make additions that render them less harmful. Also, the type of deoxidation practice has an important influence on the grain-coarsening characterisitcs of the steel. For instance, the level of S impurity can be drastically reduced (from 0.02 to 0.002%) by treatment of the liquid steel with Ca. Also, additions of rare-earth elements (La, Pr) can react with S to form hard $(RE)_2O_2S$ inclusions that do not deform during rolling and remain spherical in contrast to soft MnS inclusions. Such treatments can result in more nearly isotropic and higher impact energies. Other harmful impurities such as hydrogen can be removed by special vacuum degassing treatments. Additions of Al for deoxidation purposes during melting are beneficial also because of formation of AlN during solidification and subsequent cooling. As discussed above, fine AlN precipitates prevent austenite grain growth during reheating. Al-killed steels are thus finer grained than Si-killed steels.

(*b*) *Casting.* Casting processes can also affect steel properties primarily because of segregation during solidification. Faster casting speeds, as in continuous casting, can lead to less segregation during casting. This will generally result in more uniform properties in the final product than ingot casting. Extremely rapid solidification can be achieved by gas-jet cooling of liquid droplets. The solidified droplets must then be compacted and sintered to form a larger part, but the final product is uniform and free of segregation.

(*c*) *Hot rolling.* Hot rolling of steel ingots is used to produce useful shapes such as structural shapes, plates, bars and rods. Generally, this requires reheating the ingot to high temperatures (1200 °C) followed by a series of roughing or breakdown passes and then a series of finishing operations. The hot-rolling process also causes some rather important metallurgical changes. The austenite phase undergoes repeated recrystallization during hot rolling and the grain size becomes increasingly smaller as the finishing temperature is lowered. Also, soft MnS inclusions and segregated zones are aligned in the rolling direction into stringered inclusions or bands, respectively. These latter two effects lead to a directionality of properties, such as ductility or toughness, with minimum values occurring in the through-thickness direction, intermediate values in the transverse direction, and maximum values in the longitudinal direction.

(d) Cold rolling and annealing. Cold-rolling operations are used to produce thinner sheet products from hot-rolled slabs. In contrast to hot rolling, cold rolling is performed at or near room temperature where the steel has a ferritic structure and where no recrystallization occurs during the working operation. As a result the dislocation density of the ferrite and the strength increase during cold-working operations. Also, the ferrite grains are rotated and deformed to develop a cold-worked texture. In general, the surface finish of cold-rolled products is superior to that of hot-rolled products because of the absence of oxidation. After cold rolling, steel products are usually annealed to soften them and increase their ductility. In the annealing operation the cold-worked ferrite recrystallizes to produce an equiaxed, dislocation-free ferrite grain structure. A crystallographic texture is still present in the annealed product, although it will be different from that of the cold-worked texture. Control of the annealed texture is important in design of steels for deep-drawing operations.

(e) Thermomechanical treatment. This term is used to describe steel processing operations that combine a mechanical deformation and heating operation. Hot rolling can be considered as one form of thermomechanical treatment, but in general the latter term is applied to temperature ranges much lower than those used for hot rolling. Controlled rolling is a special type of thermomechanical treatment in which the hot-rolling operations occur at lower temperatures in the austenite range where recrystallization does not occur, and the time–temperature deformation steps are carefully controlled to produce flattened austenite grains. Controlled rolling is used to produce ultrafine-grained steels for applications requiring high toughness and high strength. Other thermomechanical treatments are ausforming and isoforming but these have found only limited applications.

(f) Heat treatment. Heat treatment of steels is used to improve their strength or toughness. It involves heating steels to elevated temperatures in the austenite range followed by air cooling or quenching and tempering. If air cooling is employed, the operation is called normalizing, and it results in a marked refinement in the grain size and some increase in strength. Normalizing operations are used to lower the ductile-to-brittle transition temperature of many hot-rolled steels. If quenching operations are used, the austenite transforms to a hard martensite phase that results in very high strength products. Normally, quenched steels are subsequently tempered to increase their ductility and toughness with a slight decrease in hardness usually being unavoidable.

To produce higher surface hardness for increased wear resistance, the surface of steel parts may be locally heat treated by induction, flame or laser hardening methods. In this case there is no change in the surface chemistry of the steel. Alternatively, gas nitriding or gas carburizing methods may be used to increase the C or N content of the surface. Carburizing is performed at high temperatures in the austenite range and is generally followed by quenching and tempering. Nitriding is performed at lower temperatures in the ferrite range and is not followed by a heat-treatment operation.

5. Steel Properties

Steel properties that are subject to control are physical, chemical, mechanical and manufacturing-related. Physical properties include magnetic behavior, electrical resistivity and thermal conductivity. Chemical properties include resistance to corrosion and to high-temperature oxidation. Mechanical properties include elastic moduli, strength, ductility and toughness. Manufacturing-related properties include weldability, formability and machinability.

5.1 Physical Properties

(a) Magnetic behavior. Magnetic behavior of steels is sensitive to crystallographic texture because the intensity of magnetization of Fe is a minimum in the $\langle 100 \rangle$ direction. Special hot- and cold-rolling and annealing procedures are used to develop (110) [100] textures to produce sheet steels with high permeabilities. High permeabilities are also favored by low C contents ($<0.02\%$) since Fe_3C particles tend to pin domain walls and prevent their movement during the magnetization cycle. Finally, core loss in magnetic steels is decreased by addition of Si (1–3%) to increase the electrical resistivity and decrease the I^2R loss caused by eddy currents.

(b) Electrical resistivity and thermal conductivity. In certain applications, the electrical resistivity and thermal conductivity of steels are important. For instance, low thermal conductivity is desired in steels used for containers for liquid gases. High electrical resistivity is desired in some transformer steels to decrease I^2R losses.

Several factors determine the electrical resistivity. These include thermal, impurity and magnetic scattering of electrons. The first two factors are considered to be additive and can be analyzed by use of Mathiessen's rule. The impurity scattering is specific to each element in solid solutions and values for many of the common alloying elements are available. The impurity scattering is assumed to be temperature independent, with the thermal scattering term accounting for all the temperature dependence of the electrical resistivity. The magnetic scattering term is very large and, as the temperature is increased to the vicinity of the Curie temperature (770 °C), a rapid increase in the electrical resistivity occurs because of the decrease in magnetic ordering. In fcc austenitic steels which are paramagnetic, the electrical resistivity is higher than in bcc ferromagnetic steels because of the absence of magnetic ordering. Also, the temperature dependence of the electrical resistivity of fcc steels is lower because of the absence of a paramagnetic–ferromagnetic transition.

Heat is conducted by both atomic vibrations (phonons) and by conduction electrons. For good electrical conductors like steels, heat transfer by conduction electrons is more important than heat transfer by phonons. This leads to a parallelism between thermal and electrical conduction as expressed by the Lorentz function. Thus, many of the conclusions reached for electrical conductivity (reciprocal of resistivity) also apply to thermal conductivity. Since electrical resistivity can be measured more easily than thermal conductivity, the Lorentz function can be used to calculate the thermal conductivity from electrical resistivity measurements. The Lorentz constant in steels is not, however, equal to the Lorentz number as predicted by the Wiedemann–Franz law. Rather, the Lorentz constant is equal to $1.16 \times$ Lorentz number.

5.2 Chemical Properties

(a) Corrosion resistance. The corrosion resistance of hot-rolled carbon structural steels exposed to atmospheric conditions may be increased two to four times by additions of small amounts of copper (0.2–0.5%). The Cu may be added in conjunction with other elements such as P, Mn or Mo, which not only further increase the corrosion resistance but also increase the strength. For resistance to more severe corrosive environments such as acidic solutions, larger amounts of alloying elements must be added. Cr is the most effective alloying addition for giving adequate corrosion resistance in acidic solutions. Steels containing substantial amounts of chromium (at least 5.5% and generally more than 10%) are designated as stainless steels, and are resistant to corrosion in many severely corrosive environments (see *Steels: Classification*). Of course, the corrosion resistance of many steel products is improved by simply putting a protective layer of another metal (Zn, Pb or Sn) on the surface to protect the metal against attack by the corrosive media. This can be accomplished by electroplating (Zn or Sn) or by hot dipping (Zn or Pb).

(b) Oxidation resistance. Oxidation resistance is an important consideration in steels designed for high-temperature service as in steam and gas turbines, exhaust valves and catalytic converter housings. Cr is again the element that is used to improve oxidation resistance by formation of a thin oxide layer that protects the base metal from being exposed to the oxidizing atmosphere. As a result, most steels for high-temperature service contain 5–20 % Cr.

5.3 Mechanical Properties

(a) Elastic moduli. The elastic moduli of polycrystalline steels are dependent on the degree of crystallographic texture because the elastic constants of single crystals of bcc Fe are anisotropic ($C_{11}=23.7$, $C_{12}=14.1$, $C_{44}=11.6 \times 10^{10}$ Pa). The maximum elastic modulus, $E\langle 111 \rangle = 28.4 \times 10^{10}$ Pa, occurs in the $\langle 111 \rangle$ direction whereas the minimum, $E\langle 100 \rangle = 11.6 \times 10^{10}$ Pa, occurs in the $\langle 100 \rangle$ direction. By suitable hot- and cold-rolling and annealing practices, it is possible to produce sheet steels with a preponderance of (111) planes in the plane of the sheet, and thus a high E value in a direction perpendicular to the plane sheet. Such steels are more resistant to thinning and fracture during deep drawing than steels with a random texture.

In general, alloying additions have much less effect on the elastic moduli than texture because the atomic binding energies of the iron lattice cannot be altered significantly by alloying additions. The maximum effect for elements in solid solution is about 1% for each 1 at. % alloying element addition. For example, the isotropic E value for Fe is lowered from 20.82 to 19.25×10^{10} Pa by addition of 9 at. % Ni.

The elastic moduli of bcc Fe decrease slowly with temperature up to temperatures near the Curie temperature (770 °C) where a rapid decrease in E occurs as the ferromagnetic bcc Fe undergoes magnetic disordering. This results from the effect of magnetic ordering on the atomic binding energies. At 910 °C, where bcc Fe transforms into fcc Fe, a small increase in the value of E is observed.

(b) Strength. The strength of steels is dependent on many factors including grain size, solid-solution hardening effects of individual elements, dislocation density, and the size, spacing and distribution of precipitate particles. In general, several of these factors are operative in any steel alloy so that rigid theoretical analysis or prediction of the strength from first principles is difficult. However, by assuming simple additivity of the various strengthening factors, useful empirical equations based on linear regression analysis of hundreds of steels have been developed. These can be used to analyze or predict effects of various changes in heat treatment or alloying additions.

(c) Ductility. The ductility of steels generally refers to the reduction-of-area or total elongation values of a tensile test specimen. It is dependent on both strength level and inclusion content and decreases with increasing tensile strength and with increasing inclusion content. Ductility values are sensitive to testing direction in rolled or extruded products because of the elongation of sulfide inclusions in the primary deformation direction.

(d) Toughness. Toughness refers to the energy absorbed in breaking notched impact specimens. The impact energy is sensitive to tensile strength, inclusion content, and testing direction in a manner similar to tensile ductility. Toughness is drastically reduced by lowering the temperature of testing because of the

ductile-to-brittle transition exhibited by all bcc metals. The high-temperature ductile fracture surface has a fibrous appearance and the energy value absorbed in the 100% fibrous region is called the shelf energy. The low-temperature brittle fracture surface has a shiny appearance with cleavage facets. The temperature at which the steel exhibits a 50% brittle appearance is called the fracture appearance transition temperature (FATT). The FATT is sensitive to many variables including grain size and various embrittlement phenomena such as temper embrittlement.

5.4 *Manufacturing-Related Properties*

(*a*) *Weldability.* The weldability depends on both carbon and alloy content. High carbon and high alloy content promote cracking in the weld bead or weld heat-affected zone. For many steels, a limit of 0.20% C is considered adequate to assure good welding characteristics because the hardenability of low-carbon steels is low enough so that martensite is not formed in the weld heat-affected zone. However, alloying elements also play a role because they too affect hardenability. As a result, the weldability of steels is generally considered in terms of a carbon equivalent equation which includes terms both for C and for other alloying elements. The weldability may also be affected by hydrogen generated from water present in the welding electrodes or in the air because of hydrogen embrittlement mechanisms.

(*b*) *Formability.* Formability is an important consideration in deep drawing operations such as stamping of automobile parts. Steels for these applications are designed on the basis of r_m values which are a measure of the relative plastic strains in the plane of the sheet and in the thickness direction in a tension-test specimen. Steels with high r_m values (near 2.0) resist thinning during drawing operations and exhibit good drawability, whereas steels with low r_m values (near 1.0) exhibit poorer drawability. The r_m value is sensitive to crystallographic texture with the $\{111\}\langle 110\rangle$ texture resulting in high r_m values. Careful control of cold-rolling and annealing cycles can be used to produce special deep-drawing steels with this texture.

In other forming operations, control of stretchability is more important. Steels for stretching operations require high work-hardening rates so that necking is delayed. The work-hardening rate is measured in terms of n, which is the exponent in the Hollomon work-hardening power law. High n values are promoted by increasing the grain size of the steel. However, grain sizes coarser than ASTM No. 6 (45 μm) are not desirable as they produce an "orange-peel" effect.

(*c*) *Machinability.* The machinability of steels can be increased by increasing the sulfur content of the steel beyond the normal limits. Sch "free-machining" steels may have S contents as high as 0.15%. The high S content makes chip breaking easier because of the low ductility of the steel caused by a high volume fraction of MnS inclusions. The sulfides probably also act as a lubricant between the tool and the workpiece. Other alloy additions for machinability include Pb which occurs as inert Pb particles and also serves as a lubricant. Bi and Te additions are also used to increase machinability. In general, Al_2O_3 particles present in Al-killed steels are detrimental to machinability because of excessive tool wear.

6. *Cast Irons*

Fe–C–Si alloys, commonly called cast irons, and the liquid forming processes of metal casting, are an ideal combination of material and process that has been used for centuries to provide mankind with metal in desired shapes. Cast iron is of relatively low cost, can be readily melted and cast into intricate shapes, and has very useful properties. It is an excellent engineering material with good strength and hardness, and is readily machined. It has sufficient heat and corrosion resistance for many common applications such as furnaces and the drainage grating in streets. Critical components of the internal combustion engine such as the motor block, head, exhaust manifold, connecting rods and flywheel are made of iron castings.

All cast irons contain more than 2% C in contrast to steels which contain less than 2% C and usually less than 1% C. Since 2% C is the maximum carbon content at which Fe can solidify as a single phase alloy with all the C in solution, all cast irons solidify as heterogeneous alloys and have more than one constituent in their microstructure. In addition to C, cast irons usually contain from 1 to 3% Si. The high C and Si make cast iron an excellent casting alloy. It is readily melted in a simple furnace using charcoal or coke and ambient temperature air. The molten iron is very fluid and allows the metal to be handled and cast effectively. Also, the precipitation of graphite in the solidifying iron reduces the normal liquid to solid contraction so that very little volume change occurs on solidification. This allows the casting of complex shapes.

Cast iron designates a family of five different types: white, malleable (cast white), gray, ductile and compacted graphite.

(*a*) *White cast iron.* This consists of a mixture of Fe_3C and austenite upon solidification. The austenite transforms to pearlite during cooling. White iron is hard and brittle because of the large amount of Fe_3C it contains. Its name derives from the white crystalline surface which appears when it is fractured. White iron can be produced in selected areas at the surface of a casting by chill casting of these regions. White iron provides very high compressive strength, good wear resistance, and retains its hardness for limited periods up to red heat. Typical compositions are 2.0–3.6% C, 0.5–1.9% Si, 0.25–0.8% Mn, 0.06–0.2% S and 0.06–0.2% P.

(*b*) *Malleable cast iron.* Malleable cast iron is iron in which carbon occurs in the microstructure as irregularly shaped nodes of graphite. This form of graphite is called temper carbon because it is formed in the solid state during heat treatment. The iron is cast as white iron of a suitable chemical composition. After the castings are removed from the mold, they are given an extended heat treatment starting at a temperature above 885 °C. This causes the iron carbide to decompose and the free carbon precipitates in the solid iron as graphite. Typical chemical compositions are 2.2–2.9% C, 0.9–1.9% Si, 0.15–1.2% Mn, 0.02–0.2% S and 0.02–0.2% P.

(*c*) *Gray iron.* Gray iron is iron in which the carbon separates during solidification to form graphite flakes that are interconnected within each eutectic cell. This graphite grows into the liquid and forms the characteristic flake shape. When this iron is subsequently fractured, most of the fracturing occurs along the planes established by the graphite, thereby accounting for the characteristic gray color of the fracture surface. Because the large majority of the iron castings produced are of gray iron, the generic term, cast iron, is often improperly used specifically to mean gray iron. The properties of gray iron are influenced by the size, shape and distribution of the graphite flakes, and by the relative amounts of ferrite and pearlite in the matrix. Low cooling rates and high C and Si contents tend to produce more and larger graphite flakes in a matrix with more ferrite and less pearlite. The flake graphite provides gray iron with some unique properties such as excellent machinability at hardness levels that produce superior wear-resisting characteristics, the ability to resist galling, and good vibration damping from a nonlinear stress–strain relationship at relatively low stresses. The damping capacity increases as the graphite flakes become coarser. Typical chemical compositions are 2.5–4.0% C, 1.0–3.0% Si, 0.2–1.0% Mn, 0.02–0.25% S and 0.02–1.0% P.

(*d*) *Ductile iron.* This material was developed in the 1940s. It is sometimes referred to as nodular iron and is called spherulitic graphite iron in Europe. It had a phenomenal ninefold increase in use as an engineering material during the 1960s. An unusual combination of properties is obtained in ductile iron because the graphite occurs as spheroids or spherulites rather than as individual flakes as in gray iron. This mode of solidification is obtained by adding a very small, but definite amount of Mg to molten iron of a proper composition. The added Mg reacts with the S and O in the melt and changes the way the graphite is formed. Typical chemical compositions are 3.0–4.0% C, 1.8–2.8% Si, 0.1–1.0% Mn, 0.01–0.03% S and 0.01–0.1% P.

(*e*) *Compacted graphite iron.* Compacted graphite iron is characterized by a graphite structure which is intermediate between that in gray and in ductile irons. The graphite exists as blunt flakes which are interconnected. Compacted graphite iron has been referred to as quasiflake, aggregated flake, seminodular iron, and by other names. Compacted graphite is formed by additions of Ce or other rare-earth elements with the presence of about 0.06 to 0.13% Ti, and a limiting S content of 0.03%. The tensile and yield strengths of compacted graphite iron are comparable to the lower values of ductile iron, while the ductility is limited to a maximum of about 6% elongation in 50 mm. The physical properties of compacted graphite such as thermal conductivity and damping capacity are closer to those of gray iron because the graphite is interconnected. The chemical composition ranges of these irons are 2.5–4.0% C, 1.0–3.0% Si, 0.2–1.0% Mn, 0.01–0.02% S and 0.01–0.1% P.

See also: Metals Processing and Fabrication: An Overview

Bibliography

American Society for Metals 1978 *Metals Handbook*, 9th edn., Vols. 1, 3 and 4. ASM, Metals Park, Ohio
Angus H T 1976 *Cast Iron: Physical and Engineering Properties.* Butterworth, London
Honeycombe R W K 1981 *Steels: Microstructure and Properties.* Edward Arnold, London
Leslie W C 1981 *The Physical Metallurgy of Steels.* McGraw-Hill, New York
Pickering F B 1978 *Physical Metallurgy and the Design of Steels.* Applied Science, Barking, UK
US Steel Corporation 1985 *The Marking, Shaping, and Treating of Steel*, 10th edn. Association of Iron and Steel Engineers, Warrenville, Pennsylvania
Walton C F 1981 *Iron Castings Handbook.* Iron Castings Society, Des Plaines, Illinois

G. R. Speich
[Illinois Institute of Technology, Chicago, Illinois, USA]

Fiber-Reinforced Cements

Cement mortar and concrete differ from most other matrices used in fiber composites in that their essential characteristics are those of low tensile strength (generally less than 7 MPa) and very low strain to failure in tension (generally less than 0.05%) which result in materials which are rather brittle when subjected to impact loads.

The binder in the matrix is cement paste produced by mixing cement powder and water which harden after an exothermic reaction to form complex chemical compounds that are dimensionally unstable due to water movement at the molecular level within the crystal structure. Inert mineral fillers known as aggregates are added at up to 70% by volume to increase stability and reduce the cost, and therefore the space available for fiber inclusion may be severely limited.

Fiber volumes in excess of 10% of the composite

volume are difficult to include in mortar and, in the case of concrete, 2% by volume of fiber is considered to be a high proportion which is generally insufficient to permit significant reinforcement of the matrix before cracking occurs, particularly where individual fibers may be spaced up to 40 mm apart by the aggregate particles.

The type of fiber composite described in this article is therefore one in which the fiber has little effect on the properties of the relatively stiff matrix until cracking has occurred at the microlevel. The fibers then become effective, resulting in a composite with reduced stiffness and increased strain to failure providing increased toughness compared with the unreinforced matrix.

Notable exceptions to this type of behavior are asbestos-cement and the cellulose fiber reinforced thin sheets introduced to the market in the mid-1980s. In these materials the fibers are used to increase the matrix cracking strain to more than 1000×10^{-6}.

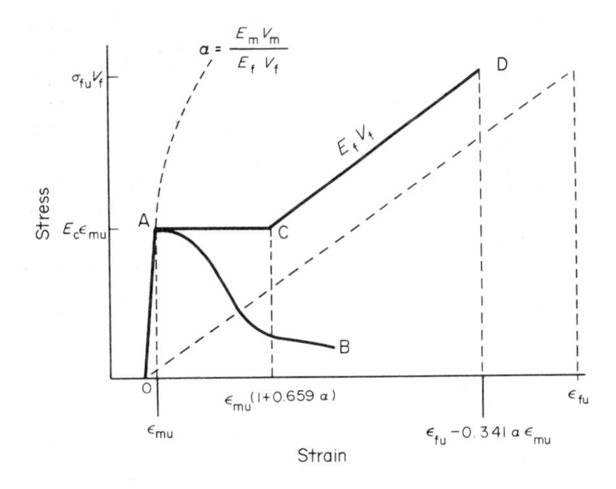

Figure 1
Idealized tensile stress–strain curves for a brittle cement matrix composite

1. Principles of Reinforcement in Tension

Owing to the relatively stiff matrix and its low strain to failure, fibers are not normally included in cement matrices to increase the matrix cracking stress. Therefore the merits of fiber inclusion lie in the load carrying ability of the fibers after microcracking of the matrix has been initiated.

1.1 Critical Fiber Volume

The critical fiber volume, $V_{f(crit)}$, is defined as the volume of fibers which, after matrix cracking, will carry the load which the composite sustained before cracking. For the simplest case of continuous aligned fibers with frictional bond, the critical fiber volume can be expressed in terms of the elastic modulus of the composite, E_c, the ultimate strain of the matrix ε_{mu} and the fiber strength σ_{fu}:

$$V_{f(crit)} = \frac{E_c \varepsilon_{mu}}{\sigma_{fu}} \tag{1}$$

Figure 1 shows stress–strain curves for composites with less than the critical volume (OAB) and greater than the critical volume of fiber (OACD).

Efficiency factors for fiber orientation and fiber length increase $V_{f(crit)}$ considerably above that calculated from Eqn. (1). The result is that most concretes are unable to contain the critical volume of short chopped fibers and still maintain adequate workability for full compaction to be achieved. However, many techniques are available which enable the inclusion of more than the critical volume of fiber in fine grained cement mortars.

The dashed curve in Fig. 1 results when high volumes of very small diameter, stiff, well bonded fibers are included. The existing flaws are reinforced and may be restrained from propagating at the stress at which they would propagate in an unreinforced matrix, thus leading to an increased cracking strain in the matrix. Typical examples are asbestos-cement and cellulose fibers in autoclaved calcium silicate.

1.2 Stress–Strain Curve, Multiple Cracking and Ultimate Strength

Figure 1 shows the idealized stress–strain curve (OACD) for a fiber-reinforced brittle matrix composite with more than the critical volume of fiber. The assumptions used to calculate the curve are that the fibers are long, aligned with the stress, and that the bond τ is the purely frictional. The matrix is assumed to crack at A at its normal failure strain ε_{mu} and the material breaks down into cracks in region A–C with a final average spacing S given by

$$S = 1.364 \frac{V_m}{V_f} \frac{\sigma_{mu}}{\tau} \frac{A_f}{P_f} \tag{2}$$

where V_m, V_f are the volume fraction of the matrix and fiber respectively; A_f, P_f are the cross-sectional area and perimeter of the fiber; and σ_{mu} is the cracking stress of the matrix. It will be noted that crack spacing is not dependent on the elastic modulus of the fiber. Region C–D represents extension of the fibers as they slip through the matrix without further matrix cracking.

The ultimate strength of the composite depends only on the fiber strength σ_{fu} and volume V_f whereas the ultimate composite strain is less than the fiber failure strain by a factor depending on the volume and elastic modulus of the matrix and of the fiber, and the matrix failure strain.

The multiple cracking represented in Fig. 1 is a desirable situation because it changes a basically brittle material with a single fracture surface and low energy requirement to fracture into a pseudoductile material which can absorb transient minor overloads and shocks with little visible damage.

1.3 Efficiency Factors for Fiber Length and Orientation

The efficiency of fibers in reinforcing the matrix in practice may be very much less than that assumed in the calculation of the idealized stress–strain curve. Efficiency depends on many factors including the orientation of the fibers relative to the direction of stress, whether the matrix is cracked or uncracked and the fiber length relative to the critical length (defined as twice the length of fiber embedment which would cause fiber failure in a pullout test).

For short fibers in a three-dimensional orientation, the combined efficiency could be less than one-fifth of the efficiency in the aligned fiber case (Hannant 1978).

2. *Principles of Reinforcement in Flexure*

The critical fiber volume for flexural strengthening can be less than half that required for tensile strengthening so that the load which a beam of the composite can carry in bending can be usefully increased, even though the tensile strength of the materials has not been altered by the addition of fibers. In fact, the flexural strength is commonly between two and three times that predicted from the tensile strength using an elastic analysis. The main reason for this is that the postcracking stress–strain curve in tension (AB or ACD in Fig. 1) is very different from that in compression (similar slope to OA in Fig. 1). As a result, the neutral axis moves towards the compression surface of the beam and conventional beam theory becomes inadequate because it does not allow for a plastic stress block in tension combined with an elastic stress block in compression (Hannant 1978).

The relationship between idealized uniaxial tensile stress–strain curves and flexural stress–strain curves is shown in Fig. 2.

The importance of achieving adequate postcracking tensile strain capacity in maintaining a high flexural strength is demonstrated in Fig. 2. For instance, curve OAC is a tensile stress–strain curve and OAC′ is the associated bending curve. If the tensile strain reduces from C to A at a constant stress, then the bending strength will reduce from C′ to A with an increasing rate of reduction as A is approached and the material becomes essentially elastic. Thus, tensile strength on its own cannot be used to predict bending strength and composites which suffer a reduction in strain to failure as a result of natural weathering will also have a reduced bending strength.

3. *Asbestos-Cement*

Since 1900, the most important example of a fiber cement composite has been asbestos-cement. The proportion by weight of asbestos fiber is normally between 9 and 12% for flat or corrugated sheet, 11–14% for pressure pipes and 20–30% for fire-resistant boards and the binder is normally a Portland cement. Fillers such as finely ground silica at about 40% by weight may also be included in autoclaved processes where the temperature may reach 180 °C.

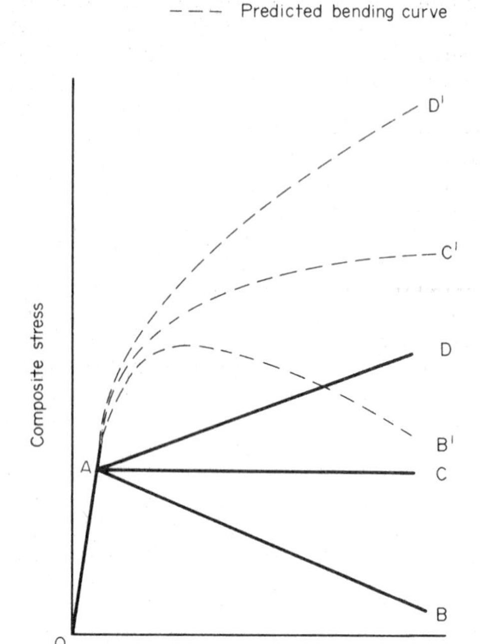

Figure 2
Apparent bending (flexure) curves predicted for assumed direct tensile curves (after Laws and Ali 1977)

The most widely used method of manufacture for asbestos-cement was developed from papermaking principles about 1900 and is known as the Hatschek process. A slurry, or suspension of asbestos fiber and cement in water at about 6% by weight of solids, is continuously agitated and allowed to filter out on a fine screen cylinder. The filtration rate is critical and coarser cement than normal (typically 280 $m^2 kg^{-1}$) is used to minimize filtration losses. Also, although chrysotile fibers form the bulk of the fiber, most formulations have included a smaller proportion of the amphibole fiber wherever possible. This is because of the dewatering characteristics of crocidolite and amosite. Mixtures of cement and chrysotile drain very slowly and production rates are consequently slow. By using a blend of chrysotile and amphibole fibers (usually between 20% and 40%), considerable improvements in the drainage rate have been achieved, plus the additional bonus of enhanced reinforcement.

Other types of fiber such as cellulose derived from wood pulp or newsprint have also been added to the slurry to produce different effects in the wet or hardened sheet.

There are several other manufacturing processes such as the Mazza process for pipes and the Magnani process for corrugated sheets. Also the material may be extruded and injection molded.

The tensile strength varies between about 10 MPa and 25 MPa depending on fiber content, density and direction of test whereas the flexural strengths may be between 30 MPa and 60 MPa.

Durability is exceptionally good, a lifetime in excess of 50 years under external weathering being possible. A disadvantage is the brittle nature of the material under impact situations due to its relatively low strain to failure.

Major applications have included corrugated sheeting and cladding for low-cost agricultural and industrial buildings, and flat sheeting for internal and external applications has also proved very successful.

Pressure pipes are another worldwide market and the ease with which the material can be formed into complex shapes has enabled a wide range of special purpose products to be supplied.

The future for asbestos-cement products depends largely on the attitudes taken by Government Health and Safety organizations to the health problems associated with asbestos fibers.

4. *Glass-Reinforced Cement*

The early development of glass fibers in low alkali cements took place in China and Russia in the 1960s but because of the susceptibility to attack of E-glass fibers by the higher alkali Portland cements used in the West, considerable investment in the UK took place in the 1970s to produce alkali-resistant zirconia glass on a commercial scale.

The fibers are commonly included in the matrix in strands consisting of 204 filaments of 10 μm diameter fibers.

The three basic techniques are premixing, which gives a random three-dimensional array unless the mix is subsequently extruded or pressed, the spray method which gives a random two-dimensional arrangement, and filament winding which gives a one-dimensional array or a multi-angled oriented array. The volume of fibers is generally greater than the critical volume for tensile strengthening.

The spray method is the most popular and consists of leading a continuous roving up to a compressed-air operated gun which chops the roving into lengths of between 10 mm and 50 mm and blows the cut lengths at high speed simultaneously with a cement paste spray onto a forming surface which may contain a filter sheet and vacuum water extraction in certain applications.

The fiber-cement sheet can be built up to the required thickness and demolded immediately, an additional advantage being that it has sufficient wet strength to be bent round radiused corners to give a variety of product shapes such as corrugated or folded sheets, pipes, ducts or tubes. Coloured fine aggregate can be trowelled into the surface to produce a decorative finish.

The process can be readily automated for standard sheets. The optimum fiber volume for flexural strength using the spray technique is about 6% because of increase in porosity of the composite at higher fiber volumes.

One of the problems with this type of glass-reinforced cement has been the reduction in strain to failure with time under water storage or natural weathering conditions. This effect results in the material becoming essentially brittle under these conditions but air storage in a dry atmosphere causes little change in the initial properties even after periods of ten years.

Continuing developments in glass–cement systems show promise of reducing the rate of embrittlement under natural weathering conditions.

Typical properties are tensile strengths between 7 MPa and 15 MPa and bending strengths between 13 MPa and 35 MPa.

Applications include cladding panels, permanent formwork for concrete construction, pipes, flat sheeting, sewer linings and surface coatings for dry blockwork.

5. *Steel-Fiber Concrete*

Steel fibers have been used mainly in bulk concrete applications as opposed to cement mortars.

It has been found from experience that, for traditional mixing and compaction techniques to be applicable, the maximum aggregate size should not exceed 10 mm and there should be a high content of fine material. In these circumstances 1–1.5% by volume of fibers typically 50 mm long by 0.5 mm diameter can be mixed with little modification of existing plant.

The fibers are generally deformed to give better "keying" to the matrix but, as the fibers are arranged randomly in two and three dimensions, the critical fiber volume for tensile strengthening is rarely exceeded. However, the flexural strength can be increased up to about 15 MPa and toughness is greatly increased.

Higher volumes of longer fibers can be included by spray techniques known as "gunite" or "shotcrete" and these materials are often used as tunnel linings or for rock slope stabilization.

The increased toughness is utilized in highway and airfield pavement overlays which have reduced thickness and greater impact resistance than traditional construction. The use of stainless steel has proved especially advantageous in refractory applications.

Fiber production processes include spinning from the melt which has reduced the cost of the fiber thus increasing its economic viability.

6. *Organic Fibers in Cement*

6.1 Man-Made Fibers

Polymer fibers used in cement matrices cover the complete range of fiber elastic moduli. The moduli range from 2 GPa to 12 GPa for polypropylene to 200 GPa to 300 GPa for carbon fibers but the basic principles of reinforcement remain the same for all the fiber types.

Polypropylene, because of its high strength combined with excellent alkali resistance and low price, has had the greatest application in cement and concrete. It is generally used as fibrillated film between 50 μm and 100 μm thick made from isotactic polypropylene with draw ratios between 5 and 20 to produce a high degree of molecular alignment. The stretched fibrillated film with strengths between 300 MPa and 500 MPa may be used in concrete as chopped twine in volumes generally less than 1% or as layers of flat opened networks in fine grained mortar at volumes between 5% and 20%.

The chopped twine is mixed with the concrete in standard mixers but as the volume of fiber is well below the critical volume for tensile strengthening the major benefits are in handling the fresh material or in impact situations such as piling or pontoons where the fibers serve to hold the cracked material together when damaged.

The manufacture of thin cement sheets containing many layers of film requires special production machinery. The sheets, which are intended as alternatives to asbestos-cement, are characterized by high toughness and flexural strength (up to 40 MPa) resulting from the closely spaced multiple cracking.

The higher elastic modulus fibers such as carbon fibers and Kevlar show considerable promise as cement reinforcements but are too expensive for bulk applications. Cheaper, high-modulus polymers, such as high-modulus polyethylene (20 GPa to 100 GPa) will fibrillate in a similar manner to polypropylene and show potential should commercial production become available.

6.2 Natural Fibers

Cellulose fibers produced from wood pulp have been used for many years as additives in asbestos-cement products but their precise physical properties seem to vary widely depending on the authority quoted and this variability may also depend on the pulping techniques used. Fiber lengths between 1 mm and 4 mm have been quoted and the helical structure of the fibers gives a variable diameter between 10 μm and 50 μm. Tensile strengths for the fiber between 50 MPa and 600 MPa have been quoted with moduli between 10 GPa and 30 GPa.

Flat and corrugated sheets containing cellulose fibers in autoclaved calcium silicate have been marketed as alternatives to asbestos-cement. The strength of these products is lower in the wet than the dry state due to the sensitivity of cellulose fibers to water absorption.

Other natural fibers such as sisal, agave, piassava, coconut and bagasse fibers have been used in cement in an attempt to produce cheap cladding sheets for houses in developing countries but although some success has been achieved, the process is labor intensive and there are many problems still to be overcome.

See also: Composite Materials: An Overview; New Cement-Based Materials

Bibliography

Building Research Establishment 1979 *Properties of GRC: 10 Year Results*, Information Paper IP 36/79. Building Research Station, Watford

Cook D J 1980 Concrete and cement composites reinforced with natural fibres. *Fibrous Concrete—Concrete International 1980*. Construction Press, London, pp. 99–114

Hannant D J 1978 *Fibre Cements and Fibre Concretes*. Wiley, Chichester

Laws V, Ali M A 1977 The tensile stress–strain curve of brittle matrices reinforced with glass-fibre. *Conf. Proc. Fibre Reinforced Materials*. Institute of Civil Engineers, London, pp. 101–9

D. J. Hannant
[University of Surrey,
Guildford, UK]

Float Glass Process

As predicted by its inventor, Sir Alastair Pilkington, the float glass process has become established as the preferred method for the continuous manufacture of transparent flat glass. Almost all of the units producing high-quality plate glass by grinding and polishing and a high proportion of the plants making lower-quality sheet glass by upward drawing have disappeared. By 1987, more than 100 float installations existed throughout the world, with a designed annual productive capacity approaching 20 Mt, equivalent to 2×10^9 m^2 of glass of 4 mm thickness.

A deceptively simple description of this elegant process, masking all the effort needed to perfect it, has been given by Sir Alastair.

> In the float process, a continuous ribbon of glass moves out of the melting furnace and floats along the surface of an enclosed bath of molten tin. The ribbon is held in a chemically controlled atmosphere at a high enough temperature and for a long enough time for the irregularities to melt out and for the surfaces to become flat and parallel. Because the surface of the molten tin is dead flat, the glass also becomes flat.

The ribbon is then cooled down while still advancing across the molten tin until the surfaces are hard enough for it to be taken out of the bath without the rollers marking the bottom surface; so a ribbon is produced with uniform thickness and bright fire polished surfaces without any need for grinding and polishing.

These essential elements of the process provide the framework for the more detailed exposition that follows.

1. *Ribbon Formation*

Since the commercial objective is to continuously produce a flat ribbon of glass of uniform thickness, free of discrete faults or variations in optical thickness which can distort or obscure the view through the glass, a prime requirement is a supply of molten glass from a continuous melting unit which provides the defined output of appropriate composition (Hynd 1984) at the highest possible quality for solid inclusions, bubble and homogeneity. The molten glass supply must then be formed into a continuous ribbon, which can be achieved by using the float bath itself to form a ribbon, or by a rolling process prior to the float bath.

1.1 Rolling

It seemed appropriate at first that the developed expertise in forming a size ribbon of glass between water-cooled steel rolls should be used as a primary forming process, and that the float bath should be used to reheat the ribbon and to remove both the surface irregularities and thickness variations introduced by the rollers. In practice, as already described, (Pilkington 1969) irremovable surface defects are introduced by the rollers and, more fundamentally, as the glass ribbon is heated sufficiently to achieve flatness and parallelism, the glass thickness changes towards a defined value where gravitational and surface tension forces are in equilibrium, independently of thickness of the rolled ribbon.

For such a nonspreading liquid–liquid system, the central portion of the ribbon can be considered to be of uniform thickness T if the width is greater than 150 mm and the system is in an equilibrium state.

Equating gravitational and surface tensional forces, this "equilibrium" thickness is given by the relation

$$T^2 = (S_g + S_{gt} - S_t)2\rho_t / g\rho_g(\rho_t - \rho_g) \qquad (1)$$

where S_g, S_{gt} and S_t are the glass–atmosphere, glass–tin and tin–atmosphere surface tensions, respectively, g is the gravitational acceleration and ρ_t and ρ_g are the densities of molten tin and molten glass, respectively.

In a sealed float system using a nitrogen–hydrogen atmosphere, S_{gt} and S_t are about equal and the value of T is approximately 7 mm, fortunately close to the $\frac{1}{4}$ inch thickness which formed a significant proportion of sales when float development and early production were carried out. The timescale for the production of a high-quality fire-finished surface on a rolled glass of soda lime–silica composition was found to be approximately 1 min at 1050 °C, corresponding to a glass viscosity of 1 kPa s, and, inherently, the ribbon moves towards equilibrium thickness with the same timescale (Charnock 1970). An analytical evaluation of this time constant shows it to be exactly proportional to the glass viscosity (Hynd 1984) and to have a value of 40 s at 1 k Pa s.

The use of a rolling process to supply a ribbon to the float bath was thus important in establishing the physical parameters for a successful process and is used here to indicate some of those parameters.

1.2 Direct Pour Process

A consequence of the nonspreading of molten glass over molten tin was the development of a ribbon-forming process in which molten glass is poured directly onto molten tin to form the continuous ribbon within the float bath itself. This is the process developed by Pilkington and now used in float plants throughout the world.

In essence the molten glass flows from the melting furnace to the float bath through a canal, the flow being controlled by a movable refractory tweel as shown in vertical cross section in Fig. 1a and in plan view in Fig. 1b.

The glass then flows down an inclined refractory spout before falling freely onto the surface of the molten tin, flowing forward and spreading to form a

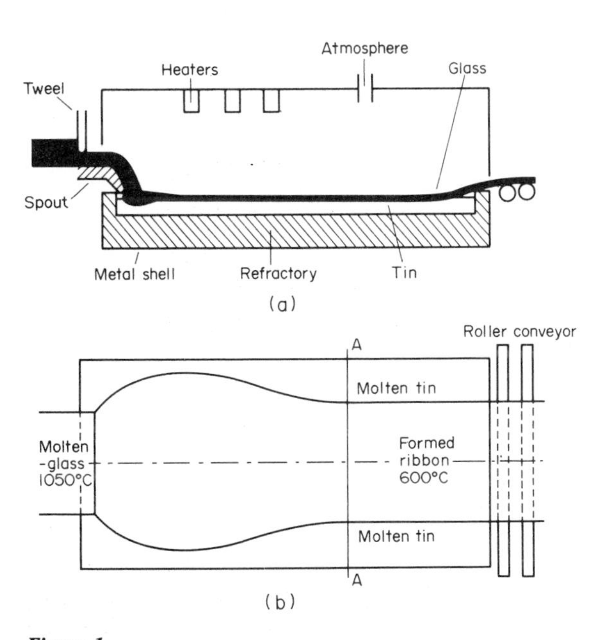

Figure 1
Direct pour process: (a) vertical longitudinal cross section, and (b) plan view

ribbon, the glass temperature being approximately 1050 °C.

The float bath itself is a steel container lined with refractory material to contain the molten tin and enclosed above glass level to provide a nitrogen–hydrogen atmosphere which prevents oxidation of the tin.

As described in detail by Pilkington (1969), it was necessary to devise an appropriate flow system to divert any refractory contamination of the bottom surface of the glass into the extreme edges of the ribbon.

The lateral spread of the glass over the surface of the molten tin and the formation of a continuous ribbon is illustrated in Fig. 1b. The indefinite lateral flow of the glass is prevented by surface tension as the thickness falls to 7 mm, and the tractive force applied to the ribbon to transport it along the bath may cause further loss of width, as shown, and some reduction in thickness. The glass is maintained at a temperature sufficiently high to remove surface imperfections or thickness variations within the required time, as described above. As the ribbon is advanced along the bath it is cooled, which eventually restricts further dimensional changes as the viscosity rises to, say, 10 MPa s, corresponding to a temperature of about 720 °C. As the temperature falls to 600 °C, its viscosity of 10 GPa s is sufficiently high to prevent surface damage as it is lifted onto a roller conveyor system, thence to be annealed and cooled before cutting to give the required product.

1.3 PPG-Industries-Modified Process

A modified ribbon forming system was announced by PPG Industries in 1975 and has been described by McCauley (1980). The main structural differences from the Pilkington process are that, first, the molten glass flows from the melting unit over a full-width lipstone which dips into the molten tin, replacing the Pilkington spout and free-fall glass flow system. Second, the molten glass in the PPG modification remains in contact with special guides which form the bath sides at the hot end of the bath, in contrast to the Pilkington process which allows the unhindered flow of the glass at the hot end of the bath.

2. *Removal of Surface Irregularity*

Since the prime objective of the float process is to produce flat glass which is substantially free of optical distortion, the mechanics of the removal of surface irregularities which are the main cause of such distortion is very relevant. When the molten glass supply is free of any inhomogeneity which can cause variations in refractive index, the variations in optical path which cause distortion are due only to changes in thickness.

Although thickness variations in a floating fluid will be reflected in both top and bottom surfaces, the essential features can be appreciated by considering an idealized irregularity in the top surface of the glass only, as illustrated in Fig. 2.

Since any form of thickness variation can be analyzed into its periodic components, we assume that the top surface of the glass is represented by a displacement $y = a \sin(2\pi x/\lambda)$ from the mean level $y = 0$, where a is the amplitude and λ the wavelength and the features extend indefinitely in a direction normal to this cross section. When $a \ll \lambda$, the curvature at any position is very close to d^2y/dx^2. Within the glass near to the surface there is a fluctuating change in hydrostatic pressure due to the surface distortion, first due to gravity and of magnitude $g\rho_g y$, and second due to the surface tension S_g, and of magnitude approximately $S_g d^2y/dx^2$ or $S_g 4\pi^2 y/\lambda^2$. This gives a combined gravitational and surface-tensional fluctuating pressure of magnitude $(g\rho_g + S_g 4\pi^2/\lambda^2)y$ attributable to the surface distortion, giving pressure gradients in the glass which will produce viscous flow and so reduce the amplitude of the distortion.

Substituting a value of 0.35 N m^{-1} for the glass–atmosphere surface tension S_g and 2400 kg m^{-3} for the density of molten glass in the bath, this relation shows that the surface-tensional and gravitational forces are equal when λ is 24 mm and the total force is then double the value of either, hence the time constant for the removal of a surface distortion of this wavelength is half that for irregularities of much longer wavelengths, such as an overall change in ribbon thickness as described above. When the associated wavelength decreases below 24 mm the surface-tensional restoring force increases rapidly, proportionally to $1/\lambda^2$, and the rate of removal of such features is correspondingly rapid.

Referring again to Fig. 2, the thicker parts of the ribbon approximate to convex lenses of optical power $(n-1)d^2y/dx^2$, where n is the refractive index of the glass, and the thinner parts of the ribbon to concave lenses of similar power. The view through the glass thus presents adjacent areas of positive and negative magnification, constituting obvious optical distortion. However, as described above, the surface-tensional restoring forces are of magnitude $S_g d^2y/dx^2$, directly proportional to the power of the optical distortion,

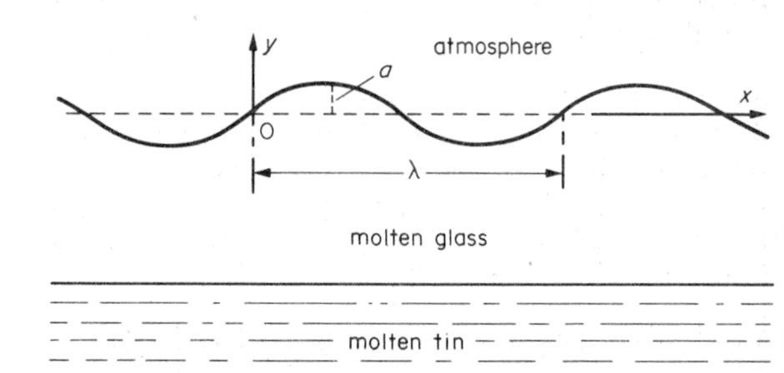

Figure 2
Idealized irregularity in top surface of ribbon

hence the inherent potential of this process to produce glass of high optical quality and of uniform thickness can be appreciated.

3. Materials Problems

From the viewpoint of the materials scientist or chemical engineer, the development of the float process presented unique problems in the high-temperature interactions between molten glass, molten metal, the atmosphere and the refractory/steel container, and the viability of the process for commercial production depended on the resolution of these problems (Pilkington 1969, 1976).

3.1 Bloom

Tin was chosen as the metal to support the glass for a variety of reasons: its density is higher than that of the glass (6500 kg m^{-3} at 1050 °C), the molten range from 232 °C to 2623 °C is more than adequate and its vapor pressure is thus relatively low at 1050 °C, its chemical interation with the glass is low and it is readily available. Nevertheless it was found that molten tin could chemically interact with molten glass in the presence of oxygen in the tin and produce a thin layer on the bottom surface of the glass which is rich in stannous tin. No adverse effect is produced unless the glass is reheated during further processing, under which conditions oxidation to the stannic form may take place, increasing the volume of the surface layer to give a microscopically wrinkled surface which shows a slight bluish haze known as "bloom." Operation of the process with good seals against the ingress of oxygen and the maintenance of a positive pressure in the nitrogen–hydrogen atmosphere system are both required to produce bloom-free glass.

3.2 Oxide Formation

A further problem related to the presence of oxygen is that the solubility of oxygen in molten tin at the lower bath temperature of 600 °C is little more than 5 ppm by weight and so solid oxide may come out of solution if the oxygen level exceeds this value, with consequent damage to the underside of the ribbon. Again, this problem is overcome by minimizing oxygen ingress into the system.

3.3 Sulfur Contamination of Tin

A different form of chemical problem encountered relates to the contamination of the molten tin by sulfur. At a level of 10 ppm by weight of sulfur in the tin, the level of tin in the saturated stannous sulfide vapor (100 mg m^{-3}) is more than 300 times higher than that above clean molten tin at 1027 °C (0.3 mg m^{-3}). The stannous sulfide vapor carried by the atmosphere may condense in cooler parts of the structure, then be reduced, and so deposit specks of tin-rich compounds onto the upper surface of the ribbon. The apparent size of these specks is enhanced by a halo due to interaction with the glass and so may cause a visible discrete fault. Occasional cleaning of the roof of the bath minimizes the incidence of this fault (Hynd 1984).

It is not possible to completely eliminate the oxygen or sulfur content of the float system, since the molten glass itself continuously brings in small quantities of dissolved oxygen and sulfur compounds. The mass balance of these materials have been analyzed in detail (Pilkington 1969) and so operational techniques have been developed to minimize any adverse effects.

4. Control of Ribbon Dimensions

The process has been developed to provide a range of glass thickness from 2 to 25 mm to meet general market demands, while glass of lesser thickness has been made to meet special product requirements. Ribbon width may be selected to minimize loss when the ribbon is cut into plates of the required length and width, a process that is now often automated. Typically, widths are in the range 2 m to over 3 m. Thus a manufacturing unit of medium size, with a throughput of 3000 t per week, will produce a ribbon of 4 mm thickness and 3 m width at a speed of nearly 600 m h^{-1}.

The variants of the process developed to provide ribbons of defined width and thickness in the widely used Pilkington process have been described in detail by Hynd (1984). Ribbons of thickness less than the equilibrium value, say 4–6 mm and at required widths, can be made using the basic process illustrated in Fig. 1 when the unit is at least of medium size. Under these conditions the molten glass spreads to form a ribbon of near-equilibrium thickness and is subsequently attenuated by tractive force applied to the ribbon by the roller conveyor. The reaction to these tractive forces is provided by the drag of the molten tin on the glass which increases as the square of ribbon speed, hence, the total tractive force on the ribbon increases rapidly along the length of the bath. Control of the degree of attenuation is provided by the final ribbon speed and by regulating the rate of increase of viscosity along the bath by controlling the cooling rate. The variation of the width, thickness and speed of the ribbon with distance along the bath can be modelled mathematically if the viscosity regime is known (Narayanaswamy 1981).

Further control of width can be obtained by introducing an array of driven toothed disks, called top rolls, on either side of the ribbon, which may be operated to reduce loss of width during the production of thinner ribbons. This additional control is useful in matching ribbon widths to market requirements.

Glass of thickness greater than the equilibrium value is produced by placing a pair of symmetrical nonwetting longitudinal guides to restrain the outward flow of molten glass at the hot end of the bath.

These guides are set to define the width of ribbon required and are long enough to allow the glass to be cooled sufficiently to inhibit any further dimensional change. The ribbon thickness is clearly defined by throughput and conveyor speed, and high-quality float glass up to 25 mm thickness has been produced by this process.

5. Solar Control Glass

The lightweight curtain clad construction of larger buildings such as office blocks and schools has led to large increases in the glazed area of such structures. When clear glass is used, the "greenhouse effect" can produce undesirable increases in internal temperatures during summer months or impose excessive loads on air-conditioning systems; hence there is a need to reduce the heat input to the building through the glazing. A solution now widely adopted is to use tinted heat-absorbing or reflecting glass to reduce both solar heat and glare.

A range of absorbing glasses are made, for example, in green, blue, bronze and gray colors, by adding appropriate materials to the glass composition, melting the glass and using the float process to form a ribbon. A disadvantage of these products is that it may be uneconomic to change the composition of the glass in the melting unit, containing over 1000 t of glass, unless a very large quantity is required. Hence the float process has been developed to meet the requirement for solar control glass in novel ways.

5.1 Electrofloat Process

The exploitation of the particular features of the float system to produce small quantities of tinted glass by an electrochemical technique has been described in detail by Robinson (1970). Within glass-forming technology the environment is unique: the glass is held in the molten or plastic state for a period of minutes at temperatures where it displays a useful electrical conductivity; it rests on an electrical conductor, molten tin; it is horizontal and so electrical contact can be made with the top surface via molten metal; and the reducing atmosphere gives the possibility of transforming the metallic ions introduced into the glass surface to a colloidal metallic form. The embodiment of these features into a production process is illustrated in Fig. 3.

A direct voltage is applied to the ribbon by means of an anode consisting of a metal pool resting on top of the glass at the position AA in Fig. 1b, the molten tin acting as the cathode. Under the influence of the applied field, sodium ions migrate downwards through the glass into the molten tin, current being carried into the top surface of the glass by metallic ions from the pool. Under the reducing atmosphere and at temperatures well above 600 °C these ions are reduced to colloidal metallic particles within the top surface of the glass, which can both absorb and reflect solar energy. The color imparted to the glass is characteristic of the metal electrolyzed and the radiant energy transmitted is varied by changing the current flow, thus controlling the concentration of metal in the surface. Since this colloidal metal is within the glass, the product is both chemically durable and abrasion-resistant. Because the introduction of metallic ions into the surfaces increases the electrical resistance, current flows more readily to areas less heavily treated, which is important in achieving uniform coloring of a ribbon which may be more than 3 m in width, and where some center-to-edge temperature difference can occur.

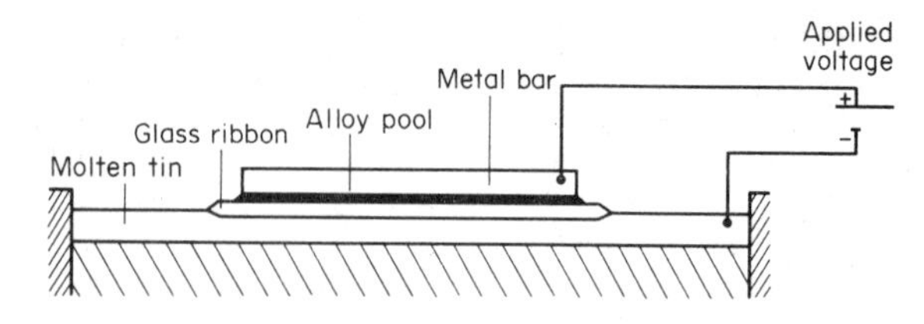

Figure 3
Vertical lateral cross section (AA in Fig. 1b) of electrofloat process

The upper molten metal pool is defined and held over the moving ribbon by causing it to wet the solid metal bar positioned just above the glass surface, as illustrated in Fig. 3.

Copper, silver and lead are examples of metals that can be introduced into the glass by this process. It has been found that molten bismuth (melting point 271 °C) is a suitable solvent metal for both copper and silver, and the bar itself can be used as the source of copper or silver which will maintain the pool at a constant composition.

This process is used commercially to manufacture a bronze-tinted glass using a copper–lead alloy pool with a copper bar and continuous addition of lead. It is ideal for short production runs, minimizing warehousing costs, and the capital cost of the equipment is very low.

5.2 Vapor Coating Process

A solar control glass is also made by exploiting the unique features of the float process. The availability of a flat horizontal glass surface at elevated temperature within a reducing atmosphere has provided the vehicle for the online application of a reflective coating to the glass by chemical vapor deposition.

Means are provided to pass a gas over the hot glass, and the gas decomposes to deposit a durable uniform film onto the upper surface of the float ribbon. The glass has the appearance of a semisilvered mirror and because it reflects both the visible and infrared components of solar radiation it is a very effective product for solar control.

6. Commercial Advantages

The float process meets all the requirements of a modern flat glass production process (Hynd 1984).

(a) Flat glass of high quality in thicknesses of 2–25 mm in widths up to over 3 m can be produced.

(b) No upper limit on output is perceived; present plants manufacture in the range 1000–6000 t per week.

(c) The process is efficient in respect of energy usage and glass utilization.

(d) Continuous operation, apart from minor repairs, is possible for the whole life of the melting unit, which now extends up to ten years.

(e) The production of a horizontal ribbon of defined width lends itself to high standards of annealing and automated glass-cutting, and the whole manufacturing line lends itself to coordination or control by computer.

(f) The labor requirement is low.

(g) The unique features of the float process can be exploited to produce glass for solar control.

Presently it is very difficult to envisage a process that will supersede the float process for the supply of transparent flat glass to the massive architectural and automotive markets.

Bibliography

Charnock H 1970 The float glass process. *Phys. Bull.* 21: 153–56

Hynd W C 1984 Flat glass manufacturing processes. In: Uhlmann D R, Kreidl N J (eds.) 1984 *Glass: Science and Technology*, Vol. 2. Academic Press, New York, pp. 45–106

McCauley R A 1980 Float glass production: Pilkington vs PPG. *Glass Ind.* 61: 18–23

Narayanaswamy O S 1981 Computer simulation of float glass stretching. *J. Am. Ceram. Soc.* 64: 666–70

Pilkingon L A B 1969 The float glass process. *Proc. R. Soc. London, Ser. A* 314: 1–25

Pilkington L A B 1976 Flat glass: Evolution and revolution over 60 years. *Glass Technol.* 17: 182–93

Robinson A S 1970 The float glass process. *Educ. Chem.* 7: 144–45

H. Charnock
[Pilkington, Ormskirk, UK]

Fracture and Fatigue of Glass

The design strength of glass is low because it is brittle, and becomes weaker with time. The prediction of usable lifetime from knowledge of glass fatigue is quite uncertain, and necessitates large safety factors in design. Better knowledge of fracture criteria and fatigue life could lead to higher design strengths and much wider application of glass. Special applications of glass, such as in fiber optics, require improved techniques for the prediction of fracture limits and fatigue life.

Until the 1950s, there was little concern given to detailed analytical treatments of fracture and fatigue in brittle solids. The reduction of strength by surface flaws was accepted, but the emphasis in strength studies was on experimental measurements of the effects of sample shape, size and composition on strength. More recently, as a result of the recognition of the importance of the strength of brittle solids, a vast amount of analytical and interpretive work has been done on fracture.

1. Fracture Testing Methods

In brittle materials, failure usually takes place at much lower tensile stresses than compressive stresses. In principle, the simplest test is to subject a rod to uniform tensile stress. In practice, there are difficulties in gripping the sample, and except for fine fibers, one must make the central length of the rod smaller in diameter to make certain that failure does not occur at the grips. Making such samples is difficult and expensive, so other tests are used in which the stress distribution is not as simple as in a tensile test. For long fibers, the direct tensile test is the simplest and most used, since gripping problems can be overcome in a variety of ways.

The most common tests involve bending (flexure), either with three or four points of stress application. Rod samples are better than bars, because chips at the corners of the bars can initiate fracture. Bend tests favor failure starting at the sample surface, where the tensile stress is high. It is always wise to examine the specimen after testing, to make certain that failure did not take place at the contact points, where the stress is uncertain. The four-point bend test is preferable to the three-point test because the former samples more of the specimen surface and is less prone to failure at the contact points.

The diametral compression test provides an attractive alternative to bend tests for some materials. In this test, a right circular cylinder is compressed across its diameter by two flat plates, giving a maximum tensile stress normal to the loading direction across the loading diameter. This stress is present across the diameter of both flat ends of the cylinder, and also across a plane from one end of the cylinder to another, so that both surface and volume failures are possible. A pad of material softer than the compressing plates must be inserted between the sample and plates, to prevent failure at the plates.

Compressive tests are sometimes used for glsss, but failure usually takes place at a corner contact of the sample with the compressing plates.

2. Strength Distributions

When the failure stress of a number of glass samples is measured under identical conditions of sample preparation and loading, the failure stresses vary over a wide range, as shown in Fig. 1. It is desirable to express this distribution of strengths in statistical terms. The strengths are often symmetrically distributed, so they can be described by a normal (Gaussian) distribution:

$$P = [1/d(2\pi)^{1/2}]\exp[-(S-S_m)^2/2d^2] \qquad (1)$$

where P is the probability of finding a sample of strength (failure stress) S, S_m is the strength of greatest probability $[P_m = 1/d(2\pi)^{1/2}]$ and d is the standard deviation. The integral of Eqn. (1) gives the fraction F of samples that break below the stress S.

Recently another distribution function called the Weibull distribution has become quite popular in describing strength distributions:

$$F = 1 - \exp[-(S/S_0)^m] \qquad (2)$$

where S_0 is a scaling factor and m is a measure of the spread of the distribution—a broader distribution has a smaller m. For $m \geqslant 3$, the Weibull distribution is nearly symmetrical and closely resembles the normal distribution.

Strengths of carefully drawn glass fibers are much higher than those of ordinary glass; E-glass can have a strength of 1–4 GN m^{-2} at room temperature, and silica fibers have even higher strengths (Proctor et al. 1967). (Compositions of glasses are given in the article *Glass: An Overview*.) The strength distributions of short E-glass fibers can be quite narrow, with a coefficient of variation of 1–2%. On the other hand, high-strength flame-polished or etched glass rods can have very broad and nonuniform distributions, as shown in Fig. 2.

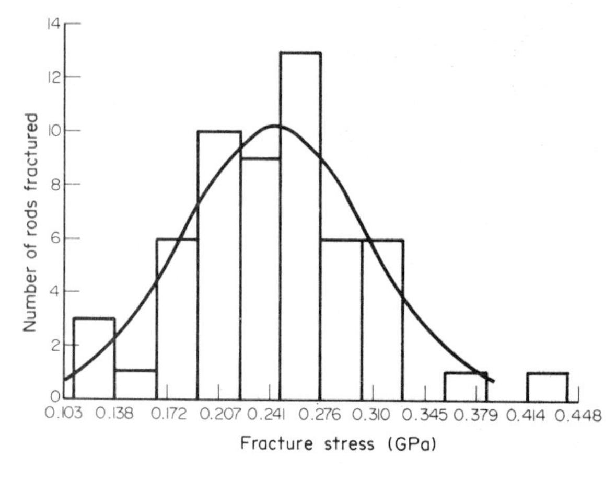

Figure 1
Distribution of fracture strengths of pyrex borosilicate glass rods in four-point bending at $-196\,°C$: line from Eqn. (1) with $S_m = 0.245$ GPa and $d = 0.061$ GPa; 3 mm diameter rods

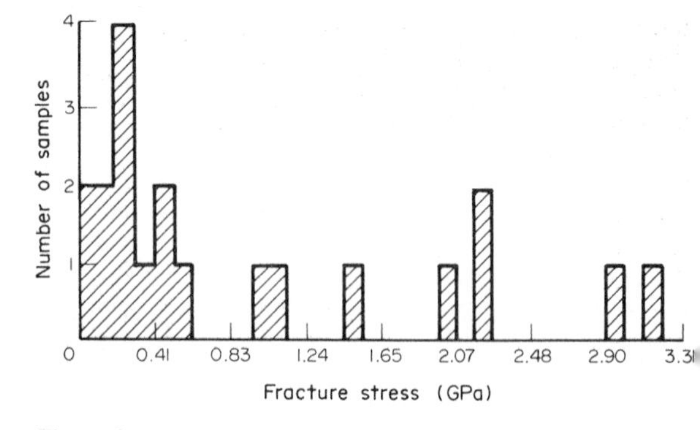

Figure 2
Distribution of fracture strengths of rods of soda-lime glass in four-point bending at $-196\,°C$: 3 mm diameter rods, after removing 9.4 μm by etching with dilute HF

3. Effect of Surface Area and Treatment on Strength

Many investigators have confirmed that as the stressed area of a glass surface increases, the strength decreases. When the data of Griffith (1920) are plotted as the log of breaking strength S against the log of diameter D, the relation is linear. Fused-silica fibers also show a linear relation between log S and log D below 25 μm diameter, with higher than expected strengths at larger diameters. This size dependence of strength can be understood in terms of strength and flaw distributions.

If glass samples are carefully prepared and kept pristine, they will have very high strengths ($\leqslant 10$ GN m^{-2}). Hillig (1962) measured strengths of up to 15 GN m^{-2} at $-196\,°C$ for 1 mm diameter fused-silica rods. These samples were flame-drawn from larger rods, and carefully protected from contact with solids.

The strength of ordinary or abraded glass can be substantially increased by etching with hydrofluoric acid (HF). The distribution of strength of soda-lime glass rods after etching is shown in Fig. 2. Etching gives some samples with very high strengths, but because samples etched in HF are highly sensitive to damage by handling, many samples have strengths lower than before etching.

The strength of an abraded soda-lime glass can be increased by a factor of up 1.5 by standing in water for a few hours after abrasion.

The strengths of strong glasses increase at lower temperatures. However, abraded glass of low strength often shows little temperature dependence of strength at short loading times. It seems likely that these temperature effects are caused by the same mechanisms as delayed failure. Hillig (1962) found no difference in glass strength between −196 and −268 °C, at which temperatures delayed failure is absent.

Glass composition can influence the strength indirectly through surface chemical reactions, through the introduction of surface stresses and through other secondary effects. However, there is no evidence for direct effects of glass composition on strength (e.g., as a result of changes in bonding in the silicon–oxygen network).

4. Fracture Surfaces and Origins

A glass fracture surface shows several characteristic regions—an origin (see Fig. 3), a smooth region called the mirror, a misty region outside the mirror and finally a region of surface roughness called hackle. The distance r from the fracture origin to the mirror–mist boundary is related to the fracture strength S by the empirical equation $Sr^{1/2}=A$, where A is a constant for a particular material and temperature. This constant has a value of about $2\,\mathrm{MN\,m^{-3/2}}$ for glass at room temperature. Apparently, A is almost independent of temperature or glass composition. This relation is of considerable practical value, because it allows accurate estimates of fracture stress to be made without direct measurement.

A variety of chemical treatments, such as sodium vapor or a molten lithium salt, give visible crack patterns on a glass surface. However, these cracks result from the treatment, and do not represent prior surface damage. Etching a glass surface with HF reveals elongated pits resulting from surface cracks.

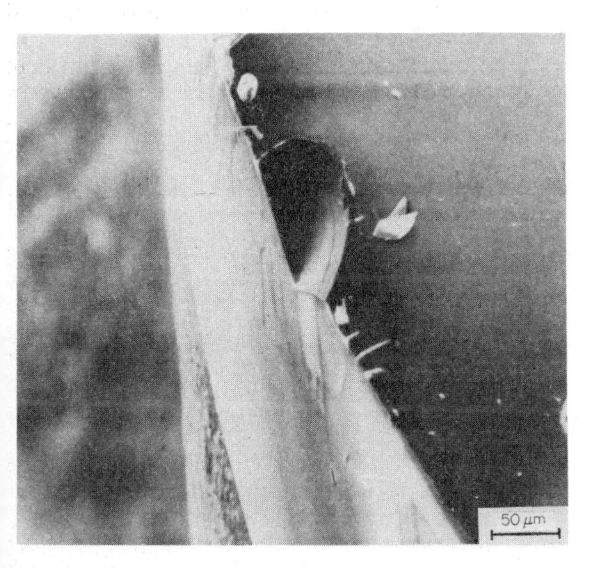

Figure 3
Scanning electron micrograph showing a fracture origin in a soda-lime glass

If a hard sphere is pressed against a glass surface, a crack develops around the contact and then grows into the glass in a conical shape. If the stress is applied with a sharp rather than spherical object, a median vent crack propagates into the glass, directly under the contact area. When the load is released, winglike cracks (lateral vents) propagate radially from the contact region. The impact of a hard spherical projectile causes damage with features of both spherical and sharp indenters. At low impact velocities cone cracks are formed, but at higher velocities the cone cracks resemble radial winglike cracks; at high velocities median cracks also propagate down into the glass.

When a hard object such as a diamond indenter is dragged across a glass surface, median and lateral cracks form along its track in much the same way as described above. Abrasion of a hard surface against the glass also gives this kind of linear damage.

5. Velocity of Fracture Propagation

The velocity of fracture propagation in glass can be measured from line markings on the mirror surface known as Wallner lines. Fracture velocities can also be measured from ripple marks on the fracture surface, which result from ultrasonic waves introduced perpendicular to the fracture face. Various techniques of high-speed photography, such as rotating mirror and multiple-spark cameras, can be used to measure crack velocity. Another method is to measure the electrical conductivity of metal strips painted on the glass as they are progressively broken by the fracture front.

6. Fracture Criteria and Theoretical Strength

The experimental observations described above provide strong evidence that the surface condition of glass determines its strength, and that flaws in the surface reduce the strength. This lead to a model of surface cracks as the origin of brittle fracture. Unfortunately it is difficult to observe these cracks directly, and there still is uncertainty about their size, depth and shape that requires more experimental investigation.

The stress σ at the tip of an elliptical crack of depth c and tip radius ρ subjected to a tensile stress S is (Inglis 1913):

$$\sigma/S=1+2(c/\rho)^{1/2}\approx 2(c/\rho)^{1/2} \tag{3}$$

Thus, if the crack is deep and has a sharp tip, the stress at the tip can be much higher than the applied stress. It seems reasonable that various types of surface damage on glass could give rise to cracks with different depths and tip radii, resulting in the influence of surface treatment on strength found experimentally. Experiments show that glass samples with cracks of the same length but with different treatment (e.g., freshly abraded or abraded and aged in water) have different strengths, showing the importance of the crack tip radius.

The theoretical strength σ_t of glass is approximately given by:

$$\sigma_t^2 = E\gamma/4a \quad (4)$$

where E is Young's modulus, γ is the surface energy and a the average interatomic distance in the material. This equation gives a strength for vitreous silica of about $24\,\mathrm{GN\,m^{-2}}$, or about 60% greater than the highest measured value.

7. *Plastic Deformation in Glass*

Tensile and bend tests on glasses to stresses close to the theoretical strength show no evidence of any nonrecoverable deformation. However, indentation of glass with a sharp point leads to some regions of deformation that do not recover when the load is removed. In fused silica, this deformation is apparently a densification that recovers completely on heating. In glasses with modifiers (alkali and alkaline earth oxides), this deformation is partly a densification, but only a small part of the deformation recovers on heating. More detailed examination shows that the deformed region is made up of flow or fault lines between which there are only elastic strains. These deformations are very different from plastic deformation in metals and other ductile materials, since they arise only from compressive loads, and do not occur with large-scale stressing in tensile and bending modes. Since crack propagation under normal conditions occurs as a result of applied tensile stress, and high tensile stress at the crack tip, there is no reason to expect densification or faulting under these conditions.

8. *Cracks and Energy*

From the strain energy change in a stressed material containing a crack, Griffith (1920) derived the equation:

$$S_f^2 = 2E\gamma\pi c \quad (5)$$

for the fracture stress S_f in a material with a crack of depth c, surface energy γ and Young's modulus E. In a treatment of crack energies and fracture criteria in terms of thermodynamic laws, the Griffith equation provides a necessary but not sufficient condition for fracture, the correct criterion for fracture being provided by Eqn. (3). The second law of thermodynamics shows that the fracture stress cannot be less than that given by Eqn. (5), but can be greater. It also predicts that fracture will only occur when the tip radius is larger than some critical minimum value ρ_m. A measure of ρ_m can be gained by comparing Eqn. (3) and Eqn. (5) for the fracture stress. Then,

$$\rho_m = 8E\gamma/\pi\sigma_t^2 \quad (6)$$

is the minimum tip radius for fracture. From Eqn. (4) and Eqn. (6), $\rho_m = 32a/\pi$. Thus the minimum tip radius for fracture is about ten times the interatomic spacing and is independent of other parameters.

9. *Nucleation of Cracks*

How do cracks get into a material? As described above, pressing a hard sphere against a glass surface causes a ring crack and then a cone crack. In some cases, the ring cracks appear to start at preexisting flaws, and then propagate. However, indenting a flaw-free surface of fused silica can nucleate a ring crack if the load is high enough. In soda-lime glass, sharp indenters and projectiles give median and radial cracks, and scribing gives median and lateral cracks. These types of damage are probably common in a glass surface as a result of ordinary handling and contact with other solids. Thus, nucleation of flaws in a pristine glass surface can be understood qualitatively as resulting from the stresses and deformation caused by colliding projectiles, indentations and scribing that are involved in handling.

10. *Strengthening of Glass*

Etching is not a practical way of strengthening glass because the etched surface is very sensitive to new damage, even if it is protected. Glass can be strengthened by introducing a compressive stress at its surface; then the total tensile stress needed for failure is increased by the amount of the compressive stress. Compressive stress is introduced by either quenching or ion exchange (chemical strengthening). Quenching is simpler and cheaper, but harder to control, and results in a region of high tensile stress inside the glass. Exchange of large ions for small ones already in the glass gives a thin region of high compressive stress in the glass surface, and only a low tensile stress inside the glass.

11. *Fatigue*

In principle, the simplest method of investigating fatigue in glass is a static test in which a constant load is imposed on a set of samples and the time of failure of each sample is recorded. Such tests can be done in bending, simple tension or diametral loading; bending is easiest for rods and tubes, and requires simple test equipment. Although static tests are the closest to practical situations and are the least complicated in principle, they have some disadvantages. They require a large number of samples for each of several stresses to define the stress–failure time relationship, because of the large spread in failure times at a particular stress. It is often difficult or expensive to obtain the required number of samples. Static tests must also be carried out over a large spread of times, leading to long delays in obtaining results.

These disadvantages have led to a search for alternative methods to test fatigue in glass and other brittle oxides. Two methods have been widely used. In the first, samples are tested at different loading rates; the dependence of fracture strength on loading rate is then related to the stress–failure time relationship in static loading. This method requires fewer samples and less time, but a testing machine with a linear load application is needed. The main disadvantage of this technique is that it can be used for only a narrow range of failure stresses, because of limits in the loading rates available. Thus, this method is good for surveys of materials, and rapid or preliminary tests, but static fatigue tests are needed for complete definition of strength–failure time relations.

A second method is to measure directly the velocity at which large (~1 cm) cracks propagate under different applied loads. This method has been widely used, but experimental and theoretical treatments suggest that the results it provides are not always directly comparable with the fatigue behavior of materials with much smaller flaws (a few micrometers long).

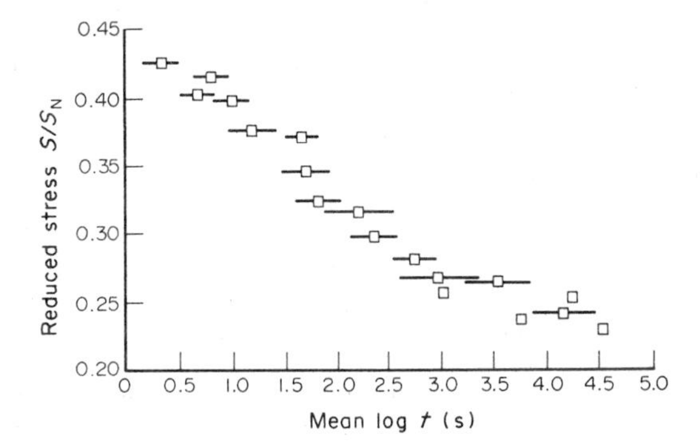

Figure 4
Mean log failure time at 100% relative humidity against relative stress for soda-lime glass rods abraded and then aged 24 h in water before testing; lines show 95% limits for the spread of log t values

In Fig. 4, static fatigue results for soda-lime glass in four-point bending are shown. Each point represents the mean of at least twenty samples. The results are plotted as reduced stress S/S_N against the log of fracture time, where S is the maximum stress in four-point bending and S_N is the mean fracture stress in the same test at liquid nitrogen temperature, where there is no fatigue.

The influence of pretreatment and humidity on the time for failure $t_{1/2}$ at $S/S_N = 0.5$ for soda-lime glass is summarized in Table 1. The results are taken from different studies which used many samples and included measurements at $-196\,°C$ for each condition; each $t_{1/2}$ value is based on measurements at many (five to twenty) stresses, and at each stress twelve to twenty samples were broken.

The results in Table 1 show that fatigue times increase strongly as the ambient concentration of water is reduced. When the samples are not aged after abrasion, their fatigue times are up to one hundred times longer than after soaking in water for 24 h with other conditions the same. Flaw type and severity and direction of abrasion also have a strong effect on both inert fracture strength and fatigue times.

Charles (1958) measured the static fatigue of soda-lime glass at different temperatures from $-50\,°C$ to $242\,°C$ in saturated water vapor; he found that failure times decreased with temperature increase. At temperatures above about $50\,°C$, the failure times no longer fit a simple Arrhenius line, and become longer than expected. At temperatures below about $-100\,°C$, the fatigue becomes very slow compared with experimental time.

Table 1
Fatigue times for abraded soda-lime silicate glass

Test	Relative humidity (%)	Average fracture stress at $-196\,°C$ ($MN\,m^{-2}$)	$\log t_{1/2}$ (s)	Remarks[a]
1	100	118	1.67	not aged
2	50	118	2.46	not aged
3	100	160	−0.66	
4	50	160	0.48	
5	water	75	0.94	aged 20–60 s in water
6	water	86	0.94	
7	43	86	2.30	
8	0.5	86	3.54	
9	water	90	2.30	heated at 470 °C for 3 h in a vacuum of 10^{-5} mm Hg, stored in dry nitrogen, then tested in water
10	50	187	1.47	heated to 400 °C for 1 h after abrasion, then held 24 h in water

[a] All samples were aged 24 h in water after abrading and before testing, unless otherwise noted. Tests 1–4 and 10 from Mould and Southwick (1959); tests 5–9 from Pavelchek and Doremus (1976)

12. Life Prediction

An important practical problem is to predict the lifetime of a piece of glass under stress. Several studies have shown that there is no separate effect of cyclical stress on the fatigue of glass, so the fatigue results from a cumulative process depending on the time at different applied stresses. Thus the problem of predicting the lifetime is to find a reliable function that can be extrapolated to longer times and lower stresses than are convenient experimentally.

Since the inert (liquid nitrogen) strengths are distributed, failure times at a constant stress are also distributed. Several different functional dependencies between failure time and stress have been suggested:

$$\log t = a - b\, S/S_N \qquad (7)$$

$$\log t = c - n \log S/S_N \qquad (8)$$

$$\log t = d + g\, S_N/S \qquad (9)$$

where the values of b, n and g are positive.

The simplest way to examine the equations is to calculate a regression equation by the least-squares method in the form of Eqns. (7–9), and compare the correlation coefficient R^2 from the regression. For pyrex glass, R^2 values for Eqns. (7–9) have been calculated as 0.920, 0.947 and 0.976, respectively. Although the correlation coefficients are all close to one, there is a clear increase in the closeness of fit from Eqn. (7) to Eqn. (9), especially when $1 - R^2$ values are compared.

Often data are not extensive or reliable enough to find any differences between correlation coefficients; a more sensitive method is needed. Pavelchek and Doremus (1976) suggested that the variation of the spread (standard deviation) of $\log t$ values with stress is a sensitive measure of the functional dependence of failure time on stress. If the differential of $\log t$ is considered to equal the standard deviation of $\log t$ then the following relations are found. For Eqn. (7) the spread in $\log t$ values should decrease as the stress decreases, for the power law of Eqn. (8) the spread in $\log t$ should remain constant with changes in stress, and for the inverse exponential of Eqn. (9) the spread in $\log t$ should increase as the stress decreases.

For every set of static fatigue data on glass that are reliable, the spread in $\log t$ values increases as the stress decreases (Pavelchek and Doremus 1976). Figure 4 also shows this increase. Thus Eqn. (9) again fits best.

13. Theories and Interpretations of Static Fatigue in Glass

The most widely accepted theory of static fatigue (Hillig and Charles 1965) is that a stress-accelerated chemical reaction of water with the glass leads to a sharpening of the crack tip. Thus, as stress at the tip increases, the reaction rate becomes faster, and finally the stress at the crack tip becomes high enough to cause rapid propagation of the crack. A comparison of the Hillig–Charles equations with experimental fatigue data shows that one expects tip sharpening rather than crack lengthening to take place during fatigue in glass.

Further evidence to support the conclusion that cracks sharpen rather than lengthen during fatigue comes from observations of crack initiation sites (Fig. 3). These sites have the same appearance after fracture at −196 °C and at 25 °C. A comparison of sizes of these sites shows no statistically significant difference between sites formed at −196 °C and at 25 °C. If fatigue at 25 °C caused crack lengthening, one would expect larger sites at this temperature.

There are striking differences between failure times of samples not aged after abrasion and those aged in water (see Table 1). From the Hillig–Charles theory, one would expect only small differences in $t_{1/2}$ for these different samples. Thus this theory, and also other theories of fatigue of glass, give no explanation for this large difference in fatigue time.

Heating the glass to above 300 °C after abrasion can anneal out residual stresses resulting from abrasion, and also decompose hydronium ions. These ions are introduced into the glass as a result of their exchange and interdiffusion with alkali ions in the glass. The decomposition of the hydronium ions leaves a proton in place of the larger sodium ion, so a tensile stress develops in the glass and it becomes weaker. The net result of these two processes can give a $t_{1/2}$ between that of samples not heated and aged in water and that of those not aged, as shown in Table 1.

14. Proof Testing

Brittle parts can be stressed to a higher level than the maximum stress to which they will be subjected in practice to eliminate weak samples. This procedure is called proof testing. If the proof test is made under conditions where fatigue occurs, the surviving samples will be weaker. To reduce this effect, the proof test can be made in a dry atmosphere and/or at a temperature below ambient.

Bibliography

Charles R J 1958 Static fatigue of glass. *J. Appl. Phys.* 29: 1549–60

Doremus R H 1976 Cracks and energy—Criteria for brittle fracture. *J. Appl. Phys.* 47: 1833–36

Doremus R H 1982 Fracture and fatigue of glass. In: Doremus R H, Tomozawa M (eds.) 1982 *Glass III*. Academic, New York, pp. 169–239

Ernsberger F M 1960 Detection of strength-impairing surface flaws in glass. *Proc. R. Soc. London, Ser. A* 257: 213–23

Griffith A A 1920 The phenomena of rupture and flow in solids. *Philos. Trans. R. Soc. London, Ser. A* 221: 163–98

Hagan J T 1980 Shear deformation under pyramidal indentations in soda-lime glass. *J. Mater. Sci.* 15: 1417–24

Hillig W B 1962 Sources of weakness and the ultimate strength of brittle amorphous solids. In: Mackenzie J D (ed.) 1962 *Modern Aspects of the Vitreous State*, Vol. 2. Butterworth, Washington, pp. 152–94
Hillig W B, Charles R J 1965 Surfaces, stress-independent reactions and strength. In: Zackey V F (ed.) 1965 *High Strength Materials*. Wiley, New York, pp. 682–93
Inglis C E 1913 Stresses in a plate due to the presence of cracks and sharp corners. *Trans. Inst. Naval Arch.* 55: 219–30
La Course W C 1972 The strength of glass. In: Pye L D, Stevens H J, La Course W C (eds.) 1972 *Introduction to Glass Science*. Plenum, New York, pp. 451–512
Mould R E, Southwick R D 1959 Strength and static fatigue of abraded glass under controlled ambient conditions: II. Effect of various abrasions and the universal fatigue curve. *J. Am. Ceram. Soc.* 42: 582–92
Pavelchek E K, Doremus R H 1976 Static fatigue in glass—A reappraisal. *J. Non-cryst. Solids* 20: 305–21
Proctor B A, Whitney I, Johnson J W 1967 The strength of fused silica. *Proc. R. Soc. London, Ser. A.* 297: 534–57

R. H. Doremus
[Rensselaer Polytechnic Institute, Troy, New York, USA]

G

Glass: An Overview

Glass is an essential material in modern technological society. In addition to traditional uses such as windows and containers, glass is used in a host of specialized ways in lamps, optics, composites and electronics. Recent developments of fiber-optic waveguides, laser optics for initiating fusion reactions, and repositories for radioactive waste—all of glass—demonstrate the versatility, importance and promise of glass for future applications.

1. History

Archaeological evidence shows that natural glasses such as obsidian have been used by man from the earliest times. Synthetic glass and glazes were made far back in human history. The earliest known glaze (coating of glass) dated from 12 000 BC, and the earliest solid glass from about 7000 BC; both were found in Egypt, which continued to be a center of glass and glaze manufacture throughout ancient history. At first, glass was used only for decorative purposes, but later it was molded or pressed into vessels. The invention of glass blowing in about the first century BC greatly increased the use of glass for practical purposes in Roman times, mainly for vessels but later for windows also. In Europe, glassmaking dispersed to isolated sites after the fall of the Roman Empire, but was continued in the Byzantine Empire (Syria especially), and then by the Arabs in Syria, Egypt and Persia. Venice became the center of a resurgent glass industry in Europe after about 1300. Artistic development paralleled that of the Renaissance in the fifteenth century, the most significant advance in Venice being the manufacture of a clear, colorless glass called cristallo. With this glass Venice became the premier center of glass manufacture and export until the seventeenth century, when centers sprang up throughout Europe, especially in Bohemia, Nuremberg and the UK.

Of special interest in this period was the publication of *L'Arte Vetraria* by Neri in Florence in 1612; this work released the art and technology of glassmaking from the secret hold of the Venetian artisans and made it available to the world.

After 1600 the development of glassmaking was rapid. Cut glass was first made about 1600, and the use of coal instead of wood for fuel gave higher temperatures and a more reliable fire. Late in the seventeenth century, lead oxide was added to glass in England, resulting in lower melting temperatures, a wider temperature range for working the glass, and "sparkle" from a higher refractive index. Better selection and some purification of raw materials in the eighteenth century gave clearer glasses and better control of melting and working conditions and of the color of the glass.

In the nineteenth century, the quality of glass produced in Europe improved tremendously in response to the increased affluence of industrial society. At the end of the nineteenth century, automatic methods of making glass were developed, such as the bottle-blowing machine.

Traditionally, window glass was made by hand by either the crown or cylinder process. In the crown process a glob of glass was blown out and one side of the resulting globe flattened. A solid iron rod was attached to the flat part and the blowing pipe detached. Then the globe was reheated and rotated until it formed a flat disk about three feet in diameter. Panes of glass were cut from the disk after it was slowly cooled. The part attached to the rod was the "bull's-eye"—still seen in some older windows. In the cylinder process the blower made a large cylinder, which was then split open and flattened. Cylinder glass was also made by machine.

In the early part of the twentieth century, processes for drawing sheet glass directly from the glass melt were developed (the Fourcault and Colburn processes). When linked with a continuous-glass-melting furnace, this process is capable of producing large quantities of flat glass of reasonable quality. Plate glass of the highest quality was, until recently, made by flowing the glass from the furnace through rollers, which gives a rough surface, and then grinding and polishing it by large automatic machines. This process requires a large capital investment, but is fairly economical because of the large quantity of glass that is drawn directly from the melt. The sheet-glass surface has a fine fire-polished finish, but shows some surface distortion because of variations in processing conditions as the glass is drawn from the melt.

In the 1950s, an ingenious new method of making relatively inexpensive flat glass of high quality was developed by Alistair Pilkington of the Pilkington Glass Company of England. In this float process, a continuous strip of glass from the melting furnace floats onto the surface of a molten metal, usually tin, at a carefully controlled temperature. The flat surface of the molten metal gives a smooth, undistorted surface to the glass as it cools. After sufficient cooling, the glass is rigid and can be handled on rollers without damaging the surface finish. High forming speeds are possible, and the cost is much less than for similar quality glass made by grinding and polishing. As a result, one glass manufacturer after another has converted to the float process and today most flat glass is made by this process.

Glass containers such as bottles and jars are made by blowing hot glass into a mold on a continuous machine. Lamp bulbs are made in a similar way by blowing hot glass into a mold; however, in this case the glass is fed from the melting furnace as a ribbon, rather than blowing out an individual glob of glass as in the container machines. A special nozzle blows glass in the ribbon into a mold. The high-speed "ribbon machine" for lamp bulbs is a spectacular sight as lamp bulbs pop out from it at a rate of 2000 per minute.

Certain glass objects such as plates, tumblers and vases can be made inexpensively by pressing the hot glass in a mold. This pressed glass was especially popular in the USA in the nineteenth century because it was much cheaper than cut-crystal glass imported from Europe. In this method, a glob of hot glass is placed into a metallic mold, and a metallic plunger is forced into the mold to shape the glass in the desired way. Patterns in the mold surface are imprinted in the glass and make it look like an expensive article. Pressed glass can also be made continuously using automatic feeders for molds on a rotating bed.

The continuous manufacture of glass in a large furnace, called a "tank," was developed in response to the demand for greater, cheaper and more uniform production of glass. This furnace requires improved refractory materials to line it, and is fuelled with natural gas or fuel oil.

A variety of new methods for making special types and forms of glass have been developed. Glass fibers as fine as 10 μm in diameter are drawn and coated at high speeds. Fused-silica tubing is drawn at temperatures above 1800 °C. Highly pure and clear glass is made for a number of optical purposes. A large variety of glass compositions have been developed for diverse applications. The infinite variability of glass compositions leads to a great assortment of possible properties which are only beginning to be exploited.

2. Definition

Glass is an amorphous solid. A material is amorphous when it has no long-range order, that is, when there is no regularity in the arrangement of its molecular constituents on a scale larger than a few times the size of these groups. For example, the average distance between silicon atoms in vitreous silica (SiO_2) is about 3.6 Å, and there is no order between these atoms at distances above about 10 Å. A solid is a rigid material; it does not flow when it is subjected to moderate forces. More quantitatively, a solid can be defined as a material with a viscosity of more than about 10^{15} P (poise).

Many glass technologists object to the above definiton of a glass. These workers prepare a glass by cooling a liquid in such a way that it does not crystallize, and feel that this process is an essential characteristic of a glass. Many earlier writers insist on this criterion: "glass is an inorganic product of fusion which has been cooled to a rigid condition without crystallization" as taken from the ASTM standards for glass. The difficulty with this view is that glasses can be prepared without cooling from the liquid state. Glass coatings are deposited from the vapor or liquid solution, sometimes with chemical reactions. Thus sodium-silicate glass can be made by evaporating an aqueous solution of sodium silicate (water glass) and baking the deposit to remove water. The product of this process is indistinguishable from sodium-silicate glass of the same composition made by cooling from the liquid. It seems wise to use the same name for materials with the same molecular structure and properties no matter how they are made; consequently, the broader definition given here is preferred.

3. Uses

The original use of glass for decorative and artistic objects continues to this day. One need only visit the Stueben exhibit in New York City, the Corning museum in Corning, New York, or glass exhibits in Czechoslovakia to sense the originality and excitement of present-day artistic work in glass.

The second use of glass was for containers for food, perfume, oils and many other substances; this application still uses the largest quantity of glass today. The production of flat glass mainly for windows in buildings and vehicles, is now the second largest item of glass manufacture. Lamp envelopes and seals are another major area of use. There are many special applications of glass, some in small quantity but of high value, most of which have been developed in the last few decades. Some of these special applications are glass ceramics and surface-strengthened glass for higher strength and chemical durability; lightweight composites of fiberglass in polymer matrixes; glass for laser hosts and optic waveguides for long-distance communication; fused silica for melting semiconductors, telescope mirrors and arc lamps; amorphous silica layers on silicon in electronic devices; encapsulation of these devices; and as a medium for consolidating radioactive wastes. New compositions of silicate glasses, and even non-oxide glasses, are being rapidly developed for a variety of new applications.

These uses of glass are based on a variety of desirable properties, such as ease of forming into many different shapes, cheap and widely available raw materials, chemical durability, transparency, high-temperature durability, wide "solubility" of constituent oxides and low electrical conductivity.

4. Main Types of Glass

The possibility of incorporating a large number of different oxides in a silicate glass has led to a wide variety of commercial glasses. Nevertheless, there are a relatively small number of glasses that make up the large majority of glass production. Some of these glasses are listed in Table 1, and their approximate compositions are given in Table 2. The softening temperature in Table 1 is either the temperature at which a rod 25 cm long and 0.76 mm in diameter elongates by 1 mm min^{-1}, or the temperature above which the glass can be easily deformed and worked.

By far the most common glass is based on the soda-lime—silicate (sodium calcium silicate) system. All ancient glasses contained the oxides of sodium, calcium and silicon. These glasses are cheap, chemically durable and relatively easily melted and formed. Many minor additions to the basic composition (not far from 70% silica, 15% soda, 10% calcium oxide and magnesium oxide, and 5% other oxides) are made to improve properties of melting and forming: alumina for improved chemical durability and reduced devitrification (crystallization), borates for easier working and lower thermal expansion, zinc oxide for lower melting temperatures, and arsenic and antimony oxides for fining (removal of bubbles). Soda-lime glass is often termed "soft" glass because of its relatively low softening temperatures.

Pyrex-borosilicate glass was developed by the Corning Glass Works to be more resistant to thermal shock and more chemically durable than soda-lime glass, yet still melt at a similar temperature. The borate in this glass reduces its viscosity and the coefficient of thermal expansion, and allows low sodium content, which increases chemical durability. Pyrex-borosilicate glass is often called "hard" glass because of its higher softening temperature compared with soda-lime glass, and is somewhat more expensive than soda-lime glass because of its higher melting temperature and more expensive borate raw material. The mirror for the Mount Palomar telescope was made of Pyrex-borosilicate glass because of its low thermal expansion; nevertheless, it is necessary to correct minute distortions of the mirror surface caused by temperature differences.

Vitreous silica is made of pure silica, SiO_2, giving it excellent high-temperature stability, optical properties and thermal-shock resistance.

A variety of lead glasses are important as low melting reading and solder glasses with a wide working-temperature range. Lead glass for fine "crystal" contains much more lead than these glasses.

Some nonoxide glasses find special applications and many others are being studied for new uses. Fluoride glasses based on beryllium fluoride have low values of refractive index, Abbé number and nonlinear index and therefore are the best glass compositions for high-

Table 1
Some types of silicate glasses and their uses

Glass	Use	Approximate softening temperature (°C)	Coefficient of thermal expansion (per °C × 10^6)
Soda lime	Containers, windows, lamp bulbs	700	9.0
Pyrex borosilicate	Headlamps, cookware, laboratory ware	830	3.2
Vitreous silica	Semiconductor crucibles, lamps, optical components, fiber optics	1600	0.5
Alkali lead	Lamp tubing, sealing	620	9.3
"E" lime aluminosilicate	Fibers	830	6.0
Lime-magnesia aluminosilicate	High temperatures, cookware	900	4.6

Table 2
Approximate compositions (wt%) of the glasses in Table 1

Glass	SiO_2	B_2O_3	Al_2O_3	CaO	MgO	PbO	Na_2O	K_2O
Soda lime	72	1	2	5	4		15	
Pyrex borosilicate	81	13	2				4	
Vitreous silica	100							
Alkali lead	77			1		8	9	5
"E" lime aluminosilicate	55	7	15	21			1	
Lime-magnesia aluminosilicate	64	5	10	9	10		1	1

power laser optics. However, beryllium fluoride is highly toxic and the glasses are sensitive to moisture.

A new series of fluoride glasses based on zirconium fluoride and other heavy metal fluorides are optically transparent in the infrared below about 8 μm. Thus they have potential for application as optical fibers for transmission of information and as infrared optical components and sensors.

Calcogenide glasses, based on sulfur, selenium and tellurium compounds rather than oxides, also transmit in the infrared and are also electronic conductors (semiconductors). Various kinds of switches and devices can be made of these glasses.

See also: Float Glass Process; Fracture and Fatigue of Glass; Glass as a Building Material; Glazing

Bibliography

History of glass

Charleston R J 1980 *Masterpieces of Glass: A World History from the Corning Museum of Glass.* Abrams, New York

Douglas R W, Frank S 1972 *A History of Glass Making.* Foulis, Henley-on-Thames

Morey G W 1954 *The Properties of Glass*, 2nd edn. Reinhold, New York, Chap. 1

Neri A 1612 *L'Arte Vetraria.* Florence, Italy, and many subsequent editions and translations

Polak A B 1975 *Glass: Its Tradition and Its Makers.* Putnam, New York

Zerwick C 1980 *A Short History of Glass.* Corning Museum of Glass, Corning, New York

Monographs on glass

Babcock C L 1977 *Silicate Glass Technology Methods.* Wiley, New York

Doremus R H 1973 *Glass Science.* Wiley, New York

Dunken H H 1981 *Physikalische Chemie der Glasoberfläche.* VEB Deutscher Verlag für Grundstoffindustrie, Leipzig

Holloway D G 1973 *The Physical Properties of Glass.* Wykeham, London

Izumitani T S 1984 *Optical Glass.* Kyoritsu Shuppan, Tokyo (Translation to English in preparation)

McMillan P W 1979 *Glass-Ceramics*, 2nd edn. Academic Press, New York

Morey G W 1954 *The Properties of Glass*, 2nd edn. Reinhold, New York

Mott N F and Davis E A 1979 *Electronic Processes in Non-Crystalline Materials*, 2nd edn. Clarendon Press, Oxford

Scholze H 1965 *Glas: Natur, Struktur, und Eigenschaften.* Vieweg, Braunschweig

Vogel W 1979 *Glaschemie.* VEB Deutscher Verlag für Grundstoffindustrie, Leipzig

Volf W B 1961 *Technical Glasses.* Pitman, London

Wong J, Angell C A 1976 *Glass Structure by Spectroscopy.* Dekker, New York

Zallen R 1983 *The Physics of Amorphous Solids.* Wiley, New York

Zarzycki J 1982 *Les Verres et L'Etat Vitreux.* Masson, Paris

Collections of review articles

Mackenzie J D (ed.) 1960, 1962, 1964 *Modern Aspects of the Vitreous State*, Vols. 1–3 Butterworth, London

Pye L D, Stevens H J, LaCourse W C (eds.) 1972 *Introduction to Glass Science.* Plenum, New York

Tomozawa M, Doremus R H (eds.) 1977, 1979, 1982, 1985 *Treatise on Materials Science and Technology*, Vols. 12, 17, 23, 26, *Glass I, II, III, IV.* Academic Press, New York

Uhlmann D R, Kreidl N J (eds.) 1980–1987 *Glass: Science and Technology*, Vols. 1–5. Academic Press, New York

Reference works

Bansal N P, Doremus R H 1986 *Handbook of Glass Properties.* Academic Press, Orlando, Florida

Mazurin O V, Streltsina M V, Shvaiko-Shvaikovskaya T P 1983, 1985 *Handbook of Glass Data*, Parts A, B. Elsevier, Amsterdam

Mazurin O V 1971–1980 *Properties of Glass and Glassy Melts*, Vols. 1–4 (in Russian). Science, Leningrad

Journals on glass

Glass Technology and Physics and Chemistry of Glasses. Society of Glass Technology, Sheffield

Glastechnische Berichte (in German). Deutsche Glastechniche Gesellschaft, Frankfurt Main

Journal of the American Ceramic Society. American Ceramic Society, Columbus, Ohio

Journal of Noncrystalline Solids. North Holland, Amsterdam

Soviet Journal of Glass Physics and Chemistry (in Russian, also translation into English). Consultants Bureau, New York

Yogyo Kyokaishi, Journal of the Ceramic Association of Japan (mostly in Japanese, with English abstracts). Ceramic Association of Japan, Tokyo

R. H. Doremus
[Rensselaer Polytechnic Institute, Troy, New York, USA]

Glass as a Building Material

The window, a word believed to be derived from the Norse meaning "eye of the wind," was originally just an open hole, but over the centuries it has become a barrier between the internal, artificial environment and the weather. However, it is only in the twentieth century that architecture and glass manufacturing technology have developed glasses which can control and enhance many aspects of the environment. Large and transparent windows can now connect living spaces to the outside and also insulate those spaces. The worst effects of solar heat gain can be mitigated by glasses that either absorb or reflect the heat. Silent internal environments can be created.

Windows can be made fire-resistant as well as bulletproof. In recent years, they have proved able to conserve energy and reduce the need for some internal energy sources. The twentieth century has also seen developments that are not related to the basic manufacturing method for glass. The future will bring greater development of processed glass, coatings and combinations of glasses with different properties, and greater use of the natural properties of glass; all of these will improve the standard of living.

1. Types of Glass

The major glasses available today for architectural use are as follows.

(a) Clear transparent glass. In the first part of the twentieth century, manufacturers concentrated on more economical ways of producing flat, transparent glass. First came the crown process, then the cylindrical process, various methods of drawing glass vertically (the sheet process), grinding and polishing rolled glass (the polished plate process) and, most recently, floating glass on molten metal (the float process). Each of these processes has successively improved the quality of glass. Clear transparent glass can be manufactured up to 25 mm thick, and thicker polished plate glass is still available. The size of glass is determined by the building design, energy conservation considerations and loading requirements.

(b) Clear translucent glass. This glass bears an impressed pattern on one or both surfaces. Such textured glass is available in a variety of designs and provides various degrees of privacy and internal natural lighting.

(c) Colored transparent glass. This glass, colored during manufacture by very small modifications to the basic composition, is usually made as float glass. The most popular colors are gray, bronze, and green; the precise color depends on the composition of the glass.

(d) Colored translucent glass.

(e) Transparent polished, wired glass. In this glass, wire, in a mesh or in another configuration, is incorporated during manufacture. Such wire is centered in the thickness of the glass; this makes the product fire and impact resistant. It is made as cast glass and the resulting pattern is polished away to give transparency.

(f) Translucent wired glass. Patterned, wired glass offers privacy as well as the attributes described above.

(g) Colored translucent wired glass. Such glass is intended for internal use, because it is particularly weak when subjected to thermal stress.

(h) Colored surface-modified transparent glass. Metallic ions are injected near one surface to produce color.

(i) Colored, coated transparent glass. While types (a–h) are manufactured "on-line," this type can also be produced when glass already cut from the basic manufacturing line is subsequently coated. In the future, most technological advances will affect surface treatments. Initially the enhanced reflection properties of the coatings were merely an additional method of controlling solar heat gain. However, recent developments have led to new coatings which modify the surface coefficient of convective and radiative heat transfer. Such coated glasses can influence the insulation of the facade of the building.

In addition to the basic products, additional types of glass have been developed using three secondary manufacturing processes.

(a) Toughening (tempering). Glass is heated; under controlled cooling, residual stress is frozen in. The result is a stronger, safer product which, when broken, fractures into small, relatively harmless pieces.

(b) Laminating. Glasses are bonded together with plastic interlayers (or more recently, special inorganic interlayers) to provide glass that is difficult to penetrate.

(c) Insulation glazing. Glass is incorporated into a hermetically sealed, factory-made multiple glazing unit which provides thermal insulation. Special low-emissivity coatings or particular gases enhance the insulation ability. Gases can also give increased acoustic insulation.

The range of functions which can be provided by these secondary processes is immense. Almost any requirement can be met.

A third group of glasses offers particular attributes not found in ordinary flat glass. Included in this group are opal glasses, antique glasses, borosilicate glasses, bullions, channel glass, copper lights, glass blocks, glass domes, leaded lights and lenses (roof and pavement).

2. Performance of Glass

Glass must meet a range of design specifications appropriate to the particular building. Often an analysis of the needs leads to contradictory requirements, and compromises are required. No one requirement is usually sufficient to specify the glass to be used.

Research in many diverse fields is continuing, so that relatively simple design rules can be formulated. Each subsection below has its own field of research. Many publications, including national standards and regulations, assist in understanding local requirements for glass performance.

2.1 Natural Lighting

This is one of the most primitive reasons for using glass in windows. Although artificial lighting has been relied upon within buildings, concern for energy conservation is encouraging a return to natural lighting and a better integration of this with artificial lighting.

There are well-established methods of predicting the natural light that will enter a building, and recommendations for relating lighting levels to the building type and the task to be performed.

The design procedures for natural light are often separated into considerations of sunlight and of daylight. Glare also needs to be accounted for and separate consideration is usually given to (a) direct glare from the sun, (b) sky glare and (c) reflected glare from adjacent water or other highly reflecting surfaces. Although sky glare and reflected glare can be controlled by glass with special transmittance properties, direct glare can be reduced only by permanent shading devices.

2.2 Thermal Performance

Heat passes through glass by conduction and by radiation in both the visible region and the near infrared. However, glass is opaque to far infrared radiation produced by low-temperature objects. Some of the incident solar radiation is absorbed and reradiated both inwards and outwards; the proportions depend on the wind speed over the outer surface.

2.3 Thermal Insulation

Glass is a relatively poor conductor of heat and it is usually used in buildings as a thin membrane. Nowadays it is more often integrated into insulating glass units which provide greater insulation. Figure 1 indicates how the insulation changes with the width of the air gap; it also shows the benefits of low-emissivity metallic coatings and insulating gases.

The degree of exposure to wind affects the actual thermal insulation obtained, and design guidelines usually divide exposure into three categories indicating sheltered, normal and exposed building sites.

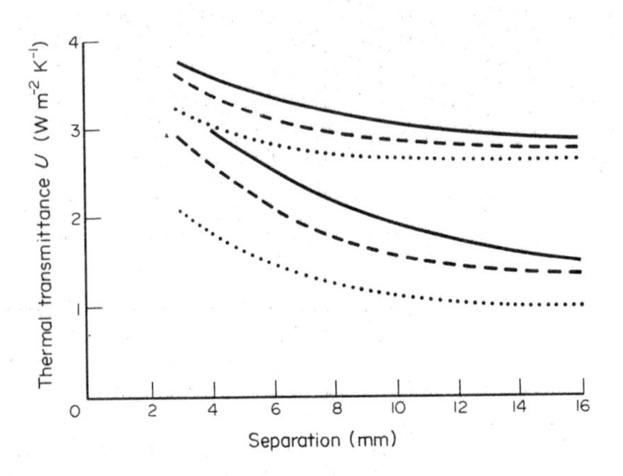

Figure 1
Effect of distance between glass sheets, low-conductivity gases and low-emissivity metallic coating on thermal transmittance of double glazing: —— air; --- argon; · · · krypton; lower set of curves is for low-emissivity coated glass

2.4 Solar Heat Gain

The transmission properties, including reradiation from the glass itself, can create a significant amount of heat within buildings. This solar heat gain is the reason for moves to replace heat absorbing glasses with heat reflecting glasses. Methods of designing the glass and determining its effects on air conditioning loads in terms of glass size, location and thermal performance are available.

2.5 Energy Balance of Windows

Until recently, the thermal insulation aspects and solar heat gain aspects were rarely considered complementary benefits of windows. However, with the growing interest in energy conservation, computer techniques have been developed to assess the energy consumption of buildings based on the type of glass, window orientation and building size for any specific form of construction.

2.6 Assessment of Thermal Safety of Glass

In certain special, and infrequent, climatic conditions, all glass types, except toughened glass, are vulnerable to thermal breakage. Such breakage occurs because of the tensile stress from the cooling effect of the window frame on the edge of the glass. Prediction methods are available from glass manufacturers, so that building designers can avoid risks by selecting a suitable glazing system (see Fig. 2).

2.7 Thermal Expansion

Soda-lime glass of all colors, toughened or not, has a thermal expansion of about $8.0 \times 10^{-6} K^{-1}$. The effects of expansion are allowed for by selecting an appropriate gap between the glass and the frame.

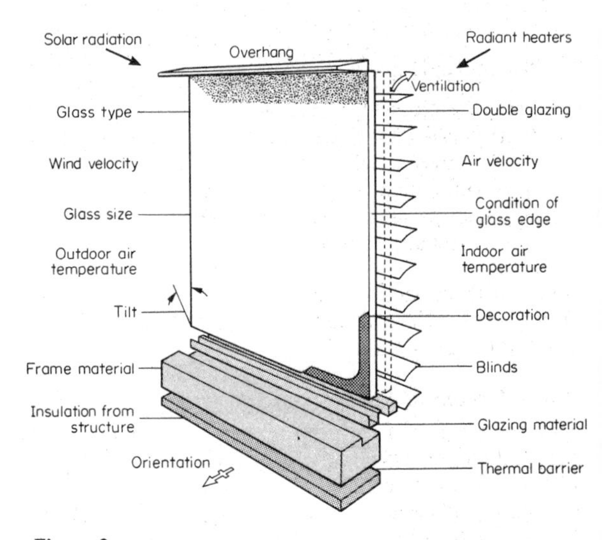

Figure 2
Factors considered in assessing the thermal safety of a glazing system

2.8 Condensation

Condensation on windows is a problem of considerable importance because it can damage building components and thus increase maintenance costs. Methods have been evolved for predicting the risk of condensation as a function of the temperature of the glass, the surrounding air temperatures and the relative humidity.

Because their inner pane is at a higher temperature, hermetically sealed insulating glass units reduce the risk of condensation. Only recently has the added benefit of energy conservation been emphasized.

2.9 Sound Insulation

The reduction of transmitted noise requires an assessment of

(a) the noise source,

(b) the human response and

(c) the glass and its surround.

Each of these elements is frequency dependent; some sounds are more objectionable than others.

Glass can be used in several ways to provide acoustic insulation. Some of the factors for assessing the characteristics of different glasses are:

(a) for single glazing:

- (i) mass per unit area, in other words, glass thickness;
- (ii) coincidence effect;
- (iii) closure or sealing (one of the most important);
- (iv) edge support conditions;
- (v) lamination (an acoustic absorbing interlayer); and
- (vi) glass dimensions;

(b) for multiple glazing, in addition:

- (vii) the pane separation;
- (viii) the acoustic insulation of the surroundings between the glass panes;
- (ix) differences in glass thickness;
- (x) nonparallel panes (not usually practical); and
- (xi) mechanical separation.

Some of these factors operate over a narrow band of frequencies, while others can effect the whole frequency range (see Fig. 3).

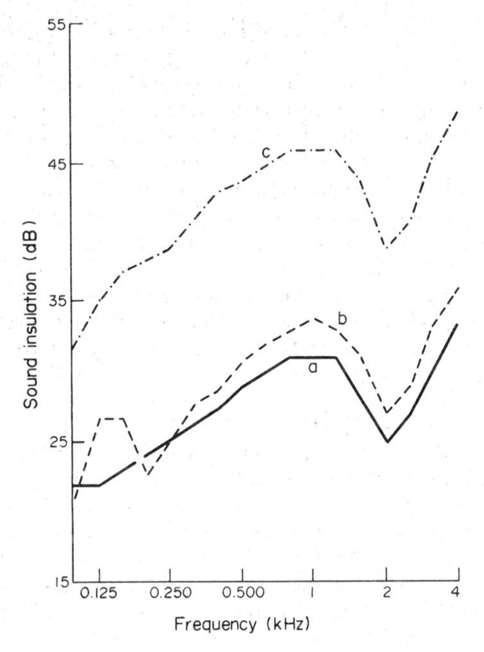

Figure 3
Variation in sound insulation of glazing with frequency, showing effect of double glazing and width of air space: (a) 6 mm single glazing; (b) 6–12–6 mm double glazing; (c) 6–200–6 mm double glazing

2.10 Mechanical Strength of Glass

Simply, glass can be considered as a pure brittle material whose surface is covered with small fissures—Griffith flaws. There has been much research on the strength of glass as a material, but relatively little has been published in relation to its use in windows. The behavior of glass in glazing has only recently begun to be understood. Because window glass is flimsy, it is subjected, even under moderate wind loads, to membrane stressing. Although much is now understood of how glass behaves under the short duration pressure loading experienced in a wind storm, most published wind loading design charts are based on an approximation. These charts avoid the need for designers to resort to complex computer techniques.

The mechanical behavior of glass in a window depends on:

(a) the magnitude of the loading;

(b) the size and thickness of the glass;

(c) the duration of the load and the time for which any part of the pane is subjected to a particular stress; and

(d) the support conditions (although most windows are considered to contain glass which is simply supported, some resistance to rotation exists at the edges and some edge deflection occurs).

With glasses combined in insulating glass units, or laminated glass, a more complex partition of loading occurs. Toughened glass behaves in a similar way to other forms of single glass, but it is up to five times as strong as normal float glass.

For building design, the thickness of glass depends on the predicted maximum wind load. However, the precise relationship between load and glass area varies from country to country, and depends on the basis of the national wind loading code—in particular, the choice of load duration and the statistical return period of such loads.

In many building uses, the maximum deflection of the glass needs to be controlled, not for structural integrity but to allow the glass to appear safe to onlookers. Glass can be used safely in large thin panes, particularly where the wind loading is low, but a certain thickness is needed to cope with any accidental impacts.

2.11 Safety Glass

In some specifically defined risk areas, glass in buildings is subject to accidental human impact. In such situations, glass which either will not break under the expected impacts or breaks safely when hit is recommended for use in many national building regulations or standards.

Throughout the world, the major risk area is in the home (about 2.5% of all home accidents can be attributed to flat glass). The major hazardous situations are:

(a) glass in sliding or hinged doors;

(b) glass in adjacent panels which can be mistaken for doors;

(c) glass fitted near ground level, particularly where many people use the adjacent areas;

(d) glass in balustrade panels; and

(e) glass near swimming pools, where people can slip on wet surfaces.

The appropriate glass for such areas is usually defined by its ability to perform satisfactorily under a simulated human impact. The US test method known as "Z97" has been accepted in slightly modified forms throughout almost the whole world; this test defines glasses that break safely under prescribed impact conditions. The prime examples of safety glasses are toughened glass and laminated glass, but other glass types have also met the particular national standards, for example wired glass, the prime function of which is to provide fire resistance.

2.12 Fire Resistance

In almost all buildings, glass is required to be used in areas where resistance to the passage of flames is required. Although too much glass can increase the fire risk, in some situations the use of glass is advantageous. For instance, a transparent area in a fire door can give advance warning of the presence of a fire. In such areas, glass must also be able to contain the fire for a specified period (30 minutes being a typical requirement).

Traditionally, wired glass has been used as fire-resistant glass, as a consequence of its success in meeting the appropriate specification over many decades. However, like all transparent glass, it transmits radiation created by fires, and this in itself can spread the fire. To overcome this drawback, special laminated glasses have recently been developed. They are normally transparent but become opaque to radiation when subjected to the high temperatures which occur near fires.

All soda-lime glasses are incombustible and will not themselves contribute to a fire. Hence, fires cannot spread by burning of the glass surface; this is particularly important on the outside of buildings and on roofs.

2.13 Security

Security glass is required more today than ever before. Such glass, which is almost always complex laminated glass, is classified in terms of its ability to withstand impacts from bullets, malicious attack or explosions.

2.14 Glass Durability

Glass made from the standard soda-lime composition is highly resistant to deterioration. On site, glass surfaces can be attacked if allowed to remain wet, and the effects are very difficult to remove except by polishing the surface. Glass should therefore be stored in dry, warm conditions. Glass is attacked by alkaline solutions, such as the effluent from rain water draining over concrete, and by welding "splatter."

2.15 Control of Reflections

Reflections can reduce a clear, uninterrupted view through a shop window. All glass reflects light, and the degree of reflection depends on the angle at which the glass is viewed. By positioning the glass at an angle to the general direction of viewing, or by using horizontally curved glass, such reflections can be avoided. The extent of the problem also depends on the relative lighting levels on the two sides of the glass. Wherever possible, the lighting level should be higher in the area to be viewed.

In other buildings, highly reflective glass creates privacy, as long as the lighting level on the outside of the building is greater than the level inside. Similarly, privacy can be obtained by lowering the lighting level in a private area. Security viewing windows work in this way.

2.16 Antifade Properties of Solar Control Glasses

Although fading of colors is attributed to the effects of ultraviolet radiation, modern organic pigments can also be affected by wavelengths in the visible spectrum and by heat. Only 3% of solar radiation reaching the earth's surface is contained within the uv range (10–380 nm) and normal clear float glass rejects about 50% of this; a polyvinyl butyral interlayer in laminated glass rejects about 95%.

Some research has been carried out on how tinted glass can reduce the risks of fading, but for irreplaceable works of art, windows need to be located so that direct solar radiation does not fall on the object.

2.17 Cleaning and Maintenance of Glass in Buildings

To preserve its surface, glass should be cleaned regularly. Dust can reduce the thermal and visual performance of glass, increase its thermal absorption and reduce its life. In extreme cases, the glass may become discolored. Glass should be cleaned with nonalkaline solutions.

3. Installation of Glass

Glass is installed in numerous ways, some being suitable only for certain climates. In areas of severe weather, more sophisticated methods of glazing are required. In some parts of the world, different factors, such as temperature rather than driving rain, are more important.

The current methods have evolved from tradition, and only in the last 20 years have internationally accepted methods of fixing glass become prevalent. Changes in glazing systems have also resulted from the changes in the materials used for window frames. First wood, then aluminum and steel were common. Now that plastic frames are gaining in use, coping with expansion has become a major problem.

With the introduction of adhesive glazing compounds which have considerable strength as well as an ability to survive on buildings for more than 15 years, interest is growing in locating glass in preformed openings without any mechanical fixings. Structural integrity can be maintained solely with these compounds.

Most countries have published guidelines to ensure that glazing is carried out to a sufficiently high quality.

3.1 Glazing of Single Glass

The major requirements of the frames are:

(a) to cope with the quality of glass cutting and the resulting variations in size and shape;

(b) to prevent glass–hard metal contact, as this could lead to glass fracture;

(c) to allow for the differences between the thermal expansions of the glass and the frame material;

(d) to allow the glazing compound to survive (many compounds require a considerable thickness of fillets to withstand temperature and humidity); and

(e) to withstand the weight of the glass and the structural effects on the glass of the many environmental factors (e.g., wind loading).

These requirements have been translated into simple glazing rules which have been almost universally adopted:

(a) The glass size should be about 3 mm less than the dimensions of the space within the window frame. This difference is increased as the glass size and thickness increase.

(b) The glass should rest centrally on two blocks set at about the quarter points. The width of these blocks corresponds to the glass thickness; their height is adjustable. As frames have become lighter in construction, the position of the "setting blocks" has been adjusted and now they can be positioned anywhere away from the corner of the glass, but they should preferably be close to the quarter points.

(c) In the past, no location pieces were used between the vertical edges of the glass and the frame. Now that glass is transported already glazed, side location pieces are necessary to prevent glass–frame contact.

(d) Clearance is ensured between both the front and back faces of the glass and the frame, by using distance pieces. The gap allowed depends on the glazing methods but is rarely less than 3 mm.

Nowadays there must be additional considerations. Traditionally, a very hard glazing compound, linseed oil putty, has been used to hold the glass in position; over many years, its suitability has been proved in countless buildings. However, putty sets very hard and is not able to cope with glass expansion. With the increasing use of, for example, heat-absorbing solar control glasses, glazing compounds have been introduced. The same requirements apply, but the importance of distance pieces is increased.

3.2 Glazing of Sealed Insulating Glass Units

For hermetically sealed, insulating glass units, a new, additional set of glazing parameters has arisen. First, a further source of glass movement was introduced. The sealed airspace in the unit changes volume as the external pressure and temperature change. Suitable glazing compounds have been developed to cope with the resulting edge rotation.

An even more important design requirement soon became evident. In almost every case, the life of sealed insulating glass units depends critically on diversion of water from the edge of the sealed unit. At first, full bedding of the unit was carried out, but the creation of a satisfactory long-life seal depends on choosing suitable materials for such thick volumes of glazing compound and achieving the necessary glazing quality and workmanship. However, at about the same time it was found that the accuracy with which aluminum frames could be extruded allowed for the incorporation of channels to hold preformed rubber-based gaskets. The use of such dry methods opened up a new era of glazing design. On-site work could be reduced to a minimum and consistent glazing quality was easier to achieve.

3.3 Drained Glazing Systems

Such systems are now almost universally accepted as the best means of glazing. Although the principles are easy to describe, achievement in practice is difficult to perfect. The system is designed so that any water entering behind the outside glass–frame seal is allowed to drain out of the frame rather than being forced into the building. Because the chamber beneath the glass must always remain at external air pressure, the seal between the inside face of the glass and the frame must be pressure-proof. Thus, no water enters the building and any water getting past the outer seal drains away. The glass edge is not subjected to continuous immersion in water, a situation which would considerably reduce the life of insulating glass units.

The two most popular methods for drainage are (a) beads which directly connect to the outside and (b) internal channels. However, experience shows that the problems associated with drainage should not be underestimated. Not only is good initial design necessary, which takes into account erection tolerances, but also a high quality of erection skill is required, particularly in terms of achieving the essential inner waterproof seal.

See also: Glass: An Overview; Glazing, Structural Sealant Glazing

Bibliography

American Society of Heating, Refrigeration and Air Conditioning Engineers *ASHRAE Guide*. ASHRAE, Atlanta, Georgia

Charles R J 1959 The strength of silicate glasses and some crystalline oxides. In: Averbach B L, Felbeck D K, Hahn T T, Thomas D A L (eds.) 1959 *Fracture*. Wiley, New York

Chartered Institution of Building Services *CIBS Guide*. CIBS, London

Collins B L 1975 *Windows and People: A Literature Survey—Psychological Reaction to Environments with and without Windows*, Building Science Series 70. National Bureau of Standards, Washington, DC

Glass and Glazing Federation 1978 *Glazing Manual*. GGF, London

Hastings S R, Crenshaw R W 1977 *Window Design Strategies to Conserve Energy*, Building Science Series 104. National Bureau of Standards, Washington, DC

Hopkinson R G, Petherbridge P, Longmore J 1966 *Daylighting*. Heinemann, London

Lynes J A 1968 *Principles of Natural Lighting*, Elsevier Architectural Science Series. Elsevier, New York

McGrath R, Frost A C 1961 *Glass in Architecture and Decoration*. Architectural Press, London

Pilkington Environmental Advisory Service 1976 *Glass and Noise Control*. Pilkington Flat Glass, St Helens

Pilkington Environmental Advisory Service 1981 *Calculating Thermal Transmittance*. Pilkington Flat Glass, St Helens

Pilkington Technical Advisory Service 1980 *Glass and Thermal Safety*. Pilkington Flat Glass, St Helens

Turner D P (ed.) 1977 *Window Glass Design Guide*. Architectural Press, London

Walsh J W T 1961 *The Science of Daylight*. MacDonald, London

D. W. Armstrong
[Pilkington Flat Glass, St Helens, UK]

Glass-Reinforced Plastics: Thermosetting Resins

Glass-reinforced plastics (GRP) consist of fibers of low-alkali E glass embedded in a thermosetting plastics matrix, such as polyester or epoxide resin. During fabrication, glass fibers are incorporated into the liquid thermosetting resin to which catalyst and hardener have been added. These cause the resin to polymerize and after curing give a solid plastic matrix reinforced with glass fibers. The fibers enhance the low stiffness and strength of the resin, whose main purpose is to transmit the load into stiffer, brittle fibers and to protect them from damage.

Single E glass fibers, typically 8–15 μm in diameter, are easily damaged and difficult to handle. For protection, they are coated with a silane or polyvinyl acetate size, which binds the fibers into strands containing about 200 filaments and acts as a coupling agent to provide a good fiber resin bond. To facilitate fabrication of GRP components, glass strands are incorporated into rovings, mats and fabrics. Glass fiber rovings consist of up to 120 untwisted strands, usually supplied wound together on a spool and suitable for the unidirectional (UD) fiber reinforcement of resins. Woven rovings (WR) are glass fiber rovings woven into a coarse fabric, usually with a balanced square weave. Glass fabrics are woven on textile machinery from twisted glass fibers (glass yarn) and are available in several weaves such as plain, square, twill and satin. Chopped strand mat (CSM) consists of chopped glass strands about 30 mm long held randomly orientated in a mat by a small amount of resin binder.

1. Glass-Fiber-Reinforced Resins

The principal resins used in GRP materials are unsaturated polyester resins, with epoxides being used in more limited quantities for high-technology applications, and phenolics, furanes and vinyl esters finding use for specialist applications. Polyesters are less dangerous to handle, easier to catalyze and much cheaper than the other resins. They may be cold cured and are suitable for low-pressure, ambient-temperature fabrication techniques. Epoxide resins have better mechanical properties than polyesters, with higher stiffness and strength, particularly at high temperatures, and lower shrinkage on curing. Phenolics and vinyl esters are used in high-temperature applications, with continuous temperatures up to 260 °C, and furanes are noted for their chemical resistance at high temperatures. These specialist resins must be cured at high temperature and GRP components require hot-press molding. The higher performance resins are generally used with UD glass fibers and glass fabric reinforcement, while CSM and WR reinforcements are used almost exclusively with polyester resins. Recent surveys show that polyesters accounted for 92% of resin consumption in the USA and 99% in the UK, with the remainder being mainly epoxide. Thus the important GRP materials in high-volume applications are CSM/polyester, WR/polyester and, to a lesser extent, UD glass/polyester. Glass fabric/epoxide and UD glass/epoxide composites are used in low-volume, high-technology applications.

Glass fiber content is related to reinforcement type and is shown, by weight, in Table 1 for the main GRP materials. A high fraction of aligned fibers (60–80% by weight) can be packed into a UD composite, while more resin is needed to impregnate all the fibers in CSM and WR laminates, giving glass contents as low as 25–35% by weight in CSM/polyester materials. In addition to the fibers, resin, catalyst and hardener, glass/resin composites may contain fire-retardant additives to reduce surface flame spread, uv absorbers to improve outdoor durability, pigments, dyestuffs and thixotropic additives which thicken resins to aid fabrication on inclined surfaces.

2. Glass-Reinforced-Plastic Molding Compounds

Several GRP molding compounds are now available in which the resin and glass fibers are premixed before molding. Component fabrication is achieved by placing the uncured compound in a closed mold and hot-press molding. The three main types of molding compound are sheet molding compound (SMC), dough molding compound (DMC), sometimes referred to as bulk molding compound (BMC), and preimpregnated glass fiber sheet (prepreg).

Sheet molding compound consists of E glass fibers reinforcing a mixture of catalyzed polyester resin and a mineral filler, such as particulate, chalk, limestone or clay. The filler thickens the resin sufficiently for uncured sheets of SMC to be stored, handled and cut to the correct size for molding. Sheet molding compound usually contains chopped glass strands 20–50 mm in length randomly orientated in the plane of the sheet and is supplied in several nominal glass weights, containing 20, 25, 30 or 35% glass fiber, with filler contents of 30–50% by weight. Fillers reduce material costs and modify the handling characteristics of the SMC sheet and the flow properties in the mold. High-performance SMC materials may also contain aligned chopped glass fibers or continuous fibers, at glass contents up to 70% by weight. These high-performance materials, sometimes termed HMC or XMC, are being used in automotive and other demanding high-volume applications. Their further development and a rapid growth in their usage are expected.

Table 1
Typical short-term mechanical property data for GRP materials

		DMC	SMC	CSM/ polyester	WR/ polyester	UD glass/epoxide longtitudinal	UD glass/epoxide transverse
Glass content (wt%)							
	average	20	30	30	50	70	
	range	15–25	20–40	25–35	45–60	60–80	
Modulus (GPa)							
tensile	average	9	13	7.7	16	42	12
	range	8–11	9–16	6–9	12–22	30–55	8–20
flexural	average	8	11	6.3	13		
	range	7–9	7–14	5–8	10–18		
shear		3	3	3	4	5	
Strength (MPa)							
tensile	average	45	85	95	250	750	50
	range	35–60	50–120	60–150	195–350	600–1000	25–75
flexural	average	100	180	170	290	1200	
	range	85–120	90–240	100–300	160–500	1000–1500	
in-plane shear	average			80	95	65	
	range			60–100	80–120	50–80	
interlaminar shear	average		15	25	20	40	
	range		12–20	20–30	10–30	30–50	
Impact strength[a] ($kJ\,m^{-2}$)		20–40	50–75	40–80	100–200		

[a] Measured by Izod-type test, unnotched specimen, according to British Standard BS2782, Sect. 306A

Dough molding compound contains catalyzed polyester resin, up to 50% mineral filler and 15–25% short glass fibers, 3–12 mm in length. The usual filler is chalk though some resins are chemically thickened with alkali metal oxides. Shorter fibers and less glass than SMC means that DMC flows more easily and is thus used to make small articles of intricate shape by hot-press or injection molding.

Prepregs consist of sheets of glass fiber reinforcement about 1 mm thick preimpregnated with catalyzed resin and ready for hot-press molding. The reinforcement is usually UD fibers or glass fabric with epoxide rather than polyester resin and component thickness is achieved by laminating together several sheets in the mold. The glass fiber content is high, typically 50–80% by weight. Accurate alignment of fibers is possible with prepregs, because there is minimal flow during molding and hence they are used mainly in high technology applications. Prepreg tapes are also available for selective stiffening of complex moldings and for the fabrication of GRP tubes and vessels by a dry filament winding process.

3. *Sandwich Materials and Laminates*

Different types of GRP material are frequently combined together in laminates or with metals, wood or foamed plastics as sandwich materials. Glass fibers may also be combined with carbon or polyamide fibers in a hybrid composite. A composite structure may thus be built up with material properties tailored to design loads. A widely used GRP laminate consists of CSM combined with WR or UD fibers. The CSM/polyester plies improve the interlaminar shear strength and provide a corrosion barrier to the WR or UD fibers which carry the main loads. To simplify fabrication, glass reinforcement may be supplied in a combination mat, consisting of CSM and WR or UD fibers stitched together. With UD reinforcement such as prepreg, or in filament wound structures, fibers may be orientated in different directions through the thickness. A common construction is a cross-ply laminate, with alternate plies orientated at 0° and 90° to a fixed axis, or an angle-ply laminate with plies at $\pm\alpha°$ for some fixed angle α.

Glass-reinforced-plastic sandwich materials are used widely in panel applications where the main loading is flexural. An ideal sandwich panel should consist of thin stiff skins with a thick low-density or low-cost core. For low and medium technology applications, GRP skins of CSM/polyester or WR/polyester are combined with cores of end-grain balsa wood or foamed plastic such as PVC or polyurethane. A typical construction for a meter square cladding panel might be CSM/polyester skins 3 mm thick with a balsa core 25 mm thick. For high-technology applications where weight saving is a priority, special low-density honeycomb core materials have been developed. These consist of a hexagonal cell structure arranged in a honeycomb pattern as shown in Fig. 1. The main honeycomb core materials are aluminum foil and a thin resin-impregnated paper. High-performance GRP skin materials are used with honeycomb cores, such as glass fabric/epoxide and cross-ply or angle-ply UD glass/epoxide laminates. A typical construction for a meter-square aircraft floor panel might be cross-plied skins 1 mm thick, with an aluminum honeycomb core 10 mm thick.

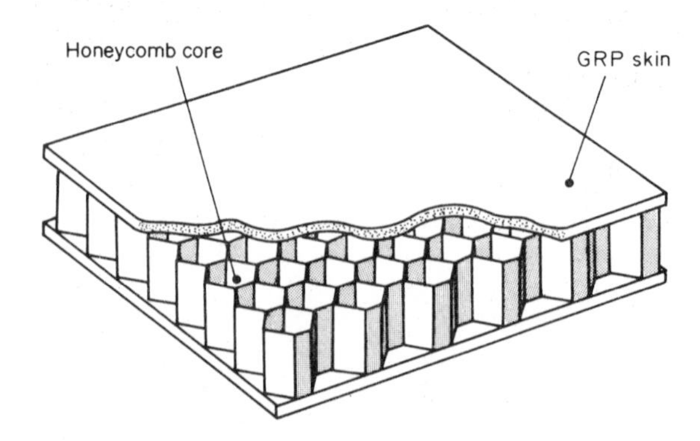

Figure 1
Schematic diagram of GRP/honeycomb sandwich material

4. *Applications of Glass-Reinforced Plastics*

Glass-reinforced-plastic materials are used in low-, medium- and high-stress applications in industries as diverse as boat building, chemical plant, motor vehicles and aerospace. Important properties of these materials which have led to their wide usage are ease of fabrication into complex shapes, high strength to weight ratio, excellent corrosion resistance, good weathering properties, low taint and toxicity for foodstuffs and good thermal and electrical insulation. The selection of a particular material is determined mainly by material properties, costs and the suitability of fabrication and quality-control procedures for use in a particular industry. For large-scale production automated fabrication methods based on filament winding, pultrusion and SMC press molding are used, with hand lay-up and semiautomated methods such as resin injection with preforms retained for smaller numbers of large moldings.

In low-stress applications such as cladding panels, shower cabinets, small boats and meter boxes, GRP materials act as space-filling panels, supporting their own weight but not subjected to any significant external loads. Stress analysis is not usually carried out and the main design requirements are a good surface finish, good weathering properties and the ability to withstand accidental damage. The most widely used material is CSM/polyester with fabrication by hand lay-up, spray-up or vacuum-bag molding. Sheet molding compound materials are used for longer production runs of larger moldings such as business-machine cases and cladding panels, and DMC for smaller components such as electrical switchgear, lampholders and meter boxes. Glass-reinforced-plastic sandwich materials containing balsa or foamed-plastic core with CSM/polyester skins are used for boats and for some of the larger panel applications.

In medium-stress applications, GRP materials may be subjected to significant loads of short duration or low sustained loads. Stress analysis is usually based on the short-term behavior of the material and product design may be governed by codes of practice or standards. Large design factors are imposed so that stresses in the material are often less than 10% of the short-term strength. Applications in this category include: pressure vessels and storage tanks with associated pipework, for use in chemical plant and the food processing industry; larger marine applications such as workboats, small hovercraft, minesweepers and small submersibles; and land-transport applications such as rail coachwork and larger automotive moldings for car bumpers, body panels and truck cabs. Resins are usually polyester, reinforced by WR or UD fibers with CSM/polyester surfaces for improved weather and chemical resistance. Most of the vessels, tanks and the marine applications are fabricated by hand lay-up, with filament winding becoming increasingly important for chemical plant pipework, sewage pipes and cylindrical tanks and pressure vessels. Sheet molding compound is used for the volume production of such items as car bumpers, lorry cab panels and rail seat shells. Structural GRP sections are becoming increasingly important; for example, box beams, channels and tubes fabricated by pultrusion for use in frame structures, ladders, walkways, ducting, etc., and hollow rotationally-molded tapered sections for the volume production of GRP street lighting columns. Where low weight is required, as in storage tank covers, hovercraft body panels and rail carriage doors, GRP/foam sandwich materials are used with fabrication by hand lay-up or vacuum-bag methods.

The high-stress applications of GRP materials are in the aerospace and defense industries. Here, low weight is important and hence materials are highly stressed and used efficiently. Design factors are kept as low as 1.5 in some applications and detailed stress analysis is thus required. Typical applications include power boats, gliders, which are fabricated by hand lay-up using WR/polyester or UD glass/polyester, and helicopter rotor blades and rocket motor casings, which are filament wound usually with epoxide resin. Sandwich materials consisting of angle-ply prepreg GRP skins with foam or honeycomb cores find use in lightweight structures such as aircraft flooring, radomes and radar antennae.

5. *Fabrication by Hand Laminating*

Hand laminating or contact molding is the traditional fabrication technique for GRP materials. It is the method used for about 50% of the UK and 30% of the US consumption of GRP materials. Glass fiber reinforcement, in the form of CSM, WR, fabric or combination mat, is placed on a mold, impregnated with catalyzed resin and consolidated by hand rolling or brushing. The required thickness is built up by adding further layers of reinforcement and resin. Alternatively, chopped glass rovings are sprayed with a controlled quantity of resin from a special gun onto the mold and then consolidated by hand. The resulting "spray-up" material has a random distribution of fibers in the plane of the molding, giving similar mechanical properties to CSM/polyester. After an initial gelation stage, the GRP molding may be removed from the mold and left to cure, during which time the material attains its full mechanical properties. Curing takes about a week at 20 °C, but is accelerated by postcuring at higher temperatures. Typical postcure times vary from 3 h at 80 °C to 30 h at 40 °C, and depend on the resin used.

Contact-molded laminates have a single molded surface usually protected by a 0.5 mm thick resin layer termed the gel coat, which is brushed or sprayed onto the mold before the main laminate. For additional protection, the gel coat may be backed by a thin glass surfacing tissue. The main advantage of hand laminating is its versatility. There is no size limitation and metal inserts or additional glass reinforcement can be added where they are needed. Because it is a low-pressure, ambient-temperature process, inexpensive molds of wood, plaster or GRP may be used, which makes it suitable for small production runs. The main disadvantages are that moldings usually only have one good surface and the process is labor intensive with quality dependent on the laminator. Control of glass content, thickness and the cure schedule is poor, which can lead to variability in mechanical properties in a single molding or between moldings in a batch.

6. *Automated Fabrication Techniques*

To eliminate the variability in hand laminating and to permit longer production runs, several automated and semiautomated fabrication techniques have been developed. The semiautomated methods are based on matched die molding which permits critical control of thickness and glass content, and produces a molding with two good surfaces. Cold-press molding makes use of relatively inexpensive tools in a hydraulic press with low pressures up to about 200 kPa. Glass fiber reinforcement such as CSM is cut or preformed to shape and placed by hand in the mold with the required quantity of catalyzed polyester resin. The mold is closed and the resin allowed to cure for about 5 min.

Two variations on cold-press molding are vacuum-bag molding and resin injection. In the vacuum-bag method, the matched molds, or an open mold, are sealed inside a rubber bag, which is evacuated so that atmospheric pressure is exerted over the surface of the molding. Resin injection or transfer molding involves the injection of catalyzed resin into the closed mold containing a preformed glass fiber mat. Injection pressures are up to 200 kPa and may be combined with vacuum assistance to improve resin flow. This fabrication method is becoming increasingly important, since it allows accurate placement of reinforcing fibers in the molding and is also suitable for the production of large moldings such as car body shells.

In hot-press molding, matched metal dies are used and a hot curing polyester resin. Pressures are 1 MPa and temperatures in the range 100–130 °C, with a mold cycle time of 2–5 min. Because of the higher pressures, entrapped air is forced out of the mold giving lower void content than with cold-press molding. In conventional hot-press molding, the operator must still handle the glass fibers and the liquid resin, but the process can be made cleaner by using molding compounds. These have a high viscosity and thus require higher molding pressures, typically 1–7 MPa, with temperatures in the range 120–170 °C. Prepregs are cut accurately and each ply is positioned in the mold to have the required orientation. Because of the short fibers, SMC, and more particularly DMC, are able to flow into the mold. With these materials, a weighed charge is positioned by hand centrally in the mold and on closing flows into the mold causing significant fiber reorientation. Sheet molding compound has the better mechanical properties and is used in larger panel applications, with DMC retained for smaller, bulkier components. The use of in-mold coatings to improve surface finish and fast curing resins giving cycle times as low as $1\frac{1}{2}$ min, has meant that SMC is now widely used for the volume production of automotive components such as body panels and bumpers. Because of the good flow characteristics, DMC components may also be fabricated by injection molding.

Fully automated methods have been developed for the large volume production of GRP pipe, sheet and rod or tube stock. In the filament winding process, glass fiber rovings are fed under tension through a resin bath onto a revolving mandrel. Several layers of reinforcement, with the same or different winding patterns, may be placed on the mandrel in a controlled way. The mandrel is removed after curing, which usually takes place at elevated temperatures. A variation, which eliminates the resin bath, is dry winding using UD prepreg tape. Filament winding is used for GRP pipes and cylindrical tanks and enables high glass contents to be achieved with close control of fiber orientation. In order to improve chemical resistance, the glass fiber reinforcement may be wound over a chemically resistant liner of CSM/polyester or a thermoplastic such as PVC or polypropylene.

Continuous laminating processes have been developed for the manufacture of GRP sheet for applications such as electrical insulating panels, corrugated panels and translucent roof sheets. A moving sheet of CSM or WR reinforcement is impregnated with catalyzed polyester resin and sandwiched between layers of plastic film. This is then passed continuously through a die, to impart the required shape, and a curing oven before being cut into lengths. Pultrusion is a similar process suitable for the production of long lengths of GRP in such forms as rods, tubes or I beams. In pultrusion, glass rovings are impregnated with resin from a bath or by injection and pulled through a shaped, heated die. Other forms of glass reinforcement such as WR or continuous-filament mat may also be pulled through the die with the UD rovings to improve the transverse mechanical properties of the pultruded section. Another recently developed automated process is rotational molding, which is used for hollow lighting and telegraph poles. Here the laminating pressure is derived from centrifugal force in the rotating mold. Methods are also being developed to automate hand lamination, for example, by machine-controlled spray-up.

Automated fabrication techniques are suitable for volume production of good quality components through careful control of glass and resin contents, fiber orientation and the cure cycle. Many of these fabrication methods are suitable for control by microprocessor, which can lead to further cost savings and help overcome the styrene emission problem by eliminating manual operations. The main disadvantage is the high capital cost of the processing equipment. Some of the versatility of GRP materials is also lost, since these methods produce GRP stock in the form of pipes or sheets which require bonding or jointing to make into a component, whereas contact or press moldings are often integral GRP structures.

7. *Glass-Reinforced-Plastic Material Properties*

On loading a GRP material such as CSM/polyester in a short-term tensile test, it is found that the behavior is approximately linearly elastic to failure, which occurs by a brittle fracture at a strain of 1.5–2.5%. The short-term mechanical properties may thus be characterized by an elastic modulus and a failure stress or strength. A more detailed study of failure (see *Composite Materials: Strength*) shows that there is progressive damage to the material at stresses below the ultimate. The first sign of damage to the material at stresses below the ultimate. The first sign of damage under tensile loading is transverse fiber debonding, that is, separation takes place between the resin and those fibers perpendicular to the load direction. In CSM/polyester and WR/polyester, this occurs typically at 0.3% strain and at a stress level of about 30% of the tensile strength. Under increased loading, the debonds initiate resin cracks and as these cracks spread, more of the load is transferred to the fibers which eventually fracture or pull out at ultimate failure. Resin cracking occurs at approximately 50–70% of the tensile strength and is observed as crazing and by a change of slope in the stress–strain curve. In this article, ultimate failure stress data are given although it should be noted that, for many applications, failure data based on alternative criteria such as resin cracking or fiber debonding may be necessary. Limited data of this type are given, for example by Johnson (1978), or alternatively a suitable design factor may be applied to the ultimate strength data.

With a knowledge of the appropriate moduli and strengths, elastic design formulae may be used for stress and deformation analysis of GRP materials under short-term loads. For CSM/polyester and SMC materials, conventional isotropic design formulae may be used, while for WR, fabric or UD reinforcement, account must be taken of the anisotropy in stiffness and strength properties arising from fiber orientation. For GRP materials, it is not usually possible to quote a single value of modulus and strength because of the inhomogeneous structure of the material. Thus, in addition to the effects of anisotropy, mechanical properties also depend on the quantity and type of glass reinforcement and on the fabrication method used.

Comparative short-term mechanical property data are summarized in Table 1 for the principal GRP materials and typical physical properties are given in Table 2. The tables are abstracted from Johnson, who has collated material property data on several types of GRP material from UK resin and glass fiber suppliers, GRP fabricators and research laboratories. Table 1 gives mean values of modulus and strength for DMC, SMC, CSM/polyester, WR/polyester and UD glass/epoxide at a typical glass content for each type of material. Below each mean value approximate bounds are given for the quantities. These bounds are usually wide because they refer to a full range of glass weight fractions, to different resins and glass fibers, fabrication techniques and test methods.

Reference to Table 1 shows that DMC, SMC and CSM/polyester, which all contain chopped fibers at low glass contents, have corresponding low moduli and strength values. The higher moduli and lower strength values for the molding compounds, compared with CSM/polyester at similar glass contents, are explained by the mineral filler which enhances the modulus but lowers the strength of the molding compound. Higher glass contents are found in WR/polyester laminates with average tensile modulus and strength about twice as high as the corresponding values for CSM/polyester. The data given in Table 1 refer to a fiber direction in a plain weave WR, with an equal number of rovings in the warp and weft directions. The highest glass contents are found in UD glass/epoxide composites and these materials show significant anisotropy in properties. Average tensile moduli range from 42 GPa along the fibers to 12 GPa transverse to them, with corresponding tensile strengths of 750 MPa and 50 MPa. Data on other GRP materials lie within the spectrum of properties shown in the table. Glass fabric reinforced resins and UD glass/polyester, fabricated by pultrusion or filament winding, usually have glass contents and mechanical properties intermediate between the WR/polyester and UD glass/epoxide values shown. Other points to note from Table 1 are the low interlaminar shear strength of all GRP materials and the differences observed between tensile and flexural properties, with flexural modulus usually below, and flexural strength above, the tensile values, as a result of the inhomogeneous structure of the material.

Under long-term loads, fatigue loads and when subjected to high temperature or a wet environment, GRP stiffness and strength properties are below the short-term values given in Table 1.

8. *Calculation of GRP and Sandwich Properties*

A wide range of theoretical formulae are available for predicting the mechanical properties of composite

Table 2
Typical physical property data for GRP materials

	SMC and CSM/ polyester	Fabric and WR polyester	UD glass/ epoxide
Density (kg m^{-3})	1300–1600 (SMC, 1600–1900)	1500–1900	1800–2000
Linear thermal expansion coefficient (10^{-6} K^{-1})	18–35	10–16	5–15
Thermal conductivity (W m^{-1} K^{-1})	0.16–0.26	0.2–0.3	0.28–0.35
Specific heat (J kg^{-1} K^{-1})	1200–1400		950
Maximum heat distortion temperature (BS2782, 102 G) (°C)	175	250	300
Dielectric strength (BS2782, 201 A, C) (kV mm^{-1})	9–12	13–16	
Permittivity at 1 MHz (BS2782, 207 B)	4.3–4.7	4.1–5.2	
Power factor at 1 MHz (BS2782, 205 B)	0.015	0.016	
Dry insulation resistance (MΩ) (BS2782, 204 A)	10^6	10^6	

materials in terms of microstructure and fiber and matrix properties (see, for example, Ashton et al. 1969, Jones 1975). Although these methods give satisfactory predictions of GRP properties, particularly moduli, they are usually only applicable to UD fiber reinforcement. The more rigorous results are also difficult to apply and often require correlation coefficients to be determined. It is therefore of interest to consider a simpler "rule-of-mixtures" approach which accounts for the main trends in the GRP properties reported here. Let E denote the Young's modulus of the composite and E_g and E_r the Young's moduli of the glass fibers and resin matrix. If v_g, v_r are the volume fractions of glass fiber and resin in the composite then the Halpin–Tsai equation may be written in the modified form

$$E = \alpha E_g v_g + E_r v_r \tag{1}$$

where α is a parameter which depends on the efficiency of the reinforcement. Suitable values are $\alpha = 1$ for UD reinforcement, $\alpha = 0.5$ for WR, with E then referring to the modulus along the fiber directions, and $\alpha = 0.3$ for CSM, which has a random fiber distribution. Taking $E_g = 75$ GPa, $E_r = 3$ GPa as typical modulus values for glass fiber and polyester resin and assuming fiber volume fractions v_g are 0.2 for CSM laminates, 0.35 for WR laminates and 0.5 for UD fibers, the following estimates for the GRP modulus are obtained from Eqn. (1): $E = 6.9$ GPa for CSM/polyester, $E = 15.1$ GPa for WR/polyester and $E = 39$ GPa along the fibers in UD glass/polyester. Reference to Table 1 shows that these predicted values agree well with typical GRP data.

An analogous expression to Eqn. (1) may be used for estimating GRP strength properties, in the form

$$\sigma = \alpha \sigma_g v_g + \sigma_r v_r \tag{2}$$

Here σ is the tensile strength of the composite, σ_g that of the glass fiber and σ_r the tensile stress in the resin at the fiber failure strain. Because of damage, the strength of a glass fiber in a composite may be only half the fresh drawn fiber strength (Parkyn 1970, Chap. 10). Taking as a nominal fiber strength $\sigma_g = 1500$ MPa, with $\sigma_r = 30$ MPa and the v_g, v_r values used above, Eqn. (2) gives the following estimates of GRP tensile strength properties: $\sigma = 114$ MPa for CSM/polyester, $\sigma = 278$ MPa for WR/polyester and $\sigma = 765$ MPa along the fiber direction in UD glass/polyester. Again, these values are seen to be in reasonable agreement with measured data.

The mechanical property data and the prediction methods described above refer to GRP materials containing a single type of reinforcement. In practice, GRP structures often contain several types of reinforcement laminated together or consist of angleply or sandwich materials and it is not feasible to provide detailed mechanical property data on each material combination. However, theoretical formulae are available which enable the designer to calculate the stiffness and strength properties of combined materials and these have been found to be in good agreement with measured properties. Laminated plate theory permits the calculation of the full anisotropic stiffness and strength properties of laminated composite materials, consisting of plies of different materials or the same material oriented in different directions. The analysis method is rigorous and requires a complete knowledge of the anisotropic stiffness and strength properties of each lamina. The calculations are lengthy and best carried out by a small computer program. A number of inexpensive microcomputer programs for laminate analysis are now commercially available (see, for example, ESDU (1985, 1987), Tsai (1987)). A simpler, strength of materials, method for estimating combined laminate tensile and flexural properties is described by Johnson (1978) and by Allen (1969) for calculating the properties of sandwich structures. When the flexural rigidity of the core material is negligible compared with that of the stiffer skin materials, a simple design formula may be derived for the flexural rigidity per unit width, D, of a sandwich panel, which is an important quantity for the design of sandwich structures. For a symmetric sandwich with thin skins

$$D = \tfrac{1}{2} E h (h + c)^2 \tag{3}$$

where E is Young's modulus of the skin material in the load direction, h is the skin thickness and c the core thickness. Similar design formulae are available for calculating strength properties and the minimum weight design of sandwich panels (see, for example, Johnson and Sims 1986).

See also: Composite Materials: An Overview

Bibliography

Allen H G 1969 *Analysis and Design of Structural Sandwich Panels.* Pergamon, Oxford

Anon 1981 Materials survey. *Med. Plast. Int.* 11(1): 34–35

ESDU 1985 *Software for Engineers 2022, Stiffness and Properties of Laminated Plates.* ESDU International, London

ESDU 1987 *Software for Engineers 2033, Failure of Composite Laminates.* ESDU International, London

Holmes M, Just D J 1983 *GRP in Structural Engineering.* Elsevier, London

Hull D 1981 *An Introduction to Composite Materials.* Cambridge University Press, Cambridge

Johnson A F 1978 *Engineering Design Properties of GRP.* British Plastics Federation, London

Johnson A F, Sims G D 1986 Mechanical properties and design of sandwich materials. *Composites* 17(4): 321–28

Parkyn B (ed.) 1970 *Glass Reinforced Plastics.* Iliffe, London

Rubber and Plastics Research Association 1977 *Survey of Fibre-Reinforced Plastics Markets and Needs.* RAPRA, Shawbury

Tsai S W 1987 *Think Composites Software: User Manual "Composites Design,"* 3rd edn. Think Composites, Dayton, Ohio

Whitney J M 1987 *Structural Analysis of Laminated Anisotropic Plates.* Technomic, Basel

A. F. Johnson
[National Physical Laboratory,
Teddington, UK]

Glazing

Glazing is the trade practised by glaziers. A glazier is a craftsman who prepares and places glass sealants or gaskets, positioning shims and setting blocks in sash or curtailwall rabbets to ensure desired long-term appearance, weather-tightness and structural integrity. When practices of the Construction Specifications Institute are followed, glass and glazing specifications are included in contract documents.

Glazing systems must satisfy a remarkable combination of owner, occupant and public expectations. Glazing is subject to critical, close-up inspection, to hands-on contact with windows and doors, to aesthetic and comfort responses to appearance, and to glare, drafts, noise, condensation, heat gain, heat loss and dirt accumulation. Reliable long-term performance requires design materials, installation and maintenance that address conservatively and cope practically with a multitude of significant changes in details and relationships. Changes occur from minute to minute (wind, solar radiation, rain) as well as from year to year (live loads, concrete curing, building settlement).

Knowledgeable and skilled craftsmanship by each glazier and careful planning by the owner are essential if design and performance goals are to be achieved. The most important considerations for proper glazing are listed in Table 1.

Table 1
Glazing design and planning considerations for proper glazing

Barometric pressure differentials	Mock-up tests
Buffeting	Post-installation inspection and tests
Cascading of broken glass	Roof gravel abrasion
Concrete curing	Settling
	Stains and etching alkali
Earthquake resistance	Storage
Explosions	Thermal expansion/contraction
	Tolerances
Falling objects	
Human safety	Ultraviolet radiation disintegration at adhesive interface
Impact	Vandalism
Maintenance pressures and chemicals	Wind and stack pressure

1. Weather Tightness

Corners and other rabbet fabrication joints must be sealed in the fabrication plant or by other trades, since this operation is not normally included in glazing contracts.

Weather tightness requires that moisture penetration be limited, by a long-lasting, watertight seal between glass perimeter and sash rabbet, to an amount that will flow by gravity from the building through weep holes. On a tightly sealed rabbet, usually three 1 cm diameter weep holes in each sill rabbet which lead outdoors from under the glass provide adequate relief. Adequate seals can be achieved and maintained with gaskets, glazing tape and wet sealants. Effective glazing gaskets provide weather tightness by establishing continuous "window-wiper" contact between gasket tip and smooth glass surface and by maintaining compression of 7–18 N per linear cm of tip contact. Lock strip gaskets that comply with ASTM standard C542 provide this performance (Fig. 1).

Wet sealants (Fig. 2) and tape sealants require that glass and sash sealing surfaces be absolutely clean, dry, free of oily deposits and not colder than 4.5 °C during application. Some sealants require that glass and rabbet surfaces be treated with special primers after careful cleaning before application.

Usually sealant sections must be controlled in the 3–6 × 6–13 mm range by rabbet design, proper placement of glass and proper shimming. Resilient shims positioned at regular intervals within the rabbet at indoor and outdoor glass surfaces near glass edges limit movement between glass and rabbet (Fig. 3). This restricts sealant squeezeout and gives the sealant an opportunity to perform.

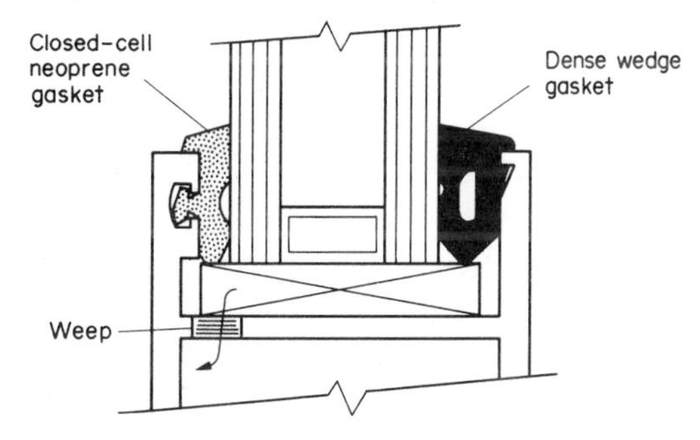

Figure 1
Typical dry glazing system

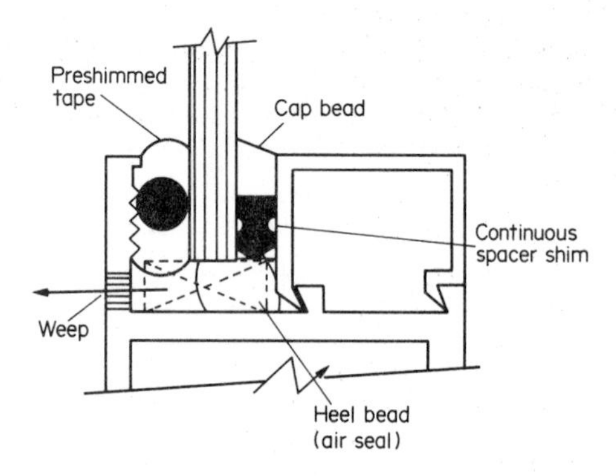

Figure 2
Typical wet glazing system

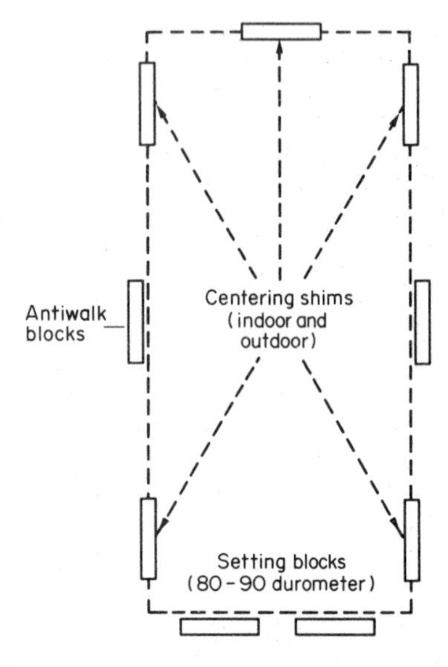

Figure 3
Positioning of centering shims and setting blocks

Resilient shims may be continuous or on 45–65 cm centers. Structural glazing systems require extra care, including temporary structural stabilization during a 2–3 week curing period.

2. Structural Reliability

Annealed glass manufactured by the continuous float process is a brittle material. In practical use, wind resistance may be halved by even mild surface abrasion. Slight damage to a skillfully clean-cut edge, caused by contact with hard materials during installations, may reduce thermal break resistance by 80% or more.

As shown in Fig. 4, breaking strength can be doubled by heat strengthening or quadrupled by tempering. However, where significant safety or economic losses may result from spontaneous failure, tempered glass should be utilized in laminates with annealed or heat-strengthened lites. A laminated lite (or piece of flat glass) has about 60% of the load-carrying ability of a monolithic lite of the same total thickness. A sealed insulating glass unit (Fig. 5) will carry ~80% more wind load than the thinner lite of which it is fabricated.

Usually, architects rely on manufacturers' charts for selecting glass thickness. To select proper glass type and thickness required to meet load and safety requirements, the architect must specify job goals and regulatory requirements. If a lite is to perform up to its nominal structural capacity, edges must be resiliently and uniformly supported in a stable glazing rabbet. Minimum clearance between lite and rabbet must be 3 mm at every point. For good performance where earthquakes may occur, continuous clearance of at least 6 mm at the perimeter, as required by the Uniform Building Code, is recommended.

Only two setting blocks of 80–90 durometer neoprene or ethylene–propylene terpolymer meeting ASTM C542 should be placed at quarter points of the sill (not closer than 15 cm to the nearest corner). Each setting block must be at least 10 cm long, equal to full rabbet width, and thick enough to prevent glass contact with the sill under all circumstances. 20–40 durometer centering shims placed on 45–65 cm centers against indoor and outdoor glass surfaces and soft, antiwalk blocks placed in the perimeter space between glass and rabbet are essential.

Shims, spacers and sealants must remain resilient over a long period of time, especially at corners.

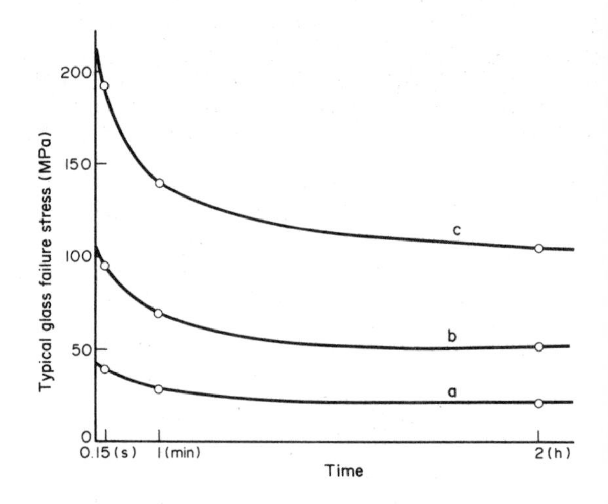

Figure 4
Typical failure stresses for glasses: (a) annealed; (b) heat-strengthened; (c) tempered

Under positive and negative wind loads, lites must be free to "curl" in or out at corners. If flexing of corners is not permitted, stress reversals may lead to breaks originating near corners.

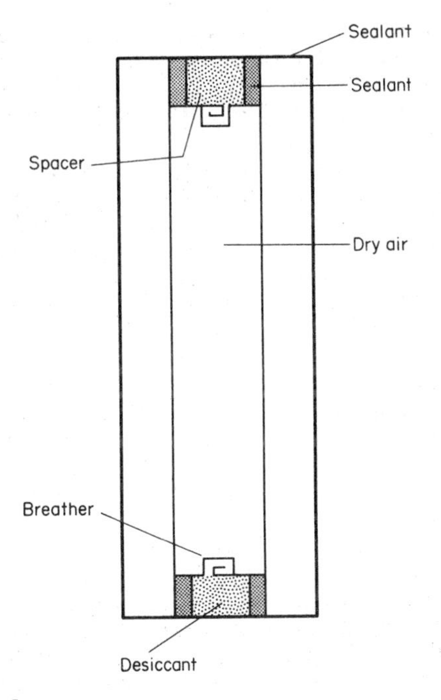

Figure 5
Sealed insulating glass unit

To reduce thermal stress, especially in tinted and reflective lites and in insulating units, edges should be insulated to prevent heat loss to adjacent heat sinks such as outdoor concrete or aluminum fins (Fig. 6).

3. *Satisfactory Appearance*

When lites are exposed to sunlight at grazing angles, glass surfaces are viewed critically and close up by building occupants. Even minute surface dirt, scratches, abrasions or etchings may become visible and unattractive. For this reason, glass surfaces must be protected carefully during storage, handling, installation and service. Glass surfaces may be pitted by welding splatter or etched by alkaline or fluoride deposits from adjacent masonry or concrete surfaces after rainstorms or maintenance washing. Strength may be reduced significantly. In situ surface repairs are not economically feasible.

Like all manufactured products, glass has subtle appearance characteristics related to its production line and fabrication orientation. To minimize "checkerboarding" in a facade, all lites should be cut and installed to achieve the same up–down, in–out, right–left relationships. All blocks, shims and so on must be uniform in dimension and installed in uniform positions and with uniform tightness of fit.

Facade appearance may vary significantly in color and flatness depending upon lighting and viewing location as well as upon color and geometry of images reflected. With barometric and temperature changes, sealed insulating units bow in or out and may cause

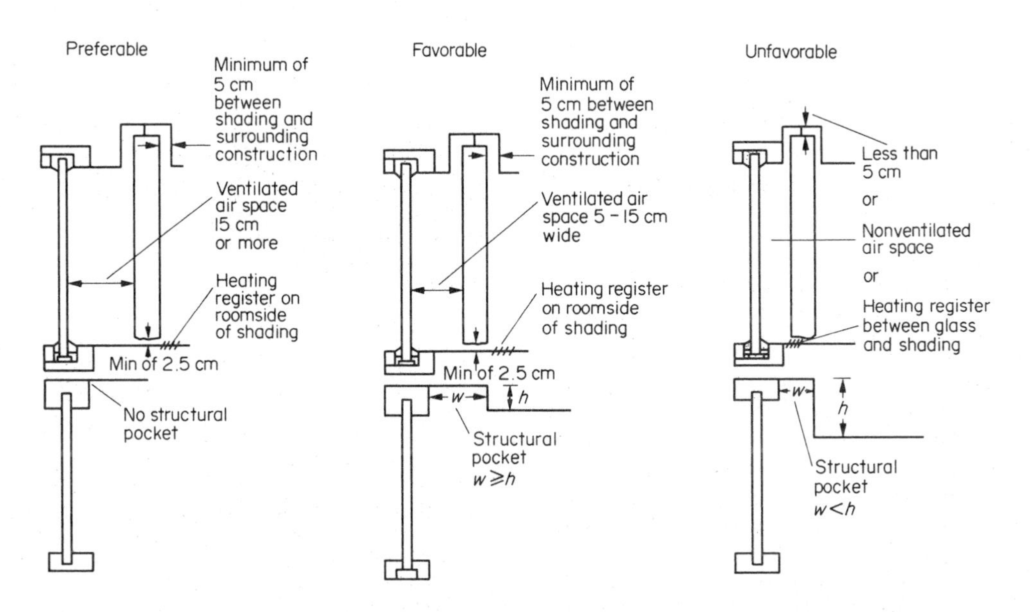

Figure 6
Preferable, favorable and unfavorable conditions of indoor shading

significant differences in reflected images from day to day.

All heat-strengthened and tempered glass is distorted to some degree by the heating process that imparts increased strength. Also, under some north-sky (partly polarized) viewing conditions, quench patterns related to processing may become visible. These characteristics vary between processes but cannot be eliminated.

4. Lessons from Experience

The following considerations should always be borne in mind by glaziers:

(a) explain risks to the owner so that he understands the design and agrees with cost–performance compromises;

(b) show glazing details in shop drawings two to four times full scale;

(c) label each item and material;

(d) explain the function of each;

(e) dimension each item and show the maximum permissible tolerance;

(f) protect glass from impact and abrasion. against alkalis in fluoride washes;

(g) replace damaged glass promptly; and

(h) clean glass regularly.

See also: Glass as a Building Material; Structural Sealant Glazing

Bibliography

Adams F W 1942 *Behavior of Glazing Materials Subject to Explosion*, ASTM Bulletin No. 122. American Society for Testing and Materials, Philadelphia, Pennsylvania

American Society for Testing and Materials. *Recommended Practices and Standards on Glass and Glazing*, ASTM Committees C24 and C26. ASTM, Philadelphia, Pennsylvania

Architectural Aluminum Manufacturers' Association 1971 *Lock Strip Gaskets*, Curtainwall Monograph No. 4. AMMA, Chicago

Architectural Aluminum Manufacturers' Association 1972 *Joint Sealants in Glass and Glazing*, Curtainwall Monograph No. 6. AMMA, Chicago

Beason W L 1980 *A Failure Prediction Model for Window Glass*. Institute for Disaster Research, Lubbock, Texas

Cook J P 1970 *Construction Sealants and Adhesives*. Wiley, New York, pp. 121–25

Ernsberger F N 1960 Detection of strength impairing surface flaws in glass. *Proc. R. Soc., London* 257: 213–23

Flat Glass for Glazing, Federal Specification DDG 451D. US General Services Administration, Washington, DC

Heat Strength and Fully Tempered Flat Glass, Federal Specification DDG 001403. US General Services Administration, Washington, DC

Flat Glass Marketing Association 1978 *Glazing Manual*. Glass and Glazing Federation, London

Frownfelter 1959 *Structural Testing of Large Glass*, ASTM Special Technical Publication No. 251. American Society for Testing and Materials, Philadelphia, Pennsylvania

Orr L 1972 Practical analysis of fractures in glass windows. *Mater. Res. Stand.* 12: 21–47

Panek J 1976 *Building Seals and Sealants*. American Society for Testing and Materials, Philadelphia, Pennsylvania, pp. 81, 134–150

Purchase S 1981 You can become an expert glazer. *Boston Sunday Globe*, Oct. 18th Issue

Further information on glazing can be found in the following publications issued by PPG Industries, Pittsburgh, Pennsylvania:

Butt Joint Glazing System, PPG Brochure

Care and Storage of Glass, PPG Technical Service Memo, Feb. 11, 1980

Design Windloads, ANSI A58.1, PPG Technical Service Memo, June 21, 1974

Framing and Glazing–Does it Float? PPG Technical Service Memo, Sept. 30, 1977

Installation Recommendations for Tinted and Reflective Glass, PPG Technical Service Report No. 130

Installation Recommendations for Windows, PPG Technical Service Report No. 230

Mock-Up Test Requirements versus Glass Design Loads, PPG Technical Service Memo, May 8, 1981

Recommended Glazing Practices for Insulating Units over 20 square feet in Area, PPG Recommended Practice

Recommended Glazing Practices for LHR Spandrelite and Hestron Glasses, PPG Technical Service Memo, Aug. 18, 1980

R. W. McKinley
[Hancock, New Hampshire, USA]

Glued Joints in Wood

Glued joints are used in wood construction to increase the size of available materials and also for product assembly. Natural adhesives have been used for furniture and other nonstructural products throughout most of recorded history (US Forest Products Laboratory 1981). Development of durable synthetic resin-based adhesives prior to World War II allowed the use of adhesives in demanding structural products and applications.

Three types of joints are readily defined with respect to the grain directions of the members joined together: edge-to-edge grain, end-to-end grain and end-to-side grain (where edge grain, also called side grain, refers to radial or tangential surfaces, and end grain to the cross section). Edge-to-edge- and end-to-end-grain joints are widely used in glued, laminated beams and other structural products. The use of adhesives in other types of building construction is expanding. The greatest variety of joints occurs in furniture construction, where not only edge-to-edge- and end-to-end-grain joints are used, but also end-to-side-grain joints. Most of the following discussion will deal with furniture joints; for applications of glued joints in building construction and laminated beams, see *Design with Wood* and *Glued Laminated Timber*.

1. Edge-Grain Joints

Edge-to-edge-grain joints are used for furniture parts such as solid wood seats and tops, and lumber cores for panels. The material is laid up in large, revolving glue-clamp carriers called reels, which hold the glued stock under pressure while the adhesive dries. High-frequency electronic glue presses are also used. Pressure is used to secure proper alignment of the parts and to ensure intimate contact between the wood and the glue. Animal glues, cold- and hot-setting urea formaldehyde, and poly(vinyl acetate) adhesives are commonly used to bond the wood together (see *Adhesives for Wood*). Boards are carefully machined to mate smoothly along their entire lengths without gaps, to prevent uneven glue lines and nonuniform distribution of pressure. Machined edges are preferred to sanded. Machining is carried out just prior to gluing, to remove surface resins and surface oxidation products. Pressures of 0.35–2 kPa are used. Wide boards are placed towards the outsides of the panels to help distribute pressures uniformly throughout the entire assembly. Temperature of the glue, the wood and the surroundings are controlled to produce joints with thin but adequate glue lines which retain sufficient adhesive to prevent starved joints. Assembly times are held as short as possible to reduce subsequent cure times. Hold-downs are used to prevent panels from bowing while being pressed. Pneumatic torque wrenches lessen the time required to tighten clamps, and ensure that clamping forces are applied uniformly. The wood is conditioned to a uniform moisture content prior to gluing to prevent subsequent problems with unequal shrinking and swelling. Glued panels are allowed to set until the swollen glue lines have dried sufficiently to allow machining.

2. End-Grain Joints

Furniture joints which require the connection of end-to-end-grain or end-to-side-grain surfaces are termed assembly joints. Included are dowel, mortise-and-tenon, finger, corner-block and gusset-plate joints (see Fig. 1). An illustration of finger joints is given in the article *Glued Laminated Timber*. Several of these assembly joints are also used in light frame building construction.

2.1 Dowel Joints

Owing to their simplicity, dowel joints are the most common connection employed in furniture construction. Withdrawal strength of a dowel from the side grain of a wooden part is directly proportional to the diameter of the dowel to the shear strengths parallel to the grain of the dowel and of the materials joined together; withdrawal strength is slightly less than proportional to depth of embedment of the dowel. For highest strengths, both the dowel and the walls of the hole are coated with glue; a close fit between dowel and hole is essential. The in-plane bending strength of joints constructed with two symmetrically spaced dowels is directly related to the withdrawal strength of the dowels and to the width of the spacing between them. The out-of-plane bending strength of dowel joints is much less than in-plane strength; thickness of the end-grain member is the dominant strength parameter (Eckelman 1979). Dowel joints also carry shear loads in furniture. Strength is limited largely by the splitting strength of the wood. Typical joints constructed with two dowels 1 cm in diameter carry loads in excess of 2.5 kN. Methods of calculating the strengths of most of the dowel joints described have been developed (Eckelman 1978); therefore, these joints can be designed to meet specified in-service requirements.

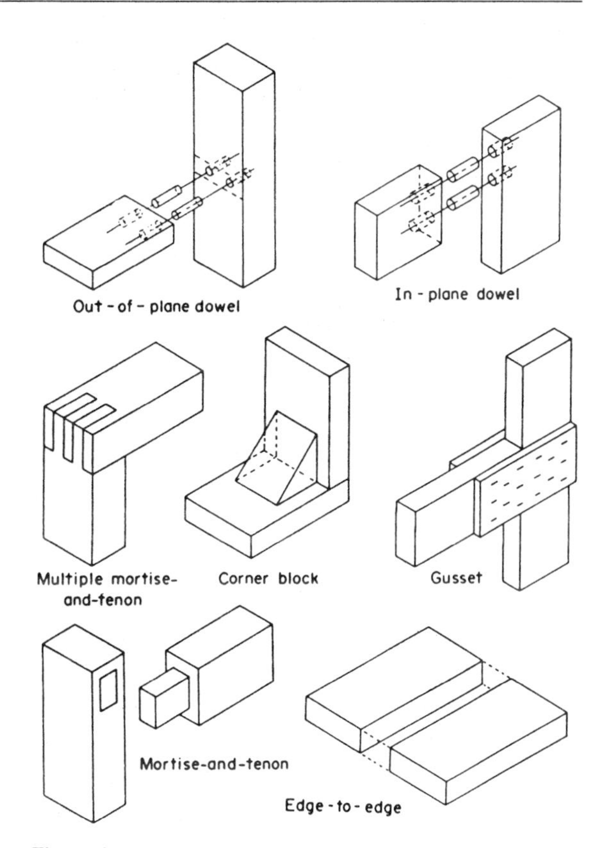

Figure 1
Common types of assembly joints

2.2 Mortise-and-Tenon Joints

Mortise-and-tenon joints have largely been displaced by dowel joints in furniture because of the relative ease of construction of the latter. Adhesives are used in nearly all mortise-and-tenon furniture joints. Well-made joints develop somewhat more strength than comparable dowel joints (Sparkes 1968). Bending strength is related to the thickness, width and length of the tenon, and the closeness of fit of the tenon into the mortise (Eckelman 1988). Bending strength of the tenon itself provides the ultimate constraint on strength.

2.3 Corner Blocks

Corner blocks are used in furniture construction to reinforce points of high shear and bending stress. Ordinarily, corner blocks are cut with each of the two glue faces at angles of 45° to the grain; as a result, both faces form bonds intermediate to side grain and end grain. Blocks may also be cut with one face parallel to the grain to allow a side-grain-to-side-grain bond on one face. Properly fitted glue blocks can effectively resist bending loads which tend to close a joint, but may be less effective in resisting bending loads which tend to open it (Eckelman 1978, 1988).

2.4 Finger Joints

End-grain finger joints are used both to increase the longest length of lumber available and to salvage material by creating long members from short scraps (Blomquist et al. 1983). Strongest joints occur when the fingers have a small slope and are sharply pointed. Corner finger joints are more difficult to cut and clamp than end joints, although joints with slight angles are readily fabricated since they are essentially end-grain joints. Such joints are regularly used in items such as the back posts of chairs. Techniques for cutting geometrically complex high-strength finger joints have been developed (Richards 1962), but are little used commercially.

2.5 Multiple Mortise-and-Tenon Joints

Multiple mortise-and-tenon joints consist of a number of tenons with parallel sides, which are cut into the ends of two mating members. A common example of this construction is the box joint. They are also used to form visible decorative joints in fine furniture. Well-made multiple mortise-and-tenon corner joints may develop up to 50% of the strength of the wood (Richards 1962). Dovetail joints are basically similar in design, although they are not usually thought of as multiple mortise-and-tenon joints. Primary use of dovetail joints is in fine-furniture construction, where they are used for drawer construction. Close fit of the parts and use of adequate glue are necessary for high strength.

2.6 Staple Glued Plywood Gusset Joints

Staple glued plywood gussets are used in hidden areas in furniture to reinforce heavily stressed joints (Eckelman 1978, 1988). Such joints were once widely used in the construction of roof trusses, but have largely been supplanted by metal-tooth connector plates (see *Design with Wood*). Strength of such joints is dependent on the size of gusset used and the rolling shear strength of the plywood.

See also: Wood: An Overview

Bibliography

Blomquist R F, Christiansen A W, Gillespie R H, Myers G E 1983 *Adhesive Bonding of Wood and Other Structural Materials*. Pennsylvania State University, University Park, Pennsylvania

Eckelman C A 1978 *Strength Design of Furniture*. Tim Tech, West Lafayette, Indiana

Eckelman C A 1979 Out of plane strength and stiffness of dowel joints. *For. Prod. J.* 29: 32–38

Eckelman C A 1988 *Product Engineering and Strength Design Manual for Furniture*. Tim Tech, West Lafayette. Indiana

Feirer J 1983 *Cabinetmaking and Millwork*. revised edn. C A Bennett, Peoria, Illinois

Gillespie R H, Countryman D, Blomquist R F 1978 *Adhesives in Building Construction*, US Department of Agriculture Handbook 516. US Government Printing Office, Washington, DC

Pound J 1973 *Radio Frequency Heating in the Timber Industry*, 2nd edn. Clowes, London

Richards D B 1962 High strength corner joints for wood. *For. Prod. J.* 12: 413–18

Selbo M L 1975 *Adhesive Bonding of Wood*, US Department of Agriculture Bulletin No. 1512. US Government Printing Office, Washington, DC

Slaats M A 1979 Glue bond quality in wood. *Int. J. Furniture Res.* 1: 24–26

Snider R 1975 *Gluing and Furniture Design*. Franklin Chemical Industries, Columbus, Ohio

Sparkes A J 1968 *The Strength of Mortise and Tenon Joints*, Report 33. Furniture Industry Research Association, Stevenage

US Forest Products Laboratory 1981 *Proc. 1980 Symp. Wood Adhesives—Research, Application, and Needs*. US Forest Products Laboratory, Madison, Wisconsin

US Forest Products Laboratory 1987 *Wood Handbook*, US Department of Agriculture Handbook 72. US Government Printing Office, Washington, DC

C. A. Eckelman
[Purdue University, West Lafayette, Indiana, USA]

Glued Laminated Timber

Glued laminated timber used for structural purposes is an engineered stress-rated product comprising assemblies of suitably selected and prepared wood laminations securely bonded together with adhesives.

'he grain of all laminations is approximately parallel ›ngitudinally. The individual laminations are of ımber thicknesses. Laminations may be composed of ieces end-joined to form any length, of pieces placed r glued edge-to-edge to make wider components, or f pieces bent to curved form during gluing.

Glued laminated timber (often referred to by the eneric term glulam) was first used as early as 1893 in :urope, where laminated arches (probably glued with asein adhesives) were erected for an auditorium in witzerland. Improvements in casein adhesives during Vorld War I created further interest in the use of lued laminated timber structural members for air-raft and, later, as framing members for buildings. 'he development of durable synthetic-resin adhesives uring World War II permitted the use of glued ıminated timber members in bridges, truck beds and ıarine construction, where a high degree of resistance o severe service conditions is required. Today, ynthetic-resin adhesives—primarily of the phenol–esorcinol and melamine types—are the principal ad-ıesives used for structural laminating (see *Adhesives or Wood*).

Typically, glued laminated structural timbers are ectangular in cross section. They may be either traight or curved between supports. Straight beams :an be designed and manufactured with horizontal aminations (load applied perpendicular to wide faces of laminations) or vertical laminations (load applied parallel to wide faces of laminations), as shown in Fig. 1. Horizontally laminated beams are the most widely used. Curved members are horizontally laminated to permit bending of the laminations to a curved form during gluing.

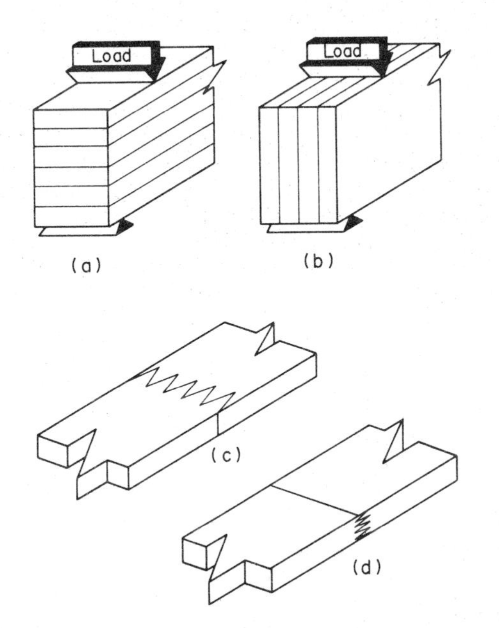

Figure 1
Glued laminated timber forms: (a) beam, horizontally laminated; (b) beam, vertically laminated; (c) end joint, vertical finger joint; (d) end joint, horizontal finger joint

The advantages of glued laminated timber construction include the provision of the ability to manufacture larger structural elements from smaller commercial sizes of lumber, giving better utilization of available lumber resources; the ability to achieve architectural effects through the use of curved shapes; and the ability to design structural elements varying in cross section between supports, and in accordance with strength requirements. Advantages also include minimization of checking or other seasoning defects associated with large sawn-wood members, because the laminations can be dried before being glued, thus permitting designs to be based on the strength of seasoned wood. In addition, lower grade lumber can be used for less highly stressed laminations without adversely affecting the structural integrity of the member.

Factors determining the assigned strength values of structural lumber will also affect glued laminated timber, since glulam is made up of individual pieces of structurally graded lumber. The mere gluing together of pieces of wood does not, of itself, improve strength properties in laminating material of lumber thicknesses over those of comparable sawn timbers. Three of the benefits inherent in the laminating process do, however, form the bases for assigning higher glued laminated timber design values: (a) drying or seasoning of the laminations; (b) positioning or placement of the laminations in the member; and (c) distribution of growth characteristics.

As wood seasons or dries under controlled conditions, it increases in strength. It is difficult to thoroughly dry pieces of wood with least dimensions greater than about 10 cm within a reasonable time. But laminated timbers are fabricated of individual laminations less than 5 cm thick, dried to a moisture content of 16% or less; therefore, regardless of the size of the finished laminated timber, it is dry, and design values can be based upon the strength of dry wood.

The bending strength of horizontally laminated timbers depends on the positioning of the various grades of lumber used as laminations. High grade laminations are placed in the outer portions of the member where high strength is effectively used, and lower grade laminations are placed in the inner portion, where low strength will not greatly affect the overall strength of the member. By selective placement of the laminations, knots are scattered and improved strength can be obtained. Studies have indicated that knots are unlikely to occur one above another in several adjacent laminations.

A comparison of the allowable design values for a typical laminated-timber grade for bending members,

Table 1
Comparison of design values, Douglas-fir

Design values	Bending (MPa)	Compression perpendicular to grain (MPa)	Shear (MPa)	Modulus of elasticity (MPa)
Glulam, typical grade	16.55	4.48	1.14	12410
Sawn timber, No. 1 "beams and stringers"	9.31	4.31	0.59	11030

and a No. 1 grade of sawn timber of the size classification "beams and stringers," both made from Douglas fir, is shown in Table 1.

Principal steps in the manufacture of glulam members are as follows.

(a) Selection and preparation of laminations. Lumber used for structural laminating is commonly of the same softwood species used for other construction purposes. In the USA, for example, the softwood species most commonly used are Douglas-fir, Southern pine and hem-fir (a marketing combination of western hemlock and true firs). The lumber is graded for knot size and location, rate of growth, density, slope of grain and other characteristics that affect its use as a structural material. The laminations are surfaced prior to gluing, to assure that the wide faces can be brought into close contact for adhesive bonding.

(b) End joining of laminations. Lumber in the long lengths necessary for large laminated timbers is not normally available; therefore, the fabrication of structural end joints is an important part of the laminating process. The most common type of end joint used by the laminating industry consists of a series of intermeshed tapered "fingers" cut into the ends of laminations. These end joints, known as finger joints, can be well bonded and develop the high strengths needed for structural laminating (Fig. 1).

(c) Spreading of adhesives. Adhesives are spread on the laminations by means of rollers through which the laminations pass, or, more commonly, by an extruder which spreads beads of adhesive along the length of the lamination as it passes beneath.

(d) Clamping. After the laminations have been spread with adhesives, pressure is applied in forms that have been preset to the shape required in the finished glulam product. Pressure is applied, usually by means of a series of evenly spaced clamps, to bring the laminations into close contact, to pull the member to its final shape, to force out excess adhesive and to hold the laminations together during the adhesive curing period. After this curing period, which may range from 8 to 24 h, the glulam members are removed from the forms and are surfaced by large planing equipment to the specified width dimension.

(e) Finishing. In the finishing area of a laminating plant the completed glulam members are cut to the proper length, appropriate taper cuts are made, holes and daps for connectors are formed, end sealers are applied and wrapping material is applied as specified.

(f) Quality control. Throughout the laminating process, samples of production are regularly checked to be sure that the products are in conformance with established laminating standards. Quality control is an essential part of the laminating process, since the strength of the glulam members is dependent upon the quality of the glue joints. Special equipment, plant facilities and manufacturing skills are needed to assure that this high quality is maintained.

The advent of techniques for glue laminating of wood provided a practical means for manufacturing wood structural members not limited by the size and shape of a tree. The use of glued laminated timber permits the construction of timber buildings with long clear spans and a variety of shapes.

See also: Glued Joints in Wood; Wood: An Overview

Bibliography

American Institute of Timber Construction 1983 *Structural Glued Laminated Timber*, ANSI/AITC 190. 1. American Institute of Timber Construction, Englewood, Colorado

American Institute of Timber Construction 1987 *Standard Specifications for Structural Glued Laminated Timber of Softwood Species*, AITC 117–87. American Institute of Timber Construction, Englewood, Colorado

American Society for Testing and Materials 1983 Standard method for establishing stresses for structural glued laminated timber (glulam) manufactured from visually graded lumber, ASTM D3737–83. *1983 Annual Book of ASTM Standards*, Vol. 4:09, *Wood*. American Society for

Testing and Materials. Philadelphia, Pennsylvania, ASTM D3737-83, pp. 645–63
Chugg W A 1964 *Glulam: The Theory and Practice of the Manufacture of Glued Laminated Timber Structures.* Benn, London
National Forest Products Association 1986 *National Design Specification for Wood Construction.* National Forest Products Association, Washington, DC

R. P. Wibbens
[Littleton, Colorado, USA]

H

Hardboard and Insulation Board

Hardboard and insulation board are rigid or semirigid sheet materials with thicknesses of less than 1 in. (~25 mm). Summarily known as fiberboards they are a category of the large class of wood composition-boards, made by various methods of manufacture from wood elements ranging from veneers to fibers (Fig. 1). The term fiber refers to any wood or other lignocellulosic element possessing the size and shape of the wood cell, the biological building block of all woody material.

In the terminology of the American Hardboard Association the boundary between insulation board and hardboard is a board density of 480 kg m^{-3} (30 lb ft^{-3}). In the trade, however, a distinction is made between medium-density fiberboard, often referred to as MDF, and higher density fiberboard, called hardboard in the narrower sense.

In the manufacturing process solid wood is first reduced to fiberlike elements (pulping) which are subsequently recombined in sheet form with or without the addition of binders.

Fiberboard processes utilize the lower quality end of the available range of raw wood material. Owing to the small size of the elements, these processes achieve a higher degree of homogenization and randomization of the anisotropic raw material than any other wood utilization process, with the exception of paper.

The most important fiberboard markets are for interior and exterior cladding, and industrial applications in the furniture and automotive industries.

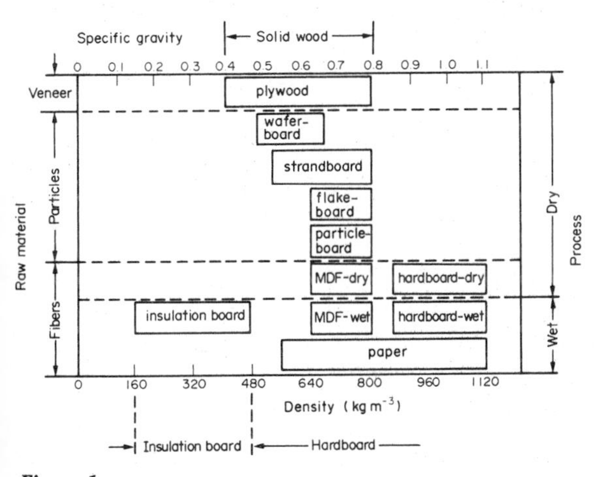

Figure 1
Classification of wood composition-boards (after Suchsland and Woodson 1986)

1. Bonds in Fiberboard

Although the tensile strength of the wood fiber is very high, it is generally not utilized to its fullest potential in the structural configuration of a paper sheet or a fiberboard. This is due to the total bonding area between fibers being incapable of transmitting the shear stresses required to stress the fiber to its limit, either because of the limited size of the total shear area or because of the quality of the bond. In low- and medium-density fiberboard, stress failures occur predominantly in the bond. At high board densities failures do occur in the fiber, reflecting both more intimate contact between fibers and possible modification of fiber characteristics under the severe conditions in the hot press.

The establishment of maximum quality fiber bonds is, therefore, one of the predominant goals of process design. Three different types of bonding occur in fiberboard, either individually or simultaneously, depending on the process (Back 1968).

1.1 Hydrogen Bonding

Hydrogen bonds depend on extremely short-range attractive forces associated with surface molecules, and develop between positively charged hydrogen atoms and any negatively charged atoms such as oxygen. This type of bonding occurs primarily in wet fiberboard processes, where the surface tension of the evaporating water pulls the fibers together within the range of molecular forces.

1.2 Lignin Bonding

Lignin bonds are attributed to the thermal softening of the lignin, a major wood component concentrated in the outer region of the cell wall, and a fusing together of lignin-rich surfaces under the conditions existing in the hot press (Spalt 1977). It has also been proposed that such bonds are formed by the condensation of phenolic substances derived from lignin. Lignin bonds are limited to wet-process fiberboards densified in hot presses, although they may occur in dry-process hardboard as a result of severe pulping conditions (Suchsland et al. 1983, 1987). Like the hydrogen bond, the lignin bond is unavailable to any of the other wood composition-board products.

1.3 Adhesive Bonding

Adhesive bonds formed by added binders are essential in the manufacture of dry-process fiberboard and all other wood composite products, and are used to enhance board quality in wet processes. The most common binders are phenol–formaldehyde (dry-process hardboard) and urea–formaldehyde resins (dry-process medium-density fiberboard). Other adhesives include drying oils and thermoplastic resins (Table 1).

2. *Fiberboard Processing*

Figure 2 gives a simplified outline of the entire range of fiberboard processes. Primary defiberizing of the raw material occurs by one of three mechanical pulping processes: (a) the Masonite process, subjecting the wood chips to high steam pressure (7 MPa), which upon sudden release explodes them into pulp; (b) the pressurized refiner, separating fibers between rotating grinding disks while under elevated steam pressure; and (c) the atmospheric refiner, grinding pretreated (steamed or water cooked) or untreated chips at atmospheric pressure. Chemical pulping is not used in fiberboard manufacture.

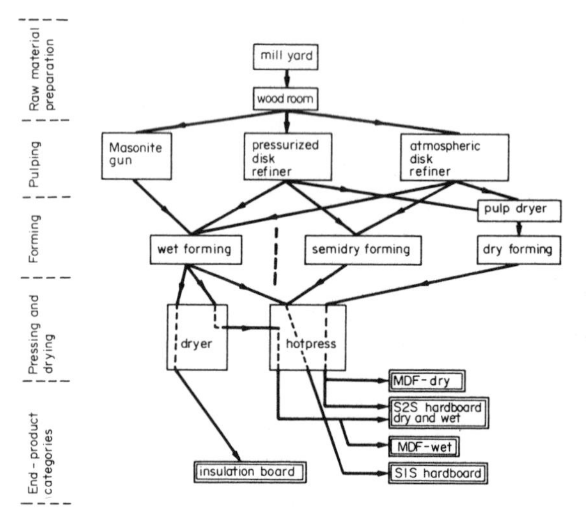

Figure 2
Simplified flowchart of fiberboard processes (after Suchsland and Woodson 1986)

The fibers are either diluted with water and the mat formed on fourdrinier machines (wet process), or dried and formed by air-felting machines (dry process). Binders and sizing (chemicals to reduce water absorption) are added to liquid furnish and precipitated on fibers by pH reduction (wet process), or are distributed on fibers by spraying or by mechanical action (dry process).

Insulation board is made by drying the wet-formed mat. It relies almost entirely on hydrogen bonding. All other fiberboards are hot pressed. The hot press densifies the mat, provides the condition for bond development (plasticizing of lignin, curing of adhesives) and reduces the water content of the wet mat. Water and steam removal in the hot press from the wet and moist (semidry) mats is facilitated by inserting a screen between the bottom mat surface and press platen. The screen pattern is embossed on the back of the finished board (S1S—smooth one side). Dry-formed mats are pressed without screens (S2S—smooth two sides); S2S boards can also be made from dried wet-formed mats. The design of press cycles (pressure–, temperature–time relationships) influences density distribution over the board cross section and thereby affects strength properties. Thicker boards (dry medium-density boards) are particularly sensitive to such effects (Suchsland and Woodson 1976).

Heat treatment (baking at 260 °C) of pressed boards with or without prior application of drying oils (tempering) increases bending strength and reduces water absorption and swelling. Bending strength improvement is believed to be due to fiber bond enhancement at high temperatures. Reduction of water absorption follows conversion of hemicelluloses into less hygroscopic furfural polymers (Stamm 1964). The surface application of tempering oils and subsequent heat treatment increases surface hardness

Table 1
Binder systems used in fiberboard manufacture

Fiberboard	Bonding (listed in increasing order of amount of binder used) Primary	Secondary
Wet process		
insulation board	hydrogen	starch, asphalt
S1S masonite	lignin	
S1S	lignin	phenolic thermoplastics
S2S	lignin	thermoplastics, drying oils
MDF	lignin	thermoplastics, drying oils
Dry process		
S2S	phenolic	
MDF	urea	

substantially, a property of great importance in the manufacture of premium quality factory-finished wall panelling. Excessive heat treatment causes pyrolytic breakdown of the cell wall. Insulation board and thick medium-density fiberboard are neither heat treated nor tempered.

Table 2
Classification of hardboard by thickness and physical properties[a]

		Water resistance (maximum average per panel)			Tensile strength (minimum average per panel)	
Class	Nominal thickness (in.)	Water absorption based on weight (%)	Thickness swelling (%)	Modulus of rupture (minimum average per panel) (MPa)	Parallel to surface (MPa)	Perpendicular to surface (MPa)
Tempered	1/10					
	1/8	25	20			
	3/16			41.3	20.7	0.9
	1/4	20	15			
	5/16	15	10			
	3/8	10	9			
Standard	1/12	40	30			
	1/10					
	1/8	35	25			
	3/16			31.0	15.2	0.6
	1/4	25	20			
	5/16	20	15			
	3/8	15	10			
Service-tempered	1/8	35	30			
	3/16	30	30	35	13.8	0.5
	1/4	30	25			
	3/8	20	15			
Service	1/8	45	35			
	3/16	40	35			
	1/4	40	30			
	3/8	35	25			
	7/16	35	25			
	1/2	30	20			
	5/8	25	20	20.7	10.3	0.3
	11/16	25	20			
	3/4					
	13/16					
	7/8	20	15			
	1					
	1–1/8					
Industrialite	1/4	50	30			
	3/8	40	25			
	7/16	40	25			
	1/2	35	25			
	5/8	30	20			
	11/16	30	20	13.8	6.9	0.2
	3/4					
	13/16					
	7/18	25	20			
	1					
	1–1/8					

[a] Source: US Department of Commerce (1980a)

Table 3
Property requirements for medium-density fiberboard for interior use[a]

Nominal thickness (in.)	Modulus of rupture (MPa)	Modulus of elasticity (MPa)	Internal bond (tensile strength perpendicular to surface) (MPa)	Linear expansion (%)	Screwholding Face (kg)	Screwholding Edge (kg)
13/16 and below	20	2000	0.62	0.30[b]	148	125
7/8 and above	19	1723	0.55	0.30	136	102

[a] Source: US Department of commerce (1980) [b] For boards having nominal thicknesses of $\frac{3}{8}$ in. or less, the linear expansion value is 0.35%

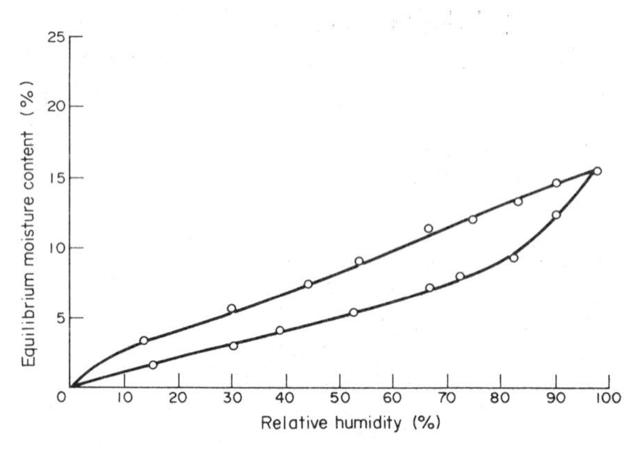

Figure 3
Adsorption and desorption isotherms between 0 and 98% RH at 20 °C for heat-treated hardboard (after McNatt 1974)

3. Fiberboard Properties

Most strength properties of fiberboard are strongly affected by raw material and manufacturing variables such as fiber characteristics, board density, binder content and heat treatment, and are, therefore, controllable within considerable margins. Real fiberboard properties are compromises between market demands and economic limitations, and are often adjusted to the requirements of particular applications. Table 2 lists minimum and maximum hardboard properties prescribed by the commercial standard.

Properties related to the hygroscopicity of the wood cell wall are much more difficult to control. All fiberboards, by absorbing and giving off water, maintain a moisture content in equilibrium with the surrounding air (Fig. 3). Important consequences of moisture content fluctuations are dimensional changes (thickness swelling and linear expansion), of which the linear expansion is critical in fiberboard exterior siding. Also affected by moisture content changes are most strength properties, but most severely the interlaminar shear modulus. The shear modulus of $\frac{1}{4}$in. (~6.25 mm) tempered hardboard at 80% relative humidity (RH) average only 70% of the modulus at 65% RH, and only slightly more than half the value at 30% RH. The modulus of elasticity in bending varied less than 20% of the value at 65% RH over the range 0–90% RH. Strength properties are similarly affected by increased temperature (Haygreen and Sauer 1969). Fiberboard, like solid wood, is viscoelastic, and the time-dependent responses to stress of fiberboard require special consideration in structural applications (McNatt 1970, Moslemi 1964, Suchsland 1965).

Thick medium-density fiberboard properties as recommended by the commercial standard are listed in Table 3. Mechanical properties are less critical for insulation board, with the exception of insulation sheathing, which requires tensile strength values up to 2.07 MPa and racking loads of at least 2360 kg. Minimum thermal conductivity values for insulation board vary from 0.0548 to 0.0692 $W m^{-2} K^{-1}$.

Bibliography

Back E L 1986 The bonding mechanism in hardboard manufacture—A summary. *18th IUFRO World Congr.*, Ljubljana, Sept. 7–13. Preprint
Haygreen J, Sauer D 1969 Prediction of flexural creep and stress rupture in hardboard by use of time–temperature relationship. *Wood Sci.* 1: 241–49
McNatt J D 1970 Design stresses for hardboard—Effect of rate, duration and repeated loading. *For. Prod. J.* 20: 53–60
McNatt J D 1974 Effects of equilibrium moisture content changes on hardboard properties. *For. Prod. J.* 24: 29–35
Moslemi A A 1964 Some aspects of visco-elastic behavior of hardboard. *For. Prod. J.* 14: 337–42
Spalt H A 1977 Chemical changes in wood associated with wood fiberboard manufacture. In: Goldstein I S (ed.) 1977 *Wood Technology: Chemical Aspects*, ACS Symposium Series 43. American Chemical Society, Washington, DC, pp. 193–219
Stamm A J 1964 *Wood and Cellulose Science*. Ronald, New York
Suchsland O 1965 Swelling stresses and swelling deformation in hardboard. *Mich. Agric. Exp. St., Q. Bull.* 47(4): 591–605

Suchsland O, Woodson G E 1976 *Properties of Medium Density Fiberboard Produced in an Oil Heated Laboratory Press*, US Department of Agriculture Forest Service Research Paper SO-116. US Government Printing Office, Washington, DC

Suchsland O, Woodson G E 1986 *Fiberboard Manufacturing Practices in the United States*, USDA Agriculture Handbook No. 640. US Government Printing Office, Washington, DC

Suchsland O, Woodson G E, McMillin C W 1983 Effects of hardboard process variables on fiberbonding. *For. Prod. J.* 33(4): 58–64

Suchsland O, Woodson G E, McMillin C W 1987 Effect of cooking conditions on fiber bonding in dry-formed binderless hardboard. *For. Prod. J.* 37 (11/12): 65–69

US Department of Commerce 1973 *Cellulosic Fiber Insulating Board*, Voluntary Product Standard PS57-73. US Government Printing Office, Washington, DC

US Department of Commerce 1980a *Basic Hardboard*, American National Standard ANSI/AHA A135.4. US Government Printing Office, Washington, DC

US Department of Commerce 1980b *Medium Density Fiberboard for Interior Use*. US Government Printing Office, Washington, DC

US Department of Commerce 1982 *Prefinished Hardboard Paneling*, American National Standard ANSI/AHA A135.5. US Government Printing Office, Washington, DC

US Department of Commerce 1983 *Hardboard Siding*, American National Standard ANSI/AHA A135.6. US Government Printing Office, Washington, DC

O. Suchsland
[Michigan State University, East Lansing, Michigan, USA]

I

Ice and Frozen Earth as Construction Materials

The cold regions of the world, characterized by the presence of ice and frozen earth, are centered around the poles and extend to about the 40th parallel in both the northern and southern hemispheres. Ice covers between 9% and 15% of the earth's surface during the year, while nearly one-fifth of all land surface of the earth is underlain by perennially frozen earth or permafrost. The polar zone consists of continuous masses of perennial ice and frozen earth, while the sub-polar zone consists of discontinuous or broken masses of ice and frozen earth, both perennial and seasonal in nature. The temperate zone consists mainly of seasonal ice and frozen earth.

Development and construction, particularly in the polar and extreme subpolar zones, require innovative engineering solutions. Transportation of personnel, materials and equipment as well as communications pose complex and expensive logistical problems. The environmental impact of development on people and wildlife is a problem of public concern. These problems have been compounded in recent years by the rapid pace of development imposed by society's needs for exploiting scarce natural resources and for maintaining the security of sovereign nations.

The ability to work with ice and frozen earth as construction materials is central to cold-regions engineering. Both materials are plentiful in supply and inexpensive. Specialized knowledge concerning their many unique mechanical and thermal characteristics is essential for developing safe and economical, yet innovative, engineering solutions. Examples of innovations in recent decades include the development of:

(a) ice roads and bridges for aircraft and ground transport;

(b) ice platforms/islands for working space offshore;

(c) ice barriers for protecting offshore structures from moving ice;

(d) small embankment dams built with permafrost;

(e) large dams and dams in warm permafrost built with thaw-stable and artificially thawed soil; and

(f) constructed facilities with passive and active methods of control for the frozen earth in foundations.

1. Structure and Forms of Ice

Ice can be either a naturally occurring or an "engineered" construction material. Naturally occurring ice is found in many forms and varying morphology, e.g., as river and lake ice, ice islands, glacier ice and icebergs, as well as sea ice sheets, floes and ridges. Engineered ice may be formed either by controlled flooding of an area and waiting for the water to freeze, or by spraying water into freezing air so that the vapor instantly crystallizes and falls back to accumulate as a sheet or mound. In certain applications, such as in the construction of ice roads, natural ice may be strengthened by compaction and by additives such as wood shavings and sawdust.

Naturally occurring and engineered ice are generally crystalline in nature. This type of ice is known to possess ten polymorphic forms or stable structures under differing temperature and pressure conditions. The most prevalent form is ice I_h, which has a hexagonal lattice structure. The oxygen atoms are all concentrated along basal planes that are perpendicular to the principal hexagonal axis, called the *c* axis. The H_2O molecule consists of an oxygen atom covalently bonded to two hydrogen atoms. Each H_2O molecule is hydrogen-bonded to four other molecules in an approximately tetrahedral coordination.

Grain boundaries develop when growing ice crystals intermingle and achieve an equilibrium configuration (associated with a minimum in surface energy). Grain sizes are typically in the 1–20 mm range, although much larger sizes can exist. During the growth and formation of ice masses, the crystals may be granular, columnar, or in one of many generally less important forms. The *c* axes of different crystals in an ice mass can exhibit various degrees of alignment, thereby imparting a definite crystallographic texture or fabric to the solid.

Ice crystals reject impurities such as gas bubbles, salts and organic or inorganic matter during the freezing process. However, large amounts of impurities get trapped within the ice, especially when freezing is fast. Consequently, most ice has a porous structure. In freshwater ice, the impurities tend to concentrate along grain boundaries, together with a thin water film. In saltwater ice, on the other hand, the brine is mostly contained in platelets or cells within the ice crystals and normal to the *c* axis. The platelets, with typical spacings of 0.1–1.0 mm, impart a cellular substructure to the material. Gravitational forces cause the formation of large tubular brine drainage channels in natural sea ice features. Typically, the brine channels have a diameter of 10 mm and a spacing of 200 mm.

2. Structure and Forms of Frozen Earth

The thermal regime of earth in cold regions depends on its surface temperature, thermal properties and the geothermal gradient. At the top there is the active layer, typically extending from 5 to 15 m, where seasonal freezing and thawing occurs. Heat and moisture movements between the atmosphere and the earth below take place in this layer.

Below the active layer, the temperature increases steadily under the influence of heat generated deep in the earth. This heat flows upward at a rate dependent on the geothermal gradient, typically 2–3 °C per 100 m, and achieves equilibrium with the mean ground surface temperature. At this depth, the earth is either thawed or may include a layer of permafrost. Sometimes, layers of unfrozen ground may exist within the permafrost or between active layer and permafrost. The depth of permafrost ranges from a few centimeters at the edge of the subpolar zone to about 60–100 m at the boundary with the polar zone. The thickness of the permafrost layer can reach several hundreds of meters in the polar zone. Ice wedge polygons and other frost features may be present in permafrost regions. The wedges form due to the infiltration of cracks in frozen earth by snow and frozen melt-water. Such cracks can endure as long as there is no thawing.

Frozen earth is a multiphase material consisting of solid mineral particles, gaseous inclusions (vapors and gases), liquid (unfrozen and strongly bound) water and ice inclusions. When earth is frozen, its complexity is increased by the appearance of the new solid phase, by the formation of new interfaces and by the random appearance of air bubbles and brine pockets. Ice is the most important component of frozen earth. Frozen water in earth or constitutional ice may exist in the form of large crystals and lenses, or in the form of pore ice which bonds the mineral particles. The frozen pore moisture is termed ice-cement. Ice-cement bonds depend on many factors, including the degree of negative temperature, the total volume content of ice or iciness, the structure and coarseness of ice inclusions and the content of unfrozen water.

If freezing is rapid, ice formation occurs only in the pores and is characterized by a fairly uniform distribution. This type of constitutional ice imparts a fused or massive cryogenic texture to frozen earth. If freezing is slower, continuous ice interlayers and lenses can form which additionally impart laminar and cellular textures to frozen earth.

Unfrozen water, which is present in frozen earth at temperatures down to about −70 °C, generally exists in two states: (a) water strongly bonded by the surfaces of the mineral particles that cannot crystallize even at very low temperatures due to the electromolecular forces of the surface; and (b) loosely bound water of variable-phase composition, between the layer of bound water and free water, that freezes at temperatures below 0 °C. The unfrozen water separates ice from the mineral grains in frozen earth. The amount of unfrozen water in frozen earth decreases as the temperature falls below 0 °C. Depending on the unfrozen water content, frozen earth can be classified as either hard-frozen or plastic-frozen. Hard-frozen earth is firmly cemented by pore ice and contains very little unfrozen water. In most cases this occurs at temperatures below −0.3 °C to −1.5 °C, although for certain very fine mineral particles the limit can be as low as −5 °C to −7 °C. Plastic-frozen earth, which occurs at higher temperatures, has a larger unfrozen water content, e.g., as much as half the total pore water.

3. Structure–Property Considerations for Ice

The structure of polycrystalline ice governs its macroscopic behavior, both when used as a construction material and when present as the most important constituent of frozen earth.

The hexagonal lattice structure makes ice an intrinsically anisotropic material. Single ice crystals are transversely isotropic because of their hexagonal symmetry. Five independent constants are necessary to characterize their elastic behavior. In spite of its open hexagonal structure, the H_2O molecules are relatively close packed in the basal plane. Consequently, ice crystals deform (slip) easily, and may also crack, between basal planes when subjected to shear stresses. The resistance against deformation and cracking is much higher on other planes.

The behavior of polycrystalline ice is complicated by the presence of grain boundaries and by the degree of grain alignment or orientation. Grain boundaries can impede slip and cause dislocation pile-ups. Since grain-boundary planes are weakened by the concentration of impurities during freezing and the resulting increase in defect density, intergranular sliding may occur in polycrystalline ice. In addition, stress concentrations can cause crack nucleation at irregularities in the grain boundary and at grain-boundary junctions or triple points.

The degree of grain alignment is generally expressed in terms of the relative orientation of the *c* axis in ice crystals. If the *c* axes are randomly oriented, the resulting ice is isotropic. On the other hand if the *c* axes lie along a plane but are randomly oriented, the resulting ice is transversely isotropic. If the *c* axes are all aligned in one direction, the resulting ice is truly anisotropic. The latter two types of mechanical behavior represent texture or material anisotropy.

Ice is generally very close to its melting temperature in construction applications, i.e., it occurs at a homologous temperature (ratio of ambient to melting temperature in Kelvin) greater than about 0.8. The mechanical behavior of crystalline solids at such temperatures is controlled by thermally activated rate processes. Thus, ice behavior is very sensitive to rate and temperature variations.

4. Deformation and Failure of Ice

The deformation and progressive failure behavior of ice is governed by three primary mechanisms: flow, distributed cracking and localized cracking. In particular, ice may display purely ductile, purely brittle or combined behavior, depending on the temperature and conditions of loading. The primary mechanism associated with flow and the constitutive framework of rate theory are appropriate for characterizing purely ductile behavior. The primary mechanism associated with distributed cracking and the constitutive framework of damage theory are appropriate for characterizing deformations during ductile-to-brittle transition or when the material is purely brittle. Such deformations are accompanied by the formation and stable growth of multiple cracks and/or voids. The primary mechanism associated with localized cracking and fracture mechanics theory are appropriate for predicting the failure strength corresponding to the onset of material instability.

The effect of the evolution of material structure on the macroscopic flow behavior of polycrystalline ice is only beginning to be understood at the present time. There is general agreement, based on theoretical and experimental work, that at least two thermally activated deformation systems, a "soft" system and a "hard" system, must be present for flow to occur. They may be either grain-boundary sliding (with diffusional accommodation) and basal slip or basal slip and slip on a nonbasal plane. A combination of these processes could be present as well.

Initially, the solid resists the applied stresses in an elastic manner and then flow begins on the soft and hard systems. However, flow, particularly on the easy soft system, causes the build-up of internal elastic stresses. This may occur as a result of grain-boundary sliding next to grains poorly aligned for deformation or dislocation pile-ups at the boundaries of such grains. Dislocation pile-ups at grain boundaries have been observed in ice through scanning electron microscopy. The internal elastic stresses, termed back or rest stresses, resist flow. In addition, internal drag stresses which resist dislocation fluxes are generated in annealed materials undergoing flow. These drag stresses are the outcome of creep-resistant substructures, e.g., subgrains and cells, formed by grain-boundary sliding, and of dislocation entanglement, dipole formation and kink band formation during slip (particularly on the basal plane).

An increasing characteristic drag stress contributes to isotropic hardening, while an increasing rest stress contributes to kinematic hardening. In isotropic hardening, material properties are independent of the direction of straining. On the other hand, kinematic hardening induces directionally dependent material properties, referred to as deformation or stress-induced anisotropy. The Bauschinger effect in metals is an example of kinematic hardening.

The deformations resulting from the interactions between the soft and hard systems can be decomposed into two components: a transient-flow component and a steady-state-flow component. Steady-state flow, representing a balance between work-hardening and recovery processes, is associated with viscous (irrecoverable) strains and obeys an incompressible Norton-type power law. Isotropic and kinematic hardening phenomena are active during transient flow and give rise to anelastic strains. These strains are recoverable on unloading since equilibrium requires the internal elastic back stress to reduce to zero. The transient strain rate can be taken to follow an incompressible Norton-type power law driven by a reduced stress equal to the applied stress minus the rest stress and with the viscous resistance defined in terms of the drag stress. The time-dependent elastic strains defining transient deformation represent the phenomenon of delayed elasticity or anelasticity. The time-temperature equivalence for both deformation systems is given by the Arrhenius law, with similar activation energies for each system.

As the rate of deformation is increased, flow is accompanied by crack formation in ice that is initially crack free. Sudden changes in temperature, i.e., thermal shocks, can also induce cracking. Less is known about cracking in ice than is known about its flow behavior. Cracks generally tend to nucleate in regions of high defect density, which for polycrystalline freshwater ice occur at grain boundaries. Elastic stress concentrations are severe at irregularities in grain boundaries and particularly at grain-boundary junctions, where the lenticular equilibrium shape of a triple point void is similar to the shape of a stressed crack. Once growth initiation occurs, the crack may follow either an intergranular or transgranular path.

At fast loading rates or under thermal shocks, transgranular cracking dominates. Observations indicate that these cracks are characterized by the smooth and often striated surface features of cleavage cracks. In addition they often change direction abruptly on intersecting a boundary in order to be parallel or perpendicular to the basal plane. On the other hand, at slower loading rates, intergranular cracking tends to dominate. The back stresses associated with anelasticity significantly contribute to the (recoverable) strain energy for crack growth at these rates of deformation. These stresses tend to concentrate on grain-boundary planes associated with crystals poorly oriented for deformation, particularly on the soft system; hence, the preference for intergranular cracking. In addition, the stress required to cause growth initiation (or the stress for crack nucleation) is smaller when anelasticity is present since at any given stress level more elastic strain energy is stored in the material.

Commencement of crack growth does not necessarily lead to instability if toughening mechanisms are active. Crack deflection at grain boundaries and voids provides one form of toughening. If the applied loading is inadequate to overcome these barriers, further growth is arrested. This is typically the case under stress states involving pure compression. The length of the arrested cracks is typically a fraction (0.6) times the grain diameter. The first crack which forms under compressive loading tends to be stable. The material sustains additional compressive stress prior to reaching its ultimate strength. The development of multiple stable cracks leads to a distributed cracking phenomenon which governs the ductile-to-brittle transition or purely brittle behavior in ice. This type of cracking weakens ice; under compressive creep stresses the phenomenon leads to "tertiary" or accelerated creep and under constant strain-rate loading in compression to "strain softening." At relatively fast rates of loading involving brittle behavior, ultimate failure is by splitting or slabbing, while in the transition range of loading rates a shear mode of failure dominates. The application of a low to moderate level of confining pressure tends to suppress the development of stable cracks, which in turn allows a higher shear/distortional stress to be sustained. Ultimate failure generally is in the shearing mode. As the confining pressure is further increased the material displays pressure-insensitive ductile flow, and eventually at very high confining pressures pressure-melting is induced and the material can no longer sustain shear/distortional stress. Theoretical models which integrate the flow and damage behavior of ice under multiaxial loading histories are under active development.

States of stress involving tension typically provide adequate energy for unstable crack propagation. In ice with grain diameters larger than a critical value (typical of most engineering applications), the first crack which forms (nucleates) propagates in an unstable manner. The stress at which this crack forms defines the tensile strength. On the other hand, when the grain diameter is smaller than the critical value, crack growth is arrested. Additional stress is necessary for unstable crack propagation. Although viscous flow often accompanies ice deformation prior to the onset of instability or localization of cracking, it is the stored elastic energy in the material which defines the onset of instability.

The energy or stress intensity required for growth initiation and unstable crack propagation in ice are not the same. Growth initiation is associated with a preexisting defect which travels along an intergranular or transgranular path. The resistance to its growth is provided by the surface energy (or energies) appropriate for the particular crack path (or paths). Crack propagation in ice is governed by toughening mechanisms, one of which is crack deflection. Since crack tip plasticity is generally absent in ice, the development of a process zone containing discontiguous cracks can also toughen the material. The amount of toughening clearly depends on the structure and form of the polycrystal. Experimental determination of the critical fracture energy involving such toughening mechanisms is sensitive to specimen geometry and the amount of crack-tip damage introduced when precracking the specimen.

5. *Deformation and Failure of Frozen Earth*

Engineering problems related to frozen earth are caused by the freezing process, thawing, and the steady-state frozen condition. The freezing of earth causes frost heaving, a phenomenon of special importance in the temperate zone. Three conditions are required for frost heaving to occur: (a) a cold surface must exist to propagate freezing; (b) a source of water must exist to feed ice growth; and (c) the physical composition of the soil must promote the migration of moisture to the freezing front. When all three conditions are satisfied, free moisture from the earth below migrates along the thermal gradient toward the colder surface on top. The moisture freezes preferentially to existing ice grains when it reaches the frost line, forming small ice lenses. The lenses form normal to the direction of heat flow and are therefore approximately parallel to the ground surface. As these lenses grow and expand, the surface of the earth moves upward. This movement often is nonuniform due to variations in the soil profile, drainage pattern, surface cover or soil conductivity. These differential movements produce undesirable effects on constructed facilities.

Although the ice segregation process is quite complex, it is known that growth occurs as long as moisture can be attracted from the adjacent soil at the rate of freezing. When the heat removal rate exceeds the moisture supply, the freezing front advances and a new ice lens forms. The total amount of heave is often close to the thickness of segregated ice layers in noncompressible earth. Between 2 and 5 cm of heaving in a single season is common and movements of 15–20 cm are not unusual. The temperate and subpolar regions characterized by seasonal frost have greater potential for frost heaving, unlike polar regions with permafrost, due to the almost unlimited availability of free moisture from groundwater or earth near the freezing front.

Experimental studies show that particle size provides a valid criterion of frost susceptibility in frozen earth. Although there is no sharp dividing line between frost-susceptible and nonsusceptible materials, smaller particles seem to have a dominating influence on the maximum heaving pressures that are developed in a particular material. Both the rate of heave and the heaving pressure increase with constraint provided by confining pressure.

Two basic countermeasures against frost heaving are available: the first involves earth improvement measures, while the second involves stabilization of constructed facilities. Earth improvement measures include: mechanical techniques such as changes in the composition and density of earth or loads on earth; thermophysical techniques such as changes to the temperature and humidity conditions to control migration of moisture; and physicochemical techniques such as artificial salinization to prevent freezing, introduction of inorganic compounds to change filtration and capillary properties, introduction of less wettable compounds, and electrochemical treatment. Stabilization measures include: improved ground drainage for preventing moisture migration to the freezing front, heating of the earth in constructed facilities to prevent freezing using heat insulated screens or artificial methods of heating, application of counteracting forces, and use of non-frost-susceptible materials in construction.

Thawing of earth in the active layer causes melting of the ice lenses. The resulting water escapes and voids are left, increasing the porosity of the soil. As melting progresses downward, the melt water cannot penetrate the frozen earth below and often is unable to dissipate laterally. The trapped water induces excess pore pressures and as a consequence reduces the effective strength of the earth. This phenomenon is termed thaw weakening. If consolidation occurs subsequently, pore pressures dissipate and the material recovers part of its strength. The dissipation of excess pore pressures is termed thaw consolidation, while additional dissipation of pore pressure represents ordinary consolidation. Part of the recovery may also be associated with desaturation of the earth. Thawing causes uniform and differential settlements as well as stability problems in earth masses. The movement of mineral particles together with water on thawing can give rise to serious erosion problems. Thawing may be accompanied by a lowering of the permafrost table. Although some variation can occur due to changes in the natural weather pattern, larger shifts are caused by modifications to the surface cover or introduction of artificial heating sources. In ice-rich earth, the potential for settlement associated with the thawing of permafrost is substantial.

The initial effective stress in earth thawed under undrained conditions, called the residual stress, governs the magnitude of settlements in the earth. When the residual stress is high, the subsequent consolidation settlements and pore pressures generated during thaw are smaller while the undrained shear strength is higher.

Three major principles of construction are available for dealing with thawing earth. They are: (a) to preserve the frozen state of the earth; (b) to design for settlements due to thawing earth; and (c) to improve the earth by preconstruction thawing. The first principle is appropriate in the polar and extreme latitudes of the subpolar zones and when the constructed facilities do not release substantial amounts of heat and do not occupy large areas. The second principle requires a predictive theory of settlements in thawing earth. If the settlements are excessive, an alternative principle must be adopted. The last principle is adopted when it is necessary to reduce (a) future settlements due to thawing and thawed earth, and (b) differential settlements in earth of highly nonuniform compressibility in the frozen and thawed states. Methods for preserving the frozen state of earth attempt to remove heat. One approach is to provide ventilation between the heat source and the frozen earth through physical separation. A second approach involves the use of special pipes with natural or forced ventilation or coarse-pored ventilated rock fill between the heat source and frozen earth. Artificial cooling of thaw-susceptible soils may also be considered. Preconstruction thawing methods rely on natural solar heat or on artificial heat sources. Examples of artificial heat sources include: hydraulic sources such as hot and cold water; steam; electric heating with alternating current; and thermochemical methods. Preconstruction thawing can be augmented by mechanical, physical and chemical methods for compacting and artificially strengthening the earth.

Frozen earth in its steady state, particularly permafrost, is both stable and strong due to the development of ice-cement bonds. However, the stresses and strains in frozen earth vary with time since the ice cement and ice interlayers flow when loaded. Thus frozen earth displays creep, stress relaxation and recovery behavior. When the degree of ice saturation is large, the mechanical behavior may approximate that for ice and the long-term strength is very small. On the other hand, when the degree of ice saturation is small and interparticle forces begin to contribute, the mechanical behavior may approximate that for unfrozen soil. However, the presence of the unfrozen water film surrounding the mineral particles which restricts interparticle contact complicates mechanical behavior. In addition, pressure melting occurs due to stress concentrations on the ice component between soil particles and from hydrostatic pressure on the ice. The migration of unfrozen water under stress gradients also influences the mechanical behavior of frozen earth.

Two general approaches exist for predicting the deformation and compressibility of frozen earth: (a) consider frozen earth to be a quasi-single-phase medium with mathematically well-defined properties, neglecting the fact that one portion of the deformation is due to volume changes; or (b) consider consolidation and flow as two simultaneous but separate phenomena, whose relative amounts of deformation depend on the applied loads and elapsed time. The first approach is easier and more practical, while the second approach is complex but more realistic.

Flow of frozen earth can be divided into three stages: transient flow, steady flow and accelerating flow. Transient flow is characterized by the closure of microcracks, healing of structural defects by moisture that has been pressed out of overstressed zones and refrozen, a decrease in free porosity due to particle dislocation, and partial closure or size reduction of macrocracks. All of these factors cause rheological compacting of the frozen earth. Compressibility of frozen earth can be significant during transient flow, although the actual amount depends on the pressure. This may occur due to instantaneous compression of the gaseous phase, creep of the ice-cement due to shear stresses at the grain contacts, and hydrodynamic consolidation due to the expulsion of unfrozen water under stress. The transition from transient flow to steady flow is characterized by microcrack closure, but the decrease in free porosity is offset by the formation of new (mainly microscopic) cracks. Essentially incompressible steady flow results when equilibrium is established between healing of existing structural defects and the generation of new defects. Finally, during accelerated flow new microcracks are generated at an increasing rate which eventually develop into macrocracks. In addition, the recrystallization and reorientation of ice inclusions causes a decrease in their shear resistance and, consequently, of the frozen earth. In practical applications, it is often sufficient to consider only transient flow in ice-poor earth and only steady flow when excessive ice is present.

Bibliography

Andersland O B, Anderson D M 1978 *Geotechnical Engineering for Cold Regions.* McGraw-Hill, New York

Ashton G D 1986 *River and Lake Ice Engineering.* Water Resources Publications, Littleton, Colorado

Barnes P, Tabor D, Walker J C F 1971 The friction and creep of polycrystalline ice. *Proc. R. Soc. London, Ser. A* 324: 127–55

Chamberlain E J 1981 *Frost Susceptibility of Soil: Review of Index Tests,* Monograph 81-2. US Army Cold Regions Research and Engineering Laboratory, Hanover, New Hampshire

Chung J S, Hallam S D, Maatanen M, Sinha N K, Sodhi D S 1987 *Advances in Ice Mechanics—1987.* American Society of Mechanical Engineers, New York

Eranti E, Lee G C 1986 *Cold Region Structural Engineering.* McGraw-Hill, New York

Farouki O T 1981 *Thermal Properties of Soils,* Monograph 81-1. US Army Cold Regions Research and Engineering Laboratory, Hanover, New Hampshire

Gandhi C, Ashby M F 1979 Fracture-mechanism maps for materials which cleave: F.C.C., B.C.C and H.C.P. metals and ceramics. *Acta Metall.* 27: 1565–602

Glen J W 1975 *The Mechanics of Ice,* Monograph 11-C 2b. US Army Cold Regions Research and Engineering Laboratory, Hanover, New Hampshire

Gold L W 1963 Crack formation in ice plates by thermal shock. *Can. J. Phys.* 41: 1712–28

Goodman D J, Frost H J, Ashby M F 1981 The plasticity of polycrystalline ice. *Philos. Mag. A* 43(3): 665–95

Hobbs P V 1974 *Ice Physics.* Clarendon, Oxford

Johnston G H (ed.) 1981 *Permafrost Engineering Design and Construction.* Wiley, Toronto

Ketcham W M, Hobbs P V 1969 An experimental determination of the surface energies of ice. *Philos. Mag.* 19(162): 1161–73

Mellor M 1983 *Mechanical Behavior of Sea Ice,* Monograph 83-1. US Army Cold Regions Research and Engineering Laboratory, Hanover, New Hampshire

Michel B 1978 *Ice Mechanics.* University of Laval Press, Quebec City

Schulson E M 1988 Ice: Mechanical properties. In: Cahn R (ed.) 1988 *Encyclopedia of Materials Science and Engineering,* Supplementary Volume 1. Pergamon, Oxford

Shyam Sunder S, Wu M S 1989 A multiaxial differential flow model for polycrystalline ice. *Cold Regions Sci. Technol.* 16: 45–62

Tsytovich N A 1975 *The Mechanics of Frozen Ground.* Scripta Book Company, Washington, DC

Weeks W F, Ackley S F 1982 *The Growth, Structure, and Properties of Sea Ice,* Monograph 82-1. US Army Cold Regions Research and Engineering Laboratory, Hanover, New Hampshire

Weertman J 1983 Creep deformation of ice. *Annu. Rev. Earth Planet. Sci.* 11: 215–40

In addition, several scientific journals and international conferences, listed below, emphasize the study of these materials.

Scientific journals:
Annals of Glaciology
Canadian Geotechnical Journal
Cold Regions Science and Technology
Journal of Glaciology
Journal of Offshore Mechanics and Arctic Engineering

Proceedings of international conferences:
IAHR Symposium on Ice
International Permafrost Conference
International Symposium on Ground Freezing (ISGF)
IUTAM Symposium on Ice
Offshore Mechanics and Arctic Engineering (OMAE)
Port and Ocean Engineering Under Arctic Conditions (POAC)

S. Shyam Sunder
[Massachusetts Institute of Technology, Cambridge, Massachusetts, USA]

Indigenous Building Materials

Indigenous building materials have two inherent advantages over other materials: low cost and local availability. As such, they have long been employed throughout the world. Centuries of use have resulted in the attainment of high levels of skill in the use of these materials, as attested by the great pyramids of Egypt, the Greek Parthenon and the Taj Mahal of India.

Developing countries today continue to rely on indigenous materials. In hot and humid lands, which support the growth of plants and forests, a variety of materials such as grasses, leaves, straws, reeds, bamboo, timber and allied materials are found. Hot and dry areas yield inorganic materials like clays, laterite, sand and stone for building purposes.

An example from India illustrates the widespread reliance on mud as a building material. According to the 1981 census, there was a total housing stock of 90 million dwelling units in India. Of these, 57% or 51.2 million employed mud as a principal building material, particularly for walling.

The wide range in type and use of indigenous materials and their corresponding importance in sheltering much of the world's population is evident from the following discussion.

1. *Inorganic Materials*

1.1 Soil—An Overview

The type of soil available in a given area depends on the climatic region and topography. Silty soils, composed of very fine sand and clay, are found in river deltas. Such soils are well suited for construction of walls and production of mud bricks, burnt clay bricks and roofing tiles, for instance. Muram types of soil, containing gritty material, are common in mountainous plateaus. These soils are suitable for construction of road surfaces, embankments and dams. Laterite soils which contain gravel in compressed formation are also used for building. When laterite soils are stabilized with lime, good quality blocks can be produced.

1.2 Adobe

Adobe bricks are one of humankind's oldest building materials. These unburnt clay bricks are formed from highly plastic soils containing >70% clay and dried in the sun for 2–3 weeks.

Mud bricks are made in the following manner. Soil is collected in heaps and water is poured into it. The soil is then puddled into plastic consistency by trampling with bare feet. Straw or chopped hay is spread over the top in order to allow the bricks to "breathe" while curing or drying. The entire mass is then kneaded to distribute the binder.

1.3 Pisse-de-Terra

The conventional method of construction of mud walls is to convert the soil available at the site into mud with the addition of water. Lumps of mud are then piled one over another to form a wall. For protection and finishing, the mud wall is plastered with a paste of the same soil mixed with small quantities of cow dung and straw.

To make mud walls strong and durable, techniques of soil stabilization have been evolved using materials such as lime, cement and emulsified bitumen. Nearly any soil can be made into a better building material with the addition of the correct stabilizer. Briefly, stabilizers perform the following actions:

(a) they cement the particles of soil together so that they are stronger and more durable;

(b) they increase resistance to weathering from sun and wind;

(c) they render the mud wall watertight, thus protecting it from rain and erosion; and

(d) they protect soil from shrinkage and swelling.

Stabilization thus eliminates the need for frequent repair or complete rebuilding of walls or roofs after rain. Stabilizers have also been developed to lengthen the life of bricks.

1.4 Water-Repellent Mud Plaster

The application of water-repellent mud plaster can also increase the life of soil-based walls and roofs. To prepare this plaster, bitumen emulsion composed of 80/100 grade bitumen, kerosene and creosote is mixed with ordinary mud plaster prepared with cow dung and straw. This treatment lasts for 4–5 years.

1.5 Laterite Blocks

A final type of sun-dried brick is the laterite block. Laterite is an abundant subtropical soil or rock formed by the weathering of igneous rocks. When combined with other locally available materials such as lime, it can be processed to obtain good-quality building bricks with a wet compressive strength of ~5–6 MPa.

1.6 Burnt Clay Bricks

Not all soil-based materials rely on the sun to dry them to their finished state. Many soil products such as clay roofing tiles require high-temperature burning.

The alluvial soils, such as those of the Indo-Gangetic plain, are ideally suited for the manufacture of very good quality burnt bricks and tiles. To produce such products, the soil is dug out and mixed with water and kneaded well to make the soil mass homogeneous. After the soil has been kept wet for several days, bricks are molded in 230 × 115 × 75 mm wooden molds, although other brick sizes can be produced. The wet bricks are left in the open to dry in the sun for about a week. After drying, they are fired in a clamp or kiln, depending upon the scale of production and types of bricks required, using firewood and coal as fuel. For regular production of bricks, large kilns called Bulls' Trench kilns are adopted using coal or oil firing.

Many improvements have been made in the preparation of soil, molding of bricks and firing in kilns to improve the quality of bricks and to reduce the production cost. For production of bricks of high compressive strength, mechanical methods are now used, such as table molding using simple machines and extrusion. A variety of structural clay products have also been produced for use in construction, such as hollow clay blocks, perforated bricks and flooring tiles.

1.7 Clay Roofing Tiles

Plastic clays are used for producing half-round country tiles with the help of a potter's wheel. After drying, the tiles are burnt in a clamp using firewood, cow dung, straw or other agricultural waste as fuel. These tiles are used extensively for covering sloping roofs, mostly in rural areas. However, country tiles are very fragile and are easily displaced or broken. For this reason, roofing is often made of bamboo or local timber rather than tile.

Improved types of clay roofing tiles have been evolved. One example is pantiles, which are formed in a wooden mold, dried and then burnt in clamps or kilns. Even more durable clay tiles are produced by an extrusion process. They are dried and then burnt in kilns. These tiles interlock into place.

In addition to roofing tiles, burnt clay bricks and tiles are also manufactured with dimensions 230 × 115 × 38 mm. They are used for masonry wall construction as well as for paving floors. Square clay tiles of 60 mm size are also produced and used for flooring in low-cost houses. Finally, vitrified clay roofing tiles having high compressive and abrasive strength are used for paving of footpaths and streets.

Research has been undertaken to upgrade the quality of clay roofing tiles by adopting better methods of molding and improving the design.

1.8 Clay Pozzolana (Surkhi)

To make clay pozzolana, burnt brickbats and overburnt bricks are ground to powder and mixed with lime for preparation of mortars, plasters and foundation concrete. Because the pozzolana prepared in this manner has a low lime reactivity, suitable clay deposits for producing reactive pozzolana have been sought. When calcined at optimum temperature, these specially selected clays produce high pozzolanicity. If used in conjunction with lime for mortars, the reactive pozzolana gives strength similar to that of cement mortars. Development of economical processes for reactive clay pozzolana has hastened the spread of this material within the developing world.

The lime—clay pozzolana mixture can replace cement in mortars, plasters and foundation concrete and is better than the cement—sand mortars. In rural areas, lime pozzolana mixtures could completely replace cement in single and double storey constructions.

1.9 Building Lime

In addition to soil bricks and blocks, rock and stone of various types make a valuable contribution to the building industry. Limestone is often available in the form of rock in the hills or as small pieces of limestone in the soil called Kankar lime. When burnt, limestone provides quicklime, the source of building lime, widely used as a binding material for preparation of mortars, plasters and concrete, with additives such as sand, pozzolanic material obtained from clay, fly ash and other materials.

Lime is produced in small kilns of 3–5 tonnes capacity with wood or coal generally used as fuel. For regular production, continuous kilns of higher capacity are adopted. However, nowadays dry hydrated lime is available in ready-to-use form.

1.10 Gypsum

Large deposits of gypsum of various grades are indigenously available in some developing countries. This material is commonly used in preparation of plasters and the production of cement. It can also be calcined to obtain gypsum plaster, which is used for internal decorative work. In addition, gypsum blocks can be manufactured and used for non-load-bearing internal partitions with good thermal insulation. However, because gypsum absorbs water, it cannot be used for external applications.

Fibrous gypsum plasterboard reinforced with jute, reeds, glass, wood, galvanized wire mesh or other materials is used for floors, ceilings and light partitions. Manufacture of this product could be undertaken on a cottage industry basis.

Acoustic tiles of gypsum are also produced and used in auditoria and other types of room where reduction in reflected sound is required.

1.11 Sand

Sand is made of crushed particles of rock of all sizes. Although it is usually obtained from river beds, rock can be crushed to form sand of required sizes.

Sand is a siliceous material with widely varying characteristics which depend on its place of origin. It is used as an admixture for preparation of mortars and plasters for masonry work and for production of concrete in conjunction with lime or cement in specified proportions. The quality of the local sand determines the amount needed in a particular mortar or cement.

Sand is also used for production of building components such as solid and hollow concrete blocks and other types of structural products. Cement–sand blocks as well as sand–lime bricks produced by autoclaving are often used for loadbearing masonry construction.

Lightweight concrete blocks and other structural building components are produced from sand, cement or lime with the aid of air-entraining agents. These prefabricated products are used for construction of walls, roofs and floors.

1.12 Building Stone

A variety of stone is used in construction, including igneous rocks such as granite, basalt and trap; sedimentary rocks like sandstone, limestone and laterite; and metamorphic rocks such as slate and marble.

Stone is used for construction of thick and massive walls, of which random rubble masonry walls are an example. For fine work, ashlar masonry walls are built. In olden days, stone arches formed the support system for the roof. Also, flat stone slabs from materials such as slate were adopted for roofing.

For stone masonry work, lime-based mortars and plasters are generally adopted. Lime concrete with stone aggregates, for example, is used for foundation work. For flooring and paving, stone is also used to provide a hard durable surface. One such example is marble, a costly stone which finds use in buildings of architectural character.

Nowadays stone is used in the production of cement concrete, a popular building material. Stone spalls can be used with cement concrete to produce stone blocks used for masonry work. Using molds, these blocks can easily be produced at the construction site or factory by skilled workers. Stone masonry blocks are made of loose cement concrete and stone pieces of 50–250 mm, which should be sound, durable and free from defects. The concrete mix used is generally 1:5:8 (cement, sand and concrete aggregate, respectively) to obtain blocks with a compressive strength of >6–7 MPa. The use of stone masonry blocks results in a saving of 15–20% in the cost of walls compared with random rubble masonry.

2. Organic Materials

2.1 Grass Thatch

In rural areas, grasses from forests and fields are extensively used for thatch roofing, formed by binding the grasses to a bamboo or timber roof frame. Yet, despite its widespread usage, thatch is highly vulnerable to fungi, insects and fire.

A number of preservatives have been developed in response to these threats. To protect thatch from fungi and insects, a solution of copper sulfate, sodium dichromate and acetic acid is often coated on the thatch. For protection from fire as well as decay, an application of a solution of boric acid, copper sulfate, zinc chloride and sodium dichromate is used. However, although the service life of thatch can be extended by 3–5 years with these treatments, the required chemicals are costly and often not available locally, so these treatments have not found wide application. A cheap alternative treatment can be made from mud mixed with 5% bitumen emulsion.

2.2 Palm Leaves

Palm leaves and midribs of dead palms are often used for roofing. Palm leaves are used for roof thatch, whereas palm midribs are used for reinforcing concrete in floors, roof coverings, walls and columns of low-cost housing. However, water-repellent and fire-retardant treatments should be applied to extend the life of palm thatch.

2.3 Husks and Straw

Husks from rice or wheat are available in large quantities after the threshing season. These husks are commonly used as an admixture in mud plaster to provide strength and prevent cracking. They can also be used with cement to form roofing sheets for low-cost housing.

Rice-husk ash provides siliceous material of pozzolanic properties. It is therefore used in conjunction with lime or cement to provide a low-cost binding material used in mortars and plasters.

The straw from rice and wheat is also useful, both as a component of building materials and as a fuel. Often, chopped straw is an admixture for mud plaster. Straw mixed with mud is also used to produce small roofing tiles. At the same time, straw and other agricultural wastes are used as fuel for burning bricks, tiles and lime.

2.4 Jute Stalk

Numerous building materials come from the jute stalk: building boards for low-cost roofing as well as strings and ropes for tying bamboo and timber frames. In addition, jute bags are commonly used for packaging of cement.

2.5 Coir

Like jute, coconut fiber, called coir, is used for making strings and ropes used for tying bamboo and timber frames. In addition, cement-bonded coir can be employed to manufacture building boards used for internal application, such as for acoustic treatment. Cement-bonded coir and roofing sheets can also be produced to provide low-cost roofing material.

2.6 Cow Dung

Cow dung is commonly used as an admixture with soils that are clayey and silty, to reduce shrinkage in preparation of mortars and plasters used to repair cracks in mud walls.

2.7 Reeds

The main types of reed used for building—cane (*Phragmites karka trin*) and papyrus (*Typha latifolia*)—are found in abundance on the edge of marshes. Both of these types have been used as building materials since time immemorial, in part because few if any tools are needed to fashion them into a shelter or building component.

Reeds are most commonly used in the form of sheets and boards. Reed sheets, made of reeds bound with string, are commonly used for roofing over timber or bamboo rafters, despite their high susceptibility to rain and high winds. Reed sheets are also used to enclose spaces in the form of party walls, while reed panels are used to enclose spaces between timber frames. Both sheets and panels are normally finished with mud or cement–sand plaster.

A recent innovation in the use of reeds is the reed board, formed by compressing a small bunch of reeds together and wiring them tightly. These boards, varying from 25 to 50 mm in thickness, are used between timber framework for walling or for roofing. They are coated with mud, lime or cement plaster.

Reed sheets as well as reed boards are vulnerable to termites and thus require antitermite treatment.

2.8 Bamboo

Bamboo is one of nature's most versatile products. By virtue of its ready availability, comparative cheapness, high strength–weight ratio and easy workability, it has continued to play a vital role in rural housing. Bamboo is an excellent material for the framework of houses particularly in seismic areas. It is also used for foundations, flooring, walls, partitions, ceilings, roofing, doors, windows and drainage systems. Because of its high tensile strength, bamboo has also been used for manufacturing a variety of boards and as a reinforcement for concrete, although its bond strength is low.

Despite its many attributes as a building material, bamboo also has some shortcomings which limit its wider application in the field of building construction. One of the most serious disadvantages of bamboo is its low durability; another is its extreme fissility. Except for certain thick-walled types, most bamboos have a tendency to split easily. This prevents the use of normal joints or even the use of nails. However, cutting a notch or mortice in a bamboo drastically reduces its ultimate strength. The only remedy is the use of nodes as places of support and joints, and the use of lashing materials in place of nails.

As an organic material, bamboo is subject to decay and can also be destroyed by insects, particularly borers. However, research shows that bamboo treated with copper–chrome–arsenic and copper–chrome–acetic-acid compositions are adequately preserved. Methods of treatment include the Boucherie process and dipping. It is estimated that the expected service life of treated bamboo is 15–20 years, compared with 2–3 years for untreated bamboo.

2.9 Wood

Wood is one of the oldest building materials used considerably in construction. There is a general botanical classification of trees into two groups: exogenous or outward-growing, and endogenous or inward-growing. The former includes all the commercial timber used for building purposes and the latter includes the plants and bamboos of the tropics. The exogenous trees are divided into two main groups: hardwoods and softwoods. This division rests on a purely botanical distinction and not upon the hardness or softness of the timber; many hardwoods are softer than certain softwoods.

Woods that are particularly important to the construction industry include teak, deodar (a member of the cedar family) and sal. If these woods are in short supply, there are more than 120 varieties of secondary species which could be used for construction if properly seasoned and given preservation treatment.

The process of drying timber under controlled conditions is called seasoning. It is especially important for hard, heavy timbers located in countries with wet climates, such as India. Once seasoned, timber should be stored under cover, protected from direct sun and rain.

Timber deteriorates from a number of causes, among which are fungi, insects and marine borers. It will also break down quickly if subjected to alternate dry and wet conditions or if used in a dark, damp, unventilated area. However, timber kept in a dry, airy place or continuously submerged under fresh water will last indefinitely. The use of preservatives to protect against various organic pests further ensures the longevity of the wood.

Various types of wood preservatives are available: oils, organic-solvent-based compositions as well as water-soluble "leachable" or "fixed" preservatives. The timber can be treated by surface application, soaking treatment, the hot and cold process, the pressure process or the Boucherie process.

It is not only sawn timber which finds an application in construction. Branches and twigs are used in conjunction with mud to build low-cost houses, particularly in the construction of walls. A variety of agricultural wastes such as timber refuse, bagasse and sugar cane also find use in building.

See also: Roofing Materials; Structural Clay Products; Wood as a Building Material

Bibliography

Central Building Research Institute 1983 *Gypsum as a Building Material*, Building Research Note 14. Central Building Research Institute, Uttar Predash, India

Central Building Research Institute 1983 *Precast Stone Block Masonry*, Building Research Note 7. Central Building Research Institute, Uttar Predash, India

Government of India 1976 *Status Report on Rural Building Materials.* Department of Science and Technology, Government of India, New Delhi

Mathur G C 1976 *Survey of Rural Housing and Related Community Facilities in Developing Countries of ESCAP Region.* United Nations Economic and Social Commission for Asia and the Pacific, New Delhi

Mathur G C 1983 *Development and Promotion of Appropriate Technologies in the Field of Construction and Building Materials Industries in India.* United Nations Industrial Development Organization, Vienna

United Nations 1976 *Use of Agricultural and Industrial Wastes in Low Cost Construction.* United Nations, New York

G. C. Mathur
[National Building Organization,
New Delhi, India]

International Trade in Materials

The international trade of materials has reached a new level of importance in the economic and political relations between nations. Factors which have long affected materials trade patterns have also received new prominence. Resource depletion in the industrialized countries has increased their materials imports from the rest of the world, particularly the developing countries. The unequal geographic distribution of these resources has now made geopolitics a critical factor in national materials decision making. Multinational materials companies have also expanded their activities, as have their counterparts, in the form of state trading organizations. Institutions also have arisen which would influence the competitive or free trade characteristic of materials flows. These range from producer organizations such as the Intergovernmental Council on Copper (CIPEC) to complex commodity arrangements such as those recently discussed within the Integrated Program for Commodities (IPC) of the United Nations Conference on Trade and Development.

1. *Structure of Trade*

The starting point for analyzing the structure of materials trade is the commodity composition of that trade; that is, the materials that are most important among world exports and how this importance has changed over time. Table 1 provides this information for the major commodity groups traded. Although the fuels are shown to dominate total trade, the main concern here is with the agricultural materials and minerals commodities. Among these, the agricultural raw materials and iron and steel are the most important. Exports of both of these materials groups have more than doubled since 1973, with iron and steel alone increasing from US$28.46 billion in 1973 to US$76.50 billion in 1980. Next in importance are the nonferrous metals with an export total of US$52.00 billion in 1980. Ores and minerals exports also have increased sharply, reaching some US$39.00 billion in 1980, but represent a smaller fraction of minerals exports because of the high transport costs relative to those for refined metals. Nonmetals surprisingly are not very significant in world minerals trade. Since the magnitude of their transport costs is similar to that of ores, the low unit values of the nonmetals do not justify their transportation over long distances.

Moving from trade composition to trading areas, it is obvious that these areas can be grouped only in a very general way. Different groupings are possible, but the one considered here is based on United Nations data for (a) industrial or developed market economy countries, (b) developing market economy countries, and (c) centrally planned economy countries. Table 2 provides the geographical patterns of materials trade over these areas. The materials data available for this purpose divide total materials trade into agricultural raw materials and into ores and metals. The percentage distributions reported show the destination of exports and sources of imports for each area.

Table 1
Value of world export trade in major materials

Materials	Export trade (billion US$)		
	1973	1978	1980
Agricultural raw materials	34.61	52.05	75.00
Ores and minerals	14.91	24.50	39.00
Nonferrous metals	17.28	27.80	52.00
Nonmetals (crude)	5.90	7.72	10.00
Iron and steel	28.46	57.15	76.50
Fuels	63.48	223.60	468.00
Total commodity exports	574.30	1303.00	1973.00

Source: *International Trade, 1980-81*. General Agreement on Tariffs and Trade (GATT), Geneva, 1981

The more important materials trade patterns both in terms of exports and imports are among the industrial economies. These countries, which basically form the Organization for Economic Cooperation and Development (OECD), dominate world trade. The developing countries export materials mostly to the industrial countries. They also import materials from them, most of which have been processed in some way. The centrally planned economies of Eastern Europe export and import mostly to and from each other, with the exception of agricultural raw materials exported to the industrial countries.

What Table 2 fails to enumerate in describing materials trade is the nature of primary material, ore and metals trade, compared with trade in materials that are processed or semiprocessed. It also fails to indicate trading patterns between individual countries. Although Table 2 cannot be disaggregated completely to show these more detailed trade flows, an attempt has been made in Table 3 to perform a disaggregation for the more important materials and countries. For each material the five most important countries have been selected and ranked according to percent market share.

Examining the agricultural raw materials first, one finds a similar trade pattern for cotton, jute, lumber and rubber. These materials are mostly exported from developing countries and are imported by the industrial countries. Some exceptions are the USA as a major lumber exporter and Pakistan as a major jute importer. A similar, though more mixed trade pattern emerges for the metals. Australia, for example, is a major exporter of bauxite, iron ore and lead, and Canada is a major exporter of lead and copper. All of the major importers of these metals are industrialized countries. While these trade patterns become less

Table 2
Gross exports and imports of agricultural and minerals materials 1978

	Total value (billion US$)	Distribution (%) World total	Industrial market countries	Developing countries	Centrally planned countries
Exports (fob)					
Industrial market countries					
Agricultural materials	31.39	100.0	80.6	13.6	5.8
Ores and metals	81.97	100.0	68.2	20.7	11.1
Total	871.99	100.0	70.8	23.8	5.4
Developing countries					
Agricultural materials	14.44	100.0	61.8	26.4	11.8
Ores and metals	16.24	100.0	73.3	20.9	5.8
Total	300.80	100.0	70.6	24.4	5.0
Centrally planned countries[a]					
Agricultural materials	4.83	100.0	55.9	10.0	34.1
Ores and metals	10.79	100.0	27.5	7.8	64.7
Total	112.43	100.0	26.1	15.2	58.7
Imports (cif)					
Industrial market countries					
Agricultural materials	37.95	100.0	67.6	23.4	9.0
Ores and metals	70.81	100.0	78.6	16.9	4.5
Total	863.54	100.0	71.5	24.6	3.9
Developing countries					
Agricultural materials	8.80	100.0	49.0	43.0	8.0
Ores and metals	21.24	100.0	79.2	16.0	4.8
Total	303.47	100.0	68.3	24.2	7.5
Centrally planned countries[a]					
Agricultural materials	4.31	100.0	31.4	26.6	42.0
Ores and metals	13.36	100.0	41.6	5.4	53.0
Total	110.45	100.0	31.2	10.3	58.5

Source: *Handbook of International Trade and Development Statistics*, Supplement, 1980. UNCTAD, United Nations, Geneva
[a] Includes Eastern Europe only

significant when viewed in terms of Table 2, which reflects the total mix of primary, semiprocessed and processed materials, they still suggest a flow of primary materials from the developing to the industrialized countries. This structure of trade has led to the so-called "North–South Debate" between materials producing and consuming countries. We will return to this important materials trade issue later.

2. *Determinants of Trade*

A number of different determinants underlie the trade patterns described above. Geographic determinants relate to the geological distribution of resources; political determinants relate to particular strategic blocs such as the COMECON countries; and social determinants relate to the need to accelerate development in different parts of the world. However, economic determinants are the primary concern here.

These determinants may be understood by first abstracting trade flows from the above material by organizing them into a general trade table or matrix (Table 4). Its elements T_{ij} indicate the quantity of the material shipped from country i to country j during a period under consideration. Total exports of the ith country (T_i^X) can be calculated from this matrix simply by summing the ith row ($\Sigma_{j=1}^{m} T_{ij}$), and the total imports of the jth country (T_j^M) by summing the jth column ($\Sigma_{i=1}^{n} T_{ij}$). Total world trade T in the material is the sum of all elements ($\Sigma_{i=1}^{n} \Sigma_{j=1}^{m} T_{ij}$). Economic analysis of this trade is normally divided in three ways: (a) imports into a country T_j^M, (b) exports from a country T_i^X, and (c) flows between any pair of countries T_{ij}.

Materials import demands depend on a number of factors which range from a country's general economic level and its military goals to its relative prices and costs of producing that material domestically. Steel imports may thus depend on the size of the domestic automobile industry, the extent of the armaments industry, and the costs of producing steel domestically. Economic demand theory would refine this explanation by stating that the quantity of material imports purchased is based on real domestic income or prod-

Table 3
Geographical distribution of major materials exporters and importers (and percent market share in 1978)[a]

Materials	Major exporting nations	Major importing nations
Rubber	Malaysia (25.7), Indonesia (11.9), Singapore (18.1), Sri Lanka (1.2), Thailand (6.6)	USA (17.4), Singapore (13.3), Federal Republic of Germany (8.4), Japan (7.6), France (6.5)
Lumber	USA (27), Indonesia (20), Malaysia (16), Ivory Coast (5.2), Philippines (3.2)	Japan (54.6), Korea (10.1), Italy (6.3), Federal Republic of Germany (5.5), France (3.4)
Bauxite	Australia (38.3), Jamaica (19.1), Surinam (9.5), Guinea (4.4), USA (5.2)	USA (40.1), Canada (8.1), Japan (7.7), Norway (7.0), Federal Republic of Germany (7.0)
Copper	Canada (5.3), Chile (16.0), Zambia (8.8), Zaire (6.3), Peru (4.6)	Federal Republic of Germany (13.8), USA (13.3), Belgium (10.3), France (10.2), UK (8.8)
Iron ore	Australia (21.2), Brazil (20.9), Canada (13.4), India (7.6), Sweden (6.4)	Japan (38.9), Federal Republic of Germany (15.6), USA (15.8), UK (5.2), Belgium–Luxembourg (5.1)
Tin	Malaysia (39.6), Thailand (16.4), Bolivia (15.7), Indonesia (12.7), UK (4.9)	USA (31.2), Japan (18.6), Federal Republic of Germany (11.1), France (7.6), UK (5.3)
Cotton	USA (36.3), Egypt (7.3), Turkey (7.2), Sudan (5.6), Mexico (2.3)	Japan (22.1), Korea (9.3), Italy (7.9), France (6.7), Federal Republic of Germany (7.4)
Lead	Australia (23.6), UK (10.6), Federal Republic of Germany (12.0), Canada (9.1), Mexico (5.6)	UK (19.4), USA (17.8), Federal Republic of Germany (10.3), Italy (9.6), Netherlands (5.8)
Jute	Bangladesh (66.7), Thailand (12.3), Nepal (7.7), Burma (4.6), India (4.2)	UK (13.5), Pakistan (13.4), Egypt (7.0), Japan (5.9), France (5.7)
Tungsten	Australia (18.3), Thailand (16.5), Bolivia (13.0), Canada (11.7), Korea (7.5)	USA (23.7), Australia (16.6), UK (12.5), Sweden (9.0), Federal Republic of Germany (10.3)

Source: *Yearbook of International Trade Statistics*, Vol. 2, 1980: *Trade by Commodity*. United Nations, New York [a] A single country may be classified as both importer and exporter because stages of process have been aggregated

Table 4
Generalized trade matrix

T_{11}	T_{12}	$\cdots$	T_{1j}	$\cdots$	T_{1m}		T_1^X
T_{21}	T_{22}	$\cdots$	T_{2j}	$\cdots$	T_{2m}		T_2^X
$\vdots$	$\vdots$		$\vdots$		$\vdots$		$\vdots$
T_{i1}	T_{i2}	$\cdots$	T_{ij}	$\cdots$	T_{im}		T_i^X
$\vdots$	$\vdots$		$\vdots$		$\vdots$		$\vdots$
T_{n1}	T_{n2}	$\cdots$	T_{nj}	$\cdots$	T_{nm}		T_n^X
T_1^M	T_2^M		T_j^M	$\cdots$	T_m^M		T

uct, the price level of that import relative to the price level of an equivalent domestic material and other factors, such as foreign exchange reserves, credit availability and so-called dummy variables which reflect unusual import-adjusting events or institutions. The substitutability of the material imports against other imported or domestic goods is also important (see *Substitution: Economics*).

Materials exports depend on the demand for a material in other countries as well as the domestic capacity to produce that material. Economic trade theory normally considers exports in the demand context; they are primarily the demand for imports in other countries or areas of the world. A country's material exports are thus dependent on world income or product (or national income in the respective importing countries) and its export prices relative to that of the world or other countries.

Whether or not a country performs well in providing exports is often considered in terms of its ability to increase its market share. Since this depends on its relative export price level, export analysis must inevitably deal also with the conditions underlying the domestic supply of that material. These often include differential growth rates of available productive factors and the responsiveness of export supply to the domestic supply of these factors, differential rates of productivity increases, and the extent to which the country is concentrated in exports to very rapidly growing markets. Behind these factors is the quality of the materials resource base, whether it be the richness of the mineral deposits or the high yields from agricultural production.

Instead of dealing just with total imports or exports, economic analysis also attempts to explain the individual trade flows between countries, as given in the above matrix. In this context, a country's choice of trading partners is seen as depending on a number of complex factors in addition to prices and incomes. The material trade flow between countries i and j is thus seen as depending on: (a) the potential of country i to export the material, (b) the potential of the country j to import the material, and (c) the resistance to trade between country i and country j. The distance between the two countries is seen as a major factor determining resistance to trade, since transportation costs rise with distance. However, other factors can be equally if not more important. Membership of the same political bloc such as the European Economic Community could be expected to stimulate trade. Also in this category is the proximity of being a neighboring country. Similarly, when multinational companies based in the importing country own production facilities in the exporting country, ownership ties may promote trade. Production differentiation such as that arising when mineral concentrates contain particular impurities that only certain smelters can remove may also influence trade. Finally, the nature of purchase agreements such as long-term mineral contracts can also be significant.

An analysis of the relative importance of these determinants of trade flows in certain metals has been made by Demler and Tilton (1980). Materials studied included bauxite, alumina, aluminum, copper concentrate, blister copper, refined copper, iron ore and refined nickel. The results of the analysis generally support ownership ties and the multinational companies that create them as greatly influencing trade, at least at one stage of production. For aluminum and steel they are very important at early stages of processing but they are not so important for refined copper. Political blocs also are significant for trade in refined products but not for trade in ores and concentrates. Sharing a common border has also been found to stimulate trade in refined products such as aluminum, nickel and copper, but not in ores other than iron ore. Finally, long-term contracts have also been found to be an important determinant, particularly in more recent years.

An alternative view for explaining trade flows or why countries are net material importers or exporters concentrates on the concept of comparative advantage. This concept assumes international markets to be freely competitive and that trade improves in those countries which have lower production costs due to more efficient factors of production. If the factors of production can be extended to include resources as well as labor and capital, then countries with a higher quality (or lower cost) materials resource base will eventually become the major exporters. However, a number of imperfections in the world materials markets would complicate this simple concept. These include the stage of process of the material and the growth in demand for it, transport problems and costs, integrated and oligopolistic market conditions, and tariff and nontariff barriers levied on processed or semiprocessed materials as they move to other markets.

The latter are of particular importance here. In theory, they are seen as altering levels of economic welfare in particular countries. In practice, they are seen as influencing the benefit that can be derived from producing materials or having access to them. Thus tariffs on imports provide a form of domestic subsidy that would protect and foster growth of an infant materials industry. Domestic production of such materials might additionally be encouraged for security reasons. Tariffs could also preserve an inefficient yet active domestic mining industry to sustain exploration and to stimulate employment. Both arguments also seem to be influenced by industrialized countries having higher trade barriers for processed as compared with unprocessed materials. These tariffs are seen as being harmful for a number of reasons: materials cannot be obtained from lowest cost sources; rapid exhaustion of resources may occur, sometimes under premature technology; and trade patterns can develop that will lead to less than optimal growth in the long run.

3. *Important Trade Issues*

Several issues have tended to reappear frequently in recent discussions of international materials trade. In this section, they are briefly reviewed and their significance assessed.

3.1 Commodity Arrangements

Materials prices are known to fluctuate more frequently and widely than prices of semiprocessed or processed goods. The explanation normally given for this phenomenon is that materials markets experience variations in demand caused by changing industrial and economic activity in various nations. Given that materials producers cannot change their output rapidly to meet such demand variations in the short run, materials prices will then have to adjust or fluctuate considerably to move a particular market towards equilibrium. This instability has thus been seen as harmful to materials producers, particularly those located in developing countries where materials earnings dominate total export or foreign exchange earnings.

As a consequence, considerable international deliberations and negotiations arose during the 1970s. The focus of these negotiations has been the United Nations Conference on Trade and Development (UNCTAD). In particular, and at the request of the developing countries, it has launched the Integrated Program for Commodities (IPC). The set of policy initiatives contained therein relate to a set of corrective commodity arrangements as well as a so-called "Common Fund."

These commodity arrangements would attempt to maintain commodity prices at levels that in real terms provide producing countries with favorable terms of trade (i.e., maintain material export purchasing power in relation to manufactured goods import requirements). In addition they would also attempt to reduce price and volume fluctuations that cause fluctuations in export earnings and balance of payments. The issue of raising prices has tended to be met with disagreement because of the lack of international institutional structures that would accomplish this task efficiently.

Price stabilization would seem to be easier to accomplish and could benefit consumers as well as producers. However, the possibilities of a favorable outcome from enacting stabilization arrangements are difficult to assess and forecast (see *Prices of Materials: Stabilization*). The tin buffer stock stabilization mechanism associated with the International Tin Agreement is the only materials arrangement of this type which has come to fruition in the past. However, it has not been notably successful in its attempt to stabilize international tin prices.

The Common Fund is seen as a capital fund useful for financing buffer-stock stabilization as well as for assisting nonstocking efforts among developing countries such as diversification and marketing effects. While these countries can be viewed as benefiting from these activities, the financial viability of the Fund, particularly that of pooling buffer stock financing for as many as eight or ten commodities, has been questioned (Brown 1981). Several nations have agreed to make donations to the Fund, but no operational agreements have been set in motion.

3.2 Market Structure and Bargaining Power

A trade issue related to commodity arrangements is whether producers in individual materials markets can develop sufficient market power to control international materials supplies and prices. To a large extent, this issue has been raised because of the success of the Organization of Petroleum Exporting Countries (OPEC) in raising the levels of crude oil prices. The possibility of market control can be seen in the context of a recent study showing the relatively high degree of concentration in materials markets (Labys 1980). A small number of countries or multinational firms have the capacity to dominate the supplies of materials in a number of individual materials markets. This has given rise to two institutions in particular. First, materials producers have organized into producer associations such as CIPEC for copper, BAUPEC for bauxite, and PHOPEC for phosphate. Second, an increasing number of countries have concentrated their materials selling and buying power into state trading organizations (STOs).

There is considerable question, however, whether such organizations can attain price control power in the form of a cartel. Such prospects are also further dimmed by the need of materials exporting developing countries for sizable foreign investments in project development.

3.3 Import Dependence

In the above explanation of materials trade patterns, attention was drawn to the fact that many of the industrial or OECD countries import a considerable portion of their materials needs from other countries. Some of these other countries also come within the OECD bloc, but others are developing countries or fall within the COMECON bloc. This import dependence has raised problems concerning the availability of strategic and critical materials and the need for stockpiling them. Also relevant is the possibility of increased recycling and increased substitution.

The extent of materials import dependence varies from nation to nation. In the case of the USA, Japan and the UK this dependence is relatively high, particularly in the ferroalloying minerals such as manganese, chromium and nickel. At present, such dependence is not seen as particularly serious. New mineral investments are taking place in different parts of the world which tend to diversify sources of supply. In addition, the maintenance of a competitive or free trade situation in raw materials can assure ready access to low-cost supplies, at least under normal conditions.

3.4 Tariffs and Processing

Materials trade between nations has been shown to be influenced by a variety of tariff and nontariff barriers. Among these are specific and advalorem tariffs, fixed and variable quotas, and price and export controls. Import tariffs or duties tend to be lower per unit of material contained in the commodity to be imported. For example, duties are lower per unit of metal contained in ores and concentrates than on smelter or refinery products. While such a differential allows in part for metallurgical losses, it also tends to encourage the importation of crude rather than refined materials. This would give a nation the benefit of the economic activity resulting from the processing of foreign crude materials for domestic consumption.

In recent years, developing countries have suggested that existing world tariff structures have not worked to their benefit. Nations are ready to purchase the materials they export, but impose relatively high tariffs on semiprocessed or processed materials they might purchase. This issue has received consideration in the context of the UNCTAD IPC, where the IPC would attempt to improve market access for processed materials sold to industrialized nations. It has also received attention in the international tariff negotiations, particularly in the context of the General Agreement on Tariffs and Trade (GATT). Some tariff concessions have taken place which would permit an expansion of materials processing in developing countries. However, the negotiations of tariff and nontariff barriers is an arduous and difficult process. In many cases the industrial nations have a comparative advantage (greater efficiency) in processing these materials. In other cases, the industrial nations have large and long-established materials processing industries which cannot easily be displaced.

4. *Materials Trade and Development*

Primary commodity or materials exports have long been recognized as a base for importing goods and technology which can then serve to accelerate domestic economic growth. The argument for export-led growth stemming from a materials base applies principally to developing countries where materials are abundant and the manufacturing sector is not very strong (Bosson and Varon 1977). In this case materials exported are often those which reflect the natural resource base of a country and those in which it should be a reasonably efficient producer. By concentrating production in one or two commodities, the country should also be able to take advantage of productive economies of scale.

However, an even stronger rationale is that the export industry would feature certain linkages which can stimulate growth in the related national economy. Backward linkages exist because the materials export industry purchases other materials, energy and other inputs from the rest of the economy. Forward linkages exist because other export industries or domestic industries can process the material produced for export, thus increasing added value and employment. Final demand linkages exist because expenditures by the materials export industry in the form of salaries or wages create further demands for domestic goods. Fiscal linkages exist because the export industry through taxation provides a major source of government revenues, and technological linkages can exist in the form of knowledge, skills and other externalities passed on to other sectors of the economy.

As a consequence of this rationale, a number of developing countries have seized on their materials base as a way of accelerating their economic growth. For example, Bolivia has attempted to exploit its tin base, Chile its copper base, Indonesia its timber base, Jamaica its bauxite base, Liberia its iron ore base, Malayasia its rubber base, and Zaire and Zambia their copper bases. For a number of reasons, however, this opportunity base has not always led concomitantly to rapid growth. Firstly, materials prices which fluctuate considerably have often led to similar fluctuations in export earnings with a subsequent stop–go effect on development plans. This effect is particularly severe where one or only a few materials dominate the total export earnings of a country.

Secondly, dependence on one or two export materials is often coincident with a situation where the high costs of investment require that the country depends on foreign capital sources, such as that provided by a multinational company. The host country may then view the multinational company as a sole and major source of fiscal revenues. This unfortunately has led to an "obsolescing bargain" whereby the initial dependence lessens and the host country increases taxation and capitalization until nationalization or expropriation occurs. In this case, the host country may lose the required operational skills as well as possible sources of further investment.

Thirdly, the economic linkages obtained have not always been as strong as hoped for. For example, mining activity does not always produce strong backward or final demand linkages since its technology is generally highly capital and skill intensive, and hence it has input requirements that are highly divergent from domestic factor supplies. The fiscal linkage, though often stronger, has often not been combined with an ability of the host government to invest productively.

The developing countries with a strong materials base thus view themselves as being in a dilemma. They possess materials representing a potential source of wealth, yet they cannot develop their materials export industries sufficiently rapidly to capture this wealth. As a consequence, there has been to some extent a polarization of relations between the developing countries which export materials and the industrialized countries which import them. The results of this polarization have been discussed above as trade issues relating to commodity arrangements, market power, and tariffs and processing.

This dilemma, however, is not insoluble. Multinational materials companies have in many cases achieved a degree of cooperation with host governments and will continue to do so. The former can provide the stimulus for materials-based industrialization by helping to overcome the uncertainty of exploration and related investment decisions, by providing access to markets together with increased materials processing, and by organizing consortia to provide the large amounts of capital necessary for materials investment projects.

Bibliography

Bosson R, Varon B 1977 *The Mining Industry and the Developing Countries.* Oxford University Press, Oxford

Brown C P 1980 *The Political and Social Economy of Commodity Control.* Macmillan, London

Demler F R, Tilton J E 1980 Modeling international trade flows in mineral markets. In: Labys W C, Nadiri I M, Núñez del Arco J (eds.) 1980 *Commodity Markets and Latin American Development: A Modelling Approach.* Ballinger, Cambridge, Massachusetts

Kindleberger C 1968 *International Economics*, 4th edn. Irwin, Homewood, Illinois

Labys W C 1980 *Market Structure, Bargaining Power and Resource Price Formation.* Heath Lexington, Lexington, Massachusetts

L'Huillier J 1968 *Les Organisations Internationales du Coopération Economique et le Commerce Extérieure des Pays en Voie de Développement.* Institut des Hautes Etudes Internationales, Geneva

Pöyhönen P 1963 A tentative model for the volume of trade between countries. *Weltwirtschaftliches Archiv* 90: 93–100

W. C. Labys

[University of West Virginia, Morgantown, West Virginia, USA]

Investigation and Characterization of Materials

Present-day materials science depends heavily on understanding how the properties of a material relate to its composition and microstructure. However, this approach is relatively recent, and for centuries before Bohr's model of the atom, skilled craftsmen and builders produced many objects of remarkable utility and beauty without any knowledge of what is now considered basic chemistry or metallurgy. They started with materials found in their native state or those requiring minimal processing. Desired properties were achieved by trial and error experimentation or the utilization of experience passed down through generations.

For thousands of years certain metals were extracted from their ores by smelting, with variations in the resulting properties often being attributed to supernatural intervention rather than to subtle differences in the materials or processes used. As time passed and printing developed, practical experience was written down. For example, in the sixteenth century Agricola's *De re metallica* described many of the methods used for the extraction, measurement and classification of metals. At about the same time, empirical knowledge existed of how to use distillation to prepare a variety of simple compounds including various alcohols and oils.

Until the eighteenth century only a small number of elements were known, and physical attributes like density, color and hardness were sufficient to tell one from another. As chemistry and physics developed in the nineteenth century, the growth of more basic understanding led to the discovery of a number of elements and the extraction or synthesis of thousands of compounds. These new substances were applied to many areas including medicine, construction, transportation, weaponry and agriculture. By the mid-nineteenth century the amount of scientific information available had grown so large that various scientific disciplines began to split into specialties. In chemistry, general chemists were replaced by organic, inorganic, physical, physiological, agricultural and analytical chemists. In 1848 Fresenius founded the first analytical laboratory which, in addition to teaching, provided government agencies and industry with checks on material specifications and the analysis of water, food and physiological specimens. The idea caught on and today there are hundreds of independent analytical laboratories, as well as those associated with hospitals, industrial concerns, universities and government agencies. Following the specialization of analytical chemistry there has remained a strong interplay between the analysis and development of materials. Better analytical methods lead to both more systematic and complete classification of existing materials and insight required for the development of new ones.

1. Historical Background

The most significant event in laying the foundation for all phases of modern materials science took place at the very beginning of the nineteenth century when Dalton postulated that matter consisted of aggregates of "ultimate particles," or atoms. Atoms of any given element were indistinguishable from other atoms of the same element with respect to mass and all properties, while atoms of different substances could combine to form "atoms" (molecules) of a third substance. Dalton further recognized that elemental atoms combine only in certain ratios, so as to form products with unique stoichiometry. Knowing that he could not weigh individual atoms, he used this concept to determine the relative weights of atoms and molecules based on hydrogen. Thus, the formation of a given amount of water (assumed to be HO) with an atomic weight of 6.5 required the combination of oxygen with an atomic weight of 5.5 with hydrogen with an atomic weight of 1.0. Although Dalton did not recognize that two atoms of hydrogen combine with one atom of oxygen to form one molecule of water (a point which was later clarified by Avogadro) and his atomic weights were not particularly accurate, his ideas are of fundamental importance to modern chemistry.

The concept of definite proportions and atomic weights combined with a fairly consistent symbolic nomenclature for the elements, introduced by Berzelius, led to the description of matter and chemical reactions in the form with which we are now familiar. This approach also led to the emergence of analysis techniques based on chemical behavior rather than physical property determinations. For example, the method known as gravimetric analysis was widely used in the nineteenth century. A chemical reaction between a solution of the material to be analyzed and a second reagent results in the formation of an insoluble compound containing the element of interest. The precipitate is isolated, dried and weighed, and the quantity of the element is determined from its stoichiometric fraction.

A second technique, known as volumetric analysis (titration), developed at about the same time. It involves measuring the volume of a reagent necessary to react completely with a fixed quantity of a sample in a predetermined chemical reaction. At first it was not as commonly used as gravimetric analysis, because of difficulty in measuring reaction end points, but as more indicators became available the method grew in popularity. A third technique, known as colorimetry, was also known in the nineteenth century. Selected reactions are performed to produce colored compounds containing the element of interest. Color intensity is then measured, relative to standard solutions, to estimate the amount of that element present. Although this approach was initially limited to a few elements and quantitative analysis was fairly subjective, its use increased as more colorimetric reagents were identified and instrumentation was developed to make measurements more accurate.

The nineteenth century also saw the initial development of electroanalytical chemistry and the use of polarized light to discriminate between materials, but the most significant instrumental method developed was the spectroscope. Its operation is based on the ability of a prism to disperse white light into its various constituents, a fact demonstrated by Newton in the seventeenth century. In 1814 Fraunhofer, a lens manufacturer, recognized that a number of both bright and dark lines could be observed in both sunlight and the flames produced by burning different fuels. The idea of using absorption spectra for more general chemical analysis was proposed by Brewster a few years later. In the 1850s Bunsen and Kirchhoff developed a spectroscope with an improved source and viewing optics which was the forerunner of instruments used today. With it they discovered the elements cesium and rubidium.

At the beginning of the nineteenth century only about thirty elements were known; by the end this number had more than doubled. During this period it also became apparent that some type of systematic classification of the elements was needed. The concept of the periodic table was independently developed by Mendeleev in Russia and Meyer in Germany using schemes based on atomic weights and various other physical properties. Although ordering of the elements is now related to atomic number and the electronic structure of the elements, most of the changes made over the years have been to add new elements rather than to reorder known elements.

At the end of the nineteenth century and the beginning of the twentieth century a series of major events took place which revolutionized understanding of both the atom and electromagnetic radiation. Studies of the electrical conductivity of gases at reduced pressure led to the discovery of cathode rays by Hittorf in 1869, confirmed by Goldstein in 1876. Although initially the nature of these rays was unclear, work by Thomson, published in 1897, showed that they consisted of negatively charged particles, from their deflection in magnetic and electrostatic fields. He also demonstrated that these particles (later called electrons) had the same charge to mass ratio as those given off from hot filaments or released during the exposure of certain metals to electromagnetic radiation (the photoelectric effect, discovered by Hertz in 1887). Thus they appeared to be a key component of all matter, and when removed would leave atoms as positive ions.

The nature of this positive ion core was studied in the first decade of the twentieth century by Rutherford, who experimented on the scattering of α particles from thin metal foils. From this work came a model of the atom in which electrons revolve around a positively charged nucleus. Although this model was consistent with his observations, it was unacceptable from the point of view of classical physics, since the electrons would continuously radiate energy and should, therefore, spiral into the nucleus. However, the dilemma was soon reconciled by Bohr, who postulated that the electron orbits corresponded to discrete energy states and as long as an electron remained in its orbit no radiation would occur. The idea of discrete energy states was based on the principles of quantum mechanics introduced by Planck in 1900 to describe radiation emitted from a black body. Using these states, Bohr was able to describe correctly the line spectra of hydrogen observed in the 1880s by Balmer and mathematically formulated by Rydberg in 1890. Although the Bohr model was significantly modified by Sommerfeld and Schrödinger to describe multielectron atoms more accurately, the concept of electron transitions between discrete levels and the accompanying release or absorption of energy is of central importance to many of the different types of photon and electron spectrometries used today.

Thomson's use of magnetic and electric fields to determine charge to mass ratios was also applied to the characterization of positive ion beams (called Kanalstrahlen by Goldstein), which were also known to exist along with cathode rays, as shown in experiments on the electrical conductivity of gases. Thomson's apparatus, known as a parabolic mass spectrometer, produced a series of photographically recorded parabolas, one for each charge to mass ratio. The system lacked mass resolution, however, because it had no energy selectivity. This difficulty was overcome several years later (1919) when Aston developed a velocity focusing instrument which showed that the spectrum of neon actually contained two peaks, one at mass 20 and the other at mass 22. This demonstrated that an element of a given number can have more than one form (isotopes) and thus revealed the power of the mass spectrometer as an important analytical instrument. In the decades that followed, resolution and sensitivity continued to improve as a result of the efforts of Dempster, Mattauch and Herzog, and Nier and Johnson. Some commercially available instruments today have mass resolutions of 1 part in 100 000, compared with the 1 part in 12 of the first parabolic spectrometers.

Experiments on the electrical conductivity of gases also led to the accidental discovery of x rays by Roentgen in 1896. He recognized that their unusual penetrating power was influenced by the tube operating voltage as well as the thickness and type of absorbers used. Since x rays were not deflected by magnetic or electrostatic fields, it was believed that they were part of the electromagnetic spectrum, but that they must have a very short wavelength. Determining their wavelength distribution presented something of a problem, because although diffraction gratings had previously been developed for dispersing visible spectra, they were not fine enough to separate x rays. The solution was proposed by Laue in 1912, who believed that the interatomic spacing of crystals should be of the right order of magnitude to act as suitable diffraction gratings. He proved this hypothesis by passing a beam of x rays through a crystal of zinc sulfide and obtaining a pattern of spots on a photographic plate.

This work stimulated further studies by W. H. and W. L. Bragg (father and son), who built the first crystal diffraction spectrometer. Using a crystal of rock salt with a known interplanar spacing, they were able to characterize the spectral distribution of a platinum x ray target and also develop the simple equation which relates interplanar spacing, x ray wavelength and diffraction angle. The spectra they observed consisted of both a broad band of general radiation (the x ray continuum) and discrete lines characteristic of the tube anode. Investigations by Moseley described in 1913 showed that there also existed a simple relationship between the wavelengths of the characteristic lines emitted from an element and its atomic number. This unique relationship forms the basis for elemental analysis by a variety of x ray, electron and ion-induced x ray spectroscopies. The work of the Braggs also provided the key to crystal structure determination by using a selected characteristic x ray wavelength to determine a set of interplanar spacings for a single crystal.

The numerous scientific discoveries that took place at the turn of this century provided the physical foundation for the hundreds of instrumentally oriented analytical methods which are widely used today. It is astonishing that most of what we now consider standard laboratory instruments were developed within the past 50 years. The following techniques and key contributors are examples:

transmission electron microscopy, Ruska (1934)
scanning electron microscopy, von Ardenne (1938)
liquid chromatography, Martin and Synge (1941)
carbon-14 dating, Libby (1946)
nuclear magnetic resonance, Block and Purcell (1946)
electron microprobe analysis, Castaing (1951)
gas chromatography, Martin and James (1952)
secondary-ion mass spectrometry, Castaing and Slodzian (1960)

The names, events and dates mentioned so far tell only a small part of the chronological development of some of the methods described. Practical refinement of many of these instrumental methods often took many years and considerable ingenuity on the part of other key contributors. For example, the first commercially available scanning electron microscope did not appear until 27 years after von Ardenne's first instrument (1965), and it has only been during the last few years that relatively easy-to-use computerized complete single-crystal structure determination equipment has been readily available. At present one of the most active areas of analysis is the study of surfaces by x ray photoelectron spectroscopy but, as mentioned previously, the photoelectric effect was known before the turn of the century. Relatively recent developments in electronics, computers and high-vacuum systems have all had an enormous impact on analytical instrumentation in terms of ease of operation, sophistication of the analysis, quality of results and speed. Industrial laboratories now routinely undertake types of analysis that would formerly have required the effort needed for a PhD thesis.

With the great diversity of instrumentation available, the role of the traditional analytical laboratory has now been expanded to include determinations of morphology, microstructure, crystallography and a variety of physical properties. The term often associated with the combination of analytical chemistry and these other methods of analysis is materials characterization. The concept of a materials characterization group goes hand in hand with the evolution of interdisciplinary materials science and engineering centers which bring together experts in such areas as metallurgy, electronic materials, ceramics, solid state physics, polymers and surface chemistry. Not only have the variety of materials and the types of instrumentation available increased, but the reasons for looking at these materials have also multiplied. Materials characterization specialists must now be concerned with basic research, product and process development, failure analysis, regulatory compliance, patent issues, quality control, manufacturing productivity and the examination of competitive products.

It is important also to recognize the degree of analytical specialization which has been taking place. Individuals who are experts in nuclear magnetic resonance spectroscopy may not have expertise in single-crystal x ray structure determination, even though information gained by the two techniques may be highly complementary and important to a polymer chemist. An expert in transmission electron microscopy may have only an incidental knowledge of secondary-ion mass spectrometry, even though both chemical and structural information may be the key to understanding the behavior of an electronic device. As the demand for materials with unique properties grows, so also does the complexity of the materials themselves and the methods needed to analyze them. Examples are the chemical identification of submicrometer second-phase particles which can significantly influence the properties of high-temperature alloys, the determination of pollutants at the parts per billion level and the characterization of a few atomic layers on the surface of a catalyst. Clearly, scientists and engineers have enough difficulty in keeping up with advances in their own fields without having to be materials characterization experts. However, it is essential to have enough basic understanding of currently used analytical methods to be able to interact effectively with such experts.

2. Methods Available in the Modern Materials Characterization Laboratory

As described above, the structure and composition of matter can be studied by examining how it interacts with radiation, electrons and ions. A summary of some of these interactions and the related analytical techniques is given in Fig. 1. The approach that is best suited for a particular problem depends on a variety of factors, many of which are strongly interrelated.

Chemical or microstructural analysis in most laboratories of any size can be carried out by a centralized analytical facility which maintains core analytical instruments staffed by personnel who are experts in each area. The following methods are fairly essential in any type of multidisciplinary research or engineering laboratory:

classical chemical analysis techniques
thermal analysis techniques
optical microscopy and metallography
ultraviolet, visible and infrared spectroscopy
mass spectrometry
gas and liquid chromatography
scanning and transmission electron microscopy
nuclear magnetic resonance spectrometry
optical emission and absorption spectrometry
x ray diffraction and fluroescence spectrometry
Auger and x ray photoelectron spectrometry

Other techniques employing commercially available equipment which are somewhat less commonly used although often important include:

low-energy electron diffraction
secondary-ion mass spectrometry
Raman spectrometry
low-energy ion scattering spectrometry
field-ion microscopy and atom probe analysis
electron spin resonance spectrometry
Mössbauer spectrometry
laser microprobe analysis
acoustic microscopy
quantitative image analysis

What must be recognized also is that even if a well-equipped analytical laboratory has all the specialties listed in the first category, each of the methods can be substantially subdivided, requiring separate instruments and often separate skills. Taking x ray diffraction as an example, the following subdivision can easily be made:

conventional diffractometry
high- and low-temperature diffractometry
back reflection and transmission Laue photography
residual stress measurement
single-crystal structure determination
pole figure determination
double-crystal diffractometry
microdiffraction
Debye–Scherrer and Guinier powder photography
Berg–Barrett and Lang topography measurements

Services not readily available in-house can often be purchased externally from general analytical laboratories or specialists in some of the above methods. This approach is particularly useful when the number of samples does not justify setting up an in-house facility or if in-house facilities are severely overloaded. No single laboratory can possess everything and the use of external analysts often provides an effective way of providing needed resources. In some cases, the use of external laboratories is the only approach when the instrumentation required is in the multimillion dollar price range and large support staffs are needed. Examples of such specialized facilities include:

extended x ray absorption fine structure measurements (using synchrotron radiation)
neutron activation analysis
carbon-14 dating
picosecond laser fluorescence spectrometry
neutron activation spectrometry
high-voltage transmission electron microscopy
Rutherford backscattering and ion channelling spectrometry
proton-excited x ray fluorescence

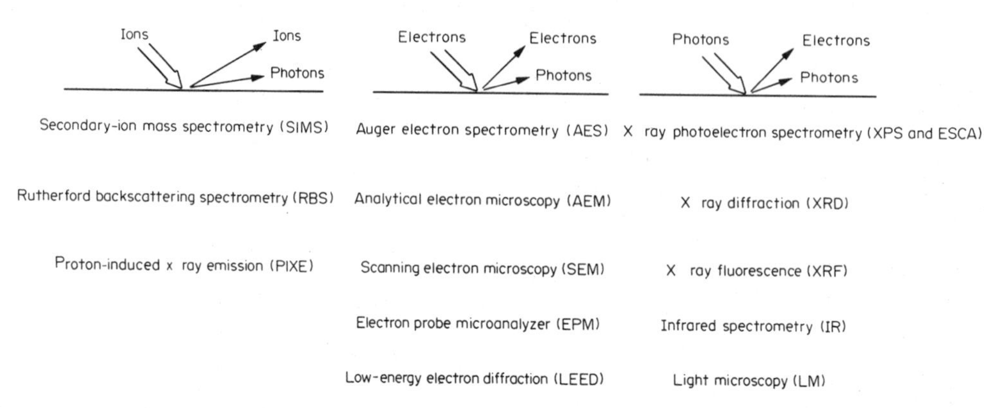

Figure 1
Analytical methods: probes and responses

There are other factors in addition to cost and speed of analysis that should be considered in the decision to use in-house or external facilities. In an industrial environment, proprietary concerns often limit sample distribution. Another important factor is the nature of the interaction between the analyst and the research scientist or engineer. In those cases where the specific type of analysis needed is known and the right sample has been selected, it may be quite satisfactory to send a sample several thousand kilometers and receive the results in a few weeks. However, in other cases a much higher degree of interaction may lead more rapidly to meaningful results. This is particularly true if the technique or the sample selected has to be changed because the initial approach is unsuccessful. It can also be true if detailed data interpretation relative to the objectives of the overall investigation is required.

Sometimes the question of who should do the analysis becomes an internal one. Should scientists or engineers have their own characterization tools or should they use a centralized service? While there is no single answer to this question, it is clearly an issue of optimizing the utilization of resources. It is not uncommon for every chemist in a laboratory to have his own gas or liquid chromatograph and every metallurgist to have his own light optical microscope. In such cases an investment of a few thousand dollars could improve research productivity by providing quick answers expediting the next experiment. It is highly unlikely, however, that each chemist would have a private nuclear magnetic resonance spectrometer and each metallurgist a separate transmission electron microscope. These require very large initial capital as well as the expenditure of both money and time to acquire the needed expertise and to maintain a piece of equipment that may not even have long-term use. Obviously the overall impact could be a serious decrease in research productivity.

Occasions do arise, however, when it is important to have fairly sophisticated analytical instruments closely integrated with other types of experimental apparatus. For example, samples may be altered by being transferred to separate instrumentation, as in the characterization of thin films formed by molecular beam epitaxy by the use of integral low-energy electron diffraction, Auger spectrometry and residual gas analysis instrumentation. In a manufacturing environment a process may require rapid on-line monitoring to maintain quality control. Chemical reactors frequently have in-line gas chromatographs and infrared analyzers. Dedicated analytical equipment outside a centralized facility may also be required if it serves only the objectives of a single individual's research interest and both the degree of interaction needed and the duration of the project justify the associated cost. An example would be a laser Raman spectrometer uniquely set up to study combustion chemistry and dynamics.

3. Interaction with the Analyst

Given hundreds of analytical methods, it is obvious to ask how is an appropriate method selected for solving a given problem? Before the recent proliferation of so many new types of analytical instruments, a scientist or engineer would go to a designated analytical chemist in his organization and describe the information needed. The analytical chemist would either go to suitable references or know immediately from personal experience which technique to use. Today this dialogue could be considerably extended and would undoubtedly involve experts in several fields. The scientist or engineer initiating the study should be prepared to answer a number of questions which will aid in establishing the method or methods most suitable for solving the problem. The following is a list of representative questions frequently asked, separated into three general categories. In addition some comments are added which should help to understand the role of these categories in establishing an analytical strategy. If the scientist or engineer and the materials analyst are, in fact, the same person the discussion is replaced by introspection but the questions remain the same.

(a) Questions relating to the type and detail of information sought:
Is there a need to know what elements are present and/or how much of each?
Is there a need to know all the elements or only some of them?
Down to what level?
What degree of accuracy is required in the results?
Is crystal structure or microstructural information needed?

(b) Questions relating to what is already known about the sample:
What is the physical state of the sample?
Is the sample a metal, ceramic, semiconductor, chemical compound or polymer?
Does the sample consist of one or more than one phase?
Is the sample amorphous or crystalline?
Is special handling of the sample required, either because of toxicity or because the sample might transform to something different before or during analysis?
Was the sample (or its source) exposed to any unusual physical or chemical conditions which might have altered its composition or microstructure?
Does the sample consist of layers, and if it does, how many are there and how thick are they?
How much of each sample is there and how many samples are to be run?

(c) Questions relating to the working relationship

between the analyst and the scientist or engineer:
How soon are results needed?
How much can be spent to obtain the results?
Is your presence needed during the analysis?
Is a detailed report on the results needed?
Are the samples proprietary?

These three categories of questions can be thought of as screens which help to identify one or more methods of approaching a given problem. They can easily be related to three general classes of analysis requests as follows.

(i) In the simplest case, the analysis is merely to repeat a particular type of measurement made earlier on a similar sample. Therefore, the method has been preselected and the only issues to be resolved are associated with the questions of category (c). The answers are determined by laboratory policy, backlogs, urgency of the work and availability of people and equipment. An example of this class of problem would be a request for mercury analysis in water by atomic absorption as part of periodic sampling of a body of water.

(ii) More frequently, a suitable method may be suggested based on earlier work, but the characteristics of the sample may be sufficiently different that the questions asked in category (b) also become important. The new sample may be of a different size or shape. It may also contain other elements or compounds that could complicate the analysis if the same analytical method is used. As an example, a researcher develops a new polymer containing a chlorinated flame retardant. He/she wants a chlorine analysis in the 50 ppm region but has only a 5 mg sample. Previously, successful results would be obtained by x ray fluorescence.

(iii) Samples that can be classified as totally unknown occur rarely, since materials are usually not picked at random for analysis. The situation does arise, however, when one material behaves differently from another and the reasons for the differences are sought. Such samples may be the result of laboratory experiments or may be associated with a device or process that has changed or even failed for some unknown reason. The individual initiating the work may not even be aware of the right questions to ask in category (a) and must therefore work with the analyst to define a set of unique material characteristics which can be used both to classify the specimen and also to contribute to an explanation of its behavior. Although problems in this class are sometimes amenable to quick solutions, they tend more generally to be the most challenging and expensive ones to deal with, since they frequently involve the use of multiple techniques and often the samples are not optimized for those techniques. An example of this class of analysis would be a request to analyze a catalytic material obtained from a new vendor which, although nominally of the same composition as the previously used material, dramatically changes the rate of a chemical reaction. Also included in this class are problems in failure analysis such as determining why a turbine blade has failed prematurely. In this situation questions in category (b) relating to the sample environment can be extremely important.

4. Choosing a Sample and a Method

The quality of an analysis can be no better than that determined by the sample on which the analysis is performed. In trace-level determination, a sample can easily be contaminated by handling or the use of unclean glassware or other laboratory supplies. Samples can also be affected by exposure to the atmosphere, as in the case of hygroscopic materials or hydrocarbon buildup on a sample to be studied by surface analysis techniques. Materials are also not always as homogeneous as expected. If a method is used with a precision of $\pm 1\%$ but the composition of a randomly picked sample varies by $\pm 10\%$, obviously the sampling error will dominate. This effect can take place at either the macro- or the microscale. In other words, a 1 kg sample taken from a single location in a tank car can lead to the same magnitude of error in assessing the overall contents of the car as using a 1 ng microprobe measurement to determine the composition of a highly segregated alloy ingot.

Collecting samples from enough regions of a material to be analyzed to determine its homogeneity and/or average composition is an essential part of characterization measurements. If the method of analysis to be used in a particular problem is known in advance, as in the case of a research experiment, samples should be prepared so that they are truly representative of the problem and also of the proper size and geometry to be compatible with the method to be used. For most methods there is some correlation between sample size, sensitivity, accuracy and precision. Situations where the amount of sample is limited are frequently encountered, however, and in these cases the quantity of samples may strongly influence the analytical method selected and minimum detectability limits. Consider the following:

a geologist working in an inaccessible location gathering samples to map out its mineral content
a forensic scientist working with only a small piece of evidence
a metallurgical failure analysis expert working with a fragment of a broken device
a research chemist or metallurgist developing a new material in which it is important to explore hundreds of variations of composition and/or phase distribution

There are also times when an abundant amount of sample is available but either the concentration levels sought by a particular instrumental method are too low to detect because of maximum acceptable sample size limitations, or interferences from other constituents can introduce serious errors in the analysis. In such situations, determining whether the component of interest (phase or element) is dispersed homogeneously throughout the sample or is nonuniformly distributed is important not only from the sampling point of view but also as a means of selecting a technique to give enhanced localized analysis for inhomogeneous samples. Being able to detect 1 ng of material in a 1 g sample is a real challenge for any bulk analysis technique but is very simple by electron microprobe analysis if the relevant location is known. In the case of solids, this can often be done by microscopic examination of the sample either as it is, or specially prepared to enhance phase differences. When possible, conventional light microscopy should always be tried first. Low-magnification observation of an unmodified sample often reveals whether a sample is nonuniform. If the sample consists of particles observable under a low-power binocular microscope, they can often be separated for easier analysis by a variety of methods including sedimentation, magnetic separation or even hand selection with tweezers.

If magnifications above ~100 × are needed and the sample consists of discrete particles, scanning electron microscopy is usually a better approach. However, since the attribute of color is not available, differences between particles tend to be based on morphology and in some cases on atomic-number-dependent electron backscattering. Furthermore, since particles smaller than a few tens of μm are difficult to handle, the analysis may have to be done in situ using energy-dispersive x ray spectroscopy. This method primarily provides qualitative elemental analysis and can be applied to particulates using magnifications up to about 5000 × if the particles are well separated. Light microscopy is used in the magnification range of ~100–2000 × primarily to examine polished sections by reflection for opaque materials or transmission for transparent materials.

While some chemical information is obtainable by polarized light and refractive index measurements, the presence of multiple phases may indicate that the sample should be analyzed either by a high-spatial-resolution instrumental method, like electron microprobe analysis, or that the sample should be dissolved so as to isolate second-phase particles for analysis by other techniques. The transmission electron microscope extends the ability to perform combined microstructural and microchemical analysis to magnifications well above 5000 ×, but the complexities of specimen preparation and the relatively high cost and inaccessibility of this method in most laboratories, compared with light and scanning electron microscopy, tend to make it a method which is chosen only when the other methods are unsuccessful.

If by microscopic examination the sample appears to be fairly homogeneous, the advantages associated with the spatial selectivity of microanalysis cannot be realized and various methods of bulk analysis are probably more suitable. In addition to traditional wet chemical analysis, these methods include spark-source mass spectrometry, optical emission and absorption spectrometry and x ray fluorescence spectroscopy. Each has a sensitivity extending down to the parts per million level and in some cases well below. In addition, the optical and x ray techniques can both be made quantitative with relative ease. These techniques can often also be applied to inhomogeneous samples, provided that no breakdown by phases is required. Figures 2, 3 and 4 give an approximate summary of the performance of some of the methods described, with regard to their ability to determine elemental composition, crystal structure and phase morphology.

The preceding comments about determining sample homogeneity refer principally to solid inorganic samples, although all the techniques mentioned can be used on polymers as well, within certain limitations associated with vacuum and electron beam compatibility requirements of the electron microscopic methods. Furthermore, x ray measurements are primarily concerned with elemental analysis rather than compound identification. Since carbon, hydrogen, nitrogen and oxygen are ubiquitous in organic materials, qualitative elemental analysis by x ray techniques (hydrogen is undetectable by these methods) does little to aid in the identification of the material other than to show that it is organic. Determining whether

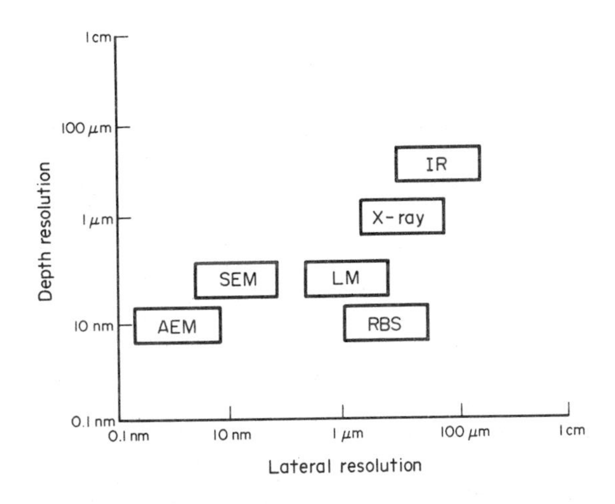

Figure 2
Lateral and depth resolution comparison for the methods indicated in Fig. 1 used for studying morphology

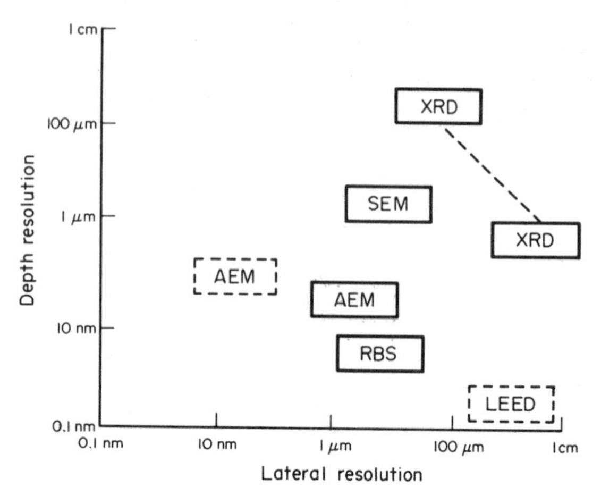

Figure 3
Lateral and depth resolution comparison for the methods indicated in Fig. 1 used for obtaining crystal structure information

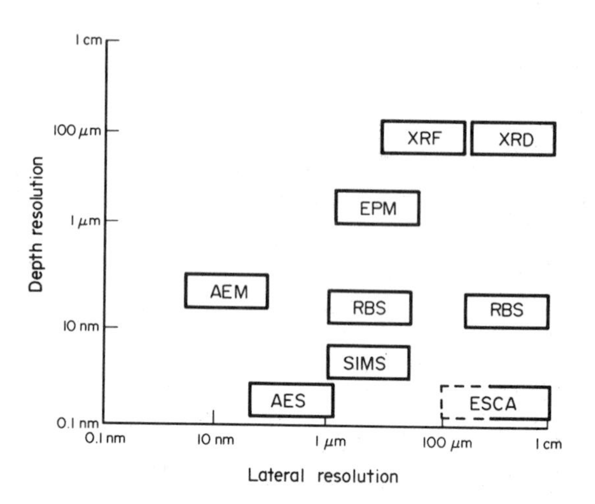

Figure 4
Lateral and depth resolution comparison for the methods indicated in Fig. 1 used for elemental analysis

an organic sample consists of one or more phases or compounds is also an important issue, however, but the types of samples, the separation techniques used and the kinds of information required can differ considerably from those associated with the examination of inorganic samples. First of all, most samples are either submitted as solutions or are taken up into solution before analysis. Separations are usually achieved by solution chemistry, solvent extraction or chromatography (gas or liquid). Separated fractions are either analyzed by direct introduction into a physically coupled instrument like a mass or infrared spectrometer or are collected and transferred to another location for analysis. If the sample is a polymer of only one type, the information sought may be molecular weight distribution and chromatography (gel permeation) may be all that is required. In other cases the information needed is the structural formula. This could be determined from information obtained by a variety of methods including mass spectrometry, infrared spectroscopy and nuclear magnetic resonance, as well as by traditional carbon, hydrogen and oxygen analysis based on combustion and gas absorption. The sample size and concentration required again become an issue, because the characteristics of the various techniques differ broadly. While nuclear magnetic resonance is excellent for structure determination, it requires larger samples than the other methods mentioned and can rarely detect compounds or functional groups at a concentration level $<0.1\%$. Organic mass spectrometry, on the other hand, can work with much smaller samples and detect compounds even down to the parts per billion level in a well-separated sample. Structural information is harder to obtain, however, and compounds with high molecular weights (>2000) are difficult to analyze by conventional means. With sufficiently high resolution systems and soft ionization sources, it is possible to obtain molecular ions and determine stoichiometry based on direct packing fraction measurements. In some cases, structure can also be determined using conventional electron impact ionization and comparing fragmentation pattern spectra with those of known compounds. Infrared spectroscopy can also have relatively high sensitivity to specific functional groups and can be used to examine relatively small samples, although larger ones are preferred for higher sensitivity. It is used more effectively in relating the spectrum of an unknown sample to a reference spectrum rather than as a means of determining the structure of a complete unknown. Of particular value is the fact that it can be routinely used to examine solid samples such as cross-linked insoluble polymers.

5. Cost and Complexity

There can be an enormous range in the costs associated with materials characterization, depending on the complexity of the problem, the amount of information sought, the number of samples to be run, how quickly the analysis has to be done and the investment in equipment and manpower. Multielement analysis for large numbers of routine samples can be effected for less than US$25 per sample, while complex surface analysis can exceed US$1000 per sample. Access to a particular type of instrumentation of the types described so far does not provide all the information needed about whether a given technique will have the sensitivity, resolution or sample-handling capability to do a given job. It is possible to purchase complete mass spectrometer systems for under US$5000 or over US$500 000. While the former may be perfectly adequate as a residual gas analyzer where unit mass resolution is sufficient, it would be totally unsatisfactory for the analysis of high-molecular-weight organic compounds. Even if the larger sum of money were available, a high-resolution mass spectrometer for isotopic analysis would be decidedly different from one used for combined gas chromatography–mass spectrometry.

The trade-off between cost and performance features is characteristic of all major pieces of analytical instrumentation. To deal effectively with this issue when selecting an instrument to purchase or use for a given application requires the development of a specification for instrument performance relative to intended use. Again taking mass spectrometry as an example, if the need is for high-sensitivity, high-resolution analysis of gas chromatograph effluents, an instrument might be specified with a highly efficient electron impact source coupled to a fast-switching, double-focusing analyzer with high-sensitivity electrical detection. Furthermore, it should also be possible to specify a generally accepted performance standard. For example, the instrument should be able to detect 10^{-9} C of ion per μg of methyl stearate at a mass resolution of 1 part in 10 000 and a source voltage of 8 kV. Generally, when selecting an instrument of any type, cost rises with performance. It is therefore incumbent on the buyer to be realistic about how many special features and what level of performance are needed for a given range of applications. It makes little sense to pay the added cost for a scanning electron microscope capable of 100 000 × magnification relative to a lower-resolution model if the details of interest are clearly observable at 5000 × or if the nature of the sample is such that pictures at higher magnifications are not possible.

5. *Trends*

The major impact of computers and microelectronics on materials characterization techniques, already alluded to, will undoubtedly continue for a number of years. In addition to instrument control, data collection and the processing of results, computers will play a more important role in method selection and the management of laboratory information. It can be expected that more general spectral data bases and efficient means for searching them will lead both to faster and to more complete results. Since specimen preparation is often a rate-limiting step in many forms of analysis, the use of laboratory robots may also improve quality and productivity.

In very general terms, most analytical instruments can be thought of as consisting of a source, an analyzer and a detector. As was shown in Fig. 1, the source can be based on the interaction of the sample with probing radiation of any wavelength or particles of different types and energies. It may even be the specimen itself, as in the case of the analysis of radioactive samples. The analyzer serves as a means of separating the signals emitted or scattered by the sample, based on differences in their physical or chemical properties, and finally the detector converts the separated signals into a comprehensible form. A major trend which has emerged during the past decade is to try all manner of combinations of sources, analyzers and detectors and thereby create instruments which are optimized for particular applications. In organic mass spectrometry, for example, ion sources are based on electron impact, field ionization, field desorption, laser ionization and both ion and neutral particle beams. In gas chromatography, detectors include flame ionization, electron capture, thermal conductivity and infrared spectrometry.

If the sample is only partly consumed or not consumed at all by a given detection process, multiple detectors can be used in series to provide complementary information. This has been done in both gas and liquid chromatography. Sometimes the detector is itself a spectrometer and thus allows separated species to be characterized more fully. Examples include gas chromatography–mass spectrometry and both gas and liquid chromatography–infrared spectrometry. An important characteristic of spectrometric detectors used in combination methods is that they must either be very fast or function in a parallel detection mode. Thus, in the examples just mentioned, a mass spectrometer capable of scanning its complete mass range in a few seconds is used in one case and Fourier transform infrared spectrometry in the other. If separation techniques are to be used with atomic emission techniques, some of the newly developed types of parallel detection optical spectrometers will prove very valuable. In other analytical situations, several detectors capable of monitoring a variety of signals are used at the same time. For example, simultaneous x ray and electron detection in scanning electron microscopes and analytical transmission electron microscopes makes it possible to relate chemical to microstructural information easily.

It is always difficult to predict where any field will go in the future, but clearly defined needs coupled with active technologies can provide some hints. All analytical methods share the common goals of higher speed, lower cost, greater sensitivity, more flexibility and increased accuracy. Over the past two decades, remarkable progress has been made in advancing localized elemental analysis capability to the limit of single-atom detection by atom microprobe analysis and ultrahigh-resolution scanning transmission electron microscopy, although only limited kinds of sample are suitable for these methods. Similar developments are needed in the high-spatial-resolution identification of compounds, particularly organics, since a knowledge of elemental constituents is usually of little value in distinguishing one material from another. Selected area diffraction, and more recently microdiffraction, are routinely used to identify inorganic compounds in transmission electron microscopes. The newer method has even given results from areas <5 nm in diameter. However, these approaches are of very limited value in the examination of organic materials, because of electron beam damage and the fact that many organics of interest are in a noncrystalline form.

The full characterization of thin organic films even when lateral spatial resolution is not an issue is complicated for other reasons. X ray photoelectron spectroscopy has already proved itself to be a powerful tool, but the chemical shift information contained in a spectrum is often difficult to extract because of the inherent energy resolution limits of the technique. Perhaps one of the most interesting and challenging problems is the atomic level characterization of interfaces within materials, where any attempt to reach those interfaces by material removal, as is done in secondary-ion mass spectrometry and Auger electron analysis, so alters them as to make the information obtained useless. For these kinds of problems, nondestructive techniques like solid state nuclear magnetic resonance and some forms of laser spectroscopy may provide valuable solutions.

Bibliography

Casper L A, Powell C J (eds.) 1982 *Industrial Applications of Surface Analysis.* American Chemical Society, Washington, DC

Czandema A W (ed.) 1975 *Methods of Surface Analysis.* Elsevier, Amsterdam

Ewing, G W 1975 *Instrumental Methods of Chemical Analysis*, 4th edn. McGraw-Hill, New York

Friend J 1951 *Man and the Chemical Elements.* Griffin, London

Ihde A J 1964 *The Development of Modern Chemistry.* Harper and Row, New York

Kane P K, Larrabee G B (eds.) 1974 *Characterization of Solid Surfaces.* Plenum, New York

Mulvey T, Webster R K (eds.) 1974 *Modern Physical Techniques in Materials Science.* Oxford University Press, Oxford

Partington J R 1962 *A History of Chemistry*, Vols. 1–4. Macmillan, London

Peters P G et al. 1976 *A Brief Introduction to Modern Chemical Analysis.* Saunders, Philadelphia

Skoog D A, West D M 1982 *Fundamentals of Analytical Chemistry*, 4th edn. CBS College Publishing, New York

E. Lifshin

[General Electric, Schenectady, New York, USA]

Iron Production

Iron is usually produced by smelting or reduction of iron ore, or by melting of ferrous scrap. A number of industrial processes have been developed for iron production, of which the blast furnace accounts for the majority of worldwide iron production. The metallic iron produced by primary extraction (hot metal or pig iron) contains substantial quantities of impurities, and must be refined to yield cast iron or steel. Small amounts of high-purity iron powders and electrolytically deposited iron are produced from high-purity iron oxide powders and aqueous solutions by more specialized processes involving hydrogen reduction and carbonyl reactions. This article covers the primary extraction of metallic iron from ores containing iron oxide.

1. Raw Materials

The principal raw materials for iron production in the blast furnace include coke derived from coal, iron oxide concentrates, pellets and agglomerates derived from iron ores, and limestone and other fluxes (Fig. 1). In addition, the blast-furnace charge may include small amounts of iron and steel scrap, steelmaking slag and consolidated waste materials such as dusts and sludges collected in gas-cleaning equipment. A small but significant portion of the fuel and reductant requirements in the blast furnace is supplied by inclusion of hydrocarbons, heavy oils and tars, and, to a lesser extent, natural gas and pulverized coal added with the preheated air blast entering the blast furnace through the tuyeres. Figures 2 and 3 provide some representative information on the design and relative consumption of raw materials in modern blast furnaces. In Fig. 3, the quantities of materials shown are based on the production of one tonne of iron of the indicated

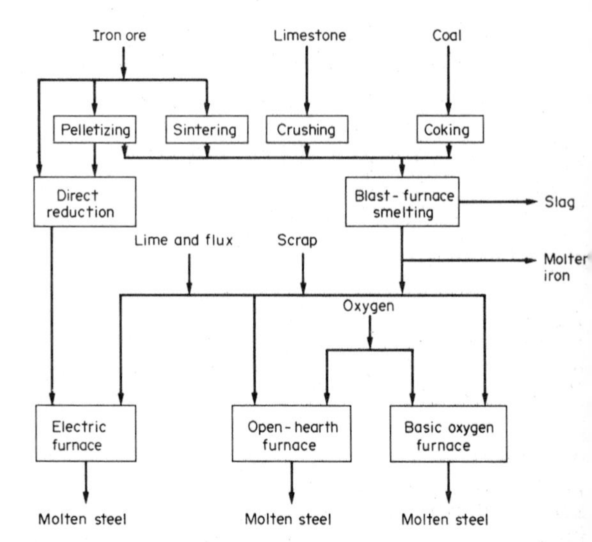

Figure 1
Flowchart showing the major steps in steelmaking

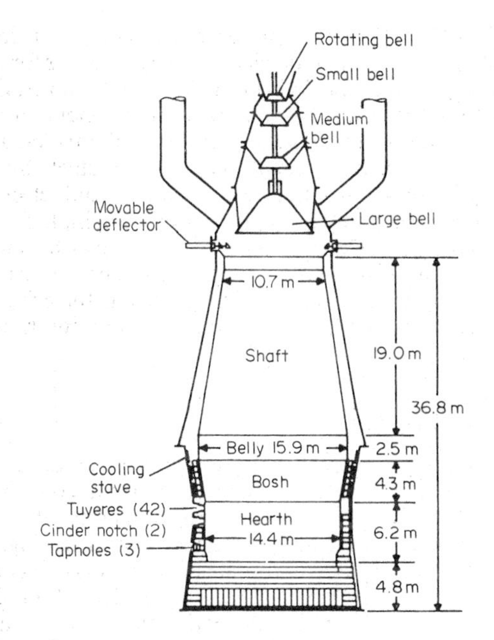

Figure 2
Features and approximate dimensions of a modern blast furnace (after Sugarawa et al. 1976)

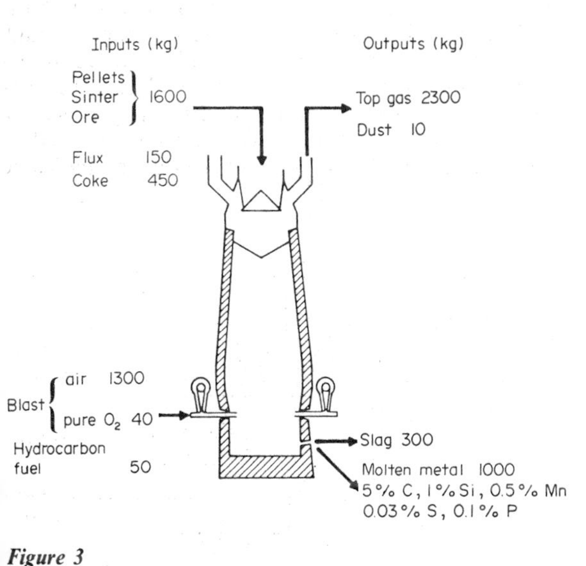

Figure 3
Representative materials balance for a modern blast furnace based on the production of 1 t of iron (after Peacey and Davenport 1979)

composition. Specific raw material consumption depends very strongly on the furnace design, composition and physical characteristics of available raw materials, and operational practices.

The principal raw materials for production of iron by direct reduction processes include iron oxide concentrates and pellets, and natural gas. A few of the direct reduction processes, particularly those that employ rotary kilns, use solid carbonaceous materials such as lignite and reactive coals. Figure 4 is a schematic diagram of the Midrex system, which is based on the use of natural gas as a reducing agent, and Fig. 5 is a schematic diagram of the SL/RN process, which is based, in part, on the use of solid carbonaceous materials as fuel and reductant. These are only two of many different direct reduction processes that exist. Continued interest in the direct reduction processes based on natural gas is expected, particularly in those countries where natural gas is abundant. The commercial development of coal-based, direct reduction processes has been slow, and significant technological progress is required before a significant competitive position in iron production

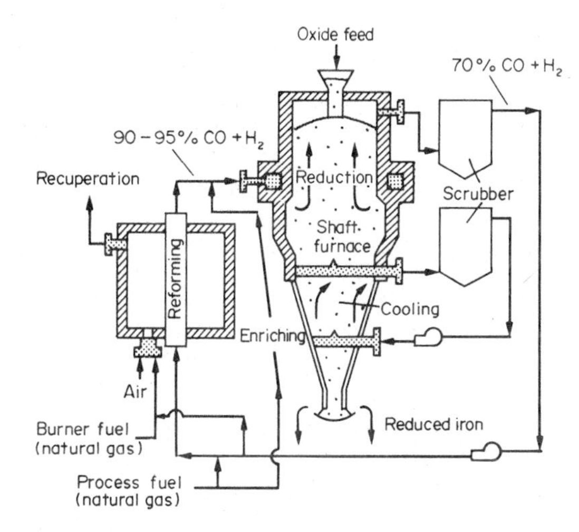

Figure 4
Simplified flowsheet of the standard Midrex direct reduction process (after Stephenson and Smailer 1980)

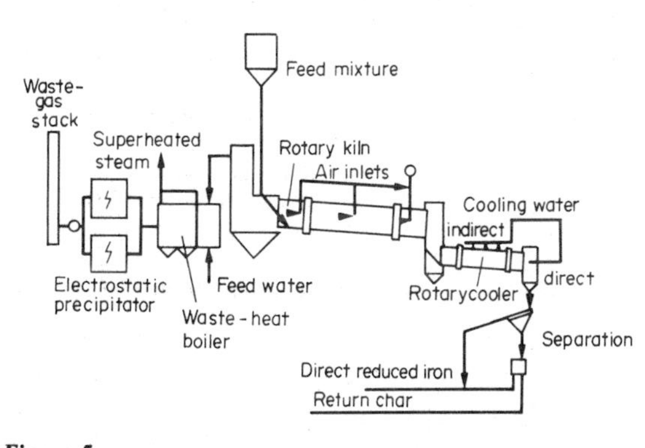

Figure 5
Schematic arrangement of SL/RN direct reduction process (after Stephenson and Smailer 1980)

can be attained. However, in specialized situations, the coal-based direct reduction processes may offer advantages.

In the production of coke, coals having both high and low volatile matter content are ground and blended to produce a suitable mixture for the destructive distillation process. Here, it is essential that the carbon formed from the plastic mass develops suitable strength, wear resistance and chemical reactivity for use in blast furnaces. The details of coke-making and the characteristics of the resulting coke, coke-oven gas and by-products may be found in several sources cited in the Bibliography. In general, coke is produced adjacent to, and as an integral part of, blast furnace iron production. In this way, the fuel value of the coke-oven gas may be used directly.

Available forms of iron oxide include high-grade lump ores, beneficiated iron ore fines, iron ore pellets, sintered iron ore, mill scale, and agglomerated and sintered iron oxide dusts. Iron oxide pellets are produced by reconsolidating iron oxide particles that are primarily finer than 325 mesh (0.044 mm). The iron oxide particles are obtained by crushing low-grade ores, typically of the taconite class, and separating the iron oxide from the siliceous gangue. The fine particles are reconsolidated into moist pellets about 1 cm in diameter which are then indurated by heating to temperatures sufficient to bring about recrystallization of the oxides. Because of the high gas velocities and abrasive conditions in most ironmaking processes, fine particles are not suitable as charge materials. They tend to be carried out with the gas streams, from which they must be separated and recirculated. The fluidized-bed direct reduction processes are an exception. Information on the types of ores and charge oxide materials that are acceptable as charge materials for ironmaking processes can be found in works cited in the Bibliography.

2. *Principles, Operations and Products*

In the simplest instance, iron production consists of removing oxygen from iron oxide with a suitable reducing agent such as carbon monoxide or hydrogen:

$$Fe_2O_3 + 3CO = 2Fe + 3CO_2 \qquad (1)$$

or

$$Fe_2O_3 + 3H_2 = 2Fe + 3H_2O \qquad (2)$$

All of the ironmaking processes depend, in part, on these equations. The manner in which hematite (Fe_2O_3) is converted to metallic iron through the intermediate oxides magnetite ($\sim Fe_3O_4$) and wustite ($\sim FeO$) is very complex. At temperatures greater than 565 °C, the overall reaction represented in Eqns. (1) and (2) requires three major structural changes. (At about 565 °C, wustite decomposes by a eutectoid reaction to magnetite and metallic iron.) The rates at which these gas–solid reactions occur and the morphological features of the products, in particular the metallic iron, are highly variable. If the starting oxide materials are very dense, the reduction usually proceeds topochemically in accord with the shrinking core model. That is, the outside of each reacting particle or pellet is covered with a layer of metallic iron beneath which there are well-defined layers of wustite and magnetite and an unreacted core of hematite. If the oxide particles are very porous, the reducing gases, H_2 and CO, penetrate easily and reduction to metallic iron proceeds throughout. Within a single particle, there may be many grains of hematite undergoing topochemical reduction and there may be many interconnected pores along which the reduction steps are proceeding. The term sponge iron is used to describe particles of oxide that have been reduced to metallic iron under conditions which lead to a highly porous structure. Under certain conditions iron ore pellets may expand dramatically during reduction (catastrophic swelling). This phenomenon is attributed to the formation of filamentary iron growing from wustite at a limited number of sites.

Each reduction step is limited by the usual mass-action principles (e.g., at 1000 °C, reduction of Fe_3O_4 to FeO and FeO to Fe can occur only when p_{CO}/p_{CO_2} is greater than about 0.40 and 2.50, respectively, so that CO and H_2 cannot be fully utilized in iron oxide reduction alone). The exhaust gas always contains considerable quantities of the useful reactants. A variety of schemes are employed for using the values of the exhaust gas. Most of these depend upon blending with richer fuel gases to produce a satisfactory fuel for auxiliary plant operations. In a few instances, a portion of the exhaust gas is chemically regenerated and recycled (e.g., the Wiberg process).

The production of sponge iron (directly reduced iron, DRI) in most of the direct reduction processes involves little more than the removal of oxygen from iron oxide as just described, although many technical problems, such as prevention of interparticle sticking, deactivation of the iron by light carburization to suppress pyrophoric qualities and subsequent transportation and handling, have to be solved. However, these do not include major chemical or physical changes in the material. DRI is charged directly to electric-arc steelmaking furnaces in the same manner as steel scrap.

On the other hand, production of iron in the blast furnace does involve a great deal more than removal of oxygen from oxide particles. As the iron oxide particles descend in the shaft, they yield oxygen (by conversion of CO to CO_2 and H_2 to H_2O) to the ascending gas stream. The CO and H_2 content of the ascending gas is regenerated by "solution-loss" reactions:

$$CO_2 + C = 2CO \qquad (3)$$

$$H_2O + C = CO + H_2 \qquad (4)$$

Optimization of blast-furnace operations depends, in

part, on proper control of the extent to which the reactions shown in Eqns. (3) and (4) occur. Coke consumed by Eqns. (3) and (4) is unavailable for combustion at the tuyeres. Furthermore, solution loss at a position too high in the furnace may lead only to some enrichment of the exhaust gas, a highly unprofitable way to use coke. The temperature of the descending solids in the blast furnace rises as shown in Fig. 6. The materials move through the upper preparation zone into the middle elaboration zone where the rate of temperature change is much less, and then into the fusion and smelting zone where the temperature rises rapidly. The reactions occurring in the smelting zone are very complex. In the vicinity of the tuyeres, temperatures as high as 2100 °C are developed in the raceways where the incoming preheated air and additives react with coke. In the bosh, around and above the raceways, fluxes, coke ash, ore gangue and unreduced iron oxide combine to form a fluid slag within the incandescent mass of coke particles. At the same time molten iron saturated with carbon is formed. Important reactions occur between the molten iron, molten slag and gas phases in this region. Sulfur is captured entirely by the slag and iron phases. Although the mechanism is not fully understood, the gaseous species SiS is an important part of the reaction sequence. As the slag and iron settle and separate in the hearth, almost all of the phosphorous and about three quarters of the manganese charged to the furnace are transferred to the iron phase. At the same time, the iron absorbs some silicon. Although it is possible to

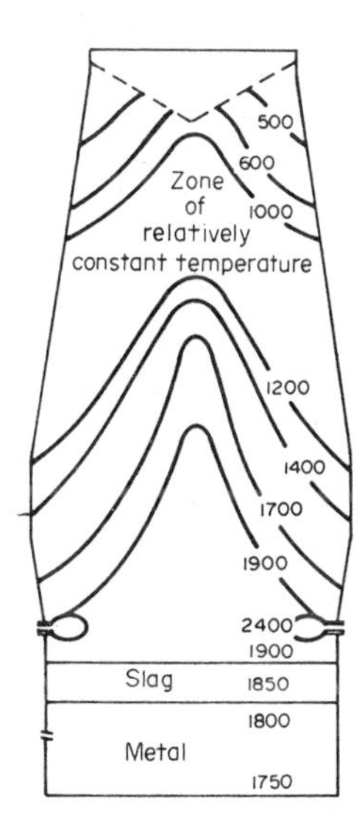

Figure 6
Representative temperature profile of a modern blast furnace, based on a quenched Nakamura furnace. Temperatures are in K (after Peacey and Davenport 1979)

Table 1
Representative compositions of blast-furnace iron and directly reduced iron

Component	Composition[c] (%)	
	Blast-furnace iron	Directly reduced iron
Total Fe	~94	>92
Metallic Fe	~94	>85
Carbon	4.0–5.0	0.8–1.7
Silicon	0.2–1.2	1.3–2.0[a]
Manganese	0.3–2.0	0.1–0.2[b]
Phosphorus	0.1–1.0	<0.05
Sulfur	~0.03	~0.005

[a] SiO_2 [b] MnO [c] Balance includes small amounts of Al_2O_3, CaO, MgO and TiO_2

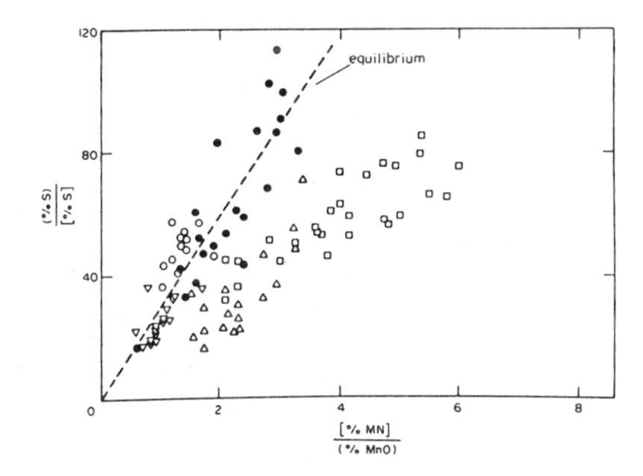

Figure 7
Representative operating data for sulfur and manganese partitioning between slag and iron at 1500 °C (after Turkdogan 1978). The different symbols indicate data for various steelworks

produce iron of high silicon content (silvery pig iron), blast-furnace iron usually contains <1 wt% silicon. Typical compositions of blast-furnace iron and DRI are shown in Table 1. Conditions in the hearth zone are highly favorable for the capture of sulfur and almost all is found either in the iron or slag phases. Representative information on the partitioning of sulfur between the two molten phases is given in Fig. 7. In recent years, advances in blast-furnace productivity and fuel consumption rates have tended to create conditions less favorable for transfer of sulfur from iron to slag. Coupled with higher sulfur content in coke and the demand for steels of much lower sulfur content, it has become desirable to adopt methods for desulfurizing blast-furnace hot metal in a separate processing unit between the blast furnace and the oxygen steelmaking processes where all hot metal is refined. Final desulfurization of steel to less than

50 ppm by ladle metallurgy injection techniques following electric or oxygen steelmaking is routine.

Typical operating data for blast furnaces are shown in Table 2. The slag from blast furnaces is used primarily as fill or road-bedding material. When properly prepared, it is suitable as a starting material for certain glass, brick, aggregate and insulation products.

The consumption of energy in the production of iron is relatively high but it must be remembered that the iron product is itself a fuel (the oxidation of 1 g of Fe to Fe_2O_3 releases 7.37 kJ, while the oxidation of 1 g of C to CO_2 releases 32.79 kJ). Heat balances on modern blast furnaces show remarkably high recovery of thermal values. Direct heat losses are usually $<10\%$ of total energy input; however, $\sim 40\%$ of total energy input is transferred as calorific value in the blast-furnace gas and there is substantial heat loss in its subsequent cleaning and reuse. About 3% of the input energy is carried off with the slag and unrecovered. Table 3 shows a simplified energy balance for a blast furnace. Care must be taken in evaluating such data because some energy is expended in preparing the charge materials (e.g., production of coke and beneficiation and induration of iron ore pellets). For comparison, the production of equal masses of copper and aluminum consume about three and twelve times as much energy respectively, as the production of iron.

The consumption of energy in direct reduction processes is highly variable. Direct comparison with blast furnaces is not valid because the products are so dissimilar. Proper comparison is made by evaluating the blast furnace–oxygen steelmaking route and the direct reduction–electric furnace steelmaking route separately. The former method uses less energy.

Table 2
Representative operating data for modern blast furnaces

Iron production rate	2–10 kt d^{-1}
Number of tuyeres	20–40
Blast temperature	1050–1280 °C
Slag composition	40% CaO, 10% MgO, 35% SiO_2, 10% Al_2O_3
Top-gas analysis	22% CO, 20% CO_2, 4% H_2, 4% H_2O, 50% N_2
Top pressure	0.3–2.2 atm (gauge)
Coke consumption	0.400–0.450 kg per kg iron
Oil injection	0.040–0.120 kg per kg iron

Table 3
Simplified representative heat balance for a blast furnace

Component	Energy (MJ per kg iron)
Input	
calorific value of coke	15.0
calorific value of injected fuel	1.0
sensible heat of air blast	2.0
Total	18.0
Output	
calorific value of top gas	7.5
calorific value of iron	7.4
sensible heat of iron	1.3
sensible heat of slag	0.5
miscellaneous losses	1.3
Total	18.0

3. Further Developments

Recent gains in productivity and fuel consumption in blast furnaces have been achieved primarily through improvements in both the chemical and physical character of the solid changes and the uniform charging, distribution and descent of the solid materials. In addition, the use of increased top pressure and improved control of tuyere conditions (e.g., adiabatic flame temperature) have been highly beneficial. Proper care in the control of size distributions in solid feeds and proper blending and distribution of the solids is receiving attention.

The direct reduction processes offer an economical route to iron and raw steel where natural gas is cheap. However, the current coal-based direct reduction processes have not established a generally competitive position and their commercial adoption will occur only in special and limited circumstances until significant technological improvements are made. Therefore, the annual demand for iron of about 500 Mt throughout the world will be met primarily by blast-furnace production through the year 2000. Gas-based direct reduction processes may assume a 10% share over that time span.

Bibliography

American Iron and Steel Institute 1981 *Annual Statistical Report 1981*. American Iron and Steel Institute, Washington, DC

Desy D H 1980 Iron and steel. *Minerals Yearbook 1980*. US Bureau of Mines, Washington, DC

Fine H A, Geiger G H 1979 *Handbook on Material and Energy Balance Calculations in Metallurgical Processes*. Metallurgical Society of AIME, Warrendale, Pennsylvania

International Iron and Steel Institute 1982 *World Steel in Figures 1982*. International Iron and Steel Institute, Brussels

McGannon H E (ed.) 1971 *The Making, Shaping and Treating of Steel*. US Steel Corporation, Pittsburgh, Pennsylvania

Peacey J G, Davenport W G 1979 *The Iron Blast Furnace: Theory and Practice*. Pergamon, Oxford

Poveromo J J, Hlinka J W 1982 Development and application of Bethlehem Steel's blast furnace charging model. *Iron Steelmaker* 9(10): 19–25

Stephenson R L, Smailer R M (eds.) 1980 *Direct Reduced Iron: Technology and Economics of Production and Use*. Iron and Steel Society of AIME, Warrendale, Pennsylvania

Sugarawa T, Ikeda M, Shimotsuma T, Higuchi M, Izuka M, Kuroda K 1976 Construction and operation of no. 5 blast furnace, Fukuyama Works, Nippon Kokan K. K. *Ironmaking Steelmaking* 3(5): 242

Szekely J, Evans J W, Sohn H Y 1976 *Gas–Solid Reactions*. Academic Press, New York

Turkdogan E T 1978 Blast furnace reactions. *Metall. Trans. B* 9: 163–79

G. R. St Pierre
[Ohio State University, Columbus, Ohio, USA]

L

Life Cycle Costing

The cost of an item to the buyer includes not only the purchase price but also the cost of ownership for its lifetime and the cost of disposal, i.e., the total life cycle cost. Durable, reliable and easily maintained items may have a lower life cycle cost, despite a higher purchase price, than cheaper, more fallible and nonrepairable items. Materials properties and costs often determine the life cycle cost characteristics of the design. An automobile fender of steel, for example, may be cheaper to manufacture than a fiberglass bonded fender, but repairs may be expensive and inferior for the steel fender and relatively inexpensive and satisfactory for the fiberglass fender. The choice between alternatives depends on the expected frequency of repairs and the buyer's assignment of the relative importance of first cost and the longer perspective of life cycle cost.

1. Principles and Areas of Application

Formally, the life cycle cost is the total cost of an item to the buyer including the development costs (if separately stated), the production costs (the usual purchase price for off-the-shelf items) and other investment costs, the operating and support costs, and the disposal costs.

In estimating or controlling life cycle cost (LCC), the following four phases of a program are used.

(a) Research and development: those activities leading to the release of manufacturing plans and drawings.

(b) Investment or production phase: the production of the prime equipment and its support equipment, initial provisioning, the initial training of operational personnel, transportation to the user, initial tests or break-in, and facility construction (e.g., fuel storage tanks or repair stations). Acquisition of the elements of production (e.g., tooling, test equipment, etc.) is also included in this phase.

(c) Operations and support phase: the operational costs such as operators, fuel or utilities, maintenance replenishment provisioning, support equipment maintenance and replacement, replacement training, facility maintenance and security. Modification or upgrading costs may be included in this phase or may be regarded as a new and distinct LCC program.

(d) Disposal phase: the disposal costs are usually relatively small and are often offset by the salvage value of the equipment. For some equipment however, such as nuclear power plants, disposal costs are significant and should be estimated and controlled.

These later costs may be discounted to correct for the time value of money (Seldon 1979), and economic adjustment factors should correct for any expected monetary inflation (or deflation). The use of "constant dollars" is recommended for LCC analysis; if decision makers require it, results may be presented in both constant and "then-year" values (incorporating any predicted monetary inflation or deflation).

LCC analysis often provides guidance in choosing between alternatives in design, replacing aging equipment, making purchasing comparisons between various equipment, contractors or suppliers, choosing operations and support concepts, developing maintenance methods, and preparing logistics support plans. In design, for example, the selection of throwaway, nonrepairable or discard-on-failure items would be discouraged if the potential consumer could be shown the advantageous lower LCC of more durable designs. As increasing scarcity in time raises raw materials costs, more durable designs will have a more attractive LCC. Rising labor costs could negate the effect of materials cost increases; quantitative economic analysis can elucidate the controlling parameters.

LCC may provide financial insight for the evaluation of alternative mining methods or sites, the choice of processing facilities, buildings and energy costs, the choice of alternative materials for design and the choice of protective maintenance methods (e.g., finishes and washes).

2. Methods of LCC analysis

LCC analysts often prepare cost estimates during the early phases of a project, as the most benefit can be gained before detailed designs or plans are available. Logistics plans and other operating support phase concepts are usually sparse or absent. In recent years, techniques have been developed for cost estimating with such rudimentary information. Even with these new techniques, often the LCC analyst must develop some exemplary plans for purposes of the baseline cost estimate; other personnel can then propose alternative plans, the cost of which will be compared with the baseline.

The cost estimates of the four phases (see Sect. 1) are combined to yield LCC and presented for analysis. The costs that comprise LCC must be combined and presented for analysis and review. The simplest model of LCC is:

LCC = Research and development
+ Investment
+ Operation and support
+ Disposal

Each of these four costs is time-phased and this variation with time is often portrayed as in Table 1. Analysts often portray LCC as a three-dimensional matrix (Fig. 1). Due to the volume of detail resulting from such a multidimensioned cost structure, computer techniques have been developed to ease the task of data manipulation, computation and structuring. The computer can generate requested output reports at whatever level of detail or generality is desired. Furthermore, the computer model can be used to convert constant dollars to inflated "real" dollars, or from one currency to another.

The computer mechanization portrayed utilizes cost-element values entered by the analyst. A further extension of the computer methodology would have the cost-estimating relationships built in to the computer program for some or all of the cost elements. For example, one of the operating and support costs might have a cost-estimating relationship as follows:

Operating and support maintenance costs per period = (number of maintenance personnel) × (labor rate per hour) × (hours per period)

Output formats of the LCC computer model should assist the analyst in understanding the details of cost contributions and show the decision makers the significant cost sensitivities of the program, as well as the overall costs. Various other formats and desirable characteristics of computer programs are discussed in the literature (Seldon 1979).

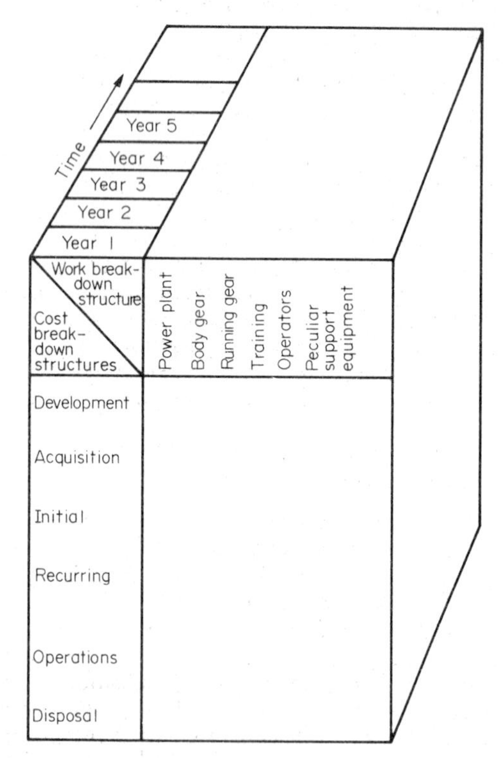

Figure 1
Three-dimensional cost matrix

3. Determinants of Ownership Costs

Life cycle costing focuses attention on the continuing ownership costs of an item (i.e., the operating and support costs). An automobile's ownership costs, for example, include operating costs (principally fuel) and support costs (principally maintenance). The calculation of operating costs, such as fuel and lubricants, is straightforward, even prior to having actual experience. Estimates of fuel consumption, for example, are routinely provided by design engineers. Maintenance costs, however, are more difficult to estimate.

Maintenance costs, particularly maintenance labor costs, usually dominate other support costs and should be given careful attention in a cost analysis. Most analysts divide maintenance costs into (a) scheduled or preventive maintenance, and (b) unscheduled

Table 1
An example of LCC estimates phased over 11 years (in US dollars or other money units)

	Year							
LCC phase	1982	1983	1984	1985	1986	1987	1992	Total
Research and development	10	15	5					30
Investment		2	20	50	5			77
Operations and support					12	18	18	120
Disposal							1	1
Total	10	17	25	50	17	18	19	228

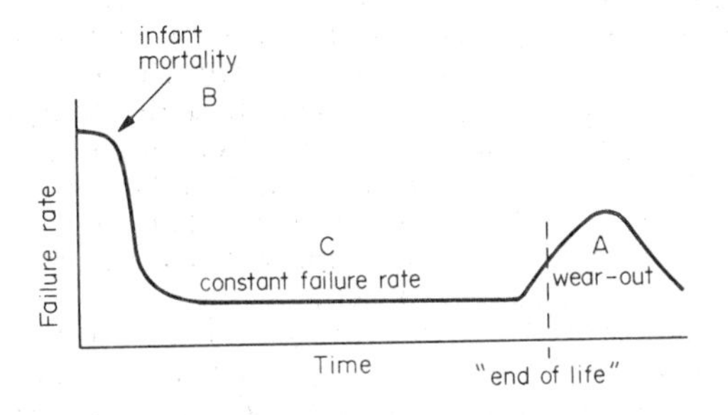

Figure 2
Diagram outlining the variation of failure rate with operating time

or corrective maintenance (repair). Preventive maintenance should reduce the requirements for corrective maintenance and cost analysts can examine expenditures for preventive maintenance to determine whether a proper balance has been achieved for minimum total cost. Changes in the preventive maintenance schedule can often reduce the total maintenance cost. For this reason, the cost analysis is often developed in conjunction with a maintenance concept.

The reliability of an item, that is, how long it will operate before failure, determines how often corrective maintenance is required. The LCC analyst must, therefore, obtain estimates of the item's reliability (with a specified preventive maintenance schedule). The frequency of necessary repair actions should be estimated by specialized reliability personnel. They must identify the frequency of required repair actions. The predicted reliability will enable the analyst to estimate parameters such as maintenance labor required, maintenance material required, the facilities, maintenance personnel training, and maintenance tooling and test equipment. If maintenance engineering personnel (logistics planning engineers) are available, their help will improve the maintenance planning and, therefore, the cost estimates.

Durability, a measure of the life of an article, also affects a planning trade-off. For many items and materials, only a limited number of repairs can be made before the costs of satisfactory repairs begin to increase sharply. Furthermore, many mechanisms exhibit a wear-out phenomenon with the failure rate increasing with operating time. The general shape of the variation of failure rate with time holds true for many kinds of equipment (Fig. 2). The end of life or durability is represented by A in this figure. It should be noted that the initial failures B are significantly influenced by the quality control of the manufacturing process, and that the random failure rate C is determined principally by design.

4. *Other Operating and Support Costs*

Replenishment provisioning identifies costs of replacing supplies and repair parts which are consumed during maintenance. Maintenance may be subdivided into prime equipment maintenance and support equipment maintenance, and further into scheduled and unscheduled maintenance. The finer breakdowns allow more precision in estimating.

The initial costs of training, such as technical manuals, training facilities, training of instructors and classroom aids, are accounted for in the investment phase, but during the operating and support phase the costs of replacement training due to attrition must be allocated. During the operating and support phase, costs will be incurred for

(a) maintenance of support equipment,

(b) maintenance facility costs,

(c) crew pay and fringe costs,

(d) fuel and other provisions,

(e) utilities (e.g., electricity, gas, telephone, water and drainage) and

(f) taxes and rent depletion allowances.

Improved logistics planning provides a major benefit of the LCC analysis. Early in a program, analysis can alert the developer and the user to the cost implications of the design and suggest design changes and logistics plans to minimize the cost of product support.

5. *Systems Engineering Using LCC with Cost–Performance Trade-Offs*

An effective LCC program should reduce the LCC of the system under consideration. LCC can show desirable changes to the system configuration, design, or the way it is produced or used.

System engineering transforms an operational need (mission requirements) into a description of system performance parameters (design specifications) and a preferred system configuration. Design engineering develops a number of feasible design concepts that could perform to the design specification. Each alternative set of performance parameters and system configuration can be evaluated for LCC. Related pairs of performance parameters, such as service life and purchase price, can be varied over a possible range. The variation in LCC as these pairs are traded-off against one another, can be evaluated and the pair with the minimum resultant LCC can be selected. This process of examining various suitable alternatives and selecting the most desirable combination of performance and cost is known as the cost–performance trade-off process or cost–benefit analysis.

6. *Marketing and Contractual Methods of Life Cycle Costing*

The buyer, the consumer, and the user, both as individuals and as organizations, must be convinced that LCC is an important criterion in the choice between alternative courses of action. For relatively simple "off-the-shelf" items, the marketing task is to convince the consumer that the particular product has a lower LCC.

In buying large, complex, one-of-a-kind systems (e.g., an ore processing plant or military weapons systems) the LCC considerations become even more important. These large complex systems are procured as a design and construction package or as an architect and engineer package followed by a construction contract.

One cannot, from first principles, specify the LCC in such circumstances, particularly as the systems are often uniquely advanced. Therefore, the buyer must use contractual incentives that will motivate the contractor(s) towards a low LCC design and a low LCC product. Military systems procurement agencies have developed a number of contractual methods of motivating contractors toward lower LCC: (a) target logistics support cost, (b) reliability improvement warranty, and (c) design to cost incentive (production costs as a surrogate for LCC).

Acknowledgement

The publishers wish to thank Mr E. L. Tinkham for having reviewed and updated this article.

Bibliography

Brown R J, Yanuck R R 1980 *Life Cycle Costing: A Practical Guide for Energy Managers.* Fairmont, Atlanta, Georgia
Earles M E 1978 *Formulas and Structures for Life Cycle Costing.* Eddins-Earles, Concord, Massachusetts
Sackman H 1975 *Delphi Critique; Expert Opinion, Forecasting, and Group Process.* Lexington Books, Lexington, Massachusetts
Seldon M R 1979 *Life Cycle Costing: A Better Method of Government Procurement.* Westview, Boulder, Colorado
US Office of Management and Budget 1976 *Major Systems Acquisition,* Circular A-109. US Government Printing Office, Washington, DC

M. R. Seldon†
[Formerly of General Dynamics, Pomona, California, USA]

Linear Viscoelasticity

Polymers are attractive because of their light weight, low cost, corrosion resistance and ease of fabrication. They show creep and relaxation, however. Short-term tests, dictated not only by the cost of testing but also by the promise of newly developed materials, must be used to predict the behavior of parts of structures over their 10–100 year lives.

Polymers are long chains of repeated molecular groups with side groups off this backbone, all attached with primary bonds. Secondary bonds act between the chains, and can be shifted by thermal motion at room temperature. Any cross-linking between the chains with chemical (primary) bonds limits the sliding of chains past each other. Glass can be thought of as a polymer with very short chains of silicon and oxygen in a three-dimensional network.

After a qualitative introduction to time-dependent deformation, the linear dependence on stress (for immediately recoverable unloading strains of 0.1–1%, which are often all that is acceptable in structures) is noted. Linearity allows the definition of a time-dependent modulus, analogous to the time-independent modulus of elasticity. The way the polymer structure affects the modulus of elasticity is described. The similarity of time and temperature effects leads to a time–temperature shift function, interpreted from a general stress- and structure-modified thermal activation. This allows the presentation of data for many polymers in a concise form, which is often adequate for predicting long-term uniaxial behavior under varying service temperatures.

For general stress analysis, the linear stress dependence of the strain allows the superimposing of results from creep, relaxation, or constant strain-rate tests to predict the behavior under arbitrary stress histories in service. In particular, a simple relation is given between creep and relaxation moduli. Further, linearity means that many two-dimensional elastic results apply to viscoelastic materials, when the time-dependent behavior is taken into account at just one convenient point in the body.

1. Physical Basis of Viscoelasticity

Common tests and kinds of viscoelastic behavior are reviewed, and their qualitative dependence on temperature and structure is noted. Molecular models provide a basis for some of the phenomena and relations.

1.1 Viscoelastic Tests and Results

Creep, relaxation, constant strain-rate and cycling tests are all used for data, depending on convenience, the data sought, and strain rates or frequencies of interest. The kinds of response include elasticity, transient and steady-state creep, delayed elasticity (elastic aftereffect on unloading) and damping.

Creep under constant stress gives transient and steady creep which may be indistinguishable from that of a metal; the typical viscoelastic response is revealed by delayed elastic recovery on unloading.

Relaxation at constant strain gives a stress falling with time, perhaps ultimately to zero.

Constant strain-rate tests, as approximated by a constant rate of crosshead motion, give a stress–time behavior similar to that of a strain-hardening metal until the crosshead is stopped, when the stress begins to fall as in relaxation.

Cyclic stress or strain data may be desired. At very high frequencies the behavior is stiff and elastic; the stress and strain are in phase. Lower frequencies allow some viscous flow, and the strain lags the stress. At still-lower frequencies in rubber, the stress and strain are again in phase. In viscous fluids (steady-state creep), the strain lags the stress by 90°. To cover the range of time response requires many frequencies.

Linearity, in the sense that doubling the applied stress or strain doubles resulting strain or stress at each time, greatly simplifies viscoelastic analysis. The limit of linearity is typically 0.1–1% of the immediately recoverable strain; it is higher for pure rubber or highly oriented crystalline polymers; it is lower for highly filled or isotropic polycrystalline polymers. The limit of linearity is determined by tests at a variety of stress or strains (or only two different constant strain rates since, except for pure viscosity, nonlinearity must eventually occur).

Spring–dashpot models give a qualitative insight into viscoelastic responses (see *Polymers: Mechanical Properties*). A spring and dashpot in series (a Maxwell element) models steady-state creep, complete relaxation, a final constant stress under constant strain rate, and damping in cyclic tests. Adding a spring in parallel with the dashpot to limit its travel (a Voigt element in series with the original spring, forming a "standard linear solid") gives transient creep or partial relaxation. Adding another dashpot in series with the other elements gives both transient and steady-state creep, and two relaxation time constants. Other four-element models can give exactly the same time response with appropriate stiffnesses and viscosities; thus, the underlying molecular structure cannot be determined from the viscoelastic response. While these models give a qualitative description, the real behavior of polymers involves a whole spectrum of time constants; stress analysis requires the use of experimental data rather than models.

1.2 Creep and Relaxation Moduli and Compliances

With linearity, a creep modulus $E_c(t)$ is defined as the ratio of constant stress σ to time-dependent strain $\varepsilon_c(t)$, and a creep compliance $J_c(t)$ is defined as its reciprocal:

$$E_c(t) \equiv \frac{\sigma}{\varepsilon_c(t)}, \qquad J_c(t) \equiv \frac{\varepsilon_c(t)}{\sigma} \tag{1a,b}$$

Likewise, relaxation data give a relaxation modulus $E_r(t)$ and a relaxation compliance $J_r(t)$:

$$E_r(t) \equiv \frac{\sigma(t)}{\varepsilon}, \qquad J_r(t) \equiv \frac{\varepsilon}{\sigma(t)} \tag{2a,b}$$

Note that the creep and relaxation moduli are not equal because they result from different stress histories (see Sect. 2.2).

For cyclically applied strain, the stress response will also be cyclic, but with an amplitude and a phase lag δ (or a phase lead $-\delta$) that both depend on the frequency (see *Polymers: Mechanical Properties*). For small damping, the fractional loss of energy $\Delta U/U$ and of amplitude $\Delta a/a$ per cycle are related by $\Delta U/U = 2\Delta a/a \approx 2\pi\delta$.

The modulus and compliance functions of time correlate a wide range of stress and strain data, allowing concentration on time and temperature.

1.3 Time and Temperature Effects

In common with all thermally activated processes creep and relaxation moduli are also very temperature dependent, as shown in Fig. 1. Often, especially for amorphous polymers, a constant horizontal shift of each isotherm (by an amount depending on the temperature) makes all the curves nearly coincide. That is, the ratio of the equivalent time dt_S at a service temperature T_S to the time dt_R at a reference (data) temperature T_R for the same physical effect, called the shift factor $a(T_S)_R$, is a unique function of the temperatures:

$$a(T_S)_R \equiv \frac{dt_S}{dt_R} \tag{3}$$

The resulting master curve shown in Fig. 1 is generated in two ways: by matching curves for successive temperatures at the highest and at the lowest common moduli. The difference between the two gives a measure of the uncertainty in extrapolating from the test data to typical service times. The corresponding range of shift factors $a(T_S)_R$ is plotted in Fig. 2, which is consistent with the slowing of processes at lower temperatures. (The term "shift factor" may come from the "shift" on a logarithmic scale and a "factor" on a linear scale of time.) Thus the master and shift-factor

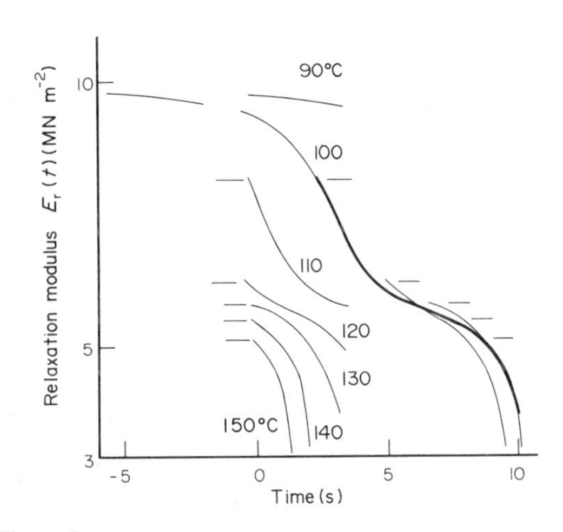

Figure 1
Relaxation data for various temperatures, and master curve at 100 °C (after Aklonis 1986)

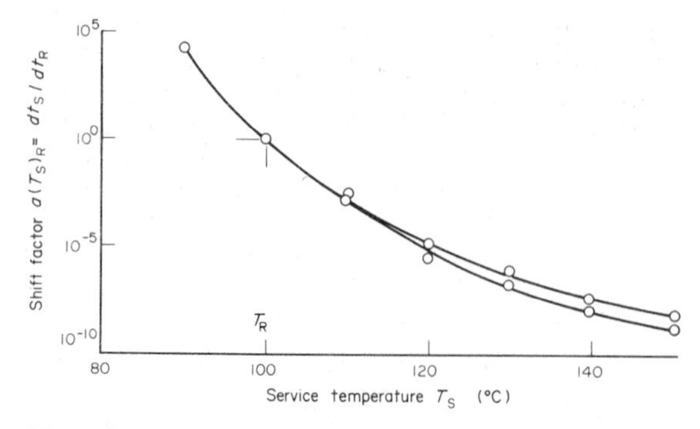

Figure 2
Shift factor $a(T_S)_R$ for data of Fig. 1. Reference temperature $T_R = 100\,°C$

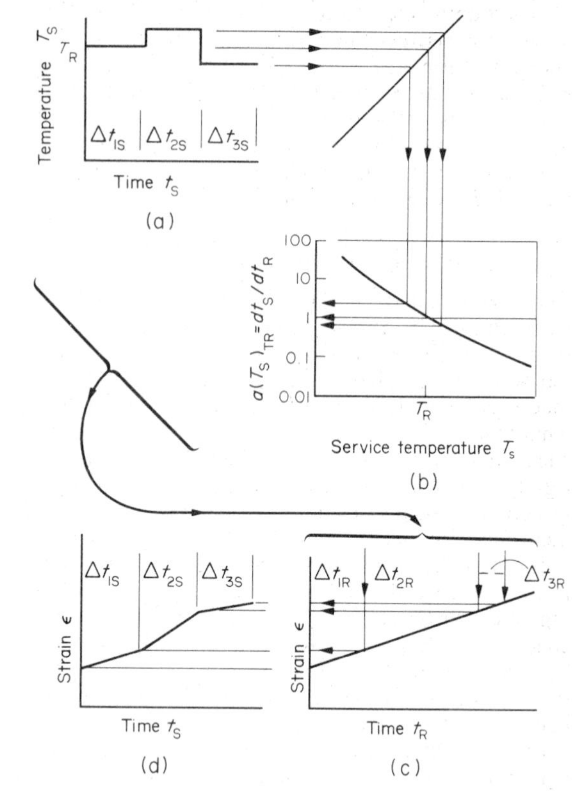

Figure 3
Maxwell creep in service at various temperatures, from shift factor and Maxwell creep at reference temperature: (a) history in service, (b) shift factor versus temperature, (c) creep in service, (d) creep at reference temperature

curves of Figs. 1, 2 summarize the complete time–temperature behavior. A small vertical shift may be added to account for the effect of temperature on rubbery elasticity.

The gradual change of the master curve over many orders of magnitude of time means that it is not practical to derive the general viscoelastic response from a model with a few time constants. Rather, it is necessary to operate directly on the data to predict service behavior. Note that polymers used as structural components should be near one of the plateau regions of Fig. 1 to avoid undue sensitivity to time or temperature. In estimating the 10–100 year behavior of a new material without waiting so long for tests that the material is obsolete, other factors must be taken into account as well, such as humidity, aging, loss of solvent, biological attack, oxidation and ultraviolet degradation.

Master curves of modulus or compliance, for either creep or relaxation and shift factor curves, $a(T_S)_R$, are still widely scattered and uncommon in the scientific and manufacturers' literature; but computer-controlled test and data-acquisition systems, along with the programs to apply the curves in practice, are making these master curves much more available.

As an example of using the master and shift curves, the prediction of creep in service from creep at a reference temperature and the shift-factor curve is shown in Fig. 3. From the temperature history in service (Fig. 3a) and the shift-factor curve (Fig. 3b), the service times are plotted onto the creep curve at the reference temperature (Fig. 3c). The strains are then plotted back on to the scale of time in service (Fig. 3d). In Sect. 2 we shall extend the history to include varying uniaxial stress and in Sect. 3 to various stress components.

1.4 Effects of Polymer Structure

In the amorphous polymer of Fig. 1, the drop in modulus by three orders of magnitude from 90 to 110 °C represents a melting of interchain forces. (The point of inflection at a given relaxation time is sometimes called the glass transition temperature.) Polymers with a regular, compact arrangement of side groups can crystallize (usually much less than the most dislocated metal crystal), in which case the drop in modulus is typically only by a factor of ten.

The plateau below the glass transition temperature is a rubbery region (or "leathery" for crystalline polymers), where the chains are locally free to slide over each other, but restrained by their entanglements, since the chains are thousands of molecules long. The rubbery modulus can be found from the tendency of the chains to assume a random configuration (see *Elasticity*). The plateau can be extended by lengthening the chains; this lengthening can even be extended to the point of decomposition of the chains themselves, by cross-linking some of the chains with primary bonds. In crystalline polymers, the modulus drops gradually through the leathery region and is less dependent on chain length.

When complete chain pullout occurs, the polymer flows as a viscous fluid. If it is still below its chain decomposition temperature, it is called thermoplastic and recovers on cooling. Otherwise it is called thermosetting. Ordinary glass can be thought of as a cross-linked polymer that is thermoplastic, because the silicon in the glass is already oxidized and the cross-link bonds reform on cooling.

1.5 Molecular Model of Viscosity

The low mass and high binding energy of carbon and hydrogen mean that they have frequencies which are too high to be excited by thermal motion at room temperature (c.f. thermal activation in *Creep*). On a macroscopic scale, their specific heats are more nearly what would be expected if all atoms except the lightest were classically activated, and the lightest atoms were rigidly bound to a neighbor. Thus a model of thermally activated deformation is more tentative for polymers than for metals.

In amorphous polymers, inelastic strain can result from a cooperative molecular rearrangement producing a distortional strain component of unity in a cluster of volume V_c. Assuming n_c such clusters of volume V_c in a total volume V gives an average strain in the total volume of $n_c V_c/V$. At high enough temperatures, thermal activation keeps generating new clusters which are available for switching, leading to continued viscous flow. Applied stress does not itself force the switch, but rather biases the activation energies of the switch in the favorable and unfavorable directions, giving a linear dependence of strain rate on stress. The resulting viscosity depends on the activation enthalpy per cluster h, which can be estimated from the elastic strain energy required to compress the molecules sufficiently to squeeze past a point of minimum spacing into the new position. In terms of Boltzmann's constant $k = 1.381 \times 10^{-23}\ \mathrm{J\,mol^{-1}\,K^{-1}}$ and the temperature T (Ward 1983, McClintock and Argon 1966)

$$\mu \equiv \frac{\sigma_{12}}{(d\gamma_{12}/dt)} = \left(\frac{V}{nV_c}\right)\left(\frac{2kT}{\nu V_c}\right)\exp(U/kT) = \mu_0 \exp(U/kT) \tag{4}$$

1.6 A Model of the Time–Temperature Shift Factor

In spite of the complexity of real polymers, the time–temperature shift factor of Eqn. (3) does seem to hold for a variety of amorphous polymers, and even some crystalline ones. Consider the approximations needed to derive this from the general equation for thermally activated deformation, in which the strain rate $\dot{\varepsilon}$ depends on a preexponential factor $\dot{\varepsilon}_p$ and an activation enthalpy h, both of which may depend on the current stress σ, the internal structure of the material s, and the temperature T (see *Creep*):

$$\dot{\varepsilon} = \dot{\varepsilon}_p(\sigma, s, T)\exp[-h(\sigma, s, T)/kT] \tag{5}$$

The softer polymers especially, are in a continuous state of thermal rearrangement, which is only biased by the stress. As in the model of viscosity, any effect of stress on the activation enthalpy is small and can be subtracted out from the exponential, becoming a linear factor in the preexponential term. Then the stress effects can be normalized by dividing by the stress. Choose the strain increments to be equal under the service and reference temperatures, then for properly scaled time increments:

$$\frac{d\varepsilon_S}{\sigma} = \frac{d\varepsilon_S/dt_S}{\sigma}dt_S = \frac{d\varepsilon_R}{\sigma} = \frac{d\varepsilon_R/dt_R}{\sigma}dt_R \tag{6}$$

Before introducing Eqn. (5) for the strain rates, assume that the structural changes are composed of two distinct parts: the particular molecular stretching and sliding due to the history of stress and strain, and the general isotropic changes associated with equilibrium at different temperatures. Assume that Eqn. (6) has applied since the start of the histories, so that the structures due to the stress and strain are the same under the service and reference conditions. Then the preexponential terms in Eqn. (5) for the service and reference conditions differ only in structural changes due to temperature $s(T)$. Thus the preexponential factor is a function only of temperature: $\dot{\varepsilon}_p(T)$. Likewise, the activation enthalpy depends only on the temperature. Now substituting Eqn. (5) into Eqn. (6) for both reference and service temperatures, with the stress divided out as described above, gives the times for equal strain increments

$$a(T_S)_R \equiv \frac{dt_S}{dt_R} = \frac{d\varepsilon/dt_R}{d\varepsilon_S/dt_S} = \frac{\dot{\varepsilon}_p(T_R)}{\dot{\varepsilon}_p(T_S)} = \frac{\exp[-h(T_R)/kT_R]}{\exp[-h(T_S)/kT_S]} \tag{7}$$

For amorphous polymers, Williams, Landell and Ferry (WLF) first found an empirical form of Eqn. (3), and later derived it from a molecular model based on the free volume between chains (see Ward 1983 pp. 149–55). Taking the reference temperature to be the glass transition temperature T_g gives the WLF form

$$\log a(T)_g = \frac{-17.4(T - T_g)}{51.6\,^{\circ}\mathrm{C} + T - T_g} \tag{8}$$

(Ferry (1980) gives data showing that the first "constant" $c_1 = 17.4$ ranges over $-10 < c_1 < 22$, and the second $c_2 = 51.6\,^{\circ}\mathrm{C}$ ranges over $30\,^{\circ}\mathrm{C} < c_2 < 100\,^{\circ}\mathrm{C}$, for 35 out of 40 polymers investigated.)

1.7 Plastic Flow in Polymers

Beyond the linear range, polymers show increasing strain per unit stress, and an increase in the deformation remaining after the release of the load. In polymers, such plastic deformation is very dependent on temperature and strain rate.

In crystalline polymers, the molecular disarrangement accompanying plastic flow may initially reduce the flow strength, leading to a drop in load and the localization of flow in a neck. Subsequent flow aligns the polymer chains and increases the strength, limiting the deformation in the neck. With continued extension of the specimen, the neck propagates along the specimen. This process, called drawing, can increase the strength and stiffness in the axial direction by factors of from five to several hundred. The stiffness, strength and ductility in the transverse direction may also increase by large factors, but remain sufficiently lower than the axial properties to give very large anisotropy. The combination of rate effects and developing anisotropy make the prediction of plastic flow in polymers very difficult, even though it is very important in forming processes and in determining crack toughness (see *Polymers: Mechanical Properties*).

2. Solving for Arbitrary Service Histories by Superposition of Data

In this section it is assumed that a history of varying temperatures has been mapped onto time at a reference temperature, to which "time" refers. When a history of stress in service is matched by repeated scaling, shifting and superposition of reference (test) stress data, applying the same operations to the reference strain history gives the desired strain history in service. For many applications, including predicting relaxation from creep data, this superposition can be carried out in one or two steps, without the need for a computer.

2.1 Linearity and the Boltzmann Superposition Principle

Viscoelasticity involves not only displacements u but also the displacement rates or velocities $\dot{u}$. The corresponding strain rates are

$$\dot{\varepsilon}_{ij}=\frac{d}{dt}\left[\frac{1}{2}\left(\frac{\partial u_i}{\partial x_j}+\frac{\partial u_j}{\partial x_i}\right)\right]=\frac{1}{2}\left(\frac{\partial \dot{u}_i}{\partial x_j}+\frac{\partial \dot{u}_j}{\partial x_i}\right) \quad (9)$$

Viscoelastic deformation involves numerous internal parameters associated with inhomogeneous chain stretching and interchain sliding. Eliminating these parameters in terms of derivatives of the macroscopic stress and strain gives

$$\sum_m p_m \frac{d^m \sigma_{ij}}{dt^m}=\sum_n q_n \frac{d^n \varepsilon_{ij}}{dt^n} \quad (10)$$

Equation (10) is far too complex for reasonable analytical modelling, so reliance has been placed on applying numerical methods directly to the data. The data are commonly obtained from creep at constant stress or relaxation at constant strain, but they may also come from stress under more or less constant strain rate, or from any other convenient experiments. Applying whatever test data are available to service histories that are often irregular is greatly facilitated by linearity.

For strains less than 0.1–1% for structural polymers, the response is linear in the sense that if a stress history $\sigma_1(t)$ and a strain history $\varepsilon_1(t)$ occur together, then multiplying both the stress and strain histories by the same weighting factor w_1 gives another pair of stress and strain histories that occur together. Because either stress or strain could be the independent variable, any pair of histories that occur together will be denoted by $\sigma_1(t)\leftrightarrow\varepsilon_1(t)$. Linearity implies $w_1\sigma_1(t)\leftrightarrow w_1\varepsilon_1(t)$. Linearity provides an even more powerful result. Assume a second stress–strain history $\sigma_2(t)\leftrightarrow\varepsilon_2(t)$, possibly completely unrelated to the first one, except that it is obtained with the same material at the same temperature and humidity. Then for any second weighting factor w_2 and for any shift in time t_2, the sum of the two scaled and shifted stress histories goes with the corresponding sum of the scaled and shifted strain histories:

$$\sigma_1(t)\leftrightarrow\varepsilon_1(t) \quad \text{and} \quad \sigma_2(t)\leftrightarrow\varepsilon_2(t)$$

imply

$$w_1\sigma_1(t-t_1)+w_2\sigma_2(t-t_2)\leftrightarrow w_1\varepsilon_1(t-t_1)+w_2\varepsilon_2(t-t_2) \quad (11)$$

Equation (11), a form of the Boltzmann (scaling, shift and) superposition principle, can be verified by substitution into Eqn. (10) and the strain–displacement and equilibrium equations, or by experiment.

2.2 Application of a Discrete Form of Boltzmann Superposition

A general, discrete form of the Boltzmann superposition principle is useful for one- or two-step hand calculations, or for a computer program.

(a) Divide the total reference time corresponding to the service life into equal intervals. Denote the time at the middle of the ith reference time interval by t_i, and the corresponding reference strain and stress by $\varepsilon_R(i),\sigma_R(i)$.

(b) Use superposition to represent the strains and stresses in service, $\varepsilon_S(i)$, $\varepsilon_S(i)$, at the midpoints of the same time intervals in terms of weighting factors w_i (to be determined later) and summations to the desired service time t_j:

$$\varepsilon_S(j)=\sum_{i=1}^{i=j} w_i\varepsilon_R(j-i+1)$$

$$\sigma_S(j)=\sum_{i=1}^{i=j} w_i\sigma_R(j-i+1) \quad (12a,b)$$

(c) Determine the weighting factors from the strain or stress, as prescribed by the service conditions. For instance, for a given service stress history, find the jth weighting factor by separating the term for $i=j$ from the sum of Eqn. (12b) and solving:

$$w_j=\left[\sigma_S(j)-\sum_{i=1}^{i=j-1} w_i\sigma_R(j-i+1)\right]\Big/\sigma_R(1) \quad (13)$$

(d) From the weighting factors, find the strains in service from Eqn. (12a):

$$\dot{\varepsilon}_S(j)=\sum_{i=1}^{i=j} w_i\dot{\varepsilon}_R(j-i+1) \quad (14)$$

The procedure can be reversed if the strain in service is given, as for an O-ring press-fitted on a shaft.

The superposition principle in the form of Eqns. (12–14) can be stated in the limit as a convolution integral, but the series form is more useful since the data for real materials and varying service loads must almost always be integrated numerically.

Is the accuracy sufficient, with a small enough number of steps to be practical? To find out, try one and then two steps to predict the creep from relaxation of a Maxwell element with elastic modulus E and time constant τ:

$$\varepsilon_R=\text{constant}\leftrightarrow\sigma_R=E\varepsilon_R\exp(-t/\tau) \quad (15)$$

or $E_R=E_r=E\exp(-t/\tau)$. The exact solution is

$$\sigma_S=\text{constant}\leftrightarrow\varepsilon_S=\frac{\sigma_S}{E}\left(1+\frac{t}{\tau}\right) \quad (16)$$

or $E_S=E_c=E/(1+t/\tau)$. As an example, find $E_S(t=\tau/2)$.

For a one-step approximation, let $\tau/2$ be the midpoint of $0\leqslant t\leqslant\tau$. From Eqn. (13), $w_1=\sigma_S/\sigma_R(\tau/2)$ and from Eqn. (12a) $\varepsilon_S(\tau/2)=w_1\varepsilon_R$. These combine to give the creep modulus $E_c(\tau/2)=E_r(\tau/2)=E\exp(-1/2)=0.607E$, compared with the exact value from Eqn. (16) of $E/(1+1/2)=0.667E$. Thus to a first approximation, the creep and relaxation moduli are equal.

For a two-step approximation, let $\tau/2$ be the midpoint of the second interval, $\tau/3\leqslant t\leqslant 2\tau/3$. Then $w_1=\sigma_s/[E\varepsilon_R\exp(-1/6)]$, $w_2=[\sigma_s-w_1E\varepsilon_R\exp(-1/2)]/E\varepsilon_R\exp(-1/6)$, giving $E_c=E\exp(-1/6)/[2-\exp(-1/3)]=0.660E$. Similarly good results are obtained with other reference and service histories unless the stress or strain drops by a large fraction during the last interval.

This example shows how the superposition principle gives the desired strain history directly from the data without finding and integrating a differential equation describing the material. This general form applies whether the service history is given in terms of strain or stress, and whatever the history that gave the reference data.

For reference data consisting of either a creep compliance or modulus (Eqn. (1)), let the pair be $\sigma_R=1$, $\varepsilon_R(t)=J_c(t)\equiv 1/E_c(t)$. For reference data consisting of either a relaxation modulus or compliance (Eqn. (2)), let the pair be $\varepsilon_R=1$, $\sigma_R(t)=E_r(t)\equiv 1/J_r(t)$.

The numerical stability of superposition is improved by using relaxation data for the reference functions, so that neither strain nor stress increases with time. If relaxation data are not available, an appropriate reference pair can be obtained by starting from a creep or constant-strain-rate pair and subtracting from them an identical pair, shifted along by the time interval used with Eqns. (12a,b). For creep or relaxation data this gives one of the reference pair as a step function in stress or strain, respectively, and greatly simplifies the superposition.

Conversion between creep and relaxation functions can be derived from Laplace transforms and Gamma-function identities if the slope of either E_c or E_r on a log–log plot of stress versus time is $-m$: $E_r(t)=E_c(t)\sin(m\pi)/m\pi$.

3. Analytical Solutions

Stress and strain distributions from plane elastic stress and deformation analysis can often be applied to viscoelastic materials when time-dependent behavior is taken into account at just one convenient point in the body. This correspondence can be extended to many more problems when Poisson's ratio can be assumed constant or almost constant.

3.1 The Correspondence Principle between Linear Elastic and Viscoelastic Solutions

The correspondence principle states that load–stress or strain–deflection relations from elasticity are material invariant if they do not involve the modulus of elasticity or Poisson's ratio. Then the corresponding viscoelastic relations will be both material-invariant and independent of time.

This correspondence principle can be proved with Laplace transforms (e.g., McClintock and Argon 1966 or Pipkin 1972). With it, material-invariant load–stress relations can be used to obtain the stress history at any point from the applied load history. The transformation from stress history to strain history need only be made at one point, often in terms of just one component of stress and strain. When the strain–deflection relations are material-invariant, elastic relations then give the time-dependent deflections from the time-dependent strain. Material-invariant relations can also be used to find the stress and strain at other points.

3.2 Applications of the Correspondence Principle to Elementary Problems

For a simply supported elastic beam with length L, depth h and moment of inertia I, subject to a central load P, the maximum stress is

$$\sigma=\frac{PL(h/2)}{4I} \quad (17)$$

Equation (17) is material-invariant (independent of E and ν). Hence, by the correspondence principle, it is valid for linear viscoelasticity. Furthermore, there are no residual stresses because the change in elastic stress on unloading is exactly equal and opposite to the (constant) viscoelastic stress distribution. The deflection v is typically found from $v=PL^3/48EI$, where E can be eliminated using $E=\sigma/\varepsilon$, then σ can be eliminated using Eqn. (17), giving:

$$v=\frac{PL^3}{48I\sigma/\varepsilon}=\frac{1}{48}\frac{4}{L(h/2)}L^3\varepsilon=\frac{L^2\varepsilon}{6h} \tag{18}$$

This relation for the deflection is material invariant, and is valid for linear viscoelasticity. To find the strain in Eqn. (18), find the stress from the load using Eqn. (17), and apply the superposition principle to find the time-dependent strain. Then substitute into Eqn. (18) to find the time-dependent deflection of the beam. Note that there are residual strains and deflections.

For pressure p inside a thin-walled, closed-end tube of radius r and thickness t, the circumferential and axial stress components are

$$\sigma_{\theta\theta}=p\frac{r}{t}, \qquad \sigma_{zz}=\frac{pr}{2t} \tag{19a,b}$$

The radial u_r and axial u_z displacements in terms of the components of strain are

$$u_r=r\varepsilon_{\theta\theta}, \qquad u_z=L\varepsilon_{zz} \tag{20a,b}$$

Here again the stress–load and the strain–displacement relations are material invariant. With biaxial stress, however, the viscoelastic calculations of the strain histories involve Poisson's ratio (see Sect. 3.3(c)).

For an O-ring $\sigma_{zz}=0$, so with an applied displacement Eqn. (20b) would give the strain. Uniaxial viscoelastic data and superposition would then provide $\sigma_{\theta\theta}$ and, through Eqn. (19a), the contact pressure.

3.3 *Applications of the Correspondence Principle to Plane Problems*

A sufficient condition for the material-invariance of the stress-load relations in plane problems is: in either plane stress or plane strain, with specified loads as boundary conditions and no net force on any closed contour (as there would be on a loaded rivet hole), the in-plane components of stress are independent of the elastic constants, and are the same for plane stress as plane strain (see, e.g., Gurtin 1984 p. 163). Several examples and corollaries follow.

(a) Stress concentrations are material-invariant. (The strain–displacement relations generally depend on Poisson's ratio and are time dependent.)

(b) There are no mechanically induced residual stresses with material invariance.

(c) Thermally induced residual stresses are not material invariant.

(d) Assuming incompressibility, $\nu=0.5$, or an elastic $\nu=0.35$, often makes the strain–deflection relations material invariant, and gives results close enough together to be practical (see (f) below).

(e) Effects of transverse stresses in superposition may also be well enough approximated by taking $\nu=0.5$ and 0.35, even in estimating shear from tensile moduli.

(f) For a thick-walled cylinder of inner and outer radii r_i, r_o, under internal pressure p_i, the in-plane stress components are

$$\sigma_{\theta\theta}, \sigma_{rr}=p_i\left[\frac{(1\pm r_i^2/r^2)}{1-r_i^2/r_o^2}-1\right] \tag{21}$$

which are material invariant, as expected. The strain components for $\sigma_{zz}=0$, in terms of the displacement u_i at the inner radius, are

$$\varepsilon_{\theta\theta}, \varepsilon_{rr}=\frac{u_i}{r_i}\left[\frac{(1-\nu)\pm(1+\nu)r_o^2/r^2}{(1-\nu)+(1+\nu)r_o^2/r_i^2}\right] \tag{22}$$

which are not material invariant. Furthermore, the superposition principle would have to deal with two components each of stress and strain, although taking $\nu=0.35$ and 0.5 might give results close enough together to solve this problem.

(g) For fracture, the relation between stress and the stress intensity factor K_I of linear elastic fracture mechanics is material invariant. Fracture will be governed not only by the stress, but also by the strain history near the tip of the crack, however. Thus crack initiation and growth will depend on a time integral of K_I over previous crack lengths, even if the material remains viscoelastic down to the size of the fracture process zone. With plastic flow the process becomes highly nonlinear and more complicated.

4. Sources of Solutions

4.1 *Analytical Methods*

Beyond using elasticity and the correspondence principle (see *Elasticity*), or creep in solids where nonlinearity is important (see *Creep*), solutions in the literature are widely scattered.

4.2 *Experimental Methods*

In photoelasticity, an optical effect reveals the differences of principal strain components under load. Along the stress-free edges of a body the results may be sufficiently material invariant for photoelasticity to apply to a viscoelastic material, but with a different fringe constant. For three-dimensional problems, strain-freezing techniques fix the viscoelastic strains when sections are cut out (even though the elastic strains are relieved), and photoelasticity may be applied to the sections. Photoplasticity can be used to approximate nonlinear problems (Kobayashi 1987).

Bibliography

Aklonis J J 1986 Viscoelasticity. In: Bever M B (ed.) 1986 *Encyclopedia of Materials Science and Engineering*. Pergamon, Oxford, pp. 5249–53

Doremus R H 1986 Viscosity and annealing of glass. In: Bever M B (ed.) 1986 *Encyclopedia of Materials Science and Engineering*. Pergamon, Oxford, pp. 5253–60

Ferry J D 1980 *Viscoelastic Properties of Polymers*, 3rd edn. Wiley, New York

Gurtin M E 1984 The linear theory of elasticity. In: Truesdell C (ed.) 1984 *Mechanics of Solids* II. Springer, Berlin, pp. 1–295

Kobayashi A S (ed.) 1987 *SEM Handbook on Experimental Mechanics*. Prentice–Hall, Englewood Cliffs, New Jersey

McClintock F A, Argon A S 1966 *Mechanical Behavior of Materials*. Addison–Wesley, Reading, Massachusetts

McCrue N G, Buckley C P, Bucknall C B 1988 *Principles of Polymer Engineering*. Oxford University Press, Oxford

Mascia L 1982 *Thermoplastics: Materials Engineering*. Applied Science, New York

Pipkin A C 1972 *Lectures on Viscoelasticity Theory*, 2nd edn., Applied Mathematical Sciences 7. Springer, Berlin

Ward I M 1983 *Mechanical Properties of Solid Polymers*, 2nd edn. Wiley, New York

F. A. McClintock
[Massachusetts Institute of Technology,
Cambridge, Massachusetts, USA]

M

Materials and Innovations

Innovations in materials have always been important to mankind, as is evidenced by application of the terms "Stone Age," "Bronze Age" and "Iron Age" to significant stages of human history. The development of metals was certainly economically important to the few people who actually smelted and worked them but of greater consequence was the fact that these new metals advanced human capabilities in other areas by making possible new equipment for farming, warfare and various daily needs. This "multiplying" effect of innovation in materials has always been important, and it is now more so than ever before, as materials are called upon to perform more and more closely to their ultimate capability.

The major motivation for innovation in materials in modern times has always been the opportunity for a materials supplier to introduce a new product with a combination of properties and cost that would compete successfully in the market place. In some cases, such as the innovations in steelmaking which occurred during the second half of the nineteenth century, loosely coupled innovations in other areas have followed. In other cases, such as the evolution of synthetic polymers, convenient, useful and inexpensive substitutes for other materials have been produced.

In a few early cases and many more recent ones, materials innovation has been necessary to permit innovation in other technological areas. The chief economic driving force for such materials innovation is the advantage to be gained from the major new product introduced. The new material is merely a means to an end. The development of such materials must be closely coupled to the development of the final product. As a result, such materials development is frequently carried out by users of the materials.

A materials innovation rarely involves only the introduction of a new composition of matter which has useful properties or cost; generally it requires a whole series of innovations in processing techniques such as compounding, melting, pressing, forming, heat-treating and finishing. In fact, some of the most important innovations in materials technology have been in technique, for instance zone refining for purification, injection molding of plastics and ceramics, float forming of plate glass and vacuum melting for superalloys.

The time between a materials innovation and its important use varies enormously. Materials whose development is essential to the development of another technological device are applied quickly. In general, however, because of corollary developments, as discussed above, a period of at least ten years passes between the initial laboratory demonstration of a new material and its important commercial application.

There was a great change in the amount and kind of work done on materials at about the time of World War II, so in this article developments up to the start of World War II are discussed separately from those which followed.

1. Innovations Preceding World War II

Table 1 presents some early materials developments made in close collaboration with the potential users, who would benefit most from their development. The use of carbon filaments in incandescent lamps by Edison is the most celebrated example. Taylor, who was motivated by the need to speed up metal cutting, carried out investigations which defined heat treatment that for the first time allowed 18–4–1 high-speed tool steels to retain their edge at a red heat. In developing ductile tungsten for incandescent lamp filaments, Coolidge was interested only in improving his company's competitive position in electric lamp sales. The development of motion pictures was a little less closely coupled than the above innovations: celluloid was developed by Hyatt in 1870, Goodwin filed a patent for a photosensitive emulsion on this support in 1887, Eastman produced such a film commercially in 1888 and Edison used that film to make his Kinetiscope in 1894.

1.1 Metals

Since the introduction of bronze, innovation in metals (see Table 2) has been important in promoting innovation in other areas. Until World War II, metals were almost the only materials which were coupled to

Table 1
Early product innovations closely related to the development of new materials

Product	New material	Developer
Incandescent lamps	carbon filaments ductile tungsten	Edison Coolidge
Cinematography	flexible support for photosensitive emulsion	Hyatt Goodwin Eastman
High-speed machining	proper heat treatment for 18–4–1 high-speed steel cobalt-cemented tungsten carbide	Taylor

Table 2
Important early innovations in metals

Period	Innovation
3500 BC	bronze developed
1500 BC	iron developed
1850–1860	Bessemer process, blast furnace. Cowper stove
1860–1870	Siemens open hearth
1870–1880	Mushet tool steel. Hadfield manganese steel
1880–1890	basic open hearth. Hall–Héroult aluminum
1890–1900	phase rule applied to alloys, other alloy steels developed
1900–1910	ductile tungsten
1910–1920	18–8 stainless steel
1920–1930	age-hardened aluminum alloys
1930–1940	commercial magnesium alloys, martensite transformation understood

major technological innovations in other areas. Between 1840 and 1940, major improvements in metals cost, availability and properties allowed significant innovations in manufacturing machinery, transportation equipment, communications methods and other areas of technology. Over this period, almost all improvements in the cost of manufacture and the properties of the product were made by metals manufacturers. In general, the demands made on the metals were not severe by modern standards, and were met by commercially available metals.

1.2 Polymers

A great number of important polymers were introduced before World War II (Table 3). In some notable early cases, for example in celluloid film and electrical wire insulation, new polymers made possible important technological innovations. However, the greater part of the volume produced was used to provide substitutes for other materials such as natural fibers, wood and metals.

Since many polymers are commonly used where extremes of temperature or electrical or mechanical stress are not encountered, there has rarely been strong motivation for polymer users to develop new materials. As a result most innovations in polymers have been made by polymer producers.

1.3 Semiconductors

Before World War II semiconductors were of little economic importance. However, some metal oxide semiconductors, such as copper oxide, were used for alternating-current rectification.

1.4 Ceramics

Until well after World War II the term ceramics referred to silicate-based products made from natural minerals. In the early eighteenth century, Europeans learned how to make porcelain which had hitherto been made only in China. During the nineteenth and early twentieth century, important advances were made in developing improved refractories for steel melting which facilitated the important advances in steelmaking of that period.

Table 3
Important early developments in organic polymers

Time period	New polymer
1870	celluloid
1900–1910	phenol–formaldehyde
1910–1920	polystyrene
1920–1930	rayon cellophane, cellulose acetate
1930–1940	acrylic resins, nylon, melamine–formaldehyde

2. The Revolution in Materials Science and Engineering

Starting at the beginning of World War II and continuing at least until the 1960s, there was a major increase in the amount of work aimed at the development of new materials, accompanied by a rapidly growing broad effort to understand the behavior of materials and to develop new materials for specific needs. In part this growth was stimulated by the success in developing specific materials for radar and the atomic bomb, and by other wartime materials efforts. Perhaps more important, the need to quickly solve the wartime problems led a number of nonmetallurgists—chemists, physicists and crystallographers—to work with metallurgists on the problems of developing the new materials. The association introduced them to a variety of the other unsolved problems in metallurgy. This in turn led to the innovative idea that a "science of metals" was possible. The high mobility of people in the immediate postwar period permitted the assembly of a number of interdisciplinary groups to work on metals at several locations. Out of this came the new field of metals science, or metal physics, which eventually became materials science.

Some support for these new activities came from the materials-producing industries, particularly in the synthetic polymers area, which developed quite independently from the other materials fields. An innovation in the other areas was that a large part of the support came from the materials-using industries interested in advanced technological developments likely to be limited by available materials and also from government agencies, particularly those concerned with nuclear energy. Guidance for the development of required new materials was expected to come from the new science.

Out of this activity five different overlapping fields associated with materials science and technology gradually developed: (a) the academic field of materials science: (b) a broadened field of metallurgy: (c) a broadened field of polymers and plastics: (d) the new field of electronic materials: and (e) a substantially new field concerned with nonmetallic inorganic materials which came to be called ceramics. Although the broad field is frequently called "materials," the separate fields continue to develop quite independently of each other.

2.1 Materials Science

One of the most important innovations in materials was the concept that it would be possible to study materials properties and behavior independently of any specific material. The approach was to avoid the complexity of real materials and to study elementary phenomena in the simplest systems experimentally available. Single crystals, high-purity materials, model systems such as alkali halides, and purely mathematical or geometrical models were widely used. Understanding of phenomena on an atomic basis was sought and usually accomplished.

An enormous range of topics has been covered. Good theoretical understanding of mechanical, electrical and magnetic properties in a wide variety of materials has been achieved. An important consequence of this was to allow the definition of theoretical limits of many of these properties. For example, the maximum possible values of flow and fracture stress, and of magnetic saturation and certain electrical properties can be stated with precision. The mechanisms and kinetics of the reactions occurring during phase transformations and heat treatment were established for many systems, and their role in determining microstructure was demonstrated. Finally, improved understanding of the relationship between microstructure and physical properties was obtained.

Since this understanding was obtained in simple systems, it can be quantitatively applied only in the simplest systems. In most real materials, many phenomena occur at the same time and the observed behavior is a result of the competition among these phenomena. The consequence is that the new understanding serves best as a qualitative guide to the empirical optimization of a new material, rather than as a direct method for the prediction of the effect of composition or treatment on properties. Much current work in materials science seeks to attack some of the more complex problems.

At the end of World War II materials characterization equipment was limited essentially to microscopes and x ray diffractometers, with measurement of thermal effects used to study some transformations. Since that time, a host of new kinds of equipment has appeared, permitting the determination of structure and composition on an ever decreasing scale (see *Investigation and Characterization of Materials*).

2.2 Semiconductors

The main factor in the development of semiconductor materials was the invention of the transistor by Bardeen, Brattain and Shockley at Bell Laboratories in 1948. It was immediately apparent that semiconductor purity was a major problem in this field, and this promoted substantial improvements in purification and characterization techniques. There was early competition between several semiconductor materials (chiefly silicon and germanium) for dominance in device production, and silicon eventually won because of its superior processing behavior and its superior native oxide which can be used as the insulating layer necessary for device processing.

Semiconductor-grade silicon is an exception to the rule that real materials are too complex to permit quantitative application of the knowledge of materials science. In a major development Pfann provided a technique for producing single crystals of silicon of extreme purity and perfection of structure. This had a profound multiplying effect on the rest of technology. In addition, because the material is so simple, it is possible to predict the effect of composition and structure on its behavior in a quantitative way. At one time it was hoped that with the development of further understanding similar predictions would be possible for other materials, but unfortunately their complexity has prevented the attainment of that goal. Moreover, in the case of silicon, present development of cheap silicon cells for solar energy requires that some of the perfection of structure be sacrificed to greatly decrease the manufacturing cost, but the trade-offs have not yet been optimized.

In recent years other semiconductor materials have challenged silicon for specific applications. However, silicon's overall dominance of the semiconductor industry appears assured for the foreseeable future. For high-speed, high-frequency and some high-power applications, certain semiconductors composed of the group III and group V elements (chiefly GaAs and related species) have higher carrier mobilities and therefore an inherent advantage. If the difficulties of processing these materials can be dealt with adequately, they may replace silicon in these applications. Other semiconductors of the III–V or II–VI variety are being developed for photodetector or photodiode applications where the magnitude of the band gap of silicon is insufficient.

Other materials work is also driven by electronic device needs. For instance, metals alloys and their combinations with semiconductors are continuously being refined for providing contacts on electronic devices with desired properties. New organic and dielectric films are being developed, and materials for packaging and protection of device components are increasingly needed. Materials purification and characterization demands continue to be much higher in electronics than in any other field and are increasing.

2.3 Metals

Since the beginning of World War II, many of the important innovations in metals have been closely coupled to and motivated by other major technological innovations. The first of these was the development of uranium and plutonium for atomic bombs. It involved close collaboration among metallurgists, physicists and chemists and set the stage for many further interdisciplinary developments. The nuclear effort continued after the war with the development of zirconium as a low-cross-section, fuel-cladding material and of beryllium as a neutron moderator in nuclear reactors.

Gas turbines are probably more clearly limited by the availability of metals which will resist high temperature than any other important technological device. Since their important introduction immediately after World War II, a major metallurgical activity has been to produce superalloys which will resist ever increasing stresses for longer times at higher temperatures. Great success has been achieved, but has often required very considerable metallurgical ingenuity. Much of this development work has been done by engine manufacturers or with very close collaboration between them and metals producers. The most advanced approaches involve powder metallurgy using rapidly cooled powders which undergo hot isostatic pressing to produce a final product of nearly the desired shape and with a very fine microstructure.

The development of techniques for making commercially useful superconducting wire has made magnets with fields of 15 T a commercial reality.

Certain alloys of iron containing boron or phosphorus can be rapidly solidified to produce a noncrystalline or amorphous product. Reasonably large amounts have been produced on a semicommercial scale. Although the product is a major innovation as far as metallurgical structures are concerned, it is speculative whether it will lead to important new products. The currently most probable application is as a low-loss magnetic material for transformer cores.

2.4 Ceramics

Working in Holland in the late 1940s, Snoek began the development of the magnetic ferrites (magnetite in which some of the iron has been replaced by other metals). The product is a magnetically soft, electrically nonconducting material in which the direction of magnetization can be rapidly changed because there are no eddy currents to oppose the change. These soft ferrites have found important application in various high-frequency electronic devices and are now important commercial materials. Perhaps more important, they were the first of a large number of nonmetallic inorganic solids which had combinations of properties not previously available in any material. Because they were made from powders and processed by pressing the powders to a final shape and then heating, they were called ceramics, although they had little in common with the silicate-based materials to which the term had previously been restricted.

A number of other ceramic materials followed, some directed at specific applications, others at generalized technological needs. Usually, the intrinsic value of the ceramics themselves is low, and their importance stems from their ability to make other devices possible. Some, such as the ferrites, the ferroelectric materials based upon barium titanate, and a group of varistors based upon zinc oxide, have become important commercially and are used in a variety of devices.

Other materials have had a limited but important impact. Pore-free, translucent aluminum oxide has made possible the high-pressure sodium vapor discharge lamp, which is exceedingly efficient. Uranium dioxide is the only important fuel for nuclear reactors, and the ability to make uranium dioxide fuel pellets with highly reproducible dimensions and controlled density has been essential to the development of modern nuclear reactors. Zirconium oxide has been known to be an ionic conductor for many years, and is now extensively used as the electrolyte in a high-temperature galvanic cell to determine the partial pressure of oxygen in gases.

Several new ceramics possess combinations of properties which would lead to most exciting innovations in other fields if certain difficulties could be overcome. One of the most studied is the group of materials based upon silicon carbide or silicon nitride. These substantially pure sintered materials have impressive strength at elevated temperatures, and much effort has been devoted to developing them to the point where they can be used as moving components in gas turbines. The problem with these materials is that their strength is not quite high enough, and their failure is catastrophic.

So-called beta-alumina, really a sodium aluminate, is a good conductor of sodium ions at only a few hundred °C. Attempts are being made to develop it as the electrolyte in galvanic cells which would react sodium and sulfur. If difficulties with lifetime and the cost of the ceramic electrolyte can be overcome, this product will permit important innovations in the method of energy storage.

2.5 Polymers

Although there were very important polymer developments before World War II, the field has blossomed since then, and now the volume of synthetic polymeric materials produced is greater than that of metals. A partial list of important polymers introduced since World War II and of their major uses is given in Table 4. The economic importance of these materials is very large. A few basic characteristics account for their explosive growth. In spite of the large increases in cost of petroleum, the raw materials for these materials are

Table 4
Important polymers introduced since the beginning of World War II

Polymer	Application
Polyethylene	sheet, bottles
Polypropylene	filaments, fibers, sheet, bottles
Poly(ethylene terephthalate)	fibers, films
Polystyrene copolymers (e.g., acrylonitrile–butadiene–styrene)	molded shapes
Poly(vinyl chloride)	sheet, textile facing
Polytetrafluorethylene	nonstick frying pans
Epoxy	adhesives, electrical insulation
Polycarbonates	impact-resistant glazing
Noryl/poly(phenylene oxide)	impact-resistant molded parts
Polyimides	high-temperature films, molded parts
Silicones	elastomers, fluids coatings
Aromatic polyamides (aramids)	high-strength fibers
Polysulfones	molded parts

still plentiful and relatively inexpensive. These raw materials may be manipulated to produce plastics of widely differing properties. A final most important characteristic is that they can be formed into complex and attractively finished shapes more inexpensively than other materials.

Bibliography

American Chemical Society 1973 *Chemistry in the Economy*. ACS, Washington, DC

Baker W D 1981 Using materials science. *Science* 211: 359–63

Burke J E 1968 Ceramics today. *Science* 161: 1205–12

DuBois J H 1972 *Plastics History USA*. Cahners, Boston, Massachusetts

Kingery W D, Bowen H K, Uhlmann D R 1976 *Introduction to Ceramics*. Wiley–Interscience. New York

Mehl R F 1948 *A Brief History of the Science of Metals*. American Institute of Mining, Metallurgical and Petroleum Engineers, New York

Pfann W G 1952 Zone refining. *J. Met.* 4: 747–49

Seitz F 1939 *The Modern Theory of Solids*. McGraw–Hill, New York

Smit J, Wijn H P J 1959 *Ferrites: Physical Properties of Ferrimagnetic Oxides in Relation to their Technical Applications*. Wiley, New York

Smith C S (ed.) 1963 *Sorby Centennial Symposium on the History of Metallurgy*. Gordon and Breach, New York

J. E. Burke
[Burnt Hills, New York, USA]

Materials and the Environment

For the purpose of this article the term material encompasses any substance used without significant chemical conversion. The notion of a materials cycle popularized by the well-known COSMAT Report of the US National Academy of Sciences (1973) is actually based on an analogy with natural biogeochemical cycles involving water, carbon and nitrogen. However, the latter are true cycles, whereas the so-called materials cycle is incomplete except in the very crude sense that materials are extracted from the environment, processed, used and finally disposed of as waste. It is important to bear in mind that materials are normally returned to the environment in a form very different from that in which they were extracted. A graphic characterization of the situation is given in Fig. 1.

The idea of recycling low-grade wastes is intellectually appealing but attractive opportunities for recycling are limited. In practice each waste flow in the economy must be regarded as a potential source of materials on an equal footing with virgin sources. If the waste stream is easily recovered and processed compared with natural sources it will generally be exploited. Not surprisingly, the more economically valuable a material is, the more likely it is to be recycled. Thus, it pays to go to considerable lengths to recover platinum from industrial catalysts or silver from film processing wastes, but to date, it is cheaper to extract aluminum from bauxite ore than from fly ash. Similarly, it is cheaper to extract sulfur from natural deposits via the Frasch process than to recover it from combustion products, even when the flue-gas desulfurization is subsidized.

It is clear that materials recycling will gradually become more widespread as high-quality virgin sources are exhausted or as increasingly strict environmental regulations force processing of combustion products or waterborne waste streams to remove hazardous substances. Nevertheless it is most likely that significant quantities of materials in forms that are not economically recoverable (such as fuels, pigments, surface coatings, flocculants, water softening agents, detergents, antifreeze, fuel or plastic additives, lubricants and solvents) will be discarded for many decades to come. Fuels are, of course, sources of energy. The others are all examples of what may be termed labor-saving, capital-saving or energy-saving—but nonstructural—materials. For example, lubricants cut down on friction enabling a machine to function with less energy input and to last longer, detergents save scrubbing, and so on (Ayres 1978). In nonstructural uses, the materials are physically or chemically transformed by use, usually into low-grade dispersed forms. Since it is unlikely that an advanced economy will dispense with these categories of uses of materials it is not feasible in principle to recover a high percentage of many materials.

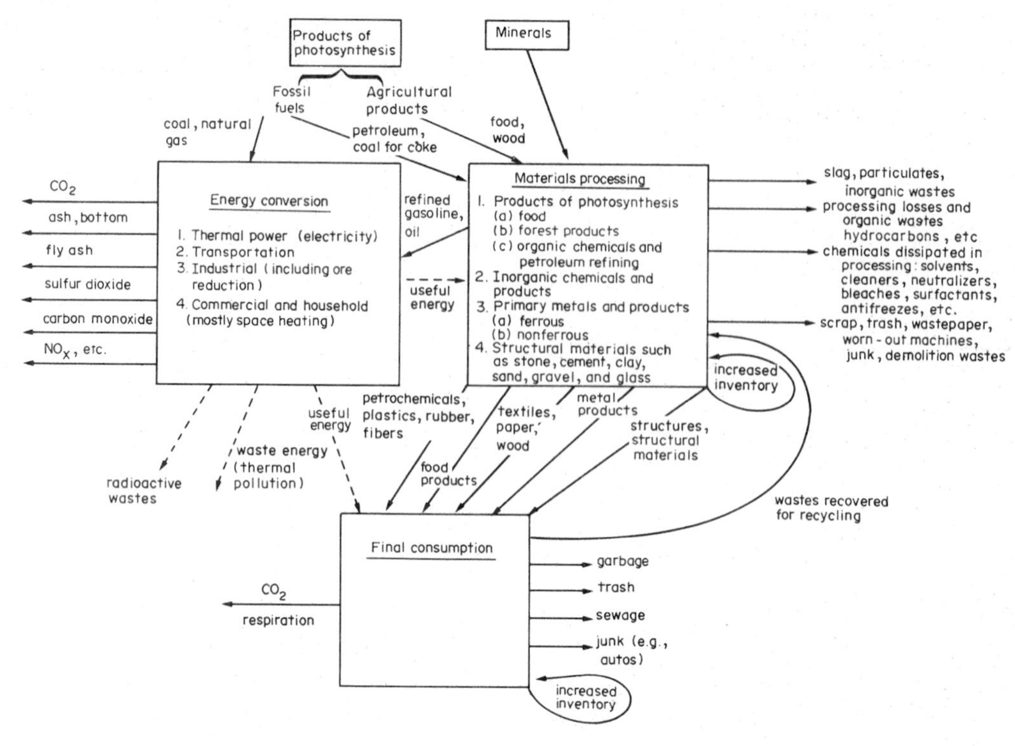

Figure 1
Materials flow in the economy (after Ayres and Kneese 1969)

The discharge of wastes to the environment creates a variety of possibilities for adverse effects on man, both direct and indirect. Major examples include:

(a) disturbance to land (from mining, materials processing, construction, or waste disposal);

(b) contamination of soil, ground or surface water used by humans and livestock or for irrigation (by acids, sludges or toxic industrial wastes), with secondary effects on fish and birds, for example;

(c) contamination of air by toxic or irritating combustion products (e.g., total solid particulates, oxides of sulfur and nitrogen) with direct effects on health;

(d) disturbance of fresh-water ecosystems by eutrophication or acid-rain deposition;

(e) disturbance of ocean ecosystems due, for example, to oil spills, ocean dumping and ocean mining; and

(f) climatic disturbance due to buildup of CO_2 and/or other chemical pollutants in the atmosphere.

Further discussion can be most usefully carried out by separately considering three major categories of materials: nonfuel minerals, fuels, and packaging materials and synthetics.

1. Minerals

It is convenient to divide this category into metals and nonmetallic minerals. Metals are worth considering separately because they are mainly used in structural applications (i.e., embodied in products), and consequently recovery and reuse is often feasible. The major tonnage metals are: iron (and steel), aluminum, copper, zinc, lead and nickel. World production of metals and ores for the years 1971–75 is summarized in Table 1.

Two important points to emphasize at the outset are that the first two metals listed above (iron and aluminum) are obtained almost exclusively from oxide ores and that their compounds are comparatively benign. Environmental problems associated with their extraction and processing are primarily due to the large scale of operations and the need to separate and dispose of significant quantities of economically useless or mildly hazardous materials. The quantities of

Table 1
Output of metals and ores (average 1971–1975)

Material	World (kt)	Non-USA (kt)
Crude steel	642000	507000
Iron ore	432000	346000
Nickel	618	458
Manganese ore	20030	18101
Chrome ore	6941	5792
Cobalt	23.4	16.2
Tungsten	40.2	33.7
Refined copper	7923	6037
Primary aluminum	12249	7861
Platinum[a]	5387	3726
Zinc	5506	4346
Tin	232	180

Source: Malenbaum 1978 [a] In units of 10^3 troy ounces

solid mining and milling wastes produced by the mining of metal ores are typically much greater than the quantities of processed metals (Table 2). Environmental controls are primarily needed, in the case of iron/steel facilities, to minimize the discharge of toxic gases from coke ovens and furnaces and waterborne acidic wastes or sludges, from pickling sheet or strip. In the case of aluminum smelting, the most serious environmental problem is to prevent the escape of fluorine from the electrolytic cells. (Fluorine originates in the breakdown of the molten cryolite bath, not in the aluminum ore.) A minor problem is the disposal of so-called red mud from bauxite processing, although recovery of iron and possibly silicon from bauxite wastes may ultimately be feasible.

Environmental problems associated with end-use consumption and disposal of iron/steel and aluminum are also primarily due to the wide variety of minor uses. Scrap (e.g., from demolished structures, pipelines, worn-out rails or rolling stock, obsolete machinery and junk cars) is easily aggregated and recycled via electric furnaces, but it is much harder to recover metal economically from small items, such as containers, wire products, foil, fasteners (e.g., nails), razor blades, mattresses and springs. These tend to be mixed with other kinds of waste as in household refuse or are scattered over the landscape as litter. In this context, aluminum cans and bottle caps constitute by far the most serious problem, since aluminum is highly corrosion resistant, whereas small iron or carbon steel objects eventually rust away.

The other major tonnage metals (copper, zinc, lead, nickel) are geochemically quite different from iron and aluminum. Firstly, being chalcophiles, they are typically found in nature as sulfides rather than oxides, although nickel is mined in both forms. The sulfides are commonly first converted to oxides by the simple process of roasting, the sulfur being oxidized and driven off as SO_2. Copper, zinc, lead and nickel smelters are major sources of air pollution despite recent efforts to improve environmental controls in some countries. For example, the large copper–nickel refineries in the Sudbury district of Ontario emit 2.7 Mt of sulfur oxides annually causing severe damage to forest over nearly 2000 km^{-2} and lesser damage over a much larger area.

A second significant characteristic of these metals is that copper, zinc, lead and nickel are typically found with coproducts or by-products. Major coproduct groups are shown in Table 3. Thus a copper, zinc, lead or nickel smelting operation will normally yield tailings, slags, ashes, flue dust or sludges that are rich in by-products (such as antimony, arsenic, bismuth, cadmium, selenium or silver) which may or may not be desirable. Several of these by-products are not worth enough to justify highly efficient recovery methods, so significant quantities of these metals are discarded or lost in refinery operations.

A third point is that many compounds of the most important metals other than iron and aluminum, including copper, zinc, lead, nickel and other transition metals (chromium, manganese, cobalt, vanadium) as

Table 2
World mineral production and land utilization (1976)

Mineral commodity	Production (Mt)	Solid waste (Mt)		Load utilized (km^2)
		Mining	Milling	
Stone	6985	544	91	934.8
Sand and gravel	6260			1396.1
Clays	526	424		291.4
Bituminous coal	3175	>424		1841.3
Iron (in ore)	895	3084	1906	344.0
Phosphate rock	107	424	272	105.2
Copper	7.3	3629	1361	408.7
Uranium	0.036	544	1361	20.2
Others	1361	NA	NA	364.2

Source: Barney 1980

well as mercury, are quite toxic to animals and plants. The extreme toxicity of lead, arsenic, cadmium and mercury, is well known, but salts of copper, zinc, chromium, bismuth and thallium are also toxic enough to have found agricultural uses as insecticides, fungicides and rodenticides.

Clearly nonferrous metal extraction, processing, and use all create possibilities of significant environmental damage. The cases of zinc and its by-product cadmium are typical. Figure 2 shows schematically the magnitudes of various flows in zinc-refining operations. Significant fractions of zinc and cadmium are lost in processing, and much of this may be accessible to biological organisms.

Environmental hazards are also associated with end uses of some metals, especially where the metal is toxic and the use is nonstructural. In the case of cadmium, for instance, almost all of the present uses are inherently dissipative. These include electroplating, pigments and stabilizers for PVC. A flow chart for cadmium is shown in Fig. 3.

Taking into account extraction, processing and end use, human activities have sharply increased the natural rates of mobilization of many metals, particularly the more toxic ones. Comparisons of worldwide natural rates of discharge into the oceans and those from mining processes are given in Table 4.

Table 3
By-product and coproduct groups

By-product groups			
Copper	Zinc	Lead	
antimony	antimony	antimony	
arsenic	cadmium	bismuth	Platinum
cobalt	gallium	selenium	iridium
rhenium	germanium	silver	osmium
selenium	indium	tellurium	palladium
silver	silver		rhodium
tellurium	thallium	Nickel	ruthenium
		cobalt	
Niobium	Zirconium	manganese	Lithium
tantalum	hafnium		rubidium
Coproduct groups			
	Brine	Rare earths	
	sodium	yttrium	
	magnesium	europium	
	chlorine	erbium	
	bromine	terbium	
		dysprosium	
		ytterbium	

Table 4
Estimates from annual discharges of metals into the oceans (kt)

Metal	By geological processes (in rivers)	From mining
Iron	25000	319000
Manganese	440	1600
Copper	375	4460
Zinc	370	3930
Nickel	300	358
Lead	180	2330
Molybdenum	13	57
Silver	5	7
Mercury	3	7
Tin	1.5	166
Antimony	1.3	40

Source: Waldichuk 1977

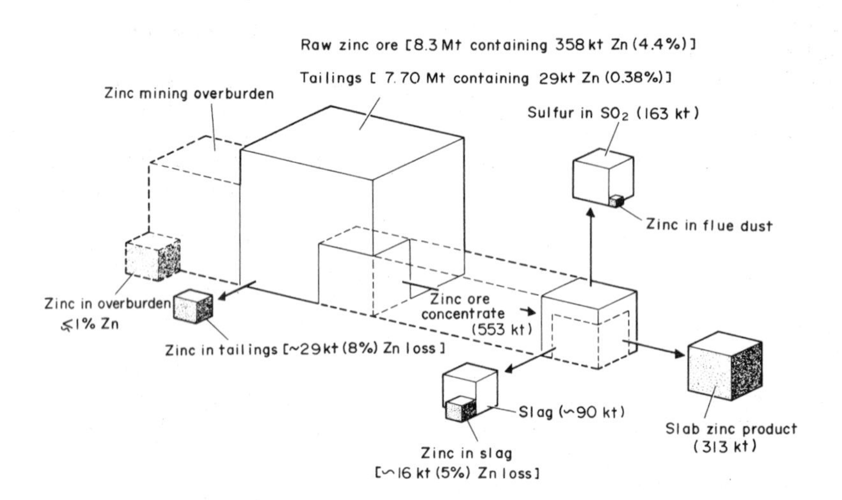

Figure 2
Zinc flow in mining and processing of zinc ores in the USA (1969) (after Fulkerson et al. 1972)

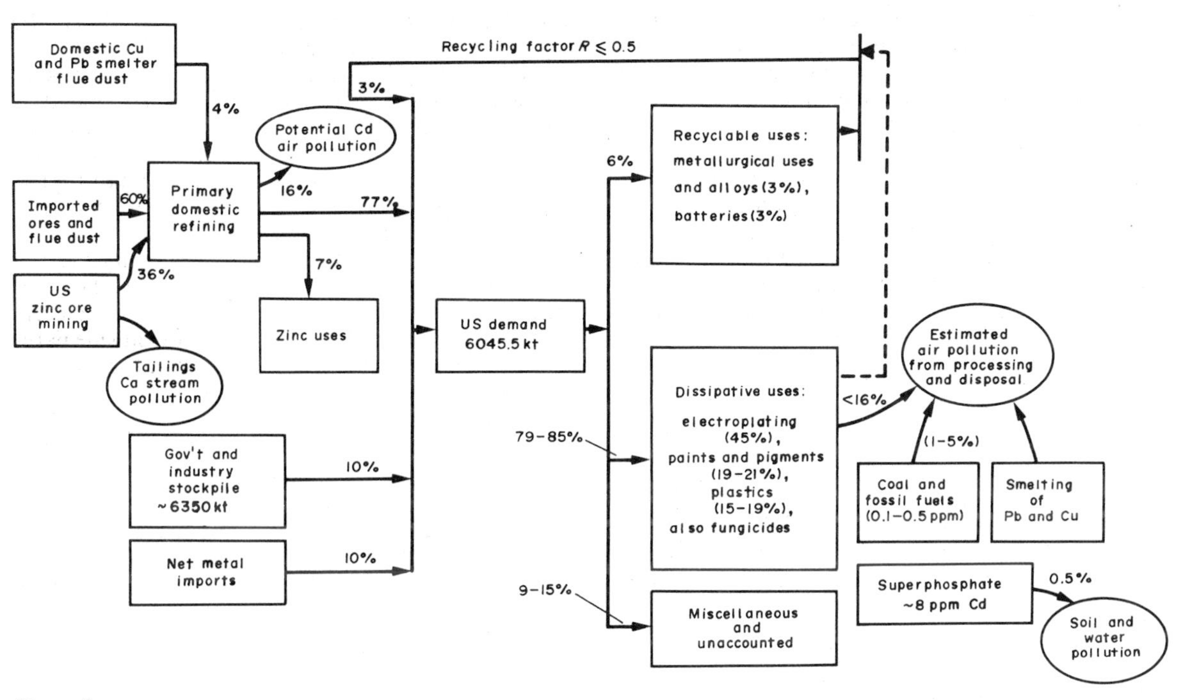

Figure 3
Cadmium flow in the USA (1968) (after Fulkerson et al. 1972)

Nonmetallic minerals such as stone, gravel, clay and sand are mined in larger quantities than metal ores, although processing losses are smaller and these materials are somewhat more benign. They are used mostly in construction. Production data for 1976 are shown in Table 2.

Environmental impacts arise from mining itself, which is mostly on the surface, and significant land disturbances, dust and erosion which often accompany such activity. Processing of phosphate rock (to produce fertilizer) also results in significant further environmental impacts of several kinds, including toxic fluoride emissions to the atmosphere and accumulation of mildly radioactive tailings. It is noteworthy that erosion rates for disturbed land, such as active strip mines, is up to ten times greater than for cropland and perhaps 200 times greater than natural grassland.

2. *Fuels*

Fossil fuels (petroleum, coal, natural gas) are used in tonnages comparable to stone, clay and sand. Land disturbance from source mining of coal is a major environmental effect. Cleaning and washing of coal also creates large piles of waste material which can later cause problems, such as mudslides and acid runoff.

However, the major environmental problems associated with fossil fuels are attributable to the combustion process and its waste products. Most air pollution is attributable to combustion of fossil fuels, the contaminants being: carbon dioxide, carbon monoxide, unburned or partially burned hydrocarbons, total solid particulates, oxides of nitrogen and oxides of sulfur. Estimated 1985 emissions are given in Table 5.

Because of the vast quantities involved, some of these effluents, notably oxides of sulfur and nitrogen and total solid particulates, can cause respiratory and other health problems in polluted areas. Unburned hydrocarbons, especially in the presence of ultraviolet radiation, can produce a mix of highly reactive, irritating and carcinogenic compounds called smog. Moreover, oxides of sulfur and nitrogen combine with atmospheric moisture and return to earth as acid rain, downwind of major power plants or smelters. This process slowly but irreversibly damages conifer forests and freshwater lakes. Airborne sulfate particles from anthropogenic sources (SO_x) are now estimated to be comparable in mass to natural sources, i.e., 130–200 Mt year^{-1} (National Academy of Sciences 1977).

It is important to note that atmospheric emissions and other wastes from the combination of natural gas and petroleum products are relatively minor

Table 5
Annual emissions: 1985 (DOE low growth case)

	European OECD countries	USA and Canada	Japan	Less developed countries	OPEC countries	Centrally planned economies	World
Carbon dioxide (Gt)	4.34	16.21	1.37	2.33	0.73	6.82	21.8
Carbon monoxide (Mt)	20.4	12.2	6.06	14.2	4.99	20.3	78.1
Sulfur dioxide (Mt)	10.7	11.8	4.10	6.41	1.21	27.0	61.1
Oxides of nitrogen (Mt)	12.3	14.2	4.27	7.45	2.16	17.9	58.2
Particulates	5.7	8.18	1.84	5.95	0.49	27.8	49.9
Hydrocarbons (Mt)	2.32	1.67	0.72	1.57	0.54	2.71	9.5
Land use (10^3 km^2)	55.4	75.7	13.6	48.2	0.016	61.1	247
Solid wastes (Mt)	72.7	198	6.74	41.4	0.53	135	453
Tritium activity (10^{13} Bq)	381	525	84.7	128	13.5	177	1310
Population exposure (10^3 man-rem)	3.98	5.44	0.88	1.33	0.14	1.84	13.6
Solid high-level wastes (10^{19} Bq)	40.7	55.9	9.03	13.6	1.44	18.9	139

Source: Barney 1980

compared with the wastes that are associated with use of coal. Natural gas has little sulfur to begin with, and it is easily removed. In the case of petroleum, refining automatically removes the sulfur from most distillate products (and concentrates it in residual fuels, tar and coke) and total sulfur removal is approaching feasibility.

The biggest difference between natural gas/petroleum fuels and coal is their ash. Gas is ash-free and even residual oil has an insignificant ash content, but coal ash content varies from a few percent to 20% or more. In conventional combustion most of this is bottom ash, which becomes a solid waste disposal problem because of the large amounts of coal involved. However, ~10% of the ash goes up the stack as fly ash and, even with efficient electrostatic precipitators, a small percentage is lost to the environment as solid particulates. Apart from the undesirable physiological effects of small particles, fly ash contains both toxic and radioactive constituents, notably zinc, vanadium, nickel, chromium and manganese (Table 6).

Table 6
Trace elements in coal

Element	Concentration (ppm)	Total amount (t year^{-1})	Percentage of 1968 US consumption of the metal
As	10	4500	21
B	100	45000	57
Ba	60	27000	4
Be	0.3	14000	46
Cd	0.1–0.5	45–230	1–4
Co	5	2300	36
Cu	15	6800	0.5
Cr	30	14000	3
Pb	5	2300	0.3
Mn	30	14000	2
Hg	0.1–0.5	45–230	2–11
Mo	1	450	0.2
Ni	50	23000	16
V	70	32000	610
Zn	100	45000	3

Source: Fulkerson et al. 1972

Assuming the world's energy needs are met increasingly from coal in the future as natural gas and petroleum become relatively scarcer, the environmental health problems associated with coal combustion—especially fly-ash emissions—will become more significant.

Finally, the buildup of atmospheric carbon dioxide from massive fossil-fuel combustion appears to be capable of affecting the heat balance of the earth itself via the so-called greenhouse effect. Although scientific uncertainties remain, evidence seems to be accumulating to support the hypothesis that CO_2 buildup will result in a general warming of the earth's surface, particularly in the polar regions. This, in turn, is likely to cause northward shifts in the average locations of the jet streams (which control precipitation patterns). The resulting climatic impact might be favorable for some sparsely populated northern countries (e.g., Canada, Siberia) but could be quite unfavorable for the more heavily populated countries in the Temperate Zone.

In addition, it has been suggested that a few degrees of general warming might trigger a fairly rapid melting of floating ice in the Arctic and Antarctic oceans. Although this appears unlikely, the effect would be to cause the sea level to rise by up to 10 m. A rise of sea level of even a few meters would cause drastic changes in the map of the world. Many island groups would disappear under water (the Bahamas, for instance), and offshore barrier islands would be swept away (including most of Long Island). Highly productive coastal and delta areas such as the Nile (Egypt), Pearl, Yellow, Yangtze (China), Mekong (Vietnam), Irrawaddy (Burma), Brahmaputra (Bangladesh), Indus (Pakistan) and Mississippi (USA) would be inundated. Many of the world's largest cities, including New York, Los Angeles, Tokyo, Osaka, London, Hamburg, Berlin, Amsterdam, Rotterdam, Bombay, Calcutta, Djakarta, Shanghai, Canton, Hong Kong and Singapore are close to sea level. Thus, it is not exaggerating to note that if significant ice melting should occur, the effects could be catastrophic.

3. Synthetics

Synthetic polymers—mostly plastics and rubber—are made largely from derivatives of petroleum. Petrochemicals already account for perhaps 5% of world petroleum consumption, and this percentage can be expected to increase. Thus, use of synthetics is inseparable from extraction and refining of petroleum, though in the future, chemicals may also be manufactured from coal.

The three most important polymers are polyethylene, polystyrene and poly(vinyl chloride) (PVC). There are, of course, a host of other resins including polyesters, phenolics and acrylics. Neglecting environmental problems associated with the paper and chemical industries there remains a significant problem of disposal. The physical volumes involved are equal to or greater than the volume of metals produced each year. Most plastics are not biodegradable: they must be disposed of in landfills or by incineration. Incineration of chlorinated plastics in particular is environmentally hazardous because of the emission of toxic or carcinogenic compounds. Tires are also very difficult to incinerate efficiently. Nondegradable litter, especially polystyrene foams and polyethylene films, is also an increasing problem in many public places, such as beaches.

4. Paper

Paper remains an important material in the economy. It is, of course, made from woodpulp by a process in which one component of the wood (the liquor) is chemically dissolved and removed. In older pulp and paper mills, this waste material was simply discarded into the nearest river, where it created biological oxygen demand (BOD), removing the dissolved oxygen and killing organisms (such as fish) that must have oxygen to survive. In recent years pulp and paper mills in most countries have substantially reduced their waterborne waste output, under pressure of environmental regulations.

5. Policy Issues

It is generally acknowledged that the environment is a resource. The fact that it has been available in the past for use as a "sink" for the disposal of industrial and consumption wastes at zero (or very low) cost is a consequence of its indivisibility and the fact that it cannot be contained and privately owned or exchanged in a market place. It is, in short, a public good, outside the market pricing mechanism. It is an axiom of economics that (in the absence of cultural restraints) the use of unpriced public goods will be excessive unless regulated by public policy. This is the nub of the problem.

Environmental services to man can be said to include the biosphere (photosynthesis, the water, carbon and nitrogen cycles, biodegradation (decay) of organic materials), the maintenance of oxygen in the atmosphere (the atmosphere of the earth prior to the evolutionary development of the blue-green algae did not contain free oxygen) and the ozone layer in the stratosphere, and a host of local "public service" functions. There is no question that these environmental functions (and others) are interfered with (e.g., by large-scale mobilization of toxic materials, erosion, silting, buildup of CO_2 and particulates in the atmosphere). Unfortunately, much remains to be learned about the specific mechanisms involved and their quantitative impact. Research is therefore the first prerequisite of intelligent environmental policy.

In the absence of the scientific knowledge that would permit us to discriminate between minor and major problems (e.g., between reversible and irreversible disturbances), the safest strategy is to limit the mobilization of materials by human activities to levels significantly below the levels found in natural processes. By this criterion, the processing and disposal of heavy toxic metals into surface waters is already excessive. Similarly, combustion and other processes are injecting certain materials into the atmosphere at rates comparable with or greater than natural processes.

Policy instruments for abating these undesirable flows include emissions regulations applicable to particular sources (e.g., mines or smelters) or on the use of particular substances (e.g., phosphates in detergents, chlorofluorocarbon propellants). More broadly, however, policies tending to promote conservation of scarce resources will also tend to reduce environmental problems (e.g., Page 1977). It is important to note that many relevant existing policy instruments have had the opposite effect of encouraging excessive use of materials, especially virgin materials. In the USA, the mineral depletion allowance (recently reduced, but still in effect) and freight rates favoring transportation of mine products as compared with recycling scrap, are two examples. Similarly regulations or indirect subsidies tending to reduce consumers' energy costs (e.g., for gasoline or home heating) also tend to encourage excessive use of energy and its associated environmental problems.

Severance taxes on mineral extraction, on the other hand, would tend to discourage use of virgin resources and encourage conservation and recycling. Similarly, effluent taxes or fees would tend to discourage the discharges of recoverable wastes into surface waters or the atmosphere. This approach is generally preferred by economists over the more common regulatory approach ("standard setting") because it allows greater discretion to the waste generator. It is frequently opposed by environmentalists on the ground that imposition of effluent charges would not guarantee meeting desired environmental standards. If a given fee does not yield the desired result, of course the answer would be for the public authority to raise the effluent charge until it is effective. In the long run, the price of virgin materials will rise as the high-quality reserves are exhausted. This will also induce conservationists' activity, though not necessarily optimum recovery of toxic by-products. A combination of approaches will certainly be needed.

See also: Recycling of Demolition Wastes; Recycling of Metals: Technology

Bibliography

Ayres R U 1978 *Resources, Environment and Economics: Applications of the Materials/Energy Balance Principle.* Wiley, New York

Ayres R U, Kneese A V 1969 Production, consumption and externalities. *Am. Econ. Rev.* 59(3): 282–97

Barney G O 1980 *The Global 2000 Report to the President: Entering the Twenty-First Century.* Pergamon, New York

Fulkerson W, Goeller H E, Phelps D H 1972 Environmental sources, sinks and chemical forms of heavy metals. *American Association for the Advancement of Science, Selected Symposium.* Westview Press, Boulder, Colorado

Malenbaum W 1978 *World Demand for Raw Materials in 1985 and 2000.* McGraw-Hill, New York

National Academy of Sciences 1973 *Materials and Man's Needs*, Report of the Committee on the Survey of Materials Science and Engineering (COSMAT) 1973. National Academy of Sciences, Washington, DC

National Academy of Sciences 1975 *Mineral Resources and the Environment*, Report of the Committee on Mineral Resources and the Environment (COMRATE) 1975. National Academy of Sciences, Washington, DC

Page T 1977 *Conservation and Economic Efficiency: An Approach to Materials Policy.* Johns Hopkins University Press, Baltimore, Maryland

Robinson G D 1977 Effluents of energy production: Particulates. In: *Energy and Climate*, Report of National Academy of Sciences Geophysics Study Committee. National Academy of Sciences, Washington, DC

Waldichuk M 1977 *Global Marine Pollution: An Overview.* United Nations Educational, Scientific and Cultural Organization, Paris

R. U. Ayres
[Carnegie Mellon University, Pittsburgh, Pennsylvania, USA]

Materials Availability

Industrialized countries depend on materials produced from minerals and other primary products to maintain their economies and sustain their living standards. As a result, concern over the future availability of materials, particularly metallic materials, arises in these countries from time to time.

Availability implies the absence of shortages, but what constitutes a shortage is open to different interpretations. To some, particularly economists, a shortage is an excess of demand over supply for a commodity at its prevailing price. With this definition, a shortage can be eliminated simply by allowing the price to rise. At some point, the higher price will stimulate supply and reduce demand sufficiently to bring the two into balance.

For many purposes, this definition is too narrow. A several-fold increase in the price of cobalt in response to an interruption of supplies from Zaire might successfully constrict demand to the available supply, yet still cause serious disruption, at least in certain sectors of the economy. For this reason, the term shortage is used here to encompass situations where material prices rise sharply in the short run or persistently in real terms over the long run, as well as those situations where demand exceeds supply at the prevailing price. Materials availability, in turn, prevails when none of these conditions exist.

Materials availability may be threatened in four ways: mineral resources may be depleted or exhausted; investment in mines and processing facilities required to extract ore from deposits and convert it into useful materials may be insufficient; cyclical booms or military buildups may cause unanticipated surges in demand; and finally, interruptions in supply or constraints on trade may arise as a result of embargoes, wars, cartels or other factors.

1. Resource Depletion

Two views of resource depletion are commonly encountered. The first, the physical view, points out that the earth is finite, and contains only so much copper, mercury, zinc and other mineral materials. As a result, the supply of such commodities is a fixed stock. Demand, on the other hand, is a flow variable that continues year in and year out. This means that sooner or later demand must consume the fixed amount of supply, at which time the world physically runs out of the commodity. Moreover, the end is often assumed to be coming sooner rather than later, because demand is perceived as growing exponentially over time.

While logically appealing, this view of depletion suffers from two serious shortcomings. First, with a few trivial exceptions, such as the fissioning of uranium, elemental minerals are not destroyed or transmuted when used. Consumption may diffuse or degrade the material, making recycling difficult and expensive, but it does not diminish the size of the fixed stock. Second, and even more important, long before all of a mineral material found in the earth's crust were consumed, the cost of mining and processing would climb sharply and stifle demand.

For these reasons, the economic view of depletion is generally favored by those who have carefully examined this issue. This second view of depletion does not see the world literally running out of mineral materials, but rather expects their real costs to rise as society is forced over time to rely on deposits which are of lower grade, more remote and more difficult to process. Eventually, higher prices cause consumers to curtail or abandon their use of materials.

While both views of depletion raise the spectre of severe material shortages, the latter are not inevitable under the economic view of depletion. Other factors, primarily technological change, can offset the cost-increasing effects of depletion.

Since forecasting the future course of technological change and its impact on material costs is fraught with difficulties, it is not surprising to find a wide range of opinion regarding the seriousness of depletion. At one end of the spectrum are the optimists or cornucopians, who maintain that real material costs are not likely to rise substantially in the foreseeable future. They cite Barnett and Morse (1963) and other studies that show the real costs of producing mineral materials have actually fallen, and fallen substantially, during the twentieth century. In addition, they point out that technological change is not totally random and fortuitous. If depletion drives material costs up, this will increase the economic incentives to develop new technologies that offset such cost increases.

At the other extreme are the pessimists, often called Malthusians after the eighteenth-century British economist and clergyman who feared the limited supply of land would keep most of humanity in a perpetual state of poverty. The pessimists claim the past is not a reliable guide to the future. In particular, they fear more stringent environmental regulations, rising energy costs and considerably higher levels of demand will in the future overwhelm the cost-reducing effects of new technology.

On the basis of identified deposits and other information, depletion does not appear to pose a serious threat to materials availability for the next 25 to 50 years. Thus, the debate between the cornucopians and the Malthusians focuses for the most part on the more distant future, the latter half of the twenty-first century and beyond. Which viewpoint eventually will prove closer to the truth will depend largely on the race between the cost-increasing effects of depletion and the cost-reducing effects of new technology. Since the information needed to determine which of these two opposing forces will ultimately dominate is lacking, the controversy over the long-term threat of depletion is not likely to be settled in the near future.

2. *Investment*

No matter how rich the world's endowment of mineral deposits, material shortages may arise if investment in mines and processing facilities is insufficient. Such shortages can, of course, be remedied by building new mines and plants, but this takes time. The expansion of existing facilities often takes two to four years, the construction of entirely new projects five years or longer. In the meantime, shortages can be severe.

Concern over the adequacy of investment may occur for several reasons. In recent years, the increased uncertainty associated with government policies is often cited as a deterrent to investment. Expropriation or sharp increases in taxation are feared in some countries, while changes in environmental regulations, energy policy and the availability of public lands for mineral development make planning difficult in others.

While such uncertainties have altered the geographic distribution of investment, apparently away from certain developing countries and towards the USA, Canada, Australia and South Africa, the effect on the overall level of investment is less clear. In theory, a rise in uncertainty should not deter investors as long as they can realize an increase in their risk premium. While this may require some reduction in overall investment, the amount is presumably small. In practice, however, investors may react more cautiously, preferring to defer investment, at least for a time, in the hope that the uncertainty will decline.

Price controls may also impede investment. For reasons discussed in the next section, the mineral industries are highly cyclical. Profits appear unusually large when the economy is strong, and low or negative when the economy is weak. Price controls imposed when the economy is booming and inflationary pressures are greatest, prevent firms from earning abnormally large profits during such periods. However, such profits are needed to allow cyclical industries to earn a rate of return over the entire business cycle sufficient to attract capital in the financial markets. For this reason, the growing use of price controls for macroeconomic stabilization in the USA and other countries falls heavily on the mineral industries, and tends to hamper their ability to compete with other industries for capital.

The cyclical nature of the mineral industries coupled with myopia may impede access to capital during economic recessions. When demand is down, profits low and stock prices depressed, firms may find it difficult to raise investment capital from retained earnings, new equity issues or borrowing, even though they may need to invest at such times to ensure adequate capacity in subsequent years. This presumably accounts in part for the decline in investment associated with depressed market conditions (Collins 1981). In addition, the firms themselves may allow current market conditions to color their expectations about the future, and thus voluntarily reduce investment during such periods.

Insufficient investment by restricting capacity eventually produces higher prices and profits, which in turn provide the incentives needed to correct the problem. However, it may be several years before the resulting new capacity comes on stream and the shortages are eliminated.

3. Surges in Demand

Unanticipated surges in material demand can occur for a number of reasons. In practice, though, most are the consequence either of military buildups or of swings in the business cycle.

Alloy steels, copper, titanium and a number of other materials are heavily used in the production of military aircraft, ships, tanks, ammunition and other armaments. In modern times, the outbreak of hostilities has produced a sharp jump in anticipated material requirements. The resulting shortages have forced governments to impose price controls and to allocate available supplies to priority uses. While such shortages are typically intense, rarely do they persist for more than several years, in part because hostilities often cease within a few years and in part because new capacity is eventually brought on stream.

Cyclical instability, which has long plagued material markets, is caused by three basic characteristics of material supply and demand. First, as output approaches the available capacity, short-run supply becomes unresponsive, or inelastic in the terms of the economist, to changes in price. New capacity can be built in the long run, a period of several years or more, but not in the short run. As a result, further increases in price do not add much to supply.

Second, the demand for most materials is similarly unresponsive or inelastic in the short run to changes in price. Materials are intermediate goods whose demand is derived from the demand for final goods and services. Since the cost of a material usually constitutes only a small portion of the selling price of final products, a change in the price of the former normally has little effect on the price or demand of the latter. It is true that a rise in a material's price may encourage manufacturers to substitute other less expensive materials. Rarely, however, are such changes made in response to short-run and perhaps temporary price increases, for they often require the ordering of new equipment, the retraining of personnel and the redesign of production processes.

Third, the demand for materials tends to be highly responsive in the short run to fluctuations in the business cycle. Most materials are heavily used in the construction, capital equipment, automobile and other transportation and consumer durable sectors of the economy. As is well known, these sectors tend to fluctuate with, and to a much greater extent than, the overall economy. Their cyclical swings are transmitted back to the demand for materials.

Cyclical instability resulting from these three characteristics of supply and demand manifests itself in somewhat different ways depending on the type of market in which a commodity is sold. In competitive markets, such as the London Metal Exchange (LME) copper market, price is free to vary as often and as widely as needed to keep short-run supply and demand in balance. As a result, cyclical instability in such markets produces frequent and sizeable fluctuations in price. Output also varies. Since output is down when prices are low and up when prices are high, competitive markets exhibit large fluctuations in revenues and profits.

The nature of cyclical instability in competitive markets is illustrated in Fig. 1a. The demand and supply curves, as capacity is approached, are drawn with steep slopes to reflect their short run inelasticity with respect to price. The three demand curves illustrate the shifting of demand over the business cycle.

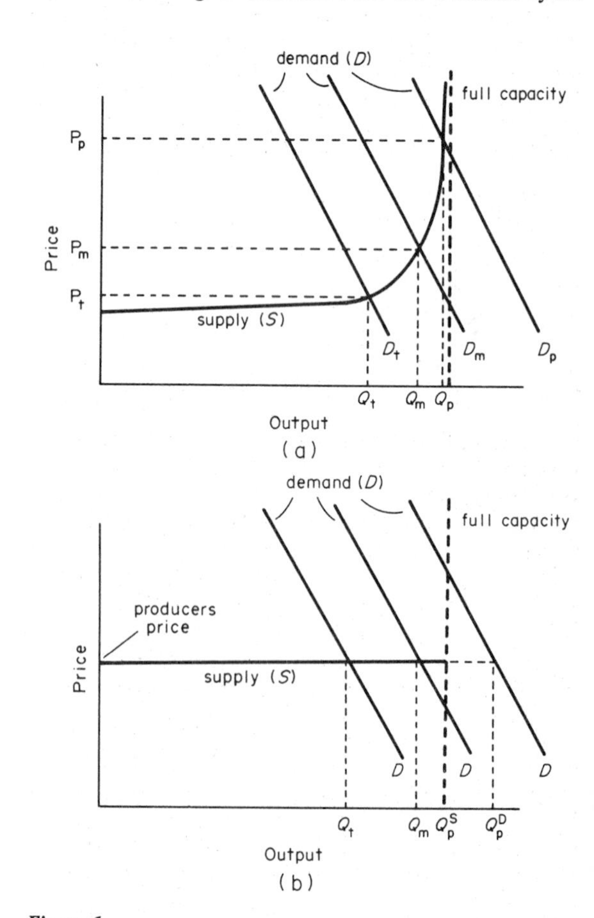

Figure 1
Implications of material supply and demand for market instability: (a) competitive markets; (b) producer markets

At the peak of the cycle, demand is strong as indicated by the curve D_p. The market price P_p is high, and output Q_p approaches capacity. Industry revenues, given by the product of the market price and industry output, are high, as are profits. At the midpoint of the business cycle, the curve D_m reflects demand. Prices, output, revenues and profits are all more modest. At the trough of the cycle, demand portrayed by the curve D_t is depressed. A significant portion of available capacity is idle, price and revenues are low and profits are small or negative.

While competitive markets are important for a number of materials, most materials are sold in markets where the major firms quote a producer price. Actual or transaction prices may deviate from the quoted price due to discounts and other factors, and the quoted price itself may at times change in response to short-run market conditions. However, prices are not allowed to fluctuate to the extent necessary to keep short-run supply and demand in continual balance. For this reason, prices are far less volatile in this type of market.

This has led some to conclude that cyclical instability is a problem only for materials sold in competitive markets. However, the same three characteristics of supply and demand that cause instability in competitive markets are shared by materials sold in producer markets. Instability simply manifests itself in different ways.

This is illustrated in Fig. 1b, which shows the consequences of cyclical instability in producer markets. Again, three demand curves are pictured, reflecting the sizeable shift in demand over the business cycle. The slopes of these curves are also steep, since short-run demand is not greatly affected by changes in price. It is the supply curve that is different. It is drawn as a horizontal line at the producer price that terminates once full capacity is reached. This is somewhat of a simplification since, as pointed out earlier, short-run market conditions may encourage some change in the producer price. They may also cause actual prices to deviate from the quoted producer price.

The figure shows that, although price instability is eliminated or considerably reduced in a producer market, output still varies greatly. When the economy is booming, it is at or near industry capacity. When the economy is depressed, much capacity stands idle. These fluctuations in output cause sizable variations in industry revenues and profits. In addition, at the peak of the business cycle, demand Q_P^D exceeds supply Q_P^S at the producer price, and physical shortages occur. Producers cope with such shortages by drawing down their inventories, lengthening their delivery times and rationing their customers. Consumers may have to do without the material they need, regardless of the price they are willing to pay.

Thus, it is at the peak of the business cycle that cyclical instability poses the greatest threat to materials availability. The resulting shortages are revealed by sharply rising prices in competitive markets, and by an excess of demand over supply in producer markets. Though often intense while they last, such shortages tend to be temporary. Within a few years new capacity can be built, and normally the business cycle declines from its peak.

Cyclical instability and the shortages it creates, however, may be growing more severe over time. Recent evidence suggests the business cycles of the major industrialized countries are becoming more synchronized (Chien 1981). Such synchronization limits the opportunities for countries experiencing an economic boom to supplement their traditional sources of material supply by importing from producers who normally serve other geographic regions. In addition, the ability or willingness of material users to vary their inventories over the business cycle in a stabilizing manner may be declining (Grubb 1981).

4. Interruptions in Supply and Constraints on Trade

Strikes, accidents, natural disasters and many other factors may interrupt the supply or constrain the trade of materials. This section focuses on those factors of most concern—cartels, embargoes, wars and civil disruptions.

A cartel is an agreement among producers to divide markets, restrict supply, maintain minimum prices or collude in some other way to raise prices above their competitive level. The objective is to earn greater profits, not to cut off supplies to consumers, although supplies must be curtailed to some extent to keep prices artificially high.

Cartels were quite common before World War II in certain industries, such as copper. Generally they lasted for only a few years, before collapsing for one reason or another. In the postwar era, the incidence of cartels declined, in part as a result of antitrust policy in the USA and other countries. The major exception is the Organization of Petroleum Exporting Countries (OPEC), which, in the early 1970s, successfully raised oil prices several fold and has managed to keep prices artificially high ever since.

The success of OPEC raised the fear that similar cartels would arise in copper, bauxite and other material industries. Research on this threat has identified several necessary conditions for a successful cartel: (a) production and exports of the material must be concentrated in a few major producing countries; (b) demand for the material must not fall greatly as its price increases; (c) supply from countries outside the cartel must not expand appreciably as the price of the material increases; and (d) the internal cohesion among cartel members must be strong enough to overcome the temptation to cheat or abandon the cartel.

The first condition is satisfied for many materials. Over half of the world mine output of iron ore comes from four countries: the same is true of bauxite, copper and nickel. The degree of concentration is even higher for chromium, cobalt, platinum and molybdenum.

The last three conditions vary, not only among materials, but over time. Material demand, as noted earlier, is generally unresponsive to price changes in the short run. However, in the long run, few if any materials are immune to the threat of material substitution in at least some of their end uses. As a result, the adverse consequences for demand grow the longer the cartel maintains artificially high prices.

Similarly, high prices are likely to encourage new sources of supply outside the cartel. However, since it takes several years to develop new mines and to construct new processing facilities, such sources will not come on stream in the short run.

Finally, the history of cartels suggests that internal cohesion is often maintained for several years, but eventually breaks down. As material substitution and the entry of new producers take place, disputes arise over how the dwindling market for cartel members should be shared. The temptation to reduce prices secretly or to cheat in other ways increases.

This review of the conditions necessary for a successful cartel suggests that a cartel attempt might succeed for a few years, but is eventually likely to collapse. This conclusion is consistent with the history of most cartels, and implies that material shortages caused by cartels are not likely to last more than a few years.

Since cartels produce only short-term benefits, while encouraging the substitution of alternative materials and the entry of new competitors over the longer run, most of the research on this topic concludes that material producers are not likely to follow OPEC's example. This view, however, assumes that producers are aware of the adverse long-run consequences, and are always willing to sacrifice the short-run benefits that cartels may offer in order to avoid them.

Embargoes, in contrast to cartels, are imposed to interrupt or restrict material trade, not to raise price and earn greater profits. They are primarily motivated by political rather than economic considerations. In some instances, the objective is to punish other countries, or to pressurize them into modifying their behavior. The United Nations embargo against chrome exports from Rhodesia during the 1970s is an example. In other instances, embargoes are the result of internal political pressure from domestic consumers who want exports restricted to ensure their access to adequate supplies at low prices. The USA has, on occasions, curtailed exports of metal scrap and soybeans for this reason.

Embargoes tend to be most disruptive in the short run. With time, countries cut off from traditional sources of supply can develop new sources and switch to alternative materials. For this reason, embargoes are rarely effective over the longer term in forcing political concessions from other countries.

Wars and civil disruptions can also interrupt material production and trade, as the invasions in 1977 and 1978 of the cobalt-rich Shaba province of Zaire so clearly illustrate. Normally, the resulting shortages are not deliberately provoked, as is the case for cartels and embargoes, but instead are an unintended by-product of other objectives and activities. There are, though, exceptions. Where mining and material processing are vital to the economic well-being of a country, an adversary may sabotage such activities. In any case, since wars and civil disruptions seldom last longer than a few years, the shortages they produce are not of long duration, even if facilities are damaged or destroyed. Even when hostilities continue over extended periods, consuming countries can, within a few years, turn to new sources of supply and alternative materials.

This examination of the threats to material availability points out several important differences between the first threat—resource depletion—and the last three—investment, demand surges and interruptions in supply and constraints on trade. The latter induce shortages that tend to arise quickly and unexpectedly. The disruption they cause can be severe. Such shortages are, however, temporary, rarely lasting more than several years. The cessation of the war, economic boom or other underlying cause often results in their termination. When this is not the case, they are eliminated by the development of new sources of supply and the substitution of more available alternative materials. Consumers, if they wish, can protect themselves against the worst effects of these shortages by maintaining sufficient stockpiles.

Shortages arising from resource depletion are very different in nature. They arrive gradually with ample warning in the form of a persistent and long-run climb in real material prices. Once in effect, they are likely to be permanent, or at least of long duration. Fortunately, such shortages appear far less likely to occur over the foreseeable future than shortages arising from other threats. Over the very long term, the outlook is less clear. The principal weapon available to society for fending off the threat of shortages from resource depletion is new technology.

Bibliography

Barnett H J, Morse C 1963 *Scarcity and Growth*. Johns Hopkins University Press, Baltimore, Maryland, pp. 1–288

Chien D 1981 Business cycles and instability in metal markets. *Mater. Soc.* 5: 257–65

Collins L S 1981 Market instability and investment. *Mater. Soc.* 5: 301–30

Fischman L L 1980 *World Mineral Trends and U.S. Supply Problems*. Resources for the Future, Washington, DC, pp. 1–535

Grubb T J 1981 Metal inventories, speculation, and stability in the U.S. copper industry. *Mater. Soc.* 5: 267–88
Smith V K (ed.) 1979 *Scarcity and Growth Reconsidered.* Johns Hopkins University Press, Baltimore, Maryland, pp. 1–298
Tilton J E 1977 *The Future of Nonfuel Minerals.* Brookings, Washington, DC, pp. 1–113

J. E. Tilton
[Pennsylvania State University, University Park, Pennsylvania, USA]

Materials Conservation: Economics

Historically, conservation has implied different things for different people. The Conservation Movement of the late nineteenth and early twentieth centuries called for the adoption of a set of values other than those implied by consumer sovereignty for judging what constituted the appropriate uses of natural resources. To the economist, by contrast, conservation has come to mean a concern over whether natural resources play a unique role in production and consumption activities. Uniqueness, in this case, is defined by judging whether it would be possible to increase, or indeed even to maintain, levels of economic well-being without a minimum supply of some or all of these natural resources. This economic interpretation of the rationale for resource conservation has not been easy to address. Full knowledge of all the earth's stocks of natural resources, including a detailed specification of their attributes, would not by itself be sufficient information to permit a judgement about the essentiality of the resources involved. Thus, to provide an economic interpretation of the problems posed by natural resource conservation requires consideration of the geological availability of those resources, of the ways the resources are used to respond to society's needs, and the relative importance of those needs requiring natural resources in comparison to other needs. This article summarizes the considerations involved in addressing each of these issues.

1. Conservationist Perspective on Materials Usage

The early conservationists maintained, as a basic assumption, that natural resources were important, not solely because of their role in production and consumption activities, but rather because nature's ecological balance had intrinsic value (Barnett and Morse 1963, pp. 77–78). Resource conservation was also warranted for the conservationists because natural resources were scarce and man's activities wasted them. Barnett and Morse (1963, pp. 80–81) identified three rules as the conservationist's proposed means for realizing an efficient, or nonwasteful, use of natural resources:

(a) the regenerative capacity of renewable resources should not be physically damaged,
(b) renewable resources should be used in place of nonrenewable resources where physically possible,
(c) plentiful nonrenewable resources should be used before the less plentiful where physically possible.

The conservationists extended their calls for maximizing the physical yield and minimizing the physical destruction of natural resources to the evaluation of the appropriate end uses of natural resources as well. These end uses, if "unwise," contribute to the scarcity of virgin materials. Thus, conservationists called for avoidance of any waste, as well as for the full recovery and reuse of extractive materials.

One would not expect that these goals would be achieved through private transactions, motivated by self-interest, and taking place within ideal markets. Indeed, the conservation movement sought an alternative to this perspective (Barnett and Morse 1963):

> The Conservationists were seriously concerned with ethical revisions of *laissez faire* and economic atomism—with values involving a holistic conception of man and nature, obligation to future generations, so-called scientific management based on physical values and rejection of the individualistic values of the classicists. They were not concerned with the professional economists' maximizing of the present value of the stream of real income over time, based on consumer sovereignty—which professional economists frequently term conservation economics.

This distinction is important because it has affected subsequent economic analyses of resource scarcity and the definition of materials policy. That is, it has very probably promoted the belief that natural resource policy can be based on judgements as to the overall availability of natural resources. Barnett and Morse, for example, called for the use of scarcity indexes to signal the importance of growing limitations on the available stocks of natural resources. Thus, while their approach recognized the dilemma posed at the outset (i.e., knowledge of resource stocks alone does not convey information as to the importance of those resources), it tended to focus attention on the development of these indexes at an aggregate level, for example, all minerals, agricultural products and forestry products.

Page (1977) continued this focus by proposing that the conservationist criterion can be interpreted as calling for the maintenance of constant (or declining) real prices for a composite of virgin materials. He argued that this objective can be considered as assuring a sustainable yield of virgin materials resources. Clearly, this objective will be different from a present-value criterion, which would call for maximization of the present value of each material resource individually.

Thus, in discussing materials conservation, it is important to distinguish between conservation as the basis for defining a criterion with which to judge the desirability of alternative allocations of natural resources versus conservation as one of several goals that might be articulated as part of the definition of an overall materials policy. In this latter role one might utilize the conservationists' three rules for avoiding waste as the basis for defining the conservation goals in the larger policy framework. However, this would not imply that conservation was the primary decision criterion in the allocation of natural resources. Equally important, this choice would not suggest that materials policy could be based on broad indexes of resource availability.

2. *Measuring Natural Resource Scarcity*

There is no reason to believe that private market allocations of natural resources will guarantee that conservationist objectives will be realized. However, it is also unreasonable to conclude that the results of these allocation decisions will be completely at odds with resource conservation. Accordingly, it is desirable to consider whether it is possible to evaluate resource availability and, if so, to judge what results are available from past efforts at this task. The first comprehensive effort to evaluate the adequacy of the US natural resource base was undertaken by Barnett and Morse (1963). Their study used scarcity indexes to signal the growing importance of any natural resource in meeting society's needs. Their primary scarcity index was called the real unit costs of extractive output. It was defined as the real quantity of homogeneous inputs required to produce one unit of a constant quality material (or extractive) output. Using index numbers for labor L, capital K and extractive output Q for each period t relative to a predefined base period B, the Barnett–Morse real-unit-cost index UC_t is given as:

$$UC_t=\left[w_{\mathrm{B}}\left(\frac{L_t}{Q_t}\right)+r_{\mathrm{B}}\left(\frac{K_t}{Q_t}\right)\right]\Bigg/\left[w_{\mathrm{B}}\left(\frac{L_{\mathrm{B}}}{Q_{\mathrm{B}}}\right)+r_{\mathrm{B}}\left(\frac{K_{\mathrm{B}}}{Q_{\mathrm{B}}}\right)\right] \quad (1)$$

where w_{B} and r_{B} are the base levels of the input prices for labor and capital, respectively (for the derivation see Smith 1980).

Their examination of the trends in this index (from 1870 to 1957) for the aggregate extractive sector and for three subsectors (i.e., minerals, agriculture and forestry) indicated that, with the exception of the forestry subsector, there was no evidence of increasing resource scarcity. This conclusion was consistently upheld under a wide variety of modifications to the ways the data were used in the index. Their analysis has had a profound effect on the methods used to consider questions associated with natural resource scarcity and on the judgements made as to long-term resource availability.

Recently this work has been subjected to detailed reevaluation (Brown and Field 1978, Fisher 1979, Smith 1978a), suggesting that there are conceptual and empirical limitations to Barnett and Morse's ideal scarcity index. Moreover, given the complexity of the issues involved in judging resource scarcity, there appears to be a need to consider disaggregate analyses of individual resources using indexes of resource scarcity that can be directly linked to the behavioral decisions of the firms involved in extracting the resources. These indexes should be evaluated with detailed supplementary information including: (a) the geological conditions governing current and prospective future deposits of each resource, (b) the market structure governing the patterns of exchange of the resource and the prospects they provide for strategic behavior on the part of market participants, and (c) an understanding of the process by which economic agents form expectations on the state of geological and market conditions.

These conclusions follow from three interrelated aspects of recent research. Firstly, the appropriate scarcity index is one which summarizes the direct and indirect sacrifices made to obtain one unit of the resource (Fisher 1979). This definition implies that attention be given to what is considered to be the resource under study, that is, the choice will be different if one considers in situ resources versus processed natural resource commodities. Recent analyses suggest that both the Barnett–Morse real unit costs and neoclassical unit costs of extractive outputs are inappropriate measures of resource scarcity. The first is intended to measure the time path of the physical inputs required to obtain a unit of the output without regard to changes in the opportunity costs of these inputs (Smith 1980). The second takes account of changes in these opportunity costs, but does not fully meet the needs of a scarcity index. Indeed, neither measure conveys information concerning the cost features of those resources remaining for extraction (Brown and Field 1978, Hanson 1980).

Secondly, the empirical evidence on the movements of scarcity indexes is not clear-cut. Replications of the Barnett–Morse analysis with updated data (Barnett 1979, Johnson et al. 1980) generally agree with the earlier findings. By contrast, evidence from relative prices (Smith 1978a) seems to cast doubt on the conclusiveness of the original findings. Recursive fitting of linear and nonlinear models using relative prices as the scarcity index indicated that the trend coefficient was quite sensitive to sample composition (Slade 1983).

Finally, theoretical models of the extractive firm's behavior suggest that the past conclusions on these firms' behavioral responses are quite sensitive to the modelling of: (a) the geological conditions governing the availability of the extractive resource (Harris and Skinner 1982); (b) the treatment of the inherent uncertainty in the exploration process (Devarajan and Fisher 1982); and (c) the character of the market structure for the firm's output, especially in cases where the most reasonable characterization of the market is one of limited competition (i.e., those market forms falling between perfect competition and monopoly).

Thus, it is not clear that the private market allocation of natural resources has succeeded in maintaining a constant supply price for these resources. Indeed, the recent analyses of the properties of scarcity indexes suggest that a much more detailed level of enquiry is needed if these indexes are to be useful signals of resource availability.

3. Economic Mechanisms for Resource Conservation

Ideally a scarcity index will reflect both the character of the demand for a resource and the incremental costs of obtaining it. Given the difficulties identified in the previous section with developing and using such indexes it is reasonable to consider alternative approaches to judging the importance of particular natural resources and to gauging the prospects for conservation. One such alternative involves investigation of how natural resources individually contribute to production activities (Brown and Field 1978). In economic models these judgements are based on: (a) measures of the prospects for substituting other inputs for nonrenewable resources within existing technologies, and (b) the pattern of nonneutral innovations to these technologies and their implications for the use of nonrenewable resources. Both of these factors can be described in either of two ways. The first of these, an engineering analysis of materials usage, would describe each of these mechanisms for conserving natural resources (i.e., input substitutions and resource-saving innovation) with examples of specific cases. For example, in iron and steel production, one might cite the presence of an electric-arc steelmaking furnace along with a basic oxygen furnace provides the basis for substituting scrap steel for iron ore in composing the set of iron-bearing materials used to produce molten steel (Russell and Vaughan 1976). For resource-saving innovations, the introduction of scrap preheating to steelmaking practices (Vaughan et al. 1976) enhances the feasible substitutions between scrap and iron ore within basic oxygen furnace production technology. Similarly, within certain electrical conductor uses, it is possible to replace copper with aluminum. However, this approach faces a significant limitation: it is unable to summarize adequately the quantitative nature of the conversion rates between the inputs over different levels of input usage, or to gauge the quantitative impact of the resource-saving innovations that would accompany a given change in relative prices of inputs.

The economic approach to summarizing input substitution is based on a neoclassical description of production activities. As such, it maintains that engineering production processes can be approximated adequately with smooth functions relating the inputs to these activities to the output(s). In this framework, the substitution between inputs permitted within a given technology is generally measured by an Allen (1938) partial elasticity of substitution. In heuristic terms, this index describes how the rate of conversion between any two inputs changes with changes in the relative usage of those inputs, while maintaining the output rate at a constant level and allowing other inputs to adjust conformably (for a discussion of the technical details see Mundlak 1968, and for a heuristic review, Smith 1978b). Large positive values for the elasticity imply that the relevant two inputs are substitutes and that each factor can readily be used in place of the other. An elasticity value of zero implies that the two inputs involved must be used in fixed proportion. Negative values suggest complementary relationships between inputs. While this index has the advantage of providing a convenient quantitative index of input conversion relationships, its descriptive ability is contingent on the authenticity of the neoclassical production model as an approximation for the underlying engineering production processes (Marsden et al. 1974). Neoclassical models also offer indexes of the pace of nonneutral technical change (Bingswanger 1974, Jorgenson and Fraumeni 1981). In this case, they generally measure the rate at which the innovation increases the effectiveness of (or augments) the existing inputs.

The available evidence of the quality of neoclassical models as descriptions of the substitution relationships present in the underlying engineering technologies suggests that the performance depends on the level of aggregation of the input measures used and on the degree of disaggregation in technologies under study (Lau 1982, Kopp and Smith 1980a, 1980b). Under ideal conditions a neoclassical cost function, used as a means for describing production activities, can provide a reasonably good description of the input associations implied by the underlying technology. The available experimental evidence indicates that the estimated substitution elasticities generally identify the correct input association (i.e., substitution or complementarity) and tend to be conservative in their appraisal of input relationships when there are conflicts in the underlying true associations at different input prices.

These findings suggest that, in evaluating the available estimated neoclassical elasticities for actual production processes, one should focus on those with disaggregated input descriptions that have been conducted within carefully defined, disaggregate technologies. Unfortunately, the available data for statistical estimation of production models have greatly limited the ability of empirical efforts to respond to these guidelines. Table 1 summarizes a representative sampling of the available estimates to date. The industries are represented by Standard Industrial Classification

Table 1
Summary of selected Allen elasticities of input substitution[a]

Study	Character of data used in study	Estimated Allen elasticities of substitution σ_{KN}[b]
Humphrey–Moroney (1975)	cross-sectional data for 4-digit SIC code industries grouped by 2-digit category in 1963	0.64 (SIC 20) 1.12 (SIC 24)
Bernt–Wood (1975)	time series (1947–1971) aggregate manufacturing	0.56
Moroney–Toevs (1977)	time series for seven 3- and 4-digit SIC-coded industries (1954–1971)	0.370 (SIC 202) 0.602 (SIC 2011) −0.320 (SIC 2041, 2043) 1.292 (SIC 3241) 0.565 (SIC 3275)
Hudson–Jorgenson (1978)	annual interindustry accounts (1947–71)	
	agriculture, nonfuel mining construction	0.61
	manufacturing	0.10
	commercial transport	0.58
	services, trade communications	0.07
Wills (1979)	time series for the primary metals industry (1947–1974)	0.48
Moroney–Trapani (1981b)[c]	time series for seven 4-digit SIC code industries (1954–1974)	0.18 (SIC 3351) 1.08 (SIC 3312) 0.47 (SIC 3334) 0.02 (SIC 3691) 0.08 (SIC 3275) −0.38 (SIC 3241)

[a] Smith (1980) [b] σ_{KN} is the partial elasticity of substitution between capital K and the natural resource input N [c] These estimates are different from those reported in Smith (1980). The present results relate to the Moroney–Trapani model without variables to measure nonneutral technical change. The previous summary reported the estimates derived from models with measures of nonneutral technical change

Table 2
Some SIC-coded industries and their primary resource inputs

SIC code	Industry	Resource input
20[a]	food and beverages	meat, grain, dairy and nondairy fluids
2011[a]	meat packing	slaughtered cattle, calves, lambs and hogs
202[a]	dairy products	cream and fluid milk
2041[a]	flour milling	wheat, rye, oats and corn sold to produce foodstuffs
2043[a]	cereals	
24[a]	lumber products	forestry products
3241	hydraulic cement	gypsum (uncalcined)
3275	gypsum products	gypsum (calcined)
3312	blast furnaces and basic steel	crude iron ore and scrap
3334	primary aluminum	bauxite
3351	copper rolling and drawing	refined copper
3691	storage batteries	lead

[a] Uses renewable natural resources

(SIC) codes. This is a way in which the US Government classifies manufacturing industries, depending on the degree of detail in the definition of the industry. Thus, for example, a two-digit specification is a highly aggregative one. Within the two-digit specification, there would be a number of different categories, and these would be specified using a three-digit code. Within each three-digit code there would be further specification up to a four-digit code. The specific codes used in Table 1 for the SIC-coded industries are defined in Table 2. Based on the criteria discussed earlier, only the Moroney–Trapani (1981a) estimates approach the conditions necessary for being judged as reasonable approximations of the underlying input associations. These results clearly suggest that the role of natural resource inputs varies greatly across production processes. Moreover, when they were used in simulation analyses (Moroney and Trapani 1981b) designed to evaluate the implications of increased resource scarcity, it was found that substitution provided very limited relief from increased resource scarcity. Indeed, Moroney and Trapani (1981b) concluded that

> Rapid increases in resources prices are translated into substantial commodity inflation, and factor substitution provides little relief . . . .From these results, we conjecture that the most promising paths for exhaustible resource conservation are likely to be found in substitution among semifinished materials and changes in final demand.

The available empirical evidence on the prospects for resource-saving technological change is even more limited. Moreover, it is important to recognize clearly the limitations on the interpretation of these studies. The existing empirical results provide a description of the pattern of technical change that has taken place over the period of analysis. It does not permit a judgement as to what innovations would be adopted under alternative patterns of input availability. Moreover, these studies have been conducted at fairly aggregate levels of analysis, both in terms of the input definitions used and the descriptions of the technologies involved. Jorgenson and Fraumeni's (1981) analysis offers the most detailed evidence available. It considered 36 industrial sectors over the period 1957–1974 and indicated that nearly all the industries experienced materials-saving and energy-using technical change. Nonetheless, the available empirical evidence remains too limited to draw conclusive judgements on the prospects that technological innovations might relax specific resource needs on an industry-by-industry basis or a specific material basis.

Thus, while economic analysis identifies, in principle, input substitution and nonneutral technical change as mechanisms for resource conservation, the available empirical evidence (from neoclassical economic models) cannot, as yet, provide detailed judgements on the magnitude of these effects for specific natural resources and production activities.

4. Definition of Natural Resources and Conservation

This discussion has implicitly maintained that natural resources and materials were synonymous. This has been conventional practice. However, it is reasonable to ask whether this definition for resources reflects all the implications of alternative patterns of materials usage for economic well-being; Krutilla (1967) has argued it does not. This limited perspective fails to recognize the contribution made to economic well-being by the amenity services provided by unspoilt natural environments. It also fails to recognize that the ways in which economic processes respond to increased materials needs can affect the quality of environmental resources over time as well. To the extent that both of these types of service contribute to society's economic well-being, it is important to consider them as an integral part of the definition and evaluation of resource-conservation policies. It is within this more general framework that the economic analyses of the importance of natural resources must be conducted. Since the framework is new little can be concluded, at this point, as to its implications for the adequacy of our natural resource base.

See also: Materials Conservation: Technology

Bibliography

Allen R G D 1938 *Mathematical Analysis for Economists.* Macmillan, London

Barnett H J 1979 Scarcity and growth revisited. In: Smith V K (ed.) 1979 *Scarcity and Growth Reconsidered.* Johns Hopkins University Press, Baltimore, Maryland, pp. 163–97

Barnett H J, Morse C 1963 *Scarcity and Growth.* Johns Hopkins University Press, Baltimore, Maryland

Berndt E R, Wood D O 1975 Technology, prices and the derived demand for energy. *Rev. Econ. Stat.* 57: 259–68

Bingswanger H P 1974 The measurement of technical change biases with many factors of production. *Am. Econ. Rev.* 64: 964–76

Brown G M Jr, Field B C 1978 Implications of alternative measures of natural resource scarcity. *J. Polit. Econ.* 86: 229–43

Devarajan S, Fisher A C 1982 On measures of natural resource scarcity under uncertainty. In: Smith and Krutilla 1982, pp. 327–46

Fisher A C 1979 Measure of natural resource scarcity. In: Smith V K (ed.) 1979 *Scarcity and Growth Reconsidered.* Johns Hopkins University Press, Baltimore, Maryland, pp. 249–71

Hall D C, Hall J V 1984 Concepts and measures of natural resource scarcity with a summary of recent trends. *J. Environ. Econ. Manage.* 11: 363–79

Hanson D A 1980 Increasing extraction costs and resource prices: Some further results. *Bell. J. Econ.* 11: 335–42

Harris D P, Skinner B J 1982 The assessment of long-term supplies of minerals. In: Smith and Krutilla 1982, pp. 247–326

Hudson E A, Jorgenson D W 1978 The economic impact of policies to reduce US energy growth. *Resour. Energy* 1: 205–29

Humphrey D B, Moroney J R 1975 Substitution among capital, labor, and natural resource products in American manufacturing. *J. Polit. Econ.* 83: 57–82

Johnson M H, Bell F W, Bennett J T 1980 Natural resource scarcity: Empirical evidence and public policy. *J. Environ. Econ. Manage.* 7: 256–71

Jorgenson D W, Fraumeni B M 1981 Relative prices and technical change. In: Berndt E R, Field B C (eds.) 1981 *Measuring and Modeling Natural Resource Substitution.* MIT Press, Cambridge, Massachusetts, pp. 17–47

Kopp R J, Smith V K 1980a Measuring factor substitution

with neoclassical models: An experimental evaluation. *Bell J. Econ.* 11: 631–55
Kopp R J, Smith V K 1980b Input substitution, aggregation, and engineering descriptions of production activities. *Econ. Lett.* 5: 289–96
Krutilla J V 1967 Conservation reconsidered. *Am. Econ. Rev.* 57: 777–86
Lau L J 1982 On the measurement of raw materials inputs. In: Smith and Krutilla 1982, pp. 167–200
Marsden J, Pingry D, Whinston A 1974 Engineering foundations of production functions. *J. Econ. Theory* 9: 124–40
Moroney J R, Toevs A 1977 Factor costs and factor use: An analysis of labor, capital and natural resource inputs. *South. Econ. J.* 44: 222–39
Moroney J R, Trapani J M 1981a Alternative models of substitution and technical change in natural resource intensive industries. In: Berndt E R, Field B C (eds.) 1981 *Measuring and Modeling Natural Resource Substitution.* MIT Press, Cambridge, Massachusetts, pp. 48–69
Moroney J R, Trapani J M 1981b Factor demand and substitution in mineral intensive industries. *Bell J. Econ.* 12: 272–84
Mundlak Y 1968 Elasticities of substitution and the theory of derived demand. *Rev. Econ. Stud.* 35: 225–36
Page T 1977 *Conservation and Economic Efficiency: An Approach to Materials Policy.* Johns Hopkins University Press, Baltimore, Maryland
Russell C S, Vaughan W J 1976 *Steel Production: Processes, Products and Residuals.* Johns Hopkins University Press, Baltimore, Maryland
Slade M E 1983 Trends in natural resource commodity prices: An analysis of the time domain. *J. Environ. Econ. Manage.* 9: 122–37
Smith V K 1978a Measuring natural resource scarcity: Theory and practice. *J. Environ. Econ. Manage.* 5: 150–71
Smith V K 1978b Duality principles and measuring the production technology: A heuristic introduction. *Socio-Econ. Plann. Sci.* 12: 161–66
Smith V K 1980 The evaluation of natural resource adequacy: Elusive quest or frontier of economic analysis? *Land Econ.* 56: 257–98
Smith V K, Krutilla J V (eds.) 1982 *Explorations in Natural Resource Economics.* Johns Hopkins University Press, Baltimore, Maryland
Vaughan W J, Russell C S, Cochrane H C 1976 *Government Policies and the Adoption of Innovation in the Integrated Iron and Steel Industry.* Resources for the Future, Washington, DC
Wills J 1979 Technical change in the US primary industry. *J. Econom.* 10: 85–98

V. K. Smith
[Vanderbit University, Nashville, Tennessee, USA]

Materials Conservation: Technology

Materials conservation comprises those strategies by means of which present or projected materials shortages are addressed by effectively reducing demand rather than increasing supply. This approach is relied upon more heavily as the limits to increasing supply, such as added energy demands and environmental damage, become more evident. The major technical strategies employed to accomplish materials conservation are: (a) effective materials utilization, (b) materials substitution, and (c) materials recycling.

Effective materials utilization is the broadest of these strategies since it addresses not only product design, development and deployment, but also manufacturing materials and processes. Materials substitution is probably the most important of these strategies, however, since it directly alters materials usage. Recycling is an obvious strategy, but one still hampered by a variety of economic, technical and institutional factors.

1. Total Materials Cycle

Materials conservation and its supporting strategies are best understood in the context of the total materials cycle shown in simplified form in Fig. 1. In this global system, as first discussed in *Materials and Man's Needs* (National Academy of Sciences 1974), natural resources extracted from the earth are refined into basic materials and then further processed into engineering materials. These, in turn, are designed and manufactured, or assembled, into the myriad products required by society. At the end of their useful life, they are either returned to the earth as waste or recycled. Materials, energy and the environment interact strongly at virtually every point in this materials cycle and events or decisions occurring at any given point may affect other points. Furthermore, every operation in the materials cycle requires energy and, at every stage in the cycle, material (and energy) losses are encountered. These losses for common metals are estimated at 50–70% of the amounts that enter the cycle and, in fact, materials conservation can be alternatively defined as the reduction of these losses from the materials cycle. The concept of the total materials cycle provides a useful framework for examining the global materials–energy–environment system, though its operation is far more complicated than that shown in Fig. 1. This complexity has been

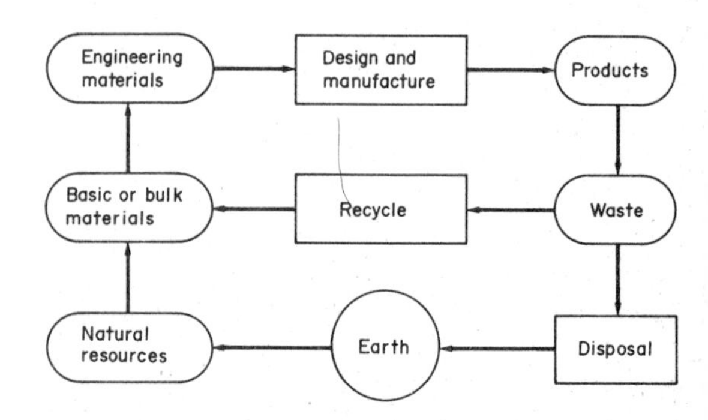

Figure 1
Simplified diagram of the total materials cycle

discussed by Bever (1978) for those parts of the materials cycle related to recycling. The economics of materials conservation is treated elsewhere (see *Materials Conservation: Economics*), while Chapman and Roberts (1983) relate the future availability of resources to the energy required to produce them.

2. *Effective Materials Utilization*

Effective materials utilization is broadly recognized as an essential materials conservation strategy. It refers to the development, selection and design of products that most efficiently meet application requirements, that have optimum durability and life, and that are recyclable. It further refers to the obtaining, processing and fabrication of materials so as to consume, waste or disperse the least amount of material (or energy) for equivalent performance. The specific tactics for achieving this fall conveniently under the headings of research and development, design, and manufacture.

Research and development are expected to provide the stockpile of technical information needed to implement any of these conservation strategies. Examples that apply to effective materials utilization include the reduction of material losses due to corrosion and wear, the development of new and improved materials and processes, and higher manufacturing yields obtained through improved flaw detection and materials characterization techniques.

Design is the preeminent tactic for achieving effective materials utilization, since it dictates not only the kinds and amounts of materials to be used in a product, but also the likely manufacturing and assembly methods to make it. These tactics include miniaturization, designing to closer tolerances, using more abundant and renewable materials, designing for longer life and easier maintenance, facilitating recycling and the use of reclaimed materials, simplifying and standardizing material and product variants, and making greater use of composites while avoiding wholly dissipative uses of materials that thwart conservation. Dissipative uses were of central concern in a NATO study of eleven potentially scarce metals (NATO 1976), which estimated that some 30% of the mine production of these metals was dissipated beyond possible recovery. They included such dissipative uses as silver in photographic prints, cobalt in paints, tin in toothpaste and zinc for galvanizing. They further estimated that another 40% of the mine production of these scarcer metals was dissipated by virtue of being technically but not economically recoverable. Such combined dissipation, which is estimated to be some two-thirds of all metals produced, can only be reduced by the substitution or recycling strategies discussed below.

Improved manufacturing processes that minimize materials (and energy) waste are the third and final tactic used in achieving effective materials utilization. Examples of the kinds of manufacturing processes to be preferred where a choice is available would include:

(a) additive over subtractive processes,

(b) solventless over solvent processes,

(c) closed-loop over open-loop processes,

(d) net-shape over material-removal processes, and

(e) feedback-controlled over sampling processes.

In each of these cases, the preferred process results in less materials waste and usually in less energy waste, since the energy content of the material itself is large in comparison with process energy.

More specific information relative to the material and energy losses associated with various metalworking processes has been collated by van den Kroonenberg (1979) among others. Ranking these in terms of decreasing material losses, the order is: machining > stamping > hot forging > cold or warm pressing, or diecasting. If the processes are ranked in terms of decreasing energy required per unit weight of chips produced or product formed, the order is: grinding > forging > hot forming > planning > turning > drilling. Such guidelines are helpful, though obviously not determinant, in selecting preferred manufacturing processes.

A necessary corollary of effective materials utilization is that the material remaining is more highly stressed and, therefore, this strategy requires higher performance materials, better processing control and understanding, better reliability information, and better materials characterization methods. These requirements can serve, in fact, as impediments in implementing the strategy of effective materials utilization.

3. *Materials Substitution*

Materials substitution is common and open to many possibilities because it focuses normally on the function(s) a material provides, rather than on the material itself. It will, however, only occur if the substitute material has properties which permit the required function to be achieved at lower cost under the prevailing circumstances.

Several kinds of substitution may be identified that can contribute to the overall goal of materials conservation. They are listed below together with selected examples.

(a) Abundant for scarcer materials (e.g., the use of aluminum for copper in electrical applications, or ceramics for high-temperature alloys in turbine engines).

(b) Nonmetals for metals (e.g., plain or reinforced plastics for zinc or aluminum in structural applications, or glass for metals in containers).

(c) Renewable for nonrenewable materials (e.g., wood for metals in construction).

(d) Synthetic for natural materials (e.g., hydrothermally grown quartz crystals for electronic use, and synthetic diamonds for cutting and polishing).

(e) Composite for monolithic materials. These would include all those materials where needed surface properties are conferred by coatings or claddings.

Although many substitutions offer the possibility of materials conservation, the time required to bring them into use can be substantial because of the need to examine alternative materials, to develop design and reliability data and to evaluate the substitute in a manufacturing environment. In addition, the proposed substitute may require novel processing or substitute material in its intended use. All of these factors make the substitution process slower than desired. Furthermore, where certain unique features of a material are involved, substitution may be difficult, if not impossible, to effect. This would include, for instance, those uses that depend on the unique catalytic properties of platinum or the thermal properties of helium. Equally difficult are those cases where the scale of use or cost cannot be met by any other material. For example, no material could replace steel in its structural uses or lead in its battery uses.

The altered material flows caused by substitution also create other problems in various parts of the materials cycle, particularly in basic materials supply and in recycling. Thus, a metal obtained solely as a by-product of another will have its own supply reduced if the latter is substituted for in any important use. The changes in recycling practices required as cars become smaller, shift to a greater use of high-strength low-alloy steels, aluminum and plastics, and use more platinum-group metals to control exhaust emissions are other examples of the problems brought about by substitution.

4. Recycling

Recycling has long been an integral part of the materials industry and is a strategy which, by its very nature, adds to material supply and modestly alleviates resource depletion. It has the advantage over effective use and substitution strategies that no product or manufacturing changes are needed to implement it, though such changes can be used to enhance recyclability. Recycling can also contribute to energy conservation, since material produced from scrap sources requires much less processing energy than that from virgin sources.

Scrap recycling occurs at three stages in the material cycle of Fig. 1. The first is at the production stage for the basic material where scrap, known as home revert or runaround scrap in the case of metals, is returned directly to the basic production process. Having known composition and little contamination, this scrap is the most desirable though it cannot be said to add to materials supply. The second stage at which recycling occurs is in manufacturing, where scrap (new, prompt-industrial or processing scrap) is collected as the next most desirable type. Like home scrap, it does not really add to supply, though it is a very important component of recycling. The third and most familiar place where recycling occurs, is at the waste or post-user stage. Here, the old or obsolete scrap recovered does add to materials supply. The availability of home and new scrap depends on current production levels, whereas that for old scrap depends on the life cycle of the products from which it is recovered and on the number of products then produced. The recovered new or old scrap is returned to the materials cycle at the basic production stage, as shown in Fig. 1, or to a secondary production stage not shown. In fact, the secondary materials industry handles the majority of all new and old scrap since it specializes in this activity.

In addition to the conventional type of scrap recycling just described, other types should be noted. The first is the closed-loop recycling of such items as paper or aluminum cans back into the same products. The second is the de facto recycling that occurs with the rebuilding and reuse of products or the dismantling or cannibalizing of them to provide useful parts.

The number of years that recycling can postpone the exhaustion of a nonrenewable resource has been developed in the NATO study (NATO 1976) and is given by the expression:

$$\log[1/(1-R)]/\log[1+(G/100)]$$

where R is the fraction currently recycled and G is the percentage annual growth rate for the resource. This relation is plotted in Fig. 2 for some common metals and shows clearly that for all but very low growth rates, recycling contributes only modestly to resource extension and that this extension is independent of resource size. In the USA only about 20% of the available old metal scrap is recycled, though it has been estimated that this figure could be easily doubled. The level of recycling for an individual material depends on the ease of collection and separation in the waste stream, the value of the scrap material, the difficulty of recovering pure material from the scrap and its degree of dependence on co-recovery. For example, 90–95% of all cars in the USA are recycled principally for the scrap iron and steel they contain but copper, lead, aluminum and other nonferrous metals can be corecovered at the same time. The level of this co-recovery will be highly dependent on the current market for old iron and steel scrap.

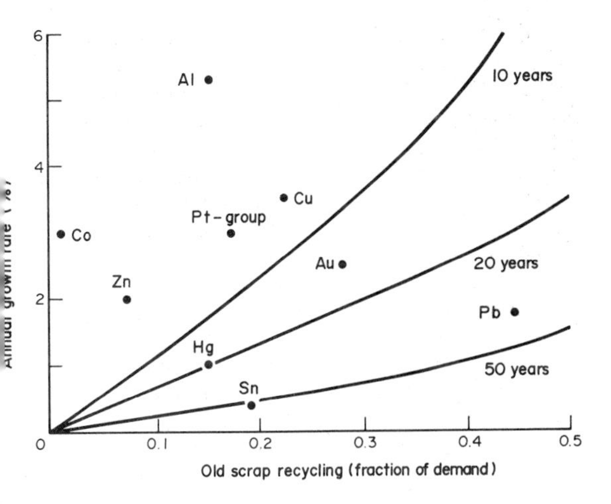

Figure 2
Number of years postponement of metal exhaustion due to recycling for selected common metals (based on US recycling and growth rates for the 1976–78 period)

5. A New Conservation Ethic

Materials conservation and its supporting strategies are part of a broader conservation ethic whose overall aim is to maximize the effectiveness of materials and energy usage. This, in turn, calls for a new philosophy of industrial design that treats a product as a materials system and seeks to optimize it in terms of the total materials cycle. This optimization requires some governmental incentives and changed attitudes on the part of society if it is to be realized, since present commercial incentives are insufficient to do so. Finally, this ethic suggests that there are certain inherent limits in our ability to continuously expand supplies and that it, therefore, is the most viable means for meeting these future needs of society.

See also: Recycling of Demoliton Wastes; Recycling of Metals: Technology; Substitution: Economics; Substitution: Technology

Bibliography

Bever M B 1978 Systems aspects of materials recycling. *Conserv. Recycling* 2: 1–17

Chapman P M, Roberts F 1983 *Metal Resources and Energy*. Butterworth, London

National Academy of Sciences 1974 *Materials and Man's Needs*. National Academy of Sciences, Washington, DC pp. 4–6

North Atlantic Treaty Organisation 1976 *Rational Use of Potentially Scarce Metals*. NATO Scientific Affairs Division, Brussels

van den Kroonenberg H H 1979 Solutions to future resource problems by design engineering. *Mater. Soc.* 3: 271–79

T. D. Schlabach
[Bell Laboratories, Murray Hill, New Jersey, USA]

Materials Economics, Policy and Management

This article presents an overview of the entire range of topics relevant to materials economics, materials policy and materials management. The first part of the article addresses materials supply, demand, markets, prices and the government policies aimed at safeguarding supplies and stabilizing prices. The second part deals with specialized aspects of materials economics and policy such as process analysis, value added and energy requirements, materials and the environment, recycling and resource recovery, and the relationships of materials to productivity and innovation. The third part discusses management issues and problems in the materials-producing and materials-consuming industries. The problems discussed are materials transportation and materials handling, materials requirements planning and materials risk management, cost estimation, value analysis and product liability.

1. General Materials Economics and Materials Policy

Materials, like all resources, are limited in supply. As a result, some mechanism is required to allocate materials among competing uses. The principal alternatives are economic markets or government directives. This part of the article reviews the economics of materials markets and the use of government to influence these markets. This requires an analysis of materials use, substitution, supply and materials availability, the structure of the supplying industry, the geographical characteristics of materials markets, and the prospects for exhaustion of materials supplies in future decades.

The analysis of materials markets begins with the very nature of these goods. Typically, they are homogeneous, fungible products. They can be produced, classified and graded in such a manner as to make brand identification relatively meaningless. Moreover, materials are rarely final products, but rather intermediate products required for the production of other goods and services. Lastly, most materials are derived in some fashion from natural resources—they are extracted, grown or synthesized from products of the land. It is this derivation from natural resources that creates the concern over the sufficiency of supply or the availability of materials, particularly during national defense emergencies.

1.1 Materials Supply

Most materials policy issues involve the current or future supply of materials. There is always a concern that the prospective demand for materials may outrun supply in future years or generations. As a result, assuring an adequate future supply of materials is often an important policy concern.

The supply of most materials depends in part upon the availability of mineral deposits. The mineral resource base is therefore an important first element in analyzing the supply of these materials. Nonfuel mineral resources are limited—in finite supply—but this supply limitation depends upon current knowledge. Over decades or centuries, the size of the identified resource base of any mineral may increase substantially despite man's depletion of existing deposits.

The resource base of any mineral does not equate with economically feasible supply. Reserves of a nonfuel mineral are that subset of the resource base that may be extracted economically with known technology at current prices. An even smaller subset is the quantity of "proved" or "demonstrated" reserves—that is, those whose existence is known with some certainty. In addition to these demonstrated reserves are inferred reserves, reflecting a geological assessment of economically recoverable deposits based upon less certain evidence.

Renewable resources may appear at first thought to be in unlimited supply, but on closer examination this is not the case. In the long run, agricultural and forestry crops are limited by the available land. In the short run, they are vulnerable to weather, disease and parasites. There has been concern about future supplies of timber, especially of certain types and in certain parts of the world. The supplies of natural fibers and any materials derived from biomass are also limited in principle.

Minerals are produced from nonfuel mineral deposits or renewable resources by a variety of different industries. Often, firms in these industries are vertically integrated, that is, they combine the extraction, smelting, refining or other processing of a nonfuel mineral or the harvesting and processing of a renewable resource such as wood. As the technology of extraction or processing changes, these industries change. For instance, direct casting of thin sheets may lead to increased integration in the copper and aluminum industries. Similarly, revolutionary changes in electric furnace technology and direct reduction of iron ore may lead to a much less concentrated iron and steel industry.

Materials supply is often augmented by the reuse or recycling of obsolete products containing the material or manufacturing trimmings, turnings and rejected products. Recycling is particularly important in the metals industries, and is likely to increase in relative importance if economic growth proceeds more slowly than in the first three decades after World War II. Recycling is also important for other materials, such as paper, and even contributes to the supply of building materials (see *Recycling of Demolition Wastes*). In recent years, the slow growth in the demand for steel has seen scrap-based electric furnace production increase its share of total steel output.

Expanding supply in most materials industries requires large lead times because of the necessary investment in mining or processing. As a result, supply tends to respond slowly to sharp increases in demand—such as the 1973–74 boom—or to sharp downturns in demand—such as in the 1930s or early 1980s. This sluggishness in the response of supply, reflecting large fixed costs of production, is one reason why materials prices tend to be extremely volatile.

1.2 The Use of Materials: Demand and Substitution

Materials are generally used as inputs into the production of more finished economic products. Most have a number of substitutes, given sufficient time to implement the necessary technology and to construct the productive capacity. Moreover, final products themselves have substitutes. The development of these substitutes in response to rising prices of a material is a natural reaction to increasing scarcity (see *Substitution: Economics*). Rising materials prices induce a search for new technologies and result in the conservation of the remaining supply of the materials (see *Substitution: Technology*).

As societies develop, they tend to use relatively smaller quantities of materials per unit of final output or consumption. Moreover, they tend to use different types of materials, such as plastics for appliances or silicon for electronic components. With rising wealth consumers do not increase their purchases of durable goods proportionately; instead, they tend to spend their income increasingly on services, such as travel, recreation and amusement. The tendency of mature economies to consume less materials per unit of gross national product is well documented by Malenbaum (1986). This declining "intensity of materials use" is important in forecasting the demand for materials.

The extent of materials use in each sector of the economy is often modelled by input–output (I–O) analysis. Most investigations by I–O analysis assume fixed coefficients in production, but these coefficients can be changed to allow for technological change and in particular for substitution. Such analyses are useful for tracing the effects of major changes in materials usage through the economic system. Carter (1986) provides a detailed analysis of the use of materials in the industrial sector of the economy, relying heavily upon I–O techniques. She demonstrates that the changes in materials intensity derive in large part from different growth rates of materials-using industries.

1.3 Materials Markets: Structure

Materials markets have a number of distinctive features. Since most materials are used as inputs to further production, demand is likely to vary substantially over the business cycle. This "derived demand" is likely to be less price sensitive than the demand for most final goods. Varying prices and the large investment and inventory requirements create the need for organized futures markets in most major materials. Firms that must hold substantial raw-material or in-process inventories often wish to hedge the risks of price fluctuation in these materials.

The degree of competition in materials markets depends upon the minimum efficient scale of production, the size of the market and the cost of entry. Materials markets have become more international in scope in recent years (see *International Trade in Materials*). As a result, few materials firms enjoy sufficiently large shares of their markets to have monopoly power over prices.

In a few cases—such as uranium—cartels of competing producers have been formed. However, these cartels are often unstable and are susceptible to collapse owing to independent actions by their members. While the OPEC cartel has succeeded in raising crude oil prices for a decade, it has shown signs of crumbling in the 1980s. The cartel of bauxite producers has never had much success in raising the price of bauxite because of the number of potential competitors in the market. Nor have copper producers been able to cartelize successfully despite their limited numbers, in part because of an ample supply of scrap.

It is always difficult to delineate a market with precision because of the constantly developing substitution possibilities among materials. Rising relative prices of steel have led users to substitute aluminum, plastics and even concrete for steel. High real copper prices in the 1970s encouraged substitution of plastics and aluminum for copper. Because of the various substitution possibilities, the boundaries of a market are to a large extent determined by judgement, not by precise measurement.

1.4 Materials Markets in Action: Prices

The interaction of supply and demand generates a market-clearing price. This price in a materials market is likely to be quite volatile over time because of the inelasticity of supply and the variability of demand over the business cycle. Nevertheless, economic theory suggests that the secular trend of materials prices should be upward. The rate at which materials prices rise over time should be the interest rate, assuming unchanging technology and information concerning deposits of the natural resources. Prices may be volatile, but their general trend should be upward in the absence of technological changes or new discoveries of reserves. However, new discoveries have generally been sufficient to depress the real price of most materials over the long run (Lovering 1969).

In fact, materials prices have not trended upward over the past century. Detailed research by Potter and Christy (1962), by Manthy (1978) and by Landsberg et al. (1963) demonstrates that most nonrenewable resources have shown a downward real price trend during this period. The reasons for this are obvious: technological change in extraction and processing as well as new resource discoveries. Were these technological breakthroughs to subside and new discoveries of resources to become less frequent, natural resource prices might begin a long upward trend as economic theory predicts.

A quadratic function appears to fit the time series data for many metals prices rather well. The general downward trend through the last decades of the nineteenth century and the early decades of the twentieth century has slowly given way to a rising trend. These U-shaped relationships may, however, be only a reflection of the sharp rise in materials prices in the early 1970s.

The prices of most "renewable" resources have shown a tendency to rise over several decades. For instance, the price of the average forest product trebled in real terms between 1875 and 1975 (Manthy 1978). Most timber products prices rose more rapidly than the general price level in the three decades after World War II. The principal exception was the pulpwood used in the manufacture of paper. Similarly, the prices of most natural fibers have increased in the inflationary 1970s, but they have begun to decline in the early 1980s as consumption growth has slowed. Paper prices, however, have been much more stable, rising very little in real terms since World War II.

If most nonfuel minerals prices begin to rise at rates exceeding the inflation rate for a considerable period of time, there will be renewed concern over minerals exhaustion, leading governments to begin to reassess their materials policies.

1.5 Materials Policies: Sufficiency of Supply

For generations there has been concern about depletion and exhaustion of mineral deposits. Generally, such concern follows periods of extended economic boom or a period of war. Soon after World War II, the Paley Commission reported on the US government's options for assuring sufficient supplies of materials (US President's Materials Policy Commission 1952). Interest waned, however, until the early 1970s when a commodity boom and the OPEC cartel once again created general concern about the future supply of materials (National Commission on Materials Policy 1973).

In the USA, federal government minerals policies involve the leasing of mineral rights on federal lands, the implementation of environmental policy and the use of tax policy to stimulate exploration and production of minerals. With little formal government ownership of minerals industries, the US government has little direct role in determining the rate at which mineral deposits are exploited. However, through its leasing policies on federal lands—particularly for coal and timber—the government can have a substantial effect upon some materials industries.

Western European countries are typically more dependent upon imports for their minerals than is the USA. In some countries, such as France, the government has become quite concerned about the growing role of imports. As a result, France has placed increased emphasis upon recycling and conservation.

Most European countries provide some protection for their minerals industries, particularly their coal industries. But in the UK and the Federal Republic of Germany, there is very little else in the nature of a formal policy toward materials supply or self-sufficiency.

Japan, which has very few mineral deposits, relies heavily upon imported raw materials. This is especially true for energy resources and metal ores. As a result, the Japanese government encourages Japanese firms to invest in foreign materials industries and participates in programs to assure the security of materials flows from foreign sources. Low tariffs on imported raw materials and encouragement of domestic stockpiling are among the more important of these policies.

Unlike Japan, Canada is a large net exporter of materials. Its mining sector accounts for about 6% of total employment and 5% of its gross national product. During the period of rapid development of its mineral industries, foreign investment played a major role. In recent decades, Canada has pursued a policy of attempting to reduce foreign control and ownership of its minerals industries, but in the 1980s, renewed concern for optimal development of the country's minerals industries has led to a somewhat relaxed view of foreign investment and a heightened concern for competing in world export markets.

Australian materials policy considers resource development and raw materials production from the viewpoint of the competing needs of the country's domestic manufacturing industries and materials exports. Foreign investments and resource ownership are attracting particular attention. Conservation and, in the case of uranium, international political considerations are also concerns. The two major political parties generally favor different policies on most of the issues. Changes of political parties in power have resulted in a materials policy characterized by several reversals in recent decades.

The 1980s have brought a changed world economic climate from that of the 1970s. Concern over sufficiency of minerals supplies appears to have diminished with declining economic growth. With slow growth in the developed countries and major debt problems in the developing countries, concern has shifted from that of rationing scarce supplies to exploiting and developing resource deposits to allow mineral-rich countries to grow. This is especially true in many developing countries that are heavily dependent upon sales of basic fuel and nonfuel minerals.

1.6 Materials Policy: Price Stabilization and Stockpiling

The decade of the 1970s was one of extreme price volatility for many materials and particularly for fuels. This had led to an increased interest in government policies to mitigate such price volatility or at least to dampen the oscillations. In addition, governments have become concerned about maintaining sufficient supplies of critical or strategic materials.

Behrman (see *Prices of Materials: Stabilization*) demonstrates that there can be substantial gains in economic efficiency if governments are able to stabilize materials prices. Among the mechanisms for achieving such stabilization are buffer stocks, direct price controls, and policies to stabilize production levels.

Public policies towards critical and strategic materials have a long history. In the USA, energy reserves were established before World War I because of security considerations. More recently the USA has attempted to maintain stockpiles of a number of strategic materials such as titanium, nickel, tungsten, cobalt and rubber. This policy became important during and immediately after the Korean War, but subsided in the 1970s and 1980s. In the mid-1970s, after the sharp rise in commodity prices in 1971–1974, the US Congress established the Commission on Supplies and Shortages to examine public policies for reducing price fluctuations in future business cycles.

Many other countries also maintain stockpiles of strategic and critical materials. France and the UK have modest public stockpiles, while Japan, Sweden and Switzerland pursue government-supported private stockpiling policies.

One of the most difficult problems in designing a stockpiling policy is identifying the truly critical and strategic materials. Most definitions of critical materials focus on the lack of substitutes for particular uses that are central to maintaining a national defense capability. However, there are substitutes for most materials, particularly with sufficient lead time. Identifying the speed at which substitution may take place is therefore very important.

Public stockpiles maintained for the purpose of limiting price volatility are even more difficult to manage. Private firms maintain large inventories of most materials to smooth out production and to ensure against price fluctuations. Public stockpiles of materials may simply reduce private inventories and provide little additional protection against supply disruptions or other sources of price volatility.

Another problem with public stockpiles that are designed to reduce price volatility involves the timing of purchases for and sales from the stockpile. Optimal decision rules for reducing price volatility are complex; moreover, political pressures often prevent public authorities from using stockpiles optimally. Sales or purchases designed to smooth out prices have obvious effects on private sellers and buyers and inventories of any material.

2. *Topics in Materials Economics and Materials Policy*

There are a number of specialized topics in materials economics and materials policy that do not fit neatly into a general analytical framework of supply–demand–prices. These include process analysis, modelling, environmental issues, recycling and resource recovery, productivity and innovation. This section surveys some of the relevant concepts and ideas and serves as an introduction to the articles involving each of them.

2.1 Process Analysis: Value Added and Energy Assessment

Materials-producing and materials-using industries differ from most other economic activities—such as agriculture, transportation, communication and finance—in that they employ physical processes in which one or more materials flow through a progressive set of linked processing stages from raw materials to final product. The processes are investigated and assessed by process analysis. Process analysis lends itself to systematic quantification of the important variables in a variety of decisions. In this way, it contributes to allocation and investment decisions as well as to process design and control at levels ranging from unit operations and single plants to the level of an entire economy.

A manufacturing plant typically produces more than one product. Similarly, many materials-processing plants produce more than one material. In this context, the distinction between by-products and co-products is relevant.

The physical progress through the various operations depicted by a flowchart is accompanied by a cumulative buildup of value embodied in the material or workpiece. The value added across all industries can be aggregated into their contribution to the gross national product.

A buildup of the cumulative energy expended in the production and conversion of materials is also possible. A complete energy assessment, however, includes the energy used in conjunction with the use of the material and its disposal after discard. Recycling of the material often results in an energy saving compared with the use of primary material. Such an analysis is a life cycle energy evaluation.

2.2 Modelling

Since World War II, there has been a considerable increase in the interest in developing models of individual materials markets, the materials sector of the economy and the entire economy. In part, this has been facilitated by the remarkable improvement in quantitative techniques and the equally rapid improvement in computers. As the cost of developing large models has declined, business and government demand for them has increased. Businesses often develop their own models in order to make production, inventory and investment decisions. Governments use models to forecast prices and supplies so as to aid in stabilization policies, sectoral policies or national defense policies.

Clark and Busch (1986) identified three types of models: econometric models, engineering models and input–output models. In principle, these models are substitutes for one another, but they are often used for different purposes. Econometric models are generally used by firms or governments to predict prices, output and investment. Engineering models are more often used to analyze the response of individual production units to changes in materials availability, prices or product demands. Input–output models are generally more aggregated than engineering models and are used to trace the impact of structural changes—such as supply stocks or shifts in demand.

In the early 1970s, a new form of dynamic model was developed, using techniques drawn from electrical engineering. This type of model involves complex feedback loops that are thought to characterize a variety of social systems from individual markets to the world economy. These models are often developed through a series of simulation exercises on high-speed computers in order to place values on the important parameters. Probably the most publicized of these exercises was the report to the Club of Rome by Meadows et al. (1972) that related scarcity of materials, congestion and pollution to the deceleration of world economic growth.

2.3 Materials and the Environment

Materials production and consumption can cause environmental degradation unless government establishes the appropriate regulatory rules of property rights. Many production processes—such as mining, extraction, refining or smelting—pollute the air, water or land. Wastes or "residuals" are discarded from both production and consumption. Discards by consumers include a wide range of materials from packaging materials and newsprint in municipal solid waste to the materials in discarded automobiles and discarded capital goods. This waste generation creates a variety of problems for public policy that require technical solutions (see *Residuals Management*).

The immediate or ultimate repository of residuals is the environment. Its absorptive capacity depends on the nature of the discarded material, but it is always limited. The environment is a public good that is usually outside the market pricing system and is therefore subject to excessive use unless it is protected or appropriately rationed by government (see *Materials and the Environment*). Protection can be brought about by regulations that reduce the immediate impact or by the appropriate use of property rights (see Coase 1960).

In recent years, economists have advocated pollution taxes or marketable pollution permits to reduce environmental pollution (see Montgomery 1972, Crandall 1983). Long-range policies reducing residuals generation and encouraging materials recovery can also contribute to environmental protection. Any policy to control environmental pollution has major impacts upon producers and consumers. Any attempt to change regimes from no control to regulation or to a system of private property rights can be disruptive to private economic agents (see *Residuals Management*).

2.4 Recycling and Resource Recovery

Recycling is the recovery of discarded materials for reuse. Complete recovery is generally not possible. It is not technically or economically feasible to recycle most materials used in dispersive and dissipative applications (see *Materials and the Environment*). As an exception, small amounts of dispersed materials can be co-recovered in the recycling of associated major metals; an example is the recovery of solder in the recycling of automobile radiators (see Alward et al. 1976).

By its very nature, recycling contributes to the conservation of materials (see *Materials Conservation: Technology*). If recycling is conducted efficiently, it generally also yields savings of equipment, pollution abatement expenditures and energy. The use of ferrous scrap in steelmaking in place of hot metal produced in a blast furnace is a case in which all three savings are realized. In addition to its contribution to conservation, the recycling of materials produced and discarded in large volumes also achieves waste disposal; this is especially true of ferrous scrap and paper and paperboard. An extreme case of waste disposal versus materials conservation is resource recovery from municipal solid waste. Often, by-products are captured that would harm the environment if discharged as residuals.

The recycling industries should be viewed like other materials-producing industries. Their raw materials are of a special kind: residuals from production and residuals from consumption. The quality of these raw materials depends on their physical form and chemical composition. Both the absolute price of the material produced and the price of scrap relative to primary raw materials determine the extent of reycling. This explains the highly developed state of nonferrous metals recycling (Bever 1978).

In contrast to the recycling of metals and other individual materials, resource recovery from municipal solid waste is driven by the necessity of waste disposal. Institutional barriers and declining materials prices have slowed the development of resource recovery. The course of its development was also affected by the rise of energy prices in the 1970s which made incineration with energy recovery viable; in this way, incineration contributes to the economics of resource recovery but also competes with it for raw materials.

2.5 Productivity and Materials

Productivity is a measure of the ratio of output to inputs of labor, capital or materials. The most common measure of productivity is output per man-hour, but this measure does not capture the full effects of changes in each of the productive factors. Total factor productivity is the ratio of output to a weighted average of all inputs used to produce it.

In most manufacturing industries, including the materials industries, labor productivity has grown more rapidly than total factor productivity because capital has been substituted for labor. However, it has been argued that decision-makers within a firm are likely to be unable to measure total factor productivity because of the difficulty of estimating the rate of change of all productive factors. As long as capital is adding rapidly to productive capacity, labor productivity is likely to grow rapidly, but once capital investments are made simply to maintain the capacity of existing plants, productivity growth slows.

Productivity is important in materials industries in several ways. Productivity growth in these industries themselves tends to reduce the real costs of materials over time. Improvements in technology have led to increases in the productivity of materials used as inputs by materials-producing industries and other industries. For example, continuous casting and improved rolling techniques have increased the ratio of finished to crude steel in the steel industry.

Materials usage has declined in many processing and downstream manufacturing industries owing to improvements in technology and production practices. Improved materials can also generate increases in productivity of other inputs in materials-producing industries and in downstream-using industries (see *Materials and Innovations*).

2.6 Materials and Innovation

Innovation increases man's ability to produce a variety of products. It may reduce the cost of producing a current product or yield an entirely new product that is useful in itself. This new product may provide a superior substitute for an existing material, such as the substitution of plastic for lead in plumbing, or it may facilitate the introduction of entirely new final products, such as memory devices used in small computers.

Through the first quarter of this century, materials innovation largely concentrated on metals. Beginning in the 1930s and throughout the next several decades, research on synthetic polymers was greatly accelerated. This led to a large number of new products. The development of synthetic rubber because of the shortage of natural rubber during World War II was an especially important achievement of polymer research. Since World War II, research on nonsilicate inorganic oxides has led to the development of a new generation of ceramics. Continued refinements in the technology of producing and using semiconductor-grade silicon have been crucial to the rapid progress of the electronics industry (see *Materials and Innovations*).

Innovation is driven by the desire to become more competitive through lower costs or improved products. Since the resulting advantage tends to be lost in time as competitors either adopt the innovation or invent new products and processes of their own, the stimulus for innovation persists.

3. *Materials Management*

Materials-related management functions include the physical movement of materials, the planning of required supplies and inventories, the estimation of materials costs, value analysis and the disposal of production wastes, such as metallic processing scrap. These functions will be surveyed in this section.

3.1 Materials Transportation and Materials Handling

The physical inputs and outputs of materials-producing and materials-converting industries must be available in the required amounts at definite times and in definite locations. Fundamentally, the locations involved in the movements of materials, both the origins and destinations, are determined by factors of economic geography. The movement of materials, especially over short distances and within plants, is the object of materials handling.

3.2 Materials Requirements Planning; Materials Risk Management

Materials requirements planning is a technique for managing inventories and controlling the replenishment of materials, components and subassemblies; it is a module in the total manufacturing control system (also referred to as manufacturing resource policy). Materials requirements planning depends in part upon the ability to process large amounts of information. For this reason, the computational speed and memory capacity of computers have been important in developing materials requirements planning in the period since 1960.

Materials requirements planning needs an accurate bill of materials and a master schedule for the final products. A combination of four steps generates planned orders for the purchases of materials and parts. As one way to provide protection for uncertainty, a minimum inventory quantity called a safety stock is specified; this is considered always to be required when calculating net requirements. It ensures materials availability under conditions of uncertainty. An alternative method employs a safety time provision. Each of these methods has its own advantages—the safety time method results in lower average inventory investment when the primary uncertainty is in timing and the safety stock method is preferable when the primary uncertainty is quantity.

Materials risk management is concerned with the strategies by which a firm can try to cope with material scarcities, especially scarcities caused by supply interference of an international nature. It is thus the private-sector equivalent of stockpiling by government and was referred to in this context earlier in this article (Sect. 1.6). Materials risk management is based on vigorous assessments of the firms' need of critical materials; various companies have developed detailed analytical procedures toward this end.

The "just-in-time" method of managing the supply primarily of components to an assembly-type manufacturing operation is related to materials requirements planning. It has received much attention in the US automobile industry's attempt to become competitive with Japan.

3.3 Cost Estimation and Value Analysis

Materials costs are one item in the estimation of production and manufacturing operations. The mineral content of ores can be an important factor in projecting their processing costs; this can be used in the parametric method of cost prediction. In the industrial engineering method, materials costs are assessed as an independent item. This should favor the use of this method for projects in which materials costs are a major consideration.

In life cycle costing, materials can be a decisive factor. As a typical example, a more expensive material, such as stainless steel, may result in lower life cycle costs because it requires less maintenance and repair than competitive materials. More expensive materials may be more economic because of the increased durability of products made from them.

In an analogy to life cycle costing, a life cycle energy assessment may favor a lighter, more energy-intensive material because of energy savings during use. This is well illustrated by metals used in transportation applications. The energy saving can be translated into a life cycle cost saving.

Value analysis measures the absolute value of a material characteristic, or other input feature, in terms of the benefit derived from it and compares it with the material's cost. This value may relate to costs of production or to advantages in the market place. In the latter case, the value may be better quality leading to larger sales; the cost is the cost of producing a higher-quality product.

3.4 Product Liability

In recent years in the USA, manufacturers have become more concerned about their exposure to product liability suits. As courts move increasingly toward a standard of strict liability and accept weaker evidence as proof of damage to plaintiffs, firms have been forced to reduce their exposure to suits by increasing product quality. In part, this is manifested through an increased demand for high-quality materials that will not expose consumers or workers to risk regardless of the nature of their use. These changes have taken place in part because of the contingency-fee system employed by lawyers and the increasing concern over risks with long latency periods.

Product liability suits have increased in number, and the magnitude of the settlements has increased even more rapidly. Large class-action suits such as those involving asbestos have caused manufacturers to become much more risk averse in offering new products.

Bibliography

Alward J R, Clark J P, Bever M B 1976 The dissipative uses of lead, *Proc. Council of Economics*. American Institute of Mining, Metallurgical and Petroleum Engineers, New York, pp. 17–26

Bever M B 1978 Systems aspects of recycling. *Conserv. Recycling* 2: 1–17

Bever M B 1986 *Encyclopedia of Materials Science and Engineering*. Pergamon, Oxford

Brooks H 1972 Materials in a steady state world, ASMTMS Distinguished Lecture. *Metall. Trans.* 3: 755–68

Carter A P 1970 *Structural Change in the American Economy*. Harvard University Press, Cambridge, Massachusetts

Carter A P 1986 Materials in the industrial system. In: Bever M B 1986, Vol. 4, pp. 2838–44

Clark J P, Busch J V 1986 Materials markets: Modelling. In: Bever M B 1986, Vol. 4, pp. 2855–61

Coase R H 1960 The problem of social cost. *J. Law Econ.* 3: 1–44

Crandall R W 1983 *Controlling Industrial Pollution*. The Brookings Institution, Washington, DC

Fischman L L 1980 *World Mineral Trends and US Supply Problems*. Johns Hopkins University Press, Baltimore, Maryland

Herfindahl O C, Kneese A V 1974 *Economic Theory of Natural Resources*. Merrill, Columbus, Ohio

Landsberg H C, Fischman L L, Fisher J A 1963 *Resources in America's Future*. Johns Hopkins University Press, Baltimore, Maryland

Lovering T S 1969 Mineral resources from the land. In: Cloud P (ed.) 1969 *Resources and Man*. Freeman, San Francisco, California, pp. 109–34

Malenbaum W 1986 Intensity-of-use of materials. In: Bever M B 1986, Vol. 3, pp. 2355–61

Manthy R S 1978 *Natural Resource Commodities—A Century of Statistics*. Johns Hopkins University Press, Baltimore, Maryland

McMains H, Wilcox L (ed.) 1978 *Alternatives for Growth: The Engineering and Economics of Natural Resources Development*. National Bureau of Economic Research (Ballinger), Cambridge, Massachusetts

Meadows D, Meadows D, Randers J, Behrens W 1972 *The Limits to Growth*. Universe Books, New York

Montgomery W D 1972 Markets in licences and efficient pollution control programs. *J. Econ. Theory*. 5: 395–418

National Commission on Materials Policy 1973 *Materials Needs and the Environment Today and Tomorrow*. US Government Printing Office, Washington, DC

National Commission on Supplies and Shortages 1976 *Government and the Nation's Resources*. US Government Printing Office, Washington, DC

Potter N, Christy F T 1962 *Trends in Natural Resource Commodities*. Johns Hopkins University Press, Baltimore, Maryland

Smith V K (ed.) 1979 *Scarcity and Growth Reconsidered*. Johns Hopkins University Press, Baltimore, Maryland

US President's Materials Policy Commission 1952 *Resources for Freedom: A Report to the President*. US Government Printing Office, Washington, DC

Vogely W A (ed.) 1975 *Mineral Materials Modeling—A State-of-the-Art Review*. Resources for the Future, Washington, DC

Vogely W A (ed.) 1976 *Economics of the Mineral Industries*, 3rd edn. American Institute of Mining, Metallurgical and Petroleum Engineers, New York

M. B. Bever
[Massachusetts Institute of Technology,
Cambridge, Massachusetts, USA]

R. W. Crandall
[Brookings Institution, Washington, DC, USA]

Materials: History Before 1800

Man has existed on earth for about 2.5 million years and has used materials for almost as long. Materials were no less important to prehistoric man than to his descendants of today. He needed materials to obtain and prepare food, for clothing and shelter, for tools and weapons (Fig. 1), and for adornment and artistic expression.* Initially man had to content himself with natural materials: wood, stone, bone, shell, feathers, straw, fibers, clays and other earthy deposits. Despite this apparent limitation, man over the millenia became enormously sophisticated in his ability to select from the available materials the best for the particular application at hand and to work these materials with the tools and forces available to convert them to his needs and in so doing to take best advantage of their internal structure and properties. In these terms we do precisely the same today but now proceed from elemental or other derivative forms of materials rather than only those forms which nature provides.

In this brief article we can no more than sketch an outline of this kaleidoscopic development. Two major limitations of scope have been applied. First, concentration is almost exclusively on the technical aspects of the subject leaving aside the equally important and fascinating economic, political and artistic history of materials. Secondly, the history is brought only to the time of about 1800, a point early in the so-called industrial revolution; the history of materials since that time is reviewed elsewhere (see *Materials: History Since 1800*). Figure 2 presents a summary chronology of the history of materials over this time period. Individual sections of the present article treat metals,

* Smith (1981) makes a major thesis of the idea that the selection and manipulation of materials for their decorative and sensual effects exerted an important influence on the development of their technology.

Figure 1
A fine hand axe from Swanscombe, England, dated at 220 000 BC, found only 3 m from the skull bone of a representative of Swanscombe Man, one of the earliest specimens of Homo Sapiens. The stone was first struck to detach a large flake leaving a rough edge. A soft hammer of wood, antler or bone had to be used to produce a very sharp and straight edge travelling all the way round the axe (after Cipolla and Birdsall 1980 from original in the British Museum)

ceramics and glass, cement and concrete, fibrous materials, paper, materials processing and the development of the understanding of materials behavior.

1. Metals

It is interesting to speculate about the very first discovery of a metal by man. It almost certainly was a native metal: gold, copper, silver and iron all occur—gold and copper rather commonly, silver rarely and iron only as a meteoritic iron–nickel alloy. This first discoverer was undoubtedly attracted by the color of this new stone, contrasting with the drab surrounding rocks if it were either gold or copper, and possibly by the luster as well if it were gold. He was even more surprised when he tried to make something of it by smashing it with another stone and found that this bright, colored stone deformed rather than shattered under the blow. From this time, somewhat over 10 000 years ago, slowly over the next six or seven thousand years, man learned that these materials—metals (from the Greek *metallam*, to search)—were uniquely different from the natural materials with which he was familiar in many, many ways.

Consider the following general qualities of metals:

(a) they are readily shapeable via such processes as forging and casting;

(b) they are durable in the sense of being relatively inert to the ravages of fire and flood;

(c) they represent high value in small weight or volume and thus can serve as a medium of exchange (coinage) or as reserve assets:

(d) they are repairable in a sense that no stone, wood or ceramic artifact can begin to approach;

(e) they are convertible in form and function (e.g., "swords into plowshares");

(f) their properties are controllable by mechanical treatment, by thermal treatment and by alloying and diffusion treatment;

(g) they can be made bright and lustrous, hence decorative or even useful as mirrors;

(h) they are sonorous, hence suitable for gongs, cymbals and bells; and

(i) they lend themselves to various types of compositing (e.g., gold-leaf, plating, weldments).

Figure 3 illustrates an early use of some of these qualities in the frieze on Im-dugud.

It is therefore no wonder that early man persevered in his search for these extraordinarily useful materials and in his experimentation with them.

Several origins have been postulated for the discovery of the smelting of metallic ores to yield the metal: the accidental heating of a colored stone (a copper ore) in an open fire under reducing conditions; the heating of malachite during a potter's glazing operation; or the knowledge that heating could alter the color of ocher (ocher has been used by man as a pigment for 300 000 years!) and hence perhaps other colored, and more readily reducible, stones. Regardless of the stimulus for the discovery, once the trick was learned and repeatable, it must have been truly awe-inspiring to witness the intentional action of transmuting a rather ordinary rock into such a totally new material.

The several aspects of metal technology mentioned thus far—the difficulty of finding appropriate ores for smelting, the special furnaces required to get consistent results and the considerable judgment and manipulative skill needed—combined to exclude metals producing and working from a part-time home occupation and to establish it as perhaps the first full-time craft occupation. It is also obvious why metallurgical developments occurred in conjunction with centers of civilization rather than in nomadic or hunter/gatherer societies. Only in settled societies could surplus food be generated from irrigation and other communal activities so as to support a full-time

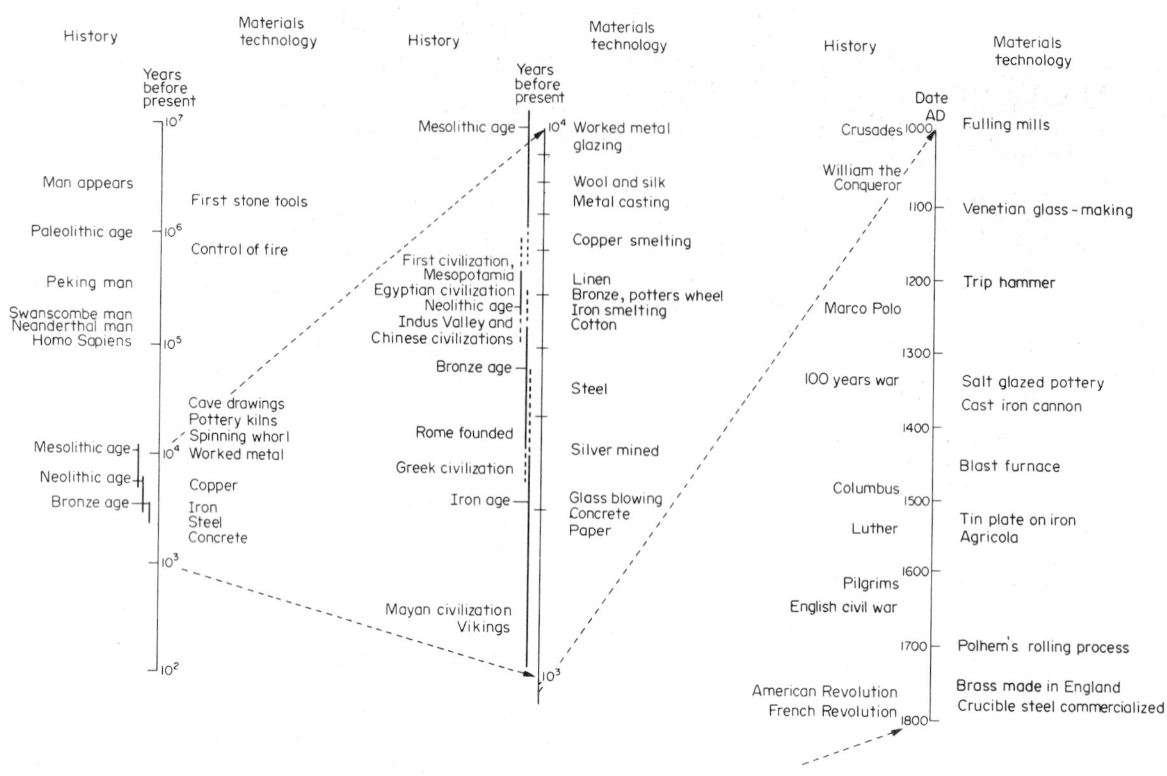

Figure 2
A chronological perspective of the history of materials. The flow of time is from top to bottom on each of the three scales. The left and middle scales are necessarily logarithmic to accommodate the exponentially increasing rapidity of technological development. The middle scale is a blow-up of the period from 10 000 to 1000 years ago, while the right hand, linear scale summarizes events over the 800-year period following 1000 AD

craftsman. The relative richness of the urban centers also provided a market for the metalworker's products and the capital necessary to support the increasingly diverse and complex equipments his trade required. In time, as the quest for ores and fuel went further afield, mining and smelting became a separate profession from metalworking or smithing. The name Smith in other languages is rendered: Kovacs, Haddad, Kowel, Gowan, Herrara, Schmidt, Fabricius, Kuznetzov and Le Fevre. The commonality of these names is indicative of how widespread and important the metalworking craft became.

The oxidic copper ores were presumably the first smelted, both because they were accessible near the surface of the ground and because such ores were frequently brightly colored; for example, cuprite, malachite and azurite. This discovery is presumed to have first been made in north-east Iran between the Caspian Sea and the Elburz mountains about 4300 BC. With the exhaustion of these oxidic near-surface ores in a given deposit, the miners would come upon the lower lying copper sulfides, frequently contaminated with the metalloid, arsenic. Although somewhat different smelting practice was required, the metalworker soon found that the new ore yielded not simply copper but a superior form of copper in that it was both more readily castable and also capable of more extensive work-hardening. Thus was born the first alloy family, The Cu–As series. At a later time, about 3000 BC, by some circumstance not yet determined, it was found that a similar result could be obtained by adding a tin ore during the copper smelting, hence the tin bronzes.

Iron is supposed to have been first produced accidentally during the smelting of copper or lead ores when iron oxide had been used as a flux, probably in Mesopotamia about 2700 BC. The discovery was much more chancy than in the case of copper because the product "bloom" would have been a spongy, impure mass. Only when this mass had been repeatedly reheated and hammered would the slag be ejected and the globules of iron consolidated into a solid lump. The use of a shaft furnace is depicted on the Greek vase shown in Fig. 4.

While copper was readily shaped by successive cold working with intermediate anneals, iron required hot

Figure 3
The frieze of Im-dugud, found on the site of the settlement of Al'Ubaid and dating from the first dynastic period of Ur, ~2700 BC. The whole frieze is covered with copper, the metal having first been hammered into sheet and afterwards beaten to fit closely on a carved, wooden foundation which was surrounded by a copper frame. The copper sheets were secured in position on the wood with copper nails. This remarkable decoration represented an eagle, having a lion's head, holding two stags by the tail. The stags' antlers, also made of wrought copper, were developed in high relief and were soldered into their sockets with lead. The complete panel is 7 ft $9\frac{1}{2}$ in. long and 3 ft 6 in. in height (after Aitchison 1960 from original in the British Museum)

Figure 4
A shaft furnace from Greece, depicted on a vase showing a smith forging an iron bloom in front with the bellows behind (after Tylecote 1976 adapted from Blümmer 1886–87 *Technologie und Terminologie der Gewerbe und Künste bei Griechen und Römern*, Vol. 4)

working to effect substantial deformation without fracture. The transition from iron to steel was first realized by the Chalybes, a subject people of the Hittites, living south-east of the Black Sea about 1400 BC. These people learned that by heating wrought iron in a charcoal fire it became much harder than bronze. Although individual steel artifacts have been found from several hundred years BC which were apparently hardened in part by a quench and temper heat-treatment, consistent results were not obtained until Roman times.

The metals of antiquity were but seven: copper, tin, gold, silver, iron, lead and mercury, although the presence of others such as nickel or arsenic (a metalloid) was recognized through the beneficial behavior of certain ore additions in copper smelting. Recognition of other base metals was slow in coming. Brass was made by the Romans by the calamine process for coinage in 45 BC and Bactrian coins of 170 BC were apparently made by smelting a mixed copper–nickel–zinc ore to yield an alloy similar to the Chinese paktong or modern nickel–silver; yet zinc was not regularly produced, at least in the Western world, until Champion in 1736. Zinc is really the only metal of commercial importance to be added to the seven of antiquity before the close of the period under review in this article (~1800). This is not to say that no progress was made in developing a fuller recognition of the metals family. However, it must be realized that identification of a new elemental metal almost always preceded isolation, and production of substantial quantities for either experimentation or application required decades more. Nonetheless, by 1800 the additional metallic elements that had been recognized were, in order of their discovery: Pt, Sb, As, Zn, Co, Ni, Bi, Mn, Mo, W, Te, U, Zr, Ti, Y, Be and Cr.

Mention has already been made of purposeful alloying of metals as with Cu–As and Cu–Sn. A few further remarks are in order. Some alloying was done not so much to improve the service behavior of the product metal but to facilitate processing. Thus we find at a very early date lead added to tin bronzes to improve castability. Mercury was used at least as early as the Greeks to recover gold and silver and, from mercury amalgams, base metals and alloys could be readily gilded with pure gold, the mercury being driven off by heating. Solders, not only Pb–Sn but also Cu–Au and Cu–Ag, were other important alloys of antiquity. It is also remarkable that the quantitative aspects of alloying to control properties were appreciated very early. For example, in the Cu–Sn system, rivets had a composition of 2–4% Sn (to be soft enough to head readily), while 8% Sn was chosen for greater strength in tools and weapons, bell metal of 18–24% Sn for castability and sonority, and 23–28% Sn for the hardness and optical qualities needed for mirrors. In all these examples the preferred compositions were well established before the beginning of the Christian era. Early alloy compositions were not confined to simple binary systems. To cite only a single, but surprising, example, the familiar Ag–Sn–Hg dental amalgam was described by Chinese writers as suitable for this purpose about 650 AD!

2. Ceramics and Glass

The term ceramics comes from the Greek *keramos* meaning potter's clay. Clay has been defined as a hydrated earthy mineral, principally consisting of various proportions of alumina and silica and showing the property of plasticity. This property derives from the platelike character of the minerals and the water of hydration. Clays are formed in nature by the decomposition of certain minerals, particular the feldspars. They may be classified into: (a) primary clays—the result of in situ decomposition by the action of carbon dioxide and water on surfaces from which air is excluded, and (b) secondary clays—those transported to another site. The secondary clays are the more widely distributed over the earth's surface, contain more impurities, and generally have a finer particle size. The primary clays usually have too little plasticity for practical use. They must be washed and purified by a sequence of settling operations and then perhaps mixed with a secondary clay and organic matter. The moldability of clay was discovered very early by prehistoric man and its hardening by fire almost simultaneously (a 25 000-year-old kiln site has been discovered in Czechoslavakia) as when a fire was built on or near clay. In this sense, clay is perhaps the first material whose properties were deliberately altered by man.

Egyptian pottery, evidently hand formed, predates the potters' wheel, the earliest example of which has been found in Mesopotamia from the Urok period (prior to 3000 BC). The key steps in the making of pottery are the selection and preparation of the clay, forming, and baking in the sun or in a furnace. Upon firing of a clay body, little change occurs other than the removal of free water until a temperature of 450 °C is obtained, at which point the chemically combined water begins to be removed and significant shrinkage occurs. Such temperatures are readily attained in an open fire. A kiln, although requiring a capital investment justifiable only by mass production for an urban market, yields more uniform heating and with draft controls can reach 1100–1300 °C. At such temperatures crystallographic transformations occur in the mineral species and some glass is formed from the silica, soda and other impurities leading to a considerable increase in strength, imperviousness and (in some materials) translucency.

A signal contribution of the Chinese to ceramic technology was the development of porcelain, a whiteware characterized by being both vitreous, sonorous and translucent. It thus stood in contrast to earthenware which was porous and opaque and to stoneware which was vitreous but not translucent. The body of hard Chinese porcelain is formed from the clays, kaolin and petuntse or china-stone, a crushed feldspar and quartz. Thus petuntse consists of aluminum silicate combined with potash, soda and lime which upon firing fuses with the kaolin to convert the whole to a (macroscopically) homogeneous, vitreous product. Impurities, especially iron, must be strictly controlled or the desired whiteness will not be achieved. The first true porcelains were introduced in the Tang dynasty (618–907 AD), but the culmination of the art was not reached until the early years of the Ming dynasty in the mid-fourteenth century. The high regard of the Western world for this product resulted in extensive imports in the seventeenth and eighteenth centuries and the application of the name *China* to such ware. European duplication of Chinese porcelain was achieved at Meissen, near Dresden, in 1709, but for the full impact of this rediscovery to spread throughout Europe took another 50 years.

Glazes were applied to pottery objects as early as 7000 BC (Egyptian faïence), either for decoration or waterproofing purposes. By the use of various mineral additions a considerable variety of colors could be produced. Cobalt compounds were used in Mesopotamia to create blue glazes and glasses by 2000 BC. Furthermore some oxide colorants, e.g., iron oxide, could even be altered by controlling the firing conditions from oxidizing to reducing to achieve different colors. (The Greeks during the fifth century BC were particularly adept.) The marvellous early Chinese glazes with deep opalescence in greenish yellow or yellowish blue hues were an attempt to reproduce the appearance of jade. The Romans developed glazes containing PbO which flowed at lower temperatures than the soda-lime–silica glasses and were applied to fired pottery and subjected to a second firing.

From glazes which flowed readily and could be produced in a variety of colors, it was only a small step to produce whole articles of glass—at first beads and other small items and then containers and decorative ware. In Egypt, glass was made from washed sand and a naturally occurring mineral, natron, a mixture of sodium carbonate and sodium bicarbonate. The Phoenicians later found that wood ashes could be substituted for the imported natron. The earliest glass-forming operations were pressing and casting in molds, but with increasing knowledge and skill and particularly with the availability of iron blowpipes, glass blowing became generally feasible about the beginning of the Christian era. The world center of glassmaking art and technology moved from Phoenicia to Rome about 100 BC. The Romans not only developed many artistic techniques but also new products: prisms, window glass and magnifying glass. China independently developed its own glassmaking industry in the third century BC but also imported glass products and glassmaking technicians. By about 1100 AD, Venice became the preeminent glass center, a position she was to hold for several centuries. Figure 5 illustrates sixteenth-century glass makers at work.

Despite the development of great skill in the handling of glass, and a wide variety of techniques, colors and compositions, glass technology remained essentially a hand operation until well into the nineteenth century. Neither power nor machine controls played any significant role (see *Glass: An Overview*).

Figure 5
Sixteenth century glass makers at work around their domed furnace. Molds, pontils and blowpipes are all clearly recognizable (after Agricola 1556 *De Re Metallica*)

3. *Cement and Concrete*

Dry-laid mudbrick or tamped earth walls obviously lacked much in both strength and durability. New techniques had to be devised for effective masonry construction. Structures of sun-dried or fired brick were bonded in ancient times with a clay slip, perhaps mixed with straw reinforcing, or occasionally with bitumen, while the earliest stone edifices could be laid up without mortar by virtue of very careful dressing of the stones at the joints. Cement by definition is a compound which, under some conditions, is plastic while under other conditions becomes strong, self-bonding and adherent to other materials. Most inorganic cements are based on various combinations of lime, alumina and silica. Selected rocks after crushing and burning are then in such a state that they will take up water and are converted to hydrated calcium silicates and aluminates and crystalline calcium hydrate. These lime mortars were first used in Hellenistic times; the Egyptians had used a mortar based on gypsum, $CaSO_4 \cdot 2H_2O$. Babylonian lime-kilns, used to prepare lime for the plastering of walls, date to ~2500 BC. Gypsum plaster was preferred for walls that were to be decorated.

Cement is important not only as a mortar but also as a water-impervious lining for cisterns, aqueducts and reservoirs, and as the bonding material for concrete—a mixture of sand, gravel and stone—used for road and building construction. The Romans were perfectors of cement masonry, both in the sense of the materials used and in the construction methods and architecture. About the third century BC they discovered that a cement, superior to the Hellenistic lime mortar, could be made from volcanic ash (pozzolana) mixed with lime. This material dried rock-hard and when dry was both fire- and water-resistant. This new mortar was sometimes spaced out with sand, small stones and bits of broken brick or tile. When it was realized that the aggregate contributed reinforcement, not just space filling, what we now know as concrete was discovered. It was at first used as filling for walls whose faces were covered with a thin layer of stone or brick (Fig. 6). Later, concrete was found to be an efficient medium for arch construction by casting over a temporary support (centering) of wood and/or brick. When set, the concrete could support superposed loads as well as itself and the temporary falsework could be removed (see *Concrete as a Building Material*).

Cement and concrete technology rested at this point for hundreds of years with widely variable results obtained as a function of the composition of locally available materials, the care used in mixing

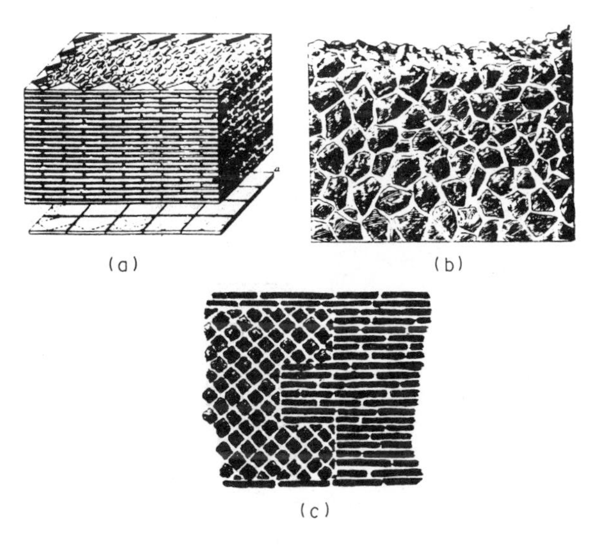

Figure 6
Roman methods of wall construction: (a) concrete wall faced with brick, *opus testaceum*, (b) *opus incertum*, and (c) *opus reticulatum* (after Schreiber T 1895 *Atlas of Classical Antiquities*. Macmillan, London)

them and the attention paid to the removal of air bubbles and free water. In particular, it was not understood why one rock produced a quick-setting cement, another slow-setting, a third quick-setting but very strong. Accordingly, the famous English engineer, John Smeaton, when planning in 1759 the third Eddystone lighthouse, determined to make a systematic investigation. By analyzing lime made from different rocks obtained all over the country, he found that the hydraulic quality of the hardened cement (its water resistance) depended on the amount of clay the original limestone contained. This finding then directed his selection of those particular rock sources. The final step was taken by Joseph Aspedin in 1824 who specified that certain proportions of clay and limestone be used and calcined together to form "Portland" cement, so called because of its supposed similarity to Portland building stone.

4. Fibrous Materials

Materials made from fibers were needed by early man not only for clothing but also for household textiles (carpets, curtains, table and bed linens), cordage, nets, sails and tenting. Animal and vegetable sources of fibers appropriate to such uses were discovered in prehistoric times, bred and cultivated, and subjected to increasingly sophisticated processing. The principal fibers used were wool, flax (linen), cotton and silk.

Wool was obtained from sheep which were domesticated in Neolithic times, probably first in Afghanistan. The fibers of wool can be short or long, coarse or fine, depending on the type. Wool is usually white but black, gray and brown are other natural colors. An important feature of the fiber is that it grows with a natural wave or crimp which imparts the desirable qualities of elasticity and resilience. The scaly surface structure of the wool fiber accounts for its excellent felting qualities. Clothing made of wool was worn by the peoples of Anatolia as early as 6000 BC, and by 3000 BC nearly all Asiatic peoples were raisers and breeders of sheep.

Linen is derived from the flax plant. The original wild species was perennial but the cultivated type is an annual. Linen cloth is known to have been made at least as early as 2700 BC in Europe and 3200 BC in Egypt. Following harvesting of the flax, several processing steps must be undertaken: rippling to separate the seeds, retting—soaking and fermentation in water—to eliminate resinous matter, grassing to dry the stems, scutching to remove the woody portion of the stem from the fibers, and heckling to separate the best of the fiber, readying it for spinning. A significant point is the use of linen by several peoples in their religious ceremonies—fine white linen was held to represent divine light and purity.

The cotton plant is a member of the mallow family, a small, dark-leaved tree which produces bolls of fine white fibers. Cotton cloth was produced in India by the Mohenjo-daro civilization as early as 2500 BC and from there spread to other parts of Asia and Europe. Columbus and other early explorers found it in use by the natives of the New World. The most important feature of raw cotton fibers is the number of times they are twisted, usually about 200 to 300 per inch, which accounts for the excellent spinning qualities of the fiber.

Silk was discovered by the Chinese as early as 6000 BC when it was discovered that a beautiful and useful fiber could be unravelled from the cocoons of a certain moth, *Bambyx mori*, which could be reared and fed on the leaves of mulberry trees. A less desirable form was made from the silk obtained from cocoons of a variety of wild moths feeding on trees. Although knowledge of the existence of silk was brought to Europe ~320 BC by Alexander the Great, the Chinese jealously guarded this secret and maintained a monopoly for hundreds of years. About 600 AD, a stock of silk worms was supposedly smuggled to Asia Minor and Byzantium in the staff of two monks who had long been resident in China, and the monopoly was broken.

Other natural fibers employed very early, although more local in use and specialized in application, include hemp (an Asiatic herb), jute (an East Indian plant), sisal (a plant of the Yucatan), bark (mulberry, breadfruit and hibiscus, from Central Africa and the South Sea) and papyrus (a sedge of the Egyptian delta). These fibers were used for such applications as sail-cloth, matting, ropes and sacking, and were often combined with the other fibers in woven goods.

Concomitant with the discovery of the fibers came the development of processes for working with them; spinning, weaving, felting, twining, plaiting, dyeing and printing all go back many thousands of years into the prehistoric period (Fig. 7). (The spinning whorl ~25 000 years BC may well be man's oldest machine.)

It is noteworthy that the four principal natural fibers: wool, cotton, linen, and silk were not supplanted by artificial fibers for thousands of years. Hooke (1664) suggested the possibility of an artificial fiber in his *Micrographia* as did Réaumur (1734) in *Histoire des Insectes*, but nothing came of these ideas. It remained for Count Hilaire de Chardonnet, who after several years research obtained a patent in 1884 for the first artificial silk, a nitrocellulose, to establish the beginning of the synthetic fiber industry.

5. Paper

Apart from monumental inscriptions in rock, the early civilizations of Asia Minor used soft clay as writing material in which indented characters (cuneiforms) could be formed with a stylus. Sun drying or firing was employed to create a permanent record. Alternatively, thinned and smoothed animal skins such as sheepskin (parchment) and calfskin (vellum) could be written on with ink and paint. The Egyptians

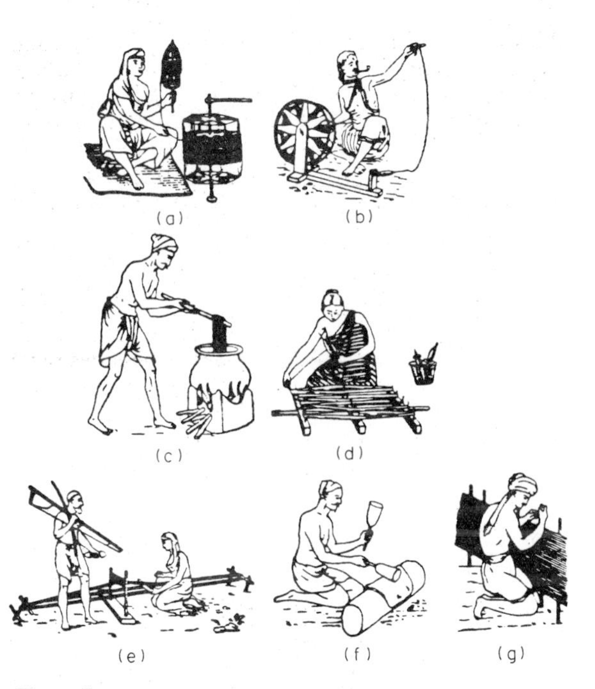

Figure 7
Indian textile operations with cotton, silk and linen: (a) winding dyed silk, (b) spinning cotton, (c) dyeing hanks of silk, (d) winding cotton thread, (e) weaving cotton (the bow is for cleaning the cotton), (f) beating canvas, a coarse cotton or composite cotton/linen fabric, and (g) making a loincloth on a vertical loom, by knotting threads (after Daumas (ed.) 1969 *A History of Technology and Invention*. Crown, New York)

very early (~2500 BC) began to make use of a locally available material, the papyrus reed, to fashion the first paper-like material. The pith of the reed was cut into slices about 42 cm × 6 cm and very thin; the slices were then laid in alternating cross-wise layers, pressed and sun-dried. Following this the sheets were beaten and polished to make them thin and smooth (Fig. 8). These sheets, approximately 50 cms × 29 cm were then glued together in a long continuous scroll. Papyrus was not supplanted as a writing material until Roman times. In China in the earliest years materials completely different from those of Egypt or Mesopotamia were employed: flat bones, turtle shells, wooden or bamboo slats, or silk.

True paper is supposed to have been invented in 105 AD in China by Tsai Lun, a eunuch in the service of the Han emperor. By beating the inner bark of the mulberry tree he obtained fibers which in a water slurry could be matted together to form a dense white sheet eminently suitable for writing. In his later researches Tsai Lun discovered that hemp, old fish nets and linen rags could be similarly decomposed to form a pulp from which paper could be made by draining an aqueous slurry spread in a thin even layer on a frame-supported mesh. The Chinese in the first few centuries

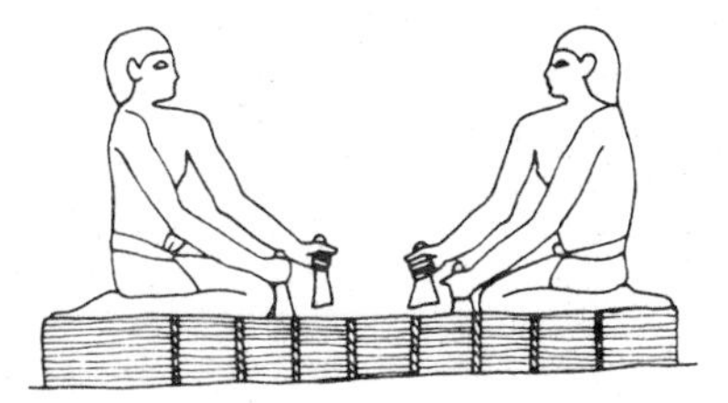

Figure 8
Two men beating out papyrus; from an Egyptian tomb of about 2500 BC (after Hodges 1970)

AD not only produced writing paper but also wrapping paper, colored paper, paper screens, napkins and even toilet paper! As with silk the secret slowly spread westward, reaching Baghdad about 800 AD. This stop on the diffusion path and a key element in the technology itself, is recorded in the word "ream" for a quantity of paper, which comes from the Arabic *risma* (bundle of clothes). Presumably the Moors brought the paper-making art to Europe via North Africa and Spain. Certainly there were water-powered paper mills in Spain and Italy by 1150 AD, in France by 1189, Germany by 1291, England by 1330 and America by 1690.

The demand for paper increased tremendously in Renaissance times, spurred in particular by the invention by Gutenberg of printing from movable metallic type (1438). Gutenberg was in fact preceded by the Koreans who introduced movable cast bronze type in the twelfth century. His contribution was a method of molding which gave type with parallel sides and constant height. The demand for paper stimulated both the development of power driven machinery and a search for a more abundant source of fibers. Water-driven stamp mills had been in use since ~1200; a power driven device for macerating rags (the Hollander) originated in the Netherlands in the mid-eighteenth century and the fourdrinier paper-making machine in 1802 in England. Although Réaumur in the early eighteenth century had conceived the idea of mechanical disintegration of wood to form pulp by watching the nest building of wasps, a practical process for pulping of wood was delayed for many years. It was not until the mid-nineteenth century that mechanical and then chemical means of pulping wood were first introduced. It is again remarkable that although today some 5000–10 000 different kinds of paper are produced, all are direct descendants of Tsai Lun's original invention, no satisfactory substitute for paper has yet been found. More paper is consumed today than ever before despite the burgeoning use of electronic recording and communicating devices which are supposedly bringing us the "paperless" office.

6. Materials Processing

Previous sections of this article describe the early development of various particular processes used in producing, forming and treating materials. In this section, some general topics that affect nearly all processing of materials are reviewed.

First physical and chemical processes are distin guished between. Physical processes, since they alter the form but not the intrinsic nature of the material, were the first introduced—for example, the flaking of flint, the shaping of wood, the casting of metal and the blowing of glass (the latest). Chemical processes, which resulted in a conversion or apparent transmutation of a single material or combination of materials to something totally new and different were much more mysterious—almost magical in character—and came later in time. Examples include the smelting of metals by oxidation and reduction, alloying, diffusion, fulling of cloth with Fuller's earth and potash, and the setting of cement by reaction with water.

A basic requirement in much materials processing is a source of heat. While the heat of the sun was used from earliest times as in the drying of mud bricks or the evaporation of brines to yield salt, most chemical processes and certain physical processes (fusion or annealing) require substantially higher temperatures that could only be realized with fire. Man's first use of fire with reference to materials is probably the hardening of wood as for spear points by charring; examples have been found dating to as early as 400 000 years ago. Fire making, either by the friction method with wood, or by the percussion method with pyrites chunks, or later flint and steel involved the careful selection of particular materials for regular success.

The capabilities of an open fire for materials processing were limited, not simply as to temperature (400–800 °C) but also with regard to the condition of oxidation or reduction and to the separation of the charge from the fuel or reaction products. Ovens and simple furnaces were known from baking and other household operations and hence were adapted to technical uses. The higher temperatures necessary for metal smelting and glazing required the natural draft of a shaft furnace. To melt copper or gold or smelt iron ore required forced draft, at first on a small scale with blowpipes but later (~2500 BC in Egypt) with bellows. Many of these furnaces were of such size and design that they had to be torn down after each heat to recover the product. An important development in the iron blast furnace came in Germany about 1300 with the so-called Stückofen. This was of brick construction, lined with clay, bulbous in the middle with a narrow top. Provision was made for tapping both metal and slag and for introducing air blast from water-powered bellows. Temperatures above the melting point of cast iron (1180 °C) and capacities of a tonne or more of metal were readily achieved.

Power was another important consideration in materials processing. For thousands of years the only power available was human power, often augmented by the use of the lever in various forms. Animal power was important in other fields such as agriculture and transportation but in the materials field it seems to have been used only in the donkey driven ore-crushers for the silver mines at Laurion (500 BC).

The most significant increase in available power came with the introduction of water power about 1000 BC. Water wheels of both horizontal and vertical design were originally employed for grinding of grain but beginning about 1000 AD were adapted to a variety of material processing operations: fulling of cloth (1000), ore-crushing stamp mills (1100), paper making (1200), tan-bark mills (1200), sawing (1200), and most importantly the trip hammer (thirteenth century, Fig. 9) and the bellows (fourteenth century). These mills, typically 10 hp or less and never more than about 25 hp, persisted almost to the present day. Not only were they important in the production and treatment of materials as just related; but, as the primary and most ubiquitous engineering mechanism for over a millenium, water-powered mills were themselves a powerful stimulus for the development of improved materials—for shafting, gearing, bearings and lubricants.

Water power was ultimately replaced largely by steam power. Although the Greek, Hero, had built a steam turbine, it was no more than a toy; no significant progress in harnessing steam power was made until Papin (1647–1712). He demonstrated a steam pressure cooker to the Royal Society, was the first to apply steam to move a piston in a cylinder, invented (1697) a steam powered pump for raising water from mines and (1711) a bellows to create "great and lasting blasts of wind to melt and refine ores." In 1698 Savery developed an improved steam engine for

Figure 9
A triphammer for forging powered by an overshot wheel. The judicious use of metal for reinforcing and at wear points can be seen as well as the application of simple bearings (after Aitchison 1960)

draining mines, serving towns with water and for working various sorts of mills. This may be rightly said to be the first practical steam engine. Subsequent improvements were made by Newcomen, and particularly by James Watt who introduced the separate condenser, a device for conversion of reciprocating to rotary motion, the throttle valve and the centrifugal governor. These innovations revolutionized the power business, and by 1786 Watt's engines drove paper mills, corn mills, cotton spinning mills, a brewery, blast furnaces, and an iron works. Power units averaged about 15 hp, and in exceptional cases 50 hp was attained. Thus the steam engine as a power source was an attractive substitute for the water-powered mill, not because it was much more powerful, but because the power source was now site-independent and season-independent. Again, as with water powered mills, the steam engine required the application of new and improved materials, cast iron and wrought iron substituted for wood, in order to achieve satisfactory performance, and the powerful synergism between machine and material was again demonstrated.

Control of an ongoing process and the function we now refer to as technology transfer were greatly inhibited by the imperfect quantification of such parameters as weight, time, and temperature. Quantitative weight determinations came first in ~5000 BC, with the balance for weighing gold dust, and much later (~2500 BC) for commercial transactions and technical purposes. By 1350 BC a weight of about 10 g could be measured with an accuracy of about 1%. Time was not measured for periods of less than a day other than by water clocks, shadow clocks and sand glasses until the mechanical clocks of the 1300s. Temperatures for thousands of years could only be judged by viewing colors, e.g. "cherry-red heat" and "white heat." Galileo (1593) made the first attempt to measure temperature with a crude gas thermometer, a device considerably improved in accuracy by Amontons a hundred years later. The key advance marking the practical beginning of scientific thermometry came with Fahrenheit's mercury thermometer (1714). The thermocouples, radiation pyrometers and ceramic cones, so important today were not introduced until the nineteenth century. A possible exception is the Wedgwood pyrometer (1780 to 1830 or so) important in the ceramics field.

Mechanization of materials processing had several important effects. It eased the drudgery of hand labor, removed the variability of hand production, increased the volume of goods produced and therefore lowered their price, and removed the limitation of recurrent labor shortages. On the other hand, products only rarely increased in scale beyond what had been possible by hand techniques. For example, our present standard paper size $8\frac{1}{2}$ in. × 11 in. is derivative from halving a quartered sheet roughly 24 in. × 38 in., which was the maximum size of the framed screen used in paper-making (Fig. 10) which a man could

Figure 10
Making paper in China, a reconstruction based on a drawing of about AD 500 (after Hodges 1970)

conveniently lift. Factors other than tradition and human dimensions also played a role. Large process equipments required considerable capital, always difficult to come by, and such projects were considerably more risky in early times because of more limited markets. Large machinery also was susceptible to easy breakage and malfunction due to the shortcomings of the available construction materials. As a result, from classical to near-modern times even large manufactories most often consisted of batteries of hand-scale process equipment even when fully mechanized.

7. *Understanding of Materials and Their Behavior*

The foregoing discussion has made clear that by 1800 a considerable sophistication had been achieved in the selection of materials and in processing them so as to alter their form and properties. This sophistication, however, was one of individual technique and knowledge of recipes for obtaining a desired result. True understanding of the nature of materials and of the reasons for the dependence of properties on structure, processing and composition was almost completely lacking.

The modern concept of the element did not replace the quartet of earth, air, fire and water used by the ancients or the principles of liquidity (mercury), fixity (salt) and combustibility (sulfur) until Lavoisier's chemical treatise in 1789. (Even he included in his list of elements heat and light, now known to be forms of energy.) Becher's phlogiston theory of the reversible reduction/oxidation reaction persisted for over a century until Lavoisier. Similarly the distinction between mixtures, compounds and alloys remained confused until well into the nineteenth century. The identification of carbon as the key ingredient distinguishing wrought iron, cast iron and steel was made only in 1781 by Bergman in Sweden.

Although quantitative density measurements of alloys had been made regularly since at least the twelfth century, and assaying (especially for precious metals) also carried out very much earlier, general quantitative measurement of physical properties of materials began only in the seventeenth and eighteenth centuries. The first measurements of the strength of materials were made by Galileo (1632), Hooke (1660) and Musschenbroek (1729). Gilbert's studies of magnetic behaviors (1600) marked the beginning in that field. With reference to electrical properties, conductors and insulators were distinguished only in the early decades of the eighteenth century. Achard (1784) suggested the close connection between electrical and thermal conductivity, and in 1788 carried out a monumental study of the physical properties of some 900 alloys. Measurements of electrical properties of matter did not begin until Ohm (1826). It is little wonder that correlations of composition and processing with properties remained qualitative and highly subjective.

The early materials scientist was also hampered by vague and conflicting notions of the structure of matter and the lack of instrumental techniques for elucidating it. Following up a centuries-old shop practice for quality control, Réaumur (1722) used fractography to follow the changes imparted by the processing of metals. It was not at all obvious, before the days of x ray diffraction, that metals were crystalline since they so rarely exhibited a regular external morphology as did many minerals and salts. The simple microscope was invented near the close of the sixteenth century and applied mostly to biological specimens. With the development of the compound microscope in Italy in the mid-seventeenth century by Divini and Campani, an instrument of much greater usefulness was at hand, but its quality was still limited by the available materials: wood, cardboard tubes and poor optical glass. A typical microscope from the early 1700s is shown in Fig. 11. Another 150 years were to pass, however, (with progressive improvements in both microscope mechanisms and materials) before Sorby showed the applicability of the microscope to study of the internal structure of solids—first of rocks with thin sections and later of metals by reflection from a polished and etched surface.

When so much of materials art and science depended on individual skills and a myriad of seemingly unrelated empirical observations, it is not surprising that progress was halting, diffusion of technology from one people to another slow, and much independent invention took place. With the development of writing, it became possible to communicate across the generations and between widely separated people who had never seen each other. Some technical writings go back very far indeed. Babylonian cuneiform tablets have been found which concern glass-making, enumerating both plain and colored glasses and imitations of precious stones. Pliny (23–79 AD) wrote a *Natural History* in 37 books, which is distinguished by its broad range of subject matter and the evident ignorance of the author of many of the technical topics on

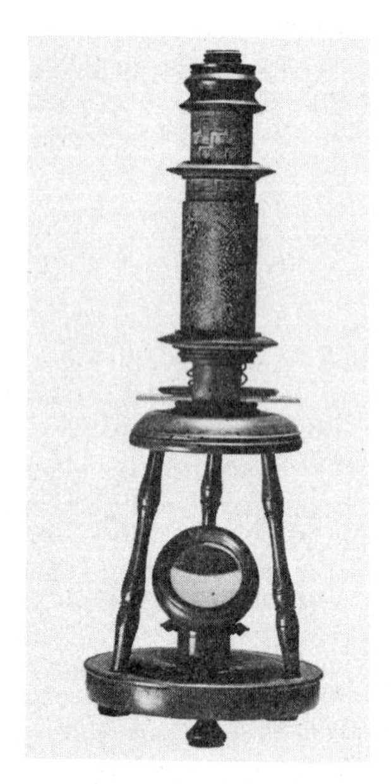

Figure 11
A Nuremberg microscope of Culpeper design (circa 1725) in which the optical axis and the axis of symmetry are coincident (courtesy of Carl Zeiss Optical Museum)

which he wrote. Vitruvius (b. 70 BC) prepared the first technical treatise by an expert about his own field. The Arab, Razi (866 AD), left a long list of equipment for melting and a systematic classification of chemical substances defined by experiment as to animal, vegetable, mineral or derivative. After 1125 AD Theophilus, a monk, wrote *Notes on Various Arts*, a collection of recipes for materials, processes and fabrication techniques for churchly ornamentation. The real breakthrough in technical books came in the sixteenth century after the introduction of the printing press. Those of special relevance to materials include Biringuccio's *Pirotechnia* (1540), Piccolpasso's *Three Books of the Potter's Art* (1550), Rossetti's *Plictho de L'Arte de Tentore* on dyeing (1540), Agricola's *De Re Metallica* (1556) and Ercker's *Treatise on Ores and Assaying* (1574). These texts and most of their illustrations convey considerable detail as to techniques used and the state of knowledge at the time. The way was now clear for the more comprehensive works, in the seventeenth century, of Bacon (1627) and Moxon (1677), and the eighteenth-century technical encyclopedias edited by Chambers (1714) and by Diderot and d'Alembert (1751–72) and the famous *Descriptions des Arts et Métiers*, sponsored by the French Academie des Sciences (1761–89). Such works contributed enormously to both the dissemination of knowledge and the acceleration of technological development.

8. Concluding Remarks

It is hoped that this brief account has left the reader with a few general impressions of the history of materials over the millenia to the beginning of the nineteenth century:

(a) the increased diversity of materials and processes man has brought into application and under control;

(b) the increased capability man has achieved with reference to the complexity of materials and processes, their reproducibility, and the volume of production;

(c) the interdependence of materials and machines, leading to improved processes, improved materials and improved machines;

(d) the interdependence of materials and instruments, leading to improved understanding of materials, improved control of processing, and improved materials;

(e) the varied contributions of different civilizations, widely separated in space and time;

(f) the nonlinearity of progress in materials technology (the achievements in the two centuries from 1600 to 1800 are enormous, compared with all prior time, but almost trivial compared with what has occurred since); and

(g) the importance of written records and publication to the acceleration of materials technology—by communication across space and time, by articulating and sharpening ideas and understanding, and by exposing each man's findings and ideas to analysis and debate.

Bibliography

Aitchison L 1960 *A History of Metals*. Interscience, New York

CBA Review (a series covering the complete history of textiles and allied crafts). CBA, Basle

Cipolla C M, Birdsall D 1980 *The Technology of Man—A Visual History*. Holt, Rinehart and Winston, New York

Davey N 1961 *A History of Building Materials*. Phoenix House, London

Douglas R W, Frank S 1972 *A History of Glassmaking*. Faulis, Henley-on-Thames

Fine M E 1973 *Notes on the History of Metals and Ceramics*. Private communication

Forbes R J 1950 *Man the Maker*. Constable, London

Forbes R J (ed.) 1955–1972 *Studies in Ancient Technology*. Vols. 1–9. Heinemann, London

Hodges H 1970 *Technology in the Ancient World*. Knopf, New York

Hunter D 1957 *Papermaking, The History and Technique of an Ancient Craft*, 2nd edn. Cresset, London

Kranzberg M, Pursell C W (eds.) 1967 *Technology in Western Civilization*, Vol. 1. Oxford University Press, Oxford

Needham J W (ed.) 1954–1970 *Science and Civilization in China*, Vols. 1–6. Cambridge University Press, Cambridge

Singer C et al. (eds.) 1954–1958 *History of Technology*, Vols. 1–5. Oxford University Press, Oxford

Smith C S 1981 *A Search for Structure*. MIT Press, Cambridge, Massachusetts

Smith C S, Kranzberg M et al. 1975 Materials and society. *Materials and Man's Needs*, COSMAT Report, National Academy of Sciences, Washington, DC

Timoshenko S P 1953 *History of the Strength of Materials*. McGraw-Hill, New York

Tylecote R F 1976 *A History of Metallurgy*. Metals Society, London

Weeks M E, Leicester H M 1968 Discovery of the elements. *J. Chem. Educ.* 45

Wertime T A and Muhly J D (eds.) 1980 *The Coming of the Age of Iron*. Yale University Press, New Haven, Connecticut

J. H. Westbrook
[Scotia, New York, USA]

Materials: History Since 1800

If a craftsman or industrialist from the time of Christ were to have been suddenly transported to 1800, he would have found little in the materials world to surprise him. Structural materials were still principally wood, brick and stone with metals used sparingly, and only in critical applications. Glass and ceramics were unchanged, fibers were still the familiar natural ones, paper was derived only from rags, and electronic materials, polymers and composites were unheard of. Perhaps the most noticeable change would have been the increased scale of materials processing operations and the enhanced degree of mechanization. Power sources were either thermal, wind or hydraulic, all of which had been in use for hundreds of years. Only steam offered a new opportunity.

How greatly and rapidly all this changed in the succeeding two centuries is the subject of this article. Metals, wood, paper, glass, ceramics, cement and concrete, fibers, electronic materials, polymers and composites will be discussed successively.

What are some of the principal factors which enabled this rapid materials advance? In the opinion of the author, the following four are among the leaders.

(a) Electrical energy. Beginning with Volta's construction of the electric pile (battery) in 1800, man had a new, powerful and versatile form of energy at his disposal for both the synthesis of materials and the control of their processing.

(b) Periodic table. Mendeleev's conception of the periodic table in 1871 both laid out the full range of the natural elements and systematized the relations between them, enormously stimulating directed materials developments.

Table 1
Types of radiation and their use in solid-state analysis

Radiation		Property to be analyzed				
Type	Wavelength	Chemical composition	Crystal structure	Microstructure	Field and domain structures	Shell, band and magnetic structure
Electromagnetic radiation						
γ rays	<5 pm	γ spectrometry		γ autoradiography		Mössbauer spectrometry
X rays	0.05–30 nm	x ray spectrometry	x ray diffraction, low-angle scattering	x ray macrostructure, x ray micrography, x ray microscopy		
Ultraviolet rays	20–400 nm	uv spectrometry		uv microscopy		
Light rays	400–800 nm	spectral analysis		microscopy (metallography, ceramography), holography	Kerr microscopy, Faraday microscopy, (ferromagnetic and superconductive domains)	
Infrared rays	0.8–30 µm	ir spectrometry	ir spectrometry	ir microscopy, thermography		
Microwaves	0.001–1 m					microwave spectrometry, spin resonance spectrometry
Radio waves	>1 m					cyclotron, helicon, plasma resonance spectrometry
Corpuscular radiation						
Electron rays	0.3–8 pm	β spectrometry		β autoradiography		
	4–10 pm	energy loss spectrometry	electron diffraction	electron microscopy	Lorentz microscopy (ferromagnetic domains), Bragg microscopy (ferroelectric domains), Coulomb microscopy (imaging of microfields)	
	40–600 pm		low-energy electron diffraction			
Neutron rays	0.01–0.1 nm		neutron diffraction			neutron diffraction
	0.3–1 nm		neutron scattering			
Atomic and molecular (ion) rays		mass spectrometry		ion microscopy		

Source: Pfisterer et al. 1967, 1968 *Siemens Rev.* 34: 279, 418; 35: 58

(c) Tools for the study of internal structure. To the microscope were added x ray diffraction, neutron diffraction, the various spectroscopies, nuclear magnetic resonance and a host of other specialized tools (see Table 1 and *Investigation and Characterization of Materials*).

(d) Scientific communication. The regularization of the reporting of new findings in science and technology through journals, monographs and public meetings did much not only to broaden the dissemination of knowledge, but also to quicken its application and the interaction between fields. Also contributory were the adoption of national patent systems in England, France, Germany and the USA, and the establishment of societies such as the Society for the Encouragement of Arts and Manufactures (1754).

As shall be seen, an interesting theme runs through all of materials development—the sensing and exploitation of the heterogeneity of matter, of structure and of composition over extremely broad ranges of scale. Segregatory and migratory habits of clusters of atoms or electrons give rise to most of the interesting properties and phenomenology of materials. These concepts, first appreciated by materials people, are now being used by chemists, physicists, astronomers and historians!

1. Metals

It is difficult today to realize how limited were the choices of metal for engineering purposes in the year 1800. Wrought iron was by far the principal construction material, with cast iron used secondarily for heavy sections and noncritical applications. Steel and the nonferrous metals were very distant also-rans.

The first event bringing about a change in the relative importance of wrought iron/cast iron compared with steel was the Kelly–Bessemer process in which an air blast through the molten pig iron oxidized carbon and many of the impurities which could then be removed with the slag. Kelly, an American, and Bessemer, an Englishman, independently invented the same process. Kelly apparently was first, but Bessemer obtained the first patent (1856) followed by Kelly. The process was not an immediate commercial success because the high phosphorus introduced by most British ores was not removed from the steel which was left brittle. This problem was solved in 1883 by two cousins, Thomas and Gilchrist, who used a basic furnace lining and basic slag to remove the phosphorus and retain it in the slag. Meanwhile, Siemens (1868) had applied the regenerative principle to steelmaking furnaces. By giving higher temperatures and with capability for much larger heats, the Siemens "open hearth" furnace was very attractive. Unfortunately, until the work of Thomas and Gilchrist, it suffered the same limitations as low phosphorus pig iron. The impact of the basic steelmaking process can be seen from the fact that whereas in 1850, no more than 60 000 tons of all kinds of steel were produced in the world, by 1900, world production of mild steel alone amounted to 28 mllion tons! The volume production of steel came just in time for the growing demands for railway rails, for steel gave ten times the life of iron rails.

It was some time before the advantages of alloying steel were appreciated. Faraday and Stodart studied steel alloys very early (1820) but perhaps the first successful commercial alloy was Mushet's tungsten steel in 1868 followed by Hadfield's manganese steel in 1887 and Riley's nickel steel in 1889. The last was particularly notable because, with only a moderate amount of nickel, the strength of steel after heat treatment was about three times as great with no loss of toughness. These steels found immediate application as naval armor plate as well as for other structural purposes. Still later (1900) came Taylor and White's famous 18-4-1 tool steel which would cut even at red heat. The stainless steels had their origin in the work of Brearley in England in 1914 (the 12% Cr grades) and in the work of Strauss and Mauer in Germany about 1910 (the Cr/Ni grades). The soft magnetic steels (0.5–4.5% Si) were introduced by Hadfield et al. in 1903 and have since become indispensable in all types of electrical machines.

A development near the beginning of the 1800s which was destined to have major effects on metallurgical processing was the availability of electrical energy. Hare of the University of Pennsylvania was one of the first to experiment with the new energy source. In 1839 he demonstrated aqueous electrolysis with the mercury cathode, arc welding and consumable electrode arc melting, all of which were reduced to commercial practice by others some decades later. Other electrometallurgical developments were Elkington's electroplating process (1840) and the electrolytic purification of copper (1865). Perhaps the most significant electroprocess of all was the simultaneous invention by Hall in America and by Héroult in France of the electrolytic reduction of cryolite to yield aluminum. Prior to this development, aluminum had been little more than an expensive curiosity; it now became a tonnage metal well before 1900 and by 1966 had passed copper in total consumption to become the world's second most important metal.

Other nonferrous process developments of note include the Pattinson process (1833) for Ag/Pb separation, later to be supplanted by the Parkes process (1850) involving the formation of Ag_2Zn_3 compound; the cyanide process (1887) by McArthur and R. and W. Forest for recovery of gold; powder metallurgy for consolidating metals by Wollaston (1801); production of tungsten wire by Coolidge (1909); and magnesium reduction of titanium tetrachloride by Kroll (1940) to form titanium.

On the physical metallurgy side, the important developments were: the discovery by Wilm in 1906 of precipitation hardening, the first new hardening process since ancient times; the full elucidation of the hardening and tempering of steel by a whole succession of workers from Sorby in 1863—Osmond, Tschernof, Charpy, Sauveur, Robarts-Austen, and others near the turn of the century—to Davenport and Bain in the 1930s; the application of the phase rule to alloy constitution by Roozeboom in 1899; and the application of x ray diffraction to alloys by von Laue, Debye and Scherrer and Hull. Also to be noted are the beginnings of systematic studies of some of the phenomena associated with structural failure: fatigue (Wöhler 1860), impact (Hodgkinson 1843), and creep (Andrade 1910) (see also *Materials and Innovations; Ferrous Physical Metallurgy: An Overview*).

2. Wood

Despite all the developments in alloys, polymers and other exotic materials synthesized by man, wood remains by far the most significant material class by volume and even on a weight basis. For most of man's history, the intelligent application of wood was largely a matter of selecting the optimum species for the job at hand and skillfully shaping the desired component. Within the last 200 years, however, several developments have occurred which have improved the utility of wood and greatly extended its range of application (see *Wood: An Overview*).

Antifungal and antibacterial treatments have had substantial effect in extending the life of railroad ties and utility poles. Even today, about 50 million railroad ties are laid each year; it is estimated that this number would be three times as great were it not for the decay protection afforded by the high-pressure creosote treatment which is universally applied. Other wood significantly exposed to the weather may receive zinc chloride or pentachlorophenol preservative treatment. Various chemicals such as ammonium phosphate or boric acid may also be pressure impregnated to impart fire-resistant qualities. Resin impregnation may be used to improve resistance to chemical attack and/or to stabilize wood dimensionally under conditions of fluctuating relative humidity (see *Chemically Modified Wood*).

Although plywood (for furniture production) was being made as early as the French Revolution, it was not until the advent of inexpensive, water-resistant glues (casein and soybean types) spurred by the needs of the fledgling aircraft industry in World War I, that plywood began to come into its own as a significant construction material. More recently the synthetic resins (phenol and urea formaldehyde, melamine and resorcinol) have led to glues which are truly waterproof, albeit requiring heat curing. In 1925, plywood produced in the USA amounted to some $14 \times 10^6 \, m^2$ while by 1980 more than $1.9 \times 10^9 \, m^2$ were produced.

Glues also were responsible for making feasible laminated wooden arches and beams (enabling spans and clear heights greater than the trees from which the basic material came) as well as particleboard ($> 2 \times 10^8 \, m^2$ annually) and the latest innovation, the highly efficient parallel laminated veneer lumber. Thus even the oldest material has been subject to some substantial developments (see *Glued Laminated Timber*).

3. Paper

Paper technology essentially stood still for over 1500 years. Two basic inventions were needed to really expand production: a continuous papermaking process and a more abundant source of raw material. Robert in France in 1799 invented a papermaking machine but it was H. and S. Fourdrinier of England who financed the brilliant engineer Bryan Donkin and built the first perfected machine of this kind in 1812. In today's machines of the same basic design the finished product comes off at $9 \, m \, s^{-1}$ in widths up to 20 ft (6.10 m).

Wood fiber had been proposed early on as a raw material source for paper by Réaumur in France and Schäffer in Germany in the eighteenth century, but it remained for F. G. Keller in 1840 to obtain the first patent for a practical wood pulp process—a mechanical grinding. Today's wood pulping processes supplement the mechanical processes (grinding and the new refiner mechanical pulping (1950) and thermomechanical pulping) with chemical ones—sulfite, soda or sulfate—the first of which was patented by Tilghmann of Philadelphia in 1867. Choice among all of these is determined by the species of wood used, the nature of the end product desired and bleaching characteristics. The sulfate (kraft) process, the last introduced ($\sim$1910), has gradually become dominant, but all are still in use. The quality of paper was also enhanced by chlorine-based bleaching processes developed by Tennant and others at the beginning of the nineteenth century.

Table 2
Chronology of the introduction of significant paper products

Year	Product
1829	Paper with threads for banknotes and stamps
1830	Sand paper
1844	Paper boxes
1849	Photographic paper
1850	Paper bags
1856	Corrugated paper
1861	Papier maché for stereotyping
1862	Tracing paper for draftsmen
1869	Paper coffins, clothing, barrels, etc.
1871	First use of roll toilet paper in the USA
1900	Paper for electrical insulation

A brief chronology of the introduction of significant paper products is summarized in Table 2.

4. *Glass*

The glass industry in the nineteenth century changed very slowly because of the complex problems associated with the properties of the material. One of the first significant developments was in glass-melting furnaces. The first application of the regenerative heating principle of using hot exhaust gases to preheat incoming fuel gas and air, embodied in Sir William Siemens' patent of 1856, was in an English glass works. The higher temperatures afforded by the new type of furnace reduced the life of the fire-clay pots containing the glass but on the other hand permitted the substitution of salt cake for soda and the melting of the more refractory borosilicate glasses (the basis of today's Pyrex). The pot problem was eventually solved by the introduction of the tank furnace by Siemens in 1872.

Mechanical engineering developments gradually converted the glass industry from a hand to a fully mechanized operation: equipment for plate-glass casting ($\sim$1850); pressed glass, a cheap but commercially successful imitation of cut glass (begun crudely in 1827 but outstandingly successful with the use of chilled cast iron molds in 1866); blown glass (begun in 1866 with Arnall and Ashley's "plank" machine for bottle making and culminating in 1926 in the famous Corning ribbon machine for blowing light bulbs); and drawn sheet glass (the British Clark patent of 1857 followed by Foucault of Belgium in 1902 and by Colburn of the USA in 1903). These achievements were overshadowed in modern times by the ingenious development of the float glass process by Pilkington in 1959 which directly yielded a finished plate glass, a product requiring no subsequent flattening, grinding and polishing. In this process a continuous ribbon of glass is poured out on the surface of a molten bath of tin on which it floats. The glass ribbon cools while still resting on the molten tin and is then passed directly through the annealing oven.

Continued experimentation and the availability of higher temperature furnaces gradually widened the range of compositions from which glass could be made. The first systematic studies of glass composition were made by E. Abbé and O. Schott at the famous Zeiss–Jena glassworks in the late nineteenth century. Now that the effects of superhigh quenching rates are appreciated it has been possible to extend the glassy state much further; even certain metals can now be obtained in the fully amorphous condition!

Optical glasses remained crude and unsatisfactory for hundreds of years because of chemical inhomogenities and bubbles. P. L. Guinand, a Swiss, demonstrated in 1798 that these faults could be overcome by prolonged stirring with a burnt fireclay cylinder. When the glass was frozen, crushed and the most homogeneous pieces selected and remelted, a truly superior optical glass was obtained.

Control of the color of the glass improved gradually. Soda-lime–silica glasses prepared from the purest materials varied from water-white to pale greenish. Cobalt was known from ancient times to impart a beautiful deep blue color. Sodium uranate was used following 1789 identification of uranium to give yellow. Browns, black, and dark green resulted from admixtures of iron, manganese, and other impurities while a clear medium green was obtained with the use of chromium since the early nineteenth century. Ambers, yellows, reds and purples resulted from antimony and reds from tin and especially gold.

Various methods of treating glass were developed which led to entirely new applications: wire glass incorporating a reinforcement of wire netting, safety glass with an intermediate layer of transparent plastic (originally cellulose nitrate but now polyvinyl butyral), glass fiber, nonreflective glass, tempered or pre-strained glass, chemically strengthened glass, divitrified glass ceramics (Pyroceram), photosensitive glass, phase-separated glass and fused silica glass (leached and shrunk from glasses of special composition). The production of fiberglass is illustrated in Fig. 1.

5. *Ceramics*

Materials and techniques in the nineteenth century for producing decorative and household wares changed hardly at all from those practised hundreds of years earlier, apart from increasing mechanization and improved quality control. The major changes arose from industrial applications—as furnace refractories, as electrical components and as synthetic abrasives and cutting tools. These new uses focused attention on new base materials and properties other than those previously considered.

Early furnace linings were made from natural stone, quarried blocks of mica schist, sandstone or other siliceous rock. While these materials performed relatively satisfactorily, supplies were localized and difficult to extract and transport. Such considerations led to the development of fireclay refractory brick made from kaolin deposits. Such bricks were used in furnaces for steam boilers, in steelmaking, the nonferrous industry, glass pots and furnaces, ceramic kilns and cement kilns. Superior bricks resulted from finding a superior deposit and double-firing of the brick to obtain a dense hard brick. Later improvements were effected by purification through practising a kind of mineral dressing prior to forming and firing. Silica bricks were first made in the USA in 1866 and high alumina types in 1888. Basic magnesite brick, which assumed great importance with the advent of the basic open-hearth steel process, was introduced from Austria in 1880. Chromite refractory brick, used by steel makers as a neutral layer between acid and basic courses, was applied as early as 1896.

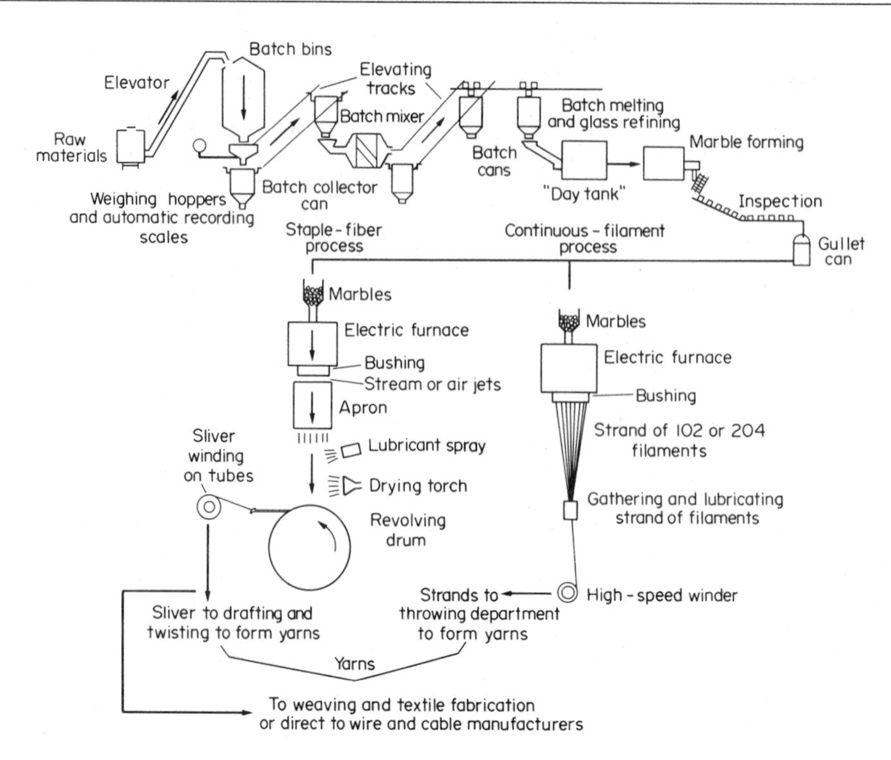

Figure 1
Flow sheet for the production of fiberglass

The burgeoning electrical industry in the mid-1800s put new demands on ceramics which continue to this day. Porcelain insulators were used on telegraph poles as early as 1850 and ceramic spark-plug insulators were introduced by 1888. In the electronics field, the molded carbon resistor was patented by Bradley in 1885 and the ceramic capacitor by Lombardi of Italy in 1900. Materials with a high negative temperature coefficient of resistivity were known since the days of Faraday, but a satisfactory commercial product for sensors and controls (mixed oxides of manganese and nickel) emerged from the Bell Laboratories only in the 1930s. Oxide magnets or ferrites were first proposed in 1909 by Hilpert in Germany, but first successful commercialization required years of work by Snoek and associates at the Philips Laboratories in the Netherlands in the 1940s.

The natural materials—quartz, corundum, garnet, and diamond—were used as abrasives for hundreds, if not thousands, of years before the appearance of the first artificial abrasive, silicon carbide. This material was discovered in 1891 by E. G. Acheson who produced it by reacting sand with carbon and named it carborundum. Abrasive-grade aluminum oxide was produced in that same year by Jacobs by the electric furnace fusion of bauxite. The next novel abrasive material to appear was boron carbide introduced in 1934 by Ridgeway of the Norton Company. A tremendous breakthrough occurred in 1960 when a team of workers from General Electric announced the successful synthesis of diamond from graphite in a specially designed high-temperature, high-pressure apparatus. Later, using the same equipment, Wentorf synthesized cubic boron nitride, a super-hard material not found in nature. Both developments have had considerable commercial success.

Efficient metal removal requires cutting tools, not grinding materials. For centuries this requirement had been filled by specially developed steels. In 1923, Schröter introduced a revolutionary new cutting material—cemented tungsten carbide, a powder metallurgy product formed by liquid phase sintering with a cobalt metal binder. Metal removal rates, hundreds of times better than with the best steel tool, were readily demonstrated. Since then, a host of similar simple and mixed carbide cutting tool compositions have evolved.

6. Cement and Concrete

Although cement and concrete had been used extensively in construction, at least from Roman times, the exclusive use of natural materials and poor control over firing of the cement and mixing of materials gave highly variable results. In 1806, Vicat in France demonstrated that natural cement rock was not necessary; he obtained a good hydraulic cement by burning finely ground chalk and a clay mixed into a paste. He used this cement in constructing works at Cherbourg Harbor.

In 1824, Joseph Aspdin obtained a patent for making Portland cement by firing clay and lime and subsequently grinding the clinker. J. C. Johnson proved the importance of a high firing temperature to obtain the highest strength in the cement when set. Foundations, sewers, tunnels, and harbor works were the principal applications of the new material; the great expansion in canal building during the early nineteenth century also intensified demand. The French in the period 1833–1840 rebuilt the mole at Algiers harbor with a considerable saving in money over the ineffectual efforts of the Moors over prior centuries using stone construction. The rotary kiln, an American development of about 1892, increased the production capacity of the cement mills enormously, and in addition, yielded a more uniform product.

The utility of concrete as a construction material was tremendously increased by the introduction of steel reinforcing. J. L. Lambot of France built a reinforced concrete rowboat in 1848, but credit for the invention and promotion of steel reinforcing is usually given to Joseph Monier who built some reinforced concrete tubs for orange trees in 1849 and went on to elaborate designs for reinforced beams and columns. The steel and concrete combination works well for three reasons: there is good bonding between the matrix and the reinforcement; the coefficients of thermal expansion are nearly equal; and the compressive strength of the concrete is complemented by the good tensile strength of the steel. In addition, the concrete protects the steel from the ravages of water and fire.

Major contributions to the design and use of the new material were made by German engineers. Reinforced concrete was employed in road construction beginning in the 1850s. The Swiss were building bridges with 128 ft (39 m) spans in the 1870s, while 1903 saw the construction of the first fully-formed reinforced concrete building in the USA. Throughout this period there was continued attention to the effects of cement:sand:gravel ratio and water:cement ratio on the setting characteristics and final strength of the concrete. Developments in modern times include the substitution of fine mesh for reinforcing rods (ferroconcrete), foamed concrete and polymer-faced concrete.

7. Fibers

At the turn of the century in 1800 and for the next 50 years or so, the textile fibers available to man were still the natural fibers—wool, cotton, linen and silk. The technological advances which were made during this period related to the mechanization of the spinning and weaving processes and to chemical treatments, e.g., starch applied to cotton to increase strength for machine handling, mordants used to improve the acceptance of dyes, and the mercerizing of cotton with caustic soda, (developed by Henry Mercer in 1850) to render it more elastic and readily dyed. Although Hooke and Réaumur had speculated about synthetic fibers, nothing was accomplished until the work of de Chardonnet in the period 1878–1889 produced a crude nitrocellulose fiber. In 1890, production was confined to one European company with a total output of 14 t. With continued development, four different processes evolved to produce what was later known as rayon: the viscose, acetate, cupra-ammonium and nitrocellulose processes. Not until 1910 was this new technology transferred to the USA, later (1919) to become the world's largest producer (> 140 kt annually in 1942 and nearly 450 kt in 1984). Common to all of these processes is the concept of dissolving natural cellulose or modified cellulose in an appropriate solvent, extruding (spinning) the solution through a multiple-hole spinneret and removing the carrier to leave fine solid threads of cellulosic material.

The manufacture of truly synthetic fibers, as opposed to regenerated natural material, began with Carothers' invention of nylon (polyamide) at the du Pont Company and its commercialization via melt spinning in 1937–39. The success of this new fiber and the growing understanding of structural organic chemistry led to the development of many others with the result, not only that fibers are now a significant fraction (25%) of total polymer production, but also that synthetic fibers show increasing domination of the fiber field, illustrated by the statistics of Table 3. Despite this amazing trend, the absolute consumption of cotton has remained about constant. Another development of modern times has been the introduction (~1938) of nonwoven fiber products for such applications as carpeting, automotive fabrics and disposables. In this process, the spun fibers are immediately bonded together either by applied weaving or knitting. By 1977, nonwoven fiber products amounted to 300×10^3 t and represented 10% of all textile products versus 25% knitted and 65% woven. The diversity of today's synthetic fibers may be seen from Table 4.

Table 3
A comparison of the US production of cotton and synthetic fibers as a percentage of all fibers, 1940–1973

Year	Cotton (percentage of all fibers)	Synthetic fibers (percentage of all fibers)
1940	72	9
1960	65	29
1973	29	70

Table 4
Fiber type, trade name and 1977 world production for various fibers

Fiber type	Trade name	1977 world production (10^6 t)
Cellulosic	Rayon	2.3
	Acetate	1.2
Polyamide	Nylon	2.9
Polyester	Dacron, Fortrel, Terylene, Kodel	4.2
Olefins (polyethylene, polypropylene)		0.9
Acrylic	Acrilan, Orlon	1.8
Aromatic polyamide	Kevlar, Arenka, Nomex, Conex	0.1
Vinylidene chloride	Vinyon, Saran	
Polyurethane	Spandex	
Novoloid	Kynol	

From a structural chemistry point of view, it is now recognized that three levels of molecular structure are of importance in understanding the formation and behavior of fibers:

(a) organochemical structure—the repeating unit of the polymer;

(b) macromolecular structure—the length, distribution of length and stiffness of the polymer chains; and

(c) supermolecular structure—the arrangement of the chains.

Thus, all synthetic fibers are to be regarded as semicrystalline, irreversibly oriented polymers whose properties are controlled largely by controlling the three levels of structure. Some of this control is exercised through the base chemistry, some through the synthesis path and some by post processing of the spun filament (drawing and annealing). Other modern treatments are applied to the finished goods, e.g., high-temperature pressing of fabric padded with a cross-linking formulation to form permanent creases (Permapress).

8. Electronic Materials

It is often assumed that the age of electronic materials began with the 1947 invention of the transistor by Bardeen, Brattain and Shockley. Actually its roots go back much further. Faraday (1833) observed negative thermal coefficient of resistance behavior in silver sulfide. Rosenschold (1835) found certain solids exhibited rectifying action on ac currents. This phenomenon was rediscovered in 1874 by Braun, who observed the difference in conductivity of metal contacts on sulfides and oxides with the direction of current flow, and by Schuster with a comparison of the conductivities of copper and tarnished copper. The rectifying action of selenium was observed as early as 1876, but dry power rectifiers did not become commercially available until the 1920s. Grondahl (1926) introduced the Cu/Cu_2O rectifier. Germanium rectifiers date from about 1925.

Bose patented in 1904 the diode as a radio detector which competed with Fleming's oscillation valve (vacuum diode) in that it was smaller, cheaper and more reliable. Unfortunately, Bose failed to realize the essentiality of the semiconducting behavior of his crystals. Pickard (1906) invented the silicon radio detector calling it the "Perikon" detector. Losser (1924) also experimented with the "cat's whisker" point contact rectifiers, especially zincite with carbon or steel. Eccles introduced the crystal diode oscillator in 1909.

These early experimenters used natural crystals such as galena. The Bell Laboratories' invention of the transistor relied on laboratory-grown crystals and emphasized three features of the new solid state electronics: the need for ultrapurity, the use of single crystals with a high degree of structural perfection, and controlled doping with selected additive elements. Initial work with these electronic semi-conductors was with germanium with a 0.6 eV band gap. Major activity now is with silicon with 1.1 eV band gap and there is developing interest in GaAs with still higher band gap (1.45 eV) and much higher carrier mobility. GaAs is a member of the group of so called "isoelectronic" compounds between III–V and II–VI element pairs which were first synthesized and studied by Welker in the early 1950s.

Other developments in electronic materials have followed two paths. In one, research has been directed toward the development of synthetics to replace natural materials in critical applications, e.g., potassium niobate for mica, lithium tantalate for crystal quartz, and barium titanate for Rochelle salt. Such work has not only freed the industry from dependence on a rare material, but has also produced a more consistent quality of material. In the other path, the aim has been to find or create a material whose properties are

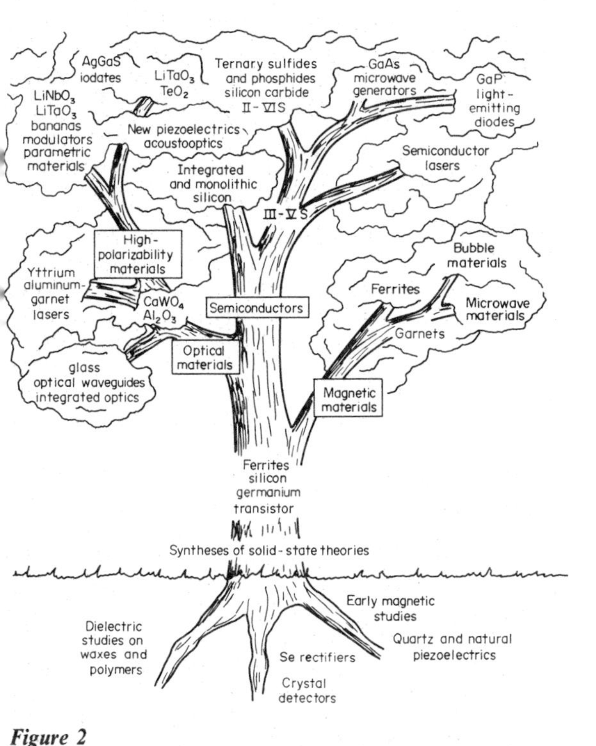

Figure 2
The electronic materials tree. Representing progress in electronic materials, the roots lie in the basic sciences and in early studies of materials, the trunk signifies the development of the transitor, the branches represent those areas of current electronic materials progress—semiconducting materials more complex than those used in early transistors, magnetic materials now being used in computer memories, and optical materials for use in light-emitting, transmitting, and modulating devices (after Chynoweth A G 1976 *Science* 191: 726)

optimized for a particular application. Examples are antimony sesquisulfide for photoconductors for light-to-electrical interconversion, specially doped phosphors for color TV screens, and ferrites as high-frequency beam deflection magnets. The developments of electronic materials is illustrated in Fig. 2.

9. Polymers

Natural polymers—ivory, tortoiseshell, bone, shellac, wool, for example—were recognized and used for hundreds of years before any attempts were made to synthesize them. Perhaps the first was Schönbein's, who, in 1845, made nitrocellulose, or "explosive cotton wool" as he called it by reacting cellulose with nitric acid. Then in the early 1850s, Schönbein—experimenting with solvents on nitrocellulose—found that when a mixture of ether and alcohol containing dissolved nitrocellulose was evaporated, a liquid glue was left which could be spread into thin films. These were suggested for surgical application but found their first real application in photography. Parkes' attempts to make a moldable material from nitrocellulose led first to "Parksine" (1862) and eventually through J. W. Hyatt to the first commercially successful plastic, celluloid (1870). Curiously, the applications Hyatt had in mind were billiard balls and dental plates. Work with regenerated cellulose for fiber production has already been discussed in Sect. 7. Baekeland, in 1901, introduced a new moldable thermosetting resin as a condensation product of the reaction between phenol and formaldehyde (later known as Bakelite) to replace the highly flammable cellulose nitrate derivatives as an insulating plastic.

So far, there had been little study or knowledge of the composition or structure of either the natural or synthetic polymers. About 1900 this situation began to change with the synthesis studies of Fischer, Leucks and Landis in 1906 on the polypeptides, and the accidental preparation of polyethylene by Bamberger in 1900. A significant breakthrough was made by Staudinger (1920) who, from study of systematic viscosity measurements, postulated a chain structure for all polymers, natural or synthetic. Though not fully accepted for many years, accumulating physical and structural evidence as well as the success of novel synthesis techniques based on the concept, finally established the chain as the basic structural feature of polymers. Cross-linking, the formation of copolymers (graft, random, alternating, block and network) and polyblends, today all practical methods of molecular engineering, all derive from the chain motif.

A chronology of the leading commercial types of plastics is presented in Table 5 (see also *Polymers: Technical Properties and Applications*).

Elastomers are another polymer family with an interesting history. Early explorers of the new world in the sixteenth century were surprised to find Amazonian natives with clothing, toys and footwear of natural rubber, then called caoutchouc. A rubber industry was begun in 1824 by Macintosh with the establishment of a raincoat factory. The shortcomings of rubber, embrittlement with cold and stickiness with summer humidity, were overcome by Goodyear in 1839 with his process of hot vulcanization using sulfur. Later important rubber products were Galloway's flooring material, linoleum (ground cork and rubber bonded by cross-linked linseed oil) in 1844 and Thomson's pneumatic tire in 1845. The growing understanding of polymer structures contributed to the Germans' success in their initial efforts to synthesize rubber, first with methyl rubber and then with Buna (polybutadiene–styrene), and American developments just prior to World War II brought out neoprene and butyl rubber. The threat of the cutoff of Far Eastern natural rubber supplies by the Japanese caused the US government to fund a forced draft development of a synthetic rubber industry based on

Table 5
Chronology of plastics development

Year	Family	Tradenames
1909	Phenolics	Bakelite
1925	Polystyrene	Styron, Styrofoam
1926	Alkyds	
1931	Poly(vinyl chloride)	PVC
1932	Poly(methyl methacrylate)	Lucite, Perspex, Plexiglas
1935	Polyethylene, low density	Polythene
1938	Polytetrafluoroethylene	Teflon
1941	Polyamides	Nylon
1943	Silicones	
1948	Acrylonitrile–butadiene–styrene	ABS
1940s	Epoxies	
1954	Polyurethane	
1954	Polyethylene, high density	
1954	Polypropylene	
1955	Acrylics	
1959	Polyformaldehyde	Delrin, acetals
1959	Polycarbonate	Lexan, Merlon
1964	Poly(phenylene oxide)	Noryl, PPO
1965	Polysulfone	Udel
1970	Poly(butylene terephthalate)	PBT, Valox, Celanex
1972	Poly(phenylene sulfide)	PPS, Ryton
1977	Poly(ethylene terephthalate)	PET

the styrene–butadiene formulation. By 1957 an annual production of over 10^6 t was achieved, far surpassing natural rubber and the other synthetics.

10. Composites

Structural composites date back at least to the days of the Mongol nomads with their bows composited of wood, sinew and animal glues. Significant use of composites in the modern age was initiated by the availability of fiberglass in the 1930s. Today's composites range from high-performance automobile tires to subway seating to exotic aerospace structures. Matrices may be plastics, ceramics or metals and the reinforcement filaments may be polymer, glass, metal, carbon or ceramic. Some of the properties of the leading reinforcements are shown in Table 6.

The attractions of the composite approach are the ability to synthesize materials with combinations of properties unavailable in any homogeneous material, to predict the properties and to tailor the properties to the specific part and application. Reinforcement geometries have been principally fibers, either chopped or continuous filament, and occasionally flakes or laminates. Initial markets have often been found in sports equipment (fishing poles, golf clubs and recreational boats) and military applications (aircraft, space structures and weapons), but as costs were reduced and experience gained, large civilian markets were entered, particularly transportation. In 1980, more than 1 Mt of fiberglass-reinforced plastics were used.

Table 6
Properties of reinforcing fibers

Fiber	Modulus of elasticity (GPa)	Tensile strength (MPa)
Graphite 25	172	1200
Graphite 40	276	1700
Carbon	365	2500
Boron	400	2750
Ceramic (Al_2O_3–B_2O_3–SiO_2)	152	1700
Fiberglass S	82	4800
Fiberglass E	72	1700
Kevlar 49 (aromatic polyamide)	124	2300
Nylon	2.8	6620
Silicon carbide	427	2800
Asbestos	145	3710

Some of the outstanding properties of fiber-reinforced composites are shown in Table 7, and a section of a metal–ceramic natural composite is shown in Fig. 3.

11. Concluding Remarks

The foregoing discussion of materials developments over the past two centuries allows a projection of what may be reasonably expected in future years:

Table 7
Typical values of some properties of fiber-reinforced composites compared with steel, aluminum and Ti6A14V

Material	Weight of fiber (%)	Tensile strength (MPa)	Modulus of elasticity (GPa)
Graphite-I–epoxy	65 continuous	1380	124
Graphite-II–epoxy	65 continuous	1100	180
S-glass–epoxy	65 continuous	1660	48
E-glass–epoxy	65 continuous	1100	35
Kevlar–epoxy	65 continuous	1240	62
SMC-30	30 chopped	140	10
SMC-65	65 chopped	280	13
BE–Ti 6-4	50	700–1400	185–220
SI–Ti 6-4	28	980	200
Borsic–Ti 6-4	50	965‖, 290⊥	250‖, 195⊥
Steel (5160)	0	1380	200
Aluminum	0	560	70
Ti6Al4V	0	1000	110

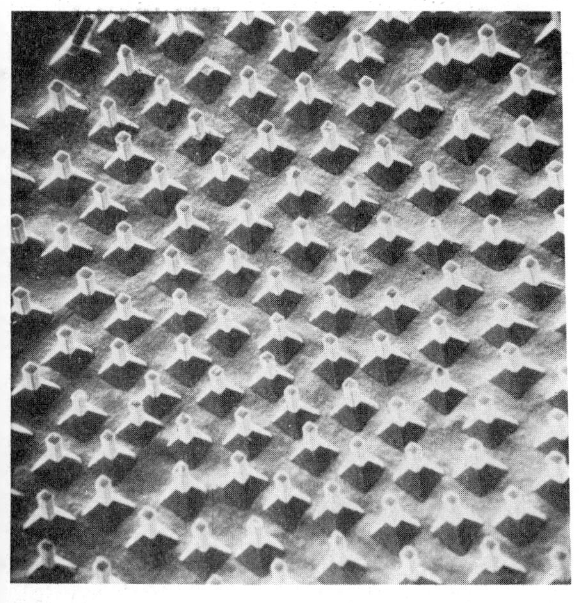

Figure 3
Section of a metal–ceramic natural composite formed by unidirectional solidification of eutectic Ni, Cr solid solution and 1.5 μm TaC fibers with 8 μm spacing (courtesy of M F Henry, General Electric)

(a) new compositions of matter with properties substantially beyond anything we now know;

(b) new structures achieved even with conventional compositions, e.g., theoretically dense poly-crystalline ceramics or glassy metals;

(c) new processes capable of more precise control, greater productivity and more continuous in operation;

(d) control of both composition and structure virtually on an atomic scale rather than a grain size scale as is mostly the case today, e.g., ion implantation, molecular beam epitaxy;

(e) wider and more rapid advance in polymer chemistry relative to the other major materials fields; and

(f) the beginnings of a technology of biomaterials, both the application of synthetic materials to living systems and also direct intervention in the control of structure and composition in living matter.

See also: Materials: History Before 1800

Bibliography

Aitchison L 1960 *A History of Metals.* Interscience, New York

Brown E MacDonald S 1978 *Revolution in Miniature: The History and Impact of Semiconductor Electronics.* Cambridge University Press, Cambridge

Craver J K, Tess R W (eds.) 1975 *Applied Polymer Science.* American Chemical Society, Washington, DC

Davey L 1961 *A History of Building Materials.* Phoenix House, London

Dummer G W A 1977 *Electronic Inventions 1745–1976.* Pergamon, Oxford.

Holonyak N 1979 The evolution of silicon semiconductor technology 1952–1977. *J. Electrochem. Soc.* 126: 20

Holstead P E 1962 The early history of Portland cement. *Trans. Newcomen Soc. Study Hist. Eng. Technol.* 34: 37–54

Hunter D 1978 *Papermaking, The History and Technique of an Ancient Craft*. Dover, New York
Kaufman M 1963 *The First Century of Plastics: Celluloid and its Sequel*. Plastics Institute, London
Kranzberg M, Pursell C W (eds.) 1967 *Technology in Western Civilization*, Vols. 1 and 2. Oxford University Press, Oxford
Mehl R F 1948 *A Brief History of the Science of Metals*. American Institute of Mining, Metallurgical and Petroleum Engineers, New York
Phillips C J 1941 *Glass: The Miracle Maker, Its History, Technology, and Applications*. Pitman, New York.
Rebenfeld L 1975 Chemistry and technology of fibers. In Craver H K, Tess R W (eds.) 1975 *Applied Polymer Science*. American Chemical Society, Washington, DC
Science (Feb. 20 1976): Issue devoted to Materials
Science (May 23 1980): Issue devoted to Advanced Technology Materials
Singer C et al. (eds) 1958 *A History of Technology*, Vols. 4 and 5. Oxford University Press, Oxford
Skempton A W 1963 Portland cements 1843–1887. *Trans. Newcomen Soc. Study Hist. Eng. Technol.* 35: 117–52
Smith C S 1965 *The Sorby Centennial Symposium on the History of Metallurgy*. Gordon and Breach, New York
Tylecote R F 1976 *A History of Metallurgy*. Metals Society, London
Westbrook J H 1970 Materials for tomorrow. In: Bronwell A B (ed.) 1970 *Science and Technology in the World of the Future*. Wiley, New York

J. H. Westbrook
[Scotia, New York, USA]

Mechanics of Materials: An Overview

Reliability in load-bearing applications is generally limited by a material's inelastic response. This response may involve plastic deformation, crack growth or some combination of deformation and cracking. The primary objective of studies of the mechanical behavior of a material is the development of relations between the deformation and crack growth in the material and the loading applied to the system. The loading may be mechanical or thermal in origin, monotonic or cyclic. The resulting inelastic behavior may involve time (rate) invariant or time-dependent deformation or cracking. Various material responses to applied loading thus obtain, depending on the material, the temperature and the loading conditions. It is convenient to classify the material response into two major types: brittle and ductile. However, it should be recognized that the distinction is often vague.

Brittle materials in the absence of cracking are subject only to elastic deformation at low and intermediate temperatures (less than $\sim T_m/3$, where T_m is the melting temperature), while ductile materials typically exhibit plastic deformation by dislocation motion. Nevertheless, substantial time-dependent deformation can occur at elevated temperatures in both material types.

In the presence of cracks, permanent deformation can occur in the highly stressed region near the crack tip in both brittle and ductile materials. However, the relatively low crack propagation resistance that characterizes brittle materials may be attributed to a restricted permanent strain capacity (associated with, for example, phase transformations and microcracking); whereas, substantial crack-tip strains can be generated, by dislocation motion, in more ductile materials.

The description of a material's response to applied load has been conducted at both the continuum and microstructural levels. The ultimate objective of mechanical property studies is to provide constitutive relations for the macroscopic load response based on considerations of microscopic material behavior. This association has not yet been well developed in most regimes of material response.

It is essential to recognize that the application of material properties, especially fracture parameters, to reliability assessment is probabilistic in nature.

1. Fracture

The assessment of fracture is contingent upon the loading conditions and the temperature. The brittle/ductile distinction is used to describe the fracture resistance of materials at low/intermediate temperatures. In this temperature range, both brittle and ductile materials are susceptible to environmentally assisted cracking (stress-corrosion cracking), but only ductile materials exhibit accelerated crack growth under cyclic loading conditions (fatigue crack growth), as shown in Fig. 1. At higher temperature

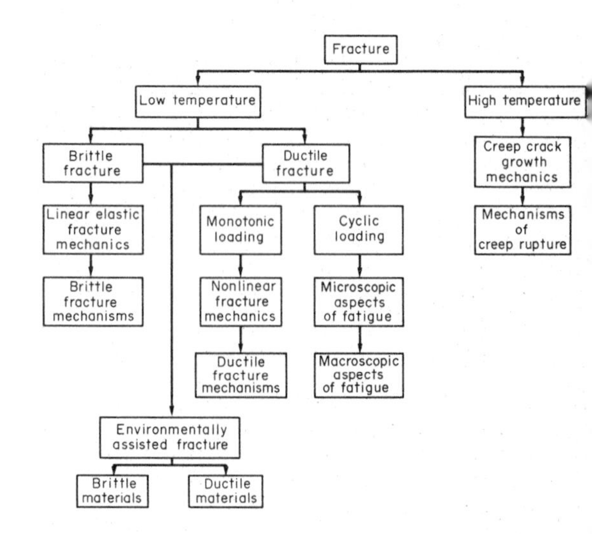

Figure 1
The various modes of fracture in materials and their association

all materials are subject to inelastic deformation, and failure is invariably accompanied by some permanent strain. Coupled effects of the environment and cyclic loading also emerge at high temperatures. Some important considerations pertinent to each of these behavioral regimes are summarized in this article.

The continuum aspects of brittle crack propagation are relatively straightforward, because the macroscopic response is linear elastic (Fig. 2a). Crack growth can thus be fully characterized by a stress intensity factor K_I. Similarly, the essential linearity of the material at both large and small strains (Fig. 2a) has permitted the development of advanced micromechanics descriptions of crack growth. These descriptions include relatively complete models of microstructural interactions involving crack deflection, stress-induced martensite transformations and stress-induced microcracking. However, the brittle fracture problem is subject to concerns regarding the flaws that initiate failure and their relative severity, because flaws of very small dimensions are responsible for catastrophic breakage. The micromechanics of crack initiation from flaws in brittle materials, although of major importance, are as yet incompletely understood. Flaws produced by machining are most thoroughly comprehended. However, considerable additional

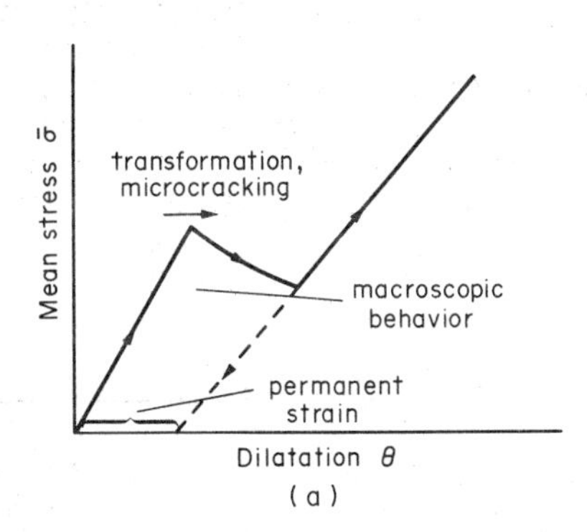

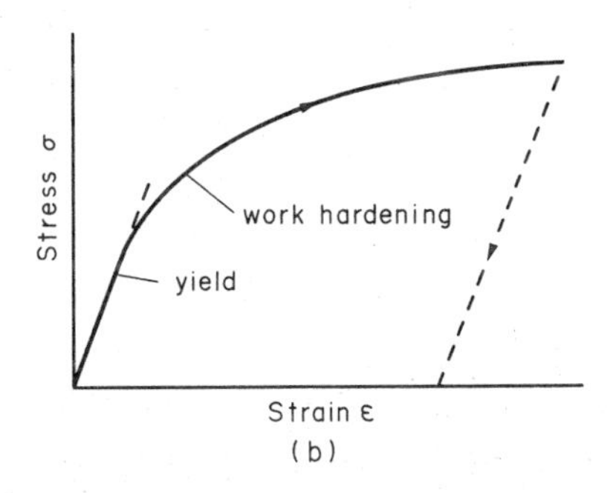

Figure 2
Stress–strain relations for materials: (a) brittle materials; (b) ductile materials

work is needed to provide a quantitative understanding of the role of processing flaws. Consequently, statistical considerations presently provide the most useful working characterization of reliability. Other major issues regarding brittle fracture are discussed in the articles *Brittle Fracture: Linear Elastic Fracture Mechanics; Brittle Fracture: Micromechanics.*

The characterization of ductile crack growth is much more complex, both at the continuum and microstructural levels, because of the history-dependent, nonlinear behavior of the material (Fig. 2b). This complexity is reflected in the non-uniqueness of the crack driving force parameters used to characterize crack extension and the consequent restrictions imposed on the use of these parameters. For example, although intial crack propagation can be described by a critical value of a nonlinear energy-release-based parameter, J_{IC} characterization of the subsequent growth depends on the crack geometry and the associated plasticity. The most useful parameter, J_R, provides a limited potential for describing the crack growth resistance. This resistance may be many times larger than J_{IC}.

The micromechanics descriptions of ductile crack growth are based on the evolution of cleavage cracks, or holes from inclusions or second-phase particles in regions around the crack tip (see *Ductile Fracture: Micromechanics*). The stress and strain fields ahead of the crack are used as the basis for models of crack growth. However, the final results are generally expressed in terms of a length parameter that, in many instances, does not have an obvious association with predominant microstructural features. Additional understanding is thus needed to describe fully the crack growth process. Nevertheless, important material trends can be deduced from available crack models.

Ductile materials are susceptible to accelerated failure, by fatigue under cyclic loading conditions (see *Fatigue Crack Growth: Macroscopic Aspects; Fatigue Crack Growth: Mechanistic Aspects*). The fatigue propagation of cracks constitutes a major concern for the structural longevity of these materials. The continuum description on crack growth in cyclically loaded systems is based on the same crack driving force parameters used to characterize crack growth during monotonic loading, but expressed in cyclic form. Certain difficulties, especially with regard to short cracks, derive from this premise. Fatigue crack propagation is generally regarded as microstructure insensitive. However, important microstructural influences prevail in the vicinity of a driving force threshold, where closure of the crack surfaces during the unloading phase of the load cycle exerts a dominant influence. Environmental effects near the threshold are also often explicable in terms of crack closure considerations. Preliminary micromechanics models provide insight into some of these effects, such as the beneficial effects of large grains. However, quantitative descriptions have yet to be developed. Some microstructural effects also occur at high growth rates where ductile rupture contributes to the growth rate. However, these effects are of less practical significance.

Crack growth in both brittle and ductile materials can be influenced by the environment: the stress-corrosion cracking phenomenon. Materials, when subject to stress-corrosion cracking, behave in a relatively brittle manner and the crack growth rate is directly characterized in terms of linear elastic crack driving force parameters, such as K_I. However, the specific mechanisms of crack growth are not thoroughly comprehended. In very brittle materials, such as glasses and ceramics, the active environmental species (usually OH^-) appears to interact electronically with the crack tip bond to cause a local reduction in crack propagation resistance. In more ductile materials, the weakening process often involves a chemical reaction, as well as diffusion of the environmental species into the material in advance of the crack tip. The rate-limiting process then depends explicitly upon the material composition and the active environmental species, but general rules governing cracking susceptibility have not yet been developed. Substantial environmental effects can also obtain under cyclic loading conditions and at elevated temperatures.

Fracture at elevated temperatures, creep rupture, is time dependent and occurs by similar mechanisms in materials that are either ductile or brittle at lower temperatures. Crack growth at high temperature is governed by time-dependent crack driving force parameters, such as C^*, determined by the fundamental growth rate. The characterization of crack growth is thus appreciably more complex than at lower temperatures, and careful consideration must be given to the loading conditions. The mechanism of cracking in crystalline materials involves the diffusive nucleation and growth of cavities on grain boundaries in a damage zone ahead of the crack. Alternative damage mechanisms occur in noncrystalline solids, such as crazing in polymers. The growth of the damage, and its coalescense, to induce crack propagation, is relatively well modelled. However, damage nucleation is still the subject of extensive debate. In practical systems, much of the creep rupture life is consumed before discrete cracks can be identified within the component. Substantial research has thus been devoted to the problem of creep damage and crack nucleation. Creep damage, in the form of cavities, develops primarily on grain boundaries in crystalline solids. The cavities grow and coalesce, by diffusive processes, at a rate dictated by the straining of the surrounding solid, subject to creep deformation. The associated mechanisms of cavitation are relatively well understood at the quantitative level, but again, the damage nucleation process is less well comprehended. It is now known that the damage cannot nucleate homogeneously; and furthermore, there is circumstantial evidence that stress concentrations produced by grain-boundary sliding transients are responsible for continuous nucleation throughout the creep lifetime. However, certain key details of this process remain to be elucidated. The extensive coalescence of cavities into discrete cracks has not been thoroughly investigated, and is not dealt with here in any detail.

2. *Deformation*

The mechanics of deformation are comprehensively developed. Sophisticated continuum theories of deformation have been derived that take account of the complex history dependence of the plastic flow stress, for example, by employing incremental theories of plasticity and associated yield criteria. Limit theorems that closely bound the stress-strain response of materials have been developed and are widely employed (see *Time-Independent Plasticity*). Dilatational effects, and the related pressure-dependent plastic flow, have also been addressed by considering the plastic expansion of voids. However, understanding of such phenomena is still incomplete.

Direct associations between descriptions of deformation based on dislocation arrays and the continuum theories have been attempted but are not thoroughly developed. Dislocation mechanics provide a rationale for the observed trends in flow stress with microstructure, but do not yet permit the *a priori* prediction of continuum flow laws. The theory of dislocations has been most satisfactory in providing a description of trends in flow stress with alloy additions, in the form of solutes or precipitates. Of particular importance in this context is the recognition of the role of the precipitate penetrability on both flow stress and work hardening rate. Notably, low-penetrability precipitates produce concurrent increases in flow stress and work hardening rate. Dislocation theory, involving forest dislocation, has also provided a means of rationalizing general trends in work hardening. However, more complex phenomena, such as the effects of non-proportional loading, are not yet well explained by dislocation models.

3. *Reliability*

The structural reliability of components based on deformation requirements can be generally analyzed using deterministic criteria, for example, employing limit theorems. Statistical considerations are not normally required. However, reliability analyses for fracture-susceptible components require probabilistic criteria, because the toughness and flaw size parameters that dictate failure exhibit statistical variability. Substantial measurement uncertainties are also involved, especially in flaw size estimation. Considerable recent attention has been devoted to this topic. In particular, reliability estimates based on crack growth data and proof testing are important. In the absence of proof testing, statistical considerations related to preexistent flaw distributions are required. Relatively sophisticated descriptions, based on weakest link statistics, and fracture mechanics, have been developed for brittle materials. Size effects are of particular importance in this development. These descriptions have been adequate for materials to which the weakest link assumption clearly applies and provided that the extrapolation beyond the data range is not excessive. Models that allow confident size effect extrapolation to large components from data obtained on small test specimens have not yet been derived. Statistical methods are also required in conjunction with nondestructive techniques of initial flaw characterization, especially when the critical flaws are relatively small, as in high-strength materials. However, much work has yet to be done in characterizing measurement errors and in prescribing the statistics of crack growth from naturally occurring defects, such as inclusions and weld defects.

Ultimate assurance of reliability involves the analysis of failures experienced in the field. Such analysis identifies errors in design, materials selection and materials characterization. Techniques of failure analysis pertinent to structural alloys, ceramics and plastics are discussed in the article *Failure Analysis: General Procedures and Applications to Metals*. Failure analysis requires knowledge of fracture markings on failed components discerned from fractographic techniques, as well as stress computations, fracture mechanics formulations and materials data regarding crack initiation and crack growth. Experience is also an invaluable asset in successful failure analysis.

Bibliography

Bradt R C, Hasselman D P H, Lange F F (eds.) 1978 *Fracture Mechanics of Ceramics*. Plenum Press, New York.

Erdogan F (ed.) 1976 *The Mechanics of Fracture*. American Society for Mechanical Engineers, Warrendale, Pennsylvania

Hirth J P, Lothe J 1982 *Theory of Dislocations*, 2nd edn. Wiley, New York

Lawn B R, Wilshaw T R 1975 *Fracture of Brittle Solids*. Cambridge University Press, Cambridge

Nauman F K 1983 *Failure Analysis: Case Histories and Methodology*. American Society for Metals, Metals Park, Ohio

Nemat-Nasser S (ed.) 1978 *Mechanics Today*. Pergamon, Oxford

A. G. Evans
[University of California, Berkeley, California, USA]

Metals Processing and Fabrication: An Overview

Metals processing and fabrication is concerned with producing useful shapes with economically acceptable properties. In general, processing has the objective not only of shape generation but also of controlling the metallurgical structure, and hence the material properties. Shape generation and property control are often carried out in separate operations, as when steel bar is hot forged into a shaft, machined, and then heat treated to obtain the required final properties. Much current emphasis is on economical metals processing that will produce the shape and the required properties in a single operation. Thus, the steel shaft discussed above could be cold or warm extruded from a bar with the properties needed for use in service.

The term processing refers to a change in the characteristics of an object. These characteristics may include geometry, state of the matter, strength level and information content (data on form). Fabrication involves the assembly of simply shaped pieces into more complex shapes or structures; for example, sheet metal produced by the rolling process may be fabricated into a television chassis containing intricate cutouts and contours. Manufacturing is a term which encompasses both processing and fabrication but is most often applied to operations which are closer to the user end of the materials cycle than the production of the metal from its ore. It should be noted that much of the following discussion, although explicitly concerned with metals, also applies to other materials.

1. Classification of Manufacturing Processes

It is not a simple task to classify the tremendous variety of manufacturing processes. A start can be made by considering the simplified hierachial classification of business and industry shown in Fig. 1. The service industries consist of enterprises like banking, communication and education that provide important services for modern society but do not create wealth by converting raw materials. The producing industries process raw materials into products needed by society and so create real wealth. The distribution industries, like merchandising and transportation, make these products available to the general public. One of the characteristics of a modern industrialized society is that an increasingly smaller percentage of the population produces the wealth that makes an affluent society possible. In the USA, the distribution of gross national product in 1980 was about 68% to service industries, 23% to producing industries and 9% to distribution industries. In 1947 about 30% of the work force was in manufacturing, whereas in 1980 it was about 22%. It is predicted that by the year 2000 less than 10% of US workers will be engaged in manufacturing.

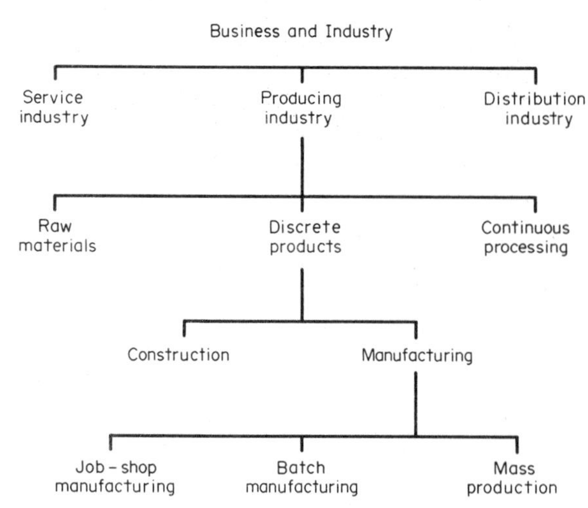

Figure 1
Simplified hierarchical classification of business and industry

The producing industries can be conveniently divided into raw-materials producers (e.g., mining, petroleum, agriculture), producers of discrete products (e.g., automobiles, computers) and industries engaged in continuous processing (e.g., gasoline, wood products, metals, chemicals). Two major categories of discrete products are construction (e.g., buildings, bridges) and manufacturing.

Under manufacturing, three classes can be identified based on the volume of production: job-shop, batch and mass production. Although this classification is usually associated with discrete product manufacturing, it may also apply to the processing industries; for example, some chemicals are produced by batch production, whereas others are produced by continuous-flow processes (mass production).

Low production volume is the distinguishing feature of job-shop production. There is great variety in the work that is done. Often, one-off parts are made to the specific design of the customer. Production equipment must be flexible and general purpose to provide this product diversity, and highly skilled workers are required to allow for a range of work assignments. Typical products made in a job shop are machine tools, specialized machinery, spacecraft, aircraft and prototypes of new products.

Batch production is concerned with the manufacture of medium-size lots of the same product, made only once or at regular intervals. Since batch production plants are capable of a production rate which exceeds the demand rate, an inventory is established and then the plant changes over to making other parts. The equipment used in batch production is general purpose but may be modified with special jigs and fixtures to increase the output rate. Typical products made in a batch-type shop are components for assembled consumer products, furniture, many types of industrial equipment and textbooks. Many machine shops, foundries and pressworking shops operate in the batch mode. It has been estimated that nearly 75% of all parts are manufactured in lot sizes of 50 units or less.

Mass production is the continuous specialized manufacture of identical parts. There is a very high production rate coupled with a high demand rate for the product. Special purpose equipment completely dedicated to the manufacture of one product is used, requiring a heavy capital investment but reduced operator skill. Typical examples of mass production are automobiles, household appliances, light bulbs and hardware items.

2. *Morphological Model of Manufacturing Processes*

All manufacturing processes can be described generically by a general morphological model containing the fundamental elements of material, energy and information flow, as shown in Fig. 2 (Alting 1982).

Manufacturing processes can be divided into three categories with respect to material flow:

(a) mass-conserving processes (through flow)—shape change without metal removal;

(b) mass-reducing processes (diverging flow)—shape change with metal removal;

(c) assembly or joining processes (converging flow)—shape is built up with components manufactured by method (a) or (b).

These categories are detailed in Table 1. Auxiliary materials, such as lubricants, cooling fluids and filler material, may be required in specific processes.

The material being processed may be solid, liquid, gaseous or particulate (granular). In processing, changes in the geometry and/or properties of the materials are effected by mechanical, thermal or chemical interaction with the material. Examples of each category are shown in Table 2.

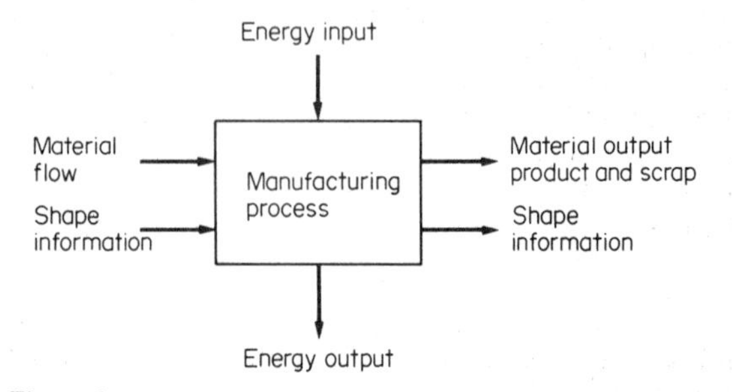

Figure 2
Model of manufacturing processes

Table 1
Classification of manufacturing processes

Type of flow process	State of material	Category of basic process	Primary basic process	Manufacturing process examples
Mass-conserving processes	solid	mechanical	plastic deformation	forging and rolling
	fluid	mechanical	flow	casting
	particulate	mechanical	flow and plastic deformation	powder compaction
Mass-reducing processes	solid	mechanical	fracture	turning, milling, drilling
		thermal	melting and evaporation	electrical discharge machining (EDM)
		chemical	dissolution	electrochemical machining (ECM)
Joining processes	fluid	thermal	melting and solidification	fusion welding
	solid (fluid filler material)	mechanical	flow	brazing

Table 2
Examples of processes generated by choosing between the "morphological box" described by state of material, basic processes and type of flow process

Mechanical	Thermal	Chemical
Elastic deformation	Heating	Solution/dissolution
Plastic deformation	Cooling	Combustion
Brittle fracture	Melting	Phase transformation
Ductile fracture	Solidification	Precipitation
Mixing	Evaporation	Diffusion

To carry out the basic processes energy must be provided to the workpiece material through a transmission medium. This energy flow system may be divided into two subsystems: (a) the tool–die subsystem and (b) the equipment subsystem. The tool–die subsystem describes how energy is supplied and transmitted to the material whereas the equipment subsystem describes the characteristics of the energy supplied from the metalworking equipment. For a mechanical process the energy flow can be provided through (a) relative motion between a transfer medium and the work material; (b) pressure differences across the work material; or (c) mass forces generated in the work material. When the energy is supplied through active motion the transfer medium can be rigid, fluid or granular, depending on the process. When pressure differences are used to supply the energy the transfer medium can be elastic, plastic, granular or gaseous (including vacuum). Mass forces generated in the workpiece are created primarily through gravity, accelerations or magnetic fields; the medium interposed between the field generator and the workpiece is not important so long as it does not interfere with energy transmission. In addition to mechanical energy sources, energy may be provided by electrical, chemical or thermal sources (Alting 1982).

The term "information flow" is used to describe the impressing of the shape information on the workpiece. Combining the geometric possibilities derived from information flow with the material and energy flow defines various possible manufacturing systems. There are two components to the information flow: the surface creation principle and the pattern of motion which creates the shape. There are four methods of surface creation, as follows.

(a) Free forming—the transfer medium does not constrain the desired geometry.

(b) Two-dimensional forming—the transfer medium contains a point or a surface element of the part geometry. Two relative motions are required to produce the surface, as in lathe turning.

(c) One-dimensional forming—the transfer medium contains a line or surface area along the line of the desired surface. Only a single relative motion is required to produce the surface, as in rolling.

(d) Total forming—the transfer medium contains the whole surface of the desired geometry. Thus, no relative motion is needed.

The advantage of the morphological approach advanced by Alting is that it clearly identifies the characteristic features of a process without becoming enmeshed in technical detail. By concentrating on such factors as the nature of the basic process, the energy source and transmission medium, and the surface creation features, it is easier to concentrate on the suitability of a process with regard to factors such as material utilization, tolerance conformance and surface finish generation. Another important aspect of the morphological approach is that when it is fully and carefully implemented all processing possibilities are considered; no unique or unconventional processes are ignored.

3. *Categories of Manufacturing Processes*

The common metal processing and fabrication processes are listed in Fig. 3. The form of the starting materials for each group of processes can vary considerably. Casting processes all start with a molten metal or alloy, and solidification is the metallurgical process involved. Some common casting processes are listed. Bulk deformation processes may start with an ingot which has been hot rolled into a billet of smaller cross section. Other processes, such as hot extrusion, can start with a cast billet. Yet other processes, such as precision closed-die forging, start with a hot-rolled or extruded bar. Sheer forming processes use hot- or cold-rolled sheet or light-gauge plate, as the starting material. Powder metallurgy processing starts with a particulate (granular) form of metal produced by atomization of liquid metal or some other process.

Most shapes produced by casting or bulk deformation processes require further processing to achieve the final shape. A myriad of machining processes are available, involving controlled cutting (fracturing), grinding or the use of unconventional energy sources (electrical or chemical). Parts may also have to be joined to produce the final geometric configuration. Common processing steps beyond this involve heat treatment, to increase properties by phase transformation or precipitation processes, or to decrease hardness (annealing) or remove unfavorable residual stresses. Finishing operations to remove unwanted oxidation or scale, or to impart a corrosion-resistant or decorative coating, may also be required. The final step is to assemble the finished parts into components and subassemblies, and to combine these to make the complete product.

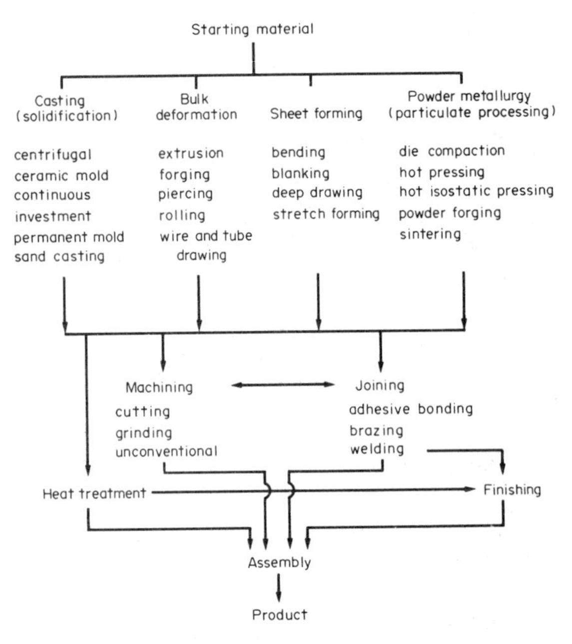

Figure 3
Major categories of metals processing and fabrication

Selecting the best manufacturing process for each part is not an easy task. Most frequently a part can be made using more than one process. As a general rule, parts with similar geometric features will be produced by similar manufacturing processes unless there is a major discontinuity in material properties or tolerance and surface finish requirements. Group technology is a methodology for classifying shapes in the manufacturing context. The specific factors which influence the choice of the manufacturing process are discussed in the article *Design and Materials Selection and Processing*. The overall decision will be based on the economics of competing processes. The process which will produce a part having acceptable properties at the lowest unit cost will be selected.

4. *Economics of Manufacturing*

The selection of the workpiece material has a strong influence on the cost of manufacturing. The factors requiring consideration are: (a) grade or composition of alloy or metal, (b) uniformity of characteristics, (c) workability or weldability, (d) form of material (e.g., bar, wire, tube, plate), (e) size (i.e., dimensions and tolerances), (f) heat-treated condition, (g) surface finish, (h) quality level, and (i) quantity available.

In general, the lower the alloy content of a material the lower its cost. Therefore, the least expensive alloy consistent with the required properties should be used. In order to reduce purchasing and inventory costs, the number of commonly used alloys should be restricted to a modest number of standard materials.

The workability of the material can have an important influence on cost through the scrap or reject rate. There is an inherent conflict between high-performance properties and high workability. Frequently the solution to this dilemma is a change in manufacturing process, for example, from forging to precision casting, or from drop forging to isothermal forging.

For simply shaped parts, such as shafts or bolts, the form in which the material should be obtained and the method of manufacture are readily apparent. However, more complex shapes offer more possibilities for choice. For example, a small gear could be machined completely from bar stock, or it could be precision forged into a rough gear blank and the gear teeth finished by machining. The most economical process will depend on the quantity level beyond which the unit cost of a gear is lowered by making the extra investment in tooling and special equipment. However, for small quantities it will be cheaper to machine each gear from bar stock.

To minimize the cost, the workpiece should be just enough larger than the final part to allow for removal of surface defects. Thus, extra material in a casting may be necessary to allow machining of the surface to the required finish, or a heat-treated steel part may be made oversize to allow for removal of a decarburized layer. Frequently the manufacturing process used dictates the use of extra material, such as the sprues and risers in castings and flash in forgings, which must be removed. Sometimes extra material must be provided for purposes of handling, positioning or testing the part. For expensive aerospace materials there is a major cost incentive to develop advanced processing technology to produce parts to net shape so that they do not require machining. However, even though extra material removal is costly, it is usually cheaper to purchase a slightly larger workpiece than to pay for a scrapped part.

Many engineering alloys can be purchased from the supplier in a variety of metallurgical conditions other than the annealed (soft) condition. Examples are quenched and tempered steel, solution-treated and aged aluminum alloys, and cold-drawn and stress-relieved brass rod. It may be more economical to have the metallurgical strengthening produced in the workpiece by the supplier of the material than to heat treat each part separately after it has been manufactured.

Cold-worked metals have a better surface finish than hot-worked metals. Although cold-worked bars and sheets are more expensive than hot-worked products, in some applications it may be more economical to start with a workpiece with the necessary surface finish than to clean and finish each individual manufactured part.

Materials produced to special quality limits (e.g., lower inclusion content, tighter hardenability limits) carry a premium price. This extra quality is often well justified by superior performance. However, each situation requires special consideration. There is an economic balance between extra testing in the development phase to confirm that the premium material is not needed and the higher cost of the quality material.

Each manufacturing process has the capability to produce a part to a given range of dimensional tolerance. To attempt to produce a design with tolerances outside this range will incur unusual manufacturing costs. Moreover, there is a general relationship between the surface roughness generated by a process and the tolerance which the process allows: the lower the surface roughness, the lower the tolerances that can be maintained. For the most economical design, the loosest possible tolerances and the coarsest surface finish that will fulfill the function of the part should be specified. Processing cost increases nearly exponentially as the surface finish goes from a cast or forged finish (tolerance ± 0.030 in.) to a surface which has a machined finish (± 0.00025 in.).

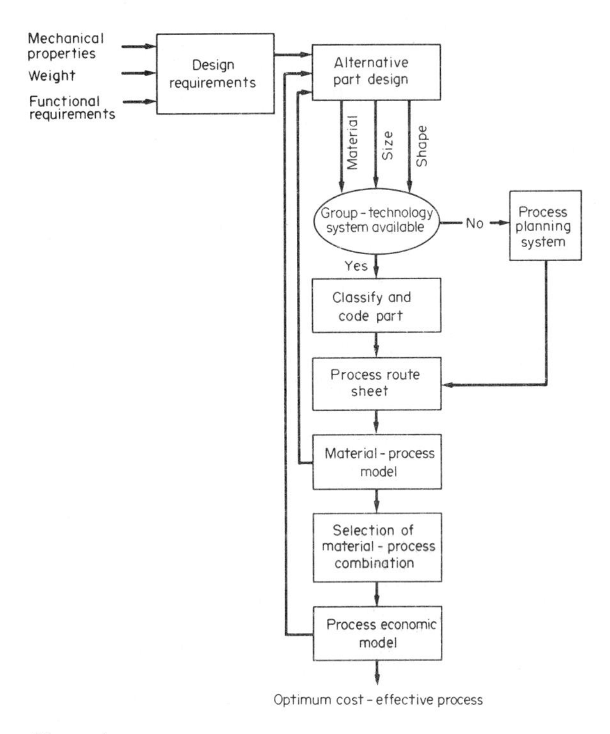

Figure 4
Flow diagram of a computer-aided method to select the least-cost manufacturing process

The relation between design details and processing is complex and not easily reduced to simple formulae. Cost information is highly proprietary and not readily available. There is, however, a growing trend for assembling cost data into computer databases so that process selection, materials selection and part design can be performed in an integrated and rational way. Figure 4 illustrates the components of such a computer-aided design process.

5. *Automation*

The economics of manufacturing are greatly influenced by the degree of automation that can be achieved. Automation is the technology which applies complex mechanical, electronic and computer-based systems to the operation and control of production. Some of the components of automation are automatic machinery for higher production rate (such as numerical control machining centers), automatic materials handling systems, automatic assembly machines, feedback control systems, computer process control systems, and computerized systems for collecting manufacturing data and aiding in the planning and decision-making activities of the manufacturing function.

Automation of manufacturing systems can be divided into fixed automation and programmable automation. Fixed automation, of which the automotive transfer line is a good example, fixes the sequence of processing operations by virtue of the configuration of the equipment. This type of "hard automation" requires a high initial investment, and therefore the volume demand must be high. In fixed automation the individual basic operations are simple but the integration of many operations into one piece of equipment produces a very complex and costly system.

In programmable automation the production equipment is designed to be flexible. The sequence of operations is controlled by a computer program and the system can be reprogrammed to change the seqence of operations. The flexibility of this type of automation makes it well suited to job-shop and batch production. While the investment cost is high, it is not as high as with a fixed automation system.

A simplified model of manufacturing leads to the following equation for the time required to process Q parts through a process consisting of n_m machines:

$$\text{manufacturing lead time} = n_m\,(T_{su} + QT_o + T_{no})$$

where T_o is the average operation time for the process averaged over n_m machines; T_{no} is the average non-operation time for the process, due to delays waiting to get on machines and in transporting incomplete work from machine to machine; and T_{su} is the average setup time for the process due to procuring tools, attaching them to the machine and checking out the setup. Also.

$$T_o = T_m + T_h + T_{th}$$

where T_m is the actual machining time, T_h is the workpiece handling time and T_{th} is the tool handling time per workpiece. The various strategies of automation that can be used are summarized in Table 3.

Special purpose machines aimed at executing a single operation with the greatest possible efficiency are a longstanding aspect of fixed automation. The automatic transfer machine is the classic example.

The objective of combining operations is to reduce the number of production machines through which the part must be routed. Typically, this is accomplished by performing more than one operation at a workstation. In a combined operations strategy the operations are performed in sequence on the same machine. The simultaneous operations strategy is a logical extension in which the machine not only performs multiple operations at the same workstation but performs these operations at the same time. A multiple-spindle drill press is a simple example.

Integration of operations is a strategy in which a series of workstations are linked together through automatic work handling devices into a single integrated system. Since there are several stations, more than one part can be processed at the same time. The overall production output is increased through reducing work in process and total throughput time. The transfer line is the classic example.

Reducing setup time, as opposed to reducing the number of setups required, can be achieved by scheduling similar parts through the machine and the use of more flexible and common fixtures for similar parts. The application of principles of group technology are important in this strategy.

Major opportunities for reducing nonproduction time exist through the use of automated materials-handling systems. Studies have shown that in batch-type manufacturing the workpiece is on the machine tool for only 5% of the time; the rest of the time is spent in nonproductive waiting and transportation.

Process control and optimization is aimed at reducing the machining time through more efficient man-

Table 3
Automation strategies (Groover 1980)

Strategy	Applicable manufacturing function	Influence on manufacturing lead time
Specialization of operations	M	reduce T_m, T_o
Combined operations	M,H	reduce T_h, T_{th}, n_m
Simultaneous operations	M,H	reduce T_h, T_m, T_{th}, n_m
Integration of operations	M,H	reduce T_h, n_m
Reduce setup time	M,C	reduce T_{su}
Improve materials handling	H	reduce T_{no}
Process control and optimization	M,C	reduce T_m
Computerized manufacturing database	D	reduce n_m, T_{no}
Computerized manufacturing control	C, D	reduce T_{no}

Manufacturing functions: M, materials processing; H, materials handling; C, process control; D, manufacturing database

agement of the equipment using feedback control and computer control. The use of the computer for process control also provides the opportunity for automating the generation of the manufacturing database. This includes process planning, work measurement and production planning and control. Much of the paperwork, which is an essential but annoying part of manufacturing, is being replaced by computer communication and analysis.

Computerized manufacturing control differs from process control in that the latter is concerned with the control of the machine–workpiece interaction, whereas the former is concerned with control at the level of the whole manufacturing plant. This strategy involves collecting data, through a computer network, from throughout the factory and using the data to aid factory management.

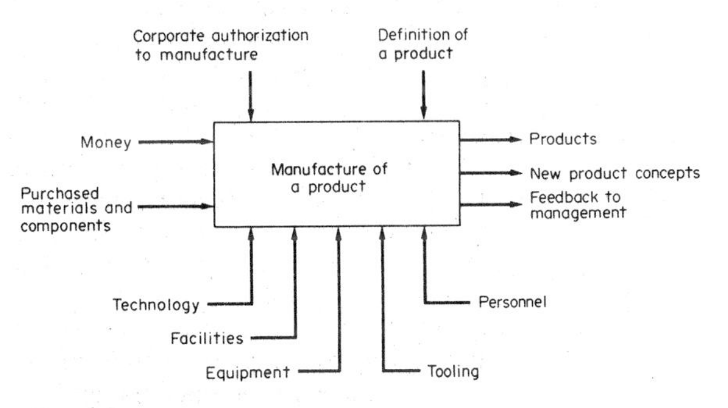

Figure 5
Resources needed for a successful management enterprise

6. The Manufacturing System

Manufacturing is the vital link between product conception and product delivery. Since it lies at the center of many business enterprises, it is a complex interactive function. Here, an attempt is made to define the many complex engineering and business functions which must combine to produce a viable manufacturing system.

To manufacture a product successfully requires the resources shown in Fig. 5. Facilities, equipment, tooling and people must be combined with the appropriate technology to achieve the goal of manufacturing the product. To accomplish this requires funding from the corporation to pay for these resources. These funds also provide the materials and components used in the manufacturing process which are purchased outside of the corporation. All this relies on corporate authorization to provide for the manufacturing operation, and thus there needs to be an approved contract of sale for an existing product, or an agreed design for a new product. The output of this manufacturing operation is a stream of saleable products and a minimum of scrap. The process may also generate ideas for new products or the modification of existing products.

The manufacturing functions can be divided into the four functions shown in Fig. 6. This figure de-

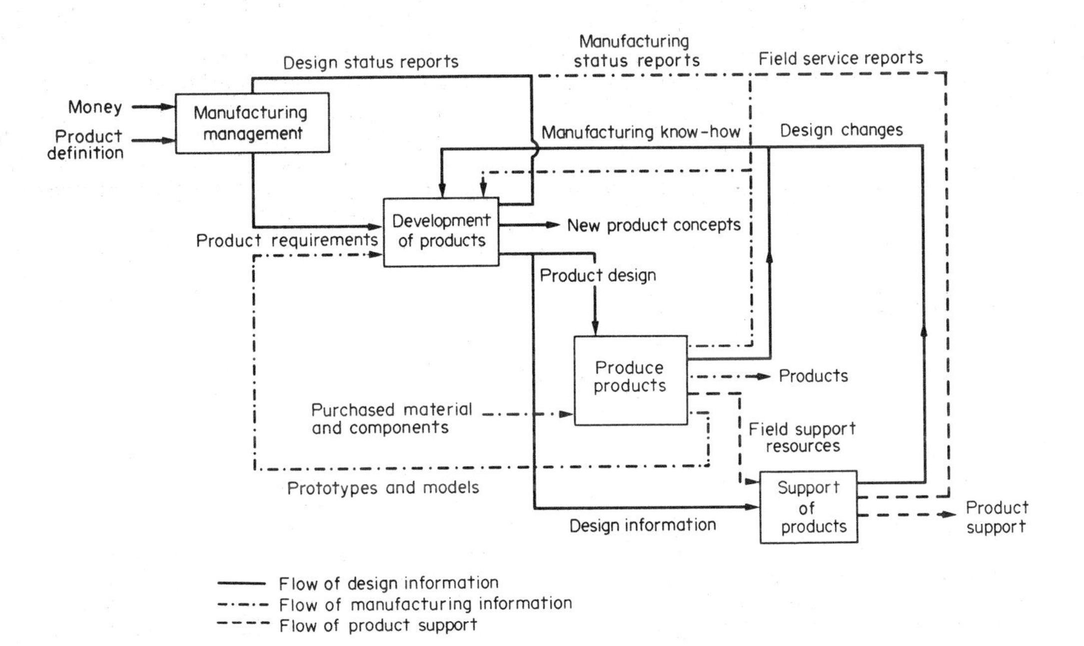

Figure 6
Schematic representation of a complete manufacturing system

lineates the complete manufacturing system and emphasizes the close interaction which must occur between manufacturing management, the design group which develops new products, the manufacturing department which produces the products, and the field-service support group which maintains the products. Both manufacturing and field service provide design changes based on their expertise, while all groups keep the management informed of current status.

Bibliography

Alting L 1982 *Manufacturing Engineering Processes.* Dekker, New York

Boltz R W 1981 *Production Processes: The Productivity Handbook*. Industrial Press, New York

De Garmo E P 1979 *Materials and Processes in Manufacturing.* Macmillan, New York

Groover M P 1980 *Automation, Production Systems and Computer-aided Manufacturing.* Prentice–Hall, Englewood Cliffs, New Jersey

Hitomi K 1979 *Manufacturing Systems Engineering: A Unified Approach to Manufacturing Technology and Production Management.* Taylor and Francis, London

Niebel B W, Draper A B 1974 *Product Design and Process Engineering*. McGraw-Hill, New York

Schey J A 1977 *Introduction to Manufacturing Processes.* McGraw-Hill, New York

G. E. Dieter
[University of Maryland, College Park, Maryland, USA]

N

New Cement-Based Materials

This article describes recent advances in the production of strong materials from hydraulic cements such as Portland cement and the calcium aluminate cements. The hydraulic cements form castable or moldable pastes when mixed with water and these harden at ambient temperature. This ease of forming shaped objects under mild conditions would be of great utility if it were not for the very low tensile properties of the hardened material compared with those of common metals, polymers and ceramics used in fabrication (see Table 1).

Apart from an ability to form shaped objects under mild conditions, an added attraction of the use of hydraulic cements (in particular Portland cement) is the low total energy cost relative to common metals and plastics. Broadly, the energies required to produce 1 m^3 of sand, cement, a synthetic organic polymer and aluminum are, respectively, 0.5, 10, 100 and 1000 GJ.

Recently, very considerable progress has been made in (a) understanding the microstructural features of cement paste responsible for poor tensile strength and (b) discovering means by which cement paste can be formulated and processed to give material having a flexural strength that can be >150 MPa, a Young's modulus >40 GPa and a fracture toughness of >300 Jm^{-2}. Such material is thus comparable with some polymers, metals and ceramics.

1. Porosity and Strength

Hardened cement paste as conventionally prepared contains 25–30 vol% porosity, the pore sizes ranging from a few nm to ~1 mm. The pores of small size are within the structure of the gel-like products of cement hydration, whereas the large pores result from poor particle packing and entrained air bubbles. It has been known for many years that the strength of cement paste—usually compressive strength—decreases as the water–cement ratio of the mix is increased and it has been common to relate strength to total volume porosity by the equation

$$\sigma_c = \sigma_0 \exp(-bp) \tag{1}$$

in which σ_0 is the strength at zero porosity, p is the porosity (volume fraction) and b is a constant.

However, recent work has shown that the tensile or bend strength of cement paste can be correlated with the length c of crack-like pores in accordance with the Griffith relationship:

$$\sigma = (ER/\pi c)^{0.5} \tag{2}$$

in which E is Young's modulus and R is the fracture surface energy. Thus, if the maximum size of pore is reduced, the strength should rise significantly. The low bend strength (5–10 MPa) of conventionally prepared hydraulic cement pastes was found to be due to the presence of pores of effective length around 1 mm.

Total porosity volume has a secondary effect on tensile or bend strength because E and R fall as total porosity volume is increased. A more accurate description of the effect of both porosity volume fraction p and pore size c on the tensile or bend strength of cement paste is given by

$$\sigma = \left(\frac{E_0(1-p)^3 R_0 \exp(-bp)}{\pi c}\right)^{0.5} \tag{3}$$

in which E_0 and R_0 are Young's modulus and the fracture surface energy at zero porosity.

2. Macrodefect-Free (MDF) Cement

In an ideal cement paste, the cement grains should be packed as closely as possible so that hydration products unite the grains and efficiently fill what interstitial space remains. There should be no voids due to entrapped air. Such a microstructure can be brought about by the incorporation of a small proportion of a water-soluble polymer in a water–cement mix having a low water–cement ratio—say <0.2. Polymers such as cellulose derivatives, polyacrylamide and poly(vinyl alcohol) can be used.

Intense mixing under conditions of high shear transforms such a water–cement–polymer mix to a plastic dough in which the volume fraction of particles is ~0.6. Such doughs can be shaped by extrusion, press

Table 1
Mechanical properties of hardened Portland cement paste and other common materials

Material	Density ($g\,cm^{-3}$)	Young's modulus (GPa)	Flexural strength (MPa)	Fracture energy ($J\,m^{-2}$)
Portland cement paste	2.4	20–25	5–10	20
Wood	1.0	10	100	10^4–10^5
Glass	2.5	70	70	10
Polystyrene	1.1	3	100	1000
Aluminum	2.7	70	100–300	10^5–10^6
Steel	7.8	200	500	10^5–10^6
Asbestos cement	2.5	30	50	300
Unglazed pot	2.5	40	50	60

molding or other methods and their deformability allows the collapse of any air bubbles, which do not reappear after the hardening of the paste. Air elimination can improve the volume fraction of particles to ~0.7 and a subsequent 10% volume contraction on drying increases it to 0.8. The remaining space is filled by hydration products. In this way, materials have been prepared in which the total porosity volume is <1% and there are no pores larger than a few μm across—hence these materials have been termed macrodefect-free cements.

3. *Application of MDF Cement Materials*

The ability to form complex cement-based shapes (sheets, pipes, rods and so on) by polymer processing technology such as extrusion or press molding and the mechanical properties of the materials suggest applications ranging from building materials (e.g., tiles, pipes) to electrical components, in which the arc resistance, volume resistivity (10^9–10^{11} Ω cm) and electrical strength (8.7 kV mm^{-1}) are attractive. Cryogenic applications are suggested by an increased flexural strength at −196 °C. These materials have good acoustic damping and hi-fi speaker boxes are being made from MDF cement sheet.

The plastic dough can be filled with a variety of particulate materials; metal powder is of special interest, for the manufacture of electromagnetic interference screens. Since low temperatures (<100 °C) are involved in fabrication, the materials may be reinforced with organic fibers such as nylon to give cement-based materials having fracture energies of 10^4–10^5 Jm^{-2} (see Table 1).

Overall, MDF cements show considerable promise as low-energy alternatives to certain plastics, metals and ceramics.

See also: Calcium Aluminate Cements; Cements, Specialty; Portland Cements, Blended Cements and Mortars

Bibliography

Alford N McN, Birchall J D, Howard A J, Kendall K 1984 Macro-defect-free cement—Strong solids made easy. *Proc. 1st Conf. on Materials Engineering*. Institute of Metallurgists, London

Birchall J D 1983 Cement in the context of new materials for an energy-expensive future. *Philos. Trans. R. Soc. London, Ser. A* 310: 31–42

Birchall J D, Howard A J. Kendall K 1981 Flexural strength and porosity of cements. *Nature (London)* 289: 388

Kendall K, Howard A J, Birchall J D 1983 The relationship between porosity, microstructure and strength and the approach to advanced cement-based materials. *Philos. Trans. R. Soc. London, Ser. A* 310: 139–53

J. D. Birchall
[ICI, Runcorn, UK]

O

Offshore Structure Materials

There are several important characteristics and operations of many offshore drilling units that require the use of different materials, or restrictions on the application of similar materials, from those used for ships. Much offshore exploration by mobile drilling units (those possessing positive buoyancy) and production by fixed (founded in the seabed) or mobile units is carried out in areas where ambient air temperatures remain below 0 °C for long periods, such as in the Arctic. Toughness for these applications must be considered on the basis of the lowest expected steel temperature because ferritic steels undergo a ductile-to-brittle fracture transition. The lowest steel temperature expected in service is estimated from meteorological data representing the coldest region in which the structure is designed to operate.

1. Toughness Versus Structural Application

In general, toughness testing of structural steel for offshore structures is conducted at a temperature below the minimum expected service temperature, the extent of the temperature increment being based on certain design criteria such as anticipated stresses, stress concentrations and redundancy of the structural member. For example, the American Bureau of Shipping defines three structural categories to which different levels of toughness are assigned (Table 1). The three categories are assigned to structural components based on the consideration of the following: (a) consequence of failure, (b) redundancy of component, (c) service experience with similar design, (d) stress level, including stress concentration and restraint, (e) fatigue, (f) loading rate and (g) capability of inspection and repair in service.

Table 1
Charpy V-notch test temperature vs structural category for steels up to 415 N mm^{-2} yield strength

Structural category	Charpy V-notch test temperature
Secondary: least critical—redundant	at service temperature
Primary: main structural members that sustain primary tensile stress and whose failure would jeopardize the safety of the structure	10 °C below service temperature
Special: most critical—fracture critical members which could experience rapid loading at points of stress concentration	30 °C below service temperature

Absorbed energy obtained by a Charpy V-notch test is the most commonly used criterion for grading steels according to toughness. This criterion cannot be related directly to design, but is used as an expression of a toughness quality level. Attempts are being made to relate fracture mechanics tests to structural performance. However, extrapolation of small-scale test data to large complex structures leads inevitably to overly conservative or overly liberal implications. In order to permit the use of materials with toughness test temperatures close to or above the design temperature, the allowable maximum thickness must be restricted in order to reduce the likelihood of plane strain fractures.

Structural steel elements for offshore structures are often fabricated from cold-formed sections. As most structural steels are strain-age sensitive, it is important to evaluate the extent to which a given steel loses toughness after being cold strained. Outer fiber strains less than 3% do not generally reduce toughness significantly. In general, above 3% strain the material is either heat treated to restore the toughness, or tested to determine that the post-strain toughness has not degraded below minimum acceptable values. To simulate the long-term aging effect after cold straining, strained test pieces are heated to 285 °C for 1 h.

2. Lamellar Tearing

Many offshore structural designs, which often resemble civil engineering structures rather than ships, consist of lattices of tubular steel members. Where these lattices come together in node-type connections, lamellar tearing is a potential hazard. Lamellar tearing is the through-thickness separation of steel plates along planes of nonmetallic inclusions, usually manganese sulfides. To reduce the likelihood of lamellar tearing, steels with improved through-thickness ductility (often called Z steels) are frequently specified. Z steels have guaranteed through-thickness reduction of area values (usually 20% minimum). The improvement in through-thickness properties is brought about by lowering the permitted sulfur contents (typically 0.010% maximum) and treating the molten steel with rare earth elements to combine with sulfur and form refractory sulfides that resist deformation at hot-rolling temperatures. Another approach to eliminate lamellar tearing, and to simplify fabrication, is to use steel castings for the node connections. Centrifugally cast straight tubulars have also been proposed for offshore structural applications.

3. *Quenched and Tempered Steels*

Certain offshore drilling unit designs, especially of the self-elevating type, make weight saving in the upper structure and the legs an important consideration, in order to reduce the size and power requirements for lifting equipment, and to aid stability. Therefore, high-strength quenched and tempered martensitic steels of approximately $700\,\mathrm{N\,mm^{-2}}$ yield strength find wider application in offshore structures than in ships. Areas subject to concentrated loads, such as openings for leg structures, cantilever overhangs for drilling booms, helicopter decks and crane pedestals, are other locations in offshore drilling structures in which high-strength steels are commonly used.

4. *Concrete*

Both reinforced and prestressed concrete have been used successfully for fixed offshore structures in the North Sea. The knowledge gained from the long-term behavior of these concrete structures in the sea could lead to an expansion of the use of concrete for large, fixed sea structures.

Bibliography

American Bureau of Shipping 1980 *Rules for the Construction and Operation of Mobile Offshore Drilling Units*. American Bureau of Shipping, New York.

Federation Internationale de la Précontrainte 1977 *Recommendations for the Design and Construction of Concrete Sea Structures*. Cement and Concrete Association, Slough, UK

Sommella J 1979 *Significance and Control of Lamellar Tearing of Steel Plate in the Shipping Industry*, Ship Structure Committee Report SSC 290. US Coast Guard, Washington, DC

UK Department of Energy 1981 *Guidance for the Design and Construction of Offshore Installations*. HMSO, London

B. L. Alia
[American Bureau of Shipping,
New York, USA]

P

Paints, Varnishes and Lacquers

A thin film applied to protect or decorate an inanimate surface is known as a surface coating—a term covering paints, varnishes and lacquers. These materials differ not so much in their function as in their mode of drying or curing.

To be effective, coatings must, regardless of the mode of drying, be applied in the proper manner by brushing, rolling, spraying or rubbing, and must go over surfaces that have been suitably prepared. Surface preparation is as important as selection of the coating material. Generally, results are better when a coating of lesser quality is used over a properly prepared surface than when a superb coating material is applied over a poorly prepared surface.

Paints are usually pigmented vehicles—derived from petrochemicals or vegetable oils—that dry or cure by oxidation or some other chemical reaction. To facilitate spreading they must be reduced by solvent or some other liquid. They also contain one or several of the following additives: driers, which are catalysts to hasten or complete oxidation; anti-skinning agents; ultraviolet inhibitors; microbiocides to prevent mildew and protect against bacterial action in storage; thickening agents; and rheological agents to aid flow and levelling.

Varnishes are transparent or nearly transparent vehicles with additives that aid drying and stabilize color. Lack of pigmentation distinguishes them from paint, although transparent colors, mainly dyes, can be added to produce wood stains.

Lacquers differ from paints and varnishes in that they dry by evaporation of their solvent.

1. Paints

Paints are categorized in a number of ways. One categorization is by relative gloss: flat paints, semigloss enamels and high-gloss enamels. Another categorization is by function: architectural, heavy-duty maintenance and factory-applied. A third designates paints by the vehicle family on which they are based, which often but not always relates to their functional category: a poly(vinyl acetate), poly(vinyl acrylic), acrylic, linseed oil or alkyd paint is usually an architectural coating; a urethane, epoxy or chlorinated rubber is likely to be maintenance paint for use in a difficult environment; and urea melamines, cellulose acetate butyrates, vinyl chlorides, and certain urethanes and acrylics are used for factory-applied or product finishes.

1.1 Relative Gloss

Flats and enamels are found in all the functional and vehicle categories. Flats are used mostly on the interior and exterior walls of residences, factories or institutions, and also on product finishes; semigloss enamels are used mainly for architectural trim, such as molding, and wherever a dressy appearance is desired without excessive shine. This may include metal, wood or cementitious surfaces—either painted on-site or in a factory. High-gloss paints are frequently found on factory-applied finishes such as appliances and furniture, on architectural areas subject to frequent cleaning and where durability is demanded.

Flat paints differ from the enamels in probable durability as well as gloss, because they usually, but not always, contain less vehicle than enamels and more hiding pigment and extender. For low-cost paints, excessive amounts of extenders—clay, calcium carbonate and talc—are used to reduce raw-material expense.

The more of a given vehicle in a paint, the greater its ability to withstand scrubbing and impact and to avoid cracking and blistering. Therefore, flats are less durable, as a rule, than semigloss and high-gloss enamels. Exceptions are the so-called flat enamels, which are paints with sufficient vehicle content to qualify as enamels but whose gloss is minimized by a flatting agent.

A flat paint, except as noted, has a high pigment volume concentration (or relatively low vehicle content) and an enamel has a low pigment volume concentration (relatively high vehicle content). A semi-gloss enamel is midway in pigment volume concentration between the two.

1.2 Function

Coatings serve several general functions. As architectural coatings, they usually provide a balance of protection and decoration, mostly in mild environments. Maintenance coatings usually function in harsher environments, some of them extreme, such as in industrial areas where airborne chemicals are present.

Both architectural and maintenance coatings are normally applied in at least two coats: primer and topcoat. Primers differ considerably when used on metal, wood or cementitious surfaces, whereas topcoats may not differ at all if the environment is the same. Where necessary, a third, or intermediate, coat is used to provide extra protection. This is often the same as the topcoat.

2. Primers and Topcoats

Primers are basically of two kinds: barrier or reactive. Barrier primers are used to block off environmental influences such as wind, sun, rain and chemicals. They contain the usual vehicles and pigments. Reactive primers contain pigments that combine chemically with materials in the vehicle to provide the substrate with some form of specialized protection.

In the case of steel, this specialized protection reduces corrosion if water reaches the substrate; in the case of natural dye-containing wood, the reactive substance combines with the natural dye and forms an insoluble substance that will not migrate to the surface and stain the paint.

Reactive primers can also serve as barriers. Zinc oxide, the reactive pigment, increases the barrier effect of linseed oil primers by reacting with the oil's fatty acids to enhance the primer's impermeability to water. Lead pigments act similarly in blocking moisture.

Steel protection, in particular, calls for reactive primers. Finely powdered zinc dust is used in coatings to protect steel substrates. Dispersed in organic vehicles, such as epoxies or phenoxies, or in inorganic vehicles, such as ethyl silicate, zinc protects when water is present by sacrificing itself in a physical reaction known as cathodic protection.

This reaction occurs because zinc in water is more chemically active than iron and is preferentially consumed in the tiny electric cell that results when water reaches two metals. Zinc becomes the anode and is slowly consumed.

Other pigments also change the susceptibility of steel to corrosion. These include zinc phosphate, calcium molybdate and barium metaborate. Protection against corrosion is also enhanced by passivating the surface before painting. This is seldom done on structural metal, but is used extensively on home appliances and other devices whose original manufacturer is zealous about providing satisfactory service life.

Passivation is a procedure whereby phosphate or chromate compounds are applied in solution to the metal surface. There they form metal salts that are less active electrically than the original metal in the presence of moisture and, therefore, the resulting surface is less likely to be corroded.

Primers for cementitious surfaces, and for wood surfaces of some types, seal off pores or cells that could absorb too large a portion of vehicle in non-specialized coatings. These are called primer-sealers because they block pores so that the vehicle in topcoats will not be excessively absorbed.

Proper formulation of primers to be used with specific topcoats is important for three reasons: the topcoat must be able to adhere; the primer, when dry, must not degrade a topcoat by absorbing part of its vehicle, thus weakening it and leaving an unsightly surface; and the primer must not contain a vehicle which, when dry, can be lifted by the solvent in the topcoat. For these reasons, when buying primer—topcoat systems it is important that one manufacturer's products be selected.

Some primer formulations can also be used as topcoats or as intermediate coats. More often, though, coatings with qualifications for particular sets of circumstances are selected for topcoats. Where circumstances are only mildly taxing, relatively inexpensive latex emulsions or acrylics may be used with suitable pigments and additives for decorative effects.

For architectural exteriors, where rain, wind, sun, atmospheric pollutants and microbial attack must be endured, more flexible and durable primers and topcoats are required. These may consist of water-reducible poly(vinyl acetate) copolymers or acrylics, but with a greater percentage of flexible monomers than is used for interior coatings.

The most effective architectural-exterior water-reducible coatings are made of acrylic resins, which use a tough monomer, methyl methacrylate and soft flexible monomers, such as ethyl or butyl acrylate. Acrylics have the advantage of wet adhesion, which means they are less likely than poly(vinyl acetate) copolymers to be harmed if rain or high humidity is present shortly after application.

Older, solvent-reduced, architectural-exterior paints based on linseed oil or alkyd resins are still used, but they are losing ground to the more convenient water-thinned coatings which most experts believe to be about as good.

For harsh environments, high-performance or heavy-duty topcoats are needed, often with special-purpose primers, as noted. For vehicles, these topcoats use materials derived from somewhat exotic sources, often more difficult to make than those used for mass-produced coatings. Selection should be determined after site conditions are thoroughly understood. Temperature changes, wind velocity, proximity to industrial or marine atmospheres, and resulting contaminants must all be considered when selecting a high-performance coating system.

3. Vehicle Systems

Vehicle families used for coating systems—including primers and topcoats and intermediate coats when necessary—are described below. Selection of systems rather than choice of individual primers and topcoats is clarified.

3.1 General Systems

(*a*) *Oils and alkyds.* Older vehicles still in use are linseed oil, used exclusively for exteriors, and alkyd resins—both solvent-reduced. Alkyds are used for exterior flats, for enamel-trim paints and for some high-quality interior paints. Their main use in architectural coatings is for enamels, but water-reducible versions have been gaining slowly.

Alkyds are really upgraded vegetable oils. Soybean, linseed or tall-oil fatty acids are reacted with a fatty alcohol, usually glycerine or pentaerythritol, and a dicarboxylic acid, usually phthalic anhydride. Drying characteristics and performance of the oil are improved.

Virtually all exterior latex primers require the addition of either a vegetable oil or an alkyd—usually about twenty percent of the vehicle—to improve adhesion, particularly if it is to go over an old paint with unseen surface breakdown in the form of chalk.

(*b*) *Poly(vinyl acetate) and acrylate monomers*. Water-reduced architectural paints are based mostly on poly(vinyl acetate) copolymerized with an acrylate monomer, usually ethyl or butyl acrylate. They are used extensively in exterior water-reducible formulations and to a lesser extent on interiors, where their higher cost is a disadvantage. Their superior color retention and ability to resist moisture dilution soon after application are advantages over competing vehicles.

The acrylate monomers soften and flexibilize the poly(vinyl acetate) and the acrylic monomers. Because they are more costly than the harder constituents, they are sometimes slighted as to quantity used. While this is not very significant on interior coatings, it may result in cracking and shorter life expectancy in exterior products.

(*c*) *Terpolymers*. Terpolymers which utilize vinyl acetate, an acrylate and a third monomer have been introduced. One version uses vinyl chloride as the third monomer and approaches acrylic emulsions in performance but at a lower cost. A second version uses low-cost ethylene with some success.

3.2 High-Performance Systems

For harsh environments where coatings may have to protect against airborne acids, alkali or salt spray, or may have to withstand rough handling or the impact of sand, rock or battering by shod feet or machines, coatings based on more durable vehicle families are selected. These include the following, which, except as noted, are solvent-thinned.

(*a*) *Acrylics*. Acrylics, which are chemically related to the water-reducible versions, are more durable. Factory-applied versions are used for curtainwalls and for automobile enamels. For metal coatings under harsh conditions where solvents cannot be used, for example on delicate machinery, water-reducible acrylic latexes of special kinds are available, usually with acrylonitrile or some other additive that improves durability.

(*b*) *Chlorinated rubber*. Chlorinated rubber is used where low water permeability is required for concrete or steel surfaces. It is also desirable if microorganisms are to be encountered, or if the surface may encounter radioactive materials that must be cleaned off. Effective use of this material in metal primers has been achieved with zinc dust.

(*c*) *Epoxies*. Epoxies are tough chemically-resistant vehicles made by combining various proportions of epichlorohydrin and bisphenol A, and various modifiers. Some liquid epoxies can cure at a temperature as low as 10 °C and some types have been cured in ocean water. Up to 100 mils (2.5 mm) of epoxy have been applied in a single coat; in a suitable solvent, epoxies can be sprayed. Blended with coal tar, epoxies form thick coatings for surface protection underground or underwater. One low-melting-point epoxy, a solid, liquefies between 65 and 75 °C. It is thinned with solvents.

Two-component epoxy systems are used for heavy-duty maintenance and marine and floor coatings. Two subtypes are identified by the resin selected for curing. When cured by a polyamide resin, the system is quite flexible and water resistant; a polyamine curing resin is selected if hardness and chemical resistance are more important.

A third type, epoxy ester, is made by precuring reactive portions of the epoxy with a fatty acid such as those found in soybean oil, dehydrated castor oil or tall-oil fatty acids. They dry and perform somewhat like alkyds, but are more water resistant. Their main use is to fortify acrylics, polyesters or phenolics.

Epoxies in water-reducible forms have shown promise and are used in some cases where complete elimination of moisture is difficult. In recent years, epoxies in powder form have proved effective when near-vitreous, yet moderately flexible, protection is required in environments where salt or corrosive alkali incursions occur. Steel reinforcement bars embedded in concrete bridge decks, for example, and subjected to road salt have been protected by these fused powders, following techniques developed with the support of the US Federal Highway Administration and the National Bureau of Standards.

Although they provide long-term protection, epoxies have limited color stability: they fade and chalk. Some success in stabilizing color has been achieved by using a hydantoin modifier.

Epoxies have been combined with polyesters, acrylics and silicones to yield improved coatings. Epoxy polyesters have added hardness and gloss, and acrylic and silicone modifications improve weatherability and chalk resistance.

(*d*) *Fluorocarbons*. Fluorocarbons, which are among the most expensive coatings, are used where repainting is difficult and costly, such as on metal curtainwalls or skyscrapers and for remote powerline towers, or where chemical exposure is severe, as in pulp mills or water-treatment plants. Shop application, closely controlled and done on meticulously prepared surfaces, is necessary, although field touch-up by specially trained applicators is possible. Fluorocarbons are prepared by combining fluorine with hydrocarbons to form a highly stable, weather-resistant polymer with a life expectancy of forty years.

(*e*) *Formaldehyde resins.* Formaldehyde resins are nitrogen-containing polymers (amines) that form when formaldehyde is reacted with urea or melamine. The resulting urea–formaldehyde or melamine–formaldehyde is added to alkyd baking resins to impart hardness and color stability when used for architectural metals or other factory-applied finishes.

(*f*) *Polyesters in combinations.* Polyesters are often combined with other vehicles to impart characteristics not otherwise obtainable. For example, an epoxy polyester provides a glazelike coating resistant to many chemicals and much abuse; silicone polyesters are chemical resistant and highly durable and heat resistant. Silicones improve chemical and heat resistance when added to epoxies, alkyds or acrylics.

(*g*) *Urethanes.* Urethanes are of two basic types: aliphatic and aromatic. The most effective as to chemical and environmental resistance is the aliphatic, which must be cured by a polyester, an epoxy or an acrylic; hence they are known as two-package systems.

Aromatic urethanes lack the weatherability and color stability of the aliphatic. They are usually cured by a polyester or a polyether component, packaged separately and added just before use.

A third general type is known as moisture-cured, a single-package material cured by atmospheric moisture after application. The three types may be applied in the field. In addition, certain elastomeric urethanes are compounded to yield rubbery films for use as tank linings exposed to acids, alkalis or abrasives.

(*h*) *Neoprene.* Neoprene (polychloroprene) is a synthetic rubber applied as a coating if the surface is exposed to ozone, strong chemicals or abrasion. Modified neoprenes are supplied for various circumstances, in solvent-reducible or water-thinnable forms, and in cured or uncured sheets. If neoprene is to be exposed to sunlight it is usually coated with Hypalon (chlorosulfonated polyethylene).

(*i*) *Hypalon.* Hypalon provides chemical-resistant, high-build coatings useful in resisting chemicals and ozone. The selection of curing agents must be related to the material to which Hypalon is to be exposed. Highest chemical resistance is imparted by a litharge (lead monoxide) curing agent, whereas magnesium oxide imparts resistance to strong sulfuric acid. Other curing agents enable the coating to function in other circumstances.

Ethyl silicate is used with zinc dust as an effective primer for steel.

(*j*) *Vinyl resins.* Vinyl resins have been effective as intermediate and topcoats in protecting steel in marine environments. They may be applied in very thin coats or in conventional film build. Vinyl resins for coatings consists of a vinyl chloride monomer with a vinyl acetate monomer to provide strength.

(*k*) *Phenoxy resins.* Phenoxy resins are closely related to epoxies, since they are made of the same basic materials, bisphenol A and epichlorohydrin, but they are thermoplastic, whereas epoxies are thermosetting. They have the advantage, however, of being as thermally stable as many thermosetting resins and have a long shelf life. They also have the advantage of requiring no curing agents. Their acceptance has not been great, but they have been used with some success in anticorrosive painting systems as a primer with zinc dust for vinyl systems in marine environments.

4. Varnishes

These transparent or semitransparent coatings consist of any one of several resins, or may consist of one of these resins dissolved in vegetable oil. Varnishes are thinned with hydrocarbon solvents; driers are added to aid curing. Resins used include alkyds, phenolics, coumarone–indene, vinyl toluene, limed rosin, phenol formaldehyde, ester gum, epoxy ester and urethane. Oils used as varnish components are linseed oil, castor oil and tung oil.

The resins in a varnish usually provide the hard portion, whereas oils impart softness and flexibility. The relative amounts of oil and resin are often expressed as oil length, with softer, more flexible varnishes having larger amounts of oil. Those described as 10-gallon lengths or less are known as short-oil varnishes and are hard. Medium-oil varnishes with 20- to 25-gallon lengths (meaning 20–25 gallons of oil per 100 gallons of resin) are not as flexible as spar varnishes, or long-oil versions, with 33- to 50-gallon lengths.

In addition to hardness and flexibility, desirable properties are water resistance, reasonable speed of drying and freedom from wrinkling. Dyes or transparent brown or red pigments may be added to varnishes to provide stains.

5. Lacquers

Advantages of lacquers are rapid drying resulting from solvent evaporation, consistent results since the product does not react or cure, reduced pickup of dirt, and receptivity to polishing that removes imperfections.

Lacquers are used mainly for factory-applied finishes; also, a considerable quantity is used for automobile and furniture refinishing.

Nitrocellulose, modified by plasticizers and other resins, is used extensively for wood coatings, but petrochemically derived synthetics have been growing in popularity because of better abrasion and heat resistance, and higher solids content, which, in some cases, reduces the number of coats to be applied.

These synthetics include: polyurethanes, cellulose acetate butyrate and melamine–alkyds. Other important lacquer resins include poly(vinyl chloride), poly(vinyl acetate) and acrylics.

See also: Corrosion Protective Coatings for Metals

Bibliography

Banov A 1979 *Paints and Coatings Handbook for Contractors, Architects, Builders, and Engineers.* Structures, Farmington, Michigan

Burns R M, Bradley W 1967 *Protective Coatings for Metals*, 3rd edn. Reinhold, New York

Myers R R, Long J S (eds.) 1967–76 *Treatise On Coatings.* Vols. 1–5. Dekker, New York

Nylén P, Sunderland E 1965 *Modern Surface Coatings: A Textbook of the Chemistry and Technology of Paints, Varnishes, and Lacquers.* Wiley, New York

A. Banov
[American Paint Company, New York, USA]

Particulate Composites

Particulate composites are defined in terms of the size, shape and concentration of the constituents and their roles in load bearing. A large number of composites can be considered as belonging to this class if due account is taken of scale effects. In this article, two categories are considered: metal-bonded composites and resin-bonded composites. They are classified by composition, method of preparation and use.

Mechanical properties, thermal and electrical conductivity, thermal expansion and dielectric constant, are discussed here and, wherever possible, the theory for the property, which is applicable to the entire class of composites, is given.

1. Definitions

Particulate composites may be defined as follows.

(a) They have components which are not noticeably one- or two- dimensional (as fibers of lamellae would be) but have similar dimensions in all directions (and in some cases are spheres).

(b) All phases carry a proportion of the load.

Particulate composites therefore form an intermediate class between dispersion-hardened materials and fiber-reinforced ones, for in the former the particles are very small, their proportion is a few percent, and they act by blocking dislocation movement and therefore yield in the matrix. In fiber reinforcement, by contrast, the proportion of fibers is high and they are the main load bearers. Particulate composites have proportions of the hard phase from a few percent up to 70%.

In complete generality we may include in the class of particulate composites materials such as rock, concrete, ceramics such as porcelain and alumina–silicate (fire-brick), hard metals such as tungsten carbide–cobalt, and filled resins such as reinforced phenolics, epoxies and thermoplastics.

Although we shall not treat the subject in such generality, it is as well to point out that many of the physical properties of particulate composites (elastic moduli, yield and fracture properties, thermal and electrical conductivity, thermal expansion) may be described by theories which are completely general and depend only on the physical properties of the components and their concentrations. The only problem then to be considered in each case is the size of the "representative volume element," that is, the unit of scale above which the material can be considered as a continuum but below which the microstructure must be taken into account. Hill (1963) has defined the representative volume element in precise terms in a paper dealing with theoretical principles which apply to all reinforced solids. Holliday (1966) included the concept of the "representative cell" in a comprehensive discussion of geometrical considerations and phase relationships in composite materials.

2. Metal- and Resin-Bonded Composites

We shall be concerned in this article mainly with metal- and resin-bonded composites rather than with the broader class outlined above. "Metal-bonded" composites include structural parts made by powder metallurgy, electrical contact materials, metal-cutting and rock-drilling tools, and magnet materials. Several of these have been in use for 50 years or more, but a thorough theoretical understanding of their behavior has only been achieved since the 1970s following the upsurge of work on composites in general and fiber composites in particular. Details of some metal-bonded composites are given in Table 1.

In parallel with the metal-bonded composites is the class of resin-bonded ones of which the earliest examples were the filled phenolics (e.g., bakelite and formica) in which the filler particles served primarily as an adulterant to reduce the cost of the finished molding but were later found to play a significant part in reinforcement and enhancement of other properties.

In resin-bonded composites a great range of different filler types, concentrations and sizes is used depending upon the properties desired and the mode of preparation. The size may vary from submicrometer, in the case of silica flour, to several millimeters (e.g., glass spheres or sand), and the concentration from a few percent up to 70%. Typical fillers are silica flour, wood flour, mica (for insulators), china clay, carbon, cellulose, glass spheres, chalk, kaolin, talc, ground slate, cork and metal powders.

Table 1
Metal-bonded composites

Type	Particle size and concentration	Method of preparation	Uses	Reference
Cemented carbides	0.8–5 μm particles of WC, TiC or TaC in cobalt or other metal of iron group; up to 94% carbide	powder metallurgy and liquid-phase sintering	cutting tools, dies, weights, rock-drilling bits	Jones (1960)
Reinforced electrical conductors	300–400 μm particles of W or Mo in Ag or Cu matrix 30–80% refractory	powder metallurgy and impregnation with conducting matrix	heavy-duty electrical contacts, electrodes for spark erosion machining	Jones (1960)
Electrically conducting composites	50–150 μm particles of graphite as soft phase dispersed in copper, bronze or silver; 5–70% graphite	powder metallurgy	electrical contacts (brushes on motors and other moving parts carrying electric current)	Jones (1960)
Magnetic composites (e.g., Alnico 5)	0.1 μm particles of FeCo precipitate in Al–Ni matrix phase; 75% Fe–Co phase	casting or powder metallurgy	permanent magnets of high coercivity	de Vos (1969)
Bearing alloys				
(a) White metal	10–100 μm particles of hard intermetallic (Cu_6Sn_5 or SbSn) in soft Sn- or Pb-rich matrix; up to 20% hard phase	casting	plain bearings; particularly superior as regards accomodation of foreign bodies	Neale (1973)
(b) Copper base	10–100 μm particles of soft (Sn/Pb) phase in Cu-base matrix; up to 30% soft phase	casting or sintering; plating	plain bearings; inferior to (a) as regards accommodation, but better high-temperature faitgue properties	Neale (1973)
(c) Aluminum base	10–100 μm "islands" of Sn in Al; 20% of soft (Sn) phase	casting and cold work	plain bearings; similar to (b) but better corrosion resistance	Neale (1973)
(d) PTFE-filled porous metal	Up to 100 μm particles of soft PTFE impregnated into porous metal matrix, typically bronze; up to 30% PTFE	sintering	plain bearings	Neale (1973)
(e) Graphite-filled metal	Similar to PTFE impregnation, 8–15% graphite	sintering	plain bearings	Neale (1973)

The resins used as the matrix fall into three groups.

(a) Phenol, urea or melamine formaldehydes—these are molded by pressure bonding of powders and the curing involves a condensation reaction. Typical uses are for electrical insulators, durable kitchenware and general-purpose moldings.

(b) Epoxy or polyester resins—these are cured with a hardener in a two-part (usually liquid) mixture in an addition reaction which does not involve high pressures in the mold. Typical uses are for electrical insulation at moderate temperatures, molded goods, the building and construction industry and tooling.

(c) Thermoplastic (heat deformable) resins—these include poly(methyl methacrylate), poly(acrylic acid), polypropylene, polyimide, nylon and PTFE. Such resins with filler included may be extruded into rods, tubes or films as well as being injection molded into intricate shapes. Block copolymers, which are two-phase polymer composites, may be included within this class.

For a comprehensive account of resin-bonded composites, see Lubin (1982).

3. Discussion of Technology

The technology used in each of the cases listed above depends upon the materials employed, the ease of mixing to ensure a uniform dispersion, and the end use. In some cases solid particles of one component may be successfully dispersed in a liquid phase of the other, whereas in other cases both components are or need to be in powder form before they can be mixed and combined to form a composite. In some cases (e.g., with magnet alloys) both sintering and casting can be used and the microstructure of the resulting material differs accordingly, with appropriate end uses for each material. In the case of many polymer-based composites, the polymer is a viscous liquid and powder methods are not possible. With phenolics, however, powders are again available and mixing of resin and filler powders is carried out before pressure molding and curing. The technology may allow control of a critical grain size; for example, in magnet materials the final grain size needs to be of the order of the critical domain size. This is best achieved by powder techniques.

Impregnation of a powder compact by molten metal is used for tungsten- or molybdenum-reinforced copper, whereas solid- or liquid-phase sintering (a situation where the molten phase is formed by reaction) is used for tungsten carbide–cobalt composites.

In deciding upon the appropriate technology many factors other than the physical properties of the composite need to be taken into account (Jones 1960, Lenel 1980, Davidge 1979, Spriggs 1974, Broutman and Krock 1967, Lubin 1982).

4. Mechanical Properties

4.1 Elastic Properties

The mechanical properties of particulate composites in the elastic region are well described by the Hashin–Shtrikman bounds as follows. Let G_c, G_1, G_2 and K_c, K_1, K_2 denote the shear and bulk moduli, respectively, where c, 1 and 2 indicate the composite, phase 1 and phase 2, then the following inequalities hold:

$$\frac{c_1}{1+(K_1-K_2)c_2/(K_2+K_1^*)} \leqslant \frac{K_c-K_2}{K_1-K_2} \leqslant \frac{c_1}{1+(K_1-K_2)c_2/(K_2+K_g^*)}$$

and

$$\frac{c_1}{1+(G_1-G_2)c_2/(G_2+G_1^*)} \leqslant \frac{G_c-G_2}{G_1-G_2} \leqslant \frac{c_1}{1+(G_1-G_2)c_2/(G_2+G_g^*)}$$

In these expressions, c_1 is the concentration by volume of phase 1, c_2 that of phase 2, and $c_1+c_2=1$. In addition,

$$K_g^*=\frac{4}{3}G_g, \quad K_1^*=\frac{4}{3}G_1$$

$$G_1^*=\frac{3}{2}\left(\frac{1}{G_1}+\frac{10}{9K_1+8G_1}\right)^{-1}$$

$$G_g^*=\frac{3}{2}\left(\frac{1}{G_g}+\frac{10}{9K_g+8G_g}\right)^{-1}$$

and G_g, K_g are the greater of G_i, K_i and G_l, K_l are the lesser of G_i, K_i in the mixture. These bounds were proved by Walpole (1966) and are a more general form of those found by Hashin and Shtrikman (1963).

Figure 1, from the latter reference, illustrates the bounds in the case of a tungsten carbide–cobalt composite, while Fig. 2, from Crowson and Arridge (1977), shows their relevance to experimental results on glass-bond-reinforced epoxy resin.

The bounds are completely general and apply whatever the phase geometry provided each phase is isotropic. Modified bounds for anisotropic phases are to be found in Walpole (1969) and Kantor and Bergman (1984).

For many purposes, however, the simpler "rule of mixture" formulae may be used though they lead to bounds which lie outisde the Hashin–Shtrikman bounds. The rule-of-mixture bounds assuming uniform strain throughout the composite give $M_c=c_1M_1+c_2M_2$, where M is any modulus (Young's, shear or bulk), whereas the bound assuming uniform stress involves the compliances M^{-1}; thus $M_c^{-1}=c_1M_1^{-1}+c_2M_2^{-1}$. For comprehensive reviews see Hashin (1983) and Hale (1976).

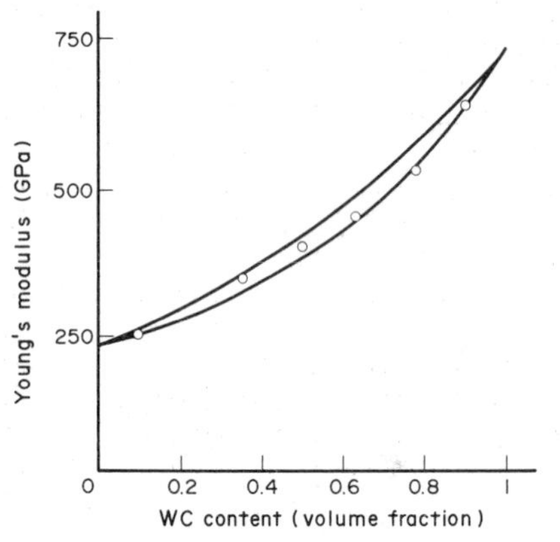

Figure 1
Hashin–Shtrikman bounds for Young's modulus of a WC–Co alloy (after Hashin and Shtrikman 1963)

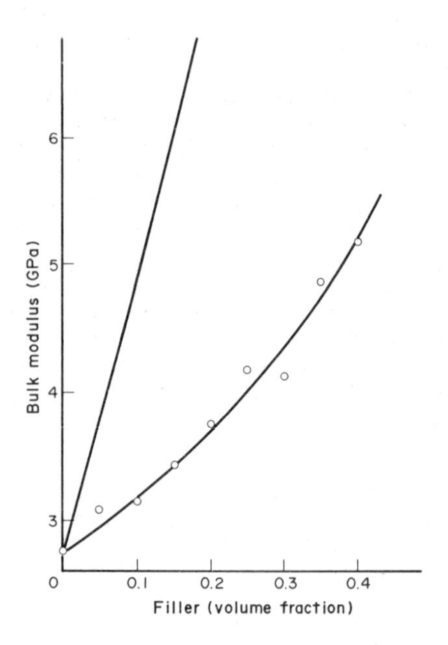

Figure 2
Hashin–Shtrikman bounds for bulk modulus of glass-filled epoxy resin (after Crowson and Arridge 1977)

4.2 Yield and Hardness

When the particles are not deformable by the matrix, theory suggests that the yield strength should be proportional to (mean interparticle spacing)$^{1/2}$, and this is observed for steels (considered as particle-reinforced composites) and for cemented carbides (Fig. 3) at least until the interparticle spacing falls below 0.5 μm (Broutman and Krock 1967). By contrast, when the particles are deformable by the matrix

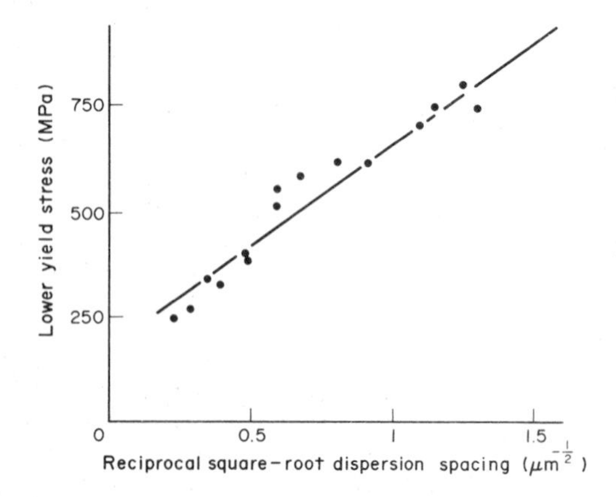

Figure 3
Lower yield points of several hypoeutectoid, eutectoid and hypereutectoid steels as a function of reciprocal square root of dispersion spacing (after Broutman and Krock 1967)

the yield stress is independent of interparticle spacing, and this is observed to be the case in W–Ni–Fe particulate composites. The deformability or otherwise of the dispersed particles thus determines the nature of the composite. If they are deformable the entire flow characteristics of the composite are determined by those of the dispersed phase (i.e., flow stress, work-hardening rate and ultimate elongation are independent of the concentration of the reinforcement). By contrast, where the dispersed phase is nondeformable, the composite is brittle and of low impact strength although the hardness may be very high, leading to uses such as cutting tools. Yield strength, tensile strength and hardness are then all directly proportional to the concentration of the hard phase.

4.3 Fracture

It is found that the fracture toughness K_{Ic} depends upon the concentration of the reinforcing elements in a particulate composite in a manner consistent with a linear variation of the fracture energy $\mathcal{G}$ with volume percent reinforcement. But $K_{Ic}=(E\mathcal{G})^{1/2}$, where E is Young's modulus, which increases linearly with volume percent reinforcement. This would then explain the linear dependence of K_{Ic} upon concentration as is found for example for WC–Co (Chermant and Osterstock 1976) and for TiC-strengthened alumina (Wahi and Ilschner 1980).

5. Thermal Expansion

The rule-of-mixtures law, $\alpha_c=c_1\alpha_1+c_2\alpha_2$, for the expansion coefficient α is close to Kerner's formula $\alpha=c_1\alpha_1+c_2\alpha_2-c_1c_2(\alpha_1-\alpha_2)$, and the experiments of Crowson and Arridge (1977) on glass-reinforced epoxy resin confirm the rule-of-mixtures law as a reasonable approximation at least for this system. Fahmy and Ragai (1970) and Hartwig et al. (1976) using a composite sphere model give the formula

$$\alpha_c = \alpha_m - \frac{3(\alpha_m - \alpha_i)(1-\nu_m)c_i}{2E_m/E_i(1-2\nu_i)(1-c_i)+2c_i(1-2\nu_m)+1+\nu_m}$$

where E is Young's modulus and ν Poisson's ratio, and the subscripts i and m refer, respectively, to inclusion and matrix. c_i is the volume concentration of inclusion. The relation has been tested to concentrations of about 40% with good agreement between theory and experiment (see also Hashin 1983). Ishibashi et al. (1979) found that combinations of different fillers and different resin-curing agents enabled the thermal contraction of composites to be matched to that of any metal. This property is also exploited in dental cements composed of poly(acrylic acid) with added alumina.

6. Other Properties

The formulae for dielectric properties of a particulate composite are the same as those for the elastic moduli (Hashin and Shtrikman 1963) if dielectric constants are exchanged for the elastic constants.

For the electrical conductivity σ of binary metallic mixtures, Landauer (1952) found that the relation

$$c_1\frac{\sigma_1-\sigma_c}{\sigma_1+2\sigma_c}+c_2\frac{\sigma_2-\sigma_c}{\sigma_2+2\sigma_c}=0$$

gave a good fit to experiment.

See also: Composite Materials: An Overview

Bibliography

Broutman L J, Krock R H 1967 *Modern Composite Materials.* Addison-Wesley, Reading, Massachusetts
Chermant J L, Osterstock F 1976 Fracture toughness and fracture of WC–Co composites. *J. Mater. Sci.* 11: 1939–51
Crowson R J, Arridge R G C 1977 The elastic properties in bulk and shear of a glass bead reinforced epoxy resin composite. *J. Mater. Sci.* 12: 2154–64
Davidge R W 1979 *Mechanical Behaviour of Ceramics.* Cambridge University Press, Cambridge
de Vos K J 1969 Alnico permanent magnet alloys. In: Berkovitz A E, Kneller E (eds.) 1969 *Magnetism and Metallurgy.* Academic Press, New York, pp. 473–512
Fahmy A A, Ragai A N 1970 Thermal expansion behavior of two-phase solids. *J. Appl. Phys.* 41: 5108
Hale D K 1976 The physical properties of composite materials. *J. Mater. Sci.* 11: 2105–41
Hartwig G, Weiss W, Puck A 1976 Thermal expansion of powder filled epoxy resin. *Mater. Sci. Eng.* 22: 261–64
Hashin Z 1983 Analysis of composite materials—a survey. *Trans. ASME, J. Appl. Mech.* 50: 481
Hashin Z, Shtrikman S 1963 A variational approach to the theory of the elastic behaviour of multiphase materials. *J. Mech. Phys. Solids* 11: 127–40
Hill R 1963 Elastic properties of reinforced solids: Some theoretical principles. *J. Mech. Phys. Solids* 11: 357–72
Holliday L 1966 Geometrical considerations and phase relationships. In: Holliday L (ed.) 1966 *Composite Materials.* Elsevier, Amsterdam, pp. 1–27
Ishibashi K, Wake M, Kobayashi M, Katase A 1979 Powder-filled epoxy resin composites of adjustable thermal contraction. In: Clark A F, Reed R P, Hartwig G (eds.) 1979 *Non-Metallic Materials and Composites at Low Temperatures.* Plenum, New York
Jones W D 1960 *Fundamental Principles of Powder Metallurgy.* Arnold, London
Kantor Y, Bergman D J 1984 Improved rigorous bounds on the effective elastic moduli of a composite material. *J. Mech. Phys. Solids* 32: 41–62
Landauer R 1952 The electrical resistance of binary metallic mixtures. *J. Appl. Phys.* 23: 779–84
Lenel F V 1980 *Powder Metallurgy: Principles and Applications.* Metal Powder Industries, Princeton, New Jersey
Lubin G (ed.) 1982 *Handbook of Composites.* Van Nostrand Rheinhold, New York
Neale M J (ed.) 1973 *Tribology Handbook.* Butterworth, London
Spriggs G E 1974 Cemented carbides: Why do they work? *Practical Metallic Composites.* Institution of Metallurgists, London
Wahi R P, Ilschner B 1980 Fracture behaviour of composites based on Al_2O_3–TiC. *J. Mater. Sci.* 15: 875–85
Walpole L J 1966 On bounds for the overall elastic moduli of inhomogeneous systems. *J. Mech. Phys. Solids* 14: 151–62; 289–301
Walpole L J 1969 On the overall elastic moduli of composite materials. *J. Mech. Phys. Solids* 17: 235–51

R. G. C. Arridge
[University of Bristol, Bristol, UK]

Passive Solar Energy Materials

Until recently, flat plate collectors were thought of whenever solar space heating was considered. These collectors, composed of flat glass coverplates over black solar absorbers, convert solar energy to transportable heat when either water or air is pumped over (or through) the metal absorber. This type of solar energy collection is known as active solar heating, since it takes energy to accomplish the pumping.

Now, however, passive solar heating is the preponderant residential form, since it is much less expensive and more reliable and aesthetically attractive. Passive solar heating requires no pumping or parasitic energy to collect thermal energy, since the dwelling itself becomes the cover plate (the windows) and the absorber (the building interior finishes).

The materials used for passive solar space heating affect thermal performance just as critically as the materials used in active collector systems. The entire building acts as a solar collector and thermal storage element when heating passively, so the solar materials also must behave as architectural materials. Although common building materials such as glass, masonry and wood have been used for passive solar heating applications, their thermal performance is not always satisfactory. Some problems are: overheating of room air during sunny winter days; poor thermal resistance of glazing products; and high-glare interiors. Many classes of material now exist which overcome these problems by enhancing the thermophysical properties while maintaining the architectural qualities. Among these are the high solar gain, superinsulating windows and the low-cost phase change materials for storing solar energy.

1. Glazing Alternatives

Glass, plastic sheet and other similar glazing products have been characterized erroneously as solar radiation traps because they are opaque to longwave infrared (ir), or thermal, radiation. The thermal radiation is produced when the transmitted shortwave solar radiation is absorbed by the objects in the interior of a building. The objects emit the lower quality energy as longwave ir. Although glass does not transmit this ir, it does not trap appreciable amounts of thermal radiation either; since it is opaque to longwave ir, it absorbs the energy readily, only to conduct the heat to the outside. Careful measurements have shown that the net radiation inside a greenhouse glazed with commercial glass or sheet plastic is slightly less than just outside the greenhouse. This is because the average solar transmission of single glazing is typically <80%, and as much as 60% of the thermal radiation generated in the hothouse still leaves the structure through absorption, conduction and emission in the glass. Greenhouses get warm because they are convection traps, not radiation traps; the heat accumulates convectively and cannot be dispersed as it is outside. Glazing products are typically very poor thermal insulators.

1.1 Selective Transmitters

It is possible to coat various glazing products so that they behave as true radiation traps, which means that their thermal resistance can be more than doubled without appreciable lessening of the solar gain. The ideal behavior of such a coating is shown in Fig. 1a. Nearly all the solar spectrum is transmitted. The coating changes its radiation characteristics at ~4 μm and behaves like a nearly perfect reflector to thermal radiation centered at ~10 μm. The popular name for this transparent coating is "heat mirror," since it reflects heat so well. Real selective transmitters coated on double strength glass exhibit overall normal solar heat gains up to 81% (ordinary double glass gain up to 88% of the incident solar energy) with thermal reflectances of ~95% (an emissivity of 5%). Typical performances for some coatings are shown in Table 1. For coatings that exhibit ir emissivities of ~5%, the thermal resistances are essentially doubled. For example, a double-glazed unit with a 12 mm airspace and a low-emissivity coating on the third surface from the outside has a k value of $1.8\ \mathrm{W\,m^{-2}\,K^{-1}}$.

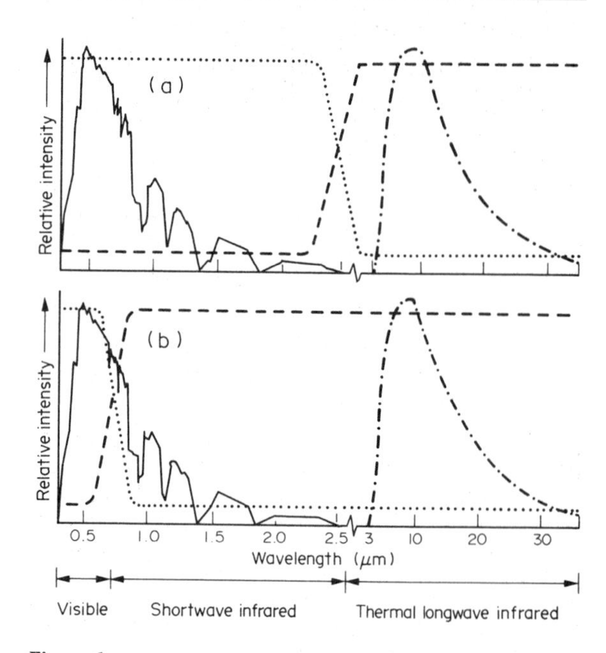

Figure 1
Ideal heat mirror transmission characteristics: (a) residential; (b) commercial; —— solar spectrum; –·– spectrum of a source at 300 K; · · · ideal transmittance; – – – ideal reflectance

The coatings can be applied to a variety of glass and plastic film or sheet glazing products by either vacuum sputtering or ion-beam deposition techniques. A single coating machine is capable of producing several million m² per year of coated product measuring up to 2 m in width. Coated, low-emissivity plastic films enhance the thermal performance when they are affixed with a water release adhesive to existing single or double pane windows. Low-emissivity coatings are normally corrosion protected by laminating a thin film of polypropylene over the coating. This raises the emissivity to 25–30%.

The selective transmitters can be classified into two film categories: (a) multilayer films composed of a metal film overcoated with an antireflection layer(s), or (b) relatively thick single layer films of highly doped semiconductors.

The metal layer in a multilayer system is used to reflect ir. Coatings transparent to ir such as TiO_2, Al_2O_3, SiO_2 and ZnO are used to antireflect one or both sides of the metal layer, a typical system for residential applications being $TiO_2/Ag/TiO_2$. This type of film is inert to water vapor and stable up to 200 °C. The film on glass has a solar transmission of over 70% and a far ir reflectivity of 95%. Since this system can be deposited at low substrate temperatures, it can also be roll coated on plastic films such as polyester.

Table 1
Thermal performances of various transmitters on double strength glass

Coating	Solar heat gain (%)	Infrared emissivity (%)	Corrosion resistance	Color
Indium–tin oxide	81	15–20	good	pale green
Copper–tin oxide	75	5–10	poor, must be hermetically sealed	pale brown
Silver–titanium oxide	80	5–10	poor, must be hermetically sealed	neutral to pale blue

The thicknesses of the film sandwiches must be carefully controlled to achieve the desired optical qualities and to avoid unwanted color distortions. Transparent conductors can be physically deposited on glass and plastic substrates by vacuum evaporative sputtering, rf sputtering or chemical deposition. Deposition on plastic tends to be more difficult due to its poor resistance to elevated temperatures.

Doped semiconductors can be used to avoid the quality control problems of thin films. Certain highly doped semiconductors exhibit high ir reflectance due to their controlled concentration of charge carriers. Transparent semiconductors include SnO_2:F, SnO_2:Sb, In_2O_3:Sn and $CdSnO_4$. The solar transmission can be increased by antireflection coatings and etched micrgrids. In_2O_3:Sn shows the best performance, and also has extremely good resistance to water and abrasion. Unfortunately, indium is currently too expensive for the thick layers ($\geq 0.3\,\mu m$) commonly used in this approach.

These thick films are usually laid down by spray hydro processes or rf sputtering. Pyrolitic processes have also been used, but color distortions can result with the large variations in thickness this process tends to generate, although recent low-cost products in this area show acceptable color properties. The emissivity tends to be around 40%, so some performance is sacrificed. The product does not have to be hermetically sealed and therefore can be used in storm windows.

Denoting the surfaces of a double glazing system by numbering them from the outside in, the overall thermal resistance is doubled by coating surface no. 2. Solar gain is maximized by coating surface no. 3, with only a 4% reduction in thermal resistance. Comfort is also increased by coating surface no. 3, since surface no. 4 will run warmest under this condition. If surface no. 4 is coated, it will run the coldest, since the ir is reflected back into the room, and thereby minimizes comfort. Coating surface no. 1 does not affect the thermal resistance noticeably.

Once one of the inside surfaces in a double-glazed unit is coated, it makes sense to reduce the remaining convection between the plates by using noble gases with lower thermal conductances than air. For instance, when argon is substituted for air, the k value falls from 1.8 to 1.2 $W\,m^{-2}\,K^{-1}$. Edge seals must be improved, however, to prevent the gas from diffusing out over a 10 year period.

The transmission point for a selective transmitter can be moved by proper doping of the coating. Figure 1b shows a curve for a so-called office or commercial heat mirror where the transmission point has been moved into the solar spectrum. Here the visible portion of the solar spectrum (daylight) is transmitted, while the larger fraction of the heat contained in the solar spectrum is reflected to the outside. This gives a cool daylight so that air-conditioning loads can be reduced greatly where natural lighting is desired. This form of lighting is much cooler than fluorescent lighting or any other form of artificial lighting. A double-glazed window coated with this material can admit up to 50% of the daylight, while rejecting 75% of the solar heat gain. Typical office aluminized glass rejects 60% of the solar gain while transmitting only 20% of the daylight. Although this type of coating is still under commercial development, the trend is clear: there will be a spectrum of coatings that will be tailored to each region's heating or cooling needs.

1.2 Antireflective Coatings

Another way to increase thermal performance of glazing products is to increase solar transmission with antireflection coatings or surface treatments. Thus the introduction of multiple, spaced layers of low-reflection glazing film or sheet in a window product increases thermal insulation without significantly decreasing solar heat gain.

Dipping glass in a fluosilicic acid bath saturated with silica results in a microscopically etched or dendritic surface that effectively lowers the index of refraction and therefore the reflection losses of the glass. This type of glass surface treatment, known as a graded index film, has been used for several years for prototype collector covers without any loss of transmission. Care must be taken, however, to avoid fouling the surface with greases or waxes. The process is expensive, mostly because of the glass handling difficulties. Costs have been lowered by generating similar dendritic surfaces on plastic films, such as polyester, by steam oxidizing an aluminum film. This processing forms needle-like structures of aluminum hydroxide which reduce the reflection losses considerably. Processing costs are significantly lower because entire rolls of the film can be processed continuously.

Alternatively, polymers with a low refractive index such as Teflon FEP can be adhered to glass as films or as colloidal dispersions.

Thin film coatings are used to increase the solar transmission of bulk glazing materials or other thin films such as low-emissivity surfaces. Inorganic thin films such as MgF_2, SiO, SiO_2 and TiO_2, are used for single or multiple interference applications. The films are usually applied by vapor deposition.

Currently, a commercial ultraviolet resistant polyester film is offered with a mechanical surface treatment which results in an overall solar transmission of 95%, rather than the 90% associated with untreated polyester. Normally, additional glazing layers decrease the solar gain more than the thermal resistance increases after the third layer is added. With film transmissions of 95%, the glazing thermal resistance will always increase faster than the reduction in solar gain, giving improved solar heating performance. The films are either heat shrunk or mechanically tensed on rigid frames cladded with glass to protect the inner films. The surfaces of plate glass can also be treated similarly.

2. *Thermal Storage by Phase Change Materials*

Phase change materials can be used when relatively inexpensive and dense thermal energy storage is required that is still in radiation contact with the room. Here the thermal energy leaves the storage mostly as ir, because the material is at a different temperature from the surroundings. Heating or cooling by ir radiation is always more comfortable and energy conserving than space conditioning with heated (or cooled) air. This is because the occupants and the furniture are conditioned directly by radiation, so the air can run cooler while maintaining normal comfort levels. Radiant heating or cooling also avoids energy-wasting drafts and air temperature stratification.

Room air temperature excursions can be limited substantially by placing the phase change material (PCM) in the room, since the material can absorb or liberate heat at a nearly constant temperature, as opposed to masonry materials, which store heat in sensible form and must rise in temperature, when exposed to solar energy. Generally, materials which change state between 23 and 27 °C are favored for radiant heating applications; for cooling, transition temperatures of 16–18 °C are used when the material is also in radiation contact with the room. For well-insulated spaces, the area of PCM exposed to the room is 30–50% of the floor area. It is difficult to find PCMs which change state in these ranges, so additives are usually used to adjust the transition point. The most common solid-to-liquid PCM materials used for storing heat latently are the salt hydrates, since they are inexpensive, nonflammable, nontoxic and nonbiodegradable. Most organic PCMs, such as the waxes, have just the opposite qualities.

Calcium chloride 6-hydrate is used when its expansion and contraction difficulties can be overcome, because it has a superior heat content of 163 $kJ\,kg^{-1}$ and a convenient phase change temperature of 27 °C. Its density is 1603 $kg\,m^{-3}$. There are additives to prevent separation of compounds during incongruent melting. Typically, the PCM is packaged in semirigid 25–50 mm diameter cylinders or 25 mm thick trays formed from 1.5 mm thick polyethylene. These containers can be stacked to form heat-absorbing walls. Larger diameter packages will not change phase fully when placed in sunshine because of poor heating conduction to the core, though this can be overcome by accelerating heat transfer with forced convection.

Sodium sulfate decahydrate is also used, because packaging is easier due to its dimensional stability during a change in state. Practical versions of the material with nucleators and suspension media typically hold up to 140 $kJ\,kg^{-1}$ and change phase at 31 °C. The density is 1666 $kg\,m^{-3}$. Even in the presence of thickeners, the denser anhydrous sodium sulfate will irreversibly separate from the lighter fluids after several freeze-thaw cycles unless it is packaged with vertical dimensions <10 mm so that crystallization can occur by diffusion. This is accomplished either by using thin pouches or an architectural matrix to lock the PCM in a closed cell structure with small vertical dimensions.

The thin, impermeable pouches have been used inside ceiling tiles, in floor tiles made of heat-conducting polymer concrete (20% polyester resin, 80% fine aggregate by weight) or laid on a variety of metal deck or cassette ceiling systems as shown in Fig. 2. The pouches are fabricated from a laminate of metal foil, extremely tough plastics and heat-sealable film. The foil contains the water and its vapor, the plastic material arrests the crystal punctures and the outer film permits the heat sealing of the product.

Solid–solid phase changes are intriguing because packaging becomes easier since the material is self-supporting in all phases. Although there are many candidates, most have phase-change temperatures well above 40 °C. One important exception is neopentyl glycol, which changes phase at 40 °C with a latent heat of transition of 119 $J\,g^{-1}$. The cost of the material is high, over ten times the cost of the salt hydrates, and the material burns. Since the container price is usually the predominant price, however, the simplified container system for this material might

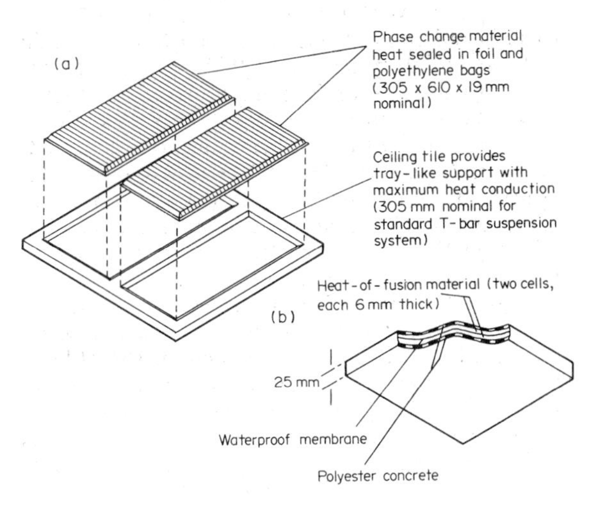

Figure 2
Diagrams of (a) PCM bags with ceiling support tile; (b) floor tile system

give a lower overall cost. Supercooling is also exhibited by the solid–solid PCMs. Powdered graphite has been used as a nucleating agent, but the undercooling cannot be entirely eliminated. The cast-in-place nucleator cannot migrate as is sometimes the case with solid–liquid PCMs, so the material remains stable. Many of the important physical parameters of the material remain unstudied, including thermal stability, thermal expansion, vapor pressure and flammability.

Various architectural matrices are being designed for containing PCMs while providing the appearance of an architectural finish element. Certain polymeric and inorganic foams are being used to suspend the incongruently melting materials in a closed cell structure. The matrices have to be sealed to prevent the water from leaving as vapor. Metal and rigid plastic envelopes are currently used, but other impermeable materials can be used if delamination can be overcome. What can be expected in the near future is a variety of architectural finish materials such as wall and ceiling boards, bricks and blocks that behave as constant-temperature thermal storage elements.

3. *Masonry*

Thermal energy can be stored sensibly in masonry. This requires either high fan power to transfer the heat (or cold) to the mass by convection or, in the case of direct solar gain schemes, diffusion of the sunlight over large areas of 10 cm thick masonry so that the solar energy will not elevate the surface temperature of the masonry and thus cause overheating of the room air. Usually, the masonry area must be five to seven times the south window area.

The thermal storage effectiveness of masonry varies widely among materials. The amount of heat that penetrates the mass and is subsequently stored is a function of the density ρ, thermal conductivity λ, specific heat c, and thickness x of the mass. Figure 3 shows as successive approximations the progression of a thermal wave through a mass where the right side is insulated and the left side is exposed to a sudden change in temperature. The mass is initially at the temperature indicated by the horizontal line labelled 0. The line labelled 1 is the temperature distribution through the mass after a time t; the line labelled 2 shows the temperature after two such time periods have expired, and so on. Notice that it takes five time periods for a change in temperature on the left side to reach the right side.

The length of the time period is related to the properties of the materials by the equation:

$$t = \frac{x^2 c \rho}{2\lambda} \quad (1)$$

where $x = b/5$ and b is the width of the masonry slab. Most building materials have the same specific heat, so the controlling parameters are conductivity and density. A short time period t means that more of the material participates in a given time frame. The value of t is 1.73 h for 15 cm thick wood, and 0.27 h for the same thickness of concrete. Thus a given temperature pulse will penetrate concrete 6.5 times as fast as in wood. The amount of heat stored in the material is given by $Ac\rho lh$, where A is the area under any given temperature profile (relative to the initial temperature distribution) and l and h are the length and height of the slab. Table 2 gives the thermal properties of common masonry materials.

The choice of material makes a very large difference to the thermal performance of passively heated (or cooled) structures. Fortunately the choice is becoming

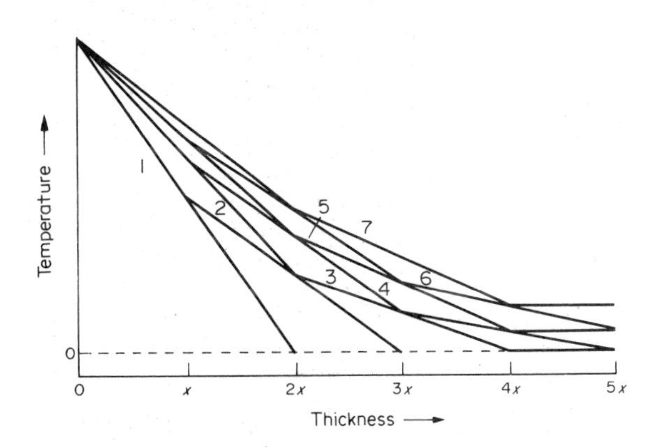

Figure 3
Approximation of temperature distribution in a solid heated from one side (after Kreith 1960)

Table 2
Thermal properties of some building materials

Material	Specific heat ($kJ\,kg^{-1}\,K^{-1}$)	Density ($kg\,m^{-3}$)	Conductivity at 20 °C ($W\,m^{-1}\,K^{-1}$)
Brick			
Fireclay	0.84	2307	1.00
Masonry	0.84	1699	1.64
Concrete			
Stone	0.84	2307	0.93
10% moisture	0.84	2244	1.21
Glass	0.84	2724	0.78
Limestone	0.84	1683	0.69
Wood (perp. to grain)			
Oak	2.38	817	0.21
Fir	2.80	497	0.10
Gypsum	1.08	875–1250	—

wider with the emergence of materials with enhanced thermophysical properties.

See also: Solar Thermal Materials

Bibliography

Johnson T E 1977 Lightweight thermal storage for solar heated buildings. *Sol. Energy* 19: 669–75

Johnson T E 1981 *Solar Architecture: The Direct Gain Approach.* McGraw-Hill, New York

Kreith F 1960 *Principles of Heat Transfer.* International Textbook Co., Scranton, Ohio

Lampert C M 1982 *Solar Optical Materials for Innovative Window Design*, LBL 14694. Lawrence Berkeley Laboratory, University of California, Berkeley, California

Rosen J E 1981 Natural daylighting and energy conservation. Thesis, Massachusetts Institute of Technology

Telkes M 1974 Storage of solar heating/cooling. *ASHRAE Trans.* 80 (II)

Ya Kondrat'yev K 1965 *Radiative Heat Exchange in the Atmosphere.* Pergamon, Oxford

T. E. Johnson
[Massachusetts Institute of Technology,
Cambridge, Massachusetts, USA]

Plasticity

Small plastic strains reduce stress concentrations, which can lead to cleavage fracture in structures, but often increase strain concentrations, which promote fracture by ductile void growth. Plasticity also provides a limit to elastic behavior by determining the onset of large-scale inelastic deformation in structures.

Plasticity solutions, in common with all continuum statics, must satisfy the strain–displacement and equilibrium equations (see *Small Strain, Stress and Work*). Plasticity is distinguished by the time-independent deformation that remains on release of the load (see *Time-Independent Plasticity*), expressed through the strain–stress relations. After discussing the physical basis of these relations and typical values of the yield strength, we turn to the form of the relations, to the general character of solutions using them, and to sources of solutions using analytical, approximate, numerical or experimental methods.

1. Physical Basis of Plasticity

Plastic deformation is inhomogeneous on a scale hundreds of times larger than the atomic scale, often up to that of the crystals in an aggregate. A major source of this inhomogeneity is the slip by discrete atomic displacements of parts of lattice planes over each other. The edges of the slipped regions are the line imperfections discussed in the article *Dislocations*. Their motion helps to explain plasticity at low (compared to the ideal) shear strengths. Dislocations help to suggest the form of the relations between stress and strain increments, and also how to strengthen metals.

1.1 Dislocations as Sources of Low Shear Strength

Under uniform stress, a perfect crystal would only deform inelastically by simultaneous slip of entire lattice planes over each other, in directions and by amounts corresponding to the minimum lattice spacing b, called the Burgers vector. Approximating the interplanar force law as sinusoidal shows that the ideal shear strength τ_i is a small fraction of the linear elastic shear modulus G:

$$\frac{\tau_i}{G} \approx \frac{1}{2\pi} \approx 0.16 \tag{1}$$

The actual yield strengths τ_y are less than ideal by many orders of magnitude for pure single crystals, and by one order of magnitude for even the hardest alloys. For example, for aluminum and iron,

$$\left.\begin{aligned}\left(\frac{\tau_y}{G}\right)_{Al} &\approx 0.000\,04\text{–}0.012\\ \left(\frac{\tau_y}{G}\right)_{Fe} &\approx 0.0005\text{–}0.016\end{aligned}\right\} \quad (2)$$

1.2 Dislocation Mechanics

Those aspects of dislocations especially helpful for plasticity are summarized here; for further information, see *Dislocations*, Argon (1986) and Honeycombe (1984).

The strain in a crystal of volume V due to dislocations with Burgers vector b sweeping out a total area ΔA, whether by a few dislocations moving a long distance or by many travelling a short one, is

$$\Delta\gamma = \frac{b\Delta A}{V} \quad (3)$$

Dislocations can be observed through etch pits, or from transmission electron micrographs or x-ray diffraction topographs of thin films. In terms of Burgers vectors b, typically 0.3 nm, dislocation spacings range from

$$\begin{aligned}&l_d \approx 10^5 b \approx 30\ \mu\text{m for well-annealed crystals, to}\\ &l_d \approx 10^2 b \approx 0.03\ \mu\text{m for heavily cold-worked alloys.}\end{aligned} \quad (4)$$

A resolved shear stress τ on the slip plane in the slip (Burgers vector) direction tends to increase the slipped area and hence to move the dislocations at its edge. Equating the applied work by τ to the work done by the force on a nearly straight dislocation shows that the force per unit length on the dislocation, f, is normal to the dislocation and

$$f = \tau b \quad (5)$$

An applied stress thus tends to move any initial dislocations out of a crystal. The resulting plastic strain before exhaustion of dislocations and the end of plasticity would be far less than observed, so there must be a multiplication mechanism to increase the dislocation density.

The Frank–Read mechanism of dislocation multiplication arises from a dislocation segment pinned at its ends by a junction with inactive dislocations on other slip planes, or by rigid second-phase particles. A resolved shear stress bows the segment into an arc which eventually becomes unstable. The slipped area spreads out in all directions, eventually doubling back on itself, regenerating a dislocation segment between the pinning points, and throwing out a dislocation loop along its growing outer boundary (see Fig. 2 of *Dislocations*). The process can be repeated until the leading dislocations are blocked by obstacles or other dislocations. Estimating the critical stress for multiplication requires some further results from dislocation mechanics.

Solving the isotropic elasticity equations around a dislocation, subject to the boundary condition of a displacement b on following any closed circuit around a dislocation (the Burgers circuit), shows that the shear stresses at a radius r from the dislocation, in terms of the shear modulus G, are of the order of

$$\tau \approx G\frac{b}{2\pi r} \quad (6)$$

Such shear stresses produce forces on nearby dislocations (Eqn. 5), tending to attract or repel them, or to form stable dislocation networks.

The (elastic) stress field of a dislocation gives it a strain energy per unit length, dU/dl. This means that work must be done as the dislocation is lengthened. In effect, it has a line tension T. For a straight dislocation in a field of dislocations, integrating the strain energy from the stress field of Eqn. (6) shows that the total energy is nearly independent of the distance to nearby dislocations, and is given roughly by

$$\frac{dU}{dl} = T \approx \frac{Gb^2}{2} \quad (7)$$

Applying a shear stress τ to a Frank–Read dislocation segment tends to bow it out, with the force per unit length balanced by the line tension (as with a stretched rubber band being bowed out by a distributed lateral force). The peak stress is reached with the minimum curvature, i.e., with the segment bowed into a semicircle whose diameter is the spacing of the pinning points, l_d. The critical stress τ_d to activate dislocation multiplication is then found by equating the tension to the result of the distributed force from the applied stress:

$$\left.\begin{aligned}2T &= f l_d\\ Gb^2 &\approx \tau_d b l_d\\ \frac{\tau_d}{G} &\approx \frac{b}{l_d}\end{aligned}\right\} \quad (8)$$

The above results show a number of aspects of plastic strain (i.e., dislocation motion). These are listed below.

(a) Elastic (interatomic) and plastic strains are additive.

(b) Plastic strain is incompressible.

(c) If the applied stress is removed after a segment is partly bowed out, the segment tends to contract against lattice friction, giving reversed plastic flow (anelasticity and damping), small compared to the elastic strain.

(d) After dislocations break away from their original pinning points, they interact with other dislocations and remain nearly stationary on release of load: plastic strain is permanent.

(e) A critical resolved shear stress (on a slip plane in the slip direction) gives the associated component of plastic strain increment.

(f) Temperature and hydrostatic pressure have rather small effects on the stress for plastic flow according to their effects on the shear modulus G of Eqn. (8). The energies to lengthen dislocations are far too large to be supplied by thermal motion. The small rate and further temperature effects that do exist are due more to breakaway from pinning points than to dislocation stretching (see also *Creep*).

(g) As the dislocations multiply, the dislocation density increases; l_d in Eqn. (8) decreases; and the resistance to further plastic flow increases. Such strain-hardening is discussed further below.

1.3 Hardening Mechanisms

Hardening is achieved by changing the crystal structure, adding alloying elements, precipitating new phases, and by plastic deformation.

The lattice shear resistance (friction stress) varies widely with the directionality of atomic bonding, from $\tau_l/G = 10^{-9}$ in Cu through 10^{-6} in (fcc) Al and Ni, 1–5 $\times 10^{-3}$ in the (bcc) transition metals Fe, Cr, V and Mo, to 0.01–0.1 (nearly the ideal strength) in carbides, oxides, some intermetallic compounds, and especially diamond.

Adding carbon to steel allows the formation of the hard tetragonal phase martensite on quenching the high-temperature soft (fcc) phase. Otherwise, only the soft (bcc) ferrite, surrounding the hard intermetallic compound cementite (Fe_3C), is formed. Adding nickel delays the formation of cementite so that martensite can form on quenching parts of up to several centimeters in thickness, which cool more slowly.

Solution hardening is due to solute atoms changing the energies of dislocations and hence exerting forces on them through differences in stress, elastic modulus, electronic structure and charge. The shear resistance of single crystals with one species of solute, τ_s, is at most

$$\tau_s/G = 0.0005\text{–}0.001 \tag{9}$$

but several species of solutes can further increase the strength.

Precipitation hardening consists of heating sufficiently to put a second phase into solution, quenching to a supersaturated solid solution, and then holding at a moderate temperature (age hardening) to nucleate and grow the second phase as precipitate particles. At first the precipitates are so small that they are easily sheared by dislocations. At the other extreme, they agglomerate to very large sizes that are more nearly in thermal equilibrium, and become so far apart that dislocations can easily move between them (the alloy overages). For a volume fraction f_v of precipitates of strength per unit width $dK/dw \approx Gb/100$, the optimum width and peak strength are roughly given by:

$$\frac{w_{peak}}{b} = \frac{Gb}{dK/dw}, \qquad \frac{\tau_p}{G} = \frac{dK/dw}{Gb} f_v^{1/2} \tag{10}$$

w_{peak} is typically only $100b$, far too small to be seen with an ordinary optical microscope. f_v is limited by the amount of alloy that can be dissolved in order to precipitate it later as the fine particles needed for peak strength. Using a small amount of a variety of alloying elements helps to overcome this limitation.

Strain hardening can be estimated from the Frank–Read model, with dislocations thrown off by a mean free path R_{mfp} before stopping:

$$\frac{1}{G}\frac{d\tau_d}{d\gamma} \approx \frac{0.1\text{–}0.2}{R_{mfp}/l_d} \tag{11}$$

R_{mfp}/l_d must be of the order of 50 to get the rate of strain hardening down to the observed value of $(d\tau/d\gamma)/G = 0.002\text{–}0.004$. (See Honeycombe 1984 pp. 129–44 for further models.) Equation (11) is consistent with the observation that, normalized by the shear modulus, the steady-state rate of strain hardening is of the same order of magnitude in a wide variety of single crystals and alloys. The dislocation motion of $R_{mfp} = 50l_d = (5000\text{–}50\,000)b$ is also consistent with plastic flow occurring in slip lines or bands that are visible in an optical microscope, and hence must be thousands or millions of times larger than the Burgers vector b. Another source of large bands is the fact that a series of dislocations from one source causes a stress concentration that triggers further sources on the same general plane.

If the above resistances to dislocations are all present and occur at successively and distinctly larger scales, they are more or less additive:

$$\tau_t = \tau_l + \tau_s + \tau_p + \tau_d \tag{12}$$

Platelet and grain size strengthening occur when concentrations of dislocations are required at a phase or grain boundary to nucleate plastic flow on the other side. An empirical equation is

$$\tau_{tgs} = \tau_t\left[1 + \left(\frac{l_{gs}}{D/2}\right)^{1/2}\right] \tag{13}$$

l_{gs} depends strongly on the alloy in an unexplained way, but in steel the effect is so strong that, in combination with most of the above mechanisms, it plays a large role in the thermomechanical processing of microalloyed, fine-grain, high-strength constructional steels.

The tensile yield strength Y of a polycrystal is about 3 times that of the resolved shear stress τ_t due to the necessary interaction of slip systems. Together these mechanisms give yield strengths $Y/E \approx 0.001$ for soft alloys, ≈ 0.01 for hard ones, where E is Young's modulus.

2. Plastic Stress–Strain Relations

For an isotropic polycrystal under multiple stress components, plastic strain requires that a function of the stress components (an equivalent stress $\bar{\sigma}(\sigma_{ij})$) reach the current flow strength (yield stress) in tension, Y. Many experiments, isotropy, independence of pressure, and some mechanics models of polycrystals are approximated or satisfied by the Mises yield locus in terms of stress deviator or stress: i.e., plastic flow requires

$$\left.\begin{aligned} Y^2 = \bar{\sigma}^2 &\equiv \sum_i \sum_j \frac{3}{2}\sigma'_{ij}\sigma'_{ij} \\ &= \frac{1}{2}[(\sigma_{11}-\sigma_{22})^2+(\sigma_{22}-\sigma_{33})^2 \\ &\quad +(\sigma_{33}-\sigma_{11})^2]+3[\sigma_{23}^2+\sigma_{31}^2+\sigma_{12}^2] \end{aligned}\right\} \quad (14)$$

Alternative yield loci are found by expressing Eqn. (14) in terms of principal components of stress and using higher exponents than two. In the limit as the exponents approach infinity, the equivalent stress depends only on the maximum difference of principal stress components (the Tresca equivalent stress), which may be more appropriate in normalized steel, with both upper and lower yield points.

For strain increments greater than about 1%, with constant ratios of the stress components, the flow strength is a scalar function of an integrated equivalent plastic-strain increment, defined so its product with the equivalent stress is the increment of plastic work. Note that since plastic flow is incompressible, the plastic strain deviator is the plastic strain itself:

$$\left.\begin{aligned} d\bar{\varepsilon}^{p^2} &\equiv \sum_i \sum_j \frac{2}{3} d\varepsilon^p_{ij}\, d\varepsilon^p_{ij} \\ &= \frac{2}{3}(d\varepsilon^{p^2}_{11}+d\varepsilon^{p^2}_{22}+d\varepsilon^{p^2}_{33})+\frac{4}{3}(d\varepsilon^{p^2}_{23}+d\varepsilon^{p^2}_{31}+d\varepsilon^{p^2}_{12}) \\ Y &= Y\left(\int d\bar{\varepsilon}^p\right) \end{aligned}\right\} \quad (15)$$

The apportionment of the equivalent strain among the different components is governed by the associated flow rule, integrated up from the critical resolved shear-stress law for dislocation motion by the principle of virtual work (Bishop and Hill 1951, McClintock and Argon 1966). In fact the rule also applies to structures under generalized loads Q_i and corresponding deformation increments dq^p_i such that their dot product is the plastic work: $dW^p = \Sigma_i Q_j\, dq^p_i$. Then if the equivalent strength or yield locus is $\bar{Q}(Q_i)$ and $d\lambda$ is a scalar,

$$dq^p_i = \frac{\partial \bar{Q}(Q_i)}{\partial Q_i}\, d\lambda \quad (16)$$

That is, the deformation increments are normal to the yield locus. Applying the associated flow rule, Eqn. (16), to the Mises yield locus, Eqn. (14), gives the components of macroscopic strain increments in terms of the deviatoric stress components:

$$d\varepsilon^p_{ij} = \frac{3}{2}\sigma'_{ij}\frac{d\bar{\varepsilon}^p}{\bar{\sigma}} \quad (17a)$$

where the scalar $d\lambda$ has become $d\bar{\varepsilon}^p/\bar{\sigma}$ to satisfy $dW^p = \bar{\sigma}\, d\bar{\varepsilon}^p$. Alternatively,

$$\left.\begin{aligned} d\varepsilon^p_{11} &= [\sigma_{11}-\frac{1}{2}(\sigma_{22}+\sigma_{33})]\frac{d\bar{\varepsilon}^p}{\bar{\varepsilon}}, \ldots, \ldots \\ d\gamma^p_{23} &= 3\sigma_{23}\frac{d\bar{\varepsilon}^p}{\bar{\sigma}}, \ldots, \ldots \end{aligned}\right\} \quad (17b)$$

Equations (17b) may be contrasted with the corresponding elastic equations where the strain has become the strain increment, Poisson's ratio has become 1/2, and $1/E$ has become $d\bar{\varepsilon}^p/\bar{\sigma}$.

The function $Y(\bar{\varepsilon}^p)$ of Eqn. (15) is often approximated in terms of a flow strength at unit strain Y_1, a prestrain ε_0, and a strain-hardening exponent n:

$$Y = Y_1(\bar{\varepsilon}^p + \varepsilon_0)^n \quad (18)$$

Typically $\varepsilon_0 = 0.0$–0.1 and $n = 0.01$–0.25. The initial flow (yield) strength is $Y_0 = Y_1\varepsilon_0^n$.

For plastic strain reversals or increments after a change in ratios of components of less than about 1%, the locus of initial yield can be thought of as being distorted, shrunk and embedded within that of Eqn. (14), which for a time grows less rapidly than implied by Eqns. (15) and (18). This is known as the Bauschinger effect (see *Time-Independent Plasticity*, Szczepiński 1963 and Hecker 1976). In dislocation terms, for a volume l_d^3 corresponding to a typical dislocation spacing, from Eqns. (3, 8) the ratio of plastic strain increment to yield strain is proportional to the area swept out by dislocations in that volume:

$$\frac{\Delta\gamma^p}{\gamma^e} = \frac{\Delta A}{l_d^2} \quad (19)$$

Thus the Bauschinger strains tend to move the tails of the dislocation distribution, rather than reworking the entire dislocation structure as in the strain hardening of Eqns. (15) or (18). For such strain reversals, typically of interest in low-cycle fatigue of structures, a cyclic stress–strain curve is often in the form of Eqn. (18) but with different coefficients, giving a 25–50% lower curve than the monotonic one for most alloys, especially if cold worked, but a 20–40% higher cyclic curve for some hard, heat-treated alloys (Fuchs and Stephens 1980, Boller and Seeger 1987). When all strain increments are reversed, the cyclic stress–strain curve is laid off from the reversal point, with the stress and strain increments both double those of the cyclic stress–strain curve. Eventually such a curve will intersect a curve from a prior cycle or osculate with the original monotonic curve in that direction. The previous curve is then followed.

Especially in materials that cyclically soften, the Bauschinger effect can lead to a relaxation of mean stress under cycling between strain limits with a nonzero mean strain (see, e.g., Rie 1987). Such cyclic relaxation (or creep) tends to arise in a region of local yielding within a generally elastic structure under nonzero mean loads.

3. *Character of Solutions*

While the concepts of strain–displacement relations, equilibrium with stress gradients, and boundary conditions are the same as for elasticity, the presence of a yield criterion and the incremental nature of the flow equations make solutions very different for plasticity, especially with vanishingly small strain hardening. Consider first axial tension and compression, then bending, and finally more general results.

3.1 Axial Loading

For uniaxial loading with elastic strains neglected, define the integral of the plastic strain increments, based on the current lengths, as $\varepsilon \equiv \int dl/l$, or from incompressibility as $\varepsilon = -\int dA/A$ and the stresses in terms of true stress (force per unit actual area). When a specimen is extended, the load increases because of strain hardening but decreases owing to a reduction of the area, reaching a maximum for the strain hardening of Eqn. (18) at the "uniform" strain ε_u:

$$\left.\begin{aligned} dP &= 0 = A\,dY + Y\,dA \\ \frac{dY}{Y} &= -\frac{dA}{A} = d\varepsilon \\ \varepsilon_u &= n - \varepsilon_0 \end{aligned}\right\} \qquad (20)$$

The maximum load per unit original area is called the tensile strength, $TS = P_{max}/A_0 = Y_1(n)^n \exp(\varepsilon_u)$. It gives the strength of the specimen in tension, but not of the material, for which already $Y(\varepsilon_u) > \mathrm{TS}$. Furthermore, if there is no fracture so the tensile strength is set by Eqn. (20), strain hardening continues.

Further extension decreases the load in the first section reaching its maximum. Any nonuniformity in the bar (some of which has already spread from the constraint by the shoulders) causes the rest of the bar to unload elastically. Such nonuniform elongation is called necking. It means that the usually reported percentage elongation at fracture is dependent on the proportions of the specimen and, while used for standardization, has no direct physical interpretation. The fractional reduction of area at fracture (*RA*) does give the equivalent strain at fracture, $\varepsilon_f = ln\,[1/(1-RA)]$. Finding the flow strength after necking involves correcting the load per unit current area P/A for the effect of triaxiality in the neck. In terms of the ratio of the radius of the minimum sections and the longitudinal radius of curvature of the profile, a/R (which, in the absence of data, can be estimated empirically from the strain beyond the uniform at the beginning of necking), the flow strength is approximately given by the Bridgman relation (McClintock and Argon 1966 pp. 323–24)

$$\left.\begin{aligned} \frac{Y(\varepsilon)}{P/A} &= \frac{1}{(1+2R/a)\ln(1+a/2R)} \\ \frac{a}{R} &\approx \varepsilon - \varepsilon_u \end{aligned}\right\} \qquad (21)$$

Additional reporting of the fracture strain $\varepsilon_f = -\ln(A_f/A_0)$ and the flow strength at fracture $Y(\varepsilon_f)$ from Eqn. (21) would, with *RA*, *TS*, $Y(\varepsilon_0)$ and the condition that *TS* occurs at maximum load, allow determination of the constants in Eqn. (18). Even so, the fit for loads before *TS*, which are of most concern in structures, is usually not very good. Also, a stress–strain curve must be taken with an extensometer short enough to avoid the nonuniform strains near the shoulders, or else a diametral extensometer must be used.

Such necking is a lack of uniqueness: in a perfectly uniform bar (with no shoulders) the neck could form anywhere along the length. The solution is said to bifurcate. Note that under given end loads, necking is unstable, but with given end displacements it is stable: in general, bifurcation may or may not be stable, depending on boundary conditions.

Buckling is another form of bifurcation, occurring under compression as in elasticity. Here, however, the process is history-dependent. In normal service the bar is not loaded and then released to buckle to either side. Rather, a small eccentricity increases with increasing load, both sides being in compression. The buckling load is given by the usual elastic equations, with the modulus of elasticity being replaced by the slope of the engineering (nominal) stress–strain curve. Again stability depends on end conditions: in a stiff enough structure a buckled member is stable, carrying a nearly constant load if elastic, but a decreasing load with progressive compression if plastic.

3.2 Bending, Residual Stress, and Beam–Cable Transitions

In pure bending of a beam into the plastic regime, the strain still varies in proportion to the distance from some neutral plane (or axis), but the stress varies nonlinearly. The condition of zero axial force on integrating the stress across the section sets the location of the neutral plane. The bending moment is found by integrating the stress multiplied by the distance from the neutral plane, over all elements of area.

The residual stress on unloading is found in general by adding the elastic stress due to an equal and opposite load. This superposition is valid as long as no yielding occurs under the resulting residual stresses (taking into account the Bauschinger effect, if need be). If the unloading is elastic, the reloading is elastic: residual stresses reduce the tendency to plastic deformation under the original load. In a beam of nonhardening material, the moment–curvature relation under reversals large enough for yielding follows exactly the cyclic stress–strain rule mentioned at the end of Sect. 2.

The same part of a structure may be modelled in different ways at different stages in loading. For a bar built-in between two rigid supports, the effect of a transverse load is initially that on an elastic beam. With a thin enough bar and high enough loads and yield strength, modelling the system as an extensible cable becomes appropriate. With a lower yield strength and a thicker section, the best models may change from an elastic beam to a plastic beam to a plastic cable. Similar transitions occur with plates and with columns in compression.

3.3 General Results

There are a number of striking differences between elastic and plastic distributions of stress and strain.

(a) Uniqueness of stress is absent in the rigid region of a rigid-plastic material not only because of the possibility of residual stress, as in elasticity, but also because of various possible loading histories. A more striking lack of uniqueness is that the strain in the deforming region is not always unique. For example, the deformation increment is not unique in a long straight bar loaded just to the tensile strength, because necking could begin anywhere along the length. A higher rate of strain hardening would mean that necking would not begin at this strain and thus would provide uniqueness.

Two uniqueness theorems hold even for nonhardening plasticity, however. Both apply to complete solutions (ones that satisfy equilibrium, yield, strain–displacement and stress–strain relations in any deforming region, and satisfy equilibrium without exceeding yield in any rigid region in which no further strain is occurring): (i) for a rigid-plastic nonhardening material, the stress distribution is unique in any region that deforms in any complete solution; and (ii) any region that is necessarily rigid in one complete solution is rigid in all solutions.

(b) The governing differential equations are hyperbolic for incompressible, nonhardening, rigid–plastic, plane-strain deformation, as well as for plane stress when the ratio of in-plane principal stresses is greater than two. This is in contrast to the elliptic equations that apply to strain-hardening or elastic materials. Elliptic versus hyperbolic behavior can be visualized in terms of a tensile bar with a transverse hole. In elasticity St. Venant's principle indicates that the effect of the hole will become negligible in a few diameters. In nonhardening plasticity, the hole localizes the strain into a narrow band that extends all the way to the next-nearest free boundary. While the plasticity becomes elliptic with some strain hardening, the tendency toward hyperbolic behavior is still present. As a result, plastic strain concentrations are often, but not always, much larger than elastic ones.

(c) Notch and crack behavior. Neuber (See Fuchs and Stephens 1980 p. 136) has pointed out that often with constrained yielding the square root of the product of the plastic strain- and stress-concentration factors is the elastic stress-concentration factor. A counterexample, as general yield is approached, occurs at the shoulders of a tensile specimen, where the plastic stress concentration factor is of order unity and the constraint of the shoulders ultimately produces a strain deconcentration. Further counterexamples are provided by the stress and strain distributions around notches and cracks.

For internal and external notches in both plane stress and plane strain, the in-plane elastic distributions of stress and strain are the same, within the scalar K_I. The plastic distributions are very different (Fig. 1). As in the tensile test, sufficient strain hardening provides uniqueness for static cracks (the HRR singularity), but growing cracks show yet other singularities, due to nonradial loading as a crack approaches, and residual plastic strains left in its wake (see, e.g., McClintock 1989). Finally the region in which the singularities dominate may be smaller than the fracture process zone size, which is the smallest region to which homogeneous continuum mechanics can be applied.

Alternating sliding off from fans of concentrated shear gives sharp grooves with faces of freshly bared

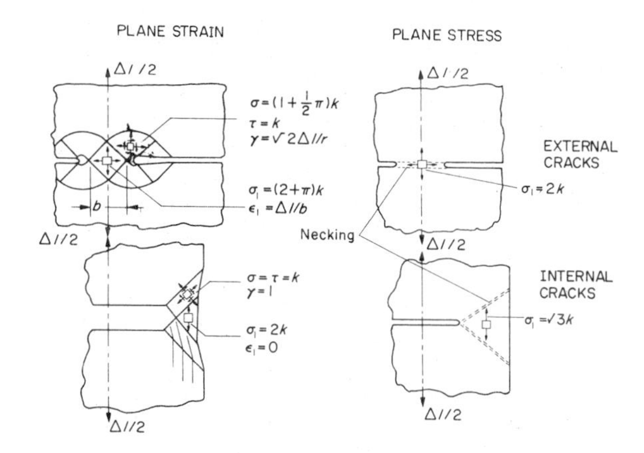

Figure 1
Contrast between four nonhardening, fully plastic flow fields: plane stress and plane strain with double (external) and single (internal) cracks (after McClintock) 1965, courtesy of the Royal Society of London

material, another form of bifurcation (McClintock 1971, pp. 155–62). This plays a role in fatigue crack growth and is not modelled by conventional finite-element analysis.

(d) The nonlinearity of the yield criterion prevents the superposition of the stress fields of two solutions to obtain a third. Plastic flow fields can be superposed if they have the same stress fields. In contrast, the linearity of the elastic stress–strain relations does allow superposition of any two solutions to obtain a third.

(e) For structural metals in general, the slope of the strain-hardening curve is less than Young's modulus E by a factor of 100 to 1000. If such low strain hardening, changes in geometry, and fracture are neglected, the load–deformation curve of a part is horizontal at a load called the limit load.

The concept of limit loads is illustrated by the load–deformation curves of two bars in tension, one without and one with a transverse hole, such as an oil hole. In the elastic–plastic region, the overall load–deformation curve is at first limited by a continous elastic region from one end to the other. Relatively little further deformation allows unlimited plastic deformation to occur across a short section of the bar, after which the load–deformation curve is nearly horizontal, at the limit load. If the bar is sufficiently stiff that the elastic deformation is negligible, then the plastic deformation before general yielding is also negligible. For many ductile structures subject to monotonic loading it is, therefore, sufficient to design to some large fraction, say 85–90%, of the limit load. For other structures and service conditions, elastic deflection or cracking from monotonic or repeated loading changes the original shape significantly before the limit load is reached, and the concept of a limit load is not useful.

4. *Sources of Solutions*

4.1 Bounds to the Limit Load

An advantage of designing to the limit load is that one can find upper and lower bounds to it (values below and above which the limit load must be), by satisfying only some of the fundamental equations for an exact solution: equilibrium, strain–displacement, yield, and strain–increment versus stress. Thus, practical results can often be obtained from relatively simple approximations. For instance, by guessing several different incompressible deformation fields, several upper bounds to the limit load can be obtained. The least of these upper bounds is closest to the limit load and hence is the most useful. Likewise, several different stress distributions satisfying yield and equilibrium conditions give several different lower bounds to the limit load. The greatest of these lower bounds is the most useful. If the least upper bound and the greatest lower bound are close enough for practical purposes, a satisfactory estimate of the limit load has been found without ever treating all the equations simultaneously. A lower bound to the limit load is given by any stress distribution which:

(a) satisfies any stress boundary conditions,

(b) everywhere satisfies the equations for equilibrium of stress gradients, and

(c) nowhere violates the yield criterion.

An upper bound to the limit load is a load for which a field of displacement increments $\delta u_i(x_j)$ can be found that:

(a) satisfies any displacement boundary conditions,

(b) gives no change in volume anywhere, and

(c) gives an integral of the plastic work increment throughout the body that is the upper bound P_{ub} multiplied by the corresponding plastic-displacement component under the load:

$$P_{ub}\delta u^p = \int Y\delta\bar{\varepsilon}^p \, dV \qquad (22)$$

A number of useful corollaries of these theorems are described below.

(a) Elastic or plastic solutions based on the net sections of parts containing cracks, holes, shoulders or other stress concentrators give lower bounds to the limit load.

(b) The bounding theorems are applicable even with strain hardening, if the current local flow strength is appropriately bounded at each point. An alternative is the loading of rigid–plastic power-law hardening material with constant stress ratios, and using the analogous theorems from the article *Creep*.

(c) In nonhardening material, the limit load is the same for elastic–plastic as for rigid–plastic material, assuming the same current shape. This is because at the constant limit load there are no further stress changes and no further elastic strain increments, so the further strain increments are rigid–plastic in either event.

(d) Residual stresses do not affect the limit load for the current geometry.

(e) Adding weightless material to a part never lowers its limit load, and removing weightless material never increases it. This result may seem obvious, but in fact it does not apply to stresses in elasticity, where adding square shoulders to a part of uniform cross section will produce a stress concentration. Likewise, removing shoulders to smooth out the specimen will decrease the elastic stress concentration.

There are also some restrictions or conditions on the theorems for bounds to the limit load. These are discussed in the following paragraphs.

(a) The current geometry must be used for a number of reasons.

(i) In tension, if there has been straightening, the limit load will be higher than that for the original geometry; if there has been thinning, the limit load will be less.

(ii) In compression, bending will increase any misalignment and decrease the limit load; in long columns, elastic buckling may occur before any plastic flow at all.

(iii) If the part contains a sharp crack, local plastic deformation may lead to crack growth (one form of shape change) before the limit load is reached.

(iv) If fracture (an extreme shape change) occurs under elastic strains (as in glass) or under small plastic strains in rapidly strain-hardening materials (as in cast iron), a limit load will not be reached.

(b) Friction never decreases the limit load, but the amount of any increase depends on the problem (Chakrabarty 1987, pp. 98–99). When the friction is directly proportional to the limit load, valid bounds are obtained.

A convenient way of generating incompressible, plane-strain displacement-increment fields for upper bounds to the limit load is to consider the sliding of triangular blocks over each other to accommodate a displacement under the applied load. This technique is especially convenient for problems involving inhomogeneous deformation, such as cracked plates in tension, hardness indentations, and drawing or extruding thin sheets (see, e.g., Johnson and Mellor 1983, pp. 421–51). (For bending, it is often convenient to consider sliding across cylindrical surfaces, which can accommodate relative rotation between rigid regions.)

Lower bounds can be established by constant stress patches, perhaps visualized by smearing out parallel arrays of bars forming trusses (see Johnson and Mellor 1983, 451–63, Chakrabarty 1987, pp. 421–25 and, especially for bars and for plates with bolt holes, Szczepiński 1972).

4.2 Shakedown

Under repeated loads, an elastic nonhardening body will develop residual stresses that provide completely elastic deformation if some pair of residual and loading stress fields can be found satisfying equilibrium and together nowhere exceeding yield (Chakrabarty 1987, pp. 99–101). There is an analogous upper bound to the shakedown load.

4.3 Plane Solutions

Where available, a plane slip-line solution for nonhardening material provides a compact representation of the effects of variables in a wide variety of problems, as well as providing limiting cases to check numerical solutions for strain-hardening materials. Roughly speaking, solutions for strain-hardening materials lie between nonhardening and linear elastic solutions. For plane strain, slip-line solutions are based on expressing the in-plane components of stress in terms of the radius, center and angular coordinate of the x-axis on Mohr's circle, with the radius constant at the yield strength in shear k. Along the directions of maximum shear stress (the slip lines), the partial differential equations of equilibrium can be integrated completely. Parts of the slip-line mesh can be found by integrating from a deforming free boundary, but some interior parts, with deformation rather than stress boundary conditions, must be found by experiment, experience, intuition or finite-element analysis.

The incompressibility equations for the displacements along the slip lines become ordinary differential equations (the Geiringer equations) along the slip-lines, and are integrated from the rigid regions as initial conditions. Unless the rigid regions are proved to be so, using lower bounds, the slip-line solutions are only upper bounds. The strain–displacement equations of Green are found in relatively few places (see McClintock 1971, p. 130 or Chakrabarty 1987, p. 419). For pressure-dependent yield criteria, as in soil mechanics, see Collins (1989).

In plane stress, the slip lines (characteristics of the hyperbolic equations) are real only when the ratio of maximum to minimum in-plane stress exceeds 2:1.

Figure 1 provides examples of slip-line fields. The doubly grooved, plane strain field, cut across the ligament and loaded in compression, is a basis for the empirical rule that the hardness as P/A is three times the tensile strength (in consistent units). For collections of solutions, see Chakrabarty (1987), Johnson and Mellor (1983), McClintock (1971) and their respective bibliographies.

4.4 Finite-Element Solutions

Finite-element methods require either very small steps or iterative solutions at each step to match the current slope of the stress–strain curve. For the limiting case of nonhardening material, surprisingly few checks have been made with slip-line solutions. There are difficulties:

(a) lack of unique strains with some boundary conditions;

(b) the need for fine meshes to approximate displacement and strain discontinuities, as well as the relatively small slip-line patches in some solutions;

(c) the need for finite deformations in dealing with incipient necking;

(d) the sliding off at crack tips, as in the singly grooved plane-strain example of Fig. 1, wherein adjoining particles become widely separated;

(e) algorithms which become indeterminate with zero slope (stiffness). (This difficulty should be avoided, since it will cause increasing trouble as the slope of the stress–strain curve drops from E to 10^{-2}–10^{-3} of E.)

The advantages of the finite-element method are the ability to match empirical stress–strain curves and to include elastic strains and rate effects (Owen and Hinton 1980, Pittman *et al.* 1984, Nakazawa *et al.* 1987).

4.5 Experimental Methods

Scribing grids on central planes of axisymmetric or plane-strain specimens subject to compression during interrupted or steady-state deformation provides displacement fields that can be differentiated to give strains. The relations between stress and strain-increment, along with some traction-free or overall force boundary conditions, then provide the stress (Thomsen *et al.* 1965). In plane stress, particularly sheet forming, photodeposited circles on the sheet provide a similar starting point.

Many of the early slip-line fields were suggested by annealing high-nitrogen steel, deforming, aging, sectioning and etching. Ordinary steel can be nitrided after sectioning and then aged and etched. Other techniques include photoplasticity, modelling with lead or clay, and observing the flaking of rust. For references to laser-optic surface observations see *Elasticity*.

See also: Creep; Dislocations; Elasticity; Small Strain, Stress and Work; Time-Independent Plasticity

Bibliography

Argon A S 1986 Plastic deformation: Time-independent dislocation mechanisms. In: Bever M B (ed.) 1986 *Encyclopedia of Materials Science and Engineering*. Pergamon, Oxford, pp. 3543–54

Bishop J F W, Hill R 1951 A theory of the plastic distortion of a polycrystalline aggregate under combined stress. *Philos. Mag.* 42: 414–27

Boller C, Seeger T 1987 *Materials Data for Cyclic Loading* A, B, C, D, E. Elsevier, Amsterdam

Chakrabarty J 1987 *Theory of Plasticity*. McGraw–Hill, New York

Collins I F 1989 Plane strain characteristics theory for density dependent soils and granular materials. *J. Mech. Phys. Solids* (in press)

Fuchs H O, Stephens R I 1980 *Metal Fatigue in Engineering*. Wiley, New York

Hecker S S 1976 Experimental studies of yield phenomena in biaxially loaded metals. In: Stricklin J A, Saczalski K J (eds.) *Constitutive Equations in Viscoplasticity: Computational and Engineering Aspects*, AMD 20. American Society of Mechanical Engineers, New York, pp. 1–33

Honeycombe R W K 1984 *The Plastic Deformation of Metals*, 2nd edn. Edward Arnold, London, pp. 129–44

Johnson W, Mellor P B 1983 *Engineering Plasticity*, 2nd edn. Van Nostrand, London

McClintock F A 1965 Effects of root radius, stress, crack growth and rate on fracture instability. *Proc. R. Soc. London, Ser. A.* 285: 58–72

McClintock F A 1971 Plasticity aspects of fracture. In: Liebowitz D (ed.) *Fracture* 3, Academic Press, New York, pp. 47–225

McClintock F A 1989 Reduced crack growth ductility due to assymetric configurations. *J. Mech. Phys. Solids* (in press)

McClintock F A, Argon A S (eds.) 1966 *Mechanical Behavior of Materials*. Addison–Wesley, Reading, Massachusetts

Nakazawa S, Willam K, Rebelo N (eds.) 1987 *Advances in Inelastic Analysis*, AMD 88 PED 28. American Society of Mechanical Engineers, New York

Owen D R J, Hinton E 1980 *Finite Elements in Plasticity*. Pineridge, Swansea, UK

Pittman J F T, Zienkiewicz O C, Wood R D, Alexander J M (eds.) 1984 *Numerical Analysis of Forming Processes*. Wiley, New York

Rie K-T (ed.) 1987 *Low Cycle Fatigue and Elasto-Plastic Behaviour of Materials*. Elsevier, London

Szczepiński W 1963 On the effect of plastic deformation on the yield condition. *Arch. Mech. Stosow.* 15: 275–94

Szczepiński W 1972 Limit analysis and plastic design of structural elements of complex shape. In: Kuchemann D (ed.) *Progress in Aerospace Sciences*. Pergamon, Oxford, pp. 1–47

Thomsen E G, Yang C T, Kobayashi S 1965 *Mechanics of Plastic Deformation in Metal Processing*. Macmillan, New York

F. A. McClintock
[Massachusetts Institute of Technology, Cambridge, Massachusetts, USA]

Plastics and Composites as Building Materials

Plastics and composites of plastics combined with other materials are relative newcomers to building but their use is growing rapidly. They possess a number of attributes applicable to building, among them.

(a) Formability—plastics have no inherent form, so they can often be shaped efficiently for a given application. Form can also help to overcome limitations such as low stiffness.

(b) Consolidation of parts—one-piece forming can often eliminate separate pieces, joints, seams, fasteners and sealants.

(c) Light weight and good strength-to-weight ratio—the low specific gravity of most plastics and reinforcing materials results in high ratios of strength to weight (see *Polymers: Mechanical*

Properties). The toughness of many plastics and composites often allows them to be used in thin sections, unlike some conventional hard building materials. By taking advantage of formability, light-weight structural shapes such as shells, folded plates and tension structures can be achieved. The extremely low specific gravity of plastics foams (0.015 and up) contributes to low weight in such structures as foam-cored building sandwich panels (Fig. 1).

(d) Corrosion resistance—because they are largely immune to electrochemical corrosion, plastics and composites can often be used in such applications as tanks to handle corrosive chemicals, water treatment plants and piping.

(e) Transparency and light transmission—plastics may be highly transparent, translucent or opaque. Examples include dome-shaped sky-lights, breakage-resistant glazing, lighting fixtures, luminous ceilings, security screens and cover sheets for solar collectors.

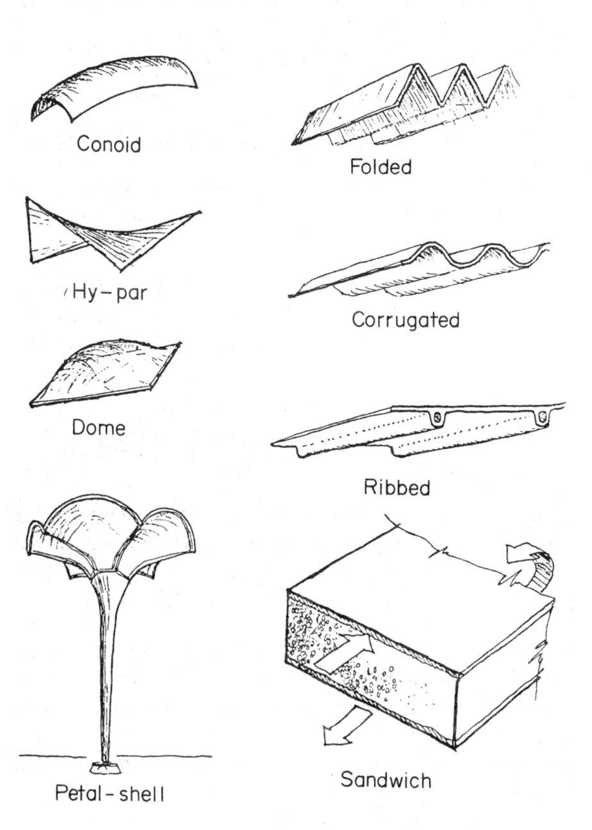

Figure 1
Examples of structural shapes utilizing the formability of plastics. In sandwiches, facings form a moment-resistant couple; the core provides shear resistance and stabilizes facings

(f) Integral color—plastics integrally colored with dyes and pigments can be cut and otherwise worked without loss of color. Examples are flooring and building panels. Permanence is largely a matter of the permanence of the pigments and dyes. Surface colors may be applied as paints or gel coats.

(g) Wear resistance and smoothness—although they may be scratched and indented more easily than glass and other harder materials, plastics often provide good to excellent resistance to wear or abrasion. Nylon hardware parts frequently exhibit outstanding wear resistance and quiet operation. Vinyl flooring is resistant to wear and abrasion. The smooth inner surfaces of extruded pipes result in low resistance to flow of liquids.

(h) Energy absorption and impact resistance—some plastics shatter easily; others are tough and resilient. Examples include acrylic and polycarbonate glazing and lighting fixtures; poly(vinyl chloride), acrylonitrile–butadiene–styrene and polypropylene piping; and glass-fiber-reinforced plastic concrete forms.

(i) Flexibility—some plastics are inherently soft and flexible; others become so by additives and copolymerization. Applications include electrical wire insulation, membranes for air-supported structures, vapor barriers, pond liners and flexible pipe.

(j) Thermal expansion—plastics generally have much higher coefficients of thermal expansion than the other common building materials such as glass, masonry, concrete, metal and wood. Plastic building components must allow for expansion and contraction (Fig. 2). Incorporation of reinforcement such as glass fiber greatly reduces thermal expansion coefficients, but even so, reinforced plastics typically expand and contract more than conventional structural materials.

1. Major Uses of Plastics in Building

The applications of plastics in building may be classified as (a) auxiliaries to other materials, (b) nonstructural and (c) structural and semistructural.

1.1 Auxiliaries to Other Materials

Among the principal uses of plastics as auxiliaries to other building materials are (a) protective and decorative coatings (see *Paints, Varnishes and Lacquers*); (b) adhesives and resin-bonding agents, especially for wood products (see *Adhesives for Wood; Structural Adhesives*); and (c) building sealants in the form of caulking compounds and gaskets. Contemporary sealants augment the older oil- and resin-based varieties, and may be two-component (mixed on the

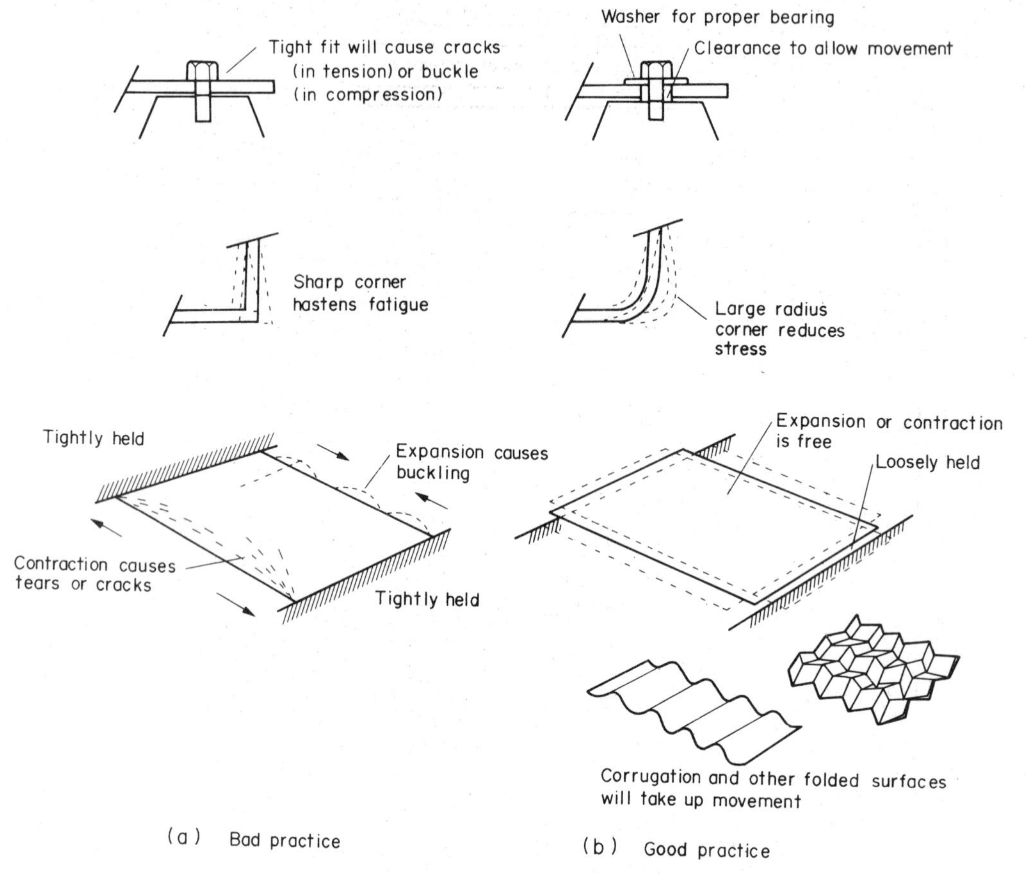

Figure 2
Thermal expansion: good and bad practice

job) or single-component (ready to use, often as cartridges for sealant guns). Polymers employed include polysulfide rubber, acrylics, silicones, chlorosulfonated polyethylene and urethanes. Gaskets, similarly, draw upon various polymers including natural and synthetic rubbers and poly(vinyl chloride).

1.2 Nonstructural Uses

These constitute by far the major applications in building. In general, they are not called upon to carry significant loads, and failure does not entail appreciable danger to occupants of buildings. Among the most important of such uses are the following.

(a) Pipe, fittings and conduit—this, the largest single class of uses, includes pipe and fittings mainly for drains, vents, water and underground lines. Poly(vinyl chloride) is most widely used, followed by high-density and low-density polyethylene, acrylonitrile–butadiene–styrene, reinforced polyester, polypropylene and polystyrene.

(b) Thermal insulation—efficient thermal insulating materials are provided by prefoamed or foamed-in-place plastics. Prefoamed plastics, such as polystyrene and polyurethane, are typically blown as large masses cut into slabs and blocks for use under concrete slabs, in cavity walls and as sheathing and sandwich cores. Two foamed-in-place types are common. In the first, for example polystyrene, pellets containing gas-forming ingredients are placed in the space to be filled and are heated. As gas is formed the pellets soften and expand, filling the space and coalescing. In the second, the liquid components of the foam, for example polyurethane and urea–formaldehyde, are mixed and poured into the space to be filled. Gas is formed as the reaction takes place; the foaming mass rises, fills the space and hardens. Depending upon the foaming agent and density, the thermal insulation achievable is one of the most favorable among building materials. Some highly efficient gases, such as fluorocarbon, unless sealed, as in refrigerator walls, may gradually be replaced by air, decreasing the thermal value of the insulation. Cells may be open or closed, that is, interconnected or not. Closed-cell foams may be impervious to passage of water vapor and constitute vapor seals; open-cell foams are not. In some foams, such as urea–formaldehyde, water is produced as a by-product of the foaming reaction and must be allowed to escape. Some foamed-in-place materials may shrink upon curing and allow heat to escape through the cracks. Precautions may be necessary to avoid residual irritating gases (see Sect. 3).

(c) Flooring—poly(vinyl chloride) is compounded with fillers, fiber and additives to provide resilient sheet and tile flooring. Urethane foams are employed as underlayments for rugs. Epoxies provide paving materials.

(d) Plumbing—bathtubs, shower stalls and other plumbing fixtures utilize large quantities of reinforced polyesters in addition to acrylics, polyacetals, thermoplastic polyesters and polystyrene.

(e) Glazing and skylights—flat, corrugated and thermoformed acrylic and polycarbonate sheets are widely used, as are flat and corrugated reinforced-polyester sheets. Flat reinforced-plastic sheets are bonded to metal grids to form translucent structural wall and roof panels.

(f) Lighting fixtures—acrylics, cellulosics, polycarbonates, polystyrene and poly(vinyl chloride) provide a multiplicity of indoor and outdoor lighting fixtures.

(g) Moisture barriers—low-density polyethylene and poly(vinyl chloride) find extensive use as barriers to prevent moisture migration through concrete slabs on the ground and vapor passage through walls and roofs, and as liners for swimming pools.

(h) Profiles—such as window parts, rainwater systems, moldings and many others are chiefly extruded poly(vinyl chloride) and polyethylene, solid or foamed.

(i) Wall coverings—poly(vinyl chloride) and polystyrene account for most wall coverings, including flexible sheets for walls subject to considerable wear.

(j) Panels and siding—these are mainly poly(vinyl chloride) and reinforced polyesters, with acrylics and cellulose acetate–butyrate also employed.

1.3 Structural and Semistructural Uses

Plain unmodified plastics are moderate in strength, but generally too low in stiffness (having low moduli of elasticity) to be useful for structural and semistructural applications. These may be defined as those in which failure is dangerous to life and property. For such uses, composites are structural; the decorative high-pressure laminates (see Sect. 2.3), for example, are usually not employed structurally. Many fibrous and particulate composite parts (see below) are nonstructural.

2. Composites

Among the most significant uses of plastics in building are the composites, combinations of materials whose properties transcend those of the individual materials acting alone. Plastics-based composites may be classified as particulate, laminar or fibrous (see *Composite Materials: An Overview*).

2.1 Particulate Composites

Portland-cement concrete may be modified with polymers to produce polymer-impregnated concrete, polymer-cement concrete and polymer concrete.

Polymer-impregnated concrete is produced by drying cured standard concrete, evacuating the pores and impregnating with a monomer subsequently polymerized in place by heat, chemical action or high-energy radiation. Compressive strengths may be increased fourfold, creep drastically diminished, and resistance to sulfate attack and freezing and thawing greatly enhanced. Road surfaces have been dried at high temperatures in temporary enclosures, then soaked with monomer which is subsequently cured by heating.

Polymer-cement concrete is standard concrete with a monomer or polymer latex added during mixing. It is subsequently handled like standard concrete. Results have ranged from disappointing to several-fold increases of compressive and tensile strength.

Polymer concrete consists of regular mineral aggregate in a polymer matrix, customarily polyester, instead of Portland cement paste. The ingredients are mixed and cast into forms or molds where the polymer hardens, in hours at room temperature or minutes with heat. Unlike standard concrete, no curing time is needed. A common application is in wall panels with thin polymer-concrete facings surrounding a foamed plastic core. Polymer concrete is used for rapid repair of holes in concrete roads.

2.2 Fibrous Composites

For building purposes, the most commonly used fibrous composite consists of E-glass fibers in a polyester matrix. Other fibers, natural and synthetic, are used occasionally, as are other matrices such as epoxy, phenolic and acrylic. Glass fibers are employed as continuous-filament and chopped-fiber mats, in various fabrics woven of rovings or twisted yarns and in filament winding. Of these, chopped-fiber is the most common, employed in layup, sprayup, bulk-molding compounds and sheet-molding compounds.

Features of fibrous composites favorable for building applications are increased strength and stiffness relative to plain plastics, toughness, reduced coefficients of expansion and lightness compared to some other building materials. As noted above, curved shells, folded plates, sandwiches, ribs, corrugations and similar features serve to utilize the material to the fullest extent (see Fig. 1).

Membranes are a special case of fibrous composites. These are usually woven fabric, glass or other synthetic, coated with a plastic such as poly(vinyl chloride) or polytetrafluoroethylene. Such membranes are employed in air-supported structures and in tension structures commonly supported and held taut by cables.

2.3 Laminar Composites

For buildings, high-pressure laminates and structural sandwiches are most important.

High-pressure laminates are most commonly decorative. They consist of layers of paper impregnated with phenolic resin, faced with printed paper (sometimes fabric or wood veneer) impregnated with melamine–formaldehyde and overlaid with a melamine–formaldehyde-impregnated veil, all simultaneously pressed into a decorative laminate, usually subsequently bonded to plywood, particle board or other substrates for counter and table tops, door facings, furniture and cabinetwork. Such laminates are resistant to most household solvents and are harder than common varnishes and lacquers, but not as hard as glass.

Structural sandwiches consist of relatively thin facings of dense, strong, stiff materials bonded to relatively thick, less dense, less strong and less stiff cores to provide lightweight combinations that are strong and stiff because of their geometry (see Fig. 1). In building panels, the combination is also expected to provide thermal insulation, acoustical properties, fire resistance, weather resistance and resistance to normal wear and tear.

Facings most commonly employed include plywood, fiberboard, particle board, cement–asbestos board, gypsum board, metals such as steel and aluminum, high-pressure laminates and glass-fiber-reinforced plastics.

Cores include foamed plastics, honeycombs of paper impregnated with phenolic resin, fiberboards, plywood, particle board, cement–asbestos board, gypsum board, foamed concrete and various wood and metal grids.

Adhesives to bond facings and core are generally based on synthetic resins. Some foamed-in-place plastics, for example polyurethanes, are self-adhering. The adhesive must withstand shear and tensile stresses between facing and core.

3. Properties of Building Plastics

Some properties of plastics and composites compared with other building materials are shown in Fig. 3. The higher values of strengths and elastic moduli are achieved by fibrous composites, whereas the lower values are those of plain unreinforced plastics. Although glass has an elastic modulus of about one third that of steel, when combined with a low-modulus plastic matrix in the usual proportions, the combined elastic modulus seldom is more than half that of glass alone, and is frequently less than that of wood parallel to the grain or of unreinforced concrete.

The high values of thermal expansion shown in Fig. 3 are those of plain plastics, whereas the low values are caused by the constraining effects of reinforcements, especially glass with its low thermal expansion and high elastic modulus.

The thermal conductivity of molded and reinforced plastics is comparable to wood, concrete and glass. Foamed plastics, on the other hand, are among the best insulators for buildings.

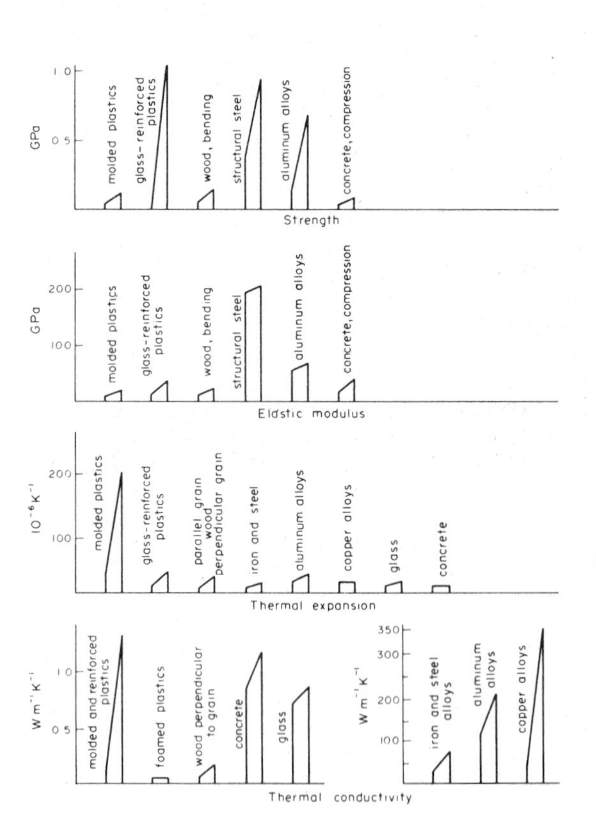

Figure 3
Comparison of the properties of plastics and other materials

3.1 Durability and Weathering

Depending upon the compositions and upon the conditions of exposure, plastics may be highly durable or may disintegrate rapidly. Plain polyethylene deteriorates quickly outdoors, but when compounded with carbon black provides highly durable cable covering and irrigation pipe. Acrylics and PVC have given many years of good performance. Glass-fiber-reinforced polyesters exposed to weathering have performed well or poorly depending upon the type of polyester employed, the care with which the fiber has been thoroughly wetted and embedded in the matrix and the surface treatments such as veils and gel coats. Once exposed by erosion, fibers tend to act as wicks to draw moisture inward, leading to deterioration. Localized stress, as in sharply curved transparent sheets, often results in localized crazing and eventual cracking. Stress may also be caused by ignoring the typically high expansion coefficients of plastics and confining them too rigidly.

Plastics may perform well under conditions such as moisture and polluted atmospheres that cause deterioration of metal, stone and other susceptible materials. They provide some of the best protective coatings for such conditions and may themselves be protected from attack, or have eroded surfaces rejuvenated, by such coatings.

3.2 Flammability

Like other organic building materials, such as wood, plastics can be destroyed by fire. Some burn readily, some with difficulty, and others do not support combustion upon removal of flame, but none withstand the temperatures that can be encountered in building fires.

Smoke and gases may be more hazardous than flame. Some gases, such as water and carbon dioxide, may be innocuous, but carbon monoxide is likely to be generated copiously when oxygen is deficient. Depending upon the chemical structure, gases may be noxious or toxic, and dense smoke may be generated. A choice may therefore have to be made between flame and smoke in a given situation.

Some of the most effective fire retardants diminish the durability of plastics in uses such as outdoor exposure. A choice or balancing of requirements may be necessary.

4. Building Codes and Standards

There are no Federal building codes in the USA. Many states have building codes but they may not be completely binding upon municipalities. Local codes govern. Other countries often have national codes. Codes may be prescriptive, that is, specify materials and installation in detail, or they may specify the performance to be expected of buildings and the components, but leave the choice of materials and methods open.

Several organizations in the USA write recommended uniform codes that are frequently adopted by municipalities. Their provisions for plastics are similar, and may be taken as representative, realizing that the municipal codes, although often based upon the recommended codes, prevail.

In addition to general measures applicable to all materials, the recommended codes have separate sections devoted to light-transmitting plastics and to foamed plastics. Plastics are classified as thermoplastic, thermosetting and reinforced. Both sections refer to standard tests and specifications of the American Society for Testing and Materials respecting strength, durability, sanitary properties, flammability and smoke.

4.1 Light-Transmitting Plastic Construction

Self-ignition temperatures must be 350 °C or above (ASTM D1929) and smoke density not greater than 450 (ASTM E84). Two classes of plastics are based upon burning rates. Types of applications are glazing, plastic wall panels, roof panels, skylights and light-diffusing systems. Glazing of unprotected openings is limited according to occupancy, percentage of wall area, height above ground and presence or absence of automatic fire-suppressant systems such as sprinklers. Exterior wall panels are subject to similar provisions. Roof panels are limited according to occupancy (prohibited in some), requirements for roof coverings, automatic fire-suppressant systems, horizontal spacing and proximity to exterior walls. Skylight assemblies must conform to requirements respecting curbs, edge protection (incombustible strips), maximum area, minimum rise of dome, horizontal separation and proximity of exterior walls. If automatic fire-suppressant systems are present or if skylights are used as fire vents, areas may be more generous. Light-diffusing systems (illuminated ceilings) must comply with interior finish requirements or fall from their supports at specified temperature ranges. Sprinklers may be needed above or below the installation. Areas are limited. Partitions, bathroom accessories, awnings and greenhouses are subject to special provisions.

4.2 Foamed Plastics

Similar language applies to formed plastics. Generally, flame-spread ratings are limited to 75 and smoke developed to 450 (ASTM E84) unless specifically exempted.

Foamed plastics are permitted within the cavities of masonry or concrete walls and, generally, on room surfaces or as elements of non-fire-resistive walls if protected by a barrier with at least a 15 minute fire rating (ASTM 119), such as 13 mm gypsum board or Portland-cement stucco, 0.81 mm aluminum or 0.45 mm (26 gauge) galvanized steel sheet.

With protective surfaces at least equal to the above, foamed plastics are permitted as cores of sandwich panels. When such panels are load bearing, as in roof panels, special requirements may be imposed. Automatic sprinklers may be required, depending upon occupancy.

See also: Composite Materials: An Overview; Design with Polymers

Bibliography

Agranoff J (ed.) 1984 *Modern Plastics Encyclopedia.* McGraw-Hill, New York

American Society of Civil Engineers 1984 *Structural Plastics Design Manual.* American Society of Civil Engineers, New York

Basic Building Code 1975 Building Officials and Code Administrators International, Chicago, Illinois

Dietz A G H 1969 *Plastics for Architects and Builders.* MIT Press, Cambridge, Massachusetts

Saechtling H (ed.) 1973 *Bauen mit Kunststoffe.* Carl Hanser, Munich

Skeist I (ed.) 1966 *Plastics in Building.* Reinhold, New York

Standard Building Code 1977 Southern Building Code Congress International, Birmingham, Alabama

Uniform Building Code 1978 International Conference of Building Officials, Whittier, California

A. G. H. Dietz
[Massachusetts Institute of Technology, Cambridge, Massachusetts, USA]

Plywood

Plywood comprises a number of thin layers of wood called veneers which are bonded together with an adhesive. Each layer is placed so that its grain direction is at right angles to that of the adjacent layer (Fig. 1). This cross-lamination gives plywood its characteristics and makes it a versatile building material. Plywood has been used for centuries. It has been found in Egyptian tombs and was in use during the height of the Greek and Roman civilizations.

The modern plywood industry is divided into the so-called hardwood and softwood plywood industries. While the names are truly misnomers, they are in common use, the former referring to industry manufacturing decorative panelling and the latter referring to that which serves building construction and industrial uses. This article is concerned primarily with the plywood manufactured for construction and for industrial uses in North America.

1. Manufacture

The fundamentals of plywood manufacture are the same for decorative and for construction grades, the primary differences being in the visual quality of the veneer faces.

1.1 Log Preparation

Logs received at the plywood plant are either stored in water or stacked in tiers in the mill yard. From there they are transported to a deck where they are debarked and cut into suitable lengths. The cut sections of logs are referred to as blocks. Debarking a log is most often accomplished by conveying it through an enclosure that contains a power-driven "ring" of scraper knives. The whirling "ring" has the flexibility to adjust to varying log diameters and scrapes away the bark.

Before transforming the blocks into veneer, many manufacturers find it helpful to soften the wood fibers by steaming or soaking the blocks in hot-water vats. The block conditioning process is usually performed after debarking. The time and temperature requirements for softening the wood fibers will vary according to wood species and desired heat penetration. Block conditioning usually results in smoother and higher-quality veneer.

1.2 Conversion into Veneer

The most common method of producing veneer is by rotary-cutting the blocks on a lathe. Rotary lathes are equipped with chucks attacked to spindles and are capable of revolving the blocks against a knife which is bolted to a movable carriage. New "spindleless" lathes are being introduced which drive the block

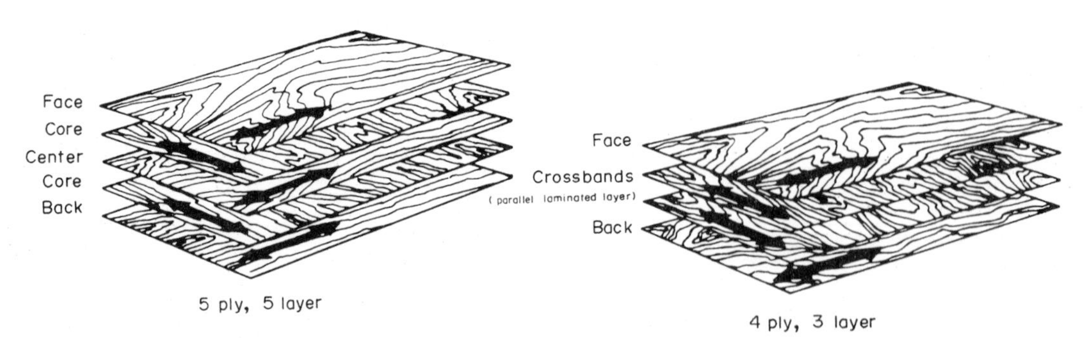

Figure 1
Ply-layer construction of plywood (courtesy of the American Plywood Association)

against a fixed knife with exterior power-driven rollers. Once in place, the chucks revolve the block against the knife and the process of "peeling" veneer begins. The first contact with the lathe knife produces veneer which is not full length. This is due to the fact that the blocks are never perfectly round and are usually tapered to some extent. It is therefore necessary to "round up" the block before full veneer yield is realized. Once this step is complete, the veneer comes away in a continuous sheet, in much the same manner as paper is unwound from a roll. Veneer of the highest grade and quality is usually obtained at the early cutting, knots becoming more frequent when the block diameter is reduced. A knot is a branch base embedded in the trunk of a tree and is usually considered undesirable.

All veneer has a tight and a loose side. The side which is opposite to that of the knife is the tight side because of pressure from the machinery holding it against the knife. The knife side is where the lathe checks occur and is called the loose side. The loose side is always checked to some degree and the depths of the lathe checks will depend upon the pressure exerted by the pressure bar opposite the knife. Various thicknesses of veneer are peeled according to the type and grade of plywood to be produced.

Wood blocks are cut at high speed, and for this reason veneer is fed into a system of conveyors. At the opposite end of the conveyor is a machine called the clipper. Veneer coming from the lathe through the system is passed under a clipper where it is cut into prescribed widths. Defective or unusable veneer can be cut away. The clipper usually has a long knife attached to air cylinders. The knife action is a rapid up and down movement, cutting through the veneer quickly. Undried veneer must be cut oversized to allow for shrinkage and eventual trimming. Up to this point the wood has retained a great deal of its natural moisture and is referred to as green. Accordingly, this whole process before drying is called the green end of the mill.

1.3 Veneer Drying and Grading

The function of a veneer dryer is to reduce the moisture content of the stock to a predetermined percentage and to produce flat and pliable veneer. Most veneer dryers carry the veneer stock through the dryer by a series of rolls. The rolls operate in pairs, one above the other with the veneer between, in contact with each roll. The amount of moisture in the dryers is controlled by dampers and venting stacks. Moisture inside the dryer is essential to uniform drying, since this keeps the surface pores of the wood open. The air mixture is kept in constant circulation by powerful fans. The temperature and speed of travel through the dryer is controlled by the operator and is dependent upon the thickness of the veneer species and other important factors.

Dry veneer must be graded, then stacked according to width and grade. Veneer is visually graded by individuals who have been trained to gauge the size of defects, the number of defects and the grain characteristics of various veneer pieces. Veneer grades depend upon the standard under which they are graded, but in North America follow a letter designation A, B, C and D, where A has the fewest growth characteristics and D the most.

1.4 Veneer Joining and Repair

For some grades of plywood, and especially where large panels may be necessary, pieces are cut with a specialized machine, glue is applied and the veneer strips are edge-glued together. The machine usually utilizes a series of chains which crowd the edges of the veneer tightly together, thus creating a long continuous sheet. Other methods of joining veneer may also be employed, such as splicing, stringing and stitching. Splicing is an edge-gluing process in which only two pieces of veneer are joined in a single operation. Stringing is similar to the edge-gluing process except the crowder applies string coated with a hot melt adhesive to hold the veneers temporarily together before pressing. Stitching is a variation of stringing in which industrial-type sewing machines are used to stitch rows of string through the veneer. This system has the advantage of being suitable for either green or dry stock.

The appearance of various grades of veneer may be improved by eliminating knots, pitch pockets and other defects by replacing them with sound veneer of similar color and texture. Machines can cut a patch hole in the veneer and simultaneously replace it with a sound piece of veneer. While the size and shape varies, boat-shaped and dog-bone-shaped veneer repairs are most common.

1.5 Adhesives

The most common adhesives in the plywood industry are based on phenolic resins, blood or soybeans. Other adhesives such as urea-based resorcinol, polyvinyl and melamine are used to a lesser degree for operations such as edge gluing, panel patching and scarfing.

Phenolic resin is synthetically produced from phenol and formaldehyde. It hardens or cures under heat and must be hot pressed. When curing, phenolic resin goes through chemical changes which make it waterproof and impervious to attack by microorganisms. Phenolic resin is used in the production of interior plywood, but is capable in certain mixtures of being subjected to permanent exterior exposure. Over 90% of all softwood plywood produced in the USA is manufactured with this type of adhesive. Another class of phenolics is called extended resins, which have been extended with various substances to reduce the resin solids content. This procedure reduces the cost of the glueline and results in a gluebond suitable for interior or protected-use plywood.

Soybean glue is a protein type made from soybean meal. It is often blended with blood and used in both cold pressing and hot pressing.

Blood glue is another protein type made from animal blood collected at slaughter houses. The blood is spray dried and applied to the plywood and supplied in powder form. Blood and blood–soybean blends may be cold pressed or hot pressed. Protein-based glues (blood or soybean) are not waterproof and therefore are used in the production of interior-use plywood. Protein glues are not currently in common use.

The glue is applied to the veneers by a variety of techniques including roller spreaders, spraylines, curtain coaters and a more recent development, the foam glue extruder. Each of these techniques has its advantages and disadvantages, depending upon the type of manufacturing operation under consideration (see *Adhesives for Wood*).

1.6 Assembly, Pressing and Panel Construction

Assembly of veneers into plywood panels takes place immediately after adhesive application. Workmanship and panel assembly must be rapid, yet careful. Speed in assembly is necessary because adhesives must be placed under pressure with the wood within certain time limits or they will dry out and become ineffective. Careful workmanship also avoids excessive gaps between veneers and veneers which lap and cause a ridge in the panel. Prior to hot pressing, many mills pre-press assembled panels. This is performed in a cold press which consists of a stationary platten and one connected to hydraulic rams. The load is held under pressure for several minutes to develop consolidation of the veneers. The purpose of pre-pressing is to allow the wet adhesive to tack the veneers together, which provides easier press loading and eliminates breaking and shifting of veneers when loaded into the hot press. The hot press comprises heated plattens with spaces between them known as openings. The number of openings is the guide to press capacity, with 20 or 30 openings being the most common, although presses with as many as 50 openings are in use. When the press is loaded, hydraulic rams push the plattens, together exerting a pressure of 1.2–1.4 MPa. The temperature of the plattens is set at a certain level, usually in the range 100–165 °C.

2. Standards

Construction plywood in the USA is manufactured under US Product Standard PS 1 for Construction and Industrial Plywood (US Department of Commerce 1983) or a Performance Standard such as those promulgated by the American Plywood Association. The most current edition of the Product Standard for plywood includes provisions for performance rating as well. The standard for construction and industrial plywood includes over 70 species of wood which are separated into five groups based on their mechanical properties. This grouping results in fewer grades and a simpler procedure.

Certain panel grades are usually classified as sheathing and include C-D, C-C, and Structural I C-D and C-C. These panels are sold and used in the unsanded or "rough" state, and are typically used in framed construction, rough carpentry and many industrial applications. The two-letter system refers to the face and back veneer grade. A classification system provides for application as roof sheathing or subflooring. A fraction denotes the maximum allowable support spacing for rafters (in inches) and the maximum joist spacing for subfloors (in inches) under standard conditions. Thus, a 32/16 panel would be suitable over roof rafter supports at a maximum spacing of 32 in. (800 mm) or as a subfloor panel over joists spaced 16 in. (400 mm) on center. Common "span ratings' include 24/0, 24/16, 32/16, 40/20 and 48/24.

Many grades of panel must be sanded on the face and/or back to fulfill the requirements of their end use. Sanding most often takes place on plywood panels with A- or B-grade faces and some C "plugged" faces are sanded as well. Sanded panels are graded according to face veneer grade, such as A-B, A-C or B-C. Panels intended for concrete forming are most often sanded as well and are usually B-B panels.

Panel durability for plywood is most often designated as Exterior or Exposure 1. Exterior plywood consists of a higher solids-content glueline and a minimum veneer grade of C in any ply of the panel. Such panels can be permanently exposed to exterior exposure without fear of deterioration of the glueline. Exposure 1 panels differ in the fact that D-grade veneers are allowed. Exposure 1 panels are most often made with an extended phenolic glueline and are suitable for protected exposure and short exterior exposure during construction delays.

3. Physical Properties

Plywood described here is assumed to be manufactured in accordance with US Product Standard PS 1, Construction and Industrial Plywood (US Department of Commerce 1983). The physical property data were collected over a period of years (O'Halloran 1975).

3.1 Effects of Moisture Content

Many of the physical properties of plywood are affected by the amount of moisture present in the wood. Wood is a hygroscopic material which nearly always contains a certain amount of water. When plywood is exposed to a constant relative humidity, it will eventually reach an equilibrium moisture content (EMC). The EMC of plywood is highly dependent on relative humidity, but is essentially independent of temperature between 0 °C and 85 °C. Examples of values for plywood at 25 °C include 6% EMC at 40% relative humidity (RH), 10% EMC at 70% RH and 28% EMC at 100% RH.

Plywood exhibits greater dimensional stability than most other wood-based building products. Shrinkage of solid wood along the grain with changes in moisture content is about 2.5–5% of that across the grain. The tendency of individual veneers to shrink or swell crosswise is restricted by the relative longitudinal stability of the adjacent plies, aided also by the much greater stiffness of wood parallel as compared to perpendicular to grain.

The average coefficient of hygroscopic expansion or contraction in length and width for plywood panels with about the same amount of wood in parallel and perpendicular plies is about 0.002 mm mm^{-1} for each 10% change in RH. The total change from the dry state to the fiber saturation point averages about 0.2%. Thickness swelling is independent of panel size and thickness of veneers. The average coefficient of hygroscopic expansion in thickness is about 0.003 mm mm^{-1} for each 1% change in moisture content below the fiber saturation point.

The dimensional stability of panels exposed to liquid water also varies. Tests were conducted with panels exposed to wetting on one side as would be typical of a rain-delayed construction site (Bengelsdorf 1981). Such exposure to continuous wetting on one side of the panel for 14 days resulted in about 0.13% expansion across the face grain direction and 0.07% along the panel. The worst-case situation is reflected by testing from oven-dry to soaking in water under vacuum and pressure conditions. Results for a set of plywood similar to those tested on one side showed approximately 0.3% expansion across the panel face grain and 0.15% expansion along the panel. These can be considered the theoretical maximum that any panel could experience.

3.2 Thermal Properties

Heat has a number of important effects on plywood. Temperature affects the equilibrium moisture content and the rate of absorption and desorption of water. Heat below 90 °C has limited long-term effect on the mechanical properties of wood. Very high temperature, on the other hand, will weaken the wood.

The thermal expansion of wood is smaller than swelling due to absorption of water. Because of this, thermal expansion can be neglected in cases where wood is subject to considerable swelling and shrinking. It may be of importance only in assemblies with other materials where moisture content is maintained at a relatively constant level. The effect of temperature on plywood dimensions is related to the percentage of panel thickness in plies having grain perpendicular to the direction of expansion or contraction. The average coefficient of linear thermal expansion is about 6.1 $\times 10^{-6}$ m m^{-1} $°C^{-1}$ for a plywood panel with 60% of the plies or less running perpendicular to the direction of expansion. The coefficient of thermal expansion in panel thickness is approximately 28.8×10^{-6} m m^{-1} $°C^{-1}$.

The thermal conductivity k of plywood is about 0.11–0.15 W m^{-1} K^{-1}, depending on species. This compares to values (in W m^{-1} K^{-1}) of 391 for copper (heat conductor), 60 for window glass and 0.04 for glass wool (heat insulator).

From an appearance and structural standpoint, unprotected plywood should not be used in temperatures exceeding 100 °C. Exposure to sustained temperatures higher than 100 °C will result in charring, weight loss and permanent strength loss.

3.3 Permeability

The permeability of plywood is different from solid wood in several ways. The veneers from which plywood is made generally contain lathe checks from the manufacturing process. These small cracks provide pathways for fluids to pass by entering through the panel edge. Typical values for untreated or uncoated plywood of 3/8 inch (9.5 mm) thickness lie in the range 0.014–0.038 g m^{-2} h^{-1} $mmHg^{-1}$.

Exterior-type plywood is a relatively efficient gas barrier. Gas transmission (cm^3 s^{-1} cm^{-2} cm^{-1}) for 3/8 in. exterior-type plywood is as follows:

oxygen	0.000 029
carbon dioxide	0.000 026
nitrogen	0.000 021

4. Mechanical Properties

The current design document for plywood is published by the American Plywood Association (1986), and entitled *Plywood Design Specification*. The woods used to manufacture plywood under US Product Standard PS 1 are classified into five groups based on elastic modulus, and bending and other important strength properties. The 70 species are grouped according to procedures set forth in ASTM D2555 (*Establishing Clearwood Strength Values*, American Society for Testing and Materials 1987). Design stresses are presently published only for groups 1–4 since group 5 is a provisional group with little or no actual production. Currently the *Plywood Design Specification* provides for development of plywood sectional properties based on the geometry of the layup and the species, and combining those with the design stresses for the appropriate species group.

4.1 Section Properties

Plywood section properties are computed according to the concept of transformed sections to account for the difference in stiffness parallel and perpendicular to the grain of any given ply. Published data take into account all possible manufacturing options under the appropriate standard, and consequently the resulting published value tends to reflect the minimum configuration. These "effective" section properties computed by the transformed section technique take into account the orthotropic nature of wood, the species group used in the outer and inner plies, and the manufacturing variables provided for each grade. The section properties presented are generally the minimums that can be expected.

Because of the philosophy of using minimums, section properties perpendicular to the face grain direction are usually based on a different configuration than those along the face grain direction. This compounding of minimum sections typically results in conservative designs. Information is available for optimum designs where required.

4.2 Design Stresses

Design stresses include values for each of four species groups and one of three grade stress levels. Grade stress levels are based on the fact that bending, tension and compression design stresses depend on the grade of the veneers. Since veneer grades A and natural C are the strongest, panels composed entirely of these grades are allowed higher design stresses than those of veneer grades B, C-plugged or D. Although grades B and C-plugged are superior in appearance to C, they rate a lower stress level because the "plugs" and "patches" which improve their appearance reduce their strength somewhat. Panel type (interior or exterior) is therefore important for bending, tension and compression stresses, since panel type determines the grade of the inner plies.

Stiffness and bearing strengths do not depend on either glue or veneer grade but on species group alone. Shear stresses, on the other hand, do not depend on grade, but vary with the type of glue. In addition to grade stress level, service moisture conditions are also typically presented—for dry conditions, typically moisture contents less than 16%, and for wet conditions at higher moisture contents.

Allowable stresses for plywood typically fall in the same range as for common softwood lumber, and when combined with the appropriate section property, result in an effective section capacity. Some comments on the major mechanical properties with special consideration for the nature of plywood are given below.

Bending modulus of elasticity values include an allowance for an average shear deflection of about 10%. Values for plywood bending stress assume flat panel bending as opposed to bending on edge which may be considered in a different manner. For tension or compression parallel or perpendicular to the face grain, section properties are usually adjusted so that allowable stress for the species group may be applied to the given cross-sectional area. Adjustments must be made in tension or compression when the stress is applied at an angle to the face grain.

Shear-through-the-thickness stresses are based on common structural applications such as plywood mechanically fastened to framing. Additional options include plywood panels used as the webs of I-beams. Another unique shear property is that termed rolling shear (see *Wood: Strength*). Since all of the plies in plywood are at right angles to their neighbors, certain types of loads subject them to stresses which tend to make them roll, as a rolling shear stress is induced. For instance, a three-layer panel with framing glued on both faces could cause a cross-ply to roll across the lathe checks. This property must be taken into account with such applications as stressed-skin panels.

Bibliography

American Plywood Association 1986 *Plywood Design Specification*, Form Y510. API, Tacoma, Washington

American Plywood Association 1987a *Grades and Specifications*, Form J20. API, Tacoma, Washington

American Plywood Association 1987b 303 *Plywood Siding*, Form E300. API, Tacoma, Washington

American Society for Testing and Materials 1987 *Annual Book of ASTM Standards*, Vol. 4.09: *Wood*. ASTM Philadelphia, Pennsylvania

O'Halloran M R 1975 *Plywood in Hostile Environments*, Form Z820G. API, Tacoma, Washington

Sellers T Jr 1985 *Plywood and Adhesive Technology*. Dekker, New York

US Department of Commerce 1983 *Product Standard for Construction and Industrial Plywood*, PS 1. USDC, Washington, DC (available from the American Plywood Association, Tacoma, Washington)

Wood A D, Johnston W, Johnston A K, Bacon G W 1963 *Plywoods of the World: Their Development, Manufacture and Application*. Morrison and Gibb, London

M. R. O'Halloran
[American Plywood Association, Tacoma, Washington, USA]

Polymers: Mechanical Properties

Most applications of polymers require load-bearing capacity. The geometrical responses to loading lead to a wide range of mechanical properties. However, nearly all can be grouped under stress–strain properties, viscoelasticity and failure properties. These properties are in turn largely determined by the polymer structure (i.e., molecular weight, cross-linking, branching, segmental structure and morphology) and the nature and extent of compounding. In addition, external conditions such as temperature, loading rate and environment must be considered when characterizing the mechanical behavior of a polymeric system.

1. Stress–Strain Properties

The most common technique for measuring mechanical properties is tensile or compression testing. The results of such tests are stress–strain diagrams. Polymers exhibit three basic stress–strain behaviors: brittle, ductile and elastomeric (Fig. 1). There are a number of factors that affect the stress–strain curves of a particular polymer, such as strain rate, compression or tension, and orientation. Strain rate is important because polymers are generally viscoelastic and thus their behavior is time dependent. Figure 2 illustrates the effect of strain rate on tensile properties. Note that increasing temperature has the same effect as decreasing the strain rate. In glassy polymers, compression differs from tension because surface flaws are detrimental to tensile properties (Fig. 3). Finally, orientation of the chains in the direction of stress substantially increases the properties in that direction, but properties in other directions decrease for both crystalline and amorphous polymers (Fig. 4).

Modulus is a function of temperature and is generally divided into three regions—rigid, leathery and rubbery—as temperature increases. The actual shape of modulus versus temperature curves is dependent on molecular weight and morphology. Increasing molecular weight increases the length of the rubbery region because of increased molecular entanglements (Fig. 5). Cross-linking broadens the leathery region because the random cross-links hinder molecular motion, increasing and broadening the glass-transition temperature range. Also, the rubbery region remains constant and increases in modulus with increased cross-linking (Fig. 6). Crystallinity raises the melting point and increases the modulus of the rubbery region because crystallites act as cross-links in the amorphous region. The negative slope in the rubbery region is due to the melting of imperfect crystallites and thermal expansion (Fig. 7). Finally, plasticizers, copolymerization and pressure also affect the modulus versus temperature curves.

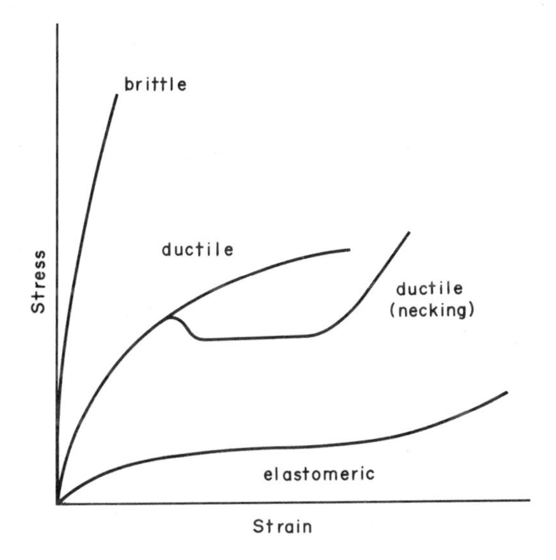

Figure 1
Typical stress–strain curves of brittle, ductile and elastomeric polymers

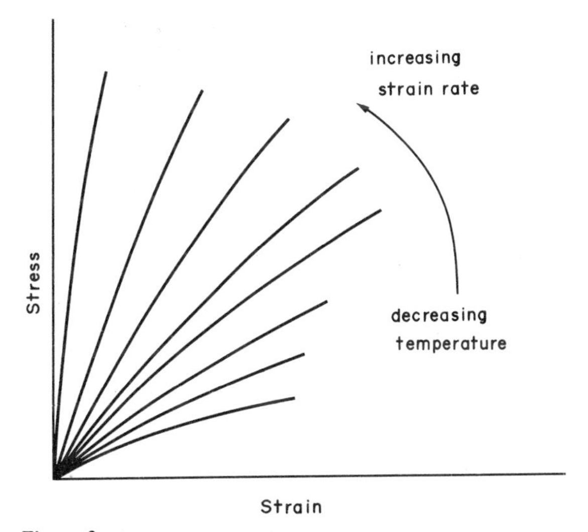

Figure 2
Stress–strain curves as affected by strain rate and temperature

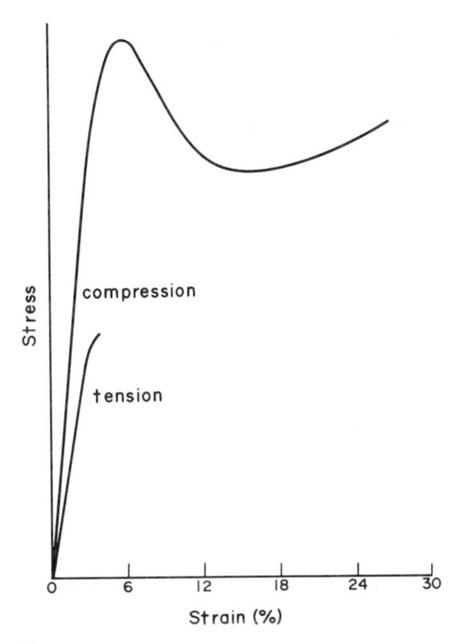

Figure 3
The stress–strain behavior of a glassy polymer showing the difference between tension and compression

2. *Viscoelasticity*

2.1 Creep and Stress Relaxation

Polymers are generally viscoelastic and hence exhibit creep and stress relaxation behavior. Creep and stress

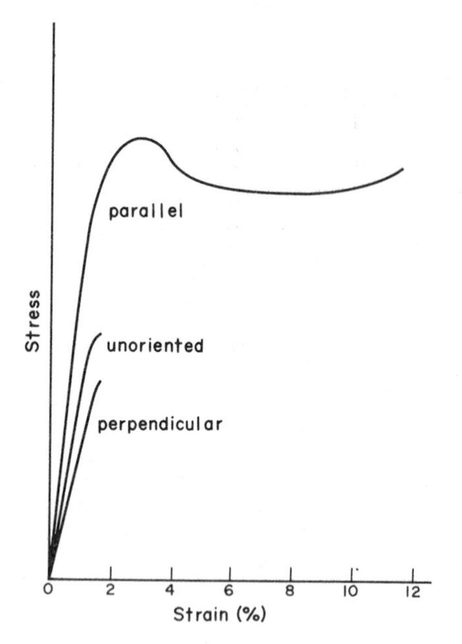

Figure 4
The stress–strain behavior of an oriented, glassy polymer tested in tension parallel and perpendicular to the orientation compared with that of an unoriented polymer

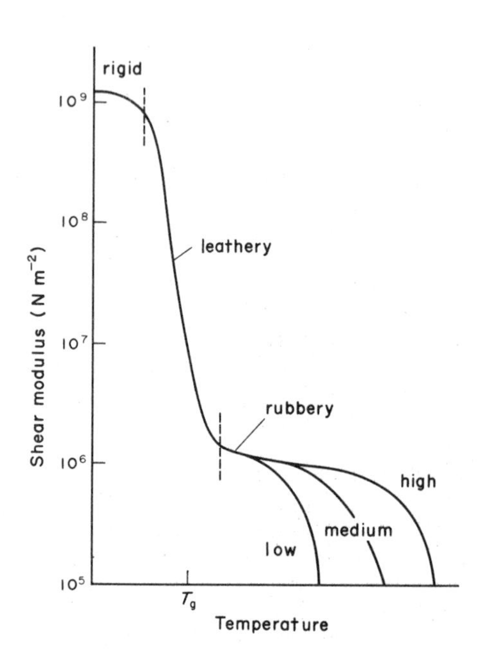

Figure 5
Shear modulus versus temperature for an amorphous polymer of relatively low, medium and high molecular weight

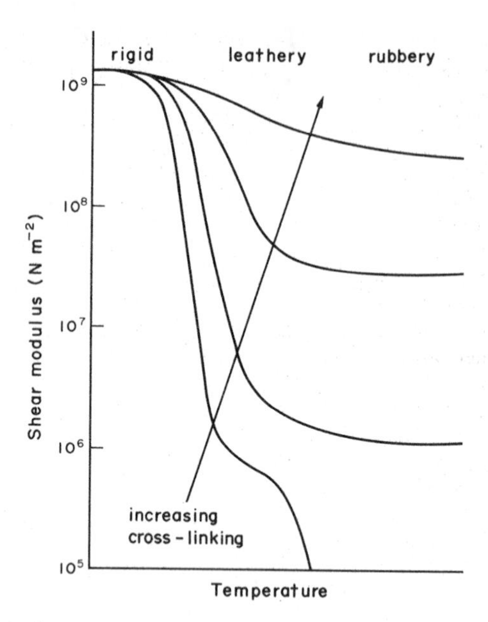

Figure 6
Shear modulus versus temperature of an amorphous polymer as the amount of cross-linking increases

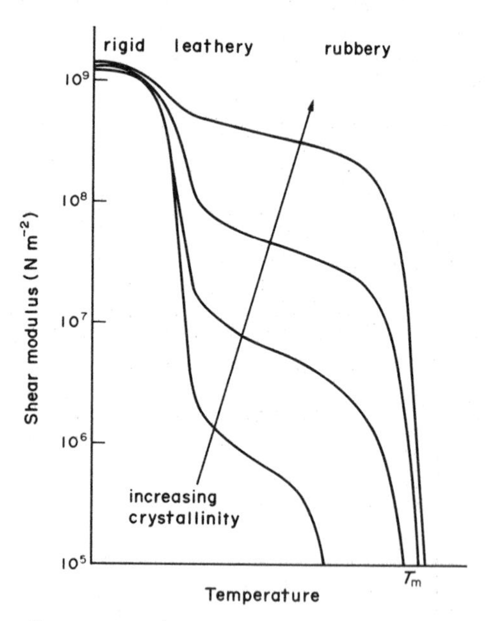

Figure 7
Shear modulus versus temperature as a function of increasing crystallinity

relaxation tests measure the viscoelastic behavior by monitoring elongational changes at a constant stress and by measuring stress reduction at constant elongation, respectively. Both tests permit calculation of the same viscoelastic constants. Most of the examples given here involve creep phenomena but can be applied to stress relaxation.

Several models have been developed to describe the viscoelastic nature of polymers. The models consist of Hookean springs and viscous Newtonian dashpots (Fig. 8). A Hookean spring elongates instantaneously to an equilibrium elongation ε which is directly proportional to the applied stress σ (i.e., $\varepsilon=\sigma/E$ where E is Young's modulus) and instantaneously recovers when the stress is removed. A dashpot is similar to a shock absorber in that an equilibrium elongation is never reached and it does not recover elongation already incurred. Mathematically, the applied stress σ is equal to the rate of elongation through a proportionality constant η known as the viscosity (i.e. $\varepsilon=(\sigma/\eta)t$) effect of an applied and removed stress as a function of time.

Three relatively simple models (Maxwell, Voigt–Kelvin and the four-parameter) and their creep behavior are shown in Fig. 9. The creep behavior of the Maxwell model adequately represents the instantaneous elongation or recovery and permanent deformation due to stress of an actual polymer. However, it does not characterize the exponential decrease in elongation rate or recovery demonstrated by most polymers. The Voigt–Kelvin model demonstrates the exponential creep and recovery of polymers, but is deficient in near instantaneous elongation or recovery. Also, this model reaches an equilibrium elongation not indicative of a true polymer and shows no permanent set.

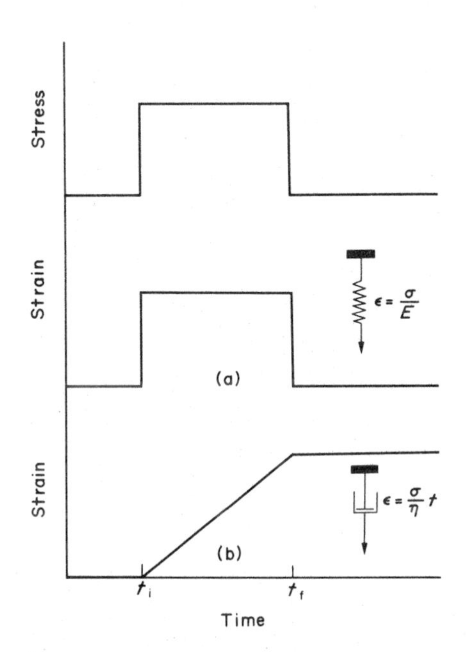

Figure 8
The creep behavior of a spring and dashpot when a stress is instantaneously applied and removed

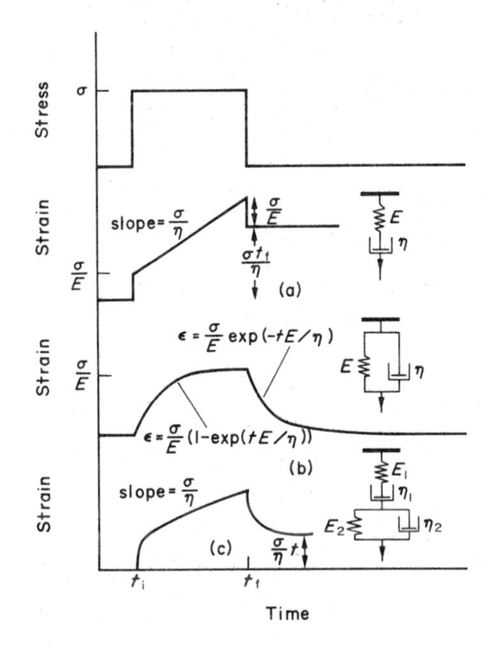

Figure 9
The creep behavior of (a) a Maxwell element, (b) a Voigt–Kelvin element and (c) a four-parameter model when stress is instantaneously applied and removed

The best features of both models are combined into the four-parameter model and thus the creep response of this model adequately matches the creep behavior of polymers. However, this model cannot fully represent polymeric creep behavior because of the distribution in molecular weight and other complex morphological structures associated with polymers. Thus, several models consisting of many Maxwell or Voigt–Kelvin elements with a distribution of relaxation times have been introduced, although the equations are complicated, as shown for Voigt–Kelvin series:

$$J(t)=\int_0^{\infty} L(\lambda)(1-\exp(-t/\lambda_i))\,d\lambda \qquad (1)$$

where J is the compliance and $L(\lambda)$ is the continuous relaxation time distribution.

2.2 Superposition Principles

In order to describe the creep response as a function of loading history, the Boltzmann superposition principle states that the sum of the creep responses due to each stress is the total creep response. Thus, the individual creep response from each stress is independent of other creep responses (Fig. 10) and is mathematically represented by

$$E(t)=J(t-t_1)(\sigma_1-\sigma_0)+J(t-t_2)(\sigma_2-\sigma_1) + J(t-t_3)(\sigma_3-\sigma_2)+\dots \qquad (2)$$

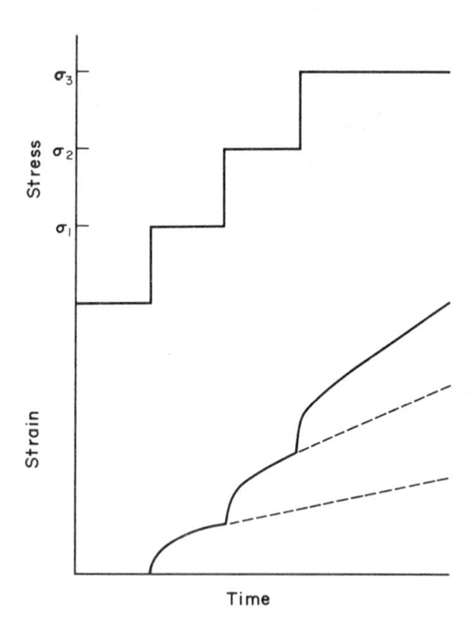

Figure 10
The creep behavior due to the subsequent addition of stresses as predicted by the Boltzmann superposition principle. Dotted lines indicate creep behavior boundaries for each stress applied

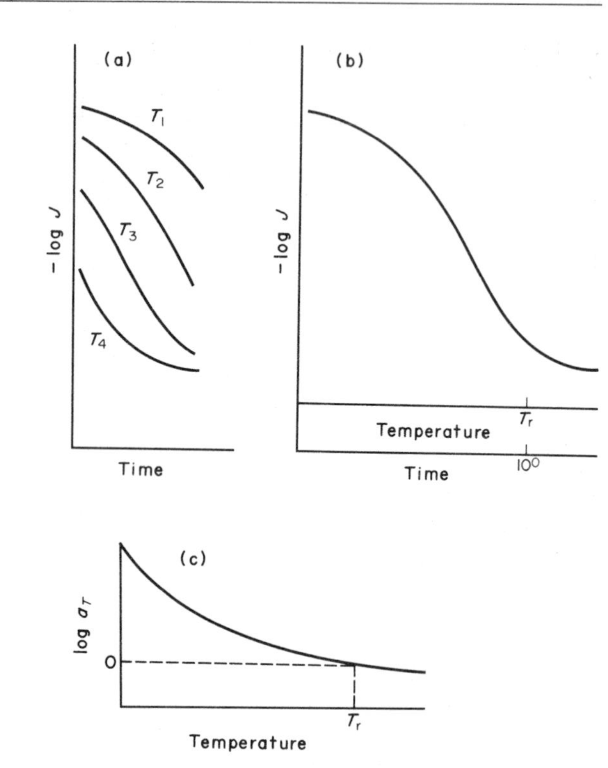

Figure 11
(a) Stress relaxation data for different temperatures. (b) The resulting master curve at reference temperature T_r. (c) Shift factor versus temperature

Unfortunately, the Boltzmann superposition principle is only valid for small strains (<2%) and is thus extremely limited. The Nutting equation takes into account the nonlinear creep behavior when the large stresses are applied:

$$E = K\sigma^{\beta}t^{n} \tag{3}$$

Nonlinearity occurs when the stress exponent β exceeds unity.

The time–temperature superposition principle establishes an equivalence between time and temperature which allows the prediction of long-term properties from short-term, high-temperature tests. This is accomplished by smoothly shifting the log of the creep compliance versus the log of the time curves (Fig. 11a) along the log of the time axis to form a master curve (Fig. 11b) in reference to a curve conducted at the desired temperature T_r. Unfortunately, the "shift factor" itself is dependent on temperature (Fig. 11c). The WLF equation relates the shift factor a_T to the glass-transition temperature T_g:

$$\log a_T = -17.4(T - T_g)/(51.6 + T - T_g) \tag{4}$$

However, this equation is only applicable to amorphous polymers and to the temperature range from T_g to 100 °C above T_g.

2.3 Dynamic Mechanical Properties

Dynamic mechanical properties indicate the ability of a polymer to dissipate energy. Owing to their visco-

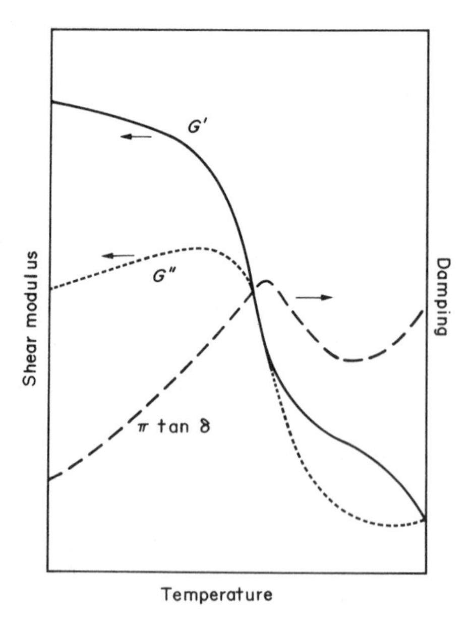

Figure 12
Dynamic mechanical properties as a function of temperature

elastic nature, dynamic mechanical properties measure the in-phase and out-of-phase moduli with an applied oscillatory stress function. More importantly, $\tan\delta$, the "damping factor" (a measure of the phase angle between the applied stress and subsequent strain), is also obtained. This allows a qualitative estimation of the ability to dissipate energy. Figure 12 shows shear storage modulus (G', in phase), shear loss modulus (G'', out of phase) and $\tan\delta$ for a typical polymer as a function of temperature, where

$$G''/G' = \tan\delta \tag{5}$$

3. *Deformation and Failure*

There are many types of mechanical failure. At the microscopic level, most of these processes involve bond stretching and bond rotation leading to a reduction in secondary bonding, and to a lesser extent, primary bonding. Other influences such as chemical reactions or radiation usually serve primary bonds, thereby affecting the material properties.

Since most products are designed so as to avoid plastic deformation and failure, it is necessary to establish the yield point, where plastic deformation begins. This may be done by taking the value from a stress–strain curve. However, this assumes uniaxial tension and is insufficient for most cases where three-dimensional stresses need to be considered. In these cases a yield criterion is used. This is a general condition that must be satisfied for yield to occur. It may be defined in terms of strain or stress and is represented by a surface in space surrounding the origin. As stress is increased from the origin, yield does not occur until a point on the surface is reached.

The Tresca yield criterion is based on a simple concept which states that yield will occur when the resolved shear stress on any plane reaches a critical value. Figure 13 shows the yield criteria in a plane. The actual yield surface in three dimensions is a hexagonal prism.

A further developed model which works better for polymers is the von Mises yield criterion, which was originally developed because the discontinuities at the corners of the Tresca model were difficult to deal with mathematically. In this model yield begins when the elastic shear strain energy density reaches a critical value. The three-dimensional yield surface in this case is cylindrical.

There are a number of more fully developed, more complex yield criteria. A model that fits real polymer data reasonably well is the modified von Mises model, which includes a correction for hydrostatic pressure.

The ability of the polymeric material to yield is an indication of ductility and usually implies that the polymer is tough. A brittle polymer will generally fail before the onset of yielding while a ductile polymer will yield and then elongate with little or no increase in stress. The amount of elongation is related to the ductility of the polymer (Table 1). The interaction of bulk yielding and brittle fracture processes determines whether or not brittle fracture occurs. Figure 14 shows how each of these modes varies with temperature for a glassy polymer. Although the yield strength generally decreases with increasing temperature, the toughness increases significantly above the ductile–brittle transition.

Many polymers exhibit a transition from ductile or high-energy failure to brittle or low-energy failure as the temperature is lowered. Generally, this transition occurs just below T_g when T_g is measured at the frequency of testing, although some polymers such as polycarbonate have ductile–brittle transitions as

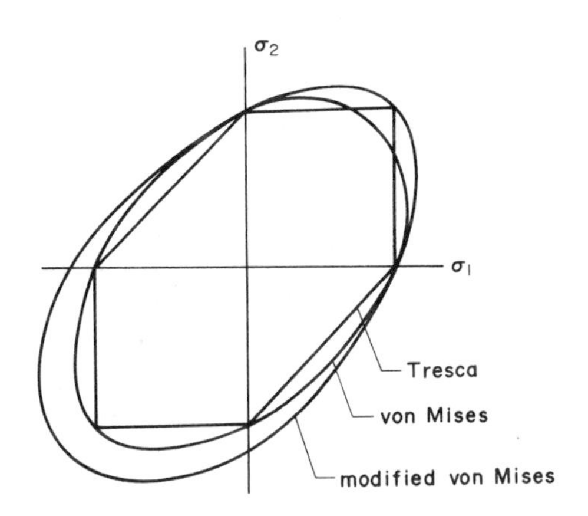

Figure 13
Yield criteria for Tresca, von Mises and pressure-modified von Mises models for biaxial stress conditions

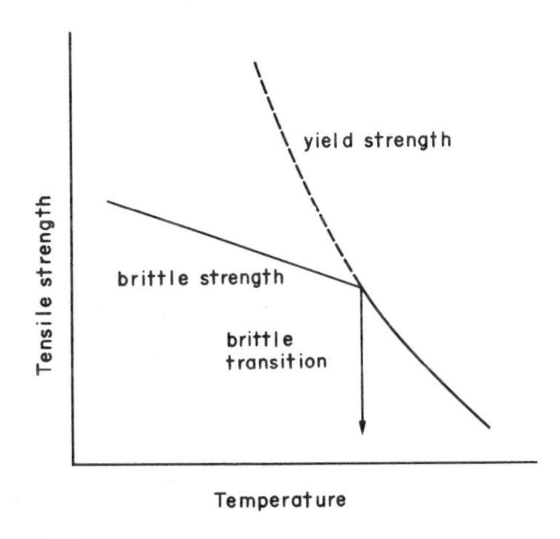

Figure 14
Resultant tensile strength from interaction of the brittle strength and the yield strength

Table 1
Typical mechanical properties of polymers

Material	Tensile modulus (10^9 N m^{-2})	Tensile strength (10^6 N m^{-2})		Elongation (%)		Izod impact strength (kJ m^{-1} notch)
		Yield	Ultimate	Yield	Break	
Acrylonitrile–butadiene–styrene	2.4	50	55	10	15	17.7
Acetal	3.2	55	68	10	50	10.2
Nylon 66	2.0	59	80	18	150	7.6
Phenolic	3.5		56		1	1.6
Polycarbonate	2.6	66	64	25	100	76
Polypropylene	1.35	32	34	12	400	7.6
Polyethylene						
(high density)	1.0	29	30	10	500	5–100
(low density)	0.2	7.5	10	20	1000	>100
Poly(methyl methacrylate)	3.1	65	65	10	10	2.1
Polystyrene	3.4	51	51	2	2	1.8
Poly(vinyl chloride)	2.5	49	50	3	29	5.1
Rubber (natural)			8		>1000	no break

much as 200 °C below T_g. Although there is some disagreement, this behavior is attributed to secondary relaxations below T_g. These secondary transitions are associated with very localized chain motions compared with the long-range segmental motions associated with T_g. Generally, in order to affect the ductile–brittle transition temperature, a secondary transition must be associated with a main chain motion.

The brittle nature of a polymeric material may be characterized by the use of fracture mechanics. Fracture mechanics theory presupposes the existence of a crack of length $2c$ subjected to a stress σ. Failure will occur when

$$\sigma^2 \pi c = K_c^2 \quad (6)$$

where K_c is the fracture toughness. An additional parameter is the energy release rate G_c necessary to initiate a crack:

$$G_c = K_c^2 / E \quad (7)$$

where E is Young's modulus. The values of K_c and G_c are dependent on sample thickness as the state of stress at the crack tip varies from plane stress in very thin sections to plane strain near the center of thick sections.

3.1 Mechanisms of Deformation

There are two different types of deformation in amorphous polymers. The deformation of brittle amorphous polymers is characterized by crazing, a voiding process; ductile deformation is characterized by shear yielding.

Crazes are narrow zones of highly deformed and voided polymer resembling a crack. They are initiated in localized regions where inhomogeneities exist. As the stress near the craze tip increases, coiled molecules align in the direction of the applied stress and form fibrils connecting the sides of the craze. This allows an increase in thickness without lateral contraction of the material. As stress increases the craze grows perpendicular to the stress with constant thickness (usually 1–3 μm) until the fibrils break resulting in crack formation and fracture (Fig. 15).

Crazing occurs to some extent in ductile amorphous plastics but deformation here is dominated by shear yielding. Shear yielding includes both diffuse shear yielding and localized shear band formation. These shear bands are thin planar regions of high shear bands are thin planar regions of high shear strain formed by a cooperative movement of molecular segments of the chains. They allow the material to deform for some time without loss of intermolecular cohesion. The bands form at about 45° to the stress and are composed of highly oriented material.

The deformation of semicrystalline polymers containing inhomogeneous spherulites involves a more complex mode of deformation. The spherulite in uniaxial tension undergoes a transformation to a fibrillar

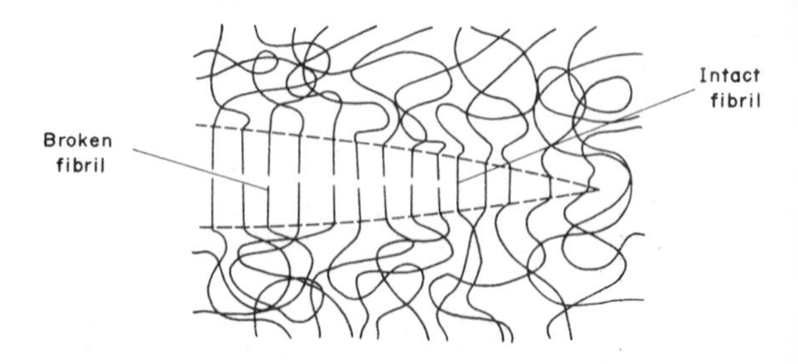

Figure 15
Schematic representation of a craze

structure. As the material is strained the crystalline lamellae undergo small-scale changes (phase changes, twinning, tilt) which allow the parallel packed chains to slip apart near the macroscopic yield point. The breakup of these lamellae and subsequent fibril formation then takes place as blocks of folded chains from individual lamellae are drawn into microfibrils. The fibrils have higher strength than the spherulitic material and contain folded chain blocks with chain segments and tie molecules oriented in the axial direction. The yielding phenomena in ductile amorphous and semicrystalline polymers absorb a considerable amount of energy, which accounts for the higher impact strengths of these materials.

3.2 Types of Failure

It is convenient to examine the mechanical behavior of polymers by observing stress–strain relationships. Even though most mechanical failures of polymer products involve more than simple tension and compression, understanding the mechanisms inolved in tension and compression contributes to the understanding of these other, more complex modes of failure. However, factors such as time, impact loading, repetitious loading and environment must also be considered.

The effect of a constant load over a long period of time (creep) is a very important design consideration. Creep basically involves the strain varying as a function of time for a given load. Initially the strain increases instantaneously by an amount determined by the modulus of the material. Then it gradually increases with time owing to the viscoelastic nature of the polymer, with a strain rate which decreases with time. The rate of increase is highly dependent on temperature and markedly increases as the temperature nears T_g. The effect of stress on the creep properties of a typical polymer can be seen in Fig. 16.

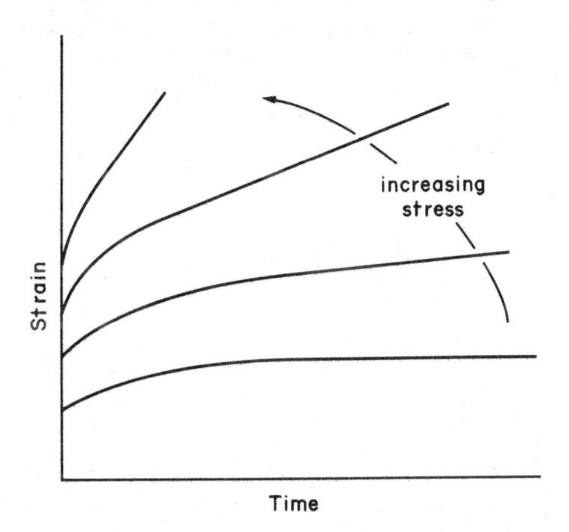

Figure 16
The effect of increasing stress on the creep properties

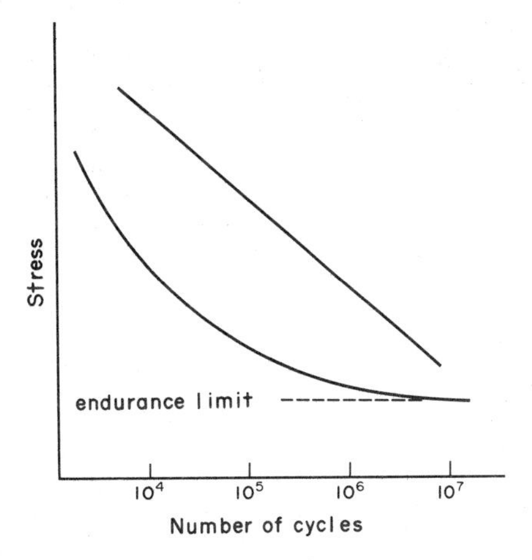

Figure 17
Fatigue life curves of typical polymers

The response of polymeric materials to high rates of loading is related to the toughness of the polymer. The amount of energy needed to rupture the polymer in an impact loading is the impact energy, which is used as a measure of the toughness. As noted in the previous section, ductile yielding in response to such a loading absorbs substantial energy and is indicative of a tough material. Temperature is very important as many polymers undergo a ductile–brittle transition near T_g. Sample geometry, if and how the sample is notched and the test method itself are all significant factors. Most impact tests use some form of swinging pendulum (as in the Izod impact strength) or falling weight to supply the impact energy.

Cyclic loading can often lead to fatigue failure at lower stresses than expected. Samples are usually tested while cycling the stress between zero and some fixed maximum value, although other cycles may be used. The number of cycles before failure is plotted for several stresses to obtain fatigue curves. The lower the applied stress, the greater the number of cycles that are needed to cause failure. Many polymers exhibit an endurance or fatigue limit: an applied stress below which failure will not occur in a measurable number of cycles. Temperature and cycle frequency have important effects on the fatigue properties. High cycle frequencies may cause a significant temperature rise in the polymer sample. A sufficient increase in temperature may lead to T_g or the melting point being reached, resulting in softening of the material and rapid failure. Fatigue life curves for a polymer with and a polymer without an endurance limit are given in Fig. 17.

Other factors such as machining, abrasion and environmental effects (e.g., solvents, oxidation and radiation) also affect mechanical properties.

See also: Crazing and Cracking of Polymers

Bibliography

Andrews E H 1973 Cracking and crazing in polymeric glasses. In: Haward R N (ed.) 1973 *The Physics of Glassy Polymers.* Wiley, New York, pp. 394–453

Bowden P B 1973 The yield behaviour of glassy polymers. In: Haward R N (ed.) 1973 *The Physics of Glassy Polymers.* Wiley, New York, pp. 279–339

Bucknall C B 1977 *Toughened Plastics.* Applied Science, Barking, UK

Hartmann B, Lee G F 1979 Impact resistance and secondary transitions. *J. Appl. Polym. Sci.* 23: 3639–56

Hertzberg R W 1976 *Deformation and Fracture Mechanics of Engineering Materials.* Wiley, New York

Hertzberg R W, Manson J A 1980 *Fatigue of Engineering Plastics.* Academic Press, New York

Kaufman H S, Falcetta J A (eds.) 1977 *Introduction to Polymer Science and Technology.* Wiley, New York

Nielsen L E 1962 *Mechanical Properties of Polymers.* Reinhold, New York

Peterlin A 1977 Plastic deformation of crystalline polymers. *Polym. Eng. Sci.* 17: 183–93

Rodriguez F 1970 *Principles of Polymer Systems.* McGraw-Hill, New York

Ward I M 1971 *Mechanical Properties of Solid Polymers.* Wiley, New York

Williams J G 1977 Fracture mechanics of polymers. *Polym. Eng. Sci.* 17: 144–49

R. D. Corneliussen, T. Frederick and R. J. Gordon
[Drexel University, Philadelphia, Pennsylvania, USA]

Polymers: Structure, Properties and Structure–Property Relations

The versatility and utility of macromolecular materials, or polymers, is manifested by their growing consumption in existing applications and expansion into new ones. From a polymer scientists' perspective, this satisfying position and the intriguing potential for polymeric materials have arisen in part as a direct consequence of the ability to elucidate structure and properties and correlate the complex relations that exist between them for many different natural and synthetic polymers. The voluminous information on the nature and concepts of polymer structure, properties and structure–property relations is conceivably material for an encyclopedia itself. This article seeks to present a concise summary that will allow the materials scientist or engineer to form a clear idea of the elements and principles involved.

1. Historical

The application of any material requires knowledge of its macroscopic response, in both the short term and the long term, to many external influences, physical, chemical or mechanical. This behavior is obviously primarily determined by the fundamental microscopic structure at the molecular level. However, the ability to decipher in total the exact properties of a polymer from the nature of its constituent macromolecules may be elusive for various reasons. Indeed, polymer scientists have only relatively recently begun to understand how long-chain molecules behave and organize themselves to confer unique properties on this class of materials. For example, the macromolecular requirements necessary to impart the once perplexing attributes of rubberlike elasticity are now clearly understood and can be interpreted and reconciled using appropriate fundamental scientific principles.

The acceptance of the idea of very high-molecular-weight covalently bonded long-chain molecules, in a historical sense, is a relatively recent occurrence. Although long-chain synthetic macromolecules had often been encountered in the laboratory in the nineteenth century, the concept of such structures met considerable resistance among the mass of the scientific community even up to the 1940s. This attitude can be illustrated by a comment made by an acquaintance of the prolific researcher Staudinger, who suggested to him, "dear colleague, drop the idea of large molecules; organic molecules with a molecular weight of 5000 do not exist." Similarly, the great pioneering work of Carothers was prefaced by a comment written in a letter to a friend where he speculated on the possibility of "being able to beat Fischer's record of 4200 as a ceiling for high molecular weight materials." It is a tribute to these researchers, and many like them, that it is now understood that the special position of polymers as a class of materials is almost exclusively attributable to their high molecular weight.

Once the principle of high-molecular-weight long-chain molecules had been established, subsequent decades witnessed extensive research, culminating in the large-scale production of a wide range of synthetic polymers. Also, it became increasingly apparent that the building blocks of natural materials were also composed of high-molecular-weight long-chain molecules, although of far greater complexity. These successful developments were founded on the ability to understand polymers in terms of their basic structure and the structural factors influencing properties. Major contributions to these objectives were the result of capitalizing on developments in sophisticated analytical techniques which allowed detailed and definitive structural information to be obtained. Similarly, synthetic skills in the laboratory and in chemical engineering became more refined, allowing the controlled fabrication of a considerable volume and variety of macromolecules. These innovations and the invaluable contributions from many theoreticians permitted the physicochemical structure and properties of many polymeric materials to be assimilated and correlated with the composition and structure of their constituent macromolecules.

Figure 1 presents a synopsis of the elements of structure and various properties of polymers, suggesting the many possible relations between the two.

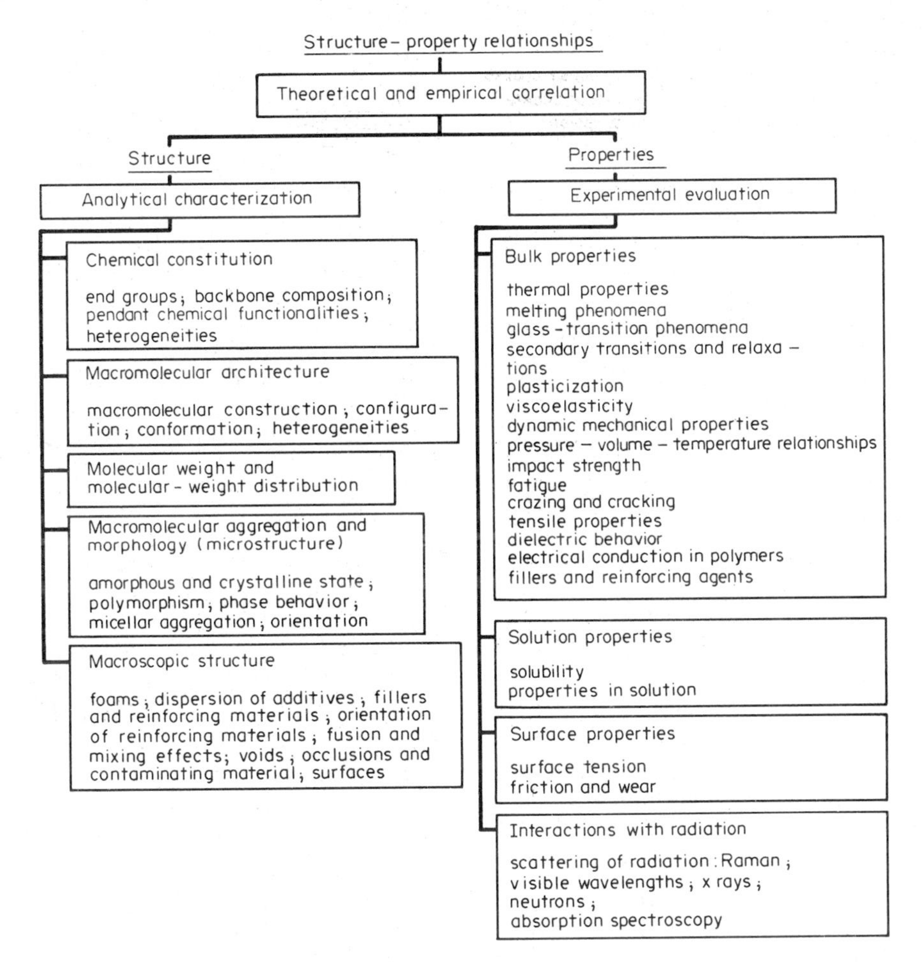

Figure 1
Synopsis of polymer structure and properties

2. Polymer Structure

Polymer structure is often rationalized in terms of microscopic and macroscopic elements. However, although such a division is often useful for descriptive purposes, perhaps a more beneficial description from a materials science viewpoint is to divide structure into that which is defined by the macromolecules themselves and that which is determined by compounding, processing and fabrication treatments. The latter is strictly beyond the scope of this article, but these treatments are nonetheless important modifiers of structure which have the potential to influence strongly the properties and therefore the performance of a polymeric material.

It is invariably hazardous to make discrete divisions when dealing with polymers, since there are always exceptions and contradictions to be found. However, the elements of structure which can be considered to be defined by the macromolecules themselves can be summarized as (a) chemical constitution, (b) macromolecular architecture, (c) molecular weight and molecular-weight distribution and (d) macromolecular aggregation and morphology.

The chemical constitution or composition of simple or complex macromolecules confers chemical properties which depend on the substance of the constituent chemical functionalities. Assigning a macromolecule to a particular family membership is often based on a chemical nomenclature represented by end groups, for example poly(ethylene glycol), backbone composition, as in polyphosphazenes and polyesters, or pendant chemical functionalities, as in polyacrylates and cellulose ethers. A further subdivision of chemical constitution, although often overlooked, difficult to identify and for that matter to quantify, may be termed chemical heterogeneities. Arising from a number of sources, such as monomer impurities, the incorporation of oxygen during polymerization and energetically permissible isomeric variations in polymerization mechanisms, they may also result in structural heterogeneities, but their inclusion as a component of chemical constitution is plainly justified.

If an analogy is made between macromolecular architecture and engineering construction, such that the preceding description has concentrated on the composition of the basic building materials, then the topographical nature of a macromolecule, defined here as macromolecular architecture, may be divided into three separate components. Macromolecular construction delineates the sequencing of the basic building blocks and consequently is determined primarily by the chemical structure and functionality of the monomer and the polymerization mechanism. Terms commonly encountered to describe macromolecular construction include linear homopolymer, branched homopolymer, random copolymer, graft copolymer, block copolymer, network polymer and ladder polymer. These terms should be regarded only as a general framework for construction nomenclature as there are many permutations possible, particularly with the advent of careful synthetic tailoring, which can result in novel macromolecules, such as star di-block copolymer.

By extension of the analogy between civil engineering and macromolecular architecture, the geometric relationships between the individual building blocks and how they are joined together are often termed the configuration of macromolecules. Modern interpretation of the configurational aspects of macromolecules is restricted to the architecture that is ordained by the same structural, geometric and stereoisometric states as are usually encountered in simple organic molecules and are therefore determined by normal covalent bonding. The configurational variant is a dramatic example of substantial property differences between polymers in which, although the sequential arrangement of chemically identical groups is the same, the geometrical or stereoisometric relation to each other is different, as in *cis*- and *trans*-polyisoprene and atactic and isotactic polypropylene.

The conformation of a macromolecule is concerned with the overall shape of the completed assembly and in the ideal case of a single macromolecule is dictated by the spatial arrangement of atoms or groups of atoms as determined by their rotation about σ bonds. In reality, the conformation of macromolecules in solution or bulk phase is also influenced by factors such as inter- and intra-molecular interactions. Macromolecules can assume a great variety of conformations—random, helical, coil, planar zigzag and rigid rod-like geometries—which will ultimately determine the manner in which they aggregate in the bulk phase as well as their solution behavior.

The formation of structural heterogeneities which may stem from architectural or configurational sources has been alluded to in the preceding description of chemical constitution. Such heterogeneities should also never be discounted as a possible prime factor in affecting the properties of a polymeric material.

The structural parameters of molecular weight and molecular weight distribution overshadow all others as fundamental sources of influence on polymer properties. The molecular weight of a polymer chain is obviously related to the chemical structure and number of monomer units, but since polymeric materials are composed of chains of different length, molecular weight averages and the statistical distribution of individual molecules about their averages must be relied on as useful measures of the molecular weight of a polymeric material. Molecular weight and molecular-weight distribution are dictated by the chemical reactivity of the monomer, polymerization mechanism and polymerization conditions and in many cases can be carefully controlled to produce a polymeric material with the desired characteristics.

A further reduction of magnifying power brings into focus the subject of macromolecular aggregation and morphology. The cumbersome dimensions of macromolecules often dictate a random conformation, which leads to an amorphous aggregation, sometimes pictured as a bowl of spaghetti. This state of aggregation is impossible to avoid, even when macromolecules obey certain configurational and local conformational requirements in order to form ordered crystalline units. The extent to which crystallinity develops in a semicrystalline polymer and the morphology of the crystalline or amorphous domains are very complex and influenced by many structural and environmental variables. Many materials with the ability to form crystalline aggregates also exhibit polymorphism, which is the presence of different types of crystalline packing.

These two fundamental states of aggregation determine the microstructure of all macromolecular materials. In the case of polymer blends and block copolymers, there is also the potential to form separate phases whose composition is chemically rich in one of the components present. Both phases may be completely amorphous or one or both of them may be semicrystalline. This morphological patchwork is often a desirable feature of microstructure. The converse may also be true in the case of polymer alloys or blends which form a completely homogeneous mass and do not exhibit any phase separation. Ionomers can also develop a micellar aggregation, rich in the ionic species present along the polymer backbone, which is a facet of microstructure also capable of inducing novel properties. A further component of microstructure attributable to the properties of high-molecular-weight long-chain molecules is the potential for orientation or alignment of molecules and crystalline aggregates in one or more directions. Although arising strictly from processing effects, the anisotropic characteristics imparted to a material by orientation are often utilized to develop enhanced mechanical performance.

This brief roll call of macromolecular structure has been conducted to emphasize one important prerequisite for the study and comprehension of structure–property relations. Any attempt to explain or correlate the properties of a macromolecular material requires a thorough knowledge and careful characterization of all the features outlined here, before meaningful conclusions can be reached.

The macroscopic structure of a polymeric material has been loosely defined as that resulting from processing and fabrication variables. In a sense, the fine detail surrounding the microstructure of a material is also strongly influenced by processing variables, but the intention here is to restrict the description of macroscopic structure as follows. There are many different processes and procedures in the forming of a polymeric material, once the base material has been synthesized; they depend upon the material in question and the application for which it is intended. The processing route and conditions, as well as the final physical form, can all influence the properties and performance of the final assembly. Examples of possible variations in macroscopic structure produced intentionally or not by processing include: dispersion of additives, fillers and reinforcing agents; fusion and mixing; foaming, voids, occlusions and contaminating substances; differential thermal and shrinkage effects; and surfaces. There have been many studies on structure–property characterization of materials which have failed to take into account the effects of processing on the macroscopic structure and therefore have made meaningful comparisons very difficult.

3. Properties and Structure–Property Relationships

A rudimentary description of a material property can be summarized as the measurement of the response of the material to an external stress or stimulus in the form of heat, electromagnetic radiation, chemical agents and mechanical or electromagnetic force fields. Property evaluations often use heat as an environmental variable so that a property may be monitored over a temperature range. Similarly, some properties are a composite response of chemical and physical elements and therefore do not lend themselves easily to any orderly classification system in the same way as that adopted for the description of structure in Sect. 2. A further difficulty with respect to classification is that a material property or measurement assumes different levels of significance depending upon the observer, the reason for a particular observation or measurement and the information sought. On the surface, a property evaluation provides necessary information pertaining to the applicability of a specific material, which in some instances is all that may be required. This database approach to property measurement also contains a subcutaneous layer of information which can be utilized to define the chemical and physical state of a material. The latter concept forms the basis for the evaluation of structure–property relationships which is concerned with the qualitative and quantitative assessment of chemical and physical structure—molecular, microscopic and macroscopic—and how it controls a particular property or group of properties.

The foregoing can be illustrated by reference to the amorphous and crystalline states of polymeric materials. On an increasing temperature scale, the amorphous state can be divided phenomenologically into the glassy and rubber regimes, separated by the glass transition temperature; the crystalline state is converted to a viscous liquid state at the melting point. The temperatures at which these transitions occur in either totally linear amorphous or semicrystalline materials mark important property changes which are of paramount importance from an applications or processing perspective. Pursuing the subcutaneous information in these characteristic transition temperatures, their theoretical description and expression in fundamental thermodynamic, compositional and structural parameters allow simple structure–property correlations to be made.

This type of description also emphasizes two crucial points regarding the significance of a property measurement. Property measurement for the purpose of estimating usefulness and applicability of a material is a reciprocal process, in that it can also be used as an analytical tool to probe polymer structure. Secondly, it demonstrates that it would be naive to assume that a unilateral approach to structure–property correlations hinges on experimental and empirical evaluation of material properties without the involvement of any theoretical foundation. The complementary advances in experimental interpretation and theoretical description of the behavior of macromolecules have provided a solid base from which to substantiate all levels of structure with properties. Consequently, for a materials scientist fully to comprehend and appreciate the relationships involved, a considerable flexibility and a complete interdisciplinary approach is required, engaging many principles of chemistry, physics, mathematics and engineering.

Inevitably, almost without exception, the properties of a macromolecular material are dominated by the physical state, and in any structure–property evaluation it is the macromolecular architecture which is the determining factor in the manifestation of the physical state. This in turn is reflected in the macromolecular aggregation and morphology of a material, often termed the microstructure. It is with respect to these components of structure that macromolecules display a rich variety of behavior. Some properties, however, can be expressly related to the chemical constitution of the macromolecules of which a material is composed.

The response of a polymeric material to thermal stress or attack by chemical agents and certain frequencies of electromagnetic radiation is dominated by bond strengths, chemical reactivity of free radicals and intermediates, functional groups and charged species. However, diffusion processes and hence kinetic factors are obviously influenced by the macromolecular and morphological environment. The solvation of macromolecules in the amorphous state is primarily determined by solvent–solute interactions which are unequivocally related to the chemical state of the participants. Chemical cross-linking logically inhibits solubility, despite considerable driving forces to the contrary leading to solvent absorption and extensive swelling. The encroachment of physical state in the solubility process, however, can also play a decisive role. For example, the presence of crystallinity demands a solvent capable of overcoming strong intermolecular forces within the crystalline lattice before individual molecules can be dispersed in the solvent.

The physical state and the bulk properties of a polymeric material arising from the unique characteristics of macromolecules in an amorphous aggregation perhaps deserve special mention. The nonequilibrium glassy regime and the combination of the elastic and viscous nature of the amorphous phase above the glass transition temperature serve to provide a rewarding complexity of property response in the study of structure–property relationships. The modification of morphology and therefore bulk properties by the presence of crystalline aggregations or the introduction of chemical cross-links into the amorphous phase also serve to influence the properties of a polymeric material dramatically. At this point the reader must be directed to the Bibliography for more explicit accounts of the influence of aggregation and morphology on the bulk properties of a material.

The foreseeable climate for future improvement of the understanding of the structure–property relations of polymers can be regarded as healthy. Modern theoretical and experimental developments now encourage more careful scrutiny of the solid state nature of polymers in the areas of macromolecular conformation, polymer chain dynamics, aggregation, morphology and microstructure. Experimentally these investigations sometimes require sophisticated and expensive equipment and instrumentation such as that required for neutron scattering, solid state nuclear magnetic resonance spectroscopy and scanning and transmission electron microscopy (SEM and TEM). Though they have cost and access disadvantages for many researchers, these are very valuable tools for structure characterization. At present there are growing demands for new specialty polymers fabricated for specific needs, such as high thermal resistance, electrical conductivity, superior mechanical integrity and carefully controlled chemical reactivity and functionality, for example, in biochemical and biomedical applications. These demands present a challenge to polymer scientists for the effective design and synthesis of macromolecules whose properties can be predicted and manipulated to meet the required characteristics. Success in such ventures will be achieved only by implementation of, and strict adherence to, the principles of structure–property relations.

Bibliography

Bailey R T, North A M, Pethrick R A 1981 *Molecular Motion in High Polymers*, International Series of Monographs on Chemistry. Oxford University Press, Oxford

Bovey F A, Winslow F H (eds.) 1979 *Macromolecules: An Introduction to Polymer Science*. Academic Press, New York

Cowie J M G 1973 *Chemistry and Physics of Modern Materials*. International Textbook Company, Aylesbury, UK

Elias H G 1977 *Macromolecules*, Vols. 1 and 2. Plenum, New York

Rabek J F 1980 *Experimental Methods in Polymer Chemistry: Physical Principles and Applications*. Wiley, Chichester

Rodriguez F 1982 *Principles of Polymer Systems*, 2nd edn., McGraw-Hill Chemical Engineering Series. Hemisphere Publishing, Washington, DC

Schultz J M 1974 *Polymer Materials Science*, Prentice-Hall International Series in the Physical and Engineering Sciences. Prentice-Hall, Englewood Cliffs, New Jersey

Van Krevelen D W 1976 *Properties of Polymers—Their Estimation and Correlation with Chemical Structure*, 2nd edn. Elsevier, Amsterdam

F. E. Karasz
[University of Massachusetts, Amherst, Massachusetts, USA]

T. S. Ellis
[General Motors Research Laboratories, Warren, Michigan, USA]

Polymers: Technical Properties and Applications

The number and variety of applications in which polymers are used have increased at a phenomenal rate since the first totally synthetic polymer (bakelite) became available commercially about 80 years ago. Applications of the naturally occurring polymers, cellulose and rubber, can be traced back to the earliest history of mankind. Coupling the many applications of synthetic polymers with the historical use of the naturally occurring polymers, it is not surprising that these complex macromolecules now constitute an important class of materials which have become a major factor in improving the standard of living.

The remarkable versatility of polymers allows them to compete in many applications including transportation, construction, wiring, agriculture and medical devices. As decorative materials, polymers have a distinct advantage over competitive materials as a result of their low density, moderate cost, ease of pigmentation and the low energy required for their fabrication into finished products.

1. Economic Considerations

Growth of the synthetic polymer industry is illustrated in Fig. 1. Spurred by the need to develop synthetic rubber during World War II, the industry expanded rapidly and continues to grow. Table 1 shows production figures for the USA. Much of the growth can be attributed to the introduction of a wide variety of new polymers. Polymers are now classified as commercial, engineering and specialty types. Polymers produced in large quantity are classified as commercial or commodity products. Engineering polymers are those having exceptional mechanical strength which are therefore suitable for applications in which strength is an essential requirement. Several engineering polymers are now produced in quantities comparable with the commodity polymers. Although not strictly an engineering application, packaging is a major application for synthetic polymers, consuming about half the annual production.

Since almost all synthetic polymers are made from fossil fuel, costs of these materials have increased sharply with the rise in cost of crude oil and natural gas. However, the use of inexpensive fillers as extenders has helped to hold down the costs of products made from polymers.

It is difficult to predict the point at which production of synthetic polymers will level off or when prices will stabilize. The industry is still in its infancy and further development of new polymers can be expected. In engineering applications, the introduction of reinforcing fillers has enabled polymers to compete effectively with metals in certain areas. A typical example is the composites using carbon fibers (see *Carbon-Fiber-Reinforced Plastics*). Advances in reinforcement and induced orientation of fibrous fillers may be expected to widen still further the scope of applications for polymer composites.

Table 1
Polymer sales in the USA (10^3 t)[a]

	1981	1982
Low-density polyethylene	3416	3431
High-density polyethylene	2129	2181
Polypropylene and copolymers	1766	1638
Poly(vinyl chloride) and copolymers	2548	2431
Polystyrene	1638	1472
ABS resins	441	328
Phenolic resins	1048	883
Saturated polyesters	294	310
Unsaturated polyesters	450	393
Polyurethane foams	785	674
Polycarbonate	109	96
Epoxy resins	150	142
Acrylics	255	219

[a] From *Modern Plastics*, January 1983

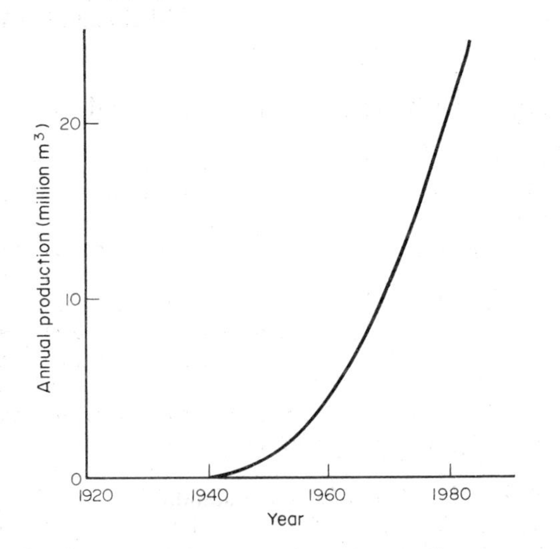

Figure 1
Growth of the synthetic polymer industry

Interest is growing in the use of polymer blends and laminate structures. Blending of two or more different polymers aims to take advantage of the unique properties of each component. For example, poly(methyl methacrylate) (PMMA) is used as a glazing material despite its vulnerability to abrasion. Lamination of an abrasion-resistant polymer over the surface would allow PMMA to compete more effectively with conventional glazing materials.

As the technology for compatibilization of polymer blends advances, these composites may find much broader application. Combinations of polymers with other materials will also contribute to growth of the synthetic polymer industry. The lightweight but exceptionally strong polymer concretes are an example. Glass-reinforced polymers are widely used in the automotive and other industries.

2. Properties Influencing Engineering Applications

Several key properties are important to the suitability of polymers for engineering applications. A brief summary is presented here with emphasis on comparison of corresponding properties with other, competitive materials.

2.1 Density and Ultimate Strength

The low density of polymers provides an important advantage over metals and ceramics in many applications (see *Polymers: Mechanical Properties*). The density of most engineering polymers lies between 0.9 and 1.9. These values compare with a range of 2.7–3.0 for the lightweight metals and 7.8–8.0 for steel. Therefore, polymers are very useful in those applications in which the weight–volume ratio is critical. The ultimate strength of polymers, on the other hand, does not compare favorably with that of metals. For example, polyesters have a Young's modulus in tension in the region of 3 GPa, in contrast to 200 GPa for steel and 10 GPa for wood. It should be noted, however, that some of the new polymer composites, such as carbon-fiber-reinforced polyimides, have strengths well within the competitive range. Because of the trade-off between low density and high strength, polymers are usually materials of choice when very high strength is not required but where advantage can be taken of their low density and ease of fabrication.

2.2 Morphology

Many polymers exist in a semicrystalline state. This complex morphology has the advantage of strength in the ordered crystalline regions and flexibility in the disordered amorphous regions. Thus polymers find application when strength must be combined with flexibility, as in fibers, electrical insulation and some pipe applications. Orientation at the macro level has been used effectively to maximize strength in polymers. Monofilaments owe their great strength per unit cross-sectional area to orientation induced in the drawing process. The single thread spun by a silkworm has a tensile strength comparable with that of metals of the same cross-sectional area. In general, polymers exhibit much higher strength in tension than in compression.

2.3 Dielectric Properties

Polymers as a class have little competition in nonrigid dielectric applications. Their strength and flexibility provide an ideal combination of properties for wire insulation and cable jacketing. Being less fragile than ceramics, polymers also have the advantage of resistance to breakage when used as rigid insulators. However, the electrical breakdown strength of polymers does not compare favorably with that of ceramics, so they are less widely used in high-voltage applications. Polymers are ideal materials for the manufacture of printed circuit boards. The ease of fabrication and flexibility are important factors favoring the use of polymers in these applications.

Polymers have excellent dielectric properties. Polyethylene has a dielectric constant about twice that of dry air at 1000 Hz. Polytetrafluoroethylene has an even lower value, while poly(vinyl chloride) represents a compromise with respect to resistance to burning; it has a dielectric constant between 4 and 8, depending on the level of plasticization. Both the dielectric constant and the surface resistivity of polymers are influenced by moisture. Since most polymers are hydrophobic, moisture absorption must be attributed to additives such as the fillers used in thermosetting resins. Large amounts of fillers such as wood fibers or cotton are responsible for dielectric losses in filled phenol resins. For use in dielectric applications, a water-resistant filler such as mica is preferred.

2.4 Processability

Polymers are fabricated into final products by a variety of processes including calendering, extrusion and a number of molding techniques. Because of their low melting temperatures, polymers can be formed into useful shapes at temperatures in the region of 200 °C. In contrast, if melting is required, metals must be fabricated at much higher temperatures.

Despite the fact that almost all synthetic polymers use fossil fuel as feedstock, the overall energy requirement for fabricating the same article from a polymer is still much less than from metals. It has been estimated that 0.1 kWh would be required to fabricate a 350 g container from polyethylene, whereas 3 kWh would be needed to form a similar container from aluminum.

Polymers have a further advantage in fabrication, as a result of the ease with which their melts flow. Thus it is relatively easy to fill complex and intricate molds to form products with a high degree of detail.

Reaction injection molding is a new fabrication technique for polymers in which monomers are polymerized within the mold. Relatively lightweight molds can be used. Since monomers flow quite freely, considerable detail can be obtained in the final polymeric product. This process is expected to lead to an extended use of polymers in automobiles, where they contribute to weight reduction and increased mileage.

2.5 Resistance to Deterioration

All polymers degrade when exposed to high temperatures or ultraviolet radiation, although there is considerable variation between individual polymers. However, major advances have been made in the stabilization of polymers, so the required service life is now attainable in many applications. Polyethylene, for example, can be stabilized to resist outdoor weathering for > 50 years by addition of carbon black and an effective thermal antioxidant.

Those polymers that are formed by addition reactions are very resistant to hydrolytic breakdown. In contrast to many metals which corrode on exposure to moisture, this class of polymers is ideally suited for applications in which high humidity or moisture is present. Synthetic polymers are also very resistant to biodegradation.

2.6 Reuse of Scrap Polymeric Material

The difficulty of reusing polymer scrap is a serious restriction on the expanding application of these materials. There are well established processes for recycling metals and ceramics, but the recycling of polymer scrap has not developed to a comparable extent. The higher densities of metals and ceramics makes it easier to separate these materials from other components of an industrial or municipal waste stream. Polymer scrap, on the other hand, is difficult to recover once it has been mixed into a waste stream. There are, however, several successful processes for the recycling of source-separated polymeric scrap.

High temperatures can be used in recycling metals and ceramics. Because of their lower decomposition temperatures, polymers are likely to decompose under the severe conditions required for recycling. In many instances, clean polymer scrap is recycled in fabrication processes. However, the danger of decomposition must always be taken into account.

3. Applications for Engineering Polymers

Although the use of polymers in packaging is not generally considered to be an engineering application, protective coating and encapsulation are important in certain engineering designs or processes. Furthermore, the volume of polymeric materials used for packaging is large.

3.1 Polymers in Packaging

Lower cost and ease of fabrication account for the increased use of polymer protective coatings in engineering applications. Foamed cushioning to protect instruments during shipping and encapsulation of solar cells are important applications. The preservation of fresh and frozen food products by polymeric materials has been responsible for extending the food supply by avoiding contamination and spoilage. The impermeable nature of polymer films tends to exclude both moisture and microorganisms.

Polymers are finding increasing application in the manufacture of beverage bottles, where they have an advantage over glass in resistance to breakage as well as lighter weight. When polymeric materials are used for packaging food or medicine, it is very important that no additives or residual monomer are present which could be absorbed by the contents of the container. A further problem with polymers in beverage containers is the difficulty encountered in washing and sterilizing plastic containers. The opacity of many plastic bottles prevents adequate visual inspection after washing. Moreover, most polymeric materials would distort at sterilization temperatures.

Many household chemicals are packaged in containers made from polymers. In these applications, resistance to breakage is an important advantage when the contents are corrosive or toxic.

Disposal of waste products in plastic bags is another important application of polymers in packaging. Plastic garbage bags have the advantage of light weight, strength and the capability to retain odors as a deterrent to rodents. Plastic garbage cans are not only lighter in weight than those fabricated from metals but they also resist damage effectively, when the correct polymer is selected. Leaf bags of thin polymeric films are also popular because of the ease of disposal.

3.2 Polymers in Construction

Polymers are used in a wide variety of construction applications where advantage can be taken of their excellent insulation capability (both thermal and electrical), light weight, design flexibility and ease of pigmentation. Even in applications where moderate strength is required, polymers are finding wide use. Such items as extruded gutters, siding, imitation wood beams, room dividers and window or door frames are growing in popularity. Since colorants can be dispersed throughout the polymer mass, repeated painting is not required. Resistance to biodegradation also favors polymers in those applications where attack by organisms is likely to occur.

Poly(vinyl chloride) (PVC) is used widely as siding for houses, competing with wood and aluminum. Color permanence, resistance to impact, insulating value and relatively low cost are all in favor of polymers in this application. PVC can be extruded as siding in long, uniform lengths as required and either applied directly over sheathing in new construction or over deteriorated wood siding. Installation is at least as simple as for competitive materials. Resilience of PVC siding minimizes damage by impact, a distinct advantage over aluminum siding. Stability to biogradation, especially in humid areas, is a further advantage of PVC siding.

Pipe fabricated from PVC, ABS molding resin, high-density polyethylene and other polymers is popular with "do-it-yourself" homeowners. Perforated drainage pipe is not as fragile as ceramic tile used for this purpose, and long lengths can be extruded easily. Cutting into desired lengths is easy and joining is relatively simple. However, in those applications in which the pipe must withstand high pressure, metal pipe is superior.

Polyolefins, paper and rubber are used as wiring insulation, but PVC is favored because of its greater resistance to burning. Codes covering inside wiring often restrict the use of polymers, but new polymers have been developed which provide outstanding resistance to burning. These new materials belong to the fluorocarbon class. Because of their improved flame resistance, these materials compete effectively as insulation for inside wiring, particularly in aircraft and marine construction, where weight is an important factor.

Decorative items constitute a major application area for polymers in construction. Decorative panels, room dividers, light diffusers, imitation beams, doors, moldings, various hardware items, bathroom fixtures and other components are easily fabricated from polymers. Only moderate strength is required in most of these applications, and attractive colors can be obtained. Ceiling panels are fabricated from a variety of polymers, or a moisture-resistant polymer can be used as a protective film over conventional ceiling tiles. Bathroom wall tiles made from polymers have the advantage that they are not cold to the touch.

3.3 Polymers in Transportation

Several factors have contributed to the expanded use of polymers in transportation. The basic property of low density provides the main reason for most of these applications. This property has always been very important in aircraft construction. More recently, the use of light polymers in automobile design has attracted considerable attention as one method for reducing fuel consumption. Marine applications, particularly in hull construction, take advantage of the resistance of polymers to attack by marine organisms.

Polymers are well adapted to fabrication of complex decorative parts in automobiles, both exterior and interior. Polymer foams are used extensively as seat cushions and in impact-resistant dashboards. Ductwork for heaters and air-conditioning units is fabricated from polymers, as are most under-the-hood containers. Battery cases are now made almost entirely of polypropylene, which has adequate strength for this application while providing a significant weight advantage over bakelite or hard rubber. As reaction injection molding technology expands in the automotive and other industries, thermosetting polymers are expected to become a major fraction of the polymers used in automobile construction. Complete body sections are envisioned, and longer range predictions include drive shafts and even engine blocks with metal inserts.

Polymers as adhesives are widely used in aircraft construction. Laminated structural units are especially adaptable to small aircraft. Thermoforming of polymers is ideally suited to fabrication of hulls for small boats. Larger hulls have been made using glass-reinforced thermosetting polymers. Though barnacles may attach themselves to plastic hulls, they do not degrade the material and are easily removed.

3.4 Polymers in the Communications Industry

The introduction of polyethylene in 1942 revolutionized the construction of wire and cable. Before the discovery of this synthetic polymer, communication cable used lead as the outer protective cover. Replacement of lead by polyethylene greatly reduced the weight while still providing adequate strength and abrasion resistance. Cost savings were substantial, and sheath fabrication was greatly simplified. Although the original polyethylene was rapidly degraded by heat and ultraviolet radiation, modern formulations of this polymer provide more than adequate service life.

The introduction of polyethylene and other synthetic polymers as a replacement for paper insulation on primary conductors has provided many improvements in cable construction. All of the synthetics have very good insulating properties and in addition they are hydrophobic. Several synthetic polymers are now used as expanded insulation to reduce the weight still further while providing even better dielectric quality. The semicrystalline morphology of most polymers used in this application provides the desired properties of strength and flexibility. Even in the advanced technology of transmission through optical fibers, a polymeric coating must be applied immediately to the glass fibers, to prevent development of surface cracks which would lead to rapid embrittlement of the glass.

3.5 Polymers in Biomedical Applications

Although the volume used is small, polymers have assumed an important role in medical applications. In most of these applications, polymers have little or no competition from other types of materials. Their unique properties of flexibility, resistance to biochemical attack and light weight make polymers the primary, if not the only, choice for the following applications.

From the familiar use of plastic tubing for intravenous infusion, biomedical applications of polymers have now been extended into complex and sophisticated techniques. A wide variety of products fabricated from polymers are used as surgical implants. The inert nature of these materials reduces the possibility of rejection by the host which is often encountered with implants of living tissue. The artificial hip joint has been used successfully for many years to restore mobility to arthritic patients.

Synthetic polymers are also fabricated into thin-wall tubing as replacements in the circulatory system. The value of breakage-resistant containers for medicine and whole blood is very important in these applications. The transport of blood plasma to soldiers in combat zones by poly(vinyl chloride) bags has been responsible for the saving of many lives. The artificial heart, which is receiving much attention, can be expected to make use of synthetic polymers in its final design. The list of biomedical products fabricated from polymers is certain to expand as medicine advances.

See also: Design with Polymers

Bibliography

Hall C 1981 *Polymer Materials: An Introduction for Technologists and Scientists*. Wiley, New York

Kaufman H S, Falcetta J J 1977 *Introduction to Polymer Science and Technology*. Wiley, New York

Kaufman M 1968 *Giant Molecules: The Technology of Plastics, Fibers and Rubber*. Doubleday, Garden City, New York

Patton W J 1976 *Plastics Technology*. Reston Publishing Company, Reston, Virginia

Rogers D C 1964 Applications. In: Raff R A V, Doak K W (eds.) 1964 *Crystalline Olefin Polymers*. Interscience, New York

Seymour R B 1975 *Modern Plastics Technology*. Reston Publishing Company, Reston, Virginia

Williams H L 1975 *Polymer Engineering*. Elsevier, Amsterdam

W. L. Hawkins
[Montclair, New Jersey, USA]

Portland Cements, Blended Cements and Mortars

Portland and blended cements are vital to civil engineering and construction. The world's annual production of these elements is about 1 billion tonnes (10^{12} kg) and they are the binding ingredient for about 10 billion tonnes of concrete. Portland cements are water-reactive inorganic powders formed by intergrinding Portland cement clinker and ~5% of gypsum. Blended cements are blends of Portland cements with other finely divided inorganic materials such as fly ash and granulated blast furnace slags. When mixed with 1 to 2 times their volume of water, cements give pastes which harden to stonelike products. Their essential ingredients are anhydrous calcium silicates.

1. Cement Industry Nomenclature

Because the compositions of their raw materials and products are usually expressed in terms of oxides, cement chemists have adopted a shorthand notation using a single letter to represent a unit of an oxide. The commonly used symbols are:

$A=Al_2O_3$	$f=FeO$	$N=Na_2O$
$C=CaO$	$H=H_2O$	$S=SiO_2$
$\bar{C}=CO_2$	$K=K_2O$	$\bar{S}=SO_3$
$F=Fe_2O_3$	$M=MgO$	$T=TiO_2$

Using this notation, tricalcium silicate, Ca_3SiO_5, is written C_3S, and calcium hydroxide, $Ca(OH)_2$, is written CH. Phases which cannot be written solely in oxide notation may be written in a mixed notation. For example, the compound with the formula $Ca_{11}Al_{14}O_{32}.CaF_2$ is writen $C_{11}A_7.CaF_2$.

2. Mechanisms of Cementing Action and the Microstructure of Hardened Cement Pastes

The setting and hardening of cement–water pastes results from the formation of hydrated solid phases of greater volume than the anhydrous cement. The hydrated phases grow into the available water-filled space to give a rigid porous structure. For both Portland and blended cements, the main hydration products at normal ambient temperatures are a poorly crystallized calcium silicate hydrate (usually designated as C–S–H, although its composition is approximately $C_3S_2H_3$) and calcium hydroxide. While these products generally adapt well to the available space, some hydration products grow as large crystal which displace other particles to accommodate their growth. Examples are brucite, $Mg(OH)_2$, formed by hydration of MgO, and ettringite, $C_3A.3C\bar{S}.H_{32}$.

The ability of a cement paste to develop strength depends largely upon the formation of a low-porosity solid; the exact composition and morphology of the hydration products appear to be less important. Intrinsic factors affecting the engineering performance of a cement are the particle size distribution, the compositions and quantities of the phases present and the distribution of phases between particles of different sizes and shapes. However, the ability to predict the performance of cements in concrete reliably from the characteristics of the cements and the conditions of use has not been attained.

3. Chemical and Mineralogical Compositions of Ingredients of Portland and Blended Cements

3.1 Portland Cement Clinker

Portland cement clinker is a synthetic product composed essentially (see Table 1) of impure tricalcium silicate (C_3S) and β-dicalcium silicate (β-C_2S) together with lesser quantities of impure tricalcium aluminate (C_3A) and a ferrite solid solution (fss) approximating to the composition of tetracalcium aluminoferrite (C_4AF). Small quantities of other phases may also be present. The phases in the clinker are similar to those which would be expected from the Portland cement area of the C–A–S phase diagram (Fig. 1), allowance being made for the effects of F and other oxides, of which M, N, K and $\bar{S}$ are the most common. Other factors which may affect the nature of the clinker and the phases present are nonattainment of high-temperature equilibrium and the rate of cooling.

The approximate compound composition of a clinker can be calculated using the Bogue potential compound composition formulae, which, for the case where $A/F \geqslant 0.64$, are:

$$\begin{aligned}\%C_3S &= 4.071(\%C) - 7.600(\%S) - 5.718(\%A) \\ &\quad - 1.430(\%F) - 2.852(\%\bar{S}) \\ \%C_2S &= 2.867(\%S) - 0.7544(\%C_3S) \\ \%C_3A &= 2.650(\%A) - 1.692 \\ \%C_4AF &= 3.043(\%F)\end{aligned}$$

Similar formulae have been deduced for the less common case where $A/F < 0.64$.

3.2 Gypsum

Gypsum ($CaSO_4.2H_2O$ or $C\bar{S}H_2$) is interground with most cements containing Portland cement clinker to control an otherwise too rapid reaction of the tricalcium aluminate with water. Because gypsum affects strength development and drying shrinkage, the quantity added should be close to the optimum for the proposed application of the cement. Natural gypsum is usually preferred, but less hydrated calcium sulfates and certain by-product gypsums have been used.

Table 1
Some phases commonly found (in impure form) in Portland cement clinkers

Chemical name of pure phase	Common name for impure phase	Chemical composition	Industry designation	Frequent impurities	Number of crystalline forms known in clinker
Tricalcium silicate	alite	Ca_3SiO_5	C_3S	A, M	3
β-Dicalcium silicate	belite	Ca_2SiO_4	β-C_2S	A, M, K	3
Tricalcium aluminate	C_3A	$Ca_3Al_2O_6$	C_3A	M, N, S	2
Tetracalcium aluminoferrite	C_4AF[a]	$Ca_4Al_2Fe_2O_{10}$	C_4AF	M, N	2
Magnesium oxide	periclase	MgO	M	f	1
Calcium oxide	free lime	CaO	C	—	1
Calcium sulfate	anhydrite	$CaSO_4$	$C\bar{S}$	—	1
Sodium potassium sulfate	aphthitalite	$(Na, K)_2SO_4$	$(N, K)\bar{S}$	—	—

[a] Frequently referred to as brown millerite or the ferrite solid solution (fss) phase; compositions between C_2F and C_4A_2F have been identified in Portland cement clinkers

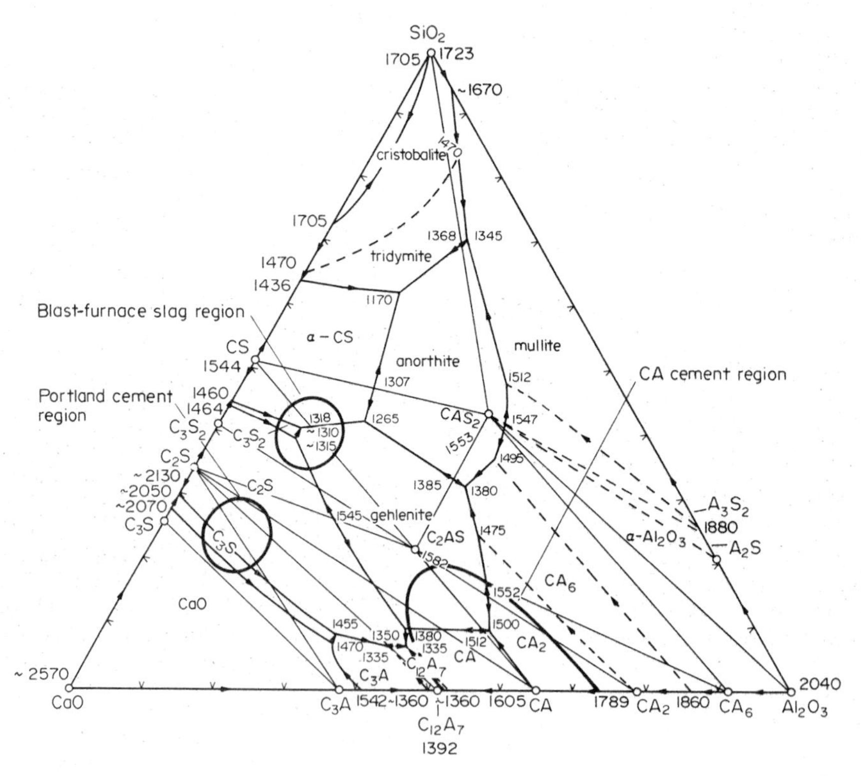

Figure 1
C–A–S phase diagram showing cement regions (after Taylor 1964; courtesy of Academic Press)

3.3 Granulated Blast Furnace Slag

Slags used with Portland cements are obtained by rapid quenching (granulation) of the molten slag from iron blast furnaces. They are predominantly glassy silicates or aluminosilicates with calcium and magnesium as the main cations present. Ignoring other oxides than C, A and S, their ranges of compositions are indicated in Fig. 1. Some crystallization of the slag usually takes place during cooling. As a rough guide, the more basic the slag and the higher its glass content, the more suitable it is for use in cement manufacture.

3.4 *Pozzolans*

Pozzolans are finely divided aluminosiliceous materials which, though having no cementing action with water alone, combine with calcium hydroxide in the presence of water to form cementing compounds. They are used in quantities of ~10–40% in Portland–pozzolan cements. Naturally occurring pozzolans include volcanic glasses and certain claylike rocks which require heat treatment to develop their full pozzolanic activity. Artificial pozzolans include fly ash, a by-product from the combustion of pulverized coal. Fly ash consists mostly of spherical particles, 1–50 μm in diameter, of a silicate glass with smaller quantities of refractory phases such as mullite, hematite, magnetite and wustite. Crystals of these phases may be included in the glassy particles. The fly ash glasses are usually less basic than the glassy material in granulated blast furnace slags.

4. *Chemistry of Portland Cement Hydration*

When the cement is mixed with water, the water quickly dissolves any alkali sulfates prevent. Also, as a result of reactions with C_3S and other cement compounds, the water becomes saturated or supersaturated with respect to calcium hydroxide and gypsum. The reactions then stop almost completely for a period of several hours referred to as the induction or dormant period; the causes are not fully understood, but they may be the formation of a lowpermeability coating on the cement grains or slow processes of nucleation of calcium hydroxide, a calcium silicate hydrate or both. At the end of the induction period, the reactions accelerate until, typically, about a quarter of the cement has reacted. The reactions then gradually slow down due to diffusion control by the formation of an increasingly thick layer of hydration products around the cement grains.

Some major hydration products are listed in Table 2 and the reactions which result in their formation are given in Table 3. The complexity of the system of simultaneous and consecutive reactions which take place during cement hydration is indicated by Fig. 2.

Table 3
Some reactions of Portland cement compounds with water

$2C_3S+6H$	$\rightarrow 2C_xSH_x+(6-2x)CH$ $(1.3<x<1.7)$
$2C_3S+4H$	$\rightarrow 2C_xSH_x+(4-2x)CH$
C_3A+6H	$\rightarrow C_3AH_6$
$C_3A+3C\bar{S}H_2+26H$	$\rightarrow C_3A.3C\bar{S}.H_{32}$
$C_3A+C\bar{S}H_2+10H$	$\rightarrow C_3A.C\bar{S}.H_{12}$
$C_4AF+2CH+6CSH_2+5OH$	$\rightarrow 2C_3(A,F).3C\bar{S}.H_{32}$
$2C_3A+C_3A.3C\bar{S}.H_{32}+4H$	$\rightarrow 3C_3A.\bar{S}.H_{12}$

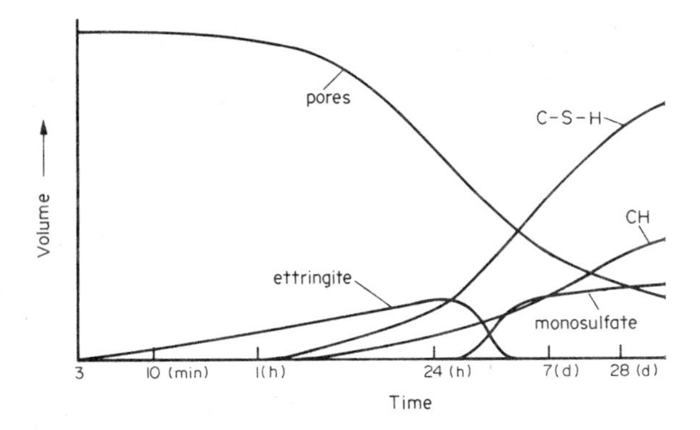

Figure 2
Schematic representation of the course of reactions of a Portland cement with water; exact volumes depend on cement composition and water–cement ratio

The main products are C–S–H and calcium hydroxide. The C–S–H does not have a well defined stoichiometry, though its C/S ratio is usually in the range 1.5 ± 0.2. It appears to have a claylike silicate layer structure and its specific surface area, as deduced from water vapor adsorption data, is $\sim1.5\times10^5\ m^2\,kg^{-1}$; the area determined by nitrogen adsorption is much lower. The highly hydrated hexacalcium aluminate trisulfate, ettringite, $C_3A.3\bar{S}.H_{32}$, which is formed at early ages, is transformed into tetracalcium aluminate

Table 2
Some hydrated phases commonly found in Portland cement pastes

Name of pure phase	Common name	Chemical composition	Cement industry designation
Calcium hydroxide	hydrated lime	$Ca(OH)_2$	CH
Calcium sulfate dihydrate	gypsum	$CaSO_4.2H_2O$	$C\bar{S}H_2$
Calcium silicate hydrate gel	C–S–H gel	$Ca_xSi_yO_{x+2y}.xH_2O$	$C_xS_yH_x$
Calcium aluminate sulfate 32-hydrate	ettringite	$Ca_3Al_2O_6.3CaSO_4.32H_2O$	$C_3A.3C\bar{S}.H_{32}$
Calcium aluminate sulfate 12-hydrate	monosulfate	$Ca_3Al_2O_6.CaSO_4.12H_2O$	$C_3A.C\bar{S}H_{12}$

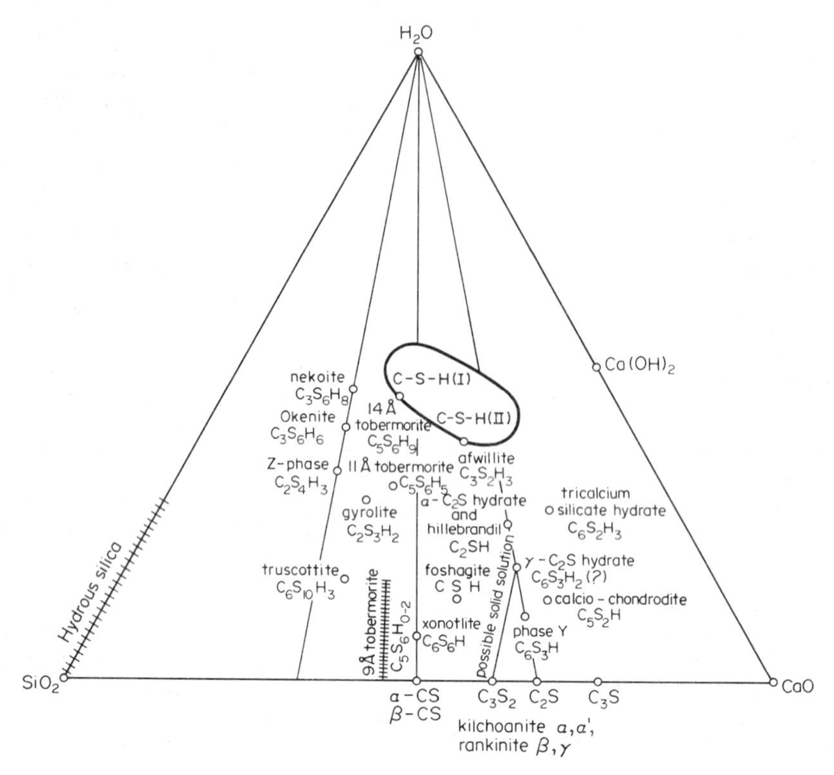

Figure 3
Phase diagram showing some calcium silicate hydrates found in hydrated Portland cements and related systems at ambient and elevated temperatures (after Taylor 1964; courtesy of Academic Press)

monosulfate hydrate, $C_3A.C\bar{S}.H_{12}$, after all the gypsum has reacted and the sulfate concentration in the aqueous phase falls.

The ferrite solid solution reacts more slowly with water than do the other cement compounds. It can form analogs of ettringite and the tetracalcium aluminate monosulfate hydrate in which at least part of the Al is replaced by Fe. It may also form $C_4(A,F)H_{13}$. The aqueous phase in a cement paste has a pH of 12.5 or higher, depending on the concentrations of sodium and potassium ions dissolved in it. The reaction rates depend upon the exact natures of the cement compounds and their states of subdivision, as well as the temperature and the concentrations of any soluble materials added to the mixing water to retard or accelerate setting. The calcium silicate hydrates formed in Portland cement and related systems at elevated temperatures (e.g., $\geqslant 100\,°C$) may be significantly different in composition and degree of crystallinity from those formed at lower temperatures (see Fig. 3).

5. Chemistry of Blended Cement Hydration

The hydration of the Portland cement component of blended cements is similar to that of the Portland cement by itself, though at a given fineness the rate of strength development is likely to be somewhat slower because of the slower reactions of the other components. The high pH and the sulfate content of the aqueous phase resulting from the Portland cement hydration reactions activate the slags or pozzolans in the cements. Once activated, slags appear to react independently to form additional quantities of calcium silicate hydrates. On the other hand, pozzolans can form calcium silicate hydrates only by reaction with the calcium hydroxide liberated from the Portland cements. The C–S–H from blended cements is similar to that from Portland cements.

6. Performance Requirements

If a cement is to perform satisfactorily in general concrete construction, most or all of the following performance requirements should be met by mortars and concretes which contain it:

(a) adequate rate of strength development,

(b) ease of mixing and placing,

(c) dimensional stability (soundness)—no detrimental volume changes,

(d) low creep and low drying shrinkage,

(e) durability—capacity to withstand normal environments above and below ground level,

(f) nonreactivity with aggregates,

(g) noncorrosivity to reinforcing steel,

(h) low heat evolution,

(i) uniform appearance, and

(j) capacity for control of setting and rheological properties by means of admixtures.

Portland and blended cements can be manufactured to meet essential requirements for most applications. The rate of strength development at early ages can be increased by finer grinding of the cement or increasing the contents of C_3S and C_3A. The ease of mixing and placing usually offers no problems, provided that the particle size distribution is broad and the gypsum additions are sufficient to prevent flash setting due to rapid hydration of C_3A. Dimensional stability is usually acceptable unless a large quantity of ettringite forms after setting or if, in products which are to be autoclave-cured, the cement contains more than a small amount ($\sim 2\%$) of periclase (MgO). Creep and drying shrinkage are related phenomena which are at a minimum when the optimum quantity of gypsum is contained in the cement.

The chemical durability of hardened cement paste is usually good in natural environments, except where sulfate groundwaters and soils are encountered. Pastes of cements with higher C_3A contents are more prone to sulfate attack as the result of the formation of large ettringite crystals within the structure. For cements which are to be used with certain siliceous sands and aggregates, a low-alkali Portland cement or a Portland–pozzolan cement is generally desirable if the possibility of disruption due to alkali–aggregate reactions is to be avoided. Hardened cement pastes form a good corrosion-protective cover for reinforcing steel because the high pH of the paste passivates the steel unless chloride is present or neutralization of the paste by atmospheric carbon dioxide has reached the level of the steel.

The heat evolution accompanying the hydration of a cement depends primarily on its composition. Cements which are high in C_3A and C_3S evolve more heat than those which are low in these constituents. The appearance of a hardened cement paste is affected by the composition of the cement and the structure of the hardened material. Differences in porosity due to different water–cement ratios and different degrees of compaction can lead to differences in appearance; also, soluble salts, such as sodium sulfate, in the pore water may be deposited on the surface due to evaporation under certain environmental conditions, giving a white incrustation (efflorescence).

The effects of intentionally added soluble materials in the mixing water on the setting and rheological properties of cement pastes are technically important. Certain organic compounds, such as lignosulfonates, can retard the setting and improve the flow of cement pastes. The effects are due to adsorption on the surfaces of the cement grains and hydration products. Acceleration of setting and hardening reactions is sometimes achieved by addition of calcium chloride to the mixing water, though this may lead to accelerated corrosion of steel reinforcement.

Frequently, a specific performance objective can be met in different ways with different cements. For example, high early strength could be achieved either by using a finely ground Portland cement or by curing the product at an elevated temperature. As another example, a concrete which is likely to be exposed to sulfate groundwaters could be made either with a Portland–pozzolan cement known to be sulfate-resistant. Because of the large number of composition and particle size variables which can be controlled in cement manufacture, the number of performance options which can be exercised is very large. In practice, the number is limited by the cost of manufacturing and storing many types of cement.

7. *Standards for Portland and Blended Cements*

In the USA the most widely used standards for cements are those established by the American Society for Testing and Materials (ASTM). The ASTM standard specification for Portland cement, ASTM C 150, covers five main types with differing performance characteristics and also recognizes some optional subcategories. The chemical and physical requirements for the five types are given in Table 4.

The ASTM standard specification for blended hydraulic cements, ASTM C 595, is newer than the Portland cement specification and rests on a less complete base of knowledge. Needs for conservation of energy and materials have increased interest in blended cements, and improvements in the standards are being brought about to facilitate substitutions between blended cements and Portland cements. The categories of blended cement recognized in ASTM C 595 are indicated in Table 5. Cements manufactured in different countries tend to be similar, but the standards may differ substantially in detail. An important factor in the selection of cements is uniformity, and a standard for uniformity, ASTM C 917, has been established.

8. *Methods of Characterization of Cements*

For specification purposes, Portland cements are characterized in terms of their oxide compositions using wet chemistry or instrumental methods such as

Table 4
Types of Portland cement defined in ASTM C150

Type	Characteristics	Limits on clinker phases	Typical potential compound composition (%)			
			C_3S	C_2S	C_3A	C_4AF
I	General purpose	—	53	23	10	8
II	Moderate heat evolution, moderate sulfate resistance	<8% C_3A	50	24	7	11
III	High early strength	—	57	18	10	8
IV	Low heat evolution	≥40% C_2S, ≤35% C_3S, ≤7% C_3A	30	44	6	12
V	Sulfate resistant	<5% C_3A	44	30	4	14

Table 5
Types of blended hydraulic cement defined in ASTM C595

Limits on % of				
Pozzolan	Slag	Type	Name	Use
<15	a	I(PM)	Pozzolan-modified Portland cement	General purpose
15–40	a	IP	Portland–pozzolan cement	General purpose
15–40	a	P	Portland–pozzolan cement	General purpose where strength at early ages is not important
	0–25	I(SM)	Slag-modified Portland cement	General purpose
	25–65	IS	Portland blast furnace slag cement	General purpose
	≥60	S	Slag cement	For use with Portland cement in concrete or with lime in masonry mortars

[a] May contain slag because the cement may be made by blending the pozzolan with either a Portland or a Portland blast furnace slag cement; in practice, slag is seldom present

atomic absorption, x-ray fluorescence or direct-coupled plasma spectrometry. Assuming that the only phases present are tricalcium silicate, dicalcium silicate, tricalcium aluminate, tetracalcium aluminoferrite, gypsum, periclase and free lime, the phase composition is calculated from the elemental analysis using a modified Bogue formula. The results reflect errors in the assumptions made.

In addition to composition, cements are characterized in terms of fineness as indicated by specific surface area, particularly as determined by the Blaine air permeability method. The characterization of blended cements is less definitive than that of Portland cements because there is less knowledge about the phases present in blast furnace slags and pozzolans. Chemical analyses of blended cements are carried out only to establish that the oxide compositions are being maintained within a defined range. It is recognized that, for purposes of quality assurance and prediction of performance, as well as for research, it would be desirable to be able to characterize cements in terms of phase composition and the distribution of the phases by particle size and shape. This is not yet practicable.

For research purposes, techniques used for determining the phases present in both Portland and blended cements include x-ray diffraction, thermal analysis, petrography and scanning electron microscopy complemented by energy-dispersive x-ray analysis. These techniques can be applied to whole cements or cements which have been separated into size fractions. Progress is being assisted by the development of techniques for selective dissolution of cement phases. For example, the calcium silicates in Portland cements can be preferentially dissolved by solutions of salicylic or maleic acids in methanol to leave the aluminate phases.

9. *Uses of Portland and Blended Cements*

Uses of Portland and blended cements (Table 6) can be divided into factory and field applications. In factory applications, the operations can be better controlled and elevated-temperature curing with steam is frequently used to accelerate strength development and shorten the production cycle. In field use, by far the greatest quantity of cement is used in ready-mixed concrete delivered to the worksite in a mixer truck.

10. *Manufacture of Portland and Blended Cements*

Manufacture of Portland cement clinkers begins with the grinding of a carefully proportioned mixture of previously crushed raw materials. Common raw materials, and the main oxides they contribute, are limestone (CaO), oyster shells (CaO), marl (CaO, SiO_2), cement rock (CaO, SiO_2, Al_2O_3), sand (SiO_2), iron ore (Fe_2O_3) and mill scale (Fe_2O_3). The raw meal is usually fed through a preheater into a rotary kiln in which it is gradually heated to ~1500 °C by a slightly oxidizing countercurrent flow of the combustion gases from a pulverized coal, oil or natural gas flame. During passage through the kiln, which rotates at 1–2 rev min^{-1}, the minerals in the raw meal are progressively dehydrated and calcined (decarbonated) and undergo phase changes, including melting, and chemical reaction before being discharged into a clinker cooler. At the highest temperature, the liquid phase makes up ~25% of the volume of the material; it is predominantly a calcium aluminate and ferrite melt and the solid phases are the calcium silicates. The viscosity of the melt depends on its composition: MgO and alkali metal oxides have large effects. The distribution of the minerals (compounds) in the clinker is controlled by the raw materials, their fineness, the homogeneity of the raw meal and the operating conditions of the kiln and cooler. The clinker is normally in the form of roughly spherical nodules ~5–50 mm in diameter. It is stored in stockpiles until needed to make the finished cement by grinding in a ball mill with ~5% of gypsum. Cement is stored in silos from which it may be shipped in bulk by truck, rail or barge. In 1981, the selling price of Portland cements in the USA was in the range \$50–60 per tonne.

The manufacture of Portland cement clinker is energy-intensive and accounts for ~0.8% of the nation's energy consumption. About 80% of the energy is used in the pyroprocessing, with lesser quantities for crushing and grinding. The theoretical energy requirement for the formation of Portland cement clinker from typical ground raw materials is ~1750 kJ kg^{-1}; the most efficient of present day plants use about twice this.

Blended cements are formed by intergrinding Portland cement clinkers with granulated blast furnace slag or fly ash and a small quantity of gypsum. They may also be formed by blending of the separately ground ingredients.

Table 6
Current uses of cements in the USA

	Usage (%) (estimate)
Cast-in-place concrete	
foundations	36
slabs	14
highways	13
columns	7
masonry mortar	4
canal linings	2.5
tunnel linings	2
other	8
Precast concrete	
concrete block	4
pipe	2
panels	2
beams	2
fiber-reinforced products	2
other precast	1.5

11. *Research Needs*

Cement research needs can be divided into two categories—those of cement manufacture and cement use. Improved knowledge of the basic properties of cements and of the mechanisms and kinetics of the hydration reactions are required to help make concrete a more predictable material. In the high-temperature reactions in which Portland cement clinker is formed, there is need for greater knowledge about effects of the raw materials and their processing on the formation and characteristics of the clinker, and the relationship between the clinker characteristics and the properties of the finished cement. With growing interest in the use of blast furnace slags, fly ash, and other waste and by-product materials in cement manufacture, there is a need for improved methods for characterizing these materials and identifying the characteristics which lead to superior performance in cement and concrete.

12. Application of Computers in Cement Manufacture and Use

The use of computers in cement science and technology is growing rapidly. Computers are being applied to control of most aspects of cement manufacturing operations and useful mathematical models of the operations have been developed. As the capabilities of computers have grown, the models have become more comprehensive and realistic. Manufacturing processes modelled include raw material proportioning and blending, grinding, and processes in kilns, including clinker formation. The models have been used for improving the manufacturing process in terms of energy use, product uniformity, reduced grinding costs and lengthened kiln lining life.

A difficulty in developing models relating the steps in cement manufacture to cement performance lies in the difficulty of determining the particle size distribution of the cement and the distribution of phases between particles of different sizes. There are similar problems in the characterization of raw meal and in the determination of the composition and structure of clinker. In application to the finished cement, computers are being used to control analytical instrumentation and process the data, and attempts are being made to develop mathematical models which can predict the performance of cements in concrete.

13. Portland and Blended Cement Mortars

Mortars are mixtures of cements with sand, or other fine aggregate, and water. The sand–cement ratio is usually ~3. Mortars are used extensively in masonry construction as the bedding materials for the masonry units and as a rendering applied to vertical surfaces for architectural purposes or to increase water tightness. The rheological properties of masonry mortars are important in their use, a "buttery" consistency usually being desired by the mason. Beneficial effects are frequently obtained by including in the mixture hydrated lime, flow modifiers, such as methylcellulose, and air-entraining agents. Important properties of masonry mortars are water retentivity of the plastic material when brought into contact with the masonry units and relatively low shrinkage after hardening. Quality assurance testing of cements for strength and dimensional stability is usually carried out on standard mortars made with a standard quartz sand.

See also: Calcium Aluminate Cements; Cement as a Building Material; Cements, Specialty; New Cement-Based Materials

Bibliography

American Society for Testing and Materials 1981 *Annual Book of Standards*, Pt. 13, *Cement; Lime; Ceilings and Walls*. American Society for Testing and Materials, Philadelphia, Pennsylvania

Duda W H 1977 *Cement Data Book*, 2nd edn. Macdonald and Evans, London

Lea F M 1970 *The Chemistry of Cement and Concrete*, 3rd edn. Chemical Publishing Company, New York

McKee H J 1973 *Introduction to Early American Masonry, Stone, Brick, Mortar and Plaster*. National Trust for Historic Preservation, Washington, DC

National Materials Advisory Board 1980 *The Status of Cement and Concrete R & D in the United States*, NMAB-361. National Materials Advisory Board, Washington, DC

Proceedings of the International Congresses (Symposia) on the Chemistry of Cements: Paris 1980, Moscow 1974, Tokyo 1968, Washington 1960, London 1952, Stockholm 1938

Ramachandran V S, Feldman R F, Beaudoin J J 1981 *Concrete Science Treatise on Current Research*. Heyden, Philadelphia, Pennsylvania

Taylor H F W (ed.) 1964 *The Chemistry of Cements*. Academic Press, London

US Bureau of Mines 1980 *Mineral Facts and Problems*, US Bureau of Mines Bulletin 671. US Bureau of Mines, Washington, DC

World Cement Standards. Cembureau, Paris, France

G. J. Frohnsdorff
[National Bureau of Standards,
Washington, DC, USA]

Prices of Materials: Stabilization

Materials prices may be valuable signals of relative scarcities. High prices tend to encourage limited use of a particular material and search for cost-reducing technological changes and cheaper substitutes, whereas low prices tend to encourage expanded use of a particular material instead of more expensive alternatives. In a world of imperfect information, data on prices which may reflect relative scarcities are often much more available and more timely than are data on quantities which directly reflect relative material scarcities. Some basic principles of how materials markets operate without price stabilization policies are discussed in Sect. 1.

Despite the possibly very useful roles of freely fluctuating prices, there have long been frequent attempts by governments and private entities to stabilize materials prices. Among the major policies that have been used for such efforts are buffer stocks; isolating a particular market and using exports from it to lessen downward price movements and imports into it to lessen upward price movements; announcing a price and rationing available quantities among users if demand at that price exceeds supplies; providing information; attempting to lessen the underlying causes of fluctuations; and intervening in futures markets for the material (Sect. 2).

The motivation for such efforts includes gains due to risk aversion (Sect. 3), macroeconomic gains regarding fuller employment and less inflation (Sect. 4), increases in consumers' and producers' surpluses (Sect. 5), and distributional gains (or losses) for particular groups often associated with changing the average level of material prices (Sect. 6).

Analysis with simple models suggests that some of these gains may indeed be attainable. But further analytical complications and empirical experience (see Behrman 1977) suggest that the costs of attaining such gains may frequently be considerable. Moreover, there is often confusion between the logically distinct efforts to stabilize materials prices and to change their expected (or average) values, usually to the benefit of some particular supplying or demanding group but with widespread efficiency costs (Sect. 7). Therefore, it is important that materials price stabilization policies should be analyzed carefully within a framework that incorporates the important features of a particular materials market, that considers the distribution of total costs and benefits, and that reflects the difficulties of predicting the appropriate expected prices—and that decisions should not be made on the basis of too simple analytical models.

1. Simple Materials Market Without Price Stabilization Policies

Figure 1 is a graphical representation of a simple materials market. Underlying the market supply curve is maximizing behavior by each of a large number of small suppliers so that no one supplier can affect the market price perceptibly, but each one responds instantaneously to the given price (from the point of view of the individual supplier) by supplying the profit-maximizing quantity (i.e., perfect competition). These quantities increase for higher prices since it is profitable to expand production at higher prices even if fixed inputs (e.g., mines and managerial capacities) have to be stretched further or higher prices paid for variable inputs (e.g., labor). By summing the quantities supplied at each price across all suppliers, the market supply curve is obtained. In Fig. 1 the market supply curve in period $t(S_t)$ is assumed to be linear, with an additive stochastic term v_t due to fluctuations in, for example, the weather, with mean value of zero (as in the solid, upward sloping supply curve) and a finite variance σ_v^2:

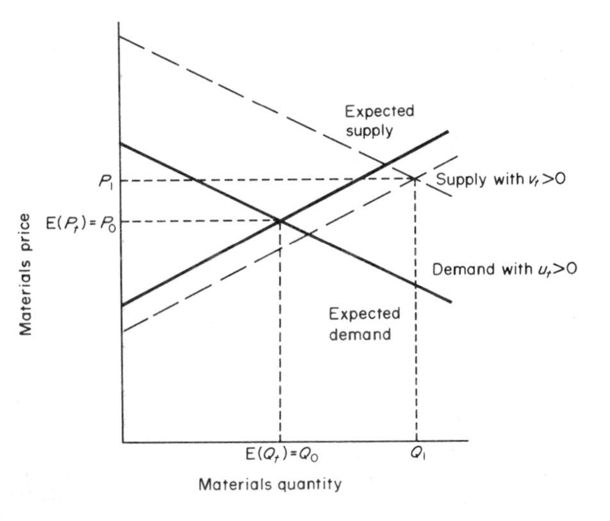

Figure 1
Simple linear materials market supply and demand curves with additive stochastic terms

$$S(P_t)=a+bP_t+v_t$$

$$b\geqslant 0, \qquad \mathrm{E}(v_t)=0, \qquad \sigma_v^2>0 \tag{1}$$

where a and b are constants, $\mathrm{E}(v_t)$ is the expected value of v_t, and P_t is the short-run equilibrium price. Likewise, underlying the market demand curve is maximizing behavior by a large number of demanders, each of which is assumed to be unable to affect the market price perceptibly, but each of which responds instantaneously to the given price (from the point of view of the individual demander) by purchasing the optimizing quantity (i.e., perfect competition again). At higher prices this optimizing quantity tends to be lower since other relatively cheaper goods or services are substituted for this material. By summing the quantities demanded across all demanders, the market demand curve is obtained. In Fig. 1 the market demand curve in period D_t is assumed to be linear, with an additive stochastic term u_t due to fluctuations in income, for example, with mean value of zero (as in the solid, downward sloping demand curve) and a finite variance (σ_u^2) and zero covariance with the stochastic term in the supply relation ($\sigma_{uv}=0$):

$$D(P_t)=c-dP_t+u_t$$

$$d>0, \quad \mathrm{E}(u_t)=0, \quad \sigma_u^2>0 \tag{2}$$

where c and d are constants and $\mathrm{E}(u_t)$ is the expected value of u_t. In the absence of any stabilization policy, in period t the short-run equilibrium price is that price at which supply equals demand. (At a higher price the desired quantity supplied exceeds the desired quantity demanded, creating pressure for the price to fall, and vice versa for a lower price.) This price P_t can be obtained by setting Eqn. (1) equal to Eqn. (2) and solving for P_t:

$$P_t=(c-a+u_t-v_t)/(b+d) \tag{3}$$

The short-run equilibrium quantity Q_t can be obtained by substituting Eqn. (3) into either Eqns. (1) or (2):

$$Q_t=S(P_t)=D(P_t)$$
$$=(cb+da+bu_t+dv_t)/(b+d) \tag{4}$$

Thus the short-run equilibrium price and quantity both are stochastic since they depend on the stochastic terms y_t and u_t in the market supply and demand relations due to fluctuations in other relevant variables like the weather and income. The dashed lines in Fig. 1 give an illustration in which both v_t and u_t are positive, with the resulting equilibrium price P_1 and quantity Q_1. From Eqns. (3) and (4) the expected values of the price $E(P_t)$ and quantity $E(Q_t)$ at which the stochastic terms take on their mean values of zero can be calculated:

$$E(P_t)=(c-a)/(b+d) \quad (5)$$

$$E(Q_t)=(cb+da)/(b+d) \quad (6)$$

In terms of Fig. 1, these expected values are just the equilibrium price P_0 and quantity Q_0 associated with the intersection of the solid supply and demand schedules (for which $v_t=u_t=0$). However, Eqns. (3) and (4) indicate that the actual price and quantity from period to period fluctuate around these expected values, with variances that depend on the size of the variances of the additive stochastic terms in the supply and demand relations (σ_v^2 and σ_u^2) and on the slopes in the nonstochastic part of these relations (b and d):

$$\sigma_P^2(\sigma_u^2+\sigma_v^2)/(b+d)^2>0 \quad (7)$$

$$\sigma_Q^2=(b^2\sigma_u^2+d^2\sigma_v^2)/(b+d)^2>0 \quad (8)$$

The reasons for these fluctuations are the underlying fluctuations in the supply and demand curves due to fluctuations in relevant factors such as weather and income. In response, the short-run equilibrium price varies so that the current quantity demanded and supplied responds to the current situation. If, for example, the weather is bad so $v_t<0$ and income is high so $u_t>0$, the short-run equilibrium price is relatively high, which induces a movement along the demand curve in that period to conserve use of the material in question and a movement along the supply curve to expand its supply. Thus the relatively high price serves to improve the situation by reducing use and encouraging supply in a period of relative shortage (and vice versa if $v_t>0$ and $u_t<0$). If both suppliers and demanders are perfect competitors and there are not externalities (e.g., products like pollution which are distributed outside of markets), the short-run equilibrium materials price tends to reflect the true social opportunity cost of materials and using it to allocate the materials results in an efficient allocation. On the other hand if there are market failures due to factors like externalities or imperfect competitors (e.g., monopolies), various policy interventions (perhaps including materials price stabilization) may make the allocation of resources more efficient. (For further considerations regarding modelling such markets and efficiency, see Mansfield 1979.)

2. Major Price Stabilization Policies

Among the most used and most analyzed of major price stabilization policies is the institution of buffer stocks which try to stabilize material prices by buying the material when otherwise the price would be relatively low and selling it when otherwise the price would be relatively high. In the situation described in Fig. 1, a buffer stock authority could completely stabilize the price at its expected value (P_0 or $E(P_t)$ in Eqn. (5)) by trading in each period a quantity equal to v_t-u_t (i.e., buying if $v_t>u_t$, selling if $v_t<u_t$). The variance in price would go from the positive value in Eqn. (7) to zero. The variance in quantity supplied would go from that in Eqn. (8) to σ_v^2, and the variance in quantity demand would go from that in Eqn. (8) to σ_u^2 (which implies an increase in at least one of these two quantity variances). The expected trading of the buffer stock is zero ($=E(v_t)-(u_t)$), so in the long run (assuming that the structure of the market is stable) the buffer stock tends to sell what it purchases at the same price. The buffer stock suffers an expected loss, therefore, equal to its expected transaction and storage costs. In a static market this cost could be covered by buying only when the price otherwise would go below some floor P below P_0 and selling only when the price otherwise would go above some ceiling $(\bar{P})$ above P_0, with the width of the price band $(\bar{P}-P)$ being a positive function of such buffer stock costs. The wider the price band, of course, the less the price stabilization and gains associated with price stabilization. In a dynamic market in which there is a secular trend in the expected price over time, the buffer stock costs may be covered by buying low and selling later at a higher price even without price bands if the secular trend in the expected price is positive enough, but the buffer stock may not be able to cover its direct costs even with very wide price bands if the secular trend in the expected price is negative enough.

A second widely used materials price stabilization policy is to isolate geographically a certain materials market (e.g., a nation or a group of nations like the European Economic Community, EEC) and regulate exports and imports of materials to offset fluctuations internal to the market. That is, if the objective is to stabilize the domestic materials price at the expected value P_0 in Fig. 1 (assuming that this figure refers only to the isolated market), let exports minus imports be equal to v_t-u_t. This strategy depends on there being some external rest-of-the-world market to which to sell exports or from which to buy imports when needed, and therefore is not applicable on a global or near-global level. This policy has the direct cost advantage over the buffer stock alternative of avoiding direct storage costs, although direct transaction costs remain (and probably are greater since direct administrative costs are incurred to isolate a particular market). The more important question regarding direct costs probably pertains to the comparison between the expected prices in the rest-of-the-world market and the market of interest. If the expected price in the rest-of-the-world market is less than that in the internal market of concern, as is apparently the case for many materials produced in the EEC, direct subsidies are required to export surpluses and internal users pay more for the material than they would if they were able to buy on the world market. If the expected price in the rest-of-the-world market is greater than in the isolated market of concern (as, for example, for major material exports from the Ivory Coast and a number of other African nations), direct subsidy costs are not likely to be a problem, but internal production is discouraged and consumption encouraged relative to the alternative of trading with the rest of the world. For such reasons it would be best to stabilize around the world price (which generally implies expected net exports or imports with variations at the margin to offset internal stochastic fluctuations) unless there are good non-efficiency reasons to aim for a different target (i.e., a higher expected internal price to encourage development of internal "strategic" materials for national defence purposes). However, it is not clear that such advice implies anything much different than integration into the world market, which has the advantage of sharing one's own shocks with the rest of the world but the disadvantage of sharing shocks that originate in the rest of the world.

A third, very common procedure for stabilizing materials prices is to set a price and ration use if the quantity demanded at that price exceeds the quantity supplied. Such policies have been followed by a wide range of governments, both socialist and mixed capitalist, for strategic materials, particularly at times of war or other large shocks (e.g., the Organization of Petroleum Exporting Countries (OPEC) oil price shocks). Also, private entities have used such a strategy at times apparently because of the perception that a stable price with rationing at times of shortages leads to a larger longer-run demand (e.g., US copper producers in the 1960s and early 1970s). If the stabilized price is below the short-term equilibrium price and allocation by rationing is effective, one major cost of such a policy is that those whose demand is fulfilled pay less than the true scarcity cost of the material and are likely to use more of it than is socially efficient: others who are willing to pay an equal amount have their demands only partially fulfilled (or not at all), and suppliers are discouraged from supplying as much of the material as they would at the equilibrium price. In addition to these induced inefficiencies, rationing is likely to encourage illegal activities and exploitation of "connections" for those who have them. Also, considerable costs are incurred in administrating any effective rationing system.

A fourth price stabilization policy is to provide information which can lead to integration of markets across space or time. If the stochastic terms are not perfectly correlated across such markets, integration resulting from such information indeed may reduce fluctuations by "pooling risks." Since information has the quality of being a "public good" (i.e., more of it for one person does not diminish what is available for others), private entities probably do not provide enough of it and public information policies possibly are efficient and are certainly undertaken. On the other hand, provision of information is not costless and it is not at all clear that governments have better information about materials markets than do private entities which also may have strong incentives to integrate markets over space and time.

A fifth materials price stabilization policy which the discussion in Sect. 1 suggests, but which is more indirect, is to attempt to lessen the underlying stochastic terms. If supply fluctuations originate in weather or labor market conditions, for example, they may be lessened by explicit private or public measures. Many such measures are undertaken, in fact, although often with materials prices stabilization as only one objective (and, at times, a very minor one). Examples include private and public water control measures to lessen supply shocks or public macroeconomic policies to lessen demand shocks.

Other policies also could be mentioned. For example, Café Mundial (a consortium of coffee producers) has recently attempted to affect current prices by trading in future markets in which their leverage is greater (since contracts can be purchased with margins much less than 100% of their value), but this logically can be considered a variant of buffer-stock type policies. As another variant of buffer stock policies, private entities could be subsidized to hold such stocks, with the subsidies related inversely to current price levels. However, the above include the major materials price stabilization policies, with the possible exception of restrictions on supply or demand (see Sect. 6).

3. Gains Due to Risk Aversion

If producers are risk averse in the sense that they are better off for a given level of expected profits (or surplus, see Sect. 4) if the variance in materials prices is lower, then there may be a gain from price stabilization even if it leaves expected profits the same and has none of the other possible effects discussed below. If users are risk averse in a similar sense, there also may be such a gain, though it probably is less important for users since any particular material is probably a much smaller proportion of the users' expenditures than it is of the producers' revenues. If a nation is dependent upon production, exports or use of a particular material for a significant share of its economic activity and risk aversion is widespread, there may be widespread gains since material price fluctuations may be amplified throughout the economy by government and private revenues and expenditures (for example, see Adams and Behrman 1982).

Risk aversion is widely accepted as a reason for which price stabilization may be advantageous. However, at least two questions occur in this connection. First, is risk aversion likely to be with regard to prices or with regard to revenues, expenditures or surpluses (see Sect. 5)? If it is with regard to suppliers' revenues, for example, the observation in Sect. 2 that buffer-stock materials price stabilization may increase variance in quantities supplied is relevant. In fact this increase may be sufficient that price stabilization actually increases the variance of suppliers' revenues. In such a situation price stabilization may make suppliers worse off, all other things being equal, if suppliers are risk averse with respect to revenues and not to prices. However, limited analysis to date suggests that the tradeoff between price and revenue stability is less likely if stabilization is partial rather than complete, the larger stochastic terms are in demand relative to those in supply and the less is the slope of the demand curve (Nguyen 1980). The available empirical evidence about materials markets suggests that probably there generally is not such a trade-off.

Second, is risk aversion empirically important? A large number of empirical studies have explored this question in recent years, albeit with measurement problems with regard to the representation of risk. The general conclusion seems to be that the effect of risk aversion as best it can be represented empirically often seems to be significant in a statistical sense, but not substantial in magnitude (e.g., see Adams and Behrman 1982). Thus, while risk aversion is a conceptually attractive reason for gains from materials price stabilization, it does not seem to be empirically important enough to warrant much economic or political cost.

4. Gains due to Macroinflation and Employment Effects

Increases in materials prices may cause upward price pressures which are costly in terms of increased inflation or negative effects of anti-inflationary policies on output. For materials price stabilization to have a gain from lessening inflationary pressures, however, there has to be a reason why there is not a symmetrical offsetting downward anti-inflationary pressure when materials prices fall. In the period after World War II, in the industrial economies at least, there seems to be an asymmetry in price movements. Prices tend to move upwards much more easily than downwards. The reasons for the asymmetry are not fully understood. Important factors probably include knowledge that governments have been committed to fairly full employment and are not likely to allow a repeat of the Great Depression, much better benefits for the unemployed than earlier and administrative pricing with an upwards bias by large (and not perfectly competitive in the sense of Sect. 1) companies. Although agreement is not universal on this matter, in this context there may be a considerable macrogain from materials price stabilization due to lessened inflationary pressure and lessened need to pursue anti-inflationary policies. One crude recent estimate calculates this gain for the USA alone to be three times as large as the gain to producers worldwide from international price stabilization of 10 major commodities: copper, tin and eight agricultural products (Behrman 1977).

5. Gains in the Form of Producers' and Consumers' Surpluses

The combination of an upward sloping supply curve and the assumption that all units are sold at the same price in period t (i.e., P_1 for the dashed lines in Fig. 1) has an important implication. For the last (or marginal) unit sold there is no producers' (or suppliers') surplus in that it would not be supplied if the price were lower, but for all other units sold there is a producers' surplus in that they would be supplied even if the price were somewhat lower. For the dashed lines in Fig. 1 this producer surplus is measured by the triangle bounded by the dashed supply line, the dashed horizontal price line at P_1 and the vertical price axis. In any period t the change in the producer surplus due to price stabilization at P_s is the integral bounded by P_t and P_s and by the supply curve and the vertical price axis. The expected value of this change in producers' surplus due to price stabilization under the assumptions of Sects. 1 and 2 is:

$$\mathrm{E}(\mathrm{SUR}_p) = [-b\sigma_u^2 + (2d+b)\sigma_v^2]/[2(d+b)^2] \quad (9)$$

This may be positive or negative, depending on the relative magnitudes of the terms in the numerator (Turnovsky 1978).

Likewise, the combination of a downward sloping demand curve and the assumption that all units are purchased at the same price in period t has a similar interpretation (with some technical qualifications that apparently are not critical empirically) for consumers' (or demanders') surplus, since demanders would be willing to purchase all but the last unit of the material even if the price were somewhat higher. For the dashed lines in Fig. 1 the consumers' surplus is measured by the triangle bounded by the dashed demand line, the dashed horizontal price line at P_1 and the vertical price axis. Under the assumptions of Sects. 1 and 2, the expected value of the change in consumers' surplus due to price stabilization is:

$$\mathrm{E}(\mathrm{SUR}_c) = [(2b+d)\sigma_u^2 + d\sigma_v^2]/[2(a+b)^2] \quad (10)$$

This, too, may be positive or negative depending on the relative signs of the terms in the numerator.

By summing Eqns. (9) and (10), the total expected surplus can be calculated:

$$E(SUR_p)+E(SUR_c)=(\sigma_u^2+\sigma_v^2)/[2(b+d)] > 0 \qquad (11)$$

This total surplus or gain is unambiguously positive.

Equations (9)–(11) lead directly to the Waugh–Oi–Massell (WOM) conclusions about the impact of materials price stabilization.

(a) Demanders gain (lose) from materials price stabilization if the only source of price instability is random shifts in demand (supply).

(b) Suppliers gain (lose) from materials price stabilization if the only source of price instability is random shifts in supply (demand).

(c) If both supply and demand have stochastic terms, the gains to suppliers alone and to demanders alone may be positive or negative, depending upon the variances of the stochastic terms (σ_u^2 and σ_v^2) and the magnitudes of the supply and demand slopes (b and d).

(d) However, the total gains to suppliers and demanders together are positve, with the gainers in principle able to compensate the losers.

Empirical estimates suggest that for many materials, the above types of gains could be considerable.

These conclusions from the WOM analyses have conditioned a widespread belief that there may be substantial gains from materials price stabilization. However, in several important respects (which are discussed in more detail by Turnovsky (1978) and Newbery and Stiglitz (1981)) they are conditional on the special assumptions of the simple model of Sect. 1. If there are multiplicative (rather than additive) stochastic terms, if supply responds to expected prices generated by some common expectational mechanisms (instead of adjusting instantaneously to current price), or if some simple partial stabilization rules are followed (instead of there being a full stabilization at P_s), only the fourth conclusion may hold strictly. Of course this is the most important conclusion since it states that total gains from materials price stabilization are positive.

However, even this conclusion is thrown into some doubt with further extensions of the analysis. First of all, the analysis ignores the transaction, administrative and storage costs of price stabilization. These may be considerable in some cases and subtracting them from expected producers' and consumers' surpluses may give a greatly reduced and even negative value. Second, the analysis presumes that the appropriate target price (P_s) is known, but there may be uncertainty in regard to this price (Sect. 6). Third, there may be dynamics in regard to both supply and demand that complicate the analysis considerably (e.g., long gestations in expanding mine capacities or in shifting from use of one material to another). Fourth, the analysis assumes that there are no private holdings of stocks. However, at least for storable commodities, why would not suppliers or demanders (depending on who would gain) reap the expected benefits (up to the point at which they equalled the expected costs) by undertaking their own storage? Fifth, this and related analyses ignore the fact that there is a basic asymmetry in some materials price stabilization policies in that the desired price (or price ceiling) may not be defensible if the buffer stock runs out or if more supplies for imports are not available from the rest-of-the-world market.

These doubts are not conclusive. There may be good reasons why, for example, private inventory behavior for materials is not socially optimal: capital market imperfections, differences between private and social discount rates to value future events, externalities such as macroeffects (Sect. 4), and distributional concerns (Sect. 6). For such reasons public buffer stock or other materials price stabilization government policies may be very sensible, although such possibilities should be evaluated carefully in light of the actual condition in a particular materials market—and the overly specialized WOM model results should not be blindly used to justify advocacy of ubiquitous materials price stabilization.

6. Distribution of Gains

Up to this point limited attention has been paid to the distribution of losses and gains from materials price stabilization policies among materials suppliers, materials demanders and the general public (through the government). Yet often such distributional effects are at the heart of the matter. If a particular group is thought to be more deserving or more needy (e.g., poorer producers of the materials), it might be socially desirable to promote price stabilization if this target group benefits enough, even if the total costs outweigh the gains if special weight is not placed on the gains of the target group. Such distributional concerns seem very important in understanding the emphasis in many societies on price stabilization of basic mass consumption commodities used relatively greatly by the poor. It also is a factor in the call by many developing country groups for price stabilization of the materials important in their exports or imports. A related concern is to use price stabilization to redistribute gains from private inventory holders (i.e., middlemen or speculators) or traders with market power (i.e., sole buyers or one of a few buyers in a market) rather than perfect competition. In certain circumstances price stabilization may help in the desired redistribution, but the effects may be somewhat arbitrary (not all of the poor or even the poorest produce or use particular materials) and have negative side effects (e.g., discouraging socially useful private inventory-holding).

Alternatively, materials price stabilization may be perceived to benefit a powerful political group (e.g., well-organized producers) at a cost to a politically weaker one (e.g., diffused and unorganized consumers). In such a case, even if the total social costs of materials price stabilization outweigh the gains and there is nothing particularly socially deserving about the beneficiaries, the political system may bow to the pressures of the politically powerful pressure group and introduce the price stabilization program.

7. *Stabilization Versus Changing Expected Materials Prices*

Recognition that materials price stabilization may be directed towards benefiting a particular group, even if the total costs outweigh the total benefits, is basic in understanding the widespread advocacy of materials price stabilization policies. A closely related fact is that materials price stabilization is often used to refer to a situation in which the real (often implicit) objective is to change the expected materials price, not just to lessen its variance. If special weight is placed on gains to suppliers (demanders), materials prices stabilization may be a misleading characterization of an effort to raise (lower) the expected materials prices. Some stabilization policies, such as curtailing supplies (demands) by production or export (import) restrictions, are oriented not towards reducing materials price variances, but towards changing the expected materials prices.

The impact of changing the expected materials price may be very different from the impact of stabilizing it, and logically should be separated from it. Unless the effort to change the expected materials price also offsets some market distortion (e.g., due to a monopoly which is restricting supplies), for example, it generally has inefficiency effects akin to those discussed in Sect. 2 with regard to isolating a particular market or rationing. The distributional impacts (as mentioned in Sect. 6) also are likely to be capricious in that the beneficiaries are not likely to be exactly those whom society considers to be particularly deserving, and these effects are likely to be larger the more the expected materials prices in fact are changed.

The frequent confusion between lessening the variances of materials prices and changing their expected values partly may reflect efforts of those who wish to change the expected prices to obscure their intent, but they are possible only because there is considerable uncertainty about what is the expected price for a particular material at a point of time or, probably more importantly, how it changes over time. This same uncertainty creates an important problem in operating a materials price pure stabilization program and this problem is ignored in most of the analysis above: how can the appropriate expected materials price be identified? Simulations of price stabilization in materials markets suggest that such identification is difficult (Behrman 1977). If the expected materials price is wrongly identified, moreover, the results may include considerable unintended distortions and large operating costs (Behrman and Tinakorn 1979).

See also: Materials Availability; Materials Economics, Policy and Management

Bibliography

Adams F G, Behrman J C (eds.) 1978 *Econometric Modeling of World Commodity Policy.* Lexington Books, Lexington, Massachusetts

Adams F G, Behrman J R 1982 *Commodity Exports and Economic Development: The Commodity Problem and Policy in Developing Countries.* Lexington-Heath, Lexington, Massachusetts

Adams F G, Klein S A (eds.) 1978 *Stabilizing World Commodity Markets: Analysis, Practice and Policy.* Lexington Books, Lexington, Massachusetts

Behrman J R 1977 International commodity agreements: An evaluation of the UNCTAD integrated commodity programme. In: Cline W R (ed.) 1979 *Policy Alternatives for a New International Economic Order: An Economic Analysis.* Praeger, New York, pp. 63–153

Behrman J R, Tinakorn P 1979 Indexation of international commodity prices through international buffer stock operations. *J. Policy Modelling* 1: 113–34

Mansfield E 1979 *Microeconomics: Theory and Applications.* Norton, New York

Materials and Society (5:3 1981): issue devoted to Market Instability in the Metal Industries

Newbery D M G, Stiglitz J E 1981 *The Theory of Commodity Price Stabilization: A Study in the Economics of Risk.* Clarendon, Oxford

Nguyen D T 1980 Partial price stabilization and export earning instability. *Oxf. Econ. Pap.* 32: 340–52

Turnovsky S J 1978 The distribution of welfare gains from price stabilization: A survey of some theoretical issues. In: Adams F G, Klein S A (eds.) 1978 *Stabilizing World Commodity Markets: Analysis, Practice and Policy.* Lexington Books, Lexington, Massachusetts, pp. 119–48

J. R. Behrman
[University of Pennsylvania, Philadelphia, Pennsylvania, USA]

Properties of Materials

The properties of materials can be classified in various ways. A simple classification scheme involves categories which relate to different generic types of physical or chemical phenomena. For example, terms such as mechanical properties, electrical properties and magnetic properties are generally well understood. Each of these terms designates a set of properties which are related to each other and are important for practical applications and hence for engineering design.

Another classification scheme establishes a hierarchy of categories of properties which is headed by rigorously defined and measurable equilibrium properties of materials, each property having its roots in thermodynamics. At the other end of the scheme are composite and qualitative properties such as the weldability of a metal or the color of a fabric. This article describes this hierarchy of properties and discusses various properties in relation to it.

Relatively simple and well-behaved properties give the response of a material to a specified change in a given set of conditions. Many properties relate independent and dependent variables in a particular process. Equilibrium properties of materials are usually defined in terms of reversible thermodynamic processes. Transport properties characterize processes that occur between parts of a system that are not in equilibrium.

Some properties have hysteretic behavior, and for a given value of an independent variable have a range of values for the dependent variables. Even more complex are properties determined by tests that are inherently irreversible, such as the yield strength of a metal.

The properties of materials have other characteristics. Certain properties vary with direction in some materials; they are termed anisotropic. A structure-sensitive property has a value that depends strongly on the density or distribution of imperfections in a material whereas a structure-insensitive property has a value that is relatively independent of variations in the structure of the material, especially of the defect structure.

1. Equilibrium Properties

The equilibrium properties of materials specify the relationship between a generalized "force" Y_i and a generalized "displacement" y_i, and are characteristic of equilibrium states of the material. Examples of generalized forces are: temperature T, electric field intensity vector $\boldsymbol{E}_i$ and stress tensor $\boldsymbol{\sigma}_{ij}$. Corresponding to these forces are the conjugate generalized displacements: entropy S, electric displacement $\boldsymbol{D}_i$ and strain $\boldsymbol{\varepsilon}_{ij}$. The force–displacement pairs are conjugate when their product is thermodynamic work. The equilibrium properties known, respectively, as heat capacity C, permittivity κ_{ij} and elastic compliance s_{ijkl} are the proportionality coefficients relating conjugate forces and displacements according to the following relations:

$$dS=(C/T)dT \tag{1}$$

$$d\boldsymbol{D}_i=\kappa_{ij}d\boldsymbol{E}_j \tag{2}$$

$$d\boldsymbol{\varepsilon}_{ij}=s_{ijkl}d\boldsymbol{\sigma}_{kl} \tag{3}$$

These linear equations are valid for small values of the generalized forces. Deviations from linearity require the definition of higher-order constants to provide correct functional relationships at large values of the generalized forces.

In addition to properties that relate displacements to their conjugate forces, certain properties may be defined that relate nonconjugate variables, such as the thermal expansion coefficient, which relates strains to temperature changes. For a system in which all three forces, T, $\boldsymbol{E}$ and $\boldsymbol{\sigma}$, can be varied independently, the following relationships between displacements and forces are obtained:

$$dS=\underbrace{\left(\frac{\partial S}{\partial \boldsymbol{\sigma}_{ij}}\right)_{E,T}}_{\text{piezocaloric effect}} d\boldsymbol{\sigma}_{ij}+\underbrace{\left(\frac{\partial S}{\partial \boldsymbol{E}_i}\right)_{\sigma,T}}_{\text{electrocaloric effect}} d\boldsymbol{E}_i+\underbrace{\left(\frac{\partial S}{\partial T}\right)_{\sigma,E}}_{\text{heat capacity}} dT \tag{4}$$

$$d\boldsymbol{D}_i=\underbrace{\left(\frac{\partial \boldsymbol{D}_i}{\partial \boldsymbol{\sigma}_{jk}}\right)_{E,T}}_{\text{direct piezoelectricity}} d\boldsymbol{\sigma}_{jk}+\underbrace{\left(\frac{\partial \boldsymbol{D}_i}{\partial \boldsymbol{E}_j}\right)_{\sigma,T}}_{\text{permittivity}} d\boldsymbol{E}_j+\underbrace{\left(\frac{\partial \boldsymbol{D}_i}{\partial T}\right)_{\sigma,E}}_{\text{pyroelectricity}} dT \tag{5}$$

$$d\boldsymbol{\varepsilon}_{ij}=\underbrace{\left(\frac{\partial \boldsymbol{\varepsilon}_{ij}}{\partial \boldsymbol{\sigma}_{kl}}\right)_{E,T}}_{\text{elasticity}} d\boldsymbol{\sigma}_{kl}+\underbrace{\left(\frac{\partial \boldsymbol{\varepsilon}_{ij}}{\partial \boldsymbol{E}_k}\right)_{\sigma,T}}_{\text{converse piezoelectricity}} d\boldsymbol{E}_k+\underbrace{\left(\frac{\partial \boldsymbol{\varepsilon}_{ij}}{\partial T}\right)_{\sigma,E}}_{\text{thermal expansion}} dT \tag{6}$$

Each of the coefficients on the right-hand side of Eqns. (4–6) is a measure of a physical response (displacement) produced by a physical stimulus (force); the designation of the physical effect for a particular coefficient is given below the relevant right-hand term. These coefficients are properties of a material, and the equations define the process by which the property can be measured. The relationships between conjugate and nonconjugate thermal, electrical and mechanical properties, the names of the properties and the variables that they relate are shown schematically in Fig. 1. Other variables, such as magnetic field intensity, could also be included in Eqns. (4–6), resulting in definitions of additional properties, as well as relationships between properties, and in multidimensional relationships.

The thermodynamic state of a material at equilibrium is determined when values of state variables are specified. In the energy representation (Callen 1960) the internal energy of a simple system is specified by values of extensive variables. An extensive variable has a value in a composite system at equilibrium equal to the sum of its values in each subsystem. For a simple system, the extensive variables are entropy, volume and mole numbers. Intensive variables, such as temperature, describe changes in internal energy with variations in extensive parameters. In simple systems at equilibrium, intensive parameters have values that are uniform throughout the system.

A free-energy function G may be defined which includes the contributions of conjugate forces and

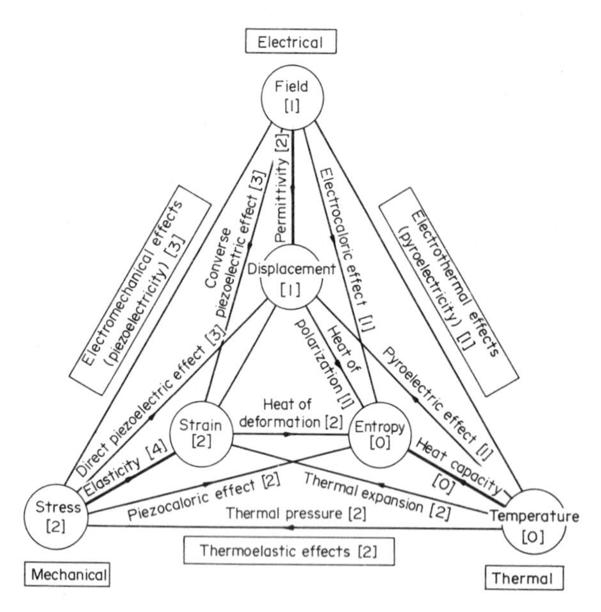

Figure 1
Relationships between forces (outer circles) and displacements (inner circles) for electrical, mechanical and thermal properties of materials. The properties that relate these variables are shown by the connecting lines. Heavy lines connect conjugate variables. Numbers shown are the tensor rank of the variables and properties (adapted from Nye 1957, p. 170)

displacements as a sum of terms of the type $y_i Y_i$. Maxwell relations can be obtained by equating mixed second derivatives of G with respect to two different independent variables (Nye 1957, pp. 178–83). The resulting expressions prove the following relations between coefficients in Eqns. (4–6):

$$\left(\frac{\partial S}{\partial \boldsymbol{E}_i}\right)_{\sigma,T} = \left(\frac{\partial \boldsymbol{D}_i}{\partial T}\right)_{\sigma,E} \qquad (7)$$

$$\left(\frac{\partial \boldsymbol{D}_k}{\partial \boldsymbol{\sigma}_{ij}}\right)_{E,T} = \left(\frac{\partial \boldsymbol{\varepsilon}_{ij}}{\partial \boldsymbol{E}_k}\right)_{\sigma,T} \qquad (8)$$

$$\left(\frac{\partial \boldsymbol{\varepsilon}_{ij}}{\partial T}\right)_{\sigma,E} = \left(\frac{\partial S}{\partial \boldsymbol{\sigma}_{ij}}\right)_{E,T} \qquad (9)$$

Relationships such as Eqns. (7–9) between properties and processes, which at first seem quite unrelated, are in fact a consequence of the nature of thermodynamic equilibrium.

2. *Transport Properties*

At equilibrium, the thermodynamic state of a material is characterized by the values of its intensive and extensive parameters. A system in which there are gradients of intensive variables is rarely at equilibrium, and in general these gradients produce fluxes of heat, electricity, matter or some other quantity. Transport properties of materials are proportionality constants that relate thermodynamic forces $\boldsymbol{X}$ to the fluxes $\boldsymbol{J}$ that they produce. The mathematical relations between forces and fluxes are usually linear when the system is not too far from equilibrium and the forces are small. Certain pairs of forces and fluxes are naturally interrelated, for example temperature gradient and heat flow; such pairs of forces and fluxes are said to be conjugate. When only one type of force and its conjugate flux are present in a material, the conductance $\boldsymbol{L}$ that relates $\boldsymbol{X}$ to $\boldsymbol{J}$ is defined by

$$\boldsymbol{J}_i = \boldsymbol{L}_{ij}\boldsymbol{X}_j \quad (i, j = 1, 2, 3) \qquad (10)$$

In Eqn. (10), $\boldsymbol{L}$ is a second-rank tensor that relates the two vectors $\boldsymbol{J}$ and $\boldsymbol{X}$, i and j represent spatial coordinates and summation of the repeated index j is implied (Einstein convention).

The situation is made more complex when more than one type of force is present, giving rise to corresponding conjugate fluxes and the possibility of cross-coupling of nonconjugate forces and fluxes. When there is a tensor relation between conjugate forces and fluxes, the component of each flux is in the general case a linear function of all the components of each force. For n different fluxes, this gives rise to equations of the form

$$\boldsymbol{J}_i = \boldsymbol{L}_{ij}\boldsymbol{X}_j \quad (i, j = 1, 2, \ldots, 3n) \qquad (11)$$

The set of equations represented by Eqn. (11) is considerably simplified when conjugate forces and fluxes are parallel, as in an isotropic material. In this instance, if two forces represented by vectors $\boldsymbol{X}_1$ and $\boldsymbol{X}_2$ are parallel and present simultaneously, both will in the general case make a contribution to each of the respective conjugate fluxes $\boldsymbol{J}_1$ and $\boldsymbol{J}_2$. That is,

$$\boldsymbol{J}_1 = L_{11}\boldsymbol{X}_1 + L_{12}\boldsymbol{X}_2 \qquad (12)$$

and

$$\boldsymbol{J}_2 = L_{21}\boldsymbol{X}_1 + L_{22}\boldsymbol{X}_2 \qquad (13)$$

where the conductances L_{11} and L_{22} relate conjugate forces and fluxes, and L_{12} and L_{21} relate nonconjugate variables. For example, equations of the type of Eqns. (12) and (13) can be used to express the fluxes of heat and electric current that arise due to the simultaneous presence of a temperature gradient and an electric field.

Principles of irreversible thermodynamics (Prigogine 1967, de Groot and Mazur 1962) have been applied to work out relations between coupling coefficients such as L_{12} and L_{21} in Eqns. (12) and (13). Onsager's principle states that provided the forces and fluxes are chosen properly, $\boldsymbol{L}_{ij} = \boldsymbol{L}_{ji}$. The necessary conditions for proper selection of force–flux pairs so that Onsager's principle is obeyed are not yet known. Some of the difficulties and inconsistencies encountered in the application of principles of irreversible thermodynamics to the theory of interface kinetics during the solidification have been elaborated by Baker and Cahn (1971).

3. Other Properties

The properties of materials discussed above are quantitative in the sense that there is a formal mathematical relation that defines the property. Using the term more loosely, there are many other "properties" of materials that, although not meeting the rigorous criteria of the equilibrium or transport properties, are nonetheless essential for determining the suitability of a material for a specific application.

The usefulness of a specific material for a given application can be indicated by "functional" properties that provide a relative measure of a material's performance under certain conditions. For example, tests have been developed to assess the durability of dental materials in service. Other functional properties of materials include flame retardancy, the ease of applying durable protective coatings like paint, the weldability of metals and the degree of toxicity on exposure through inhalation.

Some materials properties are "collective" in the sense that they may represent the resultant behavior of several individual properties in combination. Consider the biocompatibility of materials implanted in the human body. Among the factors determining biocompatibility for this application will be low corrosion rate, durability of joints between the implant and body tissues and minimal toxic reations between the implant and the body.

Some collective properties are not normally quantified, but could nevertheless be important for a material's appeal to consumers. For textiles, consumer appeal could depend collectively on fabric color, texture and moisture absorbancy. Wood for fine furniture is often selected on the basis of "figure," which refers to characteristic structures in the grain that are particularly appealing to the eye.

4. Directionality of Properties

Some properties of materials are inherently nondirectional. Since the mass and the volume of a material can be measured without reference to direction, their ratio, density, is nondirectional. On the other hand, thermal conductivity is determined by the relation between the temperature gradient (a vector) and the heat flux (a vector). For many materials the thermal conductivity depends on the direction of measurement, and is thus a directional property. The fundamental reason for the directionality of properties of materials is the fact that the densities and periodicities of the structural units that make up the material can vary with direction. The foundation of this relationship between structural symmetry and anisotropy of properties is known as Neumann's principle (e.g., see Nye 1957, pp. 20–24).

A material is said to be isotropic with respect to a certain property if the magnitude of the property is independent of the direction in which it is measured; otherwise, the material is anisotropic with respect to the property. There are a number of ways in which isotropic properties can arise. Symmetry constraints can require some properties to be isotropic as discussed below. All properties of disordered (noncrystalline) materials, like oxide glasses in the absence of any polarizing effects of drawing, are isotropic. It is also possible for some properties of a given material to be isotropic when in general, for reasons of symmetry, anisotropy is expected. Important examples are the elastic moduli of tungsten and aluminum, which are very nearly isotropic, although in principle elastic moduli of crystals are anisotropic.

Whether a property of a crystalline material is isotropic or anisotropic depends on both the property being considered and the symmetry of the structure of the material. Consider the electrical conductivity of a crystal as an example: the conductivity $\boldsymbol{\sigma}$ is in general a second-rank tensor relating the current density $\boldsymbol{J}$ to the electric field $\boldsymbol{E}$, according to

$$\boldsymbol{J}=\boldsymbol{\sigma}\boldsymbol{E} \tag{14}$$

which can also be written as

$$J_i=\sigma_{ij}E_j \tag{15}$$

where summation over $j=1-3$ is implied. It is evident from Eqn. (15) that $\boldsymbol{J}$ and $\boldsymbol{E}$ are not necessarily parallel.

Structural symmetry can impose constraints on the tensor coefficients of properties. For example, in all materials with cubic symmetry, properties describable by second-rank tensors like electrical conductivity are isotropic. Properties such as elastic constants which are fourth-rank tensors are generally anisotropic for all crystal structures. It is the combination of crystal symmetry and tensor rank of the property being considered that determines how many of the tensor coefficients are independent (Nye 1957, pp. 293–94).

It is important to note that tensors are also used to represent imposed "forces" on materials, such as a magnetic field (first-rank tensor) or a mechanical stress (second-rank tensor). These external forces and the tensors that represent them need not conform to the underlying symmetry of the material. Such tensors are called field tensors. Tensors that represent the physical properties of materials are called matter tensors.

In polycrystalline materials the structural anisotropy characteristic of the individual grains can be averaged out if the grains are oriented randomly. Randomly oriented polycrystalline materials therefore have isotropic properties and are said to be "quasi-isotropic." Many processing techniques, however, result in nonrandom grain orientations, or "textures," in crystalline materials. Polycrystalline materials may thus have anisotropic properties. Properties of polymeric materials may be anisotropic owing to the preferred alignment of structural entities.

5. Structure-Sensitive and Structure-Insensitive Properties

Broadly speaking, the "structure" of a material encompasses not only the ideal or perfect structure, but also imperfections in atomic arrangements, and the arrangements of various phases that make up the material. In an absolute or literal sense, no property of a material is completely independent of structure. Considering crystalline materials as an example, however, some properties change very little with small changes in the density and distribution of imperfections such as point defects and dislocations or even with appreciable changes in grain structure. Such properties are termed structure insensitive, and their values often depend primarily on the number of atoms of each type present in the material and the individual behavior of the atoms. Examples of structure-insensitive behavior are the small dependences of saturation magnetization of a ferromagnetic crystal on dislocation density, heat capacity of a polycrystalline material on its grain size and the compressibility of a two-phase material on the arrangement of the phases.

There are other properties of materials that can change dramatically with changes in the microstructure of the material. These are termed structure sensitive properties. For example, the electrical resistivity of a crystal at low temperatures is a sensitive function of the density of point defects. The initial permeability of a ferromagnet has a marked dependence on the sizes and arrangements of dispersed phases in a multiphase material. Permeation of a porous solid by a fluid depends not only on the volume fraction of pores, but also on their size, shape and connectivity. As a final example, the electrical conductivity of a semiconductor is extremely sensitive to dopant concentration.

6. History-Dependent Properties

Most of the properties discussed above are characteristics of materials having values that are dependent on the instantaneous values of the state variables for the material. There are other properties, however, that depend on the production history, or path, by which the material has been brought to a given state. Correspondingly, materials (and hence their properties) have been specified in terms of their origin or production process, for example wrought iron or Bessemer steel. It has been one of the goals of materials science and engineering to replace the older descriptive designations by developing methods of characterization in terms of composition, structural features and other measurable characteristics (see *Structure of Materials*).

A simple example of history-dependent behavior is provided by the direct-current magnetization curve for a ferromagnetic material (Fig. 2). The curve plots the magnetic flux density B vs the magnetizing force H for a material that is initially demagnetized and is then subjected to an increasingly positive magnetizing force (dashed curve), followed by an alternately decreasing and increasing magnetizing force (solid curve). It is seen in the figure that the response of the flux density is not a single-valued function of the magnetizing force, but depends on the path by which the material has been magnetized. The maximum value of B that is achievable is the saturation magnetization B_s. As the magnetizing force falls to zero, the flux density falls to B_r, the remanence. A reversed field of magnitude H_c, the coercive force, must be applied to result in zero flux density. If the magnetizing force is cycled repeatedly, the material will retrace the same B–H curve, thus achieving a dynamic "steady-state" behavior in which the values of B_r, H_c and B_s are constant.

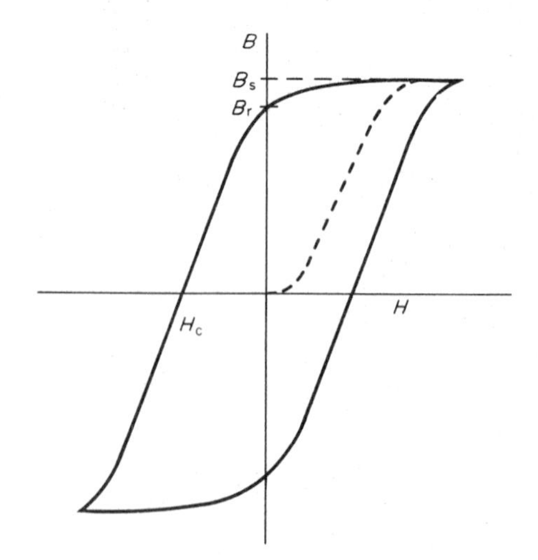

Figure 2
History-dependent behavior of a ferromagnet shown by a plot of magnetic flux density B as a function of magnetizing force H

The plastic deformation of a crystalline material can also show hysteretic behavior. Consider a well-annealed crystal cyclically deformed in tension and compression to a maximum total strain amplitude ε_T. Figure 3a shows a possible imposed straining cycle for the deformation. The resulting stress–time behavior is shown schematically in Fig. 3b. Figure 3c illustrates the combined stress–strain behavior for the cyclic deformation. Two phenomena are evident in Fig. 3c: first, the material undergoes cyclic hardening, because for a given strain the stress rises with each cycle; second, as the number of cycles becomes large, the material approaches a dynamic "steady-state" behavior in which the curves from one cycle to the next coincide. Such a steady-state behavior can be achieved for tests at elevated temperatures, provided that fracture does not preclude its occurrence.

Some properties of materials are measured by tests that are irreversible, causing structural changes to

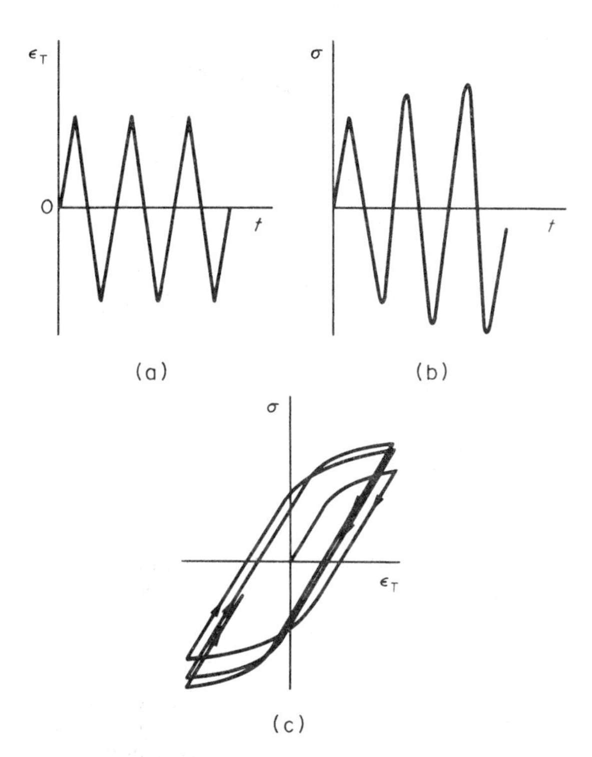

Figure 3
Mechanical hysteresis illustrated by the cyclic stress–strain behavior of a crystalline material: (a) total strain (elastic + plastic) amplitude ε_T vs time t; (b) stress σ vs time t; (c) resulting stress–strain behavior (after Landgraf R W 1970, in *Achievement of High Fatigue Resistance in Metals and Alloys*, ASTM STP 467, p. 4. © American Society for Testing and Materials, Philadelphia, Pennsylvania. Reproduced with permission)

occur in the material. Structure-sensitive properties can be altered by such tests. For example, the yield strength of a polycrystalline specimen tested in tension could be measured twice by recording its stress–strain behavior on loading, unloading and reloading, as illustrated in Fig. 4. In tests at low to intermediate temperatures, the yield strength determined on reload ing is higher. This is because the yield strength of a crystalline material is sensitive to the density of imperfections in the crystal structure, and the plastic deformation that occurs after the first yield point is reached results in a significant increase in the dislocation density in the material. In the second loading cycle, the yield strength is raised as a result of the work hardening that occurs during the plastic flow in the first loading cycle.

7. Properties and Engineering Design

Properties of materials are commonly defined, measured and compiled for use in the design of engineering

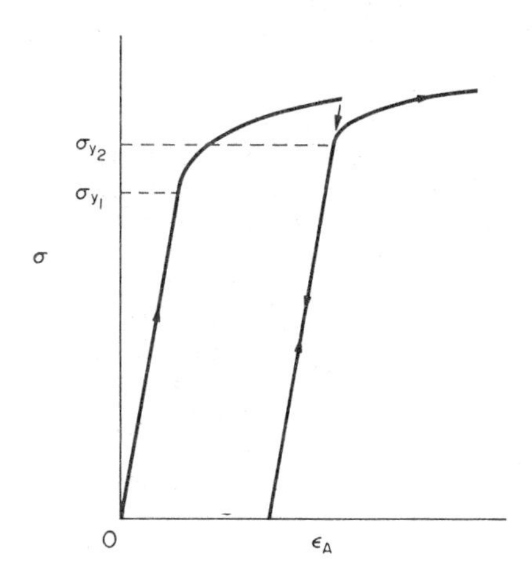

Figure 4
Stress σ vs accumulated strain ε_A behavior on repeated loading for a crystalline material

structures, devices and industrial products. The designer, both in the selection of materials and the design of specific components from those materials, must be guided by knowledge of the material's properties. Some properties of materials, such as the fracture strength of glass fibers, have a statistical distribution of values, and this fact must be taken into account in the design. Designers are also faced with the problem of extrapolating existing values of properties to specific engineering requirements which have no standardized tests for property determinations. Lacking such data, the designer must rely on experience, judgement and mathematical modelling to solve problems of this kind.

See also: Structure–Property Relations of Materials

Bibliography

Baker J C, Cahn J W 1971 Thermodynamics of solidification. In: *Solidification* 1971. American Society for Metals, Metals Park, Ohio, pp. 48–56

Billings A R 1969 *Tensor Properties of Materials—Generalized Compliance and Conductivity*. Wiley, London

Callen H B 1960 *Thermodynamics*. Wiley, New York, pp. 31–36

de Groot S R, Mazur P 1962 *Non-Equilibrium Thermodynamics*. North-Holland, Amsterdam

Mason W P 1966 *Crystal Physics of Interaction Processes*. Academic Press, New York

Nye J F 1957 *Physical Properties of Crystals*. Oxford University Press, Oxford

Prigogine I 1967 *Introduction to the Thermodynamics of Irreversible Processes*, 3rd edn. Interscience, New York

S. M. Allen and M. B. Bever
[Massachusetts Institute of Technology,
Cambridge, Massachusetts, USA]

R

Recycling of Demolition Wastes

Owing to the large volume and generally low specific value of demolition materials, the degree to which they are recycled depends not only on where and in what condition and quantity they are found, but also on the materials themselves. In this article, the origin of demolition materials is first discussed because of its strong effect on the potential for recycling. The approximate flows of different materials from different sources are given, and then the manner in which the principal materials are recycled is reviewed. The present data come from US and European studies, but the proportions of the materials involved should be broadly applicable to the demolition of similar structures elsewhere.

The origins of demolition materials can be classified into two groups: pavements and structures. Pavements include highways, sidewalks, parking lots, malls, airport runways, quays and stockyards, where a hard, strong surface has been laid over the ground to withstand the passage of motor vehicles, people or animals. Structures include buildings, bridges, tunnels, ramps and walls. The distinction is significant because pavements are usually demolished only for repair, and not for a complete change of form and use, as is often the case for structures. As energy, materials and landfill space become relatively more expensive, there is an increasing tendency for pavement materials to be recycled in situ and be incorporated into the new pavement. The proportion of pavement materials being recycled in this way is, however, still small; it is almost certainly under 10% of available wastes.

In situ recycling is more usual with asphalt pavements, which can be lifted by travelling machines, crushed, screened, mixed with new binders, and pressed or rolled back into place. Concrete, whether from pavements or structures, can also be crushed and screened, and that fraction falling within certain size limits can be incorporated into new concrete in place of aggregate (Nixon 1976, Frondistou-Yannas and Ng 1977). The strength of concrete made in this way is typically reduced to 80% of that made with crushed stone, but is normally adequate for use in pavements. However, machinery capable of crushing large pieces of concrete is not portable enough to be made part of pavement-resurfacing equipment. The presence of steel reinforcing mesh in pavement concrete, or heavier reinforcing bars in most structural concrete, is a complicating factor that only a few plants are equipped to handle, and most demolition concrete is used as coarse-fill or base material. At present in the USA and the European Economic Community (EEC), the demand for this material is such that it can be dumped on site free of charge (Wilson et al. 1979, Environmental Resources Ltd. 1980). Crushed and graded concrete can be sold. Contractors wishing to dispose of mixed demolition wastes, particularly mixtures including wood and other degradable or combustible material, must often transport them considerable distances (typically 50 km in the neighborhood of large cities) and pay a tipping charge. There is thus an increasing incentive to recycle as large a proportion of demolition materials as possible.

Demolition materials from structures can be classified into those from residential and those from nonresidential buildings. In North America, residences are usually constructed either mostly of wood or mostly of masonry. The resulting demolition flow has slightly more wood than bricks by mass for the USA (see Fig. 1). It is expected that the proportions for Canada would be similar. In the EEC, the proportion of wood in demolition materials is almost negligible compared with brick (Environmental Resources Ltd. 1980). The ready availability of wood in the USA results in very little recycling of wood as lumber, fuel or chips for building board, despite the considerable

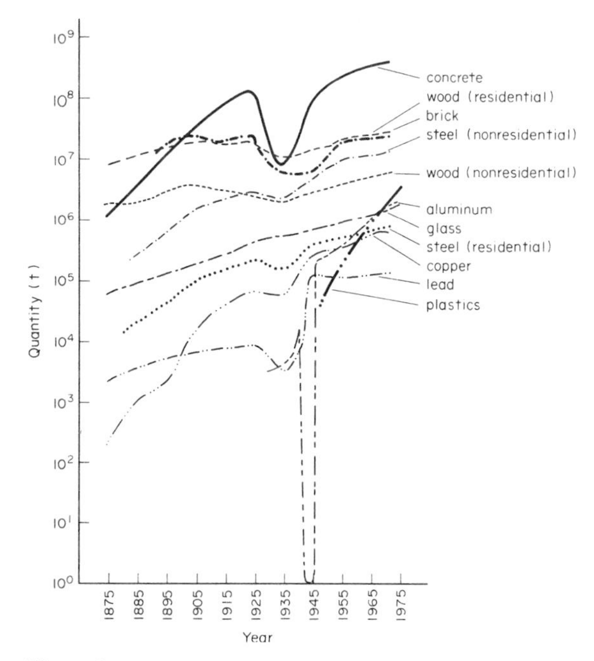

Figure 1
Quantities of materials used in building construction in the USA

proportion of wood in residential demolition materials. Companies formed to process demolition wood in the USA have ceased operations. In contrast, Mexican demolition firms operating near the US border have removed all lumber into Mexico (Wilson and Malarkey 1974).

In most countries, a high proportion of the unbroken bricks in demolition wastes is recycled. It is still profitable in the USA for contractors to employ people to remove mortar from bricks by hand, as some machines that were developed to clean bricks were found to be uneconomic (Wiesman and Wilson 1976).

Most of the "heavy" (i.e., thick-section) metals in demolition wastes, both ferrous and nonferrous, are recycled. This category includes copper pipe but not insulated copper wiring, although machinery has been developed to strip insulation from wiring. However, while there is always a market for recycled copper, brass, lead and aluminum which will repay the contractor's work in separating these metals from the overall waste stream, scrap prices can fall low enough in times of recession to make the recycling of steel and all but the heaviest nonferrous sections uneconomic. At these times the larger contractors often stockpile steel beams, for example, until market conditions improve.

Other demolition materials such as glass, plaster and plastics are recycled only to a very small extent.

The quantities of materials which have been used in the USA in building construction (excluding pavement, bridges and tunnels) are shown in Fig. 1. The average age of demolished buildings has been approximately 60–80 years for the older cities and 40–45 years for the newer cities of the west and south. The resulting quantities of materials produced in the 1970s from building demolition are estimated for the USA in Table 1 and for the countries of the EEC in Table 2. Because of the frequent immediate reuse as base or fill, there are no good data for the quantities of pavement (especially highway) demolition materials.

Table 1
Estimated quantities of materials from the demolition of buildings in the USA (mean values for the 1970s)

Material	Quantity (kt year^{-1})	Percentage share
Steel and iron	477.4	1.57
Aluminum	3.0	0.01
Copper	15.2	0.05
Lead	18.2	0.06
Glass	100.3	0.33
Brick and clay	4564	15.01
Concrete	19255	66.33
Wood	5792	19.64
Plastics	0.9	<0.01
Total	30405	100

Table 2
Estimated quantities of materials from the demolition of buildings in the EEC (mean values for the 1970s)

Material	Quantity (Mt year^{-1})	Percentage share
Masonry	40.85	57
Concrete	26.52	37
Wood	1.43	2
Steel	0.22	0.3
Other	2.65	3.7
Total	71.67	100.0

See also: Recycling of Metals: Technology

Bibliography

Davidson T A, Wilson D G 1982 US building-demolition wastes: Quantities and potential for resource recovery. *Conserv. Recycling* 5: 113–32

Environmental Resources Ltd. 1980 *Demolition Waste.* Construction Press, New York

Frondistou-Yannas S A, Ng H T S 1977 *Use of Concrete Demolition Wastes as Aggregates in Areas that have Suffered Destruction: A Feasibility Study*, NSF Report AEN 75-23221. National Science Foundation, Washington, DC

Nixon P J 1976 The use of materials from demolition in construction. *Resour. Policy* 2: 276–83

Saylak D, Gallaway B M, Epps J A 1976 Recycling old asphalt–concrete pavements. *Proc. 5th Mineral Waste Utilization Symp.* Illinois Institute of Technology Research Institute, Chicago, pp. 16–26

Wiesman R M, Wilson D G 1976 *An Investigation of the Potential for Resource Recovery from Demolition Wastes*, NSF Report AEN 75–14197. National Science Foundation, Washington, DC

Wilson D G, Davidson T A, Ng H T S 1979 *Demolition Wastes: Data Collection and Separation Studies*, NSF Report RA-790407. National Science Foundation, Washington, DC

Wilson D G, Malarkey J M 1974 *The Resource Potential of Demolition Debris*, NSF Report 74-SP-0889. National Science Foundation, Washington, DC

D. G. Wilson
[Massachusetts Institute of Technology, Cambridge, Massachusetts, USA]

Recycling of Metals: Technology

Collection and processing of scrap are the first two steps in recycling. Scrap can be classified into three categories: (a) home, revert or run around; (b) prompt industrial; and (c) postconsumer or obsolete. Purchased scrap consists of the last two categories. Since home scrap is recycled within the plant where it is generated and requires no additional processing prior to melting, only purchased scrap will be discussed. Prompt industrial scrap is usually new, clean material generated during an industrial operation (e.g., trimmings which result from the stamping of automobile fenders) or reject products (e.g., defective cans). Prompt industrial scrap is usually sold by the generator and may be purchased by either scrap processors or final consumers. Obsolete scrap can come from a variety of sources (e.g., discarded automobiles, outdated electronic gear, old buildings, railroad cars and ships).

Since scrap comes from a variety of sources in many different forms, it must be processed to facilitate efficient use. The primary roles of the scrap processor are to collect, accumulate, sort, grade, prepare, market and distribute scrap. Sorting is done following identification of the scrap. Preparation may involve any of a number of activities, such as shearing, baling or briquetting.

1. Ferrous Scrap

Ferrous scrap supplied a slowly increasing proportion of the iron consumed in the USA between 1976 and 1981 (Table 1). Almost all of this scrap was used to produce steel and cast iron. Small amounts have been sold for copper cementation (about 450 kt per year) and for production of ferroalloys.

Ferrous scrap is usually identified and sorted according to source or prior use. As ferrous scrap is low in value compared with nonferrous scrap, stainless steel and specialty alloy scrap, only simple identification techniques are economic. When mixed scrap is received, it is sold as such and commands an appropriately lower price.

Processing of ferrous scrap depends on the requirements of the ultimate consumer and the form in which it is received by the scrap processor. Oxyacetylene torches are used to cut up obsolete railroad cars and ships, structural steel from buildings, and other very large objects. Shears reduce large pieces of scrap (e.g., structural members) to a dimension suitable for furnace charging, usually 1–1.5 m long. Automobiles pressed into logs also are sheared. Two types of shears are used: guillotine and alligator shears. The guillotine shear is the larger of the two, consisting of a movable head which is hydraulically forced past a stationary anvil. The scrap to be cut rests on and protrudes beyond the anvil. Alligator shears function like giant scissors, except only the upper part is movable.

Loose, light-gauge scrap is frequently baled or briquetted. Examples of this type of material include sheet remaining after a stamping or punching operation, and turnings (metal removed in a machining operation). In addition, entire automobiles and appliances are sometimes baled. Balers consist of hydraulic cylinders that compress the scrap by pushing it against fixed walls of a cube. Steel briquetters have a pair of rotating drums with a series of cavities around the periphery of each drum into which the scrap is compressed. The scrap (e.g., turnings) assumes the shape and dimensions of the cavities.

Shredders are used mostly for preparing light scrap (e.g., automobiles and large home appliances). They consist of a series of steel hammers mounted on a rotating drum. The hammers pass by a series of fixed anvils. Feeding the car, appliance or other scrap between the hammers causes it to be quickly ripped into fist-sized chunks. Magnetic separation removes the ferrous metal from nonferrous metals and nonmetallic materials such as glass, rubber, plastics and dirt. The resulting ferrous product has the advantages of increased purity and improved consistency, handling, transportation and melting characteristics. In 1980 there were an estimated 200 automobile shredders operating in the USA with a capacity of more than 13 Mt per year.

The ferrous fraction of municipal solid waste presents a unique case. This material may be sold directly to steel mills for blending with other scrap materials, to a detinner, or to a scrap processor. The cans must be cleaned of paper and food waste by the seller or scrap processor prior to detinning. The detinner will leach (dissolve) the tin away from the steel using an $NaOH–NaNO_3$ solution. This process also removes any aluminum present. The resultant ferrous scrap has considerably increased value, and the tin can be recovered from solution by electrolysis.

Lower purity scrap usually is restricted to applications with less stringent chemical and mechanical property specifications. These include the manufacture of reinforcing bars and manhole covers. High-purity scrap can be used to make any grade of steel or cast iron.

Table 1
US consumption of ferrous scrap

Year	Consumption (kt)	Percentage of total ferrous consumption
1976	37590	31
1977	39040	34
1978	37860	36
1979	42680	35
1980	36590	37
1981	37460	35

2. Copper

Copper-bearing scrap materials are among the oldest used by man, and indications are that these materials were recycled by the earliest civilizations. Extensive copper recycling has continued up to the present time, when copper-bearing scrap accounted for almost 50% of all copper raw materials in 1979 and 1980 (Table 2). Depending on the purity of the scrap and the melting technique employed, using scrap to produce copper requires only 3–40% of the energy needed to produce pure copper from ore.

Copper-bearing scrap is divided into a number of grades. No. 1 copper scrap must be unalloyed, as it is used directly to make wirebars. No. 2 copper scrap is refined electrolytically and therefore may contain some impurities, though a copper content of at least 94–96% is normally required. Light copper also passes through an electrolytic refinery; its copper content must exceed 92%. Refinery-grade brass must contain at least 61.3% copper, while copper-bearing material generally contains less than 60% copper. These last two grades normally are melted in a blast furnace and blown in a converter prior to electrolytic refining.

Copper-bearing scrap can be used by tube mills, refiners, brass mills, ingot makers and specialty companies. The end use determines the preparation of the scrap by the processor. As tube mills do not refine electrolytically, metallic contamination of scrap to be used by tube mills is unacceptable. However, small amounts of nonmetallic contaminants may be acceptable, depending on the user's ability to fire refine. Refiners can produce wirebar directly from No. 1 copper scrap, but No. 2 and light copper scrap must first be cast into anodes and refined electrolytically before it can be used for this purpose. Brass mills usually use No. 1 copper and brass scrap in their melts in combination with pure copper.

The composition of brass scrap must be known and kept within narrow limits so as to keep the product made from the scrap within composition specifications. Brass and bronze ingot makers supply alloys of known, fixed composition to foundries. Hence, they can and do melt a wide variety of copper-bearing scrap. Specialty uses for copper-bearing scrap include chemicals, electroplating, castings (both copper and gray-iron), copper–aluminum alloying and steelmaking.

Prompt industrial copper-bearing scrap, which accounts for 60–70% of recycled copper, is generated by a variety of industrial operations. For example, scrap turnings from brass screw machines amount to up to 65% of the brass feed. Other sources include stamping plants, wire-drawing rejects, die shavings and short rod wire lengths. Postconsumer scrap includes: metal in surplus and obsolete items from communication, rail, electronics, power utility and defense operations; metal in disused automobiles and ships; and metal in spent shell casings.

Identification may be made using the material's color, spectroscopy, spot testing with acids, chemical analysis or, in some cases, magnetism. Highly trained and experienced sorters are required for the first three methods. Together with the shape of scrap and its past use, the color of a clean surface may be all that is needed to determine the alloy type. Spectroscopy offers only a semiquantitative analysis of an alloy, but this information is often sufficient for sorting scrap into alloy groups. Spot testing involves observing the reaction, or lack thereof, between various reagents and the metal.

A variety of techniques is used to prepare copper-bearing scrap materials by the scrap processor. Traditional processes include shearing to the size required by the customer, briquetting or crushing of thin and lightweight materials (e.g., turnings and borings), magnetic separation to remove iron, cleaning and degreasing.

One new technique is chopping of copper wire scrap. Large, stiff pieces of wire are presheared prior to chopping. Bales of wire must be broken down and are usually cut up with an alligator shear. The major steps then consist of chopping in three stages, with magnetic separation to remove ferrous contaminants following the first stage and sizing following the second stage. The third stage handles unliberated wire from the air separation process which follows sizing. Both the coarse and intermediate size fractions are processed by air separation to produce four fractions: (a) clean metal, (b) liberated metal with some insulation contamination, (c) unliberated wire, and (d) clean insulation.

Shredded automobiles also yield copper-bearing scrap. Following magnetic recovery of the ferrous metals, the nonferrous rejects can be processed by a number of techniques. The nonmetallics can be removed by air separation and water elutriation, and/or heavy media separation. The copper-bearing metals can then be hand picked after the aluminum and zinc are floated in a heavy medium.

Cryogenics (i.e., cooling scrap to temperatures as low as $-185\,^{\circ}\mathrm{C}$) can be used to recover copper from items such as small motors, transformers, generators, and their components. When exposed to these very low temperatures, certain materials, including steel and plastics, become brittle, whereas copper remains malleable. Therefore, crushing of an item immediately after cooling will allow relatively large pieces of copper to be separated from the other, finer materials by screening.

Table 2
US consumption of copper scrap

Year	Consumption (kt)	Percentage of total copper consumption
1976	1139	41
1977	1085	40
1978	1247	41
1979	1553	47
1980	1427	47
1981	1418	45

3. Aluminum

The increase in aluminum recycling has been dramatic. In 1960, recovery of secondary aluminum was about 300 kt but by 1976 that figure had increased to over 1 Mt, accounting for 21% of total aluminum production (Table 3). Recycling during the period 1976–1980 fluctuated slightly, ranging from 20–22% of total aluminum consumption. However, in 1981 the rate rose to 34% as production decreased and secondary recovery increased.

Prompt industrial aluminum scrap is produced by a variety of industrial operations, such as foil rolling, canmaking, airplane manufacturing and wire drawing. It is usually kept sorted and moves in bulk to the scrap processor or recycler. Prompt industrial scrap is, therefore, easy to handle.

Major suppliers and potential sources of obsolete scrap include industry, government and municipal solid waste. Products from which aluminum scrap is obtained are automobiles, household appliances and furniture, used cans and containers, trucks, aircraft, boats and buildings.

Scrap sorted by alloy class brings the highest prices, so the first step in scrap processing is identification. The first step in the identification process is to consider the nature (e.g., sheet, plate, forging, casting) and sources of the scrap. If more information is needed, a spectroscope may be used. In the case of castings, chemical spot tests are used to determine whether the alloy is based on aluminum, magnesium or zinc.

Once the scrap has been sorted, one or more processing techniques may be employed to prepare it for shipment. Borings and shredded material may be shipped loose for bulk handling and unloading. Other types of scrap, such as clippings and sheets, usually require baling or briquetting. Large pieces of aluminum are often sheared to a size specified by the customer.

In addition to the technology mentioned above, there are five other processes for recovering aluminum: wire chopping, shredding, aluminum-can recycling, sweating and dross processing. Depending on the type of wire metal, wire chopping recoveries are usually in the range 55–65%. The technology is similar to that described for copper. However, air classification of the coarse and intermediate aluminum wire size fractions gives three products, not four fractions. These products are clean metal, plastic and middlings (a mixture of some metal with insulation). The middlings are sent to a tertiary chopper or granulator to separate the metal from the insulation. The chopped middlings are then recycled to join the product of the second-stage chopper before sizing. Wire chopping produces a well-accepted product containing about 99.5% aluminum.

Most shredded aluminum scrap is obtained from three sources: shredded automobiles, scrap cans and aluminum shredded for handling in the scrapyard. For every tonne of shredded steel scrap produced from disused automobiles, about 13.5 kg of aluminum is recovered. With the potential recovery in 1979 being about 135 kt of aluminum from the 10 Mt of shredded automobile scrap, actual recovery was approximately 100 kt. Given projections for increased use of aluminum in cars, these numbers should increase.

Following automobile shredding, the material undergoes magnetic separation. A number of techniques exist to recover the aluminum from the nonmagnetic shredded material. Initially, the non-metallic materials can be recovered by air classification, water elutriation, eddy current, or a friction sliding board to produce a heavy fraction. Alternatively, a sink–float technique, in which the density of a heavy medium is controlled between that of the metallic and nonmetallic materials, can be used. Finally, a second sink–float tank, using a medium or higher density, can be used to float the aluminum and zinc from the other metals. Final separation is made by hand picking or in a sweat furnace.

Scrap aluminum cans come from both prompt industrial and postconsumer sources. Collections of cans from postconsumer sources exceeded 165 000 t in 1979, about 30% of the aluminum used for cans that year. There are three levels of aluminum-can recycling: (a) minimal, with a scale and sorting table; (b) intermediate, with scales, magnetic sorters and balers or flatteners; and (c) full-scale, with shredders, material handling and truck or rail-car loading equipment in addition to (b). New separation processes (e.g., the eddy current linear-induction motor) are the key to recovering aluminum from municipal solid waste.

Sweating is the process of separating materials based on differences in their melting points. In this process the scrap is heated to a temperature between the melting points of the higher and lower melting point materials. Sweating should be used only to recover aluminum from contaminated scrap. This scrap should: (a) be free of contamination with other low-melting metals (e.g., zinc and magnesium, which should have been sweated off first); (b) be at least

Table 3
US recovery of aluminum scrap

Year	Amount recovered (kt)	Percentage of total aluminum consumption
1976	1049	21
1977	1154	22
1978	1201	22
1979	1272	22
1980	1261	21
1981	1502	34

6 mm thick so that the particles will melt into droplets large enough to flow in the furnace; and (c) contain at least 30% (preferably 50–60%) aluminum. Ingots produced from the sweated aluminum may not contain more than 1% iron, 1% magnesium and 1–3% zinc depending upon requirements.

There are three kinds of sweating furnaces: (a) direct-fired sloping hearth, (b) direct-fired rotary, and (c) indirectly fired rotary. The direct-fired sloping hearth is the original style of sweating furnace. It consists of a refractory-brick lined box with a sloping floor and a tap hole at its lowest point. The aluminum in the scrap charge is melted and then runs down the sloping hearth to the tap hole and into an ingot mold or collection chamber. The scrap on the hearth must be manually raked and rabbled to free any trapped aluminum which cannot flow off the hearth. Manual raking is unnecessary if a rotary hearth furnace is employed. These may be heated directly, with a burner firing into a kiln from the discharge end, or indirectly, with a ceramic-lined, stainless steel tube rotating in a temperature-controlled furnace. Both styles of rotary furnaces are inclined, with weep holes at their lower (discharge) end. The aluminum runs down the hearth through the weep holes and is collected, while the remaining solid material passes over the weep holes and is discharged.

Dross recovery is the separation of metallic aluminum from the nonmetallic residues produced whenever aluminum of any type is melted. The dross is a mixture of aluminum oxide, dirt, spent fluxes and entrained aluminum. The aluminum content, which depends on the care that was taken to prevent oxidation in and after removal from the furnace, can range from zero up to 70%. When the aluminum content is sufficiently high, usually in excess of 40–45%, the dross may be sent directly to a rotary furnace where it is heated along with a flux consisting of equal proportions of NaCl and KCl, and up to 5% cryolite. However, the majority of drosses are milled and screened at about 20 mesh (0.84 mm) with the oversize processed in the rotary furnace and the undersize discarded or sold to the steel industry for use as hot-topping compound.

Wrought aluminum scrap of sufficient purity can be sold directly to a primary smelter where it is used to make new alloy. Economics are the primary incentive for the dramatic increase in this practice since the mid-1960s. For example, construction of an aluminum reduction plant with a capacity of 73 kt per year would cost about 250 million US dollars, whereas construction of the same size plant producing can stock from can scrap would cost about 16–20 million US dollars (1983). In addition, recycling aluminum scrap requires only about 5% of the energy needed to produce new metal.

Most secondary aluminum smelters produce recycled secondary ingot from both wrought and cast scrap (i.e., an ingot with a known composition). This material is used primarily for aluminum castings. In some cases notch bar or shot is produced for use as a deoxidizer or alloy addition. Scrap (e.g., from wire chopping) can also be used in these applications. Other possible uses for aluminum from wire chopping include production of powder, pharmaceuticals, and paints and coatings.

4. Zinc

The USA is the world's largest consumer of zinc. From 1976 to 1981 the fraction of domestic consumption met from scrap increased from 24 to 29% (Table 4). One reason for the relatively low level of zinc recycling is that about 45% of zinc consumption is used dissipatively (e.g., for galvanizing).

Zinc is recovered primarily as alloy scrap, which may contain zinc, copper, aluminum or magnesium as the major constituent. Zinc is also recovered from skimmings, drosses and fumes. The major sources and uses of zinc scrap materials are presented in Table 5. Galvanizing provides the largest single source of scrap zinc, in the form of skimmings, drosses, fumes and spent fluxes. These zinc-containing wastes are sold directly to recyclers. Zinc die casters also produce a dross, as well as rejected castings. The availability of fragmentized die-cast scrap has grown markedly since the early 1970s as a result of the tremendous expansion of automobile and appliance shredding. However, owing to the smaller amount of zinc used in the automobile during the last few years as a result of increased use of plastics, future zinc recovery from automobile shredding is expected to diminish considerably.

The major role of the scrap processor is in the recovery of zinc from shredded automobiles and appliances. If this scrap zinc is recovered by mechanical means (e.g., flotation and/or screening) rather than by sweating, a number of problems are created for the consumer. If the zinc content of this material is too low, operation of the recycling furnaces is inefficient, resulting in higher production costs. Excessive aluminum and finely divided zinc in the scrap increases the amount of skimmings and therefore the loss of

Table 4
US consumption of secondary zinc

Year	Consumption (kt)	Percentage of total zinc consumption
1976	338	24
1977	330	24
1978	339	23
1979	370	27
1980	304	27
1981	305	29

Table 5
Major sources and uses of zinc scrap materials

Type of scrap	Sources	Uses
Fragmentized die-cast	nonmagnetic portion of crushed automobiles and appliances	distillation
New industrial die-cast	rejected castings from die cast operations	die-cast alloy distillation
Old die-cast	old, oxidized and contaminated accumulators, mostly from automobile wrecking	distillation
Die-cast skimmings	ashes, drosses from melting die-cast scrap	distillation chemicals
Aircraft forming dies	forming dies from aircraft and automobile industry	die-cast alloy distillation
Zinc skimmings	hot-dip galvanizing kettles	distillation chemicals (fluxes) agricultural compounds
Sal-ammoniac skimmings	hot-dip galvanizing kettles	chemicals (fluxes)
Zinc dross (top)	top of continuous galvanizing kettles	distillation
Zinc dross (bottom)	bottom of continuous and hot-dip galvanizing kettles	distillation
Brass ingot makers fume	baghouse collectors	chemicals (agricultural compounds)
New zinc, painted, oxidized, corroded	printing plates, battery casings, addressograph plates, punchings from rolled zinc	remelt zinc brass ingot makers galvanizing
Old zinc, painted, oxidized, corroded	roofing, boat anodes, jar tops, spent plating anodes	remelt zinc distillation
Prime die casters dross	die-casters kettle (high metallic skim)	die-cast alloy distillation
Remelt die-cast (blocks)	sweating iron-containing and fragmented die cast	distillation

zinc. The recovery of zinc is decreased by the presence of glass, rubber, rocks and trash. The degree of purity required is determined by agreement between the scrap processor and the consumer.

5. *Magnesium*

Consumption of magnesium-based scrap for 1976–1981 is shown in Table 6. In 1980, consumption of magnesium-based scrap increased significantly both in absolute terms, and as a percentage of total magnesium consumption.

Magnesium, scrap or primary, has both nonstructural and structural applications. Nonstructural uses are as alloying agents, as a reducing agent for the production of other metals, as an addition to nodular cast iron, as anodes for the cathodic protection of steel structures and as a scavenger for sulfur in blast-furnace iron destined for primary steel production. Except when used as an alloying agent, the composition of magnesium scrap does not need to be closely controlled. However, when magnesium scrap is used for structural purposes, composition of the scrap is critical. For example, high-grade casting alloys may contain no more than 0.05% copper and 0.05% nickel, with some alloys limited to 0.002% nickel. Therefore, scrap processors must take great care to remove these metals or their alloys from the magnesium scrap. Removal is usually done manually.

Because magnesium and aluminum are very similar in appearance, sorting by visual identification of these two metals, or their alloys, is extremely difficult. A reliable method of distinguishing between magnesium and aluminum is to apply a 5% silver nitrate solution to a scratched surface of the metal in question. The presence of magnesium is indicated by the surface turning black. Aluminum is not affected.

Table 6
US consumption of magnesium scrap[a]

Year	Consumption (t)	Percentage of total magnesium consumption
1976	7570	8
1977	7825	8
1978	9220	9
1979	8900	9
1980	10175	12
1981	7643	9

[a] Not including magnesium in aluminum-based alloy scrap or magnesium used to make aluminum-based alloys

Magnesium-bearing wastes, such as skimmings, generally have not been recycled owing to the low prices offered by magnesium smelters for them.

6. *Lead*

More than half the lead consumed in the USA is recycled from scrap (Table 7). In 1980, 86% of lead scrap was obsolete and 83% of the obsolete scrap was obtained from recycled batteries.

Table 7
US consumption of lead scrap

Year	Consumption (kt)	Percentage of total lead consumption
1976	659	49
1977	758	53
1978	769	54
1979	801	59
1980	676	63
1981	641	55

Lead scrap is of three types: whole battery scrap (both prompt and depleted batteries); prompt industrial scrap (such as drosses, skimmings and lugs); and other scrap (such as cable sheathings, bearings, babbits and solders). Of these types, only batteries require significant processing to recover the lead.

Batteries are processed to drain the contained sulfuric acid and to liberate the lead. Four techniques are used to accomplish this, of which two, sawing and guillotining, are similar. Both remove the top of the battery, including the posts and connectors of the battery section, from the rest of the case and the plates. The remainder of the battery is hit against a striker bar to allow plates, separators and acid to fall into a conveyor. Acid drains off the conveyor into a sump where it is neutralized with lime or ammonia. The battery top is crushed and the metallic posts and connectors are separated from the crushed casing material by air classification. A more recent concept in processing scrap batteries is to break the whole battery and separate the battery plates from the casing, followed by heavy media and hydrometallurgical techniques to separate the grid metal from the paste. An alternative method involves breaking the battery to drain the acid, then charging the entire battery to the smelting furnace.

Although about 60–65% of the lead used in batteries is recycled, the same is not true of lead used in other products. For example, only about 10% of the lead used in solder, 25% of the lead consumed by the cable industry and 30% of the lead in bearings is recycled. The reasons for this lower recycling rate are diverse. Most solder applications are dissipative (i.e., the solder is used in very small quantities bonded to other metals). Cable sheathing, while easy to recover, has an average life of 40 years. Recovery of lead in bearings frequently requires disassembly of a large piece of machinery, which is uneconomic.

Virtually all lead scrap is sold to a secondary smelter and/or refiner. The resulting product finds its principal use in the production of new batteries. However, the resulting lead also may be used to produce other alloys for type metals, solders, shot and bearing metals.

7. Nickel Alloy and Stainless Steel Scraps

Although treated together because of the importance of nickel, nickel alloys and stainless steels are fundamentally different. Scrap nickel alloys can be used in stainless steel production. However, because of the high iron content, stainless steel scrap cannot be used as a feed material for making nickel alloys. Table 8 presents data on the rate of nickel recycling.

Nickel alloys are usually classified as follows.

(a) Superalloys possess high strength at high temperatures. A typical application of these alloys, such as the Inconels, Hastelloys and Waspaloys, is in gas turbine engines. Nickel content can vary from 20 to 80%. Most of the prompt industrial scrap is generated by the aerospace industry in the form of turnings, rejected turbine components, and casting stubs and sprues. A major source of obsolete scrap is jet aircraft engines.

(b) High-nickel alloys and heating alloys are used in corrosion- or oxidation-resistant applications. Nickel contents vary from 20 to 100%. As usual, prompt industrial scrap results from metal forming and fabrication. Obsolete scrap is recovered from discarded industrial equipment.

(c) Pure nickel scrap is recovered as depleted anodes from electroplating. This represents only a small portion of the nickel available for recycling.

(d) Other nickel-bearing wastes include Monels, small amounts of special alloys, grinding swarfs, sludges and spent catalysts. Their importance is not sufficiently significant to warrant further treatment.

Stainless steel alloys are classified into three broad categories: 200, 300 and 400 series. As the 400 series contains only chromium it will not be treated. The 300 series contains about 18% chromium and about 8% nickel. Other elements such as niobium and molybdenum may also be present. Applications of the 300 series stainless steels are found in kitchen utensils,

Table 8
US consumption of nickel scrap

Year	Consumption (kt)	Percentage of total nickel consumption
1976	47.4	22
1977	45.8	23
1978	40.1	16
1979	52.1	25
1980	44.8	22
1981	41.4	27

hospital equipment, chemical plants and the aerospace industry. In the 200 series, 5–10% manganese is added to the alloy to promote the formation of austenite. Because nickel serves the same function, the nickel content can be reduced to under 4%.

The users of superalloys and stainless steels have rigid specifications for these alloys, and so reliable segregation and identification of this type of alloy scrap is essential to its recycling and composition. The source or form of the scrap is often a good indication of its identity. In addition, four relatively simple procedures (magnetic testing, spark testing, chemical spot testing and x ray fluorescence) can be used for initial identification. If more reliable information or a guaranteed composition is required, a sample is taken for chemical analysis.

The magnet test is conducted with a small permanent magnet. Unless it is used to separate two alloys of known composition, it can provide only an approximate or initial classification of alloys. Spark testing can be used to differentiate among iron- and nickel-based alloys. However, the reliability of this procedure depends strongly on the skill of the operator. Spot tests include susceptibilty to acid attack and reaction of test reagents to form specific colors.

Following identification of the scrap, it is processed for sale. Processing steps may include cleaning and or sizing. Turnings are usually crushed in a ring-type or hammer mill crusher and then sampled for chemical analysis. Superalloy scrap is degreased with trichloroethylene, screened to remove fines and processed through a magnetic separator. The screened fines are sold to a nickel refinery, while the prepared turnings are sold to a superalloy manufacturer or a stainless steel mill or are exported, depending on their composition. Stainless steel scrap may be crushed and, if series 200 or 300, passed through a magnetic separator, but degreasing is rarely economic and screening is never performed.

Superalloy solids are cut with a plasma torch, and the surface is cleaned by sand blasting. With the exception of cleaning the surface by a combination of steel and grit blasting, solid, stainless steel scrap is handled in an identical manner to solid, plain, carbon-steel scrap.

Stainless steel dusts and grinding swarfs can be pelletized or briquetted prior to charging back to the steelmaking furnace. Alternatively, service companies convert this material to alloy ingots of known composition suitable for charging to the steelmaking furnace.

8. Titanium

Titanium alloys are classified as alpha, beta, and alpha–beta. Alpha alloys include unalloyed metal or low-alloyed titanium grades. Typical alpha alloys contain aluminum and tin or palladium. Of the beta alloys, the alloy Ti–13V–11Cr–3Al was widely used during the 1960s, but has since been displaced by Ti–8V–4Mo–6Cr–4Zn–3Al. The alpha–beta alloys are the most important, with a production volume comparable to that of the alpha and beta alloys combined. Typical alpha–beta alloys are Ti–6Al–4V, Ti–4Al–3Mo–1V and Ti–8Mn.

Physically, titanium scrap is classified as chips (light) and solids (heavy). This material is recycled to the smelter in the ratio 1:2. The processing method employed by scrap processors is determined by the physical properties of the scrap.

Table 9 presents data on the tonnage and degree of recycling of titanium from 1976 to 1981. Note that 5% of total ingot weight becomes home scrap in the form of croppings, collars and turnings. About 86% of total ingot weight ends up as prompt industrial scrap. Because of restrictions on contamination, only about 35% of total ingot weight is recycled to the titanium-melting furnace. The remainder is used for alloying in the steel and aluminum industry, exported, or is lost as grinding and other wastes. A small amount of obsolete scrap, mostly from aircraft maintenance, is also recycled.

Titanium scrap must meet very strict standards for purity, especially for aerospace applications, because there is very little refining action during melting. Contaminants that must be handled by processing include surface oxides resulting from exposure to high temperatures during use, hydrocarbons from metalworking lubricants and pieces of cutting tool bits (usually tungsten carbide).

Chips are crushed in a hammermill to improve their handling and their packing density for shipping and subsequent briquetting. The crushed scrap is then solvent-vapor degreased in trichloroethylene. This also removes tramp impurities and any foreign metal pieces entrained in the grease. Next, magnetic separation is used to extract any ferrous or ferromagnetic impurities that are present. A representative sample is then taken for chemical analysis. Care shold be taken in handling and storing turnings as they are easily flammable, particularly if they are light and oily.

Table 9
US consumption of titanium scrap[a]

Year	Consumption (t)	Percentage of total titanium consumption[b]
1976	8360	41
1977	9890	40
1978	11190	38
1979	12700	37
1980	13990	36
1981	13420	34

[a] Excludes titanium scrap used as an alloying addition [b] Total consumption includes only scrap and sponge; pigment is excluded

Solids are subjected to minimal processing by the scrap processor. Following magnetic separation each piece is analyzed by x ray spectroscopy, segregated by grade and shipped to a melter. The melter cuts the heavy scrap with a plasma torch, descales the titanium by abrasion, further cleans the surface with caustic soda and pickles the surface with nitric acid containing some hydrofluoric acid.

9. Precious Metals

This section covers gold, silver and platinum-group metals recycling. With the exception of periods, such as early 1980, when the price of precious metals was exceptionally high, most secondary precious metal bearing materials have been obtained from industry. Once the secondary precious metals bearing scrap is put into a concentrated or segregated form, it is sold to an appropriate refinery where it is converted into pure metal or alloys identical to those obtained from primary sources. Consumption of gold, silver and platinum-group metals scrap is presented in Tables 10, 11 and 12, respectively.

Table 10
US consumption of gold scrap

Year	Consumption (kg)	Percentage of total gold consumption
1976	33250	23
1977	32380	21
1978	43090	29
1979	52150	35
1980	67990	68
1981	49460	57

Table 11
US consumption of silver scrap

Year	Consumption (t)	Percentage of total silver consumption
1976	3216	60
1977	3101	65
1978	2436	49
1979	2410	49
1980	3697	95
1981	2606	72

Table 12
US consumption of platinum-group metals scrap

Year	Consumption (kg)	Percentage of total platinum consumption
1976	6704	13
1977	6077	12
1978	6007	11
1979	9620	11
1980	10305	15
1981	12182	12

Precious metals are obtained from a variety of secondary sources. Film users, such as hospitals and printers, generate scrap film and hyposolutions from which silver is recovered. Electrical and electronic components may contain gold, silver and/or platinum-group metals (e.g., as contacts or solders). Catalytic converters from automobiles contain platinum and palladium, although converters on newer models also contain rhodium. Precious-metals scrap also is generated during plating and jewelry manufacture, and by scientific and dental laboratories. Aircraft and aerospace equipment contain precious metals in sparkplugs and brazing alloys.

Precious metals are sometimes found in conjunction with base metals, most frequently in an assembled part (e.g., electrical equipment and automobile catalytic converters). The function of the scrap processor is to remove as much of the base metals and nonmetals as possible.

Complicated devices may be chopped or shredded to produce a fairly uniform material. This material may be subjected to magnetic separation, screening, air classification, and/or incineration, to produce precious metal concentrates. When appropriate, hand dismantling may be practised.

The material is sampled and analyzed prior to shipment. Preparation for sampling would be by grinding to fine powder or by melting and either taking a dip sample or pouring into an ingot from which saw cuttings or drill turnings may be obtained.

Automobile catalytic converters are removed from junked cars and then sold for further processing. The processor opens the stainless steel containers and removes the platinum–palladium covered alumina balls or honeycomb that constitute the catalyst. The stainless steel is sold to a stainless steel producer and the catalyst is sold to a precious-metals refiner specializing in platinum-group metals.

See also: Materials Conservation: Technology; Materials Economics, Policy and Management

Bibliography

Alter H 1982 Toward a national policy for secondary material symposium on resource recovery and environmental issues of industrial solid wastes. *Resources and Conservation.* Elsevier, Amsterdam

Arthur D Little Inc. 1979 *Proposed Industrial Recovered Materials Utilization Targets for the Metals and Metals Products Industry.* US Department of Energy, Washington, DC

Battelle Columbus Laboratories 1972 *Identification of Opportunities for Increased Recycling of Ferrous Solid Waste.* National Technical Information Service, Springfield, Virginia
Battelle Columbus Laboratories 1977 *A Study to Identify Opportunities for Increased Solid Waste Utilization: I. General; II. Metals.* National Technical Information Service, Springfield, Virginia
Carrillo F V, Hibpshman M H, Rosenkranz R D 1974 *Recovery of Secondary Copper and Zinc in the United States.* US Bureau of Mines, Washington, DC
Fine P, Rasher H W, Wakesberg S (eds.) 1973 *Operations in the Nonferrous Scrap Metal Industry Today.* National Association of Recycling Industries, New York
Kusik C L, Kenehan C B 1978 *Energy Use Patterns For Metal Recycling.* US Bureau of Mines, Washington, DC
Rheimers G W, Rholl S A, Dahlby A B 1976 Density separations of nonferrous scrap metals with magnetic fluids. In: *Proc. 5th Mineral Waste Utilization Symposium.* US Bureau of Mines, Washington, DC
Siebert D L 1970 *Impact of Technology on the Commercial Secondary Aluminum Industry.* US Bureau of Mines, Washington, DC
Spendlove M J 1961 *Methods for Producing Secondary Copper.* US Bureau of Mines, Washington, DC
US Bureau of Mines 1980 *Minerals Yearbook 1980.* US Bureau of Mines, Washington, DC
US Bureau of Mines 1981 *Mineral Commodity Summaries.* USB, Washington, DC
US National Association of Recycling Industries 1975 *Recycled Lead in the United States: A Study of Current Market Trends and Projections.* USNARI, New York
US National Association of Recycling Industries 1977 *Recycling of Nickel Alloys and Stainless Steel Scrap.* USNARI, New York
US National Association of Recycling Industries 1977 *Wire and Cable Chopping and Recovery of Metals from Shredder Operations.* USNARI, New York
US National Association of Recycling Industries 1978 *Recycling of Zinc.* USNARI, New York
US National Association of Recycling Industries 1979 *Recycling Precious Metals.* USNARI, New York
US National Association of Recycling Industries 1980 *Recycling Copper and Brass.* USNARI, New York
US National Association of Recycling Industries 1981 *Recycling Aluminum.* USNARI, New York
US National Association of Recycling Industries 1982 *Recycled Metals in the 1980s.* USNARI, New York

R. S. Kaplan
[US Bureau of Mines, Washington, DC, USA]

H. Ness
[National Association of Recycling Industries, New York, USA]

Residuals Management

A decent quality environment is an important element in the economic and social development of all societies, rich and poor alike. Environmental degradation can affect the well-being of people in two ways: directly through exposure to pollutants and contaminants which can result in both acute and chronic health effects and in reduced recreational and other opportunities; and indirectly through damages to life-support systems, to water and food supplies, and to marketed goods and services. An important public policy issue for all nations is to strike an appropriate balance between levels of environmental quality and levels of other public and private goods and services.

Protection of the environment requires the same resources—human, land and capital—that could be put to productive use in other sectors of the economy. Trade-offs between levels of environmental quality and levels of material well-being must be made by all nations regardless of the political–economic systems used to produce and allocate goods and services. Within the environmental sector, further trade-offs must be made between the efficient use of resources employed to protect the environment and equity. That is, are the desired levels of environmental quality being purchased at least cost in terms of other opportunities foregone? Who benefits from improvements in environmental quality, and by how much? What are the costs of these improvements, and who pays for them? Are the costs and benefits of pollution control distributed equitably?

In this article, environmental pollution and its control are viewed from the perspective of a public management problem. In designing government policies for pollution control, three primary functions are distinguished. Firstly, an institutional and legal framework within which pollution control is to be conducted must be established. Secondly, environmental goals and targets—usually expressed in the form of ambient standards—must be adopted. Finally, a management strategy for achieving these goals and targets within the institutional and legal framework that has been established must be identified, adopted and implemented.

The intent of this article is to provide a brief overview of residuals management in the context of environmental pollution and its control. First, the nature of residuals is discussed. Next, the major elements of a residuals management system are delineated. Thirdly, alternative procedures for establishing environmental targets are presented. In Sects. 4 and 5, the steps to be taken in preparing a regional management plan for pollution control, steps to be taken in implementing the plan, and assessment of the plan in terms of achieving the environmental goals and targets established by government are discussed.

1. The Nature of Residuals

Residuals—sometimes referred to as wastes—are generated at virtually every stage in the production and consumption of goods and services. An example for iron and steel production is shown in Fig. 1. Residuals are of two basic types, material and energy. Most

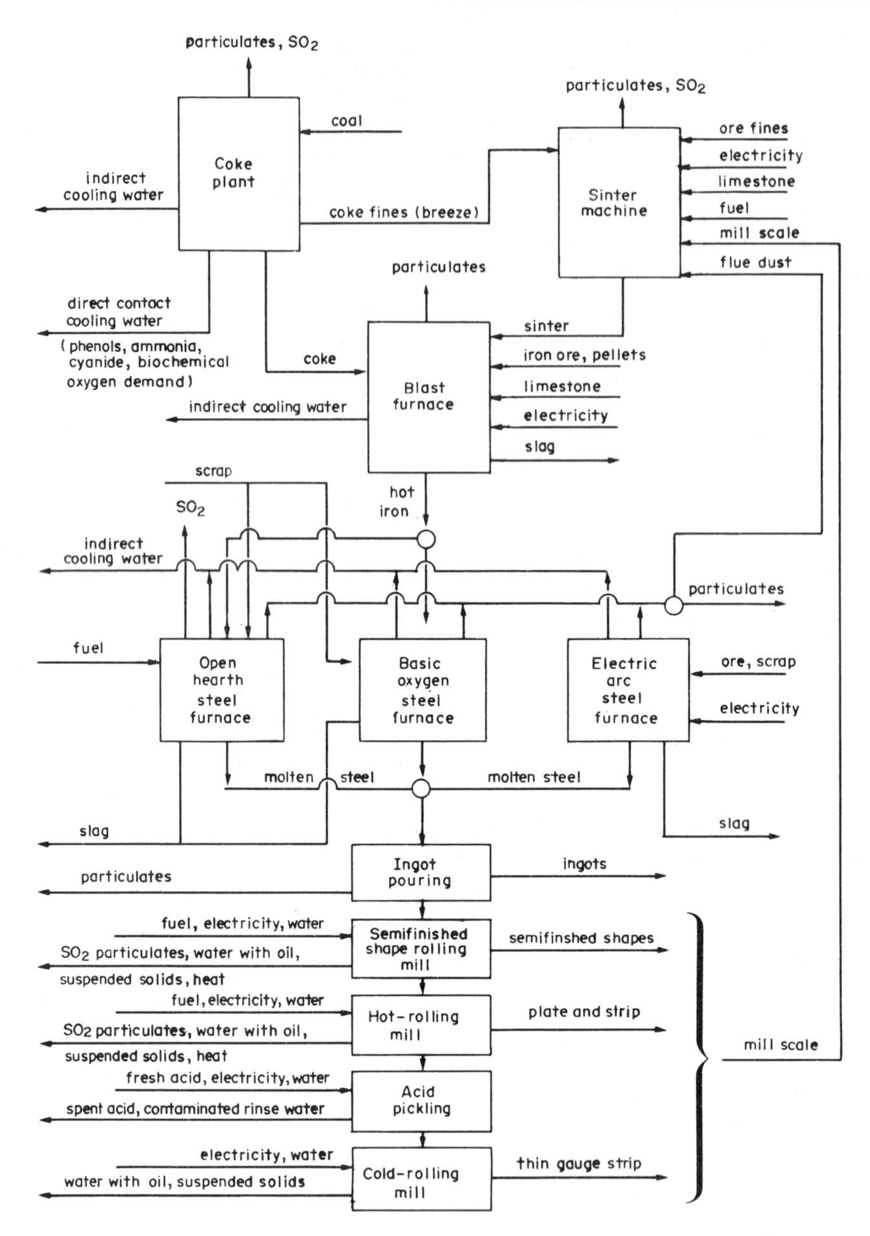

Figure 1
Schematic representation of iron and steel production indicating the generation and discharge of liquid, gaseous and solid residuals (after Russell and Vaughan 1976. © Johns Hopkins University Press, Baltimore. Reproduced with permission)

material residuals are readily transformed from one form to another in production, consumption and treatment activities. Examples include: solids to gases as in the combustion of municipal refuse and sewage sludge; gases to liquids as in the wet scrubbing of gaseous effluent streams containing SO_2; and gases to solids as in the electrostatic precipitation of gas-stack particulates. Changes in the form of residuals, however, often involve inputs of energy and the generation of additional residuals.

In the context of residuals management, the principle of mass conservation is important. Application of this principle in the analysis of environmental problems is referred to by some environmental economists as the materials–balance approach. Residuals can be changed from one form to another, but the total weight to be handled and disposed of does not change. (In some cases, it increases.) Consequently, solids are sometimes referred to as the irreducible limiting form of residuals. By the application of appropriate equipment and energy, all undesirable substances can be removed from water and air streams (except CO_2 which may be harmful in the long term). As restrictions on the discharge of gaseous and liquid residuals to the atmosphere, water courses and land become more stringent, problems of handling greater and greater quantities of bottom ash, sludge and solid wastes can be expected to gain in importance. This has already been demonstrated in countries with well-established environmental protection programs. Both air and water pollution can be reduced, and in some cases eliminated entirely, but the solid residuals problem cannot be eliminated without total materials recovery and reuse.

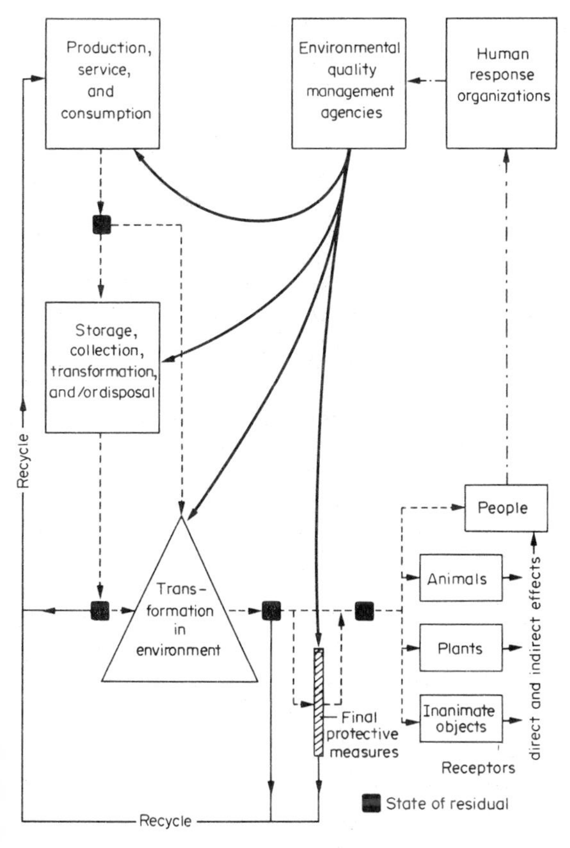

Figure 2
Schematic representation of an environmental quality management system: —— government action, -- flow of residuals,—·— information flow (from Resources for the Future 1968 *Annual Report*. © Resources for the Future, Washington, DC. Reproduced with permission)

The quantities of material residuals disposed of in the environment may be reduced by: (a) decreasing the quantities of residuals generated in production and consumption activities, and (b) increasing levels of materials recovery and reuse through recycling, by-product production and reclamation.

The generation of residuals in production and consumption activities can be reduced by: changing the nature of intermediate and final products; reducing levels of production of a particular product; using substitute raw material inputs; employing different technologies of production and consumption; and increasing the useful lifetimes of consumer goods, such as cars and household appliances.

The extent of materials recovery and reuse is a function of economic factors and public and private restrictions and incentives. The economic feasibility of materials recovery and reuse depends on: the demand for final products, the relative prices of virgin and secondary materials, the availability of a consistent supply of secondary materials of a specified quality and quantity, processing and reprocessing technologies, and product specifications. The demand for secondary materials is highly sensitive to changes in the demand for final products. The output of the secondary materials industry typically expands during periods of shortages of competitive virgin raw materials. The most notable example of this occurred during World War II.

Of the nonproduct outputs generated in production and consumption activities, those that are not recovered and reused must be disposed of, usually in the natural environment. The ability of the environment to assimilate residuals is a valuable resource. Complete elimination of all discharges of residuals to the environment would impose unnecessary cost burdens on society. However, if no restrictions were placed on the use of the environment to assimilate residuals, it would become overused, as most nations have already experienced. Controlling discharges of residuals to the environment in the most cost effective way is the essence of residuals management.

2. *Residuals Management System*

A conceptual framework for residuals management that has found considerable use in the study of pollution problems is shown in Fig. 2. This framework is applicable to residuals management in all countries. It is useful for delineating (a) the range of structural and nonstructural alternatives available to society for reducing the damaging effects of residuals in the environment, and (b) the range of policy instruments available to government for reducing discharges of residuals from individual production and consumption activities.

The management system is conceived as consisting of flows of materials and energy from production and consumption activities (reflecting decisions about production processes, raw-material input mixes, recycling and reclamation opportunities, and residuals treatment and disposal alternatives), through the environment where residuals are stored, diluted, transported, and physically, chemically and biologically transformed, to receptors—humans, animals, plants, and inanimate objects—where damages occur. Some damages are experienced by humans directly and some indirectly. In some cases, final protective measures, such as air filtration, water treatment, and noise barriers are used to reduce the impacts on receptors of pollutants and contaminants in the environment.

The flows of residuals and the damages which result from them give rise to corresponding flows of information from people who are affected by pollution to response organizations, such as conservation groups and other defenders of the environment, and local and national governments. The concerns of people ultimately reach the environmental quality management agencies that have jurisdiction. These agencies, in turn, take whatever actions seem desirable or are required by law. These actions may include requiring certain levels of treatment, issuing discharge permits, imposing effluent charges, subsidizing recycling operations, modifying the assimilative capacity of the environment, or installing measures to protect people from environmental pollution, such as water treatment. Whatever actions such agencies take will modify the damages caused by pollutants and contaminants in the environment.

3. Establishing Environmental Targets

Various procedures can be used to establish target levels of environmental quality for a region or nation. The procedure used in a given management context depends on the institutional and legal framework that has been adopted for pollution control. The three most commonly used procedures are: (a) cost–benefit analysis, where ambient standards derive directly from a comparison of the costs and benefits of increments of pollution control; (b) cost-effectiveness analysis and public deliberation, where ambient standards derive indirectly from a consensus on the willingness to pay for different levels of environmental quality; and (c) human health, ecological, or aesthetic considerations where ambient standards (and in some cases effluent standards) are based on ethical, legal, or political concerns rather than on costs.

Ideally, the objective of an environmental quality management program for a given region or river basin should be to maximize the value of expenditures in pollution control. This requires, in principle, that the benefits and costs of alternative pollution control strategies be estimated, and that the benefits and costs of increments of pollution control be compared. The most efficient pollution control strategy for the region as a whole is the one in which the net benefits (the benefits minus the costs) are maximized, or alternatively the one in which the sum of pollution control costs and environmental damages is minimized. This analysis simultaneously defines the optimal levels of the ambient standards and the optimal levels of investment in pollution control.

Although widely used in assessing other public investment projects, cost–benefit analysis has not found extensive use in environmental quality management because of problems in valuing certain kinds of benefits. Not only is it difficult to assess the physical, chemical and biological impacts of alternative pollution control strategies, but even where these impacts can be assessed with a reasonable degree of accuracy, it is often not possible to assign unambiguous values to reductions in them. This is particularly true of human health and aesthetic impacts and damages.

In cases where the benefits of environmental improvements cannot be estimated with a reasonable degree of accuracy, the next most useful information for purposes of environmental decision making is the cost of achieving alternative levels of environmental quality. Given this information, a region or nation can determine the level of environmental quality it is willing to pay for in terms of other opportunities foregone. This process involves setting ambient standards in light of the resources that will be required for pollution control. The damages associated with pollution and the benefits associated with pollution control are dealt with subjectively, in a qualitative sense, rather than objectively as is required in cost–benefit analysis. This approach to establishing ambient standards is often used, especially for those situations where the damages are primarily aesthetic.

The third procedure that has been used to establish acceptable levels of environmental quality for a region or nation is based on health, ecological, or aesthetic considerations independent of estimates of either the costs or the benefits of pollution control. Costs, to the extent that they are reflected at all in decisions, are not considered until after the pollution control program has been implemented and they are about to be incurred. At this point, some adjustments to the ambient standards based on the actual costs of achieving them may be made. For countries with well established environmental protection programs, such as Europe, North America, Australia and Japan, this is the approach that has been used to select target levels of environmental quality.

4. Regional Management Plans

4.1 Preparation

After agreement on levels of the ambient standards for a region or nation has been reached, the next step is to identify alternative residuals management strategies that can meet these standards. A management strategy for a region consists of specifying: (a) the structural and nonstructural measures that are both capable of meeting the ambient standards and available for use, and (b) regulations and/or economic incentives that will ensure that these measures will be adopted. The results of this analysis provide the basis of a management plan for pollution control in the region. A management plan consists of a management strategy, technical guidelines and administrative procedures for implementing the management strategy, and a compliance schedule for pollution control.

A large number of alternatives is available to any region for meeting the ambient standards. However, they will differ in total cost to the region, in the distribution of costs among dischargers in the region, and in other pertinent characteristics. An important consideration in residuals management is the cost of pollution control.

The potential cost savings to a region of the least-cost strategy can be substantial. Experience has shown, for example, that the total cost of meeting ambient standards by requiring a uniform percent reduction in discharges at individual sources is typically two to five times more than the cost of the regional least-cost strategy. Alternatively, in situations where there are severe limitations on the availability of capital for pollution control, as is the case in most developing nations, substantially more progress can be made in protecting the environment if the available resources are used as productively as possible. The cost savings associated with the least-cost strategy are due partly to differences in the costs of pollution control at individual discharge activities and partly to the relative locations of dischargers and of areas of low environmental quality in the region.

A residuals management strategy which may be highly desirable from the perspective of least cost to the region, however, may not be feasible because of the distribution of pollution control costs or because of concerns with equity more generally. With the least-cost strategy, dischargers who can treat their discharges most cheaply are asked to reduce them proportionally more than those whose treatment costs are greater. In a number of countries, particularly those with market economies, the least-cost strategy has not been popular because of the inability to arrange an appropriate institutional mechanism for dividing the potential cost savings among the dischargers in the region in an acceptable manner.

Because the regional least-cost strategy is often considered inequitable and also because it is generally difficult to administer, more equitable and more easily administered management strategies should also be identified and their costs assessed. Uses of uniform effluent standards, uniform effluent charges and effluent standards based on best available technology are examples of strategies that are generally considered to be more equitable and more easily administered than the least-cost strategy. Management strategies based on the use of marketable discharge permits are particularly promising because of their desirable features with respect to both the total costs of pollution control and the distribution of these costs among the dischargers in the region.

After the regional analyses have been completed, and a residuals management strategy for the region adopted, there are two final steps in the preparation of a regional management plan. Guidelines for establishing discharge standards and for issuing discharge permits, if they are to be used, and guidelines for using economic incentives, such as subsidies for construction, effluent charges, and marketable discharge permits, if they are to be employed, must be prepared. Finally, a compliance schedule for the completion of treatment facilities, for meeting permitted discharges at individual activities, or for implementing an incentive system for discharge reduction must be established. The management plan should provide for both population and economic growth, and for the additional pollution associated with this growth, for a planning period of at least 15–20 years into the future.

4.2 Implementation

Having developed a management plan for pollution control in the region that prescribes the allowable discharge levels at all sources, a pollution-control compliance schedule for individual dischargers, and the administrative steps to be taken to attain the desired levels of control, the next step is implementation of the plan.

Implementation involves three major activities: (a) construction of pollution control facilities, (b) operation of pollution control facilities, and (c) enforcement of pollution control laws by government. For those countries with well established pollution control programs, implementation has tended to focus more on the design and construction of pollution control facilities than on the operation and maintenance of the facilities once constructed. Examples exist in all countries of pollution control facilities not operating properly, or indeed at all, to reduce operating costs. Without adequate enforcement with a focus on emissions rather than on installed equipment, the management plan will not fulfill its purpose.

Enforcement requires well-trained inspectors, ambient and effluent monitoring equipment, and laboratory support, backed up with sufficiently severe penalties for offenders. The success of a pollution control program depends on frequent inspections of emissions.

5. Assessment of the Pollution Control Program

To ensure that the pollution control program achieves the environmental goals and targets established by government, it will be necessary to monitor the quality of the air, water and land throughout the region, including the toxic substance contents of plants, fish and human tissues. This information is used to ascertain whether or not the pollution control program is achieving its goals, and if not, what changes are needed in the regional management plan to ensure that these goals will be achieved.

See also: Materials and the Environment; Materials Economics, Policy and Management

Bibliography

Armstrong E L (ed.) 1976 *History of Public Works in the United States, 1776–1976.* American Public Works Association, Chicago, Illinois

Basta D J, Lounsbury J L, Bower B T 1978 *Analysis for Residuals–Environmental Quality Management: A Case Study of the Ljubljana Area of Yugoslavia.* Resources for the Future, Washington, DC

Bower B T (ed.) 1977 *Regional Residuals Environmental Quality Management Modeling.* Resources for the Future, Washington, DC

Bower B T, Löf G O G, Hearon W M 1971 Residuals management in the pulp and paper industry. *Nat. Resour. J.* 11: 605–23

Brown F L, Lebeck A O 1976 *Cars, Cans, and Dumps: Solutions for Rural Residuals.* Johns Hopkins University Press, Baltimore, Maryland

Freeman A M III 1979 *The Benefits of Environmental Improvement: Theory and Practice.* Johns Hopkins University Press, Baltimore, Maryland

Kneese A V, Ayers R U, d'Arge R C 1970 *Economics and the Environment: A Materials Balance Approach.* Johns Hopkins University Press, Baltimore, Maryland

Kneese A V, Bower B T 1968 *Managing Water Quality: Economics, Technology, Institutions.* Johns Hopkins University Press, Baltimore, Maryland

Kneese A V, Bower B T 1979 *Environmental Quality and Residuals Management: Report of a Research Program on Economic, Technological and Institutional Aspects.* Johns Hopkins University Press, Baltimore, Maryland

Mills E S (ed.) 1975 *Economic Analysis of Environmental Problems.* Columbia University Press, New York

Page T 1977 *Conservation and Economic Efficiency: An Approach to Materials Policy.* Johns Hopkins University Press, Baltimore, Maryland

Quimby T H E 1975 *Recycling, The Alternative to Disposal: A Case Study of the Potential for Increased Recycling of Newspapers and Corrugated Containers in the Washington Metropolitan Area.* Johns Hopkins University Press, Baltimore, Maryland

Russell C S 1973 *Residuals Management in Industry: A Case Study of Petroleum Refining.* Johns Hopkins University Press, Baltimore, Maryland

Russell C S, Vaughan W J 1976 *Steel Production: Processes, Products, and Residuals.* Johns Hopkins University Press, Baltimore, Maryland

Sawyer J W Jr 1974 *Automotive Scrap Recycling: Processes, Prices, and Prospects.* Johns Hopkins University Press, Baltimore, Maryland

Spofford W O Jr 1971 Solid residuals management: Some economic considerations. *Nat. Resour. J.* 11: 561–89

Spofford W O Jr 1973 Total environmental quality management models. In: Deininger R A (ed.) 1973 *Models for Environmental Pollution Control.* Ann Arbor Sciences Publishers, Ann Arbor, Michigan

Spofford W O Jr, Russell C S, Kelly R A 1976 *Environmental Quality Management: An Application to the Lower Delaware Valley.* Resources for the Future, Washington, DC

W. O. Spofford Jr.
[Resources for the Future, Washington, DC, USA]

Roofing Materials

The function of roofing is to provide shelter against the effects of the sun, rain, wind and other climatic conditions. An increasingly wide range of materials is used as roofing in two general categories of roof structures, depending on the manner in which penetration of water is prevented. Sloped roofs, conventionally used to shed water rapidly from structures, are covered with unit roofing materials arranged in an overlapping fashion to prevent water penetration through the joints. Flat or nearly flat roofs, a more recent development, contain rain and snow for longer periods of time than sloped roofs and are covered with continuous roofing materials which depend on durable waterproofing properties for their effectiveness. All roofing materials must be used in a compatible manner with supporting roof structures, and with adjoining insulation and retardant materials to control heat flow and water vapor flow between building interior and exterior environments.

1. Historical Development

The first sloping roofs developed to provide shelter from the elements made use of the watershedding properties of sloping overlapping branches, leaves and crude forms of thatch. Later, other locally available materials, such as slate and stone slabs, were found to be effective roof coverings when laid in an overlapping fashion on sloped structures. Where slate and stone were scarce, roofing tiles of clay, and later, wood shakes, shingles and metal roof coverings were developed. All of these materials continue to be used today for covering sloped roof construction. When laid, they act to prevent water penetration, despite voids or holes in the roofing at each lap joint, by sufficiently overlapping the unit materials and making the roof slopes steep enough. Thus, the gravitational forces which draw the water out of the joints and down the slopes exceed all other forces which would carry the water inward.

The modern flat roof covered with a continuous waterproof membrane was not introduced until the mid-nineteenth century. The first membranes were based on tar-saturated felts, initially pine tars and later coal tars, coal-tar pitches and asphalts. The felts were attached directly to wood and concrete decks using fasteners and bituminous materials between felt layers to achieve a continuous built-up roofing membrane capable of performing satisfactorily for twenty years or more. Following World War II, changes took place in building design and construction which affected the development of roofing technology. Larger industrial buildings with greater column spans and larger roof areas were constructed, resulting in greater thermal and structural stresses to the roofing membranes. Steel decks were used instead of concrete and wood decks, air conditioning of buildings was developed, and insulation was added between the decks and the membranes to control heat flows. The need developed in many buildings for vapor-retarding materials on the warm side of the insulation to control the occurrence of condensation on the underside of the membranes. Earlier installation practices of applying heavy quantities of asphalt to adhere the roofing membranes and the insulation and vapor-retardant materials were found to result in increased fire spread, while subsequent reductions in asphalt applications increased the exposure to wind uplift.

In the 1960s, modifications to existing materials and the development of new plastic and rubber roofing materials resulted in an increasing number of attempts to improve the performance of roofs. To distinguish them from bituminous built-up roofing systems which are often referred to as conventional systems, the new systems were termed nonconventional systems. These newly developed polymeric systems were designed to be lighter, to have fewer reinforcing fabrics or felts to provide strength, to be more elastic over a wider range of temperatures, and to be resistant to a wider variety of weather conditions, physical abuse and chemical attack. Covering a broad spectrum of materials with almost infinite variations, these materials were categorized as elastomeric systems which, after substantial deformation and release of a weak stress, quickly return to their approximate initial dimensions. The materials also were categorized as single-ply roofing because they were developed either as preformed sheet systems manufactured in factories or as liquid applied systems which solidify as a film in place by solvent or water evaporation or by chemical curing upon mixing. The wide spectrum of systems has had varying degrees of success and continues to be developed and modified.

2. Sloped Roofing Materials

2.1 Thatch

Thatch is generally considered to include any matter of vegetable origin, including grass, reed, straw, heather, sedge, leaves and similar materials depending on their local availability. Thatch roofs are designed to shed snow and rain rapidly from slopes of 50° or more. The reeds, straw or other thatch materials are arranged in parallel overlapping layers (20–40 cm thick) pointing down the slope of the roof. Water running down the roof is never expected to penetrate below the top few centimeters of the thatch materials. The bundled thatch material is securely attached along its full width to supporting rafters, purlins and roof battens by a variety of fixing techniques, including hooks, spars and twine. Thatch roofs generally have good heat and sound insulation but are combustible, although chemical solutions are available to improve the fire-resistant properties of the thatch materials. Properly maintained, thatch roofs can be serviceable for sixty years or more.

2.2 Slate

Slate is a strong, dense and durable material which is almost nonabsorbent. The weight of slate varies from 35 to 390 kg m^{-2}, and can be up to 5 cm thick. A slate roof requires a stronger supporting structure than a thatch roof. The slates are laid on roof slopes of at least 30° with minimum overlapping vertical layers of 7.5 cm. In manufacture, the quarried slate is split, resulting in variations in texture depending on its source. The manufactured slate is provided with at least two countersunk nail holes, and slating nails must be carefully driven during installation so as not to damage the overlapping slate. In addition to durability, a principal advantage of slate is its noncombustibility.

2.3 Clay Tile

Clay tile is a strong, durable and noncombustible material which is slightly more absorbent than slate. The weight of the tile varies from ~40 to 90 kg m^{-2}. Tiles are manufactured by forming plasticized shale and clay into a variety of shapes which are fired at high temperatures to develop desired weather-resistant properties. Tiles are laid on roof slopes of 45° and are interlocked by overlapping adjoining tiles vertically and horizontally. Owing to their brittleness, the tiles must be nailed snug but not tight. Among other tile types, concrete tile is manufactured from mixtures of portland cement, sand and water using high-pressure extrusion processes and high-pressure steam autoclave curing.

2.4 Wood Shakes and Shingles

These are principally manufactured from cedar or redwood having good durability. Shakes are hand or machine split from logs and are produced in various grades, thicknesses and lengths. Shingles are sawn from the logs into slightly tapered units and present a more uniform appearance than the irregularly split shakes. For long roof life, shakes and shingles need to dry quickly and are usually installed in slopes of at least 30°. They are attached to the underlying structure either with rust-resistant galvanized nails or with aluminum nails. Shakes and shingles are combustible but can be impregnated with fire-retardant materials to reduce combustibility.

2.5 Asphalt Shingles

Asphalt shingles are composed of a felt base impregnated with asphalt saturant and covered with asphalt coating in which mineral granules are embedded. The weight of the shingles varies from 5.9 to 18.6 kg m^{-2} depending on the coverage, with glass fabric shingles lighter than organic felt shingles. Galvanized nails are used to attach asphalt shingles to the roof deck. In high wind areas, shingle tabs are cemented (or otherwise fastened) to the deck or special lock-type shingles are used. Asphalt shingles have relatively good resistance to external fires, depending on their composition and weight. Asphalt roll roofing is similar to asphalt shingles but is provided in sheet form for applications of more general utility.

2.6 Metal Roof Coverings

These are manufactured in a variety of sizes and sections and include steel, aluminum, copper, lead and related alloys. Multimetals, in which two or more metals are joined in molecular bond by heat and pressure, are also used to obtain properties not available in a single metal. Steel roofing, usually available as corrugated steel, is given protection from corrosive elements during manufacture by galvanizing, terne coating with lead, enamelling or application of other protective coatings. Aluminum roofing is lighter and more resistant to atmospheric corrosion but its less favorable properties are its higher coefficient of expansion (twice that of steel) and its greater susceptibility to electrolytic corrosion and to pitting when subjected to salt spray in coastal areas. Among multimetals used on sloped roofs are terne-coated, copper-clad and brass-clad stainless steel. Attachment of the shaped metal roof coverings is made with neoprene washers and nails, clips, or other fasteners of the same metal or alloy as the roofing. Provision is made for expansion and contraction by use of several different types of seam or joint construction. Standing-seam construction uses narrow panels interconnected along their lengths by raised folded edges that move in expansion and contraction. Batten-seam construction includes a metal cap, or batten, over the raised seam. Flat-seam construction uses small sheets which are soldered on their four sides and which rely on the small amounts of expansion and contraction to be taken up by slight buckling in the center of each sheet. Corrugated metal-seam construction relies on the corrugations to absorb the thermal movements.

3. Bituminous Built-up Roofing

3.1 Hot Process Built-up Roofing

This is the largest category of roofing systems in use on modern flat roofs. The built-up roofing consists of plies or layers of roofing felts bonded together on the roof with hot bitumen to constitute a waterproof membrane. The guiding principle for the construction of flat roofs is to form a large watertight tray by turning up and sealing the roofing membrane at all corners, angles and roof projections so that the only water drainage from the tray is by scuppers or roof drains provided for the purpose.

Bitumen is a generic term which describes mixtures of complex hydrocarbons and which, for roofing materials, refers to asphalt and coal-tar pitch. In their solid state they appear alike and are characterized by low water permeability and good adhesion and cohesion. Asphalt is produced from petroleum distillation residues and, with refinement, comes in a range of viscosities and softening points allowing its use on roof slopes of up to 45°. Coal-tar pitch is a denser bitumen and is produced by distilling crude tars derived from the coking of coal. Coal-tar pitch has excellent cold-flow self-healing properties which limit its use to low roof slopes of less than 7.5°.

Roofing felts are mats of organic or inorganic fibers, saturated or coated with bitumen. Organic felts are manufactured from combinations of wood-fiber pulp, scrap paper and small amounts of rag. Inorganic felts are either asbestos felts, manufactured from asbestos mineral fibers, or glass-fiber mats. The manufacturing of organic and asbestos felts orients the fibers in the longitudinal machine or roll direction which makes the fabric stronger in that direction than across their width. Glass-fiber mats are laid up in jackstraw patterns and bonded with resin, resulting in tensile strengths more nearly the same in the two directions.

Surfacing materials are used to protect the bituminous waterproofing layer from solar rays and to provide a reflective surface to help reduce heat transfer. Surfacing materials generally include gravel, slag or crushed rock set in a flood coat of hot bitumen.

A wide variety of built-up roofing specifications, with different numbers and types of felt plies and layers of bitumen, is used in roofing construction depending on the underlying materials in the roof. The typical modern built-up roofing membrane consists of a coated base sheet nailed or embedded in bitumen on the substrate. Two to four additional layers of saturated felts are added, embedded in alternate layers of bitumen, with a heavier pouring of bitumen and surfacing material over the top felt. Application of the 91 cm wide rolls of felt is made with overlaps using bitumen temperatures which, at the point of application, provide proper viscosity and flow. Good installation practices require that the roofing materials be given protection from moisture, both during storage and during application, and that proper and complete adhesion between plies be obtained to prevent entrapment of air and vapor leading to blistering and other membrane problems. Properly applied, many conventional hot-applied built-up roofs have remained in service for 20 years or more.

3.2 Cold Process Built-up Roofing

This type of roofing uses solvent-thinned cutbacks and asphalt emulsions as adhesives which do not require heat to melt. The asphalt component in the cutbacks is dissolved in petroleum solvent which evaporates after application, causing the asphalt to become more viscous and to revert to its solid state. The asphalt emulsions solidify in place as the water in the emulsions evaporates during cure, leaving the bitumen and emulsifying agent as residue. Cold process roofing is used both as new membranes and as coatings and resaturants for reroofing or roofing repairs. Asphalt emulsions have either a bentonite-clay or a chemical emulsifying agent, with the clay type offering better weather resistance. Asphalt emulsions serve only as surfacing bitumen. Cutbacks serve as interply cement and are inferior to emulsions as surfacing bitumen. Cold process adhesives are not adaptable to aggregate surfacing and membranes are limited to smooth-surface roofs or cap sheets with roofing granules.

4. *Elastomeric Roofing*

Elastomeric roofing includes a wide variety of thermosetting and thermoplastic materials possessing different chemical, physical and mechanical properties. Thermosetting materials solidify when first heated under pressure and cannot be remelted or remolded without destroying their original characteristics. Thermoplastic materials, which will repeatedly soften when heated and harden when cooled, are not truly elastomeric materials and are separately categorized here, as are groupings for composite roofing materials and other nonconventional systems.

The elastomers consist of natural and synthetic rubbers and rubberlike materials. They are resistant to creep because of their elasticity and are tougher than thermoplastics. The vulcanized elastomers are the least affected by physical and chemical stress, are flexible at 10 °C and are virtually insoluble in petroleum distillates. In the factory, sheets are seamed together under pressure before vulcanization, to minimize the amount of field sealing required. The membranes can usually be adhered to the substrate or can be installed loose-laid with a covering of stone or other ballast weighing $\sim 60\ \mathrm{kg\,m^{-2}}$. They cannot be solvent or heat welded and field sealing typically is accomplished by contact adhesive.

The most commonly used elastomeric roofing materials are neoprene, ethylene–propylene–diene monomer and polyisobutylene.

Neoprene is the generic name for the polymers of chloroprene and exhibits good resistance to petroleum oils, solvents and weathering. It is available in liquid-applied systems and preformed sheet systems. Some formulations require a protective coating of chlorosulfonated polyethylene to retard weathering. Thickness varies from 30 to 120 mils (0.76–3.05 mm).

Ethylene–propylene–diene monomer (EPDM) is a synthesized elastomer which has good resistance to ozone and weathering, and has low water vapor and gas permeability. Although resistant to corrosive chemicals and dilute mineral acids, it is not resistant to petroleum oils and gasoline. Thickness varies from 30 to 60 mils (0.76–1.52 mm).

Polyisobutylene (PIB) is usually blended with EPDM and other synthetic rubbers and has good resistance to ozone and weathering. Although its resistance to dilute mineral acids and dilute solutions of alkalis is good, its resistance to oils and solvents is low. Thickness varies from 30 to 60 mils (0.76–1.52 mm).

5. *Thermoplastic Roofing*

Thermoplastic materials melt when heated and resolidify when cooled. At normal temperatures they are brittle unless blended with plasticizers to improve their flexibility. Chlorinated polymers improve fire resistance, solubility and control of creep. Thermoplastic sheets are sealed by heat welding or by solvent welding which fuses the seams of the somewhat soluble resin sheets in solvent cement.

Thermoplastic materials include polyethylene and polypropylene, both of which are difficult to bond and are not used as a principal material in roofing. The most commonly used thermoplastic roofing material is polyvinyl chloride. In addition, chlorinated polyethylene and chlorosulfonated polyethylene are produced as thermoplastics but these materials can be vulcanized and therefore are sometimes considered as semielastomerics.

Polyvinyl chloride (PVC), a thermoplastic polymer, exhibits good resistance to acids, alkalis and many chemicals. Through plasticizing and proper formulation, elastomeric properties and improved resistance to weathering are obtained. The loss of plasticizers by leaching, aging, solar exposure or other factors causes embrittlement and possible shrinkage. Ultraviolet inhibitors are added if the installed membrane is not to be protected from the sun by ballast. Thickness varies from 32 to 60 mils (0.81–1.52 mm) and reinforcement with nylon, fiberglass or polyester fabrics is added in some membranes to form composite sheets. Application can be loose-laid with ballast, fully adhered or mechanically fastened.

Chlorosulfonated polyethylene (CSPE) is a chlorinated polyethylene containing a small number of chlorosulfonyl groups and exhibits good resistance to ozone, heat and weathering. It also has good resistance to mineral oils and chemical attack but will swell in chlorinated solvents. The liquid-applied system is used as a coating on concrete and plywood decks and on foamed-in-place roofing, and the sheet material contains an asbestos backing to facilitate handling. Thickness is approximately 35 mils (0.89 mm) and application is by hot asphalt or contact adhesive.

6. Modified Bituminous Roofing

Modified bituminous roofing materials are composite systems which, in general, have more elongation than conventional bituminous membranes and use fewer plies. The modifications render the bitumen more flexible and less brittle at cold temperatures and also improve elasticity at high temperatures. The modifiers are generally macromolecules such as natural rubber or synthetic rubber. Because of their greater elongation, the membranes are reinforced with polyethylene films, nonwoven polypropylene and polyester mats, or similar materials with greater stretch properties than reinforcing mats used in conventional bituminous systems. Surfacing may be a thin embossed metal, special protective coating or gravel adhesive with mineral granules. The systems are adhered with adhesives or hot bitumen or in some applications are loose-laid and ballasted. Joint laps and adhesive bonding are accomplished with propane torches or with solvents or solvent-based adhesives. Thicknesses vary from 60 to 160 mils (1.52–4.06 mm), depending on the manufacturer.

7. Other Liquid-Applied Roofing

Liquid-applied roof coatings are used to provide a weathering skin on concrete and plywood decks, and on foamed-in-place roofing such as sprayed urethane foam. The coatings are generally applied by roller or airless spray equipment in two or three coats to give a minimum dry-film thickness of 20 to 30 mils (0.51–0.76 mm). In addition to the liquid-applied roofing materials already described, several other materials are commonly used.

Acrylics are single component, water based coatings that cure by evaporation of water and are the most widely used coatings.

Silicone coatings are available as single and two component types. The single component silicone cures by reacting with moisture vapor in the air following solvent evaporation. The two component silicone system cures by rapid chemical reaction.

Urethane coatings are also available as single and two component types. Again the single component types cure by reaction with moisture vapor in the air while the two component types cure by chemical reaction.

8. Relation of Roofing to the Completed Roof Structure

The wide range of available roofing materials must be used in a compatible manner with a variety of other components to complete a weathertight and durable roof structure. Flashings must be used at all roofing membrane terminations and penetrations to intercept and lead the water to drains. The manner and location of placement of components to control heat and moisture vapor flows between the outside and inside of roof structures impose different requirements on the component materials and affect the ultimate durability of the roofing.

Flashings are required to withstand differential structural and thermal movements which occur between roof membranes and independent adjoining building elements. Flashing materials include bituminous felts, glass-fiber fabrics, various metals and the newer polymer materials. Base flashings are made part of the roofing membrane to bridge the joints to the adjoining building elements. They are required to be flexible, compatible with the membrane and sag resistant where they turn up vertically. Metal counterflashings are attached to the adjoining building elements, and overlap and remain free of the base flashings. Control or movement joint flashings are deliberate discontinuities in membrane installations to allow for movements resulting from expansion and contraction.

Insulations are built into many roof structures to limit heat flow. Insulation materials include a variety of open-fibrous and closed cell materials which are produced in board form in a factory, or are mixed or foamed-in-place at the site. Their location in the completed roof system has an effect on the thermal and moisture conditions in other components. The usual location of insulation in flat roof construction is under the roofing membrane on top of the structural deck. This position subjects the membrane to wider extremes of daily and seasonal temperatures with consequent material stressing than would be experienced if no insulation were used. A new type of roofing assembly has become available which rearranges the normal elements of the roof system. Termed the protected membrane system, inverted roof or upside-down roof, the system places the roofing membrane directly on the structural deck and the preformed insulation boards outside the membrane. The extreme temperature variations experienced by the membrane are avoided in this configuration. The system requires that the insulation (usually polystyrene foam insulation boards) be non-absorptive of water, and that it be held in place against wind uplift and flotation by adhesive bonds or ballast.

Vapor retarders are incorporated into some roof structures of buildings experiencing moist air conditions or located in colder climates. Their purpose is to prevent moist inside air from reaching and condensing on cold roof surfaces, including insulation placed under the membrane. Vapor retarders are normally placed on the warm side of the insulation and are installed to be continuous, without gaps, and sealed at all joints and edges. Vapor retarders have low permeance to moisture vapor and include such materials as asphalt-saturated and coated roofing felts, and kraft-paper laminates. The use of hot-applied bituminous felts is not recommended as vapor retarders for steel roof decks because of poor resistance to internal fires.

The roof structure and deck must be designed to support all applied loads without failure. Typical modern flat roof decks, classified as nailable or non-nailable, are constructed of wood, metal, cast-in-place materials and preformed units. When selecting roofing materials, it is necessary to consider roof structure deflections and differential movements arising from loads, shrinkage, creep, moisture changes and thermal variations. Cracks and deflections, even if not structurally significant, may induce sufficient stress to tear bonded membranes and permit increased heat flow, and moisture and vapor migration. Slopes of all roofs should be designed to a minimum slope of 1.875° (1 in 48). Excessive deflections may allow ponding of water which can adversely affect the durability of the roofing materials.

Bibliography

American Society for Testing and Materials 1966 *Engineering Properties of Roofing Systems*, Special Publication No. 409. ASTM, Philadelphia, Pennsylvania

Baker M C 1980 *Roofs: Design, Application, Maintenance*. National Research Council of Canada (Multiscience), Montreal

Building Research Establishment 1970 *Built-up Felt Roofs*, BRE Digest No. 8. BRE, Watford, UK

Building Research Establishment 1979 *Flat Roof Design: The Technical Options*, BRE Digest No. 221. BRE, Watford, UK

Bushing H W, Mathey R G, Rossiter W J T, Cullen W 1978 *Effects of Moisture in Built-up Roofing—A State-of-the-Art Literature Survey*, NBS Technical Note 965. National Bureau of Standards, Washington, DC

Griffin C W Jr 1982 *Manual of Built-up Roof Systems*, 2nd edn. McGraw-Hill, New York

Handegord G O 1966 *Moisture Considerations in Roof Design*, Document No. CBD 73. National Research Council of Canada (Division of Building Research), Ottawa

Mathey R G, Cullen W C 1974 *Preliminary Performance Criteria for Bituminous Membrane Roofing*, NBS Building Science Series 55. National Bureau of Standards, Washington, DC

National Academy of Sciences/National Research Council 1974 *Roofing in Developing Countries*. NAS/NRC, Washington, DC

National Roofing Contractors Association 1981 *NRCA Roofing and Waterproofing Manual*. NRCA, Oak Park, Illinois

Rossiter W Jr, Mathey R 1979 *Elastomeric Roofing: A Survey*, Technical Note NBS 972. National Bureau of Standards, Washington, DC

Watson J A 1979 *Roofing Systems: Materials and Applications*. Reston Publishing Company, Reston, Virginia

J. D. Stelling
[Thompson and Lichtner Company,
Cambridge, Massachusetts, USA]

S

Sealants as Building Materials

Sealants represents a fraction of 1% of the cost of a building but, after the roof, represent one of the most troublesome areas of the exterior skin of a building. Air, dust and water infiltration can damage or destroy interior walls and contents. Leaking seals waste energy and are an annoyance to the occupants. However, in the last few decades the science of building sealants and the art of joint design have advanced considerably, so that failures of the sealant need not happen. No single sealant is ideal in all applications, so a knowledge of what type of sealant to use in each application and how to use it is fundamental to solving the sealing problems.

1. Factors Affecting Selection

When two or three sealants have all the necessary performance characteristics for a specific sealing job, a cost comparison will determine which one is ideal. Provision of the required performance at the lowest cost determines the ideal sealant. However, performance is the first consideration, since a failed sealant has a high labor replacement cost. Increased costs for building and from deterioration of contents due to seal failures often exceed the initial cost of the proper sealant.

The selection of a sealant, as with any building material, starts with an estimation of what kind of performance is needed. The following checklist is a guide to sealant selection, with items listed in order of descending importance.

(a) *Dynamic movement expected.* How much movement will the joint encounter due to expansion and contraction of adjacent materials, vibration, wind forces, settling of the structure and other sources? The amount of movement should be estimated as a percentage of the width of the joint to be sealed. Sealants are generally classed for use with movements of less than $\pm 5\%$, ± 5 to $\pm 12.5\%$, ± 12.5 to 25%, and ± 25 to $\pm 50\%$. The sealing joint width is decided on the basis of the total movement expected and the movement capability of the sealant chosen, or the sealant is chosen on the basis of the desired joint width and the movement expected.

(b) *Weatherability.* What environment will the sealant encounter? Outdoors requires a sunlight- and rainwater-resistant seal; indoors, less durability is generally required. Below grade or continuous water immersion provides different conditions again. The length of time that the sealant is required to perform in the environment is also a key consideration: a month, a year, many years? Every sealant performs for some time in most environments.

(c) *Extremes in environment.* Many construction seals are exposed to temperature extremes. Solar collectors can reach temperatures of 95, 140 or 185 °C when in static conditions. Sealing around hot water pipes, heating ducts and the like also presents special problems. Extreme cold causes many sealants to stiffen, but many building materials like metals and plastics shrink or contract by large amounts as they cool down. Thus, when expansion is most needed, as the temperature decreases, the sealant is least able to expand.

(d) *Adhesive and cohesive strength.* Sealants are required to adhere to the substrate; thus, specifying the substrate is fundamental in all sealing operations. Many sealants require a primer for some substrates. If the sealant also serves as a glue to hold materials in place, the strength requirement for the adhesive seal should also be noted in sealant selection.

(e) *Abuse and abrasion.* Some walkways and factory floors are sealed flush with the surface. The sealant chosen for high-traffic areas, with the flush joint design, needs to be abrasion-resistant.

(f) *Other factors.* Many seals need to be resistant to attack by oils and solvents. Perhaps radiation resistance or electrical insulation will be required. All such special features must be considered before selection.

If more than one sealant satisfies all the performance requirements, there are two further factors to be considered.

(g) *Ease of installation.* The amount of joint preparation with a given substrate can vary with the type of sealant. The temperature of installation can also be a factor. If mixing and bulk dispensing are problems, a premixed, one-package sealant is to be preferred to a two-package product that requires mixing.

(h) *Cost.* Only when all the foregoing requirements are specified is the price of the sealant a consideration. When two or more materials satisfy all requirements, then the sealant of lowest price is the obvious choice.

2. *Types of Sealant*

Building sealants currently available include oil-extended mastics, oil-based caulks, butyl sealants, latex acrylic sealants, solvent-based acrylic sealants and specialty sealants made of Neoprene–Hypalon, ethylene–propylene and polybutene–polyisobutylene copolymers and SBR. There is also a group generally referred to as high-performance sealants in the broad generic classes of polysulfides, silicones and polyurethanes. For each of these types, especially the high-performance classes, there are several variations like one-package ready-to-use versus two-package or three-package mix-on-the-job products; there are products with high, medium and low movement capability.

Most wet sealing work employs one or other of the high-performance sealants. They weather well, with outdoor lifetimes of >10 years (sometimes >20 years), and have joint movement tolerances of ±25% or more (occasionally exceeding ±50%). Their major market share is reflected in the greater detail in which they are considered in this article (see Sects. 2.5–2.7).

2.1 Oil-Extended or Oil-Based Caulks

Until 1950 these materials dominated the market. They are composed of filler(s), generally calcium carbonate and/or a clay, and silica together with one or more oils like linseed oil, fish oil, soybean oil or tung oil. Some of the oils are of the drying type; others are not, and serve as long-term plasticizers. Today, the most important attribute of this type of sealant is its low cost; it uses the least expensive materials. These sealants are most suitable for temporary structures that undergo no movement or very limited movement. Most materials in this class have a useful life expectancy of 2–10 years and a movement capability, based on joint width, of 5% or less. Failure occurs by hardening, shrinkage and cracking. The best of these materials comply with ASTM specification C570, which details initial performance and has no durability requirement.

2.2 Butyls

The butyls are generally copolymers of isobutylene and isoprene which form a cured rubber which is then swelled in solvents and filled with clay, chalk or similar filler. The butyls use a higher-molecular-weight "permanent" plasticizer for greater elasticity, together with volatile solvents which are temporary plasticizers that impart gunnability and allow for easy extrusion. The ratio of permanent to temporary plasticizer is critical, because the permanent plasticizer causes the principal problems of tackiness and staining often encountered with these sealants. These sealants cure by solvent evaporation and form a rather long-lasting seal. The expected lifetime is 10–15 years. These sealants can typically tolerate movements up to ±10% of the joint width. The relatively low cost, coupled with rather good unprimed adhesion, has made butyls the most popular sealants in the mobile-home industry and also popular with the homeowner.

This type of sealant is often used in inappropriate places: where there is more movement than the sealant can tolerate; on porous substrates where staining is most likely to occur; in highly visible positions where the tacky surface can become unsightly; or in structures whose lifetime is much greater than the expected useful lifetime of the sealant. Historically, the butyls represent an early advance in technology over the oil-based caulks. In many markets they are being replaced by sealants with greater movement capability and fewer staining problems.

2.3 Acrylics

The latex and solvent acrylics are copolymers of acrylates and substituted acrylates mixed with typical extending fillers and pigments. Both have remarkably good unprimed adhesion and good weathering characteristics. The solvent acrylics contain significant monomer, so that when extruded to cure, they exude a strong, obnoxious odor. They are not to be used in occupied dwellings but are quite acceptable outdoors. The monomer that causes the odor is also responsible for the ability to stick to most substrates, even those coated with dust or light oils. Solvent acrylics weather quite well, with a 10–15 year life expectancy outdoors. The best of the solvent acrylics tolerate movements up to ±12.5% of the joint width. They are rather soft, pick up appreciable dirt and are called self-healing, because the polymer relaxes and reorients under strain. This last property is also called compression set and is one of the reasons why these sealants tolerate only ±12.5% joint movement.

The largest use of solvent acrylics is in commercial exterior construction applications in joints with little or no movement. They are generally of intermediate cost, higher than the oil-based caulks but lower than the high-performance products. They enjoy some popularity based on their adhesion, weathering and relative cheapness, but are receiving an ever-decreasing market share as durability and performance are becoming more sought after.

The latex acrylics differ from the solvent type in that they have almost no odor. They are used extensively in some parts of the sealing trade. Since the carrier is water, installation and subsequent cleanup are easy. The elastomeric properties are usually a bit better than those of solvent acrylics because of prepolymerization to a higher molecular weight.

The ease of cleanup, good adhesion even to damp substrates, lack of odor and moderate cost make the latex acrylics the most popular sealant with the do-it-yourself trade. This type of sealant has a 15–20 year life expectancy indoors and 10–15 years outdoors. Poor low-temperature flexibility, high shrinkage and slow cure below 3 °C are the major problems. The relatively low movement capability keeps these products from taking an important share in most commercial and industrial sealant markets. Even in the DIY market the push for longer lifetimes and more water resistance after cure is starting to erode the share of the acrylics in favor of more durable products.

2.4 Specialties

The specialty systems made with materials like polychloroprene, ethylene–propylene copolymer systems, styrene–butadiene copolymer systems, polybutene and polyisobutylene copolymer systems all offer one or more material advantages that keep them out of the high-performance sealant class. Together, these specialty products represent a very small fraction of the sealing market.

2.5 Polysulfides

The polysulfides were the first of the high-performance materials to achieve broad introduction and use in the early 1950s. The polymeric backbone of the system is generally of the type $(SSCH_2CH_2OCH_2OCH_2CH_2)_n$. The system commonly cures according to the general equation:

$$2R\text{–}SH \xrightarrow{\text{catalyst}} R\text{–}SS\text{–}R + H_2O$$

where the catalyst is a peroxide. The sealant always contains one or more of the common fillers and additives as needed.

The advantage of the polysulfides over the urethanes and silicones is primarily that of cost, because the selling price averages 15–20% less than that of either of the others. The difference in total job costs is smaller since the labor, joint preparation and installation costs are similar for all three types of sealants and many times exceed the material cost. The historical factor is also important. Sealant specifiers and applicators are more familiar with this product than with the newer silicones and urethanes, and thus the polysulfides retain some of their popularity due to this familiarity and performance history.

The first polysulfides were two-package systems mixed on the job, and this is still the most popular package system for polysulfides. With the two-package system, coloring agents can be added at the site, making an almost infinite variety of colors and shades possible. With the one-package, premixed sealant, five to ten standard colors are available. The two-package system cures to a dry surface in several hours, but the one-package system can take several days. Unfortunately, this slow cure allows dust and dirt to be embedded in the sealant surface. There is therefore a trade-off between the convenience of the one-package, no-mix system and the color flexibility and cure rate advantages of the two-package system. However, the inconvenience of mixing should not be passed over lightly. Many sealant experts judge that inadequate mixing of the two-package systems is the greatest single cause of problems (with both polysulfides and polyurethane two-package systems), after adhesion and surface preparation problems.

Adhesion of the polysulfides generally requires priming on most substrates other than steel, glass and aluminum, and there are no less than ten different primers available and recommended for various substrates. This is one of the factors in the marginal acceptance of this product in the DIY market. Priming is acceptable for large jobs with trained crews that are accustomed to the task; the sealant distributor who sells the job will have the primer needed and will give advice in the primer selection process. Small stores almost never carry primers and the DIY user is often not aware of the need for them.

Polysulfides do almost everything the sealants mentioned earlier do, and they have the ability to seal a joint which moves up to $\pm 25\%$ and to do so successfully for many years.

Ongoing markets for polysulfides are the expansion joint and perimeter sealing market and the glazing market. Another large market is the insulated glass market. Two-package systems that are easily mixed in automated equipment can cure quickly with heat acceleration. Such systems are used to seal insulated glass units. The key features here are good strength, ease of application and reasonable cost. These features, combined with acceptable weathering properties and fairly low moisture vapor transmission rate (midway between silicones and butyls), are the reason for their popularity. Polysulfides allow for the production of sealed glass windows with expected trouble-free lifetimes of 10 years and often more.

2.6 Silicones

Silicone sealants did not penetrate the sealing market until the mid-1960s. They offer some unique features over the polysulfides and other sealants available. Their backbone polymer is shown in Formula (**1**), where X represents a reactive group.

$$X-\underset{X}{\overset{CH_3}{Si}}-O-\left(\underset{CH_3}{\overset{CH_3}{Si}}-O\right)_n-\underset{S}{\overset{CH_3}{Si}}-X$$

(**1**)

Silicones brought to the marketplace thermal stability and the ability to perform at 140 °C. (Previous sealants could tolerate ~85 °C and this is still the operating limit of the nonsilicone sealants.) It also brought moisture stability so that glass could be bonded and sealed to glass for frameless aquariums. Also, for the first time a clear sealant was available that was more aesthetically pleasing in some applications.

Along with the above properties, the silicones featured better weatherability and better resistance to ultraviolet degradation than any sealing material before this, for an expected lifetime of >20 years. They also cured faster than the polysulfides or the newly emerging urethanes, giving tack-free surfaces in <1 h. In addition, the first silicones had a joint movement capability equal to the best, able to take movement of $\pm 25\%$.

One of the key limitations of the first silicones was that they released acetic acid during the curing process and thus were not acceptable against concrete, marble and some metal substances. After the initial introduction and acceptance of silicones, modifications soon produced amine cures, caustic cures and neutral cures so that silicone sealants now available can be used on virtually any substrate. The initial silicones had unprimed adhesion to glass and many household-type materials, which allowed their placement into the consumer and DIY areas, but many substrates still needed priming.

Further developments in silicone sealants produced materials that adhere unprimed to most substrates. The newer developments also produced sealants with increased capability for joint movement: silicone sealants with joint movement tolerances up to $\pm 50\%$ of the joint width are available. In fact, a very wide variety is now available and care must be used in choosing the proper silicone for each application.

One important distinction in high-performance sealants and especially within the generic class of silicone sealants is the modulus, or the force required to stretch the sealant a given distance. The lower the modulus, the less stress on the bond line and the substrate. The very high movement capability mentioned above, $\pm 50\%$ of the joint width, is possible because some silicones have a very low modulus and also very high recovery values after stretching. The low-modulus sealants are ideal for expansion joints and sealing applications where high movement occurs (where sidewalks or decks meet side walls or in highways, in airfield runaway joints or for sealing large precast panels).

The low modulus is not ideal for some applications where the sealant also performs as an adhesive. A classic case of this is in structural glazing, where the glass lite is literally glued on the side of the building with the sealant. The sealant is the glue and also the weather seal. The sealant here must not be so soft or flexible that it permits excess movement of the glass lite. Only some of the higher-modulus sealants can be used in this application and then only a silicone with its ultraviolet stability and resistance to oxidation is permitted, since the bond is in full exposure to sunlight and must not deteriorate.

Thus, silicones of one type or another are being used in virtually every segment of the construction industry, from pure glazing to adhesive seals, insulated glass, expansion joints and highway seals. While the majority are of the one-package ready-to-use type, some two-package mix-on-site materials are being used.

2.7 Urethanes

While the silicones were being introduced, yet a third generic type of high-performance sealant was being developed. The urethanes and modified urethanes were introduced in the 1960s. The urethanes generally contain polyethers and urethanes in the base polymer. The polymer is generally terminated with isocyanate groups and cures by the addition of a polyol. Formula (**2**) shows the basic chemical structure.

$$\left[O(CH_2)_a - O - \overset{\overset{\large O}{\|}}{C} - \overset{\overset{\large H}{|}}{N} - R - \overset{\overset{\large H}{|}}{N} - \overset{\overset{\large O}{\|}}{C} - O \right]_n$$

(**2**)

Many commercial variations on this exist, including the termination of the base polymer with epoxy functional groups and curing by the addition of a polyamine. Thus, some differences exist in the urethanes available. Also, some come as the one-package ready-to-use system and others as the two-package mix-on-site variety. However, the urethane sealants as a generic class have less variety and more common features than the generic class of silicones. They accept fillers well and some lower-cost, low-performance materials can be made by extensive filler addition, but only the high-performance products are considered here.

Urethanes show very little creep or flow. They generally recover from stress better than the polysulfides and not as well as the silicones, but all three types are better than any of the other common sealants on the market. The urethanes weather well enough to be acceptable in most sealing jobs, although some are not recommended for glazing and no urethane is used in structural glazing. They generally cure faster than similar polysulfides and slower than similar silicones. They cure better in the cold than polysulfides but not nearly as fast as silicones. They adhere well to glass and some metals, but every urethane manufacturer recommends primed surfaces for most substrates. Thus, in adhesion too they are similar to the polysulfides and many, but not all, of the silicones.

In joint preparation, all three major sealant groups require a clean and dry surface, the urethanes being generally more susceptible to problems if any dampness remains in the surface. The urethanes, like the polysulfides, accept paint quite well. The silicones, except for specially formulated products, do not allow paint adhesion. The paintability feature, however, is not of high value, since urethanes and polysulfides are generally used in external applications where the sealing joint is not usually painted.

In heat stability, the urethanes (up to $\sim 85\,^{\circ}C$) are comparable with the polysulfides and less stable than the silicones. Thus, active solar collectors and other hot areas are more likely to use silicone seals.

In the key feature of movement capability, the urethanes are quite good, with most able to tolerate

±25% and some of the lower-modulus products accepting movement ±30% or ±35% of the joint width. They are therefore generally near the middle of the range in this respect also, with the polysulfides and the high-modulus silicones on one side and the very low-modulus silicones on the other.

In another key feature, cost, urethanes again started out as intermediate between polysulfides (cheaper) and silicones (more expensive). This position has changed somewhat with the rising cost of hydrocarbons, because premium urethanes are now equal to or only a few percent lower than silicones in cost and the latter are 15–20% more expensive than polysulfides (for materials only).

None of the premium sealants are recommended under water on concrete unless special barrier coat primers are used. Underwater applications of all types are generally tough on urethanes, and care should be taken in the selection of the sealant for such situations.

In toughness and tear resistance the urethanes are generally better than both polysulfides and silicones. This property makes urethanes very acceptable for sealing joints in high-abrasion areas and situations where cutting, picking and vandalism are likely to occur. Areas like sidewalks and industrial floors where the sealant is often flush with the wearing surface are suitable for urethanes.

2.8 Combinations

Many sealants are available which use urethanes, polysulfides and silicones as a small part of their polymeric backbone or as an additive to enhance one property or another. These sealants do not closely resemble the premium products described above. Coal-tar-extended urethanes and polysulfides are closer to coal tar than to urethanes or polysulfides in performance. Butyl sealants with an additive of silicone or siliconized acrylic latex are really butyl or acrylic sealants with all the properties of a butyl or acrylic except for adhesion, cure rate, extrudability or some other single property or couple of properties that are enhanced slightly by the use of the silicone additive. If high performance is needed, a high-performance sealant is required. The manufacturer's data sheets are the best source of data for finding the sealant that has the properties required for a given application.

3. Installation

With any sealant, for any application, proper joint design is fundamental. The manufacturer's data sheets are also the best source of instruction as to the proper joint preparation procedure and as to the necessity of priming for particular substrates. However, there are certain general procedures that are usually appropriate for most sealants. These are as follows.

(a) Choose the sealant on the basis of the properties desired.

(b) Plan the joint dimensions on the basis of the total expected movement and the movement capability of the sealant.

(c) Test the adhesion, either primed or unprimed, by putting down a bead of sealant on a few cm of a typical joint (prepared as expected to be done on the job), let it cure for 7–14 days then try pulling it off. All this implies, correctly, that any important sealing job should be planned and investigated several weeks before sealing is actually to be done. When the sealing is actually started, attention must be paid to the joint preparation. Even if no priming is needed, the joint must be clean and dry and free of any oil or detergent residue.

(d) Install a backing material that will control the depth of the sealant and provide a surface to which the sealant will not stick. Adhesion at the joint sides is fundamental, but adhesion to the back restricts the sealant's ability to expand and contract.

(e) Now install the sealant with a pushing motion, forcing it into the joint. Immediately after installation, tool the sealant with a concave instrument. This forces the sealant into contact with the joint faces—a step that should never be skipped.

If all the above steps are followed, the performance advertised by the sealant manufacturer can be expected and the job is well done. A good sealant, well installed, will then be more than a building material—it will also be a protection for the structure that will help to prevent deterioration of other building elements.

See also: Adhesives in Building; Glazing; Structural Sealant Glazing

Bibliography

Anon 1978 Weather tester "climate" doesn't faze silicone sealant. *Mater. News* November/December: 5

Baldwin B 1976 Selecting high performance building sealants. *Plant Eng.* 30(1): 58–64

Bruner L B 1963 Room temperature-vulcanizable polysiloxane elastomers. US Patent No. 3,077,465

Cahill J M 1966 Weathering of one-part silicone building materials. *Build. Res.* 3(3): 38–39

Ceyzeriat L, Dumont P 1959 Elastomers based on diorganopolysiloxanes. French Patent No. 1,188,495

Crossan I D 1977 Proper selection and use of silicone adhesives insures extended life. *J. Appl. Polym. Sci: Appl. Polym. Symp.* 32: 421–42

General Electric Company 1967 Platinum catalyst. British Patent No. 1,054,658

Gibb J F 1980 Hidden, but essential—A technical review of backer rod. *Constr. Specifier* 3(3): 40–45
Hyde J F 1951 Siloxane elastomers. US Patent No. 2,571,039
Kasperski M G, Klosowski J M 1978 Construction sealants. *Constr. Specifier* 31(10): 21–26
Klosowski J M 1981 "Premium" sealants of the 1980s. *Adhes. Age* 24(11): 32–39
Klosowski J M 1982 Byword for sealants: Durability. *Adhes. Age* 25(1): 30–32
Matherly J E 1972 Selection and use of RTV silicones as structural adhesives. *Appl. Polym. Symp.* 19: 231–55
Noll W 1968 *Chemistry and Technology of Silicones*, 2nd edn. Academic Press, New York
Rochow E G, LeClair H G 1955 On the molecular structure of methyl silicone. *J. Inorg. Nucl. Chem.* 1: 92–111
Roth W L 1947 The possibility of free rotation in the silicones. *J. Am. Chem. Soc.* 69: 474–75
Smith A H Jr 1972 Silicon dioxide-containing fillers for hardenable silicones. US Patent No. 3,635,743
Smith A L, Angelotti N C 1959 Correlation of the SiH stretching frequency with molecular structure. *Spectrochim. Acta* 15: 412–20

J. M. Klosowski
[Dow Corning Corporation, Midland, Michigan, USA]

Selection Systems for Engineering Materials: Quantitative Techniques

Decision making in the materials selection process is a very complex operation due to the large number of materials, spanning a wide range of physical and mechanical properties, which have to be compared in order to select the optimum material. To reduce the risk of overlooking possible candidate materials, a systems approach to the selection problem is necessary. This article reviews some of the quantitative techniques used in the systems approach to selecting engineering materials. All the techniques reviewed here can readily be adopted for computer-aided selection from a materials database containing specifications and properties.

1. Cost per Unit Property Method

In most materials selection exercises one property usually stands out as the most dominant service requirement. In such a case it is a simple task to find out how much a material will cost to provide that requirement (cost of function). In the past, strength has traditionally been the primary requirement for most traditional applications, and by introducing the density and selling price, the cost of buying 1 MN m^{-2} of tensile strength can be calculated via a loading test. In such a test, the rod just supports the load when

$$W = \sigma \pi d^2/4 \tag{1}$$

where W is the load (MN), σ is the tensile strength of the material (MN m^{-2}), and d is the diameter of the rod (m). The weight of the rod is $\rho L \pi d^2/4$, where L is the length of the rod (m) and ρ its density (kg m^{-3}); the cost of the rod is therefore $k\rho L\pi d^2/4$, where k is the price per unit weight of the material. Substituting from Eqn. (1), the cost per unit strength C is thus

$$C = LW(K\rho/\sigma) \tag{2}$$

The only material variables in Eqn. (2) comprise the factor $K\rho/\sigma$, which can be evaluated for a range of materials in order to compare the relative costs of providing a unit of tensile strength.

Similar equations can be derived to compare materials on the basis of other properties. The lower the value of C, the cheaper it is to buy the property.

Also, since materials selection is a comparison exercise, a basis material can be selected and various candidate materials compared with it. The relative cost (RC) between the basis material and the candidates can be expressed as

$$\mathrm{RC} = (P_\mathrm{B}/P_i)(\rho_i/\rho_\mathrm{B})(K_i/K_\mathrm{B}) \tag{3}$$

where P refers to the property and the subscripts B and i refer to the basis material and the ith candidate material from a list or database. RC less than unity indicates that the candidate material is cheaper than the basis material. Since manufacturing costs can contribute significantly to the final cost of a material, K can represent the cost of the material on the job, that is, the cost of the stock material plus the manufacturing cost.

Although this method is useful for a preliminary analysis, the limitation of allowing only one dominant property is its main drawback. In most real-world applications, the situation is more complicated and applications nearly always dictate more than one dominant property.

2. Weighted Property Factors Method

The weighted property factors (WPF) method is a technique for evaluating materials in which each parameter (e.g., strength, cost, toxicity, specific heat) is assigned a weighting which depends on its relative importance to the overall system. For properties that can be represented by numerical values the application of weighting is a simple matter, but with properties such as corrosion resistance and toxicity, numerical values are rarely available. In such cases, test data and experience are employed to devise a scale or rating from 0 to 100. For certain properties such as density and electrical resistivity, lower values are desirable and thus the lowest value rather than the highest should be rated as 100.

At present, the biggest drawback of this approach is that it is largely intuitive and when several properties must be considered it is obvious that the analysis becomes quite complex. Statistical techniques such as the Spearman Rank Coefficient or the Digital Logic approach (Farag 1979) can be employed to assign the weighting factors in a more systematic manner.

The National Aeronautics and Space Administration (NASA) Weighted-Index system is an outstanding example of this method. This system, developed by NASA for the SuperSonic Transport (SST), is one of the best attempts yet made to rationalize the process of materials selection on an industrial scale (Raring 1964).

For the SST exercise, properties such as strength, stiffness, corrosion resistance and cost are defined as being the key requirements. The first step separates these requirements into three groups, each of which performs a separate screening function. These groups are: (a) go/no go parameters, (b) nondiscriminating parameters, and (c) discriminating parameters.

Go/no go parameters, such as weldability and corrosion resistance, are those requirements which must meet a certain fixed minimum value. Any merit in excess of a minimum level is of no special advantage, nor would it make up for any lack in another parameter. Thus, go/no go parameters do not lend themselves to compromise or relative rating.

Nondiscriminating parameters, such as availability and ease of fabrication, are requirements that must be met if a material is to be considered at all. Like the previous group, they represent parameters that do not allow comparison or quantitative discrimination. Thus, if a material is not readily available or cannot be formed in the shape desired, then it does not merit further consideration.

Discriminating parameters are those characteristics to which quantitative values can be assigned. It is on the basis of these parameters that trade-offs can be made. Each discriminating parameter is assigned a weighting factor. For example, in the case of the SST, strength has a factor of five and cost a factor of unity, indicating that strength is of greater importance than cost. Next, each of the pertinent properties of the candidate materials is assigned a rating depending on how closely it meets the requirements. These rating factors are in turn multiplied by the weighting factor for each parameter. The final rating number for each candidate material is obtained by taking the sum of the relative rating numbers and dividing by the sum of the weighting factors used. If the rating process is correct in concept and execution, then the material with the highest rating number is the optimal material.

3. The Hanley–Hobson Method

This method employs an analytical approach using minimization procedures for the computer-aided selection of engineering materials (Hanley and Hobson 1973).

The algebraic approach to choosing materials from a database is to minimize the sum of the per unit deviations of the properties of candidate materials from the specified properties.

If the specified properties are designated Y_i and the properties of candidate materials X_i, then the criterion is expressed as follows:

$$\min Z = \sum_{i=1}^{n} |X_i/Y_i - 1| \tag{4}$$

With the incorporation of subjective weighting factors α_i between zero and unity, the expression becomes

$$\min Z = \sum_{i=1}^{n} \alpha_i |X_i/Y_i - 1| \tag{5}$$

The Hanley–Hobson algebraic approach to the problem of materials selection is based on this simple algorithm. For any database it is necessary to limit the number of candidate materials presented for consideration. In the example given by Hanley and Hobson, the selection process ignores materials if any property goes outside limits by a weighted value of 0.4 or if the sum of all weighted deviations exceeds 2.0.

Using a conversational programming language such as BASIC, the algorithm can be coded and run on a desktop microcomputer.

4. The Fulmer Materials Optimizer

The Fulmer Materials Optimizer is an information system for the selection and specification of engineering materials (Waterman 1974). The system consists of four volumes of readily assimilated information on the performance and current costs of commercially available metals, plastics and ceramics, and their related manufacturing processes. The Optimizer also proposes an analytical method for selecting materials based on the WPF concept.

There are two main reasons for the setting up of this information system: (a) to select and specify the materials and manufacturing route for a new product; and (b) to evaluate alternative materials or manufacturing routes for an existing product.

The quantitative basis for optimizing in this system is the well-known weighted property factors (WPF) concept. For each material, material characteristics M_i are determined along with the relevant weighting factors W_i and the processing cost C (per unit weight). The merit rating R can then be expressed as

$$R = \sum_{i=1}^{n} (M_i W_i / C) \tag{6}$$

where n is the number of required material characteristics. Materials with the highest values of R can now be considered in more detail using the Optimizer.

A materials information system very similar to the Optimizer has also been developed by the General Electric Company in the USA, called EMPIS—Engineering Materials and Processes Information Service (Westbrook 1981).

5. International Harvester CHAT System

The International Harvester Computer-Harmonized Application-Tailored (CHAT) system incorporates a computer for the design and selection of alloy and carbon steels for heat-treated applications (Breen and Walter 1972). Computer Harmonizing is the determination of the least costly combination of alloying elements that meets the property requirements called for by Application Tailoring. Application Tailoring is the translation of engineering requirements into quantitative metallurgical requirements.

Computer-Harmonized steels have a composition that is optimized with respect to the cost of alloying elements. The need for using a computer approach becomes apparent when the various variables are taken into consideration. A least-cost steel which only needs to meet a specified base-hardenability value D_{IB} could be designed fairly easily by using nomographs and tables of alloy costs and hardenability multiplying factors. When a carburizing grade is required, however, at least one additional restriction, a minimum case-hardenability value D_{IC}, is specified. Because hardenability factors for the individual alloying elements are different for the case and base compositions, a steel designed to satisfy the former requirement would not necessarily satisfy the latter. If further restrictions, such as the martensitic start temperatures M_s of the case and base are included, it becomes impossible to find, manually, the least-cost combination of alloying elements that satisfies the multiple requirements. Because of these multiple restrictions, cost optimization requires the use of linear programming (LP) methods (Gillam 1979).

The CHAT system is designed to optimize one objective while simultaneously satisfying multiple restrictions. In the work described (Breen et al. 1973), the system determines the least costly alloy combination that satisfies specified values for D_{IB}, D_{IC} and M_s. Other objectives can be optimized and other restrictions added, provided all features can be expressed as quantitative functions of the alloying elements.

6. Weibull Analysis

Weibull analysis is a statistical method for accurately predicting the safe mechanical-stressing limits of materials which exhibit wide scatter in test data. When ceramics were first considered for engineering applications, there was a problem of interpreting and designing around a wide and apparently erratic scatter in mechanical-property test data. The amount of scatter varies from ceramic to ceramic, but coefficients of variation (i.e., standard deviation/mean strength) of 20% or more are not uncommon. A statistical distribution, known as the Weibull distribution, was found to account for the test results and predict how the material would behave under specific design conditions. Justification for the use of this method is empirical rather than theoretical, but designers use the Weibull model in two ways: (a) to accurately assign mechanical properties of brittle materials in probabilistic terms, and (b) to define design requirements precisely, in terms of strength and reliability, for selection of brittle materials.

The Weibull distribution replaces the traditional safety factor—a highly unreliable estimate—because the reported strength of a material is a single number representing an average, or an otherwise subjectively selected value based on tests of many samples, some of which failed well below the value. Standard procedure has been to reduce this value by a safety factor to allow for the variations in the material. The modified Weibull distribution is written as

$$\ln \ln [1 - p(\sigma_c)]^{-1} = M \ln (\sigma_c) + C \quad (7)$$

where σ_c is the critical stress (i.e., the measured stress at failure in a strength test), $p(\sigma_c)$ is the probability of failure for σ_c, and M is the Weibull modulus, a quantitative indication of variability in test data. High values of M correspond to consistent material with less scatter in test data, and $M = \infty$ corresponds to a material characterized by a single strength value and which exhibits no size effect. A low value corresponds to wide scatter in test data. M values for ceramics vary between 5 and 25.

Thus, a low-strength material with a high modulus can be substituted for a high-strength material with a low modulus. Work at the Lucas Research Centre, UK, has shown that when considering hot-pressed silicon nitride for automotive turbine rotors, a material with a mean bend strength of 700 MN m^{-2} and an M value of 25 is preferable to a material of 900 MN m^{-2} bend strength and an M value of 11.

Computer programs have been developed to estimate M from strength-test data. Detailed and accurate stress analysis is required to make this technique useful. Also, at present Weibull moduli and probability plots are available only for very few materials.

7. The Computerized WPF–CUR System

This is an online computerized materials-selection system based on the cost per unit requirement (CUR) concept (Appoo and Alexander 1976, Appoo 1981).

The most critical aspects in any materials selection program are the mechanical properties which will be required to withstand service conditions. Since most service conditions dictate a combination of properties, the system grades these according to their importance and combines them into an overall property using the familiar WPF technique. The WPF–CUR method goes further and costs the material according to the contribution each property makes to the overall combined property. The online accessibility makes the WPF–CUR system a powerful tool in evaluating the cost-effectiveness of materials. A simple example illustrates the theory.

Table 1
Design strength (weighting 3), electrical conductance (weighting 10) and weighted property factors for the materials A and B

Material	Design strength (MN m^{-2})	Electrical conductance (% IACS)	WPF
A	500	50	154
B	1000	40	262

Table 2
Design strength (weighting 3), electrical conductance (weighting 10), scaling factors and weighted property factors for the materials A and B

Material	Design strength (MN m^{-2})	SF	Electrical conductance (%IACS)	SF	WPF
A	500 (1000)[a]	0.1	50 (50)[a]	2	88
B	1000 (1000)[a]	0.1	40 (50)[a]	2	85

[a] Maximum value

A component such as an electrical conductor has to be designed with two main requirements: (a) high design strength, and (b) high electrical conductance. Since this is primarily an electrical application, conductance is assumed to be the primary requirement and three times as important as strength. Assuming that two materials, A and B, are the candidates, the WPF of each is calculated and the choice is based on the WPF value. The higher the WPF, the more the material has to offer in overall property terms.

The WPF is calculated by multiplying each property by its weighting and dividing the sum of these products by the sum of the weightings, leading to a value of 154 for A and 262 for B (see Table 1). Hence, even though conductance is the primary property, B has a better overall property combination. This is because strength influences the WPF to a much greater extent because of its higher absolute value.

But, can unlike units be combined to give a rational result? This problem is overcome by the introduction of scaling factors. The scaling factor is a simple normalization technique used to bring all the different properties within one numerical range. The fact that one property may be best when at its largest magnitude and another may be best when at its smallest does not matter, as it is the range within a property which is compressed. Each property is scaled or normalized so that the best (highest or lowest) value, from a materials database, is given a rating of 100 and the rest are scaled accordingly:

scaling factor (SF)
= 100/(best property value in database)

scaled property value = actual property value
× scaling factor

Thus, each property is given equal importance and affects the WPF according to its weighting only: the units are dropped and the properties can be arithmetically manipulated when necessary. Like the WPFs, they become dimensionless factors.

Considering the above example again and assuming the maximum values in the database for strength and conductance are 1000 MN m^{-2} and 50% IACS respectively, and using scaled property values, then the WPF becomes 88 for material A and 85 for material B (see Table 2). Hence, the calculations now favor the material with the higher conductance in accordance with the weightings and requirements.

Generalizing the above theory, if WPF_i is the weighted property factor for material i, P_{ji} is property j of material i, W_j is the weighting for property j, SF_j is the scaling factor for property J, and n is the number of specified properties, then

$$\text{WPF}_i = \sum_{j=1}^{n} (P_{ji} W_j \text{SF}_j) \Big/ \sum_{j=1}^{n} (W_j) \tag{8}$$

Since materials selection is usually a comparison exercise, a basis material is assumed, against which the candidate materials are compared. Hence, the WPFs of the candidates are compared in turn with the WPF of the basis material, to give the relative WPF:

$$\text{RWPF}_i = \text{WPF}_i / \text{WPF}_\text{B} \tag{9}$$

The RWPF is thus a comparison between the overall combined properties of two materials. The higher the value of RWPF_i, the more superior is material i as it has more to offer in terms of properties for that particular application.

Now, from analysis of loading conditions it is known that the relative cost can be expressed as

$$\text{RC}_i = \left(\frac{P_{j\text{B}}}{P_{ji}}\right)^n \left(\frac{\rho_i}{\rho_\text{B}}\right) \left(\frac{K_i}{K_\text{B}}\right) \tag{10}$$

where ρ is density, K is cost per unit weight, and n is the loading exponent, which depends on the property and type of loading.

The relative overall property contributions to the particular application can now be evaluated against the relative costs by expressing the ratio as a cost per unit requirement (CUR):

$$\text{CUR}_i = \text{RC}_i / \text{RWPF}_i \tag{11}$$

If the ratio is less than unity, then material i is cheaper and is more economical in overall terms than the basis material.

Continuing with the example of the electrical conductor, cost and density can be incorporated in the analysis. The values are as follows: $\rho_\text{A} = 8900$ kg m^{-3}, $\rho_\text{B} = 8800$ kg m^{-3}, $K_\text{A} = 525$ \$ t^{-1} and $K_\text{B} = 450$ \$ t^{-1}. Assuming that material B is the basis material established on the market, and A is a new material challenging B for market penetration, then from Eqns. (9) and (10),

$$RWPF_A = WPF_A/WPF_B = 88/85 = 1.035$$

$$RC_A = \left(\frac{40}{50}\right)^n \left(\frac{8900}{8800}\right) \left(\frac{525}{450}\right)$$

where $n=1$ if the conductors are designed for a specified voltage gradient, and $n=\frac{2}{3}$ if the conductors are designed for a specified temperature rise.

$$\text{For } n=1, \quad RC_A = 0.944,$$

$$\therefore \; CUR_A = 0.944/1.035 = 0.91$$

$$\text{For } n=\tfrac{2}{3}, \quad RC_A = 1.016,$$

$$\therefore \; CUR_A = 1.016/1.035 = 0.98$$

Therefore, material A is more cost-effective on an overall property basis than material B.

Factors such as availability and energy cost can all be quantified to some extent and incorporated into the analysis. Materials selection is then said to be carried out on a total basis using a systems approach.

A systems approach that takes into consideration the interplay between the various factors involved and can narrow the field to a few candidate materials, which can then be subjected to experienced and detailed evaluation, would in the long term be of vital importance in the optimum use of metals and materials.

Bibliography

Appoo P M 1981 A computerised materials database for an automotive manufacturer. In: Glaeser P S (ed.) 1981 *Data for Science and Technology*. Pergamon, Oxford, pp. 431–40

Appoo P M, Alexander W O 1976 Computer analysis speeds assessment of engineering materials. *Met. Mater.* July/August: 42–45

Breen D H, Walter G H 1972 Computer-based system selects optimum cost steels. *Met. Prog.* 102: 42–45

Breen D H, Walter G H, Sponzilli J T 1973 Computer-based system selects optimum cost steels—V. *Met. Prog.* 103: 43–48

Farag M M 1979 *Materials and Process Selection in Engineering*. Applied Science, London, pp. 160–62.

Gillam E 1979 Materials selection: Principles and practice. *Metall. Mater. Technol.* 11: 521–25

Hanley D P, Hobson E 1973 Computerized materials selection. *J. Eng. Mater. Technol.* 95: 197–201

Raring R H 1964 SST Materials. *Astronaut. Aeronaut.* 2: 54–59

Waterman N A (ed.) 1974 *The Fulmer Materials Optimizer*. The Fulmer Research Institute, Slough

Westbrook J H 1981 EMPIS: A materials data program of an electrical manufacturing company. In: Glaeser P S (ed.) 1981 *Data for Science and Technology*. Pergamon, Oxford, pp. 462–68

P. M. Appoo
[Prospect, New South Wales, Australia]

Small Strain, Stress and Work

For most structural applications, strains and rotations are small compared to unity. The resulting definitions of strain in terms of displacements, along with the equations of equilibrium in terms of stress gradients, are common to various idealizations of material behavior (see *Elasticity; Plasticity; Creep*; and *Linear Viscoelasticity*). These limiting cases, and the resulting solutions for various boundary conditions, are distinguished by the relations between stress, strain and their increments.

Stress and strain are defined as properties of a continuum; that is, they apply to elements which are small compared to the size of a part or other surrounding structure but large compared to the next smallest structure (such as the fiber, grain or atomic structure), so that average properties are relatively uniform across each element.

1. Strain–Displacement Relations: Definition of Strain Components

The often invisibly small displacements in a structure vary with the coordinates as $u_i(x_j)$. The local deformation or strain in an element is defined in terms of gradients of the displacements $\partial u_i/\partial x_j$. For most structural applications, only the displacement gradients (strains and rotations) that are small compared to unity need to be considered.

The components of strain (see Sect. 5 for visualization) are

$$\varepsilon_{ij} \equiv \frac{1}{2}\left(\frac{\partial u_i}{\partial x_j} + \frac{\partial u_j}{\partial x_i}\right) \tag{1a}$$

The components of strain have two subscripts, one corresponding to a component of displacement and the other to a component of gradient. Normal components are those with $i=j$; shear components are those with $i \neq j$. Note that $\varepsilon_{ij} = \varepsilon_{ji}$. When the factor 1/2 is dropped from these tensor shear strains, the result is often called an engineering shear strain,

$$\gamma_{xy} \equiv \frac{\partial u_x}{\partial y} + \frac{\partial u_y}{\partial x} \tag{1b}$$

As a counter-example to Eqn. (1a), consider the large 180° rigid-body rotation $u_1 = -2x_1$, $u_2 = -2x_2$, for which Eqn. (1a) would give normal strain components of -2 instead of zero. Strains defined for large displacement gradients (see, for example, Gurtin 1981) may be useful in the deformation of elastomers (if they can survive at high strains over the desired service life), in the development of localized flow as in shear bands, and in finite-element calculations where they may save computing time.

The six strain components of Eqn. (1) must all be derivable from a displacement field with just three components. There exist partial derivative relations between the strain components to ensure this. These compatibility conditions (see, for example, Timoshenko and Goodier 1970) are more useful in obtaining solutions for elasticity than in the understanding of them, which is the objective here.

An important special displacement field is plane strain, with a plane normal to which there are neither displacement components nor the gradients of stress, strain or displacement.

2. *Definition of Stress Components*

For elements which are large compared with the underlying microstructure, but small enough that strains are relatively uniform, stress may be defined as the set of forces δF_i per unit area δA_j on three orthogonal sections through the element:

$$\sigma_{ij} \equiv \lim \frac{\delta F_i}{\delta A_j} \qquad (2)$$

Positive normal components are tensile, negative ones compressive. Moment equilibrium of an element requires that $\sigma_{ij}=\sigma_{ji}$. Note that since Eqn. (2) for the definition of stress involves division but not differentiation, it is the same in any orthogonal coordinate system.

3. *Equilibrium with Stress Gradients*

There are three equations (found from equilibrium of a cube) relating the gradients of stress: one for the equilibrium of forces in each of the three directions. Each equation has three terms, each term corresponding to the change in stress between one face and its opposite. Including components of a body force per unit volume ρb_i, gives

$$\sum_{j=1}^{3} \frac{\partial \sigma_{ij}}{\partial x_j} + \rho b_i = 0 \qquad (3)$$

In thin sheets there are often no stress components on, say, the x_3 faces. This, along with $\sigma_{ij}=\sigma_{ji}$, eliminates all components with a subscript 3. Furthermore, if there are no gradients in the x_3 direction, the condition is called plane stress. Note that the surface of a body which is traction-free is not necessarily in plane stress: near a small hole or crack there may be appreciable stress gradients normal to the surface, so the full Eqns. (3) must be used. With cylindrical or spherical coordinates, extra terms arise in Eqns. (3) because of the partial differentiation in curvilinear coordinates (see Sect. 5).

4. *Changes in Strain and Stress Components with Rotation of Axes*

Since components of stress or strain involve two directions, corresponding to their two subscripts, the rules for the change of components with rotation of axes are different from those for vector components, which involve only one direction. (A 180° rotation of coordinate axes leaves stress and strain components unchanged, but reverses the sign of the components of a displacement vector.)

For two-dimensional transformations, the geometrical construction called Mohr's circle is useful and assumed known. (The engineering shear strains γ must be divided by two before the Mohr's circle transformation applies, and positive shear components as defined by Eqns. (1) or (2) must be plotted along the negative y axis for the same signs of angles in Mohr's circle and the physical plane (see Marguerre 1962).) Mohr's circle shows that for either stress or strain there exist orthogonal coordinate axes, called principal axes, for which the shear components are zero and the normal components, e.g., σ_{I}, σ_{II}, are extrema, called principal components.

Improved insight into stress or strain fields can be obtained by graphically displaying, at a number of points, the orthogonal vector pairs representing the orientations and magnitudes of the local principal components (in contrast to displaying contour lines representing only one or some combination of components).

Concise computational rules can be given for the transformations of tensors in three dimensions. Consider the set of direction cosines $l_{i'i}$ of the angles $\theta_{i'i}$ between the new i' and the old i axes (both orthogonal). The transformation rule for vector (displacement) components a_i is

$$a_{i'} = \sum_i l_{i'i} a_i \qquad (4)$$

The transformation rule for strain and stress components is an extension of Eqn. (4):

$$\varepsilon_{i'j'} = \sum_i \sum_j l_{i'i} l_{j'j} \varepsilon_{ij} \qquad (5)$$

and similarly for stress. Equations (5) can be solved to find the three principal axes, referred to which the shear components are zero and the normal components are principal (extrema) (see, for example, Marguerre 1962).

5. *Dilatational and Distortional Components of Strain and Stress*

For isotropic materials, definitions of strain components other than those of Eqns. (1) and (2) give better physical insight and are often more convenient for stress–strain relations. The resistance to pure dilatation (change in volume) is almost always elastic, governed by interatomic forces. The resistance to distortion (the remaining part of the deformation) may be governed either by interatomic forces, by the thermodynamics of long-chain molecules in rubber, or by

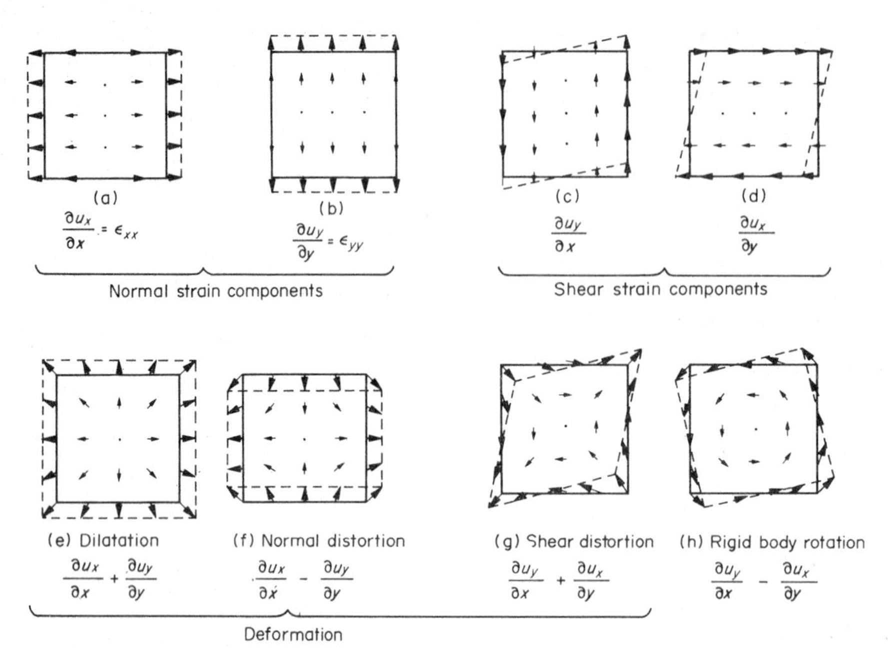

Figure 1
Combinations of displacement fields with constant gradients, giving various components of strain. In cases (a)–(d), arrows represent displacement gradients of equal magnitude. In cases (e)–(h), arrows represent the sums and differences of individual gradients

the mechanisms of inelastic behavior, as in the plastic flow of metals or the viscosity of liquids and gases.

To visualize dilatational and distortional components of strain, consider the displacement fields associated with each of four equal in-plane partial derivatives (Figs. 1a–1d) and their sums and differences (Figs. 1e–1h). The three combinations (Figs. 1e–1g) all involve deformation, while Fig. 1h is a rigid body rotation (as long as the derivatives are small compared to unity). Rotation is not part of deformation, so consider only the first three combinations. The first is pure area change, without distortion, called dilatation. The second is normal distortion, while the third is shear distortion.

Note that a 45° rotation of coordinate axes changes a normal component of distortion to a shear component, and conversely. Thus normal and shear distortions are not physically distinct, but are only two components of the same physical entity, distortion. (A 45° coordinate rotation would not simply interchange components of a vector, such as displacement.)

The dilatational strain is often given as the fractional change in volume, $\varepsilon \equiv \varepsilon_{11} + \varepsilon_{22} + \varepsilon_{33} = \Delta V/V$. Similarly a dilatational stress σ is defined as the mean of the normal components of stress, so that the increment of work per unit volume is $\sigma d\varepsilon$: $\sigma \equiv (\sigma_{11} + \sigma_{22} + \sigma_{33})/3$.

The distortional (or deviatoric) components of strain and stress are denoted by ($'$) and best given in terms of δ_{ij}, defined as unity when $i=j$ and 0 when $i \neq j$:

$$\varepsilon'_{ij} \equiv \varepsilon_{ij} - (\varepsilon/3)\delta_{ij}, \quad \sigma'_{ij} \equiv \sigma_{ij} - \sigma\delta_{ij} \tag{6}$$

The normal and shear deviatoric components of stress and strain differ only by a 45° rotation of coordinate axes, so their stress–strain relations are identical for an isotropic material. (This is not evident from the γ_{xy} definition of the shear strain.)

6. Cylindrical (Polar) and Spherical Coordinates

When the shape of the body is a circular cylinder or a sphere, polar or spherical coordinates may be convenient. Relations involving partial derivatives then have more terms, because of the changing directions of the unit vectors in curvilinear coordinates.

6.1 Strain–Displacement Relations

Using cylindrical coordinates, Equations (1a,b) become

$$\varepsilon_{rr} = \frac{\partial u_r}{\partial r}, \qquad \varepsilon_{\theta\theta} = \frac{\partial u_\theta}{r\partial\theta} + \frac{u_r}{r}, \qquad \varepsilon_{zz} = \frac{\partial u_z}{\partial z} \tag{7a}$$

$$\left.\begin{aligned} \gamma_{r\theta} &= \frac{\partial u_\theta}{\partial r} + \frac{\partial u_r}{r\partial\theta} - \frac{u_\theta}{r}, \qquad \gamma_{\theta z} = \frac{\partial u_z}{r\partial\theta} + \frac{\partial u_\theta}{\partial z} \\ \gamma_{zr} &= \frac{\partial u_r}{\partial z} + \frac{\partial u_z}{\partial r} \end{aligned}\right\} \tag{7b}$$

For spherical pressure vessels, both the shape of the body and the loading have spherical symmetry, so the displacements are purely radial. The strain–displacement relations then become

$$\left.\begin{aligned}\varepsilon_{rr}=\frac{\partial u_r}{\partial r}, \qquad \varepsilon_{\theta\theta}=\varepsilon_{\phi\phi}=\frac{u_r}{r}\\ \gamma_{r\theta}=\gamma_{\theta\phi}=\gamma_{\phi r}=0\end{aligned}\right\} \quad (8)$$

More general loadings of spheres, as in ball bearings, require more general forms of Eqns. (8) (see, for example, Timoshenko and Goodier 1970).

6.2 Equilibrium Equations

In polar cylindrical coordinates

$$\left.\begin{aligned}\frac{\partial\sigma_{rr}}{\partial r}+\frac{\partial\sigma_{r\theta}}{r\partial\theta}+\frac{\partial\sigma_{rz}}{\partial z}+\frac{\sigma_{rr}-\sigma_{\theta\theta}}{r}+\rho b_r=0\\ \frac{\partial\sigma_{\theta r}}{\partial r}+\frac{\partial\sigma_{\theta\theta}}{r\partial\theta}+\frac{\partial\sigma_{\theta z}}{\partial z}+\frac{2\sigma_{r\theta}}{r}+\rho b_\theta=0\\ \frac{\partial\sigma_{zr}}{\partial r}+\frac{\partial\sigma_{z\theta}}{r\partial\theta}+\frac{\partial\sigma_{zz}}{\partial z}+\frac{\sigma_{zr}}{r}+\rho b_z=0\end{aligned}\right\} \quad (9)$$

With spherical symmetry and no body forces, Eqns (3) for equilibrium become

$$\left.\begin{aligned}\frac{\partial\sigma_{rr}}{\partial r}+\frac{2}{r}\left(\sigma_{rr}-\sigma_{\theta\theta}\right)=0\\ \frac{\partial}{\partial\theta}=\frac{\partial}{\partial\phi}=0, \quad \sigma_{\theta\theta}=\sigma_{\phi\phi}, \quad \sigma_{\theta z}=\sigma_{zr}=\sigma_{r\theta}=0\end{aligned}\right\} \quad (10)$$

7. Work per Unit Volume

The incremental work per unit volume is useful in approximate stress analysis, as well as in theoretical proofs:

$$\left.\begin{aligned}(dW)/V=\sigma_{xx}d\varepsilon_{xx}+\sigma_{yy}d\varepsilon_{yy}+\sigma_{zz}d\varepsilon_{zz}+\sigma_{yz}d\gamma_{yz}\\ +\sigma_{zx}d\gamma_{zx}+\sigma_{xy}d\gamma_{xy}\\ =\sum\sum\sigma_{ij}d\varepsilon_{ij}=\sum\sum\sigma'_{ij}d\varepsilon'_{ij}+\sigma d\varepsilon\end{aligned}\right\} \quad (11)$$

Bibliography

Gurtin M E 1981 *An Introduction to Continuum Mechanics.* Academic Press, New York

Marguerre K 1962 Basic concepts (in elasticity). In: Flügge W (ed.) 1962 *Handbook of Engineering Mechanics.* McGraw–Hill, New York, pp. 33–3–33–14

Timoshenko S P, Goodier J N 1970 *Theory of Elasticity*, 3rd edn. McGraw–Hill, New York

F. A. McClintock
[Massachusetts Institute of Technology, Cambridge, Massachusetts, USA]

Solar Thermal Materials

Solar-to-thermal energy conversion systems utilize optical elements to collect solar radiation and convert it into heat. Heat-transport systems take the heat from the collector to the point of utilization. Basic physical properties of materials play an important role in determining the cost and configuration of the optical elements and transport systems. Typically available concentrating and nonconcentrating collectors make use of different combinations of basic optical elements and heat-transport systems, as shown in Fig. 1. To maximize system performance, the requirements are: reflectors with 100% reflectance, transmitters with 100% transmittance, absorbers with 100% absorptance and no thermal emittance, and coolants and containment materials which have high-energy densities and low corrosion rates. Typical optical properties achievable with new materials are 95% reflectance using silvered glass, 92% transmittance using glass and 95% absorptance with 10% emittance for black chrome, but these properties tend to degrade with time.

1. Solar Collector Areas

The diffuse nature of solar energy requires the use of relatively large areas of solar collectors in order to

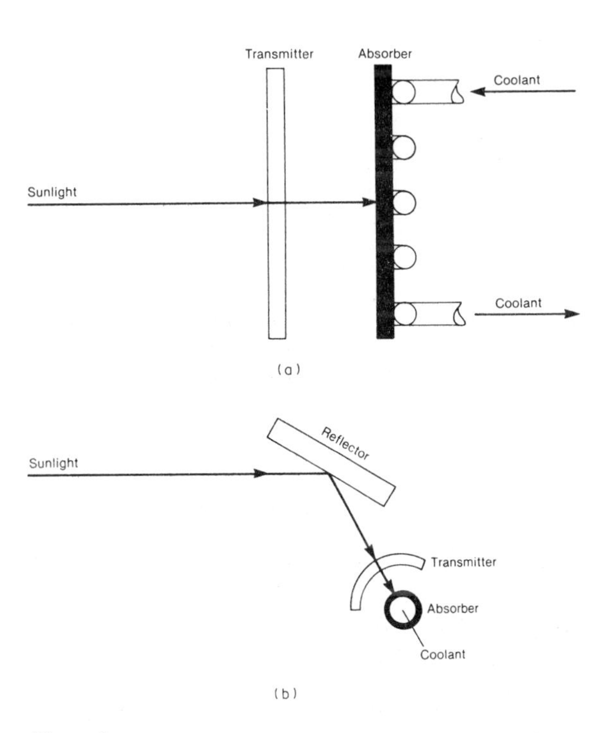

Figure 1
Optical paths for typical solar-energy systems: (a) nonconcentrator and (b) concentrator

capture a reasonable amount of solar energy. The sun's terrestrially measured output can peak at 1000 W m^{-2} on a clear day at noon in the southwestern USA. With a typical system efficiency of 50%, 500 W m^{-2} can be collected. The daily average solar intensity, under these same conditions, is approximately 850 W m^{-2}, which would give a daily average thermal energy collectable as 3.4 kWh m^{-2}. The average family uses about 30–80 k Wh per day for heating and cooling a home, which requires tens of square meters of collectors (US Department of Energy 1978, Fleck 1975, Kreider and Kreith 1981). The requirement that large active areas be used has a different impact if concentrating instead of nonconcentrating collectors are used. Nonconcentrating collectors tend to require very large areas of transmitters and absorber plates, while concentrators tend to use large areas of reflecting surface and smaller areas of transmitters and absorbers. A 100 × concentrator would have a reflector aperture (i.e., the projection of the reflector surface on the plane normal to the sun's rays) which would be 100 times the surface area of the absorber (Klein 1970). Owing to the large areas of materials required, collector designers select thin-film materials as reflectors, transmitters and absorbers, in order to minimize material weight and cost.

2. Solar Reflector Materials

Solar energy is distributed in a wavelength band between 250 and 2500 nm with the peak occurring at approximately 500 nm. In order to have a high solar reflectance, a mirror must reflect all wavelengths from the ultraviolet through the near infrared. The reflectivity R, defined by Eqn. (1), is dependent on the angle of incidence (Klein 1970):

$$R=\frac{1}{2}\frac{\sin^2(\theta_2-\theta_1)}{\sin^2(\theta_2+\theta_1)}+\frac{\tan^2(\theta_2-\theta_1)}{\tan^2(\theta_2+\theta_1)} \quad (1)$$

where θ_2 and θ_1 are the angles of incidence and refraction, respectively, as related by Snell's law ($n_1 \sin\theta_1 = n_2 \sin\theta_2$, where n_1 and n_2 are the refractive indices of the media).

Many metals exhibit high reflectances in various parts of the solar spectrum, but only a few have a high reflectance across the full range. Only aluminum and silver in their metallic form possess solar reflectance that is high enough to be of interest. Since aluminum and silver films in thicknesses of 60–100 nm develop the full reflectance of the bulk material, the most common mirror configuration is a transmitting film superstrate, backed by a thin film of the reflecting metal (silver or aluminum) which is then coated to protect the metal from the corrosive atmospheric environment. Silver is reactive with the environment and tarnishes easily by forming sulfides and chlorides which decrease the solar reflectance dramatically, so protection of the reflective layer is important. Aluminum can be used in sheet form as reflectors which have on their surfaces very thin aluminum-oxide films, which are very stable; with frequent washing they can maintain their reflectance when exposed to outdoor environments for long periods of time (between 10 and 30 years) without substantial degradation of the solar reflectance. Aluminum, because of its relatively low cost compared with silver, finds use in two forms of solar reflectors: (a) as a bulk polished sheet of aluminum, which is then anodized to create a thin surface oxide to maintain high reflectance; and (b) as a vacuum vapor-deposited film on a thin or thick transmitting polymer superstrate. Typically, polymer superstrates are of the order 0.1–6.0 mm thick and are used to support the thin film of aluminum. Acrylics are the most widely used of the transmitting superstrate polymers. The properties of these aluminum mirrors are given in Table 1. Owing to the high cost of the metal and its reactivity with the environment, silver reflectors must be protected from environmental moisture and contaminants by the optically transmitting superstrate.

As most polymers are permeable to the atmospheric gases, glass superstrates are widely used to support thin-film silver reflectors. At present no viable polished and overcoated silver mirrors are available for large-area applications. A new silvered polymer film is now available for outdoor use. Additional research on tarnish prevention for silver mirrors is being continued. The glasses used as superstrates for solar reflection layers range from 0.5 to 6.0 mm in thickness, and in the thin form they are flexible for contouring to parabolic-trough shapes. The aluminum sheet, aluminized and silvered acrylics and silvered glass composites represent the currently available choice of materials for high solar reflectance (Butler and Claassen 1980).

Table 1
Properties of common mirror types[a]

	Solar reflectance (%)		
Material	Specular, 4–18 mrad cone angle	Total hemispherical, 2π rad	Durability
Aluminum polished sheet	70–80	85	good
Aluminized polymer films	70–85	85–88	good
Silvered polymer films	90–93	93–95	good
Silvered glass (back surface)	83–95	83–95	excellent

[a] Approximate property values. Actual properties of materials may vary from manufacturer to manufacturer and lot to lot

3. Solar Transmitting Materials

The solar transmittance of materials is defined as the ratio of solar energy transmitted to the solar energy incident on the surface of the transmitting panel. There are two causes for losses in transmittance: one is the front and back surface reflection losses due to the refractive index mismatch between air and the transmitting material; and the second is the absorption loss, which is characterized by an absorption coefficient α and optical path length x, as defined in Eqn. (2) (Fowles 1968):

$$I_0 - I = I_0[1 - \exp(\alpha x)] \quad (2)$$

where I_0 and I are the incident and transmitted intensities, respectively.

The values given in Table 2 are solar transmittances for flux incident normal to the plane of the transmitting material. A wide variety of glass and plastic sheets as well as thin plastic films are available with high solar transmittance. The high bonding energies and inherent stability of glass make it an excellent choice for solar transmitters. Common float-window glass has a solar transmittance that is lowered by the inclusion in the glass of ferrous ions, which absorb energy from the solar spectrum, giving the glass a greenish cast. By removing this ferrous iron either by oxidizing it into the ferric state or by decreasing the total iron content of the glass, the solar transmittance can be increased from 84 to 90% in an equivalent thickness of glass. Antireflection etching processes have been developed to minimize the front and back surface reflections from these glasses which represent approximately an 8% effective transmission loss. These processes do not affect the optical clarity of the glass, and produce an antireflection layer of glass with an intermediate refractive index providing 96–98% transmittance when used on glass with low ferrous-iron content. Antireflection coatings based on the deposition of thin films which provide index matching and interference effects have also been developed. Surface etchings and coatings are being developed for polymer sheets and films, but are not yet widely available.

Self-supporting sheet material can be used as a structural element and represents a thermal and infrared barrier in addition to providing weather isolation. Thin films, however, need to be combined with another structural element and must have a frame to support them. Plastic films tend to be less effective as infrared and thermal barriers. The selection of these materials must take into account their mechanical properties, thermal properties and solar transmittance, all of which can be altered by ultraviolet ex-

Table 2
Optical properties of common transmitting materials[a]

Material	Solar transmittance (%)	Infrared transmittance (%)	Photodegradation resistance[b]
Sheets			
Float-glass sheet (3.2 mm thick)	84	2	excellent
Low-iron glass sheet (3.2 mm thick)	90	2	excellent
Acrylic sheet (3.2 mm thick)	90	2	excellent
Polycarbonate sheet (3.2 mm thick)	84	2	fair
Fiberglass-reinforced polyester sheet (0.6 mm thick)	87	8	fair
Thin Films			
Polyethylene film (0.1 mm)	92	80	poor
Polyester film (0.13 mm)	87	20	poor
Poly(vinyl fluoride) film (0.1 mm)	92	20	good–excellent
Polytetrafluoroethylene (0.05 mm)	96	26	excellent
Poly(vinylidene fluoride) (0.1 mm)	93	23	excellent

[a] Approximate property values. Actual properties of materials may vary from manufacturer to manufacturer and lot to lot [b] Degradation is a subjective consensus based on similar uses in glazing applications

posure. Degradation and durability must also be considered for the specific application (Gupta et al. 1980).

4. Solar Absorptance

Solar absorptance is the amount of solar energy absorbed per unit area in the solar wavelengths, divided by the total energy intensity impinging on the absorbing surface per unit area. The absorptances reported in Table 3 are direct normal, that is, the absorbing plane is normal to the incident radiation. A variety of materials possess very high solar absorptance; these include mostly thin films of black pigments (paints) or electrochemically deposited films such as black chrome, black cobalt and Tabor black (nickel and zinc sulfides).

The amount ϕn of solar energy retained for useful purposes per unit area of absorber plate in solar collectors is dependent not only on the amount of solar energy absorbed, but also on the amount reradiated as expressed in Eqn. (3) (Edwards et al. 1962):

$$\phi\eta = \phi\alpha_s - \sigma\varepsilon_\tau(T_1^4 - T_2^4) \qquad (3)$$

where η is the effective thermal efficiency per unit area, α_s is the solar absorptance, ε_τ is the thermal emittance of the solar absorber, ϕ is the solar irradiance on the absorber (i.e., after passing through the system optics), T_1 is the surface temperature of the absorber, T_2 is the temperature of material or matter to which the absorber is radiating and σ is the Stefan–Boltzmann constant ($=5.67\times10^{-8}$ W m^{-2} K^{-4}). It can be seen that, for a given solar input, if the only heat loss from the plate is its radiation, then the maximum temperature the plate will reach depends on its solar absorptance and its thermal emittance; the higher the emittance value, the lower the stagnation temperature of the absorbing plate or tube. The benefit of reducing emittance generally increases dramatically as the operating temperature of the solar-energy system increases. The benefit of paying higher prices for coatings with high solar absorptance and low thermal emittance is a compromise that the system designer must make.

Nonselective coatings such as organic-binder black-pigment paints are inexpensive and provide good durability and temperature stability. High-temperature black paints based on pigments bound by inorganic binders are considerably more expensive, but because they do not have the organic binder systems, they are stable to much higher temperatures and are widely used. The selective coatings based on electrodeposition require a base-metal substrate with a low emissivity, such as nickel, which must then be overcoated with an electrodeposited surface that has the absorbing solar properties. These coatings require the high solar absorptance of the surface coat teamed with the low thermal emittance of the undercoat to achieve the selective absorbing properties. Thus, these coatings tend to be inherently more costly than paint coatings which are simply applied by spray techniques. Recently, Pt–Al_2O_3 cermet materials have been developed which are fabricated by advanced reactive-sputtering techniques. These coatings offer high solar absorptance and low thermal emittance based on a similar principle to the electrodeposited coatings, but they are stable at much higher operating temperatures. A number of advanced concepts based on thin films using interference are being researched with a variety of thin-film layers and configurations to determine whether further advances in selective-absorber technology are possible.

Table 3
Properties of common absorber coatings[a]

Material	Solar absorptance (%)	Thermal emittance at 25 °C (%)	Stability maximum (°C)
Black paints (low temperature organic binders)	92–96	90–92	100
Black paints (high temperature inorganic binders)	95–96	84–85	600–800
Black chrome (electrodeposited)	94–96	5–10	300
Black cobalt (electrodeposited)	93–96	14–18	700
Tabor black (electrodeposited NiS–ZnS)	91	14	100
Platinum–aluminum-oxide cermets	95	8–15	500

[a] Approximate property values. Actual properties of materials may vary from manufacturer to manufacturer and lot to lot

5. Conclusions

There are a variety of existing materials that satisfy the solar optical-property requirements for solar collection and utilization. It is possible to make some moderate improvement in the solar reflectance, transmittance and absorptance–emittance properties of solar optical materials. However, major improvements in these materials can come from the use of less material to serve the same function and/or increasing the life expectancy of these optical properties in real solar collecting environments. All of the optical properties quoted are for clean, new materials; a variety of data exist indicating that these optical properties degrade as a function of environmental exposure and operating duration. Thus, the maintenance of optical properties becomes the key area of concern maintaining the optical performance of a solar collection system. Ongoing research activities are expected to increase the materials options while decreasing their susceptibility to environmental degradation.

See also: Passive Solar Energy Materials

Acknowledgement

The work of the author was sponsored by the US Department of Energy under Contract No. EG-77-C-01-4042 to the Solar Energy Research Institute.

Bibliography

Butler B L, Claassen R S 1980 Survey of solar materials. *J. Sol. Energy Eng.* 2: 175–87
Call P J 1979 Applications of passive thin films. In: Kazmerski L L (ed.) 1979 *Polycrystalline and Amorphous Thin Films and Devices.* Academic Press, New York, pp. 257–91
Edwards D K, Gier J T, Nelson K E, Roddick R E 1962 Spectral and directional thermal radiation characteristics of selective surfaces for solar collectors. *Sol. Energy* 6: 1–8
Fleck P A (ed.) 1975 *Solar Energy Handbook.* Time Wise Publications, Pasadena, California
Fowles G R 1968 *Introduction to Modern Optics.* Holt, Rinehart and Winston, New York
Gupta A, Liang R, Morcanin J, Goldbeck R, Kliger D 1980 Photochemical processes in polymeric systems—II. Photochemistry of a polycarbonate of bisphenol A in solution and in the solid state. *Macromolecules* 13: 262–67
Klein M V 1970 *Optics.* Wiley, New York
Kreider J F, Krieth F (eds.) 1981 *Solar Energy Handbook.* McGraw-Hill, New York
US Department of Energy, Division of Solar Technology 1978 *Solar Thermal Power Systems Program: Program Summary January 1978*, DOE/ET-0018-1. US Government Printing Office, Washington, DC

B. L. Butler
[Solar Energy Research Institute, Golden, Colorado, USA]

Steel for Construction

Man began using iron for tools and weapons about 3500 years ago, but the first structural application, the bridge over the River Severn at Ironbridge, England, is only 200 years old. Cast iron, and later wrought iron, were used in structures until they were gradually replaced for this purpose by steel in the second half of the nineteenth century. It was the higher strength and competitive cost of steel which led to the abandonment of iron in construction.

The first bridge to incorporate steel as a structural material was the 100 m suspension bridge over the Danube Canal in Vienna (1828). Three 30 m steel truss bridges were built in Holland in 1862, but subsequent applications of steel in bridges were unsatisfactory. Indeed, the British Board of Trade prohibited the use of steel in bridges until 1877. Steel was introduced to American construction practice in the tubular arches of the Eads Bridge at St. Louis (1868–74) and the bridge at Glasgow, Missouri (1874–79) included five 100 m all-steel trusses. The Brooklyn Bridge (1869–83), the first suspension bridge with steel wire cables, was followed by the Forth Bridge (1882–87), an all-steel cantilever truss, in Scotland.

Steel was introduced into building construction in the Home Insurance Building, Chicago (1884). This building was the first with the features of a high-rise building—an internal frame and curtain walls. In early high-rise buildings many of the structural members were riveted plates and angles. The Gray mill, devised in 1897 in Duluth, Minnesota, and capable of rolling wide-flange shapes, was put into use by Bethlehem Steel corporation in 1907. These shapes were more efficient in both weight and strength than riveted members and they ushered in the skyscraper era.

The structural applications of steel have expanded from bridges and buildings to include many other structures. Today, a wide range of steel grades in many forms—plates, rolled shapes and cold-formed sheet—offer a spectrum of strength, weldability and corrosion resistance. Also, the replacement of rivets by welds and high-strength bolts has produced lighter, more efficient structures.

1. Types of Steel in Construction

Structural steels are available in many grades that cover a wide range of properties and may be classified into four major categories: carbon steels; high-strength low-alloy (HSLA) steels; heat-treated carbon steels; and heat-treated alloy steels. Chemical compositions and mechanical properties of selected structural steels used in the form of plates and rolled shapes are listed in Table 1.

Carbon steels consist almost entirely of the element iron but also contain small quantities of carbon, manganese, silicon, phosphorus, sulfur and sometimes copper. Carbon steels are the least expensive structural steels on a cost per unit weight basis.

HSLA steels are carbon steels to which additional alloying elements have been added. Manganese, often present at higher levels than in carbon steels, increases strength and toughness. Nickel improves low-temperature toughness; copper improves atmospheric corrosion resistance. Vanadium or niobium is added to increase strength by grain refinement and precipitation strengthening. Chromium improves resistance to atmospheric corrosion and both chromium and molybdenum increase strength and heat resistance.

Heat-treated carbon and alloy steels may be normalized or quenched and tempered to improve mechanical properties. Normalizing produces essentially the same ferrite–pearlite microstructure as that of carbon and HSLA steels, except that the heat treatment produces a finer grain size. This makes the steel tougher and more uniform throughout its length. Quenching and tempering produce a martensite or

Table 1
Chemical compositions and mechanical properties of selected structural steels

Steel	Chemical composition (%)						Minimum yield strength (MPa)	Minimum tensile strength (MPa)	Minimum ductility (elongation in 50 mm, %)
	C (max.)	Mn	P (max.)	S (max.)	Si	V (min.)			
Carbon[a]	0.29	0.60–1.35	0.04	0.05	0.15–0.40	—	170–250	310–410	23–30
HSLA[a]	0.26	0.50–1.65	0.04	0.05	0.15–0.40	0.02	280–450	410–550	18–24
Heat-treated carbon									
normalized[a]	0.36	0.90 max.	0.04	0.05	0.15–0.40	—	200	420	24
quenched and tempered[b]	0.20	1.50 max.	0.04	0.05	0.15–0.30	—	550–690	660–760	18
Heat-treated alloy									
normalized[c]	0.20	0.70–1.60	0.04	0.05	0.15–0.50	—	320	350	23
quenched and tempered[d]	0.21	0.45–0.70	0.035	0.05	0.20–0.35	—	620–690	720–800	17–18

[a] If copper is specified, the minimum is 0.20% [b] Boron, 0.0005% minimum [c] Copper, 0.35% maximum; nickel, 0.25% maximum; chromium, 0.25% maximum; molybdenum, 0.008% maximum [d] Molybdenum, 0.45–0.65%; nickel, 1.50% maximum; chromium, 1.20% maximum; boron, 0.001–0.005%

martensite–bainite microstructure in both carbon and alloy steels. The alloying elements (boron, manganese, molybdenum, chromium, silicon and nickel) make appreciable contributions to the hardening of alloy steels.

Structural steels are produced in many forms. Plates may be hot rolled, control rolled, normalized or quenched and tempered. Rolled shapes, bars and sheet piling are hot-rolled and the shapes and bars may be normalized. Sheet may be either hot or cold rolled.

Bolting materials include carbon steels with minimum tensile strengths of 410 MPa, quenched and tempered carbon and low-alloy steels with minimum tensile strengths of 720 MPa, and quenched and tempered low-alloy steels with tensile strengths of 1.0–1.2 GPa.

Structural cables (wire rope and strand) are constructed from carbon steel wire that is strengthened by cold-drawing rods to reductions of 70–90%. Tensile strengths vary from 1.3 to 2.2 GPa, depending on the carbon content (0.40–0.85%).

Welding rods and wire used in welded structures may be either carbon or low-alloy steel. As a general rule, since the rod or wire contains less carbon than, and does not harden as easily as, the steel in the members being welded, the strength of the rapidly cooled weld is similar to that of the member. Nickel can be added if the weld needs to be tougher.

2. *Important Properties of Steel in Construction*

A structural steel member must resist one or more modes of failure: excessive deformation, plastic collapse, fatigue, fracture or corrosion. The properties of a steel grade selected for a particular structural member must be adequate to resist failure of the member within a stipulated service life.

One of the most important properties of structural steel is ductility, that is, the ability to deform plastically. Ductility makes it possible to construct highly complex structures which can accommodate, without cracking, large deformations due to lack of fit during erection, nonuniform settling of supports, high cyclic loads such as earthquakes and loads applied at very high rates such as blasts. Ductility also permits steel to be fabricated by bending, cutting, shearing, punching and so on.

The load and deformation of a structural steel member increase in a linear, proportional (elastic) fashion until the yield strength is reached. When it is loaded beyond its yield strength, the structural steel member deforms plastically and sheds its load to less highly stressed parallel members (if present in the structure), unless its ductility is significantly inhibited. Ductility of steel may be inhibited by an increased thickness of the member, sudden changes in shape (notches), multiaxial loading, low temperatures, high loading rates, cracks or cracklike discontinuities or a combination of these conditions. A steel grade with adequate fracture resistance (toughness), when combined with proper design, detailing and fabrication, ensures a sufficient load redistribution in spite of inhibited ductility. Inadequate toughness, on the other hand, may result in sudden fracture of the member.

If a structural steel member is subjected to a cyclically varying load of sufficient amplitude, it may fail after a certain number of load repetitions, even though the maximum load in a single cycle is much less than that required to cause yielding or fracture. A crack will initiate at a notch and, with successive load repetitions, will propagate until the member loses its ability to carry load. This failure mode is known as fatigue. In general, the fatigue resistance of steel increases with tensile strength for long-lifetime, low-load cases and with ductility for short-lifetime, high-load cases. However, the sensitivity of fatigue resistance to tensile strength and ductility decreases with an increase in notch severity. Because the fatigue behavior of steel is a function of many parameters, typical values of fatigue resistance cannot be cited. Instead, fatigue tests of the steel member or connection are performed to ensure adequate resistance to fatigue failure. The fatigue criteria of most current structural design specifications are based on such tests.

In the elastic range, the ratio of stress to strain is the material stiffness or modulus of elasticity. This property is essentially the same for all structural steels—200 GPa. If a structural steel member is exposed to fire for a period of time, the stiffness and yield strength of the steel decrease and excessive deformations result unless the steel member is protected.

Corrosion resistance is the ability of a steel to resist the attack of gases and liquids. Loss of material due to corrosion may eventually reduce the load-carrying ability of a structural member to the point where its strength is exceeded. Corrosion pits on the surface of a steel member may initiate a fatigue crack, and subsequent crack growth may be accelerated by the corrosive environment. Certain alloys and protective coatings enhance the corrosion resistance of steel.

3. *Protection of Steel*

Frequently, organic coatings are used to protect steel from corrosion by isolating it from oxygen, moisture and other corrosives or by inhibiting electrochemical action. Paints containing inhibiting agents, such as zinc chromate and red lead, in an oil vehicle provide protection when the environment is not too severe. Longer durability and protection in severe environments can be obtained with paints based on synthetic resins—alkyds, epoxies, vinyls, zinc-rich primers, for example. These paints require more careful control of surface cleanliness, paint application and thickness than the oil-based paints.

Metallic coatings used on structural steels for corrosion protection can be applied by hot dipping the steel in a bath of molten metal, by electroplating or by spraying. Hot dipping in molten zinc (galvanizing) is the most widely used metallic coating process. A mixture of aluminum and zinc as a replacement for zinc alone is gaining acceptance because it offers improved corrosion protection. The specific coating material and method of application depend on the degree of corrosion resistance required, the costs and the processing requirements.

Weathering steels denote a class of HSLA steels that, under atmospheric conditions, develop a dense, adherent, protective rust layer. These steels do not require painting when exposed openly and allowed to weather naturally. The corrosion resistance of the weathering steels is derived from a combination of alloying elements (chromium, copper, nickel, silicon, manganese and possibly phosphorus, vanadium and molybdenum). Tests in a severe industrial environment have shown that after 10 years the average corrosion penetration of the weathering steels is only $\sim 25\%$ of that of carbon steels.

Cathodic protection reduces the corrosion rate of steel by lowering its electrochemical potential by one of two methods: (a) electrical coupling to a more active metal, such as zinc coating; or (b) impressing a dc current through the steel–environment interface. Cathodic protection by impressed current uses an external power source and an inert anode. For instance, a buried steel structure is connected to the negative terminal of the power supply and the anode, for example graphite, buried nearby is connected to the positive terminal. Current passes through the soil and corrosion is inhibited.

4. *Structural Steel Building Elements*

Buildings usually have a skeleton of steel shapes: beams and girders with cross sections shaped for optimum bending resistance, and columns with cross sections designed to carry compressive loads. These shapes may be hot rolled, welded from plates or sheet, or cold rolled from sheet (see Fig. 1). Hot-rolled shapes

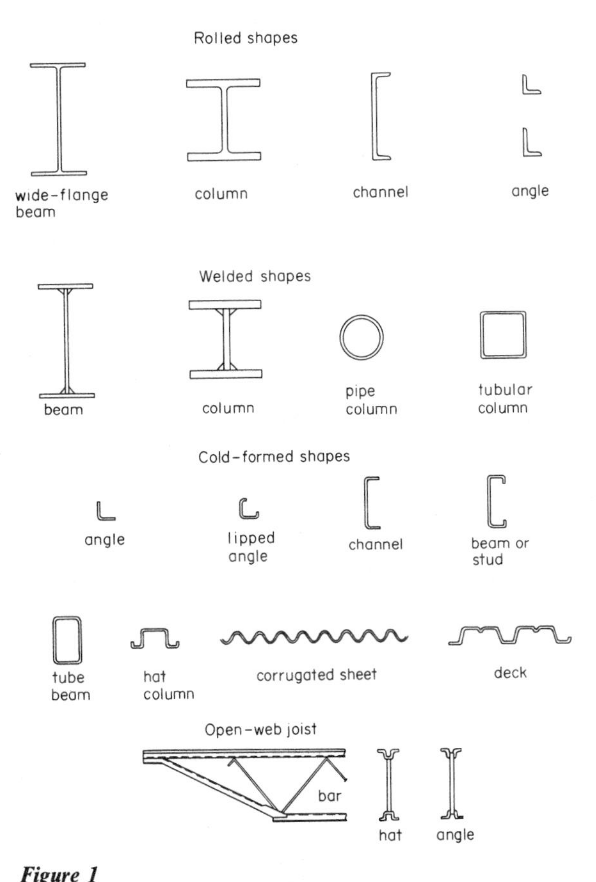

Figure 1
Typical structural steel shapes

are limited to 1 m depth by the capabilities of the rolling mill, whereas welded shapes can be manufactured in virtually any size.

Cold-formed shapes are used in many structural applications, such as concrete forms, roof and floor decks, purlins, farm silos, culverts, highway sign and lighting supports, highway guard rails, flanges of open-web joists and roof trusses.

Pipe and square or rectangular tubing can be either hot or cold formed. These sections are used where an architecturally pleasing appearance is important.

Open-web joists are used as flexural members spaced at close intervals to support roofs and floors of buildings. Such joists are small, welded trusses, fabricated from rolled angles, bars or cold-formed members (Fig. 1). Larger trusses built up from hot-rolled shapes are used for longer spans, as in power houses and mill buildings.

Steel plates can be curved and welded to form storage tanks, bins, hoppers, pressure vessels and other containers.

Steel H-piles are generally wide-flange shapes with flanges and webs of equal thickness and a width approximately equal to the depth. These carry loads either as end-bearing or as friction piles in foundations of structures. The ductility and toughness of steel allow the H pile to penetrate hard soil or dense material. Sheet piles are rolled shapes with interlocking edges. When interlocked and driven into soil, sheet piling forms a continuous wall for retaining earth or water. Sheet piling is widely used for temporary structures such as cofferdams and for permanent installations such as seawalls, bulkheads and retaining walls.

Wire ropes and strands are manufactured from cold-drawn carbon steel wires. Wire rope is relatively flexible and is used in elevators, crane hoists, slings and tramways and for stationary guys in construction. Wire strand is used in suspension structures, cable-supported roofs, masts and permanent guys on radio and TV towers. The main cables of a suspension bridge consist of many individual steel wires bundled together in a protective sheath. The bridge deck is hung from the main cables by strand suspenders or suspended directly from towers by wire-strand stays. The latter structures are cable-stayed bridges.

Steel reinforcing bars, prestressing strands and posttensioned strands, wire or bars are used to provide tensile and shear strength to concrete beams, columns and slabs. The surface of steel reinforcing bars has raised projections that create a bond with the concrete. This locking action ensures that the concrete and steel act together as a composite structural material.

5. *Steel Structures*

Steel is used as a major load-carrying material in a variety of bridge types. Short-span girder bridges use rolled, wide-flange shapes up to 1 m deep, while girders 6 m or deeper built from welded plates are capable of spans of >200 m. Also, suspension bridges (Fig. 2) and trusses fabricated of many rolled or welded elements are used for long spans. Because of their relatively high strength–weight ratio, suspension bridges make efficient use of steel in tension and are capable of spans exceeding 1.5 km. Bridge members are fabricated by welding and are field-assembled with high-strength bolts.

Figure 2
Verrazano Narrows suspension bridge, New York

The main components of a steel building are beams and girders (horizontal members), columns (vertical members) and bracing (to resist lateral loads). Horizontal members can also be trusses or open-web joists. Attached to the underside of roof and floor systems is a fire-resistant ceiling. Fire protection can also be provided for the beams and columns by an insulating material sprayed onto the steel elements.

Other important steel structures are tanks for liquid storage, and bins and hoppers for the storage of dry materials. Steel pipelines transport liquids and slurries over long distances. Electric power transmission lines are usually carried by towers of structural steel in the form of either trusses or tubes.

See also: Steels: Classification

Bibliography

American Society for Metals 1978 *Metals Handbook*, Vol. 1, *Properties and Selection: Iron and Steels*, 9th edn. ASM, Metals Park, Ohio

American Society for Testing and Materials 1984 *Annual Book of Standards.* ASTM, Philadelphia, Pennsylvania
Beckett D 1969 *Great Buildings of the World—Bridges.* Hamlyn, London
Bethlehem Steel Corporation 1977 *Modern Steels for Construction.* Bethlehem Steel Corporation, Bethlehem, Pennsylvania
Condit C W 1968 *American Building.* University of Chicago Press, Chicago, Illinois
Gordon J E 1976 *The New Science of Strong Materials,* 2nd edn. Pelican, London
Gordon J E 1978 *Structures or Why Things Don't Fall Down.* Pelican, London
Hayden M 1976 *The Book of Bridges.* Cavendish, London
Jefferson T B, Woods G 1962 *Metals and How to Weld Them,* 2nd edn. James F Lincoln Arc Welding Foundation, Cleveland, Ohio
McGannon H E (ed.) 1971 *The Making, Shaping and Treating of Steel,* 9th edn. United States Steel Corporation, Pittsburgh, Pennsylvania
Mainstone R J 1975 *Developments in Structural Form.* MIT Press, Cambridge, Massachusetts
Plowden D 1974 *Bridges—The Spans of North America.* Viking, New York
Steel Structures Painting Council 1955 *Steel Structures Painting Manual.* SSPC, Pittsburgh, Pennsylvania
Tall L (ed.) 1974 *Structural Steel Design,* 2nd edn. Roland Press, New York
Winter J A 1969 *Great Buildings of the World—Modern Buildings.* Hamlyn, London

H. S. Reemsnyder
[Bethlehem Steel Corporation, Bethlehem, Pennsylvania, USA]

Steels: Classification

Steel forms probably the most widely used category of material, and occurs over a wide range of compositions. Many of these compositions are detailed in the numerous American, British, German, French, Swedish and Japanese national specifications, while others are listed under the proprietary compositions and names of individual steel producing companies. It is clearly impossible to list the multitude of steel compositions within the framework of this article. Rather, the intention is to classify steels into generic types which are applicable to all countries.

1. Carbon and Carbon–Manganese Steels

The largest category of this class of steel is the low-carbon mild steels used for forming and packaging. These are predominantly produced as flat rolled products in the cold rolled or subcritically annealed conditions, often with a final temper rolling treatment. The carbon content is low, <0.08% (all percentages in wt%), with up to 0.4% Mn, and in many cases the steels are rimming qualities. Strain aging is minimized in some steels by aluminum additions which remove the nitrogen as aluminum nitride; with correct processing this can also optimize the crystallographic texture for improved deep drawing. Typical uses are in automobile bodies, tin plate and wire products, e.g., for galvanizing. For rolled steel structural plates and sections, the carbon content may be increased to ~0.25%, with higher manganese up to 1.5% and aluminum if improved toughness is required. These materials may also be used for stampings, forgings, seamless tubes and boiler plate, in which case aluminum is not used. A major type of this classification of steel is the low-carbon free-cutting steels which contain a maximum of 0.15% C with up to 1.2% Mn. These free-cutting steels contain a minimum of silicon and up to 0.35% S, with or without 0.30% Pb. They are never aluminum killed and are used for many products made by automatic mass-production machining methods.

Most of the lower-carbon steels are used in the as-rolled, forged or annealed conditions, and are seldom quenched and tempered. Increasing the carbon content to about 0.5%, usually accompanied by an increase in manganese, also allows the steels to be used in the quenched and tempered condition. The low hardenability, however, restricts the section sizes for such treatments. The uses of these higher carbon–manganese steels include shafting, couplings, crankshafts, axles, gears and many types of forgings. Some can be made free-cutting by lead and/or sulfur additions and they may be aluminum grain refined to improve the toughness. Steels in the 0.40–0.60% C range are also used as rails, railway wheels, tires and axles, whilst higher carbon contents can be used as laminated spring materials, often with silicon–manganese or chromium–vanadium additions. The higher carbon steels can also be used for high-strength wire, and over a range of carbon contents for reinforcing bar. Tool steels will be considered later.

2. High-Strength Low-Alloy Steels

The high-strength low-alloy steels, used for weldable and formable applications, are an ever-increasing and sophisticated class of steel. They are invariably of low carbon content to give adequate weldability and formability, some of the more modern steels being as low as 0.05% C. The manganese content is up to 1.8%. They are grain refined by aluminum, and usually contain one or more of the microalloying additions niobium, vanadium (sometimes with enhanced nitrogen contents) or titanium. Under appropriate processing conditions these can introduce precipitation strengthening. The steels have a predominantly ferrite–pearlite structure, and are produced usually as strip or plate by controlled rolling and cooling techniques. Two recent processing innovations have been:

(a) the use of inclusion shape control by zirconium, titanium, rare-earth or calcium treatment to improve through-thickness toughness and ductility; and

(b) treatments designed to produce a ferrite/austenite–martensite dual phase structure for improved formability and strength.

Within this class of material are the bainitic low-carbon steels, usually containing combinations of such elements as manganese, nickel, molybdenum, chromium, copper and vanadium, often with a boron addition. These give bainitic structures in the air-cooled or as-rolled condition over a wide range of section sizes, and a significant improvement in ductility and toughness has been obtained by the use of a low carbon content, <0.06%, to produce the acicular ferrite steels, some of which contain up to 3.5% Mn with microalloying additions. Uses of these high-strength low-alloy steels include oil and gas pipelines, drilling rigs, pressure vessels, structural members and bridges.

3. Engineering Quenched and Tempered Steels

These steels are designed to have high strength with good toughness, and with hardenability sufficient to give thorough or adequate hardening in a wide range of section sizes. Their carbon contents are in the range 0.10–0.45% with alloy contents, either singly or in combination, of up to 1.5% Mn, 5% Ni, 3% Cr, 1% Mo and 0.5% V. Generally the higher the alloy content, the greater the hardenability, and the higher the carbon content, the greater the available strength. Many of the steels are aluminum grain refined to promote toughness, whilst some contain boron to enhance hardenability and conserve alloy additions. Alloying elements such as molybdenum and vanadium increase the tempering resistance and also introduce secondary hardening, while chromium increases the tempering resistance and imparts some oxidation and corrosion resistance but does not secondary harden.

The lower-carbon (0.10–0.25%) steels are often used for case carburizing, while the 0.30–0.40% C steels with up to 3% Cr, 0.5% Mo, 0.25% V and sometimes 1% Al are used as nitriding steels. Occasionally, plain carbon–manganese steels are used for carburizing, but the section size is limited and the Ni–Mo types are perhaps the major carburizing grades. Also in the lower carbon category the Cr–Mo and Cr–Mo–V steels, which are secondary hardening, are often used as ferritic creep-resisting steels for steam generation and power plant as both superheater tubing and steam pipe. In general, however, these alloy steels are used in the quenched and tempered condition to produce a range of strengths depending on the tempering temperature, and in a range of section sizes depending on the alloy content and hardenability. Often a small molybdenum addition is made more to minimize the temper embrittlement so often encountered in Ni–Cr steels than to impart hardenability or tempering resistance. The uses of these steels are wide ranging, from engine components and transmission assemblies in automobiles to gears and many machine parts. A particularly interesting property is the increased cryogenic toughness as the nickel content is increased, leading eventually to the 9% Ni steels which are tempered within the critical range to produce a two-phase ferrite–austenite structure and excellent toughness at low temperatures.

Also within this category are the ultrahigh-strength quenched and tempered steels with proof stress values >1500 MN m^{-2}. A particularly common type is the 1% C–Cr ball and roller bearing steel which is quenched and tempered at ~150 °C. This same material can be used for cold-rolling rolls. In general, however, higher tempering temperatures are used, and the ultrahigh-strength steels fall into two major categories.

(a) Steels tempered at low temperatures, containing 0.30–0.40% C and of the order of 2% Ni, 1% Cr and 0.3% Mo. To overcome an embrittlement effect during tempering at ~300 °C, the tempering resistance is increased by additions of up to 2% Si. The appropriate strength, with enhanced toughness, can then be achieved by tempering at or above 350 °C.

(b) Steels which are tempered at 550–650°C and which are of the secondary hardening type. These steels contain 0.3–0.4% C with 3–5% Cr and are secondary hardened by up to 2.5% Mo and 0.5% V. Tempering is usually at 600°C or 650°C, i.e., above the temperature for maximum secondary hardening, and the tempering resistance can be further increased by lowering the chromium content (if permitted by the hardenability requirements) and by adding about 1% Si.

A different type of ultrahigh-strength steel, based upon the precipitation hardening of a very low carbon iron–nickel martensite, is maraging steel. A minimum of 18% Ni is employed, together with precipitation hardening by cobalt–molybdenum, titanium or aluminum additions. The high-nickel, low-carbon martensitic structure in which the precipitation occurs results in a steel with maximum toughness.

4. Tool Steels

Tool steels comprise many types, varying from the plain carbon steels to the high-speed steels. The water-hardening tool steels contain essentially 0.6–1.4% C, sometimes with 0.25% V. They have low hardenability, which may be taken advantage of to produce a tough core. Cracking and distortion may be minimized by oil quenching in the smaller section sizes, and the steels are tempered at ~200 °C. These materials are used for cutting tools, including dies, drills, taps and punches, and for razors and surgical instruments.

The shock-resistant tools, e.g., hammers, chisels and shear blades, are essentially made of 0.5% C steels with additions of up to 1.5% Cr, 2% Si, 1.5% Mo, 2.5% W and 0.3% V. Their higher hardenability enables them usually to be oil quenched and tempered at up to 350 °C. The cold work steels on the other hand, which are used for blanking and forming dies, are essentially high-carbon steels with up to 0.5% Cr, 0.5% W and 1.75% Mo. They are oil hardened and tempered to the desired hardness, but a 5% Cr steel can be air hardened. Other cold work steels with greater wear resistance contain 1.4–2.4% C, 12% Cr and up to 1% of either vanadium or molybdenum, the lower carbon levels imparting improved toughness. The structures of these steels resemble to some extent the high-speed steels, with carbide networks in the cast condition and banded carbide segregation in the wrought condition. However, they have good dimensional stability. A special wear-resistant type of tool steel contains ~2% C, 5% Cr, 4% V with up to 1.3% of molybdenum or tungsten. These steels are essentially air hardening and, because of their low M_s temperature, must be stabilized between −75 °C and −196 °C before tempering to the desired hardness. A tool steel used for plastic molding and zinc die casting of complex shapes, and operating at heavy pressures up to 200 °C, has a low carbon content in the range 0.10–0.60% with up to 5% Cr, 0.75% Mo and 0.3% V. It may also contain up to 4% of nickel or cobalt, and range in hardenability from water to air hardenable. In some cases they can be replaced by the low-carbon 12% Cr steels where extra corrosion resistance is required to handle certain plastics.

The hot work die steels are used for hot metal forming such as drop forging, stamping and extruding. They must possess a combination of hot strength, wear resistance and toughness, and are conventionally classified into three types. The chromium steels are similar to the 5% Cr–Mo–V ultrahigh-strength steels but sometimes may contain up to 1.5% W. The molybdenum steels contain rather higher carbon contents of ~0.6% with 4% Cr, 5–8% Mo and 1–2% V, while the tungsten grades contain various carbon contents in the range 0.25–0.65% with 3–4% Cr, 9–18% W and up to 1% V. Normal working hardness values are 400–600 HV, and the steels retain high hardness to temperatures of 550 °C. They are air hardenable and are tempered usually around 500 °C when they exhibit secondary hardening. For optimum toughness, however, it is advisable to avoid maximum secondary hardening.

The high-speed tool steels are used as machine tools for high cutting speeds and there are two main types. The tungsten steels contain 0.70–0.85% C, 4% Cr, 18% W and 1–2% V. On the other hand, the molybdenum grades contain similar carbon contents but 4% Cr, 1.5% W, 8–9% Mo and 1–2% V. A combined tungsten–molybdenum type is popular, containing 0.8–1.3% C, 4% Cr, 5–6% W, 4–6% Mo and 1.5–4% V. There are other types in which up to 9% Cr is added to the tungsten steels, and up to 8% Co can be added to tungsten–molybdenum steels containing 4% Cr, 3–9% Mo, 2–6% of tungsten and vanadium. Some of the steels also contain 0.15% S to improve machinability in the annealed condition, and also to improve the cutting properties of the tool itself.

Since these steels are prone to carbide segregation and banding, much attention is paid to homogenization and hot working. After machining, the steels are stress relieved at ~700 °C and finish-ground to size. Austenitizing at up to 1290 °C (depending on the composition) requires great care as the molybdenum steels especially are prone to decarburization. Quenching is carried out at a rate sufficient to form martensite. The high carbon and alloy content, however, produces retained austenite which can be minimized by subzero treatments and/or a multiple tempering treatment. All the high-speed steels show pronounced secondary hardening.

5. *Transformable Stainless Steels*

In order to provide the required corrosion and oxidation resistance, these steels contain a minimum of 11–12% Cr with 0.1–1.0% C. The higher-carbon steels are the cutlery, stainless razor and tool steels, which in some cases contain more than 12% Cr. The lower-carbon (0.1%) steels are widely used in petrochemical, chemical, gas turbine and electrical generation applications. They are fully austenitic at 950–1050 °C and are air hardenable to form martensite in quite large sections. They can be tempered either below 400 °C to produce high strength and limited toughness, or at 600–650 °C to give lower strength but good toughness and stress-corrosion resistance. In order to produce increased strength, the tempering resistance may be increased by additions of up to 2% Mo and 0.3% V, or alternatively the nitrogen content may be slightly increased. The addition of ferrite-forming elements tends, however, to introduce delta ferrite which lowers the potential strength, and this effect is often offset by additions of up to 2% Ni. Cobalt, however, can be used, and is beneficial as it does not depress the M_s temperature. Up to 0.5% Nb may be added in some cases to increase the tempering resistance still further by accentuating secondary hardening and retarding overaging, and at the same time refining the austenite grain size to improve the toughness.

A somewhat higher-chromium steel type, containing 14–17% Cr with 3–7% Ni or 2–3% Mn and additions of varying amounts of Mo, Cu, V, Co, Al or Ti in varying combinations was developed, which would be austenite at room temperature and thus capable of cold forming prior to subsequent transformation by either (a) refrigeration −78 °C or −196 °C; or (b) precipitating $M_{23}C_6$ carbide at 700 °C which so increased the M_s temperature that the austenite transformed to martensite on cooling from 700 °C to room temperature. These so-called controlled transformation steels could then be age hardened at 450 °C. However, control over the constitution and transformation characteristics was so critical that consistent response to heat treatment was difficult to achieve. Consequently, many of the basic compositions have been modified to dispense with the controlled transformation behavior so as to produce a fully transformable precipitation hardening steel for use in more corrosive environments.

6. *Ferritic Stainless Steels*

The ferritic stainless steels are presenting an ever-increasing challenge to the well-established austenitic steels for other than the most severe service requirements. They contain 11–26% Cr, usually in the range 17–26%, and in the presence of 0.1% C contain a little austenite at temperatures of about 1000 °C. This austenite transforms to martensite on cooling. The steels usually contain no nickel as this would increase the amount of austenite, but with the currently available very low carbon and nitrogen contents, or with the fairly common additions of molybdenum, titanium or niobium, they are fully ferritic up to the melting point. This gives problems with welding, grain growth and toughness, although the general formability is acceptable for most applications with the possible exception of stretch forming.

The normal alloy additions used are 1% Mo to increase the general and pitting corrosion resistance, and 0.5% of either niobium or titanium to provide stabilization against intergranular corrosion. A problem with this type of steel is the poor toughness, but the recently developed very low interstitial (carbon and nitrogen) steels help to relieve this problem despite an increased tendency for grain growth. Low interstitial contents also improve the corrosion resistance. Recent developments include an increased chromium content, up to 26–30%, with higher molybdenum contents of up to 4%, at the lowest possible carbon and nitrogen levels. Some of these higher-alloy steels contain a few percent of nickel but still remain fully ferritic. A major advantage of the ferritic steels is their considerable immunity to chloride-induced stress-corrosion cracking, but they can suffer from sigma phase and 485 °C embrittlement. The steels also can be used as heat-resisting alloys, when they normally contain 20–25% Cr with additions of aluminum and silicon.

7. *Austenitic Stainless Steels*

The standard austenitic stainless steels contain 16–25% Cr, with 7–22% Ni. The normal compositions are 18% Cr, 8–10% Ni. The lower nickel contents give enhanced stretch formability due to the formation of strain-induced martensite. Normal carbon contents are 0.06–0.08%, but extra low carbon grades can contain a maximum of 0.03% C. There are three main alloy additions:

(a) molybdenum (2–4%) which imparts improved general and pitting corrosion resistance, especially in chloride environments;

(b) titanium in amounts of five times the carbon and nitrogen contents to inhibit intergranular corrosion; and

(c) niobium in amounts of ten times the carbon and nitrogen contents, also to stabilize against intergranular corrosion.

All these alloy additions are ferrite-forming elements, and especially with the very low carbon contents in current steels, have to be compensated by increased nickel contents. This prevents the formation of delta ferrite which is detrimental to hot workability and also can impair corrosion resistance. The only safe way of preventing intergranular corrosion, however, is to use very low carbon contents.

In some steels, nickel is partially replaced by high (0.25%) nitrogen and by up to 9% Mn, which also enables the solubility for nitrogen to be increased. These replacement nickel steels have rather poorer corrosion resistance and formability than the standard austenitic steels.

The austenitic stainless steels are prone to stress-corrosion cracking, but this can to some extent be minimized by low interstitial contents and increased nickel and silicon additions. Owing to their toughness and high work hardening rates, the austenitic stainless steels are not easily machinable, but additions of 0.15–0.25% S or a similar selenium addition are made to improve machinability, albeit at the possible expense of some corrosion resistance. The steels are readily weldable, particularly if the weld metal contains a controlled delta ferrite content, and the presence of delta ferrite in the steel improves the resistance to stress-corrosion cracking. This has led to the recent development of a series of proprietary duplex stainless steels, containing approximately equal proportions of austenite and delta ferrite, for use in chemical plant. The delta ferrite also increases the strength and is achieved by careful balance of the ferrite and austenite forming elements in the steel. These duplex stainless steels are excellent casting alloys but may have somewhat decreased hot workability.

In view of their high chromium contents the austenitic steels form the basis of heat-resisting steels, when their chromium content is increased to 22–26% and balanced by increased nickel contents of 15–22%. The standard steels which contain molybdenum, niobium or titanium also form well-known types of austenitic creep-resisting steels, some of which contain additions of two of these elements. Higher-strength austenitic steels, containing nickel contents up to 25%, may be precipitation hardened by titanium and aluminum additions, when they also form the basis of the age hardening of iron-base superalloys.

Bibliography

American Iron and Steel Institute 1966 *Stainless Steel and the Chemical Industry*. American Iron and Steel Institute, New York

American Society for Metals 1984 *High Strength Low Alloy Steels: Technology and Applications*. American Society for Metals, Metals Park, Ohio

Briggs J Z, Parket T D 1965 *The Super 12% Cr Steels*. Climax Molybdenum Company, New York

British Steel Corporation 1980 *Iron and Steel Specifications*, 5th edn. British Steel Corporation, Sheffield

Climax Molybdenum Company 1977 *Proc. Int. Conf. Stainless 77, London*. Climax Molybdenum Company, New York

Institute of Metals 1985 *Stainless Steels 84*. Institute of Metals. London

Peckner D, Bernstein I M (eds.) 1977 *Handbook of Stainless Steels*. McGraw-Hill, New York

Pickering F B 1976 The physical metallurgy of stainless steel development. *Int. Met. Rev.* 21: 227–68

Pickering F B 1978 *Physical Metallurgy and the Design of Steels*. Elsevier. London

Roberts G A, Hamaker J C, Johnson A R 1971 *Tool Steels*, 2nd edn. American Society for Metals, Metals Park, Ohio

Schmidt W, Jarleborg O 1974 *Ferric Stainless Steels with 17% Cr*. Climax Molybdenum Company, New York

Union Carbide Corporation 1977 *Proc. Int. Sym. Microalloying 75, Washington*. Union Carbide Corporation, New York

F. B. Pickering
[Sheffield City Polytechnic, Sheffield, UK]

Structural Adhesives

The first important structural adhesive, a vinyl phenolic resin composition, was developed for the bonding of aircraft fuselage structures during World War II (de Bruyne 1950). Shortly after, the adhesive qualities and processing advantages of epoxy resins were recognized, and the foundations for structural bonding formed.

For the purpose of this article structural adhesives are materials used to join rigid load-bearing adherends. They must be able to distribute stresses uniformly throughout the bonded area, have sufficient cohesive strength to sustain applied loads, and be able to withstand service environments for a period of time much greater than the expected lifetime of the structure. The principal materials bonded are metals, although in recent years a need has developed to bond fiber-reinforced composite structures.

The bonding of rigid substrates is of importance to a wide variety of manufactured products. Although most adhesives were originally designed for aerospace applications, their use has gradually expanded to other industries. This article deals primarily with metal-to-metal bonding as applied to the aerospace, appliance, automotive, electrical and electronics industries. However, even in the construction field some combinations of materials, such as the epoxy–polysulfides and epoxy–polyamides, originally developed for aircraft applications, have become important commercial products.

The physical characteristics of the substrate dictate the types of materials (chemically) best suited for bonding. Proper substrate preparation, interfacial wetting and interfacial attractions are prerequisite to good bonding but beyond the scope of this article; an excellent review of the theories of bonding, however, can be found in Zisman (1977). Much of the thinking encompassing these phenomena can be expressed in a simple rule of adhesion published by de Bruyne (1939): "Provided we use pure or simple substances as adhesives then there is a good deal of evidence that strong joints can never be made by polar adherends with nonpolar adhesives or by non-polar adherends with polar adhesives."

Since most structural bonds involve polar substrates (metal oxide surfaces), it follows that polar resins are the basis for most adhesive formulations. Thermoset resins are favoured over thermoplastics. They resist creep under sustained loads and are highly fluid at some point in the bond formation process. The latter condition provides better wetting of substrates and superior interfacial bonding. In keeping with these concepts, the preferred adhesive resins are epoxies, phenolics and blends of these products with other polymers. Vinyl-based polymers including cyanoacrylates are used where rapid bond formation is important, and polyimides or other special polymers are employed where elevated temperature performance is a requirement. Urea and melamine formaldehyde resins are not widely used as structural adhesives as ureas have poor water resistance and melamine resins are outperformed by the phenolics on metal substrates. Isocyanates or polyurethanes likewise find only limited use in structural bonding. They tend to depolymerize above 120 °C, are subject to hydrolysis, and autoxidize on thermal or ultraviolet light exposure (Frisch and Reegen 1972).

The sections that follow include a description of the more popular resins and their chemistry, together with a brief discussion of the important bond test methods, formulation, bond properties, end-use patterns and future trends. Throughout the text various resins and curing agents will be referred to by chemical name and all are commercially available from chemical suppliers.

1. Polymer Chemistry of Structural Adhesives

Undoubtedly the most versatile materials used in the formulation of structural adhesives are epoxy resins. Besides metal-to-metal bonding, they are excellent adhesives for fiber-reinforced plastics, they are excellent adhesives for fiber-reinforced plastics, paper, wood, ceramics and rubber. They cure without the evolution of volatile by-products, and are highly resistant to moisture, a variety of solvents, alkalis and acids, and natural environments. Service temperatures range from the cryogenic regions to beyond 200 °C in specialized formulations. In combination with phenolics, polyamides, nitrile elastomers, polysulfides, vinyl polymers and other structural resins, properties can be tailored to fit a broad spectrum of applications.

Historically the first mention of the basic epoxy resin, the diglycidyl ether of bisphenol A (**1**) was by Schlack (1938, 1939). He disclosed that this resin could be hardened with equivalent amounts of organic amines, and showed strong adhesion. Parallel development activities started a few years later: patents were issued to Castan (1944) in Switzerland on the use of the resin as a denture material and to Greenlee (1944) in the USA as an intermediate for surface coatings. The development in adhesives started around 1948, and has since moved forward to the wide range of products available today.

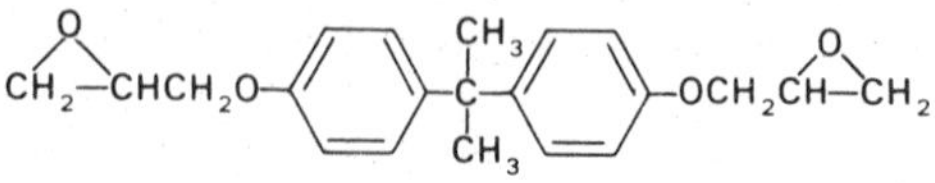

(**1**)

The versatility of epoxy resins stems from the many variations in chemical structures available from both the resins and the curing agents. Although a majority of the epoxy adhesives are based on the diglycidyl ether of bisphenol A (**1**), if resins with higher temperature performance are required the bisphenol part of the molecule can be replaced with phenolic aldehyde condensation products which increase the number of functional groups per molecule. Typical products of this class are glycidyl ethers based on novolacs (**2**, $n=2$, 3), tetraphenol ethane (**3**) and bisphenol A novolacs (**4**, $n=3, 4$).

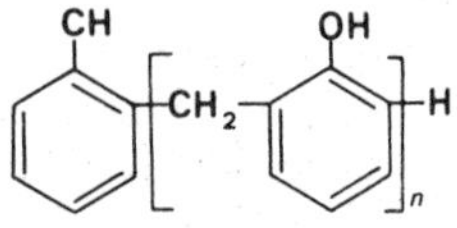

(**2**)

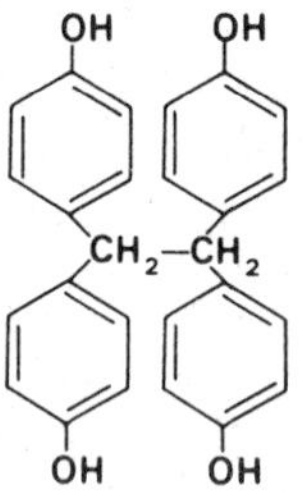

(**3**)

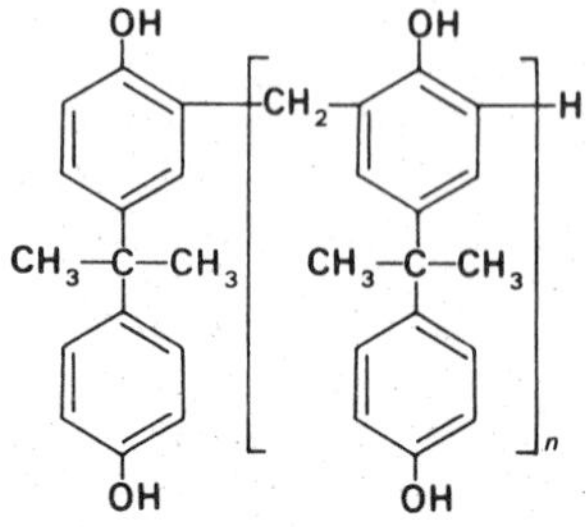

(**4**)

There are a number of low-viscosity epoxy resins which can be used to control the viscosity and tack of epoxy formulations. Most common are the monofunctional diluents such as butyl glycidyl ether and phenyl glycidyl ether. However, these materials reduce elevated temperature performance, and vinylcyclohexene diepoxide, which has a minimal effect on the glass-transition temperature, is often used as an alternative. A cycloaliphatic diepoxy carboxylate (**5**) although primarily used in electrical castings, has also been demonstrated to be an excellent diluent for use in high-temperature epoxy adhesives (May and Nixon 1961).

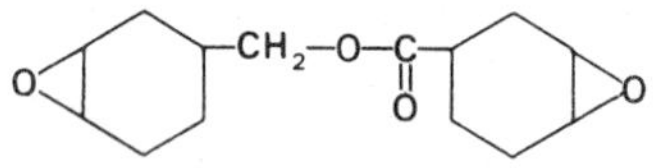

(**5**)

More recently, a number of adhesive formulations based on glycidyl amines have become availabe. Bonds prepared from products such as tetraglycidyl methylenedianiline (**6**) and triglycidyl *p*-aminophenol (**7**) display excellent bond strengths at both ambient and elevated temperatures. Also available to the adhesive formulator are flexibilizing resins where greater toughness or peel strength is required, halogenated resins which afford fire retardancy, and a variety of specialty resins designed for specific applications.

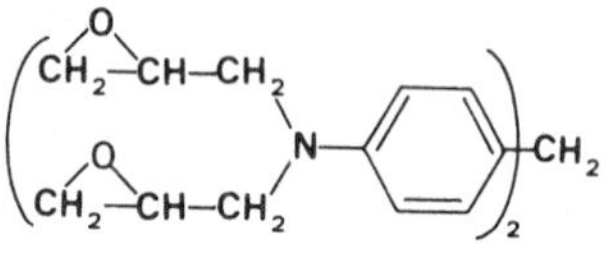

(**6**)

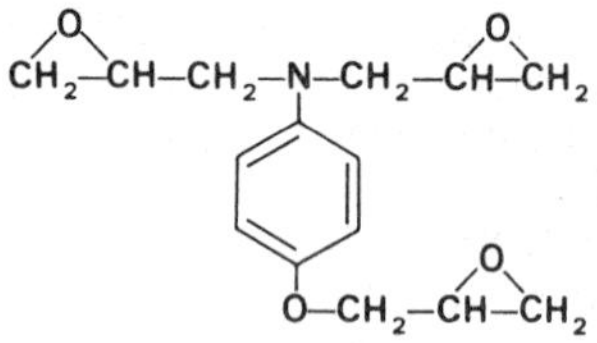

(**7**)

Epoxy resins developed their three-dimensional cross-linked structure by chemical reactions with curing agents. Table 1 shows the bond strengths that might be expected when the major curing agents are used in combination with compound (**1**) (Dannenberg and May 1969). With a wide variety of resins and curing agents available, it is readily apparent that the adhesive formulator has an almost infinite number of material combinations to choose from.

The most important curing agents are the aromatic and aliphatic amines. Thus, the curing reactions of greatest importance are as illustrated in Scheme 1. It is

Table 1
Bond strengths expected from major curing agents

Curing agent	Typical cure temperature (°C)	Expected tensile shear strength on aluminum (MPa)					
		(−57 °C)	(24 °C)	(82 °C)	(121 °C)	(149 °C)	(260 °C)
Polyamides, polysulfides	24	20.7	27.6	6.9			
Aliphatic amines	24	13.8	20.7	17.2			
	80–99	20.7	24.1	17.2	3.4		
Mixed aliphatic tertiary amines	80–99	20.7	24.1	24.1	6.9		
Aromatic amines	135–163	20.7	24.1	27.6	24.1	13.8	
Insoluble latent amines	149–177	17.2	24.1	24.1	20.7	10.3	
Carboxylic acid anhydrides	93–149	13.8	17.2	20.7	17.2	10.3	
Phenolic resole resins	166–177	17.2	17.2	13.8	13.8	13.8	10.3

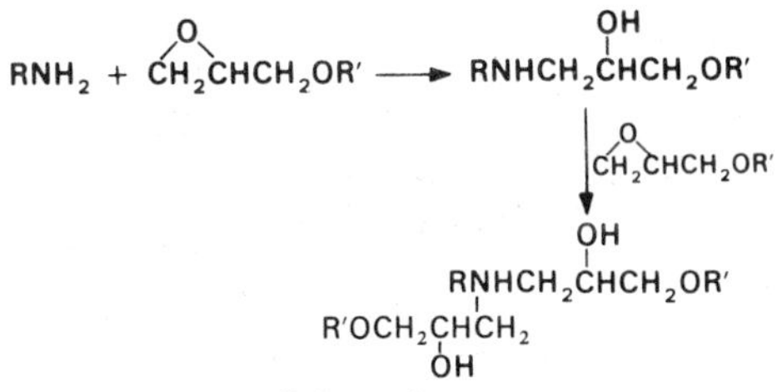

$$RNH_2 + \overset{O}{CH_2CHCH_2OR'} \longrightarrow RNHCH_2\overset{OH}{C}HCH_2OR'$$

$$\xrightarrow{\overset{O}{CH_2CHCH_2OR'}} RNHCH_2\overset{OH}{C}HCH_2OR' \;/\; R'OCH_2\underset{OH}{C}HCH_2$$

Scheme 1

also possible to cure epoxy resins with organic carboxylic acids or anhydrides, and by catalytic homopolymerization of the epoxide group using Lewis acids or bases. However, these types of curing agents are not widely used in epoxy adhesive formulations for a variety of reasons.

Phenol–formaldehyde resins are useful in structural adhesives because they are highly resistant to hostile chemical environments and form strong bonds at elevated temperatures. Service temperatures exceed those of the epoxies. The main disadvantages are the evolution of volatiles during cure and susceptibility towards atmospheric oxidation, which limit the useful life of the resins at elevated temperatures. Most structural adhesive formulations involve combinations with high-molecular-weight thermoplastics, elastomers or epoxy resins. The major use of phenolic resins is in the manufacture of plywood; major structural applications are in the aircraft and automotive fields.

There are two types of phenolic resins used in structural adhesives, the resoles and the novolacs. The resoles are generally made using an alkaline catalyst and an excess of formaldehyde. Because of the excess formaldehyde, the resole structure (**8**) is dominated by oxymethyl groups as illustrated. The predominance of oxymethyl groups dictates a highly reactive and chemically complex cure mechanism. Curing involves the elimination of water and formaldehyde. The ratio of the benzyl ether linkages x to the methylene linkages y is dependent on the phenol–formaldehyde ratio, and the reaction conditions under which the resin is prepared and cured. Formaldehyde is eliminated during the cure, but not in proportion to the number of benzyl ether linkages (Zinke et al. 1939). This suggests that the formaldehyde reacts further with the phenolic nucleus (both *meta* and *para* positions are active sites) or the methylene bridges (Waterman and Veldman 1936).

The novolac resins (**2**, $n=2$, 3) are prepared using a molar deficiency of formaldehyde. An acidic catalyst is generally preferred. Since these resins are less reactive, a catalyst or curing agent is generally added to promote cross-linking. The most common is hexa methylenetetraamine or "hexa." The curing agent serves as a latent source of formaldehyde. This combination has the advantage that minimal water of reaction is evolved during the cure. As in the case of the resoles, the chemistry of the novolac cure is highly complex and beyond the scope of this article. The chemistry of phenolic resins is discussed further in Martin (1956) and Megson (1958).

The isocyanates or urethanes, in spite of the previously mentioned drawbacks, find use in the bonding of fiber-reinforced plastics for the automotive industry (Kimball 1981) and in the bonding of elastomers to themselves or metals (Schollenberger 1977). The reactions of isocyanates with alcohols and amines form urethanes (**9**) and ureas (**10**).

There are two types of urethane adhesives. The two component systems utilize the chemistry described above. One component contains the isocyanate, whilst

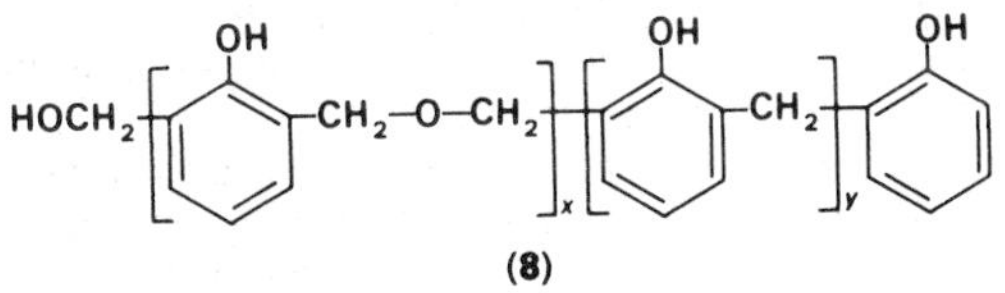

$$HOCH_2\text{–}\Big[C_6H_3(OH)\text{–}CH_2\text{–}O\text{–}CH_2\Big]_x\Big[C_6H_3(OH)\text{–}CH_2\Big]_y\text{–}C_6H_4OH$$

(**8**)

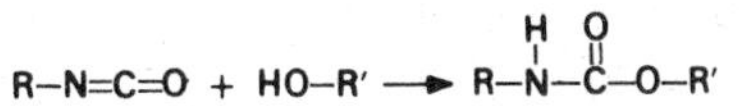

$$R\text{–}N{=}C{=}O + HO\text{–}R' \longrightarrow R\text{–}\overset{H}{N}\text{–}\overset{O}{\overset{\|}{C}}\text{–}O\text{–}R'$$

(**9**)

$$R\text{–}N{=}C{=}O + H_2N\text{–}R' \longrightarrow R\text{–}\overset{H}{N}\text{–}\overset{O}{\overset{\|}{C}}\text{–}\overset{H}{N}\text{–}R'$$

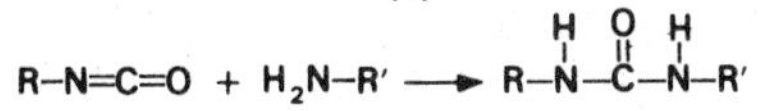

(**10**)

the other is an amine, an alcohol-containing compound, or a polymer. The one-component urethanes rely on atmospheric moisture for cure. Water reacts with the isocyanate group to form a carboxylate group, which splits off CO_2 and NH_2^-; the latter catalyzing the polymerization of the isocyanate. The rapid cure of the urethanes is of importance for assembly line applications.

Two different monomeric species are of importance to the chemistry of the vinyl structural adhesives, the glycol diacrylates or methacrylates (**11**, R=H, CH_3) and the alkyl 2-cyanoacrylates (**12**). Glycol acrylates are used to formulate products known as anerobic adhesives. They are single-component liquids that can be stored at room temperature in the presence of oxygen, and will polymerize rapidly to a solid when confined between two surfaces which exclude air. They also cure rapidly when exposed to ultraviolet light. Because of the rapid cure, products of this type are also readily adaptable to assembly line operations in the automotive and electronics industries. Second generation acrylics are now appearing, which are less affected by surface contamination and are capable of bonding a greater variety of surfaces.

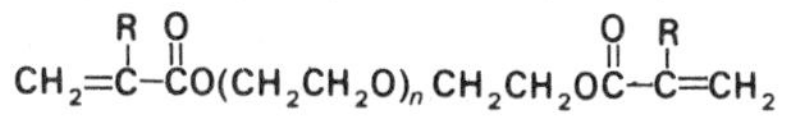

$$CH_2{=}C(R){-}C(O)O(CH_2CH_2O)_n\,CH_2CH_2OC(O){-}C(R){=}CH_2$$

(11)

$$CH_2{=}C(CN){-}C(O){-}O{-}R$$

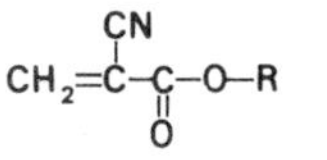

(12)

Cyanoacrylate adhesives display the same room-temperature storage and cure characteristics as the anerobics. However, in this case the cure mechanism is an anionic polymerization catalyzed by a weak base. The formulated products thus contain inhibitors such as sulfur dioxide for stability against anionic polymerization and quinone derivatives as free-radical inhibitors. These products adhere well to a variety of substrates, and are also candidates for assembly line manufacturing. Excellent reviews, with numerous references, can be found for both the anerobic (Murray et al. 1977) and cyanoacrylate (Cover and McIntire 1977) types of products.

Another important class of structural adhesives are those which yield useful bonds above 260 °C. The classic high-temperature adhesive is a phenolic–epoxy combination (Naps 1953) which displays useful bond properties up to 425 °C. Aromatic polyimides, made primarily by the reactions of aromatic diamines with tetrafunctional aromatic carboxylic acid diahydrides, have useful bond strengths at about 370 °C, and have better high-temperature aging characteristics than the epoxy–phenolics. Other polymers which appear of interest for this purpose are the polybenzimidazoles, polybenzothiazoles, polyquinoxalines, polyphenylquinoxalines and the polybenzoxazoles. Useful bond strengths have been reported up to 540 °C; however, the useful lifetime of the bonds drops to 1–10 h above 425 °C (Aponyi 1977).

2. *Testing of Adhesives*

Important adhesive-bond properties to be considered are strength in tension, shear and peel, toughness, environmental resistance and fatigue strength.

The most common test is the lap or tensile shear test. The specimens are inexpensive and easy to make, and the procedure is readily reproducible. The test is used for preliminary evaluation of adhesives and for quality assurance. The forces involved in bond rupture are a combination of the tensile and shear strengths of the adhesive and peel forces acting primarily at the adhesive–adherend interface. Testing in pure tension merely reflects the tensile properties of the polymer involved and pure-shear specimens are difficult to prepare.

The toughness of adhesive bonds can be measured in a number of ways. Most common are the T-peel strength and floating roller peel tests. In peel testing at least one of the adherends is flexible, and is pulled away from the bondline in a manner which concentrates the stress at the peel front. Another measure of toughness is the impact or block-shear test, whereby the bond between two blocks of adherend is sheared by a single, calibrated blow of a test hammer. The difficulty with this test is the need for perfect alignment between the test specimen and the hammer. Recently, toughness of adhesive bonds has been evaluated using a double-cantilever-beam specimen. This is widely recognized as an appropriate measure of bond toughness. The measurement is called the fracture energy (G_{IC}) and is the energy required to initiate the propagation of a crack in the bondline.

Another important characteristic of a bonded structure is its ability to resist hostile environments. The most common method of evaluation is to expose lap-shear specimens, either under tension or in the relaxed state, to the environment in question. The environmental effects most commonly studied are those of moisture, salt spray, selected organic solvents and temperature. One of the more important tests developed is the wedge-crack test, which serves to evaluate the entire bonding process. The substrate is bonded in a manner whereby a wedge of fixed dimension can be forced into the end of the bondline to initiate a crack. The rate at which the crack growth progresses with time is then measured at 49 °C in an atmosphere of 95% relative humidity. The rate is not only dependent on the adhesive used, but also the quality of the substrate preparation (cleaning) prior to bonding.

A further bond property tested is the resistance to fatigue. The test requires a machine which can supply a cyclic load to bonded specimens in tension, compression, or both. Since the test involves expensive equipment, and quality adhesives usually require numerous test cycles, fatigue data are time consuming and expensive to obtain. Consequently, only a few commercial adhesives have fully characterized fatigue-behaviour.

Finally, testing is also important for quality assurance. Until recently, the acceptability of purchased products was based on mechanical testing of bonds using the methods described above. However, because of the wide range of formulative components available to the manufacturer, changes can occur in formulations, but go undetected by mechanical tests. This is a serious problem. Whereas a formulation change may not affect short-term mechanical strengths, long-term effects, such as reduced environmental or fatigue resistance, could result. Fortunately, recent efforts have done much to alleviate this problem; for example, procedures have been evolved for the chemical characterization of epoxy formulations. Both analytical procedures (May et al. 1976) and characterization methods (May et al. 1979) are documented in the literature.

3. *Structural Adhesive Formulation*

The lap shear strengths of some typical structural adhesives are shown as a function of temperature in Fig. 1. Adhesive formulations consist of a matrix–resin system, and in general, a filler or mixtures thereof. The resin system can be highly complex consisting of thermoset resins, curing agents, catalysts, polymers added to achieve special properties, and other additives such as heat stabilizers. The technology of adhesive formulation is the art of compromise; the formulator must add the correct concentration of a component to achieve the desired effect without unwittingly destroying some other needed property. For example, a thermoplastic may be added to a thermoset for toughening, but too much could reduce the elevated temperature performance of the bond to undesirable levels.

Representative formulations of structural adhesives are given in Table 2. The general-purpose epoxy polyamide formulation is typical of the twin-tube kits for home use. Satisfactory cures can be obtained at room temperature; however, mild heat cures (up to 100 °C) will result in better bond performance. Fumed silica is added to one-part general-purpose epoxy adhesives to give thixotropy. This reduces sag of the adhesive on vertical surfaces. The curing agent is highly stable until heated. The normal curing temperature is 177 °C; however, accelerators can be added to the formulation which permit cures in the 115 °C range.

Aromatic amine-cured epoxy formulations are versatile products with good elevated temperature per-

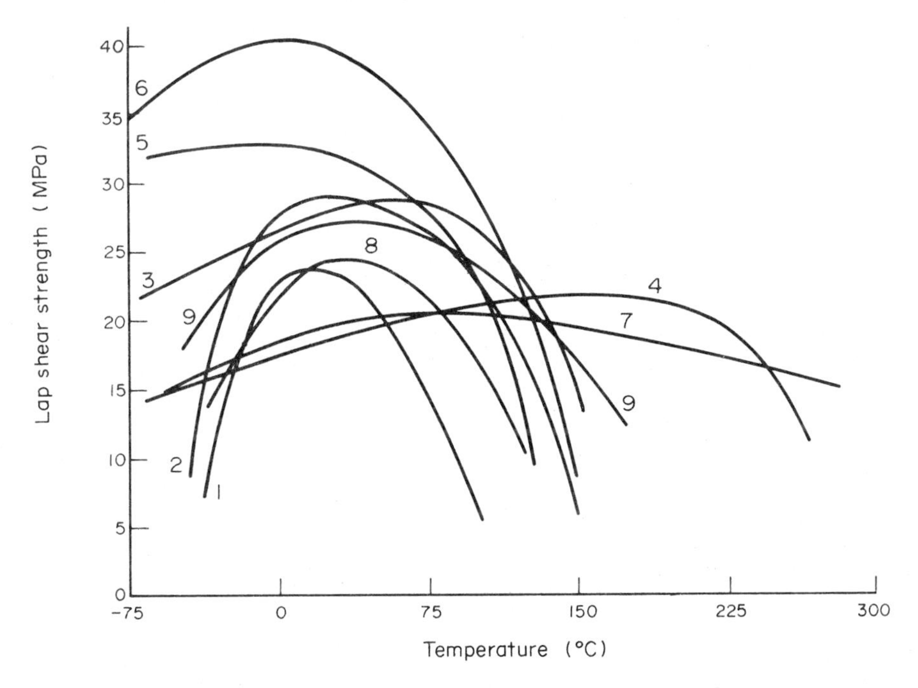

Figure 1
Lap shear strengths of typical structural adhesives: 1, general-purpose epoxy polyamide; 2, one-part general-purpose epoxy; 3, aromatic amine-cured epoxy; 4, epoxy phenolic; 5, rubber-modified epoxy; 6, nylon epoxy; 7, polyimide; 8, vinyl phenolic; 9, nitrile phenolic

Table 2
Representative formulations of structural adhesives

Adhesive	Formulation	Parts by weight
General-purpose epoxy polyamide	liquid diglycidyl BPA[a] resin	100
	polyamide curing agent	75
One-part general-purpose epoxy	liquid diglycidyl BPA resin	100
	alumina	20
	fumed silica	10
	dicyandiamide curing agent	6
Aromatic amine-cured epoxy	liquid diglycidyl BPA resin	100
	silica flour	80
	liquid aromatic amine	20
Epoxy phenolic	solid diglycidyl BPA	33
	phenolic resole	67
	aluminum dust	100
	dicyandiamide curing agent	6
	Cu 8-hydroxyquinoline	1
Rubber-modified epoxy	rubber-modified resin	100
	melamine resin	2
	fumed silica	5
	dicyandiamide	8
	3-(3,4-dichlorophenyl)-1,1-dimethylurea (Diuron)	1.5
	aluminum dust	50
Nylon epoxy	liquid diglycidyl BPA resin	25
	Nylon	75
	dicyandiamide	10
Polyimide	linear condensation polyimide	100
	aluminum dust	80
	fumed silica	5
	arsenic thioarsenate	2
Nitrile phenolic film	nitrile rubber	100
	phenolic resin	100
	sulfur	2
	zinc oxide	5
	benzothiazyl disulfide	1.5
	carbon black	20

[a] BPA, bisphenol A

formance up to 150 °C. There is a wide variety of aromatic amines available which will give roughly the same strengths when used in an amount chemically equivalent to the epoxide content of the resin.

Copper 8-hydroxyquinoline is added to epoxy phenolic formulations to improve the high-temperature aging of bonded structures. Because of the presence of the phenolic resole, water is evolved during the cure. This results in a foamy bondline, but the consistency of the bond performance is unaffected.

The rubber-modified resin, used in formulation of rubber-modified epoxy adhesives, a 10:1 mixture of liquid diglycidyl bisphenol A resin and carboxy-terminated nitrile rubber with an epoxide–carbonyl reaction catalyst at about 100 °C. The exact reaction conditions determine the morphology of the rubber as a dispersed phase and accordingly govern bond performance. Addition of Diuron permits cures in the range of 110–120 °C.

Of the other formulations given in Table 2, nylon epoxy has a very high peel strength and fracture resistance. It is fabricated as a film cast from a solution of the components in an alcohol–water or an alcohol–chlorinated-hydrocarbon solvent.

Vinyl phenolic was the first structural adhesive. It was used in the UK during World War II to assemble the DeHavilland aircraft. Bonds were made by spreading a solution of a phenolic resole on the parts to be bonded and sprinkling it with a powdered polyvinyl formal resin. The solvent was allowed to evaporate and the resultant assembly was subsequently cured with heat and pressure. An exact formulation is thus not possible. Although the technique was crude, metal-to-metal bonds proved to be consistently strong and durable.

Formulations for isocyanate and vinyl structural adhesives cover a wide variety of combinations of materials and are usually tailored for very specific applications. The main factors to consider in selection of a particular formulation are ease of application and speed of cure.

4. *Current and Future Uses of Structural Adhesives*

Presently, the major use of structural adhesives is in the fabrication of aircraft. Adhesively bonded primary structures have been in service in Europe for many years. In the USA, the approach has been more conservative, there are essentially no bonded primary aircraft structures in service today. However, in the US Air Force PABST (Primary Adhesively Bonded Structure Technology) Program, tests proved that bonding technology is ready for use on the next aircraft to be designed (Thrall 1979). That is not to say that adhesive bonding is not a part of aircraft manufacture in the US: many aircraft use adhesives to bond secondary structures. The Lockheed L-1011 TriStar, for example, uses adhesives for the bonding of skin panels, riveted and bonded doublers, triplers and failsafe straps (Petrino 1972). The US Air Force makes extensive use of adhesive bonding. The C-5 aircraft, for example, has a total of over 2300 m^2 of bonded honeycomb structure and about 900 m^2 of metal–metal bonded structure (Reinhart 1981). Suffice to say that under the impetus of programs such as PABST, structural bonding in the aerospace industry still represents a major growth area for the future.

For other industrial uses of adhesive bonding, it is difficult to define single operations which use large volumes of structural adhesives. In the automotive field, although there is a considerable amount of adhesive bonding, the material types and their use vary widely. Urethanes find widespread application in bonding fiberglass-reinforced-plastic body components (Kimball 1981). Epoxies are used for bonding numerous small parts, such as quartz headlamps, underhood electronic devices, automatic chokes, voltage regulators and ignition control modules. Weld and rivet bonding of sheet metal structures is increasing in importance. Epoxies are replacing the traditional lead body solders on trucks, buses and taxicabs (Bolger 1980). As the emphasis on fuel economy through reduced weight continues to become more important, adhesive bonding in the automotive industry will continue to grow.

Structural adhesives are also found in refrigerators, freezers, food disposers, air conditioners, dishwashers, washing machines, driers, ovens and numerous other small appliances. At one factory, over 40 000 electric motor adhesive assembly operations take place each day (Huntress 1979). The electrical and electronic industries use adhesive bonding extensively in the manufacture of motors, generators, transformers, coils and microcircuitry, and for assembling the various components into products such as computers, radios, television sets and microprocessors. For these latter purposes ease of application and speed of cure are important, and hence cyanoacrylates and acrylic type adhesives find widespread use.

As structural adhesives continue to broaden their range of physical properties and methods of application, the use of these products will continue to rise. The primary areas in need of improvement are higher temperature performance and increased toughness. Analytical methods for assuring the proper chemical composition of the adhesive and the proper cure chemistry are realizable goals. The future depends heavily on the reliability and durability of bonded structures.

See also: Adhesives in Building

Bibliography

Aponyi T J 1977 Other high temperature adhesives. In: Skeist 1977, Chap. 38, pp. 619–27

Bolger J C 1980 Epoxies for manufacturing cars, buses and trucks. *Adhes. Age* 23(12): 14–20

Borstell H 1974 *Chemical Composition and Processing Specifications for Air Force Advanced Composite Matrix Materials.* Grumman Aerospace, Bethpage, New York

de Bruyne N A 1939 The nature of adhesion. *Aircraft Eng.* 18(12): 52–54

de Bruyne N A 1950 (to Rohm & Haas Co) Adhesion between surfaces. US Patent No. 2,499,134 (28 February 1950)

Carpenter J F 1978 *Test Program Evaluation of Hercules 3501-6 Resin.* Report No. N00019-77-6-0155. McDonnell Douglas Aircraft, St Louis, Missouri

Castan P 1944 Synthetic resins. US Patent No. 2,324,954 (29 July 1944)

Cover H W Jr, McIntire J M 1977 Cyanoacrylate adhesives. In: Skeist 1977, Chap. 34, pp. 569–80

Dannenberg H, May C A 1969 Epoxide adhesives. In: Patrick R L (ed.) 1969 *Treatise on Adhesion and Adhesives,* Vol. 2. Dekker, New York, Chap. 2, pp. 3–76

Flickinger J H 1974 Elastomer modified epoxy resins. *Repr. Adhes. Seal. Counc. Spring Meeting* 3: 84–102

Floyd D E 1966 *Polyamide Resins,* 2nd edn. Van Nostrand-Reinhold, New York

Frisch K C, Reegen S L (eds.) 1972 *Urethane Science and Technology,* Vol. 1. Technomic, Westport, Connecticut

Greenlee S O 1944 (to Devoe and Raynolds Co) Synthetic surface coatings. US Patent No. 2,456,408 (14 December 1948)

Huntress E A 1979 Adhesive assembly. *Am. Mach.* 123(10): 145–60

Kimball M E 1981 Polymethane adhesives: Properties and bonding procedures. *Adhes. Age* 24(6): 21–26

Martin R W 1956 *The Chemistry of Phenolic Resins.* Wiley, New York

May C A 1977 Composite quality assurance—An art becomes a science. *Natl. SAMPE Symp. Exhib.* 24: 390–403

May C A, Fritzen J S, Wearty D K 1976 *Exploratory Development of Chemical Quality Assurance and Composition of Epoxy Formulations.* AFML Report TR-76-112. US Air Force Materials Laboratory, Dayton, Ohio

May C A, Hadad D K, Browning C E 1979 Physiochemical quality assurance methods for composite matrix resins. *Polym. Eng. Sci.* 19: 545–51

May C A, Helminiak T E, Newey H A 1976 Chemical characterization plan for advanced composite prepregs. *Natl. SAMPE Tech. Conf.* 8: 274–94

May C A, Nixon A C 1961 Reactive diluents in aromatic amine cured epoxy adhesives. *J. Chem. Eng. Data* 6(2): 290–93

Megson N J L 1958 *Phenolic Resin Chemistry.* Academic Press, New York

Murray B D, Hauser M, Elliott J B 1977 Anerobic adhesives. In: Skeist 1977, Chap. 33, pp. 560–69

Naps M 1953 *Elevated Temperature Resistant Modified Epoxide Resin Adhesives for Metals.* AFML Report TR 53-126. Air Force Materials Laboratory, Dayton, Ohio

Petrino D A 1972 L-1011 Tristar places new demands on aircraft sealants and adhesives. *Adhes. Age* 15(2): 15–19

Reinhart T J 1981 Use of structural adhesives by the US Air Force. *Adhes. Age* 24(8): 20–23

Schlack P 1938 Resinous amines rich in nitrogen and suitable for coating paper, cloth, etc. US Patent No. 2,136,928 (15 November 1938)

Schlack P 1939 (to I G F Farbenindustrie) Nitrogenous condensation products. German Patent No. 676,117 (26 May 1939)

Schollenberger C S 1977 Polyurethane and isocyanate based adhesives. In: Skeist 1977, Chap. 27, pp. 446–64

Skeist I (ed.) 1977 *Handbook of Adhesives,* 2nd edn. Van Nostrand-Reinhold, New York

Thrall E W Jr 1979 PABST program test results. *Adhes. Age* 22(10): 22–33

Waterman H I, Veldman A R 1936 Destructive hydrogenization of resins from phenol and formaldehyde I, II. *Br. Plast.* 8: 125–28, 182–84

Zinke A, Hanus F, Ziegler E 1939 Beitrag zur Kenntnis des Härtungsprozessen von Phenol-Formaldehyd-Harzen. *J. Prakt. Chem.* 152: 126–44

Zisman W A 1977 Influence of constitution on adhesion. In: Skeist 1977, Chap. 3, pp. 33–71

C. A. May
[Lockheed Missiles and Space Company, Sunnyvale, California, USA]

Structural Clay Products

The unique nature of the structural clay products industry can be explained in social and economic terms. This industry now produces a large number of individually identifiable products under the general categories of bricks, tiles, pipes and industrial specialties. The large variety of raw materials and forming methods explains the wide range of physical properties found among these products.

1. Structural Clay Products Industry

The nature of the structural clay products industry today is profoundly affected by the traditions established over a period of at least 100 000 years. Building bricks were first made when the urbanization of society began. The making of tiles and pipes for the construction of cities followed as soon as skills in the working of clays developed. Up to the nineteenth century, and in some places much later, these products were manufactured as needed near sites of building construction. An industry for the purpose of supplying structural clay products to all who required them did not exist in early periods of civilization, but the art and manual skills for making these products developed. When the industrial revolution came along in the eighteenth century, the manufacture of structural clay products in factories was a gradual and natural transition from unorganized manual skills to an industry using mechanical and electrical power. In making this transition, much of the manual arts and skills were simply transferred to machine and equipment design. Today, the scientific revolution is gradually intruding into the structural clay products industry in the form of a better understanding of raw materials and products and in the form of automation.

Partly because of its historical development and partly because of its close association with the building industry, the structural clay products industry is very conservative. It resists change in type and quality of its products and methods of manufacture as long as there is a market for them and profits are being made. Many factories have simply gone out of business and closed their doors rather than modify a product line or develop new techniques of manufacture. In fact, whole sections of the industry have disappeared over the years when a product went bad or the market was taken over by a substitute material. The building industry, which is supplied by the structural clay products industry, is characteristically conservative also. This association affects the structural clay products industry in the type of market available. Buildings, whether residential, industrial or public, are large investments for their owners; therefore, they are usually unwilling to try new materials and tend to stay with that which has proven satisfactory in the past.

This conservatism causes the structural clay products industry to be engineering oriented rather than research oriented. The industry is concerned with efficiency and process control from the mining of raw materials to the packaging of products for shipment. In order to keep the unit price low, there is an intense interest in eliminating labor costs in the manufacturing process by automation. The other major concern of this industry is the ability to use various kinds of hydrocarbon fuels and to use them inexpensively and efficiently. All of these production problems are approached from an engineering point of view, and are mostly handled on the production line.

Unlike our new-technology industries that have their origins in the research laboratory, the structural clay products industry is not research oriented. The industry as a whole does not put much of its profits into basic and applied research because it has no expectation of the development of new products, and it sees little need for the improvement of the products produced currently. In addition, the industry has little confidence in the outcome of research and not enough patience to allow a research program to be productive.

The large variation in the sizes of the companies producing structural clay products in the USA is also not conducive to proper research endeavor. The efforts that small companies can make is often too small to be effective, and large companies are unwilling to make a considerable investment that would assist their competitors as well as themselves. Cooperative efforts have been tried on a national scale in the highly diverse brick section of the industry, but they have not been successful.

Several well-developed countries meet the research needs of the structural clay products industry by government sponsorship of the pertinent programs. Such an effort would be beneficial in the USA because there are problems in this industry that should be addressed. Some of these problems are related to a better characterization of raw materials, a better understanding of clay plasticity, an identification of the phases in the final products that contribute to strength and durability, a precise description of pore structure, the adsorption of molecular species on internal surfaces and the phenomenon of secondary moisture exapansion.

The structural clay products industry could be classified as energy intensive as are most branches of the classical ceramic and metallurgical industries. A large portion of the total cost of the product is in the form of energy use and fuels consumed. Electric power is used to run the heavy machinery required for mineral dressing and forming the products. In addition, electricity operates the automation and air circulating fans connected with the dryers and kilns. A large part of the total energy consumed is in the form of heat from the burning of natural gas, oil and coal that is required to dry the products from the plastic state and to fire them to maturity. Temperatures between 980 °C and 1260 °C are obtained in the kilns in order to convert the raw materials into durable products with specified physical properties.

As in the case of the producers of most materials, the structural clay products industry has little control of the end use of its products. Once bricks and structural tiles are placed in the walls of buildings, it is an extremely complex task to trace the causes of successes or failures. When failures do occur, it is difficult to determine whether the clay products, the design of the building, the customer use or the quality of construction were at fault. The best defense this industry has against complaints is to make products good enough to withstand the most serious stress situations that can be anticipated.

2. *Size and Economics of the Industry*

The size of the structural clay products industry in the USA is relatively small, but it is an important supplier to the building construction industry. The Statistical Abstract of the USA published by the US Bureau of the Census lists 47 000 workers employed in this industry in 1976. The same source gives the dollar value of the products shipped in 1976 as $1 595 000 000 and the value added by manufacturing as $930 000 000.

The structural clay products industry is economically unstable in that the volume of business fluctuates widely with the economic climate of the country. During periods of recession, many plants have to be shut down to prevent intolerable inventories, but in periods of economic growth these same factories may be back ordered up to a year and a half. These trends follow similar fluctuations in the building construction industry. Figure 1 shows these relationships and trends over a period of 26 years. The quantity of facing bricks shipped was selected as an indicator of the activity in the whole structural clay products industry because facing bricks constitute a little more than half of the dollar value of all structural clay products used. New housing units started include private, corporate and government programs, and it was selected as a measure of the total building construction of the period.

Although the average output of facing bricks has been relatively constant over the past 30 years, some other types of structural clay products have had different histories. The production of structural facing and partition tiles has decreased steadily over this period. This decline was probably due to the substitution of lower-cost materials of a different kind. Sewer

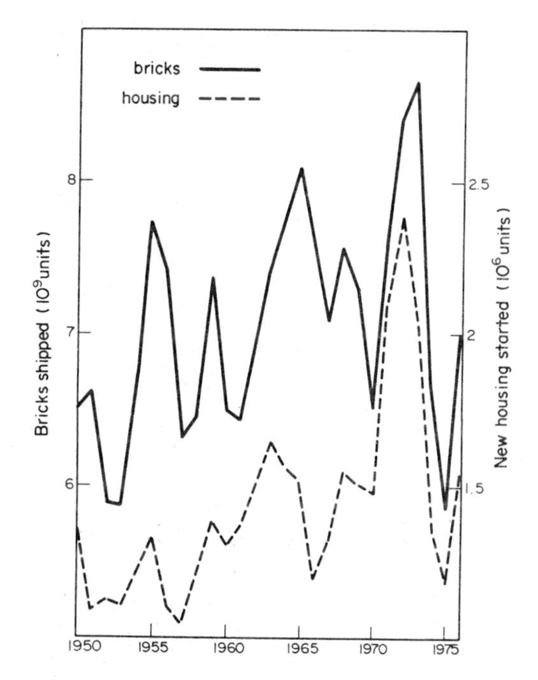

Figure 1
Fluctuations in the business activity of the structural clay products industry and the building construction industry

and drain pipe production reached a peak in 1964 and has been in a decline since that year; however, it still remains the third largest section of the structural clay products industry with an annual dollar value of shipments in 1976 of $123 000 000. The second largest portion of the whole industry in dollar value of shipments is the production of floor and wall tiles. In 1976, this value stood at $186 000 000. The actual production of these tiles has remained reasonably constant over the past 20 years.

Perhaps the large initial capital investment for new factories to produce such low-cost products discourages growth in the structural clay products industry. The cost of a new, single-line production factory is from 6 to 10 million dollars. Brick plants must produce 80 000–100 000 standard brick equivalents per day for each production line in order to make a satisfactory profit.

3. *Classification of Types of Products*

The various identifiable types of structural clay products are outlined in Table 1. The largest sections of this classification are bricks, tiles and pipes. Under each of these sections there are many varieties that have their own unique properties appropriate to the intended use. The forming methods used are a part of this classification because the appearance and physical properties are different for the various methods.

Table 1
Types of structural clay products

I Bricks
- A Facing (C62-75a, C126-76, C216-79)[a]
  - 1 Extruded
    - a Solid, smooth, textured or glazed
    - b Cored, smooth, textured or glazed
    - c Panel, smooth, textured or glazed (C652-77)
  - 2 Sand molded, natural, colored or glazed
  - 3 Dry pressed, solid
- B Paving, solid—extruded
  - 1 Sidewalk, smooth or textured (C902-79a)
  - 2 Street or highway, smooth or textured
- C Industrial floor, solid, smooth or textured—extruded (C410-78)
- D Chimney—extruded
- E Industrial chimney lining—extruded
- F Sewer and manhole—extruded (C32-79)

II. Tiles
- A Structural for walls—extruded
  - 1 Non-load-bearing, smooth or textured (C56-76)
  - 2 Load-bearing, smooth or textured (C34-75)
  - 3 Facing (C212-75)
    - a Smooth
    - b Textured
    - c Glazed (C126-76)
    - d Acoustical
    - e Lightweight
    - f Through-the-wall
  - 4 Screen, smooth or textured (C530-75)
- B Floor
  - 1 Large-cored, smooth or textured—extruded (C57-78)
  - 2 Quarry—plastic pressed
  - 3 Mosaic, unglazed and glazed—dry pressed
- C Wall, unglazed and glazed—dry pressed
- D Roofing—extruded and plastic pressed
- E Flue linings—extruded (C315-78c)

III. Pipes—extruded
- A Sewer and drain, plain or glazed (c700-78a)
  - 1 Bell and spigot end
  - 2 Plain end
- B Perforated subdrainage (C700-78a)
  - 1 Bell and spigot end
  - 2 Plain end
- C Chemical resistant, plain or glazed
- D Conduit, plain or glazed
- E Drain tile
  - 1 Plain (C4-75)
  - 2 Perforated (C498-75)

IV Liner plates, curved or flat, plain or glazed—extruded (C479-77)

V. Filter blocks for trickling filters, plain or glazed—extruded (C159-78)

VI. Chemical resistant tower packings (C515-76)

[a] Designations in parentheses refer to the appropriate specifications of the American Society for Testing and Materials

4. Forming Methods

The forming of structural clay products by extrusion induces inherently unique physical properties. The most common extrusion machinery uses an auger in a vacuum chamber to force the plastic, clay-containing raw materials through a shaping die. Because of the flaky morphology of the clay minerals and the frictional resistance of the clay going through the die under considerable pressure, laminations are induced into the extruded column. These laminations are caused by the extensive orientation of the clay particles under the pressure of the auger flights, and slip planes which develop between the center of the extruded column and the inner die surfaces. Such preferred orientation and concentration of clay particles in the vicinity of the laminations give a directional character to the crystalline microstructure and the pore structure of the products. These directional characteristics are often reflected in directional strength and adsorption. The laminations can be planes of weakness and planes of open pores. In the best industrial practice, such laminations are completely healed by maintaining adequate vacuum on the auger chamber, optimum particle-size distribution of the coarse filler grains and the proper water content in the plastic mix. Some large sewer pipes are extruded by a ram instead of an auger; therefore, laminations are less well developed and often completely unnoticeable in these products.

The sand molding of plastic clay is generally referred to as the soft-mud process. The clay material is made up in a softer plastic condition than for extrusion; then, bricks are formed by ramming the soft clay into sand-coated, wooden molds. Although this is a very ancient method for making face bricks, it is a highly automated process in our modern factories. This method has the advantage of a random distribution of crystalline phases and pores, and there is no tendency toward laminations. Bricks made in this way are intentionally less perfect in size and shape than those made by extrusion.

Clay materials for plastic pressing are prepared to the state of maximum plasticity or maximum plastic strength, and the tempered clay is vacuum extruded and cut into regular bats in preparation for pressing. In relation to the other forming methods, the plastic state is intermediate between that for extrusion on the drier side and sand molding on the wetter side. The forming of the ware is done by pressing the plastic clay bat between upper and lower molds made from plaster of paris. This method produces some distortion of the pore structure, but the product is relatively free of laminations and is of uniform strength.

Thin floor and wall tiles are dry pressed to shape in large hydraulic presses under a pressure of about 34.47 MPa (5000 psi). Some small amount of water containing a binder is added to the clay batch; so the forming operation should more accurately be termed damp pressing. This method of forming produces a different type of pore structure from that produced by plastic methods. Laminations are possible, but they are usually avoided by the proper selection of clays and compacting pressure. The major advantage of dry pressing is the more uniform size and shape of the product, since all drying shrinkage has been eliminated.

The porosity of the final product is more dependent on the raw materials used and the heat treatment than on the forming method. Firing of clay products shifts the pore-size distribution towards larger sizes and reduces the total pore volume. This change in pore structure is accompanied by shrinkage.

5. *Raw Materials*

The wide variation in the types of clay raw materials used for the manufacture of structural clay products multiplies the classification by a very large number if the aesthetics of the products are considered. Both clay and argillaceous shale deposits, located near the factories, are used as nature provided them with little or no beneficiation. The exception is in the floor and wall tile industry. Here, the clay bodies are prepared in exact compositions from several beneficiated raw materials. This industry uses mixtures of specific clays, feldspar, flint (pulverized quartz), pyrophyllite and talc.

Regardless of whether natural or beneficiated raw materials are used, there is a general range of mineral compositions that must be adhered to for satisfactory production. There are three types of minerals that must be present in proper amounts for all clay products. The clay content can range from 35% to 55%. Inert fillers, such as quartz, will generally make up from 25% to 45% of mixture. Finally, reactive fillers or fluxing materials are required in the range from 25% to 55%. Natural raw materials must contain these elements in acceptable proportions in order to be useful to the structural clay products industry. For this reason many plants elect to blend two or more naturally occurring raw materials to achieve the desired physical properties and visible appearance.

A list of the mineralogical–geological types of natural raw materials used in the USA indicates the variations that can occur in each product line. This can be illustrated by reference to fired color:

(a) crude, residual, kaolinitic clays—white;

(b) kaolinitic–illitic fireclays—buff;

(c) illitic–kaolinitic shales—red;

(d) illitic clays—red;

(e) illitic shales—red;

(f) illitic–chloritic shales—red;

(g) argillaceous–calcareous clays—pink to buff; and

(h) argillaceous–calcareous shales—pink to buff.

The clay minerals in the above list are kaolinite and illite. Kaolinite is a refractory clay that requires other reactive fillers or fluxes to produce satisfactory products; however, illite acts as its own flux on firing to 1050 °C and higher temperatures. A third clay mineral, montmorillonite, is sometimes found in small quantities in clay deposits, but it is not an essential ingredient, and it does present some special problems in production.

6. *Physical Properties*

As one might expect from the variations in forming methods and raw materials, the physical properties of structural clay products vary widely but still remain within certain broad limits. Table 2 gives the ranges of values for strength and water absorption for five general types. The ranges of values mean that in some locations these products can be found with average values in these ranges. Any specific product will have a much narrower spread of values around its mean. The compressive strength of some bricks will be as high as

Table 2
Physical properties of structural clay products

Product	Compressive strength (lb in.$^{-2}$)	Compressive strength (MPa)	Modulus of rupture (lb in.$^{-2}$)	Modulus of rupture (MPa)	Absorption (%)
Bricks, facing	1250–20000	3.62–137.90	500–2000	3.45–13.79	2–25
Bricks, paving	2500–20000	17.24–137.90	1500–2500	10.34–17.24	1–6
Tiles, structural	500–10000	3.45– 68.95			2–28
	Crushing strength (lb ft^{-1})	Crushing strength kN m^{-1})			
Pipes, sewer	1200–10000	17.5–145.8			1–8
Pipes, drain tile	680–5000	9.9–72.9			5–16

137.9 MPa (20 000 psi), and the modulus of rupture is usually around 6.9 MPa (1000 psi). The absorption of water by building bricks is preferred to be in the 5% to 8% range. Structural tiles have compressive strengths below that of bricks made from the same material because of their difference in geometry. Sewer pipes can have crushing strengths up to 68.9 kN m^{-1} (10 000 lb ft^{-1}), but drain pipes are usually weaker and have higher absorptions because of their different end use.

7. *Specifications*

The generally accepted standard specifications for the physical properties of bricks and structural tiles are too broad and relatively ineffective. For these reasons, many organizations rely on the past performances of specific products as a guide to reliability, and this practice is subject to many dangers, including undetected changes in the products and purchasing frauds. The main reason for the broadness of current specifications is that an attempt was made to write one set of specifications for all products regardless of their types of raw materials or methods of manufacture. An important secondary explanation for the broadness is that the appropriate relationships between physical properties and actual service conditions are not known. This ignorance reflects the lack of research in the field of structural clay products. As it is, the specifications are not much protection for the customer and only a vague guide for manufacturers.

Bibliography

American Society for Testing and Materials 1980 *Annual Book of ASTM Standards.* American Society for Testing and Materials, Philadelphia, Pennsylvania

Bloor E C 1959 Plasticity in theory and practice. *Trans. Brit. Ceram. Soc.* 58: 429–53

Brownell W E 1976 *Structural Clay Products.* Springer, New York

Emrich E W 1973 History and development of ceramic wall tile bodies in the United States. *Am. Ceram. Soc. Bull.* 52(9): 687–88

Grim R E 1968 *Clay Mineralogy*, 2nd edn. McGraw-Hill, New York

Grimshaw R W 1971 *The Chemistry and Physics of Clays*, 4th edn. Wiley–Interscience, New York

National Clay Pipe Institute 1972 *Clay Pipe Engineering Manual.* National Clay Pipe Institute, Washington, DC

van Olphen H 1963 *An Introduction to Clay Colloid Chemistry.* Interscience, New York

W. E. Brownell
[New York State College of Ceramics,
New York, USA]

Structural Sealant Glazing

Structural sealant glazing is the architectural concept in which glass panels are glued to the side of a building. This type of construction allows for "glass cube" buildings and continuous runs of glass as in strip windows.

This concept of structural glazing was tested extensively in sealant development laboratories in the late 1960s and the early 1970s. A few buildings were constructed in this manner each year from 1968. In 1978, after 10 years of testing both in the laboratory and in mockups, the concept had become accepted and it has since grown rapidly in popularity. Routine inspections of early structural glazing projects continue to provide confidence in the performance of this technique. The sealant that glues the glass to the building also performs the task of weathersealing the building against the elements. Besides being aesthetically pleasing and giving new freedom to the architect, glazing of this design is easy to clean and maintain due to the uncomplicated exterior.

The heart of the structural system is the sealant that transfers the wind load from the glass panel to the support mullions. The sealant, or glue, must be sufficiently strong to counteract the negative wind load yet remain flexible enough to accommodate the movement from differential thermal expansion experienced in metal curtainwall construction.

Since this adhesive or sealant must not stiffen and lose its elastic properties in extreme cold (−35 to −40 °C is experienced at some installations) or soften and lose its adhesion strength in the heat (dark glass and dark metal systems can experience 80–85 °C on a southern exposure in southern climates), only silicone adhesives or sealants are at present allowed in this application. Also the bond line sees the full radiation of the sun either directly (at the edge) or through the glass (back-sealed system) and must not deteriorate. At present only certain silicones can perform in this high-stress environment. Structural glazing is in fact a concept made possible only by the existence of a durable, elastomeric glue.

There are two different types of structural glazing design. The first, and the one with the longest history, can be classified as "two-sided" silicone glazing. This design concept incorporates silicone sealant as the structural adhesive on only two opposite sides (edges) of the glass panel. The other two sides are typically held in a metal glazing channel at the head and sill. The glazing channel supports the weight of the glass and provides outside capture of the glass.

The surface contact of the silicone sealant on the verticals is calculated on the basis of both the specified wind load and the glass width. This calculation utilizes the trapezoidal distribution of stress on the glass to approximate the force transmitted through the silicone sealant.

The sealant most often specified for this application is a high or medium modulus silicone. The design stress for acetoxy sealant is generally accepted as 140 kPa. Based on this design stress, the calculated silicone sealant contact surface allows for a two and a half- to seven-fold safety factor based on perfect laboratory specimens.

The second and newer type of structural silicone glazing is referred to as "four-sided" sealant glazing. This type incorporates silicone sealant as the only structural link between the glass panel and the metal curtainwall. Some designs specify a metal fin at the horizontals. These fins accept the weight of the glass through the setting blocks placed at the quarter points. The silicone sealant provides both the structural fastening and the weathersealing of the panels.

The most innovative utilization of "four-sided" structural sealant glazing eliminates the fin from the design. The metal framing is literally glued to the back side of the glass with the silicone sealant. In this design, the sealant constantly bears the weight of the glass. The sealant also counteracts the negative wind load as well as weathersealing the building.

The successful utilization of silicone sealant glazing demands that all accessory spacers, setting blocks and gaskets be totally compatible with the sealant. Since the ultraviolet exposure of the spacers and setting blocks is much higher than in typical wet or dry glazing in a glazing channel, newer, more stable compounds must be used in the fabrication of these rubber extrusions. Use of silicone rubber extrusions for all the accessory materials assures complete compatibility. Selected organic accessory materials can also be incorporated into the design, provided that they have been evaluated by the sealant manufacturer.

The laboratory test procedure for determining compatibility involves a 21-day exposure period under concentrated ultraviolet exposure. The criterion for determining compatibility depends on the sealant manufacturer. To determine the compatibility of spacers, a clear silicone, attached to the spacer, is exposed to accelerated ultraviolet weathering. No color change should be observed after 21 days. The result of extreme incompatibility could be total loss of adhesion of the silicone sealant.

The ultimate success of structural sealant glazing lies in the long-term durability of the adhesive bond of the sealant to both the glass and the metal mullion. The industry presently accepts ASTM C794-80 (*Standard Test for Adhesion-in-Peel of Elastomeric Joint Sealants*) as a method for determining the adhesion of structural sealants. This test method makes use of a metal screen embedded in a strip of sealant. The test also includes a 7-day water immersion cycle.

ASTM Committee C-24 on Building Seals and Sealants is currently developing a more realistic method for testing adhesion. This test method utilizes a sealant specimen 13 × 13 × 51 mm that more closely approximates the typical structural sealant joint.

Thus structural sealant glazing is a systems concept in which compatible glass (or other panels), metal mullions, setting blocks, spacers and high-strength, high-modulus silicone adhesives or sealants all work together. To make such a system successful demands competent workmanship during installation. In particular, the sealant needs to be applied to a clean, dry surface according to the manufacturer's recommendations.

See also: Adhesives in Building; Glass as a Building Material; Glazing; Sealants as Building Materials

Bibliography

Brower J R 1980 Considerations and advantages of silicone sealants in curtainwall applications. *US Glass, Met. Glazing* July/August: 45–53

Hilliard J R, Parise C J, Peterson C O Jr 1977 *Structural Sealant Glazing. Sealant Technology in Glazing Systems*, ASTM STP-6-1977. American Society for Testing and Materials, Philadelphia, Pennsylvania, pp. 67–99

Kasperski M G, Klosowski J M 1978 Construction sealants. *Constr. Specifier* 31 (10): 21–26

Klosowski J M, Gant G A L 1980 *The Chemistry of Silicone Room Temperature Vulcanizing Sealants*. Reprinted from ACS Symposium Series No. 113, *Plastic Mortar, Sealants and Caulking Compounds*, pp. 113–27. American Chemical Society, Washington, DC

Peterson C O Jr 1980 Structural silicone glazing. *US Glass, Met. Glazing* July/August: 38

Romig C A, Klosowski J M 1977 Sealants. *Glass Dig*. 56: 27, 33

J. M. Klosowski and C. M. Schmidt
[Dow Corning Corporation, Midland, Michigan, USA]

Structure of Materials

All materials are aggregate of atoms. An understanding of their structure thus starts with the description of positions of atoms in space within a material. Since an object of macroscopic size is likely to contain in excess of 10^{20} atoms, it is impossible to describe the detailed atomic positions. Descriptions of the structure of real materials, therefore, tend to focus on both the ideal or perfect structure and the imperfections in the structure. For a description of the structure of materials, it is rarely helpful to consider any material to be homogeneous. Because of the heterogeneous nature of materials, an important aspect of structure is the relationship between various parts of a material.

This article presents a survey of the structure of materials with attention to the arrangements and interrelations of the component parts of a material. It is a general treatment of the subject that is intended to be independent of any particular class of materials.

1. Structural Units and Interactions

For many purposes it is useful to consider materials as being built up from small structural units, which form a macroscopic body by joining together. In liquids, the atoms or molecules are arranged in dense random-packed structures. Such structures have been shown to be describable by a packing of polyhedral units in which atoms or molecules occupy vertices of the polyhedra.

Silica glass is composed of randomly joined structural units, called coordination tetrahedra, made up of a central silicon atom surrounded by four oxygen atoms which lie at vertices of a regular tetrahedron. The polymer polyethylene is formed by joining ethylene molecules end-to-end in a process known as vinyl polymerization. In this instance, the structural unit is a "mer" that can be represented schematically by the formula $-CH_2-CH_2-$. For crystalline materials, the unit cell is frequently considered to be the structural unit. From a crystallographic viewpoint, however, the most basic element of a crystal structure is the asymmetric unit that is repeated by all the symmetry operations of the space group of the crystal.

Interactions between the atoms in a material result in the formation of chemical bonds (Kittel 1976). Many aspects of the structure and properties of materials derive directly from the strength and directionality of these bonds. Covalently bonded materials have highly directional bonds which are generally very strong. The directionality of the bonds between carbon atoms in polymers allows for manipulation of polymer structure and properties by varying the degree of cross-linking. In materials with bonding that is largely ionic or metallic, the bonds are less directional. In such materials, arrangements on a local scale are determined mainly by atomic packing into structures of high density.

The intensity and directionality of the interaction between structural units are closely related to the mode of aggregation of the units into a macroscopic body. In metals, the metallic bond provides the primary interaction. Since the bond is nondirectional, aggregates of metal atoms have close-packed or nearly close-packed structures and are generally crystalline in the solid state. The same is true of ionically bonded materials, but because anions and cations generally differ in size, more complex packing arrangements, and hence crystal structures, occur. In long-chain polymers, the interactions may be due to strong covalent bonds when the polymer molecules are cross-linked, or due to relatively weak van der Waals forces in the absence of cross-linking.

The structure and configuration of polymer molecules govern the modes of aggregation that are possible in polymeric materials (Pine et al. 1980). If the degree of internal order, or tacticity, of the polymer molecule is high, molecules can form regular intermolecular bonds and crystalline polymeric structures may result. Atactic molecules, on the other hand, have little internal order and form only glassy structures.

In metallic alloys, different configurational arrangements of the elements are known to form depending not only on the relative strengths of interactions between neighboring pairs of atoms but also on the temperature of the alloy. Consider two species confined to a square lattice, with the nearest neighbor interactions favoring bonds between unlike pairs. At low temperatures, the equilibrium state will be an ordered arrangement that minimizes nearest neighbor bonds between like atoms. At high temperatures, the free energy is lowest for a disordered state, since this state maximizes the entropy. Schematic representations of the two states are presented in Fig. 1.

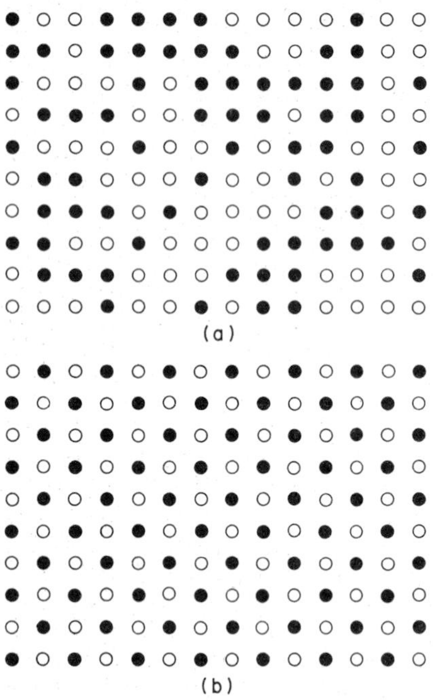

(a)

(b)

Figure 1
A square lattice occupied by two kinds of atoms, in which nearest neighbor interactions favor bonds between unlike species: (a) high-temperature equilibrium state is disordered; (b) low-temperature equilibrium state is ordered

2. *Homogeneous and Heterogeneous Materials*

Interpreted rigorously, no material is truly homogeneous. When viewed on a small enough scale, an apparently homogeneous structure like that of a molten pure element would be found to consist of dense atomic nuclei separated by a large volume of empty space. Conversely, when viewed on a scale much larger than the structural units, imperfections in the structure, which themselves comprise heterogeneities in the material, may become apparent (see Sect. 4).

Single-phase materials may exhibit a wide range of imperfections, as summarized in Fig. 2. Local configurations may be present around some of the structural units that are atypical of the average structure. Other defects become apparent only at a much larger

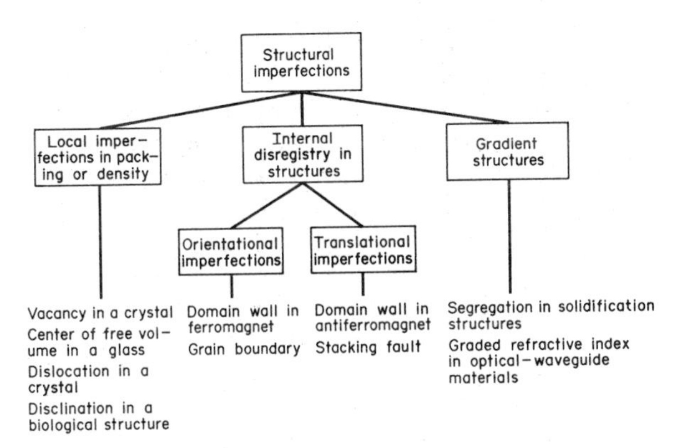

Figure 2
An outline of the types of imperfections possible in single-phase materials, with examples

scale. For example, internal disregistry between adjacent regions in a material is only evident when considering the relationship between these regions.

Multiphase materials are obviously heterogeneous with respect to thermodynamic properties. The structural features characterizing these materials are many. They include all of the features presented in Fig. 2 for single-phase materials and several additional ones: the amount of each phase present, the spatial distribution of the phases, the size distribution of dispersed phases, connectivity in the microstructure, the shapes of the regions occupied by each phase and the degree of their alignment.

Several of these latter features are illustrated in the micrograph of a hypoeutectoid steel, Fig. 3. The microstructure consists of only two phases: ferrite, which is light, and cementite, which is dark. Large volumes of ferrite are free of cementite, and in the rest of the sample ferrite and cementite alternate in a structure of lamellar morphology called pearlite. The tendency toward local alignment of the adjacent cementite lamellae is evident in Fig. 3. The spatial distribution of the phases in the sample is nonuniform. The extent of connectivity of the phases in lamellar morphologies is of special interest. In pearlities, it is believed generally that a highly interconnected morphology is produced by branching of each phase during growth. In lead–tin alloys, the lamellar two-phase morphology which results from cellular precipitation is apparently formed by repeated nucleation of one phase, resulting in little connectivity between phases.

The structural features that are pertinent to the characterization of multiphase materials are equally relevant to the characterization of porosity in materials. In this respect, the pores may be considered as a separate "phase," and structural features such as pore size, shape and connectivity may be studied.

Both single-phase and multiphase materials may contain gradients in their composition or structure.

Figure 3
Photomicrograph of polished and etched hypoeutectoid steel

During solidification of multicomponent systems, the phases in contact at the solid–liquid interface generally differ in composition. Frequently, the compositions at the interface change gradually during solidification. If diffusion in the solid phase is slow, the resulting solid will have gradients of composition.

Structural gradients in materials may be introduced by special processing steps. In general, many of the structural characteristics listed in Fig. 2 can vary spatially in a material. Case-hardened steel provides one example. In nitriding, nitrogen is diffused at elevated temperature into the steel from the surface, forming precipitates that harden the surface layer. Figure 4 shows a dark case-hardened layer at the surface of a nitrided alloy steel, the lighter region being the softer interior of the specimen that is free of nitrides and etches less deeply. In this example, gradients of composition, microstructure and mechanical properties exist normal to the surface of the steel as a result of the case-hardening treatment. The pointed black features visible in the micrograph are Knoop-microhardness indentations produced by a diamond indenter. The hardness gradient is evident from the change in length of the indentations.

Figure 4
Photomicrograph of nitrided alloy steel, polished and etched; showing Knoop-microhardness indentations made with a 300 g load

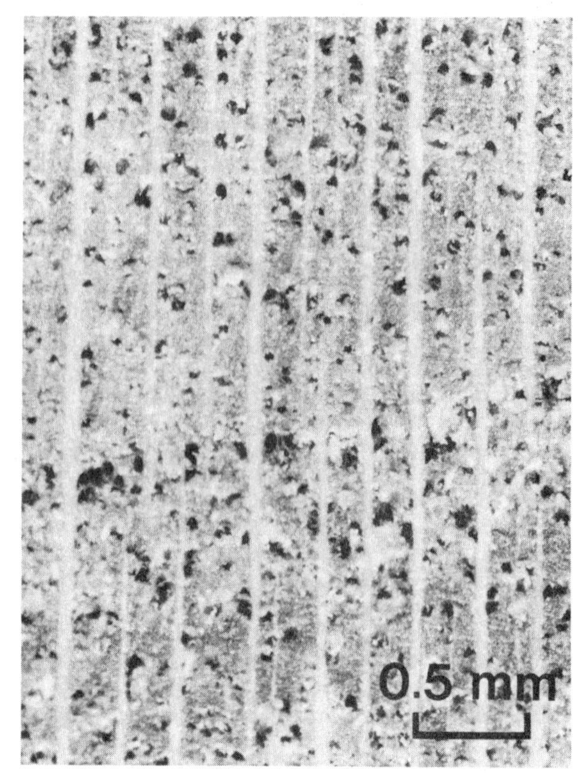

Figure 5
Photomicrograph of cherry wood in cross section: the dark features are vessels; the long light features are rays aligned in the radial direction of the tree (dark-field illumination)

Composite materials can also be designed to incorporate gradients in the relative amounts, distribution and orientation of the component structural elements. Various possibilities of structural gradients in composites have been discussed (Bever and Duwez 1972).

Materials of biological origin contain gradients in their structures. Annual growth rings in some species of wood arise from changes in the diameter of newly grown cells due to seasonal changes in sunlight, temperature and moisture levels. The rings develop because new cells are grown only in the thin layer, called the cambium, that separates the wood from the bark. Gradients in cell diameter in the circumferential direction are in principle minimal, whereas gradients in the radial directions can be appreciable.

Some wood microstructures are suggestive of reinforced composite materials. The microstructure of cherry wood is shown in Fig. 5 in cross section (perpendicular to the tree axis, or longitudinal direction). The dark features are vessels, tubular openings in the wood that allow transport of sap along the tree axis. Vessels are surrounded by much smaller cells, called tracheids, that are also aligned longitudinally. The wood is reinforced by ray cells which are aligned in the radial direction. Ray cells are the long light features visible in Fig. 5. The combination of these structural elements results in a material suited to both biological function (e.g., growth) and mechanical strength.

3. Structures and Space Filling

The assembly of structural units into a macroscopic body is ordered by geometric and topological constraints, in addition to those imposed by physical forces between the units. An exposition of diverse examples of the action of constraints on two-dimensional structures is given by Gombrich (1979). For crystalline materials, the principles of repetition theory which underlie crystallography limit the total number of distinct arrangements of a structural unit. Also, the structures of all materials, whether crystalline or not, are subject to topological laws.

In a crystalline material, a motif or structural unit is repeated periodically to produce the structure. The periodic repetition of primary interest in materials occurs in three dimensions. Crystallography is concerned with the delimiting of allowable patterns of crystals based on the simple requirement that the pattern must be periodic. There are only two ways of generating crystal structures if the crystal is one-dimensional. For two-dimensional crystals, there are 17 different arrangements. In three-dimensional crystals, 230 possible arrangements of a motif exist, corresponding to the 230 crystallographic space groups. For magnetically ordered three-dimensional crystals, the number of distinct arrangements is 1651, corresponding to the number of magnetic space groups. The total number of distinct crystalline arrangements in materials is thus determined by the dimensionality of the repetition.

The topological principles important for an understanding of materials are those that provide relationships between microstructural features of different dimensionality (Smith 1964). In polycrystalline materials, topological principles can be applied to the grain structure. The polycrystal is composed of individual crystals (called grains) which are separated from each other by grain boundaries, grain edges and grain corners. The grains are three-dimensional (volumes), the grain boundaries are two-dimensional (surfaces), the grain edges are one-dimensional (space curves), and the grain corners are zero-dimensional (points). Euler's law states the relationship between the number of volumes, surfaces, curves, and points (n_3, n_2, n_1 and n_0, respectively):

$$n_0 - n_1 + n_2 - n_3 = 1$$

This equation applies equally well to any multiply connected network; for example, it applies to the network formed by interlinking of tetrahedra in a silicate-glass structure and to the cell structure of biological materials.

The effect of capillarity at interfacial junctions provides further restrictions on grain and cell shapes. In three-dimensional structures, an analysis of interfacial junctions leads to the conclusion that for stability four edges meet at a point and three boundaries meet along an edge. These constraints result in a simplification of Euler's law (Smith 1964, Barrett and Yust 1970) which allows conclusions about grain shapes to be drawn.

A two-dimensional cell structure of soap bubbles confined between glass plates is shown in Fig. 6. In such a structure, equilibrium of capillary forces at soap-film junctions requires that angles between cell edges at a junction equal 120°. Thus, only hexagonal cells can have straight sides; cells with fewer sides tend to be convex when viewed from the outside, and those with more sides concave. In a two-dimensional structure, when the number of cells is large, the average number of edges per cell is six (Smith 1964).

Curvature of a cell boundary may lead to boundary migration, for example, by diffusion of gases through soap films or by solid state diffusion across grain

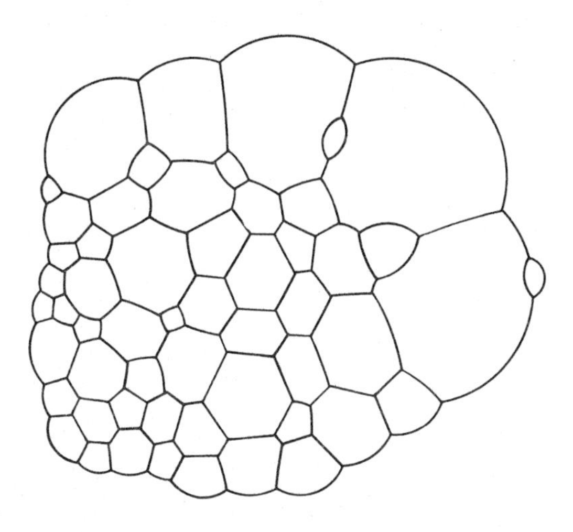

Figure 6
An aggregate of soap bubbles of irregular size placed between parallel glass plates (after Smith 1952 *Metal Interfaces*, p. 70. © American Society for Metals, Metals Park, Ohio. Reproduced with permission)

boundaries in polycrystals. Such boundaries migrate toward their center of curvature, resulting in coarsening of the cell structure. For two-dimensional cell structures, irregularities in the number of sides bounding the polygons are maintained during coarsening. As a result, coarsening would continue indefinitely in an infinitely large polygonal structure.

Similar principles apply in three-dimensional structures. The α-tetrakaidecahedron is a polyhedron with fourteen faces that in some respects plays a similar role to the hexagon in two dimensions. Identical polyhedra can be joined together to fill space. Its faces have zero mean curvature. When joined together to fill space, capillary forces at interfacial junctions are balanced.

Another aspect of space filling, alluded to earlier, is the physical problem of packing atoms into dense structures. Materials with nondirectional bonding tend to be densely packed, and many structures of such materials can be understood by considering the packing of rigid spheres. For spheres of equal size, two basic structures of maximum possible density exist: face centered cubic (fcc) and close-packed hexagonal. Many elements have these structures in the solid state. In many ionically bonded materials, however, radii of the ions differ. The coordination number for the structures of such a material is the number of ions of one element that surround a smaller ion of another element. The only possible coordination numbers are 1, 2, 3, 4, 6, 8 and 12 (Buerger 1956). For each coordination number, a range of ionic-radius ratios can be computed over which the coordination number is expected to be stable. Agreement between expected and observed coordination for ionically bonded atoms is generally good.

The structures of materials with directional bonding are strongly influenced by this directionality. A good example is diamond. Hybridization of electron wave functions of carbon atoms results in four bonds of equal strength between all carbon atoms in diamond. The four bonds around a single carbon nucleus make equal angles with each other; that is, they are directed toward the vertices of a tetrahedron. A diamond crystal has a fc cubic Bravais lattice which is perfectly compatible with the directionality of the covalent bonds between carbon atoms.

4. *Size Scales and Hierarchical Structures*

Size scales of interest in various branches of materials science and engineering range from the atomic level (~0.1 nm) to sizes that may approach the dimensions of an engineering structure (~1 m). This range spans ten orders of magnitude and has several implications. A variety of experimental techniques with different resolving powers are required to investigate structural features of materials. The materials scientist and materials engineer must often characterize microstructural features that are present in a material over this entire size range, and in addition must understand the connections between structural features on all size scales on one hand, and macroscopic properties on the other.

The general concept of structural hierarchies has been considered in a variety of contexts (Whyte et al. 1969). When a material is formed through the aggregation of structural units, there may be new aspects of the structure of the aggregate that are not present in the structure of the units. Such hierarchical features are frequently present in the structure of materials. Provided that a material is neither perfectly ordered nor perfectly random, imperfections in the packing of structural units of the material may lead to additional levels of structure on different size scales. Several hierarchical "levels" of structure may exist simultaneously. A simple hierarchical structure is the network of grain boundaries, edges and corners that results when isolated crystallites form in a liquid and grow to impingement. The boundaries exist because of misorientations between adjacent crystals. They have their own topological features distinct from those linking atoms within a particular crystallite. The size scale of the boundary structure differs from that of the atomic arrangements in the crystals.

More complex hierarchical structures occur in a polycrystalline long-range-ordered binary alloy, where in addition to grain-boundary structure a substructure of interfaces known as antiphase boundaries may exist within single grains. In a structure like β-brass with only two kinds of possible antiphase domains, the topological characteristics of the antiphase-boundary structure differ markedly from those of the grain-boundary structure, the former having a high degree of connectivity (English 1966).

The hierarchical nature of the structure of materials has important consequences for how structure–property relations are conceived. Since structural hierarchies encompass several different size scales in a material, experimental techniques with different resolutions must generally be employed to investigate and characterize the structure at all levels. A direct comparison of the size of some common structural features in materials, with the limit of resolution of various techniques for their observation, is presented in Fig. 7.

Structural characteristics at several different levels must frequently be considered to understand the properties of a material. For example, the plastic-deformation behavior of a polycrystalline material depends on the interatomic potential in the crystal, the response of lattice defects (dislocations) to the applied stress, and the grain-boundary structure since grain boundaries may act both as dislocation sources and as barriers to dislocation motion.

5. *Interfaces in Materials*

Even though interfacial regions usually occupy only a minute volume fraction of a material, their characteristics and behavior have important effects on the structure and on kinetic processes for structural changes in materials.

Most structural changes in materials take place through the motion of an interface. In transformations that are initiated by a nucleation event, the reaction can proceed toward completion by migration of the interface between the parent and daughter phases. Some transformations, like spinodal decomposition and certain order–disorder transformations, proceed through the creation of a very high density of interfaces in an initially homogeneous material.

An important property of interfaces is their excess free energy. Any readjustment of an interface that reduces the total interfacial free energy of the system will occur if the rate of readjustment is rapid enough. Readjustments of two types are allowed for interfaces with anisotropic free energies, that is, free energies that vary with orientation of the interface. Migration of a curved interface toward its center of curvature reduces interfacial area. Rotation of an interface with anisotropic interfacial free energy is possible as well, provided that the orientation change lowers the interfacial free energy. An example of the latter process is the faceting of an initially smooth polished surface that can occur when a crystal is annealed at high temperatures. Figure 8 shows faceted surface structures developed by annealing an initially smooth polycrystalline Al_2O_3 specimen at 1580 K. Each of the three surface orientations of the grains shown in the figure behaves differently: the lower grain is unfaceted; the grain on the right has developed two facet orientations, and that on the left three. Interfacial free energy also plays an important role in the kinetics of structural changes

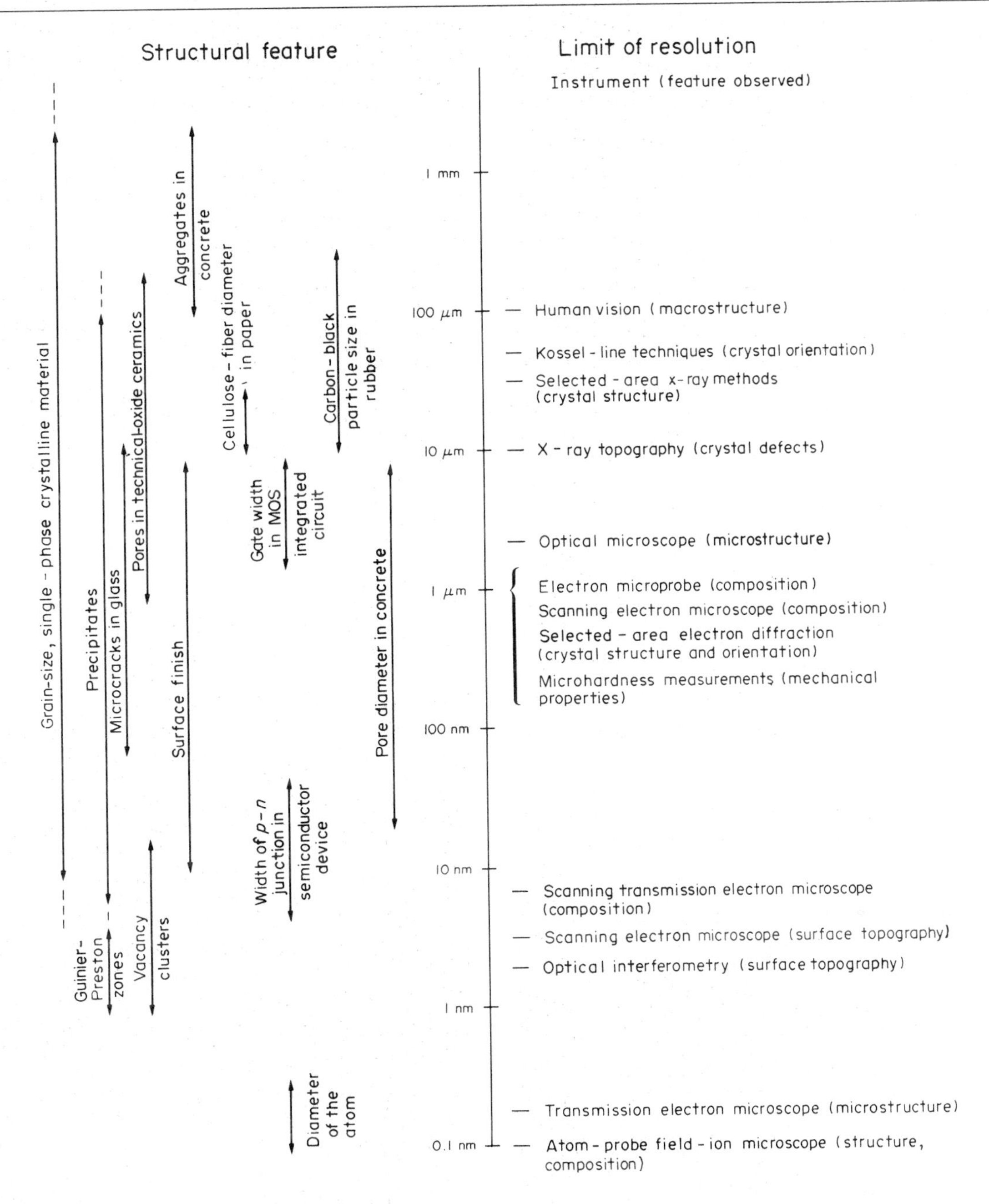

Figure 7
Size scale relating structural features of materials to experimental techniques for structural observation

that are initiated by nucleation, because for a nucleus of subcritical size interfacial free energy provides the dominant force for reversion.

Equilibrium at interfacial junctions determines the dihedral angles that are formed between interfaces at a junction, as shown in Fig. 9. In a multiphase material, the magnitudes of these angles can exert an important influence on the processing and properties of materials. In liquid-phase sintering of powders, densification is assisted if a system can be found in which the liquid phase perfectly wets the solid-powder particles. For this purpose, the overall composition of the system is chosen so that the dihedral angle shown in Fig. 9 is zero.

Figure 8
Scanning electron micrograph of faceted structures formed on the surface of polycrystalline Al_2O_3, near a three-grain junction. The number of facet orientations formed depends on the crystallographic orientation of the surface (after Dynys 1982 Sintering mechanisms and surface diffusion in aluminum oxide, Ph.D Thesis, p. 100. © Massachusetts Institute of Technology, Cambridge, Massachusetts. Reproduced with permission)

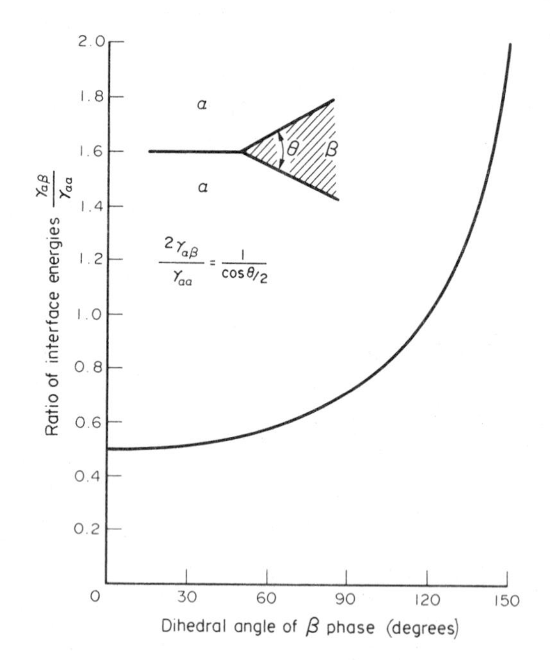

Figure 9
Graph showing the relation between equilibrium dihedral angle and ratio of interfacial free energies in a two-phase material (after Smith 1953 *Trans. Am. Soc. Metals* 45: 542. © American Society for Metals, Metals Park, Ohio. Reproduced with permission)

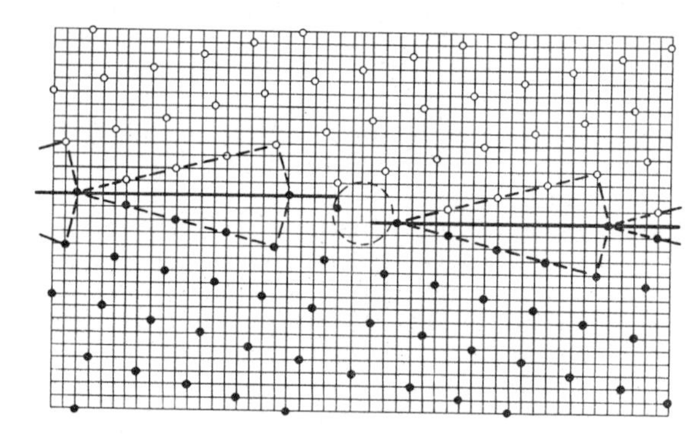

Figure 10
Atomic model of a tilt boundary (heavy line) formed by rotating two simple cubic lattices by 28.1° about a ⟨100⟩ axis. Open and filled circles are atomic sites in the two grains. A grain-boundary dislocation (encircled) appears as an extra half plane of the DSC lattice (light lines). Associated with the dislocation is a step which is a shift of the boundary plane (after Balluffi 1979. © American Society for Metals, Metals Park, Ohio. Reproduced with permission)

Many analyses of the structure of interfaces in crystalline materials make use of geometric models at the atomic level (Balluffi 1979). An example of this approach is the grain-boundary model illustrated in Fig. 10. In general, the grain-boundary region has a substructure of defects described as steps and grain-boundary dislocations. The grain-boundary dislocations appear as terminations of planes in the DSC lattice (the lattice of vectors by which one crystal can be displaced relative to another without changing the local atomic structure of the interface). Grain-boundary migration occurs through the lateral migration of these defects in the plane of the interface. "Open" sites (shown as dashed quadrilaterals) in the grain-boundary structure are believed to be favored locations for impurity atoms. Models similar to that shown in Fig. 10 also exist for interphase interfaces. The defect structures of interphase boundaries are especially relevant to the understanding of martensitic phase transformations that occur in steels and other materials (Olson and Cohen 1979).

6. Materials Engineering and Structure

The possibility of deliberate structure control lies at the root of materials engineering. It is the understanding of the relations between the structure and properties of materials that guides this aspect of materials engineering. The applications of materials-engineering principles that follow provide classic examples.

Materials engineering has been used to great advantage in developing processing methods for iron–silicon alloy sheet used in power-transformer laminations. Because the magnetic permeability of crystals is generally anisotropic, a material is most easily magnetized and demagnetized parallel to a specific crystallographic direction. In the most common applications of iron–silicon alloys, however, polycrystalline sheet is used. Special processing procedures involving deformation and recrystallization treatments have been worked out to impart specific textures in which grains are preferentially aligned. These processing steps result in alignment of the directions of easy magnetization in most grains, close to the direction of the applied magnetic field in the engineering device.

The processing of fibers made from high-molecular-weight long-chain polymers by drawing or extrusion can yield a product with exceptionally high tensile strength. In melts of such material, molecular chains coil in a random fashion. Cooling reduces the mobility of the chains, and when cooled the material deforms mostly by a combination of uncoiling of chains and by relative sliding motion of adjacent molecules. Extrusion serves to reduce greatly the capacity for coiling of molecules, inducing preferred alignment of molecular chains parallel to the extrusion axis. Mechanical properties of fibers produced by extrusion are thus more closely related to the elasticity and strength of primary carbon–carbon bonds in the molecules than to the mechanics of uncoiling.

The strengthening of glass by the ion-exchange process is another example of structure control. For instance, a sodium-rich glass can be reacted chemically with a potassium salt so that larger potassium ions replace smaller sodium ions in a reacted zone at the surface of the glass. A state of compressive stress in the surface layer results. The glass is strengthened because the level of internal tensile stress necessary to make a surface crack grow into the interior of the glass is raised.

Materials engineering is also used to improve certain properties in materials of natural origin. Natural rubber, *cis*-polyisoprene, is a very viscous liquid at room temperature. By a process known as vulcanization, sulfur is incorporated into the rubber, forming cross-links between adjacent polyisoprene molecules. Cross-linking is necessary to promote reversible elastic behavior at room temperature. Plywood is another example of an "engineered" natural material. It generally consists of a "sandwich" structure in which adjacent thin sheets of wood are bonded with the grain of the wood perpendicular. Because of the anisotropic cell structure of wood, moisture absorption causes dimensional changes that are anisotropic, these changes being minimal along the axis of the grain. In plywood, the composite structure that is produced averages the behavior in the plane of the sheet, leading to more nearly isotropic dimensional changes with moisture content than in a piece of solid wood.

These simple examples of materials engineering illustrate the ways in which the structure of a material can be modified to result in improved properties and performance. Through an understanding of the essential elements of structure and the principles by which structural changes occur, many possibilities exist not only for improving the performance of existing materials, but for developing new materials as well.

See also: Polymers: Structure, Properties and Structure–Property Relations; Structure–Property Relations of Materials

Bibliography

Balluffi R W 1979 Grain boundary structure and segregation. In: Johnson W C, Blakely J M (eds.) 1979 *Interfacial Segregation*, American Society for Metals, Metals Park, Ohio, pp. 193–236

Barrett L K, Yust C S 1970 Some fundamental ideas in topology and their application to problems in metallography. *Metallography* 3: 1–33

Bever M B, Duwez P E 1972 Gradients in composite materials. *Mater. Sci. Eng.* 10: 1–8

Bever M B, Shen M 1973 The morphology of polymeric alloys. *Mater. Sci. Eng.* 15: 145–57

Buerger M J 1956 *Elementary Crystallography: An Introduction to the Fundamental Geometrical Features of Crystals.* Wiley, New York

Cahn J W 1966 A model for connectivity in multiphase structures. *Acta Metall.* 14: 477–80

English A T 1966 Long-range ordering and domain-coalescence kinetics in Fe–Co–2V. *Trans. Metall. Soc. AIME* 236: 14–18

Gombrich E H J 1979 *The Sense of Order: A Study in the Psychology of Decorative Art.* Cornell University Press, Ithaca, New York, pp. 63–94

Holliday L (ed.) 1966 *Composite Materials.* Elsevier, Amsterdam

Kelly A, Groves G W 1970 *Crystallography and Crystal Defects.* Addison-Wesley, Reading, Massachusetts

Kittel C 1976 *Introduction to Solid State Physics*, 5th edn. Wiley, New York, Chap. 3

Olson G B, Cohen M 1979 Interphase boundaries and the concept of coherency. *Acta Metall.* 27: 1907–18

Pearson W B 1972 *The Crystal Chemistry and Physics of Metals and Alloys.* Wiley-Interscience, New York

Pine S H, Hendrickson J B, Cram D J, Hammond G S 1980 *Organic Chemistry*, 4th edn. McGraw-Hill, New York, pp. 89–129

Rostoker W, Dvorak J R 1977 *Interpretation of Metallographic Structures*, 2nd edn. Academic Press, New York

Smith C S 1964 Some elementary principles of polycrystalline microstructure. *Metall. Rev.* 9: 1–48

Smith C S 1981 *A Search for Structure: Selected Essays on Science, Art, and History.* MIT Press, Cambridge, Massachusetts
Thompson D W 1942 *On Growth and Form.* Cambridge University Press, Cambridge, UK
Wells A F 1956 *The Third Dimension in Chemistry.* Oxford University Press, Oxford
Whyte L L, Wilson A, Wilson D (eds.) 1969 *Hierarchical Structures.* Elsevier, New York
Williams R 1972 *Natural Structure.* Eudaemon Press, Moorpark, California

S. M. Allen and M. B. Bever
[Massachusetts Institute of Technology, Cambridge, Massachusetts, USA]

Structure–Property Relations of Materials

There is ample historical evidence suggesting that craftsmen and artisans of past ages knew techniques by which materials could be deliberately modified to enhance their physical usefulness and aesthetic appeal. Yet it is comparatively recently that techniques for chemical and structural analysis of materials were developed. Once tools for investigating the structure of materials became available, the stage was set for the study of the relations between the structure of materials and their properties.

Structure–property relations of materials describe the way in which a change in some structural feature of a material affects its properties. The term as generally used expresses the magnitude of changes in properties caused by a change in structure, rather than allowing absolute magnitudes of properties to be predicted. Thus structure–property relations deal with correlations and trends. Some properties of materials have values that change markedly with changes in certain structural variables. These so-called structure-sensitive properties provide the most striking examples of structure–property relations.

The field of materials science and engineering is often said to encompass the processing–structure–property–performance spectrum. Structure–property relations have remained at the center of the field because they provide the main link between a material and its production and processing on the one hand, and the use to which it will be put on the other hand.

This article is a survey of structure–property relations of materials, organized around tables that illustrate the degree of correlation between typical structural variables and the magnitude of changes of specific properties. Tables are presented for four different classes of materials: crystalline materials, polymeric materials, composites and cellular solids. The classes are not defined on the basis of consistent criteria, nor are they mutually exclusive or exhaustive. For example, polyphase crystalline materials can be considered to be composite materials and polymeric materials can be crystalline. Another limitation is that only a selection of recognized structure–property relations is presented. The tables demonstrate that structural variations over a wide range of size scales have a significant effect on different properties; at all levels of structural hierarchy, there exist properties having variations that are strongly correlated with structure.

1. Structure–Property Relations in Crystalline Materials

An overview of structure–property relations in crystalline materials is given in Table 1. In this and subsequent tables, the structural variables considered appear across the top of the table and tend to relate to features of increasing size from left to right. Properties considered appear in the left-hand column in the tables; in Table 1 they are grouped into mechanical, magnetic and electrical properties. Four intensities of structure–property correlation are used in the tables: strong, medium, weak and none. Assigning values of intensity in the tables requires a subjective judgement, because conflicting examples often exist. Entries in the tables indicate the intensity of structure–property correlations that are typically found in materials.

In the case of crystalline materials, six structural variables are considered. Electronic structure is varied with alloying additions that can change electron energy levels, electron densities and electron density distributions in materials. Point defect structure can be varied by varying temperature and hence the equilibrium concentration of point defects, by quenching-in nonequilibrium concentrations of point defects and even more drastically by irradiation. Dislocation structure can be varied most appreciably by plastic deformation. Grain size, grain crystallographic orientation anisotropy (texture) and grain shape anisotropy can be varied both by altering crystal growth conditions and by thermomechanical treatment.

1.1 Elastic Moduli

The elastic moduli of crystalline materials depend strongly on the bonding in the material and to a lesser extent on the continuous variations of electronic structure that are possible through alloying. Point defects produce local softening in a lattice but are present in such small concentrations that their effect on moduli is slight. Marked changes in elastic moduli are introduced by textures in materials that are elastically anisotropic. Other parameters such as grain size and shape have a negligible effect on elastic moduli whether or not textures are present.

1.2 Yield Strength

The yield strength of a crystalline material is sensitive to all structural features that affect the generation and propagation of dislocations through the material.

Table 1
Structure–property relations for crystalline materials

	Electronic structure (through alloying)	Point defect density	Dislocation density and structure	Grain size	Grain orientation anisotropy	Grain shape anisotropy
Mechanical						
Elastic moduli	Medium	Weak	None	None	Strong	None
Yield strength	Strong	Medium	Strong	Strong	Strong	Medium
Creep strength	Medium	Strong	Medium	Strong	Strong	Strong
Electrical						
Conductivity	Strong	Strong	Weak	Weak	Strong[a]	None
Magnetic						
Saturation magnetization	Strong	None	None	None	None	None
Permeability	Strong	Weak	Weak	Strong	Strong	Strong
Coercivity	Strong	Weak	Weak	Strong	Strong	Strong

[a] If crystallographically allowed

Strong dependence of yield strength on alloying, dislocation structure (especially forest dislocations), grain size and texture can generally be expected. Point defects can also affect yield strength, but the magnitude of the effect is smaller than that usually observed with alloying additions. Grain shape can produce significant changes in yield strength, for instance if a small dimension of the grains lies in a plane with high shear stress.

For brittle crystalline solids such as alumina and diamond, the stress at which cracks propagate is smaller than the yield strength. The fracture stress at which such materials fail depends more on the size and size distribution of flaws than on the structural features controlling yield strength discussed above.

1.3 Creep Strength

Creep of crystalline materials at elevated temperatures may involve one or more mechanisms that are structure-sensitive. Nabarro–Herring creep occurs by stress-directed diffusion and depends strongly on point defect densities (particularly vacancy concentrations in metals) and grain size. The effects of grain boundary sliding may be appreciably reduced by controlling grain shape, as in directionally solidified nickel-base superalloy turbine blades which have columnar grains aligned in the direction of tensile loading. When grain boundaries are eliminated entirely, creep occurs by dislocation motion. Monocrystalline turbine blades have stress-rupture lifetimes that strongly depend on crystal orientation. Of lesser importance for creep strength are electronic structure (through solid–solution strengthening, for example) and dislocation structure, since during creep, steady-state dislocation arrangements are developed that are not systematically variable.

1.4 Electrical Conductivity

The conductivity of electronic conductors strongly depends on the degree of perfection of local atomic arrangements. Among the factors affecting electronic structure, and hence electrical conductivity, are alloying additions, long- and short-range ordering and the density of point defects. The effects of texture on conductivity can also be marked for materials with crystal structures that allow anisotropy of second-rank tensor properties (Nye 1957). Under normal conditions (especially at room temperature), the grain size and grain shape of metals do not affect electrical conductivity, but in some varistor materials, such as zinc oxide and barium titanates, special electrical properties of grain boundaries make possible applications of the materials that require certain unusual electrical resistivity characteristics. Ionic conductors have conductivities that depend strongly on point defect densities, because charges are transported by ionic diffusion. The conductivity of zirconia, for example, is greatly enhanced by addition of calcia or yttria, since these form oxygen vacancies to maintain electrical neutrality.

1.5 Saturation Magnetization

The saturation magnetization of a material is solely determined by electronic structure and hence by composition and local atomic arrangements. A very high concentration of point defects can have a slight influence on saturation magnetization; none of the other structural variables has any significant effect in the high applied field used in the measurement.

1.6 Magnetic Permeability

The permeability of a ferromagnet depends on a number of structural variables. Electronic structure and therefore alloy composition play an important role, owing to the composition dependence of the magnetization itself as well as of the various anisotropy energies. In soft iron–nickel permalloys, for instance, additional elements such as Mo, Cu and Cr can be introduced so that both the magnetostrictive and magnetocrystalline energies are minimized simultaneously. Other structural variables that may affect permeability are those with a size scale comparable with the width of a magnetic domain wall, typically tens of nm in permanent magnets and several hundred nm in soft ferromagnets. This can cause a grain size effect on permeability in ultrafine-grain material and also allows the possibility of significant grain shape effect. The presence of texture can also increase permeability in materials with significant magnetocrystalline anisotropy.

1.7 Coercivity

Like permeability, the coercive force of a ferromagnet depends on both the local atomic arrangements in the material and structural features that have characteristic sizes on the same scale as the width of a magnetic domain wall. In hard Co–Sm alloys, high crystalline anisotropy energy associated with local electronic structure results in high coercivity. Reducing the grain size in these alloys raises the coercive force. In addition, textures can be used to enhance coercivities, as has been found in Alnicos. Large increases in coercivity can also be achieved by precipitation treatment in a magnetic field, resulting in preferred orientations of magnetic phase precipitates. The coercivity of soft ferromagnets is kept low by eliminating as far as possible fine dispersions of precipitates or inclusions that could contribute to magnetic domain wall pinning.

2. *Polymeric Materials*

The structure of polymeric materials is complex. High-molecular-weight macromolecules, usually in the form of long chains with strong internal covalent bonds, can form aggregates that are amorphous, crystalline or a mixture of the two. The interactions between macromolecules on which many properties depend are determined both by relatively weak chemical bonds between macromolecules and by strong covalent bonds when the macromolecules are entangled or cross-linked. Thus the spatial arrangements of the macromolecules in polymeric materials are more complex than those of atoms or ions found in crystalline metals or ceramics.

The properties of polymeric materials are sensitive to a wide range of structural variables (see *Polymers: Structure, Properties and Structure–Property Relations*). Some of the structure–property relations for polymeric materials are presented in Table 2. Of the structural variables considered in this table, macromolecular construction refers to the sequencing of the basic components of the polymer ("mers") and hence includes compositional as well as some topological aspects of polymeric structure (e.g., branching and cross-linking). Conformation refers to the overall shape of a macromolecule and can be described by terms such as helical, planar zigzag and random. The other structural variables listed in Table 2 have their usual meaning.

2.1 Glass Transition Temperature

The glass transition temperature of an amorphous polymer marks the transition from a glassy solid to either a viscoelastic liquid (disordered amorphous polymer) or a rubbery solid (cross-linked amorphous polymer). It depends strongly on macromolecular construction and conformation and also on molecular weight if this is low. The degree of molecular orientation can also affect the glass transition temperature. If the molecules tend to be aligned, molecular motions become frozen-in more easily on cooling to the glass transition temperature, resulting in an increase in glass transition temperature with degree of molecular alignment.

Table 2
Structure–property relations for polymeric materials

	Macromolecular construction	Molecular weight	Conformation	Degree of molecular orientation	Degree of crystallinity	Phase morphology and distribution
Glass transition temperature (amorphous polymers)	Strong	Strong[a]	Medium	Weak	None	None
Melting temperature (crystalline polymers)	Strong	Strong[a]	Medium	Medium	Weak	Medium[b]
Elastic moduli	Strong	Weak	Strong	Strong	Medium	Medium
Tensile strength	Strong	Strong	Medium	Strong	Strong	Strong
Ductility of amorphous polymers	Strong	Medium	Strong	Strong	None	Strong
Ductility of crystalline polymers	Weak	Medium	Medium	Strong	Strong	Strong

[a] Diminishes as molecular weight increases [b] Especially dependent on lamellar thickness

2.2 Melting Temperature

Crystalline polymers melt at rather well-defined temperatures, at which ordered arrangements of macromolecules in crystalline regions come apart and the material becomes a viscous liquid. The transition occurs without breaking strong covalent bonds. The primary factor influencing crystallization and melting is macromolecular construction, as crystallization is facilitated when macromolecules have chemically and geometrically regular structures. Molecular weight and conformation also play significant roles in determining the melting temperature. Since polymer melting involves the transfer of macromolecules from the crystalline to the liquid state at the crystal–melt interface, the degree of molecular orientation in the amorphous regions between crystallites can have a moderate effect on the melting temperature. The primary effects of degree of crystallinity and phase morphology on the melting transition are an influence on melting and crystallization kinetics; the effect of these variables on melting temperature is negligible.

2.3 Elastic Moduli

The elastic moduli of all polymeric materials depend most strongly on macromolecular construction and conformation. The degree of molecular alignment can also strongly affect elastic moduli, but molecular weight has no significant effect. The moduli of unoriented crystalline polymers depend on the behavior of the most compliant disordered regions of the material; hence the glass transition temperature of these regions is important in determining elastic moduli. Of lesser importance for the moduli of crystalline polymers is the degree of crystallinity and the morphology and distribution of crystallites, provided that the polymer is unoriented.

2.4 Tensile Strength

The tensile strength of polymeric materials depends on polymer structure at all levels. Macromolecular structure plays a role, as ultimately the tensile strength depends on the intrinsic strength of covalent bonds in the macromolecules. The degree of molecular orientation of macromolecules along directions of tensile loading greatly enhances tensile strength. In crystalline polymers, crystallite morphology and distribution as well as the degree of crystallinity affect the tensile strength. At small strains, deformation is confined to dsordered regions between crystallites. Deformation behavior at larger strains strongly depends on the orientations of the lamellar crystallites with respect to applied stresses.

2.5 Ductility

Polymers behave elastically at small strains and short loading times. At long loading times and elevated temperatures, they behave in a viscous manner. The entire range of behavior between these two extremes is referred to as viscoelastic (see *Viscoelasticity*). Viscoelastic behavior is strongly dependent on the temperature and on the time scale. The continuum plastic flow of polymers can be modelled by various combinations of mechanical elements whose equations of motion simulate the viscoelastic behavior of materials.

The ductility of a polymeric material depends on the ability of the macromolecules to flow relative to one another and also on the localized deformation processes that often precede fracture. In many amorphous polymers, significant ductility occurs only above the glass transition temperature. Flow in this regime is determined by macromolecular construction, conformation and molecular weight. Below the glass transition temperature, amorphous polymers can undergo large plastic flow in compression and shear. Some amorphous polymers, such as polycarbonate and many polyimides, can undergo plastic deformation under tensile loading conditions. Tensile plastic deformation in amorphous polymers is restricted by a localized deformation process called crazing. Although the atomic-level mechanism of crazing is still not well understood, there is much evidence that the more disordered and random the molecular packing, the more craze-prone the polymer becomes. Flexible-chain polymers such as polystyrene and poly(methyl methacrylate) fracture as a result of crazing before large-scale plastic flow can occur. Inflexible chain polymers with considerable liquid-crystal-like molecular order are resistant to crazing and usually undergo plastic deformation in tension.

In semicrystalline and crystalline polymers, flow strongly depends on preferred molecular orientation, degree of crystallinity and the morphology and distribution of crystallites. Plastic flow of semicrystalline polymers can involve interlamellar slip (between crystallites), yielding of the lamellar crystallites and subsequent drawing down into microfibrils, and finally strain hardening and fracture of the microfibrils. Each of these deformation stages is sensitive to the initial microstructure of the polymer.

3. Composite Materials

Composites are heterogeneous materials made up of distinct phases that are separated by interfaces which may or may not display good cohesion. Numerous materials found in nature are composites, for example granite and wood. Manmade materials are often composites that have been engineered to optimize a combination of properties that none of the component phases alone could achieve. The designer of composite materials has not only a wide variety of homogeneous materials from which to select but also an appreciable freedom of choice as to how the phases will be assembled into three-dimensional structures.

Structure–property relations in composite materials are illustrated in Table 3. A wide range of structural characteristics can be varied in composites

Table 3
Structure–property relations for composite materials

	Volume fraction of phases	Size and size distribution of phases	Phase shapes	Phase orientations	Connectivity of phases
Thermal conductivity	Strong	Weak	Strong	Strong	Strong
Elastic constants	Strong	Weak	Weak	Strong	Weak
Fracture strength	Strong	Medium	Strong	Strong	Weak
Optical properties	Strong	Strong	Medium	Medium	Medium

(Holliday 1966, pp. 1–27). A detailed understanding of structure–property relations in composites usually requires that physical models developed to describe the behavior of homogeneous materials be adapted to describe the properties of an idealized heterogeneous structure (Holliday 1966, pp. 28–62). It has been found generally that composite materials have properties that fall into two categories: those that depend on phase dimensions and those that depend on phase periodicities (see *Composite Materials: An Overview*). Two examples of each type of behavior are discussed below.

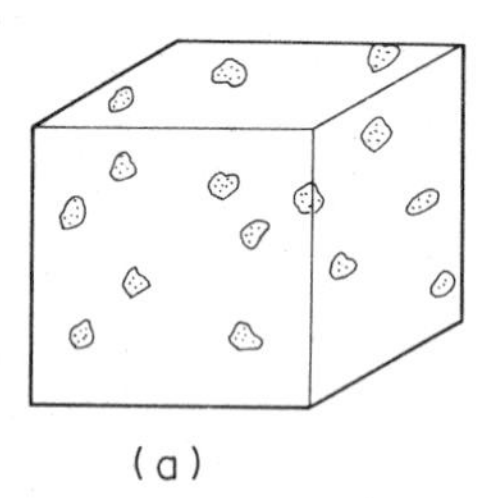

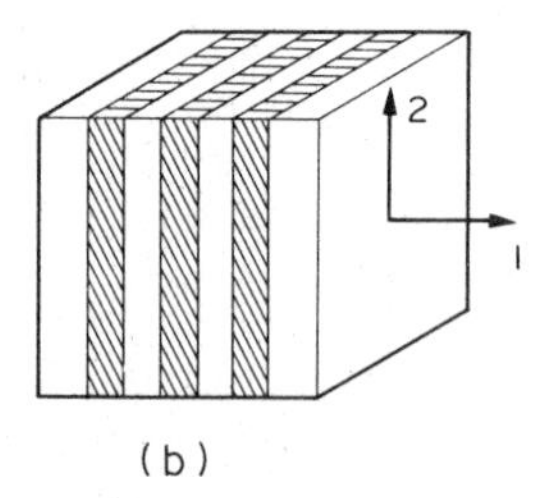

Figure 1
Examples of possible phase arrangements in composite materials: (a) phase randomly dispersed in matrix; (b) lamellar structure

3.1 Thermal Conductivity

The thermal conductivity of composites depends strongly on phase geometries, orientations and volume fractions. Here, simple two-phase composite structures are considered, as shown in Fig. 1, the two phases having different conductivities. When one phase consists of dispersed particles (Fig. 1a), the composite has a conductivity approximately equal to the volume-fraction-weighted average. When the phases are in the form of parallel slabs (Fig. 1b), the magnitude of the conductivity in directions parallel and perpendicular to the slabs varies in an analogous way to the flow of current in a resistive electrical circuit. Parallel to the slabs, both phases contribute independently to the heat flow, similar to current flow through resistors in parallel, and the resultant conductivity is always higher than that contributed by the phase with the larger conductivity. Perpendicular to the slabs, heat must flow sequentially through each of the slabs and is more effectively impeded by the lower-conductivity phase, analogous to current flowing through resistors in series. In this latter case, the conductivity is always lower than that contributed by the lower-conductivity phase alone.

3.2 Elastic Constants

The elastic constants of a composite material have very similar structure–property relations to those discussed for thermal conductivity. They are sensitive to phase geometries but not in general to phase dimensions. An interesting exception has been investigated by Tsakalakos and Hilliard (1983), who found considerable modulus enhancement in compositionally modulated thin metal films having modulation wavelengths in the region of 2 nm. The moduli of the modulated films were found to exceed significantly those of the pure elements making up the film or those of any homogeneous intermediate alloy compositions.

3.3 Fracture Strength

Numerous composites have been developed for applications in which high strength- and stiffness-to-weight ratios are desired. Among the strongest and lightest materials available for fiber reinforcement are covalently bonded nonmetallic fibers such as boron and graphite. Such materials are inherently brittle when loaded in tension, but they can be used in applications at stresses close to the ultimate fiber strength when they are embedded in a softer plastically deformable matrix (Kelly 1973, pp. 157–226). The strength of fiber-reinforced composites is markedly dependent on the volume fraction of fibers as well as the fiber orientation with respect to the loading axis and the distribution of fibers in the matrix (i.e., fiber–fiber spacing and its uniformity).

Especially important for mechanical strength is the ratio of fiber length to diameter—the aspect ratio. For a stressed fiber-reinforced composite having embedded fibers of finite length, loaded in tension along the fiber axis, it can be shown that the tensile stress in the fiber rises to a maximum midway along the fiber length (Kelly 1973, pp. 175–82). A maximum tensile strength of fiber-reinforced composites is attained when the stress in the fibers reaches the fracture stress of the fiber. This maximum strength level is achieved only if the fiber aspect ratio exceeds a certain value. For metal-matrix composites, this value depends on the ratio of fiber fracture strength to the shear yield stress of the matrix. For plastic–resin-matrix composites, the critical aspect ratio depends on the ratio of fiber fracture strength to a frictional stress, governing sliding between fiber and matrix (Kelly 1973, pp. 183–88).

Within the range of fiber diameters usually encountered in engineering composites, 5–50 μm, fiber diameter can be at most an important variable for controlling fracture strength (e.g., in glass–thermoplastic polymer composites). Because of the statistical distribution of individual fiber strengths, the actual number of fibers supporting a load is important. If the number of fibers is very small, failure of one fiber can lead to a large stress increment on each of the fibers that remain, fast fracture of the remaining fibers being the result. If the number of fibers is very large, there is a high probability that a large number of weak sections of fibers will be present, resulting in the formation of a critical crack at relatively low loads (Argon 1974, pp. 176–79).

3.4 Optical Properties

Important examples of the relations between composite structures and optical properties occur in ceramic and glass–ceramic systems. Translucency, the degree to which light is transmitted by a material, is enhanced when the density of light-scattering sites in the material is minimized. Translucency is sensitive to all the structural variables listed in Table 3. In addition to the light-transmission properties of each phase, light is scattered at interphase boundaries in a composite when the refractive indexes of the phases differ. Fine pores are thus very deleterious to the translucency of ceramics. Not only the relative volume fractions of the phases but also their dimensions are important. Light is scattered most effectively when the dimensions of the phases are comparable with the wavelength of the radiation being scattered. Phase shapes and orientations can also be important. Composites in which phases are aligned parallel to one axis can polarize light transmitted by the composite and can exhibit birefringence even when both phases are optically isotropic.

4. Cellular Solids

Cellular solids consist of a solid matrix in which a high volume fraction of gaseous phase is dispersed. There are many naturally occuring cellular solids, including wood, bone and sponge. Numerous manmade cellular materials also exist, for example foamed thermoplastics, foamed elastomers and aluminum honeycomb.

Structure–property relations in cellular solids are presented in Table 4. Important structural variables are the cell size, cell shape and cell orientation, and the relative density, which is the ratio of the density of the cellular solid as a whole to the density of the cell wall material. The relations between structure and mechanical properties considered in Table 4 can be derived by applying principles of mechanics to idealized geometries of cellular solids (Ashby 1983).

4.1 Young's Modulus

Young's modulus is the constant that appears in the linear relation between stress and strain for small-strain elastic deformation. In cellular solids such deformation occurs by the elastic bending of cell walls. For a wide variety of cellular materials, Young's modulus is found to depend on relative density and to be independent of cell geometry (Ashby 1983, pp. 1758–59).

4.2 Compressive Crushing Strength

Brittle cellular solids fail in compression by brittle fracture of cell wall material due to stresses arising from cell wall bending. The failure stress, called the crushing strength, strongly depends on relative density and, like Young's modulus, is independent of cell geometry (Ashby 1983, pp. 1761–62).

Table 4
Structure–property relations for cellular solids

	Relative density ρ_{solid}/ρ_{wall}	Cell size	Cell shape	Cell orientation
Young's modulus	Strong	None	None	None
Compressive crushing strength	Strong	None	None	None
Tensile fracture strength	Strong	Strong	Strong	Strong
Permeability to fluids	Strong	Strong	Strong	Strong
Thermal conductivity	Strong	Medium	Weak	Weak

4.3 Tensile Fracture Strength

Brittle foams loaded in tension fail by sequential failure of cell wall material at the tip of a macroscopic crack. Principles of fracture structural variables that control the fracture strength. As for most other mechanical properties of cellular materials, relative density is an important factor. In addition, cell size plays a major role in determining the fracture strength. If cells are not equiaxed, cell shape and cell orientation affect the fracture strength in tension. This is relevant for wood, in which the cell shape is highly anisotropic.

4.4 Permeability to Fluids

Foremost in determining the permeability of a cellular solid is the extent to which there are open channels between adjacent cells, that is, the connectivity of the cells. Provided that such channels exist, liquids can pass through a cellular solid by capillary flow. The primary structural variables in this process are the cell dimensions normal to the direction of fluid flow as well as the relative density, as this determines the relative volume fraction of the channels. Gases can also be transported through closed-cell structures, but the process is slower than for open-cell structures, since diffusion of the gases through the cell walls is required. Gas transport depends on relative density, as this is related to the total thickness of cell wall through which diffusion must occur, as well as to the cell size when gas–solid interface resistances to diffusion are important.

4.5 Thermal Conductivity

Both cell wall material and the gas in the void space can contribute to thermal conductivity of a cellular material (Holliday 1966, pp. 282–84). The cell wall material generally has a higher conductivity than that of a gas and is thus the major contributor to conductivity at high relative densities. At low relative densities, conductivity in the gas is of primary importance. In closed-cell structures, convective heat transport can occur provided that the cell size is larger than ~1 mm. For smaller cells, the thermal conductivity of the gas controls the conductivity.

5. Other Structure–Property Relations

In the foregoing, the properties of four major classes of materials were considered in relation to their intrinsic structural features. Typically these features are restricted in their size to the microstructural domain. The relations discussed between properties and structural features are characteristically dependent on the classes of materials. For testing and behavior in service, additional structural features and features related to structure can also be important. Among them are the size and shape of the body, surface features, residual stresses, porosity and macrostructure.

The size and shape of specimens or finished objects have significant effects on certain types of behavior. The size effect on mechanical behavior, in particular the strength of wires, is well-known. Configuration in the extreme case of notches affects static and fatigue strength. In inherently brittle materials, surface defects can be important to fracture initiation and thus to the problem of the reliability of these materials. The presence of oxide films on the surfaces of metal crystals also affects their mechanical properties. Residual stresses can be present in a wide range of materials. They can have beneficial effects on properties, as in tempered plate glass.

There are several structural features which can be present from the micro to the macro scale. An especially important example is porosity. In metals, macrostructural features are voids (such as blowholes or shrinkage cavities in metal castings) and flow lines (with a possible effect on orientation-sensitive properties). Composites with gradients in the distribution of phases provide another example of macrostructure (Bever and Duwez 1972).

Composition, which is another characteristic that can vary on both the micro and macro scales, affects structure and properties. Modulated structures have already been mentioned in connection with the enhancement of the elastic modulus (see Sect. 3). Over the entire range of composition variation in metals—from Guinier–Preston zones and grain boundary regions through microsegregation to macrosegregation—various properties are affected by composition variations. A gradient through the thickness of a body can be of special interest. Examples are composition gradients in polymers (Shen and Bever 1972) and composition gradients in glass fibers for optical transmission applications (Chynoweth 1976).

See also: Properties of Materials; Structure of Materials

Bibliography

Argon A S 1974 Statistical aspects of fracture. In: Broutman L J (ed.) 1974 *Composite Materials*, Vol. 5, *Fracture and Fatigue*. Academic Press, New York, pp. 153–90
Ashby M F 1983 The mechanical properties of cellular solids. *Metall. Trans. A* 14: 1755–69
Bever M B, Duwez P E 1972 Gradients in composite materials. *Mater. Sci. Eng.* 10: 1–8
Billmeyer F J Jr 1984 *Textbook of Polymer Science*, 3rd edn. Wiley–Interscience, New York
Broutman L J, Krock R H (eds). 1967 *Modern Composite Materials*. Addison-Wesley, Reading, Massachusetts
Cahn R W, Haasen P (eds.) 1983 *Physical Metallurgy*, Pt. 2. North–Holland Physics Publishing, Amsterdam
Chynoweth A G 1976 Fiber lightguides for optical communications. *Mater. Sci. Eng.* 25: 5–11
Cottrell A H 1955 *Theoretical Structural Metallurgy*, 2nd edn. Arnold, London
Holliday L 1966 *Composite Materials*. Elsevier, New York
Kelly A 1973 *Strong Solids*, 2nd edn. Clarendon, Oxford
Kingery W D, Bowen H K, Uhlmann D R 1976 *Introduction to Ceramics*, 2nd edn. Wiley, New York
Newnham R E 1975 *Structure–Property Relations*. Springer, Berlin

Nye J F 1957 *Physical Properties of Crystals.* Clarendon, Oxford
Seymour R B, Carraher C E 1984 *Structure–Property Relationships in Polymers.* Plenum, New York
Shen M, Bever M B 1972 Gradients in polymeric materials. *J. Mater. Sci.* 7: 741–46
Tsakalakos T, Hilliard J E 1983 Elastic modulus in composition-modulated copper–nickel foils. *J. Appl. Phys.* 54: 734–37
Van Krevelen D W, Hoftyzer P J 1976 *Properties of Polymers, their Estimation and Correlation with Chemical Structure*, 2nd edn. Elsevier, Amsterdam

S. M. Allen and M. B. Bever
[Massachusetts Institute of Technology, Cambridge, Massachusetts, USA]

Substitution: Economics

Material substitution comprises several different types of events that may substantially alter material usage. Material-for-material substitution entails the replacement of one material for another in the production of a particular good. Aluminum and plastic, for example, have been increasingly used in place of wood siding in residential construction in recent years. Other-factors-for-material substitution diminishes material use by increasing other inputs into the production process, such as labor, capital and energy. The hand soldering of many electronic components, for example, requires less solder but more labor than more automated procedures using printed circuit boards. Quality-for-material substitution occurs when material savings are produced by a reduction in the quality of the final product. For example, the thinner glass beverage containers produced during World War II consumed less glass but also had a higher breakage rate. Similarly, the amount of nickel, molybdenum and other alloying elements used in alloy steels can be reduced if lower levels of performance are acceptable.

Inter-product substitution produces a change in the mix of goods used to achieve a particular function or to satisfy a given need. Public transportation can be substituted for private automobiles, a telephone call for a visit or letter, frozen vegetables for fresh or canned produce. Unlike the other types of substitution, inter-product substitution does not alter the manufacturing processes and the use of materials in the production of individual goods, but rather affects material use by changing the composition of goods and services. This type of substitution includes what Chynoweth (1976) and others call functional substitution. The latter occurs when a product (along with the materials it contains) is replaced by an entirely different means of achieving a desired function, such as the use of satellites instead of underground cables to transmit long-distance telephone messages.

Finally, if material substitution is defined broadly, it includes technological substitution, whereby an advance in technology reduces the amount of material needed to produce a particular product. The development of electrolytic tinplating, which greatly decreased the tin required per tinplate can, is a classic example of this type of substitution. New technology can also lead to the introduction of new products, such as the aluminum can, and thus increase the possibilities for inter-product substitution.

In short, material substitution may result from the introduction of new technology, from shifts in the composition or quality of final goods, and from changes in the mix of factor inputs used in producing these goods. The technological aspects of substitution are examined in the article *Substitution: Technology*; this article concentrates on the economic aspects.

1. Importance of Material Substitution

For economists, material substitution is of interest for several reasons. First, by allowing society to replace expensive and scarce resources with cheaper and more abundant resources, it has historically helped offset the upward pressure on resource costs resulting from the depletion of high-grade, readily accessible, and easily processed mineral deposits (Barnett and Morse 1963, Barnett 1979). As for the future, there are those who believe material substitution will have to play an even greater role if the adverse effects of resource depletion are to be held at bay (Goeller and Weiberg 1976, Skinner, 1976).

Second, material substitution may in some instances help alleviate shortages caused by war, embargoes, labor unrest, natural disasters and cartels. Such shortages rarely last for more than a few years. Still, they often arise quickly with little warning, and as a result can be quite disruptive while they last.

Third, the threat of material substitution forces producers to behave in a more competitive manner. Even in highly concentrated industries where dominant firms charge a uniform producer price, material substitution may prevent high prices and monopoly profits. In a similar manner, substitution reduces the likelihood that major producing countries will join together to form a cartel, and in the event they do, that it will be successful.

Finally, economists are often called upon to forecast future material requirements. Since substitution may greatly alter the demand for materials, it must be considered in making such predictions.

2. The Economic View of Substitution

In analyzing material substitution, or more generally the substitution of any factor of production, economists rely on the branch of microeconomics called production theory. An important conceptual relationship found in this theory is the production function, which shows how much output of any particular good can be derived from given quantities of labor, capital, energy, materials and other inputs. This relationship is governed by the existing state of technology, and so changes as new innovations are introduced. For this reason, the production function shown below includes not only the output Q of the good being produced and the inputs $(A, B, \ldots)$ of all factors of production but also a variable T reflecting current technology:

$$Q = f(a, b, \ldots ; T)$$

The extent to which any two inputs, such as A and B, can be substituted for each other in producing Q is portrayed by isoquant curves. Three such curves, identified as Q_1, Q_2 and Q_3, are shown in Fig. 1. Each of these curves indicates the various quantities of A and B that are just sufficient to produce the specified output, assuming all other inputs and technology remain fixed at some particular level. The further an isoquant is from the origin, the larger the output level it reflects and the greater the inputs of A and B required to produce it.

The isoquant curves shown in Fig. 1 are downward sloping, indicating that a specified level of output can be maintained while one factor input is reduced if the other is increased. That is, some substitution between A and B is presumed possible. The curves are also drawn as continuous functions convex to the origin. Convexity implies that, as more of one factor is substituted for another, greater quantities of the factor being added are required for each unit of the factor being replaced. The presumption here is that the most economical substitutions, those requiring the least amount of the factor being added, will be made first.

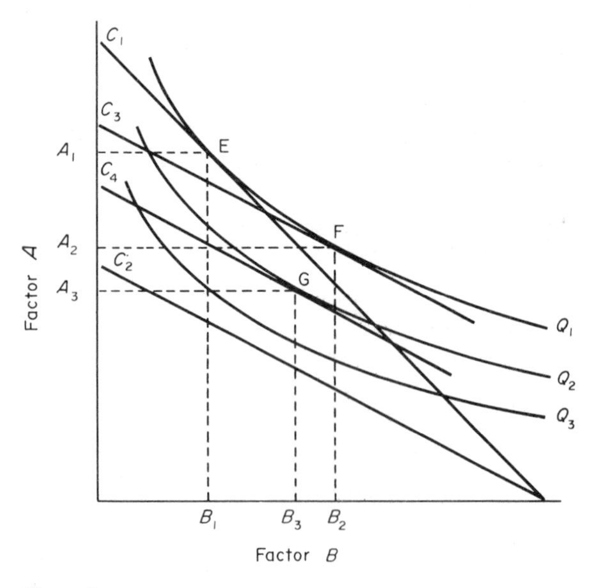

Figure 1
Factor demand determined by isoquant and isocost curves

While isoquant curves are commonly drawn as shown in Fig. 1, the implicit assumptions in such constructions can be relaxed. For example, if two inputs must be used in fixed proportions and no substitution between them is possible, the curve collapses to a point representing their requisite quantities. Alternatively, if three different production processes are feasible each using A and B in fixed but different proportions, the isoquant curve is composed of two line segments connecting the three points determined by the fixed quantities of A and B required by each process. (This assumes that all three processes are economic at some input price ratio; otherwise, the resulting isoquant is simply one line segment connecting the points determined by the two processes that can be economic.) As the number of possible processes grows, each using fixed but different proportions of A and B, the isoquant curve is defined by an even larger number of line segments and so increasingly approximates a continuous curve, such as those shown in Fig. 1.

The determination of just where on an isoquant curve a firm will operate, and hence how much of A and B it will use, requires the introduction of the isocost curve. This shows the various quantities of A and B a firm can purchase for a given expenditure. Its slope is determined by the ratio of the price of B to the price of A, and its distance from the origin by the level of expenditure. Figure 1 depicts four isocost curves—C_1, C_2, C_3 and C_4.

If a firm produces the output represented by the isoquant Q_1 and if the slope of the isocost curve C_1 reflects the existing ratio of input prices, the firm minimizes its costs by purchasing A_1 of factor A and B_1 of factor B. Any other combination of A and B on isoquant Q_1 lies on an isocost curve further from the origin and hence involves a greater level of expenditure for factors A and B.

Now let the price of factor A increase, causing the isocost curve to rotate from C_1 to C_2. This shift reflects the reduction in the amount of factor A the firm can purchase for a given expenditure. If the firm still wants to produce the same level of output, it minimizes its costs, which have increased as a result of the rise in price of factor A, by moving from point E to point F on the isoquant Q_1. In the process, it reduces its consumption of factor A to A_2 and increases its consumption of factor B to B_2.

Since the rise in the price of factor A increases the firm's costs, it may decide to reduce its output. If other firms do likewise, this would produce a rise in prices to cover all or part of the increase in costs. This implies a shift to a lower isoquant curve, such as Q_2. In this case, the firm would end up producing at point G, consuming A_3 of factor A and B_3 of factor B. The decline in the consumption of factor A can be separated into two parts: the reduction from A_1 to A_2 which reflects the substitution of factor B for factor A motivated by the rise in the latter's price; and the further reduction from A_2 to A_3 caused by the fall in output. In the case of factor B, the substitution effect increases consumption from B_1 to B_2, which is to be expected since its price has fallen relative to that for factor A. The output effect, however, reduces consumption from B_2 to B_3. The net result as shown in Fig. 1 is an increase in consumption of factor B, though this need not always be the case.

This conceptual view of material substitution reflects the economists' concern with the role of prices in influencing material demand, and has little to say about other factors affecting substitution. It is most appropriate for examining material-for-material and other-factors-for-material substitution. It can accommodate technological substitution when the latter reflects a reduction in material requirements used to produce existing goods, for such innovations simply result in a shift towards the origin of the isoquant curves. It is of little use in examining quality-for-material substitution, inter-product substitution, or technological substitution resulting from the introduction of new goods.

3. Economic Research on Substitution

Economic research on substitution falls into three general categories. The first covers studies that estimate aggregate production functions (or the corresponding information in the form of cost functions) for entire industries, economic sectors, or even the economy as a whole. These efforts, which Slade (1981) and Field and Berndt (1981) have reviewed, generally treat materials as a single factor of production, and focus on the extent to which they can be replaced by other factors, such as labor, capital and energy. Not surprisingly, their findings vary greatly depending on how they specify the aggregate production function, how they take account of technological change, and how they resolve other issues. Many of these studies also suffer from the use of aggregated input data and from their static nature (Kopp and Smith 1980, Slade 1981).

The second category of research consists of more micro investigations that differentiate individual materials and often focus on specific end uses. These studies involve the estimation of formal models, the most common of which are econometrically estimated supply-and-demand models, such as the model for copper of Fisher et al. (1972) and the model for aluminum of Woods and Burrows (1980).

The demand equations in these models typically take account of material substitution by including price variables for close substitutes as well as the material whose demand is being estimated. Thus, they assume that substitution is primarily driven by material prices. In addition, the model specification generally assumes that the relationship between demand and material prices is continuous and reversible. The latter implies that the demand lost when a material's price increases can be completely recaptured when price returns to its original level. These assumptions are all consistent with the economic view of substitution described earlier.

The short-run and long-run effects of material prices can be differentiated by introducing a lag structure, such as lagged price variables, into the demand equations. When this is done, material demand is almost always found more responsive or elastic to prices in the long run. This, as may be expected, implies that a change in material prices will effect more substitution the longer the response time.

The third category of research encompasses case studies of material substitution in particular end uses, such as beverage containers (Demler 1983), automobile radiator solder (Canavan 1983) and stabilizers for the production of PVC plastic pipe (Gill 1983). Such studies do not normally entail formal modelling, but rather attempt to determine more about the nature of material substitution by analyzing how and why the mix of materials used in certain applications has changed over time.

While the findings vary depending on the end uses examined, several conclusions are fairly common (Tilton 1983). First, material substitution may, and often does, substantially alter material requirements in specific uses, particularly over the long run. Second, a major determinant, perhaps the major determinant, of substitution is technological change. Third, changes in material prices often have little effect on the material mix in the short run as producers are constrained by existing technology and equipment. In the long run, both technology and equipment can change, and prices have more of an influence. Indeed, the major effect of material prices may occur indirectly over the long run by altering the incentives to conduct research and development on new material-saving technologies. Finally, material substitution is often motivated by changes in government activities and regulations, such as building codes or worker health and safety standards.

These findings have a number of implications. The possibility that prices may influence material substitution most by inducing technological change suggests that the relationship between price and material demand may be discrete and irreversible, rather than continuous and reversible. This coupled with the finding that other factors, such as government regulation, are often important determinants of substitution raises reservations about the traditional economic view of substitution as described earlier.

Forecasts of future mineral requirements, particularly over the long run, need to take account of substitution, for the latter can greatly alter material consumption. This, however, is not easy to do, for substitution is often the result of technological change, which by its nature is uncertain and difficult to predict.

The findings also suggest that substitution can play an important role in alleviating material shortages caused by the depletion of high-quality mineral deposits. Such shortages tend to arise slowly and persist for extended periods, providing ample time for rising material prices to induce material substitution. For other types of shortages, those caused by war, cartels, embargoes and strikes, material substitution is likely to provide much less relief. Such shortages tend to arise abruptly and unexpectedly, and the resulting disruption is likely to be greatest in the short run. By the time sharply higher prices stimulate material substitution by inducing changes in technology and equipment, these shortages are likely to have passed.

See also: Substitution: Technology

Bibliography

Barnett H J 1979 Scarcity and growth revisited. In: Smith V K (ed.) 1979 *Scarcity and Growth Reconsidered.* Johns Hopkins University Press, Baltimore, Maryland, pp. 163–217

Barnett H J, Morse C 1963 *Scarcity and Growth.* Johns Hopkins University Press, Baltimore, Maryland, pp. 1–288

Canavan P D 1983 Solder. In: Tilton 1983, pp. 36–75.

Chynoweth A G 1976 Electronic materials: Functional substitutions. *Science* 191: 725–32

Demler F R 1983 Beverage containers. In: Tilton 1983, pp. 15–35

Field B C, Berndt E R 1981 An introductory review of research on the economics of natural resource substitution. In: Berndt E R, Field B C (eds.) 1981 *Modeling and Measuring Natural Resource Substitution.* MIT Press, Cambridge, Massachusetts, pp. 1–14

Fisher F M, Cootner P H, Baily M N 1972 An econometric model of the world copper industry. *Bell J. Econ.* 3: 568–609

Gill D G 1983 Tin chemical stabilizers and the pipe industry. In: Tilton 1983, pp. 76–112

Goeller H E, Weinberg A M 1976 The age of substitutability. *Science* 191: 683–89

Kopp R J, Smith V K 1980 Measuring factor substitution with neoclassical models: An experimental evaluation. *Bell J. Econ.* 11: 631–55

Skinner B J 1976 A second iron age ahead? *Am. Sci.* 63: 258–69

Slade M E 1981 Recent advances in econometric estimation of materials substitution. *Resour. Policy* 7: 103–09

Tilton J E (ed.) 1983 *Material Substitutions: Lessons from Tin-Using Industries.* Resources for the Future, Washington, DC, pp. 1–118

Woods D W, Burrows J C 1980 *The World Aluminum–Bauxite Market.* Praeger, New York, pp. 184–230

J. E. Tilton

[Pennsylvania State University, University Park, Pennsylvania, USA]

Substitution: Technology

Substitution as a technological process is the replacement of one material, process, design or technology by another. These types of substitution occur at varying levels of complexity and are the principal responsibility of a design engineer.

In general, substitution is an evolutionary process and one motivated by cost reduction, functional advantage or supply considerations. To these historical factors are now added regulatory constraints, strategic concerns, environmental issues and public attitudes. These, together with a quickened pace of innovation, have led to more abrupt substitution and a growing emphasis on the technological planning for substitution.

Substitution is one facet of a broad conservation effort aimed at conserving scarce materials, reducing energy use and eliminating waste. By its very nature, substitution can have a substantial impact on related areas of technology.

1. Types of Substitution

There are four basic types of substitution: material, process, design and functional. They are closely related in that a substitution of one type may well cause others. Thus, for example, a material substitution often requires a modification of the process used and design changes almost invariably lead to material and process changes. Similarly, substitution stimulates competitive development and innovation in the material, process or system that it seeks to displace.

Material substitution is probably the most familiar type and of greatest concern because of its relation to the supply of critical and strategic resources. Under this heading, three categories of materials substitution may be distinguished: physical, quantitative and invisible.

Physical substitution is the replacement of one material for another and countless examples could be cited; aluminum for copper in electrical conductors, glass for steel in containers and synthetic detergents for soap to name but a few. Such substitutions only occur if the substitute material has properties which permit the function to be achieved at lower costs under the prevailing circumstances.

Quantitative substitution involves a quantitative reduction in the amount of material used per unit of output. Examples would include using a thinner tin coating on tin-plated steel, less gold on an electrical contact or the use of a composite to replace a thicker, monolithic material.

Invisible substitution is less obvious and occurs when a new product entering the market uses material(s) other than those traditionally used. This type is more difficult to trace although some examples are clear such as when a new car model or piece of sporting equipment is introduced.

Substitution in the area of processing and fabrication is responsible for the majority of ongoing cost reductions realized in product manufacture. It can range from eliminating a manufacturing step to adopting a wholly new manufacturing process as in going from a machined to a powder-metallurgy part. This type of substitution is closely correlated with technological development and materials property improvement through processing.

Design substitution involves alteration of the product. Miniaturization is one example. Others include redesigning a product for easier maintenance or to operate on a different mechanical or electrical principle.

In functional substitution, a completely different approach to performing a needed function is found. A now classic example is the substitution of transistors for vacuum tubes. Jet engines replacing piston engines and propellers in aircraft and nuclear reactors replacing boilers fired by fossil fuels are other examples. Functional substitution can lead to the profound revision of consumption patterns for materials and energy and can inspire the creation of entirely new industries (Chynoweth 1976).

2. *Levels of Substitution*

It is possible to identify several levels of substitution depending on the complexity of the substitution process involved and to attach typical development times for their implementation. These levels are adapted from Jantsch (1967) and are listed in Table 1.

Noninteractive materials substitution is one that can be implemented easily while retaining product performance. It could be as simple as changing suppliers for a given material but typically involves a different material. The development of a new material or chemical process is more common and interactive materials substitutions are so named because they often involve associated processing and design modification.

The majority of substitution activity takes place in the first four levels listed which happen to correspond to the time frames for much of present corporate planning. Higher levels of substitution induce significant lower-level changes.

Table 1
Levels of substitution and associated development times

Level	Development time (years)
1 Noninteractive material substitutions	1–3
Assembly or component change	
2 Development of new material or chemical process	3–4
3 Interactive materials substitution	4–5
Electronic technology change	
Subsystem or small system development	
4 Systems of reasonable complexity	~7
5 Complex weapon systems	~10
6 Telephone exchange technology	~13
7 Time for scientific discoveries to find large-scale technical application	~15

3. *The Substitution Process*

Once a substitution need or opportunity has been identified and defined in terms of cost and performance objectives, the steps in its implementation are:

(a) draw on the available research and development stockpile;

(b) begin design and development;

(c) conduct screening tests of viable alternatives;

(d) develop hard design and reliability data; and

(e) evaluate on a pilot scale in the manufacturing environment.

These steps are the same as those for an original design process, differing only in being more narrowly focused for low-level substitution.

The design engineer has the principal responsibility for the substitution process and the critical nature of his role can be seen by considering Fig. 1. At each stage, alternatives exist for the material and the part showing that there is no unique path to a material for use nor to the behavior of a part to meet its intended use. The choices made, then, at these various stages determine the kinds and amounts of materials (and energy) used, attendant pollution, reuse and recycling possibilities and the related technologies to be impacted. Design guides are seldom available to help make wise choices for resources and the environment. It is useful to remember that design engineers are conservative in their technical judgements and are usually rewarded on the basis of the initial effectiveness of their designs and not on end-of-life considerations such as ease of recycling.

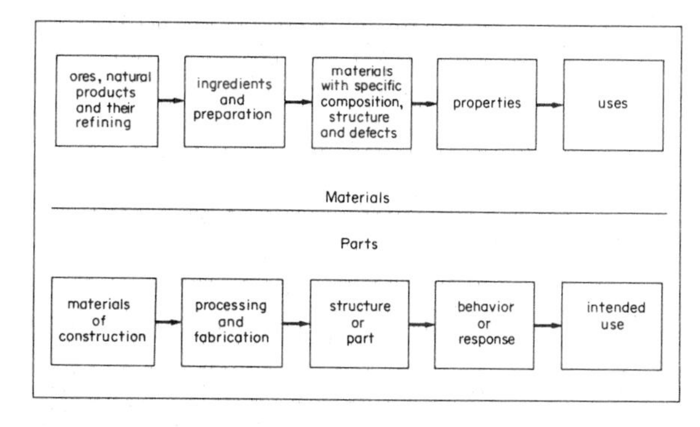

Figure 1
Schematic relationship between starting materials and their final uses

4. *Substitution and Technology Planning*

The critical element in facilitating future substitution is to ensure that an adequate technology stockpile exists on which to draw. This first step in the substitution process allows for the widest range of options to be considered. On a narrower basis, technological forecasting is used to identify likely future directions and products, and the substitution opportunities they present. A substantial methodology exists for doing this (Linstone and Sahal 1976). On yet a narrower level, the materials and components most critical to the product(s) are identified and specific substitution strategies developed to deal with their potential unavailability.

5. *Substitution Constraints*

Many factors act to inhibit substitution. At the design level, there may be a lack of property or reliability data. In manufacturing, there are retooling and start-up costs to consider. Within the organization, there is inertia to continue the past and delay in converting to the new. Government, society and individuals can also act to block or delay substitution.

A particular constraint arises when a design depends on a unique property or material. For example, superconducting systems uniquely require liquid helium. In this, and less stringent examples, design lock-in occurs that prevents substitution. Lock-in can also occur with dedicated manufacturing facilities inflexible to change. For these reasons, substitution offers little hope for mitigating short-term, unanticipated shortages. These are constrained by the in-place technology and must be a direct fit in order to work.

Substitution can also be inhibited because of its impact on related areas of technology. For example, demand for a substitute material could exceed known supply. In addition, the substitution may require developments in several related areas before it itself could be implemented.

6. *Aluminum and Copper: Cost-Driven Substitution*

These last four sections illustrate some of the principles cited and identify certain important driving forces for substitution.

The mutual substitutability of aluminum and copper in electrical applications receives wide study and close monitoring. It is a substitution driven principally by cost considerations that begins with the relative cost required for equivalent electrical conductivity. As seen in Table 2, aluminum is more attractive than copper on that basis and sodium is more attractive than either. The choice, however, is dictated by many other factors including relative availability, ease of manufacture, size, connector technology, performance in the field and customer acceptance. On the basis of these, sodium is rejected for all but a very few special cases while aluminum finds many uses.

Table 2
Electrical conductor materials

	Copper (ETP grade)	Aluminum (EC grade)	Sodium
Resistivity ($\mu\Omega$ cm)	1.71	2.83	4.88
Specific gravity	8.89	2.70	0.97
Price (US$ per kg)[a]	1.92	1.67	1.19
Relative cost per mho	1.00	0.44	0.19

[a] American Metal Market, March 1981

In the specific instance of substituting aluminum for copper in communications cables, it required some five years to accomplish because new alloys, connectors, cable constructions and manufacturing methods had to be developed and tested. This is a case where now either could be used depending on overall economics, but there are other cases where only copper or only aluminum can be used for technical reasons.

7. *Cobalt: Availability-Driven Substitution*

In 1978, due to production difficulties in Zaire, the producer price for cobalt quadrupled and a user allocation of 70% was imposed temporarily. Since Zaire supplies about 40% of the world's cobalt and since most countries are heavily dependent on imports for cobalt, these events sparked a wide effort to find substitutes. As seen in Table 3, and discussed by Chin et al. (1979), magnetic alloys were one area where several commercially viable substitutes were found. These and other substitutes served to markedly alter the pattern of cobalt consumption between 1977 and 1980 in the USA. The continued high use of cobalt in superalloys reflects the difficulty of substitution (Tien et al. 1980).

8. *Cars: Regulatory-Driven Substitution*

Government-mandated, fuel-economy requirements in the USA are acting to substantially reduce the weight of cars and to alter their pattern of materials use. The weight reduction required is being achieved through smaller cars and greater use of lightweight, high-strength materials. This has meant an increased use of high-strength steels, aluminum and plastics as seen in Table 4. By 1985, some 30% of the car's weight will be in these materials selected according to cost per unit fuel saved.

Since the automotive industry is a large user of materials, this changing pattern has a major impact on materials suppliers and on recycling. In addition, these newer materials require significant changes in manufacturing and assembly.

Table 3
US cobalt consumption (%)[a]

	1977	1980
Superalloys	26	45
Magnetic alloys	25	15
Cutting and wear tools	16	15
Catalysts and driers	24	20
Other	9	5

[a] US Bureau of Mines, Mineral Commodity Summaries

Table 4
Projected material usage in US passenger cars (%)

Material	1977	1981	1985
Cast iron	17	13	10
Steel:			
high-strength	2	9	12
other	60	52	46
Aluminum	3	4	8
Plastics	4	7	9
Other	14	15	15

9. Light-Wave Communication: Technology-Driven Substitution

In the 1970s, optical fibers having very low optical signal attenuation and good mechanical strength were developed and offered commercially. Reliable, long-lived, light-emitting and laser diode sources also became available at about the same time. Combining this transmission medium with reliable sources and detectors led to light-wave communication systems of high information-carrying capacity or bandwidth.

Such systems joined with current integrated chip technology have created a communications–computer environment that promises to have major commercial and social impact. The availability and anticipated low cost of these systems facilitates electronic mail, teleconferencing and electronic meetings, electronic journals and books, robotics and computer-aided design, and the wired city. Functional substitution of this breadth and depth profoundly alters our use of materials an energy.

See also: Substitution: Economics; Materials Economics, Policy and Management

Bibliography

Altenpohl D G 1980 *Materials in World Perspective.* Springer, New York

Chin G Y, Sibley S, Betts J C, Schlabach T D, Werner F E, Martin D L 1979 Impact of recent cobalt supply situation on magnetic materials and applications. *IEEE Trans. Magn.* MAG-15: 1685–91

Chynoweth A G 1976 Electronic materials: Functional substitution. In: Abelson P H, Hammond A L (eds.) 1976 *Materials: Renewable and Nonrenewable Resources.* American Association for the Advancement of Science, Washington, DC, pp. 123–30

Harwood J J, Boiling G F 1980 Planning for changing materials utilization patterns. *Mater. Soc.* 4: 239–46

Jantsch E 1967 *Technological Forecasting in Perspective.* Organization for Economic Cooperation and Development, Paris, pp. 39–48

Linstone H A, Sahal D (eds.) 1976 *Technological Substitution.* Elsevier–North-Holland, New York

National Materials Advisory Board 1972 *Mutual Substitutability of Aluminum and Copper*, NTIS PB-211-727. National Technical Information Service, Springfield, Virginia

Tien J K, Howson T E, Chen G L, Xie X S 1980 Cobalt availability and superalloys. *J. Met.* 32(10): 12–20

T. D. Schlabach
[Bell Laboratories, Murray Hill, New Jersey, USA]

T

Thermosetting Resins

In addition to the major thermosetting resin systems such as polyurethanes, phenol–formaldehyde resins and epoxy resins, there are several other commercially available resin systems that will be discussed here. These resins vary widely in economic importance, from production rates of only several hundred thousand dollars per year for Friedel–Crafts resins, to over two billion dollars per year for unsaturated polyester resins. Many are notable for specific properties or processing characteristics not available in the common thermosets. Some of the main characteristics of these resins are summarized in Table 1. Only the more important resins are discussed in detail here.

1. Unsaturated Polyester Resins

There are three main classes of resins that will be classified as unsaturated polyesters for the purposes of this discussion: i.e., standard unsaturated polyesters, alkyd molding compounds and vinyl esters. The latter two are sufficiently similar to the primary unsaturated polyester compositions to be included here.

1.1 Standard Polyester Resins

The most widely used unsaturated polyester resins are prepared by thermal condensation of a mixture of an unsaturated dicarboxylic acid, a glycol or mixture of glycols, and a modifying saturated dicarboxylic acid or mixture of acids. This first-stage low-molecular-weight resin is then blended with a liquid cross-linking monomer to yield a viscous liquid, which can then be thermoset by the addition of appropriate free-radical catalysts. For economic reasons, the most common unsaturated acid is maleic acid, which is usually charged as the anhydride. Common glycols are ethylene and propylene glycols or the dimeric glycols, diethylene and dipropylene glycols. Phthalic anhydride is perhaps the most common acid modifier because of its low cost. By far the most common vinyl monomer is styrene, although vinyl toluene has a significant use in fast-curing systems. The chemical reactions involved in these processing steps are summarized in Scheme 1.

It is important that the maleic-acid groups in the polyester be converted to fumarate because of the superior copolymerizability of the latter with styrene. This is generally accomplished thermally during the condensation step.

Because of the brittleness of most of these cured resins and for economic reasons, the products are almost universally compounded with fillers and fibrous reinforcements prior to curing. This requirement, together with the large number of monomers, catalysts, fillers and reinforcements available commercially, has led to the development of an extensive body of formulation and reinforcement technology that can be used to vary widely the processing characteristics depending upon the requirements of the application. Examples of the compounding versatility possible with these systems are the use of chlorendic acid and isophthalic modifying acids to improve corrosion resistance, use of chlorendic acid and/or dibromoneopentyl glycol to reduce flammability, use of inexpensive fillers such as clay and calcium carbonate to reduce the cost, and use of glass fiber and/or glass cloth in the composition to increase rigidity and impact strength. Metal oxides and high-temperature catalysts are used to prepare sheet molding compounds, while benzoyl peroxide and promoters and accelerators are used for

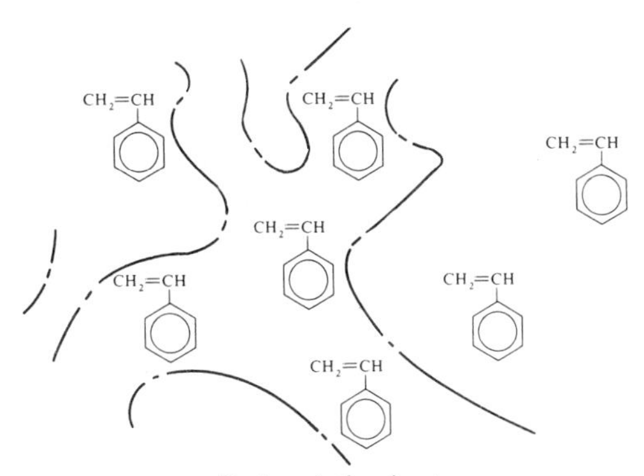

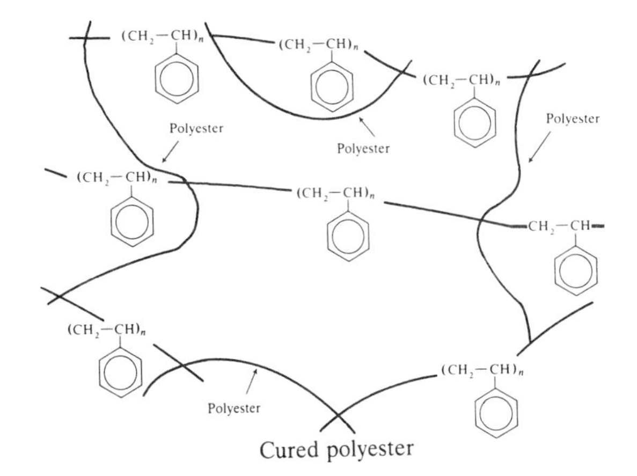

Scheme 1

Table 1
Miscellaneous thermosetting resins

Resin	Common monomer	Typical chemical structure of resin	Curing conditions	Important properties	Major applications
Furan	furan ring–CH_2OH	⁅(furan ring with CH_2)–CH_2⁆$_x$	acidic catalysts room temperature and above	heat and corrosion resistance	reinforced laminates, binders
Friedel–Crafts	phenol (OH on benzene ring)	–CH_2–(phenol ring, OH) CH_2–(benzene ring)–CH_2–(phenol ring, OH)	elevated temperatures	thermal stability	reinforced laminates and moldings
				high heat resistance	
	CH_3OCH_2–(benzene ring)–CH_2OCH_3	+ hexamethylene tetramine or an epoxy resin			
Allyls	benzene ring with two –C(=O)–$OCH_2CH{=}CH_2$ groups	–CH–CH_2– / CH_2 / C=O / (benzene ring) / C=O / O / CH_2 / –CH_2CH–	free-radical catalysts at elevated temperatures	excellent electrical properties and heat resistance	reinforced laminates and moldings for electrical applications

Unsaturated polyesters	polyglycolfumarate and styrene	$-R-O-C(=O)-CH-CH-C(=O)-O-$ / $-CH(C_6H_5)-CH_2-$ / $-R-O-C(=O)-CH-CH-C(=O)-O-$	free-radical catalysts room temperature and above	processing versatility, require fibrous reinforcement for best properties	reinforced laminates and moldings for a wide variety of business areas such as transportation and construction
Alkyds	phthalic anhydride, triglyceride vegetable oils, modifying acids and alcohols	$-R-C(=O)-O-CH_2-CH(-O-C(=O)-R'-CH-CH-R'')-CH_2-O-C(=O)-R-C-C(=O)-$ / $-R-C(=O)-O-CH_2-CH(-O-C(=O)-R'-CH-CH-R'')-CH_2-O-$	air oxidation at room temperature	low-cost raw materials, versatile formulation, low-cost production processes	paints and coatings
Acrylics	styrene, methyl methacrylate acrylamide, hydroxyalkyl, acrylates with formaldehyde and butanol	$-C(=O)-NH-CH_2-OBu$ / $-C(=O)-NH-CH_2-NH-C(=O)-$	120–175 °C with acid catalysts	good hardness and light stability	baked enamels and coatings

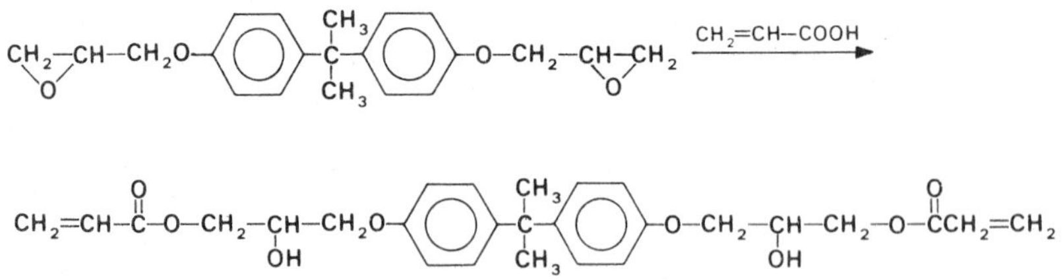

Vinyl ester resin

Scheme 2

room-temperature cure in hand-layup operations. Hydrated alumina is a commonly used fire-retardant filler in shower stalls and bathroom fixtures. Other formulation changes can be used to prepare bulk molding compounds (BMC) and reaction injection molding (RIM) compositions.

1.2 Alkyd Molding Compounds

Alkyd molding compounds are unsaturated polyesters that are formulated to form low-melting-point solids, and then compounded with fibrous reinforcements, fillers, catalysts and high-boiling-point or solid cross-linking agents such as diallyl phthalate. The ingredients are generally mixed on heated rolls and extruded into granular or rope form for molding. Such dry molding compounds are used primarily in electrical applications, because of their superior electrical properties, high impact strength, good dimensional stability, good surface appearance and high heat resistance.

1.3 Vinyl Esters

These modified polyester resins are most commonly prepared by reacting a bisphenol diepoxide or resin with acrylic acid to yield an acrylic-ester-terminated resin, which can then be dissolved in styrene or other vinyl monomers and processed as described for unsaturated polyesters above. The chemical reactions involved are summarized in Scheme 2, using the bisphenol A diepoxide for illustrative purposes.

These resins are superior to most conventional polyester resins in glass wet-out; they have greater curing speed, can take higher filler loadings and have increased resilience. They are used predominantly in corrosion resistance and electrical applications, using filament-winding and hand-layup techniques.

2. Alkyd Resins

As originally defined by Kienle, the term *alkyd* encompassed the reaction products of polyhydric alcohols and polybasic acids. This broad definition includes unsaturated polyesters, polyester polyols used in the manufacture of polyurethanes, and high-molecular-weight polyesters such as polyethylene terephthalate. In modern usage, however, the term predominantly designates polyester compounds formulated with monobasic fatty acids or vegetable oils, and the solid molding compounds described under unsaturated polyesters. It is this latter definition that will be used here.

As currently formulated, alkyd resins are the reaction products of a polybasic acid (such as phthalic anhydride), vegetable oils (linseed oil) and/or an unsaturated fatty acid, a polyhydric alcohol (glycerol or pentaerythritol) and often a saturated fatty-acid modifier such as pelargonic acid. The product is then diluted to 30–70% solids with an inert solvent such as toluene or mineral spirits. A simple example of the chemical reactions involved in such a preparative procedure is given in Scheme 3. The monobasic fatty acid (or acids) generally contains olefinic double bonds which are subsequently used in cross-linking—most commonly by reaction with atmospheric oxygen in the presence of metallic compounds called driers.

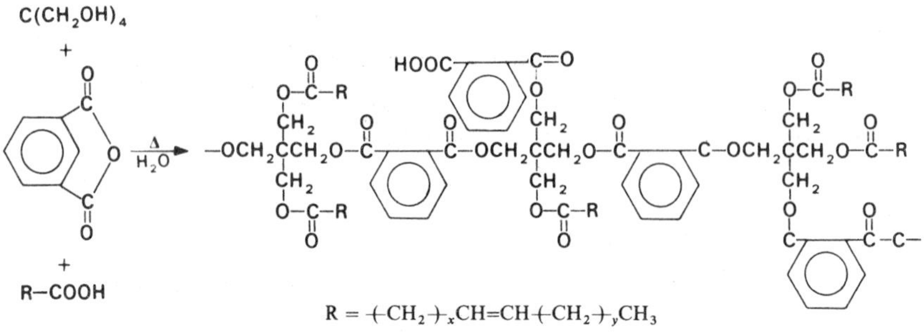

$R = -(CH_2)_x-CH{=}CH-(CH_2)_y-CH_3$

Scheme 3

Despite the apparent simplicity of the above polycondensation, the functionality and ratios of the various ingredients must be carefully calculated, using Flory's modified theory of gelation, and the degree of polycondensation carefully controlled, if a reproducible product is to be obtained and premature gelation circumvented. The use of fatty-acid triglycerides (vegetable oils) under the preparative conditions leads to a transesterification reaction between the oil, alcoholic and acid components, as indicated in Scheme 4. Oxygen must be carefully excluded during the polycondensation to prevent uncontrolled oxidation of the olefinic double bonds of the fatty acid, if the formation of undesirable gel particles is to be prevented. The large number of monomers available for use in the above reaction allows the preparation of a nearly infinite number of compositions. A wide variety of properties is thus possible in the final product. Some of the more common commercial monomers available for these compositions are listed in Table 2.

A complete discussion of the many variations commonly used in alkyd manufacture is beyond the scope of this article. A few of the more important possible variations will be mentioned for completeness.

The ratio of oil or unsaturated fatty acid used in the formulation is commonly identified by the terms long-oil and short-oil alkyds. The former contain the higher oil ratios and the shorter drying times (drying in this sense includes both the evaporation of the solvent and oxidative cross-linking to the thermoset film). The term oil-free alkyds refers to those containing no fatty acids, such as the glycol maleate ester of pentaerythritol. Modified alkyds are usually those that have been chemically reacted with other components such as rosin or abietic acid.

```
      O                                 O
      ||                                ||
CH2—O—C—R   —COOH             CH2—O—C—
|                              |
|     O                        |     O                 O
|     ||                       |     ||                ||
CH—O—C—R'  +          ——→      CH—O—C—R'  +  —CH2—O—C—R
|                              |
|     O                        CH2OH
|     ||
CH2—O—C—R   —CH2OH
```

Scheme 4

The alkyd can also be thermally polymerized during preparation by a Diels–Alder mechanism, as indicated in Scheme 4. This process is called heat bodying, and is used to alter drying time and viscosity.

Alkyd resins are often mixed with other resins such as epoxies and cellulose nitrate to yield an alkyd blend with desirable properties not easily obtainable by direct formulation.

2.1 Commercial Uses

The alkyds are predominantly used in coatings and paints, although some limited utility has also developed in printing inks, adhesives and fabric binders.

3. Less-Common Thermosets

A variety of thermoset resins of lesser importance are also produced commercially. Furan polymers and allyl resins are reviewed briefly here.

3.1 Furan Polymers

Furan resins are the reaction products of furfuryl alcohol or furfuryl aldehyde. The materials are the reaction product of many vegetable materials, such as oat hulls and corn cobs, when appropriately treated with aqueous acid. Although furfural is used to modify phenolic novolak resins under alkaline conditions, its hydrogenated derivative, furfuryl alcohol, is a much more versatile polymer precursor. It is polymerized to a dark resinous prepolymer in the presence of heat and acidic catalysts to the desired degree of polymerization (low-viscosity liquids or brittle solids). After neutralization, the stable first-stage polymer can be cured by the addition of such acidic catalysts as maleic anhydride, phosphoric acid or *p*-toluene sulfonic acid. The stronger acids are used to effect rapid cures at room temperature. Although the stable intermediate resins are complex mixtures and little understood, the main polymerization reaction is depicted in Scheme 5, with

Table 2
Most-common alkyd ingredients

Polybasic acids	Oils	Polyhydric alcohols	Monobasic acids
Phthalic anhydride	Linseed	Glycerol	Tall-oil fatty acids
Isophthalic acid	Soya	Pentaerythritol	Fatty acids from oils
Maleic anhydride	Tung	Dipentaerythritol	Synthetic fatty acids
Fumaric acid	Fish	Trimethylolethane	Pelargonic acid
Azelaic acid	Safflower	Sorbitol	Isodecanoic acid
Succinic acid	Cottonseed	Ethylene glycol	Isooctanoic acid
Adipic acid	Dehydrated castor oil	Propylene glycol	2-Ethylhexanoic acid
Sebacic acid		Neopentyl glycol	
		Dipropylene glycol	

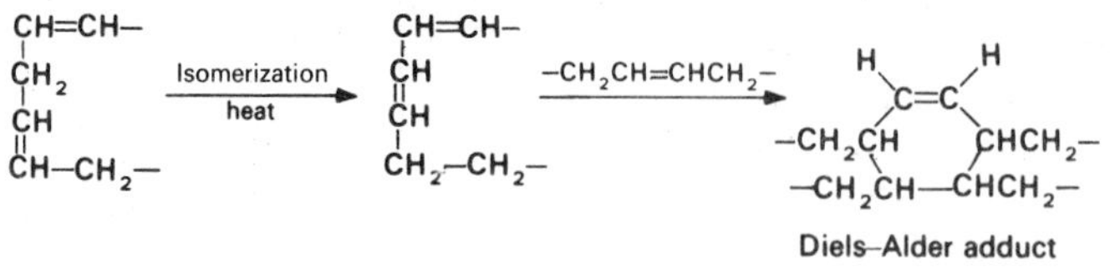

$-CH=CH-CH_2-CH=CH-CH_2- \xrightarrow[\text{heat}]{\text{Isomerization}} -CH=CH-CH=CH-CH_2-CH_2- \xrightarrow{-CH_2CH=CHCH_2-}$ Diels–Alder adduct

Scheme 5

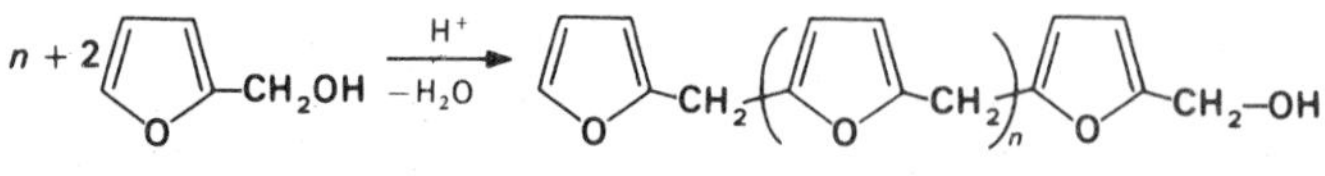

$(n+2)$ furfuryl alcohol ($C_4H_3O-CH_2OH$) $\xrightarrow[-H_2O]{H^+}$ $C_4H_3O-CH_2-(C_4H_2O-CH_2)_n-C_4H_2O-CH_2-OH$

Scheme 6

reaction indicated only at the more active α hydrogen atoms for simplicity.

The reaction is very exothermic and must be carefully controlled to prevent overheating. Thermosetting is the result of reaction at the β positions on the furan ring. Often, the above reaction is modified by the addition of formaldehyde and/or urea. The cured resins are renowned for the excellent heat and chemical resistance which dictate their major application areas.

3.2 Allyl Resins

These resins are prepared from polyallyl esters such as diallyl phthalate and diethylene glycol bis(allyl carbonate) by partially polymerizing the monomer, in the presence of free-radical catalysts, to low-molecular-weight solid resins (Scheme 6). The resins are subsequently compounded with additional free-radical catalysts, fillers and fiber reinforcements into molding compounds, whose principal utility is the preparation of high-performance composites for electrical and aerospace applications. The polyfunctional nature of the monomers ensures highly cross-linked materials after final cure.

The composites show superior electrical properties and dimensional stability, high heat and chemical resistance, and low shrinkage. Other monomers used in these molding compounds for special properties are diallyl isophthalate, triallyl cyanurate and diallyl chlorendate.

See also: Epoxy Resins

Bibliography

Brusie J P 1964 Acrylic ester polymers. In: Bikales N M (ed.) 1964 *Encyclopedia of Polymer Science and Technology*. Wiley, New York, pp. 273–317

Mraz R G, Silver R P 1964 Alkyd resins. In: Bikales N M (ed.) 1964 *Encyclopedia of Polymer Science and Technology*. Wiley, New York, pp. 663–730

Siegfried K J 1964 Furan polymers. In: Bikales N M (ed.) 1964 *Encyclopedia of Polymer Science and Technology*. Wiley, New York, pp. 432–45

Whelan A, Brydson J A 1975 *Developments with Thermosetting Plastics*. Elsevier, London

R. R. Hindersinn
[Occidental Chemical Corporation,
Niagara Falls, New York, USA]

Time-Independent Plasticity

In most efforts to cope with the physical aspects of the real world, it is helpful to think in highly compartmentalized modes. Very special cases are considered and very drastic idealizations are made of the geometry of the structures, machines and devices, and of the materials of which they are constructed. Sharp distinctions are drawn between elastic and inelastic behavior. Inelastic behavior is divided into such categories as viscous, elastic–plastic, viscoelastic and viscoplastic. Within each category, the attempt is often, and very appropriately, to be as general in approach as possible. Frequently, however, further subcategories are found useful or necessary. Idealizations of idealizations are created in attempts to capture the essence of reality relevant to the class of problems for which answers are sought.

Proposals are often made to adopt a completely opposite point of view, to seek to describe reality itself through statements of fundamental laws of universal validity and concepts applicable to all materials. Certainly, the more that can be done along these lines, the greater the understanding of material behavior will be and the sounder the idealizations will be at each level of idealization. However, a complete description of reality is no description at all. The laws of thermodynamics must be obeyed, but within that framework the possible modes of behavior of materials are infinitely complex in an infinity of ways: Bridgman spoke of a sea of irreversibility.

1. Behavior and Idealization in One Dimension

The simple stress–strain curve in tension, compression or shear (Fig. 1) is not in fact all that simple. The higher the rate of straining or stressing, the higher the curve. Most often, but not always, the higher the temperature at which the test is performed the lower the curve. Homogeneity of material, even on the scale of conventional laboratory specimens, is unlikely, so that nominally identical specimens of the same batch of material will often differ appreciably in their response. Differences in the machining or other means of preparation of the specimen also cause significant effects: small specimens are more likely to show appreciable differences than large ones. Nevertheless, for most of the common structural metals and alloys over quasi-static rates of loading of practical interest and modest temperature ranges, the different curves obtained will fall into a reasonably narrow band for carefully controlled material and carefully prepared specimens. Reference to *the* stress–strain curve is an extremely useful time-independent idealization for analysis and design, although it ignores a host of fascinating features of behavior that are the essential aspects of the response in the creep range on the one hand or the high-velocity impact range on the other. Similarly, the idealization of purely elastic response for partial unloading and subsequent reloading ignores completely the dislocation motion that does occur and contributes damping capacity in the nominally elastic range of response. Yield points and plastic deformation are matters of definition. If the motion of a single dislocation is considered to be plastic deformation, there is no purely elastic range of response. If plastic deformation is defined to be ignorable below a strain of the order of 0.001, the elastic range so defined is quite a bit larger than the limits shown by the horizontal lines in Fig. 1.

The replacement of a fairly narrow band of stress–strain curves by a single curve—and correspondingly the band of load–deflection curves for a structure by a single curve—is consistent both in computation and concept. The laboratory specimens from which the stress–strain curves are taken are inhomogeneous on the microscale. Inclusions, voids and grains of differing orientation are quite analogous to the elements of a structure or other body under load. Initial or residual stresses are present in both. The variation in stress from region to region of a loaded body or structure is normally far greater than in a laboratory specimen under nominally homogeneous stress. Departure from linearity, therefore, usually is visible on a load—deflection plot at a much earlier stage than on a stress–strain plot, and the transition region from dominantly elastic behavior to dominantly plastic behavior is proportionally greater, as shown in Fig. 1.

At strains or deflections that are a small multiple of the computed purely elastic response, the slope of the curves becomes small compared with the elastic slope for most ductile metals. This small positive slope, or work hardening, is a stable response of great practical importance. However, the further idealization that replaces the upward sloping curve by a line of zero slope, the idealization of perfect plasticity (Fig. 2), certainly gives the proper level of stress or of load for a wide range of strains or deflections about any chosen reference value. This would be true for a single work-hardening curve, and becomes even more reasonable as an approximation to the band of curves.

Monotonic loading interrupted by partial unloading and reloading is only a small part of the total picture. Reversed loading as in Fig. 1 causes plastic deformation of the opposite sign at a much smaller

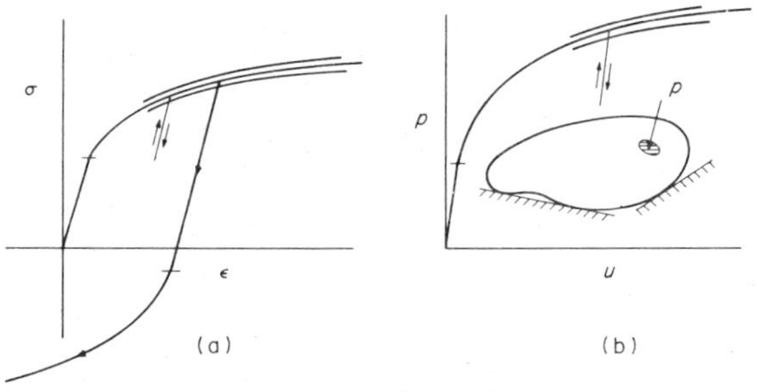

Figure 1
(a) The stress–strain (ε–σ) curve and (b) the load–deflection (u–p) curve: analogous idealizations

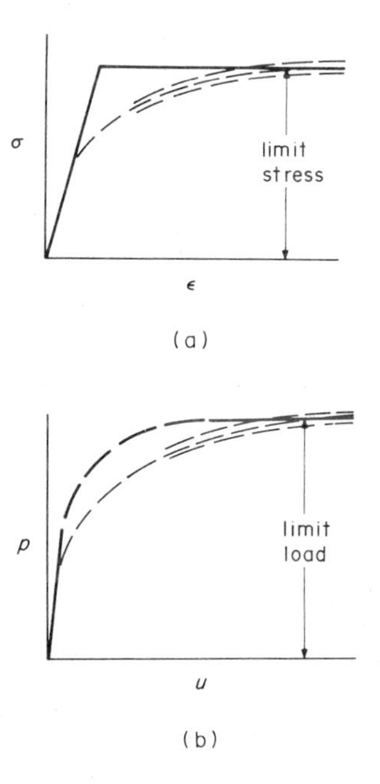

Figure 2
Further idealization from Fig. 1 to perfect plasticity

magnitude of stress than for continued plastic deformation, and often at less than the initial yield value for the forward loading. This Bauschinger effect is large and of great importance in cyclic loading and many other practical situations. However, continued reversed loading tends to reach stress levels about equal to those for forward loading once the stress–strain curve flattens out at large negative strains. This flattening out occurs when the reversed plastic strain becomes a moderate multiple of the reversed elastic response.

2. Multiaxial Behavior and Idealization

Of course, these considerations of response to one component of stress or to one load must be generalized to a six- or nine-dimensional stress space and to an N-dimensional load space and the corrresponding strain or displacement response. In one sense, we know very little about the general inelastic response of ductile metals and alloys, even when time-dependence can be ignored over the range of strain rates considered. Tests other than simple tension, shear, compression or their equivalent have been run on a small sample of materials, but they are very few indeed in comparison with the enormous complexity in detail of the response of any one material to the application of load. A statically determinate test specimen is required for direct determination of stress–strain relations, a specimen in which knowledge of the geometry and the loads give the stresses unambiguously, however they are defined, and the strains can be determined by measuring the geometry change.

When there is axial symmetry and homogeneity of material along the length, the standard thin-walled cylindrical tube specimen (Fig. 3) under combinations of twisting moment M_T, axial force F and interior pressure p (or combined interior and exterior pressure) is uniquely suitable. Almost all the detailed experimental information on multiaxial stress–strain relations comes from such tests. Yet there are severe limitations: compression is difficult; tube walls that are too thin will buckle too early in torsion; the most careful machining of specimens affects the properties of material somewhat, and annealing generally is not feasible when the structural alloys are to be examined in their heat-treated condition. Furthermore, the paths of loading that can be followed are limited to plane stress with or without superposed hydrostatic pressure. Unfortunately, also, few alloys as produced permit fabrication of a tube with axisymmetric properties.

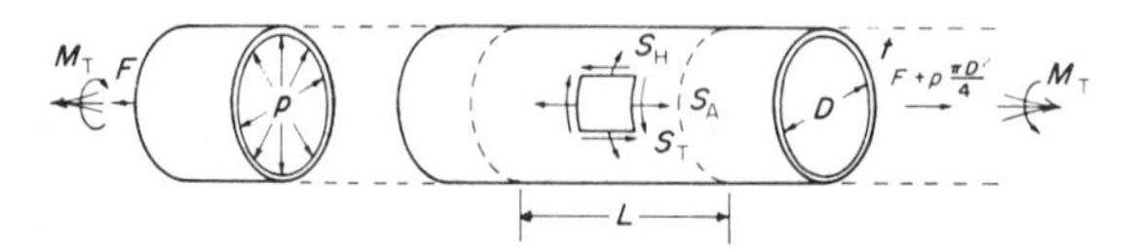

Figure 3
Axial symmetry and homogeneity along the length L required for direct determination of stress–strain relations

Why then is so much trouble taken for a few rather special materials under restricted states of stress? It is more than the inability to do better, it is the very reasonable hope that the essential features of response of a wide class of materials can be discovered. Knowledge and fundamental understanding of the stress–strain relation for any one material should carry over in principle, although not in detail, to materials of comparable qualities. Mathematical generalizations of the basic characteristics found experimentally should be applicable to initially anisotropic as well as isotropic materials, and to general states of stress, not to plane stress alone. However, it is not to be expected that the details of behavior within the established broad framework can be determined. Experimental information alone can provide quantitative values for coefficients A_{ijkm} in the relation between increments of stress $\delta\sigma_{ij}$ and of strain $\delta\varepsilon_{ij}$:

$$\delta\varepsilon_{ij} = A_{ijkm}\delta\sigma_{km}$$

or more precisely their rates:

$$\dot{\varepsilon}_{ij} = \dot{A}_{ijkm}\dot{\sigma}_{km}$$

3. Stress–Strain Relations from Tube Tests

Suppose M_T, F and p are applied to explore stress–strain relations (Fig. 3). Working displacements for the test section initially of length L_0 and diameter D_0 (volume $V_0 = \pi D_0^2 L_0/4$) are the relative angle of twist $\delta\phi$ corresponding to M_T, the change in length δL corresponding to F, and the change in interior volume δV of the gauge section corresponding to p, all of which can be viewed as measured quantities. The change in thickness δt needed to specify the geometry completely does not yet enter, and may be very troublesome to measure accurately; imagine that difficulty to be overcome.

Many methods of analyzing the data can be devised, but the most direct is simply to present results in terms of the applied generalized forces (M_T, F and p) and corresponding displacements ($\delta\phi$, δL and δV), and so avoid any detailed interpretation at the start. Suppose the response to be reasonably time-dependent over a considerable range, as indeed it will be for most useful structural metals. Suppose further that the system is in the usual work-hardening range, with its response stable for small changes in displacement in the strict sense of this definition: i.e., the work done by the increments of M_T, F and p on the displacement increments produced is positive for all equilibrium paths or cycles of loading (including unloading) involving small plastic strains. The initial yield and each subsequent loading surface in (M_T, F, p) space enclosing the current domain of purely elastic response must then be convex (Fig. 4). The infinitesimal

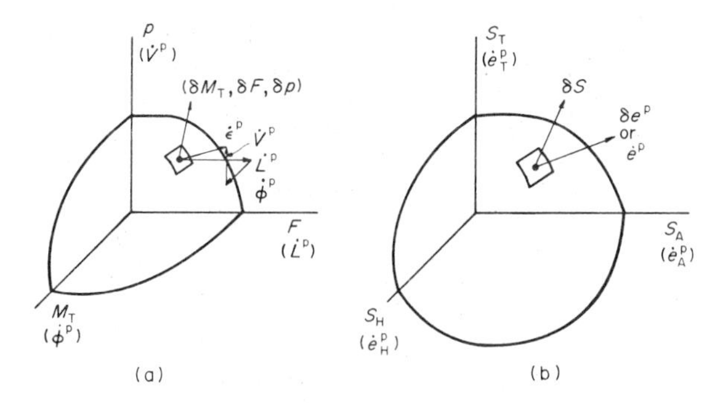

Figure 4
Convexity of each yield or loading surface and normality of the plastic strain rate (or increment) in load (a) and conventional stress space (b)

plastic displacement vector ($\delta\phi^p$, δL^p, δV^p) or the rate vector $\dot{\varepsilon}^p(\dot{\phi}^p, \dot{L}^p, \dot{V}^p)$ must be normal to the loading surface $H(M_T, F, p)=K^2$ at the current load point (M_T, F, p) (normality is to be interpreted in the extended sense at a corner or vertex):

$$\dot{\phi}^p=\Lambda\partial H/\partial M_T,\ \dot{L}^p=\Lambda\partial H/\partial F,\ \dot{V}^p=\Lambda\partial H/\partial p$$

The plastic displacement rate as defined here is the total rate minus the elastic or recoverable displacement rate when infinitesimal δM_T, δF, δp are added to (M_T, F, p) and then removed. In the small-strain range the elastic response at each stage is indistinguishable from the initial elastic response.

When the initial geometry or any later geometry is taken as reference and subsequent changes in geometry are treated as small, convexity then follows in the resulting conventional stress space, and the conventional plastic strain rate obeys normality (Fig. 4). The word conventional is used here for the shear stress S_T and the axial stress S_A on the cross section, and the hoop stress S_H along with the corresponding increments of shear, axial and hoop strain (δe_T, δe_A, δe_H) [or rates ($\dot{e}_T$, $\dot{e}_A$, $\dot{e}_H$)] and stress rates ($\dot{S}_T$, $\dot{S}_A$, $\dot{S}_H$). Normality and convexity carry over from (M_T, F, p) space with ($\dot{\phi}^p$, $\dot{L}^p$, $\dot{V}^p$) superposed, to (S_T, S_A, S_H) space with ($\dot{e}_T^p$, $\dot{e}_A^p$, $\dot{e}_H^p$) superposed, if the same geometry is used for the computation of stress and strain and their rates. Nominal stress and strain is the common terminology when the dimensions D_0, L_0, t_0 are employed; true stress and strain when the current dimensions D, L, t are used.

4. *General and Special Forms for Small Displacement*

In the usual small-displacement theory with stress and strain denoted by σ_{ij} and ε_{ij}, the elastic and plastic components of strain rate $\dot{\varepsilon}$ are separable:

$$\dot{\varepsilon}_{ij}=\dot{\varepsilon}_{ij}^e+\dot{\varepsilon}_{ij}^p$$

The plastic strain rate is normal to the current yield surface $f(\sigma_{ij})=k^2$ at the current stress point:

$$\dot{\varepsilon}_{ij}^p=\lambda\partial f/\partial\sigma_{ij}=G\frac{\partial f}{\partial\sigma_{ij}}\frac{\partial f}{\partial\sigma_{mn}}\dot{\sigma}_{mn}$$

where f and G are functions of stress, and in general depend upon the prior history of plastic strain. The simplest of all analytic forms is the J_2 isotropic hardening form associated with the names of Mises, Huber, Prandtl, Reuss and Prager:

$$f=J_2=\tfrac{1}{2}s_{ij}s_{ij}=\tfrac{1}{2}(\sigma_{ij}-\tfrac{1}{3}\sigma_{kk}\delta_{ij})(\sigma_{ij}-\tfrac{1}{3}\sigma_{kk}\delta_{ij})$$

$$G=G(J_2)\ (\text{e.g., } cJ_2^n)$$

$$\dot{\varepsilon}_{ij}^p=Gs_{ij}\dot{J}_2$$

Many other mathematical forms have been proposed and are in use, and the reader is referred to the Bibliography for further discussion on these, as well as on the notation and conventions used here.

Stress–strain relations are available which do not employ a yield surface. However, calculations of even moderate accuracy for complex loading paths cannot be made without the equivalent "book keeping": a record must be kept of the extent and direction of previous loading and taken into account properly if a reasonable estimate is to be made of the response to a radically new direction and extent of loading, whether or not the yield surface is shrunk to zero.

5. *Geometry Change*

When changes in geometry must be considered more fully in the calculation of stresses, strains and their rates, many problems arise. There is no difficulty with overall rigid-body rotation, because the axis of the tube serves as reference. However, the increments or rates of "conventional" stress are not "proper" rates, because the rotation of material associated with the shear deformation is not taken into account. Large plastic deformations are not especially troublesome in principle, because the reference coordinate system may be continuously reconfigured. Large elastic deformations do lead to difficulties of definition of plastic-strain increments, of reference directions in the material, and consequently of appropriate stress rates. Fortunately, elastic strains are small for most structural metals.

Rice has pointed out that it is not permissible in considerations of bulk stability to ignore terms of magnitude $\sigma\delta\varepsilon^p$, because they are comparable to $\delta\sigma$ when plastic moduli are comparable to stress in magnitude—a common situation. Destabilizing as such terms can be, they are already included implicitly in the conventional stress–strain relation. It is not appropriate to relabel the conventional stress–conventional strain relations (as, for example, Cauchy stress–strain relations for rotated axes) and then calculate the effective shear modulus for shear localization.

The tube test gives that modulus directly; stability in the tube test carries over to material stability elsewhere under the same state of stress, increment of stress and history of strain.

6. *Initial- and Subsequent-Yield Surfaces*

Figure 5 is a schematic of initial-yield and subsequent-yield or loading surfaces in a stress or load space. As loading proceeds along any path, radial or otherwise, and plastic deformation takes place, the subsequent-yield surfaces (surfaces enclosing states of stress or load reachable along purely elastic paths) may shrink, expand, translate or rotate. Once again, the idealization of reality may be moderate or drastic. The elastic-response range or size of the yield surface will go to zero with the definition that any dislocation motion constitutes plastic deformation, and it will be very large in extent with a 0.002 strain offset definition of yield. Figure 6 is a plot of response for loading along one of the axes of Fig. 5. For any such radial or proportional loading, or for any generally outward loading as illustrated in Fig. 5, the generalized strain response corresponding to the generalized stress imposed has the appearance of Fig. 6. If inelastic strains or displacements of the order of a few times the computed total elastic response can be ignored, all the complexity of plasticity represented by moving and changing yield surfaces disappears. In the resulting elastic–perfectly plastic idealization, the single-limit surface is the analog of the single level of stress in Fig. 2. If some of the reality of work hardening is to be included in the idealization in the form of isotropic hardening, then a nested set of loading surfaces may be suitable for generally outward loading paths (Fig. 5). If a wide variety of loading paths must be accommodated in some detail, the complex but still highly idealized picture of moving and changing yield surfaces is needed.

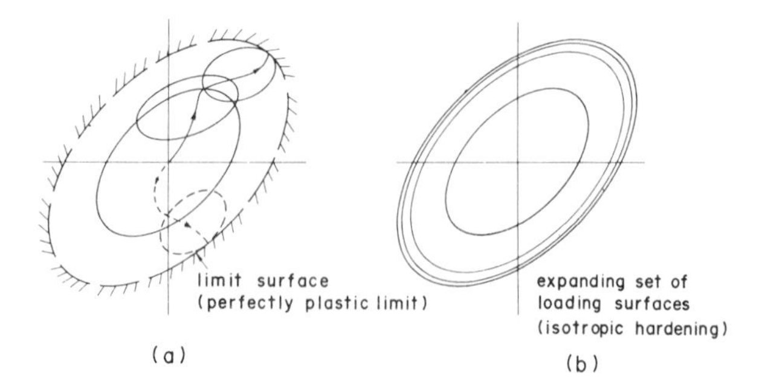

Figure 5
Idealizations of initial- and subsequent-yield surfaces in (a) stress or (b) load space

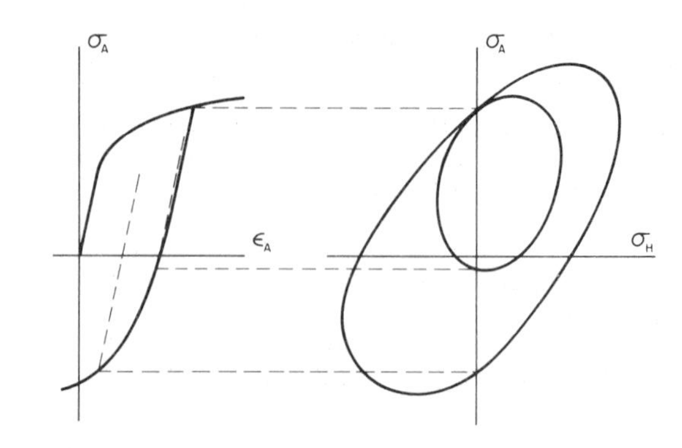

Figure 6
The yield surface is a matter of choice

7. *Limit Theorems*

The limit theorems of perfect plasticity, which derive from stability, have essential validity for amply ductile work-hardening materials when there is stability of both material and structure, and fracture does not intervene early.

(a) Lower-Bound Theorem: If an equilibrium distribution of stress can be found lying within the limit surface everywhere in the body, the body will not collapse. The translation of this idealization to reality is that if such a safe equilibrium distribution of stress can be found for the given loads, the body will not deform plastically by more than a few times the computed elastic deformation. Initial stresses wash out.

(b) Upper-Bound Theorem: If for any set of kinetically admissible deformations taken as purely plastic the work done by the applied loads exceeds the energy dissipated, collapse will occur. In translation to the real world of work-hardening materials, the applied loads will cause the body to undergo plastic deformations that are extremely large compared with the elastic, and may well be out of the range of small displacement theory.

The limit theorems on which engineers often depend consciously or unconsciously are not valid for frictional systems, or for pressure-dependent materials that do not follow an associated-flow rule (normality of $\dot{\varepsilon}^p$); a perfectly reasonable safe distribution of stress may be possible and yet the system will collapse. Initial states of stress are important and may well play a crucial role. The local inelastic strain rate does not determine the local state of stress; the rate of energy dissipated depends upon the existing stress as well as the inelastic strain rate.

8. Application of the Limit Theorems on the Microscale

Many microstructures of interest contain brittle elastic inclusions or particles in a ductile matrix on a scale of hundredths of micrometers to hundreds of micrometers. The macroscopic behavior may be very brittle in appearance, as in sintered carbides, or very ductile, as in pearlitic or spheroidized steels above their transition temperatures.

A first step in the applications of continuum plasticity theory on the microscale is to ignore the size scale and treat the microscopic constituents as though they are macroelements of elastic and elastic–plastic material. The response to applied load or stress of such inhomogeneous assemblages of material is complicated, but understandable in principle and computable in detail until fracture intervenes. However, such detailed distributions of stress and strain in each component are not generally required. Many of the major features of response when the matrix is well into the plastic range can be determined adequately by the limit theorems of perfect plasticity applied to simple representations of a volume fraction f_v of inclusions (Figs. 7 and 8). Work hardening may be included simply by choosing the yield stress for the perfectly plastic idealization, or the equivalent stress for multiaxial states of stress, in accord with the order of magnitude of the local strain levels of interest (Fig. 9).

Results of several limit analyses show that inclusion shape is relatively unimportant and that until f_v exceeds $\frac{1}{3}$, the plastic constraint factor σ_0^c/σ_0 (ratio of flow stress of assemblage to flow stress of the ductile matrix) introduced by the inclusions remains close to unity (Table 1). Consequently, in spheroidized steel, with its small volume fraction of cementite particles, the flow stress or stress–strain curve of the steel is not much different from the corresponding value of its ferrite matrix. The same is true for pearlitic steels up to at least 0.2% C or 25 vol% pearlite, but at carbon contents close to the eutectoid the volume of free ferrite is negligibly small. Then the flow stress of the randomly oriented pearlite is governed by the ferrite between the cementite platelets. The constraint factor is low. The computed increase of less than 25% in the yield strength of steel in the pearlitic state over that in the spheroidized state agrees with experimental data and confirms the applicability of continuum plasticity.

Sintered carbides are at the other end of the spectrum, with their very high volume fraction of carbides. In what can be termed the ductile domain, large plastic strains in the very thin layers of ductile matrix cause little if any noticeable deviation from linearity in overall stress–strain response, as true limit load conditions terminate in brittle fracture. Yet in this ductile domain of response under limit load conditions, a bend specimen at rupture will have an almost uniform tensile stress σ_u on the tensile half of the beam, while a tensile specimen will have an almost uniform distribution σ_u over the entire cross section. Consequently, if, as is often done, the apparent linear brittle behavior is taken to imply a linear elastic distribution of stress in

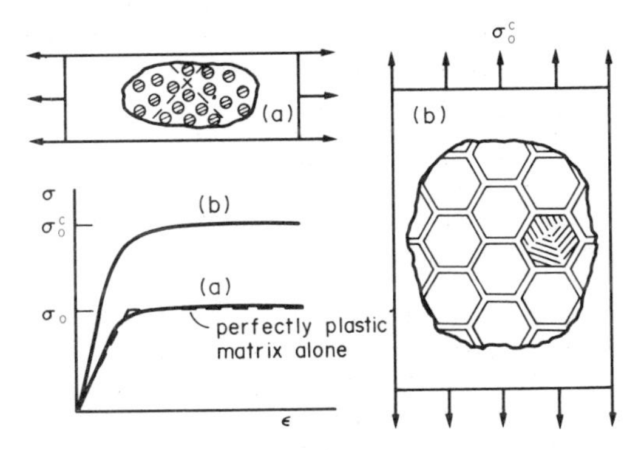

Figure 7
Plastic constraint is small until f_v is large (after Butler and Drucker 1973)

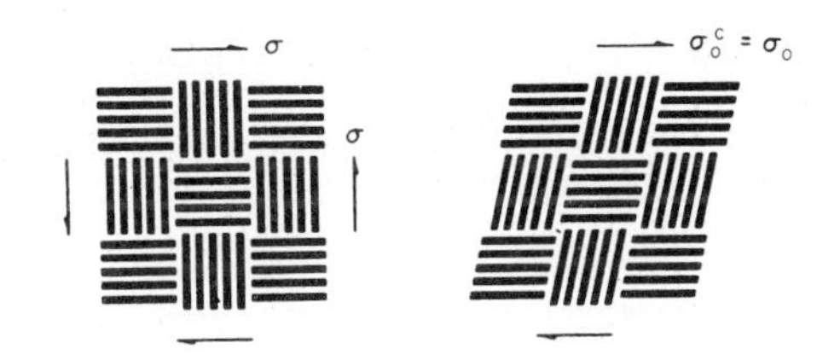

Figure 8
Large plastic deformation with no clear slip path (after Butler and Drucker 1973)

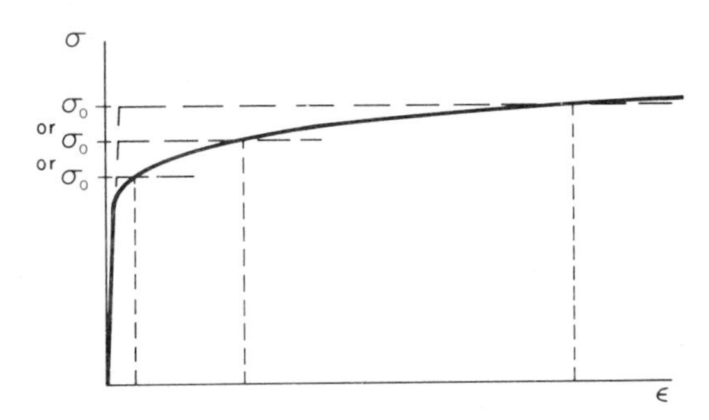

Figure 9
Choice of yield stress in accord with strain level of interest

Table 1
Approximate plastic constraint factors σ_0^c/σ_0 (Fig. 7)

f_v	0.9	0.8	0.7	0.6	0.5	0.4	0.3	0.0
σ_0^c/σ_0	5.9	3.1	2.1	1.7	1.4	1.2	1.05	1.00

bending, the bend strength or modulus of rupture will be calculated as $1.5\sigma_u$ instead of the proper value σ_u. The strength in bending will be misinterpreted as 50% greater than the strength in tension, and arguments based on statistics of strength distribution valid for genuinely brittle behavior will be misapplied. Observed rather modest variations of tensile strength from specimen to specimen, and remarkably little scatter in the correlation of strength solely with $d^{-0.5}$ (where d is the dimension of the carbide particles), lend confirmation to the validity of the limit load approach in the "ductile" domain.

When the ductile layers or regions are of very small submicrometer dimension, the number of mobile dislocations that are present or can be generated will be far too low for the ordinary bulk properties to apply. A crude correction that increases the yield stress with decrease in dimension seems to permit useful application of continuum plasticity very much farther down in the microscale than would be expected.

Continuum theory cannot give the flow properties of the ductile matrix itself, properties given by solution hardening and by dispersion hardening due to particles not visible in the best optical microscope. However, the theory provided an early indication of the strong likelihood that correlations of mechanical properties of ductile alloys with dimensions such as grain size and interparticle spacing of inclusions clearly visible in the optical microscope were found only because there were invisible controlling features that were a scaled image of those optically visible.

See also: Mechanics of Materials: An Overview

Bibliography

Asaro R J 1983 *Micromechanics of Crystals and Polycrystals*, Advances in Applied Mechanics Vol. 23. Academic Press, New York

Butler T W, Drucker D C 1973 Yield strength and microstructural scale: A continuum study of pearlitic versus spheroidized steel. *J. Appl. Mech. Ser. E* 40: 780–84

Chen W F 1975 *Limit Analysis and Soil Plasticity*, Developments in Geotechnical Engineering, Vol. 7. Elsevier, Amsterdam

Drucker D C 1966 The continuum theory of plasticity on the macroscale and the microscale. *J. Mater.* 1: 873–910

Drucker D C 1967 *Introduction to Mechanics of Deformable Solids*. McGraw-Hill, New York

Drucker D C 1984a From limited experimental information to appropriately idealized stress–strain relations. In: Desai C S, Gallagher R H (eds.) 1984 *Mechanics of Engineering Materials*. Wiley, Chichester

Drucker D C 1984b Material response and continuum relations: Or from microscales to macroscales. *J. Eng. Mater. Technol.* 106: 286–89

Hill R 1950 *The Mathematical Theory of Plasticity*. Clarendon Press, Oxford

Lee E H, Symonds P S (eds.) 1960 *Plasticity, Proc. 2nd Symp. Naval Structural Mechanics*. Pergamon, New York

McMeeking R M, Rice J R 1975 Finite-element formulations for problems of large elastic–plastic deformation. *Int. J. Solids Struct.* 11: 601–16

Martin J B 1975 *Plasticity: Fundamentals and General Results*. MIT Press, Cambridge, Massachusetts

Palgen L, Drucker D C 1983 The structure of stress–strain relations in finite elasto-plasticity. *Int. J. Solids Struct.* 19: 519–31

Prager W, Hodge P G Jr 1951 *Theory of Perfectly Plastic Solids*. Wiley, New York

Save M, Massonnet C E 1972 *Plastic Analysis and Design of Plates, Shells and Disks*. North-Holland, Amsterdam

D. C. Drucker
[University of Florida, Gainsville, Florida, USA]

W

Welding: An Overview

If joining processes are considered in a very broad sense—the joining of any type of material instead of just the welding of metals—it is apparent that joining is very basic to our everyday life. When viewed in this broad sense, it is obvious that joining is important in the fabrication and maintenance of all household appliances, clothing, furniture, buildings, machines and other objects composed of two or more parts joined in some way. This article discusses a somewhat narrower subject: those technologies related to welding. Welding processes are processes for joining materials, primarily metals, by applying heat, pressure or both, with or without filler metals, to produce a localized union through fusion or recrystallization across the interface.

1. Historical Development of Welding Technologies and Their Applications

1.1 Development of Welding Technologies

Historians have traced welding technologies back to prehistoric times. Copper–gold and lead–tin alloys were being soldered before 3000 BC and iron and steel were being welded into composite tools and weapons as early as 1000 BC. However, the only sources of heat for welding at that time were wood and coal. The relatively low temperatures available severely limited the types of welding processes and their application.

Development of modern welding technologies began in the latter half of the nineteenth century when electricity became readily available. Many of the important discoveries leading to modern welding processes were made between 1880 and 1900. Figure 1 traces some important developments in welding technologies since 1880. For example, in 1881, a Frenchman, Moissan, first used a carbon arc for melting metals. Bernardos, a Russian, carried this step further and took out a German patent for carbon arc welding in 1885. Shortly afterwards, another Russian, Slavianoff, conducted experiments on metal arc welding. The first metal arc welding patents were issued in the USA to Coffin in 1889. Meanwhile, another American, Thompson, applied for the basic patents for resistance welding in 1886. In 1895, Le Chatelier, a Frenchman, invented the oxyacetylene blowpipe.

This last invention gave oxyacetylene welding a leading position in the field. Ferrous welds made with oxyacetylene blowpipes were so superior to carbon and bare-electrode arc welds that the latter processes were nearly forgotten for joining metals subjected to high stresses. Then in 1907, Kjellberg, a Swede, revolutionized arc welding when he introduced covered electrodes. It was soon found that their use greatly improved the properties of weld metals and allowed many new welding applications. Today, shielded metal arc welding, using covered electrodes, is the most commonly used welding process for fabricating steel structures, including bridges, ships, pressure vessels and pipelines.

The oxyacetylene flame, which has the highest burning temperature of all the many industrial flames, is still widely used for various joining and cutting applications. Although many metals can be welded by an oxyacetylene torch, its uses are generally limited to noncritical applications. Oxyacetylene torches are also widely used for cutting steel plates.

Efforts to mechanize arc welding have led to various automatic processes, including submerged arc welding, which was developed in the USA during the 1930s. In submerged arc welding, the arc is maintained between the workpiece and a continuous electrode

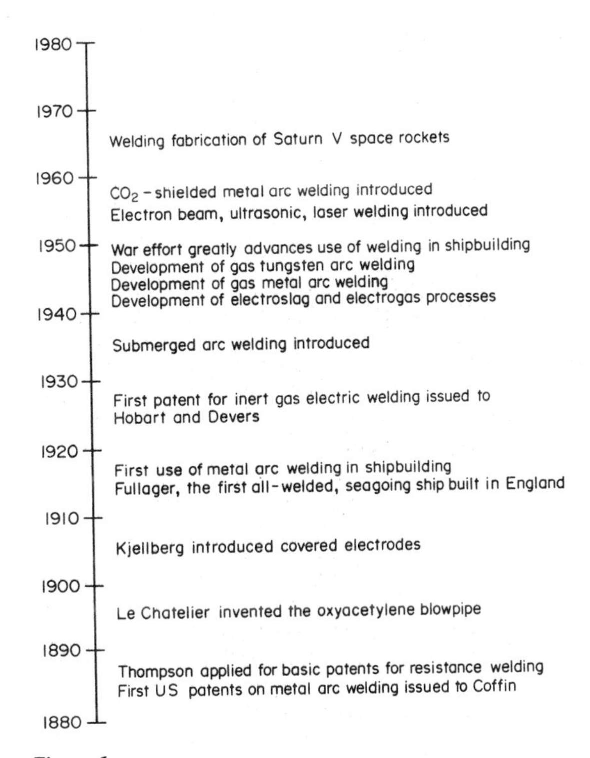

Figure 1
Chronological development of welding processes and their applications

that is mechanically fed as it melts, while the weld zone is covered by flux.

Another important advance made about the same time was the development of arc welding processes using inert gases, such as argon and helium, as the shielding gas. In the 1930s, two Americans, Hobart and Dever, explored the possibilities of improving weld quality by shielding the arc with inert gas from the atmosphere containing nitrogen, oxygen and hydrogen. Their studies led to the development of the tungsten inert gas (TIG) are welding process, in which the electric arc is maintained between the workpiece and a tungsten electrode that is virtually nonconsumable. A filler wire may be used as needed. The TIG process was first used commercially during early World War II for the welding of magnesium airplane seats. Later, metal inert gas (MIG) arc welding was developed using a continuously fed consumable metal electrode. The TIG and MIG processes proved to be very useful for welding aluminum and magnesium alloys, which had been difficult to weld using covered electrodes and oxyacetylene torches.

In the 1950s, efforts were made to use the gas-shielded metal arc process for joining stainless steel and carbon steel. It was discovered that a small percentage of oxygen added to the inert shielding gas would greatly improve the arcing and general welding characteristics without adversely affecting stainless steel or carbon steel weld metal properties to any great extent. It was also found that carbon dioxide, which is much cheaper than argon and helium, could be used as the shielding gas for welding carbon steel. With the development of new processes using shielding gases containing oxygen and carbon dioxide, it became apparent that the technical terms TIG and MIG were no longer adequate. The American Welding Society recommends that the terms gas tungsten arc welding (GTAW) and gas metal arc welding (GMAW) be used to designate these arc welding processes.

Significant advances have also been made over the years in electric resistance welding processes. Since large-capacity resistance machines tend to be expensive, their extensive use has been limited to special applications, such as multiple-spot welding machines used in automobile manufacturing plants and high-frequency resistance welding machines for welding steel pipelines.

Many other welding processes have been developed during the last 40 years. They include electroslag and electrogas welding, friction welding, plasma arc welding, electron beam welding and laser beam welding. Today there are over 100 different welding processes and process variations. As a result, almost all metals used in present-day applications can be welded.

1.2 Development of Applications of Welding

In the early days of arc welding, its application to metal fabrication was quite limited. However, as the process improved, the fields of application greatly widened. During World War I, metal arc welding was used for the first time in ship construction; however, its uses were limited to repairs and joining unimportant parts.

Welded fabrication was used for a 13 m launch, *Dorothea M. Geary*, built in Ashtabula, Ohio, in 1915. In 1921, the first all-welded oceangoing ship, *Fullarger*, was built in a British shipyard. Around this time the technologies of welding became recognized by engineers and scientists. For example, the American Welding Society was established in 1919. From these beginnings, applications of welding increased steadily until World War II.

A dramatic increase in the use of welding for building ships occurred during World War II in the USA. To meet the urgent demand for the large number of ships needed for the war, the USA entered into large-scale production of welded rather than riveted ships for the first time in history. Demand for light-metal aircraft parts also accelerated the development of inert-gas arc welding processes, as previously described.

The sudden increase in application of welding in shipbuilding created difficult problems. By that time, the techniques of welding steel plates had been well established, but insufficient knowledge and experience had been amassed regarding design and fabrication of large welded structure and their fracture characteristics. Of the 5000 or so merchant ships built in the USA during World War II, some 1000 experienced structural failures. Twenty or so broke in two or were abandoned because of structural failures. Brittle fractures also occurred in other welded structures, including bridges built in Europe just before World War II.

An interesting historical fact is that the USA persisted in building welded ships despite these brittle fractures, rather than returning to riveted ships. Instead, an enormous research effort was devoted to brittle fracture and other topics related to the design and fabrication of welded structures. Through this research effort, the technology of deisgn and fabrication of welded structures was fairly well established by the early 1950s.

A major change in postwar construction was the tremendous increase in use of welding for almost all kinds of steel structurès. It could be said that welding technology greatly contributed to the rebuilding of war-torn Europe and Japan. Another result of the increased use of welding was the founding in 1948 of the International Institute of Welding, an organization composed of major welding societies throughout the world.

The many applications of welding have extended to numerous kinds of structures, including aerospace structures, automotive products, bridges, buildings, pressure vessels and boilers, storage tanks, nuclear power plants, industrial piping, transmission pipelines, ships, offshore pipelines and platforms, as well as rails and railroad equipment. Many of the structures presently being built—space rockets, deep-diving submersibles and very thick containment vessels for nuclear reactors—could not have been constructed without the proper application of welding technology. For example, the huge tanks containing the fuel (kerosene and liquid hydrogen) and the oxidizer (liquid oxygen) of the Saturn V rocket used for the Apollo lunar program were built of welded aluminum alloys.

2. *Advantages and Problems of Welding Fabrication*

Welding has replaced riveting, forging and casting in many applications because it has advantages over these processes. However, welding fabrication is by no means free of problems. To design and fabricate soundly welded structures it is essential to have (a) adequate design, (b) proper selection of materials, (c) adequate equipment and proper welding procedures, (d) good workmanship and (e) proper postweld inspection.

2.1 Advantages of Welding Fabrication

(*a*) *High joint efficiency.* The joint efficiency is defined as

$$\frac{\text{fracture strength of joint}}{\text{fracture strength of base plate}} \times 100\ (\%)$$

The values of joint efficiency of welded joints are higher than those of most riveted joints. For example, the joint efficiency of a sound butt weld can be as high as 100%. The joint efficiency of a riveted joint varies depending upon factors such as the diameter and the spacing of the rivets, but it is never possible to obtain 100% joint efficiency.

(*b*) *Ease of obtaining water and air tightness.* It is very difficult to maintain complete water and air tightness in a riveted structure in service. Welding is more fundamentally suited to the fabrication of structures that require water and air tightness, such as submarine hulls and pressure vessels containing gases and liquids.

(*c*) *Simple structural design.* Joint design for welded structures can be much simpler than for riveted structures. In welded structures, members can be simply butted together or fillet-welded. In riveted structures, complex joints are required.

(*d*) *Weight saving.* As a combined result of the above-mentioned factors, the weight of a welded structure can be as much as 10–20% less than a comparable riveted structure.

(*e*) *No limit on thickness.* It is very difficult to obtain sound riveted joints using plates >50 mm thick. In welded structures, on the other hand, there is virtually no limit to the thickness that may be used.

(*f*) *Reduction in fabrication time and cost.* By using modular construction techniques, in which many subassemblies are prefabricated in-plant and later assembled on-site, a welded structure can be fabricated in a short time. For example, in a modern shipyard, a 200 000 t (dead weight) welded oil tanker can be launched in less than three months, while more than a year would be needed for the same ship fabricated with rivets.

2.2 Problems with Welding Fabrication

(*a*) *Difficult-to-arrest fracture.* Once a crack starts to propagate in a welded structure, it is very difficult to arrest it; therefore, the study of fracture in welded structures is very important. If a crack occurs in a riveted structure, on the other hand, it propagates to the end of the plate and stops. Although a new crack may be initiated in the second plate, the fracture has been at least temporarily arrested. It is for this reason that riveted joints are sometimes used as crack arresters in welded structures.

(*b*) *Possible defects and lack of reliable NDT techniques.* Welds often contain various types of defects, including cracks, porosity, slag inclusion, incomplete fusion and undercut. Many nondestructive testing (NDT) techniques, including radiography, magnetic particle tests, ultrasonic inspection and liquid penetrant inspection, have been developed and used. Although these techniques can provide valuable information on the sizes and distributions of weld defects, none is yet developed sufficiently to be completely satisfactory in terms of cost and reliability.

(*c*) *Sensitivity to materials.* Some materials are more difficult to weld than others. In general, materials with higher strengths tend to be more difficult to weld without cracking than similar materials with lower strengths. Furthermore, as the strength level increases, a material tends to become more sensitive to even smaller cracks. In other words, even a small crack cannot be tolerated when a structure is fabricated with high-strength materials.

(*d*) *Residual stresses and distortion.* Due to local heating, complex thermal stresses occur during welding, and residual stresses and distortion remain after welding is completed. Transient thermal stresses, as well as residual stresses and distortion, often cause cracking and joint mismatching; high tensile residual stresses produced in areas near the weld may cause fractures of weldments under certain conditions; and out-of-plane distortion and compressive residual stresses in the base plate may reduce the buckling strength of a structural member subjected to compressive loading.

(*e*) *Potential health and safety problems.* There are some health and safety problems associated with welding. For example, in arc welding, workers must be protected from electric shock, possible eye damage from the intense light generated by the welding arc, possible burns from sparks and potential health hazards from fumes.

(*f*) *Potential problems associated with welding dissimilar metals.* Great care must be taken in welding dissimilar metals, since some metals are extremely difficult to weld to certain others. For example, aluminum alloys cannot be successfully fusion-welded to steel and titanium alloys, owing to the formation of brittle intermetallic compounds. Techniques often used in joining difficult-to-weld metals include: (a) use of a third metal that can be successfully welded to both of the metals to be welded; (b) use of joining techniques that involve minimum melting such as cold welding; and (c) use of nonmetallic adhesives such as epoxy resin.

3. Recent Advances in Welding Technology and Future Possibilities

Significant advances in welding technology have been made in recent years, and it is expected that further major advances will be made in the future. Since present-day welding technology is very complex, current advances are being made in many different processes to achieve various objectives. These advances may be classified into the following categories:

(a) those which can be regarded as improvements of existing welding processes and procedures;

(b) those in other fields of science and technology which can have application to welding; and

(c) those which have been made to meet new demands or changes of priorities of society.

There have of course, been many advances in welding technology which fall into more than one category.

3.1 Improvement of Existing Welding Processes and Procedures

There have been numerous research and development efforts to improve existing welding processes and procedures. These have had various objectives: to improve the performance of weldments, to increase reliability and/or to reduce fabrication costs. For example, the development of the narrow-gap metal arc process can result in (a) reduction in fabrication time, (b) reduction in distortion and (c) reduction in the heat effect (or improved metallurgical properties).

3.2 Applications of Advances in Other Fields of Science and Technology to Welding

The electron-beam and laser-beam welding processes can be regarded as applications of advances in physics to welding. Welding robots can also be regarded as applications of robotic technology. In fact, the tremendously rapid and vast advances in electronics and computer technology will undoubtedly have complex and far-reaching effects on the welding technology of the future. Three examples among the many applications of electronics and computer technologies to welding are discussed here.

(*a*) *Automatic adaptive control.* Many automatic welding machines have already been developed and used. Most of these, except some of those developed very recently, use welding conditions that have been determined by previous experiments or experience. Therefore, they should be called "preprogrammed" or "mechanized" machines. To the best knowledge of the author, no existing welding machine has complete in-process sensing and control capability in real time. With the use of modern electronic and computer technologies, it should be possible to develop "smart" welding machines equipped with adaptive control systems. During the last few years, significant research and development efforts have been made in developing automatic welding machines with adaptive control capabilities. It is even possible to conduct inspections immediately after welding.

(*b*) *Information storage and retrieval.* As technology advances, an increasing number of different materials are being dealt with, some of which are difficult to weld. There are also a large number of joining processes. To weld different materials successfully in different thicknesses using different joining processes, different welding parameters must be selected, including proper joint configurations, welding current and arc travel speed. Computers can be used to cope with this problem by storing all the necessary data and supplying them to engineers as needed. Several investigators have developed computerized systems for selecting optimum welding conditions. Such efforts will be continued and expanded in future. In fact such efforts are essential for developing "expert" welding systems.

(*c*) *Computer simulation of welding phenomena.* Welding research has been, and still is, primarily empirical, relying on experimental data. The major reason for the relative lack of analytical studies is the fact that welding processes, especially arc welding, involve complex, transient phenomena including rapid solidification of the weld metal, transformation in the heat-affected zone, cracking in various portions of weldments and transient thermal stresses and metal movement during welding. Some investigators, mainly in universities, have conducted analytical studies on these subjects. However, the analyses using manual calculations have been limited to very simple cases that are hardly representative of complex, practical weldments widely used in industry. With the development of modern computers and such techniques as the finite-element method, the computation capability has been increased tremendously. It should be possible to develop the capability for computer simulation of various phenomena involved in welding.

3.3 Advances to Meet New Demands and Changes of Priorities of Society

A good example here is the recent advance in underwater welding technology to meet the increased demand for construction and repair of offshore structures. This demand has been created as a result of rapidly expanding oil-drilling activities. Oil exploration activities have also extended to very cold regions. Another new demand is for welding in space. Some welding experiments have already been performed aboard spacecraft. It is quite possible that there will be a need for developing techniques for welding fabrication and repair of space stations. There is also a need for developing techniques to repair damaged components of nuclear reactors.

These examples can be regarded as extensions of the ability to carry out welding in remote places and in hostile environments. It is quite possible that the welding technologies developed for these applications will extensively employ remote manipulation techniques using robots. Some of these new welding technologies could well be adopted in automated fabrication shops in the twenty-first century.

Bibliography

Masubuchi K 1980a *Analysis of Welded Structures.* Pergamon, Oxford

Masubuchi K 1980b New directions in welding research and development. Special lecture presented at the Int. Conf. Welding Research in the 1980's. October 1980, Osaka, Japan

Masubuchi K, Terai K 1976 Future trends of materials and welding technology for marine structures. Paper to Spring Meeting, Society of Naval Architects and Marine Engineers, New York

Shank M E (ed.) 1957 *Control of Steel Construction to Avoid Brittle Fracture.* Welding Research Council, New York

Weck R 1977 The need for new welding processes. *Proc. R. Soc. London, Ser. A* 356: 1–23

K. Masubuchi
[Massachusetts Institute of Technology, Cambridge, Massachusetts, USA]

Welding of Bridges

The Interstate Highway System, with its requirement for some 300 000 highway bridges, was the stimulus that brought welding of bridges into prominence in the USA. Early in the program, it was established that welded construction led to a saving of approximately 25% in the usage of steel. As a result, a technology that had promise quickly became a preferred method for building structures to span roadways and streams. Old-style cumbersome truss structures gave way to the plate girder bridge, its aesthetics often enhanced by the graceful curves of haunched design.

Most of the welding on a bridge is done in the welding shop, where automatic or semiautomatic welding equipment is used to join cut plate into girders and bracing and stiffening components. Welding is also used to convert standard rolled shapes into structural members meeting the engineering requirements of the job. Thus, a cover plate welded to a wide-flange rolled beam gives the beam the strength to serve as a primary load-carrying member. Other uses of welding in shop construction of bridge members include the fabrication of bents, floor panels, diaphragms, bearings, expansion devices and shear attachments.

During erection, welding may be used to splice beams and girders, although bolts are still used extensively for this purpose. Girders to be field-spliced by welding are specially prepared for the groove welds joining the flanges. Special procedures are defined for the critical splice joints. Low-hydrogen, shielded metal arc-welding electrodes are commonly used in field welding, although the self-shielded flux-cored semiautomatic process can reduce welding costs.

As in the case of multistory buildings (see *Welding of Buildings*), a welded bridge structure should be designed as a welded structure from the first sketches on the drawing board for maximum efficiency. The full advantages of welded design are only achieved when the bridge is conceived as a continuous structure with all segments fused into a unit programmed to carry, transfer and resist impinging forces. Yet, even when a bridge is not weld-designed, substantial economies in labor, materials and time can be realized by using weld-joining where applicable.

The girder bridge is practically synonymous with the welded bridge, but this does not mean that welding is limited to girder bridges. In fact, the technology can be applied to all types of bridges, if only for the efficient fabrication of components. Welded connections are as strong or stronger than the steel they join. Material size does not have to be increased to allow for breaks in continuity of the material, such as occurs with bolt holes. Whatever physical property the steel has, the welded joint can be made to match.

The aesthetics of the welded bridge is a feature appealing to planners and to the public. Since there are no limitations placed on the form or mechanics of a welded connection, roadways may be curved horizontally or vertically, supported on narrower piers or artistically designed bents, and the final structure fashioned into a virtual sculpture in concrete and steel. The designer can give free rein to his imagination to produce a structure that is serviceable and aesthetically pleasing. With girder bridges, even such engineering components as horizontal and vertical stiffeners are used not only for their engineering function but also as lines to relieve the monotony of masses, and thus have artistic function.

In the USA, bridge design is governed by the regulations of the American Association of State Highway and Transportation Officials (AASHTO). The welding must conform to the American Welding Society Structural Welding Code D1.1. Various square-butt, V, bevel, J and U joints are prequalified by the American Welding Society for use in welded bridges.

See also: Steel for Construction; Welding: An Overview

Bibliography

American Welding Society 1984 *Specifications for Welded Highway and Railway Bridges*, AWS D2.0–84. American Welding Society, Miami, Florida

O. W. Blodgett
[Lincoln Electric Company, Cleveland, Ohio, USA]

Welding of Buildings

Welding, assisted by bolting for positioning purposes, has become the primary method for joining the structural members of multistory buildings. Most of the welding for a building frame is conducted in the welding shop, where automatic and semiautomatic welding can be used to reduce fabrication costs. Plate steel is used extensively for the weld fabrication of columns, beams and other members, but, even when rolled sections are primary components, they are often fitted with shop-welded connecting members. Column sections, weld-built from plate, may even have beam half-lengths shop welded to them for bolt and weld splicing during erection. The trend is to do as much work as possible in the shop, where sophisticated jigging and automatic welding equipment can minimize welding costs. During erection the trend is to use semiautomatic welding processes, such as self-shielded flux-cored arc welding in place of shielded metal arc welding, again to minimize costs.

Welding is also the primary method for joining the steel materials used in small building structures such as warehouses, sheds, farm buildings and storage structures. Prefabrication of members is the key to economies in such buildings. Beams, girders, side walls, roof panels, joists and other members are shop-welded and delivered to the building precut and marked for bolt assembly. Prefabrication not only gives the economies of shop welding with automated equipment and ease of erection, but also reduces material usage. Thus, a fabricated beam correctly designed for its load may weigh considerably less than a rolled beam selected from the steelmaker's catalog. Bar joists, built from plate and bent rod, are examples of material savings through weld design.

Efficiency in welded framing reaches its peak when the structure is weld-designed—in other words, from its very inception welding is considered as the method of joining the structural members. Although it is possible to select welding as the connection method after the design has been finalized, with subsequent economies, the full advantages of using steel construction in competition with other materials can only be realized when the structure is created as a welded design, and fabricators and erectors use modern techniques of welding and production control.

Shop fabrication of members using automatic equipment is standardized and applicable to all types of weldments. The submerged-arc process is a favored method of shop joining, because of the excellent quality of welds and its ease in adaptability to mechanization. It is in erection welding that unique approaches are required and where much progress has been made in recent years.

Moment and shear-reaction transfer are the two main considerations in erection welding. In a beam, the bending forces from the end moment lie almost entirely within the flanges of the beam. The most effective and direct method to transfer these forces is some type of welded flange connection. Several different designs accomplish this.

A preferred method of connecting flanges to the column is a groove-welded tee joint. This transfers forces directly and requires a minimum amount of welding. Use of backing strips below the flanges allows the weld to be made with reasonable tolerance limits, as long as there is sufficient root opening to assure sound welds.

Shear forces lie almost entirely within the web of the beam. They must be transferred directly to the supporting column by means of a connecting weld on the web, or directly down to a supporting seat. The web connection must have sufficient length of weld so as not to overstress the beam web in shear. Similarly, the seat connection must have sufficient length to prevent overstressing the beam web in compression.

Shop welding during fabrication of members usually eliminates vertical and overhead welding. However, some vertical and overhead welding may be required during erection. When this is the case, the need for changing welding current and electrode size with varying welding positions can be eliminated by having one or more operators specializing in out-of-position work using lower currents, and other operators handling the flat welding with higher currents.

The iron powder-covered electrode has become the shielded metal arc welding electrode of choice for erection welding because of its operating properties and the quality of its welds. The self-shielded flux-cored semiautomatic process is also used, gaining popularity because of its ability to produce high-quality welds even under the severe wind conditions frequently encountered in high-rise field welding.

In the USA, multistory buildings are usually designed according to the specifications of the American Institute of Steel Construction (AISC). The welding on such buildings is governed by the American Welding Society (AWS) Structural Welding Code D1.1.

See also: Steel for Construction; Welding: An Overview

Bibliography

American Welding Society 1984 *Code for Welding in Building Construction*, AWS D1.0–84. American Welding Society, Miami, Florida

O. W. Blodgett
[Lincoln Electric Company, Cleveland, Ohio, USA]

Wood: An Overview

Wood is the hard, fibrous tissue that comprises the major part of stems, branches and roots of trees (and shrubs) belonging to the plant groups known as the gymnosperms and the dicotyledonous angiosperms. Its function in living trees is to transport liquids, provide mechanical support, store food and produce secretions. Virtually all of the wood of economic significance is derived from trees and, as timber, it becomes a versatile material with a multitude of uses. In addition, there are woody materials found in the stems of tree forms in another plant group, the monocotyledonous angiosperms, including the bamboos, rattans and coconut palms. These woody materials are similar to wood in being lignocellulosic in composition, but they differ substantially in their anatomic structure.

1. Wood as a Renewable Natural Resource

The wood used for purposes other than fuel, that is, wood which is either cut into lumber or used as a raw material for such wood-based materials as plywood, fiberboard or particleboard, is largely derived from the stems of trees.

Woods are classified as either hardwoods or softwoods; the distinction is botanical and is reflected in details for wood structure. Most (but not all) softwoods are needle-bearing evergreen trees, whereas the hardwoods are broad-leaved trees. This classification is not related to the actual hardness of wood, since woods with the least and the greatest known hardness are both found among the hardwoods. Woody materials of the monocotyledons consist of vascular bundles, which are thin strands incorporating elements of both wood and bark, embedded in a mass of soft ground tissue and thus have a structure substantially different from wood. These woody materials are therefore specialty items with properties and uses distinct from wood and lumber.

The annual world production of wood is about 2.4×10^9 t, with 10^9 t used as industrial roundwood (e.g., sawlogs, veneer logs, poles and pilings, and pulpwood) and the remainder (about 53%) used as fuelwood and charcoal. Developed and developing countries differ sharply in the ratio of industrial roundwood to fuelwood produced: in developed countries, the percentage of fuelwood and charcoal is, on average, 18% of total wood production, in developing countries it is 80%. In recent years, as energy sources other than petroleum have received increased attention, considerable interest has developed in the use of wood not only as a fuel directly but also as raw material for the production of alcohol and similar sources of energy. The forest products industry is in a unique position to use its own wood residue for generating at least part of its energy needs, and is already extensively taking advantage of this possibility. How far the use of wood as an energy source can be developed will depend on economic factors. As the technology for manufacturing wood and wood-based materials continues to develop, the tendency is for a reduction of wood residue in manufacturing operations, which will ultimately affect the availability of such residue in the factories. However, large quantities of presently unused wood that may only be suitable for such uses as fuel are available in the forest. Economic factors will determine if there will ever be conflicting demands for wood resources for fuel on the one hand and as industrial material on the other, but this is not likely in the foreseeable future.

The world production of industrial roundwood is of the same order of magnitude (by weight) as the production of steel and iron. Unlike steel, however, wood represents a renewable resource that can be (and is) constantly replenished as it is being used. Trees can be grown for harvesting in as little as ten years, but may require a hundred years or more, depending on climate, species and the purpose the wood is to serve. As long as the rate of removal does not exceed the rate of growth, forests will be a perpetual source of wood and lumber. The composition of the forest with respect to tree species may change, as may the size of trees at the time of harvesting, but under proper management the resource will always be there. Increasing the intensity of management (tree farming) can lead to substantial increases in productivity on a given area of forest land.

2. Principal Characteristics of Wood

As a natural product of biological origin, wood is characterized by a high degree of diversity and variability in its properties. No one knows exactly how many species of trees there are in the world, but estimates run from 30 000 upwards. A large proportion of these are to be found in the tropical regions, which frequently have an incompletely explored flora. The number of commercial timbers is only a small fraction of the total number of tree species. Since most wood properties are related to wood density, the fact that balsa, the lightest commercial species, has a density of about 160 kg m^{-3} as compared with 1280 kg

m^{-3} for lignum vitae, one of the heaviest, illustrates the diversity that can arise from species differences. In addition, considerable variability is found within species due to genetic and environmental factors that influence tree growth. Such variability is found not only from one tree to the next, but also within trees, in part due to natural patterns of growth that make wood properties dependent on radial position and height within a stem. This variability must be taken into consideration if wood is to be used properly. Where wood is used as a load-bearing element, for instance, predictions as to its resistance must be based on near-minimum rather than average values for safety. The coefficient of variation (standard deviation expressed as a percentage of the mean) of clear wood strength properties is of the order of 20% (see *Wood: Strength*). Further variability is introduced by the presence of strength-reducing characteristics such as knots and cross grain in lumber, major factors which determine the grade of structural lumber. The allowable bending stress of visually graded lumber of mixed Douglas-fir and larch may be as little as 1.9 MPa or as much as 16.9 MPa, depending on grade.

Wood is a highly anisotropic material; that is, its properties depend significantly on the direction in which they are measured. The principal directions in wood are the longitudinal, radial and tangential directions. The longitudinal direction is parallel to the cylindrical axis of the tree stem, and is also referred to as the direction parallel to grain since the majority of the constituent cells of wood are aligned parallel to it. If a tree stem is cut at right angles to its long axis, a series of concentric rings can often be seen on the cross section. These rings are markings produced by annual growth increments. The radial and tangential directions are those which are normal and tangential, respectively, to the growth rings on the cross section. The radial and tangential directions are also referred to collectively as directions perpendicular to grain. The tensile strength and stiffness of wood is highest parallel to grain and very low perpendicular to grain (see *Wood: Strength; Wood: Deformation Under Load*). Conversely, the shrinkage of wood that accompanies loss of moisture is very small parallel to grain and very much higher perpendicular to grain. In each of the properties of tensile strength, Young's modulus and shrinkage, the largest and smallest values are in the ratio of about 25:1. The degree of anisotropy is thus extremely high. Wood is often modelled as an orthotropic material by neglecting the curvature of the growth rings. This approximation is most appropriate for pieces of wood cut at some distance from the center of the tree stem.

Wood can be a very durable material. Elaborately inlaid wood furniture from the tombs of the Pharaohs has survived for nearly 4000 years. Being sealed in a tomb in a very dry climate provides almost ideal conditions for preservation of wood. Even in Japan, which has a humid climate, there are wooden temple buildings still in existence after 1300 years of service, and in the harsh climate of the mountains of Norway all-wood stave churches can be found that were built about 800 years ago. The key to durability is proper use and understanding of the factors that destroy wood. Wood is biodegradable, which is essential in nature and can be an advantage when wood in use becomes unserviceable and must be disposed of. The same natural process becomes a severe disadvantage when it occurs as biodeterioration of wood in use. The key to preservation is to create conditions unfavorable to the organisms that cause biodeterioration, principally decay fungi and termites. In the case of fungi this means keeping wood dry. Even wood exposed to the weather will last if the detailing is such that water will run off the surface without being trapped inside joints and interior spaces that cannot dry off readily. In the case of termites, contact of wood members with the soil must be avoided. Alternatively, preservative treatments can be effective against either fungi or termites (see *Wood: Decay During Use*).

Wood is susceptible to damage by fire. In matchstick sizes wood burns readily but it is much more resistant in larger sizes. With suitable construction methods, wood can provide a high degree of fire safety in the early stages of a building fire, giving time for occupants to egress and the fire to be brought under control. This is possible because wood is a poor conductor of heat, and the char layer formed beneath a burning wood surface an even poorer one. Wood members can therefore retain a major part of their original strength during the early stages of fire exposure, and thus resist sudden collapse of the building (see *Wood and Fire*).

As a hygroscopic material, wood will take up or give off moisture depending on the temperature and relative humidity of the surrounding atmosphere. Changes in moisture content below the fiber saturation point, which is the state where the cell walls are saturated with adsorbed water and no free water is present in the cell cavities, have an effect on virtually all properties of wood. Wood also shrinks as it dries, and conversely swells when it is wetted again. The total shrinkage in volume of wood from a freshly felled tree dried in an oven will range from about 6 to 20%, depending on species. About two thirds of the shrinkage will be in the tangential direction and one third in the radial direction, the longitudinal shrinkage being almost negligible. Wood in use will not ordinarily change moisture content sufficiently to undergo more than a fraction of the total possible shrinkage, but even then the dimensional changes are likely to be significant. The effects of shrinking and swelling can be minimized by drying wood to the approximate moisture content it will reach in service before it is used or installed. Coatings can be used that will retard moisture changes resulting from short-term fluctuations of temperature and relative humidity, but there are no coatings that would prevent moisture movement entirely. Alternatively, wood can be chemically treated to reduce its hygroscopicity, but such treatments are too expensive for most applications (see *Chemically Modified Wood*). The best method of dealing with shrinking and swelling in wood is to allow for it in proper detailing, by avoiding large dimensions perpendicular to grain, and, where this cannot be done, using floating construction so that dimensional movement perpendicular to grain can occur without constraint.

3. *Processing of Wood and Production of Wood-Based Materials*

Energy requirements to produce a unit mass of lumber (1000 kWh t^{-1}) are rather low compared with those for most materials. Steel requires about 4000 and aluminum about 70 000 kWh t^{-1}. Processing can be done with very simple handtools such as axes, saws and hammers, or on an industrial scale with sophisticated machinery. The principal industrial processes used to obtain lumber are extraction (logging), sawing, drying and surfacing (planing or sanding). Pressure impregnation is used to treat wood with such chemicals as preservatives or fire retardants. Production of wood-based materials usually involves gluing and pressing, as in the manufacture of plywood, particleboard and glued laminated timber (see *Adhesives for Wood*). Members of wood and wood-based materials are given shape mainly by either some type of cutting or by using adhesives to create shapes from smaller pieces. Forming processes by imparting permanent deformation of the material play a minor role, but are possible to a limited extent.

Traditionally, wood has not been a subject of intensive study by materials scientists and engineers. Probably the most compelling reason for this is that even if the relationship between various levels of wood structure and properties were perfectly known, the next step, namely that of altering the structure to create properties for certain use requirements, would be very difficult to realize. The quality of wood in trees can be influenced by genetic manipulation (tree breeding) and by certain silvicultural practices that affect conditions of growth, but in a practical sense the opportunities are severely limited. The many factors involved interact in complex ways, use requirements are diverse and sometimes conflicting, and time scales are measured in generations. However, since there are very diverse species of wood available, it is possible to choose the species (and possibly the grade) best suited for a particular application. For example, Douglas-fir and southern pine are the best structural timbers in North America, lignum vitae is uniquely suited for bearing blocks for steamship propeller shafts, teak is excellent for ship's decking, and willow has the right combination of light weight and shock resistance to make it very suitable for artificial limbs.

A further broadening of the range of available wood material is possible by combining wood with various plastics introduced into the wood structure as monomers and polymerized in situ, or by chemical treatments to modify the wood substance (see *Chemically Modified Wood*). By far the most widely practised method, however, is the breaking down of the log into elements that may be as small as a single fiber 2 or 3 mm long or as large as a piece of lumber 40 mm × 300 mm × 6 m, and combining these elements, usually by addition of an adhesive or binder and the application of pressure, into some type of wood-based material. By varying the size and shape of the elements, the way they are oriented with respect to each other, the type of adhesive used and the method by which the product is assembled, a large spectrum of wood-based materials can be created. Single wood fibers or fiber bundles are used as the basic unit for making such sheet or panel products as paper, hardboard, fiberboard or insulation board (see *Hardboard and Insulation Board*). Since the fibers themselves are anisotropic, panels may be isotropic in the plane of the sheet or have directional properties, depending on the degree of orientation of the constituent fibers. Larger elements, collectively referred to as particles, include chips, slivers, strands and flakes. They are generally up to 30 mm long, and sometimes longer. These are assembled into panel products known as particleboards, which again may or may not have a preferred orientation of the elements. Thin sheets of wood of just a few millimeters thickness can be assembled either into plywood where the grain direction changes by 90° from one layer to the next (see *Plywood*), or into laminated veneer lumber where the grain direction of all layers is parallel. Finally, dimension lumber, generally about 40 mm in thickness, can be glued up to form much larger structural members, such as beams and arches, than could be produced from solid wood (see *Glued Laminated Timber*).

At present, about 1 m^3 of wood-based panels (fiberboard, particleboard and plywood) is produced for each 4 m^3 of sawn wood on a worldwide basis, which illustrates how important the wood-based materials are. No doubt their relative importance will increase, as inevitable changes in the resource base to smaller logs and advances in technology will lead to further shifts from sawn lumber to wood-based materials.

Finally, it may be pointed out that in addition to the various scientific and technological tangible aspects discussed here, wood has a natural warmth and beauty that are a significant part of its overall character and at times may even be the determining factors in its selection for a particular use.

Bibliography

Forest Products Laboratory 1987 *Wood Handbook: Wood as an Engineering Material*, Agriculture Handbook No. 72. US Department of Agriculture, Washington, DC

Haygreen J G, Bowyer J L 1988 *Forest Products and Wood Science: An Introduction*, 2nd edn. Iowa State University Press, Ames, Iowa
Klass D L (ed.) 1981 *Biomass as a Nonfossil Fuel Source*, ACS Symposium Series No. 144. American Chemical Society, Washington, DC
Koch P 1972 *Utilization of the Southern Pines*: Vol. 1, *The Raw Material*; Vol. 2, *Processing*, Agriculture Handbook No. 42. US Department of Agriculture, Washington, DC
Koch P 1985 *Utilization of Hardwoods Growing on Southern Pine Sites*: Vol. 1, *The Raw Material*; Vol. 2, *Processing*; Vol. 3, *Products and Prospective*, Agriculture Handbook No. 605. US Department of Agriculture, Washington, DC
Kollmann F F P, Côté W A Jr 1968 *Principles of Wood Science and Technology*, Vol. 1. Springer, Berlin
Kollmann F F P, Kuenzi E W, Stamm A J 1975 *Principles of Wood Science and Technology*, Vol. 2, *Wood Based Materials*. Springer, Berlin
Marra G G 1981 Overview of wood as a material. In: Wangaard F F (ed.) 1981 *Wood: Its Structure and Properties*, Educational Modules for Materials Science and Engineering Project. Pennsylvania State University, University Park, Pennsylvania
Meyer R W, Kellogg R M (eds.) 1982 *Structural Uses of Wood in Adverse Environments*. Van Nostrand Reinhold, New York
Panshin A J, de Zeeuw C 1980 *Textbook of Wood Technology: Structure, Identification, Properties, and Uses of the Commercial Woods of the United States and Canada*, 4th edn. McGraw–Hill, New York

A. P. Schniewind
[University of California, Berkeley, California, USA]

Wood and Fire

The fire performance of wood depends on many variables. These include the nature of the wood specimen itself (density, moisture content, etc.), specific fire exposure, (flaming or smoldering) and geometry (i.e., thick or thin specimen).

Although combustible, wood has numerous fire-performance advantages when compared with many synthetic polymeric materials. These include:

(a) lower calorific value on a weight basis and lower rates of heat release in similar fire exposures;

(b) the presence of absorbed moisture, which tends to retard ignition;

(c) predictable charring/ablation rates; and

(d) minimal excess pyrolyzates (unburned fuel gases) which are produced when wood burns.

Extending these general properties to the use of wood in construction, the material has other advantages from a fire-safety standpoint. Amongst these are the fact that wood members are nonyielding in the early stages of a fire. This is in contrast to unprotected metal structural systems which yield early (and suddenly) in fully developed fires. In addition, wood members exhibit a predictable decrease in cross-sectional area during fires. This property, coupled with design philosophies (prescribed in building codes) which call for the use of repetitive wood members, provide a system with a substantial margin of safety in structural fire-performance applications. Finally, as compared with most synthetic polymers, absorbed moisture present in wood retards ignition of thick wood members, and the spread of fires from wood sources due to convection is minimal since excess pyrolyzates are rarely produced. Thus, physical barriers must usually be disrupted before wood-fuelled fires will spread.

1. Fire Performance of Wood

Chemical composition, macrostructural details (e.g., grain orientation, density, moisture content) and microstructure all affect the fire performance of wood. Heat-transfer processes which occur during fires result in the development of steep thermal gradients in a burning member from exposed surface to unaffected core. At the same time, mass transport of moisture caused by combustion creates steep moisture-content gradients, ranging from zero moisture content at a burning surface to the prefire equilibrium moisture content at the central, unaffected core. Such moisture-content gradients are in turn accompanied by stress gradients and stress concentrations due to the shrinkage associated with dehydration. These lead to the formation of fissures which are familar in burned wood.

In addition to the substantial moisture and stress gradients from burning wood surfaces to unaffected core areas, similar transitions and variations exist in the chemical processes occurring during fire exposure. Specifically, exothermic oxidation of carbonaceous chars created by pyrolysis predominate at or near burning wood surfaces. Simultaneously, the underlying char base undergoes complex thermal degradation reactions known collectively as pyrolysis. Successively lower levels of oxygen are present, moving from the wood surface to the base of this thermal-degradation zone. Eventually, a sharp boundary can be seen between thermally degraded wood and wood which is substantially unaffected and able to resist loads. Cracks through the char layer terminate in this zone as can be seen in Fig. 1.

Four broad temperature ranges have classically been associated with thermal degradation and fast combustion of wood in air. These are as follows.

(a) Ambient to 200 °C: generation of noncombustible gases such as CO_2, formic acid, acetic acid and H_2O occurs.

(b) 200 °C to 280 °C: endothermic heating and surface darkening occur. Gases listed above, as well as CO, continue to evolve.

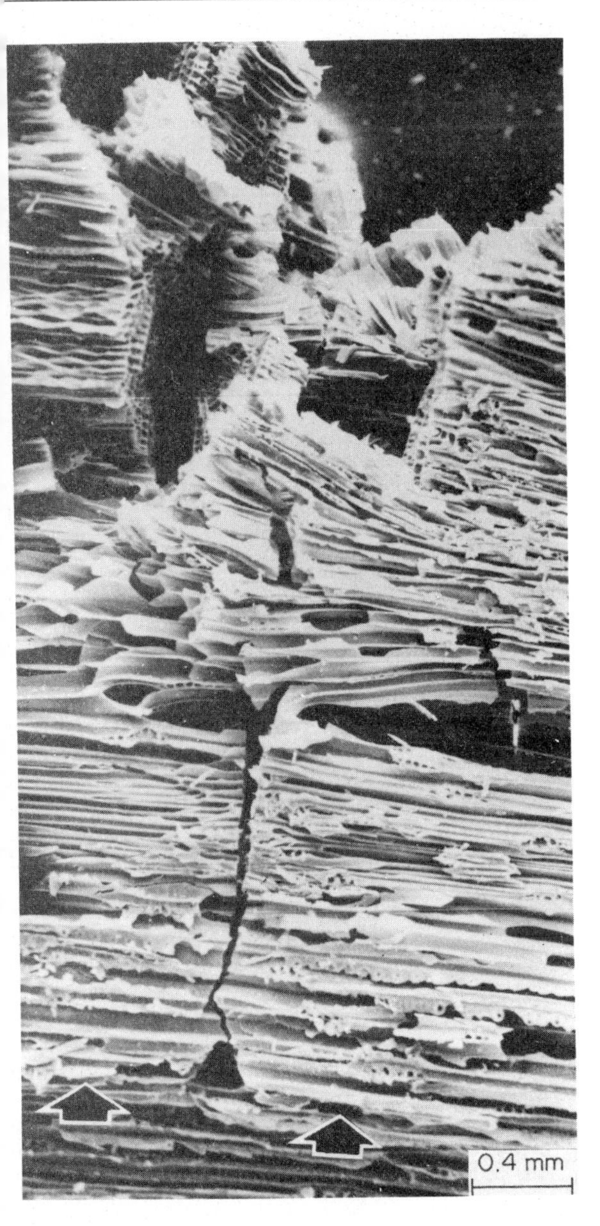

Figure 1
Scanning electron micrograph of a fissure tip in the transition zone between whole and burned wood. The arrows indicate a longitudinal crack along the major grain direction caused by mechanical property differences in burned and unburned areas. The vertical crack originated at the initial wood surface

(c) 280 °C to 500 °C: exothermic heating begins, accompanied by significant production of combustible fuel gases.

(d) Above 500 °C: sustained gas-phase combustion of pyrolyzed gases occurs, and solid state surface combustion consumes the charcoal residues remaining after fuel gases have been liberated. In this temperature range, the char front moves rapidly into the substrate.

Just as important as the temperature ranges listed above are time–temperature relationships. For example, although listed ignition temperatures of wood are usually at least 150 °C, long-term exposure of wood to temperatures in excess of 90 °C can result in thermal degradation conditions which have been observed to lead to ignition in the field. The ignition of such thermally degraded wood (sometimes referred to generically as pyrophoric carbon) may be a simple function of its extended exposure to moderately elevated temperatures, creating thermally active chars. It may also, however, be a function of low thermal inertia and conductivity (often expressed as the product "kpc") or the form of the wood subjected to heating (as with wood fiberboard and cellulose insulation which self-heat exothermically).

Chemically, wood is composed of three major polymers: cellulose, hemicellulose and lignin, each of which have their own characteristic responses to thermal degradation conditions which are based on their polymer composition (i.e., chain order, crystallization, unsaturation, ring structure, etc.).

Two reaction pathways (each having characteristic reaction rates) exist when wood is subjected to high temperatures. These compete in the combustion process for available fuel and depend on composition, density and presence or absence of treatments (such as fire retardants) in the wood. One such pathway involves polymer cleavage or fission. This yields low-molecular-weight, gaseous pyrolyzate fuels which burn through gas-phase combustion. The second pathway involves dehydration of cellulose and yields chars and tars in the substrate. This pathway favors solid state or glowing combustion. A shift in reaction rate of either of these competing reactions will change the character of the combustion and the char yields observed.

"Catalysts" such as fire retardants, metallic preservatives, fungal activity (or residue) or unusual mineral contents will alter the balance or ratio between these possible reaction pathways. Usually cell-wall polysaccharides (cellulose and hemicellulose) favor the cleavage, gas-phase combustion pathway. Lignins and extractive chemicals which contain a wide variety of cyclic rings and unsaturation follow pathways favoring solid state combustion. Thermal analysis data show lignin to be the most stable with regard to oxidative degradation followed by cellulose. Hemicellulose (coincidentally less ordered than cellulose) shows the least resistance to oxidative degradation and thus will participate in combustion earlier.

Microstructure, an expression of the cellular, three-dimensional nature of wood, affects the observed kinetics of wood combustion. For this reason, isolated individual chemical components of wood behave differently in a combustion environment than when they are intimately combined into the complex physical and chemical arrangement found in wood cells. For example, observed combustion reactions in wood vary from exothermic to endothermic, depending on the surface exposed to thermal radiation (i.e., end-grain vs. side-grain) or gross form (whole wood vs. wood meal). As a result of these structural effects, confusion has developed in reconciling research results from small, isolated specimens of wood with observations on larger members which possess intact wood structure. Data from the small, isolated specimens of specific, isolated chemical groups (such as alpha-cellulose) indicate that combustion of wood follows first- or second-order reaction kinetics. Conversely, results that take wood structure into account show zero-order kinetics, implying diffusion-controlled reaction mechanisms. In fact, diffusion control of combustion reactions in wood is consistent with both microstructural features in wood and char, and is also suggested by research linking moisture movement, pressure-driven flow and thermal conductivity effects with microstructural details.

Ignition depends on chemical composition, physical factors (especially thermal conductivity) and exposure geometry. Smoldering ignition, in which solid state combustion predominates, occurs in wood-based materials of low thermal conductivity. This takes place when wood is exposed over time in situations where thermal inputs from the environment and heat produced by exothermic oxidative degradation are greater than heat losses from the material. In such cases, heat from the environment plus the exothermic heat are fed back into the substrate and increase decomposition rates. Eventually, smoldering combustion occurs if the available fuel is not consumed before actual ignition occurs. Such processes may also have biological origins (i.e., thermogenic bacteria). Examples are smoldering fires in raw material storage areas of particleboard or paper plants, improperly installed cellulose or fiberboard insulations, or fiberboards shipped in boxcars without being allowed to cool following manufacturing. In general the phenomena of smoldering ignition in wood substrates are less well understood than those of flaming ignition.

For flaming ignition to occur, wood substrates must be raised to temperatures sufficient for production of an adequate supply of fuel gases to allow sustained, gas-phase combustion. Assignment of a single, unique ignition temperature for wood (either piloted or nonpiloted), however, is ill-advised. Variables such as specimen thickness, moisture content, rate of heating and mode of thermal input (i.e., radiative, convective or conductive) are all important and will affect the temperature required. Preheating and thermal history will also affect an observed "ignition temperature."

2. Wood Used in Construction

Contemporary references occasionally refer to historical aspects of fire performance of wood in events like the Great Fire of London (1666), the Chicago Fire or other conflagrations. Following such events it has been common to see new fire safety design criteria expressed through building code changes in countries or areas where such events occur. Historically, the USA has had an abundance of wood available for construction. The advent of the textile industry in prerevolutionary New England led to a style of building known as "mill-type" construction. This involved wood posts and beams of large dimension used with floors of heavy timber or planks, three or more inches in thickness. Interestingly, such mill-type construction is still recognized today as a fire-resistant method of building in all US building codes.

Today wood joist, floor/ceiling assemblies and stud wall constructions which utilize a variety of surface finish materials afford predictable levels of fire performance for given design situations and building use.

3. Fires Affecting Structures

Fire performance of wood is related to the type of fire exposure which occurs. Structural fires can occur both within buildings and on a building's exterior. In both cases factors of importance are ignition source and form of surface materials present.

In exterior use, ignition sources are typically either (a) low intensity, such as grass or trash fires, or (b) high intensity, such as wild fires or fires caused by exposure to a fire of an adjoining structure. An exposure which involves both interior and exterior fire impact would occur when a fire, originating in the interior of a building, extends to siding materials via a window opening. At greatest risk in exterior fire exposures are wood roofing materials which have not been treated with durable fire retardants. Their impact on fire spread has been known for decades, especially in the western USA.

4. Fires Within Structures/Compartment Fires

Interior fires typically cause a greater structural impact than those originating outside a wood-framed structure and are also more common. Such incidents usually begin with ignition at one (often small) item, device or location. If such an incident is to pose a serious threat to the structure, the fire must grow to involve fully the room of origin which is often called a compartment. (Whether or not the fire does grow to this extent can be expressed through probability treatments which lend themselves to modelling. Fault-tree analyses may be conducted or state transition techniques applied. Such analytical methods assign probabilities in response to a series of "what-if" questions to determine the compound probability that a small fire will grow and spread in a given environment.)

Growth of the fire to where it threatens the room at large is divided into two regimes or growth periods. These are the preflashover and postflashover periods. In the former, boundaries of the affected room (walls, doors, windows, etc.) are not threatened and will contain the small, growing preflashover fire. Occupants of such preflashover rooms will not usually be at risk during this period of fire growth, which is in constrast to the postflashover period.

4.1 Preflashover Fires

A substantial amount of literature exists on preflashover fires which has, as a common thread, the impact of finish materials, furnishings and room geometry on fire growth. Room size, ventilation, rapidity of flamespread over surfaces and ability of furnishings to sustain combustion and liberate heat after ignition are crucial variables. The performance of combustible finish materials such as wood panelling can be gauged by their rate of heat release as determined under various test conditions. Specific variables such as material thickness, sensitivity to ignition and moisture content are particularly important.

In compartment fires, completion of the flamespread state (which must be preceded by sustained ignition) signals the end of the preflashover stage. During this stage the fire will have spread from the point of localized ignition over combustible surfaces liberating heat and generating further gases for combustion throughout the compartment. Before this, the fire was essentially a two-dimensional phenomenon of limted intensity within a three-dimensional space. As the fire continues to grow hot gases accumulate in the upper levels of the compartment and radiative heat transfer increases.

In the preflashover stage, variables important to fire performance of wood are (a) susceptibility to ignition and (b) rate of flamespread following sustained ignition. Ease of ignition, though ill-defined from a quantitative standpoint, is an expression of the minimum flame or heat source required to ignite a sample of specified size and geometric configuration.

In terms of ease of ignition, two types of wood fuels classified as thermally thick or thermally thin are recognized. Plywoods less than 6 mm thick, for example, are typical of thermally thin fuels. Their thin cross section allows rapid attainment of thermal equilibrium with a potential ignition source and consequently ignition will occur readily. Conversely, a thermally thick fuel of the same wood species (such as a piece of dimension lumber) displays a substantial temperature gradient from exposed surface to the unexposed (back) face of the member when heated. For this reason, the thick member is much less susceptible to ignition from a similar source. Additionally, because of the effects of thermal inertia on thick wood fuels, susbstantial levels of radiative feedback from other burning members are necessary for sustained ignition. Thin fuels, in addition to showing easier ignition, will generally show much higher rates of flamespread in a given use-configuration than thick members of similar species.

From a practical standpoint, the use of thin plywood (because of its ease of ignition and high flamespread) is regulated in building codes by requiring that such materials be backed by a gypsum wallboard membrane, which essentially converts it to a thick fuel. Conversely, thicker layers of wood products (i.e., 15–16 mm), such as plywood or tongue-and-groove panelling, may be used without a backing over supporting framing because such materials do not show the ease of ignition or high rates of flamespread common to thin wood members.

4.2 Postflashover Fires

Eventually, the nature of a fire will change from the two-dimensional phenomenon discussed above to a three-dimensional phenomenon with volumetric involvement of the compartment. When this occurs, the affected room is said to go to "flashover" and the postflashover fire stage begins. This stage of a fire is characterized by:

(a) the compartment becoming untenable for human life;

(b) gas temperatures in excess of 550 °C; and

(c) exposed combustible surfaces appearing to ignite more or less simultaneously.

Interestingly, the postflashover process can be described using the kinetics of a constantly stirred tank reactor.

In the postflashover fire, one of the properties that becomes important when considering the fire performance of wood is the ability to produce a char layer. This layer regulates the flow of heat and reactants into the pyrolysis zone as wood burns and will limit its rate of heat release in a postflashover fire. The charring occurs at a rate of about 0.8 mm min^{-1} after an initial period of rapid burning. Research has shown that wood composites such as particleboard, thick plywood, laminated beams and lumber have charring rates similar to whole wood. Heat release, which is interrelated with charring, is a function of the thermal inertia of the material in a general sense. Thus, products such as plywood which have density and thermal conductivity similar to whole wood can be expected to show similar rates of heat release when exposed at identical radiant flux levels. Wood products which have substantially different materials properties (such as cellulose insulation or wood fiber sound-deadening board) can be expected to show charring rates and rates of heat release consistent with those differences.

5. Wood Composites

Forms of wood found in construction have expanded from timbers, lumber and decking to a wide variety of wood-based materials. Products such as plywood, laminated veneer lumber and more recently composites of plywood and particleboard (such as wafer board and strand board) are now commonly used. These products behave similarly to whole wood of equal dimensions with respect to fire performance, provided voids are not present and thermosetting adhesives are used in their fabrication.

Exceptions to these generalizations exist in several notable cases. For example, "comply" studs and joists composed of veneer faces over a particleboard core can be expected to show similar charring rates but reduced load-carrying ability if the veneer on a fire-exposed face is allowed to burned through since the veneer carries a disproportionately large share of the load. Likewise, when engineered wood structural elements with reduced cross-sectional areas are used (such as wood-composite I-beams or composite beams with repetitive metal tubing for web materials), load-bearing fire endurance can be expected to be reduced as compared with members of equal initial load-bearing capacity of either whole wood or laminated design. If an adequate thermal barrier is included in the design to protect such beams, they can be expected to function safely, however.

Other examples include truss designs where metal connectors transfer stresses between members. These cannot be expected to perform as well as the older style mill-type or post-and-beam constructions they replace. Similarly, nailed, metal truss plates can be expected to have poorer fire endurance than component member sizes alone would predict if a protective membrane is not included to reduce heat transfer to such plates in a fire incident. In these situations, field experience has shown that the metal connectors act as a heat sink, leading to early dehydration and charring at joints with subsequent early failures to sustain impose load.

6. Fire Retardants

Fire retardants are applied to wood in two ways. One method involves surface coating to protect the underlying wood members. Such treatments increase time to ignition and reduce flame spread following ignition. Typically, such materials are of little benefit in a postflashover fire, since they do little to increase fire endurance.

The second application is by pressure treatment. Most of these are based on mixtures of waterborne salts such as borates, sulfates and phosphates. Such treatments both reduce flamespread and produce wood elements which are accepted in the building codes as having improved fire endurance. Since they are water-soluble, however, such treatments are not suitable for exterior exposure. They must also be carefully formulated and/or applied to avoid corrosion of fasteners, pipes and associated metal in contact with treated wood. In addition, their appearance is not suitable for architectural uses.

The functioning of waterborne salts is complex, and a substantial literature exists describing how such chemicals act as retardants, reducing either gas-phase flaming combustion or increasing smoldering (glowing or solid state) combustion. In most cases, it is believed that a catalysis occurs due to the presence of the fire retardant, which promotes char development through increased glowing (or solid state) combustion. This decreases gas-phase combustion since the enhanced char layer reduces heat and oxygen transfer at the pyrolysis zone. Release of fuel gases formed in the pyrolysis zone (which would normally burn in the gas phase) is also reduced and is expressed by reduced flamespread values. In practice, formulations of such treatments must result in reduced flamespread without adverse smoldering tendencies, corrosive properties, or high levels of hygroscopicity.

In the past decade, a number of patented, exterior, durable fire-retardant formulations have been developed. These are based on pressure treatment of wood with water-soluble, thermosetting polymer systems which are heat cured following pressure treatment. Contents include melamines, urea formaldehydes, dicyandiamide, borates and other materials. Developed in both the USA and Canada, these are used mostly for fire-retardant treatment of wood roofing materials but they can also be applied to siding products which will have acceptable appearance for architectural uses.

Many other specialized wood products rely on fire retardants for their successful, safe utilization. Treatment of low-density fiberboard acoustical tiles, for example, drastically reduces their flamespread. Medium- and higher-density fiberboard products achieve low flamespread ratings by the inclusion of fire retardants such as aluminum trihydrate (which includes relatively large amounts of water of hydration) to reduce flammability. In both these cases, the fire retardant can be added to the furnish (fiber feedstock) to obtain integral treatment through the entire thickness of the product. Likewise, low flamespread particleboards are available for subsequent overlaying with fine veneers or high-pressure laminates for panelling or work surfaces.

Fire-retardant treated, loose-fill and spray-on insulations based on recycled newspaper or wood pulp are widely used in the USA. These systems are especially susceptible to smoldering ignition if not formulated properly because of their low thermal conductivity and high surface area. Use of proper fire-retarding and manufacturing techniques and proper installation practices are necessary for their safe use. Likewise, insulation boards based on wood fibers perform well but must be protected from exposure to heat sources such as torches or hot surfaces which can lead to either smoldering fires or rapid flamespread. While exists for fire-retardant treatment of such products, little practical application has occurred.

See also: Wood as a Building Material

Bibliography

Atreya A 1983 Pyrolysis, ignition and fire spread on horizontal surfaces of wood. Ph.D. dissertation, Harvard University

Browne F L 1958 *Theories of the Combustion of Wood and its Control*, USDA Forest Products Laboratory Report 2136. USDA Forest Products Laboratory, Madison, Wisconsin

Goldstein I S 1973 Degradation and protection of wood from thermal attack. In: Nicholas D D (ed.) 1973 *Wood Deterioration and Its Prevention by Preservative Treatments*. Syracuse University Press, Syracuse, New York, pp. 307–39

Kanury A M 1972 Ignition of cellulosic solids—A review. *Fire Res. Abstr. Rev.* 14(1): 24–52

Meyer R W (ed.) 1977 Trends in fire protection. Symposium proceedings. *Wood Fiber* 9: 1–170

Parker W J 1985 Prediction of heat release rate of wood. *Proc. 1st Int. Fire Safety Symp.* Hemisphere, New York

Parker W J 1988 Prediction of heat release rate of wood. Ph.D. dissertation, George Washington University (also available from University Microfilms, Ann Arbor, Michigan)

Roberts A F 1971 Problems associated with the theoretical analysis of the burning of wood. *13th Symp. (Int.) Combustion*. The Combustion Institute, Pittsburgh, Pennsylvania, pp. 893–903

Zicherman J B, Williamson R B 1981 Microstructure of wood char, Part I: Whole wood. *Wood Sci. Technol.* 15: 237–49

Zicherman J B, Williamson R B 1982 Microstructure of wood char, Part II: Fire retardant treated wood. *Wood Sci. Technol.* 16: 19–34

J.B. Zicherman
[University of California, Berkeley, California, USA]

Wood as a Building Material

Wood has a long history of use in construction, dating back for many centuries if not into prehistory. Its applications have virtually covered the range of structural utilization except for buildings beyond a few stories in height. Engineered timber is used widely in such diversified construction as schools, churches, commercial buildings, industrial buildings, residences and farm buildings, highway and railway bridges, towers, theater screens, ships, and military and marine installations. Timber structures were at one time limited in the spans over which they could be used, because of the limited lengths of timber available and the limited cross-sectional dimensions. Modern technology, including glued laminated construction and modern methods of connection, has largely removed many of these limitations. Now, longspan arches and beams, trusses and dome-type structures provide large clear areas for recreational and athletics buildings, churches, auditoriums and bridges. A number of favorable characteristics enhance the suitability of timber for structural use; these will be dealt with in the following sections.

1. Important Characteristics

1.1 Durability

Wood, because it is an organic material, is commonly considered to be shortlived. In truth, its useful life may be measured in centuries if it is not subjected to seriously adverse conditions in service. There are numerous examples of structures several centuries old and, in the USA, examples of houses that date back to the early days of its history, together with covered bridges almost as old. One of the best-known timber structures in the USA is the Tabernacle of the Church of Jesus Christ of Latter Day Saints in Salt Lake City. This structure, whose roof is supported by wood trussed arches, is still in regular use after 100 years with no problems.

The critical question for the engineer is whether or not the structural properties of wood are altered with time. A number of studies, including one on girders from a 100-year-old cathedral in Madison, Wisconsin, have shown that there is no measurable loss in strength or stiffness as a result of age alone. It is known, however, that wood structural members can sustain higher loads for short rather than for long periods, and higher design stresses are permitted for short-term loading. Since the loading history of a structure, particularly one which has been in service over many years, is not likely to be known, caution suggests that for an old structure, design stresses for evaluation of load-carrying capacity should be those applicable to long-term loading (see *Design with Wood*).

Wood is, however, subject to a number of deteriorating influences including decay fungi, insects (especially termites), bacteria and marine organisms. Fungi, bacteria and many insects require moisture to be active, so their effects can be prevented if moisture is kept from reaching a critical level by careful design of details, by using moisture barriers and by properly maintaining the structure to prevent the intrusion of water from condensation, roof leaks, plumbing or other sources. Where this is not possible, substantial increases in useful life can be obtained by treatment with preservative chemicals.

1.2 Behavior in Fire

Wood is a combustible material. It is also a good insulator, as is the char formed when wood burns. Thus the wood only a short distance from the char–

wood interface is at a temperature well below the ignition temperature (see *Wood and Fire*). As a consequence, the rate of charring is low and wood members of large cross section retain significant load-carrying capacity for long periods when exposed to fire. This capability was recognized long ago in buildings with masonry walls and heavy columns and beams—so-called mill-type construction—and led to the acknowledgement that heavy timber construction performed in a manner similar to noncombustible constructions; thus it may be given a one-hour fire rating. Many building codes in the USA permit the use of heavy timber structures in all fire districts with all types of occupancies.

Exposure of wood to elevated temperature results in a loss of strength; long-term exposure causes a permanent loss. Because wood just below the char–wood interface has been subject to only a moderate rise in temperature, the largest part of the unburned portion of a wood member retains its original strength. Thus, removal of the char, and perhaps a quarter inch of wood below the char on each fire-exposed face, yields a member whose load-carrying capacity can in some cases be determined by applying original allowable stresses to the section properties determined from its new (reduced) size. In laminated bending members, however, the grade of individual laminations commonly varies within the cross section, the higher grades being used in the areas of highest stress. Laminations at the tensile side, particularly, are commonly required to be of especially high grade. Thus the loss of a large proportion of a tensile lamination may change the overall design stress greatly and this effect, as well as the effect of changes in cross-sectional dimensions, must be considered. In some instances, char removal may be all that is needed to renovate burned structural members.

The effects of fire on wood may be markedly changed by the application of fire retardants. They may be applied in the form of coatings or of chemicals such as mixtures of waterborne salts impregnated into the wood under pressure. Fire retardants are described in some detail in *Wood and Fire*.

1.3 Rate of Load Application; Duration of Load

It has long been known that wood can sustain higher loads for short periods than for long periods and this characteristic has been accounted for by adjustment of design stresses. Similarly, very rapidly applied loads can be sustained at higher levels than can more slowly applied loads. Design stresses, for example, may be doubled for impact loads. The factors commonly used to account for the effects of load duration or rate of load application have been derived from tests on material that is free from strength-reducing characteristics. Recent data suggest that these effects are different for lumber containing the usual characteristics of knots, cross grain and the like, and that the duration effects may be much less in the lower grades. Work on this problem is continuing, but at the time of writing, design-stress adjustments have not been changed.

1.4 Anisotropy

Wood is not isotropic and has different mechanical and other properties in different directions with respect to the grain of the wood (see *Wood: Strength*). This characteristic must be considered in design and special care must be taken to avoid design details that result in large stresses applied at right angles to the grain of the wood.

1.5 Exposure to Extremes of Temperature

Long-term exposure to elevated temperature may result in marked loss in strength, the degree of loss depending on the exposure temperature and the duration of the exposure. Thus the use of wood structures in applications where it is known that the wood will be continuously exposed to markedly elevated temperatures should be approached with caution.

Conversely, wood becomes stronger at very low temperatures and this characteristic has led to its use in vessels designed for transportation of liquefied gas.

1.6 Dimensional Stability

Wood will not change dimension with changes in moisture content unless it is at or below the fiber saturation point, commonly considered to be a moisture content of $\sim 30\%$. Below that level, wood shrinks as the moisture content decreases and swells as it increases. These changes may have structural effects. For example, if a connection involves steel plates attached to a wood member by widely separated bolts, a change in moisture content may cause splitting because the bolts are fixed in position by the steel plates and, as the wood attempts to shrink, it is restrained between the bolts by their inability to move. Thus tensile stress across the grain develops; if this stress becomes higher than the perpendicular-to-grain strength of the wood, the wood will split.

An unseasoned member of large cross section may develop deep checks because the outer layers of wood reach moisture contents below the fiber saturation point before the center of the member does. Since shrinkage does not occur until the fiber saturation point is reached, the interior will remain at its original dimension while the surface layer attempts to shrink. Thus checking will occur if the shrinkage tendency is great enough to create stresses beyond the across-the-grain strength of the wood.

Laminated members (see *Glued Laminated Timber*) are made from laminations which are generally not > 5 cm (nominal) in thickness, so that they can readily be dried to a suitable moisture content without serious drying defects. Thus the high initial moisture contents that are normal in a large solid sawn member are not present in a laminated member of large cross section; problems such as splitting and checking are therefore not to be expected in the same degree.

Change in dimension along the grain with a change in moisture content is, for normal wood, only a fraction of that across the grain. However, certain types of wood such as compression wood or juvenile wood (wood from near the pith of the tree) may show substantial dimensional changes along the grain with changes in moisture content. Thus, although it is generally true that problems from changes in length of a member with changes in moisture content are not to be expected, there may be instances when it can cause significant problems.

Coefficients of thermal expansion for wood are in the region of only one-tenth to one-third those of other structural materials. This difference must be taken into account when other materials are used in conjunction with wood.

1.7 Resistance to Chemicals

While wood can be degraded by exposure to certain chemicals, it may generally be considered highly resistant to many others. Wood is superior to cast iron and ordinary steel in resistance to mild acids and solutions of acidic salts. Thus, in the chemical processing industry, it is the preferred material for tanks and other containers and for structures adjacent to or housing chemical equipment. Wood is used widely in cooling towers where the hot water to be cooled contains boiler-conditioning chemicals as well as dissolved chlorine for suppression of algae. It is also used widely in the construction of buildings for bulk chemical storage, where it may be in direct contact with chemicals.

Highly acidic salts tend to hydrolyze wood and cause embrittlement when present at high concentrations. Alkaline solutions are more destructive than acidic solutions. Iron salts develop at points of contact with tie plates, bolts and the like. They may cause softening and discoloration of the wood around corroded iron fastenings. This is particularly likely in acidic woods such as oak and in woods such as redwood that contain tannin and related compounds.

The natural resistance of wood to degradation may be improved by treatment with a variety of substances. Pressure impregnation with a viscous coke-oven coal tar may be used to retard liquid penetration. Resistance to acids may be increased by impregnation with phenolic resin solutions followed by appropriate curing. Resistance to alkaline solutions may be increased by treatment with furfuryl alcohol. Impregnation by a monomeric resin followed by polymerization by radiation or other means is a relatively new development.

1.8 Strength and Structural Design

Strength, resistance to deformation, design and related factors are discussed in other articles (see *Wood: Strength; Wood: Deformation Under Load; Design with Wood*), but it seems desirable to point out here that wood generally has a favorable ratio of strength to weight and that this is one of the characteristics that has led to its use in a wide variety of structural applications.

Each species (or species group) has its own characteristic mechanical properties and not all species are classified according to the same set of grade descriptions. Design allowables, therefore, are presented by species or species groups and by grades under each species. In the USA and Canada, the dimension lumber used in light frame construction is commonly graded under a single rule called the National Grading Rule for Dimension Lumber. That is, the characteristics such as knots or cross grain permitted or limited in the National Grading Rule are the same regardless of species, although design stresses differ.

Design stresses for lumber, plywood and glued laminated timber are commonly developed by either of two methods. In one, fundamental mechanical properties based on data from tests of clear, straight-grained specimens are modified by a series of factors. For example, a property at a near-minimum level is used as a starting point. This is then modified by factors to account for moisture level, the effect of duration of load and the effect of such things as knots and cross grain permitted by the grade; finally, a safety factor is applied. The second method starts with data from full-size pieces representative of the various grades. The base data are again modified by a series of factors except that relating to grade. In North America, such methods are contained in standards of the American Society for Testing and Materials.

Design stresses developed as above may be modified to account for the individual circumstances associated with a particular structure. For example, for loads of short duration, such as snow, somewhat higher stresses may be used than with structures on which the full design load may be expected to continue for long periods. On the other hand, adverse circumstances such as certain treatments of wood, exposure to long-continued high temperature or long-term use at high moisture content may dictate modification to lower design stresses. Design stresses and their modification are presented in design and other handbooks.

Structural design is characteristically based on the use of allowable stresses. Increased attention, however, is being given to the development of reliability-based techniques for application to wood structures.

2. *Applications*

Typical examples of the wide range of applications of wood are discussed here to indicate some of the possibilities, but a complete review is not attempted.

2.1 Foundations

Wood piles have long been used to support buildings, bridges and waterfront structures. Because of the conditions to which most piles are exposed in service and the cost and other problems incident upon replacement, most piles are treated to provide protection against decay and termite attack and, when used in salt water, against attack by marine organisms.

Round wood poles or square wood posts set in the ground and supported on a footing are used with supporting beams to provide the substructure for buildings, frequently in basementless structures. In this type of application, the posts extend far enough above ground to minimize the possibility of decay in the beams. In a similar type of construction, the poles or posts are longer and form not only the foundation but part of the framing of the building as well, the other structural members being fastened directly or indirectly to the poles or posts. The latter type of construction is especially suitable on sloping sites, since a minimum of disturbance to the soil is required—far less than would be the case for a typical foundation. The integral foundation–framing type of construction has advantages where the structure is subject to high wind and wave action, as in seacoast structures in a hurricane area.

A wall-type foundation system, commonly called the all-weather wood foundation, is made from pressure-treated dimension lumber and plywood. Essentially a wood frame wall, the panels are placed on a wood footing over a gravel bed. This type of foundation, widely accepted in North America, has a number of advantages. It can, for example, be installed during weather that precludes the pouring of a concrete foundation, hence its name. In addition, the installation time is less than for a conventional foundation and the wall may be insulated, providing a more comfortable basement than one with concrete walls.

2.2 Light Frame Construction

In North America and possibly to a lesser degree elsewhere in the world, the majority of residences and many farm and light industrial buildings are of light frame construction. The framing members are usually of dimension lumber 2 in. (nominal) (5 cm) in thickness by widths varying from 4 in. (10 cm) to ~12 in. (30 cm) depending upon the type of load applied. Typically, light frame structures have been "stick built," that is, made of individual pieces assembled on the site. Increasingly, however, prefabricated units in the form either of panels or three-dimensional modules are being utilized. In the past, light frame construction has made only limited use of engineering analytical methods; this is changing as the cost and availability of materials become ever more critical and as the requirements of building codes are refined.

Floor loads are carried by joists spanning from the exterior walls to a central girder or load-carrying partition. The width of the joists depends upon the span but is generally in the range 8–12 in. (20–30 cm). Roof loads are commonly carried by rafters resting on the exterior walls and abutting at a ridge board. Vertical loads are carried by studs normally 2 × 4 in. (5 × 10 cm) in cross section. Increasingly, the common rafter is being replaced by a truss (frequently called a trussed rafter), with the various truss components connected by metal plates attached to the members by teeth punched from the plate or by nails driven through the plate.

Flat trusses with similar jointing methods are becoming more common to support floors, instead of joists. A new technology termed the truss-framed system has recently been introduced. It consists essentially of a trussed rafter and a floor truss connected by wall studs framed into both the trusses, the whole constituting a rigid frame requiring no load-carrying framing in the interior. Not only does this construction use less structural lumber than conventional light framing, but all framing can be done with a single size, such as 2 × 4 in. (5 × 10 cm), rather than the more costly wide dimension lumber needed for solid joists and rafters. Fabrication in a plant rather than on the site permits better control of quality.

2.3 Panelized Roof Systems

A variety of panelized roof systems are available. One common in California uses a basic unit consisting of a 4 × 8 ft (1.2 × 2.4 m) panel of plywood to which are attached stiffeners (commonly 2 × 4 in. dimension lumber). These basic units are attached to purlins which span between primary supporting beams. Although the basic panels can be put into place individually after purlins are erected, efficiency can be gained by assembling the panels and purlins on the ground, lifting the whole unit into place by machine and installing the purlins into hangers previously attached to the primary beams. This system promotes safety in that only one person is required to be on the roof to set panels and there is always a roof surface on which to stand; walking on beams or purlins is not necessary.

2.4 Heavy Timber Structures

Heavy timber construction is a type included in nearly all major building codes. As mentioned in Sect. 1.2, its use is based on the inherent fire resistance of timber members of large cross section. Codes specify the minimum dimensions for various members in order to ensure the desired resistance to fire. Heavy timber construction was at one time based solely on the use of solid timbers. These are becoming increasingly difficult and costly to obtain, so glued laminated members are now more commonly used.

2.5 Beams and Columns

Beams and columns made from solid-sawn timbers are typically straight members of rectangular cross section. Their cross section and length are limited by the size of the log from which they are cut. Thus beam spans and column heights are similarly limited. The advent of glued laminated structural timbers (glulam) has largely removed size limitations in terms of both cross section and length. Thus glulam beams of 100 ft (30 m) span or more and of 5–7 ft (1.5–2 m) depth are becoming more common. Greater spans may be achieved through the use of cantilevered or continuous beams. Further, beam form is no longer necessarily straight and rectangular. Glulam beams may be straight with single or double taper, or they may be pitched and tapered with a curved lower face over part of the span. Columns may be tapered or straight and may be either solid or spaced.

2.6 Curved Members

The advent of glulam has removed also, the limits on curvature which had existed for so long. Glulam arches may be (and are) used to support both roof and bridge structures. Although both two-hinged and three-hinged arches are feasible with glulam, two-hinged arches are less common because of difficulties in treatment and transportation. The curved forms possible are almost limitless, but one of the most common is the Gothic type so frequently found in church architecture. Gothic arch spans are commonly up to 80 ft (25 m).

Arches may also be used in domes for athletics or other facilities requiring large clear floor space. A typical form employs large arches radiating from a compression ring at the top of a dome to a tension ring at the base. Domes of this type (termed radial rib) have been constructed with clear spans up to 350 ft (105 m).

Even greater clear spans have been achieved with domes framed with wood members, largely laminated, to produce the needed curvature. Framing configurations differ and special connections between members are required. Two principal framing methods are employed, termed the Varax and the Triax domes; they represent patented systems. Structural analysis of such domes requires complex computer programs. A triangulated dome housing an athletics facility in Portland, Oregon has a clear span in excess of 500 ft (150 m).

2.7 Open Web Trusses

Open web roof and floor trusses offer a number of advantages including ease of erection because of light weight and the use of the open space for ductwork, plumbing, electrical lines and lines for fire sprinkler systems. One such open web system widely available uses steel tubing for the web members and machine stress-rated lumber for the top and bottom chords. Open web roof systems may span up to ~120 ft (35 m) and floor systems up to ~50 ft (15 m).

2.8 Plywood Components

Plywood is frequently used as roof, floor and wall sheathing in structures where the primary loads are carried by the framing. Not only does the plywood serve the normal sheathing function, but because of its high resistance to shear loads it may be designed to act as a diaphragm that will enhance the resistance of the structure to wind and earthquake loads. Stressed-skin panels, consisting of framing members with plywood bonded to one or both faces of the framing, provide efficient elements for carrying roof, floor or wall loads. Sandwich panels involving plywood bonded to a continuous lightweight core such as a foam or honeycomb are also efficient structural elements for roof, floor or wall. Stressed-skin or other types of panels may be used in folded plate roof systems designed to act as diaphragms. The plates are designed as deep intersecting beams acting together and supported at the ends.

In recent years, plywood nail-glued to joists has been used to provide semicomposite T-beam action resulting in reduced deflection and less squeaking of floors. Composite action is ignored in designing for strength. The special construction adhesives are generally suitable for on-site use. I-beams with plywood webs are used as structural members. Most have dimension lumber for flanges, but one manufacturer uses a specially made laminated material built up from veneer bonded with adhesive.

2.9 Wood Shells

Shell roofs of several forms may be constructed from layers of wood fastened either with nails or with adhesives. Only limited numbers of such structures have been made of wood.

2.10 Bridges

Wood has had wide application in bridge construction and has given long service; there are wood bridges more than 100 years old in North America and even older in Europe. The secret of long service is protective design to avoid entry of water (a tight deck with a water-resistant wearing surface, for instance), coupled with preservative treatment.

Bridges of a number of types have been built with wood. The type for a particular situation depends, of course, on such factors as the span, the type of traffic, the intended service life and aesthetics. Wood bridges are, for example, particularly suited to use in rural locations because their appearance can easily be made compatible with the surroundings.

The trestle-type bridge is common, particularly on railroads. In this type, girders or stringers with an appropriate deck system are supported on a substructure of piles or on a frame bent.

Girder bridges may be built with either solid sawn or glued laminated girders supporting a deck system. Solid sawn girders are, of course, limited in length and cross-sectional dimensions to the dimensions which can be cut from available timber; the maximum span for highway loads is ~30 ft (9 m). This limitation is largely eliminated if glued laminated members are used (see *Glued Laminated Timber*). Laminated girders up to 84 in. (2.1 m) in depth and 150 ft (45 m) long have been used. Deck systems vary, but nailed laminated decks of dimension lumber are common. These decks tend to become loose with changes in moisture content, with resulting water leakage and potential decay in the girders. Glued laminated deck sections obviate this difficulty. Various methods of attachment and joining, including special brackets and clips and dowels to transmit shear at the ends of the panels, have been developed and are in use.

A number of arch-supported bridges have been built. This type may be the most economical, depending upon the site conditions, because of savings in substructure. It does, however, require substantial height between foundations and roadway, and the site must be capable of resisting horizontal thrust in the abutments. A three-level interchange on an interstate highway near Mount Rushmore, South Dakota, involved, in the top bridge, an arch spanning 155 ft (47 m).

Truss-type bridges are becoming less common. They are practical for spans up to ~250 ft (75 m). Parallel-chord and bowstring trusses are the most common. The bowstring type may be the most economical because of the lower stresses in the web members. The advent of glued laminated members offers the possibility of continuous chords, reducing some of the connection problems found with shorter members. Improved fasteners can develop strong, rigid, concentric joints.

The Canadian Forest Service has developed designs for inverted kingpost bridges in which the stringers are designed to carry only the dead load while cables prestressed over the post carry the live load. Kingpost designs have been used for spans up to 125 ft (38 m) and queenpost trusses to 160 ft (50 m).

Suspension bridges are unusual, although the US Forest Service developed designs for spans up to 400 ft (120 m) in the 1930s. A suspension bridge in Switzerland is reported to span more than 540 ft (165 m).

Composite timber–concrete bridge decks have been built in numbers. The T-beam type consists of timber stringers forming the stem of the T and a concrete slab forming the flange or crossbar. Provision must be made for resisting horizontal shear, for example by cutting notches in the top of the timber, and uplift must be resisted by such means as driving mechanical fasteners into the top of the stringer. Composite slabs consist of 2 in. (50 mm) lumber nail-laminated with the wide surfaces vertical. The concrete is cast monolithically on top of the timber. Again, provision must be made to resist shear and uplift.

3. Wood-Based Materials

Wood is the base for a wide variety of materials for use in many different applications. Some of those most likely to find use in construction are discussed briefly in this section. Many of these materials are discussed in detail in other articles (see *Plywood; Hardboard and Insulation Board; Glued Laminated Timber*). Depending upon the hazards which may be expected in service, these materials may have to be treated to resist fungal attack, marine borer attack, or to improve resistance to fire.

Softwood lumber is most commonly used to provide the structural framework of buildings in the form of dimension lumber or of timbers. In such applications as joists, girders, posts, studs and rafters, it is used as it comes from the sawmill or planing mill. In the form of trusses, dimension lumber may be used in relatively short lengths connected by metal truss plates. In addition, it may form the chords of I-beams or box beams in combination with a panel material such as plywood. It may, in addition, be used as the stringers in stressed-skin panels having faces of plywood or another panel material. In all such structural uses, it is preferable that the lumber be stress-graded to ensure sufficient strength and stiffness to support service loads. In most cases, also, it should be at a relatively low moisture content. Lumber may also serve a protective function as house sliding in various forms. In such applications, appearance and dimensional stability are important characteristics.

Panel products such as plywood, fiberboard and hardboard may serve structural or other functions in buildings. As mentioned earlier, the high shear resistance of plywood leads it into floor and wall sheathing, particularly in structures expected to be subjected to high wind or earthquake loads and, because of its availability in large sheets, into use as the facing of stressed-skin panels. Fiberboards generally are used to provide insulation, but some forms are used as wall sheathing in light-frame buildings. Hardboards find a great deal of use as a base (or underlayment) for tile or linoleum floorings.

A product which is commonly termed laminated veneer lumber but which is marketed under a number of trade names is made from layers of veneer bonded together with the grain of each layer in the same direction as in glulam. Laminated veneer lumber (LVL) could serve any of the functions of sawn lumber, but is generally limited to special applications such as truss chords, scaffold planks and ladder rails.

Wood in round form finds use as piling foundations, as poles in pole frame buildings and as utility structures, either as single poles or as pole frames. In most uses, treatment to resist decay or other hazards is common.

See also: Building Systems and Materials; Wood: An Overview

Bibliography

American Institute of Timber Construction 1984 *Standard Specifications for Structural Glued Laminated Timber of Softwood Species*, AITC Standard 117-84, *Manufacturing*; AITC Standard 117-84, *Design*. American Institute of Timber Construction, Englewood, Colorado

American Institute of Timber Construction 1985 *Timber Construction Manual*. 3rd edn. Wiley, New York

American Society of Civil Engineers 1975 *Wood Structures: A Design Guide and Commentary*. American Society of Civil Engineers. New York

Freas A D 1971 Wood products and their use in construction. *Unasylva* 25 (101–03): 53–68
Freas A D (ed.) 1982 *Evaluation, Maintenance and Upgrading of Wood Structures.* American Society of Civil Engineers, New York
Freas A D, Moody R C, Sottish (eds.) 1987 *Wood: Engineering Design Concepts,* Vol. IV: Clark C Heritage Memorial Series on Wood. Pennsylvania State University, University Park, Pennsylvania
National Forest Products Association 1986a *National Design Specification for Wood Construction.* National Forest Products Association, Washington, DC
National Forest Products Association 1986b *Design Values for Wood Construction.* A supplement to the 1986 edition of National Design Specification. National Forest Products Association, Washington, DC
US Department of Agriculture 1987 *Wood Handbook: Wood as an Engineering Material,* Agriculture Handbook No. 72. US Department of Agriculture. Washington, DC

A. D. Freas
[Madison, Wisconsin, USA]

Wood–Cement Boards

Wood–cement boards, also known as mineral-bonded wood composites, are molded panels or boards which contain about 10–70 wt% of wood particles or fibers and about 30–90 wt% of mineral binder. The properties of the composites are significantly influenced by the matrix and also by the amount and nature of the woody material and the density of the composites. Mineral binders used are Portland cement, magnesia cement (Sorel), gypsum and mixed mineral binders which contain at least two components of the aforementioned binders. The composites are classified according to their density and their binder (Tables 1, 2 and 3).

1. Manufacturing Methods

The various stages in the manufacturing process are: preparation of the woody materials (Kollmann 1955, Maloney 1977); mixing these materials with the binder components, water and additives; mat forming; pressing; hardening or curing; demolding; further curing; drying; and sizing.

Three processes—the wet, the casting, and the semi-dry method—are used to manufacture the composites. The classification is based on the amount of water used for preparing the furnish, and also on the method employed to compact it to a coherent material.

In the wet process the thin slurry made of wood fibers, the binder, the aggregates, additives and water are fed to a sieve which is dewatered using vacuum

Table 1
Some characteristics and material requirements of medium-density mineral-bonded wood composites

Characteristic	Medium-density composites	
Structure of material	coarse-pored	porous
Wood particle/fiber	wood wool	flakes, shavings, slivers, sawdust
Particle dimensions		
length (cm)	50	0.4–7
width (mm)	3–6	2–15
thickness (mm)	0.2–0.5	up to 6
Binders	Portland cement, magnesium oxysulfate	Portland cement
Products	boards	boards, molding
Density ($g\,cm^{-3}$)	0.36–0.57	0.45–0.80
Bending strength (MPa)	0.4–1.7	0.4–2.0
Elastic modulus (MPa)	300	
Compression strength parallel to surface (MPa)	0.4–1.2	0.5–2.5
Thickness swelling after 24 h in water (%)	(3.5)	1.2–1.8
Length swelling by humidity, RH 35–90% (%)	(0.7)	
Thermal conductivity ($W\,m^{-1}\,K^{-1}$)	(0.09–0.14	0.1–0.20
Sound insulation by DIN 52210 (dB)	31[a]	31[a]
Resistance to wood-destroying fungi and insects	resistant	resistant
Fire test by DIN4102[b]	Bl	Bl
Material requirement per m^3		
wood (kg)	120–180	200–300
binder (kg)	180–250	200–300
additives (kg)	6–8	5–10
water (1)	200–290	150–300
Trade names	wood-wool boards	Arbolit (USSR) Velox (Austria) Durisol (Switzerland)

[a] 5 cm thick boards and 0.4 cm plaster on each side [b] A2, incombustible, B1, low flammability

Table 2
Some characteristics and material requirements of medium-high-density mineral-bonded wood composites

Characteristic	Medium-high-density composites				
Structure of material	compact	compact	compact	compact	compact
Wood particle/ fiber	flakes, shavings strands	flakes, shavings	wood pulp (recycled pulp) + inorg. fibers	flakes	wood pulp (recycled pulp)
Particle dimensions					
length (cm)	1–3	same as resin-bonded particle-board		same as resin-bonded particle-board	
width (mm)	1–3				
thickness (mm)	nearly 0.3; fine particles for surface				
Binders	Portland cement	magnesium oxysulfate, magnesium oxychloride	magnesium oxycarbonate	gypsum hemihydrate	gypsum hemihydrate
Products	boards, moldings	boards	boards	boards	boards
Density ($g\,cm^{-3}$)	1.0–1.35	0.9–1.25	0.7–1.1	1.0–1.2	1.0
Bending strength (MPa)	6–15	7–14	8–10	6–9	4–7
Elastic modulus (MPa)	3000–5000	3000	3000	3000–5000	4000–5000
Compression strength (MPa) parallel to surface	10–15	10			
Thickness swelling after 24 h in water (%)	1.2–1.8	4–8		(destroyed)	(destroyed)
Length swelling by humidity, RH 35–90% (%)	0.12–0.25 . . . 0.4	0.25	0.10	0.05–0.07	0.04
Thermal conductivity ($W\,m^{-1}\,k^{-1}$)	0.19–0.26	0.2–0.25			0.04
Sound insulation by DIN 52210 (db)	32 (12 mm)	30 (16 mm)	29[a]	41 (2 × 10 mm)	31 (10 mm)
Resistance to wood-destroying fungi + insects	resistant	resistant	resistant	—	—
Fire test by DIN 4102[b]	B1 A2 (spec. boards)	B1 A2 (spec. boards)	A2	B1 A2 (spec. boards)	A2
Material requirement per m^3					
wood (kg)	280	340	100	200	190
binder (kg)	800	500 MgO	500	800	870
additives (kg)	45	50 $MgSO_4$ or $MgCl_2$		5	500
water (l)	300	200		200	500
Trade names	Century board (Japan) Duripanel (Switzerland) Golden Board (Japan)	Ilves mineral (Finland)	Golden Siding Board (Japan)	Sasmox (Finland) Arborex (Norway)	Fermacel (Germany)

[a] 5 cm thick boards and 0.4 cm plaster on each side [b] A2 incombustible; B1 low flammability

Table 3
Some characteristics and material requirements of high-density mineral-bonded wood composites

Characteristic	High-density composites	
Structure of material	compact	compact
Wood particle/fiber	cellulose fiber	flakes, shavings 0.2–2 cm
Particle dimensions		
length (cm)		
width (mm)		
thickness (mm)		
Binders	cement and quartz	cement, fly ash
Products	boards, moldings	hollow core panels[a]
Density ($g\,cm^{-3}$)	1.35	1.4
Bending strength (MPa)	21	3–6
Elastic modulus (MPa)	6000	3400
Compression strength parallel to surface (MPa)		10–15
Thickness swelling after 24 h in water (%)	0.25	0.2
Length swelling by humidity, RH 35–90% (%)	0.15	0.2
Thermal conductivity ($W\,m^{-1}\,K^{-1}$)	0.35	0.40
Sound insulation by DIN 62210 (dB)		high
Resistance to wood-destroying fungi and insects	resistant	resistant
Fire test by DIN4102[b]	A2	A2
Material requirement per m^3		
wood (kg)	100	100
binder (kg)	1000	390 cement 300 fly ash 430 sand 10 water glass 320 water
Trade names	e.g., Internit (Germany)	Fibrecon (Singapore) Ecopanel (Finland)

[a] extrusion process [b] A2 incombustible

chambers. The process can be continuous or discontinuous. In the continuous process a thin web is formed, e.g., using a Hatschek machine, which is wound on a cylinder until the desired thickness is reached. The web is then split and transferred to a steel caul. The green boards are compacted by pressure before the binder is allowed to harden. Since springback forces of the refined wood fibers are negligible after pressing, the boards are hardened without further compression. This method is used to manufacture flat and corrugated boards and molded products.

In the casting process, all the water needed to prepare the furnish is applied to the mixture. The furnish is a thick and fluid mixture and is compacted by gravitation or vibration or by a slight pressure, before it is allowed to harden. According to Joklik (1983), a mixture of wood chips, water containing a foaming agent, cement and sand is cast into wall elements using wooden frames as casting molds. The total water–binder ratio in this process is about 0.8. It is desirable to keep the water–binder ratio as low as possible in the manufacture of mineral-bonded wood composites, because excess water will leave the binder after hardening and curing, creating micropores which reduce the mechanical strength of the material. The extra water also prolongs the curing time needed to reach the desired strength. The total amount of water added exceeds the theoretical amount for the hydration (hardening) process, as water is also needed as a homogenizing agent. Portland cement requires about 25%, magnesia cement 20–30% and gypsum hemihydrate 17% of water. The influence of the water–binder ratio is demonstrated by the following example: hardened Portland cement of water/cement ratios of 0.4 and 0.8 resulted in a density of 2.0 and 1.6 $g\,cm^{-3}$, respectively.

In the semidry process all the water is applied to the wood furnish. However, the water becomes available to the binder after pressing (at 1–3 MPa) the furnish to the desired thickness, and forming a coherent material. The total water–binder ratio is about 0.4, and the woody material acts as a water reservoir. The furnish is spreadable, and can be formed into a mat, applying forming devices commonly used in the manufacturing of particleboards. After pressing to the desired thickness, the pressure can be released or sustained until the boards gain enough strength and can be demolded without damage. Mineral-board fiberboards made according to the semidry method are hardened without further pressure, whereas particleboards have to be clamped. In a newly developed extrusion process, wood particles, fly ash, cement and water-containing waterglass are formed into hollow core panels. Owing to the low content of wood particles and also to the addition of sand, such panels can be piled after the extrusion process (Anon 1987e, f). The hardening time of most of the inorganic binders is long compared with those of synthetic resins commonly used to manufacture particleboards and medium-density fiberboards. Portland cement requires about 8–24 h, magnesia cement 4–6 h and gypsum hemihydrate 15–30 minutes at ambient temperatures. The final strengths are attained after curing for 28 days. New developments aim to reduce the hardening time of mineral-bonded composites to minutes and enable the manufacture of composites, akin to that of particleboards and medium density fiberboards (Anon 1987a, b, c).

Another feature of mineral-bonded composites is the high ratio (1–10) of binder to woody material, whereas wood–polymer composites have a ratio of 0.08–0.12. Although the prices of inorganic binders per kilogram are rather low, due to the high amounts needed, the total cost of such binders per cubic meter of end product can be high.

1.1 Portland-Cement-Bonded Composites

Portland cement is a hydraulic binder and the bonding agent in hydrated cement is a calcium silicate. Cement is standardized in many countries. The classification is based mainly on the minimum compression strength (MPa) of specified specimens made of cement, sand and water. Generally three classes of cements are distinguished, e.g., according to German Industrial Standards PZ 35, PZ 45 and PZ 55. Wood–cement composites made of the same wood and with cement of different strength classes exhibit the same trends in strength development as shown by concrete. The hydration of Portland cement is retarded by about 0.5% of wood starch or 0.25% of free monosaccharides or oligosaccharides or phenolic compounds (based on the weight of cement), making demolding within a reasonable length of time impracticable. Wood with a high alkali-soluble content (e.g., a high hemicellulose content) or wood deteriorated by fungi or excess of light or heat, may have the same effect. In general softwoods are more suitable than hardwoods. Methods to improve the suitability of woods are: natural seasoning (2–3 months); addition of accelerators; extraction of wood particles with water; addition of diffusion-retarding compounds; drying of the wood particles; use of rapid-hardening cement; and addition of fumed silica or rice hull ash (Simatupang et al. 1987). Accelerators used in the casting and semidry processes are: calcium or magnesium chloride; a mixture of aluminum sulfate and calcium hydroxide or waterglass; waterglass; calcium formate or acetate; sodium or potassium silicofluoride; and sodium fluoride. In the wet process only waterglass is effective (Kitani 1985). The amount of accelerator used is in the range of 0.5–3% (based on the weight of cement). Curing of cement can be performed at ambient temperature or at higher temperature and higher pressure, e.g., hydrothermal treatment of cellulose cement fiberboards (Coutts and Michel 1983).

1.2 Magnesia-Cement-Bonded Composites

The bonding agent is the basic salt of magnesium hydroxide and the magnesium salt of hydrochloric, sulfuric, nitric, carbonic, or phosphoric acid with the general formula

$$x\mathrm{Mg(OH)_2} \cdot y\mathrm{MgAR} \cdot z\mathrm{H_2O}$$

where x, y and z are integers and AR denotes an acid radical. The various basic salts have different properties. Magnesia-cement or Sorel cement composed of magnesium salts of hydrochloric, sulfuric or carbonic acids are mostly used. The cement is hard and tough, but is not weather-resistant. It is prepared by mixing calcined magnesium carbonate or magnesium oxide with a concentrated solution (12–25%) of magnesium chloride or magnesium sulfate (Mortl and Leopold 1978). Owing to the rapid hardening of magnesiun oxysulfate at higher temperatures, even in the presence of a higher concentration of wood extractives, boards bonded with such cements can be manufactured continuously, e.g., Heraklithboards (Kollmann 1955, Maloney 1977). Magnesia-bonded particleboards are manufactured in Finland in a slightly modified particleboard plant, using the process described by Simatupang (1985).

Magnesium oxycarbonate is prepared by heating magnesium trihydrate. The latter is obtained by treating a magnesium hydroxide slurry with flue gases (free of SO_2) containing carbon dioxide (Kirk and Othmer 1978). Magnesium-oxycarbonate-bonded fiberboards are manufactured in Japan using a wet process (Anon 1987).

1.3 Gypsum-Bonded Composites

The binder used is gypsum hemihydrate ($CaSO_4 \cdot 1/2H_2O$) which is obtained by the calcination of naturally occuring gypsum, waste gypsum from chemical processes e.g., phosphogypsum or gypsum from desulfurizing of flue gases (Wirsching 1978, 1984). There exist two phases (α and β) of gypsum hemihydrate. The β-hemihydrate, plaster of Paris, is mostly used as binder for composites. The binder components are rehydrated gypsum dihydrate crystals

($CaSO_4 \cdot 2H_2O$) which crystallize from a supersaturated solution. Wood extractives generally inhibit the formation of perfect crystals, and cause a reduction of the shear strength between wood and hardened gypsum (Seddig and Simatupang 1988). The gypsum bond is not water resistant. Owing to the short hardening time, plaster of Paris is used for the continuous manufacture of cardboards (Wirsching 1978) and fiberboards (Williams 1987). Gypsum-bonded particleboards according to the Kossatz process are currently manufactured in two plants in Finland and Norway (Hübner 1985, Wilke 1988).

New developments aim to extend this semidry process to gypsum fiberboards using recycled paper as reinforcing material (Anon 1988).

2. *Utilization*

The medium-density composites have been used extensively for insulation purposes and as fire-resistant materials. However, because of the availability of newer building materials, production has decreased worldwide. Recently the concrete casing method using boards or precast wood–cement moldings has found increasing application.

Cement-bonded particleboards, made according to Elmendorf (1966), are used as wall elements and siding materials in prefabricated houses. The properties and durability of such materials have been reviewed (Dinwoodie and Paxton 1983, Simatupang 1987). It is feasible to construct such buildings entirely from wood–cement panels by gluing the panels into folded shapes that form load-bearing structural elements (Anon 1978). Worldwide, 27 cement particleboard plants exist. Of these, 12 are in the USSR and 5 in Japan. The rest are distributed in 9 other countries.

The replacement of asbestos fibers in cement-bonded fiberboards with less harmful materials is making great progress. In Australia, New Zealand (Sharman and Vautier 1986) and Finland, cellulose fibers are used for such purposes. In Germany such fiberboards are used indoors, whereas outdoors a combination of synthetic polymer and cellulose fibers are utilized. Alkali-resistant glass fibers are also used as asbestos substitutes. Cellulose fibers have to be added, because the synthetic fibers are too coarse to retain the cement particles from the slurry during manufacturing. The nature of this natural fiber also has an influence on the properties of such boards (Simatupang and Lange 1987).

Magnesia particleboards, mostly overlaid with veneer, are used as low-flammable ceilings and walls in theaters and hotels. Gypsum fiberboards and particleboards have the same utilization as gypsum cardboards, with which they compete.

Bibliography

Anon 1978 The Bison folding system (BFS). *Build. Prefab.* 3: 18–19

Anon 1987a *Neue, umweltfreundliche Technologien zur Herstellung zementgebundener Holzspanplatten.* Bison Information, Forschung und Entwickelung. Bisonsystem, Springe, FRG

Anon 1987b *Rauma-Repola and Falco Cement Flake Board Technology.* Rauma-Repola Engineering Works, Finland

Anon 1987c *The Ecocem Process for fast and economic production of cement bonded particleboards.* RAUTE Wood Processing Machinery, Raute Oy, Lahti, Finland

Anon 1987d *Golden Siding Board.* Nippon Hardboard, Nagoya, Japan

Anon 1987e *Natural Fibre Concrete's Mechanical and Physical Properties.* ACOTEC (Advanced Construction Technology). Salpakangas, Finland

Anon 1987f *ECOPANEL.* Raute Oy, Lahti, Finland

Anon 1988 *Information about Gypsum Fibre Board System Würtex.* Wurtex Maschinebau, Uhingen, FRG

Coutts R S P, Michel A J 1983 Wood pulp fiber-cement composites. *J. Appl. Polym. Sci. Appl. Polym. Symp.* 37: 829–44

Dinwoodie J M, Paxton B H 1983 *Wood Cement Particleboards—A Technical Assessment,* Building Research Establishment Information Paper IP 4/83. Building Research Establishment, Princes Risborough, UK

Elemendorf A 1966 Method of making a non-porous board composed of strands of wood and Portland cement. US Patent No. 3,271,492

Hübner J E 1985 The industrial production of gypsum boards with reinforcing wood flakes. *TIZ-Fachberichte* 109(129): 908–16

Kirk R E, Othmer DF 1978 *Encyclopedia of Chemical Technology,* Vol. 14. Wiley, New York, pp. 619–34.

Kitani Y 1985 Inhibition of pulp cement board hardening by pulp and its relieving methods. *Mokuzai Kogyo* 40(5): 19–23

Kollmann F F P 1955 *Technologie des Holzes und der Holzwerkstoffe,* Vol. 2. Springer, Berlin, pp. 468–88

Maloney T M 1977 *Modern Particleboard and Dry-Process Fiberboard Manufacturing.* Miller Freeman, San Francisco, California

Mörtl G, Leopold F 1978 Caustic magnesia for the building industry. In: Cooper B M (ed.) 1978 *Proc. 3rd Industrial Minerals Int. Congr. Metals Bulletin,* London

Seddig N, Simatupang M H 1988 A shear method for testing the suitability of various wood species for gypsum-bonded wood composites and SEM-microscopy of the shear areas. *Holz Roh-Werkst.* 46: 9–13

Sharman W R, Vautier B P 1986 Accelerated durability testing of autoclaved wood-fiber-reinforced cement sheet composites. *Durability Build. Mater.* 3: 255–75

Simatupang M H 1985 Caustic magnesia as a binder for particleboards. *Proc. Symp. Forest Products Research International Achievements,* Vol. 6. SCIR Centre, Pretoria, pp. 1–19

Simatupang M H 1987 Manufacturing process and durability of cement-bonded wood composites. *Proc. 4th Int. Conf. on Durability of Building Materials and Building Components,* Vol. 1. Pergamon, Oxford, pp. 128–35

Simatupang M H, Lange L 1987 Lignocellulosic and plastic fibres for manufacturing of fibre cement boards. *Int. Cement Compos. Light. Concr.* 9(2): 109–11

Simatupang, M H, Lange H, Neubauer A 1987 Influence of seasoning of poplar, birch, oak and larch and the addition of condensed silica fume on the bending strength of cement-bonded particleboards. *Holz Roh-Werkstoff.* 45: 131–36

Williams W 1987 Gypsum fiberboard proves success for Germany's Fels-Werke. *Wood Based Panels N. Am.* 7(2): 26–29

Wilke D 1988 Die Gipsplatte kommt voran-nunmehr auch in Norwegen. *Holz-Zentralbl.* 114 (33): 502–3

Wirsching F 1978 Gypsum. In: *Ullmanns Encyklopädie der Technischen Chemie*, Vol. 12. Verlag Chemie, Weinheim

Wirsching F 1984 Drying and agglomeration of flue gas gypsum. In: Kuntze R A (ed. 1984) *The Chemistry and Technology of Gypsum*, ASTM STP 861. American Society for Testing and Materials, Philadelphia, Pennsylvania, pp. 160–73

M. H. Simatupang
[Institut für Holzchemie und Chemische Technologie des Holzes, Hamburg, FRG]

Wood: Decay During Use

Decay of wood during use is a serious problem, causing significant losses in strength in stages of decay so early as to be difficult to diagnose even microscopically. The key to preventing decay in wooden buildings is the exclusion of water. Although emphasis in this article is placed on experience with wood in structures, the principles of wood decay are equally applicable to wood in any use.

1. Causal Fungi and their Requirements for Growth

Decay of wood is caused by fungi which are members of the group of higher fungi called Basidiomycetes related to the common edible mushrooms. These fungi are primitive plants which lack chlorophyll and therefore cannot manufacture their own food, relying instead on the organic matter produced by plants possessing chlorophyll such as the wood produced by trees. The decay fungi, as scavengers, play an important role in the forest, returning organic matter to the soil, returning nutrients for reuse by other plants, preventing buildup of material which would make access for wildlife impossible and reducing the fire hazard produced by combustible material on the forest floor. However, the same natural process occurring in wood in structures is considered undesirable and so preventive measures are necessary.

Decay fungi have four basic requirements for growth and production of decay in wood: air (they are aerobic organisms), water, a favorable temperature and a food source. The food is of course the wood itself. This food may be made unavailable to the fungus by poisoning it with a preservative, but normal use of wood in above-ground portions of structures does not usually involve such treatment. Decay fungi have a wider tolerance to temperature than people who occupy wooden structures. Freezing temperatures will stop the progress of decay but will not kill the decay fungi. Temperatures must usually exceed about 60 °C before thermal death occurs. Decay fungi require so little oxygen that it is usually not possible to control decay by limiting this factor in structures; water submersion or deep burial, however, excludes oxygen sufficiently that wood in such conditions is not attacked by decay fungi. Thus, the remaining factor—water—is the key to controlling decay in buildings. Throughout history a major goal of the design and construction of structures has been to shed water effectively. If this is done, and no water leaks are allowed, there is no reason why above-ground portions of wooden buildings should decay. Wood is a hygroscopic material and takes up moisture from the surrounding air, but the moisture content attained by wood in 100% relative humidity is not sufficient to support decay; a source of liquid water such as rain, plumbing leaks or ground contact is required.

There are other fungi which inhabit wood, namely those which produce mold and blue stain, but they do not cause the extensive loss in wood strength which results from the action of decay fungi. These fungi live principally on storage materials in the wood, such as starch, and therefore do not break down the wood cell walls as do decay fungi. Since they have pigmented cell walls or spores, their main effect on wood is discoloration. However, although they tend to grow faster, they have generally the same requirements for growth as decay fungi and therefore their presence may indicate that the wood has been exposed to conditions conducive to decay.

2. Sources of Inoculum

Any living cells of a fungus, when transferred to a new location favorable for growth, have the potential for reproducing the fungus in this new site. Decay fungi produce spores which act like seeds of higher plants in that they are capable of starting a new growth of fungus in a new area where conditions are favorable. The fungi produce great numbers of such spores light enough to be carried by the winds, thereby assuring that fungi are present wherever conditions conducive to decay exist. Moreover, the careless manner in which wood products are often handled tends to ensure that wood in structures is handled/infected before being placed in service.

Wood may be shipped and stored in the green condition, thereby allowing time for it to become infected. Wood products may be delivered to the job site and set on the soil; since decay fungi live in the soil this provides an opportunity for them to move into the wood before it is placed in the structure. If the wood is allowed to dry and remains dry throughout the life of the structure, such preinfection may be of no consequence. However, if conditions later become conducive to decay, the necessary fungi may already be present.

3. Types of Decay

There are two major categories of decay produced by Basidiomycetes: brown rot and white rot. The brown-rot fungi decompose only the carbohydrate fraction of wood, leaving the lignin modified but not metabolized. The residual lignin is darker in color than the cellulose, leading to the brown coloration of the rot, and has very little strength, leading to the excessive shrinkage and cubic checking pattern typical of brown-rotted wood.

White-rot fungi decompose all the major components of wood and presumably may lead to 100% loss in cell wall material. Some of the white-rot fungi appear to attack lignin preferentially in the early stages of decay, while others, called simultaneous rots by some workers, decompose both the lignin and the carbohydrates at a rate approximately proportional to their original relative concentration in the wood.

A third category of severe wood deterioration—soft rot—is not caused by Basidiomycetes and therefore should not be included under the term decay. Nevertheless, soft rot can cause such serious decomposition of wood and loss of strength in certain situations that it must be considered in any treatise concerned with the effects of wood decomposition. Soft rot is caused by fungi of the mold and stain group (Ascomycetes and Fungi Imperfecti). It is not known whether fungi which normally inhabit wood as mold or stain producers are also capable of producing soft rot under the right conditions, or if only special strains of these organisms are capable of producing soft rot when present. In any event, soft rot occurs in environmental conditions under which growth of decay fungi would not be expected, such as in preservative-treated wood, and very wet or water-submerged wood. The hyphae of soft-rot fungi grow within the S2 layer of the secondary wall of wood cells, producing cavities parallel to the microfibrils making up that layer. After removal of the S2 layer there is so little left of the cell walls that the softened wood either sloughs off or fails. Soft-rot fungi produce deterioration similar to that of brown-rot fungi in that they primarily remove the carbohydrates from the cellulose-rich S2 cell wall layer, but, unlike brown-rot fungi, they metabolize some lignin as well.

Some bacteria are capable of decomposing wood also, but their attack is usually very slow and occurs only in conditions of high moisture, such as saturation or submersion. Because many of these bacteria can grow under anaerobic conditions, their attack is often the sole cause of degradation in submerged or buried wood. Bacterial attack would not be expected in buildings.

4. Importance of Water

As stated above, the key to decay in wood structures is water. Wood kept dry will never decay. Furthermore, since the major goal of structural design and construction is the shedding of water, keeping wood dry in a structure should be the easiest, cheapest and most effective method of assuring that the wood does not decay. Building leaks, design features which lead to water infiltration, features which tend to trap water and leaks in or condensation on interior plumbing should be carefully avoided to preserve the life of a wooden structure.

5. Microscopic and Chemical Effects of Decay

Not all decay fungi attack wood in the same way, and a given fungus may produce different effects in different woods. Therefore, not all the features discussed below will necessarily be found in any given piece of decayed wood. However, the following microscopic features may frequently be required for diagnosing the presence of early stages of decay; the more they are present, the more confidence may be placed in the diagnosis.

A fungus can be thought of as being like a ball of cotton with filaments running in all directions. The filaments that make up the body of a fungus are called hyphae. Individual hyphae are too small to be seen easily without a microscope, but massed together they form what is known as a mycelium, which can readily be seen on the wood surface. The hyphae of decay fungi may contain very distinctive loop-like structures called clamp connections. Only Basidiomycetes have them, and the only Basidiomycetes which would be present in wood would be decay fungi. Therefore, if microscopic examination reveals hyphae with clamp connections, then the fungus is a decay fungus. Unfortunately, many decay fungi do not produce clamp connections and some which produce them in culture do not produce them in wood. Clamp connections, then, are a valuable diagnostic tool if present, but their absence supplies no useful diagnostic information.

The hyphae of decay fungi may invade wood in early stages by growth through ray cells and large longitudinal elements, such as resin ducts or vessels. This is true of mold and stain fungi too, but unlike these fungi the decay fungi normally leave these elements and ramify throughout the wood. In some cases decay-fungal hyphae may be most numerous in the strength-providing longitudinal elements (fibers or tracheids). Penetration may be through pits or by means of boreholes dissolved through the wood cell walls by the enzymes from the hyphae. The boreholes of decay fungi are usually numerous. They enlarge to several times the hyphal diameter, tend to run in series (serial boreholes) through a number of wood cell walls in a relatively straight line (as seen in radial section) and bear little relationship to pits. The boreholes produced by mold and stain fungi are infrequent and tend to remain at the diameter of the penetrating hypha, resulting in a "necked-down" appearance to the large hypha penetrating a small hole. They tend to be solitary, with the penetrating hypha changing direction after penetration, and penetration is frequently through pits with no borehole formation at all. The hyphae of soft-rot fungi may produce serial boreholes, but they normally consist of large, frequently pigmented hyphae which "neckdown" considerably to pass through the wall and are therefore not too difficult to differentiate from those of decay fungi.

Early stages of attack on the cell walls may be characterized by separations within or between cells, and shallow depressions or gouges at the wall–lumen interface. In later stages of decay, white-rotten wood cell walls may appear thinned in a uniform fashion from cell to cell, whereas brown-rotted wood cells appear shrunken and distorted on a very irregular basis from cell to cell. Soft rot results in cavities within the S2 layer of the secondary wall which parallel the microfibrils in this layer and tend to be diamond-shaped or elongated with pointed ends, making soft rot easily distinguishable from brown rot or white rot under the microscope.

White-rot fungi decompose all the major cell-wall components and some remove them at rates approximately proportional to the amounts present in sound wood, thereby leading to the observed progressive thinning of the cell walls.

Brown-rot fungi decompose primarily the carbohydrate fraction of the wood. Lignin is altered by their attack, but little of it is removed. In early stages of brown rot the wood constituents may be depolymerized more rapidly than the decomposition products are metabolized by the fungus, leading to a buildup of low-molecular-weight molecules which constitute mass but provide no strength.

6. *Effects of Decay on Strength Properties*

Resistance to impact loading is the mechanical property of wood most sensitive to the presence of decay. Toughness, for example, may be largely destroyed at a stage of decay so early that it may be hard to detect even microscopically (Table 1). Those strength properties relied upon in the use of wood as a beam, a common mode of service, are also severely diminished in very early stages of decay. In such early decay stages there is little difference noted between brown rot and white rot with regard to the magnitude of strength losses. However, as decay advances, further effects on strength are much more severe in brown rot than in white rot.

Because of the drastic reduction in most strength properties caused by decay at such an early stage as to be difficult to diagnose, it is easy to see that it is considerably safer to prevent decay in the first place than to rely upon ability to detect and diagnose it accurately in sufficient time to effect a remedy before failure occurs.

Table 1
Expected strength loss at an early stage of decay (5–10% weight loss) in brown-rotted softwoods (after Wilcox 1978)

Strength property	Expected strength loss (% of strength of sound wood)
Toughness	≥80
Impact bending	80
Static bending (MOR and MOE)	70
Compression perpendicular to grain	60
Tension parallel to grain	60
Compression parallel to grain	45
Shear	20
Hardness	20

7. *Natural Decay Resistance*

Some woods have what is called natural decay resistance, that is their heartwoods are infiltrated with chemicals at the time of heartwood formation which make that wood unpalatable to decay fungi. Through both laboratory testing and field performance, various woods which appear in the US commercial market have been rated according to the natural decay resistance of their heartwoods (Tables 2 and 3). Decay-susceptible wood will perform adequately in structures if kept dry. Moderately decay-resistant wood may perform adequately, even if occasionally wetted but then allowed to dry out. Highly decay-resistant wood may perform adequately in above-ground exposure even if allowed to remain wet for substantial periods of time. Most woods, even if highly decay-resistant, will provide only limited service in contact with the soil, which is the most hazardous terrestrial exposure to which wood can be put. To provide adequate service in ground contact, most woods must be pressure-treated with an effective preservative.

8. *Detailing for Shedding Water*

The first part of a building to intercept rainwater is the roof. Flat roofs have become popular in some parts of the world, no doubt primarily for economic reasons, but they are dangerous for a wooden structure in that the performance of the entire structure depends solely upon the effectiveness of the membrane placed on top of the flat roof. One pinhole or small crack in this barrier may jeopardize the safety of the entire structure. It is better to provide a slope and drainage so that the natural force of gravity is working for rather than against the building and assists in the shedding of water. Roof overhangs are useful in removing the drip line from the rest of the structure and in protecting the walls and wall penetrations from direct contact with rainwater. Gutters and downspouts are also important since they collect water from the roof and dispose

Table 2
Relative grouping of some US domestic woods according to heartwood decay susceptibility based upon laboratory tests and field performance (after US Department of Agriculture 1974)

Resistant or very resistant	Moderately resistant	Slightly or nonresistant
Baldcypress (old growth)[a]	Baldcypress (young growth)[a]	Alder
Catalpa	Douglas fir	Ashes
Cedars	Honeylocust	Aspens
Cherry (black)	Larch (western)	Basswood
Chestnut	Oak (swamp)	Beech
Cypress (Arizona)	Chestnut	Birches
Junipers	Pine	Buckeye
Locust (black)[b]	Eastern white[a]	Butternut
Mesquite	Southern	Cottonwood
Mulberry (red)[b]	Longleaf[a]	Elms
Oak	Slash[a]	Hackberry
Bur	Tamarack	Hemlocks
Chestnut		Hickories
Gambel		Magnolia
Oregon white		Maples
Post		Oak (red and black species)
White		Pines (other than longleaf, slash and eastern white)
Osage orange[b]		Poplars
Redwood		Spruces
Sassafras		Sweetgum
Walnut (black)		True firs (western and eastern)
Yew (Pacific)[b]		Willows
		Yellow poplar

[a] The southern and eastern pines and baldcypress are now largely second growth with a large proportion of sapwood. Consequently, substantial quantities of heartwood lumber of these species are not available
[b] These woods have exceptionally high decay resistance

Table 3
Relative grouping of some woods imported into the USA according to heartwood decay susceptibility based upon laboratory tests and field performance (after US Department of Agriculture 1974)

Resistant or very resistant	Moderately resistant	Slightly resistant or nonresistant
Angelique	Andiroba[a]	Balsa
Apamate	Apitong[a]	Banak
Brazilian rosewood	Avodire	Cativo
Caribbean pine	Capirona	Ceiba
Courbaril	European walnut	Jelutong
Encino	Gola	Limba
Goncalo alves	Khaya	Lupuna
Greenheart	Laurel	Mahogany, Philippine:
Guijo	Mahogany, Philippine:	Mayapis
Iroko	Almon	White lauan
Jarrah	Bagtikan	Obeche
Kapur	Red lauan	Parana pine
Karri	Tanguile	Ramin
Kokrodua (Afrormosia)	Ocote pine	Sande
Lapacho	Palosapis	Virola
Lignum vitae	Sapele	
Mahogany, American		
Meranti[a]		
Peroba de campos		
Primavera		
Santa Maria		
Spanish cedar		
Teak		

[a] More than one species included, some of which may vary in resistance from that indicated

of it safely away from the structure rather than allowing it to drip, splash or run down various portions. If no roof overhang is provided, the flashing and detailing at the intersection between the roof and the vertical walls becomes critical. Flashings and moisture barriers behind exterior wall coverings should be designed and assembled so that, at each joint, the tendency is to force water out, away from the structure. It is surprising how many structural failures due to decay in wooden buildings can be traced to careless reversal of the flashing or shingling of moisture barriers so that water coming in contact with these structures is drawn into the interior of the wall rather than being shed. Careful attention to such apparently minor features may be major factors in attaining adequate service from a wooden building.

Being a hygroscopic material, wood will adjust its moisture content to the humidity of the air around it. This means that if the air becomes more moist, the wood will increase in moisture content, resulting in swelling. If the surrounding air becomes drier, the wood will lose moisture and shrink. Wood undergoes normal fluctuation in size and shape with the annual change in seasons and the concomitant differences in relative humidity. Such changes must be taken into account in the design and construction of wooden buildings. If wood is put into a structure at a higher moisture content than it is expected to reach when at equilibrium with its surroundings, shrinkage must be expected, and this overall reduction in the size of each member must be taken into account during construction. On the other hand, dry wood products which will be expected to take on moisture in service must be provided with sufficient gaps between them to allow for the resultant swelling. Both wood shrinkage and swelling with the crushing which can accompany it may result in gaps which can lead to water infiltration.

9. Ventilation

Most processes associated with habitation of structures, such as kitchen, bathroom and laundryroom operations, may introduce significant amounts of water vapor into the living space of the structure which must be vented to the outside. Furthermore, moisture leaving the soil on which a wooden structure is built may condense on wood surfaces or on plumbing in contact with wood and provide sufficient water to induce decay problems. The solution to both these problems is adequate ventilation.

All activities within the living space which release large amounts of water vapor should be vented mechanically (using fans if necessary) to the outside to prevent the necessity for such water to make its way through the walls or ceiling of the structure. Substructure areas where the soil is likely to be moist should have sufficient cross ventilation supplied through or above the foundation to keep the relative humidity of the air in the subspace below the dew point at all seasons of the year. If the soil is so wet that the amount of ventilation required becomes excessive, a vapor barrier should be placed over the soil surface in the sub-area to prevent moisture from leaving the soil.

10. Hazards of Ground Contact

Soil is the most hazardous terrestrial environment in which wood can be placed. Soil is usually moist and, if so, has the capacity for supplying moisture to the wood at a perfect rate to be conducive to decay. In addition, decay fungi live in the soil on cellulosic residue such as wood scraps, roots, stumps and other plant debris. Therefore, placing wood in contact with the soil may provide both an actively growing decay fungus and the one missing factor necessary to produce decay—water. In many parts of the world, ground contact has the additional hazard of exposing wood to direct entry by subterranean termites.

11. Insulation and Vapor Barriers

As a result of the need to reduce heating cost, great emphasis is currently being placed on providing adequate insulation in structures. In some cases the addition of insulation to existing wooden structures is advocated. Some insulations, such as the closed-cell plastic foams, may also act as effective vapor barriers. Other insulation materials, like fiberglass placed on a metal foil, have vapor barriers incorporated in them to improve their performance as insulation materials. Both types of materials when added to wooden structures may be extremely dangerous if they impede the natural flow of moisture vapor out of the living area of the structure or change the location of the dew point in such a way as to cause condensation in walls or ceilings. At present many construction and design professionals are inadequately informed in this area. Such insulation practices, if not based on adequate knowledge and theory, though well-meaning in their intent, may end up costing more energy than they save in terms of replacement of wood which they have made vulnerable to rot.

See also: Wood as a Building Material

Bibliography

Eslyn W E, Clark J W 1979 *Wood Bridges: Decay Inspection and Control*, USDA Agriculture Handbook No. 557. US Government Printing Office, Washington, DC

Graham R D, Helsing G G 1979 *Wood Pole Maintenance Manual: Inspection and Supplemental Treatment of Douglas-Fir and Western Redcedar Poles*, Research Bulletin 24. Oregon State University, Corvallis, Oregon

Meyer R W, Kellogg R M (eds.) 1982 *Structural Uses of Wood in Adverse Environments*. Van Nostrand Reinhold, New York

Nicholas D D (ed.) 1973 *Wood Deterioration and its Prevention by Preservative Treatments*. Syracuse University Press, New York

Rosenberg A F, Wilcox W W 1982 How to keep your award-winning building from rotting. *Wood Fiber* 14: 70–84

Scheffer T C, Verrall A F 1973 *Principles for Protecting Wood Buildings from Decay*, USDA Forest Service Research Paper FPL 190. US Forest Products Laboratory, Madison, Wisconsin

US Department of Agriculture 1974 *Wood Handbook: Wood as an Engineering Material*, USDA Agriculture Handbook No. 72. US Government Printing Office. Washington, DC

Wilcox W W 1978 Review of literature on the effects of early stages of decay on wood strength. *Wood Fiber* 9: 252–57

W. W. Wilcox
[University of California, Berkeley, California, USA]

Wood: Deformation Under Load

The two outstanding features of the deformation behavior of wood under load are that it is highly anisotropic and, due to the biological nature of wood, very variable. Furthermore, the deformation of wood is time dependent even under normal ambient conditions. For many applications, however, the time-dependent deformation constitutes such a small fraction of the total deformation that the former may be safely neglected. Wood is then treated, to a first approximation, as a linearly elastic material. In this article wood will also be considered as a viscoelastic material.

The nature of the deformation behavior of wood can be clarified by reference to the glass transition temperature T_g of its components. Lignin and hemicellulose as wholly or largely amorphous polymers and the amorphous portion of cellulose all exhibit glass transition temperatures. In the dry state, values of T_g are about 230 °C for cellulose, 180 °C for lignin and 200 °C for hemicellulose. At room temperature, wood is therefore clearly in a glassy state. Adsorbed moisture, however, acts as a plasticizer and lowers T_g. Lignin, being less hygroscopic, is not affected as much as cellulose and hemicellulose. The T_g of the latter two falls below room temperature at the fiber saturation point, so that wet wood will have some components that are in a rubbery state.

1. Wood as a Linearly Elastic Material

Figure 1 shows that what might be considered typical stress–strain curves for wood in tension and compression. They are for wood loaded in a direction parallel to the grain, in other words, parallel to the long axis of the majority of the cells. Both curves have an initial portion which is more or less rectilinear, followed by a curvilinear portion to the point of maximum stress. The stress at the junction of the rectilinear and curvilinear portions is referred to as the fiber stress at proportional limit. The slope of the stress–strain curve below the proportional limit represents the modulus of elasticity parallel to the grain. Although it has been suggested that the modulus of elasticity is slightly higher in tension than in compression, the difference, if it exists at all, is so small that it can be neglected for all practical purposes.

Beyond the proportional limit the deformation becomes nonlinear, and the degree to which this takes place is much greater in compression than in tension. In tension the inelastic deformation is restricted to molecular mechanisms which are not well understood.

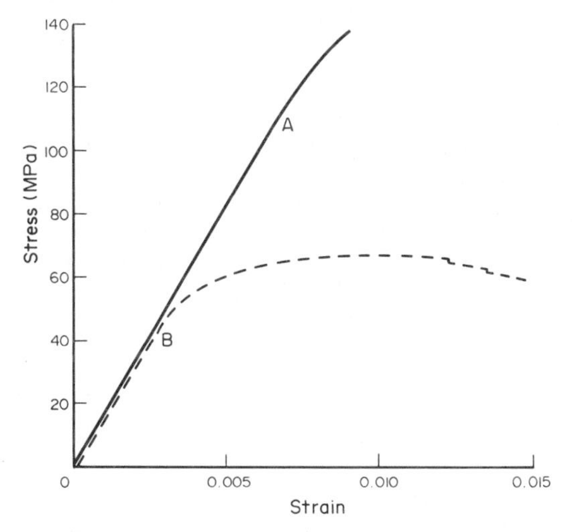

Figure 1
Stress–strain curves in tension and compression parallel to grain for air–dry samples of Bishop pine (*Pinus muricata*): ——— tension; – – – compression; A, B proportional limits in tension and compression, respectively

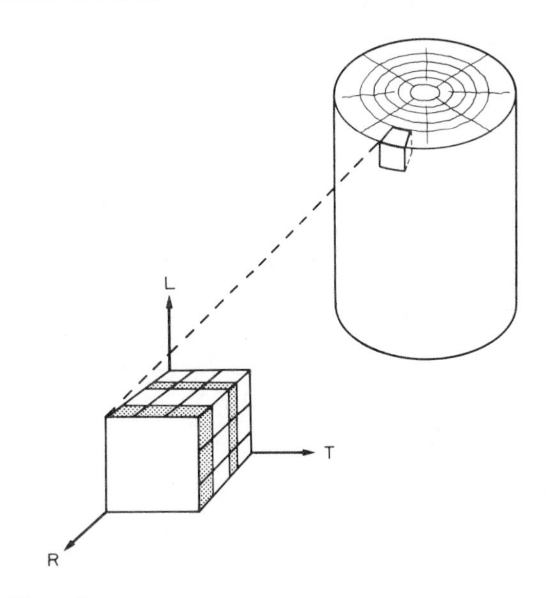

Figure 2
Wood element showing principal structural axes; longitudinal (L), radial (R) and tangential (T), together with the tree trunk section from which it was taken

Since wood is a porous material, deformation in compression can additionally take place by crinkling and folding of cell walls and localized buckling into the cell cavities. The inelastic deformation of wood is largely permanent, but wood does not exhibit plasticity in the same sense as do, for instance, metals. Figure 1 also shows that the strength is greater in tension than in compression (see *Wood: Strength*).

At any point in wood there are three mutually orthogonal planes of structural symmetry defined by three orthogonal directions, as shown in Fig. 2. The directions are longitudinal (L), which is parallel to the axis of the tree trunk and to the majority of wood cells; tangential (T), which is tangential to the annual or growth rings; and radial (R), which is orthogonal to the other two. Cylindrical anisotropy therefore seems to be the most suitable model for wood. However, since most pieces of wood are cut from positions away from the center of the tree trunk, it is customary to neglect the curvature of the growth rings and to treat wood as an orthotropic material. The matrix of elastic compliances (S_{ijkl}) for orthotropic materials is:

$$\begin{bmatrix} S_{1111} & S_{1122} & S_{1133} & 0 & 0 & 0 \\ S_{2211} & S_{2222} & S_{2233} & 0 & 0 & 0 \\ S_{3311} & S_{3322} & S_{3333} & 0 & 0 & 0 \\ 0 & 0 & 0 & 4S_{2323} & 0 & 0 \\ 0 & 0 & 0 & 0 & 4S_{1313} & 0 \\ 0 & 0 & 0 & 0 & 0 & 4S_{1212} \end{bmatrix}$$

Since the matrix is symmetric, we have nine independent elastic constants. Equating the x_1 and L directions, the x_2 and T directions and the x_3 and R directions, we can express the compliances in terms of three Young's moduli, E_L, E_R and E_T, three shear moduli, G_{LT}, G_{LR} and G_{TR} and six Poisson ratios, ν_{LT}, ν_{TL}, ν_{LR}, ν_{RL}, ν_{TR} and ν_{RT}. Thus, $S_{1111} = 1/E_L \ldots$, $S_{1122} = -\nu_{TL}/E_T \ldots$, and $4S_{1212} = 1/G_{LT} \ldots$. The first subscript in the Poisson ratios indicates the direction of applied stress and the second the direction of lateral expansion or contraction.

Some typical values for the elastic constants are shown in Table 1. Note that E_L is an order of magnitude larger than either E_T or E_R; compared with this difference, that between the two latter values, although substantial of itself, is relatively small, which explains why it is possible to neglect the curvature of the growth rings and consider wood to be orthotropic. Note also that some Poisson ratios are greater than 0.5, which is not contradictory as long as they are complemented by appropriately smaller ratios in other directions. Some of the values are less than 0.05 and very difficult to measure. They are therefore often calculated from the symmetry relations: $\nu_{LT}/E_L = \nu_{TL}/E_T$, $\nu_{LR}/E_L = \nu_{RL}/E_R$ and $\nu_{TR}/E_T = \nu_{RT}/E_R$.

The matrix of compliances given above applies only when the principal directions of structural symmetry coincide with the coordinate axes. If they do not, some or all of the zero terms in the original matrix may become nonzero in the transformed matrix, so that for instance normal stresses give rise to shear strains and, conversely, shear stresses give rise to normal strains. In lumber such an "off-angle" condition is known as cross grain, meaning that the grain or fiber direction is not parallel to the geometric long axis of the piece. This also leads to a coupling between bending and torsion, that is, a cross-grained piece of lumber tends to twist as well as bend if subjected to bending moments, or bends as well as twists if subjected to torsional moments.

2. Factors Influencing the Elastic Constants of Wood

The elastic moduli, both Young's moduli and shear moduli, are subject to various influencing factors, much as are the strength properties of wood (see *Wood: Strength*). First there is a difference between species, which can amount to an order of magnitude, as shown by balsa and birch in Table 1. A major part of the species difference can be explained on the basis of wood density, which is a measure of porosity and thus of the amount of wood substance present per unit volume. The Poisson ratios, on the other hand, are relatively unaffected by density.

Above the fiber saturation point (about 28%) moisture content has no effect on elastic properties, but as moisture content decreases below this level, the elastic moduli increase. The Young's modulus E_L parallel to the grain increases about 30% when wood initially above the fiber saturation point is dried to 12% moisture content. The Young's moduli E_T and E_R and all three shear moduli are more sensitive to moisture content than is E_L, whereas particular Poisson ratios may either increase or decrease with decreases in moisture content.

Increases in temperature result in decreases in the elastic moduli. The effect depends on moisture content, dry wood being less sensitive than wood of high moisture content. In the range from -100 to $+100\,°C$ the relationship between the elastic moduli and temperature is linear. Taking the room temperature values as a base, E_L decreases (increases) by 0.1–0.7% for each 1 °C increase (decrease) in temperature, depending on moisture content. For the two other Young's moduli and the shear moduli, the effect of temperature is more pronounced, ranging from 0.5 to 1.5% per 1 °C, again depending on moisture content. The effect of temperature on Poisson ratios is not known.

As a natural product of biological origin, even clear, straight-grained wood of the same species is variable in its properties. For the modulus of elasticity E_L as determined from bending tests, the coefficient of variation (standard deviation expressed as a percentage of the mean) is about 20%. Assuming a normal distribution, this means that about 68% of all values are within 20% of the mean.

In most practical applications it is a matter not of clear, straight-grained wood but of lumber which has a number of growth characteristics or defects that affect strength and stiffness. Since lumber is usually used in the form of linear elements, such as beams or columns,

Table 1
Elastic constants of some woods[a]

Species	Density ($kg\,m^{-3}$)	E_L (MPa)	E_T (MPa)	E_R (MPa)	G_{LT} (MPa)	G_{LR} (MPa)	G_{TR} (MPa)	ν_{LT}	ν_{TL}	ν_{LR}	ν_{RL}	ν_{TR}	ν_{RT}
Balsa	100	2440	38	114	85	124	14	0.49	0.01	0.23	0.02	0.24	0.66
Birch	620	16 300	620	1110	910	1180	190	0.43	0.02	0.49	0.03	0.38	0.78
Scots pine	550	16 300	570	1100	680	1160	66	0.51	0.02	0.42	0.04	0.31	0.68
Sitka spruce	390	11 600	500	900	720	750	39	0.47	0.02	0.37	0.03	0.25	0.43

a Hearmon (1948)

the effective modulus of elasticity E_{eff} along the axis of the piece is of much more concern than any of the other elastic constants. The two main defects of interest are cross grain and knots. Figure 3 shows the variation of E with grain angle and illustrates clearly that small amounts of cross grain (small angles of deviation) can lead to substantial reductions in E_{eff}. Knots similarly lead to reductions in stiffness. Although the knot itself may be harder and denser than the surrounding wood, the directions of its fibers are more or less perpendicular to the long axis of the piece of lumber. Furthermore, grain distortions around the knot have a detrimental effect. Reductions in E_{eff} are proportional to the severity of cross grain and the size of knots. These variations in stiffness are in turn correlated with strength, so that measurements of the axial modulus of elasticity of lumber can be used in the nondestructive evaluation of structural lumber grades.

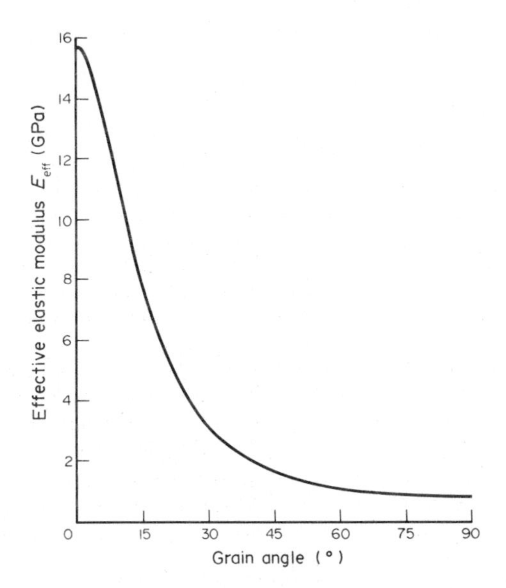

Figure 3
Effective modulus of elasticity as a function of grain angle for air-dry Douglas fir (*Pseudotsuga menziesii*) (calculated from data in Hearmon 1948)

3. *Wood as a Viscoelastic Material*

While the linearly elastic model is adequate for many applications, wood under load does in fact undergo time-dependent deformation in addition to the instantaneously elastic deformation. The additional deformation or creep occurs at a rate which decreases with time, as shown in the curves of Fig. 4. If stresses are low, the deformation can reach virtual equilibrium after a period of loading of 2–3 years. At high stresses, however, a point may be reached where an inflection appears in the creep curve and the creep rate increases again. Once the creep rate does increase again, failure is inevitable unless the load is removed. This process of delayed failure or static fatigue is known as the duration-of-load effect in wood (see *Wood: Strength*).

The creep deformation can be divided into two parts, one of which is recoverable after unloading and the other is not. They are also referred to as the delayed elastic response and the flow, respectively. Recovery of creep is the inverse process, being rapid at first and proceeding at a decreasing rate, asymptotically approaching its final value. Both forms of creep can occur at all stress levels.

Within limits, wood can be considered a linear viscoelastic material to which the Boltzmann superposition principle applies. One limiting factor is stress level, which in dry wood at room temperature generally should not exceed 60% of the short-term static strength. At elevated temperature and moisture content, nonlinear behavior may become evident at lower stress levels. Linear behavior persists at higher stress levels in tension than in compression. Another limiting factor is time, because the flow is only approximately linear for short times and occurs at a decreasing rate which virtually vanishes after a few years of loading.

Creep under constant load and steady conditions can be represented by equations of the type

$$\varepsilon_t = \varepsilon_0 + mt^n \tag{1}$$

where ε_t is the total creep strain, ε_0 is the initial strain, t is time and m and n are constants. For limited timescales it is also possible to use the equations of the Burgers model (Kelvin and Maxwell models in series):

$$\varepsilon_t = \sigma_0\{J_0 + J_\infty[1-\exp(-t/\tau)]+t/\eta\} \tag{2}$$

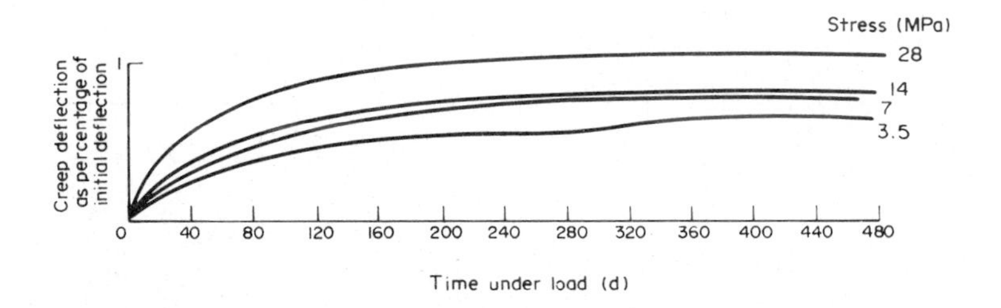

Figure 4
Creep of air-dry beams of mountain ash (*Eucalyptus regnans*) at 40 °C (after Kingston 1962)

where σ_0 is the applied stress and J_0, J_∞, τ and η are parameters of the spring and dashpot elements of the model. The three terms in braces represent the instantaneous elastic, delayed elastic and flow components, respectively. For an extended timescale, the model can be generalized by adding terms of delayed elastic response with different retardation times τ_i.

Creep deformation is often expressed as a fraction of the initial—presumably instantaneously elastic—deformation. As a general rule of thumb, fractional creep of up to 1.0 can be expected with long periods of loading under moderate conditions; that is, the additional time-dependent deformation can reach about the same order of magnitude as the initial elastic deformation. Fractional creep tends to be greater across the grain than parallel to the grain and greater in compression than in tension.

Creep in wood can be greatly increased if there is simultaneous moisture sorption, as for instance in a green beam drying under load. Even more pronounced effects are obtained if there is moisture cycling, as illustrated in Fig. 5. It should be pointed out, however, that the results in Fig. 5 are based on beams only 2 × 2 mm in cross section; in larger members the effect of humidity cycling is much reduced, since the resulting changes in moisture content tend to be confined to the surface layers.

During dynamic loading (vibrational loading), the resulting deformation of the viscoelastic material is subject to damping. In forced vibration this manifests itself by a phase difference between stress and strain, and in free vibration by a gradual decay of amplitude. In either case the damping capacity can be expressed in terms of the logarithmic decrement, which in the case of free vibration is defined as the natural logarithm of the ratio of two successive amplitudes. The value of the logarithmic decrement depends on grain direction, moisture content, temperature and frequency. For dry wood at room temperature it is in the region of 0.02–0.05.

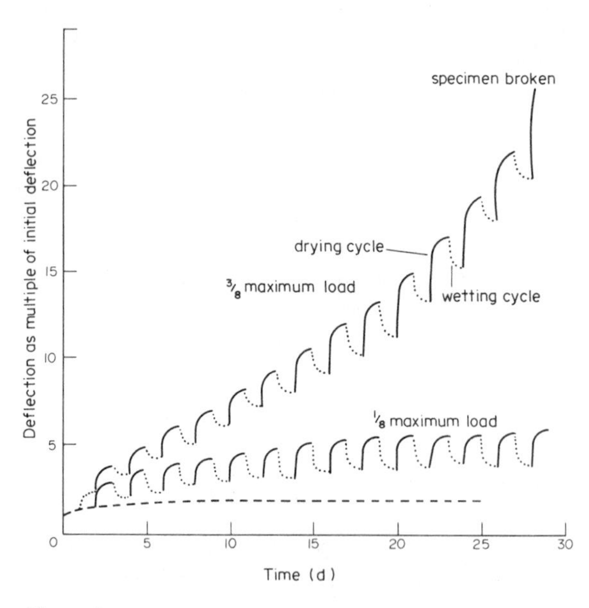

Figure 5
Creep of 2 × 2 × 60 mm beech beams: --- constant conditions ($\frac{3}{8}$ maximum load, 93% relative humidity); and alternative drying (——) and wetting (. . . .) cycles (after Hearmon and Paton 1964)

See also: Wood: An Overview

Bibliography

Back E L, Salmén N L 1982 Glass transitions of wood components hold implications for molding and pulping processes. *Tappi* 65(7): 107–10

Bodig J, Goodman J R 1973 Prediction of elastic parameters for wood. *Wood Sci.* 5: 249–64

Bodig J, Jayne B A 1982 *Mechanics of Wood and Wood Composites.* Van Nostrand Reinhold, New York

Dinwoodie J M 1981 *Timber: Its Nature and Behavior.* Van Nostrand Reinhold, New York

Hearmon R F S 1948 *The Elasticity of Wood and Plywood*, Forest Products Research Special Report No. 7. Department of Scientific and Industrial Research, London

Hearmon R F S, Paton J M 1964 Moisture content changes and creep of wood. *For. Prod. J.* 14: 357–59

Jayne B A (ed.) 1972 *Theory and Design of Wood and Fiber Composite Materials.* Syracuse University Press, New York

Kelley S S, Rials T G, Glasser W G 1987 Relaxation behavior of the amorphous components of wood. *J. Mater. Sci.* 22: 617–24

Kingston R S T 1962 Creep, relaxation and failure of wood. *Res. Appl. Ind.* 15: 164–70

Kollmann F F P, Côté W A Jr 1968 *Principles of Wood Science and Technology*, Vol. 1. Springer, Berlin

Schniewind A P 1968 Recent progress in the study of the rheology of wood. *Wood Sci. Technol.* 2: 188–206

Schniewind A P 1981 Mechanical behavior and properties of wood. In: Wangaard F F (ed.) 1981 *Wood: Its Structure and Properties*, Educational Modules for Materials Science and Engineering Project. Pennsylvania State University, University Park, Pennsylvania

Schniewind A P, Barrett J D 1972 Wood as a linear orthotropic viscoelastic material. *Wood Sci. Technol.* 6: 43–57

US Forest Products Laboratory 1987 *Wood Handbook: Wood as an Engineering Material*, USDA Handbook No. 72. US Government Printing Office, Washington, DC

A. P. Schniewind
[University of California, Berkeley, California, USA]

Wood: Strength

In the living tree, wood has several functions, one of which is structural support. The wood of the tree trunk has to withstand the compression loads from the weight of the crown above, and it has to resist bending moments resulting from wind forces acting on the entire tree. The ability of wood to resist loads (i.e., its strength) depends on a host of factors. These factors include the type of load (tension, compression, shear), its direction, and wood species. Ambient conditions of moisture content and temperature are important factors, as are past histories of load and temperature. Strength of a wood sample also depends on whether it is a small, clear piece free of defects or a piece of lumber with knots, splits and the like. This article will be concerned with the strength of clear wood and the major factors that influence it. Resistance and design values of commercial lumber are discussed elsewhere.

1. The Anisotropic Nature of Wood Strength

The cellular structure of wood and the physical organization of the cellulose-chain molecules within the cell wall make wood highly anisotropic in its strength properties. On the basis of the structural organization of wood, three orthogonal directions of symmetry can be distinguished. The longitudinal (L) direction, also referred to as the direction parallel to grain, is parallel to the cylindrical axis of the tree trunk and also to the long axis of the majority of the constituent cells. This direction has the highest proportion of primary bonds resisting applied loads, and is therefore the direction of greatest strength. The other two directions are the radial (R) and tangential (T) directions, which are perpendicular and parallel, respectively, to the circumference of the tree trunk. The latter two directions are also known collectively as directions perpendicular to grain. The strength perpendicular to grain is low because loads are resisted predominantly by secondary bonds.

The tensile strength of air-dry (~12% moisture content) softwoods parallel to grain is of the order of 70–140 MPa. For hardwoods it is often greater but may be less, depending on the particular species. The tensile strengths in the tangential and radial directions are about 3–5% and 5–8%, respectively, of the strength parallel to grain. The tensile strength perpendicular to grain of wood is thus only a small fraction of what it is parallel to grain, and in practical applications tensile stresses perpendicular to grain are avoided to the largest extent possible.

The compression strength (strength as a short column with a slenderness ratio of 11 or less) of air-dry softwoods parallel to grain is of the order of 30–60 MPa, which is much less than the tensile strength. The degree of anisotropy is also less, the compression strength perpendicular to grain being about 8–25% of the value parallel to grain. In contrast to tensile strength, which is always higher radially than tangentially, the compression strength may be either higher or lower, depending on species. Furthermore, compressive strength perpendicular to grain is often a minimum at 45° to the growth rings, that is, intermediate to the radial and tangential directions. The degree of anisotropy in compression depends on species, and in general is less in species of higher density. This is because compression failure of wood as a porous material occurs principally by instability at several levels. In compression parallel to grain, failure of cellulose microfibrils by folding can occur at stresses as low as one half of the ultimate failure stress. At higher stresses, a similar pattern of folding takes place on the cell-wall level, and these eventually aggregate into massive compression failures. In compression perpendicular to grain, failure occurs by collapse and flattening of the cells. Species with higher density have less porosity and thicker cell walls, which makes the cells much more resistant to collapse.

The shear strength of wood is characterized by six principal modes of shear failure, as illustrated in Fig. 1. The shear plane or failure plane may be in one of the three principal planes of wood (the LT, LR and TR planes), and failure in each shear plane may occur by sliding in one of two principal directions. These six modes may be divided into three groups: shear perpendicular to grain, shear parallel to grain, and rolling shear. In shear parallel to grain, both the shear plane and the sliding direction are parallel to grain. In air-dry softwoods, the shear strength parallel to grain is of the order of 5–12 MPa. In shear perpendicular to grain, where the shear plane and the sliding direction

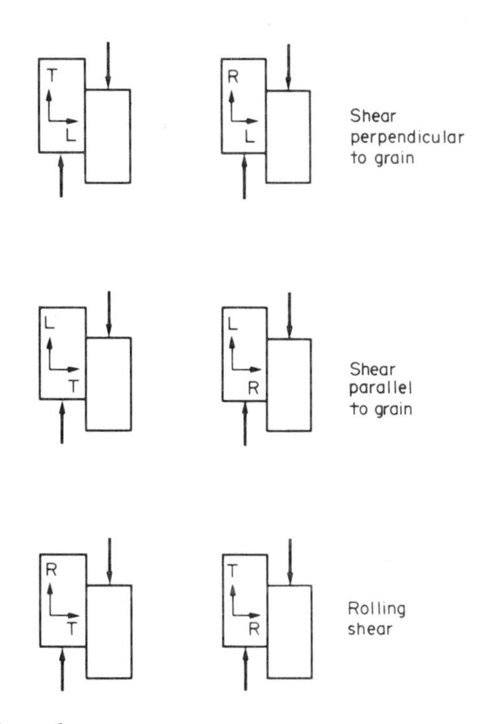

Figure 1
Schematic representation of the six principal modes of shear failure in wood, divided into three groups as shown

are both perpendicular to grain, wood is most resistant, but because of its high degree of anisotropy often fails first in some other mode such as compression perpendicular to grain. Limited data indicate that the shear strength perpendicular to grain is of the order of 2.5–3 times the shear strength parallel to grain. In rolling shear, so-called because the wood fibers can be thought of as rolling over each other as failure occurs in this mode, the strength is least, amounting to 10–30% of the shear strength parallel to grain. There are no consistent or major differences between modes within each of the three groups of shear modes.

Certain problems relating to the strength of wood members can be dealt with by using linear elastic fracture mechanics. The required material parameters, the values of the critical stress intensity factor K_c, also vary because of the anisotropic nature of wood. In the opening mode of crack loading (mode I), there are six principal systems which are somewhat analogous to the six modes of shear failure. The crack plane may be located in one of three principal planes of wood structure, and in each of these the crack may propagate in one of two principal directions. Similarly, six systems are found in the other two modes of crack loading, the forward shear mode (mode II) and the transverse shear mode (mode III), and thus 18 fracture parameters are required to characterize wood completely. Since mode I is the most important, most of the available data are for K_{Ic}. Particular systems are denoted by two subscripts, the first representing the direction normal to the crack plane, and the second the direction of crack propagation. The RT, TR, RL and TL systems (i.e., the four systems where the crack plane is parallel to grain) have very similar K_{Ic} values. Those for the remaining systems, namely LT and LR, where the crack plane is perpendicular to grain, are again similar to each other but almost an order of magnitude greater than the other four. Thus, "weak" and "strong" systems can be distinguished. K_{Ic} values in the weak systems of air-dry softwoods are of the order of 150–500 kPa m$^{1/2}$, and those in the strong systems are 6–7 times greater. Weak system K_{IIc} values are of the order of 1000–2500 kPa m$^{1/2}$.

2. *Methods of Determining Wood Strength*

Many countries have established national standards for testing the mechanical properties of wood. By and large, the objectives and methods are similar. In the USA the methods are given in ASTM Designation D143–83, *Standard Methods of Testing Small Clear Specimens of Timber*. This standard has provisions both for sampling and for the tests themselves. The strength tests included are static bending, two types of impact bending, compression parallel and perpendicular to grain, tension parallel and perpendicular to grain, shear parallel to grain, hardness, and cleavage. All tests must be made at controlled (room) temperature and moisture content, which is usually either green (moisture content above the fiber saturation point) or 12%, and at prescribed speeds of testing. Test specimens are cut from sticks 50 × 50 mm in cross section with the grain parallel to the long dimension. European systems are usually based on sticks 20 × 20 mm in cross section.

The static-bending test is the most important single test. The test specimen is 50 × 50 × 760 mm and is center loaded as a simply supported beam over a span of 710 mm. Load-deflection data are taken so that an elastic modulus of bending may be calculated. The maximum bending moment is used to calculate the modulus of rupture, which is the presumed stress at the extreme fiber of the beam assuming that stress varies linearly through the depth of the beam. Actually, since the compression strength of wood is much less than its tensile strength, there is initial yielding on the compression side, followed by development of visible compression failures and enlargement of the compression zone. The neutral surface shifts toward the tension side as tensile stresses continue to increase. Maximum moment is reached when there is failure in tension. The modulus of rupture is the characteristic bending strength of wood, and its value is intermediate between the tensile and compressive strengths.

The dimensions of compression parallel to grain specimens are 50 × 50 × 200 mm. The maximum load is used to calculate the stress at failure, which is usually referred to as maximum crushing strength. For compression perpendicular to grain specimens, the dimensions are 50 × 50 × 150 mm. The specimen is laid flat, and the load applied through a metal loading block to the middle third of its upper surface (Fig. 2d). This is intended to simulate the type of loading found in a stud or column bearing on a wood sill, so that the edges of the area under load are laterally supported by the adjacent unloaded area. In this test, a clearly defined maximum load cannot usually be obtained and only a fiber stress at proportional limit is computed.

Shear tests are generally made only parallel to grain. The specimen is 50 × 50 × 63 mm, with a 13 × 20 mm notch (Fig. 2c). The notch is cut so that the failure plane is the LT or LR plane in alternate specimens, and the results are averaged.

The four tests described so far are the most significant tests, since allowable stress values for visually graded structural lumber are based on their results. Data obtained from the remaining tests are useful for special purposes and for comparing relative strength of different species for particular applications. This includes tensile tests, because allowable stresses in tension parallel to grain are based on modulus of rupture data and there are no allowable stresses in tension perpendicular to grain. Tensile tests parallel to grain are difficult to make because of the high degree of anisotropy. Specimens are 460 mm long, have a

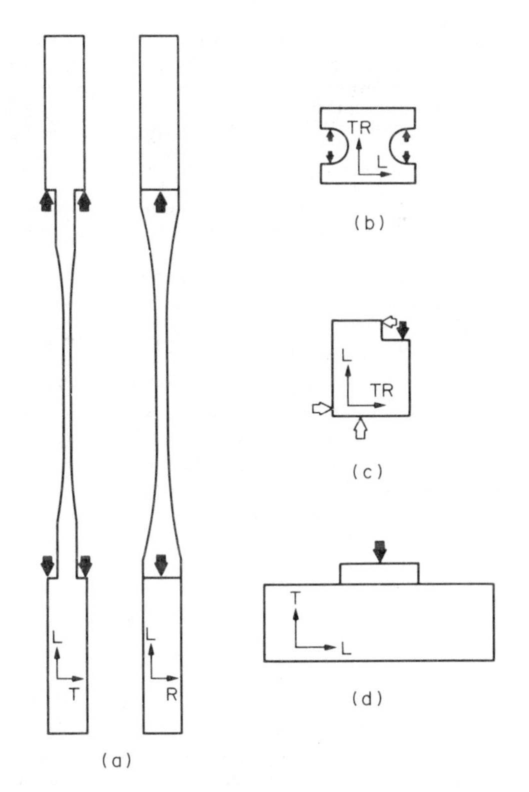

Figure 2
Standard test specimens for testing clear wood: (a) tension parallel to grain; (b) tension perpendicular to grain; (c) shear parallel to grain; and (d) compression perpendicular to grain. Application of loads is indicated schematically

25×25 mm cross section at the ends, and are necked down to a minimum cross section of 4.8×9.5 mm (Fig. 2a). Load is applied to the shoulders of notches introduced near the ends. Load transfer from machine to specimen is therefore through shear over an area of 50 cm^2 at each end. Data for tensile strength parallel to grain are not available for most species and are not usually included in tables of strength data. Tensile tests perpendicular to grain are made on specimens with circular notches at each end, and loads are applied (through suitable fixtures) to the inside surface of the notches (Fig. 2b). Because of the geometry of the specimen, the tensile stresses perpendicular to grain are distributed very unevenly over the minimum cross section, so that the tensile strength is underestimated by about one-third.

The two types of impact bending tests are impact bending and toughness. The impact bending test involves dropping a hammer from successively greater heights, the maximum height to cause failure being recorded. This test is now considered to be obsolete. The toughness test is a single-blow impact test using a pendulum to break a $20 \times 20 \times 280$ mm specimen by center-loading over a 240 mm span. The energy required to break the specimen is recorded. Toughness test data are not available for more than a few species.

The cleavage test, which is intended to measure the splitting resistance of wood, is made with a specimen which resembles the tension perpendicular to grain specimen. It has the same circular notch, but only at one end, and is 95 mm long overall. The load per unit of width, as applied at the notch, which causes splitting is recorded.

Hardness in wood is measured by embedding a spherical indentor of 11.3 mm diameter to a depth of 5.6 mm, and recording the load required to do so. Indentations are made on all three principal wood planes. Values obtained on the RT plane are referred to as end hardness and those on the LT and LR planes are known collectively as side hardness. Strength values for a few species as obtained from the various tests described are shown in Table 1.

3. *Factors Affecting Wood Strength*

Various factors can strongly affect the strength of wood. The compression strength parallel to grain, for example, can more than double when drying wood from the green condition to 12% moisture content. The major factors can be divided into three groups as shown below.

3.1 Factors Related to Wood Structure

Although the nature and composition of wood substance is basically the same for all wood, the types of cells, their proportions and arrangements differ greatly form one species to the next. Balsa, a well-known lightweight wood, has an air-dry modulus of rupture of 19.3 MPa, as compared with values of over 200 MPa for a few African and South American tropical species.

Since wood is a biological material, it is subject to environmental and genetic factors that influence its formation. As a result, wood strength is very variable, not only from one species to the next, but also within a species. It will even vary depending on location within a single tree stem. Within a species, values of wood strength follow approximately a normal distribution. The variability of a particular species, as measured by the standard deviation, is generally proportional to the mean value. Therefore, it is convenient to express the standard deviation as a percentage of the mean, which is referred to as the coefficient of variation. Some strength properties tend to be more variable than others, but for most the coefficient of variation is of the order of 20%. This means that 95% of the values will fall in the range from about 60–140% of the mean value.

Much of the variation between and within species can be attributed to differences in wood density. The density of wood substance is constant at 1500 kg m^{-3},

Table 1
Strength data for four wood species at 12% moisture content

Property	Eastern white pine	Douglas fir	White oak (overcup oak)	Yellow poplar
Static bending: modulus of rupture (MPa)	59	85	87	70
Compression parallel to grain: maximum crushing strength (MPa)	33	50	43	38
Compression perpendicular to grain: fiber stress at proportional limit (MPa)	3.0	5.5	5.6	3.4
Shear parallel to grain: maximum shear strength (MPa)	6.2	7.8	13.8	8.2
Tension parallel to grain: maximum tensile strength (MPa)	78	130	101	154
Tension perpendicular to grain: maximum tensile strength (MPa)	2.1	2.3	6.5	3.7
Impact bending: height of drop causing complete failure (m)	0.46	0.79	0.97	0.61
Toughness: work to complete failure (J)	13	32	37	24
Cleavage: splitting force per unit width (kN m^{-1})	28	32		49
Hardness: force to cause 5.6 mm indentation (kN)				
side grain	1.7	3.2	5.3	2.4
end grain	2.1	4.0	6.3	3.0

so that wood density is in effect a measure of porosity. As a result, there is a moderate to high degree of correlation between wood density and strength. It can be expressed as

$$S = k\rho^n$$

where S is a strength property, ρ is the wood density, and k and n are constants depending on the particular property. Values of n range from 1.00 to 2.25. If n is larger than unity, it suggests that as density increases there is not only an increase in the amount of wood substance (which would imply $n=1$) but qualitative changes in wood structure as well.

Grain direction has a major effect on strength as already discussed in detail. In structural lumber, which is used as an essentially linear load-bearing element (tension or compression member, or beam), it is desirable that the direction of greatest strength—the direction parallel to grain—should coincide with the long geometric axis. However, in practice the grain direction is often at an angle to the edge of the piece, a condition which is referred to as cross grain. In such a case, uniaxial stress along the geometric axis will have components both parallel and perpendicular to grain. The effective strength could be predicted if a suitable criterion for failure under combined stresses were available. For wood, such a general theory has not yet been found. An empirical relation which is often used is the Hankinson equation:

$$N = PQ/(P\sin^2\theta + Q\cos^2\theta)$$

where N is the strength at an angle θ to the grain, and P and Q are the strengths parallel and perpendicular to the grain, respectively. The variation of strength with grain angle is similar to the variation in Young's modulus (see *Wood: Deformation Under Load*), so that even small angles can have a major effect on strength.

In addition to cross-grain, the major strength-reducing characteristics of structural lumber are knots. They are portions of branches which have become part of the stem through normal growth. Although generally harder and denser than the surrounding wood, their grain is more or less perpendicular, so that they contribute little to the strength parallel to the geometric axis of the piece. An additional factor is that the grain of the wood around the knot becomes distorted, representing localized cross-grain which also reduces strength.

3.2 Factors Related to the Environment

Equilibrium moisture content of wood is a function of ambient conditions of relative humidity and temperature. Above the fiber saturation point moisture content has no effect on wood strength, but below it strength increases upon drying for most properties. A notable exception is toughness, which does not change much and may even decrease upon drying. The amount of increase depends on the particular property. Shear strength parallel to grain increases by about 3% for each 1% decrease in moisture content below the fiber saturation point; for maximum crushing strength in compression parallel to grain the increase is 6% per 1% moisture-content decrease. Most tabulated data give strength values when green (S_g) and at 12% moisture content (S_{12}). The strength (S_m) at moisture content M, in the range from 8% moisture content to the fiber saturation point (for

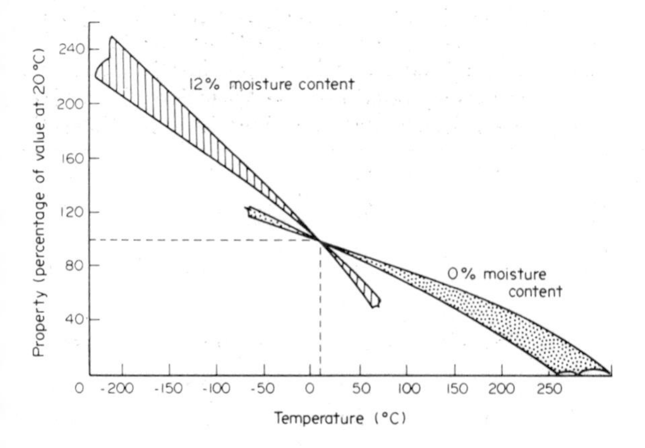

Figure 3
Immediate effect of temperature on bending strength, tensile strength perpendicular to grain, and compression strength parallel to grain. Variability is indicated by the width of the bands (adapted from *Wood Handbook*)

strength adjustments commonly assumed to be 25%), can be estimated by the equation

$$S_m = S_{12}(S_g/S_{12})^{[(M-12)/13]}$$

Below 8% moisture content this exponential relation no longer holds as the rate of strength increase becomes less, and in the case of some properties there may even be a maximum at 6 or 7% moisture content. For example, tensile strength both parallel and perpendicular to grain decreases somewhat as the moisture content is decreased below this limit. Since the moisture content of wood in use is rarely much below 6%, this is of little practical significance.

Temperature has both immediate, reversible effects on wood strength and time-dependent effects in the form of thermal degradation. One immediate effect is that strength decreases as temperature is increased. There is an interaction with moisture content, because dry wood is much less sensitive to temperature than green wood. Some properties, such as maximum crushing strength in compression parallel to grain and modulus of rupture, are more sensitive to temperature than others, such as tensile strength parallel to grain. As a general rule, the effect amounts to a decrease in strength of 0.5–1% for each 1 °C increase in temperature. This is applicable in the temperature range of approximately −20 to 65 °C. Temperature effects over a greater range for several properties are illustrated in Fig. 3.

3.3 Agents Causing Deterioration

Deterioration or degradation of wood can be caused by biological agents, by chemicals, and by certain forms of energy. The principal biological agents of concern are decay fungi and wood-destroying insects, although damage may also be sustained because of bacteria, marine borers, and woodpeckers (see *Wood: Decay During Use*). Damage by wood-destroying insects is by removal of wood substance, so that the effect on strength could be estimated by an assessment of the volume of material removed. In the case of wood decay, such estimates are more difficult because in the early stages of decay wood may appear sound but may have already suffered significant strength loss. To date, there are no satisfactory methods for estimating the residual strength of partially decayed wood members.

Chemicals which do not swell wood may have little or no effect on wood strength. Swelling chemicals, such as alcohols and some other organic solvents, may affect strength in the same way that water does, in proportion to the amount of swelling. The effect of such chemicals disappears if the chemicals are removed from the wood. Other chemicals may cause degradation of wood by such mechanisms as acid hydrolysis of cellulose. In general, wood is relatively resistant to mild acids but is degraded by strong acids or alkali. Heartwood tends to be more resistant to chemical degradation than sapwood, and softwoods are generally more resistant than hardwoods. Some chemicals used for preservative treatments do affect strength, and also the temperatures used in the treating process can have a permanent effect. Fire-retardant treatments usually result in some loss of strength. Common practice is to take a reduction of 10% in modulus of rupture for design purposes, but available data indicate that the effect on bending strength may be greater.

Wood under sustained loads is subject to static fatigue, commonly referred to as the duration of load factor. Available data show that for clear wood load capacity is linearly related to the logarithm of load duration. Load capacity decreases by 7–8% for each decade of increase on the logarithmic time scale. In structural lumber, the duration of load factor may be less pronounced, but current practice is to make adjustments for duration of loading in design. Dynamic fatigue in wood has not been studied in great detail. Results of tests on clear wood have shown that for fully reversed loading the fatigue strength after 30×10^6 cycles is about 30% of the static strength.

Wood can be degraded with large doses of nuclear radiation. Doses of γ rays in excess of 1 Mrad will cause a measurable effect on tensile strength; at a dose of 300 Mrad the residual tensile strength will be about 10%.

A more common form of degradation is by heat. The amount of degradation is a function not only of temperature but also time of exposure. Measurable effects on strength can be observed at temperatures as low as 65 °C. Degradation is more severe if the heating medium supplies water than if it does not. Softwoods are somewhat more resistant to thermal degradation than hardwoods. As in the case of other forms of degradation, including decay, toughness or shock resistance is affected first, followed by modulus of rupture, while stiffness is affected least. For softwood heated in water, a 10% reduction in modulus of rupture takes 280 days at 65 °C. At 93 °C the same reduction takes 8 days, and at 150 °C it takes only about 3 hours.

See also: Wood: An Overview

Bibliography

American Society for Testing and Materials 1987 *Annual Book of ASTM Standards*, Vol. 4.09: *Wood; Adhesives.* American Society for Testing and Materials, Philadelphia, Pennsylvania

Bodig J, Jayne B A 1982 *Mechanics of Wood and Wood Composites.* Van Nostrand Reinhold, New York

Dinwoodie J M 1981 *Timber, Its Nature and Behavior.* Van Nostrand Reinhold, New York

Forest Products Laboratory 1987 *Wood Handbook; Wood as an Engineering Material.* US Government Printing Office, Washington, DC

Gerhards C C 1982 Effect of moisture content and temperature on the mechanical properties of wood: An analysis of immediate effects. *Wood Fiber* 14: 4–36

Kollmann F F P, Côté W A Jr 1968 *Principles of Wood Science and Technology*, Vol. 1, *Solid Wood.* Springer, Berlin

Patton-Mallory M, Cramer S M 1987 Fracture mechanics: A tool for predicting wood component strength. *For. Prod. J.* 37(7/8): 39–47

Schniewind A P 1981 Mechanical behavior and properties of wood. In: Wangaard F F (ed.) 1981 *Wood: Its Structure and Properties.* Pennsylvania State University, University Park, Pennsylvania

A. P. Schniewind
[University of California, Berkeley, California, USA]

LIST OF CONTRIBUTORS

Contributors are listed in alphabetical order, together with their addresses. Titles of articles which they have authored follow in alphabetical order. Where articles are co-authored, this has been indicated by an asterisk preceding the article title.

Adams, R. D.
Department of Mechanical
Engineering
Queen's Building
University of Bristol
Bristol BS8 1TR
UK
Adhesives in Civil Engineering

Adaniya, T.
Steel Research Center
Technical Research Center
Nippon Kokan KK
1-1 Minamiwatarida-cho, Kawasaki-ku
Kawasaki-shi, Kanagawa-ken 210
JAPAN
Electrogalvanized Steel

Alia, B. L.
American Bureau of Shipping
65 Broadway
New York, NY 10006
USA
Offshore Structure Materials

Allen, S. M.
Department of Materials Science
& Engineering
Room 13-5056
Massachusetts Institute of Technology
Cambridge, MA 02139
USA
**Properties of Materials*
**Structure of Materials*
**Structure–Property Relations of Materials*

Appoo, P. M.
26 Myrtle Street
Prospect NSW 2149
AUSTRALIA
Selection Systems for Engineering Materials: Quantitative Techniques

Armstrong, D. W.
Pilkington Flat Glass Ltd
Prescott Road
St Helen's
Merseyside WA10 3TT
UK
Glass as a Building Material

Arridge, R. G. C.
H H Wills Physics Laboratory
University of Bristol
Royal Fort
Tyndall Avenue
Bristol BS8 1TL
UK
Particulate Composites

Ashby, M. F.
Department of Engineering
University of Cambridge
Trumpington Street
Cambridge CB2 1PZ
UK
Cork

Ault, C. H.
Indiana Geological Survey
611 N Walnut Grove Avenue
Bloomington, IN 47405
USA
Construction Materials: Crushed Stone

Ayres, R. U.
Department of Engineering & Public Policy
Carnegie Mellon University
Pittsburgh, PA 15213
USA
Materials and the Environment

Banov, A.
American Paint Company
Room 813
370 Lexington Avenue
New York, NY 100017
USA
Paints, Varnishes and Lacquers

Behrman, J. R.
McNeil Building, 160/CR
Department of Economics
University of Pennsylvania
3718 Locust Walk
Philadelphia, PA 19104
USA
Prices of Materials: Stabilization

Bell, J. P.
Institute of Materials Science

The University
Storrs, CT 06268
USA
Epoxy Resins

Bever, M. B.
Department of Materials Science & Engineering
Room 13-5026
Massachusetts Institute of Technology
Cambridge, MA 02139
USA
**Materials Economics, Policy and Management*
**Properties of Materials*
**Structure of Materials*
**Structure–Property Relations of Materials*

Birchall, J. D.
Solid State Science
ICI Advanced Materials
PO Box 11
The Heath
Runcorn
Cheshire WA7 4QE
UK
New Cement-Based Materials

Birley, A. W.
Institute of Polymer Technology
Loughborough University of Technology
Loughborough
Leicestershire LE11 3TU
UK
**Failure Analysis of Plastics*

Blodgett, O. W.
Design Consultant
Lincoln Electric Company
22801 St Clair Avenue
Cleveland, OH 44124
USA
Welding of Bridges
Welding of Buildings

Brach, A. M.
Department of Civil Engineering
Center for Construction Research & Education
Room 1-241
Massachusetts Institute of Technology
Cambridge, MA 02139
USA
**Ceramics as Construction Materials*

Brook, R. J.
Institut für Werkstoffwissenschaften
Max-Planck-Institut für Metallforschung
Heisenbergstrasse 5
D-7000 Stuttgart 80
FRG
Advanced Ceramics: An Overview

Brownell, W. E.
NYS College of Ceramics
Alfred University
Alfred, NY 14802
USA
Structural Clay Products

Burke, J. E.
33 Forest Road
Burnt Hills, NY 12027
USA
Materials and Innovation

Butler, B. L.
Solar Thermal and Materials Research
Solar Energy Research Institute
1617 Cole Boulevard 15/3
Golden, CO 80401
USA
Solar Thermal Materials

Carr, D. D.
Indiana Geological Survey
611 N Walnut Grove Avenue
Bloomington, IN 47405
USA
Construction Materials: Industrial Minerals

Castle, J. E.
Department of Metallurgy & Materials Technology
University of Surrey
Guildford
Surrey GU2 5XH
UK
Corrosion and Oxidation Study Techniques

Charnock, H.
R & D Laboratories
Pilkington Brothers plc
Lathom
Ormskirk
Lancashire L40 5UF
UK
Float Glass Process

Corneliussen, R. D.
College of Engineering
Drexel University
32nd & Chestnut Streets
Philadelphia, PA 19104
USA
**Polymers: Mechanical Properties*

Crandall, R. W.
Brookings Institution
1775 Massachusetts Avenue, NW
Washington, DC 20036
USA

Materials Economics, Policy and Management

Dieter, G. E.
College of Engineering,
Building 094
University of Maryland
College Park, MD 20742
USA
Design and Materials Selection and Processing
Metals Processing and Fabrication: An Overview

Dietz, A. G. H.
19 Cambridge Street
Winchester, MA 01890
USA
Plastics and Composites as Building Materials

Dluhosch, E.
Room E21-204
Massachusetts Institute of Technology
Cambridge, MA 02139
USA
Building Systems and Materials

Doremus, R. H.
Materials Science Department
Rensselaer Polytechnic Institute
Troy, NY 12181
USA
Fracture and Fatigue of Glass
Glass: An Overview

Drucker, D. C.
Department of Engineering Sciences
College of Engineering
University of Florida
231 Aerospace Engineering Building
Gainesville, FL 32611
USA
Time-Independent Plasticity

Eckelman, C. A.
Department of Forestry & Natural Resources
Purdue University
West Lafayette, IN 47907
USA
Glued Joints in Wood

Ellis, T. S.
Polymers Department
General Motors Research Laboratories
General Motors Technical Center
Warren, MI 48090
USA
**Polymers: Structure, Properties and Structure–Property Relations*

Evans, A. G.
Department of Materials Science & Engineering
Hearst Mining Building, Room 264
University of California
Berkeley, CA 94720
USA
Brittle Fracture: Micromechanics
**Failure Analysis of Ceramics*
Mechanics of Materials: An Overview

Fraser, G. S.
Indiana Geological Survey
611 N Walnut Grove Avenue
Bloomington, IN 47405
USA
Construction Materials: Sand and Gravel

Freas, A. D.
Professional Engineer
2618 Park Place
Madison, WI 53705
USA
Wood as a Building Material

Frederick, T.
Drexel University
32nd & Chestnut Streets
Philadelphia, PA 19104
USA
**Polymers: Mechanical Properties*

Frohnsdorff, G. J.
National Bureau of Standards
Center for Building Technology
Building 226, Room B348
Washington, DC 20234
USA
**Calcium Aluminate Cements*
Cements, Speciality
Portland Cements, Blended Cements and Mortars

Gent, A. N.
Institute of Polymer Science
University of Akron
Akron, OH 44325
USA
**Bond Strength: Peel*

Gibson, L. J.
Department of Civil Engineering
Center for Construction Research & Education
Room 1-232
Massachusetts Institute of Technology
Cambridge, MA 02139
USA
Cellular Materials in Construction

Goldstein, I. S.
Department of Wood & Paper Science
North Carolina State University
PO Box 5488

Raleigh, NC 27650
USA
Chemically Modified Wood

Gordon, R. J.
Drexel University
32nd & Chestnut Streets
Philadelphia, PA 19104
USA
**Polymers: Mechanical Properties*

Hamed, G. R.
University of Akron
Akron, OH 44325
USA
**Bond Strength: Peel*

Hannant, D. J.
Department of Civil Engineering
University of Surrey
Guildford
Surrey GU2 5XH
UK
Fiber-Reinforced Cements

Hawkins, W. L.
26 High Street
Montclair, NJ 07042
USA
Polymers: Technical Properties and Applications

Hemsley, D. A.
Institute of Polymer Technology
Loughborough University of Technology
Loughborough LE11 3TU
UK
**Failure Analysis of Plastics*

Hindersinn, R. R.
Occidental Chemical Corporation
Grand Island Complex
MPO Box 8
Niagara Falls, NY 14302
USA
Thermosetting Resins

Holtz, R. D.
School of Civil Engineering
Purdue University
West Lafayette, IN 47907
USA
Construction Materials: Fill and Soil

Hooper, R. A. E.
Arthur-Lee & Sons
Meadow Hall
PO Box 54
Sheffield S9 1HU
UK
Corrosion of Iron and Steel

Johnson, A. F.
Division of Materials Applications
Building 95
National Physical Laboratory
Teddington
Middlesex TW11 0LW
UK
Glass-Reinforced Plastics: Thermosetting Resins

Johnson, T. E.
Department of Materials Science & Engineering
Massachusetts Institute of Technology
Cambridge, MA 02139
USA
Passive Solar Energy Materials

Kalyoncu, R. S.
Tuscaloosa Research Center
US Bureau of Mines
Box L
Tuscaloosa, AL 34586
USA
Construction Materials: Granules

Kaplan, R. S.
US Bureau of Mines
2401 East Street NW
Washington, DC 20241
USA
**Recycling of Metals: Technology*

Karasz, F. E.
Materials Research Laboratory
Room 701, Graduate Research Laboratory
University of Massachusetts
Amsherst, MA 01003
USA
**Polymers: Structure, Properties and Structure–Property Relations*

Kelly, A.
Vice-Chancellor
University of Surrey
Guildford
Surrey GU2 5XH
Composite Materials: An Overview

Klosowski, J. M.
Dow Corning Corporation
Construction Specialist Division
3901 South Saginaw
Midland, MI 48640
USA
Sealants as Building Materials
**Structural Sealant Glazing*

Kopanda, J. E.
National Bureau of Standards
Center for Building Technology
Building 226, Room B348
Washington, DC 20234
USA
**Calcium Aluminate Cements*

Kramer, E. J.
Department of Materials Science
Cornell University
Ithaca, NY 14853
USA
Crazing and Cracking of Polymers

Kruger, J.
Department of Materials Science
Johns Hopkins University
Baltimore, MD 21218
USA
Corrosion of Metals: An Overview

Labys, W. C.
Department of Mineral & Energy Resource Economics
213 White Hall-Comer
West Virginia University
Morgantown, West VA 26506
USA
International Trade in Materials

Lawn, B. R.
Inorganic Materials Division
Mechanical Properties Group
US Department of Commerce
National Bureau of Standards
Washington, DC 20234
USA
Brittle Fracture: Linear Elastic Fracture Mechanics

Leidheiser Jr, H.
Director, Corrosion Laboratory
Center for Surface & Coating Protection
Francis Macdonald Sinclair Memorial Lab
Lehigh University
Bethlehem, PA 18015
USA
Corrosion Protective Coatings for Metals

Lifshin, E.
General Electric Company
Corporate Research and Development
Chemical and Structural Analysis Branch
PO Box 8
Schenectady, NY 12301
USA
Investigation and Characterization of Materials

Markow, M. J.
Department of Civil Engineering
Center for Construction Research & Education
Room 1-250
Massachusetts Institute of Technology
Cambridge, MA 02139
USA
**Ceramics as Construction Materials*

Mason, B. H.
The France Stone Company
PO Box 49
Waterville, OH 43566
USA
Construction Materials: Lightweight Aggregate

Masubuchi, K.
Department of Ocean Engineering
Room 5-219
Massachusetts Institute of Technology
Cambridge, MA 02139
USA
Welding: An Overview

Mathur, G. C.
National Building Organization
"G" Wing
Nirman Bhawan
New Delhi
INDIA
Indigenous Building Materials

May, C. A.
Manufacturing Research
Lockheed Missiles & Space Co
PO Box 504, B/170, 0/86-54
Sunnyvale, CA 94086
USA
Structural Adhesives

McClintock, F. A.
Mechanical Engineering Department
Center for Construction Research & Education
Massachusetts Institute of Technology
Cambridge, MA 02139
USA
Creep
Elasticity
Linear Viscoelasticity
Plasticity
Small Strain, Stress and Work

McKinley, R. W.
RFD 1
Box 742
Hancock, NH 03449
USA
Glazing

Mills, P. J.
Department of Materials Science & Engineering

University of Surrey
Guildford
Surrey GU2 5XH
UK
*Carbon-Fiber-Reinforced Plastics

Morganroth, D. E.
27880 White Road
Perrysburg, OH 43551
USA
Energy Conservation: Insulation Materials

Murray, H. H.
Department of Geology
Indiana University
1005 E 10th Street
Bloomington, IN 47405
USA
Binders: Clay Minerals

Nabarro, F. R. N.
Deputy Vice-Chancellor's Office
University of the Witwatersrand
1 Jan Smuts Avenue
Johannesburg 2001
SOUTH AFRICA
Dislocations

Ness, H.
National Association of Recycling Industries Inc
330 Madison Avenue
New York, NY 10017
USA
*Recycling of Metals: Technology

Newcomb, R.
Department of Mining & Geological Engineering
College of Mines and Engineering
University of Arizona
Tucson, AZ 85721
USA
Economics of Materials, Energy and Growth

O'Halloran, M. R.
American Plywood Association
7011 South 19th Street
Tacoma, WA 98411
USA
Plywood

Patton, J. B.
[Deceased; late of Indiana Geological Survey, Bloomington, IN, USA]
Construction Materials: Dimension Stone

Pelloux, R.
Department of Materials Science & Engineering
Room 8-237
Massachusetts Institute of Technology
Cambridge, MA 02139
USA
Failure Analysis: General Procedures and Applications to Metals

Pickering, F. B.
Department of Metallurgy
Sheffield City Polytechnic
Pond Street
Sheffield S1 1WB
UK
Steels: Classification

Potter, R. T.
Materials & Structures Department
Royal Aerospace Establishment
Farnborough
Hampshire GU14 6TD
UK
Composite Materials: Strength

Reemsnyder, H. S.
Principal Scientist
Product Research
Bethlehem Steel Corporation
Bethlehem, PA 18016
USA
Steel for Construction

Richerson, D. W.
Garrett Turbine Engine Co.
111 South 34th St
PO Box 527
Phoenix, AZ 85010
USA
*Failure Analysis of Ceramics

Ritchie, R. O.
Department of Materials Science & Mineral Engineering
Hearst Mining Building
University of California
Berkeley, CA 94720
USA
Ductile Fracture: Micromechanics
Fatigue Crack Growth: Macroscopic Aspects
Fatigue Crack Growth: Mechanistic Aspects

St Pierre, G. R.
2041 North College Road
The Ohio State University
Columbus, OH 43210
USA
Iron Production

Schlabach, T. D.
Bell Laboratories
600 Mountain Avenue

Murray Hill, NJ 07974
USA
Materials Conservation: Technology
Substitution: Technology

Schmidt, C. M.
Dow Corning Corporation
Sealant Division
3901 South Saginaw Road
Midland, MI 48640
USA
**Structural Sealant Glazing*

Schniewind, A. P.
Forest Products Laboratory
University of California
1301 South 46th Street
Richmond, CA 94804
USA
Wood: An Overview
Wood: Deformation Under Load
Wood: Strength

Schodek, D. L.
Department of Architecture
Room 201
Harvard School of Design
George Gund Hall
48 Quincy Street
Cambridge, MA 02138
USA
Architectural Design and Materials

Seldon, M. R.
Manager
Life Cycle Costing
Pomona Division
General Dynamics Corporation
PO Box 2507
Pomona, CA 91766
USA
Life Cycle Costing

Sunder, S.
Department of Civil Engineering
Center for Construction Research & Education
Room 1-274
Massachusetts Institute of Technology
Cambridge, MA 02139
USA
Ice and Frozen Earth as Construction Materials

Simatupang, M. H.
Institut für Holzchemie und Chemische Technologie des Holzes
Leuschnerstrasse 9LB
205 Hamburg 80
FRG
Wood Cement Boards

Smith, P. A.
Department of Materials Science & Engineering
University of Surrey
Guildford
Surrey GU2 5XH
UK
**Carbon-Fiber-Reinforced Plastics*

Smith, V. K.
Centennial Professor of Economics
Department of Economics & Business Administration
Vanderbilt University
Nashville, TN 37235
USA
Materials Conservation: Economics

Smith, W. R.
College of Forest Resources
University of Washington
Seattle, WA 98195
USA
Acoustic Properties of Wood

Spaak, A.
Plastics Institute of America
Stevens Institute of Technology
Hoboken, NJ 07030
USA
Design with Polymers

Speich, G. R.
Chairman
Department of Metallurgical Engineering and Materials Science
Illinois Institute of Technology
IIT Center
Chicago, IL 60616
USA
Ferrous Physical Metallurgy: An Overview

Spofford Jr, W. O.
Resources for the Future
1755 Massachusetts Avenue NW
Washington, DC 20036
USA
Residuals Management

Steiner, P. R.
Faculty of Forestry
University of British Columbia
2357 Main Hall
Vancouver, BC V6T 1W5
CANADA
Adhesives for Wood

Stelling, J. D.
Thompson and Lichtner Company
111 First Street

Cambridge, MA 02141
USA
Roofing Materials

Suchsland, O.
Department of Forestry
Natural Resources Building, Room 210
Michigan State University
East Lansing, MI 48824
USA
Hardboard and Insulation Board

Suddarth, S. K.
Department of Forestry & Natural Resources
Purdue University
West Lafayette, IN 47909
USA
Design with Wood

Tilton, J. E.
Department of Mineral Economics
221 Eric A Walker Building
Pennsylvania State University
University Park, PA 16802
USA
Materials Availability
Substitution: Economics

Tomkins, B
Risley Nuclear Power Development Laboratories
United Kingdom Atomic Energy Authority
(Northern Division)
Risley
Warrington WA3 6AT
UK
Fatigue: Engineering Aspects

Vick, C. B.
USDA Forest Service
Forestry Sciences Laboratory
Carlton Street
Athens, GA 30601
USA
Adhesives in Building

Westbrook, J. H.
133 Saratoga Road
Scotia, NY 12302
USA
Materials: History Before 1800
Materials: History Since 1800

Wibbens, R. P.
American Institute of Timber Construction
333 West Hampden Avenue
Englewood, CO 80110
USA
Glued Laminated Timber

Wilcox, W. W.
Forest Products Laboratory
University of California, Berkeley
1301 South 46th Street
Richmond, CA 94804
USA
Wood: Decay During Use

Wilson, D. G.
Room 3-447
Massachusetts Institute of Technology
Cambridge, MA 02139
USA
Recycling of Demolition Wastes

Yahalom, J.
Department of Materials Engineering
Technion-Israel Institute of Technology
Haifa
ISRAEL
Corrosion Protection Methods

Young, J. F.
Civil Engineering & Ceramic Engineering
University of Illinois at Urbana-Champaign
Urbana, IL 61801
USA
Cement as a Building Material
Concrete as a Building Material

Zicherman, J. B.
IFT Technical Services Inc
2550 Ninth Street
Suite 112
Berkeley, CA 94710
USA
Wood and Fire

SUBJECT INDEX

The Subject Index has been compiled to assist the reader in locating all references to a particular topic in the Encyclopedia. Entries may have up to three levels of heading. Where there is a substantive discussion of the topic, the page numbers appear in ***bold italic*** type. As a further aid to the reader, cross-references have also been given to terms of related interest. These can be found at the bottom of the entry for the first-level term to which they apply. Every effort has been made to make the index as comprehensive as possible and to standardize the terms used.